PALEOBIOLOGY
OF THE INVERTEBRATES

Paleobiology
of the Invertebrates
DATA RETRIEVAL FROM THE FOSSIL RECORD

Paul Tasch

Professor of Geology
W. S. U. Endowment Association
Distinguished Professor of Natural Sciences
Department of Geology
Wichita State University

JOHN WILEY & SONS New York ■ Chichester ■ Brisbane ■ Toronto

Library of Congress Cataloging in Publication Data:

Tasch, Paul.
  Paleobiology of the invertebrates.

  Includes indexes.
  1. Invertebrates, Fossil. I. Title.

QE770.T32   1980   562   79-14929
ISBN 0-471-05272-8

Printed in the United States of America

10 9 8 7 6 5 4 3 2 1

plan of this book

Paleobiology is this book's concern — systematic study of bioprogrammed creatures without backbones that have left a fossil trace. The objective, a more dimensioned presentation — places invertebrate fossils in the context of their periods of life and activity. This requires linking the fossil to the vast range of available knowledge, from molecular biology and biogeochemistry to information theory and modern ecology.

Among other innovative features not previously contained in a single volume are *identification to species level*, including reasoning behind decisions; *detailed study of neglected groups* — archeocyathids, stromatoporoids, hydrozoans, polychaetes, and others, *as well as all major groups*; the inclusion, after each chapter, of a section on *Questions* expected of the student and extensive *References* (where to find further information); *case histories* (Appendix); *labeled illustrations* of soft-part as well as hard-part anatomy, with *definitions in legends and text as they are needed;* emphasis on *information retrieval* — how to apply fossil data to larger problems; *introduction to the origins of life and micropaleontology.*

Levels of difficulty are built in. The instructor can use this text at any level he or she chooses, undergraduate to graduate. For beginning courses, more technical details and discussions can be omitted; for more advanced undergraduates and graduates, probing of the many problems raised and exploration of the extensive References can be undertaken.

Parts of the material of this book, continually updated, have been used in my undergraduate and graduate classes. Finding that students at all levels responded to the challenge of this type of presentation and material encouraged me to assemble this volume.

Chiefly, this book is meant to be an invitation for both new and more mature students to relate paleobiology to their other scientific background in biology, chemistry, physics, mathematics, engineering, and the earth sciences; and to discover the relevance of invertebrate paleobiology to most of the major problems confronting the earth sciences in particular, and science in general.

A flexible matrix of content, this book's characteristic, can serve as a permanent framework on which to place the ever-new findings as they appear.

ACKNOWLEDGMENTS

Clearly, no one can write a text in any science without drawing heavily on numerous excellent technical contributions by many investigators in all disciplines. The factual content of this book originates from the research literature, supplemented by field and laboratory observations.

D. H. Deneck, Wiley geology editor, encouraged me to undertake this effort and sustained me with friendly assistance during the years of its production. Mrs. Helen Ardrey aided in one phase of the editing. Robert Strong prepared all of the biological illustrations according to my directions. Other drawings were made by Dennis Carlton.

Many colleagues, acknowledged at appropriate places in the book, furnished published and unpublished data, illustrations, or information of a clarifying nature. Specific illustrations were provided by Drs. Rigby (sponges), Grant (brachiopods), Catala (crinoids), and Stuermer (trilobites).

The Alumni Fund of Wichita State University supplied funds for photostating and travel in connection with my researches for this book. To all colleagues, friends, and aides, and for all assistance they have given in bringing this text to completion, I am gratefully indebted and appreciative.

Wichita State University   PAUL TASCH
Wichita, Kansas
August, 1972

permissions

In addition to acknowledgment made throughout the text, the following permissions were also granted: John Wiley and Sons for Guthrie and Anderson, *General Zoology*, 1957; H. H. Ross, *A Textbook of Entomology*, 3rd ed., 1965; R. A. Boolootian, ed., *Physiology of Echinodermata*, 1966; E. T. Denton, "The Buoyancy of Marine Mollusks," in Wilbur and Young, *Physiology of Mollusca*, Vol. I. (Author acknowledgment is made in the text.) University of California Press for V. A. Zullo, "Zoogeographic Affinities of the Balanamorpha (Cirrepeida, Thoracica) of the Eastern Pacific," in R. I. Bowman, ed., *The Galápagos*, 1966, Fig. 1. American Association for the Advancement of Science for D. H. Wenrich, "Sex in Protozoa," in *Sex in Microorganisms*, AAAS Publ. #37, 1954. Pergamon Press for Charles P. Raven, *Morphogenesis*, 1958.

preface

The valuable suggestions of colleagues and students who have used this text have been incorporated in this edition, although the basic plan of the book remains the same.

The index has been significantly reorganized. All taxa are separately indexed in alphabetical order, facilitating the rapid location of any scientific name by students and other readers. The subject index incorporates updated material, making rapid inspection for a given entry more accessible.

Sections on archaeocyathids, edrio-asteroids, as well as portions of the Porifera have been completely revised, based on the latest findings and improved classifications.

New material has been added ranging from modified salinity tolerances in the Brachiopoda, to the origin of articulate crinoids; the new echinoderm class, Blastozoa, as well as the new arthropod phylum, Uniramia, are also included.

Wichita State University　　　　Paul Tasch
Wichita, Kansas
June, 1978

contents

8. The Mollis Clans: World of the Geometrical Shell   312

chapter one

BIOSPHERE AND INVERTEBRATE POPULATIONS: LIVING AND FOSSIL

ABUNDANCE OF LIVING THINGS

Our planet teems with life. When we rise above the earth in an airplane, submerge in a vessel (such as the *Trieste*) to the bottom of an oceanic trench, drill into the bedrock of the earth for oil, excavate through solid rock mining for ores, we find representatives of living things at unusual levels. Attention focuses on the entire realm of life on our planet and the spread and density of living things found in our realm.

THE BIOSPHERE

Spread

As the word "biosphere" suggests (*bios* = life), it is a name for the domain of life (Figure 1.1). This domain is visualized as a thin envelope some 21 miles thick. Presently defined, the biosphere extends to a height of 14 miles above sea level in the stratosphere; it reaches a depth of slightly more than 7 miles at the bottom of oceanic trenches.

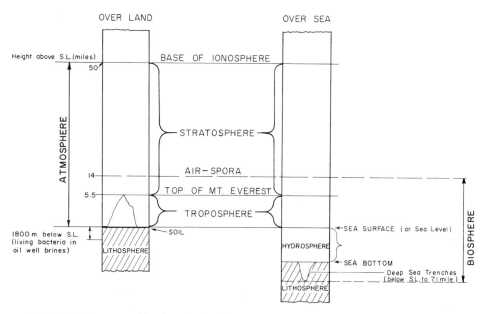

FIGURE 1.1 The present biosphere (realm of life). The extent is from the lower portion of the stratosphere to the upper lithosphere—a vertical spread of 21±miles. The interface between lithosphere and hydrosphere is the sea bottom; between ocean and atmosphere it is the sea surface (mean sea level).

Our atmosphere is subdivided into an upper unit, the stratosphere, and a lower unit, the troposphere. Transition to the hydrosphere is marked by the contact between the lowermost atmosphere and the water surface of the earth (sea level). The hydrosphere ends at the sea floor. From the sea floor downward, there is a thin veneer of superficial sediments underlain by bedrock. This begins the lithosphere (*lithos* = rock).

On land the troposphere extends downward from the top of Mt. Everest, a height of 5.5 miles above sea level, to contact at the soil surface. Below the soil there is the gradual transition to bedrock (the lithosphere). Deep drilling through this bedrock has brought to light the existence of bacteria living in salt brines at a depth of 1800 meters below sea level.

Explorations of deep sea trenches, isolated lakes at high elevations, inland waters of Antarctica, and cores from glacial and iceberg ice have shown diverse life forms widespread on our planet.

Density

Sparse viable populations of bacteria and fungi (air spora) have been obtained from the stratosphere at elevations greater than 10 km above sea level. In the lower atmosphere near the earth's surface the density of airborne fungi can reach 2000 or more individuals per cubic foot.

Beebe has estimated that in a vertical range of 600 ft depth in the sea every square mile of seawater contains 16 tons of planktonic skeletons. These microscopic skeletons contribute to the foraminiferal oozes that cover more than 35 percent of the present ocean bottom. According to Nigrelli, the sea produces from 10,000 to 14,000 tons of organic matter per square mile. Levorsen has estimated that every single year there is a potential yield of 12 million tons (80 million barrels) of hydrocarbons produced by the photosynthesis of microscopic marine plants in the ocean—algae (phytoplankton). This quantity could account for all known and future petroleum deposits.

The mud bottom of the continental shelf (Figure 1.2; 1.3) contains a rich worm-protozoan-molluscan fauna that, Harvey estimates, contributes a mass of living tissue equal to 100 grams per square meter.

Thorson (1946) showed the enormous population densities of planktonic larvae in the Danish Sound (Øresund). Sea ur-

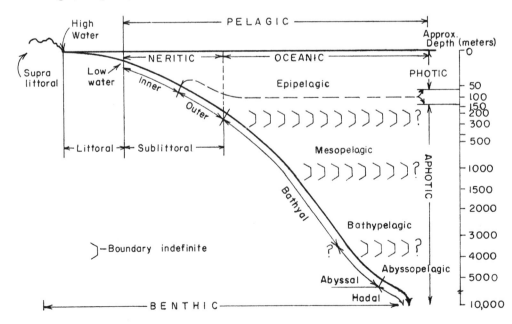

chin, *Strongylocentrotus* (Echinodermata), can reach an annual density of more than one billion individuals; pelecypod *Mytilus* (Mollusca) might attain yearly larval population counts of 4.5 billion individuals. An estimated 20 million to 1 trillion (10^{12}) larvae are contributed to the marine plankton by the intertidal barnacle *Balanus balanoides* (Arthropoda) along only 1 km of shore (H. B. Moore, 1935). Most of the larvae (spat) never settle to the bottom, being consumed by marine carnivores.

A census of only 4 sq ft of forest and meadow soil to a depth of $\frac{1}{2}$ in. led to estimates that forest soil contained over 1 million animal items and over 2 million seed-fruit items, whereas the meadow soil per acre had more than 13 million animal items and almost 34 million seed-fruit items.

If an insect in the tropics could lay 100 eggs every 30 days, it has been calculated that from a single mating pair at the end of 1 year there would be 488 trillion individuals if all survived. Fortunately, all do not.

This abbreviated account suffices to indicate that the various ecologic niches in the biosphere teem with life.

Communities—Living and Fossil

In any given sector of the biosphere, the sea, the living biota (all the plants and animals) will be organized into communities (Figure 1.3). Three main subdivisions of these communities are illustrated: the epifauna, the infauna, and nekton/plankton (swimmers and floaters). A study of Figure 1.3 shows that community distribution relates to depth of water with attendant variability in light penetration, salinity, and other parameters.

In the course of time the relationships shown in Figure 1.3 will change. This occurs as a result of the advance and retreat of the sea. Swimmers and floaters, upon demise, fall to the bottom mud and are buried under bottom sediments. The epifauna and infauna may be stranded by regressions of the sea; a freshening of the water could create intolerable conditions. Such cycles of change occur over periods of decades, centuries, millennia. Later, cores can be obtained by drilling at several sites along the mud bottom, and cycles can be deciphered from data encoded in the mineralized cores (Figure 1.4).

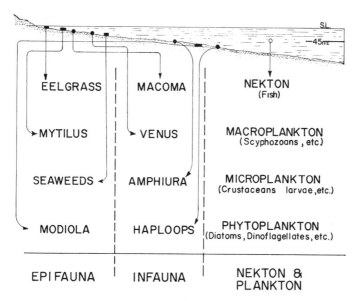

FIGURE 1.3 Modern bottom communities (Danish Sound). (Data trom Thorson et al). *Epifauna*—chiefly intertidal zone animals that live on/among vegetation, stones, pilings, and so on; very shallow water; cover 10± percent of total area of sea bottom. *Infauna*—chiefly animals living below intertidal zone; burrowers and diggers in clay/sand substrates; covers about 50± percent of earth's surface. *Amphiura*—echinoderm; *Haploops*—amphipod crustacean; *Modiola, Macoma, Venus*—pelecypods.

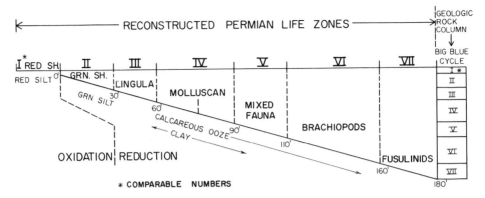

FIGURE 1.4 Ancient sea bottom communities (Permian Big Blue Cycle, Kansas). (Data from M. K. Elias.) Note depth zones, 0 to 180 ft. Mixed fauna include calcareous algae, bryozoans, corals, and calcareous brachiopods. Leathery seaweed at 30-ft depth. Red shale phase I—geologically oldest; fusulinid phase VII—youngest. (See Elias, 1964 Sympos. Cyclic. Sedimentation. Geol. Surv. Kansas Bull . 169, Vol. 1.)

In both Figures 1.3 and 1.4, *populations* are represented in each community. These populations numbering millions of individuals embrace several different species. Furthermore, each individual in a population was part of a food chain (the larger generally eating the smaller). Each individual performed the entire complement of biological functions—respiration and excretion to reproduction and conversion of nutrient intake into body substance (protoplasm and skeletal material) and body weight (biomass).

Hierarchy of Life Forms: Systems

Systems of classifying the objects of the world are devised to promote ready discourse and facilitate scientific research. The ultimate aim is to attain a classification meaningful on the genetic level.

In the history of the classification of plants and animals a few highlights have appeared (see Table 1.1). Charles Darwin's *Origin of the Species*, published in 1859, is considered a partition separating preevolutionary (Aristotle, Linnaeus)

TABLE 1.1. Ancient and Modern Systematics (Classification of Plants and Animals)

	Binomial Nomenclature		
Aristotle (348–322 B.C.) *Historia Animalia* *(Fourth century B.C.)*	*Linnaeus (1707–1778)* *Systemae naturae* *(1735; 10th ed., 1758)*	*Darwin* *Origin of the* *Species (1859)*	*Modern*
	Basis: Archetype *Preevolutionary Theory*		*Basis: Phylogenetic* *Postevolutionary Theory*
Genos—Family Eidos—Type or species	Imperium—Empire Regnum—Kingdom (3) Classis—Class Ordo—Order Genus—Genus Species—Species Varietas—Variety		Kingdom—Animalia Phylum—Chordata Class—Mammalia Order—Primate Family—Hominidae Genus—Homo Species—Sapiens Individual—Your name
Number of Species Known ca. 500	1000's		2,000,000+

from postevolutionary systems. In Aristotle's time (fourth century B.C.) about 500 species were known. By the eighteenth century thousands of species were known to Linnaeus and co-workers. Today over two million species of animals and one-third of a million of plants have been recorded. Mayr et al. (1953) observed that new species of plants are being described at the rate of 4,750 per year and new species and subspecies of animals at the rate of 10,000 per year.

MODERN SYSTEMATICS

Modern biological classification has a phylogenetic basis. It incorporates the Darwinian concept of evolutionary descent or modification of existing populations (species) through time. That concept explains the interrelatedness of apparently diverse forms. Thus the brachiopods, bryozoans, and phoronid worms have close affinities; yet, superficially, they are strikingly unlike.

The advent of biochemical genetics explained on the molecular level — and population genetics on the population level — the origin of variations in animals and plants, distribution of such variation, and the mechanics of natural selection. Which variations are to be conserved is a decision that rests with natural selection. In this way, modern systematics has much significance on the evolutionary, genetical, and molecular levels.

By contrast, Linnaeus had thought that species, although objective entities, remained fixed and immutable. Since evolution works on small variations within existing species, this concept of *unchanging* species is the essential crux of the matter. Until the possibility for change was resolved, no further advance was possible. Aristotle had no notion of the objective reality of species. On the other hand, Darwin, demonstrating that species were mutable, was unaware of the significance of contemporary Gregor Mendel's contribution to genetics. The genetical basis for speciation completely eluded him.

Unknown until recently, molecular genetics has freshly illuminated this blind spot of evolutionary theory (see Chapter 2, Figure 2.7).

BINOMIAL NOMENCLATURE

A description of any living or fossil invertebrate or vertebrate will follow the binomial system of nomenclature established since Linnaeus' time. That is, each animal will have a dual name (genus and species). Thus the fossil cryptostome bryozoan *Pachydictya ambigua* Ross, 1961, belongs to the species *ambigua*, genus *Pachydicta*, and was figured and described for the first time by J. P. Ross in 1961. The fossil brachiopod *Atrypella carinata* J. G. Johnson, 1964, similarly represents the species *carinata* and the genus *Atrypella* first described in 1964 by J. G. Johnson.

One might ask precisely what was classified in the cited examples. Ross had 90 fossil fragments of cryptostome bryozoan zoaria (Chapter 6); Johnson gave dimensions of four fossil fragments of his articulate brachiopod species (Chapter 7). In each case, only a minute sample of the total original population was available. Nevertheless the distinct morphology observed was inferred to be (*a*) characteristic of the bulk of the original population and (*b*) sufficiently unlike (that is, distinct from) any other related population. Such populations as discussed below are called *species*, and such distinctness is genetically determined (see Chapter 2 and Figure 2.7).

The respective genera (mentioned above and all others, in turn) are referred to known or newly erected families, the families to orders, orders to classes, classes to phyla (Figure 1.5; Table 1.1).

Yet, persistently, one might ask on what characters the several taxonomic groupings (species, genus, family) were erected. Essentially, that is what the rest of this book is about. We shall study the information contained in living invertebrates along with their fossilized remains and also the ways in which such data are obtained by the paleobiologist. Such pri-

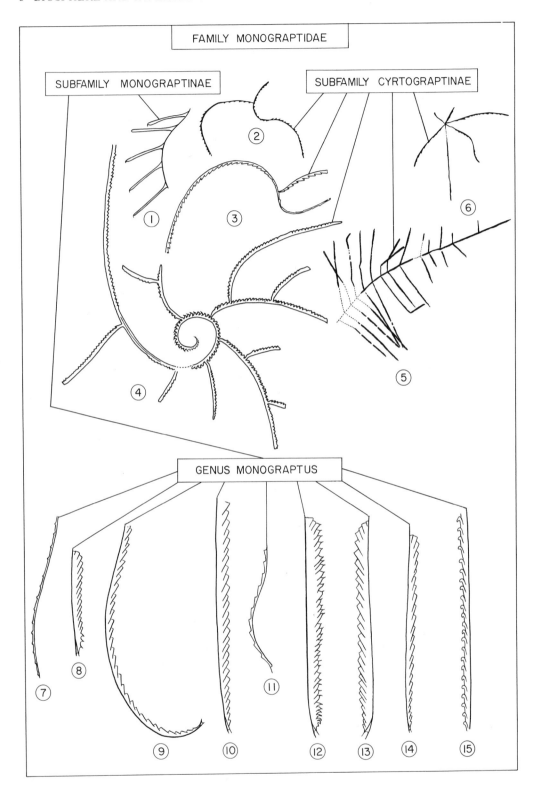

FIGURE 1.5 Classification of species, genera, and so on (class Graptolithina). Lower panel: (7–15) species of the genus *Monograptus* (which includes still other species not shown here): (7) *M. crinitus*. (8) *M. varians*. (9) *M. bohemicus*. (10) *M. dubius*. (11) *M. nilssoni*. (12) *M. colonus var. compactus*. (13) *M. tumescens*. (14) *M. forbesi*. (15) *M. scanicus* (schematic enlargements) (Data from W. B. N. Berry, 1964).[1] Each species shown may be represented by individuals—a portion of the original population. These would have a range in size and shape of theca, and so on, and yet all would be identifiable as members of a given species. Upper panel: Genera of the two subfamilies of the family Monograptidae, namely subfamily Monograptinae, and Cyrtograptinae (note subfamily ending, *-inae*, and family, *-idae*). Each genus shown includes one or several species. Subfamily Monograptinae includes only two genera (1) *Rastrites* and *Monograptus*. Subfamily Cyrtograptinae includes five genera; (2) *Diversograptus*, (3) *Barrandeograptus*, (4) *Cyrtograptus*, (5) *Abiesgraptus*, (6) *Linograptus*. (Data from Bulman, 1955, 1970.)[2] Note the morphologic distinctions of the different genera (see Chapter 13 for classification of the graptolites). J. Paleont. 1964, Vol. 38 (3).[2] Treatise on Invert. Paleont. 1955, Pt 5; 1970, same, Revised Pt 5.

mary data become the actual parts of the definition of genus, family, order. Thus the classification given in this book may be viewed as a master code in capsule form of invertebrate animals.

THE SPECIES

A sample of 859 specimens of a fossil articulate brachiopod (see Chapter 7) was collected from three successive horizons in beds of Pennsylvanian age (Tasch, 1953). Similar fossils are widely known to be numerous in the midcontinent Pennsylvanian. All specimens shared the following common characteristics: small size (width, 10 mm or less), nearly flat dorsal valve, prominent convex umbo, and nearly circular outline. Accordingly they were assigned to the same species *Crurithyris planoconvexa*. This type of species is known as a *morphological* species because it is described only on the basis of shell characteristics. In order to relate this species of the paleobiologist to the species described by biologists from living forms, a series of inferences are needed.

We may infer as follows:

1. On a given stratigraphic time plane, all fossils showing common morphology represented members of a *distinct population* (Figure 1.6).

2. In life this was an *interbreeding population* or a *deme* (Chapter 2) (Figure 1.6). Following Sylvester-Bradley two types of demes can be distinguished; *topodeme*—a geographically distinct population; *chronodeme*—interbreeding populations separated in time

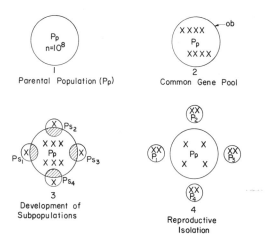

FIGURE 1.6 Simplified schematic of major steps in speciation.
(1) $n = 10^8 = 100$ million individuals.
(2) ob—original boundaries of Pp (species range). Common gene pool shared by individual members "x" of Pp, an interbreeding population.
(3) Subpopulations $(Ps_1 - Ps_4)$ develop by migration, and so on. [Note shaded areas of overlap. Individuals in the overlap areas regularly meet and mate with opposites of Pp. In unshaded areas, individuals are *geographic isolates* (subpopulations).]
(4) Individuals in populations $P_1 - P_4$ (represented by symbol "x") are now reproductively isolated from those of Pp; they constitute distinct species (that is, each has its own common gene pool).

without regard to morphological changes. It is possible for several topodemes to coexist. These would be equivalent to subspecies.

3. Although living in the same Pennsylvanian sea and region (midcontinental United States) as contemporary populations of closely related articulate brachiopods, the *Crurithyris planoconvexa* population was *reproductively isolated* from them. Since each such contemporary population likewise had its own gene pool, there could be no transfer of gene-encoded data that govern

all characters ot shell and soft-part anatomy.

4. Thus the distinctness of the *C. planoconvexa* shell morphology preserved in the fossil state contains information about its original life condition as shown in items 1–3 above.

5. It follows that when the paleobiologist describes a valid morphological species, he is *effectively* applying the biological species concept (see Mayr, 1963, p. 23).

In general, fossil specimens may be viewed as if they represented the complete living individual since recovery of a portion of the lost original information on softpart anatomy and ecology is possible. Accordingly, in this book, speciation (Mayr, 1963) of the invertebrates studied usually involved the following factors.

1. There was a population of interbreeding individuals that shared the common parental gene pool (Figure 1.6, part 2).

2. This population had a horizontal or geographic spread at any given time. The whole areal or spatial extent may be called the *species range* (Figure 1.6, part 1). Usually the boundaries of such ranges were not sharply demarcated from those of competing or other contemporary populations; rather, boundaries generally overlapped.

3. Migrants among the parent population could have extended the species range; some would have gradually become geographically isolated (Figure 1.6, part 3). (Ecologic and other spatial isolation could have achieved the same effect.)

4. In this manner the common gene pool was fractionated into subgene pools. At this stage any member of the geographic isolates (topodeme, subspecies, or subpopulation) could have interbred, with the likelihood of producing viable offspring, if it had met an opposite of the parent population (Figure 1.6, part 3).

5. Subsequently, reproductive isolation occurred. At that time, if members of the subpopulation were to meet and mate with members of the parental population, no viable offspring would have resulted. When this condition prevailed, the former subpopulation shared a gene pool distinct from

that of the parental gene pool. In other words, it now constituted a species distinct from that from which it was derived (Figure 1.6, part 4).

FOSSILIZATION

Fossils

Objects referred to as fossils include a wide variety of animal and plant remains (anatomical): biogenic items (such as fecal pellets or eggs) and animal artifacts (tracks, burrows, and nests). In general, such objects had been buried some time in the past, have undergone petrifaction or other modes of preservation, and have had to be retrieved by digging. Frequently, in the case of anatomical remains, shells have been chopped out of consolidated sediments or a crystalline rock matrix, or else released by acid treatment.

Contemporary Fossilization

In every terrestrial and marine environment presently inhabited, millions of biotic elements die daily. Which of these will be preserved in the fossil state in the future? The answer depends chiefly on how soon the carcass, skeleton, tree, or shell is removed from the effects of oxidation. That is to say, how rapidly is it buried? However, even rapid burial is no guarantee that such objects will be preserved through subsequent geological time.

The erosional cycle will remove, fragment, and often destroy vast protofossil assemblages that have been rapidly buried. Deformational events, too (as in folded belts), and attending metamorphism, volcanism, and magmatic events, can remove evidence of fossils incorporated in the original sediment. This is achieved by heat, pressure, solution, and recrystallization. Nevertheless one cannot speak of hard and fast rules; some fossils survive metamorphism (Bucher, 1953).

Fossilization is a chance event. Most biotic remains lose their identity before or after burial and never again occur in identifiable form in the rock record. Despite all this, their existence as once living populations may be inferred from their remains. For example, bituminous matter associated with evaporites as laminae in anhydrites or as inclusions in salt crystals, or disseminated in fetid brown dolomite, probably represented the organic residues of planktonic organisms. These were swept from the open sea into a gulf or inland sea before cutoff (Tasch, 1953). This example would be part of the biogeochemical cycle of carbon. Similarly other cycles—such as phosphorus, sulphur, nitrogen—contain information derived from portions of the unfossilized biotas that have come and gone through geologic time.[1]

PRIMARY AND SECONDARY ENCODED DATA

Only the most resistant parts of animals and plants survive the ravages of physicochemical agencies. Generally, soft-bodied forms or the soft-part anatomy of forms with skeletons are less likely to be preserved than organisms with endo- or exoskeletal structures.

There are, of course, exceptions to the rule, as is illustrated by such examples as the superbly preserved fossil termites found in Mexican amber in Tertiary beds (Snyder, 1960); Oligocene spiders from European Baltic ambers (Petrunkevitch, 1955, Figure 31); bacteria, probably Paleozoic, preserved dehydrated, in entrapped fluids of negative halite crystals, viable upon release (Reiser and Tasch, 1960); nematode worms hermetically sealed in the softened chitinous exocuticle of a fossil scorpion in the Lower Carboniferous of Scotland (Størmer, 1963; text Figure 30).

Newly established at the time, the crustacean branchiopod order, Lipostraca, was excellently preserved in the Devonian Rhynie Chert (Scourfield, 1926). Størmer described a new fossil scorpion from the Scottish Carboniferous in which unusual features were preserved. Some had such fine structures as the "internal 'nerve' (extension of sensory cell)" of the setae (1963, plate 14, Figures 1 and 3). In other fossil arachnids (including spiders and scorpions) are records of the heart preserved in two instances (Petrunkevitch, 1955).

In such examples noted above, the bulk, or a sizable portion, of the primary information in a given fossilized creature is retained.

Most common, however, is preservation of either shell or skeletal material. Here a large part of the original data in a given animal has been lost. What the given fossil contains is generally fragmentary information. This often suffices to permit a reconstruction of part of the lost data on soft-part anatomy. Thus muscle attachment scars on molluscan, brachiopod, or ostracod shells, in lieu of preserved muscle tissue, indicate the former presence and function of the particular muscles. Appendages also imply muscular control. In similar ways, various pieces of the soft-part anatomy may be derived from incomplete fossil remnants.

One kind of information in fossils may be viewed as secondary. It incorporates mineralogic changes in the original shell, skeleton, or tissue that occurred after death and burial. All such changes are incorporated under the term "petrifaction" (Table 1.2).

Secondary information relates to the migration of mineralizing fluids in the fossiliferous rock or sediment and to the replacement of existing minerals or organic compounds in a given fossil. Accordingly, such data, being secondary can tell us nothing about the original animal except insofar as cell outline and anatomical structure have been preserved. However, these data might have important bearing in locating mineral deposits.

CONSERVATION LAW AND FOSSILIZATION

Within the larger framework of the conservation of mass and energy, every as-

---

[1] This has opened up a new chemical branch of investigation called molecular paleontology.

TABLE 1.2. Partial Listing of the Variety of Mineral Replacements in Invertebrate Fossils (Petrifaction)

Native	*Sulfide*	*Silicate*
Silver: Mollusk (?)[a]	Galena: Foraminifer	Amphibole: Mollusk (Penn.)
Sulfur: Mollusk (Tert.)	(Perm.)	Analcite: Insect (Mio.)
Sulfate	Marcasite: Corals, etc.	Beidellite: Mollusk (Eoc.)
Barite: Mollusk (Jur.)	(Dev.)	Bentonite: Echinoid (Oligo.)
Celestite: Mollusk (Eoc.)	Pyrite: Mollusk (Dev.)	Feldspar: (?) (Miss.)
Gypsum: Brachiopod	Sphalerite: Mollusk (?)	Garnet: Mollusk (Trias.)
(Miss.)	*Oxide* (Quartz + var.)	Pyrophyllite: Graptolite (?)
Chloride	Quartz: Coral (Sil.)	Sepiolite: Mollusk (Tert.)
Halite: Brachiopod (Dev.)	Carnelian: Echinoid (?)	Wollastonite: Brachiopod (Sil.)
Fluoride	Opal: Mollusk (Cret.)	Zoisite: Echinoderm (Jur.)
Fluorite: Crinoid	*Oxide* (Iron)	*Carbonate*
(U. Paleoz.)	Hematite: Crinoids (Cret.)	Calcite: Mollusk (Mio.)
Molybdate	Limonite: Entire fauna[b]	Cerussite: Crinoid (?)
Wulfenite: Mollusk (?)	(Penn.)	Onyx: Arachnid (Tert.?)
Phosphate	Magnetite: Echinoderm	Smithsonite: Mollusk (?)
Francolite: Foraminifer	(Jur.)	
(Mio.)	*Oxide* (Other)	
Diadochite: Brachiopod	Cassiterite: Crinoids +	
(Dev.)	shell (?)	

[a]Geologic Age is given in parentheses where cited in literature.
[b]Dry shale (Penn.), Kansas.
Source. Modified after Ladd, 1957.

pect of fossilization and transformation of dead biotas can be explained. Whatever is not conserved in one state is conserved after translation or transformation into another. Thus fossil artifacts as footprints or trails retaining no anatomical parts of the animals responsible for them can be viewed as a corollary of the conservation laws—namely, conservation of form in the absence of original structure. By contrast, the contribution to the biogeochemical cycles of living planktonic and other organisms is only a translation, in part or whole, to another state, with the apparent loss of mass and/or energy accounted for by dispersion. In this way, dispersed organic matter in the sea, in coal, or as oil deposits constitutes part of the translated and/or dispersed substance of past life that left no definable fossil record. (Both coal and oil have additional, recognizable macro- and/or microfloral remains.)

When statements say that the fossil record is incomplete, representing as it does very few of the total number of indi-

viduals alive at any time, it would be wise to remember conservation of form, translation, and dispersion in the biogeochemical cycles discussed above. These last indicators surely represent a portion of this lost record of life. Paleobiologists thus study the rarest of documents, fortuitously preserved traces of former biosphere populations (see Chapter 2).

TYPES OF PRESERVATION

Unaltered

As already noted, fossils preserved, effectively unaltered, are a rarity (see discussion above).

Altered

The fossil invertebrate record consists essentially of altered remains. Such alterations are varied. Skeletal substance can be lost or gained. The entire original substance could be removed by solution, and yet a recognizable fossil might be found. If skeletal remains with their cellular organization and morphologic structures are viewed as organo-mineral systems, then element transfer into and out of systems is indicated.

Desiccation

A living creature taken from an aqueous environment, exposed to air, gradually decreases in volume by loss of water, shrivels, and stiffens. Because desiccation is a passive action, original cellularity and morphology are preserved and are always recognizable. Left exposed to weather and microbial action, desiccated corpses, being brittle, are often fragmented. Dried material, altered by mineral replacement, invasion, compaction, and/or sedimentation, might become part of the fossil record. Coquinas of molluscan shells represent heaps of living mollusks that had been tossed ashore during storm action at sea. Subsidence and subsequent compaction and mineral replacement or cementation result in a shell bed in the rock record.

For invertebrates generally, desiccation contributes in a minor way to the skeletal-remain preservation. Desiccation can contribute as merely an early intermediate factor in the process of fossilization (for example, desiccated insects in amber or dehydrated bacteria in salt crystals).

Carbonization

Volatiles are subtracted, transferred out of an organo-mineral system.

Chitinous exoskeletons or structures occur in coelenterates, annelid worms, mollusks, brachiopods, and arthropods. Under cover of water with poor bottom circulation, fouled conditions ensue after the oxygen supply has been exhausted. Volatiles (oxygen, hydrogen, nitrogen) in the chitin, chitin-protein, and protein-aceous skeletal material (see Chapter 15) are gradually driven off. This is common in crustaceans. A residue of carbon sometimes remains. Usually the thin film carbon residue retains more or less of the morphological characteristics of the original exoskeleton or structure.

Petrifaction

Mineralizing solutions are added, that is, transferred into an organo-mineral system.

Where original cellularity, internal structures, and fine details have been preserved, for each molecule of natural skeletal substance transferred out of the system, one molecule of the mineralizing fluid is substituted.

Commonly, with invertebrate skeletal remains, molecule by molecule replaces only on the surface. Compaction (after burial, flattening on bedding planes, and so on) limits the area exposed to mineralizing fluids. Relative porosity determines preferred pathways for such fluids through shell and other skeletal remains. Superficial layers often containing organic complexes, on which pores open, are likely to offer a degree of freedom for fluid passage greater than densely packed subsurface crystalline layers and internal endoskeletal structures.

Exceptions can be found to generalizations on this subject. For example, the brachidium of spiriferid brachiopods (Chapter 7) and the septa of conularids (Chapter 5), both internal structures, have been preserved by permineralization in some instances. Furthermore, completely uncompacted and/or undistorted invertebrate fossils are not rare. The articulated valves of the brachiopod *Composita* is often found uncompressed in the midcontinent Pennsylvanian-Permian. Other forms, often retrieved uncompressed, are rugose corals, Ordovician trilobites and, less frequently, three-dimensional graptolites released by acid treatment of cherts.

KINDS OF MINERALIZING FLUIDS

A glance at Table 1.2 indicates that mineralizing fluids form from almost any soluble mineral. Furthermore, these fluids can, under suitable circumstances, replace almost any invertebrate endo- or exoskeletal hard part. Occasionally replacement may lead to the preservation of delicate structures such as body hairs and appendages (Palmer et al., 1957; Scourfield, 1926).

The fossil-record evidence of invertebrates indicates that the common types of mineralizing fluids are calcareous, siliceous, and ferruginous (iron oxide, hematite, may alter to limonite or to iron sulphide, pyrite).

The chief replacement processes are calcification, silification, and pyritization (also hematization). Phosphatization, less frequent, occurs in formations such as the basal Maquoketa depauperate zone and the Permian Phosphoria formation.

Excluding silicon, the element composition of the main mineralizing fluids replacing invertebrate hard parts closely compares with the basic biological elements (Table 2.1). Calcification involves calcium, carbon, and oxygen; pyritization, iron and sulphur; and so on. As a part of one or another of the biogeochemical cycles, any given element circulates, is distributed and, by decay, solution, and so forth, is ultimately restored to circulation. The element composition of mineralizing fluids, therefore, may include any biological element contribution ensuing from the death of continental and marine invertebrates. Silicon is the exception. Still, siliceous spicules of sponges and opaline silica tests of radiolarians and diatoms do go into solution in seawater and contribute to the silica exchange (see Chapter 3).

The predominance of calcium over silicon in living invertebrate skeletons appears to insure the quantity absence of biogenic silicon in mineralizing fluids. Nonbiogenic silicon, however, is quite abundant.

Several different replacements noted may occur in the same formation, bed, or fauna. Thus, phosphatized, pyritized, hematized, and silicified fossils all occur in the basal Maquoketa fauna of Iowa although the phosphatized is the dominant type of replacement. Moreover, *partial replacement* by mineralizing fluids may occur in the same fossil; pyrite crystals, for instance, may occur in calcified or phosphatized shell walls. *Selective replacement* is also possible, crinoid stems embedded in limestone may have their hollow centers silicified while main stem body remains calcitic. Secondary cherts often contain fusulinids partially chertified, partly calcitic.

Open spaces in fossils (chambers in ammonoids, for example) may develop into geodes by growth of calcite or quartz crystals from their walls.

Permineralization often serves to preserve invertebrate skeletal remains otherwise destroyed through time. Thus silicified fossils not visible on surfaces of limestones or not represented by the calcareous fossil specimens are often released from the interior of sedimentary rocks by acid digestion. A classic example is the beautiful silicified brachiopod assemblages recovered from Permian Glass Mountain limestone blocks. Silicified trilobites, corals, and mollusks, among others, are also known. Similarly, internal and external molds may be the only fossil remains of a given invertebrate.

MOLDS AND CASTS

A shell may be thought of as having two surfaces (external and internal) separated by a wall. The features of each surface, which include ornamentation, sutures, scars, are called the *positive* (borrowing from photography). The *negative* (impressions) can be found on fossils formed by indurated material that either filled the interior (internal mold) or enclosed the exterior (external mold) of the shell. *No shell replacement is involved in mold formation*, although after its formation the shell may go into solution.

The pyritic, hematitic, or limonitic internal mold of various mollusks is often referred to as a *steinkern*, which means "rock kernel."

Material replacing the original shell (that is, the wall and its two surfaces) or the space between the internal and external molds is called a *cast*.

A single shell may leave three complete fossils as well as numerous fragments: an external mold, the shell itself or its cast, and an internal mold (Challinor, 1928). One may also find a cast of the external surface and an external mold lined with original shell material, as, for example, the Cretaceous Pierre shale mollusks. A cast of the internal surface and an internal mold is a frequent occurrence. Among the arthropods an individual in life will shed a successive series of molts as it grows. Thus, in the fossilized state in the same bed, large quantities of fossilized valves or carapaces or body parts representing relatively few individuals are not rare. The growth-stage series of an individual ostracod, for example, may be deciphered by arranging, according to size, several identical valves found on the same bedding plane.

The importance of correctly distinguishing between the different kinds of replicas or impressions of a given fossil was demonstrated by Kieslinger (1939, and see Chapter 5 and Figure 5.13*D*). He restudied the Solenhofen limestone scyphomedusans. The proliferation of distinct genera and species he found had been due to mistaken identification of internal and external molds of the ex- and subumbrella of the same scyphomedusan. Thus he gave all of these replicas the same name and placed in synonymy all previous names given to the different impressions (molds). (See *Treatise on Invertebrate Paleontology (F) Coelenterata*, p. F46. *Synonomy* consists of a chronological listing of correct and/or incorrect scientific names.)

Molds, despite their "negative" reflection of the original shell surface features, and casts contain information pertaining to the original animal. Accordingly, study of the surface characteristics of molds plus shape and size of casts permits recovery of the contained information.

Frequently at an outcrop or in samples studied in the laboratory, both the "positive" (i.e., an artificially made cast) and the internal or external impression of a given fossil skeleton are necessary for correct interpretation. Both can be made by application of modeling clay. For more permanent casts, latex can be used (Marek and Yochelson, 1964).

An entirely soft-bodied fossil fauna from southern Australia—the Ediacara fauna—of Pre-Cambrian age was preserved in quartzite. This fauna is now represented by numerous external molds and natural casts. One new genus of uncertain position, *Tribrachidum*, was described not on the actual fossil, which was an external mold, but on its "artifically produced counterfeit," that is, a cast (Glaessner and Daily, 1959) because certain details had been better seen.

FOSSIL SPOOR

Some European workers distinguish trace fossils (*Spurenfossilen*) from body fossils (*Körperfossilen*). Trace fossils include fecal pellets and castings; tracks, trails, and burrows; chambers; boreholes; eggs and miscellanea (gastroliths, pearls, and so on). (See Figure 1.7.)

Fecal Pellets and Castings

Another group of fossil objects includes mineralized fecal pellets or organic excreta of worms, mollusks, crustaceans, and so on. The term "coprolite" is restricted to larger objects of this type, generally of vertebrate origin ($\frac{1}{2}$ in. to 6 in. in length). Objects of this type, 1 mm or less in length with simple ovoid configuration, are classed as "fecal pellets" or organic excreta, generally of invertebrate origin.

The remarkable thing about such fossils is their abundance. Because they become part of the sediments after mineralization, the abundance factor is often overlooked. H. B. Moore (1955, Figure 1 for types) reviewed modern marine occurrences and illustrated sculptured and unsculptured types.

Illing (1954) showed that probable molluscan fecal pellets (generally cemen-

14

FIGURE 1.7 Fossil spoor and miscellanea.

(1) Reconstruction showing a Cambrian trilobite excavating its burrow—the latter is named *Cruziana jenningsi.* Dashed line indicates depth limit to which burrow will extend.

(2) *Lingula sp.,* L. Carboniferous of Scotland, fossilized in living position in burrow (84 percent of all lingulid shells in the formation were in this position).

(3) The horseshoe crab, *Limulus walchi,* at the end of its trail, Solenhofen limestone, reduced approximately 1/3.

(4) Acrothoracic barnacle burrows on a Pennsylvanian productid brachiopod, Texas; burrows average 1.8 mm long, 0.6 mm wide.

(5) *Arthrophycus alleghZeniensis,* Silurian, Pennsylvania, underside of sandstone; probably in-fill of trail in underlying mud, × 1/2.

(6) *Scolithus* burrows in Cambrian orthoquartzite, 1 to 15 cm long, 3 to 4 mm wide.

(7) Atokan spoor: *(a) Laevicyclus* sp. sandstone slab; concentric marking probably made by waving tentacles of a polychaete worm, diameter 5 cm; diametric ridge an anomaly unrelated to polychaete morphology; *(b) Conostichus* sp., top view, *(c)* side view showing cone-in-cone structure, diameter, 1 cm.

(8) Eggs: Eggs (e) fossilized on conchostracan left valve *(Leaia),* Permian, Antarctica, egg diameter 0.1 mm or less.

(9) Flysch spoor: two examples, schematic. *(a) Taphrhelminthopsis, (b) Helminthopsis.*

(10) Fecal pellets: *(a)* calcareous oolite with fecal pellet nucleus that had been voided by the brine shrimp *Artemia gracilis,* Great Salt Lake; *(b)* phosphatized structureless (fecal?) pellet, Permian Phosphoria formation; *(c–d)* calcified fecal pellets, Bahama Islands (a–d schematic).

(11) Crustacean gastroliths, *Wechesia pontis,* M. Eocene, Texas: *(a)* side, *(b)* exterior, maximum diameter 0.3 mm.

(12) Fossil pearls in *Ostrea* valves (Cretaceous, France), pearl is 1 cm high.

Illustrations modified or adapted from the following: (1) Fenton and Fenton, 1937; (2) Craig, 1952; (3) Richter, 1954; (4) Rodda and Fisher, 1962; (5) Hall cit Shrock, 1948; (6) Shrock, 1948; (7) Henbest, 1960; (8) Doumani and Tasch, 1964; (9) Seilacher, 1958; (10a) Eardley, 1938; (10b) Emigh, 1956; (10c–10d) Illing, 1954; (11) Frizzell and Exline, 1958; (12) Tasnadi-Kubacska, 1962.

---

ted by internal bacterial precipitation of aragonite) were a recognizable constituent of modern Bahaman calcareous sand grains. The most common bottom sediment constituent of the Great Salt Lake comes from calcified fecal pellets of the brine shrimp *Artemia gracilis* (Eardley, 1938). Borings from this area showed that similar pellets are as old as 600,000 years.

Tertiary oil-bearing sandstones and/or shales from Japan and California contain ovoid objects. Galliher referred to the California pellets as "sporbo." These pellets usually have a high phosphate content (37.7 percent, P_2O_5). Phosphatic fecal pellets and/or coprolites have been widely reported, from the basal Maquoketa shale of Iowa and from the Cretaceous phosphatic chalk beds of southeast England, among others. A structureless type of pellet—one of five types of pellet in the Permian Phosphoria formation—is interpreted to be original animal excrement, first solidified and later phosphatized (Emigh, 1956). Rod-shaped sand casts in the Jurassic Solenhofen limestone, reported by Fenton and Fenton, are pos-sibly castings of a polychaete worm (H. B. Moore, 1955).

Fish coprolites may contain parts of the undigested skeletons of invertebrates, such as conchostracan valves in Triassic coelocanth fish coprolites.

Fecal pellets, after extrusion, may be calcified, phosphatized, glauconitized, pyritized, or silicified. Termite pellets, for example, were opalized with the wood into which they had been extruded in the Pliocene of California (Ladd, 1957). Quartz grains and clay minerals may occur in fecal pellets.

Tracks, Trails, Burrows, Chambers, and Boreholes

A variety of invertebrate spoor, exclusive of organic excrement, have been found in the rock record. These may or may not identify the animal that made them; on occasion some provide indications of the probable ecologic setting in which they had been made. Study of such spoor is called palichnology (Figure 1.7).

Spurenfossilen can serve as index fossils

and paleoecologic indicators, and can aid stratigraphic correlation.

Eggs and Miscellanea

Soft, minute objects — invertebrate eggs — are rarely preserved in the fossil record. Egglike objects representing spores, excreta, or sedimentary accretions such as oölites may be found cemented or attached to fossil invertebrate valves or carapaces. A case in point is the so-called trilobite eggs — an identification most specialists have rejected.

One group of animals, however, has less debatable evidence: fossil crustacean branchiopods (Tasch, 1969). The anostracan *Branchiopodites vectensis* (Oligocene, Isle of Wight) has an egg sac and egglike objects within it preserved. The lipostracan *Lepidocaris* (Rhynie Chert, Scottish Devonian), which is related to the anostracans, also had an egg sac and egg cases fossilized. There are numerous reports of eggs attached to fossilized conchostracan valves. Carbonized cladoceran ephippia that bore two ephippial eggs are known from the *Braunkohle* (Oligocene, West Germany).

Miscellanea include crustacean gizzard stones called *gastroliths*, which are essentially small calciphosphatic objects with very fine lamination and an excavated channel. Frizzell and Exline (1958) gave the name *Wechesia pontis* to such objects from the Eocene Claiborne group of Texas (Figure 1.7, part 11). Similar gastroliths are known in Recent crayfish and other crustaceans.

Other miscellaneous objects are fossil pearls (Figure 1.7). Pearls are secreted in the outer surface of the molluscan mantle, these are called blister pearls. Those secreted within a small sac in the mantle are the true pearls of commerce, called cyst pearls. Both types were initiated by some irritant. These objects are composed of numbers of minute, overlapping flakes of aragonite arranged in thin, concentric layers that are separated by fine films of conchiolin (Russell, 1920). Widely re-

ported from Cretaceous beds in England, Europe, Africa, Japan, and the United States, they are known from the Jurassic, Eocene, and post Pliocene. Commonly they are associated with, or attached to, pelecypod shells, including *Inoceramus, Ostrea, Exogyra*, and *Pinna* (Chapter 8).

Jackson (see Russell, 1920, pp. 427 ff.) found 130 fossil pearls in a cluster in the London Clay associated with the broken shells of *Pinna* and *Ostrea*; some disappeared, leaving identifiable pittings. Many workers have identified such pittings as sites of pearly protuberances, even in the absence of fossil pearls.

PALEOPATHOLOGY

Two works catalogue and illustrate pathological conditions ("paleopathology"): Moodie, 1923; Tasnádi-Kubacska, 1962. Other data on this subject are scattered in individual papers.

Shell injury and repair is a common paleopathological indicator in fossil invertebrates and is often found in large samples of fossil molluscan or brachiopod shells. Fossil crinoids occasionally show evidence of parasitic worm infections, snails in symbiotic association, or regeneration (Hattin, 1958); fossil asteroids sometimes have regenerated arms. Responses to a pathological condition in life may have led subsequently to a distorted or deformed shell.

Although so-called dwarf fossil faunas (pyritic, limonitic, and so on) were often reported in the past, only in the last decades has more vigorous evidence been forthcoming (Tasch, 1953). Vogel (1959) found that of 700 specimens of the ammonoid species *Polyptychites pumilio*, from a restricted portion of the column, 600 were true dwarfs; the rest were juveniles or normally grown individuals. Bennison (1961) studied the small shells of the pelecypod *Naiadites obesus* (Lower Carboniferous, Fife) at two horizons. On one he found dwarfed individuals; on the other the shells were exotic and constituted a "pebble necrocoensis." Reports of living dwarf faunas are rare. Schuster-Dieterichs (1954) found, along beaches of the Pacific coast of El Salvador, both normal and dwarfed snails (*Olivella columellaris*). The dwarfed forms were about 6.0 mm smaller in height at the same growth stage as the normal forms. Magnetite sands occurred on beaches inhabited by the dwarf forms. (See Mancini, 1978.)

QUESTIONS

1. Does Figure 1.4 represent a single ancient biosphere or more than one? Explain.
2. Outline the possible steps by which the living biotas in Figure 1.3 might become a part of the rock record of the future.
3. What is the special significance of the following terms: deme, topodeme, common gene pool, geographic and reproductive isolation, species range?
4. (a) Take any Recent mollusk shell available; prepare internal and external molds using modeling clay. Secure an artificial "positive" of these impressions by filling them with moist silty clay. Compare the artificial "positive" with the original. Devise an experiment to secure a cast.
   (b) What did Keislinger (1939) find?
   (c) What information can a geologist recover from the spoor shown in Figure 1.7?

REFERENCES

Ager, D. V., 1963. Paleoecology. McGraw-Hill. 310 pp.

Asmutz, G. C., 1958. Coprolites: A review of the literature and a study of specimens from Southern Washington: J. Sed. Petrology, Vol. 28 (4), pp. 498–508, Figures 1–3.

Bennison, G. M., 1961. Small *Naiadites obesus* from the Calciferous Sandstone Series (Lower Carboniferous) of Fife: Paleontology, Vol. 4 (2), pp. 300–311, 9 text figures.

Bucher, W. H., 1953. Fossils in metamorphic rocks: A review: G.S.A. Bull., Vol. 64, pp. 275–300.

Caster, K. E., 1957. Problematica: Treat. Mar. Ecol. and Paleoecol., Part 2, G.S.A. Mem., Vol. 67, pp. 1025–1032.

Challinor, J., 1928. The terms "cast" and "mould" in paleontology: Geol. Mag., Vol. 65, pp. 410–411.

Craig, G. Y., 1952. A comparative study of the ecology and paleoecology of *Lingula*: Edinburgh Geol. Soc. Trans., Vol. 15, pp. 110–120.

Eardley, A. J., 1938. Sediments of the Great Salt Lake, Utah: Bull. A.A.P.G., Vol. 22 (10), pp. 1305–1411.

Emigh, G. D., 1956. The petrography, mineralogy, and origin of phosphate pellets in the Western Permian Formation and other sedimentary formations. Doctoral dissertation, Univ. of Arizona.

Fenton, C. L., and M. A. Fenton, 1932. Boring sponges in the Devonian of Iowa: Amer. Mid. Nat., Vol. 13, pp. 42–54.

_____, 1937. Trilobite "nests" and feeding burrows: Amer. Mid. Nat., Vol. 18, pp. 446–451.

Frizzell, D. L., and H. Exline, 1958. Crustacean gastroliths from the Claiborne Eocene of Texas: Micropaleontology, Vol. 4, pp. 273–280.

Glaessner, M. F., and B. Daily, 1959. The geology and Late Precambrian fauna of the Ediacara Fossil Reserve: Records So. Austral. Mus., Vol. 13 (3), pp. 369–401, plates 42–47, text figures 1, 2.

Harvey, H. W., 1955. The chemistry and fertility of sea water: Cambridge Univ. Press, 199 pp.

Hattin, D. E., 1958. Regeneration in a Pennsylvanian crinoid spine: J. Paleontology, Vol. 32 (4), pp. 701–702, plate 98.

Hedgpeth, J. W., 1957: Classification of marine environments: G.S.A. Mem. 67, Vol. 1, Chapters 2, 3.

Henbest, L. G., 1960. Fossil spoor and their environmental significance in Morrow and Atoka Series, Pennsylvanian, Washington County, Arkansas: U. S. Geol. Surv., Prof. Paper # 400-B, pp. B383–B385.

Higgins, C. G., 1961. Significance of some fossil wood from California: Science, Vol. 134 (3477), pp. 473–474.

Illing, L. V., 1954. Bahaman calcareous sand: Bull. A.A.P.G., v. 38, p. 1–95.

Kornicker, L. S., and E. G. Purdy, 1957. A Bahaman faecal-pellet sediment: J. Sed. Petrology, Vol. 27, pp. 126–128.

Ladd, H. S., 1957. "Introduction" and "Paleoecological evidence": G.S.A. Mem. 67, Vol. 2, Chapters 1, 2.

Marek, L., and E. L. Yochelson, 1964. Paleozoic mollusk: *Hyolithes:* Science, Vol. 146 (3652), pp. 1674–1675, figures 1–3.

Mayr, Ernst, 1963. Animal species and evolution. Harvard Univ. Press. 622 pp., glossary, refs.

_____, E. G. Linsley and R. L. Usinger, 1953. Methods and principles of systematic zoology. McGraw Hill, 284 p.

Moodie, R. L., 1923. Paleopathology. Univ. Illinois Press. 557 pp.

Moore, H. B., 1955. Faecal pellets in relation to marine deposits, in P. D. Trask, ed., Recent Marine Sediments, pp. 516–524.

Pålmer, A. R., et al., 1957. Miocene arthropods from the Mohave Desert, California: U. S. Geol. Surv., Paper 294-G, pp. 237–275, plates 30–34, text figures 83–101.

Petrunkevitch, A., 1955. Arachnida, in Treatise on Invert. Paleont., P. Arthropoda, 2, Kansas Univ. Press, pp. 43ff.

Reiser, R., and P. Tasch, 1960. Investigation of the viability of osmophile bacteria of great geological age: Trans. Kans. Acad. Sci., Vol. 63 (1), pp. 31–34.

Rodda, P. U., and W. L. Fisher, 1962. Upper Paleozoic Acrothoracic Barnacles from Texas: Texas J. Sci., Vol. 14 (4), pp. 460–479.

Russell, R. D., 1920. Fossil pearls from the Chico formation of Shasta County, California: Am. J. Sci., Ser. 5, Vol. 18, p. 417, plate 1, Figure 12.

Schafer, W., 1962. Akuto-Palaontologie nach Studien in der Nordsee. Waldemar Kramer. 666 pp.

Schuster-Dieterichs, O., 1954. Verzweigung durch Eisen?, Umschau, Vol. 21, pp. 644–646, plates 1–2.

Scourfield, D. J., 1926. On a new type of crustacean from the Old Red Sandstone (Rhynie Chert Bed, Aberdeenshire—*Lepidocaris rhyniensis* gen. et. sp. nov.: London, Philos. Trans. Roy. Soc. Ser. B, v. 214, p. 153–187, pl. 21–22, text figs 1–51.

Seilacher, A., 1953. Studien zur Palichnologie I. Ueber die Methoden der Palichnologie: Neues Jahrb. Geol. Paleont., Vol. 96, pp. 421–452.

_____, 1958. Zur ökologischen Characteristik von Flysch und Molasse: Ecolg. Geol. Helvet., Vol. 51, pp. 1062–1078.

Shrock R. R. 1948. Sequence in layered rocks. McGraw Hill, pp. 174–188.

Snyder, T. E., 1960. Fossil termites from Tertiary Amber of Chiapas, Mexico (Isoptera): J. Paleontology, Vol. 34 (3), pp. 493–494, plate 70.

Størmer, Leif, 1963. *Gigantoscorpio willsi,* a new scorpion from the Lower Carboniferous of Scotland and its associated preying microorganisms. Oslo Univ. Press. 124 pp., 22 plates, 45 text figures.

Tasch, P., 1953. Causes and paleoecological significance of dwarfed fossil marine invertebrates: J. Paleontology, Vol. 27 (3), pp. 356–444, plate 49, 6 text figures.

_____, 1961. Valve injury and repair in living and fossil conchostracans: Trans. Kans. Acad. Sci., Vol. 64 (2), pp. 144–148, Figures 1–5, appendix.

_____, 1969. Branchiopoda, in Treatise on Invert. Paleont. R. Arthropoda.

Tasnádi-Kubacska, 1962. Paläopathologie. Veb. Gustav Fischer, Verlag Jéna. 240 pp., 295 text figures, refs. (Part I for invertebrates.)

Thorson, Gunnar, 1946. Reproduction and larval development of Danish marine bottom invertebrates with special reference to the planktonic larvae in the Sound (Øresund): Meddelelser fra Komm. for Danmarks Fiskeri-Og Havundersøgelser: Plankton Vol. IV, 484 pp., refs.

Vogel, K. P., 1959. Zwergwuchs bei Polyptychiten (Ammonoidea): Geol. Jb., Vol. 76, pp. 469–540, tables, illustrations.

The following readily available references should be consulted:

Häntzschel, Walter, 1962. Trace fossils and problematica: Treatise on Invert. Paleont., Pt. W, pp. W177–W245, Figures 109–149.

Seilacher, Adolf, 1964. Biogenic sedimentary structures, in J. Imbrie and N. Newell, eds., Approaches to paleoecology, John Wiley, pp. 296–316.

(An excellent review has been provided by T. P. Grimes and T. C. Harper, eds., 1970. Trace Fossils. Liverpool Geological Society.

Walker, K. R. 1972. Trophic analysis: a method for studying ancient communities. J. Paleontology, vol. 46 (1), pp. 82–93. (Useful discussion and references. Stresses organic detritus-eaters among benthos of ancient seas). For details on a post-mortem case history see Boyd, W. D. and N. D. Newell, 1972. Taphonomy and diagenesis of a Permian fossil assemblage from Wyoming. J. Paleontology, vol. 46 (1), pp. 1–14.

Mancini, E., 1978. Origin of micromorph faunas in the geologic record. J. Paleontology, vol. 52 (1), pp. 311–312.

chapter two

ARCHIVES OF PAST LIFE AND DELAYED FEEDBACK

ENCODING, THE FOSSIL RECORD, AND FEEDBACK OF INFORMATION

Earth history is congealed in rocks. Rocks are composed of minerals. Accordingly, our planet's history is written in a mineral code (Figure 2.1).

Encoding

A group of dissimilar animals can, and often do, have the same mineral composition in their skeletal parts—for example, calcite. Clearly, the mineral code is only an arrow pointing to a more fundamental code. Ultimately dissimilarity or similarity is predetermined by the genetic code evidenced by hard parts (skeleton, shell, and/or structures). This last-named code is written in a more elementary language —the biochemical code.

Approached from a different route, the atomic code determines both mineral code and biochemical code. Consider the fossilized shell of a Tertiary mollusk. It exhibits characteristic morphology (genetic code), but the shell itself is composed of mineral calcite or aragonite (mineral code). Both the organic content of the shell (biochemical code) and the shell's mineralogy are functions of the atomic code.

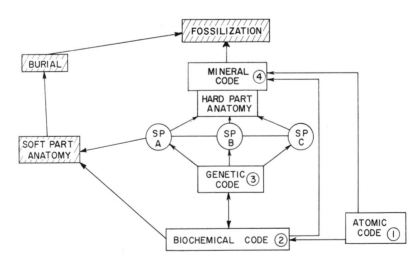

FIGURE 2.1 Encoding of data in living forms. The atomic code (1) gives rise to the biochemical (2) and mineral codes (4) written in soft- and hard-part anatomy of a given invertebrate. The genetic biochemical code (3) governs every system of an animal including its mineralized parts. (For information recovery, see Figure 2.2) (SP.A. = species A . . .).

Feedback and Data Retrieval

The cycle of life on our planet performs as a continuous system with input of information at one end—the beginning of a cycle—and output or retrieval of encoded information at the other end of the cycle.

As discussed in Chapter 1, a given living invertebrate may be viewed as a discrete organo-mineral system. A complete description of such a system would require n bits of information, where n is some undetermined large number. We can refer to the sum of the n bits as equal to S, and speak of this as encoded in the system. How much does S include? Every aspect of hard- and soft-part anatomy, shell mineralogy, organic chemistry of body fluids, biochemistry of genetic material, cell biology, evolutionary descent, the context in which the animal lives, and its biotic associates—all of this "information" is what S denotes.

The total bits of information of all kinds in the given invertebrate represent the input or data-encoding phase. In another way of speaking, S contains a *message* to be transmitted. Transmitted how, and to what receiver? (In ordinary communications—radio, television, the printed word—both the *message* and the potential receivers are apparent.)

Transmission of the *message* encoded in any living form appears elusive at first glance. Such transmission can be referred to as "spontaneous." It occurs with all the activities, functions, and interactions related to a given animal's life. A nonspontaneous transmission (the area of our concern) can be achieved only through deliberate intervention of a scanning device. A scanning device may be designated Sc; Sc represents any given investigator. For living forms, Sc may study spontaneous transmission in cultures, in the wild state, even in nonvital material such as skeletons or dried and preserved specimens. In these cases the receiving end of the given circuit is the scanner (Sc).

Upon demise, the given living form will generally undergo burial and organic decay or, decay without burial. In the common case, burial will result in some irretrievable loss of encoded information (designated as D). Why? Because microbial action will sponsor organic decay. At this stage, only a handful of years need elapse. Thus, S—the original message—is diminished by a quantity D, $(S-D)$ (Figure 2.2).

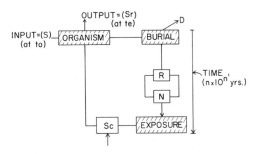

FIGURE 2.2 Feedback loop (delayed feedback): S—total bits of information encoded in an organism; R—redundancy (casts, molds, and so on): N—noise (that is, distortion of information in transmission by solution, erosion, and fragmentation): Sc—scanning device, or the energy put in by an investigator: Sr—immediate data retrieval by the investigator: D—irretrievable loss of information by organic decay. (t_o—time of origin; t_e—time of exposure.)

Other losses occur in transmission. All of these losses may be bracketed under the term "noise" (N). *Noise* may be defined as all effects that distort the original encoded message during its transmission to the decoding end of the circuit. The equivalent is static in radio transmission. Solution, erosion, fragmentation—some of the more important factors discussed in Chapter 1—constitute "noise" in the fossil record of invertebrates.

In ordinary communications there is also the factor of "redundancy" (R). *Redundancy* is a deliberate and multiple repetition of the same message in order to insure that the message arrives at the receiving end despite the effects of *noise*. With invertebrate fossils, the multiple copies of all or part of a buried animal by means of external and internal casts, molds, clues about animal's configuration and/or activity revealed by fossil spoor, the possibility of making artificial replicas or impressions in wax or latex may be interpreted as a variety of redundancy of the original message. Recall at this point that the original message comes from the complete intact living animal and its skeleton.

For the fossil record, redundancy tends to counteract information loss due to "noise" in the transmission of the original message.

Therefore, given the following units— a living form, the fossilized state, subsystems redundancy and noise, plus geologic time before the investigator (Sc) is tied into the circuit—a *delayed feedback loop* operates; output of original message information becomes possible. This output is *initial information recovery.*

Despite the fact that S—the total information possible about any given fossil—is unattainable, S'—the maximum recoverable information possible from any given fossil—is a quantity that can enlarge with time. In effect, this means that we can learn more and more from the same objects of study.

Throughout this text emphasis rests on contained and derived information and the maximization of derived information from fossil invertebrates (Figure 2.3).

Paleontology (paleobiology) is essentially the science of recovery of information about forms that lived in the past. Or, one might say, that it comprises information contained in the fossil record of life that through a delayed feedback loop, becomes available to contemporary investigators.

Atomic and Mineral Code and Fossils

The dominant composition of invertebrate fossil skeletons is calcareous (calcium carbonate). Exceptions are found among the protists such as the radiolarians, silicoflagellates, and others such as glass sponges. These forms are siliceous. Occasionally fossils are calciphosphatic— for example, inarticulate brachiopod *Lingula*, or polychete worm jaws. Arthropods are predominantly chitinous although

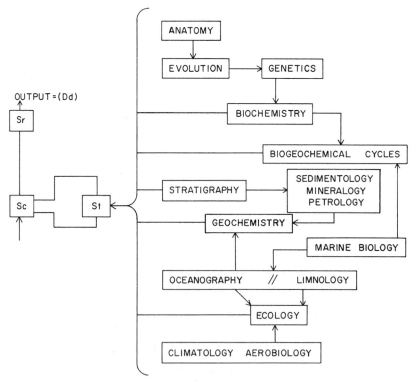

FIGURE 2.3 Derivative data (an external feedback loop): *Sc*—scanning device; *Dd*—derivative data; *St*—information storage. (Compare Figure 2.2.)

valves or carapaces may be impregnated with calcium salts. Ostracod valves exemplify this feature.

The mineral code of skeletal material in life essentially is restricted to carbonate minerals, calcite and aragonite, and opaline silica. There are also a variety of organic substances that serve as skeletal fibers, plates, shields, and complete exoskeletons.

What is the explanation for the restricted mineral code in the original skeletal material? Obviously halides (sodium chloride—common salt) would not be possible for skeletons of creatures living in water, because of ready solubility; nor can invertebrate skeletons be too ponderous, which eliminates compounds from lead, iron, and so on. These elements occasionally concentrate in given marine invertebrates, often with adverse effects on growth. Important also is availability of a mineralizing fluid, for example, calcium carbonate. Remembering the vast populations of marine mollusks and so on, a plentiful mineralizing fluid is needed. Carbonates in solution meet this requirement.

The mother liquor is critical for the ultimate crystals forming skeletal material. Because these crystals are secreted by tissues of the given animal from its nutrient intake, understandably the mother liquor must be physiologically compatible and nontoxic.

Table 2.1 shows the skeletal contribution made by the basic biological elements. These are called basic because they are found in all living things. Besides their contribution to invertebrate hard parts such as the shell and test of an animal, these elements contribute critically to soft-part anatomy (cell, tissues) as well as to body fluids, and pigments.

The skeletal remains of any living Echinodermata, such as echinoid or crinoid (see Chapter 12), will contain as its major mineral code the mineral calcite, about 80 to 95 percent of the total. Another carbonate invariably will be present also, magnesium carbonate, $10\pm$ percent. A

TABLE 2.1. Main Mineral Species and Basic Biological Elements Found in Marine Invertebrate Skeletons

Mineral Group	Mineral Species	Element Composition
Carbonates[a]	Calcite Aragonite Vaterite	Calcium, Ca[b] Carbon, C[b] Oxygen, O[b] Magnesium, Mg[b]
Phosphates	Apatites	Calcium Phosphorus, P[b] Oxygen Hydrogen, H[b] Carbon Fluorine, F
Silicates	Opals	Silicon, Si Oxygen Bound water: Hydrogen Oxygen
Sulfates	Celestite	Strontium, Sr Sulfur, S[b] Oxygen

[a]Strontianite, $SrCO_3$, belongs here although it is of limited occurrence.
[b]Basic biological elements along with potassium, K: nitrogen, N; and iron, Fe.

minor skeletal constituent will be phosphates, which, among others, are found in quantities ranging from a trace to $2\pm$ percent.

Figure 2.4 shows that magnesium concentration is temperature-controlled; it varies with temperature of habitat. Calcareous skeletons from several invertebrates indicate that, generally, skeletal magnesium concentrations varies with three factors: *mineralogy*, that is, calcite or aragonite (magnesium is lower in aragonite skeletons); *water temperature*; and *phylogenetic* level, that is, where the given invertebrate occurs on the evolutionary scale (Chave, 1954). See Figure 2.5.

A study of ten species of fossil crinoids indicated that the magnesium carbonate concentration varied from less than 1 to 3 percent. Thus it appears that fossil crinoids have 3 to 10 percent less magnesium

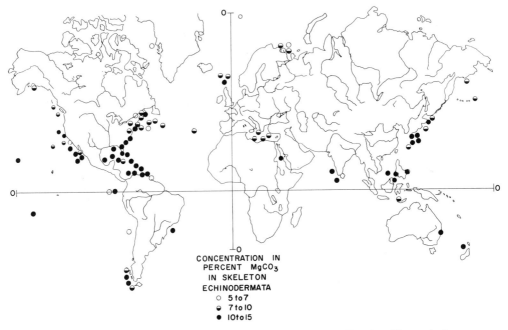

FIGURE 2.4 Magnesium carbonate in skeletal parts of echinodermata. (Data from Vinogradov.)

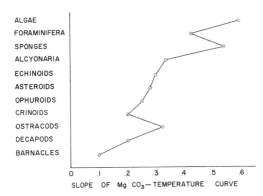

FIGURE 2.5 Magnesium concentration and phylogenetic level. (Data from Chave.) Total Mg and slope of temperature—Mg curve decrease to left; that is, towards the higher phylogenetic levels.

carbonate than living forms. What is the explanation?

More than one answer is possible. (a) The measured difference between living and fossil forms is a proven fact. From this it follows that in the geologic past a significant difference in magnesium carbonate uptake existed in crinoid metabolism. (*b*) Since the solubility of magnesium carbonate in skeletal material is high,

the difference found may be accounted for by solution. That is, the probable 10 percent magnesium carbonate in the living crinoids now represented by fossils was reduced to one-third that amount by subsequent solution effects. (*c*) The measured difference is an artifact of the less accurate methods of chemical analysis used in the past.

This problem has yet to be resolved conclusively, although Chave's researches have shed fresh light on the enigma. However, it does highlight the contained information yet to be released by cracking the atomic and mineral codes in fossil skeletal remains. Aside from the factual data derived, a student *discovers important questions to raise.* These have meaning of ever-increasing magnitude. The apparently unrelated questions raised above possibly bear on the dolomite problem vigorously pursued by sedimentologists, on the chemical history of the ocean (a vital concern to chemical oceanographers), and on the physiological evolution of echinoderms of interest to both physiologists and evolutionists.

Biochemical Code, Genetics, and Fossils

The biochemical code is written in living animals in (*a*) the organic components of skeletal material, the amino acid components of proteins (Figure 2.6), the organic compounds in tissues, body organs, blood, body fluids, and in (*b*) the genetic code—the nucleic acids, DNA and RNA (Figure 2.7).

In fossilized skeletal material it is sometimes possible to recover a portion of the biochemical code formed by the skeleton's organic compounds (chitin, pectin, cellulose, and protein). Or, one may study the amino acids in fossils' denatured proteins. Most frequent, and by far the most important, *the genetic code in a partial sense can be inferred by the determination of species differences in fossilized remains.* Each of these aspects will be considered briefly.

(a) *Organic Components of the Skeleton*

1. *Chitin.* The most frequently reported organic material in fossilized skeletons is chitin, a complex sugar $(C_{32}H_{54}N_4O_{21})_x$. It is characteristic of the arthropod skeleton (particularly in carapaces and valves), insect wings, and exuviae (molts). (See Chapters 11 and 15.) Chitin also occurs in bryozoans, brachiopods, mollusks, and hydrozoans; bristles and setae of annelid worms are chitinous.

Upon demise, the chitinous parts may remain unaltered (a rare occurrence) or experience varied changes. In the case of an Eocene coleopteran insect or that of Mesozoic insects preserved in amber, the chitin was retained unaltered (see Vallentyne, 1963). Or chitin may be attacked by bacteria and thereby decompose; or it may be carbonized by loss of the non-carbon components, especially under swamp conditions; or be decomposed by digestive enzymes such as are found in snails. In many forms during life, chitin is *sclerotized* (hardened) by either organic compounds or calcium salts. In protozoans a mucous-like substance, tectin, which is actually pseudochitin, is secreted.

This substance should not be confused with true chitin.

2. *Pectin, Cellulose, and so on.* These complex organic compounds are found in plants. In Chapter 3 it will be shown that many algae, such as the chrysophytes, have pectic substances in their walls and that the skeletal plates of dinoflagellates are cellulosic.

3. *Protein.* As supporting, covering, or foundation tissues, proteins occur in many invertebrate skeletons (Porifera, Coelenterata, Mollusca, Brachiopoda, Vermes, Echinodermata, among others). Spongin in sponges, and conchiolin in mollusks serve in this way. Proteinaceous tissues in mollusks subsequently are mineralized, generally with calcite or aragonite, and are admixed with magnesium carbonate to form the invertebrate skeleton.

Data on original skeletal proteins may be obtained from either younger or older fossilized invertebrate remains. Modern chromatographic techniques permit detection of the amino acid suite that is still present. In more than 100 species of mollusks comparative biochemistry of amino acid patterns showed that these patterns may prove useful both in classification and in clarification of evolutionary sequences (Hare and Abelson). Radiocarbon dating of the carbon in molluscan conchiolin has been effected for shells some 8000 years old (Berger et al.). From the Kansas Pleistocene, Byron Leonard and co-workers have studied the organic shell content of fossil land snails and have found some changes through time.

The stratigraphic utility of shell-protein-nitrogen (SPN) was tested. Mollusks were collected from late Cenozoic outcrops (Pliocene to Recent) in the central and southern High Plains region extending from Nebraska to Texas. The non-marine fossil molluscan shells were first decalcified and the water-insoluble shell protein or peptides subsequently isolated. Determination of the SPN in micrograms of nitrogen per gram of shell indicated that since from Pliocene to

FIGURE 2.6 The twenty amino acids. (After amino acid assembly in a chain, modification of cysteine, proline and lysine respectively yield the three amino acids below broken line.) (From M. F. Perutz. 1962. *Proteins and Nucleic Acids.* Elsevier Publishing Co. By permission of the author.)

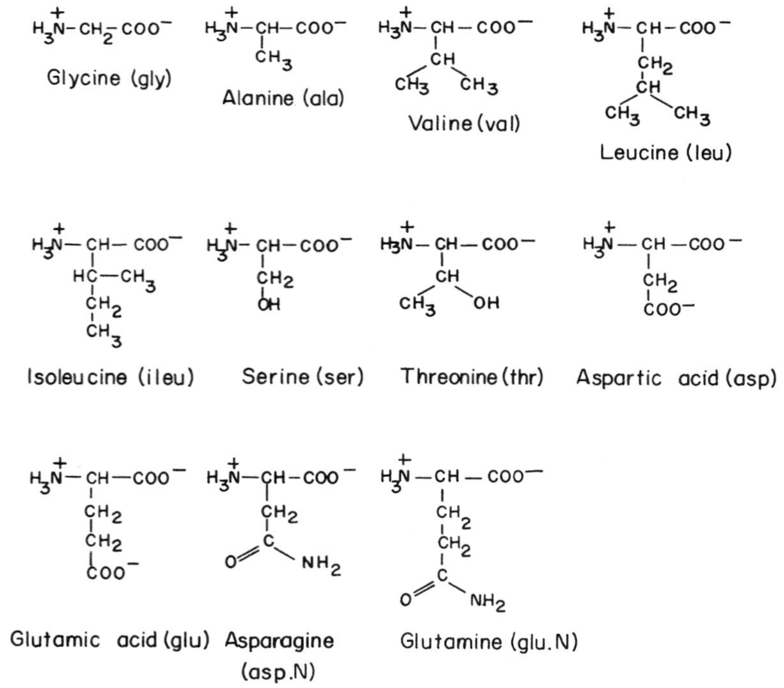

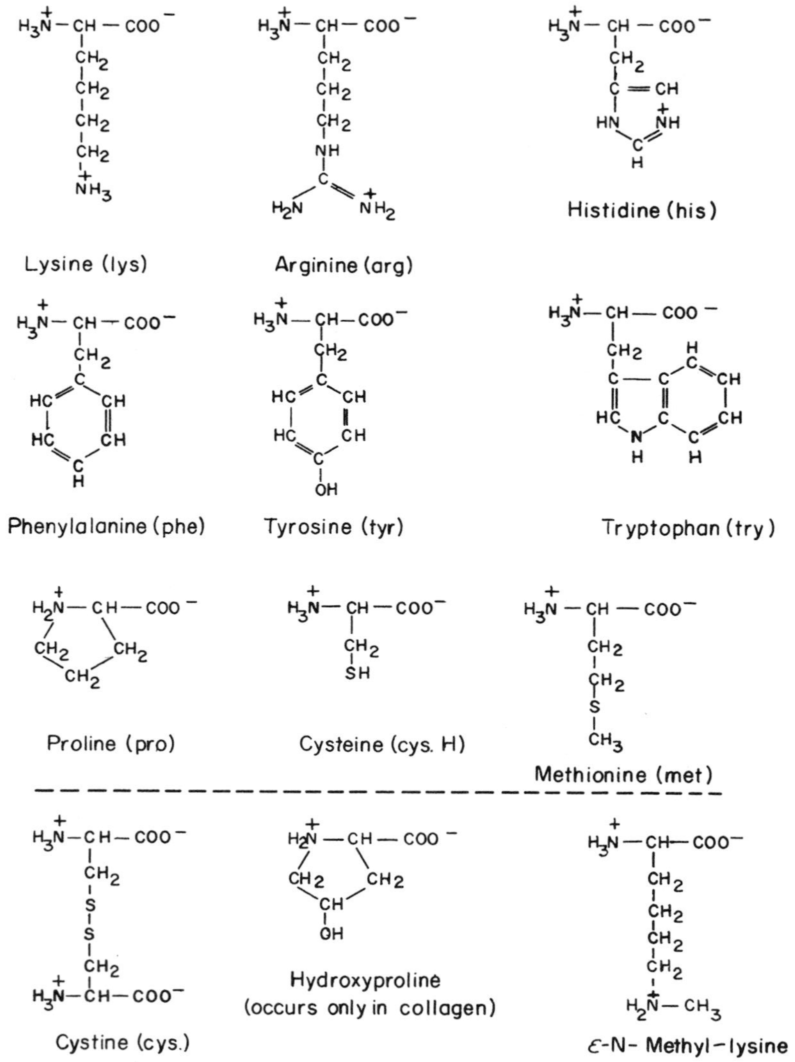

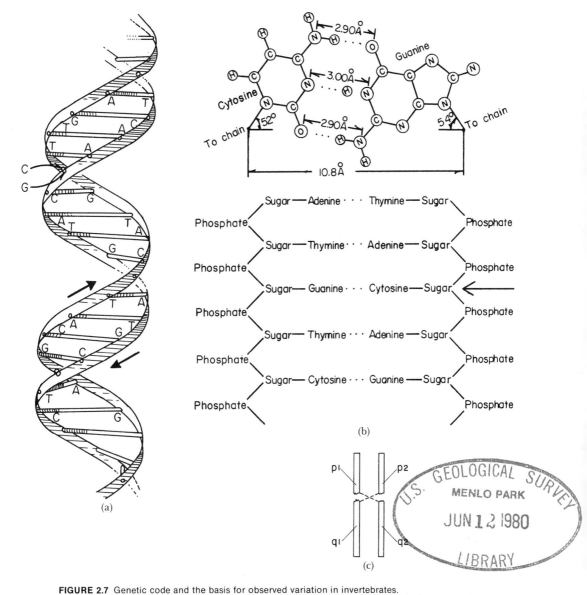

FIGURE 2.7 Genetic code and the basis for observed variation in invertebrates.
(*A*) Schematic representation of DNA molecule. Two threads arranged in double helix. Letters along each thread are bases (Adenine, A; Cytosine, C; Guanine, G; Thymine, T). Rule requires A attach only to T and C to G.
(*B*) Besides these four bases, a nucleotide has both a phosphate group and a sugar. Portion of molecule shown; distances in angstrom units.
(*C*) Recombination of parental genes. Portions of two chromosomes are shown; each with a pair of genes; respectively $p1q1$ (maternal) and $p2q2$ (paternal). Jagged ends on each chromosome denote a chromosomal break. Because of it genes can recombine $p1q2$ etc.
(*A*. From The Nature of Life by C. H. Waddington. Copyright © 1961 by George Allen & Unwin, Ltd. Reprinted by permission of Atheneum Publishers. *B*. After M. F. Perutz, by permission, see figure 2.6. *C*. After several authors.)

Recent these values increased, they were useful stratigraphic indicators for regional correlation (Ho Tong-Yan). Lowenstam has reported amino acids (a high glycine content) in the decalcified shell of a Pennsylvanian nautiloid; originally it had been composed of aragonite.

The mere presence of amino acids in fossils, according to Abelson, can serve to set an *upper* temperature limit to which the fossils have been exposed through time. In addition, applications lead to taxonomy, phylogeny, and stratigraphy (and correlations are possible). (See Table 2.2.)

cally on the molecular level. The step-by-step procedure by which this was achieved constitutes the most exciting research area in modern genetics. Simpson stressed

TABLE 2.2. Amino Acid Content of Some Tertiary Mollusks (After Abelson)

Specimen	Description	Mineralogy	Amino Acid[a] Content ($\mu M/g$)
Glycymeris parilis	Clam	Aragonite	0.80
Mercenaria mercenaria	Clam	Aragonite	0.75
Melosia staminea	Clam	Aragonite	0.90
Lyropecten madisonius	Scallop	Calcite	1.10
Eucrassatella melina	Bivalve	Aragonite	0.60
Turritella variabilis	Snail	Aragonite	0.40
Ecphora tricostata	Snail	Inner aragonite, outer calcite	1.20
Turritella indenta	Snail	Aragonite	0.50

[a]Amino acids include alanine, glycine, isoleucine, valine, glutamic acid, aspartic acid. (Fossils from Calvert Formation, Maryland; age 25×10^6 years.) $\mu M/g$ — micromoles per gram; $1 \, \mu M/g = 1$ part in 10,000.

(b) *The Genetic Code.* Only recently has more profound insight been obtained into the molecular basis for variation in animals. This clarity is critical for the study of fossils.

A single cell contains in its nuclear material *chromosomes* whose ultimate unit is the *gene*. The gene has a length of a fraction of a millimicron ($1 \, \mu = 0.001 \, \text{mm}$).

Proteins and DNA compose the chromosome. All the genetic information from both parents is transmitted to the offspring by a biochemical code (Figure 2.7). Variation within and between populations can be accounted for either by mutations or by recombination of genes during sexual mating. Recombination involves positional shifts of chemical components, called nucleotides, along the strands of DNA. At present, recombination in higher forms is thought to be quantitatively more important than mutations in speciation and evolution (Mayr, 1963, ref. Ch 1).

Every fossil either discussed in this book or encountered in laboratory or field must be studied to completion in the following terms.

1. Every characteristic, such as size, shape, structure, biochemistry, of the given fossil in life was determined geneti-

natural selection operating in populations, not individuals, as "the composer of the genetic message." This is to say that natural selection does not write the genetic code or message; rather, it selects out inadaptive portions of that code. Thus despite the primacy of mutations and recombinations in providing new genetic messages, natural selection always has the last word.

2. Even if the given fossil be a minute fragment, it represents an individual. Each such individual animal was actually one among a population of individuals. Populations numbered in the millions. (Exceptions are successive molts of the *same* individual among any of the arthropods, or colonial forms—a single coral or bryozoan consisting of many individuals.)

3. All such individuals taken together at any given time (that is, the population) consisting of larvae, juveniles, adolescents, and adults of both sexes, make up the living representatives of a given species.

4. Biochemical distinctness: Molecular biologists have performed thought-provoking experiments. They have tried to hybridize single DNA strand components (gene constituents) taken from verte-

brates differing as much as rat, fish, and man. The degree of recombination (that is, natural formation of double strands) was found to vary with the evolutionary distance between the respective animals. The method used is thought to be useful for "quantitative assessment of genetic relatedness of higher taxonomic categories" (order, class). The investigators speculated on whether there exists a persistent "particular class of nucleotide sequences" governing such vertebrate characteristics as bilateral symmetry, establishment of a notochord (precursor of spinal chord), and the presence of hemoglobin (Hoyer et al., 1964).

If such surmises prove to be correct, they will apply as well to the fossil record of vertebrates and invertebrates. The following discussion will be confined to the species level, although by inference from studies cited, one would recognize the possible applicability to higher categories [genus (many species), family (many genera), order (many families), class (many orders), phylum (many classes)].

The species level. The species is the basic unit of evolutionary change. Members of a given species population share countless characteristics in common. The morphological distinctness from all others of a given fossil species is thought to represent recombinations and/or mutations that affected limited nucleotide sites on the DNA strands of the parent population gene complex (see Figure 2.7). For example, if the resultant of recombination and/or mutation is the rearrangement of a single amino acid, this could change the character of blood hemoglobin. Other resultants of genetic modification could account for the several variations observed in skeletal anatomy.

FOSSILS AND BIOGEOCHEMICAL INFORMATION

Once the composition of a given fossil is known (mineral code) and part of the genetic information it contains is deciphered (that is, identified as to species, genus, and so on), then it is possible to obtain derivative data and useful biogeo-

TABLE 2.3. Major Kinds of Derivative Data of Ancient Biospheres Obtained from Fossil Evidence[a]

1. Geological Data
   a. Stratigraphy (correlation to establish contemporaneous time planes).
   b. Sedimentation (nature of substrates).
   c. Historical geology (the succession of biospheres).
   d. Geochronology (the position in time of a given biosphere).

2. Biological Data
   a. Evolution (speciation, genetics).
   b. Comparative anatomy (comparable structures in related and unrelated animals).
   c. Biology (marine, fresh-water, and terrestrial animals).
   d. Paleopathology (parasitism, shell repair, dwarfism, and so on).

3. Ecological Data
   a. Oceanography and limnology (physical characteristics of the aqueous environment).
   b. Climatology (paleoclimates).
   c. Zoo- and Phytogeography (lateral spread of biotas on a given time plane, see 1a above).
   d. Ecology (total interrelated network of contemporary biotas and the environmental factors affecting them).

4. Chemical and Biogeochemical Data
   a Organic geochemistry (elemental composition of invertebrates and their end products).
   b. Cycles of carbon, phosphorus, and so on.
   c. Diagenetic effects (postdepositional alteration of fossils).
   d. Trace element accumulation in skeletal remains.

[a]Cosmological Data (lunar month; earth's rotation, see pages 904, 905).[6] Geophysical Data (continental drift; sea floor spreading). Items 5 and 6 are added to indicate a vast additional potential.

chemical information. Table 2.3 lists some kinds of information that may be retrieved.

The essential for all derivative data is to obtain fossils from defined horizons in the rock column. (See appendix II.)

Program for Data Retrieval from Fossils

The program for obtaining data often varies in details, even in the order in which the items are considered. However, essentially it will follow the schedule indicated in Table 2.4.

Consider an example where repre-

sentatives of major groups of invertebrate fossils were completely absent. The Institute for Polar Studies had had a team of geologists exploring the Ohio Mountain Range, among others, in Antarctica for the past several years. From a sequence of rocks in this range only a few fossils were obtained (Figure 2.8). By study of even these fossils it was possible to retrieve part of the encoded data. The problem to be solved by study of these few fossils was the determination of the geologic age of the whole formation from which they were collected (Table 2.3).

TABLE 2.4. Abbreviated Schedule for Recovery of Information from Invertebrate Fossils

---

1. Preliminary field and laboratory work (see Appendix II and text throughout).

2. Megascopic and microscopic examination of all fossils.
    A. Segregation into groups of similar kinds of fossils. Obvious similarities are used as criteria at this stage. Gross sorting of fossils separates different phyla.
    B. Characterization of each group (apparent genera).
        i. Critical measurements (length, width, diameter, and so on).
        ii. Outstanding features (hinge line, ribs, trilobation, and so on).
    C. Further segregation using data above from B-i and B-ii.
    D. Diagnostic description of each isolate from C above (apparent species).
    E. Population analysis of each apparent species (statistical and graphic study of data from B-i; distinguish larval and juvenile from adult stages, males from females where possible).
    F. Review of similar forms described in the literature.
    G. Final designation of generic and specific names, possible assignment to a family.

3. Determination of known geologic range (time), geographic spread (space) of a given species.

4. Further detailed comparison of horizontal and vertical distribution of a given species with species selected from other formations.
    A. Stratigraphic correlation (stratigraphic paleontology).
    B. Paleogeographic reconstruction.

5. Paleoecologic analysis.
    A. Biotic associates of a given species in the rock sample being studied, or the known plant and animal associates at other localities (inferred food chain).
    B. Mineralogical and biochemical characteristics of fossils compared to the rock sample in which they occur.
    C. Population density and distribution of the given species in the rock samples under study.
    D. Evidence of transport, bottom reworking, life habitat, and so on.
    E. Ecological inferences (paleoecology). Nature of burial site (terrestrial, fresh-water, marine, or brackish); depth of water, salinity, eH, pH, and so on. (From data obtained in 5A–5D above.)

6. Summation of all data (items 1–5 above), with applications to larger problems.[a]
    A. Regional and intercontinental stratigraphic correlations.
    B. Evolution.
    C. Rate of sedimentation at burial site.
    D. Faunal migrations into or out of given area.
    E. Paleoclimatology.
    F. Dolomite and chert — origin, primary/secondary.

---

[a]Illustrative items are far from exhaustive.

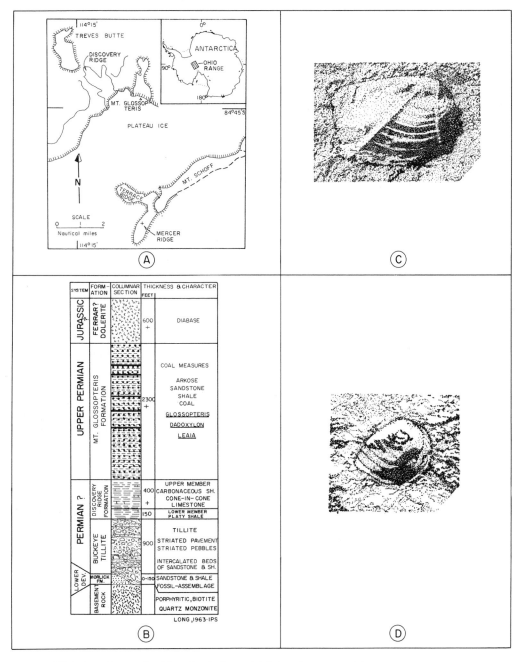

FIGURE 2.8 Data retrieval from fossils (essential data and one example).
(A) Sketch map of Ohio Range, Antarctica. The sign "X" indicates where fossil conchostracans and *Glossopteris* leaves were collected from west face, Mercer Ridge.
(B) Stratigraphic section, Ohio Range. Several species of *Glossopteris* flora also found in New South Wales.
(C)–(D) Permian conchostracans: (C) *Leaia gondwanella*, × 5; (D) *Cyzicus (Lioestheria) doumanii*.

Procedure. The schedule given follows Table 2.4.

1. Examination of all fossils (Table 2.4, Item 2—hereafter to be referred to as 2.4, 2). The fossils were found to be valves assignable to the crustacean branchiopods (order Conchostraca). (See Chapter 11.)

2. Characteristics of each kind of fossil.

Two genera were recognized (2.4, 2B). These will be referred to as genus A and genus B (2.4, 2B).

3. Determination of apparent species (2.4, 2D). One species each in genus A and genus B was recognized.

4. Paleoecologic analysis (2.4, 5A–5E).

(a) Biotic associates. Associated with the conchostracan valves on the same bedding plane were carbonized wood, *Glossopteris* leaves, and undetermined objects.

(b) Lithology and related items. The valves occurred on a black-gray, calcareous shale that had weathered white. Some of the valves of genus A were carbonized, as were some *Glossopteris* leaves.

(c) Population density (*n* = number of individuals). Genus A, species 1, *n* = 28. Genus B, species 1, *n* = 3. Distribution of valves on the slabs indicated a sparse number of individuals spottily distributed.

(d) Habitat. Conchostracan fossils described from other localities and as parts of the rock column are known to have inhabited inland water bodies grading in size from small lakes to temporary puddles. The Antarctic species are similar to these. Furthermore, the spotty, lateral, restricted, horizontal distribution suggests that the Antarctic conchostracans lived in temporary ponds or pools. Accordingly, the depth of water appears to have been very shallow.

The occurrence of carbonized valves, wood, and leaves suggests swamp conditions. The bottom sediments, as indicated by the lithology, were clay muds (2.4, 5D–5E).

(e) Other observations. One valve, fossilized eggs were attached; several valves were fragmentary (2.4, 5D–5E).

(f) Correlations. Since no other fossils of the type described are known from Antarctica, the only resource was to review similar forms from other continents (2.4, 4A). This comparison revealed that the leaiid conchostracan zone in Antarctica had equivalents in the Gondwana area (southern hemisphere). These related fossils occurred in South Africa, Australia, and Brazil and were associated again with similar species of *Glossopteris* leaves.

(g) Geologic age. The leaiid beds on other continents of the southern hemisphere had been dated—all of Middle-Upper Permian age. Since no other leaiid beds (of younger geologic age) were known[1] in the southern hemisphere, the conclusion was reached that the age of the Antarctic leaiids was also Middle-Upper Permian (2.4,3).

(h) Rate of sedimentation. This was determined in two different ways. First, one measured the smallest sediment interval separating any two successive conchostracan populations on the same slab—0.5 mm/year. The second method employed was band count along edge of the slab. Bands appeared to be regular and to represent annual deposition—0.68 mm/year. The closeness of the two independent determinations added confidence that the figure for the sedimentation annual rate was about 0.68 mm/year (2.4, 6C).

(i) Duration of the ponds. Conchostracans add a growth band to their valves approximately every three days, at which time they undergo ecdysis (molting). (See Chapter 11). Bands can be counted on valves. Such counts on the fossil valves being described indicate a life span of 1 to 1½ months for the given conchostracans; hence ponds they inhabited lasted a similar length of time. This determination can be verified independently by study of living conchostracans inhabiting temporary ponds. Such ponds, either drying up seasonally or even several times during a given season, are generally short-lived (2.4, 5E).

---

[1]Carboniferous forms from West and East Australia I have since described, do not affect the Antarctic age determination.

(j) Paleoclimatology. One can infer variable weather cycles from the evidence on the limited water bodies of short duration. Disseminated carbonized wood in the fossiliferous slabs is witness to rise and fall of a swamp water cover (2.4, 6E).

(k) Applications to still larger problems. Findings discussed here bear on the question of continental drift. This concept envisions a geologic past time when the southern-hemisphere continents were closer together or adjacent to each other. Antarctic evidence suggests that there could have been continuous swamp land during Middle to Upper Permian time, that it may have extended from Brazil across the Ohio Mountain Range area of Antarctica as well as from South Africa to New South Wales, Australia.

During Permian time transport of conchostracan eggs over long spans of salt water was not likely. These eggs will not hatch in saline waters — a conclusion verified by experiment. Furthermore, before birds, there was not a transporting agent (over oceanic distances) for eggs contained in dried muds.

The example cited merely shows possible application of purely paleobiological data to larger problems confronting geologists (2.3, 6A, 6D). Countless marine fossil examples are available (Appendix I).

(See: P. Tasch, 1971, in L. O. Quam, Ed., Research in the Antarctic. AAAS Publ. no. 93, pp. 703–716.)

QUESTIONS

1. Figure 2.5 lacks a geologic time scale. Can you provide one? Discuss.

2. If someone were to show you two identical fossils from two different localities (trilobites, for example) and to state that they are both of Lower Cambrian age, what derivative data are implied in this statement? (See Table 2.3.)

3. Outline the chain of reasoning of inferable biochemical distinctness in the genetic code from a study of a given fossil suite. (See Figure 2.8.)

4. (a) What did Simpson mean when he stated that natural selection composes the genetic message?

   (b) How can the operation of natural selection be inferred from a collection of fossil invertebrates?

5. Discuss retrieval of biochemical data from fossil material and/or sediments by recent investigators; indicate the kinds and significance of data retrieved.

6. E. Rosenberg [Science, 1964, Vol. 146 (3652), p. 1680] provided a quantitative assay of nucleic acid residues in sediments from Experimental Mohole. What can you conclude as to the source of these residues? (Consult Rittenberg et al., 1963, J. Sedimentary Petrology, Vol. 33, p. 140

7. (a) Identify "storage," "noise," "redundancy," "message," "scanning," and "irretrievable loss of data" in the fossil record; explain their significance relative to data retrieval.

   (b) Given a Devonian fossil brachiopod, referring to Figures 2.2 and 2.3, show how to maximize data retrieval.

   (c) Explain the operation of the delayed feedback loop in data retrieval from fossil invertebrates.

REFERENCES

Abelson, P. H., 1957. Organic constituents of fossils, in GSA Mem. 67, Vol. 2, pp. 87–92. (See Goldberg ref. below for complete title.)

————, 1963. Geochemistry of amino acids, in I. A. Breger, ed., Organic Geochemistry, Macmillan, pp. 431–455.

Berger, R., A. G. Herny, and W. F. Libby, 1964. Radiocarbon dating of bone and shell from their organic components: Science, Vol. 144, pp. 999–1001.

Brillouin, L., 1949. Life, thermodynamics, and cybernetics: Am. Scientist, Vol. 37 (4), pp. 554–568.

Chave, K. E., 1954. Aspects of the biogeochemistry of Magnesium. I. Calcareous marine organisms: J. Geology, Vol. 62 (3), pp. 266–283. II. Calcareous sediments and rocks: J. Geology, Vol. 62 (6), pp. 587–599.

Doumani, G. A., and P. Tasch, 1963. Leaiid conchostracan zone in Antarctica and its Gondwana equivalents: Science, Vol. 142 (3592), pp. 591–592.

Elsasser, W. M., 1958. The physical foundation of biology. Pergamon Press. 219 pp.

Goldberg, E. D., 1957. Biogeochemistry of trace metals, in Treatise on Marine Ecology and Paleoecology, GSA Mem. 67, Vol. 1, Chapter 12.

Hare, P. E., and P. H. Abelson, 1964. Comparative biochemistry of amino acids in molluscan shell structures: GSA Program, 1964 (Miami Beach), Abstracts, p. 84.

Ho Tong-Yan, 1964. Stratigraphic study on water-insoluble fraction of shell proteins in Late Cenozoic non-marine mollusks from central and southern High Plains (from central-southern Nebraska to northwestern Texas): GSA Program, 1964 (Miami Beach), Abstracts, p. 92.

Hoyer, B. H., B. J. McCarthy, and E. T. Bolton, 1964. A molecular approach in the systematics of higher organisms: Science, Vol. 144 (3621), pp. 959–967.

Perutz, M. F., 1962. Proteins and nucleic acids: structure and function: Eighth Weizmann Mem. Lect. Series, 1961. Elsevier. 181 pp.

Rankama, K., and T. G. Sahama, 1950. Geochemistry. Univ. of Chicago Press. Chapter 8, "Geochemistry of the Biosphere."

Rosenberg, E., 1964. Purine and Pyrimidines in sediments from the Experimental Mohole: Science, Vol. 146 (3652), pp. 1680–1681.

Schrödinger, Erwin, 1946. What is Life? Cambridge Univ. Press. 91 pp.

Simpson, G. G., 1964. Organisms and molecules in evolution: Science, Vol. 146 (3651), pp. 1535–1538.

Tasch, P., 1965. Communications theory and the fossil record of invertebrates: Trans. Kansas Acad. Sci., Vol. 68 (2), pp. 322–329.

Vallentyne, J. R., 1963. Geochemistry of carbohydrate, in I. A. Breger, ed., Organic Geochemistry, Macmillan, pp. 456–502.

Vinogradov, A. P., 1935–1944. The elementary chemical composition of marine organisms: Memoir Sears Foundation for Marine Research, No. 2, Yale Univ., Chapter 14.

Waddington, C. H., 1962. The nature of life. Atheneum. 125 pp. Chapter 2.

Yasushi, K., and D. W. Hood, 1961. Effect of organic material on the polymorphic forms of $CaCO_3$: GSA Program, 1961 (Cincinnati), Abstracts, p. 86a.

For further references, see Chapter 15.

(Wyckoff, R. W. G., 1972. The biochemistry of animal fossils. Williams and Wilkins Co., particularly Chapters 4 and 5, the proteinaceous residues, carbohydrates and lipids in fossils.)

chapter three

ORIGIN OF LIFE TO EMERGENCE OF THE UNICELLS

A. INTRODUCTION AND CHRYSOPHYTA

Consideration of the origin of life makes more plausible the first appearance of unicells in the Pre-Cambrian. Moreover, this chapter fills in the oldest prehistory of invertebrates.

The Origin of Terrestrial Life

Miller's Experiment. In 1952 Urey suggested that the primitive (original) earth atmosphere had the following composition: hydrogen (H_2), methane (CH_4), ammonia (NH_3), and water (H_2O). A year later Stanley L. Miller, a biochemist, set up a simple series of experiments to test Urey's thesis. Miller wanted to find out whether the components of Urey's atmosphere could yield amino acids under an electrical discharge (Figure 3.1). The result was positive. Simple amino and other acids including glycine and alanine were formed.

Abelson's Experiment. A few years later (1955), Rubey, reasoning from geochemical considerations, proposed a primitive atmosphere composed of carbon dioxide (CO_2), nitrogen (N_2), carbon monoxide (CO), and water (H_2O). The following year Abelson tested Rubey's atmosphere. Would it also yield amino acids? Again positive results ensued. If hydrogen were present, amino acids would be formed by the action of a spark discharge.

Both sets of data, the Urey/Miller and the Rubey/Abelson, indicated (*a*) that organic compounds such as the amino acids, the building blocks of proteins, could have formed in the primitive earth atmosphere; (*b*) that the atmosphere was reducing (presence of hydrogen); (*c*) that natural ultraviolet radiation could have been the source of energy instead of

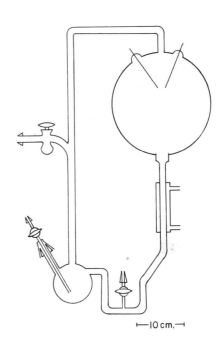

├─10 cm.─┤

FIGURE 3.1 Miller's experiment (see text for details). Spark discharge apparatus. Mixture of water and gas (small flask) is heated to boil to bring water in region of spark chamber (right, top). The volatile products condense and flow through *U* tube and are collected at spigot. Nonvolatiles accumulate in small flask.

electrical discharges. (This has been confirmed in laboratory experiments.)

These and subsequent findings reviewed at an international symposium held at Moscow (1957) and elsewhere have confirmed the pioneer work of the Russian biochemist A. I. Oparin. He had postulated that life on our planet could be traced back to the initial steps of the synthesis of amino acids and other organic compounds from simple, inorganic substances found in the primitive atmosphere and brought into reaction by natural radiation. Subsequently, Oparin reasoned, these compounds accumulated in the oceans and concentrated in coacervate drops (see Figure 3.2 for definitions and steps).

However, once organic compounds formed, there was a vast distance to cover before a single cell (a monerid) consisting of some 200 million million molecules would become possible. Further researches have led to a clearer explication of the steps involved (Bernal, 1961, 1968; Figure 3.2*A*, *B*, compare Fig. 3.2, *C*.)

From Amino Acids to a Living Cell

Let us begin the process where Miller's experiment left off. We have a suite of amino acids and other organic compounds formed by natural radiation. They had concentrated in the ocean. It is not necessary to assume equal concentrations everywhere in the ocean. In fact, recently local concentrations seemed more likely. Bernal envisions that these amino acids and other compounds were adsorbed on estuarine clays. (Estuaries are drowned river mouths.) If so, that would restrict such accumulations to a limited number of near-shore environments. Sidney Fox and co-workers have shown that amino acids could have been synthesized by heat alone. These investigators produced chains of amino acids when amino acids were placed on a dry lava bed (120°C) and simulated tides and rainwater were passed over the bed. When the water cooled, spherical particles (amino acid chains) were retrieved.

The amino acids and other compounds adsorbed on such clays would form colloidal solutions. Each colloidal particle would be surrounded by water molecules; only the inner layer gradually formed an oriented water shell around it. That would be the first step in the formation of a *coacervate drop* (Figure 3.2*A*, steps 1–3). Many such drops of all sizes would form and enlarge. Such coacervate drops have the interesting property of concentrating all the colloidal particles from the surrounding liquid. This liquid is called the equilibrium fluid to distinguish it from the drops.

Suppose that amino acids or nucleoproteins are in the original colloidal solution. Experiments by Oparin, his coworkers, and Evreinova indicate that amino acids would be concentrated in the coacervate drops when they formed (Figure 3.2*B*, step 1).

A chemical interchange between colloids inside coacervate drops and equilibrium fluid has been observed. Thus, Bernal postulates, the *original* organisms (named *eobionts* by Pirie) must have come into existence at this time, that is, when large peptide molecules concentrated inside coacervate drops and chemically reacted with the environing fluid. Present knowledge refers to the primitive non-nucleated unicell as a *prokaryote*—the oldest fossil of which is three billion years old.

Subsequently ribose nucleic acid (RNA) accumulated inside coacervate drops, forming nucleoprotein *organelles*. At this stage, according to theory, photosynthesis was possible and there was also *autoduplication* of nucleoproteins, that is, the making of one compound from the template of another as in the manufacturing of machine parts (Figure 3.2*B*, step 2). Thereafter the oxygen of the atmosphere occurred as a result of photosynthesis.

Stage 3 came into being when organelles, originally having formed inside coacervate drops from nucleoproteins, developed a lipid cover. *Lipids* are fats or related substances. DNA was formed and

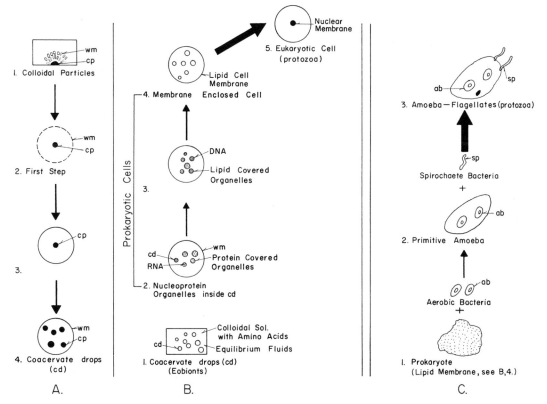

FIGURE 3.2 Simplified outline of postulated stages leading to a living unicell starting with amino acids and so on. (After Bernal, 1961, experimental data by Oparin et al., 1959; Evreinova, 1959; and Oparin, *Origin of Life.* Procaryote and eucaryote terminology added on the basis of more recent work.) Terminology: coacervate—a special type of concentrated colloidal sol that concentrates colloidal particles in minute drops; nucleoproteins—proteins and nucleic acids (DNA, RNA); lipids—fats and related substances; eobiont—original organism (in modern terminology the progenitor of oldest procaryotes); organelle—any discrete structure in cell.

(*A*) Formation of coacervate drops from colloidal solutions (after Oparin). Steps 1–4: (1) Colloidal particles (*cp*) surrounded by water molecules (*wm*). (2) First step in formation of a coacervate drop from colloidal solution. (3) Coacervate drop with *cp*. (4) Coacervate drop contains all colloidal sol. Oriented water molecules comprise an outer shell; several colloidal packets are thus within this shell. (Bernal, 1968, pp. 126–127, is inclined to place coacervate drop participation in origin of life, *after* nucleoprotein evolution.)

(*B*) One postulate of stages leading to unicells (procaryotes and eucaryotes). Steps 1–5: (1) Coacervate drops (after experiment by Evreinova, 1959). Drops will concentrate (adsorb) large peptide molecules formed from amino acids in the equilibrium fluid (*eobiont stage*—see above definition). (2) Nucleoprotein organelles inside coacervate drop. (*Photosynthesis and autoduplication of nucleoprotein.*) (3) Lipid covered organelles inside coacervate drop *(DNA and autoduplication as in step 2).* (4) Membrane enclosed cell *(organism—a procaryote, lacking a nuclear membrane); fission and conjugation.* (5) Eucaryote cells—protists *(mitosis and sexual reproduction).*

(*C*) An alternate theory on the rise of eucaryotic cells. (Data from Lynn Margulus, 1970, *Origin of Eukaryotic Cells.* Yale Univ. Press.) Starting with a prokaryote cell [see *B*(4)]: (1) Incorporate aerobic bacteria (a.b.) (endosymbionts), shown on diagram by "+". (2) Result is a primitive amoeba.[1] (3) Incorporate spirochaete bacteria (s.p.) (ectosymbionts), shown on diagram by "+". (4) Result is a flagellate-amoeboid cell, that is, a protist (protozoa) as well as fungi (not shown). (See: T. Cavalier-Smith, 1975. Nature, Vol. 256, pp. 463–468, on the limitations of the Margulus model and an alternate solution of the rise of the eucaryotes. Also, Preston, Cloud, 1976, Paleobiology, Vol. 2 (9), pp. 351–387, for update on the presently acceptable fossil data on procaryotes and eucaryotes. R. M. Schwartz and M. O. Dayhoff, 1978, Science, Vol. 199, pp. 395–403 discuss origins of procaryotes, eucaryotes, etc. based on protein and nucleic acid sequence data, called "living fossils.")

also concentrated in the organelles. Furthermore, these DNA molecules were capable of autoduplication; recall that any gene is composed of DNA (Figure 3.2*B*, step 3).

When the shell, or layer of oriented

---

[1]According to Bernal, by this stage (a protozoan), there was a protein-covered cell wall.

water molecules about the coacervate drops, had been replaced by a more permanent, resistant lipid-based cell membrane, we may speak of the *beginning of unicells*. At this stage, reproduction occurred by fission and conjugation of nucleoproteins. Bear in mind that the DNA forming the gene complex had already been present (Figure 3.2*B*, step 4).

Finally, when the lipid-based cell membrane was succeeded by a protein-covered cell wall, the cell was *nucleated*. (Nucleation of a cell occurs when a membrane separates the hereditary mechanism of the cell—the nucleoproteins—from the cytoplasm of the cell.) At this stage a eucaryotic cell is thought to have appeared on the planet. Such unicells reproduce by mitosis (simple division) and by sexual reproduction, as will be discussed elsewhere in this section (Figure 3.2*B*, step 5). Unicells with organelles (nucleus, and so on) are called *eucaryotes*. The oldest known fossils of this type (chrysophytes, and so on) came from beds that were between 1.2 and 1.4 billion years old. Prior to nucleation there were more primitive unicells than protists on the planet (Kingdom Monera).

When did all of this biochemical evolution take place? It must have occurred three or more billion years before the present. Why is this so? The oldest known fossil procaryote presently accepted by specialists is an algal species represented by stromatolites from the Pongola System of N. Zululand, South Africa that lived 3.0 to 3.1 billion years ago. Other reported Archean microbiotas are questionable (Cloud, 1976). Since the age of the earth is some five billion years, we may envision a long biochemical evolution that may have worked its way to the procaryote stage during the first two billion years of earth's formation, and a billion plus years later, to the eucaryote stage. (See Figure 3.2 part C based on Margulus, 1970.) It has been proposed that in the older Pre-Cambrian, meteorite falls could have provided the necessary preliminaries for the biochemical

sequence outlined above (Anders *et al*, 1973).

Phylum Chrysophyta
Phylum Pyrrophyta

The Protista

Starting where Bernal ended his biochemical stages and from the origin of procaryotes and subsequently protist eucaryotes (Figure 3.2*B*), we may envision the rapid radiation of newly evolved monerids and of the latter protists to different ecologic niches than that afforded by the estuarine environment; perhaps from a volcanic arc environment, if Fox and co-workers are on the right track. We may further presuppose successive generations, numerous mutations, recombinations of genes (DNA nucleotide ordering; see Chapter 2, Figure 2.7) leading to speciation and the ultimate diversity of protists.

As discussed in Chapter 1, all plants and animals are classified into species, genera, families, and so on. The distinction between species reflects differences in the DNA nucleotide ordering in their gene complexes. With this in mind, we can consider the kingdom Protista. The kingdom is the highest division possible in taxonomy. [Whittaker, 1969, proposed a five-kingdom system: (1) Monera—procaryotic cells (blue-green algae, bacteria, and so on), (2) Protista (discussed below; protists evolved from the monerids) (3) Plantae, (4) Fungi, (5) Animalia.]

Kingdom Protista. Unicell or unicell-colonial organisms; eucaryotic cells, that is, with organelles (nucleus, and so on). Range: Pre-Cambrian to Recent.

Under this kingdom ten phyla are grouped. Several of these lack a fossil record. Of the remaining phyla those treated here include the Chrysophyta, the Pyrrophyta, and several phyla once grouped under the "Protozoa." These should suffice to introduce the student to the protists.

Abbreviated Classification of Selected Protists

Kingdom Protista

   Phylum Chrysophyta

      CLASS CHRYSOPHYCEAE

      CLASS COCCOLITHOPHORIDA

      CLASS SILICOFLAGELLATA

      CLASS DIATOMACEA

   Phylum Pyrrophyta

      CLASS PERIDINIEAE (= DINOPHYCEAE)

The Chrysophyta

Forms having a fossil record in this phylum include the chrysomonads (Figure 3.3*D*), the coccolithophorids (Figure 3.3*E*), the silicoflagellates (Figure 3.3*F*), and diatoms. Algae (diatoms) constitute the so-called pasture of the sea and are the primary producers in the marine food chain (Figure 3.3*G*). Of these, it seems appropriate to stress the first three classes named above (order Chrysomonadina).

As an aid in data retrieval from fossil chrysophytes, it is necessary to determine the essential coded data in living representatives.

Living Chrysophytes

General Characteristics. Chrysomonads, coccolithophorids, and silicoflagellates share several characteristics. They can all, as do plants, produce their own food by photosynthesis. They are all capable of self-locomotion, as are animals.

Photosynthesis and Food. Living forms of each of the above possess pigmented organelles called *chromatophores* which, during photosynthesis, produce a nutritive supply for the given protists. That food is stored as a carbohydrate; the small granule or lump, a few microns in diameter, is called *leucosin.*

Motility. The motility of chrysophytes is achieved by means of flagella (and pseudopodia, in addition, in silicoflagellates and other chrysomonads). Living chrysomonads may be sorted into four series by the differences in flagellation: single, two equal, two unequal, one short and two long flagella). Coccolithophorids are biflagellate (two equal flagella), and silicoflagellates possess a single flagellum.

Reproduction. In general, reproduction occurs by simple division (fission). Although sexual reproduction is known in some forms, it is rare.

Life Cycle. A motile (flagella) stage and cyst (or dormant) stage characterize the chrysomonads. The transition to a cyst from a flagellate stage involves the loss of flagella, rounding of the cell, formation of a gelatinous coat about the cell, and secretion of a siliceous cyst within the cytoplasm (Figure 3.3*D*).

Confusion about the life cycle of the coccolithophorids in the past led to the assignment of the motile stage to one genus and species (*Crystallolithus hyalinus*) and the cyst stage to another (*Coccolithus pelagicus*). It is now known that both stages characterize the life cycle of the same animal. A similar life cycle (motile and cyst) appears to characterize silicoflagellates also, although it is still imperfectly understood.

Size. Chrysophytes are small and range in size from 5 to 25 μ (microns). Table 3.1 shows that the smallest dinoflagellates and protozoans overlap the chrysophytes in size; chrysophytes are 10 to 50 times the size of a bacterium.

TABLE 3.1. Relative Protist Size in Bacterial Units (0.5 μ = unit)[a]

	Number of Units[b]
Chrysophytes	10–50[c]
Dinoflagellates	12–200
Protozoans	20–2000

[a]Average size of a bacterium. (One micron (μ) equals 10^{-3} mm.)

[b]To be read as follows: A given chrysophyte is 10 to 50 times larger than a bacterium.

[c]Diatoms, as distinct from the other chrysophytes, which are excluded from this figure, may range in size from 20 to 2000 units.

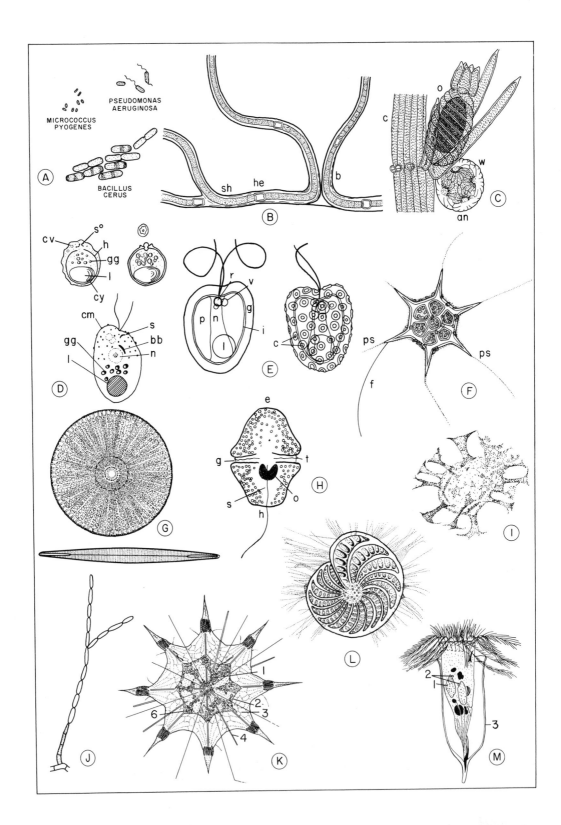

FIGURE 3.3 Variety of living monerids, protists, and so on. (A) Bacteria: (a) *Bacillus cercus*, × 1600; (b) *Micrococcus pyogenes*, var. *aureus*, × 2400; (c) *Pseudomonas aeruginosa*, greatly enlarged; note flagella. (B) Blue-green algae: *Scytomena*. (C) Grass-green algae: *Chara*. (D) Chrysophyceae. (E) Coccolithophorida: *Hymenomonas roseola*, (a) with skeletal elements (coccoliths) removed; (b) with skeletal elements (c). (F) Silicoflagellates: *Distephanus speculum*. (G) Diatoms: (a) *Arachnoidiscus*, (b) *Amphipleura*. (H) Dinoflagellates: *Glenodinium*. (I) Acritarchs (Hystrichospheres). (J) Fungi: *Fusisidium maritum*. (K) Radiolaria: *Acanthometra*. (L) Foraminifers: *Polystomella*. (M) Tintinnids: *Favela*. Illustrations from (A) several authors: (B, C, E, H) several authors; (D) after Bourrelly; (F) after Marshall; (G) adapted from photographs in Fuller and Tippo; (I) Tasch; (J) after Sutherland, cit. Johnson and Sparrow; (K,M) after Hyman and other authors; (L) after Myers.

Shape of Cysts. Cysts vary in shape from ovate to subelliptical and spherical. These configurations may be viewed as variations of a sphere. Loricae or shells of chrysomonads are rarely found as fossils. However, they are extremely varied in shape, ranging from cylinder- to vase-shaped objects. Bourrelly (1963) noted that forms like *Pseudokephyrion* have a calcareous lorica or shell, yet produce siliceous cysts.

Composition of Cysts or Skeletons. Lorica of chrysomonads (rarely fossilized) are cellulosic-pectic in composition and often have calcareous or siliceous impregnations. Cysts, on the other hand, consist of a pectic substance chemically bound to silica.

With recent biological material, one can use certain stains to detect the presence of cellulose, chitin, or pectin. With the stain Chlor-zinc-iodide cellulose turns blue; chitin, brown; and pectin, ocherous to yellow.

The envelope of a coccolithophorid cell consists of an inner membrane overlain by a gelatinous portion in which are embedded individual minute disks called coccoliths. In older individuals, disks become rigidly united when mucilage has calcified.

The older coccoliths gradually dislodged and shed, fall to the ocean bottom; new coccoliths form within the old (Figure 3.3E, part b). Coccolith size ranges from 2.0 to 30.0 μ.

Silicoflagellates have a skeleton; it is internal in early development and external in the adult. Essentially the skeleton is a latticework case of hollow, siliceous bars composed of opaline silica. This contrasts with siliceous radiolarians, which some silicoflagellates resemble, that have skeletons composed of solid bars of opaline silica.

Information Recovery

Ecology. The planktonic algae, the chrysomonads, and coccolithophorids occur in both fresh and marine waters; on the other hand, silicoflagellates are exclusively marine plankton.

Some species of coccolithophorids, such as *Coccolithus huxleyi*, are known to be responsive to changes in salinity and temperature. Thus this species (which, in cultures, shows good growth at salinities of 20 to 45 parts per thousand, and optimum growth at 20°C) is excluded from brackish waters (Braarud, 1961). A vertical size distribution of coccolithophorids at equatorial stations has been observed to have small forms abundant in the upper 50 meters. This distribution may be accounted for by the capacity of calcite coccoliths to reflect light. From this it follows that coccolithospheres would tend to the higher portions of the photic zone in the sea, whereas other protists, lacking this skeletal protection, would seek the lower (Isenberg et al., 1963, p. 63, Figure 13). From specific-gravity considerations, a coccolithophorid population (showing normal size distribution, in which all individuals seek the upper portions of the euphotic zone) would be expected to show the smaller individuals higher in that zone than the larger ones (Tasch, 1963).

Mandra (1958) plotted the data of an earlier worker that showed that the ratio of the respective abundance of each of two silicoflagellate genera (*Dictyocha* and *Distephanus*) indicates the temperature of the respective water (Table 3.2). Other studies show that silicoflagellate abundance is inversely related to diatom abundance in a given sector of the sea. Accordingly, an increase of diatoms would denote a lowered number of silicoflagellates. Whether this fluctuation in numbers is related to available silica in solution for which both diatoms and silicoflagellates may compete, has not been investigated.

TABLE 3.2. Ratio of the Relative Abundance of Two Silicoflagellate Genera (*Dictyocha* and *Distephanus*) an Indicator of Oceanic Temperature

Temperature (°C)	Abundance Ratio
0–5	$\dfrac{248^a \; D_1}{3043 \; D_2}$
5–10	$\dfrac{2088 \; D_1}{8083 \; D_2}$
10–15	$\dfrac{4853 \; D_1}{9615 \; D_2}$
15–20	$\dfrac{3367 \; D_1}{4786 \; D_2}$
20–25	$\dfrac{4678 \; D_1}{2551 \; D_2}$

[a]Number of individuals of the genus *Dictyocha* (D_1) over the number of individuals of *Distephanus* (D_2).
Source. After Mandra.

Data Lost in Fossilization. Of the three classes of chrysophytes, only the parts indicated below are fossilized.

1. Chrysomonads—entire cyst (but generally lacking plug, see Figure 3.3*D*).
2. Coccolithophorids—coccoliths.
3. Silicoflagellates—complete or broken siliceous skeleton.

From this type of fossil material, one must infer the life cycle: the motile stage, the existence of chromatophores, leucosin accumulation from photosynthesis, and secretion of the cyst wall. We need to envision the imbedded calcite coccoliths in the cell wall and hardening of the mucilage.

The data derived from living forms can also be inferred from fossil evidence, for example, a vertical size distribution of coccolithospheres in an ancient sea, or growth sponsored by suitable salinity and temperature gradients. One must infer the associated phytoplankton in the original living condition; for example, archeomonads, a family of marine chrysomonads, and/or silicoflagellates will be found fossilized with diatoms. In specific instances by counts of populations of respective species, one may infer paleotemperatures (see Table 3.2). Other derivative data from fossil chrysophytes include evolution through time.

Information in the Fossil Record

Archeomonads. The family Archeomonadae occurs in the fossil record as small, spherical, siliceous cysts. Four fossil genera are illustrated in Figure 3.4*A*. Fossil archeomonads have been recovered from diatomites, shales, marls, and silts by acid treatment. Cysts terminate in a *neck* pierced by a *pore*. In life this pore car-

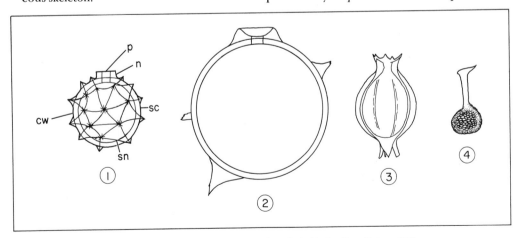

FIGURE 3.4A Fossil chrysophytes—I. Archeomonadae: Cret.; Tert.—Rec. (Schematic).
(1) *Archeomonas*, Mio. Md., *sc*-siliceous cyst; *p*-pore; *n*-neck; *cw*-cell wall; *sn*-spinous network.
(2) *Archeosphaeridium*, Mio. Md. (3) *Litharcheocystes*. 4) *Micrampulla*, U. Cret. (Figures redrawn and adapted from Tynan and from Deflandre.)

ried a cytoplasmic plug resembling a jar stopper that was not preserved.

Archeomonad cysts produce a varied *ornamentation*. Walls may have pointed spines, polygonal fields consisting of raised ridges, or other ornamentation. Some forms may lack any ornamentation. Equatorial region of cysts may or may not bear a flange (Figure 3.4*A*).

In any sample retrieved by acid treatment, archeomonads can be detected readily by the following criteria.

1. Small size (8–10 μ).
2. Siliceous wall.
3. Spherical-to-ovate configuration.
4. Termination in hollow neck (i.e., neck pierced by pore).
5. Shape and size of pore.
6. Ornamentation.

Modification in items 3 and 4 above (but sometimes items 5 and 6) distinguish one genus from another. Variations in items 5 and 6 can have specific value and thus help to isolate distinct species.

As presently known, archeomonads range from Upper Cretaceous to Recent. Archeomonads are abundant in the Miocene Calvert Formation of Maryland and the Cretaceous Moreno shale of California. They should be sought in Tertiary marine diatomites as well as in beds of Upper Mesozoic age. Van Landingham (1964) reported a chrysophyte assemblage from diatomites of the Miocene Yakima Basalt of south-central Washington.

Coccolithophorids. Fossil coccolithophorids are generally skeletal components rather than entire skeletons (Figure 3.4*B*). They constitute minute, variously shaped, calcareous disks that had once been imbedded in the investing membrane of a coccosphere. They may bear a central perforation or be imperforate, or even bear anterior or medial spinelike processes.

A variety of names have been assigned to such skeletal pieces (coccoliths): rhabdoliths (Figure 3.4*B*, part I), lopodoliths, pentaliths, and so on.

Size range of these objects is 3–15 μ.

However, some have been known to attain a size as large as 50 μ. Small size permits reworking after deposition of these pieces. Cretaceous coccoliths and discoasters (Figure 3.4*B*, part G) were eroded and redistributed widely before being intermixed with Tertiary deposits on the sea floor. An Oligocene marl bearing pelagic foraminifers, such as *Globorotalia*, contained Eocene discoasters. These associations are common. Hence the total microfauna of a given deposit needs study before one hazards an age determination.

Deflandre subdivided class Coccolithophorida into two orders on the basis of the microstructure of calcite in coccoliths and discoasters.

Class Coccolithophyceae
Order 1. Heliolithae (Range, Jurassic–Recent). The calcite of the skeletal parts shows radiating-fibrous structure. (All coccoliths are included in this order.)

This order comprises the family Syracosphaeacea, with 18 genera (se Figure 3.4*B* for examples); the family Coccolithidae, which contains the genus *Coccolithus* (Figure 3.4*B*, part C) and a few others; and the family Sphenolithacea, which includes the genus *Sphenolithus* (Figure 3.4*B*, part J).

Order 2. Ortholithae (Range, Cretaceous–Recent). The skeleton or larger skeletal parts behave as single crystals viewed in the petrographic microscope. Several families are included in this order; two common genera are *Braarudosphaera* (Figure 3.4*B*, part E) and *Discoaster* (Figure 3.4*B*, part G).

Scanning electron microscope (SEM) studies of coccoliths required a revised classification. For an illustrated key to genera (seen by SEM), see P. Reinhardt, 1972. Coccolithen, A. Zieman Verlag, Wittenberg, Lutherstadt. Compare modern SEM usage in Ellis, Lohman, and Wray, 1972, Colorado School of Mines Quarterly, vol. 67(3). (Upper Mio.-Rec., nannofossils, Gulf of Mexico).

A suite of coccoliths can be made readily available to any interested student.

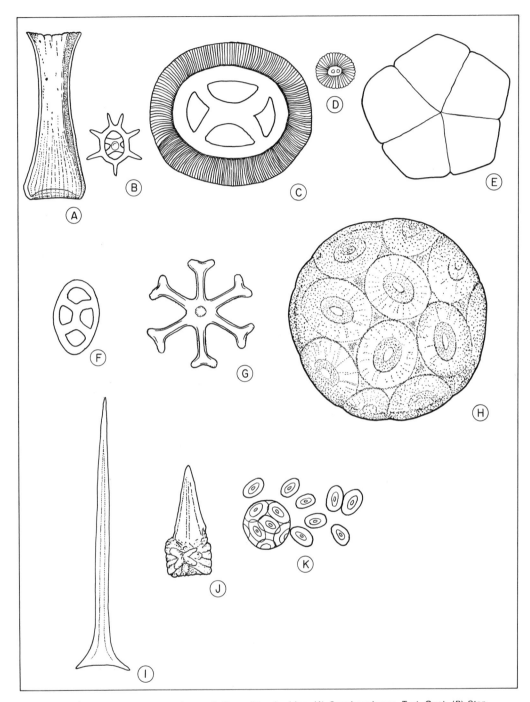

FIGURE 3.4B Fossil chrysophytes—II. Coccolithophoridae. (A) *Scyphosphaera*, Tert.-Quat. (B) *Stephanolithion*, Oxfordian. (C) *Coccolithus*, Jur.-Rec. (D) *Tremalithus*, U. Cret-Tert. (E) *Braarudosphaera*, Eoc. (F) *Neococcolithes*, Cret.-Tert. (G) *Discoaster*, Tert. (H) Complete coccolithophorid from Tertiary Chalk. (I) Rhabdolith, Tert. (J) *Sphenolithus*, Tert. (K) Microscopic appearance of a fine-textured carbonate (chalk) bearing coccoliths. (Illustrations all schematic; redrawn and adapted from Deflandre; Bramlette and Riedel; and Levine in, Isenberg *et al.*, 1963.)

Method: Take a piece of ordinary white blackboard chalk, pulverize it, mix it with distilled water in a test tube, and let it stand for about 20 minutes. A temporary wet mount may be prepared by inserting an eyedropper into the test tube and by discarding the first four or five drops; then one or two drops are put on a glass slide, and a cover slip is placed over it. At magnifications of X400–X500, numerous coccoliths and other microforms will be seen.

Information Derived from Coccoliths and Discoasters.

Mohole drill cores (consisting of gray-green clays) from the vicinity of Guadalupe Island contained numerous coccoliths and discoasters in association with diatoms, foraminifers, radiolarians, and silicoflagellates. Four biostratigraphic units were differentiated on the basis of

the calcareous nannoplankton; they ranged from Middle Miocene to Lower Pliocene (Martin and Bramlette).

The variety, density, and size of coccolithophorids permit inferences as to the probable oceanic temperatures, that is, warmer or colder. Salinities may also be inferred from known distribution of living forms today.

Coccolithophorids have provided the major constituent of Mesozoic (Jurassic and Cretaceous) and Tertiary chalks and marls. They occur in calcareous shales. Any natural chalk taken at random tests this statement.

Although coccolithophorids constitute a minor part of recent calcareous oozes

TABLE 3.3. Relative Contribution of Coccoliths, Foraminiferal Tests, and others in Calcareous Oozes and Chalk[a]

| | RECENT | TERTIARY | | | | | CRETACEOUS | |
	1.	2. Mio.	3. Olig.	4. M. Olig.	5. L. Olig.	6. M. Eoc.	7.	8.
	Globigerina	Calc.	← OOZES — Coccolith →				← CHALK →	
Total CaCO₃	86	79	87	91	84	95	93	65
Coccoliths (etc.)	10	26	35	58	48	43	18	21
Foraminifera	55	25	30	5	15	32	23*	18*
Calc–Skeletal Debris (other)	3	8	5	1	3	5	15	10
Calcite Particles (undet.)	18	20	17	27	18	15	37	16
Radiolaria + Diatoms	9	13	9	0	7	0	0	0
Non–Calc. Residues (inorg)	5	8	4	9	9	5	7	35

1—6, Deep Pacific ; 7, Maestrichtian, Denmark ; 8, Campanian, Texas.
  * =Chiefly benthonic forms

[a]Figures represent percentage of total sample.
Source. Bramlette, 1958.

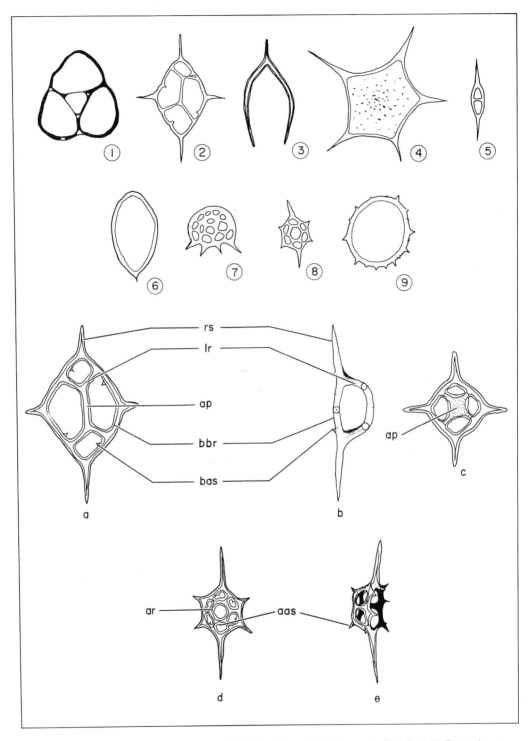

FIGURE 3.4C Fossil chrysophytes—III. Silicoflagellates. (1) *Corbisema*, U. Cret.-Eoc. (2) *Dictyocha*, Tert. (3) *Lyramula*, U. Cret. (4) *Vallacerta*, U. Cret. (5) *Naviculopsis*, Tert. (6) *Mesocena*. Tert. (7) *Cannopolis*, Tert. (8) *Distephanus*, Tert. (9) *Paradictyocha*, Tert. (after Mandra) Symbols "a–e", nomenclature of skeleton (after Tynan): *rs* — rostral spine; *lr* — lateral rod; *ab* — apical bar; *bbr* — basal body ring; *bas* — basal accessory spine; *ap* — apical plate; *ar* — apical ring; *aas* — apical accessory spine.

in the ocean, in the Cretaceous they dominated the calcareous nannoplankton. Bramlette (1958) noted that they have played a "significant role" in the whole geochemical cycle of oceanic carbonates (Table 3.3).

Table 3.3 shows how coccolith-derived data relate to the geochemistry of bottom oozes in the Pacific and in Cretaceous chalk beds.

Silicoflagellates. Complete and incomplete (broken) siliceous skeletons of silicoflagellates may be retrieved from almost any marine diatomite (Figure 3.4C). Also they are reported from siliceous shales and, although sparse, are represented in plankton hauls (Figure 3.4D), with coccolithophorids the large constituent. In size they range from 10 to 150 μ.

The internal skeleton of silicoflagellates

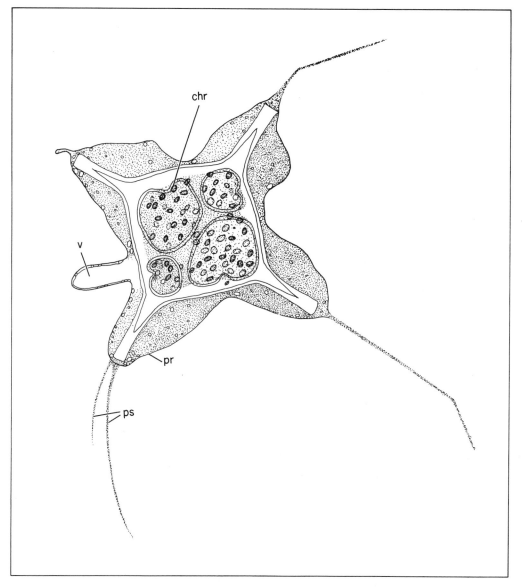

FIGURE 3.4D Fossil chrysophytes—III, continued. Living silicoflagellate, *Dictyocha fibula*, Firth of Clyde—abapical side, flagellum not seen (after Marshall). Symbols: *pr*—protoplasm; *v*—vacuole; *ps*—pseudopods; *chr*—chromatophores.

is composed of opaline silica. A lattice-work skeleton consisting of hollow bars and rods is organized about a basal body ring; radial spines generally issue from the sides (Figure 3.3F and Figure 3.4C). However the apical structure (Figure 3.4C, parts a–e) connects to the basal body ring by an apical bar. This structure may be composed of a plate or a ring (Figure 3.4C, parts c, d). A variety of basal and apical accessory spines (all very small) also occur. Finally, the skeleton may or may not be ornamented with pits or beads.

Mandra classified common shapes of silicoflagellate skeletons into 13 subdivisions. These shapes range widely from U- and Y-shaped objects to triangular and rectangular. Various polygonal configurations are also common (pentagon, hexagon, septagon, octagon, and so on). In addition, circular, elliptical, and some nongeometric forms are known. The nongeometric forms may even combine several of the other shapes.

Silicoflagellates—at least some forms—can be reminiscent of certain radiolarians and sponge spicules. However, because their bars and rods are hollow, they can be distinguished from radiolarians, which have solid bars and rods in their skeletons. To distinguish silicoflagellates from sponge spicules, Mandra suggests that one measure two diameters: the external (d_e) and the internal (d_i). For a given d_e, sponge spicules will have a smaller d_i than silicoflagellates.

Silicoflagellates range from the Lower Cretaceous, through the Pleistocene, to Recent. The longest-ranging genus is *Dictyocha*, which extends from the Upper Cretaceous to the Recent. Several extinct genera are restricted in the rock column. Thus *Vallacerta*, *Cornua*, and *Lyramula*, as presently understood, do not extend below or above the Upper Cretaceous. The genus *Cannopolis* is restricted to the Miocene.

Fossil silicoflagellates occur in the Miocene Calvert formation of Maryland and in California; they extend from the Cretaceous Moreno formation through the Miocene Monterey formation. Likewise they are known from the rock column in many parts of the world.

Preservation of Chrysophytes.   Because of their small size, complete cysts, coccoliths, or skeletons are common for chrysophyte fossils. In general, the loss of information is confined to the soft-part anatomy.

QUESTIONS

Origin of Life

1. Can you place the results obtained by Dr. Sidney Fox and co-workers in the context of Figure 3.2?

2. What data critical for protist origin are illustrated in Figure 3.2B, steps 2–3? Figure 3.2C?

3. How do the spherical particles of Dr. Fox and the eobionts illustrated in Figure 3.2B differ from a living protist?

4. Sir Julian Huxley has asserted that a simple cell will surely be created in a test tube before the twentieth century is over. Assume that this is a forecast that will be verified. (a) Where would you place such a cell (see Figure 3.3)? (b) Could such synthetic cells start a new evolutionary sequence independent of the ongoing sequence? Explain. (c) Might descendants of such cells become part of a fossil record? Justify your answer.

5. Drs. Claus, Nagy, and others reported indigenous fossil protists (resembling chrysophytes) in carbonaceous stony meteorites. The evidence was disputed; the argument stated that the alleged extraterrestrial fossils are really terrestrial contaminants. Assume for present purposes that such protists are actually extraterrestrial in origin. (a) In light of the evidence of this section of the text, can you explain the steps leading to their origin? (b) What would the existence of extraterrestrial protists prove about the origin of life?

Protista (Chrysophyta)

1. Members of the Protista share plant and animal characteristics.
   (a) What are these?
   (b) Viewed as primary data, can you trace these characteristics back to the postulated sequence of events leading to the first protists?
   (c) What is the age of the oldest known Chrysophyta?
   (d) Is it a eukaryote? Explain.

2. What data are recoverable from the cysts and/or skeletons of chrysomonads, coccolithophorids, and silicoflagellates, with reference to chemical composition, mineralogy, development and size, shape, history?

3. In what ways (be specific) can primary data bearing on each of the following items be obtained from fossil chrysophytes (that is, for the classes discussed): oceanic paleotemperature, photosynthesis in Mesozoic seas, biostratigraphy, paleosalinities?

4. (a) Discuss the geochemical cycle of oceanic carbonate and the relative contribution of coccoliths (consult Table 3.3).
   (b) What fossil chrysophytes would you expect to find in Cretaceous chalk; in Tertiary marine diatomites?

5. Some genera of silicoflagellates, for example, are long-ranging (*Dictyocha*); others are restricted in the rock column (*Cannopolis*). How can such disparate distributions be explained?

REFERENCES

Origin of Life

Bernal, J. D., 1961. Origin of life on the shores of the ocean: physical and chemical conditions determining the first appearance of biological processes, in Mary Sears, ed., Oceanography, A.A.A.S., Publ. 67, pp. 95–118.

Bernal, J. D. 1968. The origin of life. The World Publ. Co. Ohio. 333p.

Clark, F., and R. C. M. Synge, eds., 1959. The origin of life on earth. Macmillan. 656 pp. (See especially papers by Oparin, pp. 428ff., and discussion by T. N. Evreinova, pp. 493–494, Figures 1–9.)

Fox, S. W., 1972. Evolution of amino acids: Lunar occurrence of their precursors. In F. M. Johnson ed., Interstellar molecules and cosmochemistry. Ann. N.Y. Acad. Sci. vol. 194, pp. 71–85.

Miller, S. L., 1957. The formation of primitive compounds in the primitive earth, in Modern ideas on spontaneous generation: Ann. N.Y. Acad. Sci., Vol. 69 (2), pp. 260–274. (See discussion by P. H. Abelson, *idem*, at end of this paper, pp. 274–275.)

Oparin, A. I., 1938. The origin of life. (Reissued Dover Publ., S. Margulis, transl., 1953, 2nd ed., 252 pp., biblio.)

_____, 1961. Origin of life in the oceans, in Mary Sears, ed., Oceanography, A.A.A.S., Publ. 67, pp. 119–128.

Ponnamperuma, C., 1972. Organic compounds in the Murchison meteorite. (Same ref. as Fox, 1972), pp. 56–70. Cf. Anders et al, 1973, Science, 182:781.

Chrysophytes (General)

Bourrelly, P., 1963. Loricae and cysts in the Chrysophysaeae, in Life forms in meteorites (abbrev. title). Ann. N.Y. Acad. Sci., vol. 108 (2), pp. 421–429.

Fritsch, F. E., 1935. The structure and reproduction of the algae. Cambridge Univ. Press, Vol. 1, 756 pp.

Archeomonads

Deflandre, G., 1932. Note sur les Archeomonadaceés: Bull. Soc. Bot. France, Vol. 79, pp. 346–355, 38 text figures.

_____, 1933. Seconde note sur les Archeomonadaceés: *idem* pp. 79–90, 41 text figures. (Both publications are useful for fossil illustrations.)

Tynan, E. J., 1960. The Archeomonadacea of the Calvert formation (Miocene) of Maryland: Micropalentology, Vol. 6 (1), pp. 33–37, plate 1.

Van Landingham, S., 1964. Chrysophyta cysts from the Yakima Basalt (Miocene) in south-central Washington: J. Paleontology, Vol. 38 (4), pp. 729–739, plate 120, 2 text figures.

Coccolithophorids

Braarud, T., et al., 1955. Terminology, nomenclature, and systematics of the Coccolithophoridae: Micropaleontology, Vol. 1 (2), pp. 157–159.

_____, 1961. Cultivation of marine organisms as a means of understanding environmental influences on population, in Mary Sears, ed., Oceonography, A.A.A.S., Publ. 67, pp. 271–294, refs.

Bramlette, M. N., 1958. Significance of coccolithophorids on calcium carbonate deposition: Bull. GSA, Vol. 69, pp. 121–126, table 1.

————, and W. R. Riedel, 1954. Stratigraphic value of discoasters and some other microfossils related to Recent coccolithophors: J. Paleontology, Vol. 21 (4), pp. 385–402, plates 38–39, refs.

Deflandre, G., 1952a. Classe des Coccolithophoridés, in J. Piveteau, ed., Traité de Paléontologie, Vol. 1, pp. 107–115. Masson et Cie., Paris.

Deflandre, G., 1952b. *idem*, in P. P. Grasse, ed., Traité de Zoologie, Vol. 1, Fasc. 1, pp. 439–470. Masson et Cie., Paris.

Hay, W. W., and K. M. Towe, 1962. Electron microscope studies of *Braarudosphaera bigelowi* and some related coccolithophorids: Science, Vol. 137 (3528), pp. 426–428, Figures 1–6.

Isenberg, H. D., et al., 1963. Calcification in a marine coccolithophorid in Comparative biology of calcified tissue: Ann. N.Y. Acad. Sci., Vol. 109, article 1, pp. 49–64 (see Figure 13).

Lohman, H., 1902. Die Coccolithophoridae: Archiv. fur Protistenkunde, Vol. 1, pp. 1–159, 13 plates. (This paper is excellent for illustrations of living coccolithospheres.)

Silicoflagellates

Deflandre, G., 1932. Sur la systématique des Silicoflagellelés: Bull. Soc. Bot. France, Vol. 79, pp. 494-506, 42 figures.

Mandra, Y. T., 1958. Fossil silicoflagellates from California. Univ. Microfilm, Inc., Ann Arbor, Mich. 104 pp., refs., plates 1–4, text figures 1–4, Tables 1–12.

Marshall, S. M., 1934. The Silicoflagellata and Tintinnoinea: British Museum (Nat. Hist.), Great Barrier Reef Exped., 1928–1929. Scientific Reports, Vol. 4 (15), pp. 623–662, 43 text figures. (Text figure 6 shows the variation in skeletons of a single silicoflagellate species, *Dictyocha fibula* var. *hexagona*.)

Tynan, E. J., 1957. Silicoflagellates of the Calvert formation (Miocene) of Maryland: Micropaleontology, Vol. 3 (2), pp. 127–136, plate 1, text figures 1–3.

(Volumes of the "Deep Sea Drilling Project," 1961–1975, vols. 1–41, contain an excellent series of papers on coccolithophorids, silicoflagellates, and other microscopic forms. B.U. Haq and A. Boersma, 1978, Introduction to Marine Micropaleontology, Elsevier, have chapters on calcareous nannoplankton (Chapter 3) and chitinozoans, dinoflagellates, acritarchs, silicoflagellates, and others. Haq gives excellent figures and a key to coccoliths.)

B. THE PYRROPHYTA

Living Dinoflagellates

Dinoflagellates (*dino* = armored), following Chatton's succinct diagnosis, are essentially flagellates having yellow-green chloroplasts (impregnated with chlorophyll and other pigment). The two flagella — one longitudinal, the other initially perpendicular to the first — serve different functions. The longitudinal (*flagelle axial*) acts as a rudder while the perpendicular (*flagelle ondulant*) propels the individual. The *flagelle ondulant* is the anterior flagella and is contained in a helical groove. Armored forms have a cellulose plate system (Figure 3.5*A*). Unarmored forms are either naked or contained within a cellulose membrane. Many unarmored types are parasites living within other marine animals, for example, xooanthellae in Radiolaria (see Protozoans), or the intestines of marine copepods (see Crustacea, Chapter 11). Such forms exist as nonflagellate cysts.

Both asexual and sexual reproduction occur in dinoflagellates. Size range is about 6.0 to 100.0 μ.

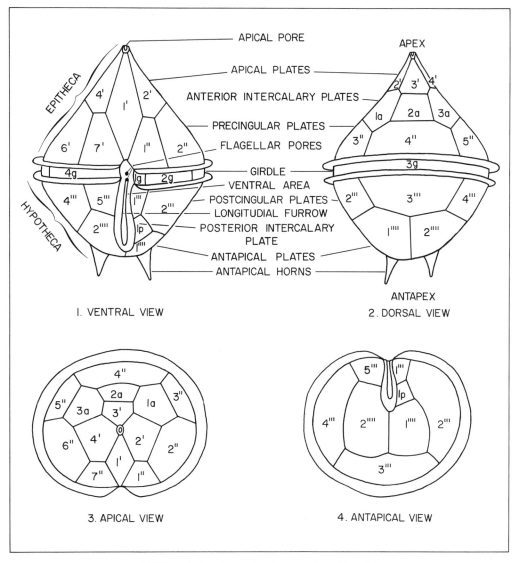

FIGURE 3.5A Terminology for dinoflagellates (after Evitt).

The cellulose plates of armored (or thecate) dinoflagellates follow a general plan (Figure 3.5A). Where the several series of plates are in contact, a suture exists containing cement. Therefore the plates are loosely bound together and the armor can expand as growth proceeds. Additionally the arrangement favors ready separation of the armor along certain sutures to permit escape of gametes.

Revised Taxonomy

The position of the hystrichosphaerids (Figure 3.5C) has been clarified. Many are dinoflagellates. Included are the well-known fossil genera *Hystrichosphaera* and *Hystrichosphaeridium*, among others. These, together with previously established dino-flagellates, belong to the class Peridinieae (= Dinophyceae). Nevertheless, other hystrichosphaere-like bodies, such as cysts, spore cysts, and vegetative stages, that belong to one or another stage in the life cycle of unicellular algae have been grouped under the *acritarchs*.[2]

Information Recovery

Food Chain and Microplankton. The microplankton serving as *primary producers* in the food chain in the sea include, in descending order, (*a*) diatoms, (*b*) dinoflagellates, (*c*) coccolithophores. Diatoms are unknown as fossils before the Cretaceous, and coccolithophores are unknown before the Mesozoic (Jurassic). Dinoflagellates alone are known to extend from the Permian on. This raises the intriguing question of whether dinoflagellates might not have played the role of primary producers during the Paleozoic.

They could have; acritarchs would have filled the second and third places in the food chain in lieu of diatoms and coccolithophores. If this view can be substantiated by accumulating evidence, then a significant change in the food chain in the sea occurred with the rise of the diatoms.

Phytoplankton Blooms and Biogeochemistry. Phytoplankton blooms are seasonal occurrences in oceans, seas, sounds, fjords, and even inland waters such as the Great Lakes. At such times the dinoflagellate population, among others, increases enormously (Figure 3.5C, part 5). In turn, such blooms are encouraged by

[2]A. R. Loeblich, Jr., 1969. Morphology, ultrastructure and distribution of Paleozoic acritarchs. In, Sympos. North Amer. Paleont. Convention. Pt. G., Allen Press, Lawrence, Kansas.

Compare D. Wall, and B. Dale, 1967. Review Paleobotany and Palynology, vol. 2, pp. 349–354.

adequate supplies of nutrients such as phosphorus and nitrogen, and a whole array of ecological factors such as temperature, pH, and salinity. Thus considerable biogeochemical and ecological data can be obtained as derivative data when a phytoplankton bloom is recorded among living microplankton. A Cretaceous phytoplankton bloom of dinoflagellates has been recorded in the Kiowa shale of Kansas (Tasch et al., 1964). And of course, various fossil occurrences of diatomites and coccolith-bearing sediments are likewise indicative.

Information Preserved and Lost in Fossilization

The resistant theca, the cellulose cuirass (or skeleton), with or without an enclosed cyst, and with or without the plate system well defined, are all that are generally preserved of dinoflagellates. While for the dinoflagellate hystrichosphaeres, the material usually found in processed samples, include a generally ovate body bearing various-sized processes that rarely terminate in an enclosing membrane.

The dinoflagellates may occur as fragmented pieces since the plate system can readily separate at the equatorial girdle (see Figure 3.5B, part 10). Thereafter the lower hypotheca bearing antapical horns and the upper epitheca, with or without an apical projection, are frequently found in processed samples.

Acritarchs often occur in samples as spherical or ovate bodies producing many small or few spines, or process-like projections. Some acritarchs such as *Membranilarnax* (Figure 3.5C, part 4) will have a reticulation or polygonal pattern on the central body.

FIGURE 3.5B Fossil dinoflagellates and acritarchs. (1) *Hystrichosphaera bentori* Rossignol (Pleist., Israel). Cycle of development: *a-h*. Starting with a cyst in the thin refringent area (*a*) through (*d*), to secretion of a protective case (which is the hystrichosphaerid with processes about it), to (*g*) where cyst is expelled through the archeopyle. (After Rossignol).[1] (2) Archeopyle (after Evitt). (3) Living *Gonyaulax* chain. Gulf of California (after Klement). (4) *Gonyaulax*. (5) *Nannoceratopsis*. (6) *Palaeosperidium* (Perm). (8) *Dellandrea* (Tert.). (9) *Pareodina*. (10) Peridinium, hypotheca with anfapical horns. (Illustrations schematic and redrawn after, Evitt, Deflandre, Jansonius and Tasch).) ([1]Compare Wall, D. and B. Dale, 1967. Review Paleobotany and Palynology, vol. 2, pp. 349–354.) [Fig. 7, deleted]

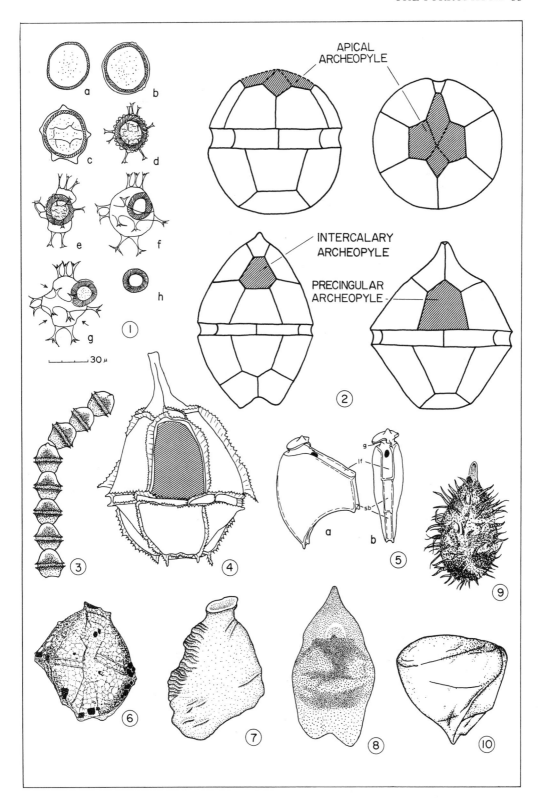

APICAL
ARCHEOPYLE

INTERCALARY
ARCHEOPYLE

PRECINGULAR
ARCHEOPYLE

30 μ

Fossil Dinoflagellates

Primary data encoded in the fossil skeleton of a dinoflagellate extend a long way, to the cellulose plate system, the two sets of grooves, the flagellar pore, enclosed cyst, general configuration, and relationship of epitheca to the hypotheca.

The plate system will establish the fossil as dinoflagellate; the grooves and pore will relate to the original flagellae that governed movement and locomotion; the cyst explains the nonmotile stage in the life cycle; the general configuration will bear on the shape adaptation for locating, remaining, and positioning in the eupho-tic zone of the sea. In turn, this configuration pertains to the operation of natural selection in the original population of which the given fossil was a member during Paleozoic, Mesozoic, or Cenozoic time. Variable size and number of antapical horns in such living forms as *Ceratium* apparently is a response to changes in oceanic and inland water temperatures and most likely, has a similar significance for fossils (see Figure 3.5*B*, part 8).

Information Recovery

Paleoecological and biogeochemical data inferred for fossils will follow that ob-

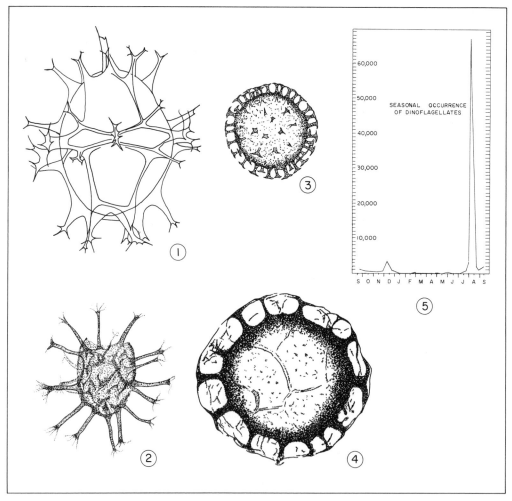

SEASONAL OCCURRENCE OF DINOFLAGELLATES

FIGURE 3.5C Dinoflagellates, Acritarchs, and Blooms. (1)–(2) Dinoflagellates: (1) *Hystricosphaera* (Tert.), with precingular archeopyle (not shaded). (2) *Hystrichosphaeridium* (U. Cret.) (After Evitt). (3)–(4) Acritarchs: (3) *Micrhystridium* (Sil.). (4) *Membranilarnax* (Jur.) (After Deflandre). (5) Seasonal occurrence (bloom) of dinoflagellates at Iglook (above Arctic Circle); data in cells per liter/per given month (after Bursa).

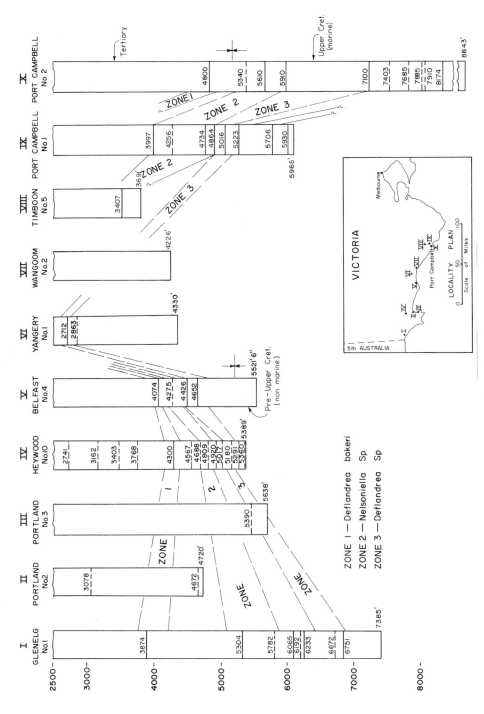

FIGURE 3.6 Microplankton horizons and Deflandreidae "zones," Western District, Deep Bores, Victoria, Australia (after Douglas, 1962).

tainable from living forms. It is also possible to mark stratigraphic horizons by zoning dinoflagellate and dinoflagellate-hystrichosphaere occurrences horizontally and vertically over a region (Douglas, 1962, and Figure 3.6).

Fossil Record

Dinoflagellate faunas have been described from the late Paleozoic (Permian of West Canada, Zechstein of Germany); from the Mesozoic around the world (Triassic, Jurassic, Cretaceous of North America, among others); from the Tertiary of India, for example; from the Pleistocene of Jerusalem. In modern seas, dinoflagellates abound not only in marine waters, but in marginal basins such as estuaries. Inland water bodies, such as the

Great Lakes, also contain abundant faunas.

Dinoflagellate-hystrichosphaeres extend back to the oldest Paleozoic. Abundant Carboniferous and Cambro-Ordovician, as well as Silurian and Devonian, hystrichosphaerid assemblages are known.

Nonmarine beds (Tertiary peat in southwest Australia) yielded some fossil dinoflagellates (Churchill and Sarjeant, *Nature*, 1962, Vol. 194, p. 1094).

R. C. Pirini reported on coccoliths from the Permian of East Turkey at the Planktonic Conference in Rome, 1970. If in place, this would put another primary producer in Late Paleozoic seas — at least in restricted areas. (See W. A. S. Sarjeant, 1974. Fossil and living dinoflagellates, Academic Press, for data on ecology, morphology, and classification of fossil families and genera.)

QUESTIONS

1. (a) Sketch and number the cellulose plate system in armored dinoflagellates (consult Figure 3.5*A*). (b) What is the significance of the longitudinal furrow, flagellar pore, equatorial girdle, archeopyle, and apical and antapical horns?

2. With reference to Figure 3.5*C*, part 5, and the text discussion, outline the biogeochemical and ecological implications of a Cretaceous phytoplankton bloom.

3. Can you distinguish between acritarchs and dinoflagellates?

4. What can you conclude about the relationship of hystrichosphaerids and dinoflagellates from Rossignol's evidence shown in Figure 3.5*B*?

5. How did Douglas (Figure 3.6) use information derived from dinoflagellate fossils for stratigraphic zonation?

6. Name several differences between fossil dinoflagellates and coccolithospheres.

REFERENCES

Classification

Deflandre, Georges, 1952. Dinoflagelles Fossiles, in P. P. Grasse, ed., Traité de Zoologie, Vol. 1, pp. 391–406.

Downie, C., W. R. Evitt, and W. A. S. Sarjeant, 1963. Dinoflagellates, hystrichosphaeres and the classification of acritarchs: Stanford Univ. Publ. (Geological Sciences), Vol. 7 (3), 16 pp.

Evitt, W. R., and S. E. Davidson, 1964. Dinoflagellate studies. 1. Dinoflagellate cysts and thecae: Stanford Univ. Publ. (Geol. Sci.), Vol. 10 (1), pp. 3–12, plate 1, text figures 1–2.

Biology

Braarud, Trygve, 1944. Morphological observations on marine dinoflagellate cultures: Avhand. Det. Norske-Videnskaps-Akademi. Oslo, No. 11, pp. 3–18. 4 plates.

Chatton, Edouard, 1959. Classe des Dinoflagelles ou Peridiniens, in P. P. Grasse, ed., Traité de Zoologie, Vol. 1, pp. 309–390.

Fritsch, F. E., 1935. The structure and reproduction of the algae. Cambridge Univ. Press. Vol. 1, 756 pp. (Dinoflagellata, pp. 680–704).

Specific Living Dinoflagellate Faunas

Bursa, Adam S., 1961. The annual oceanographic cycle at Igloolik in the Canadian Arctic. II. The Phytoplankton: J. Fish. Res. Bd., Canada, Vol. 18 (4), pp. 563–615.

Davis, C. C., 1962. The plankton of the Cleveland Harbor area of Lake Erie in 1956–57: Ecol. Monographs, Vol. 32 (3), pp. 209–247.

Graham, H. W., 1942. Studies in the morphology, taxonomy and ecology of the Peridiniales: Carnegie Inst. of Wash. Publ. 542. Biology III, 129 pp.

Kofoid, C. A., 1911. Dinoflagellata of the San Diego Region, IV. The genus *Gonyaulax* with notes on its skeletal morphology and a discussion of its generic and specific characters: U. Calif. Publ., Vol. 8 (4), pp. 187–286, plates 9–17.

Selected Fossil Faunas

Deflandre, Georges, 1944–1945. Microfossiles des Calcaries Siluriens de la Montagne Noire: Ann. Paleontologie, Vol. 31, 75 pp., 3 plates.

Douglas, J. G., 1962. Microplankton of the Deflandreidae group in Western District

sediments: Mining and Geol. Journal, Vol. 6 (4), pp. 1–16, plates 1–4.

Eisenack, Alfred, 1951. Uber Hystrichosphaeridien und andere Kleinformen aus baltischen Silur und Kambrium: Senckenbergia, Vol. 32, pp. 187–204, plates 1–4.

Evitt, W. R., 1961. The dinoflagellate *Nannoceratopsis* Deflandre: morphology, affinities and infraspecific variability: Micropaleontology, Vol. 7, pp. 305–336, plates 1–2.

Jansonius, J., 1962. Palynology of Permian and Triassic sediments, Peace River Area, Western Canada: Paleontographica Abt. B, Vol. 110, pp. 35–98, Figure 3.

Klement, K. W., 1960. Dinoflagellaten und Hystrichosphaerideen aus dem Unteren und Mittleren Malm Sudwestdeutschlands: Paleontographica: Vol. 114, Section A, pp. 1–104, 10 plates.

Sargeant, W. A. S., 1962. Upper Jurassic microplankton from Dorset, England: Micropaleontology, Vol. 8, pp. 255–268, plates 1–2.

Tasch, P., K. McClure, and Orrin Oftedahl, 1964. Biostratigraphy and taxonomy of a hystrichosphere-dinoflagellate assemblage from the Cretaceous of Kansas: Micropaleontology, Vol. 10 (2), pp. 189–206, plates 1–3.

Wall, D., and C. Downie, 1963. Permian hystrichospheres from Britain: Paleontology, Vol. 5 (4), pp. 770–784, plates 112–114.

C. THE UNICELL EUCARYOTES (PROTOZOANS)

Introduction

Our discussion of the origin of life reviewed the postulated biochemical steps leading to appearance of a protozoan or unicell (eucaryote). [A procaryote is a unicell lacking definite nucleus and other organelles; a eucaryote is a unicell with definite nucleus and other organelles. The oldest acceptable eucaryote fossils come from the Proterozoic Beck Spring Dolomite—some 1.2 to 1.4 billion years

old (P. Cloud, 1969, Nat. Acad. Sci. U.S.A., Vol. 62, No. 3, p. 623). The oldest known procaryote fossil, which is some three billion years old, comes from the Pongola System, South Africa (cf. Cloud, 1976, see ref. p. 37).

Nucleated protozoans embrace forms of geological importance including the foraminifers, radiolarians, and tintinnids. It is valuable to place fossil protozoans (eucaryotes) in the context of known living forms.

Unicells ordinarily may be separated into distinct subphyla or classes based on

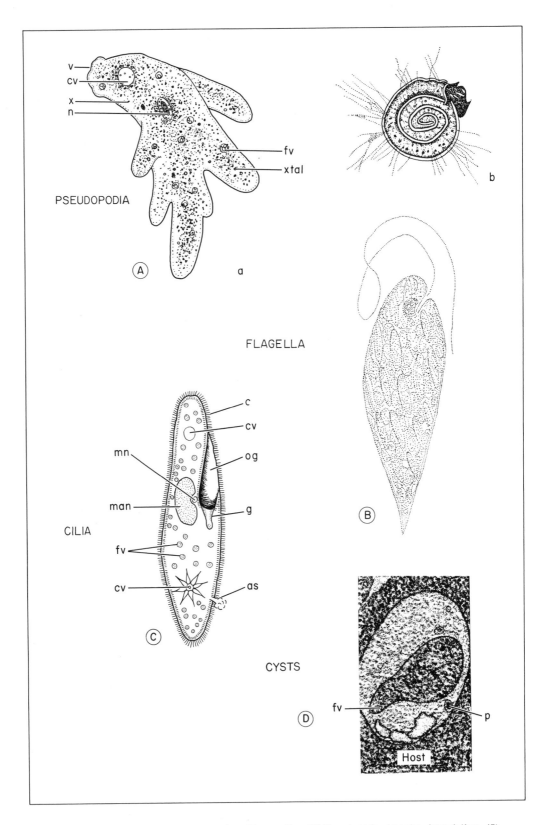

FIGURE 3.7 Living protozoans—modes of locomotion. (*A*) Pseudopods. *Amoeba;* foraminifers. (*B*) Flagella. *Euglena.* (*C*) Cilia. *Paramecium.* (*D*) Cysts. Sporozoa: *Plasmodium berghei.* Illustrations redrawn and adapted from (*A*) Monica Taylor and others; (*B*) Godjic; (*C*) Guthrie and Anderson; (*D*) from photograph by Rudzinska and Trages, 1959, cit. Pitalka.

mode of locomotion. Thus one can distinguish the flagellates that move by motion of one or more flagella (*Euglena*) from those that move by means of pseudopodia (*Amoeba*), radiolarians and foraminifers, from those that move by means of cilia (*Paramecium*). Parasitic protozoans that lack locomotor organelles (such as Plasmodium of malaria that infects the human blood stream) are assigned to the phylum Sporozoa (Figure 3.7).

Abbreviated Classification of Selected Unicells

Subphylum 1. Sarcodina (pseudopods) (=Phylum, Whittaker, 1969[3])
    Class: Rhizopoda
        SUBCLASS: FORAMINIFERA[4] (= ORDER FORAMINIFERIDA, SEE LOEBLICH 1964) (LOWER CAMBRIAN–RECENT)
        SUBCLASS: RADIOLARIA[4] (CAMBRIAN–RECENT)

Subphylum 2. Sporozoa (no locomotor organelles) (=Phylum, Whittaker, 1969)
    Class: Actinopoda

Subphylum 3. Ciliophora (cilia) (=Phylum, Whittaker, 1969)
    Class: Ciliata
        SUBORDER: TINTINNINA[4] (JURASSIC–RECENT)

The Sarcodina, Sporozoa, and Ciliophora: Eucaryotes

Reproduction, sexual or asexual (fission, budding, multiple division), and frequently an alternation of generations—asexual followed by sexual followed by asexual, and so on (Figure 3.8). Unicells are microscopic in size (see Table 3.1), may be naked (*Amoeba*) or secrete encasements, shells, and skeletons (radiolarians or foraminifers). May be solitary as with foraminifers, or colonial as with living *Volvox*. Geologic range: Pre-Cambrian to Recent.

Foraminifers (Section 1)

Order Foraminiferida (after Hyman, Pokorný, Loeblich), a subclass of some authors. Figure 3.9*A*, *B*.

Wall Composition of Test.

Wall Composition of Test. Given: a mixed sample of representative foraminifers both fossil and living. Into what large groupings would you separate the individual specimens? Inspection would soon disclose that the specimens had walls of different composition and that three major kinds of wall composition were present: membranous and pseudochitinous, agglutinated or arenaceous, and calcareous. Further study would reveal that the calcareous specimens were not all alike and that three kinds of calcareous tests were present. In this way you could isolate five separate distinct assemblages (suborders) from the original sample. These five suborders are the following.

Allogromina—membranous and pseudochitinous test.

Textulariina—agglutinated or arenaceous test.

Fusulinina—calcareous microgranular test.

Miliolina—porcelaneous calcitic test.

Rotaliina—hyaline perforate calcareous test.

Note that the name of a suborder (and of a superfamily or family) contains the stem

---

[3]*Science*, Vol. 16, pp. 150–160.
[4]The unicells considered in this chapter. R. H. Hedley, 1964, The biology of foraminifers. In, W. T. L. Felts and R. J. Harrison, Eds., International Review General and Experimental Zoology. Academic Press, vol. 1, pp. 1–45.

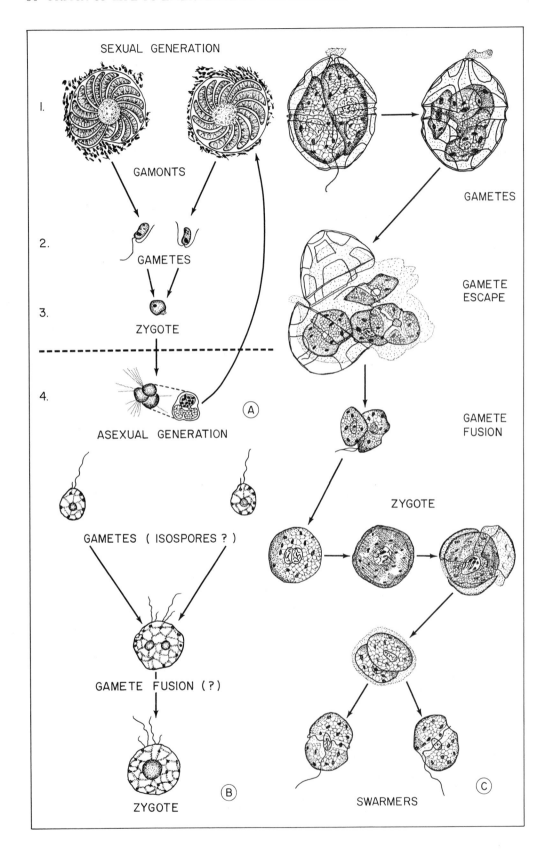

SEXUAL GENERATION

I.

GAMONTS

2.

GAMETES

3.

ZYGOTE

4.

A

ASEXUAL GENERATION

GAMETES (ISOSPORES ?)

GAMETE FUSION (?)

B

ZYGOTE

GAMETES

GAMETE ESCAPE

GAMETE FUSION

ZYGOTE

SWARMERS

C

FIGURE 3.8 Reproductive cycle in protists.

(A) *Polystomella crispa*, a foraminifer; alternation of generations: (1) Gamonts asexually produced by microspheric adult (not shown) by fission. (2) Flagellated gametes (megalospheric) from separate individuals combine to form a zygote (3). (4) Asexual generation yields new microspheric young, which upon reaching adult stage start the cycle over again.

(B) *Acanthostaurus puperascens*, an acantharian radiolarian (that is, skeleton of strontium sulphate): Gametes (thought to be "isospores") presumably fuse to form zygotes. This would be the sexual generation. Other investigators have recorded sporulation (asexual generation).

(C) *Glenodinium lubiniensiforme*, a dinoflagellate: development of gametes; escape of gametes from parent shell; fusion of gametes; nuclear fusion; encystment of zygote; excystment (that is, breaking out of cyst) of zygote; development of "swarmers"; excysted "swarmers." Illustrations schematic; redrawn and adapted from (A) Myers, 1938; (B) Schewiakoff, 1926; (C) Diwald; (A–C) all cit. Wenrich, 1954.

of an included genus. Thus the suborder name Allogromina was derived from the name *Allogromia*, a genus name, and so on.

Chambers and Wall Microstructure.

The five suborders can be further subdivided into 17 superfamilies, which in turn include 62 families. The number of chambers (one or many) and microstructure of the wall define the superfamilies. To illustrate: Suborder Textulariina; the superfamily Ammonodiscacea and superfamily Lituolicea belong to this suborder. The families of these superfamilies can be distinguished as follows.

Single-chamber test—family Saccaminidae (Figure 3.9B, part 6).

Two-chambered test—family Ammodiscidae (Figure 3.9B, part 2).

Many-chambered test—family Textulariidae (Figure 3.9B, part 1a).

Coiling, Chamber Arrangement, Aperture Position.

Within any family—for example, the Textulariidae—individual genera may be distinguished by the following characteristics.

Coiling of early test (Figure 3.9B, part 6).

And/or arrangement of chambers—biserial, uniserial, or both (Figure 3.9A, part P).

And position of aperture (Figure 3.9A, part M).

Perforation and Type of Calcareous Test.

Exclusive of the fusulinids (suborder Fusulinina), which merit separate treatment, the calcareous forms may be separated into two main groups. (The family Spirillinidae have monocrystalline walls.) The

two groups are as follows.

1. Test imperforate, porcellaneous-appearing (example: *Quinqueloculina*, Figure 3.9B, part 4).
2. Test perforate, hyaline (glassy)-appearing: (a) With radial walls (example: *Globigerina*, Figure 3.9C, part A); (b) with granular walls (example: *Anomalina*).

Fusulinaceans (Section 2)

There are important reasons why protozoan foraminifers in the superfamily Fusulinacea (suborder Fusulinina) are of unusual interest. *They have a restricted geologic range*; that is, they apparently first appeared in late Mississippian time and were extinct in the seas of the world before the close of the Permian. *They are abundant and often well preserved*, forming fusulinid limestones. *They have a wide geographic spread*, being found in the Upper Paleozoic of all continents exclusive of Antarctica and Australia. Because of features of internal structure revealed in thin sections, *their evolution can be traced* in considerable detail for a span of about 80 million years. During this time, over 1000 species evolved and these are assignable to 72 genera.

Composition of Shell.

All fusulinaceans have an original calcareous microgranular test. These tests are composed of very minute crystals of calcite. They are all about the same size, closely packed, and may include foreign matter. To check this, the student·can dissolve a few clean fusulinids in a watch glass, using dilute hydrochloric acid. Frequently this small

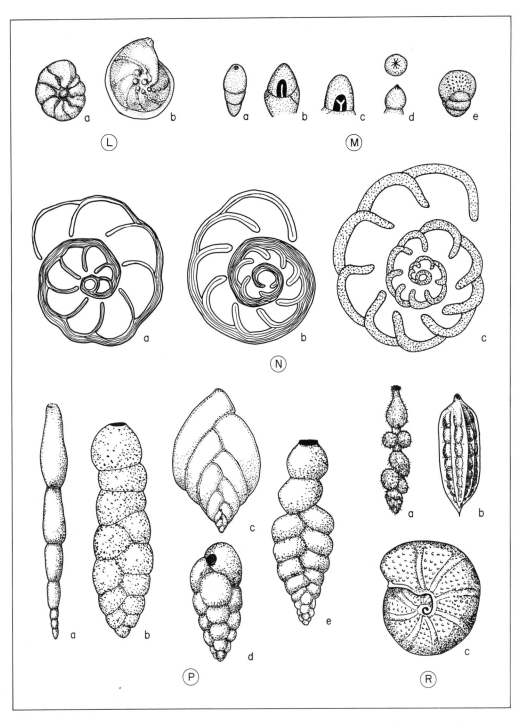

FIGURE 3.9A Information contained in foraminiferal tests: selected examples. (*L*) Structures: (*a*) umbonal plug, (*b*) keel. (*M*) Aperture, *a–e*, types (*N*) Test and wall structure, *a–c*, thin sections. (*P*) Chamber arrangement, *a–e*, uniserial, biserial–uniserial and so on. (*R*) Ornamentation, *a–c*, some types. (All figures schematic).

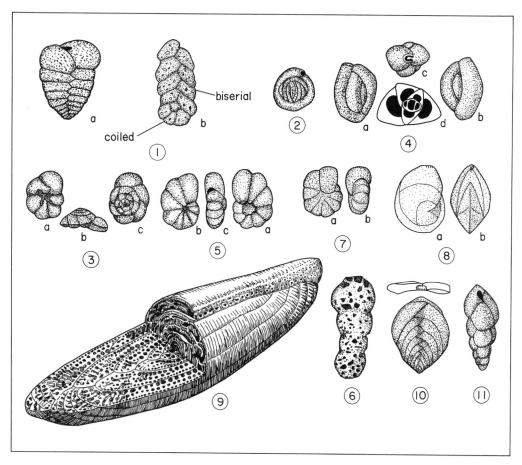

FIGURE 3.9B Some representative benthonic foraminifera (schematic). (1a) *Textularia;* (1b) *Spiroplectammina.* (2) *Ammodiscus.* (3) *Discorbis.* (4) *Quinqueloculina.* (5) *Cassidulina.* (6) *Reophax.* (7) *Endothyra.* (8) *Robulus.* (9) *Alveolina.* (10) *Polymorphina.* (11) *Bulimina.*

experiment yields a pinch of fine quartz silt, an occasional iron oxide particle (see Figure 3.10, part D), mica flakes, and other rarer mineral grains. It should be recalled that the original test of fusulinaceans may be partially or completely replaced by iron oxides or sulphides, silica (chert), phosphates (black fusulinids), and so on, and occurs so in the outcrop or rock sample.

Wall or Spirotheca. The spirotheca or wall in fusulinaceans began in primitive forms, such as *Millerella*, as a single dense layer. This layer subsequently underwent modifications in the direction of thickening in its lower portion. The lower portion is called the keriotheca.

In Figure 3.10, part D the relationship between the tectum and the upper and lower keriotheca (which is alveolar) is clearly shown. Dr. M. L. Thompson observed (1964) that the walls of all fusulinaceans have the same three layers. However, alveoli (hollow tubes) may be too small to be seen even in thin section and hence may be obscured in many forms.

Differences in wall structures, as seen in thin sections, help to distinguish between several fusulinid genera (Figure 3.10, part E).

Proloculus, Chambers, Septa. The initial chamber or proloculus contained the original protoplasmic mass that constituted a single cell. A single exit from this initial chamber is identified as the proloculus pore (Figure 3.10, part C). Out of it streamed the pseudopods of the contained

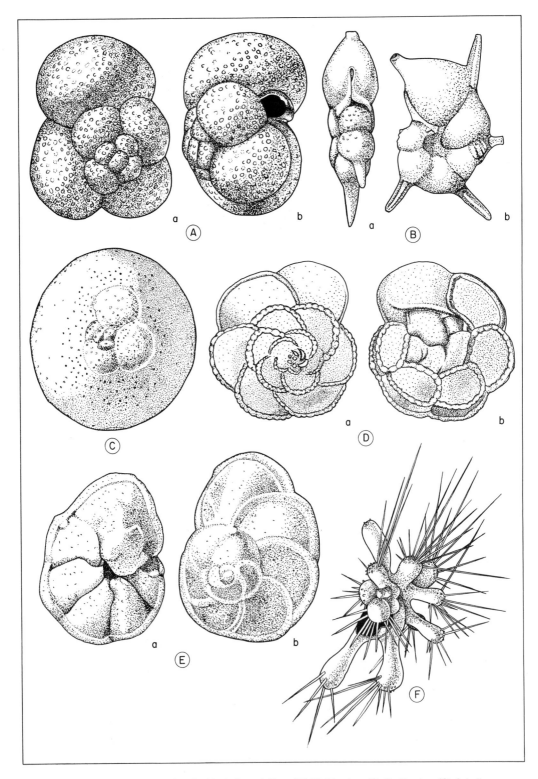

FIGURE 3.9C Representative planktonic foraminifera. (*A*) *Globigerina.* (*B*) *Hantkenina.* (*C*) *Orbulina.*
(*D*) *Globorotalia.* (*E*) *Globotruncana.* (*F*) Hastigerella. (Schematic.)

foraminifer. It was around the proloculus that subsequent chambers were added. Partitions between chambers are called septa.

In foraminifers, dimorphism is quite common (microspheric—asexual generation with a small proloculus and large test; megalospheric—sexual generation with a large proloculus and small test). Whether dimorphism also existed in the fusulinaceans is not completely established. Two sizes of test are infrequent.

Antetheca. During growth in a coiled fashion, at any stage of development the antetheca represents the last chamber's anterior wall. As a new chamber is added, the former antetheca becomes a septum separating the newly formed last chamber from the older last chamber (Figure 3.10, part A). No aperture or foramina (holes or openings) are known in the antetheca, nevertheless it does have septal pores.

Tunnel, Chomata, Septal Pores, Foramen, Axial Fillings. A variety of structures are found in different fusulinaceans. These are useful in identification. They may be classified into two major groupings: resorbed mineral calcite (tunnels, septal pores, foramen—openings at base of septa) and redeposited calcite (chomata, parachomata, axial fillings).

Structures formed by resorption apparently were related to the need of the speck of protoplasm contained in the proloculus to remain in contact with the outside aqueous media *as it grew.* Thus all types of spaces and openings from tunnels to holes and pores would permit pseudopods to stream out into the fluid environment and so receive nutriment from, and pass waste to, the water.

Those structures that are the result of redeposition (chomata, parachomata, and axial fillings) seem to have served a dual function: ready disposal sites for resorbed mineral calcite (chomata) and adjustments of specific gravity necessitated by continued growth from deposition of

axial fillings along the maximum diameter of the test.

Information Derived from Foraminifers including Fusulinaceans

. *Ecology—Environments.* Living benthonic foraminifers (as well as their fossil equivalents) occur in numerous major marine environments: estuary, coastal lagoon, beach and near-shore, continental shelf, slope, and deep sea. Characteristic suites of species occur in any such habitat although particular species may be found in more than one environment. Note that benthonic forms constitute the bulk of foraminifers. Only some 30 to 50 species are known among modern planktonic forms (Figure 3.9C).

Bathymetry. Zonation of benthonic foraminiferal faunas offshore is related to water depth. A few such bathymetric zones, according to Phleger (p. 255), must be worldwide. Thus distinct faunas occur at approximately the following depths:

  I. 70–125 meters
 II. 20–30 meters (Continental Shelf)
III. 1000 meters (Continental Slope)
 IV. 2000 meters (Continental Slope)

Fusulinaceans are associated with water depth varying from 5 to 60 ft according to some authors (Imbrie, Lane, Tasch), although others attribute maximum depths of 150–180 ft (Elias) or even up to greater than 250 ft (Bandy). Apparently they lived offshore in open seas and generally deeply penetrated continental basins only during times of marine transgressions (Thompson, 1964). Offshore, in deeper waters, they were bottom dwellers (benthonic), whereas inshore, in shallower waters, they probably lived above the bottom. The fusulinacean fauna in Pennsylvanian cyclothems and in Permian beds serve as significant environmental indicators (see Figure 3.11A, B, C).

Mixed Faunas. Foraminiferal faunas may be mixed in various ways, that is,

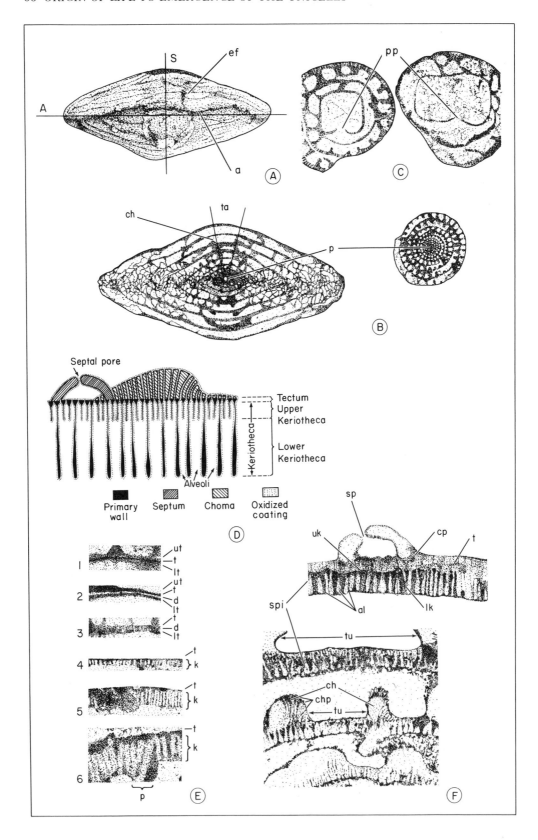

FIGURE 3.10 Morphology of fusulinaceans. (A) Fusulinid *Triticites ventricosus* (L. Perm.) showing where the two thin sections of "*B*" were obtained; slice along A (axial section); slice along S (sagittal section). (These are shown in "*B*" as they appear after slide preparation); *ef* – external furrow; *a* – antetheca (see text). (B) Same genus and species as "*A*"; *ch* – chomata, *ta* – tunnel angle, *p* – proloculus. (C) First fusulinid chamber (proloculus) and its opening, a pore (*pp*) (see text). Both right and left figures are species of *Parafusulina* from the Asiatic Permian. (D) *Schwagerina* wall structures; spirotheca (schematic), compare two main divisions with figure E, 6 below based on actual thin section. (E) Comparative spirothecal structures of four fusulinid genera: (1) *Profusulinella*; (2) *Fusulina*, with 4-layered wall; and (3) 3-layered wall; (4), (5) *Triticites*; and (6) *Schwagerina*; *t* – tectum; *d* – diaphanotheca; *k* – keriotheca; *ut* – upper tectorium; *lt* – lower tectorium. (F) Details of some schwagerinid structures seen in thin section (schematic); upper figure, sagittal section; lower figure, axial section; *sp* – septal pore; *t* – tectum; *uk, lk* – upper and lower keriotheca, *chp* – chomata pore; *al* – alveoli (compare figure D); *tu* – tunnel; *spi* – spirotheca. (Illustrations redrawn and adapted from M. L. Thompson, 1948, 1951, 1964.) (See Ross, C. A., 1972. J. Paleont., v. 46 (5), pp. 719–728.)

shells of different benthonic and/or planktonic species may occur together out of the proper context of their usual habitat. Such mixed associations provide valuable sedimentary and oceanographic data on displacement of sediments, change in sea level (rise or fall), convergence of water masses (different surface temperatures, that is, colder in higher latitudes, warmer in lower latitudes).

Morphology. Bandy (1960) found that variation in benthonic foraminiferal assemblages in modern seas was related to bathymetry (bay, shelf, and bathyal zones). Variable characteristics included overall size, shape, size of chambers, chamberlets, ratio of coiled/uncoiled forms, spinosity, and surface sculpture of the test (costa, striae). Loeblich (1957) reported that the planktonic foraminifers show a variety of morphological and structural adaptations for their floating existence. Thickness of shell varied with temperature and salinity. Increase of pore size, development of supplementary apertures, and aperture enlargment all appear to be related to reduction in the specific gravity. Similar adaptations apparently characterize *all* pelagic protists.

Zonation and Correlation. Suites of foraminiferal fossils either occur in established recognizable zones in the rock column sections, or permit regional correlation from unknown to known zones. The literature on the stratigraphic utility of foraminifers is enormous. Here we

need cite only a few selected examples. The Tertiary of California (Bandy and Kolpack, 1963) and the Gulf region have characteristic index foraminifers. So well established are such zones and others in South America and Europe ("Nummulitic," Figure 3.12*A*), that foraminifers picked from new samples frequently may be identified by inspection, and the zone, as well as the stratigraphic position of a given bed, quickly recognized.

The same applies to fusulinacean zones. Zones are defined by the dominance of a given genus or its limited vertical spread. The North American zones are the following, from older to younger: *Millerella* – the common genus in the Lower Pennsylvanian; *Profusulinella* – restricts to the early Middle Pennsylvanian; *Fusulinella* – dominates the lower Middle Pennsylvanian; *Fusulina* – found throughout the upper Middle Pennsylvanian; *Triticites* – characterizes Upper Pennsylvanian; *Pseudoschwagerina* – index of Lower Permian (Wolfcampian); *Parafusulina* – denotes the Lower Permian (Leonardian and lower Guadalupian); *Polydiexodina* – Lower Permian (late Guadalupian, may be partly equivalent to the southern European-Asiatic zone of *Verbeekina*); zone of *Yabeina* – characterizes some of the youngest fusulinacean beds in the eastern hemisphere and is known to occur above *Verbeekina* (also known from Canada and northwest United States).

The above fusulinacean genera are not restricted to the time indicated by any given zone. Thus *Millerella* ranges from the Upper Mississippian to the Upper

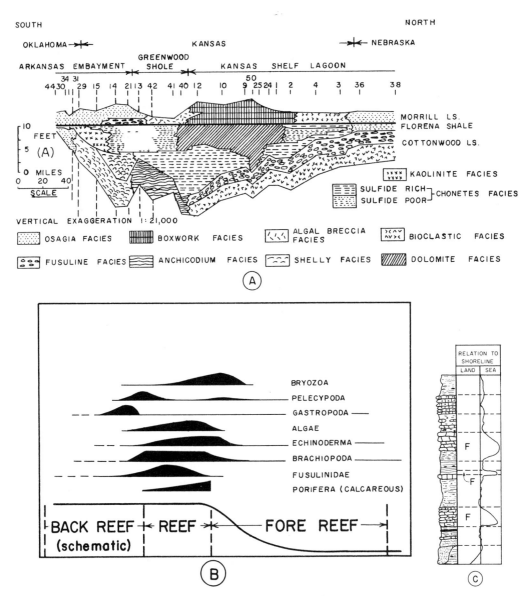

FIGURE 3.11 Fusuline phase of Pennsylvania-Permian deposits.
(A) Fusulines and other facies in the Beattie Limestone (after Imbrie et al., 1959). Shole = shoal.
(B) Inferred faunal distribution—Capitan Reef (from the Permian Complex of the Guadalupe Mountains Region, Texas and New Mexico, by N. D. Newell et al. W. H. Freeman and Company. Copyright © 1953). Note the greatest density of fusulines is on reef facies.
(C) Recurring fusuline facies (f) in a Pennsylvanian megacyclothem, (modified from H. C. Wagner, 1964).

Permian, but only in the Lower Pennsylvanian does it characterize a zone.

Radiolarians (Section 3)

Capsule. Radiolarians are distinguished from all other protozoans by the structure of a portion of their soft-part anatomy—the central body or capsule. This capsule[5] may be perforated or have one or more apertures (Figure 3.3K, Figure 3.13). The capsule separates two layers of the cytoplasm (extracapsular from intracapsular). It is seldom found as a fossil.

[5]According to E. G. Merinfeld (Dalhousie University), latest analyses (1975) show that all capsules are proteinaceous and lack chitin.

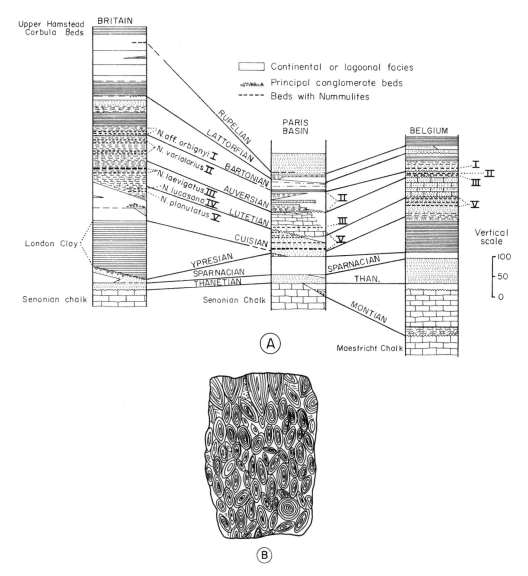

FIGURE 3.12 (A) Zonation and correlation by *Nummulites* species, in Europe and Great Britain. Paleogene (modified after Gignoux, after Wrigley and Davis). *Nummulites* species: (1) *N. planulatus,* (2) *N. lucasana,* (3) *N. laevigatus,* (4) *N. variolarus,* (5) *N. orbigny* or *N.* aff. *orbigny.* Note that of these five faunal zones, the Belgium column lacks zone 4, the Paris column lacks zones 1 and 4, whereas the Britain column is complete. However since all have zones 2, 3 and 5, and two have zone 1, interregional correlation is possible. Compare the ranges of species 2 and species 1. Explain the difference. (*B*) Nummulitic limestone with species 2 (above) in natural section on the surface of a Carpathian sample (after Zittel, 1900, and Tasch collection).

Scleracoma. Although the four radiolarian suborders are largely defined by the nature of the central capsule and its perforations or apertures, for paleobiological purposes characteristics of the scleracoma, the radiolarian skeleton, will be emphasized.

Suborder 1. Acantharia. Skeleton, centrogenous (i.e., supporting rods are generated at cell center) composed of acanthin or strontium sulphate. Range: Eocene to Recent (Figure 3.13, part *A*). (Acanthin is an organic compound of strontium sulphate.)

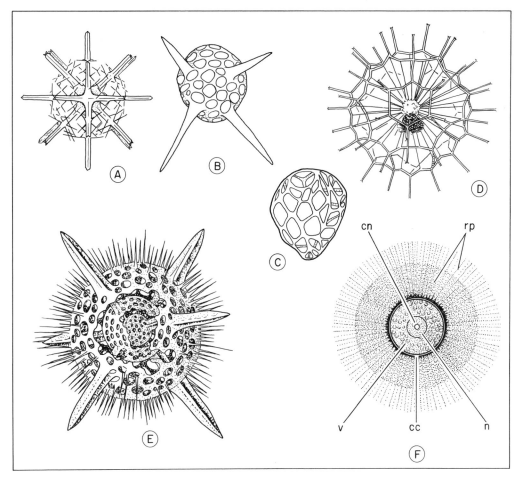

FIGURE 3.13 Genera belonging to the four radiolarian suborders, skeleton, and soft-part anatomy. (A) *Staurocantha*, Recent, an acanthine. (B) *Staurosphaera*, Ord.–Rec., a spumelline. (C) *Spongocystis*, Carb.–Jur., a nasselline. (D) *Cannosphaera*, Cret.–Rec., a phaeodarine. (E) Opaline silica skeleton, *Actinomma*. (F) Soft-part anatomy: *rp* – radial pseudopoda, *cn* – concentric nucleolus, *cc* – central capsule, *n* – nucleus, *v* – vacuoles. (Schematic, adapted from Campbell.)

Suborder 2. Spumellina. Generally spherical skeleton composed of opaline silica. According to A. S. Campbell (1954), this suborder, the most primitive, is the common stem of the other radiolarian suborders. Range: Cambrian to Recent (Figure 3.13, part B).

Suborder 3. Nassellina.[6] Skeleton, a tripod, ring, or lattice shell; opposite poles of shell disimilar; skeleton composed of opaline silica. Range: Cambrian to Recent (Figure 3.13, part C).

Suborder 4. Phaeodarina. Skeleton

[6] Suborders 2 and 3 are now united in a new subclass, Polycystina.

composed of silica-carbonate in the form of hollow or solid tubules, rods, or lattice. Range: Cretaceous to Recent (Figure 3.13, part D).

Reproduction. Seasonal or periodic reproduction in radiolarians occurs in "epidemic form" when waters are locally and temporarily high in silica. This relationship helps to explain the radiolarian cherts in the rock column.

Ecology. All radiolarians presently known live in marine environments. Generally they are free floaters among the plankton (see "Morphology," below). Representatives of suborder Phaeodarina and some Polycystina were collected from abyssal depths. Most species belonging to the Polycystina and Acantharia, by contrast, are pelagic. *Almost all fossil radiolarians are upper-zone pelagic types* — characterized, as are their Recent equivalents, by spherical and elliptical configurations and a spongy structure.

Campbell (1954) found that living radiolarians may float either at the surface or close to the sea bottom, and are found in shallow water as well as abyssal depths. Aberdeen (1940) reported that the same genus may embrace both deep- and shallow-water forms. This observation is rather important because fossil radiolarians had been generally assumed to be deep-water inhabitants. Today radiolarians occur in all seas; at the same time the Pacific Ocean yields the richest faunas.

Oozes. Radiolarians belong to the group of marine forms (planktonic diatoms, silicoflagellates, and some benthic sponges) that secrete skeletal structures of opaline silica. These skeletons are, in a majority of cases, readily dissolved upon the death of the organism. However, a portion of intact skeletons settle to the bottom, so that, in pure oozes, radiolarians may constitute as much as 75 percent or more of the total constituents (Campbell).

Radiolarian ooze is red, chocolate-brown, or straw-colored. About 6.6×10^6 km² of the Pacific floor is covered by radiolarian oozes. In contrast, the Indian Ocean floor covered by such oozes runs to about 0.3×10^6 km² (Sverdrup, Johnson, Fleming, 1946, Table 106). Elsewhere, a given deep-sea ooze may contain less than 5 percent radiolarian skeletons. Such mixed oozes are more widespread than the high radiolarian oozes of the Pacific and Indian Oceans. For example, Pessagno (1963) reported on an upper Cretaceous radiolarian fauna from Puerto Rico. This fauna contained 50–80 percent planktonic foraminifera. The author wrote that the radiolarians were masked by abundant foraminifers in the sediment and that many

were filled and coated with calcium carbonate.

Even after deposition, a continuous dissolution of siliceous skeletons may occur. Arrhenius had concluded that this is due to the "instability of silica in the interstitial water of the sediment." Dissolution of siliceous skeletons occurs in the following order: silicoflagellates, then diatoms, subsequently radiolarians and, in extreme instances, siliceous sponge spicules.

Morphology: Skeletal   Adaptations. Small-sized radiolarians are characteristic of warm surface waters, whereas large tests occur in the cold, deeper waters. A relatively delicate skeleton is found in the pelagic radiolarian when compared to the massive, solid skeleton of the abyssal type.

Pelagic radiolarians have skeletons especially adapted to remain in the euphotic zone. Globular-shaped tests and long pseudopods help to resist sinking. This is critical for forms that do not take in solid food but rather thrive on yellow cells of xooanthellae that live and multiply in their cytoplasm, depending on photosynthesis. Accordingly, such forms must remain in the euphotic zone.

Geometry. A graduate student in mathematics, Gary Crown, matched representative radiolarian test shapes to a given equation as an exercise in a micropaleontology class. When these equations were plotted on graphs, figures comparable to those of the named radiolarians were obtained by joining the plotted points. (These are illustrated in Figure 3.14.)

Information Derived From Fossil Radiolarians

Spread. A glance at Table 3.4 shows the vertical distribution of radiolarian species.[7] Reported Pre-Cambrian radiolarians are of doubtful age. Over 2000 radio-

[7] Two types of fossil distribution may be distinguished: *geologic* distribution or *vertical* spread through time; *geographic* distribution or *horizontal* spread, that is, existing at the same time.

For more current Neogene data, see Riedel and Sanfilippo, 1970, 1971, ref. p. 77.

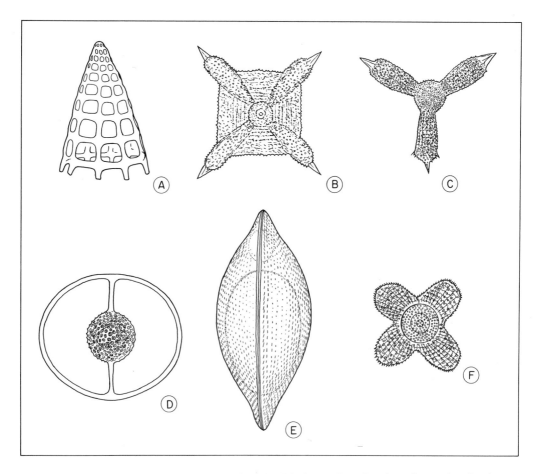

FIGURE 3.14 Some equations solved in selected radiolarian configurations (equations assigned and graphed originally by Dr. Gary Crown).
(A) *Bathropyramis*, paraboloid of revolution:

$$\frac{x^2}{a^2} + \frac{y^2}{b^2} = \frac{z}{c}.$$

(B) *Dicoccrura*, four leaf rose: $r = a \sin 2\theta$.
(C) *Trigonactura*, three leaf rose: $r = a \sin 3\theta$.
(D) *Saturnalis*, ellipse:

$$\frac{x^2}{a^2} + \frac{y^2}{b^2} = 1.$$

(E) *Conchopsis*, conchoid of Nicomedes: $(x-a)^2(x+y)^2 = b^2x^2$.
(F) *Astrococcus*, epicycloid:

$$x = (a + b) \cos \theta - a \cos [(a + b)\theta/a],$$
$$y = (a + b) \sin \theta - a \sin [(a + b)\theta/a].$$

TABLE 3.4. Distribution of Fossil Radiolarian Species in the Rock Column

Age	Number of Species	Remarks
Cenozoic	1500	Some 500 species from Barbados (U. Eocene)
Mesozoic	421	At least one-half the total number from the Alpine Jurassic
Paleozoic	261 (?)	Example: Caballos novaculite (Dev.) Texas, 24 species

Source. Campbell, 1954.

larian species are known from Cambrian to the present.

Lithologies. Fossil radiolarian skeletons may be retrieved from soft sediments or studied in thin sections in silicified material. They have been found in cherts, quartzites, siliceous shales, tuffs, silicified coprolites, soft marls and clays, limestones, and geosynclinal deformed calcareous shales and siltstones (Flysch). Association with volcanics, as in the Franciscan chert of California, is not uncommon.

Depth. Most radiolarian cherts have been interpreted to be of shallow-water origin. The common association with coarse clastics, it is thought, supports this decision. However, as Fagan (1962) pointed out, this last argument is negated by the accumulating knowledge on turbidity currents and their capacity to transport coarse shallow-water debris to deeper waters (turbidites). This is even more apparent in such examples as the Carboniferous Schooner bedded cherts of Nevada where single beds are generally 8–10 cm thick. Here the whole formation is a eugeosynclinal deposit and coarse turbidites are interbedded with clays and radiolarian cherts.

In the Caballos chert (Devonian, Texas) by contrast, preservation of some of the cytoplasm suggests rapid burial of radiolarians. Since sedimentation in deeper waters is generally slow, this type of burial would be favored by shallow depths (Aberdeen, 1940).

Laminated dark blue-gray to black siliceous shales with some thin cherts recur in the late Paleozoic Stanley-Jackfork-Atoka rocks of the Oklahoma Ouchitas. Radiolarian tests are common in association with siliceous sponge spicules. Graded bedding observed in these beds would support the interpretation of deposition in deeper waters (Cline, 1960).

Population Density, Seasonal Blooms, and Geochemistry. Some 50–60 radiolarians have been observed in a microscopic field having a diameter of 3.1 mm. This leads to the estimate of 1 million individuals per cubic inch of consolidated sediments (David and Pittman, 1899). In turn, a seasonal bloom is reflected by these data. The bloom itself was encouraged by a local increase in silica in the sea.

It is important to recall, except for local situations near centers of active volcanoes, that the origin of chert in marine sediments cannot be accounted for by inorganic precipitation (Krauskopf, 1956). Rather, siliceous skeletal remains on the sea floor may undergo partial solution, and the silica may be redeposited. Skeletal scraps can be seen not infrequently in thin sections of radiolarian cherts.

A few computations convinced the writer that in a volumetric sense there are 100 times more radiolarian skeletons (viewed as perfect spheres) that could be packed in a cubic foot of consolidated rock than David and Pittman reported. It follows that the observed population densities in thin sections of chert may represent as little as 1/100 of the original population. This appears to account for a sufficient number of absent (that is, dissolved) radiolarian skeletons to have provided the needed silica for chert formation.

Correlations. The large number of species and the extended vertical spread of radiolarians hinder their general use for widespread Paleozoic stratigraphic correlation. However, they have proven valuable for Mesozoic stratigraphy (E. A. J. Pessagno, 1977, *in* A. T. Ramsey, Ed., *Oceanic Micropaleontology,* vol. 2, pp. 913–938), and worldwide Neogene correlations (see: Riedel ref. p. 77).

Identification of a Fusulinacean to the Species Level

If the specimen being studied belongs to a *known* species, a description and classification of it will be available in print.

If it is *unknown*, that is, represents a new species, that fact cannot be established without a careful search of the literature and a comparison of critical measurements with those of known species closest to it. Useful references for this purpose are the following.

1. Protista, Loeblich et al., 1964. This is the latest and most authoritative reference on the generic and familial levels as well as for higher categories.
2. Catalogue of Foraminifera, Messina and Ellis. An encyclopedic compilation with new units issued periodically. Useful for descriptions of known species of a given genus.
3. Foraminifera, Cushman, 1950. A useful

volume at the generic and familial levels.

4. Journal publications (*Journal of Paleontology, Micropaleontology,* among others) and individual monographs. These are valuable for descriptions and illustrations of new species that may not yet have been incorporated into standard reference works. They also provide additional data on known species.

Example. A thin fusulinacean zone was discovered in the mid-Pennsylvanian Bethany Falls limestone, Missourian of Kansas and Missouri (Newell and Keroher, 1937).

Problem. To identify the fusulinaceans occurring in this bed and to establish

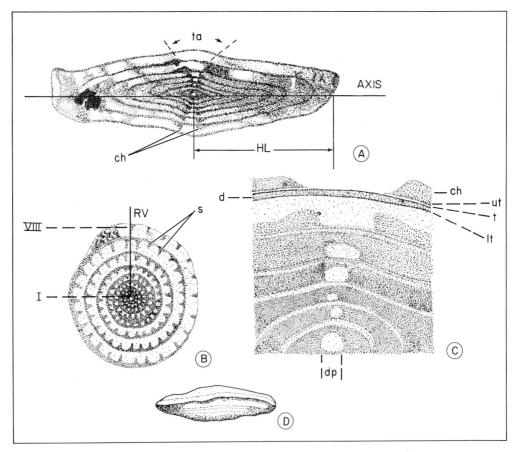

FIGURE 3.15 *Wedekindellina ultimata*—a mid-Pennsylvanian fusulinid: (*A*) Axial section; (*B*) sagittal section; (*C*) detail of wall structure; (*D*) Bethany Falls specimen before sectioning (greatly enlarged). In "*A*" *ta* represents tunnel angle. To measure, place thin section on stage of polarizing microscope. Line up one cross-hair from center of proloculum along margins of chomata and record degrees (*P*) on scale. Rotate stage until cross-hair is similarly aligned with inner margin of opposite set of chomata. Read degrees (*Q*); then *P* − *Q* = tunnel angle, *ch*—chomata; *ut*—upper tectorium; *d*—diaphanotheca; *lt*—lower tectorium; *HL*—half length (axial); *RV* (= *HD*)—half diameter; *dp*—diameter; of proloculum. Numbers I–VIII: volutions; starting at line *RV* one makes a complete circle ending at *RV* to complete one volution; *s*—septa. (After Newell and Keroher; Thompson; and Dunbar.) Tunnel angle varies a little from whorl to whorl.

relationships with equivalents reported elsewhere in beds of comparable age.

Procedure (see Figure 3.15, part A–C).

1. Observe shape of shell. Fusulinaceans are slender, highly elongate — fusiform. In other words, the shell is coiled about the long axis.
2. Prepare axial and sagittal thin sections (see Figure 3.15, part A, B).
3. The following observations can now be made:
(a) The number of layers in the spirothecal wall: The wall is four-layered. It is composed of a tectum, diaphanotheca, upper and lower tectoria (see Figure 3.15, part C).
(b) Septa are unfluted.
(c) Axial fillings are massive except for last part of the last volution.
(d) Chomata are massive and broad.
4. Consult key to fusulinacean genera (Cushman, 1950, p. 458, and check against Thompson, 1964). The only fusiform genera with a four-layered wall that meets the above criteria either lack unfluted septa or possess still other characteristics. The specimens fit the description of the genus *Wedekindellina*.
5. The problem remaining is determination of the species. In the example being analyzed, the investigators did this by obtaining measurements on 20

axial and 20 sagittal thin sections. (See Dunbar and Henbest, 1942, for the techniques of measuring fusulinaceans.) The items measured or computed included axial length (half-length, *HL*, is usually measured at present), sagittal width (half-diameter, *HD*, is usually measured at present), form ratio (*HL/HD*), tunnel angle, number of volutions, proloculus diameter, wall thickness, septal count for each volution or whorl (see Figure 3.15, part A).

6. A tabular comparison can now be made with the known American species of *Wedekindellina*. At the time of the given study ten species were known. The specimens being studied differed from all of these species in the following ways:
(a) Larger proloculum than any known species.
(b) Distinct form ratio (*HL/HD*).
(c) Septal count for each volution readily distinguished the studied specimens from five of the known species.
(d) Thicker diaphanotheca than any known species.
7. The studied material (based on data noted above) apparently represented a new American species. It was named *Wedekindellina ultimata* Newell and Keroher, 1957.

QUESTIONS

Introduction, Foraminifers, Fusulinaceans

1. On what basis are living protozoans subdivided into subphyla or classes.
2. (a) What criteria would permit you to isolate foraminifers from the other protists? (See Figure 3.3.)
   (b) Among the foraminifers, what further criterion would further allow you to distinguish the five suborders?
   (c) Once 2b is achieved, what other information would you automatically have?
3. Taxonomically, what information may be obtained from foraminiferal tests by study of each of the following: presence

or absence of radiate aperature; position of aperture; number of chambers; coiling.

4. What distinctive characters, if any, are to be found in planktonic compared to benthonic foraminifers? (See Figure 3.9*B*.)

5. From Figure 3.8*A*, draw a megalospheric and a microspheric test. What information may be obtained from a sample of fossil tests identified as one or the other, or containing both of these?

6. In fusulinaceans, what information is contained in each of the following: proloculus, prolocular pore, keriotheca, antetheca, configuration, alveoli, calcareous deposits (axial chromata), openings (foramina, tunnel, pores).

7. (a) How is a fusulinacean zone defined?

(b) What dominant fusulinacean genus would you expect to find in the early middle Pennsylvanian; the lower Permian?

8. Precisely how was the apparent worldwide extinction of fusulinaceans established?

9. Compare foraminifers to coccolithospheres as contributors to the geochemical cycle of oceanic carbonate. (See Table 3.3.)

Radiolarians[8]

1. (a) Distinguish between a fossil radiolarian and a fossil silicoflagellate. (*Hint.* See Section A of this chapter.)

   (b) Study the geologic range of the four radiolarian suborders, and rearrange them in an evolutionary sequence from most primitive to geologically youngest to appear.

2. (a) Can you find two or three plausible explanations as to why none of the suborders have become extinct?

   (b) Discuss the distribution of radiolarian oozes in the modern oceans and the implications for chemical oceanography.

   (c) Discuss the bathymetry of the four radiolarian suborders.

3. Relative to siliceous sponge spicules, diatoms, and silicoflagellates, what position do radiolarians have in the order of dissolution of silica?

4. Distinguish the existing biotic situation in the oceans indicated by pure, as contrasted to mixed, oozes.

5. (a) How do pelagic radiolarians resist sinking below the euphotic zone of the sea?

   (b) What is the generalization that applies to all pelagic protists?

6. How does Krauskopf's findings help to clarify the origin of radiolarian cherts?

[8]W. R. Riedel on Neogene radiolarian derivative data: sea floor spreading (Science, 1967, v. 157 (3788), pp. 540–542); Pliocene-Pleistocene boundary (Science, 1968, v. 140 (3592), pp. 1238–1240).

REFERENCES

Foraminiferida

Bandy, O. L., 1960. General correlation of foraminiferal structure with environment: Intern. Geol. Cong. XXI, Session. Norden, Part 22, pp. 7–19. Copenhagen.

_____, and R. L. Kolpack, 1963. Foraminiferal and Sedimentological trends in the Tertiary section of Tecolote Tunnel, California: Micropaleontology, Vol. 9 (2), pp. 117–170.

Cushman, J. A., 1950. Foraminifera. Harvard Univ. Press. 478 pp., plates 1–55, index (should be supplemented by reference to Loeblich and to Thompson).

Dunbar, C. O., and L. G. Henbest, 1942. Pennsylvanian Fusulinidae of Illinois: State Geol. Survey Bull. 67, Urbana, pp. 35 ff., 23 plates.

Elias, M. K., 1962. Comments on recent paleontological studies of late Paleozoic rocks in Kansas: 27th Field Conf., Kans. Geol. Surv. Guidebook, pp. 106–115.

Imbrie, J., L. Laporte, and D. F. Merriam, 1959. Beattie limestone facies and their bearing on cyclical sedimentation theory: 24th Field Conf. Kans. Geol. Surv. Guidebook, pp. 69–78, 6 text figures.

Loeblich, Jr., A. R., Helen Tappan, et al., 1964. Sarcodina, chiefly "Thecomoebians" and "Foraminiferida," R. C. Moore, ed., Treatise on Invertebrate Paleontology, Part C, Protista 2, 2 vols.

_____, and collaborators, 1957. Studies in Foraminifera: U.S. Nat. Mus. Bull. 215, pp. 1–235, 74 plates.

Newell, N. D., and R. P. Keroher, 1937. The fusulinid *Wedekindellina* in mid-Pennsylvanian rocks of Kansas and Missouri: J. Paleontology, Vol. 11 (8), pp. 698–705, plate 93, text figures 1–4.

Phleger, F. B., 1960. Ecology and distribution of Recent Foraminifera. Johns Hopkins Press, Baltimore. 276 pp., refs., index.

Pokorný, V., 1958. Principles of zoological micropaleontology. Trans. K. A. Allen; J. W. Neale, ed. (1963: MacMillan Co.,

New York). Vol. 1, 593 pp., index (excellent bibliography).

Ross, C. A., 1962. The evolution and dispersal of Permian fusulinid genera *Pseudoschwagerina* and *Paraschwagerina:* Evolution, Vol. 16 (3), pp. 306–315.

Skinner, J. W., and G. L. Wilde, 1954. New early Pennsylvanian fusulinids from Texas: J. Paleontology, Vol. 28 (6), pp. 796–803, plates 95, 96.

Thompson, M. L., 1948. Studies of American fusulinids: Protozoa, Art. 1: U. Kans. Publ., pp. 1–102, plates 1–38, Figures 1–7.

_____, 1951. Wall structures of fusulinid Foraminifera: Cushman Found. Foram. Research Contrib., Vol. 2, part 3, pp. 86–91, pls 9–19, 1 fig.

_____, 1964. Fusulinacea: in Loeblich and Tappan, Sarcodina. (See ref. above), v. 1, C 358 ff.

Weinrich, D. H., 1954. Sex in Protozoa, in Sex in microorganisms: A.A.A.S., 346 pp. (Dinoflagellates, pp. 139–143; Foraminifera, pp. 175–179; Radiolaria, pp. 181–187).

Zeller, E. J., 1952. Stratigraphic significance of Mississippian endothyroid Foraminifera: U. Kans. Paleont. Contrib., Protozoa, Art. 4, pp. 1–23, plates 1–6.

Radiolaria[9]

Aberdeen, Esther, 1940. Radiolarian fauna of the Caballos formation, Marathon

Basin, Texas: J. Paleontology, Vol. 14 (3), pp. 127–139, plates 20, 21.

Arrhenius, G., undated. Pelagic sediments: Contrib. Scripps Instit. Oceanography, New Series, U. Calif., Preprint.

Campbell, A. S., 1954. Radiolaria, in R. C. Moore, ed., Treatise on Invertebrate Paleontology. Pt. D. Protista 3, pp. D-11–D-163.

Cline, L. M., 1960. Late Paleozoic rocks of the Ouachita Mountains: Oklahoma Geol. Survey Bull. 85, 106 pp., index, map.

David, T. W. E., and E. F. Pittman, 1899. On the Paleozoic radiolarian rocks of New South Wales. Quart. J. Geol. Soc. London. Vol. 55, pp. 16–37.

Fagan, J. J., 1962. Carboniferous cherts, turbidites, and volcanic rocks in Northern Independence Range, Nevada: Bull. Geol. Soc. America, Vol. 73 (5), pp. 595–612.

Krauskopf, K. B., 1956. Dissolution and precipitation of silica at low temperatures: Geochemica et Cosmochemica Acta, Vol. 10, pp. 1–26.

Pessagno, E. A., 1963. Upper Cretaceous radiolaria from Puerto Rico: Micropaleontology, Vol. 9 (2), 197–214, plates 1–7. Micropaleont. Sp. Paper (1974).

Stainforth, R. M., 1948. Applied micropaleontology in coastal Ecuador: J. Paleontology, Vol. 22, pp. 113–151, plates 24–26, Figures 1, 2.

Sverdrup, H. U., M. W. Johnson, R. H. Fleming, 1946. The Oceans. Prentice-Hall, 1049 pp.

[9]For Neogene correlations, Riedel and Sanfilippo, 1970, 1971 Deep Sea Drilling Project, Vols. 4, 7 (Washington).

Tintinnids (Section 4)

Suborder Tintinnia. Range: Silurian to Cretaceous; Pleistocene to Recent. Thirteen families and 73 genera are included. Of these, 6 families embrace fossil as well as recent genera.

Specialized Protozoans. The ciliates, which embrace such diverse yet related forms as *Paramecium*, *Stentor*, and tintinnids, are perhaps the most specialized of protozoans; this can be seen for example in their food-catching organelles and pellicle differentiation, — Hyman, 1940 (Figure 3.16). Some 40 percent of all presently known marine and fresh-water infusorians are tintinnids — about 800 species (Campbell, 1954).

Figure 3.16, parts 1, 2 shows the permanent hairlike projections from the body surface (cilia) and the macronucleus (reproductive) and micronucleus (vegetative) that distinguish the ciliates from other protozoans.

Soft-Part Anatomy. The tintinnid body is conical or trumpet-shaped. It is ciliated and enclosed in a very delicate vase-shaped case or test that is called the *lorica*. Only at one place is the green-colored body attached to the test by a pedicel (Figure 3.16, parts 1, 2). Open space separates the rest of the body from its enclosing case.

The green coloration is attributed to chlorophyll that is either taken in with food (diatoms, dinoflagellates, coccolithospheres, bacteria, and smaller ciliates) or derived from zoochlorella (green algal cells).

Reproduction. Tintinnids, as other protozoans, may reproduce by sexual mating (conjugation) or asexual means (binary fission).

Locomotion. Unlike such protists as the radiolarians, tintinnids move by means of feathery locomotor membranelles (Figure 3.16 part 1). The oral end is directed backward in movement. Many of the streamlined configurations of the lorica, viewed in light of the mode of locomotion, appear to offer less frictional resistance than to forward movement (Figure 3.16, parts, 1, 5–7). Streamlined configurations would tend to confer a selective advantage on any tintinnid population.

Lorica. The gelatinous or pseudochitinous, cuplike or elongate, lorica of tintinnids is frequently agglutinated. Foreign particles encrusted or included in the delicate membranous wall may consist of fine mineral grains, coccoliths, diatoms, and organic debris (Colom, 1948, 1959).

The shape of the skeleton in living, as well as in fossil, tintinnids is extremely diverse (Figure 3.16).

FIGURE 3.16 Tintinnids: living and fossil. (1) *Tintinnopsis campanula;* after Colom, after authors; *c*—collar; *oz*—oral zone; *az*—aboral zone; *ca*—caudal appendage. (2) *Stenosenella nivalis,* after Campbell; cutaway to expose soft parts; *min*—micronucleus; *mcn*—macronucleus; *p*—pellicle; *cl*—ciliary lines; *bc*—body cilia; *l*—lorica (the part that is fossilized); *pe*—pedicel (by this means body is attached to lorica). (3)–(7) Streamlined configurations in lorica of living tintinnids (after Marshall): (3) *Rhabdonella.* (4) *Salpingella.* (5) *Tintinnus.* (6) *Xystomella.* (7) *Epiplocyclis.* (8) *Calpionellopsis.* (9) *Calpionellites.* (10) Same as 8. (11) *Stenosemellopsis.* (12) *Coxiella.* (13) *Amphorellina.* (14) *Calpionella* (elongate type, usually broad horseshoe shape). (15) *Parafavella.* (16) *Nannoconus,* an algae of uncertain position, often associated with tintinnids; as seen in thin section. (17) *Rhabdonelloides.* (18) *Favelloides.* (19) *Tintinnopsis.* (20, a–c) *Tintinnopsella carpathica,* cross-sections along aa, bb, cc (right figures) are shown as they appear (with different shapes) in thin sections (left figures, a, b, c). (Figures schematic; 16—after Trejo; others after Colom.

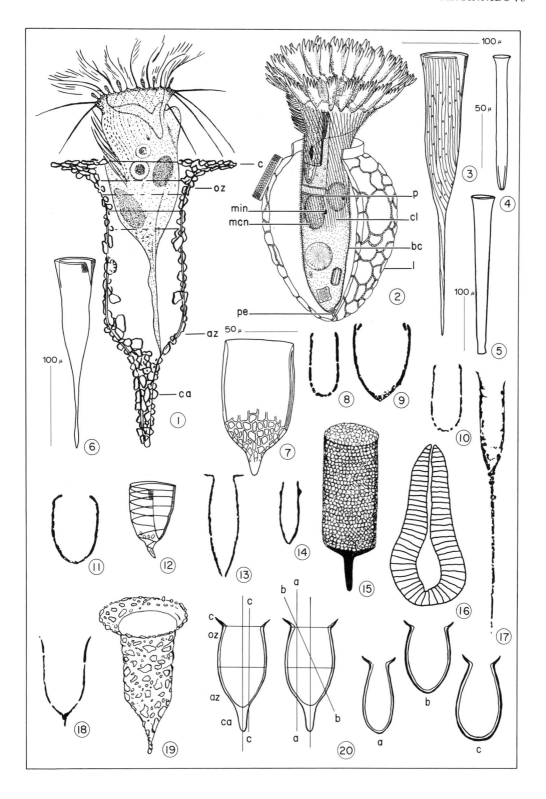

Campbell (1954) maintains that the lorica is so delicate that it could not serve a generally protective function; rather it must be related to the pelagic existence tintinnids lead. In other words, the lorica aids flotation. As with all pelagic protists the test configuration helps to slow down the animal's sinking below zones of optimum food supply—the upper portion of the euphotic zone. Flat projecting collars of some forms may serve to brake or retard vertical drop (Pokorný) and may be comparable to parachute structures in some dinoflagellates.

Food Chain. Tintinnids frequently occur in the food vacuoles of silicoflagellates, whereas diatoms, dinoflagellates, and coccolithospheres are part of the tintinnid food intake. Although the last-named forms constitute primary producers in the sea, marine biologists report that tintinnids and silicoflagellates do not occupy an important position in this economy.

Habitat, Distribution, and so on. Tintinnid species are generally marine, pelagic, and cosmopolitan. They occur at all latitudes as well as in all seas (Arctic and Antarctic waters too). Specific species are restricted to particular oceanic currents such as the Humboldt Current and the Mexican Current. Only about 2 percent of tintinnid species are known from fresh waters (estuaries, relict lakes, and some fossils from Quaternary peat bogs). Other species like *Favella* occur mainly in near-shore waters. Marshall (1934) recognized two large groups in the Great Barrier Reef collections: neritic and oceanic. The first group can tolerate lesser salinities and are found year round; the second group require relatively high salinity and are rarer in occurrence [August and September at salinity 35%. (%o = parts per thousand)].

Speciation among tintinnids, as among many other forms, is accelerated in tropical waters.

Size and Temperature. Kofoid (1930) studied 1000 individuals belonging to the same species, *Tintinnus tenue*, and found that those taken at stations with surface temperatures of 19°–23°C were predominantly larger than those from stations with temperatures of 24°–28°C.

Fossil Tintinnids

Fossil Record. Tintinnids made their apparent first appearance during the Silurian. They were first found in abundance in the European Mediterranean area (Tethys Sea). Since then, discoveries show that they also thrived elsewhere—in Mexico (Trejo, 1960) and Cuba, for example. Although abundant in modern seas, their fossil record as of this writing does not extend beyond the Cretaceous (Barremian): Table 3.5. Obviously, living forms descended from preexisting forms that must be traced back to the Barremian forms. Accordingly, one can expect that a late Cretaceous-Tertiary record will be found.

A discovery extended the range of tintinnids to the Upper Cretaceous (Cenomanian) with the single exception in the Scandinavian Quaternary. Hundreds of crushed three-dimensional fossil tintinnids have been found in Cretaceous shales, such as the Graneros shale of Wyoming, South Dakota, and Colorado (Eicher, 1964, 1965). Associated arenaceous foraminifers suggest a shallow water environment. The fossils are closer to living genera than to Mesozoic genera.

Classification. Fossil tintinnid genera are all assigned to known families erected from study of living forms (Campbell, 1954). Some ten fossil genera are known and assigned to five families. The chief features in fossils to be observed in order to make a taxonomic assignment are discussed below.

Lithology. The actual material preserved are fossilized lorica; that is, the original organic tests have been replaced

TABLE 3.5. Tintinnid Range Chart[a]

	Tithonian	Berriasian	Valan-ginian	Haute-rivian	Bar-remian	Aptian
Calpionella alpina	———	—				
Calpionella elliptica	———	—				
Calpionella undelloides	———	——				
Calpionella massutiniana	——	——				
Calpionellopsis thalmanni			———	———	——	
Calpionellopsis simplex			———	———	——	
Calpionellites darderi			———	———	——	
Calpionellites neocomiensis		——	———	— —		
Stenosemellopsis hispanica		——	———	——		
Favelloides balearica		——	———	——		
Favelloides pseudoserrata		——	———	——		
Tintinnopsella carpathica	——	———	———	———	— — —	
Tintinnopsella longa		——	———	— — —	— —	
Tintinnopsella cadischiana		——	———	— — —	— —	
Tintinnopsella batalleri		——	———	— —	—	
Amphorellina lanceolata			———	——		
Amphorellina acuta			———	——		
Rhabdonelloides ineserrata			———	——		

[a] Ranges are extended by more recent data to Cenomanian (U. Cret.) and one report in the Quaternary. *Source.* See Tappan and Loeblich Jr., *J. Paleontology* 42(6): 1378–94 for Tertiary (Eocene, Oligocene) tintinnids.

by calcite crystals and/or originally agglutinated forms that had been fossilized. Cross-sections of such tests (Colom, 1948, plates 33, 34) are embedded in a sublithographic limestone matrix. Outlines of fossil tintinnids (that can be seen only in thin sections) appear white compared to the gray microcrystalline calcareous ooze in which they are embedded. Such oozes (fine-grained pelagic calcareous muds) often contain abundant coccoliths.

A sample from Majorca (L. Neocomian), provided by Dr. G. Colom, is a white-gray sublithographic limestone. It contains the tintinnid genus *Calpionella* (with a horseshoe-shaped outline) and a probable chlorophyllous alga *Nannoconus*, which is a frequent biotic associate of fossil tintinnids.

Often tintinnids may be preserved in marls, cherts, peat (gyjtta) and shale, although most frequently they are found in sublithographic limestones that represent the deeper-water lithotopes.

The preparation of thin sections of tintinnid-bearing limestones can result in various irregular views of the given tintinnid fossil outline. Thus the same species might show several different aspects (shapes), depending on the orientation relative to the longitudinal axis—which alone reveals the true configuration (Figure 3.16, part 20).

Morphology. The most commonly observed features of the fossil tintinnid outline are the shape along the longitudinal axis, collar, caudal appendix, and oral and aboral zones (Figure 3.16, part 1). The detailed structures such as ribs, plications, flutings, reticulations, shelves observed in the lorica of living forms are not retained when they are imbedded in calcareous bottom ooze, and the test is replaced by calcite crystals, or if agglutinated, is recrystallized.

Biotic Associates. Species of the tintinnid genus *Calpionella* are frequently associated with *Nannoconus*. Coccolithospheres are not uncommon in pelagic muds in which tintinnid loricas are embedded. Various algae of uncertain affinities (*Globochaete*, *Eothrix*) occasionally occur with fossil tintinnids. Generally, radiolarians are absent in tintinnid-bearing limestones but infrequently do occur in them and in limited numbers. However, radiolarians constitute the major recurrent biofacies in Mesozoic bathyal geosynclinal deposits of the western Mediterranean (Colom, 1955, text figure 1). It is in this same sequence of deposits that the tintinnid facies has a more restricted spread.

Information Derived from Fossil Tintinnids

Morphology. The significance of the true configuration, that is, along the longitudinal axis, has been discussed for living tintinnids. The same interpretation may be made for fossil lorica.

Oceanography

Mesozoic Plankton (Tethys Area, and so on). The biotic associates of tintinnids and the tintinnids themselves reflect the plankton of the time. In turn, the abundance of such forms represented as fossils implies a whole sequence of relevant factors: nutrient salts, temperature, food chain, chemical composition of the water — all of which must have been within suitable limits to support such large plankton populations.

Alpine Geosynclinal Belt. A glimpse of bathyal sedimentation in the deeper parts of the Alpine geosynclinal belt during the Mesozoic is afforded by the condition of sedimentation that can be gleaned from tintinnid-bearing limestones. Included are such items as prevalence of fine, calcareous, microcrystalline bottom muds; calcification of the organic lorica of tintinnids; or destruction of surface and wall structures of the lorica by recrystallization after embedding, by bottom current action in fine-grain calcareous pseudo-breccia interbeds.

QUESTIONS

Tintinnids

1. (a) Of what is the lorica composed?
   (b) What protists may be found incorporated in agglutinated loricas?
2. Discuss streamlining of the lorica and the information it contains (see Figure 3.16, parts 3–7).
3. Review distribution of tintinnids in each of the following: marine and fresh water; by latitude; restricted to specific currents; neritic vs. deeper oceanic.
4. What could you conclude about the habitat if you found a fossil *Calpionella* in fine-grained calcareous indurated muds containing abundant coccoliths?
5. Study the geologic ranges of several fossil tintinnids listed in Table 3.5. List the species that are the best index fossils, and pick the top of the Mediterranean Tithonian.
6. How can Mesozoic tintinnid-bearing limestones shed light on bathyal sedimentation in a geosynclinal basin?

REFERENCES

Tintinnids

Campbell, A. S., 1951. Tintinnia, in R. C. Moore, ed., Treatise on Invertebrate Paleontology. Protista, Part B, pp. D166 ff.

Colom, G., 1948. Fossil tintinnids; loricated Infusoria of the order Oligotricha: J. Paleontology, Vol. 22, pp. 233–263, plates 33–35, Figures 1–14.

———, 1955. Jurassic-Cretaceous pelagic sediments of the western Mediterranean zone and the Atlantic area: Micropaleontology, Vol. 1 (2), pp. 109–124, plates 1–5, text figures 1–4.

———, 1965. Essais sur la biologie, la distribution géographique et stratigraphique des Tintinnoïdiens fossiles.: Ecolo-gae Geol. Helvetiae, Vol. 58 (1), pp. 319–334, 3 figs, 4 tables, 3 pls. (excellent bibliography).

Eicher, D. L., 1964. Cretaceous tintinnids from the Western Interior: G. S. A. Program (Miami Beach, Florida, November, 19–21). Abstracts, p. 54.

———, 1965. Same: Micropaleontology,

Vol. 11(4), pp. 449–456, pl. 1

Kofoid, C. A., 1930. Factors in the evolution of the pelagic ciliata, The Tintinnoinea: Contributions to Marine Biology, Stanford Univ., pp. 1–39, Figures 1–3.

————, and A. S. Campbell, 1939. The Ciliata, the Tintinnoinea: Harvard Coll. Mus. Comp. Zool., Bull., Vol. 84, pp. 1–473, plates 1–36, map.

Marshall, S. M., 1934. The Silicoflagellata and Tintinnoinea: Great Barrier Reef Exped. 1928–1929. Scientific Rpt. Brit. Mus. (Nat. Hist.), Vol. 4 (15), pp. 623–660, text figures, Table 1, refs. (Tintinnids, pp. 632 ff).

Trejo, M., 1960. La familia Nannoconidae y su Alcance Estratigrafico en America (Protozoa, Incertae Sedaedis): Bol. Assoc. Mexicana de Geól. Petroleros, Vol. 12 (9, 10), pp. 259–314, plates 1–3.

[Tappan, H. and A. R. Loeblich, Jr., 1968. J. Paleontology, vol. 42(6) on lorica composition and systematics.]

chapter four

PORE BEARERS OF MODERN AND ANCIENT SEAS

A. PHYLUM PORIFERA

In succession and over the centuries it was concluded (a) that sponges were animals, (b) that sponges were not related to corals, and (c) that sponges were distinct from other animals.

Recent researches suggest enlargement of the sponge realm to include stromatoporoids (Stearn, 1972), archaeocyathids (Balsam and Vogel, 1973; Ziegler and Rietschel, 1970), and chaetetids — usually classified as tabulate corals (see Chapter 5), (Hartman and Goreau, 1970). Only the stromatoporoids are included here since specialists are still debating the other suggested reassignments (p. 115).

Living Sponges

1. *Phylum Definition (data from Laubenfels; Hyman and others).* Phylum Porifera (Latin *porus* = pore + *fera* = to bear), sponges. The phylum has the following characteristics.

(a) Organization — cellular grade.

(b) Aquatic (dominantly marine).

(c) Metazoa.

(d) Asymmetrical or radially symmetrical.

(e) Without organs, mouth, or nervous tissue.

(f) With a body permeated by pores, canals, and chambers (inlets and exits).

(g) Water flows through openings by action of numerous flagellate cells that line the internal cavities.

(h) Larvae, free-swimming; adult, sessile.

(i) Reproduction, chiefly sexual; asexual propagation by gemmules common in freshwater sponges.

(j) Skeleton: internal, generally present, consists of crystalline spicules, organic fibers, or both; sometimes admixed with foreign particles.

(k) Geologic range: Pre-Cambrian to Recent.

Item (a) is the critical point: Organization — cellular grade. Sponges occupy a midway position between the single-celled protozoans and coelenterates: either completely independent, noncooperative individual living cells, or a colonial organization of individual cells as in *Volvox* and *Proterospongea*. (Coelenterates with tissue grade of organization exhibit united cells, those having lost their independence cooperate or act as a unit.)

2. *Cellular Grade of Organization.* It is possible to isolate individual germ and body cells or cell clumps. If prevented from reassembling to form a new individual (which any piece of sponge can do), these isolates act as protozoans and move by pseudopods and/or flagella. Unlike protozoans, they soon die if not reunited. This separability into individual protozoan-like cells illustrates the cellular grade of organization.

Two of the spongelike cells of special interest to this discussion are the collared flagellate cells, *choanocytes* (Figure 4.1B, C) and *amoebocytes*. Choanocytes contain information coming from the evolutionary

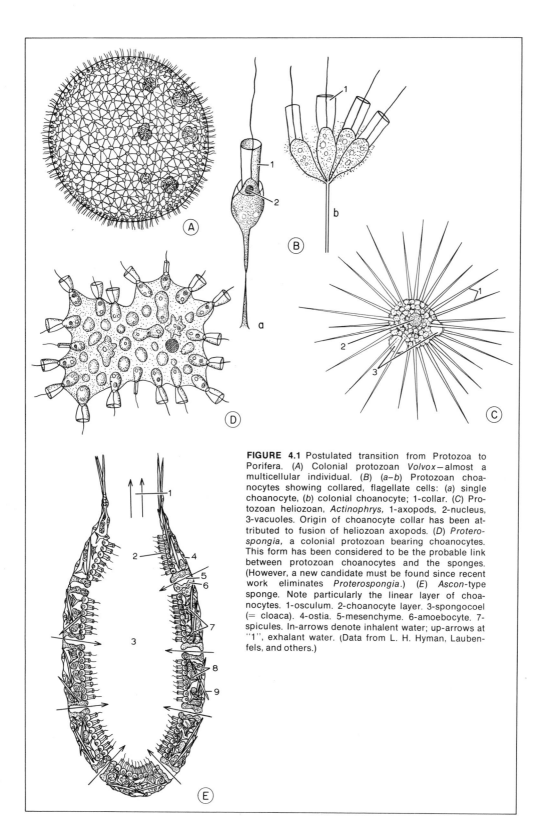

FIGURE 4.1 Postulated transition from Protozoa to Porifera. (*A*) Colonial protozoan *Volvox*—almost a multicellular individual. (*B*) (*a–b*) Protozoan choanocytes showing collared, flagellate cells: (*a*) single choanocyte, (*b*) colonial choanocyte; 1-collar. (C) Protozoan heliozoan, *Actinophrys,* 1-axopods, 2-nucleus, 3-vacuoles. Origin of choanocyte collar has been attributed to fusion of heliozoan axopods. (*D*) *Proterospongia,* a colonial protozoan bearing choanocytes. This form has been considered to be the probable link between protozoan choanocytes and the sponges. (However, a new candidate must be found since recent work eliminates *Proterospongia.*) (*E*) *Ascon*-type sponge. Note particularly the linear layer of choanocytes. 1-osculum. 2-choanocyte layer. 3-spongocoel (= cloaca). 4-ostia. 5-mesenchyme. 6-amoebocyte. 7-spicules. In-arrows denote inhalent water; up-arrows at "1", exhalant water. (Data from L. H. Hyman, Laubenfels, and others.)

history that relates sponges to an ancestral protozoan flagellate. The colonial protozoan *Proterospongea* was once thought to fill the position of type-connecting-link between protozoan choanocytes and sponges (Figure 4.1*D*). Whereas the amoebocytes, very versatile cells, serve many different functions, each separate function bears a different name. The function of greatest interest for a paleobiologist is its secretion of the spicular skeleton (Figure 4.2). When an amoebocyte is actually secreting skeletal material, it is called a *scleroblast*.

3. Sponge Architecture. Evolutionary steps leading from a flagellate colonial ancestral protozoan (Figure 4.1*D*) to the simplest sponge—Ascon grade (Figure 4.1*E*) remain unknown. Assuming a genetic step series for this evolutionary transition, it seems that the simplest sponge is simply a wall perforated by holes (pores) lined with collared flagellate cells enclosing a central cavity (*spongocoel*). In

contrast to pores taking in water (inhalent current), the *oscula* lets out water from the spongocoel (exhalent current).

Given: basic structural unit (Ascon grade). The next more complex sponge (Sycon grade, Figure 4.4 parts 3, 4) has walls built up by a repetition of Ascon-type units. The Rhagon grade of sponge architecture, being more complex, has its walls built of Sycon-type units (Figure 4.4 parts 5, 6). The problem that had to be solved here may be stated as follows: How can maximum contact with the water medium be realized as sponge volume increases? The critical site was the pore for water intake. Note that actual solution did not require anything strikingly new. Rather, more surface area was achieved by taking as a basic unit a complete Ascon sponge and by grouping many of these about a central cavity (Sycon-type). For the increased volume of the Rhagon-type over the Sycon-type, the solution repeated complete Sycon-like units grouped about

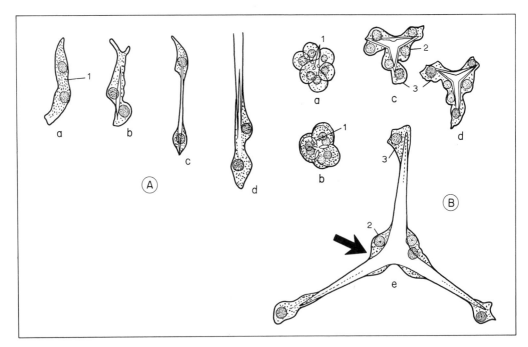

FIGURE 4.2 Secretion of spicular skeleton.
(*A*) Monaxon: (*a*) *1*—initial spicule; (*b*)–(*d*) elongation of spicules. Note increased distance separating secreting cells.
(*B*) Triradiate spicule (common in calcareous sponges—Calcispongea): (*a*) *1*—spicular grains; (*b*)–(*d*) growth of triradiate spicule. Note multiple cells involved (*2,3*) [Several stages are omitted between (*c*) and (*d*), arrow points to a fully grown spicule at (*e*).] Adapted from Woodland 1905, and others.

a central cavity. The Rhagon-type structure seems the dominant one. Furthermore, it is the most successful in an evolutionary sense. Ninety-eight percent of all individual sponges and sponge species, both living and fossil, have this structure (Laubenfels).

It appears that the essential sponge genetic code is that which specifies the Ascon-type structure.

4. Reproduction. The fertilized sponge egg gives rise to a free-living flagellated larva. This ultimately, after leaving the parent body, attaches to the bottom and grows into a new sponge. *In many invertebrates the free-living larval stage explains the wide geographic distribution of both living and fossil forms, even when the adults are attached (sessile).* Asexual reproduction may occur by budding or branching. In fresh-water forms it occurs by means of gemmules.

5. Water Circulation. A sponge 4 in. tall will pass 25 gallons of water daily through its body. In life, a sponge 200 g dry weight will circulate 1000 kg (1 metric ton) of water for each 24-hour period. *These important data help to explain some controls governing the distribution of sponges, both fossil and living.* Choanocyte flagellum move almost in unison at the rate of about 10 times per second, and create and direct the exhalent current. These cells also trap food particles — mostly very small forms, such as bacteria, (collectively called *hekistoplankton*).

6. Morphology, Size, Habitat. Shape in sponges varies from amorphous and incrusting to globular and cylindrical. The incrusting and globular types generally occur in strong currents. Crowding can distort the configuration. In relatively motionless water, Laubenfels (1957) noted, sponges may erect tall oscular tubes preventing recirculation through the pores of exhaled water-bearing wastes.

Mature sponges may be the size of a human head; others may be less than 1 cm. Incrusting types such as are found on Ordovician orthocone cephalopods may be less than 1 mm thick. Largest sponges, densest sponge populations, and most diverse species occur in equatorial waters.

In the lower reaches of the littoral zone (low tide to high tide) lives the greatest density of living sponge populations (Burton, 1948). Elsewhere on the sea floor, they are not rarities. The siliceous sponges (class Hyalospongea) are mainly known from depths of 100 fathoms (1 fathom = 6 ft). However, as Reid (1962) pointed out, in the region of Indonesia several expeditions recovered such forms in waters of half that depth.

7. Biotic Associates, Predators, and Predation. Living sponges may contain as many as three nematode worms per cc. In turn, fish seeking worms will prey on sponges. Amphipod crustaceans and shrimps (see Chapter 11, under "Arthropoda") live in the sponge pores and chambers. Sponges are eaten by such gastropods as *Patella* and *Littorina* (Burton, 1948) and some holothurians (Chapter 12). In the Chesapeake Bay area, boring sponges (family Clionidae) commonly riddle oyster and other molluscan shells with holes to reach the body fluids. (Marcus C. Old, 1942). Figure 4.3 part 4 shows an Eocene oyster with similar borings. Shallow-water forms contain numerous plant symbionts and derive nutrients from this source (Laubenfels, 1955).

Information Derived from Fossil Sponges

Fossil Record. Comparison of fossil species (that are assigned to over 1000 genera) to the more than 10,000 species of living sponges shows that less than two dozen are identical (Laubenfels). How can this be explained? The answer is probably twofold: Poor preservation of fossil sponges tends to obscure relationships to modern species; and through time there must have been a *loss of some genetic information.*

The fossil record of the Pre-Cambrian is too obscure and poorly documented as yet. More impressive is the Paleozoic record.

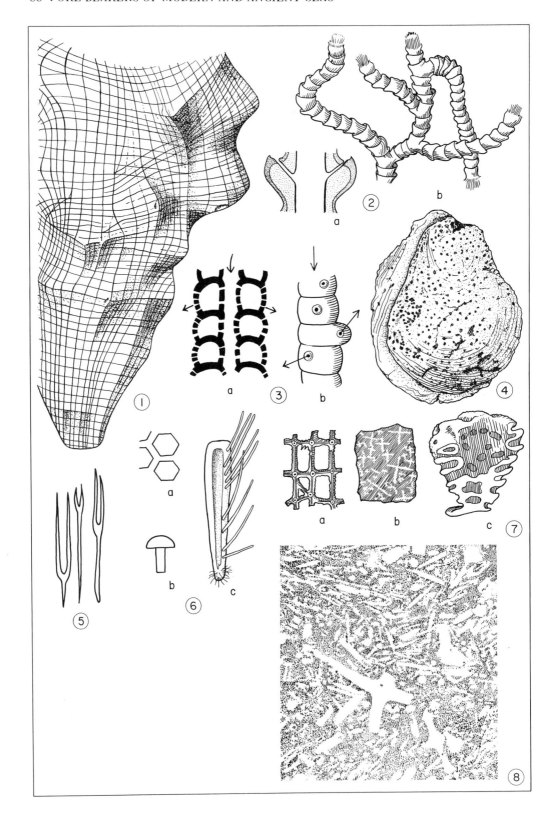

FIGURE 4.3 Some fossil sponge types. (Schematic). Hyalospongea: (1) *Hydnoceras*, U. Dev., N.Y., N.Y.S., longitudinal and transverse strands were siliceous in life. (2) *Titusvillia*, L. Carb., Pa; *a* — longitudinal thin section exposing central cavity and pores; *b* — branching colony formed by budding: note terminal bud with corona of long erect spicules. (3) Calcispongea: (*a*) *Sebargasia*, Carb., Spain, longitudinal thin section; arrows indicate inhalant and exhalant currents; internal organization of "(*b*)" similar. (*b*).*Girtycoelia*, Penn., Midcontinental U.S.A. (4) Parasitic sponge borings (black dots) in pelecypod *Ostrea* shell, L. Eoc., England. (5) Calcicpongea: *Sestrostomella*, tuning-fork-shaped pharetrone spicules, Cret., England. (6) Demospongea: *Pirania*, M. Cam., Brit. Col., *a, b,* pavement-like sponge surface (*a*) created by short spicules "*b*"; *c*, only a few tylostyle spicules (long, pointed at outer end, swollen at base) are figured; hollow central cavity exposed in longitudinal section. (7) Hyalospongea: *a, b, c, Porospongia*, U. Jur., Europe. *a* — cortex-like spicular mesh of inner sponge flesh, composed of two types of spicules: stauracts (4 rays in one plane) and hexacts (6 rays at right angles to each other). *b* — thin section through spiculated sponge flesh exposing chiefly stauracts. *c* — Myliusia, living form (Cret.–Recent), thick walled; sectioned to disclose central cavity and inner, large exhalant exits for water (oscules?). (8) Spiculitic chert (Arkansas novaculite) composed of sponge spicules, varisized and varishaped, in a siliceous matrix. (Illustrations adapted and redrawn after: (1) Hall and Clarke; (2) Caster; (3, 5–7) Laubenfels; (4) Ager; (8) Goldstein and Hendricks, 1953. G.S.A. Bull. Vol. 64).

As early as the Cambrian, demosponges are known (Figure 4.3, part 6); in the Pre-Cambrian of the Belgian Congo calcisponges may have first appeared, to become important from Devonian on (Figure 4.3, part 3). There were great assemblages of glass sponges (Hyalospongea) in the Devonian of New York–Pennsylvania (Hall and Clarke), Carboniferous and Cretaceous of Great Britain (Reid), and the Mesozoic of Europe (Gignoux). Many Paleozoic and some Mesozoic occurrences of siliceous sponges can be inferred from spiculitic cherts. cherts.

Skeleton. The four classes of the Phylum Porifera can be differentiated by the skeleton composition. This applies to both living and fossil material.

Class 1. Demospongea. Range: Cambrian to Recent. Skeletons in whole or part consisting of spongin; often with siliceous spicules whose rays meet at 60 or 120-degree angles. Architecture: compact Rhagon (Figure 4.3, part 6; Figure 4.4, parts 5, 6).

Class 2. Hyalospongea (hyalos = glass). Range: Pre-Cambrian; L. Cambrian to Recent. Siliceous spicules, skeletal spicule rays typically form at right angles. Common spicule is the stauract (a tetraxon with four rays on one plane); Figure 4.3, part 7b. Architecture: simple Rhagon type, less compact than Demospongea.

Class 3. Calcispongea (calx = lime).

Range: Cambrian to Recent. Skeletons of calcareous spicules. Three types of spicules: diacts, triacts (Figure 4.3, parts 3, 5), and tetracts (di = two, tri = three, $tetra$ = four + act = ray). Architecture: Rhagon, Sycon, and Ascon (Figure 4.4). Tuning-fork type of spicules may be interlocked or cemented together forming a rigid structure.

Class 4. Sclerospongiae. Range: Recent. Coralline sponge with compound skeleton of siliceous spicules, organic fibers and aragonite; some genera bear astrorhizal structures (see Figure 5–9,*D*). (Compare Stromatoporoids, Table 5.1. "Note.")

Sponge Fossils. At the outcrop, complete or partial sponge fossils may be collected. These will show configuration, pores, canals and, sometimes, skeletal elements. Arrangement of spicules on a bedding plane may correspond to the way in which they were arranged in the sponge flesh during life, for example, *Chancelloria*, M. Cambrian. In well cuttings, isolated calcareous and siliceous spicules may be retrieved directly or, in the case of siliceous spicules, from insoluble residues (that is, after acid digestion of limestones). Imprints, molds of spongin fibers, or root tufts may occur on bedding planes as well as in casts of canal systems. In hand specimens of sponges, spicules may become visible when studied optically in thin section.

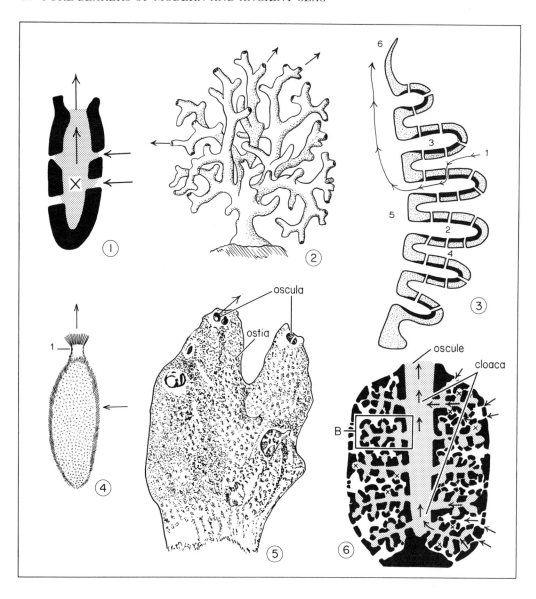

FIGURE 4.4 Sponge architecture. Arrows indicate inhalant and exhalant water flow through several sponge grades. [After Laubenfels; Hyman; Borradaile et al.]. (*1, 2*). *Ascon*-type: (*1*) black denotes sponge flesh;—many flagellate choanocytes line inner part of flesh; (*2*) Living *Ascon*-type sponge. (*3, 4*) *Sycon*-type: (*3*) Labels 1–6: 1 and 6 show inhalant aperture and water circulation route (multiplied by similar entrants at all similar pores); black-lined layer denotes flagellate choanocytes which circulate inhalant water; 2 and 3, *Ascon*-type chamber; compare 1 and envision it rotated 90° to the right; 4, entrant pore; 5-central cavity (cloaca) (Schematic). (*4*) Living form. (*5, 6*) *Rhagon*-type: (*5*) Living form; (*6*) Longitudinal section (schematic). Rectangle B details how the sponge is organized by *Sycon*-type units; compare ''3'', and envision it rotated 90° to the left with exit arrow repointed to the cloaca. Spiculated sponge flesh in black; blank area—inhalant water; gray area—exhalant water.

Taxonomy. Assignment of sponge-fossil fragments or isolated parts often can be made to a given *class* based on skeletal composition and/or composition, structure, or arrangement of spicules; to a given *family* based on such characteristics as growth, form or shape, position and arrangement of pores and canals; to a given *genus* by such characteristics as number and arrangement of oscules, the exact type of spicules (megascleres and/or microscleres), and so on.

Once such taxonomic assignment is made, a whole sequence of derivative data may be decoded.

Chemistry and Mineralogy. See Chapter 15 and discussion of spicular cherts below.

Geologic Range and Stratigraphic Problems. Limited use has been made of fossil sponges in solving field problems. However, Gutschick (Figure 4.5, parts A and C) has shown how effectively sponge spicules may be utilized in stratigraphic correlation. Some fossil sponge families and genera have restricted geologic ranges and can be helpful as indicators of geologic age. The family Barroisidae, for example, is restricted to the Mesozoic, in Europe and elsewhere; the genus *Girtycoelia* is limited to the Pennsylvanian of the midcontinental United States; and numerous glass sponge genera are confined to the Devonian of New York.

Finks et al. (1961) found that the lithistid sponge *Actinocoelia meandrina* (Figure 4.6) was a time stratigraphic marker in the Cordilleran region. Apparently the fossil is restricted in that region from late Leonardian to early Guadalupian time. Griefe and Langenheim (1963) also found that the lithistid sponge fauna of the Middle Ordovician Mazourka formation of California was of value in correlation within the Cordilleran region. These investigators found that the California sponges were equivalent to those reported from the Chazyan beds of Nevada.

Bathymetry. Although there is still much disagreement about depth indications provided by fossil Hyalospongea (Reid, 1962), useful surmises and extrapolations can be made, as Gignoux (1955) (Figure 4.7) demonstrated, because many Mesozoic and Tertiary forms have living equivalents whose distribution is known.

Reid showed that fossil siliceous sponges in the Cretaceous chalk beds of England need not imply a depth greater than 600 ft. Menzies and Imbrie (1958) noted that siliceous sponges in the modern deep sea (abyssal realm) probably came to inhabit the deeper parts of the realm relatively recently — that is, in Tertiary time. (See also "Spiculitic cherts" discussed below.)

Ecology and Habitat. For fossil reef-inhabiting sponges, there are numerous data pertaining to ecology and habitat. Newell et al. deciphered such data for the Permian Capitan Reef (Figure 3.11, part B; Figure 4.8, part 2), Lowenstam for the Niagaran (Silurian) Reefs of Illinois, and Carozzi for the Devonian of Alberta, Canada. Items have been noted, such as population density relative to other reef dwellers, through time, and the energy level (that is, the transition from quiet to rough water conditions) (Figure 4.8, part 1). Borings in Tertiary fossil oyster shells (these oysters formed reefs in life) disclose data on the ecology (food supply and biotic associates) of boring clionid sponges (Figure 4.3, part 4, compare figure 4.9), as do reports of sponge borings in brachiopod and molluscan shells. These are known from the Devonian of Iowa. Other reports disclose such borings in cephalopod interiors from the Ordovician of New York (Okulitch and Nelson).

British Cretaceous sponges (pharetrones, see Figure 4.3, part 5) characterize the littoral zone.

Studies such as those by Newell et al., Lowenstam, Carozzi, and others also give indications of the biotic associates of sponges found as fossils (Figure 3.11, part B). With these comprehensive data covering the entire biota and not sponges

		KENTUCKY	INDIANA
MISSISSIPPIAN	U. M.	Warsaw ls.	Harrodsburg ls.
LOWER OSAGE		Muldraugh fm.	Edwardsville ls.
		Floyds Knob fm. (reef)	Floyds Knob fm. (reef)
		Brodhead fm.	Carwood fm.
			Locust Point fm.
	K.	New Providence sh.	New Providence sh.

(BORDEN GROUP)

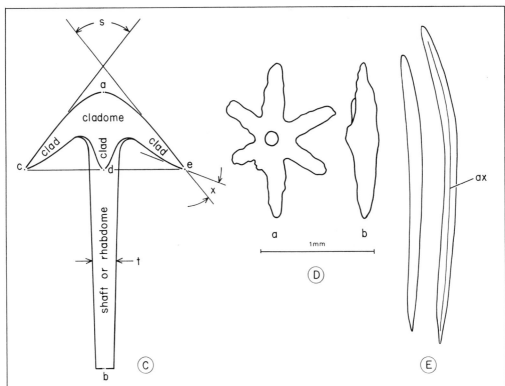

FIGURE 4.5 Fossil sponge spicules in stratigraphic correlation.

(A) Correlation by anchor-type spicules (see C).
(B) Surface outcrops containing spicules. Borden Group, Indiana/Kentucky.
(C) *Hyalostelia*, L. Miss, Indiana/Kentucky. Measurement of an anatraene spicule: a—apex; ab—length; ce—chord; ad—sagitta, t—thickness of shaft; s—dome angle; x—clad angle; ac—length of clad. (D, E) Silicified Ordovician spicules from insoluble residues. (D) a, b, Astreospongia, 8-rayed; two rays are represented by central circle, L. Ordovician, Illinois. (E) Monaxons, L. Ordovician, Wisconsin; ax—axial canal left by original organic thread around which the spicule grew. All figures schematic; redrawn after (A)–(C), Gutschick, 1954a; (D) Gutschick, 1954b, (E) Howell and Landes, 1936.

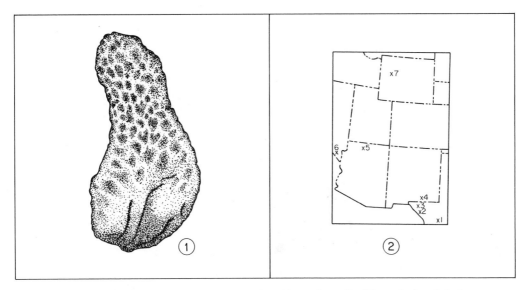

FIGURE 4.6 Permian Lithistid sponges as stratigraphic markers—Cordilleran Region, Late Leon-ardian—Early Guadalupian. (After Finks, Yochelson, Sheldon, 1961.)
(1) *Actinocoelia meandrina* Finks, sagittal section; occurs at localities 2–7, generally in shelly lime-stones.
(2) Cordilleran Region showing localities (except 1) where the above sponge species occurs. 1—Glass Mts.; 2—Sierra Diablo; 3—Southern Guadalupe Mts.; 4—Central Guadalupe Mts.; 5—Hermit Trail area; Grand Canyon; 6—Goodsprings area; 7—Dubois.

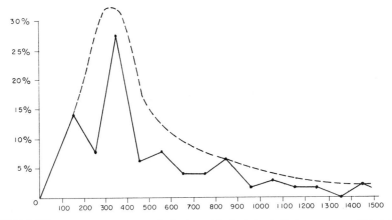

FIGURE 4.7 Siliceous sponge genera as indicators of bathymetry in Cretaceous seas. Oblug Chalk. Hannover, Germany. These deposits contain abundant siliceous sponges—Class Hyalospongea. Taking some of the same genera extant today, known depths at which they occur were applied to Cretaceous genera. Frequency of fossil genera was then plotted against probable depth. Dashed frequency curve smooths out the zigzag fossil distribution on the assumption that a larger sample would approach this limit.
(Redrawn from Stratigraphic Geology by Maurice Gignoux, W. H. Freeman and Company. Copyright © 1955)

alone, it is possible to infer the trophic levels (primary and secondary producers, consumers, etc.). See Chapter 5.

Occasionally sponges not only lived on but, importantly, helped to build reefs.

King (1934) described a sponge-algal reef of this type from the Permian of Mexico.

Permian lithistid sponges from Wyoming (Figure 4.6, part 1) were found to have overgrown many bryozoan colonies,

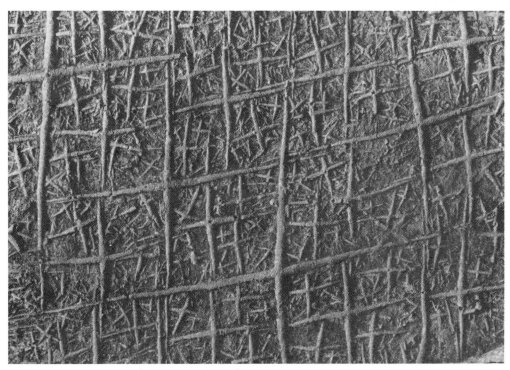

The sponge *Protospongia hicksi* Hinde (Hyalospongea), Middle Cambrian, Western Utah. Central part of silicon rubber cast, enlarged. Note the layer of rodlike spicules, and large and small four-rayed cross-type spicules (cruciform stauracts). Courtesy of Dr. J. Keith Rigby.

pelecypod and brachiopod shells, as well as crinoid stems. (See also discussion of conulatids.) Shells incorporated on opposite sides of the sponges suggest roll-over due to wave action. Numerous shelly forms found in these sponges gave indication of the substrate, depth, and salinity. Such forms lived on a firm carbonate mud bottom (shelf environment); the water was shallow and apparently of normal salinity (Finks et al., 1961). Paleoecological interpretations may be extended to the other Cordilleran localities where this sponge occurred (Figure 4.6, part 2).

Spiculitic Cherts. Valuable data on bathymetry, sedimentation, diagenesis, energy level, habitat, and associated biota may be obtained from a study of spiculitic cherts in thin section (Carozzi, 1960; Krauskopf, 1956; Finks, 1960; Newell et al., 1955; Gignoux, 1955).

Broken siliceous sponge spicules accumulated in bottom lime muds at a

given time in the geologic past. The rock formed was composed chiefly of spicules and is called a spiculite. What actually happened is that the lime mud went into solution and was replaced by opal or chalcedony—two varieties of mineral quartz. Such data can provide valuable references to the bathymetry of the sea at the time.

An important source of silica in cherts is dissolved siliceous spicules. Often only

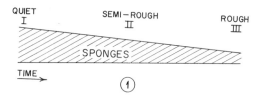

FIGURE 4.8 Paleozoic sponge reef distribution and fabric.
(1) Vertical changes in reef-dwelling sponge distribution (adapted from Lowenstam et al., 1956). I–III represent three successive stages in reef development. Note decrease in sponges through time as environment (energy level) changed; see Figure 3.11*B* which shows reef-dwelling sponge distribution at a *given* time.

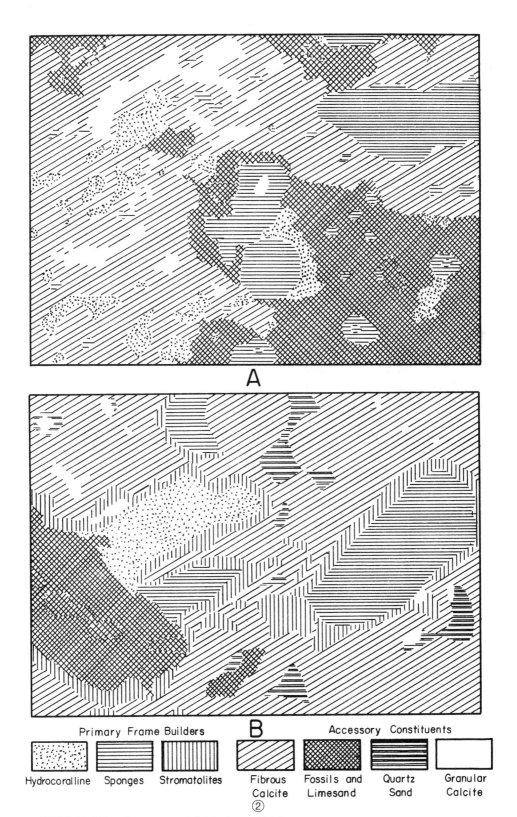

FIGURE 4.8 Paleozoic sponge reef distribution and fabric.
(2) Permian reef rock, Guadalupe Mountains. (From N. D. Newell, 1953, "Depositional Fabric in Permian Reef Limestones," *Journal of Geology*, Vol. 63. By permission of the University of Chicago Press.) (A)–(B) Reef fabric consisting of primary and accessory components, X5. Note conspicuous presence of sponges (see Figure 3.11 part B).

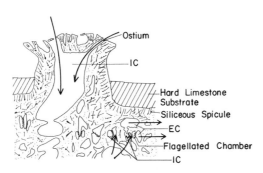

FIGURE 4.9 The boring sponge *Cliona celata* (class Demospongea) has penetrated into a hard limestone substrate which may be coralline, algal or contain molluscan shells: *ic, ec*—incurrent and excurrent channels; water flow shown by arrows. (Adapted from Goreau, T. F., and W. D. Hartman. 1963, in, Mechanisms of hard tissue destruction, R. F. Sognnaes, ed., A.A.A.S., Publ. 75, Ch. 2).

the axial canal is preserved. If definite spicule configurations such as tetracts or hexacts, are preserved, closer identification is sometimes possible.

Calcareous spicules such as monoaxons also occur in nodular cherts and may be composed of calcite, dolomite, both of these, or chalcedony. Diagenetic changes in sediments can lead not only to siliceous replacement of calcareous spicules, but to calcite replacing the opal of siliceous spicules (Delaware Basin, Permian of Texas).

FIGURE 4.10 Portion of natural outline of fossil object (Glass Mountains, Texas). Polished section: note *bs*—a series of hollow spheroidal bodies; *p*—pores; wt_1, wt_2—wall thickness; *ch*—chambers. After King, 1943; Laubenfels, 1955.

Identification of a Fossil Sponge to the Species Level

Data from R. H. King, 1943, pp. 24–26, plate 1, Figure 8, text figure 2.

1. Raw data.
(a) Silicified objects with walls, chambers, and pores embedded in a limestone matrix. Age: Permian (Leonardian). Locality: Glass Mountains, Texas.
(b) Natural sections through the object in the limestone matrix often occur. These reveal a series of spheroidal bodies having two walls and pores in the walls. [Polishing the face of such a section can bring out many details. This is especially so when the polished object (covered with a thin layer of oil, glycerine, or water) is studied microscopically.] See Figure 4.10.

2. Measurements and observations. Determine overall dimensions, diameter, thickness of each of the two cell layers, length of long axis of spheroidal bodies. Determine fine dimensions: Pores are found to be 0.2 mm in diameter (and relatively sparse). General characteristics: large, gently convex, subdiscoidal, composed of two layers of chambers.

3. Classification.
(a) Is this object a sponge? (See Figure 4.4 and phylum definition.) If so, one must next ask to which class it belongs.
(b) Some sectioned specimens were preserved as calcite. Others were silicified in a limestone matrix. *What was the original composition of this sponge?* (See earlier discussion on diagenesis.) The fact that calcitic objects and silicified objects of the same kind both

occur in a limestone matrix indicates that silicification was secondary, that is, after deposition of the *calcareous* sponge.

(c) If the object belongs to class Calcispongea, does it resemble any previously described forms in this class? The answer is yes. It resembled other forms in its cellular structure.

(d) To what genus does the object belong? In order to answer this question, one must determine how it differs from known forms that apparently are similar. A list of such known genera is given below.

description is more general, since many different species will ultimately be incorporated under it.)

(e) A new genus may be erected on the basis of a type species. Accordingly, a species description and a new species name were given, *Polyphymaspongia explanata*, to the entire collection of objects described above. One specimen, the best preserved, became the *holotype* (the type of the species). (The species description is less general and more detailed in comparison to that of the genus.)

Known Genus	Age	Differences in Characteristics of the Studied cf. Known Genus
Cystothalamia	Permian	Mode of growth; absence of ostia.
Guadalupia	Permian	Chambers larger and of different configuration.
Girtycoelia	Permian	Two layers of cells.

The investigator therefore concluded that there was no described fossil genus that fitted the characteristics of the studied objects. Accordingly, a new genus was proposed. It was given the name *Polyphymaspongia*, R. H. King, 1943. (The genus

(f) The descriptions for genus and species together with measurements, illustrations, and relevant discussion are then published. Only after publication (called an "indication") are the new names recognized as valid scientific names.

QUESTIONS

Porifera

1. Discuss the evidence for the probable transition from colonial protozoans to sponges (see Figure 4.1).

2. (a) What evolutionary problem has been solved in the development of sponge architecture?

   (b) Explain the significance of the Ascon grade.

3. Discuss information obtained from siliceous sponges from the Cretaceous chalk beds of England.

4. Relate the type of architecture of sponges to other class characteristics.

   (b) Outline the steps in the secretion of a monaxon; triradiate spicule; of spongin.

   (c) What would you measure on a

sponge spicule? (See Figure 4.5, part C.)

5. What kind of sponge faunas would you expect to find in the Devonian of New York and in the mid-continental Permian?

6. (a) How did Gutschick use sponge spicules in stratigraphic correlation?

   (b) How can insoluble residues yield data on fossil sponges?

   (c) What bathymetric inferences was Gignoux able to make about Cretaceous sponges in Germany?

7. (a) What did Lowenstam find regarding sponge population density through time in the Silurian Niagaran reefs? How is this related to the energy level? (See Figure 4.8, part 1.)

   (b) "Spiculites" are often reported from geosynclinal deposits. Can you explain how this happens?

REFERENCES

Butler, P. E., 1961. Morphologic classification of sponge spicules, with descriptions of siliceous sponge spicules from the lower Ordovician Bellefonte dolomite in central Pennsylvania: J. Paleontology, Vol. 35 (1), pp. 191–200, plate 39, 7 text figures.

Burton, Maurice, 1948. Ecology of sponges: Nature, Vol. 162 (4106), pp. 73–74.

Clarke, J. M., 1918. Devonian glass sponges: N.Y. State Mus. Bull., 196, pp. 177–186, plates 1–6.

Carozzi, A. V., 1960. Microscopic sedimentary petrography. John Wiley. 461 pp. (see Chapter 6).

Caster, K. E., 1939. Siliceous sponges from Mississippian and Devonian strata of the Penn-York embayment: J. Paleontology, Vol. 13, pp. 1–20.

de Laubenfels, M. W., 1955. Porifera, in R. C. Moore, ed., Treat. Invert. Paleontology, Part E. Archeocyatha and Porifera, pp. E21–E110.

————, 1957. Sponges of the Post-Paleozoic: Treat. Mar. Ecol. and Paleoecology, Vol. II, GSA Mem. 67, pp. 771–772.

Finks, R. M., 1960. Late Paleozoic sponge fauna of the Texas region: the siliceous sponges: Am. Mus. Nat. Hist. Bull., Vol. 120, article 1, pp. 1–160, plates 1–50.

————, E. L. Yochelson, and R. P. Sheldon, 1961. Stratigraphic implications of a Permian sponge occurrence in the Park City formation of western Wyoming: J. Paleontology, Vol. 35 (3), pp. 564–568, 3 text figures.

Girty, G. H., 1908. The Guadalupian fauna: U.S.G.S. Prof. Paper 58, pp. 1–649.

Griefe, J. L., and R. L. Langenheim, Jr., 1963. Sponges and brachiopods from the Middle Ordovician Mazourka formation, Independence Quadrangle, California: J. Paleontology, Vol. 37 (3), pp. 564–574, plates 63–65, 2 text figures.

Gutschick, R. C., 1954. Sponge spicules from the lower Mississippian of Indiana and Kentucky: Am. Mid. Nat., Vol. 52 (2), pp. 501–509, plate 1, 3 text figures.

Hall, J., and J. M. Clarke, 1900. Paleozoic reticulate sponges constituting the family Dictyospongidae: Univ. State of New York (N.Y. State Mus.) Mem. 20, pp. 1–197.

Howell, B. F., and R. W. Landes, 1936. New monactinellid sponges from the Ordovician of Wisconsin: J. Paleontology, Vol. 10 (1), pp. 53 ff.

Hyman, L. H., 1940. See ref. ch. 5.

King, R. E., 1934. The Permian of Southwestern Coahuila, Mexico: Am. J. Sci., 5th ser., Vol. 27, pp. 98–112 (sponge-algal reefs).

King, R. H., 1943. New Carboniferous and Permian sponges: Kans. Geol. Surv. Bull. Vol. 47, pp. 5–36.

Krauskopf, K. B., 1956. Dissolution and precipitation of silica at low temperatures. Geochimica et Cosmochimica Acta, Vol. 10, pp. 1–26.

Lowenstam, H. A., 1950. See ref. Ch. 5.

Menzies, R. J., and J. Imbrie, 1958. The antiquity of the deep sea bottom fauna: Oikos, Vol. 9 (2), pp. 192–208.

Newell, N. D., et al., 1955. The Permian reef complex of the Guadalupe Mountain Region, Texas and New Mexico. W. H. Freeman & Co. 209 pp., 32 plates.

Pennak, R. W., 1953. Porifera, in Fresh Water Invertebrates. Ronald Press. 726 pp. (Chapter 3, pp. 77–96 — species of the fresh water sponge family, Spongillidae).

Reid, R. E. H., 1962. Sponges and the chalk rock: Geol. Mag., Vol. 99 (3), pp. 273–278.

[The Patagonian Cretaceous has yielded the first fossil fresh-water sponge (Spongillidae) complete with gemmules, and spicules. (W. Volkheimer, 1972. N. Jb. Geol. Paläont. Abh. 140, pp. 49–63, 6 figures). Two other important additions to the literature on sponges are J. K. Rigby, 1971, Sponges and Reef and Related Facies through Time, Symp. N.A. Paleont. Conv., 1969, Part J. pp. 1374–1388; W. G. Fry, ed., 1970, *The Biology of the Porifera.* Symp. Zool. Soc., London, No. 25, Academic Press, 490 pp. (For the new sponge class Sclerospongiae, see p. 89.)]

B. PHYLUM ARCHAEOCYATHA

Introduction

The archaeocyathids are a group of extinct organisms of disputed position. At one time or another they have been placed under coelenterates or close to coelenterates (corals), foraminifers, receptaculids (algae), and sponges. Presently, there are two schools of thought: archaeocyathids are sponges structurally (Stearn, 1972) as well as functionally [Balsam and Vogel, 1973, J. Paleontology, vol. 47(5), pp. 979–984]. Or, archaeocyathids constitute a separate phylum (Zhuravleva, 1970; Hill, 1972; Hill, 1970, personal communication indicated that she considered archaeocyathids less highly organized than sponges).

Phylum Archaeocyatha. Geologic range: Lower to Upper Cambrian. (After Hill, Debrenne, and Zhuravleva.)

1. Organisms with calcareous (microgranular calcitic) skeleton; an inverted cone (= cup) that was erect or curved and held to the substrate by holdfasts.
2. Skeleton may consist of a single porous wall (Monocyathida) but more commonly has two concentric porous walls, an inner and an outer wall separated by a space (*intervallum*).
3. The inner wall surrounds a *central cavity*.
4. Intervallum may contain various radial elements including *porous septa* and *tabulae* as well as *non-porous rods, bars,* and *dissepiments.* The radial elements connect both walls. Septa subdivide the intervallar space into smaller compartments (*loculi* = space between two septa). [*Tabulae* = transverse (horizontal) porous skeletal elements of variable curvature. These connect both walls. *Septum* = radial longitudinal plate connecting walls of a two-wall cup. *Dissepiment* = non-porous thin plate, convex above (i.e., bubble-shaped) that occurs in interradial space of intervallum and/or in central cavity. *Bar* and *rod* = thin elongate skeletal elements.]

Classes. The phylum (or if reassigned to Porifera, subphylum) divides into two classes: Regulares and Irregulares. Both classes consist largely of solitary forms, but colonial types also occur.

Class Regulares. Conical, cylindrical – to saucer-shaped cups, commonly with two porous walls, enclosing a central cavity. Type of wall construction highly variable. The intervallum may contain a variety of radial skeletal structures – flat or convex tabulae (may be absent) bear round, ovate pores; septa in which pore rows show fanlike divergence; dissepiment presence variable.

Geologic range: Lower Cambrian – Middle Cambrian.

The class is divided into three orders: Order Monocyathida (*Archaeolynthus,* Figure 4.11*F*), Order Putapacyathida, and Order Ajacicyathida (*Nevadacyathus,* Figure 4.11*G*).

Class Irregulares. Conical, cylindrical – to discoid-shaped cups, commonly with two porous walls. Types of wall construction less numerous and complex than in Regulares. Intervallum with a variety of radial skeletal structures.

The class is divided into three orders: Order Thalassocyathida (one-wall cup); Order Archaeocyathia [*Archaeocyathus atlanticus* (Fig. 4.11*B*); *Archaeocyathus sellicksi* (Fig. 4.11*C*]; Order Syringocnemidida (*Syringocoscinus,* Fig. 4.11*E*).

Range: Lower Cambrian–Upper Cambrian.

Measurements cited in describing new

species include cup diameter and height and the following coefficients:

1. *RK* (Radial Coefficient) = number of radial elements/diameter of cup. (Usable where interradial space is not variable.)
2. *Rk* = number of longitudinal rows of pores per intersept*/diameter of cup. (*Intersept = that portion of outer wall defined by two successive septa.)

Inferred Biology

(a) Reproduction. Sexual and asexual reproduction (budding and fission) are both inferred. A planktonic (free-swimming) larval stage is probable in view of worldwide distribution of similar genera. Adults were attached sessile benthos (except discoid forms capable of being moved by currents).

(b) Skeletal Development. After a limited planktonic existence, the larva apparently settled (attached) to the carbonate mud bottom, to older archaeocyathid skeletons already on the bottom, or to the thallus of an alga. After attachment, amoebocyte-type cells probably secreted the skeletal units, as in sponges.

(c) Soft-Part Anatomy. The bulk of soft tissue presumably lined the straining system inside the intervallum where the life processes (digestion, reproduction, etc.) were carried on (Figure 4.11 part *A*). The exterior surface of the outer wall and the inner surface of the central cavity are thought to have been covered with tissue. Conceivably, algae also lived in this tissue in a symbiotic relationship. Food was likely taken from sea water by the incurrent flow and strained out by the complex pore system, to be digested by amoebocyte-type cells.

A mesogleal layer (as in sponges) is thought to have contained the probable amoebocytes that secreted the skeletal units.

(d) Observed Morphology, Inferred Habit. Cone, goblet, and vase shapes are most characteristic of the archaeocyathids. There are, of course, others, such as saucerlike, irregular, and crenulate.

Archaeocyathids appear to have lived along the coastlines of warm, shallow Cambrian seas as in Australia, where they formed "gardens" of sessile benthos extending as much as 400 miles. Particularly among young archaeocyathids found in the Lower Cambrian of Nevada, it appears that muddy water inhibited growth. The large number of young and immature individuals on limestone slabs led to the inference that they were suffocated or buried in bottom mud.

Optimum depths were 20–30 m to 50 m (but some forms extended to 100 m).

Extinction has been attributed (*a*) archaeocyathids being supplanted by algae (Vologdin) or (*b*) competition with proliferating sponge populations, and, in lagoonal substrates, an increase in MgO (magnesium oxide (Zhuravleva).

(e) Size. Archaeocyathids vary in size but in diameter usually lesser variation is observed (from 1.0 to 3.0 cm). The largest sizes known are 52 to 60 cm (diameter) and 25 to 30 cm (height).

(f) Biotic Associates, Reefs. Among fossils generally found in archaeocyathid limestones are trilobites (*Olenellus* and other genera), brachiopods, various mollusks, several algae, and less frequently except at certain places and times, sponges and/or sponge spicules. Some forms, such as thick-shelled inarticulate brachiopods, gastropods like *Helcionella*, and some pteropods, were originally attached to archaeocyathid reefs in Labrador. Others, such as the alga *Epiphyton*, served to cement the dead archaeocyathid skeletons into reef structures. From the helter-skelter arrangement of fossils in archaeocyathid limestones, it appears that when larval forms attached to the bottom, they often settled on the skeletons of dead archaeocyathids, although, as noted earlier, they also attached to

algal thalli. In this way, successive generations could build up a reeflike structure.

Taylor (1910, plate 5) found that the Australian archaeocyathid reefs were composed predominantly of archaeocyathids and very few sponge fossils. The Australian archaeocyathid limestones, some 200 ft thick, extend in a narrow belt for some 400 mi. In Labrador the "reefs," several yards wide, range through limestones 10 to 20 ft thick. By contrast, the Siberian archaeocyathid reefs are multiple hundreds of feet thicker.

Okulitch (1955), stressed that in life archaeocyathids did not build bioherms (topographically prominent structures rising from the sea floor as a reef); instead, a given contemporaneous population formed "gardens." Nevertheless, archaeocyathid reef structures have been reported. We may view such reefs as representing a succession of archaeocyathid "gardens." Zhuravleva (1970) related growth to depths (bioherms, for example, had smaller individuals).

Stratigraphy and Age Determination.
North American archaeocyathids do not occur above the *Olenellus* zone (Lower Cambrian). Although usually a sufficient trilobite or brachiopod–molluscan suite of fossils helps pinpoint the age of a given bed, North American archaeocyathids can prove that a given bed is Lower Cambrian in age.

In Labrador the archaeocyathid reefs (composed only of genera of the class Archaeocyathinae) occur in the Forteau Formation; they are excellently exposed in the Highlands of St. John. Zhuravleva and colleagues (1970 and see Hill, 1972) have subdivided the Lower Cambrian of the Siberian platform into eleven archaeocyathid zones and many of these zones are traceable elsewhere. Other stratigraphers (North American, Australian, European, and Chinese) have been able to correlate and date Cambrian outcrops partially by the aid of genera and species of archaeocyathids.

Kawase and Okulitch (1957) described an archaeocyathid fauna from the Yukon Territory. It occurred in dark, fine-grained, massive to oolitic carbonates. The fauna was dominated by Coscinocyathidae and Pycnoidocyathidae. These investigators found that the Archaeocyatha were instrumental in age determination of rocks underlying a large area of the Yukon Territory, that is, Lower Cambrian age. They further observed that Yukon archaeocyathids showed close relationship to those of the British Columbia Cordilleran and also to those of Siberia and Australia, but differed from the Nevada or California fauna. The latter regions are dominated by Ethymophyllidae and Ajacicyathidae.

Thus, not only is age determination possible — where otherwise puzzling — but also faunal provinces may be defined by archaeocyathid faunas.

Genetics and Evolution.
Some 700 species of archaeocyathids have been recognized (Zhuravleva, 1970). Some 79 families first appeared during Lower Cambrian time and nine new families in Middle Cambrian (Hill, 1972). The only acceptable Upper Cambrian archaeocyathids were found in Antarctica (Hill, 1972).

From these data it follows that great radiation of archaeocyathids and the time of maximum speciation occurred during the Lower Cambrian.

A particular Southern Australian archaeocyathid occurrence (Ajax Mine locality) consisted of "silicified Archaeocyathinae in limestone" 150 ft thick. Within the time represented, Taylor (1910) reported ten archaeocyathid genera, two of which had eight species (*Archaeocyathus*) and five species (*Coscinocyathus*) respectively.

There was clearly intense competition for living space between reef species, abrupt replacement of species, and in the archaeocyathid "gardens" many separated subpopulations (isolated gene

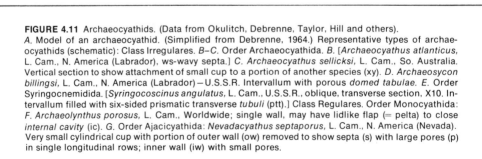

FIGURE 4.11 Archaeocyathids. (Data from Okulitch, Debrenne, Taylor, Hill and others).
A. Model of an archaeocyathid. (Simplified from Debrenne, 1964.) Representative types of archae-
ocyathids (schematic): Class Irregulares. *B–C.* Order Archaeocyathida. *B.* [*Archaeocyathus atlanticus,*
L. Cam., N. America (Labrador), ws-wavy septa.] *C. Archaeocyathus sellicksi,* L. Cam., So. Australia.
Vertical section to show attachment of small cup to a portion of another species (xy). *D. Archaeosycon
billingsi,* L. Cam., N. America (Labrador)–U.S.S.R. Intervallum with porous *domed tabulae. E.* Order
Syringocnemidida. [*Syringocoscinus angulatus,* L. Cam., U.S.S.R., oblique, transverse section, X10. In-
tervallum filled with six-sided prismatic transverse *tubuli* (ptt).] Class Regulares. Order Monocyathida:
F. Archaeolynthus porosus, L. Cam., Worldwide; single wall, may have lidlike flap (= pelta) to close
internal cavity (ic). *G.* Order Ajacicyathida: *Nevadacyathus septaporus,* L. Cam., N. America (Nevada).
Very small cylindrical cup with portion of outer wall (ow) removed to show septa (s) with large pores (p)
in single longitudinal rows; inner wall (iw) with small pores.

pools) that can explain the apparent rapid evolution.

Three other significant factors pertain to archaeocyathid speciation: (*a*) the planktonic larval stage permitted wide cosmopolitan distribution, and (*b*) entry into "open" new ecologic niches. Some larvae settled on lime mud bottoms with little competition for space. The third factor, (*c*) the unique anatomical structure, helps to account for archaeocyathid "success" in the new niches. At any rate, successive generations occupied the given area and this turnover of genetic material resulted in recombinations and mutations leading to new species.

It is noteworthy that several archaeocyathid genera and species are found in a given reef limestone and often on the same bedding plane. That is, they were coextant. In Labrador five species of the cosmopolitan genus *Pycnoidocyathus* (formerly called *Cambrocyathus*), and six genera in all, occur. In Siberia seven species of the cosmopolitan genus *Archaeocyathus*, and 14 species in all belonging to five or six genera, have been reported from the same reef complex. The thickness of the Labrador reefs range from 3 to 20 ft, suggesting that the indicated speciation during the time represented by this deposit was relatively rapid.

According to Russian investigators, the essential evolution was the complication of wall structures that led to greater efficiency in straining currents of water through wall pores.

Study of Archaeocyathids

To study archaeocyathids, it is desirable to polish a hand specimen of reef limestone. Subsequently a layer of water and a smear of vaseline or any clear lacquer may be applied so that many structure details can be studied microscopically.

Thin sections are usually needed to secure further data on fine structures. These should be from several orientations: transverse, radial, tangential, and inclined.

If specimens are silicified, careful surface etching may reveal details of structure, as with the Ajax Mine specimens from Southern Australia. Calcitic archaeocyathids found in shale, in argillaceous limestones (marls), and so on, have been studied by the peel technique. Procedure: Polish face or side to be studied. Etch with acid briefly. Wash clear of acid and apply acetone. Lay a piece of sheet acetate on acetone. Keep acetate sheet close to polished surface by a weight. When dried completely, acetate peel will bear a two-dimensional reproduction of major structures. In this way, a given fossil archaeocyathid may be cut at different places along its length and peels made that will show the developmental stages of the individual.

QUESTIONS

Archaeocyatha

1. Compare the primary data in archaeocyathid and sponge structure point by point. How are they different? In what ways are they homologous?

2. (a) What probable factors can account for the great diversity of archaeocyathid species?

   (b) How would you explain archaeocyathid extinction?

3. (a) Explain the transition from archaeocyathid "gardens" to archaeocyathid reef limestones.

   (b) Explain the skeletal organization of archaeocyathids as related to function, particularly wall and intervallum structures (see Figure 4.11*A*) and compare it to that of a sponge.

   (c) Zhuravleva related growth to depth. Discuss her findings.

   (d) Did the animal move water by inhalent and exhalent currents? Explain.

4. Can you isolate the oceanic, sedimentary, and related factors associated with, or responsible for, archaeocyathid appearance, growth, and demise?

5. Review the utilization of archaeocyathid fossils in biostratigraphic correlations. Cite several different examples.

REFERENCES

Balsam and Vogel (J. Paleont. 1973, v. 47 (5) pp. 979–984) found no functional discontinuity between sponges and archeocyathids and questioned if the latter belonged in a separate phylum.

Hill, D., 1972. Archaeocyatha. Treat. Invert. Paleontology, Part E, (Revised).

Kawase, Y., and V. J. Okulitch, 1957. Archaeocyatha from the Lower Cambrian of the Yukon Territory: J. Paleontology, Vol. 31 (5), pp. 913–930, plates 109–113.

Okulitch, V. J., 1943. North American Pleospongea: G.S.A. Sp. Paper No. 48, 86 pp., plates 1–18, 19 text figures.

_____, and M. W. de Laubenfels, 1953. The systematic position of Archeocyatha (Pleosponges): J. Paleontology, Vol. 27 (3), pp. 481–485.

_____, 1954. Archeocyatha from the Lower Cambrian of Inyo County, California: J. Paleontology, Vol. 28 (3), pp. 293–296, plate 28.

_____, 1955. Archeocyatha, in R. C. Moore, ed., Treat. Invert. Paleontology, Part E, Archeocyatha and Porifera, pp. E1–E20.

Resser, C. E., 1938. Cambrian system (restricted) of the Southern Appalachians: G.S.A. Sp. Paper No. 15, 107 pp., 16 plates. (Archeoycathid reef outcrop belt, Virginia, plate 1; plate 2, Figure 8, and p. 36.)

Schuchert, C., and C. O. Dunbar, 1934. Stratigraphy of Western Newfoundland: G.S.A. Mem. No. 1, 117 pp., 11 plates, 8 figures. (Labrador archeocyathid reefs, plate 3, Figures a, b; text figure 3.)

Taylor, T., 1910. The Archeocyathinae from the Cambrian of South Australia: Royal Soc. So. Australia, Mem. No. 2, pp. 55–183, plates 1–15. (Excellent plates show details of archeocyathids.)

Vologdin, A. G., 1937. Archeocyatha and the results of their study in the U.S.S.R., in Problemy Paleontologii (Problems of Paleontology), Vol. II, III (Publ. Lab. Paleont. Moscow Univ.), pp. 453–481 (Russian text); pp. 481–500 (English text); 4 plates, 24 text figures.

[Zhuravleva, I. V., 1970. In, W. G. Fry, Ed., The Biology of Porifera. Academic Press. Zool. Soc. Sympos. 25, pp. 41–59. B. Ziegler and S. Rietschel, 1970. Same. pp. 23–40. W. D. Hartman, and T. F. Goreau, 1970. Same. pp. 205–243. C. W. Stearn, 1972, *Lethaia*, vol. 5(4), pp. 368–385.]

chapter five

THE RADIAL SYMMETRY BIOPROGRAM IN CORAL AND OTHER SEAS (COELENTERATES)

INTRODUCTION

A remarkable diversity of forms characterizes living coelenterates, which are predominantly marine. The three classes (embracing some 9000 living species) studied in this chapter include familiar forms: the colonial hydroid *Obelia*, the fresh-water hydroid *Hydra* (Hydrozoa), the jellyfish *Aurelia* (Scyphozoa), the sea anemone *Metridium*, and the corals (Anthozoa).

In organic reef construction, coelenterates (hydroids, anthozoans) play(ed) a significant part. They either form(ed) the framework or are among the reef inhabitants. Such structures, called bioherms or biostromes (*herm* = mound; *stroma* = bed), are of first-rate geological and biological importance and will be studied in detail. (Figure 5.5*A*, *B*)

Among common fossil coelenterates encountered in outcrop or in well cores or cuttings are the anthozoans. Thus rugose corals are abundant in beds of Carboniferous age, and hermatypic (reef) scleratinian corals (Figures 5.6*A*,*C;* 5.7; 5.8) are frequent in Mesozoic to Tertiary, Pleistocene to Recent bioherms. Silurian and Devonian beds often abound in hydrozoan stromatoporoid biostromes. Tertiary reefs often contain structures built by the hydrozoan milleporines and stylasterines. Some scyphozoan conulatids are particularly characteristic of the Lower Paleozoic.

PROTOZOANS TO METAZOANS

The ancestral line that gave rise to the most primitive of the coelenterates (Figure 5.1) fades without clear traces into the Pre-Cambrian. Paleontological evidence is too sparse and incomplete to resolve the matter. Students of living forms have long deliberated about the probable ancestral coelenterate, reasoning from embryological development and other data (Hyman, 1940).

Since sponges did not give rise to any higher form, only the protozoans can have been the original source for the development of the metazoans.

A flagellate colonial protozoan such as *Volvox* (Figure 5.1*A*) resembles the early stages of metazoans (blastula stage). There appear to have been two critical steps in the evolutionary transition from a *Volvox*-like colony to the simplest metazoan:

1. *Cell Specialization.* Somatic (body) and reproductive (germ) cells are differentiated. The latter pass to the interior, leaving an outer rim of flagellate body cells (Figure 5.1*B*).

2. *Formation of Two-Layered Wall* (Diploblastic) (Figure 5.1*C*). The development of a two-layered wall could have progressed from the one-layered flagellate layer as in *Volvox*, to the *blastaea* of Haeckel (Figure 5.1*B*) by the in-wandering of ectodermal cells (see Figure 5.1*D* for cell layers). That would have provided an inner layer performing digestive functions (gastroderm or endoderm) and an external epidermis (ectoderm).

The next step in the transition was probably the development of a mouthless stereogastrula (Figure 5.1*D*). This stage

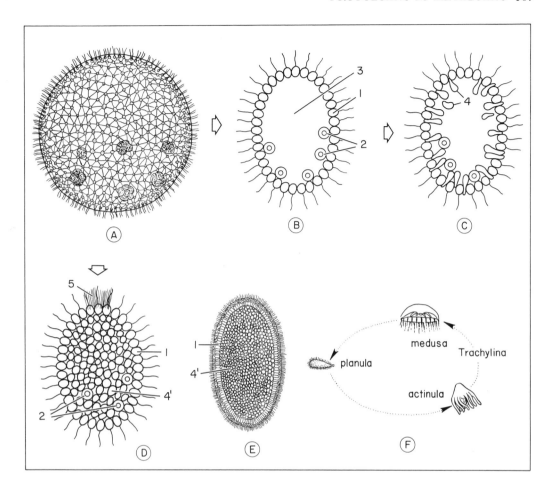

FIGURE 5.1 Probable transition from Flagellate Colonial Protozoan to Ancestral Coelenterate. (*A*) *Volvox*. Haeckel derived the Metazoa from a *Volvox*-like ancestral flagellate (see *B*). (*B*) Ancestral Blastea of Haeckel. (1) single layer of flagellated cells, (2) germ cells, (3) hollow interior. Note differentiation into body cells (1) and reproductive cells (2). (*C*) Protoendoderm stage. Multipolar ingress, that is, in-wandering of ectodermal cells forming a protoendoderm (4). (*D*) Hypothetical Stereogastrula or Planuloid ancestor of the Eumetazoa (that is, lower metazoa). (1) ectoderm, (2) germ cells, (4′) endoderm, (5) sensory bristle. Note lack of hollow interior ("3" in Figure 5.1*B*) and differentiation into two tissues, ecto- and endoderm characteristic of coelenterates. This is the postulated common ancestor of coelenterates and flatworms. (*B, C, D* from *The Invertebrates*, Vol. I, by Libbie H. Hyman. Copyright 1951 by McGraw-Hill, Inc., with permission of McGraw-Hill Book Co.). (*E*) Living hydroid planula larva. (1) cilia, (2) ectoderm (= "1" of *D*), and (3) endoderm (= 4′ of *D*). (*F*) Life cycle of the hydroid *Trachylina* (after Hill and Wells). Actinula of *Trachylina* is thought to be like the primitive medusa—the ancestral coelenterate that developed from a metagastraea (a stage beyond that in Figure 5.1*D*). In the latter stage, mouth and archenteron developed after the stereogastrula began feeding on the bottom.

strikingly resembles the coelenterate planula larval stage (Figure 5.1*E*). It is now thought to have been the common ancestor, not only of coelenterates, but also of flatworms (see Chapter 9).

Finally, to derive the three classes of cnidarians discussed below (Hydrozoa, Scyphozoa, Anthozoa), a few additional steps are postulated. After the stereogastrula (Figure 5.1*D*) settled and fed on the bottom, rearrangement (wandering) of endodermal cells would form a mouth and the precursor of the internal cavity (enteron) called the archenteron. This stage is called the *metagastraea*.

The developmental history of the living

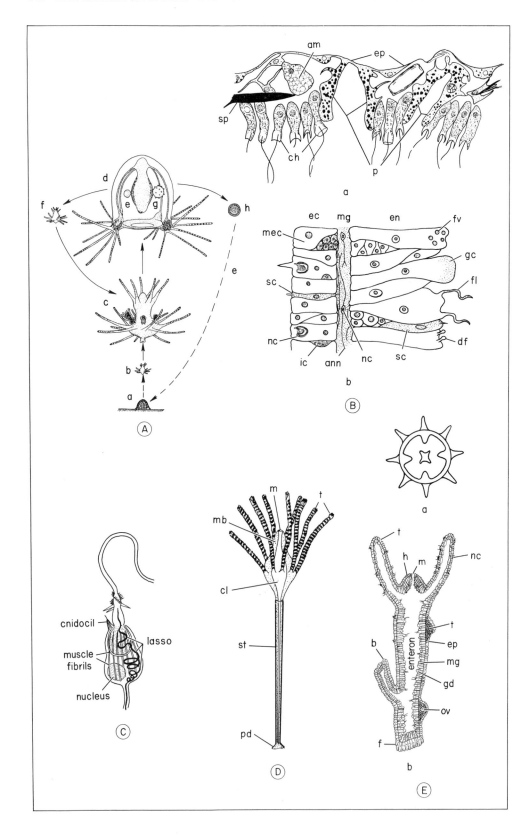

FIGURE 5.2 Distinctive coelenterate features. (A) Life cycle of a pelagic hydroid *Margelopsis haeckeli*. Hartlaub (after Werner, 1963). Reproduction in the three classes of Cnidaria emphasizes either medusoids (Scyphozoa), or polypoids (Anthozoa), or polymorphism, that is, both medusoids and polyps (Hydrozoa), as in *Margelopsis*. (B) Comparison of cellular and tissue grades of construction: (a) sponge *Leucosolenia*, cross-section-cellular grade. Note individual separable cells: choanocytes (ch), amoebocytes (am), porocytes (p); (sp — spicule); (b) coelenterate, *Hydra*, tissue grade. Here multicells are non-separable and serve different functions; they occur in a continuous tissue (ec — ectoderm, en — endoderm): sc — sense cells, gc — gland cells, nc — nerve cells, ic — interstitial cells, mec — musculo-epithelial cells, among others. (C) Nematocysts. (D) Hydroid polyp; cl — coelenteron, see *E-b*; mouth (m); tentacles (t); st — stem; pd — pedal disk, mb — manubrium (*E-a*) Medusa, transverse section (schematic) of hydrozoan (also scyphozoan); symmetry is radial and tetramerous. (*E-b*) *Hydra*, a solitary freshwater polyp: m — mouth; en — enteron (coelenteron); mg — mesogloea; t-t'-tentacles; t — testis; o — ovary; nc — compare "B" above; b — bud. (This a long view seen microscopically). Illustrations: (*B-a, D, E-a*) from *The Invertebrates*, Vol. 1, by Libbie H. Hyman (Copyright 1940 by McGraw-Hill, Inc., with permission of McGraw-Hill Book Co.): (*B-b*) and (C) after Will, from G. A. Kerkut, ed. *The Invertebrata* (3rd ed., 1959, by L. A. Borradaile and F. A. Potts with permission of Cambridge University Press); (*E-b*) after several authors.

hydroid order Trachylina (Figure 5.1*F*) indicates (*a*) that its *actinula* larval stage was preceded by a planula larval stage (metagastraea) and (*b*) that the actinula closely resembles the primitive medusa from which arose, first, the trachylinid hydroid line and, subsequently, the three classes.

LIVING COELENTERATES

Phylum Definition

Phylum Coelenterata (Greek koilos, = cavity; enteron = intestine). The phylum has the following characteristics.

1. *Multicellular* animals (Metazoa), generally *tentacle*-bearing.
2. Chief external feature is *radial symmetry*. The main axis of symmetry is the *oral-aboral axis* about which the parts are arranged (Figure 5.2*E*).
3. *Tissue grade* of construction (epithelial, muscular, and connective tissue) (Figure 5.2*B*).
4. Body wall composed of two epithelia: outer epithelium (*ectoderm*), inner epithelium (*endoderm*). An intermediate layer (*mesoglaea*) forms a kind of connective tissue or cement between the two wall layers (Figure 5.2*E*).
5. A single internal cavity, the digestive cavity (*enteron*), occupies the space within the body wall. It is closed at one end (the aboral end of the axis of symmetry) (Figure 5.2*E*).
6. The only orifice of the digestive cavity is the *mouth* (oral end of the main axis of symmetry). Mouth serves for food intake and ejection of indigestible substances.
7. *Radial folds* and *partitions* commonly divide the internal cavity (see class Anthozoa).
8. *Systems* that are *present*: digestive (see items 5 and 6 above), muscular, nervous (a simple network of cells), sensory — in an elementary stage of development (Hyman), and reproductive (sex cells aggregated into gonads) (Figure 5.12*A*).
   Systems that are *lacking*: respiratory, excretory, circulatory.
9. *Exoskeleton*: An exoskeleton is often present (in subphylum Cnidaria) but completely lacking in the subphylum Ctenophora.
10. Geologic range: Pre-Cambrian to Recent.

If the subphylum Ctenophora ("jelly combs") is eliminated from consideration because of fossil record rarity, then the given phylum definition must be supplemented to include special features of the other coelenterate subphylum, the Cnidaria.

Subphylum Cnidaria

1. Presence of stinging capsules (nematocysts) diagnostic for all living forms. Nematocysts serve for defense and food capture (Figure 5.2*C*).
2. Asexual budding can yield either solitary polyps (Figure 5.2*D*) or polypoid colonies; sexual reproduction typically produces cili-

ated larvae (planula, see Figure 5.1*F*, 5.2*A*) that attach to bottom and form a polyp.

3. Characterized by polymorphism, that is, either a fixed polypoid stage or a free-swimming medusoid stage (Figure 5.2*A*). In the Hydrozoa, both stages may occur; or the latter stage may occur exclusively.

4. Some cnidarians possess an endoskeleton (internal) or exoskeleton (external), whose composition may be calcitic or aragonitic (hydrozoans, anthozoans), or chitinophosphatic (conulatids). An endoskeleton of hornlike composition is known in some anemone-like anthozoans that are rare as fossils (see Chapter 15 for details on skeletal variation in reef corals). Some horny hydroids are also known.

5. Chiefly marine.

GENERAL DATA ENCODED IN THE THREE CNIDARIAN CLASSES

To distinguish one cnidarian from another, the subphylum is subdivided into three distinct classes. These three classes are as follows:

Class 1. Hydrozoa

Includes hydroids such as the organic reef-building and/or inhabiting milleporines, stylasterines, and stromatoporoids, as well as craspedote[1] medusae that are more rarely found as fossils.

Solitary or colonial; radial symmetry (tetrameral or polymeral, see Figure 5.2*E*); occur as polyps or craspedote medusae or both (Figure 5.2*A*); enteron lacks gullet, septal projection, and nematocysts (Figure 5.2*C*). In general, if present, the exoskeleton is chitinous, although in some orders, calcareous skeletons are important rock builders. Chiefly marine but some fresh-water forms occur. Range: Lower Cambrian to Recent.

Class 2. Scyphozoa

Includes the jellyfish or true medusae; having a vari-shaped gelatinous bell. The conulatids are the most frequently found fossils of this class; rarer are scyphomedusae.

Radial symmetry (mostly tetrameral); living forms mostly free-swimming (acraspedote medusoid); fossil forms such as conulatids attached (polypoid). Hard parts are generally lacking; chitinophosphatic periderm (Figure 5.15) present in conulatids. Four interradial endodermal septa typically divide the gastrovascular system into a central stomach and four gastric pouches (Figure 5.12). Interradial septa may be mineralized in fossil forms. Exclusively marine. Range: Cambrian to Recent.

Class 3. Anthozoa

Includes sea anemones, corals, sea fans, sea pens, sea feathers. Solitary or colonial; occur as polyps only. Oral end expanded radially into an oral disk bearing hollow tentacles (Figure 5.18*B*, *a*); mouth centrally located. *Stomodaeum* or gullet (pharynx in adults) leads from mouth into gastrovascular cavity and may have one or two flagellated grooves (siphonoglyphs) (Figure 5.18*B*, *c*).

Gastrovascular cavity biradially partitioned into compartments by complete or incomplete mesenteries.[2] (Figure 5.18*C*), which bear endodermal gonads.

Larval stage, free-swimming planula. Skeleton, if present, horny or calcareous spicular endoskeleton (that is, internal skeleton) or calcareous exoskeleton (corallum) characterizes some groups. Exclusively marine. Range: Ordovician to Recent.

ADDITIONAL INFORMATION FROM LIVING HYDROZOA

Polymorphism, kinds of polyp, and their significance. Although five living orders

---

[1]Craspedote medusae have a circular shelf called a *velum* (Figure 5.2*A*), which contains a muscular band that serves in swimming. If the velum (or craspedon) is lacking as in the scyphomedusae, the medusae are known as acraspedote.

[2]Mineralized septa occur within and between mesenteries.

of Hydrozoa can be distinguished on many small anatomical details, they differ essentially in *degree* of polymorphism (that is, medusoid generation only, polypoid only, or dominant, or both of these with one reduced or absent).

Hydrozoan Orders

Order 1. Trachylida. Medusoid dominant; most primitive of the hydrozoans. Lower Cambrian (Figure 5.3*A*, part 1*a–b*). [Only the probable oldest appearance of the orders is indicated since all these five orders are still extant.]

Order 2. Hydroida. Colonial polypoid dominant. Cambrian (Figure 5.3, part III/IV).

Order 3. Siphonophorida. Several types of polypoids and medusoids attached. The most specialized of the hydrozoans. Ordovician.

Order 4. Milleporida. Free-swimming medusae; two kinds of retractile polyps. Upper Cretaceous (Figure 5.3, part V).

Order 5. Stylasterida. Sexual generation by non-detachable gonophores; two kinds of retractile polyps. Upper Cretaceous (Figure 5.4, part XIII).

The paleontological significance of polymorphism can readily be seen in the observable morphology of fossil hydrozoans. Medusoid or polypoid generations will present correspondingly flattened, bell-shaped exumbrellar impressions (Figure 5.3) or branching or stolonal structures (Figure 5.4). On the other hand, different kinds of retractile polyps will be revealed in fossil milleporines and stylasterines by the *cyclosystems* and by the general size and spacing of the respective pores (Figure 5.3, part V). [A cyclosystem is the arrangement of the two kinds of retractile polyps (dactylozooid, gastrozooid, see Figure 5.3, part V) relative to each other. Circular rows of five to seven individual dactylozooids are arranged around a gastrozooid. See Figure 5.4, part XIV for fossil example.]

Ecology (Milleporines and Stylasterines)

Symbiosis (Millepora). The coenosarc tubes or surface system that links all parts of a milleporine colony, as in the genus *Millepora* (Figure 5.3, part V), harbors numerous unicellular algae (zooxanthellae). In the apparently symbiotic relationship the algae live by, photosynthesis yields oxygen or alga serve other needs of the milleporine colony. In turn, the depth at which living (and presumably fossil) milleporines thrive(d) is determined by this symbiotic relationship. Generally, living forms thrive in depths of 30 meters or less. At Funifuti, species range from depths of 1 to 21 ft.

Depth, Substrate, Temperature. Crossland (1928 ref. in Boschma, 1948, p. 64) found various growth forms of *Millepora* in the Tahiti reefs: (*a*) simple encrustations on reef surface, which were the first stage in all cases observed; (*b*) platelike masses with possible branching in the upper margins (surf-swept barrier edge of reef); (*c*) upright branching (in quiet water); (*d*) leafy form (in the lagoon). Yonge (1930) observed growth types (*a*) and (*b*) in somewhat similar situations in the Great Barrier Reef. (Figure 5.6*B*) However, Boschma (1948, pp. 71ff.), in the reefs of the Java Sea, found that two or more growth forms of *Millepora* existed side by side in the same environment.

An intensive study of milleporine growth and speciation showed that four major factors influenced their growth: (*a*) energy level of water (quiet or agitated); (*b*) nature of substrate to which larvae attach; (*c*) parasites such as the barnacle *Pyrgoma milleporae* (see Chapter 11) that are invariably associated with species of *Millepora* and cause "warty excrescences" on the surface; (*d*) depth of water—that is, colonies will appear stunted if they tend to grow above water into the air.

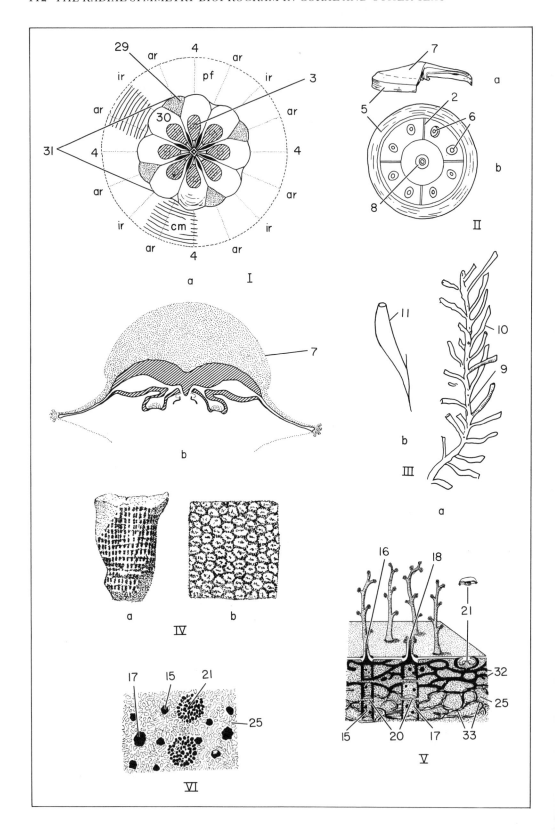

FIGURE 5.3 Fossil Hydrozoa. I. Order Trachylinida (L. Cam., and U. Jur.—Rec.): 1. *Kirklandia texana,* L. Cret., Texas: (*A*) schematic, oral or subumbrellar view; (*B*) axial section through restored disk. II. *Beltanella gilesi,* L. Cam., S. Austral., reduced. Order Hydroida (Cam.—Rec.). III. *Archaelofoea longicornis,* M. Cam. Victoria-Tasmania, enlarged. Order Sphaeractinoidea: IV. *Stromatomorpha stylifera,* Trias. Europe: (*a*) radial surface: pillars (vertical) and concentric laminae (horizontal) that bind them; (*b*) transverse section. X6. Milleporina: V. *Millepora* colony, schematic. VI. Surface of V, enlarged.
(1) Sketch of subumbrellar or oral surface. (2) Radial canals. (3) Gastro-genital pouches (?). (4) Periradius. (5) Velum (?). (6) Gonads (?). (7) Umbrella. (8) Oral structures (?). (9) Hydro-caulus (main stalk). (10) Hydrotheca. (1) Gonotheca. (12) Radial pillars. (13) Horizontal bar or perforate laminae. (28) concentric laminae. (29) Exsert lobe. (30) Insert lobe. (31) Oral or gastric disk. (32) Canal system at surface. (33) Canal system in lower layer degenerating. Illustrations redrawn and adapted from : (1) Caster, 1945; (2) Harrington and Moore, 1956; (3.4) Hill and Wells, 1956; Kuhn, 1932; (5.6) Boschma, 1948; Boschma, 1956 modified after Hickson. (For items 15 to 25 see Fig. 5.4.)

Milleporine Speciation. Boschma (1948) concluded that he could distinctly recognize ten species based on coralline form. This theory held despite high variability in form of a given species of *Millepora.*

Skeleton. Both the milleporine and the stylasterine hydrocorals possess a porous, calcareous skeleton. In the stylasterines it is less brittle and, because of narrower pores, denser and harder.

The skeleton of living milleporine species contains between 87.32 and 99.63 percent calcium carbonate = aragonite; from a trace to 2.08 percent magnesium carbonate; from a trace to a few tenths of 1 percent phosphorus pentoxide; and negligible aluminum oxide plus iron oxide and silica. Calcium sulphate ranges from 0.06 to 2.14 percent. The stylasterine skeleton is quite similar in composition. Both types of skeleton contain mineral aragonite (Vinogradov, 1953, Table 119, ref. ch. 15).

Coral Reefs (Milleporines). Vaughan (1919) saw that *Millepora alcicornis* incrusted coral reefs "nearly everywhere there are such reefs" in the West Indies, Bermuda, and Florida. In the Pacific (Raroia Atoll), incrusting *Millepora* occur in discontinuous patches that demark a narrow zone on the outer flank of the coral-algal ridges (Newell, 1954). On Eniwetok Atoll, Odum and Odum (1957) reported encrusting *Millepora* in the front encrusting zone (B in Figure 5.5*A*) and the lagoonal sand-shingle zone (E in Figure 5.5*A*).

A FOSSIL HYDROCORALLINE FAUNA

Nielsen (1919) described a fossil hydrocoralline fauna of modern aspect from a quarry in Faxe, Denmark (Upper Cretaceous, Danien Age). The hydrocorals included one species of *Millepora* (Figure 5.4) and species of eight other genera.

Biotic Associates

In the mentioned fauna, the following biotic associates occurred: octocorals, such as *Heliopora* and other corals; echinoderms, especially asteroids (Chapter 12); foraminifers; worms, such as *Serpula, Spirorbis* (Chapter 9); bryozoans, such as *Membranipora* (Chapter 6); brachiopods, such as *Terebratulina* (Chapter 7); mollusks, which included *Pecten* and numerous gastropods (Chapter 8) and crustaceans.

Lithology

The entire fauna found in the Danish quarry occurred in a fine-textured, highly porous limestone (a micrite). Fossils were readily released from sediments when washed with water.

FOSSIL HYDROZOANS

Extinct Hydrozoan Order

Order 6. Sphaeractiniida **(after Kuhn, 1939, and Galloway, 1957).** Calcareous skeletons composed of concentric, radial trabeculae and having astrohizae. Colonial. Permian to Cretaceous. *Stromatomorpha* (Figure 5.3, part IV) shows the characteristics of the order.

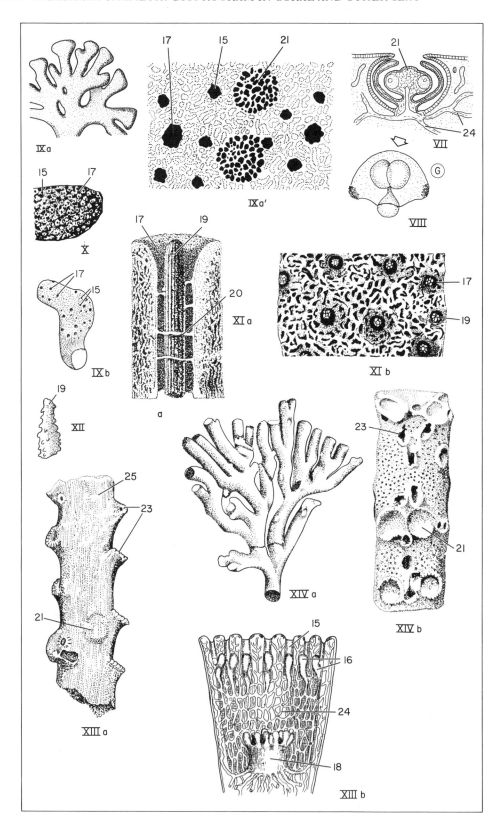

FIGURE 5.4 Fossil Hydrozoa, II. Order Milleporida continued. (All figures schematic). VII. *Millepora* sp., longitudinal section through ampulla (compare Figure 5.3, V). VIII. Medusa released from VII. IX (*a, a'*) *Millepora tenera;* (*a*) portion of a colony: (*a'*) section of surface of ''(*a*)''; (*b*) *Millepora parva.* U. Cret. (Danian) Denmark. X. *Milleaster,* Mio., Maryland. XI. *Axopora,* Eoc., France: (*a*) longitudinal section of a gastropore; (*b*) surface enlarged. XII. *Axoporella,* Eoc. Hungary; fragment of gastrostyle. XIII (*a*) *Stylaster,* Mio. Victoria-Australia; (*b*) retracted zooids (gastro- and dactylo-). XIV. *Congregopora nasiformis,* U. Cret. Denmark: (*a*) colony; (*b*) skeletal features.

(15) Dactylopore. (16) Dactylozooid. (17) Gastropore. (18) Gastrozooid. (19) Gastrostyle. (20) Tabula. (21) Ampulla. (22) Gonopore. (23) Cyclosytem. (24) Coenosarc. (25) Coenosteum. (Illustrations adapted and redrawn after Boschma, and Boschma after several authors, except IX (*b*), XIV, Nielsen, 1919; VII-VIII, XIIIb, Hickson 1891, Quart. Jour. Micros. Sci., Vol. 72 and Moseley, 1879, Philos. Trans. Roy. Soc. London, ser. B, v. 169).

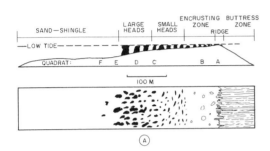

FIGURE 5.5 Eniwetok Atoll, Reef Ecosystem. (*A*) Zones.
Redrawn and adapted from Odum and Odum, 1955.)

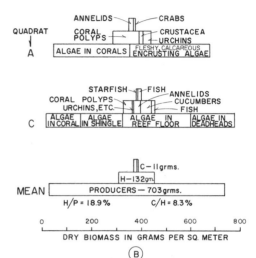

FIGURE 5.5 (*B*) Trophic levels (abbreviated) and Biomass.
(Redrawn and adapted from Odum and Odum, 1955.)

REASSIGNED FOSSIL HYDROZOA

Phylum Porifera, Subphylum Stromatoporata

Order Stromatoporida **(after Galloway, 1957; Stearn, 1972 (see p. 117, and Table 5.1).** Calcareous skeleton (*coenosteum*) with discrete or amalgamated horizontal (plates, laminae) and vertical (pillar) structures (Figure 5.9*E*). *Astrorhizae* (that is, a stellate system of outwardly branching canals) present or absent; exclusively colonial; either laminar (sheet-like), massive, cylindrical, or dendroid in growth habit. Skeletal tissue compact or spongy (minutely vesicular) with minute pores in laminae, pillars, and so on (*maculate*) or with small tubes or pores in the laminae (*tuberculate*). Lack *trabeculae* (rods). Ordovician to Devonian.

Stromatoporoid Affinities

The stromatoporoids, known only as fossils, constitute another very difficult group. The question arises, "To what phylum and class do they belong?" Many specialists have been concerned with finding an answer to this question. Some have argued that they belong to the Foraminifera; others, to the Porifera. The late Dr. Galloway had thought they might have evolved from the archaeocyathids (see Chapter 4) but, with other specialists, finally concluded that they were hydrozoans.

A glance at Figure 5.9, J–K will indicate apparent similarity in structure between living hydroid *Hydractinia* and some stromatoporoids. Vertical tabulate chambers, astrorhizae (Figure 5.9*D*), and other structures, such as latilaminae (Figure 5.9*B*), might argue for hydrozoan affinities. (Compare Sclerospongiae, Table 5.1, "Note". Sponge affinities are now preferred.)

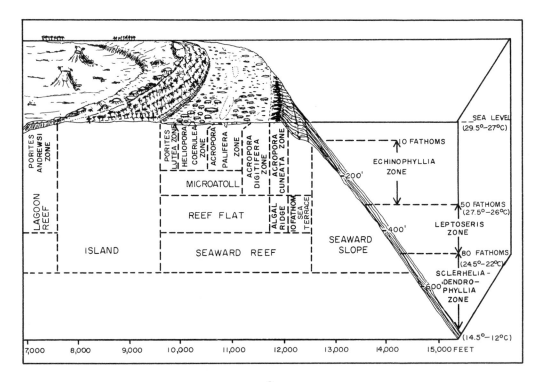

A.

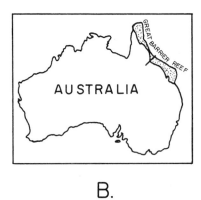

B.

FIGURE 5.6 Modern coral atolls and reefs. (*A*) Lagoon, Bikini Atoll. Coral zones (after Wells, J. W., 1954, U.S. Geol. Surv. Prof. Pap. 260-I, pp. 385–486). Figures at base show width of atoll in feet; about 3 miles across. Note deeper zone at 10, 50, 80 fathoms. (*B*) Great Barrier Reef, Australia, general location. (*C*) Low Isles, Great Barrier Reef (Traverse 1). Modified from Manton and Stephenson, U.S. Geol. Surv. Prof. Pap. 260-I, pp. 385–486.

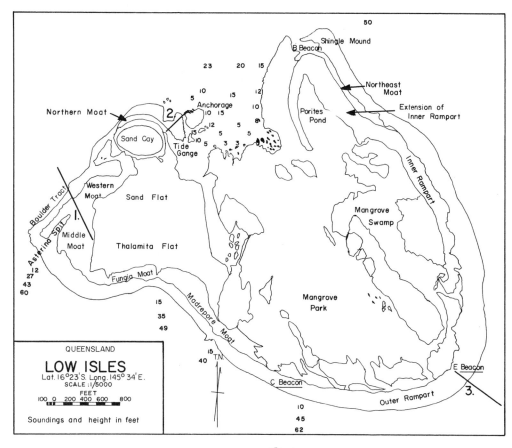

C.

Classification of Stromatoporoids

Lecompte (1956, *Treatise on Invertebrate Paleontology*), recognized eleven families. In turn, such recognition extended the range of the order from Cambrian to Cretaceous. Galloway recognized only five families. This, in turn, limited the range of the order from Ordovician through the Devonian Stearn, 1972, would include Mesozoic families (see Table 5.1).

Fossil Record of Stromatoporoids

Stromatoporoids occur abundantly in bioherms, biostromes, and "banks" that are found in Ordovician through Devonian beds in many parts of the world.

The Ordovician genus (*Cystostroma* Figure 5.10B, parts *e–f*) is considered by Galloway and St. Jean (1961) to be ancestral to all Silurian and Devonian stromatoporoids. Three types of *coenostea* appear to characterize Ordovician stromatoporoids: a thin, flattish layer (genus *Dermatostroma*); hemispherical masses (*Cystostroma*, Figure 5.10B, part *e*, *Labechia*, Figure 5.10B, part *a*, and others); and vertically erected cylindrical or branching forms (*Aulacera*, Figure 5.10B, part c and so on).

Ordovician stromatoporoids are widely distributed in North America, Asia, the Urals, and the Baltic areas (Galloway and St. Jean, 1961). In the Middle Ordovician (Chazyan–Black Riveran) of Isle La Motte, Vermont, they occur on reefs with colonial corals and algae. At this locality some protocoenostea of stromatoporoids are found intergrown with algae. The same intergrowth are also known from dolomitized Silurian reefs in Indiana.

In Gotland, Silurian reefs, ranging

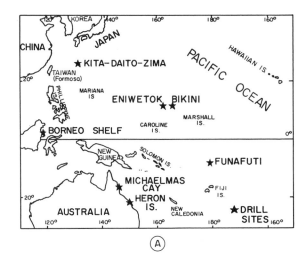

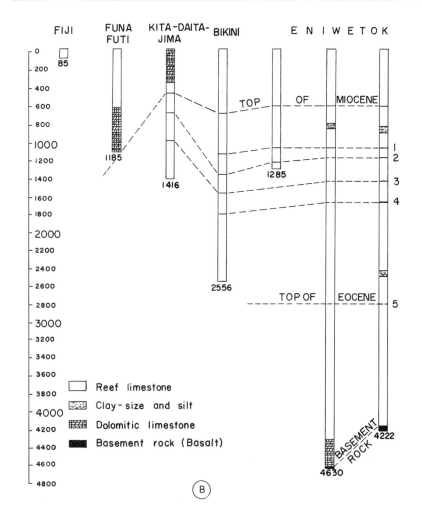

FIGURE 5.7 Coral atolls of the Pacific. (Modified from Ladd et al., 1953.) (*A*) Map of area showing drilling sites. (*B*) Coral atolls; deep borings and correlations between atolls.

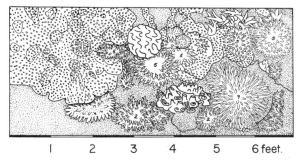

FIGURE 5.8 Great Barrier Reef: Scleratinian coral distribution in a limited reef tract (Traverse 2, seaward slope). Eleven coral species are represented in 2 square yards and include Nos. 1, 2, 5, three distinct *Acropora* species. (After Manton and Stephenson, 1935.)

from small knolls a few meters in diameter to bioherms 10 to 20 meters thick, laterally extend for several hundred meters. Stromatoporoids, corals, and calcareous algae are the chief reef-builders (Hadding, 1950). Lenticular bodies, chiefly stromatoporoid-bearing, are frequent in the Devonian of the Ardennes region of France.

The maximum distribution, abundance, and diversity of stromatoporoids were attained during Devonian time in North and Central America, Europe, Asia, North Africa, and Australia.

It is comparatively infrequent that an entire reef is composed exclusively of stromatoporoids. Nevertheless at some localities particular stromatoporoid genera and species may be dominant. In the Devonian bioherms of Alpena, Michigan, stromatoporoids are so abundant that they constitute the bulk of the reef. In Indiana a bed of limestone (biostrome) 3 ft thick is composed almost completely of the species *Labechia huronensis*. The genus *Clathrodictyon* (Figure 5.10*A*) is abundant in the Upper Ordovician–Lower Silurian of Anticosti Island, where it occurs in association with a considerable invertebrate fauna. The genus *Aulacera* (Figure 5.10*B*) abounds in the Kentucky Bardstown Reef.

Common and well represented in the Devonian formations of Indiana, Michigan, Ontario, Ohio, Kentucky, Illinois, Missouri, and Iowa, stromatoporoids have been extensively collected in quarries, roadcuts, and along stream beds.

Galloway and St. Jean found that throughout the North American Cordillera, stromatoporoids were numerous. They are known from Arizona, Wyoming, Nevada, Utah, and British Columbia. Oil-soaked stromatoporoid-bearing cores have been reported from the subsurface of eastern Montana in beds of Silurian and Devonian ages.

Lithology, Preservation

Stromatoporoids are most frequent in limestones or shaly limestones and are rarely found in calcareous shales. At places such as Gotland and Michigan, stromatoporoid heads and branches have been broken off by breaker action and the fragments incorporated in a breccia. Leaching and recrystallization destroy fine structures of the skeleton. Silicification and dolomitization, especially of exposed surfaces, are seen quite often (Parks). Some stromatoporoids have been replaced by iron oxide.

Method of Study

To study stromatoporoids, two oriented thin sections are generally needed. One section is cut vertically, that is, parallel to the pillars. Another section is cut parallel to the laminae (and is therefore a tangential section). In cylindrical and dendritic forms, an axial and a tangential section are both needed. In such cases a cross-section is also used. This section is

FIGURE 5.9 Characteristic stromatoporoid structures. (*A*) *Stromatopora polyostiolata*, M. Dev. Ger., X50. *m* — mamelons, *ar* — astrorhizal opening. (*B, E, F*) *Actinostroma:* (*B*) portion of colony, Dev., Iowa, *ll* — latilaminae; (*E*) (*a*) vertical section (*A. clathratium*, M. Dev. England), *p* — pillars, *l* — laminae; (*b*) tangential section; (*F*) Tangential section. (*A. devonense*, U. Dev. Belgium). (*C*) *Atelodictyon fallax*. M. Dev. Belgium, irregular vermiform pattern, enlarged. (*D*) *Ferrestro-matopora larocquei*, M. Dev. Ohio, astrorhizae with regularly dividing branches joining adjacent astrorhizae, X 1. (*G*) *Aulacera plummeri*, vertical section, *c* — cyst plates, *vp* — vertical pores, X 20. (*H*) Stachyodes, Dev. Germany, vertical section, enlarged. (*I, J, K*) The hydroid, *Hydractinia:* (*I*) growth of colony on small shell, schematic — (1) polyp, gastrozooid, (2) spine, (3) periderm with coenosarc covering; (*J*) vertical section through basal mat of (*I*) showing basal portion of periderm and its structure, (4–5) outer and inner epidermis (note how gastrodermal tubes (*gt*) — polyps — are related to coenosarc covering of spines); (*K* longitudinal section through (*I*) 3. (Note that entire skeleton consists of parallel narrow chambers traversed by closely spaced thin tabulae.) (*L*) The stromatoporoid *Parallelopora paucicaniculata*, M. Dev. Belgium, enlarged (compare *K* and *L*).
Illustrations redrawn and adapted, after, (*D, E, G* — Galloway, 1957b; (*B*) Tasch collection; (*I*) Hyman; (*J*) Collcut, 1897 cit. Hyman; all others Lecompte.

made without reference to parallelism to either laminae or pillars.

For determination of genus and species, thin sections are critical. By polishing two faces at right angles and studying the structures, one can determine the family. The key can then be used directly (Table 5.1).

INFORMATION DERIVED FROM STROMATOPOROIDS

Paleoecology

Biotic Associates. Stromatoporoids most often occur with tabulate corals. So common is this type of intergrowth that in thin sections of stromatoporoid skeletons, distinctive coralline tubes are often seen — caunopora. Other associates are bryozoans, algae, and crinoids. In the rock matrix between stromatoporoid fossils or fragmentary pieces of stromatoporoids, echinoids, brachiopods, and mollusks may occur.

Seasonal Growth. The distinct latilaminae, common in many forms (Figure 5.9*B*) are thought to indicate annual warm and cold seasons. The detection of seasonal changes in beds of Paleozoic age is extremely rare and hence, if the latilaminae do contain seasonal information, much more research on this subject seems useful.

Substrate. The Gotland reefs grew upward from marly limestones. The rarity of stromatoporoids in shales is indicated by the limitations to their growth imposed by a clay mud bottom and turbid conditions. Lecompte said that massive and globular stromatoporoid colonies characterize pure limestone, whereas sheetlike colonies are more likely to occur in impure limestones or marly beds.

Carozzi (1961) found a rhythmical alteration of eight carbonate microfacies in a reef sequence deciphered from a core (Beaverhill Lake formation, Upper Devonian, Alberta, Canada). The microfacies were vertically superimposed. In each, the *Amphipora*-stromatoporoid reef community grew and developed on a carbonate platform and then disappeared. In Carozzi's microfacies number 5, for example, scattered large fragments of *Amphipora* and abundant mat colonies of stromatoporoids plus mollusks, ostracodes, and sponge spicules occurred in a calcilutite matrix. This microfacies is thought to have represented a backreef environment. The reef itself, in the Carozzi study, was composed of cabbage-type stromatoporoids, with *Amphipora* appearing where the water apparently shallowed. The general type of sediment encountered was biocalcarenite.

Depth, Temperature. Stromatoporoids, by their biotic associates and the enveloping sediments, may be taken to indicate clear, warm, shallow water. They probably flourished most commonly on

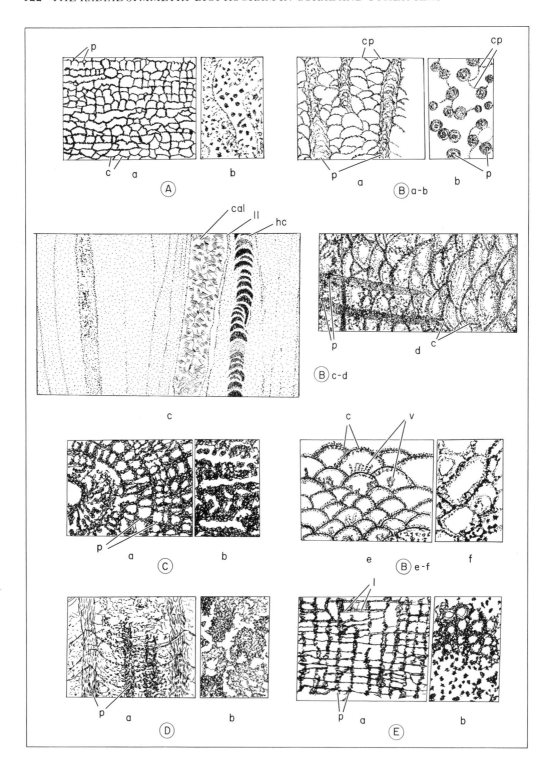

FIGURE 5.10 The five stromatoporoid families showing a representative genus in each. (Compare Figure 5.9 for other genera, and text for key.) (After Galloway; Galloway and St. Jean; Lecompte; and W. A. Parks, 1907–1910, Univ. Toronto Studies, Geol. Series. Nos. 4–7). (A) Family Clathrodict-yidae: *Clathrodictyon vesiculosum,* M. Sil. Ohio: (a) vertical section, cysts (c) end to end in place of regular laminae and pillars (p); (b) tangential section. (*Ba-b*). Family Labechiidae: *Labechia conferta,* Sil. England: (a) vertical section, pillars and arched cyst plates (cp); (b) tangential section. (*Bc-d*) *Aulacera plummeri,* U. Ord. Indiana: (c) longitudinal section showing axial column of hemi-spherical cysts (hc), large crack filled with calcite (cal) and latilaminae; (d) vertical section showing oblique pillars. (*Be-f*) *Cystostroma simplex.* Basal Trenton., M. Ord. Tennessee. This genus embraces the most primitive and oldest known stromatoporoids (e) vertical section, c — cyst plates; v — villi; (f) tangential section. (C) Family Idiostromatiidae: *Idiostroma roemeri,* Dev. Germany: (a) cross-section showing pillars superimposed; (b) tangential section. (D) Family Stromatoporidae: *Syringostroma superdensum,* M. Dev. Ohio: (a) vertical section, large pillars of maculate tissue, and laminae (1); (b) tangential section. (E) Family Actinostromatidae: *Trupetostroma warreni,* Dev. Canada: (a) vertical section, pillars superimposed; (b) tangential section.

reefs or as parasites on corals in quiet waters a few feet deep. Brecciation of these reefs suggest that such structures were subjected to breaker wave action. However, Carozzi (1961) showed reef and backreef distributions.

Size, Shape. Stromatoporoid heads range in size from a few centimeters to almost a meter in diameter. Shrock and Twenhofel described *Clathrodictyon vesiculosum* (Figure 5.10*A*) from northern Newfoundland. Specimens of this species occur as concentrically laminated, flattened, discoidal, or hemispherical masses. They were almost parallel to the bedding plane.

Stromatoporoids are particularly abundant in the Devonian of Iowa. Prominent, large, reeflike structures are well displayed in the Cedar Valley limestone. In the overlying Shellrock and younger Lime Creek formations, marked stromatoporoid zones occur. A partial section of the Shellrock formation (Figure 5.11*C*) indicates the context in which stromatoporoids grew.

Lecompte (1954) recognized three types of reef in the Devonian (Frasnian) of the Ardennes, Belgium: (*a*) reefs formed of massive stromatoporoids built up in the zone of agitation, (*b*) coral reefs constructed beneath the zone of agitation, and (*c*) mixed reefs begun by corals and ended by stromatoporoids. He noticed that two types of stromatoporoids often occurred together — lamellar and massive.

Three examples are given to illustrate further the stratigraphic distribution of stromatoporoids. The horizontal spread of *Clathrodictyon vesiculosum* in the Pike Arm formation of Newfoundland is shown in Figure 5.11*B*. Teichert (Figure 5.11*A*) saw that the older beds in the west Australian Devonian did not have reef-building stromatoporoids but only isolated occurrence of *Amphipora* species (family Idiostromatidae). However, younger Devonian beds had abundant stromatoporoid reef-builders. The same genus (*Amphipora*) was found to be abundant in the carbonate facies below the Palliser formation (Middle Devonian, Canadian Rockies).

All five of Galloway's stromatoporoid families are represented in Stearn's Canadian Rockies collections (Figure 5.11*D*) and species of genera such as *Actinostroma, Stromatopora,* and *Trupetostroma* are common. It is useful to note the geographic spread of these Devonian stromatoporoid reefs. They fall within an area defined by only three degrees of latitude (50°–53°) and longitude (115°–118°), with the greatest number of localities concentrated in an area defined by only one degree of latitude and longitude.

ADDITIONAL INFORMATION FROM LIVING SCYPHOZOANS

Symmetry

Both polyps (Figure 5.12*A,* and medusae (Figure 5.12*E*) display marked *tetra-*

TABLE 5.1. Key to Families of Stromatoporidea

A. Tissue Not Maculate[a]
1. Labechiidae: skeleton mostly of over-lapping curved plates (Figure 5.9G and Figure 5.10B).
2. Clathrodictyidae: pillars short, confined between two laminae (Figure 5.10A and Figure 5.11B).[b]
3. Actinostromatidae: pillars continuous or definitely superposed (Figure 5.9B, E, F, and Figure 5.11C).[b]
4. Idiostromatidae: coenosteum ramose, mostly with axial tube (Figure 5.9H and Figure 5.10C).

B. Tissue Maculate
5. Stromatoporidae: Pillars long, short or absent (Figure 5.9A, D, L, and Figure 5.10D).

[a]This key should be used in the following manner: First, ask whether the tissue is maculate. (Maculae are dark or light spots on a gray ground mass. They are 0.01–0.06 mm in diameter and confer a spongy texture since they constitute minute pores.) This will separate the first four families from the fifth. The second question to ask (once this separation is made, leaving only the first four families further involved) concerns the coenosteum. Is it massive, tuberose, and so on? If it is, the specimens at hand will belong to either family 2 or 3. Then the family description will be the deciding factor. If, on the other hand, the coenosteum is ramose, family 4 will be indicated. If the skeleton consists largely of overlapping curved plates, then family 1 is indicated. For genera under each family, the student can consult Galloway (1957).
[b]For families 2 and 3, the coenosteum is either massive, tuberose, or laminar.
Source. Rearranged after Galloway, 1957, p. 419.
Note. A new class of coralline sponges, Sclerospongiae, embraces living genera bearing astrorhizae. These are now considered relatives of astrorhizae-bearing genera of the stromatoporoids. The latter would be classified under the Sclerospongiae (W. D. Hartman and Thomas F. Goreau, 1970. *The Biology of the Porifera*, ed. W. G. Fry. Academic Press: Zool. Soc. London Symposia 25, pp. 205–243.) C. W. Stearn, 1972, ref. p. 105. All stromatoporoids (including Mesozoic forms) proposed as separate Porifera subphylum, Stromatoporata.

meral symmetry. That symmetry is defined by two principal planes: the peri-radial and the interradial, either of which divide the body into four structurally alike quadrants (Figure 5.12A). Body parts are arranged around the *oral-aboral* axis (which falls where the two principal planes cross at right angles, that is, tetrameral or multiples of four), compare Figure 5.12A,

and 5.12E, 1). *This symmetry is an important key in the identification of fossil remains* (Figure 5.13).

Structures

The vari-shaped colored *bells* range from goblet to saucer in configuration. These differences in shape distinctively feature the several orders. Thus the four orders that have a discoidal bell can quickly be distinguished from the two orders having a pyramidal or cubical bell. The order Coronatida alone possesses a coronal groove (Figure 5.12D, 12). In general, the bell is gelatinous, composed of more than 95.0 percent water.

A characteristic that distinguishes scyphozoan from hydrozoan medusae is the division of the gastrovascular cavity into a central stomach and four gastric pouches (Figure 5.12A, 3). Other scyphozoan structures include four *oral arms* bearing nematocysts around the mouth (Figure 5.12E, 1); the fringe of an umbrella-shaped body bearing *tentacles* (Figure 5.12E); a bell margin often scalloped into *lappets* (Figure 5.12D); marginal sensory bodies (*rhopelia*) in niches between lappets (Figure 5.12C); gelatinous basal expansions called *pedalia* (Figure 5.12C) that may bear tentacles; a *manubrium* (gullet or pharynx) that leads into the internal gastrovascular cavity; *gonads* in sacs or bands on the stomach floor (Figure 5.12E).

Scyphomedusans swim by the rhythmical contraction of the coronal muscles — a broad and strong circular muscle of the subumbrella (Figure 5.12E) that extrudes water from within the concave space. Pulsations occur regularly about 20 to 100 times per minute.

Reproduction

Sexes are separate. The male sperm migrates through water to fertilize the eggs in the female gonads. Zygotes attach on the oral arm of the female until they attain the planula stage. Planula then undergoes the transition shown in Figure 5.12B.

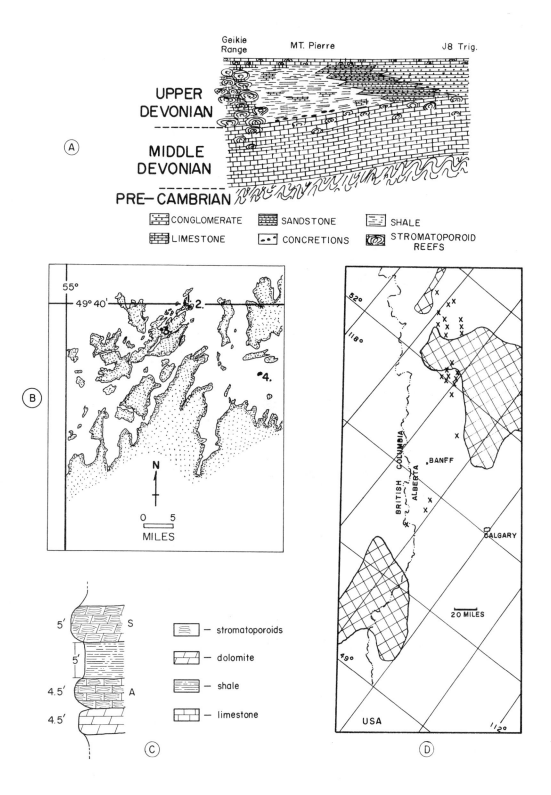

FIGURE 5.11 Stratigraphy of stromatoporoid-bearing beds. (*A*) Stromatoporoid reefs. Western Australia (after Teichert). (*B*) Distribution of *Clathrodictyon vesiculosum*. Pike Arm Formation, Silurian, Newfoundland (data from Shrock and Twenhofel, 1939). (1) Pike Arm, outer coral zone, (2) Pike Arm, inner coral zone. (3) Burnt Island, coralline argillite (4) Yellow Fox Island, coralline sandstone. (*C*) Columnar section of portion of Shell Rock Formation Dev., Iowa (see text for discussion of section). A. *Actinostroma*. S—*Stromatopora*

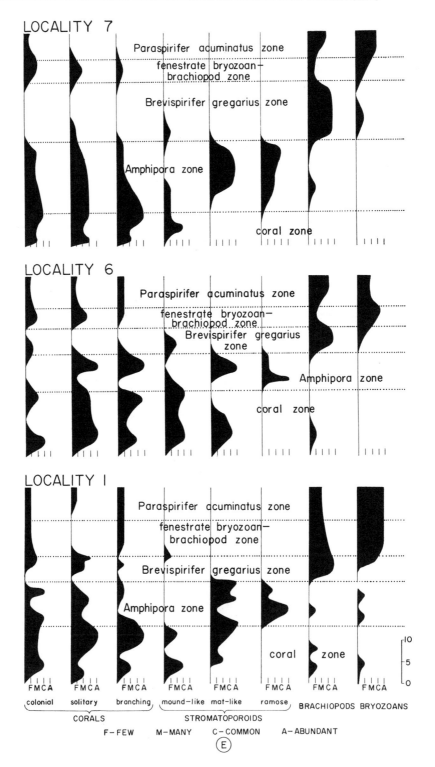

FIGURE 5.11 (continued) Stratigraphy of stromatoporoid-bearing beds (continued). (*D*) Stromatoporoid reefal dolomites of the Canadian Rockies (M. Dev.). Modified from C. W. Stearn, 1961.) Hatched area-shale, basinal facies; *x*-stromatoporoid localities, carbonate reef facies (bioherms and biostromes). (*E*) Distribution of stromatoporoid and other faunal elements of the Jeffersonville Limestone (M. Dev.), S.E. Indiana (Loc. 1. Jeffersonville, Indiana). (After R. D. Perkins, 1963.)

Ecology

Food and Biotic Associates. When food is captured by aid of the nematocysts on the oral arms, it is first conveyed to the mouth and then circulated in a ramified, *digestive canal complex.*

Generally scyphozoans are carnivorous. They have a wide range of diet that includes almost all small invertebrates. Fish is a common food.

Particular species will feed on a preferred diet. A *Haliclystis* in Puget Sound, for example, prefers amphipods (see Chapter 11); others, like *Aurelia* (Figure 5.12E), eat almost anything among the small plankton that collect under their exumbrella.

The bell of a scyphomedusan serves as a haven for various crustaceans, crabs, amphipods, and some fish. Various algae are symbionts on scyphozoans. In the Phillipines, species of the order Carybdeida are preserved in vinegar and sold as food (Mayer, 1910).

Bell Curvature. Increased salinity in *Aurelia* proportionately increases and lowered salinity proportionately decreases bell curvature. (Bell configuration, although compressed and distorted in fossil scyphomedusae, serves as an important criterion in assignment to an order.)

Habitat. (See definition of orders below.) The habitats indicated under order definitions apply to living forms but may or may not be applicable to fossil forms.

Size. Medusoids range in size from 1 in. to 7 ft; the polypoid stage, if present, is minute; tentacles, however, may exceed 100 ft in length (Berrill).

Paleosalinity and Salt Regulation. *Aurelia* and other forms have about the same total quantity of salts as the salt water medium in which they live, but medusae apparently can control salt regulation. One investigator (see Mayer, 1910) found more potassium, less magnesium, about

an equal amount of calcium, slightly less sodium, and 36 percent less sulphur dioxide in the medusae he studied than was found in the seawater in which they lived. That finding led to the possibility that *this salt content might correspond to that of an ancient sea* (Hyman, 1940). If true, this is paleosalinity and points to changed oceanic salinity through time. Since the two medusae studied belonged to the order Semontaeostomida and since the oldest known fossil of the order is Upper Jurassic age (Figure 5.13C), possibly the Upper Jurassic sea had more potassium and less sulphur dioxide and magnesium.

Mayer (1910, pp. 645–646) found that bell pulsation is stimulated by the ionic sodium of seawater and is inhibited by magnesium, calcium, and potassium. He located the control of this ionic balance in the *sense clubs* (Figure 5.12G), which hang from the margin of the medusa.

Scyphozoan Orders (the acraspedote medusae and medusa-like polypoids).

Order 1. Stauromedusida. Bell pyramidal, usually attached; eight adradial tentacle-bearing lobes. Probably degenerate. Confined to polar region and cold seas. No fossils known (Figure 5.12A).

Order 2. Carybdeida. Bell cubical, lacks lappets; four interradial tentacles. Confined to tropics and warm seas. Range: (?) Upper Jurassic–Recent (Figures 5.12C and 5.13A).

Order 3. Coronatida. Bell discoidal; coronal groove, lappets, and pedalia present; tentacles arise from clefts between lappets. Deep sea and pelagic. Range: (?) Lower Cambrian; Upper Jurassic to Reent (Figures 5.12D and 5.13B).

Order 4. Semontaeostomida. Bell discoidal, lappets present but no pedalia or coronal groove; lacks interradial septa in stomach. Mainly coastal forms; some have worldwide distribution. Range: Upper Jurassic to Recent (Figures 5.12E and 5.13C).

Order 5. Rhizostomida. Bell discoidal; lappets; tentacles absent; numerous mouths on eight adradial armlike processes. Tropical seas. Range: (?) Upper

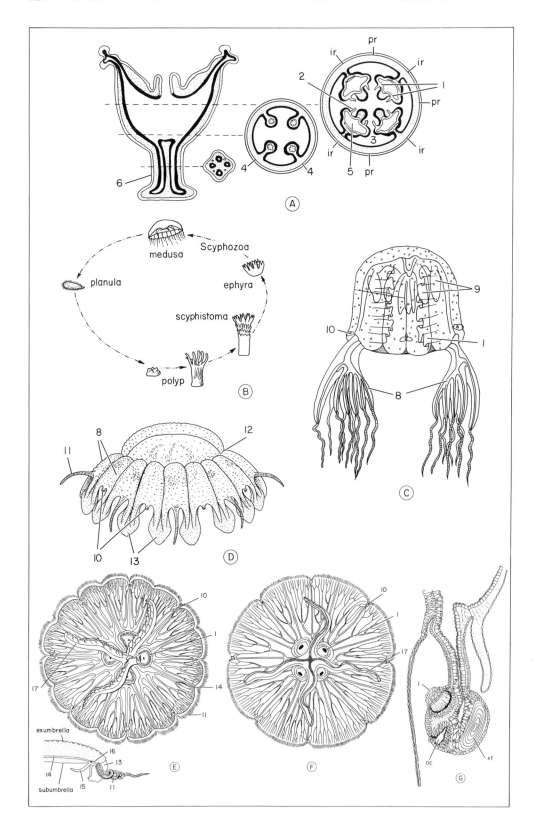

FIGURE 5.12 SCYPHOZOA (Living Types).
(*A*) Order Stauromedusida. *Haliclystis:* (1) longitudinal section showing attached sessile character of the order exclusive of one pelagic family. (Black-endoderm; outer white layer-ectoderm; dotted area between these gelatinous substance). (2) First transverse section above stalk exposing interradial septa: (3) second transverse section above stalk with reproductive and digestive elements. Planes of symmetry (tetrameral): two distinct planes, periradial (*pr*) and interradial (*ir*); between any two of these there is the adradial (*ar*) plane. (*B*) Life cycle of a scyphozoan (after Hill and Wells). (*C*) Order Carybdeida. *Chiropsalmus quadrumanus.* Beaufort, North Carolina. (*D*) Order Coronatida. *Nausithoë punctata*, Naples Zoological Station. (*E, F*) Order Semaeostomida, two species of *Aurelia*, Philippine Islands: (*E*) *Aurelia labiata* with 16 notches in bell margin and comparatively short mouth arms: (1) oral view, canals filled in for only one quadrant, one mouth arm cut off; (2) section of bell margin to expose structures. Nematocysts on exumbrella side of "11". (*F*) *Aurelia aurita*, with 8 notches in bell margin and comparatively longer arms. (*G*) *Carybdea xaymacana.* Bahama Islands, section through a rhopalium or sense club, *xt*—crystalline concretions that are composed of calcium oxalate and help maintain an excess of sodium ion. (Statoliths of calcium sulphate and calcium phosphate occur in other scyphomedusans—and the entire mass of cells serves as an organ of equilibrium.) *I*—lens of cuplike eye, *oc*—ocellus. (1) Gonads, (2) gastric filaments or cirri, (3) gastric pockets, (4) septum, (5) funnel pits, (6) stalk, (7) adhesive pads for attachment—below "6" in "A", (8) pedalia, (9) subumbrellar sacs, (10) rhopalium, (11) tentacles, (12) coronal groove, (13) lappets, (14) radial canals, (15) velum-like margin of bell, (16) circular vessel, (17) mouth arms. [(*A*)–(*C*), *G*) redrawn and adapted from Mayer, 1910] (D.M. Chapman, 1966, in The Cnidaria and their evolution, Academic Press, pp. 56–60 and Figure 3, suggested that the scyphistoma stage (cf. Figure 5.12 *B*) points to conulatid ancestry.)

Jurassic to Recent. (There is a poor specimen from the Solenhofen limestone, Upper Jurassic, Germany, *Leptobrachites trigonobrachius*).

FOSSIL SCYPHOZOANS

Fossil Scyphomedusae

One extinct order of the Scyphomedusae includes specimens found only in the Solenhofen limestone of Germany.

Order 6. Lithorhizostomatida. Dome-shaped bell; lappets; rhopalia; eight clusters of short tentacles arise from the exumbrellar surface. Range: Upper Jurassic, Germany (Figure 5.13*D*).

INFORMATION DERIVED FROM FOSSIL SCYPHOMEDUSAE

Fossil Record

The sparse record, thus far reported, for scyphomedusans can probably be accounted for by the absence of preservable hard parts. The only hard bodies are soluble, crystalline structures called statoliths. In the living condition the bell is often made stiff and rigid by fibers. Partially embedded jellies may be carried in

shore, buried again, or overturned by the tide.

The Cretaceous *Kirklandia* were, according to Caster, buried medusae. Sprigg's Lower Cambrian fauna from southern Australia were also buried pelagic forms. These fossils (Sprigg, 1949) are impressions in flaggy sandstone quartzite, which originally was a fine-grained, well-sorted sand of an intertidal flat or strand line. The grooves in these impressions bear either a ferruginous stain or an argillaceous film.

Some of the best-known scyphomedusae come from the Solenhofen calcareous slates (*Kalkschiefer*) and slab limestones (*Plattenkalk*), deposited in low-energy environments or quiet conditions. Kieslinger (1939) studied the possible modes of preservation of the Solenhofen scyphomedusans. These he incorporated into a single genus and species *Rhizostomites admirandus* (Figure 5.13*D*), finding all the impressions to be only different possible copies of the same kind of individual scyphomedusan (that is, the upper or under side of the same ex- or subumbrella).

A large collection of fossil medusoids from the Lower Cambrian of Australia led Sprigg (1949) to infer (*a*) that the uppermost Pre-Cambrian was "an age of jellyfishes," and further (*b*) that "all mod-

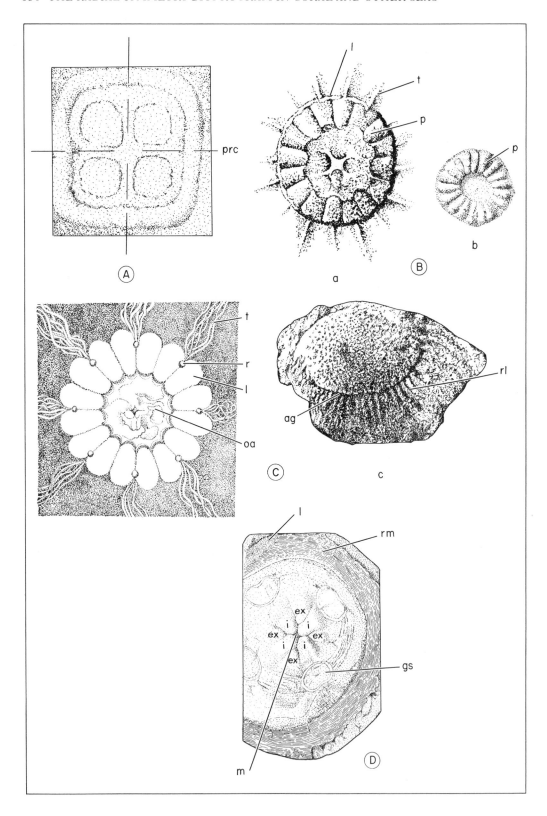

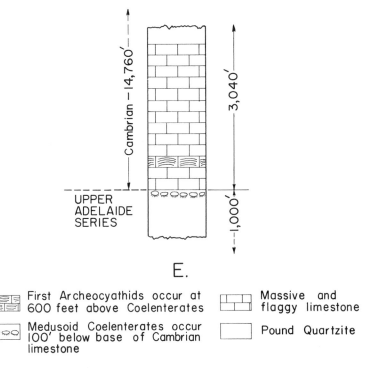

E.

First Archeocyathids occur at 600 feet above Coelenterates

Medusoid Coelenterates occur 100' below base of Cambrian limestone

Massive and flaggy limestone

Pound Quartzite

FIGURE 5.13 Subclass Scyphomedusae. Fossils of several orders.
Quadrimedusina quadrata (order Carybdeida) subumbrellar impression, U. Jur., Ger., *prc* — periradial canals (?).
(B) Order Coronatida: (a) *Epiphyllina distincta*, U. Jur., Germany; (b) (?) *Lorenzinia carpathia*, Eoc. Poland; (?) — questionably assigned to this order, *p?* — pedalia, *l* — lappet, *t* — tentacle. (c) *Camptostroma roddyi*, reassigned from Scyphozoa to Echinodermata (see Ch. 12).
(C) *Eulithota fascicolata*, U. Jur., Germany (order Semaeostomatida). Reconstruction of subumbrellar surface. Labels as in (B), *oa* — four short oral arms, *r* — rhopalium.
(D) *Rhizostomites admirandus*, U. Jur., Germany. (Order Lithorhizostomatida); *l* — lappet, *rm* — radial muscle, *I* — four insert radial lobes ("iron-cross"), *ex* — exert perradial lobes, *m* — cruciform mouth, *gs* — gonadic sacs.
(E) Stratigraphic occurrence of medusoid fossils, Flinders Range, Australia. Columnar section, generalized, showing relative position of medusoid coelenterates and archaeocyathids (modified after Sprigg, 1947, Trans. Roy, Soc. S. Australia, Vol. 71, Pt 2, pp. 212–223; pls 5–8). (Note position of medusoid fossils below base of thick Cambrian section, in upper portion of Pound Quartzite.) (A)–(D) redrawn and adapted from several authors.

ern orders" were already in existence by Lower Cambrian time. Item (*a*) requires several qualifications. (1) Several of Sprigg's medusoids are hydrozoan, that is, hydromedusae; (2) several of his genera are "medusoid problematica" or questionable scyphomedusans, and one genus, *Dickinsonia*, is a probable worm according to Dr. Glaessner; (3) Walcott's Cambrian medusoids are now referred to a distinct class of Protomedusae. Item (*b*) also needs correction since three extant scyphozoan orders have no known fossils prior to the Upper Jurassic.

Thus *various* medusoid coelenterates

(including jellyfish) were prominent indeed in late Pre-Cambrian and Lower Cambrian seas.

Fossil Conulata

Biological Affinity. Fossil objects referred to as conulatids have been variously assigned to the worms, mollusks, coelenterates and, by the Termiers, to the hemichordates (that is, the lower chordates).

Gradually, such problems are resolved. Kiderlen (1937) found that conulatids possessed quadrilateral radial symmetry

(Figure 5.15*A*) and that this symmetry bore a close resemblance to that of certain scyphomedusans.

The actual interradial septa are rarely preserved in conulatids. Their presence may be inferred from the following; the external expression of such septa is thought to be the midline furrow on each face of the pyramidal configuration. The furrows (or ridges) are arranged in a quadrilateral radial symmetry. Knight (1937) restudied *Conchopeltis* and discovered four pairs of muscle scars arranged in quadrilateral symmetry. Some specimens of this genus have tentacles preserved (Figure 5.15*B*).

Once we believe the scyphomedusan affinities of the conulatids (and some students of these forms do not accept this interpretation), then structural and other features fall into a consistent pattern. Similarly, a variety of specimens show lappet-like projections of the periderm and/or their infolding to close the aperture (compare lappets in the scyphomedusae). *Metaconularia aspersa* and *Diconularia clavus* (Figure 5.14*E* and *C*), both in the Silurian of England, have this feature preserved. In turn, these structures point to the muscles related to governing movement. Kiderlen and others have interpreted the rounded apical extremity found in some conulatids (in contrast to the more commonly occurring pointed or sharp-angled apex), the result of a transverse diaphragm (called a *schott*) sealing off this end (Figure 5.15*E*). Such forms are presumed to indicate free-swimming medusoids; those with a pointed apex supposedly represented forms apically attached to the bottom. Proof of attachment can be seen in Ruedemann's specimens from the Middle Ordovician of New York (*Sphenothallus*, Figure 5.15*D*).

There are (some) anomalies. As Sinclair (1943) observed, three-sided conulatids existed. However, these represent abnormal individuals of a species ordinarily possessing four faces. Nevertheless, over a dozen specimens of Raymond's *Conularina triangulata* examined by Sinclair showed a triangular cross-section. That led him to conclude that "this was normal for the species." If this be true, then how can this "normality" jibe with the tetrameral symmetry of the scyphomedusans?

Mesoconularia arcuata from the Middle Devonian of Algeria has its younger (adapical) end recurved (Figures 5.15*I*). If this is not a pathologic expression as is likely (see Figure 5.15*F*), it would be difficult to reconcile it with a scyphomedusan affinity.

The data from conulatids, known only as fossils, as interpreted by many specialists (Kiderlen, Knight, Bouček, and Caster et al.), suggest scyphomedusan affinity more strongly than any other.

Taxonomy

Subclass Conulata (slightly modified from Moore and Harrington, 1956; Sinclair, 1952; Kiderlen, 1937).

Definition

1. *Shape* — cone-shaped to elongate, pyramidal, or subcylindrical organisms (Figure 5.14).
2. *Periderm* — thin, chitinophosphatic, lappet-like projections infold over aperture (Figure 5.14*C*; Figure 5.14*P*).
3. *Periderm markings* — generally present (but sometimes lacking, that is, smooth); transverse and longitudinal markings include ornamentation, ribs, lirae) (Figure 5.14*B*).
4. *Oral margin* — with tentacles rarely preserved; peridermal traces considered tentacular muscle attachments (Kiderlen, 1937), leading to the inference that all conulatids bore tentacles (Figure 5.15*E*).
5. *Aboral end* — with or without attachment disk (Figure 5.15*D*); with or without adapical transverse diaphragm (schott) (Figure 5.15*E*).
6. *Tetramerous symmetry* — four interradial bifurcate septa are rarely seen. Tetramerous symmetry, however, is inferred from four pairs of muscle scars (compare *Conchopeltis*) or the four faces of the pyramid; each may bear a distinct midline inter-

preted to be the external reflection of internal septa (Figure 5.15*A*).

7. *Geologic range*—Middle Cambrian to Triassic.

The subclass definition applies to its single order, the Conulariida. The latter subdivides into two suborders, four families, as follows.

Order Conulariida

Suborder 1. Conchopeltina. Contains only the genus *Conchopeltis* in a family with the same name.

Suborder 2. Conulariina.

Family 1. Conulariellidae. Having rectilinear transverse ribs; lacking internal septa. Embraces one genus only, *Conulariella* (Figure 5.14, Part *A*). Middle Cambrian to Lower Ordovician.

Family 2. Conulariidae. Mostly with quadrilateral cross-sections. Contains the bulk of conulatids. Subdivided into three subfamilies.

Subfamily 1. Conulariinae. Corner furrows do not interrupt transverse sculpture. Upper Cambrian to Permian. Embraces the following genera: *Conularia* (Figure 5.15, part *C*), *Archaeoconularia* (Figure 5.14, part *L*), *Mesoconularia* (Figure 5.14, part *D*), *Diconularia* (Figure 5.14, part *C*), *Metaconularia* (Figure 5.14, part *E*), *Exoconularia* (Figure 5.14, part *B*), *Anaconularia* (Figure 5.14, part *M*), and *Palaenigma* (Figure 5.14, part *F*).

Subfamily 2. Paraconulariinae. Corner furrows interrupt the surface sculpture. Middle Ordovician to Lower Permian. Embraces the following genera: *Paraconularia* (Figure 5.14, *N*), *Calloconularia* (Figure 5.14, *I*), *Eoconularia* (Figure 5.15, *A*), and *Adesmoconularia* (Figure 5.14, *G*).

Subfamily 3. Ctenoconulariinae. Corners more or less furrowed and strengthened by peridermal thickenings. Middle Ordovician to Middle Devonian; (?) Lower Mississippian. Embraces the following genera: *Ctenoconularia* (Figure 5.15, *G*), *Climaconus* (Figure 5.14, *J*), *Conularina* (Figure 5.14, *O*), *Sphenothallus* (Figure 5.15, *D*).

Family 3. Conulariopsidae. Corners lack furrows. Lower Triassic, Japan. Embraces a single genus, *Conulariopsis* (Figure 5.14, *P*).

The most diagnostic feature of the conulatid shell, according to G. W. Sinclair

(1952), is the *corner of the pyramid and the nature of the structures found there*. Pyramidal corners (Figure 5.14, cg) may be indented by a longitudinal furrow or groove. This may be shallow as in *Metaconularia*, or deep as in *Adesmoconularia*. The corners may lack a distinct furrow and hence appear elevated into a rounded or sharp ridge as in *Anaconularia* or be furrowed and strengthened by a pronounced thickening of the periderm as in *Ctenoconularia*. The corner furrows may (*Paraconularia*) or may not (*Conularia*) interrupt the transverse sculpture (ribs), or they may be crossed by transverse wrinkles as in *Climaconus*. It follows that the conulatid's transverse section is most significant for classification.

A secondary feature is the *character of the midline of the faces*. This feature may be completely absent (*Neoconularia, Conulariella*) or it may be represented by a groove (*Mesoconularia*), elevation (*Palaenigma*), a slight (*Paraconularia*), or marked deflection of the ribs along it (*Ctenoconularia*).

Surface ornamentation of the faces and of the ribs (Figure 5.14*G*) or a feature such as a spinous margin or spines along the midline, taken alone, is of minor importance. However, in conjunction with the features noted previously, these characteristics can have value on the species level.

The transverse ribs can be of value on both generic and specific levels of classification. Thus in *Conulariella* the ribs are rectilinear (not curved or bent on the faces), whereas in *Paraconularia* they are bent adaperturally (away from aperture) where they terminate at the edge of the corner furrows. One of the distinguishing features that Driscoll (1963) found in two species of *Paraconularia* was the difference in the angle made by the ribs at the midline.

Other minor features include strength of definition of transverse ribs; alternation of ribs; relative steepness of the pyramid. The presence of lappet-like projections of the periderm, a *schott*, and the overall relative size may enter into discrimination of one genus from another.

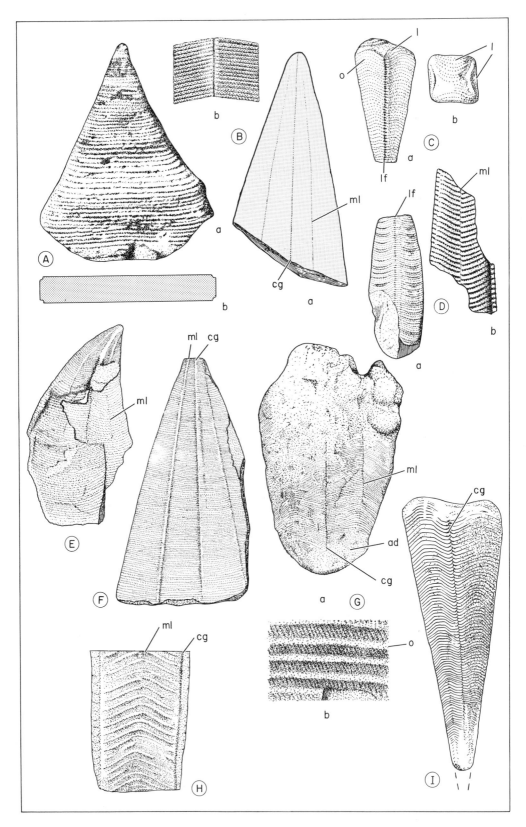

FIGURE 5.14 *A* Conulatid genera. (*A*) *Conulariella robusta*, L. Ord., Czechoslovakia, (*a*) (*b*) side and transverse section. (*B*) *Exoconularia exquisita*, M. Ord., Czechoslovakia, (*a*) side view, (*b*) facial ornamentation, apertural direction downward. (*C*) *Diconularia clavus*, M. Sil., England, (*a*) corner view,

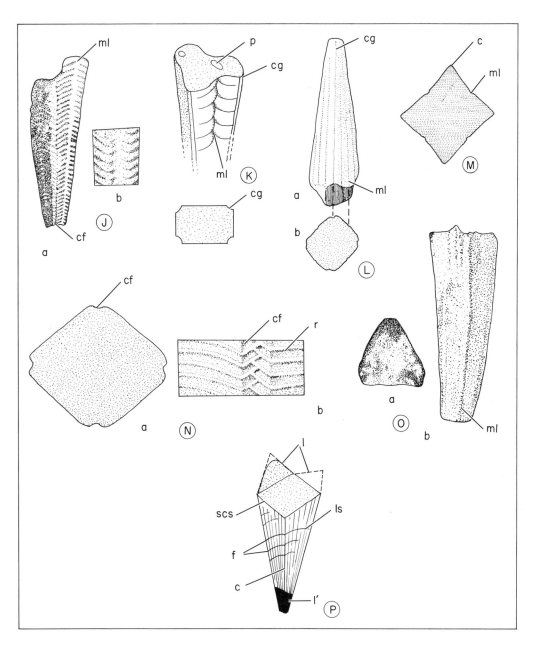

(b) apertural view. (D) *Mesoconularia fragilis*, L. Dev., Czechoslovakia, (a) corner view, (b) *M. ul-richana*, Dev., Brazil, facial ornamentation. (E) *Metaconularia solitaria*, U. Ord., Czechoslovakia, crushed specimen. (F) *Palaenigma grandissima*, Sil., Czechoslovakia. (G) *Adesmoconularia byblis*, L. Miss., Iowa, (a) corner view, (b) detail of facial (midline) view. (H) *Ctenoconularia obex*, M. Ord., Minnesota, facial view, *ml*—midline marked only by the course of the ornamentation. (I) *Call-oconularia strimplei*, U. Penn., Oklahoma, corner view X 2.5. (J) *Climaconus*, (a) *C. batteryensis*, U. Ord., Quebec, corner view; (b) *C. bottnicus*, M. Ord., detail of corner. (K) *Climaconus sp.* Basal Ma-quoketa depauperate zone; Iowa; (a) corner view, *p*—phosphate pellets, (b) transverse view. (L) *Ar-chaeoconularia fecunda*, M. Ord., Czechoslovakia, (a), (b) side and transverse views. (M) *Anaconu-laria anomala*, M. Ord., Czechoslovakia, transverse view. (N) *Paraconularia*, (a) *P. inequicostata*, L. Camb. Belgium, (a) transverse section, (b) corner view. (O) *Conularina triangulata*, M. Ord., Ver-mont, (a) imperfect transverse triangular section; (b) facial view. (P) *Conulariopsis quadrata*, L. Tri-assic, Japan, (b) corner view showing lappets; square cross-section (scs). (Note absence of corner grooves.)

k—keel, *ml*—midline, *lf* or *cf*—longitudinal furrow, *cg*—corner groove, *l*—lappets, facial view equiva-lent to side view, *o*—surface ornamentation, *ns*—normal surface, *ad*—smooth apical diaphragm (= schott), *f*—ribs, *ls*—longitudinal striae, *c*—corner without furrow.

Illustrations redrawn and adapted from (A)–(F), (M)–(N), Moore and Harrington after several au-thors; all Czechoslovakian specimens after Boucek; (G) Driscoll, (H, I) Sinclair, (K) Tasch, (P) recon-struction from Sugiyama's description and photograph.

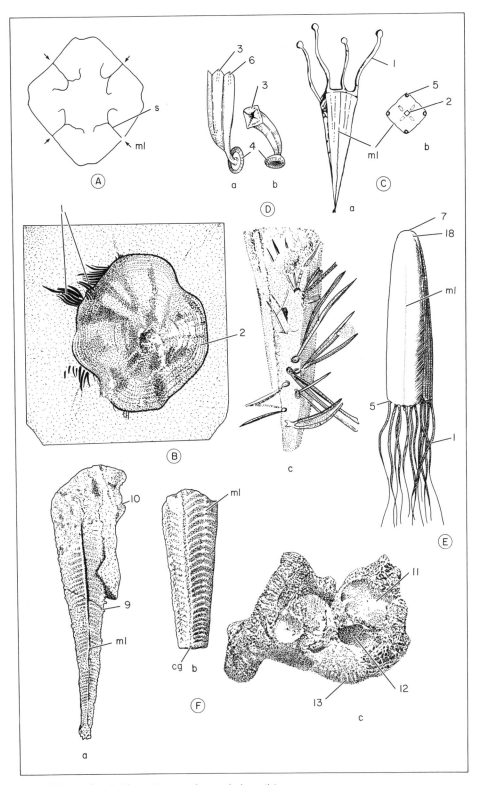

FIGURE 5.15 Conulatid structures, and so on (schematic).
(A) *Eoconularia locuta*, Sil., Gotland: transverse section; *s*—bifurcate interradial septum; *ml*—external midline that corresponds to internal septum (shown by arrows).
(B) *Conchopeltis alternata.* Apical view showing (1) tentacles; (2) lobate margins.

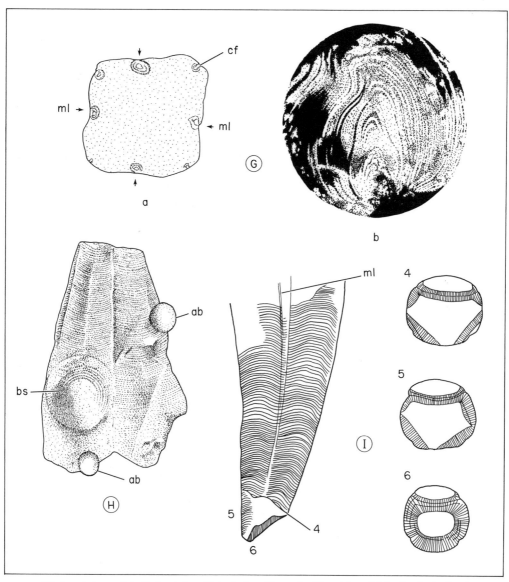

(*C*) *Conularia cambria*, U. Cambrian: (1) tentacles; (2) mouth; (5) periderm thickening for tentacular muscle attachment.

(*D*) *Sphenothallus augustifolius*, M. Ord., N.Y.S. (*a*) Juvenile; 3-lappet-like structures; 4-fixation disk; 6-muscles for underside of periderm endings. (*b*) Juvenile. 3-muscles moved lappets downwards to close aperture.

(*c*) Adult, with juveniles attached.

(*E*) *Exoconularia consobrina*, a reconstruction; medusoid: (1) tentacles; (5) periderm thickness for muscle attachment; (7) smooth or rounded proximal end of free-swimming medusoid conulatids; (18) exposed portion of diaphragm (= *schott*).

(*F a–c*) *Conularia leonardensis*, Perm., Texas; *ml* – midline; (10) bryozoans encrusting conularid; (9) sponges; (11) brachiopod *Meekella*; (12) conularid mold; (13) sponge on conularid – incurrent surface shown; (*cg*) corner groove.

(*G*) *Ctenoconularia* sp. Dev., So America (Bolivia). Thickening of periderm at midline and corners: (*a*) transverse section showing sites of thickening; (*b*) detail of one such site (note successive laminae superficially resemble a Bunsen flame that enlarge to elliptical or subspherical configurations); *ml* – midline (arrow); compare corner furrows.

(*H*) *Conularia continens*, Marcellus shale, M. Dev., New York; three individual inarticulate brachiopods attached.

(*I*) *Mesoconularia arcuata*, M. Dev., Algeria; (4) section at the contact with rib of conulatid; (5) section through bevel; (6) recurved point; *ml* – midline.

Illustrations redrawn and after (*D, H*) Hall, 1879; (*F*) Finks; (*I*) Termier and Termier, 1950; (*A, C, E*) Kiderlin, 1937; (*B*) Knight, 1937; (*G*) Knod, cit. Moore and Harrington, 1956.

Measuring the angle between any two faces, as Hall did, is of small value since some degree of compression, flattening, or distortion characterizes many conulatids.

Size

Conulatids, in general, are small megafossils even when the fragments are reconstructed to approximate the original size. Mostly incomplete specimens and fragments are found. Some (*Conularia* sp.) are generally less than 1/8 in. in length (basal Maquoketa, Iowa); others range from 1.0 to 3.6 in. (*Mesoconularia*, Morocco); this is about the most common size range. Larger specimens are found among Hall's Middle Devonian conulatids. Some of these attain as much as 6.0 in. *Anaconularia* from the Middle Ordovician of Czechoslovakia are as large as 5.5 in., and McKee's Kaibab conulatid reached 5.7 in. Larger-handsize-specimens are known in Australia, for example. The size range in any collection may include juvenile and adult individuals of the same population.

INFORMATION DERIVED FROM CONULATIDS

Periderm

Conulatids had a thin, laminate periderm in life, which was also flexible. Evidence for this last conclusion is that fossils show indented or bent periderms or flattened examples but lack any significant tears or breaks. Periderm thickness may have been as small as 0.056 mm but, on the whole, averaged 0.30 mm. The composition of the periderm is essentially chitinophosphatic. "Chitinophosphatic" is used because the chitin may be impregnated with some 70 percent calcium phosphate in Ordovician species, with as much as 96 percent in Pennsylvanian species.

Phosphatic Content of Body Fluids

The relatively high phosphatic content of the conulatid periderm contrasts with the negligible phosphate in the statoliths of living scyphomedusans.

Although they are opaque objects, conulatids can be rendered partially translucent by HF acid treatment. Experiments by the author on conulatids from the basal Maquoketa of Iowa (Figure 5.14, *K*) permitted transmission of light through the fossil after such treatment.

Paleoecology

Conulatids abound in fetid black shales (that is, dark, carbonaceous shales) but, as Caster observed, they occur in all lithofacies. Hall's Middle Devonian conulatids from New York were embedded in arenaceous and calcareous shales; rare conulatids from the Texas Permian are silicified and can be released from limestones by acid treatment. Upper Middle Ordovician conulatids were preserved in carbonaceous shales and dolomitic limestones of Minnesota. Conulatids are also found in sandstones such as St. Peter's, Oriskany, or the white sandstone member of the Kaibab limestone. Carboniferous limestone, such as the Keokuk, late Ordovician shale in the subsurface of the Williston basin (Ross, 1957), and cherts of cherty limestones (McKee, 1935) are other lithologies from which conulatids can be obtained.

Biotic Associates

Inarticulate brachiopods (see Chapter 7) are attached to the periderm of the Marcellus shale conulatids (Figure 5.15, *H*). Knight noted apparent worm tubes attached near the apex of *Conchopeltis*.

Finks found a bryozoan incrusting a conulatid that was associated with *Meekella* and productid brachiopods. He also discovered conulatid molds enclosed by the calcareous sponge *Guadalupia* (Figure 5.15, *F*). The same investigator unearthed a small, bun-shaped lithistid sponge in which two conulatid external molds occurred. On a Bohemian *Exoconularia* species, an edrioasteroid was attached (see Chapter 12). Among attached forms, smaller or juvenile conulatids attached to larger adult forms (Figure 5.15, *D*).

Harrington and Moore (1956) effectively argue that attachment scars left on conulatids by some forms prove the ectodermal origin of the periderm—a point of some importance to the scyphomedusan assignment of conulatids. Thus it is the ectoderm that invaginates to form the four septa in living scyphomedusans (see Figure 5.12) and the furrowed midline of

conulatids is thought to represent such an invagination of the periderm.

In addition to these faunal associations that point to conulatids as sites for other invertebrates, conulatids are also found with varied faunas including, among others, mollusks (especially pteropods), graptolites (*Climacograptus*), coelenterates such as the rugose coral *Streptelasma*, trilobites (*Isotelus*), serpulid worms, and the calciphosphatic inarticulate brachiopods (*Lingula* and *Orbiculoidea*) (Sinclair, 1943).

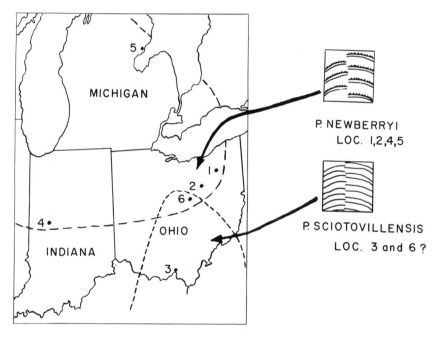

FIGURE 5.16 Reconstruction of a portion of the hypothetical species ranges for two paraconularid species during Lower Mississippian (Osagian) time (data from Driscoll, 1963). Note suggested overlap of range in vicinity of locality 6, and compare this with distinctive rib angle at midline (Table 5.2) of the two species. Further note "overlap" of this feature between species.

TABLE 5.2. Characteristic Difference between Two Species of *Paraconularia* (Data from Driscoll, 1963)

Species	Locality Number	Angle of Ribs at Midline					
		110°	115°	120°	125°	130°	135°
Paraconularia	1					xxx	
newberryi	2				xxxxxxxxxx		
	4					xxxxxxxxxxxxxx	
	5					xxx	
Paraconularia	3		xxxxxxxxxxxxxxxxxxxxxxxx				
sciotovillensis	6(?)						

Habitat

If Kiderlen's interpretations are followed (Figure 5.15, *E*), then conulatids were both sessile and free-swimming. Many conulatids from the carbonate facies (Permo-Carboniferous) and from the phosphate facies (basal Maquoketa shale) can be designated as shallow-water forms. Associated faunas support this interpretation.

Speciation

Where data are available on locality, distribution, and morphology of one or more contemporaneous or nearly contemporaneous conulatid species in a given region, it may be possible to infer range overlap, geographic, and subsequently genetic isolation of two gene pools (Table 5.2 and Figure 5.16).

Stratigraphy

Conulatid genera may be restricted to a portion of a given geologic period (Upper Pennsylvanian—Missourian, for example), or to a given region (Japan, Europe). A given species may be known from only one formation or horizon. Species of other genera can be useful in approximating the probable geologic age of a given formation. Thus *Archaeoconularia* species will be found in the Lower Ordovician to Silurian, whereas species of *Mesoconularia*, known from Europe, Africa, North and South America, will be found in Lower Devonian to Permian beds. Fossil associates will be invaluable in further confirming and pinpointing the age of a given bed.

LIVING ANTHOZOANS

Introduction

Of all the coelenterates, some anthozoans have the most impressive, extensive fossil record.

Taxonomy

The three subclasses of the class An-

thozoa may be defined according to whether or not the mesenteries (septa) are paired.

1. *Subclass Ceriantipatharia.* This subclass includes two different types of corals. One, the "black corals" (order Antipatharia), has a horny and thorny skeletal axis. The skeletal axis is composed of a horny substance, *gorgonin* (Figure 5.18*E*). "Black corals" are chiefly nonreef-dwelling, living at depths of 1 to 1900 fathoms. The Miocene of Italy has yielded one genus that has living representatives.

The second type in this subclass (order Ceriantharia) lacks a hard skeleton, although the anemone secretes a slimy protective case (Figure 5.18*A*). The protective case is buried in bottom sands to the oral disc. There is no known fossil record. However, Hill and Wells (1956) speculate that some unidentified vertical cavities (spoor) in shales and sandstones might prove to be ceriantipatharian burrows.

2. *Subclass Octocorallia.* The octocorals also have unpaired mesenteries but differ from Subclass 1 in having pinnate (that is, feathery or branching) tentacles. Several orders of this subclass have calcareous and/or horny spicular axial skeletons (Figure 5.17). Massive fibrocrystalline coralla are rare (see *Heliopora*, Figure 5.17*E*).

The name "octocoral" refers to the regular octamerous symmetry (Figure 5.19*B*).

3. *Subclass Zoantharia.* Included in this subclass are solitary or colonial forms having a calcareous exoskeleton. The preponderance of corals known as fossils are zoantharians. Paired (coupled) mesenteries (Figure 5.18*D*) are characteristic.

LIVING OCTOCORALS

Attempts to classify the octocorals on any other basis than the *spicular system* have met with failure. The classification given below follows F. M. Bayer in the *Treatise.* Spicules of several orders such as Alcyonacea (Figure 5.19*C″*) and the Stolonifera (Figure 5.19, *E*), when found as fossils, are very difficult to classify accurately with any assurance. Only three of the seven orders of the octocorals produce axial structures that permit somewhat better (but still imperfect) identification in the fossil state. Of these, one extinct order will be discussed separately. Primary attention will be directed to two extant orders.

Order Gorgonacea

This includes sea whips, sea feathers, sea fans; also called horny corals. Axial structures more or less specialized; distinct central axis of horny or calcareous material, or both; or, as a central medullar zone of calcareous sclerites (bound firmly or loosely by horny or calcareous matter). Range: Cretaceous to Recent (Figure 5.17 *A*).

Order Pennatulacea (Sea Pens)

Colonial unbranched octocorals. Spicules, smooth or three-flanged rods or needles (rarely tuberculate), or small scales or plates. [Living sea pens are classified on the basis of the arrangement of two kinds of zooids (autozooids and siphonozooids), the distribution and form of the calcareous spicules, and other non-fossilized features.] Range: (?) Silurian to Recent (Figure 5.17*C*).

Morphology

The essentials of the morphological structures are given in Figure 5.17 and in the definition of the class Anthozoa. Octocorallia do not go through a medusoid stage in their life cycle. They are exclusively polypoid. The individuals of the colony are called autozooids. Several different modifications of the basic polyp type develop in certain octocorals. These modifications generally involve loss of all or most of the eight tentacles that characterize autozooids.

The upper retractile portion of an autozooid is called the *anthocodia* (Figure 5.17). In some orders polyps arise from stolons (Figure 5.19). Generally, however,

the lower portion of the polyp is stiffened and is called the *anthostele*. It is embedded in the coenenchyme (a spicule-bearing, thick, gelatinous substance). Neighboring polyps are interconnected by a system of tubules (*solenia*) in the coenenchyme.

Spicules or Sclerites

Calcareous spicules of octocorals (Figure 5.19*E*) are "the most important single character" (Bayer) in the identification of families, and so on. In the living animal, at the base of *each* tentacle, and *in* the body wall between septa, sclerites approximate a symmetrical arrangement. The denser spiculation occurs in the anthostele and accounts for the stiffening of this portion of the polyp. It also serves a defensive function when the polyp is retracted. Spicules may be arranged in rows to form an operculum (or lid). This lid can close the aperture of the anthostele when the polyp retracts.

Attempts to distinguish octocoral families on the basis of the operculum alone have proved unreliable.

Octocoral spicules have a most varied morphology. All orders (except order Coenothecalia—"the blue corals"—which have a skeleton of crystalline, aragonite fibers fused to form lamellae) have the *basic type*, that is, the *monoaxial rod or spindle with pointed ends*. A smooth three-flanged version occurs in the Pennatulacea (Figure 5.17*C*). Spicule sculpture consists of spines, intricate warts, and other structures (Figures 5.19*C*, and 5.17*A*, *c*).

Separate calcareous spicules may be fused by a calcareous cement. This leads to the formation of such objects as plates, scales, ovals, dumbbells. In the Gorgonacea, for example, *all* of these types occur. Within the jelly-like *mesogloea*, calcareous sclerites are secreted by specialized amoeboid cells (scleroblasts, cf. Porifera); one spicule to one scleroblast. Gorgonin is also secreted by mesogloeal cells.

Specialists on living forms find the following spicule characteristics of taxonomic value: (1) arrangement on the polyps, (2) size and form, (3) distribution in vari-

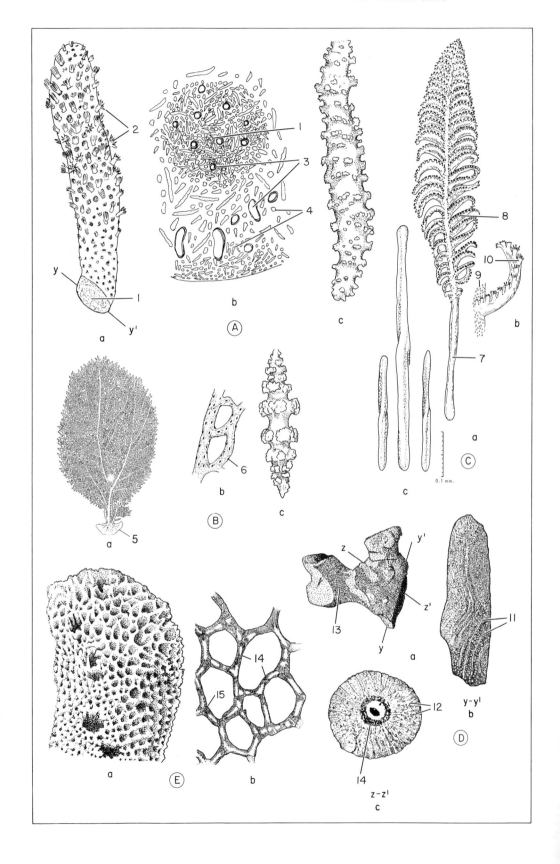

FIGURE 5.17 Octocorallia skeletal structures. Order Gorgonacea.
(A) (a) *Briareum* sp. 1, spicules of axis; 2, anthocodia, the upper retractile portion of an autozooid including tentacles. (b) Half-cross-section of (1) along y—y'; 3, canals; 4, spicules of coenenchyme (lower portion of polyps are embedded on this thick gelantinous spicule-bearing substance). (c) *Briareum asbestinum*, Rec. Florida, spicule.
(B) The common seafan *Gorgonia* sp. (a) Bermuda specimen. (b) Detail of a piece of (1)—5, expanded basal attachment plate; 6, holes for anthocodia; compare A(a), (b) above. (c) *Gorgonia flabellum*, Rec. West Indies (Nassau), spindle type spicule.
(C) Order Pennatulacea. *Pennatula*, the common seapen. (a) Complete colony. (b) Details of (a)—7, peduncle; 8, leaf; 9, siphonozooids; 10, anthocodia. (c) *Pennatula aculata*, Rec., Mass., three-flanged spicules.
(D) Order Trachypsammiacea. *Trachypsammia mediterranea*, Perm. Sicily. (a) Specimen showing sections cut from it, (b) y—y' section of (1). (c) z—z' section of (1). Extracellular tissue: 11, a system of 28–50 longitudinal septum-like lamellae that cross the concentric system shown in 12; 12, concentric system of thin laminae, finely and regularly vesicular, external expression—winding files of tubercles; 13, winding files of tubercles. Medullar region: 14, system of sinuous winding canals that frequently join and ramify.
(E) Order Coenothecalia, Cret.—Rec., (a) (b), *Heliopora coerulea*, Rec., Indo-Pacific: (a) top of massive fibrocrystalline corallum; (b) tangential section of growing edge of corallum. 15, junction of 3 calcareous laminae to form triradiate figure, and 14 (= 14'), junction of adjacent triradiate figures.
Illustrations after L. H. Hyman, *The Invertebrates*, McGraw-Hill, 1940, Vol. 1; Bayer; recent material, Gorgonacea and Pennatulacea, and data from E. Montanaro-Gallatelli.

ous layers and zones of the coenenchyme and, occasionally, (4) color. (In the Pennatulacea a fixed, that is, a nonsoluble, lipochrome pigment in the spicules colors them yellow, orange, red, or purple.)

Axial structures and spicules may both be examined under polarized light once thin sections have been prepared.

Reproduction

New polyps are produced asexually, although both sexual and asexual modes of reproduction occur.

Regeneration and Gorgonian Windrows

When branches of gorgonian colonies are broken off by severe storms, the stumps can regenerate (see Cary, 1914, Table 4). A strip of beach (Bush Key, Tortugas, Florida) 112 yards long and 1 yard wide, contained windrows of gorgonian skeletal fragments broken during a hurricane storm and heaped on shore. These fragments represented nearly 8500 colonies (Cary, 1914, pp. 85–86).

Food

The nutrition of octocorals consists chiefly of nannoplankton, larval mollusks, and small zooplankton. Some forms thrive without external food being sustained

apparently by a symbiotic relationship with the algal zooanthellae abounding in their tissues.

INFORMATION DERIVED FROM LIVING OCTOCORALS

Ecology (Distribution)

Worldwide distribution; particularly abundant in the littoral zone (between high and low tides), in coral reef environments of the tropical and subtropical belts of the Pacific, Atlantic, Indian Oceans, and the Red Sea.

Depth

Most octocorals occur on the continental shelf and slope, but some do occur in deep waters (more than 4000 meters).

Growth Form

Strong constant currents can affect growth form of any colony. Among the Gorgonacea, shallow-water forms grow at right angles to the current which explains their fanlike configuration. By contrast, deeper-water forms grow in all directions.

Carbonate Contribution to Reefs

Spicule content of a reef population of

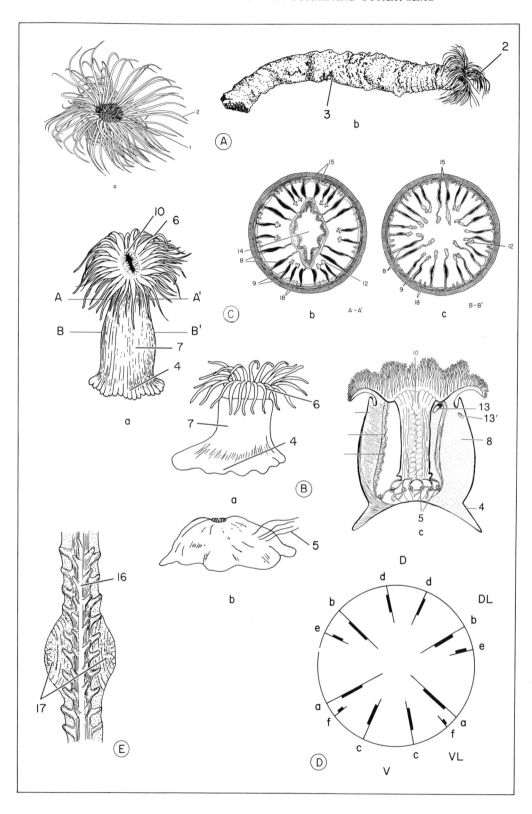

FIGURE 5.18 Anthozoans (subclass Ceriantipatharia and Zoantharia). Subclass Ceriantipatharia.
(A) Sea anemone, *Cerianthus membranaceus (?)*, Rec. Mediterranean Sea (adapted from photo-graph by S. Dunton, Turtox News, 1963, Vol. 41, No. 10). (a) Detail of tentacles. (b) Protective casing, side view, total length 28 in. (In life tube stands upright in sand or mud up to the oral disk.)
(B) Subclass Zoantharia: (a), (b), (c), sea anemone, *Metridium*: (a) Young specimen from life, side view of polyp. (b) Same specimen contracted. (c) Longitudinal section showing septa and details of anatomy.
(C) Anemone (Family Sagartiidae). (a) Polyp view. (b) Cross-section through pharynx (A–A').
(c) Cross-section below pharynx (B–B').
(D) Common order of septal appearance in anemone development, (a–f), septal couples: D–dorsal, V–ventral, DL–dorsolateral couple, VL–ventrolateral couple; first complete couple to appear is a–a', then b–b', then directives, c–c', followed by d–d'; a', b,' and so on, to left of D and V.
(E) *Dendrobrachia* (order Antipatharia) showing a portion of the thorny skeletal axis.
1, Oral tentacles encircling mouth; 2, marginal tentacles; 3, protective case, formed of hardened slimy secretions, which incorporates sand grains and other particles; 4, pedal disk; 5, acontia, the lower end of septal filaments that continue as free threads; 6, oral disk; 7, column; 8, primary sep-tum; 9, secondary septum; 10, siphonoglyph; 11, gonad; 12, septal filament; 13, oral stoma that per-mit water passage between interseptal chambers; 14, marginal stoma, same function as 13; 14, pharynx, 15, directives; 16, thorny axis of gorgonin; 17, retracted polyps; 18, tertiary septa.
Sources of illustrations (B, C, E) (after Brooks) from *The Invertebrates*, Vol. I, by Libbie H. Hyman, 1940, with permission of McGraw-Hill Book Co.; (D) (part) Wells and Hill, 1956.

octocorals (Figure 5.19C', E) has been estimated to range from 5 to 38 tons per acre. Allowing for diminishment by death of some 20 percent of the octocoral popu-lation would leave still a yield of 1 ton of calcium carbonate per acre per year. Forms like *Heliopora* with a massive ara-gonite skeleton (Figure 5.17E and Figure 5.21) that characterize distinctive reefal zones, obviously are larger carbonate con-tributors. The Bikini atoll exemplifies this source of carbonate.

Cayeux (1921, 1939) has shown that octocoral spicules contributed important-ly to the carbonate and calcareous phos-phatic rocks of the French Jurassic. Niel-sen (1917) reported *Heliopora* encrusted on corals and hydrocorals in the chalk beds (Danien) of Denmark.

Biotic Associates

Because they are attached forms hav-ing various openings, a wide spectrum of invertebrates and a few fish (egg cases of small sharks) regularly invade or live on or in living octocorals. *Millepora* larva can attach to a gorgonid stump, and the grow-ing hydroid colony that results will kill off the octocoral (Cary, 1914). Interbed-ded in or infesting the fleshy tissue of some octocorals are gastropods and vari-ous arthropods (barnacles, copepods; see Chapter 11). These lead to localized galls or swellings on a given polyp.

Further polychaete worms (Chapter 9) and copepods inhabit octocoral canals and cavities. Some serpulid worms live in the skeletal axis of alcyonarians (Cary, 1931). Some crinoids (comulatids; see Chapter 12) are found clinging to gorgonaceans as are various other echinoderms. Pro-tists, such as the protozoan foraminifers, commonly inhabit the outer cuticle of some octocorals. Sponges may grow around and directly attach themselves to octocoral colonies.

Various zoantharian corals commonly associate with octocorals but not as sym-bionts or commensals (Figure 5.21). As already noted, the zooanthellae algae live by photosynthesis and abound in the tissues of octocorals. Algae and bryozoans are also common encrusting forms (Cary, 1914).

These relationships may be important in some fossil deposits.

Speciation

Bayer (1955) affords an interesting in-sight into pennatulacean speciation. Two

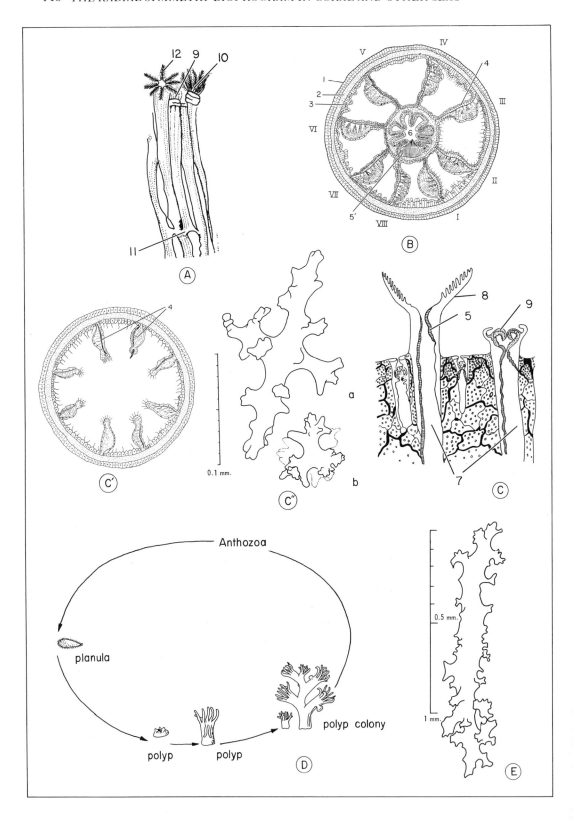

FIGURE 5.19 Living anthozoans (subclass Octocorallia). (A) *Clavularia viridis,* Rec. Indo-Pacific, schematic, portion of colony (see *F*). (B) *Alcyonium,* Rec., cross-section through pharynx in *C*. (C) Vertical section through *Alcyonium.* (C') Cross-section through *C* below pharynx to show free inner edge of septa (see *B*). (C") *Alcyonium digitatum,* Rec. North Atlantic, (a) (b), spicules.
1, epidermis; 2, mesogloea; 3, gastrodermis; 4, the eight longitudinal partitions (I–VIII), septa or mesenteries; 5, flagellate grooves that direct water currents into interior (= siphonoglyphs); 6, pharynx (stomodaeum); 7, gastrovascular cavity or tubes of polyps; 8, polyp; 9, wholly or partially contracted polyp; 10, hollow tentacles that alternate with the septa, and are extensions of interseptal chambers in *B*; 11, stolon; 12, short pointed projections (pinnules) that give a pinnate or feathery appearance to the tentacles that bear them.
(D) Polypoid generation exclusively; life cycle of anthozoans. (E) Spicule (spindle) of body wall of *Clavularia viridis,* Rec., Indo-Pacific.
Illustrations after (B, C, C') Hickson, S. J., 1895. Quart. J. Micros. Sci., Vol. 37. (D) Hill and Wells; (A, C", E) Bayer.

Recent populations (one Atlantic, one Pacific) occur on either side of the Panama isthmus. In all critical characters, these two geographic isolates cannot be distinguished from each other. Accordingly they belong to the same species. Yet the common gene pool of the original population has been fractionated (by geographic isolation established by the isthmus) since Miocene time. These observations point to negligible genetic change since the Miocene, that is, during the past 25 million years!

Awareness of stability in existing species yields paleontological value. This signifies that octocorals from the Tertiary should compare morphologically to living forms. Fossil *Virgularia presbytes* illustrates ideally this observation (Figure 5.20*A*).

INFORMATION DERIVED FROM FOSSIL OCTOCORALS

The Fossil Record: Criteria

1. The skeleton must be proved to be spicular in nature.

2. It should bear recognizable resemblances to some Recent calcareous octocorals. [Items 1 and 2 taken together led Bayer to reject the assignments of the following corals: *Favosites, Syringopora, Aulopora, Halysites* (see Tabulata, this chapter). These had been referred to the Stolonifera.]

3. External axial characters alone are inadequate to distinguish species or genera of the pennatulids (order Pennatulacea). [Thus species themselves of the Tertiary fossil *Graphularia* (Figure 5.16*H*) are of question-

able standing, whereas the genus is acceptable. Examples of unquestioned fossil sea pens are *Virgularia* (Figure 5.20*A, b*), *Pteroeides* (Figure 5.20*C*), and (?) *Cancellophycus* (Figure 5.20*I*).]

4. Numerous fossil genera casually assigned to the order Coenothecalia are actually unacceptable. [This eliminates all fossils assigned to the blue coral *Heliopora* (family Helioporidae) (Figure 5.17*E*). However, Bayer recognized two genera in the last-named family: *Octotremacis,* Miocene of Java, and *Polytremacis,* Upper Cretaceous to Lower Tertiary of Europe.]

5. It is impossible to determine with any confidence fossil material from the order Alcyonaria based on isolated spicules alone. (The only acceptable fossil alcyonarians, according to Bayer, belong to the genus *Nephthea,* Lower Jurassic, Europe.)

6. Many fossil genera and species assigned to the order Gorgonacea are unacceptable. Included in this group are *Plumulina* Hall, 1858 and *Websteria* M. Edw. H., 1850. Acceptable gorgonacean fossils include *Parisis,* Tertiary of Italy and India (Figure 5.20*F*) *Junceella* (?) Eocene (Figure 5.20*G*) *Mopsea,* Eocene (Figure 5.20*E*), *Axogaster,* Cretaceous of England, *Moltkia,* Upper Cretaceous (Danien), Denmark (Figure 5.20*B*). (See Nielsen, 1917, 1925.)

An Extinct Octocoral Order

Bayer accepted the following new order as properly belonging to the octocorals.

Order Trachypsammiacea Montanaro-Gallitelli, 1956. The corallum is dendroid (Figure 5.17*D*); new corallites arise from the medullary system (that is, central

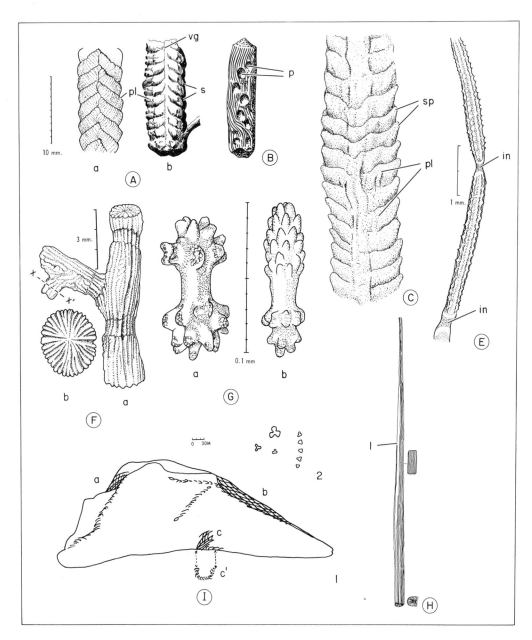

FIGURE 5.20 Fossil octocorals and/or living equivalents. (A) *Virgularia presbytes*. (a) Rec., Gulf of Mexico, ventral view; (b) mold of ventral side of colony, Cret., Trinidad, *s*—suture, *pl*—polyp leaves. (B) *Moltkia isis*, U. Cret., (Danian) Denmark, branch with impression of separate individuals (*p*), X 3. (C) *Pteroeides argentum*, Eoc. (?) Oligo. (?), S.E. Moluccas, *sp*—jagged edges indicate spicular armature of polyp leaves (*pl*), schematic. (D) figure deleted. (E) *Mopsea elongata*, Rec., Antarctica (fossils in Eocene), *in*, internodes. (F) *Parisis fructiocosa*, Rec., Sula Sea (fossils in Tertiary of Italy, India), (a) part of the axis decorticated; (b) end view of internodes. (G) *Junceella juncea*, Rec., Philippines [fossils found in Eocene (?)], (a) dumb-bell type sclerite from axial sheath layer; (b) clubtype sclerites from outer cortex. (H) *Graphularia desertorum*, Nummulitic limestone, Eocene, Libya, reduced. (*l*) *Cancellophycus* sp., Jur. (Bathonian), France. (1) Section of spiral cone (reduced), *a*–*b* mark the exterior surface of a cone; *a*—section of limb; *c*—section parallel to surface of interior spire of cone; *c'*—transverse section of same spire shown for comparison. (2) Spicules (× 150) (in section lamellae contain calcite needles, trifoliate and close to pennatulid spicules).
(A, G, H) Order Pennatulacea. (B–F) order Gorgonacea.
Illustrations after (A, C, E–G) Bayer, 1955; 1956; (B) Nielsen, 1917; (H) Zittel; (*l*) Lucas.

zone of stem; see Figure 5.17*D*, 14); there are skeletal structures of cortical region, that is, secondary thickenings oriented radially. Known only from the Upper Permian of Sicily and Timor.

The length of the fossil shown in Figure 5.17*D* is about 1 in.

Geosynclinal Indicators

Kugler (in Bayer, 1955) described the Flysch sequence of the Pointe-a-Pierre formation from the Cretaceous of Trinidad.

In graywacke block debris, at the base of a wave-cut cliff (Figure 5.21*D*), three-dimensional sandstone casts were collected – pennatulacean *Virgularia presbytes* (Figure 5.20*A*, *b*). Similar casts known from the Carpathian and Alpine "Flysch" deposits were identified as molluscan tracks (See Chapter 8) and named "*Bilobites*" or "*Palaeobullias.*" However, Bayer (1955) proved that both European and Trinidadian fossils were pennatulaceans. By preparing rubber casts of plaster molds made from living specimens of *Virgularia presbytes,* he obtained a pattern identical to that found in the fossils.

The recurrence of "*Bilobites*" markings (that is, pennatulacean octocorals) in sparsely fossiliferous Flysch can denote deeper geosynclinal environments.

Magnesium Carbonate Content

Octocoral alcyonarians from the Uteli reef, Samoa (Figure 5.21*C*) in the upper portion, had a high magnesium carbonate content; from 10 to 36 percent (Cary, 1931). Thus geologically younger reefal, high-magnesium limestones and dolomites could reveal ghosts of octocoralline spicules in petrographic thin sections.

Distribution and Reefal Zonation

The distribution of octocorals in two extant reefs (Figure 5.21, *B*) may be more or less restricted to a given zone (Figure 5.21*A*) or may be sporadically distributed from shore to coral reef with greatest areal coverage only a few hundred feet from shore (Figure 5.21*B*).

Thus, if one applies the data of Figure 5.21*B* to those of Figure 5.21*C*, it is possible to infer the relative distance from shore represented by the 54- to 105-ft interval.

ZOANTHARIANS

Introduction

General biological and morphological data on the zoantharians have been given in the subclass definition and in Figure 5.18*B*. However, more extended treatment seems desirable.

In broadest terms, the subclass consists of three orders, of which one has living representatives and lacks a fossil record (Zoanthiniaria, Corallimorpharia, and the Actiniaria); three extinct orders known only as fossils (Heterocorallia, Rugosa, Tabulata); and one order that has both a fossil record and living representatives (Scleratinia).

Essentially, subdivision of the Zoantharia into orders is based on two criteria: arrangement and development of the mesenteries, and the presence or absence of a skeleton.

Paired Mesenteries and Relationships Between Orders

Figure 5.22 and accompanying terminology set forth the essential data on the mesenteries of all the orders excepting the Tabulata whose longitudinal septation is either obscure or lacking.

The ancestral or stem zoantharian had the following mesenterial arrangement (Figure 5.22*B*):

1. Eight complete mesenteries (four pairs or couples).
2. Two additional lateral couples (that is, two pairs).
3. Of these six pairs, two pairs are dorsal and ventral *directives.*
4. Additional pairs also occur in cyclical arrangement.

Distance from Shore		*Square Feet Covered by Octo-corals*		*Character of Bottom*
0–25	–	2		
25–50	–	3		
50–75	–	0		Almost entirely covered by sand
75–100	–	0		
100–125	–	8		
125–150	–	50	–	Several *Porites* heads
150–175	–	200		
175–200	–	400		Octocoral Reefs
200–225	–	100		
225–250	–	30	–	dead *Porites*
250–275	–	80	–	branching *Porites*
275–295	–	25	–	many *Acropora*

(B)

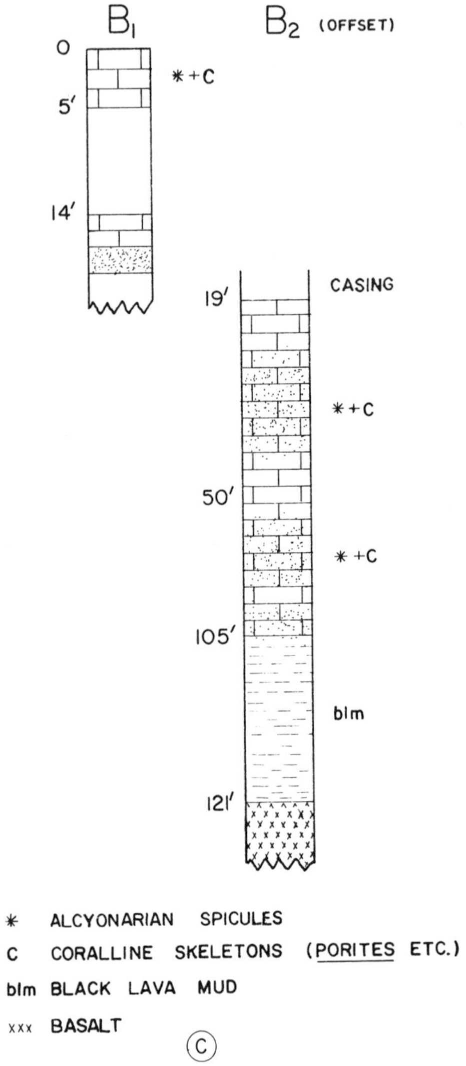
B1
B2 (OFFSET)
0
5′
14′
✱+C
19′
CASING
✱+C
50′
✱+C
105′
blm
121′
✱ ALCYONARIAN SPICULES
C CORALLINE SKELETONS (PORITES ETC.)
blm BLACK LAVA MUD
xxx BASALT
(C)

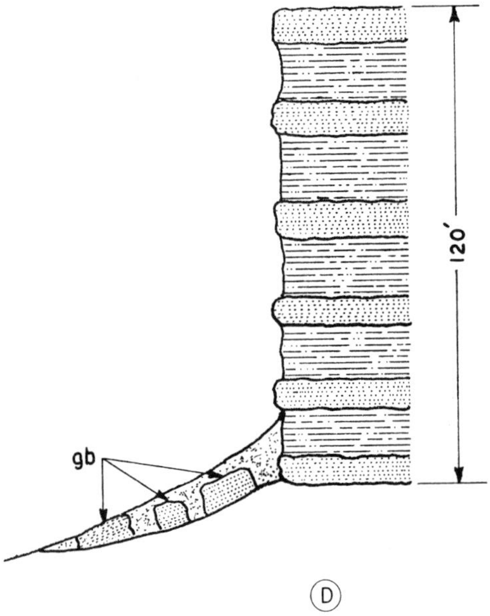
gb
120′
(D)

FIGURE 5.21 See Fig. 5.6(*A*). Note relationship of the octocoral *Heliopora* zone to that of *Porites* and *Acropora*. (*B*) Distribution of octocoral alcyonarian reef at Pago Pago, American Samoa. Along traverse 2—from Seawall, Goat Island, to reef edge. Data from L. C. Cary, 1931, Table 5. (*C*) Incorporation of reefal octocorals and zoantharian corals into rock column. Boring through Utelei Reef, Samoa. Compiled from data in Cary, 1931. (*D*) Generalized section (Flysch). Pointe-a-Pierre formation. Trinidad (prepared from description by Kugler, in Bayer, 1955). Total thickness, 120 ft. *gb*—graywacke block containing molds of pennatulaceans.

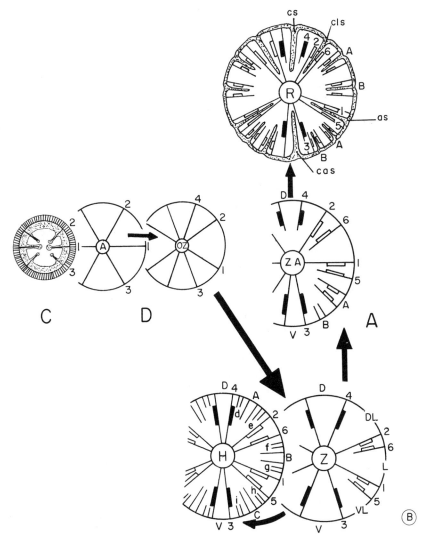

FIGURE 5.22 Terminology and symbols (after Hill and Wells, and others).
Paired mesenteries — order of appearance: 1, 2, 3 — primary (first to appear);
4, 5, 6 — later mesenteries, A, B, C — secondary mesentaries;
d, e, f — tertiary pairs.
D — dorsal, V — ventral, DL — dorsolateral, VL — ventral exocoele, L — lateral exocoele.
Complete or perfect septa or mesentery (see Figure 5.22 part A, "1,2") extends from wall to pharynx
(indicated by letter H, Z, ZA, R).
Directive (s) mesentery — 2 pairs of complete septa (see Figure 5.22 part A, "3,4") with retractor
muscle (black rectangles) on outer surface facing exocoeles.
Endocoele — the space between members of a given pair of septa.
Exocoele — the space between pairs of septa (see Figure 5.22 part B, VL, DL).
Mesentery types:
   Indefinite musculature (see B, B). Lines with no rectangles.
   Nondirective (retractor muscles — white rectangles face each other).
   Directives (see definition above).
cs — counter septum; cls — counter-lateral septum; as — alar septum; cas — cardinal septum.
Arrangement of mesenteries on ancestral or stem Zoantharian (Z) and derivative orders. (A, B, D —
after Hill and Wells).
(A) Order Zoanthiniaria (ZA) thought to be derived from Z. Order Rugosa (R) thought to be derived
from ZA.
(B) Zoantharian stem (Z), paired mesenteries (incomplete left section duplicates the right in
(A)–(D)). Hexactinarians (H) include these orders: Corallimorpharia, Actiniaria, Scleratinia. Note de-
rivation from Z.
(C) Zoanthid larva, septal development. Note the three primary couples and compare to the original
anthozoan stem of ancestral pattern in (D).
(D) (left figure) Anthozoan stem (A), or the ancestral septal pattern with three primary couples.
(right figure) Anthozoan stem with three primary couples gave rise to the octocorallian-zoantharian
stem (OZ).
(C — after Menon.)

151

Given this *basic* septal plan, under what conditions would new mesenteries be inserted to form distinctive septal plans? The basic plan was genetically determined. To alter this plan in the direction of the hexactinarians (Figure 5.22*A*) or the zoanthiniarians (Figure 5.22*B*) involved biochemical changes in DNA-nucleotide ordering and the operation of natural selection on countless generations (see Chapter 2).

Although only living representatives are known, the anemone-like zoanthiniarians lacking a skeleton are thought to be close to the most primitive zoantharians, that is, the stem line (Figure 5.22*B*). In early larval stages its septal plan is reminiscent of the ancestral anthozoan (compare Figure 5.22*C, D*). The zoantharian stem, in turn, appears to have evolved from a skeletonless polyp with eight mesenteries and eight tentacles. One evolutionary branch of this ancestral stock gave rise to the octocorals, and one to the zoantharian stem.

THREE ANEMONE ORDERS LACKING A SKELETON

1. Zoanthiniaria

The most important fact to a paleobiologist pertaining to this order is the mode of insertion of new septa. New septa are inserted exclusively in the *exocoeles* to either side of the ventral *directives* (see Figure 5.22 for terminology). This condition is also found in the Rugosa (see Figure 5.22*A* and text discussion of rugose corals). This order contains a half-dozen living genera.

2. Corallimorpharia

The order differs from the scleratinians chiefly in the lack of a skeleton (see section on Scleratinia).

3. Actiniaria

See *Metridium*, Figure 5.18*B*. There are 200 living genera and 1000 species in this order. Some Cambro-Ordovician fossils

have been assigned to it, but Wells and Hill (1956) rejected these as questionable assignments. [The questionable fossil genera include (?) *Mackenzia*, Middle Cambrian, Burgess Shale, British Columbia; (?) *Palaeactis*, Middle Ordovician of France; (?) *Palaeactinia*, Middle Ordovician of New York.]

THREE EXTINCT ORDERS

Most restricted in time, an order with but one family and two genera is the Heterocorallia.

Definition: Order Heterocorallia

Elongate coralla originally with four septa conjoined axially; new septa form in attachment to these (Figure 5.23*C, a–b*): with narrow trabeculate stereozone (i.e., marginal zone of dense tissue) (Figure 5.23*A*); tabulae complete, domes steep towards vertical edges (Figure 5.23*B*). Geologic range: Carboniferous (Visean-Namurian), Europe, Asia; Mississippian, Pacific coast region (Helen Duncan, 1965).

Morphology

The basic structural element of the heterocorals is longitudinal radial, with transverse plates. The former are the *septa*; the latter are the *tabulae*. The stereozone (a thickened margin) is the only other major structural feature.

Mineralogy of Skeleton

The two sets of plates are composed of calcium carbonate fibers. Hill observed that the fibers grow only on the upper surface of the tabulae.

Septation

There are two different interpretations of the insertion of septa in the heterocorals. According to Yabe and Hayasaka (1915), six protosepta were originally present as in the Rugosa and the Scleratinia (Figure 5.23*C*). Thereafter the new

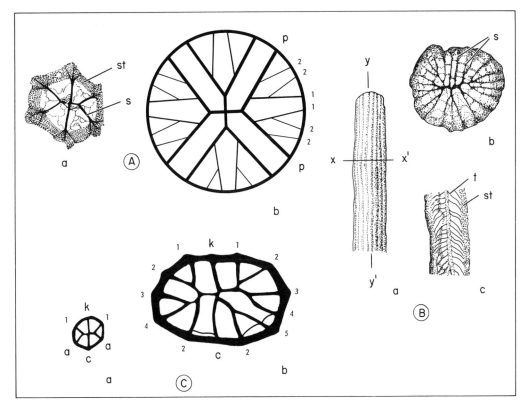

FIGURE 5.23 Order Heterocorallia
(A) (a) *Hexophyllia prismatica,* Carb. (Visean/Namurian, Europe), transverse section. This might be the more primitive of the two genera, and its septal plan might be equivalent to the young stages of *Heterophyllia* [see also C(a), and B, a–c]. (b) Septal arrangement according to Schindewolf. (= 1'), 2 (= 2^x).
(B) *Heterophyllia (Heterophylloides) reducta,* Carb., Europe.
(a) Exterior of corallite note fluting; (b) x–x' – transverse section; (c) y–y' – longitudinal section.
(C) *Heterophyllia kitakamiensis,* Carb., Asia. (a) Transverse section through juvenile portion of corallite (compare to A); (b) ephebic part of corallite (septal arrangement according to Yabe-Hayasaka); st – stereozone = zone of dense tissue, s – septa, t – tabulae, domal or saucered, the transverse skeletal element; k – counterseptum, c – cardinal septum, a – accessory (alar) septum, 1–5 = order of appearance of metasepta, p – protoseptum, 1', 2^x = metasepta in order of appearance.

septa (called *metasepta*) insertion was confined to two positions – not four as in the Rugosa (see Figure 5.23C, *a* and *b*; also see Figure 5.23A, *a*). Schindewolf (1941) stated that four protosepta were present originally. Near the periphery each of these split into two. New septa (metasepta) insertion was confined to the four spaces (quadrants) between the two split portions of each septum (Figure 5.23A, *b*).

Unique Feature

The absence of free inner edges of the septa, where each is attached to another septum, is perhaps the most distinctive heterocoral feature. (See P. K. Sutherland, 1965, Okla. Geol. Surv. Bull., Vol. 109, pp. 35–40.)

INFORMATION DERIVED FROM HETEROCORALS

Stratigraphy

Heterocorals now known from the Pacific Coast region were formerly found only in the European and Asian Carboniferous. Wherever they occur, they serve as excellent index fossils.

Pseudoplankton (?)

Schindewolf (1941) found hooks on the outer septal edges of some forms, suggesting that perhaps such heterocorals were attached to seaweeds. If so, they were pseudoplanktonic during the European-Asian Carboniferous.

Lithology

Heterocorals occur in pure limestones and calcareous shales (Stuckenberg, 1895; Yabe and Hayasaka, 1915; and Schindewolf, 1941).

Evolution

Both the geologic and the geographic restricted occurrences of heterocorals indicate an apparently unsuccessful evolutionary experiment leading to the dead-end extinction.

Since attachment of the inner edges of septa is the unique characteristic of heterocorals, gene loci in the ancestral population that governed septal insertion must have been involved. The nonpersistence of the order through time and its restricted speciation may both be related, partly, to a preferred pseudoplanktonic existence.

RUGOSA

Introduction

1. The eroded, bare, carbonate fossil skeletons once housed a living animal, a coelenterate, an anthozoan (see definitions).
2. Each morphological structure was secreted by living tissue and involved a complex biochemical sequence (see Chapter 15).
3. Differences in septation, tabulation, dissepimentation, and so on, were all genetically determined and hence may be attributed, on the molecular level, to changes in the gene DNA biochemistry.

Viewed in this light, *every structure, every term,* has biological connotations on the level of molecular and cell biology; many are significant because they designate variable characters that were being modified during growth and speciation. The individuals in which such modifications occurred often have a selective advantage conferred on them in this way.

Definition: Order Rugosa (after Hill)

1. Solitary or compound epithecate corals (Figure 5.24*A*; 5.25).
2. Septa typically in two orders (major and minor septa) alternating in length, there maybe one or three orders in some forms (Figure 5.24*B*).
3. Symmetry bilateral.
4. In solitary corallites the metasepta are inserted in four positions only (after insertion of the first six protosepta). These positions are counterside of each alar septum and on each side of the cardinal septum (see Figure 5.24*B*).
5. Marginarium (either a dissepimentarium or a peripheral stereozone, for terminology, see p. 171 and Figure 5.24*F*.
6. Tabulae present. They may be conical, domed, horizontal, sagging, or inversely conical.
7. Each tabula may be complete (Figure 5.24*G*) or incomplete.
8. Incomplete tabula consist of a number of tabellae (Figure 5.24*D*).
9. An axial structure may develop (Figure 5.24*C*).
10. Geologic range: Ordovician to Permian (333 genera).

Structures: Exoskeleton

Rugose corals represent an exoskeleton (*exo* = external; *endo* = internal). How was the skeleton formed? Visualize a planula attached to some substrate; it will develop into a polyp stage. The polyp is composed of an endoderm, mesogloea, and ectoderm. (Coelenterata defined, p. 109.) These tissues form a wall about a central gastrovascular cavity that opens only on the oral, tentaculate end. At the polyp's base a series of invaginations or upfolds of tissue occur. Given this model, the basal ectoderm is thought to have given rise to the exoskeleton in several steps:

1. Exudation of a gel.

2a. Calcium carbonate (aragonite) needles or "fibers" periodically crystallized out *within* this gel. (See Chapter 15 for chemistry and mineralogy.) The carbonate, of course, was in solution in seawater and taken in by the polyp metabolically.

2b. The needles crystallized at right angles to the gelatinous ectodermal surface (Bryan and Hill, 1941).

3. Basal epitheca — this was the site of the polyp's initial calcareous deposit.

4a. Protosepta (Figure 5.24B, a). These six major septa were the first skeletal elements to be secreted in the basal epitheca.

4b. Radial plates (septa generally) are formed by fibers crystallized as in part 2a, but located in basal tissue invaginations or upfolds. [Points or centers of calcification on the crest of the upfolds control crystallization. That, in turn, explains why septa are made of spines (*trabeculae*) (Figure 5.24H–K).]

5. Transverse plates (tabulae, dissepiments). These were formed of fibers crystallized as in part 2a but located in unfolded parts of the ectoderm. (Fibers are disposed at right angles to upper and lower surface of the plates.)

6. Parts of wall between peripheral septal edges — crystallization as in part 5 in the unfolded parts.

Solitary and Compound Corals

The essential shape of solitary corals is that of a reversed cone. However, a wide band of variation from this archeotype occurs (Figure 5.26). Such variants are found not only between different families and genera, but also in the same genus (Jeffords, 1947). At Queens Hill Quarry in Nebraska, the writer recalls a rugose coralline zone of Pennsylvanian age. This zone yielded bushels of corallites that varied in shape from ceratoid to cylindrical. Many families contain solitary corals of several shapes. In every rugose fossil population, some shape variation is to be expected.

Compound corals (that is, several individual polyps housed in a single skeleton known as a corallum) have two types of skeleton: fasciculate and massive. Fasciculate forms have corallites not in contact, and are *dendroid* (branching) or *phaceloid*. Massive forms share a common epitheca. Each corallite may have its own wall (*cerioid* arrangement), or there may be a loss of individual corallite walls without reduction in septa (*asteroid* arrangement). Still other arrangements include *thamnasteroid*, in which the septa between corallite neighbors may resemble magnetic lines of force, and *aphroid*, in which neighboring corallites are united only by dissepiments. (Figure 5.25).

Reproduction

Colonial rugose corals generally reproduce in the following manner. The founder (initial or protocorallite) was sexually generated while other individual corallites arose by asexual generation (called *increase*). The new corallites that budded from the protocorallite are *offsets*. Such increase may be *axial* (old corallite ceases to grow and two or more offsets form by extension of its septa and epitheca), *peripheral* (small offsets arise in the outer septate zone of the old calice, and old corallite growth may or may not continue), *lateral* (that is, branching or intermural — offsets arise from wall of old corallite), see Figure 5.24L, a–d, and Hill, 1956.

In a population of solitary corals, all are assumed to have been sexually produced (Oliver, 1960).

Calice

Although no soft-part preservation is known in fossil Rugosa, the bowl-shaped oral surface of a corallite provides some direct information on the polyp itself. *It is the mold of the base for the living polyp.* It is apparent that some rugose polyps could retract and close the calice by a lid (operculum). In some instances this structure has been fossilized.

Septa (Major)

The protosepta (cardinal, counter, two

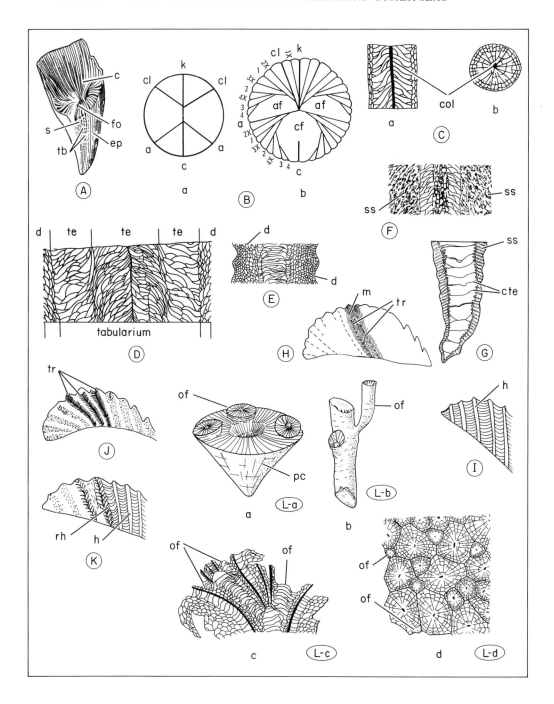

FIGURE 5.24 External and fine structural features of Rugosa (see Terminology); adapted from Hill.
(A) *Amplexizaphrentis,* complete solitary individual.
(B) (a) Protosepta—the six major septa; (b) *Zaphrenthis, K*—counterseptum, *cl*—counterlateral septum, *a*—alar septum, *c*—cardinal septum; numbers show insertion order of metasepta (later in the cycle, the minor septa are inserted).
(C) *Lithostrotion,* L. Carb., Europe. (a) Median longitudinal section; (b) transverse section; *col*—columella.
(D) *Dibunophyllum bipartium,* L. Carb., Europe. Median longitudinal section; two outer and an inner series of tabellae (= incomplete tabulae (*te*) and the dissepimentarium (*d*) occupied by dissepiments.
(E) *Entelophyllum articulatum,* Sil., Europe. Median longitudinal section showing the marginarium, which consists either of a dissepimentarium (*d*) where the tabularium occurs in other forms (see D) or a peripheral stereozone, *ss,* (see F).
(F) *Kodonophyllum truncatus,* Sil., Europe, *ss*—stereozone (= marginarium).
(G) *Tryplasma loveni,* M. Sil., Europe. Longitudinal section showing complete tabulae, *cte* (compare incomplete tabulae in D) and stereozone, *ss.*
(H)–(K) Types of trabeculae, *tr* (see terminology). (H) *Porpites porpita,* Sil., Europe. One septum in longitudinal section showing only two trabeculae—*monocanths,* or monocanthine trabeculae (*m*). (I) *Tryplasma primum,* M. Sil., England. Longitudinal section with several trabeculae shown—*holocanths* (*h*). (J) *Rhabdocyclus porpitoides,* Sil., England. Longitudinal section with three trabeculae shown; *rhabdocanths.* (K) Same as (I); at right are holocanths (*h*), to left are rhabdocanths (*rh*). The pair are called *dimorphacanthine* trabeculae.
(L) Types of increase (asexual reproduction) in compound rugose corals:
   (a) Peripheral: *Entelophyllum articulatum,* Sil., Europe.
   (b) Lateral: *Lithostrotion affine,* L. Carb., England.
   (c) Axial: *Acervularia ananas,* M. Sil, England.
   (d) Intermural: *Lithostrotion* sp., L. Carb., Scotland.
f—fossula, *ep*—epitheca, *s*—septa; *of*—offset; *pc*—protocorallite; *c*—calice.

alar, and two counterlateral) are shown in Figure 5.24*B, a*. These are the initial skeletal elements in rugose corals. Metasepta insertion occurs later and is restricted to only four points (either side of the cardinal septum and the counterside of each alar; see Figure 5.24*B, b*).

Septa (Minor)

These septa, much shorter than the major septa, alternate with them.

Septal Thickening and Rugose Coral Nutrition

If crystalline fibers of septa (formed as indicated earlier) become elongated, they are said to "thicken." Such thickening, expressed in a particular zone of the coral such as the peripheral zone, forms a *stereozone* (Figure 5.24*G*). These thickenings, seen in successive layers, suggest a periodic control. Dorothy Hill has suggested that *nutrient intake* may be such.

Septal Type

Rugose septa are commonly laminar, but variants of the continous sheet do occur. Septal length varies (long, short); thickness varies (thin, or thickened inner edges—rhopoloid or club-shaped); septa may be dilate or attenuate and are rarely perforate.

Discontinuities both vertical and radial characterize certain septal types: *amplexoid* septa are fully developed only on the upper surfaces of tabulae, for instance, *Amplexus; lonsdaleoid* septa do not continuously extend through the dissepimentarium but fade out toward the epitheca. For other structures, see terminology and illustrations.

Three Suborders

The order can be differentiated into three suborders chiefly on the basis of the differences in the *marginaria* and *tabulae.*

Suborder 1. Streptelasmatina. Solitary or colonial Rugosa; marginarium comprising either a septal stereozone or a dissepimentarium composed typically of small, globose interseptal dissepiments;

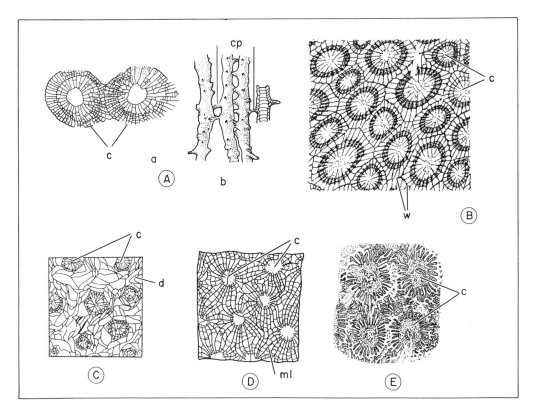

FIGURE 5.25 Compound Rugose Corals—Types. (See text for discussion).
Fasciculate: (A) PHACELOID, (a) *Eridophyllum seriale* (M. Dev., Morocco) transverse section; (b) *Eridophyllum simcoense* (Dev., North America), corallite in median longitudinal section. Massive (B–D): (B) CEROID, *Trapezophyllum coulteri* (L. Dev., New South Wales), transverse section (C–D) APHROID, (C) Orionastrea lonsdaleoides, transverse section; (D) THAMNASTEROID, *Orionastrea garwoodi* (L. Carb., England. (E) ASTEROID, *Palaearaea lobatini* (Sil., Siberia).
Symbols: *c* — corallite, *w* — wall, *d* — dissepiment, *ml* — confluent septa resembling magnetic lines of force between neighboring corallites, *cp* — connecting process (After Hill).

tabulae typically domed. Ordovician to Permian (218 genera). (See Figure 5.27.)

Suborder 2. Columnariina. Corallum compound (but less frequently, solitary). In oldest forms, marginarium absent, but later it develops as a septal stereozone that may be replaced by a dissepimentarium or as an incomplete series of elongate dissepiments. Tabulae complete and flat, or with down-turned edges or sagging medially; late forms develop axial structure. Ordovician to Permian (60 genera). (Figure 5.29.)

Suborder 3. Cystiphyllina. Solitary or compound Rugosa; septal trabeculae (Figure 5.29, parts *H–K*), large; marginarium either a stereozone in which cores of trabeculae appear, or a dissepimentarium of small, globose dissepiments; typically with septa represented by separate trabeculae based on the upper surfaces of dissepiments; tabulae flat and complete or inversely conical and incomplete. Ordovician to Devonian (30 genera).

Some 65 percent of all rugose coralline genera belong to one suborder, the Streptelasmatina. Since all three suborders seem to have appeared during Black Riveran time (M. Ord., Appalachian Geosyncline), more accelerated speciation and divergence characterized the streptelasmatids.

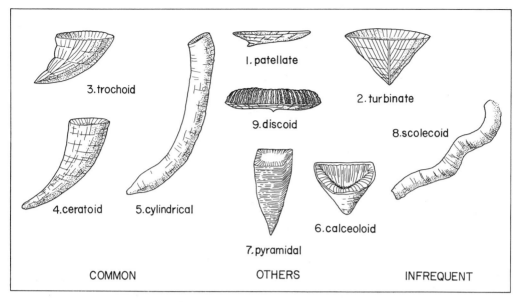

FIGURE 5.26 Rugose corals. Configurations of simple Coralla.
(1) *Patellate,* and (9) *discoid* — both types can be found in the family Tryplasmatidae, and of eight genera of family Hadrophyllidae, five are discoid, and one patellate. (6) *Calceoloid,* (7) *Pyramidal* — several genera of family Goniophyllidae are characterized by calceoloid coralla, and *Goniophyllum* is square in section (*pyramidal* coralla). (3) *Trochoid,* (4) *Ceratoid,* (5) *cylindrical* — the most common types of coralla. (8) *Scolecoid* — infrequent in occurrence (*Helminthidium mirum,* Middle to Upper Silurian, Europe).
(Adapted from Hill.)

The Streptelasmatids

The 218 genera of this suborder are grouped into two superfamilies. Cyathaxonacae and Zaphrenticae.

Included in the first superfamily, the Cyathaxonacae, are the following genera: *Lambeophyllum* (Figure 5.27*A*), *Amplexiphyllum* (Figure 5.27*B*), *Kinkadia* (Figure 5.27*C*), *Leonardophyllum* (Figure 5.27*D*), *Stereocorphya* (Figure 5.27*E*), *Cyathaxonia* (5.27*F*), *Hadrophyllum* (5.27*G*), *Zaphrentoides* (5.27*H*), *Zaphrentites* (5.27*I*), *Lophophyllidium* (5.27*J*), *Lophamplexus* (5.27*K*).

The second superfamily, the Zaphrenticae, includes *Streptelasma* (Figure 5.28*A*), *Enterolasma* (Figure 5.28*B*), *Aulocophyllum* (Figure 5.28*C*), *Heliophyllum* (Fig. 5.28*D*), *Entophyllum* (5.28*E*) *Acervularia* (5.28*F*), *Zaphrenthis* (5.28*G*), *Phillipastrea* (5.28*H*), *Eridophyllum* (5.28*I*), *Dibunophyllum* (5.28*J*), *Lithostrotion* (5.28*K*), *Meniscophyllum* (5.28*L*).

The Columnariids

Some representative genera of this sub-order include *Favistella* (Figure 5.29, part A), *Spongiophyllum* (Figure 5.29, part B), *Tabulophyllum* (Figure 5.29, part C), (?) *Diversophyllum* (5.29, *D*), *Acanthophyllum* (5.29, *E*), *Lithostrotionella* (5.29, *F*), *Lonsdaleia* (5.29, *G*).

The Cystiphyllinids

This suborder met apparent extinction by Devonian time, whereas the other two survived until the Permian. In evolution-are terms, that denotes a variety of factors. What these were can only be surmised. Perhaps some of the modifications in calcareous deposition lacked a selective advantage [for example, the coating of skeleton elements with dense calcareous tissue (sclerenchyme) and the larger volume of the coralla occupied by globose dissepiments]. Aberration in the growth cycle may have been induced by environmental influences. Thus *rejuvenescence* as a mode of coralla increase commonly occurs in genera of this suborder (Figure 5.29, parts

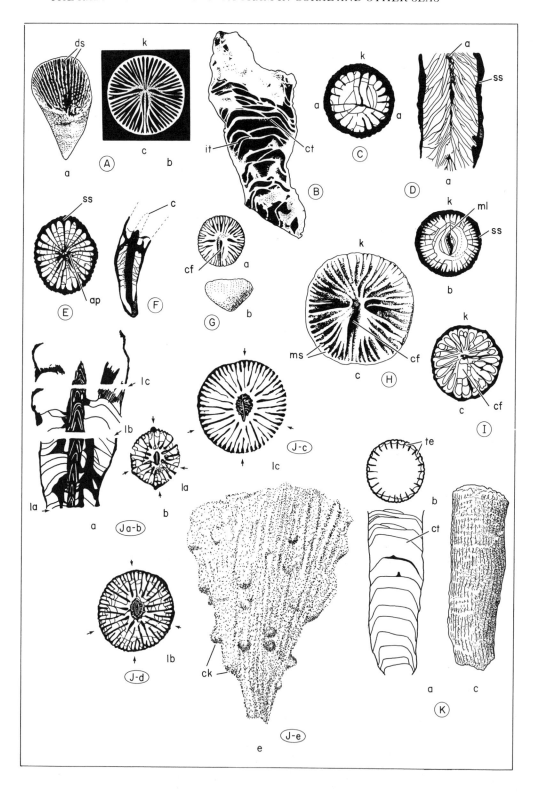

FIGURE 5.27 Streptelasmatina (Cyathaxoniicae) (schematic).
(A) *Lambeophyllum profundum.* Black River Group (M. Ord., Michigan), 30.0 mm long. (a) Side view, (b) transverse view.
(B) *Amplexiphyllum hamiltoniae.* Hamilton Group (Dev., New York), 31.0 mm long, longitudinal section.
(C) *Kinkadia trigonalis* (Miss., North America), transverse section, diam. ca. 60.0 mm.
(D) *Leonardophyllum distinctum* (L. Perm., Leonard, West Texas), length 16.0 mm. (a) Longitudinal section; (b) transverse section.
(E) *Stereocorphya annectans* (Penn., Texas), max. diam. 11.0 mm. Transverse section, adult stage.
(F) *Cyathaxonia cornu* (L. Carb., Europe), length 14.0 mm, longitudinal section.
(G) *Hadrophyllum orbignyi* (M. Dev., North America), diam. 6.0 mm. (1) Top; (2) side.
(H) *Zaphrentoides griffithi* (L. Carb., Europe), max. diam. 30.0 mm, calcial view.
(I) *Zaphrentites parallela* (L. Carb., Europe), max. diam. 6.0 mm, transverse section.
(J a–e) *Lophophyllidium spinosum,* (Perm. Texas). (a) Longitudinal section; (1a, 1b, 1c) serial sections through "(a)"; note increased complexity of septal patterns from 1a to 1c. (e) Corallite length 28.0 mm. (K) *Lophamplexus brevifolius* (Penn., Texas). (a) Longitudinal section; (b) transverse section; (c) portion of cylindrical corallite, length 24.0 mm.
Symbols: *ds*—denticulate septa formed of fused spines, *it*—incomplete tabulae, *ct*—complete tabulae, *k*—counterseptum, *a*—alar septum, *ml*—medial laminae, *a*—axial structure lacking walls; *c*—cardinal septum, *c*—columella, *cf*—cardinal fossula, *ms*—minor septa, small arrow—indicates protosepta, *ap*—axial pillar, *ck* (= *cr*—conical radicles, *ss*—stereozone.
Illustrations redrawn and adapted from (A) Stumm; (A)2, (C), (E)–(I) Hill; (B) Stumm and Watkins, (D) Jeffords and Moore; (J)–(K) Jeffords.

H, K, L, M). Since in rejuvenescence early stages are recapitulated in lieu of normal adult development, it may have been initiated by specific physico-chemical factors.

The operculate condition (Figure 5.29, part *J*) characterizes genera of the family Goniophyllidae (L. Silurian–M. Devonian) and appears to have conferred a selective advantage in a harsh or varying environment.

Representative genera of this suborder include *Tryplasma* (Figure 5.29, part *H*), *Cystiphyllum* (Figure 5.29, part *I*), *Calceola* (5.29, *J*), *Cayugaea* (5.29, *K*), *Mesophyllum (Cystiphylloides)*[3] (5.29, *L*), *Skoliophyllum* (5.29, *M*).

INFORMATION DERIVED FROM RUGOSE CORALS

Reproduction

The mode of reproduction can bear significantly on *genotypic variation* (that is,

[3]Genus *Mesophyllum*, subgenus *Cystiphylloides.* A genus is subdivided into subgenera when groups of species within it fall into natural subunits based on particular characteristics.

the extent to which individual corallites in a colony differ from the parental stock) in rugose corals. One would assume that there would be greater variation in sexually reproducing solitary Rugosa because of the possibility of genetic recombination. Oliver (1960), however, observed exceptions when two or more colonies intergrew or when several protocorallites fused either prior to or during early colony development.

Growth and Form

Wells (1957) noted that most solitary rugose corals had ceratoid (Figure 5.26), or contorted coralla. These were accounted for by (*a*) toppling over, or (*b*) negative geotrophic growth thereafter (that is, they grew upward). The well-known dual occurrence in the same bed of upright (the natural attitude) and curved coralla may be attributed to relative balance. Thus the straight coralla were well balanced in nonagitated water. As they grew and added weight, the coralla sank into a soft substrate without toppling over. By contrast, curved or contorted forms, represent individuals that toppled and possibly rolled over the bottom because of

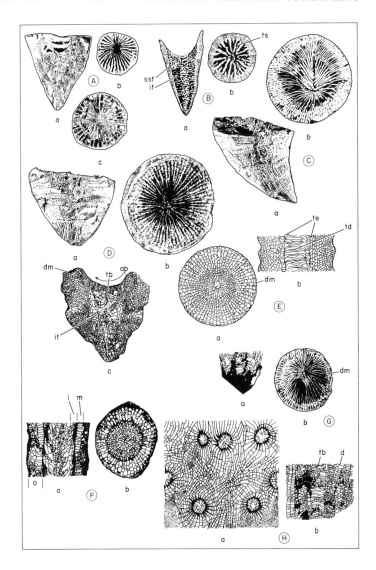

FIGURE 5.28 Streptelasmatines (Zaphrenticae).
(A) *Streptelasma breve* (Ord., Minnesota), diam. 11.0 mm. (a) Longitudinal section; (b) calyx view;
(c) transverse section.
(B) *Enterolasma strictum,* L. Dev., North America. (a) Longitudinal section, length 16.0 mm;
(b) transverse section.
(C) *Aulocophyllum scyphus,* Dev., Michigan. (a) Side view, length 48.0 mm; (b) calyx view.
(D) *Heliophyllum halli,* M. Dev., Ontario. (a) Side view; (b) calyx view, diam. 39.0 mm; (c) longitudi-
nal section.
(E) *Entophyllum articulatum,* M.-U. Sil., cosmopolitan. (a) Transverse section, diam. 16.0 mm;
(b) longitudinal section.
(F) *Acervularia ananas,* Sil., Europe, (a) Longitudinal section; (b) Transverse section, diam. 16.0 mm.
(G) *Zaphrenthis phrygia,* Dev., North America. (a) Transverse section, diam. 26.0 mm; (b) longitudi-
nal section.
(H) *Phillipastrea:* (a) *P. hennahi* (U. Dev., Europe), transverse section, diam. 24.0 mm; (b) *P. currani*
(Dev., New South Wales), longitudinal section.

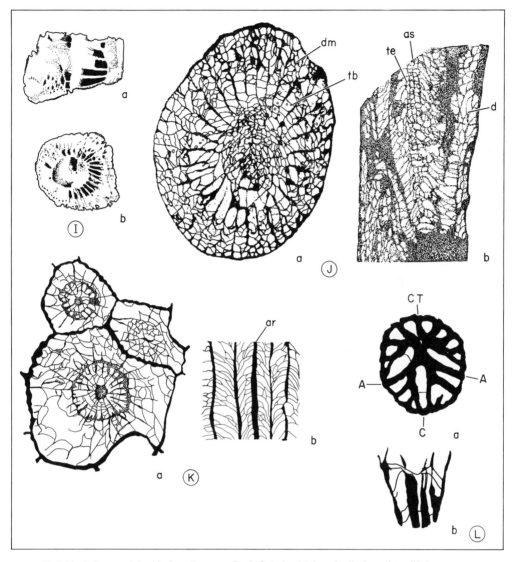

(*I*) *Eridophyllum seriale*, M. Dev. Formosa Reef, Ontario. (*a*) Longitudinal section; (*b*) transverse section.

(*J*) *Dibunophyllum uddeni*, Penn., Glass Mts., Texas. (*a*) Transverse section, diam. ca. 16.0 mm; (*b*) longitudinal section.

(*K*) *Lithostrotion* (*L*.) *circinatus*, Redwall-Limestone, Miss., Arizona. (*a*) Transverse section; (*b*) longitudinal section, length ca. 7.0 mm.

(*L*) *Meniscophyllum durhami*, Upper Tin Mt. Fmtn., L. Miss., California. (*a*) Transverse section; (*b*) longitudinal section, ca. 10.0 mm length.

Illustrations redrawn and adapted from (*A*)–(*C*) Stumm, (*B*), (*E*)–(*H*) Hill, (*D*) Stumm and Tyler, and Hill, (*I*) Fagerstrom, and Hill, (*J*) Ross and Ross, (*K*) Easton and Gutschick, (*L*) Langenheim and Tischler.

Symbols: *sst*—axial stereozone, *t*—tabula, *it*—incomplete tabulae, *A*—alar septum, *tb*—tabularium, *dm*—dissepimentarium; *te*—tabulae, *CT* (= *K*)—counterseptum, *C*—cardinal fossula, *d*—dissepiments, *td*—tabular domes, *o*, to *i*—complex dissepimentarium: outer to inner zones, *ap*—axial pit, *as*—axial structure, *ar*—axial end or columella, *au*—aulos (inner tube).

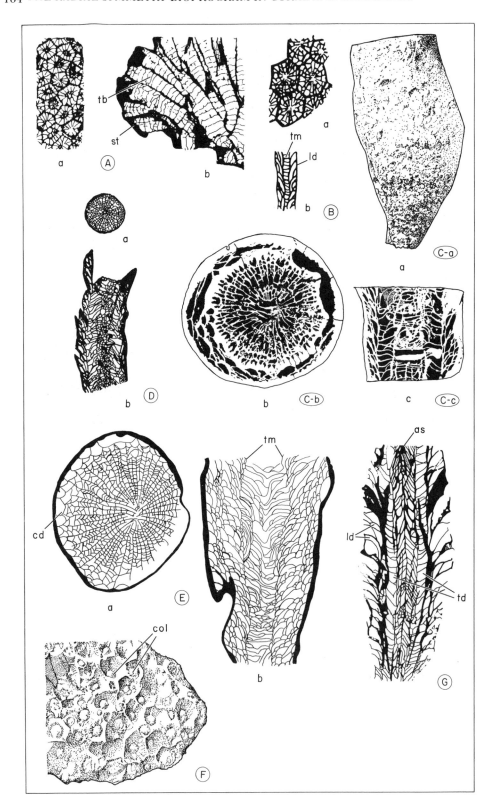

FIGURE 5.29

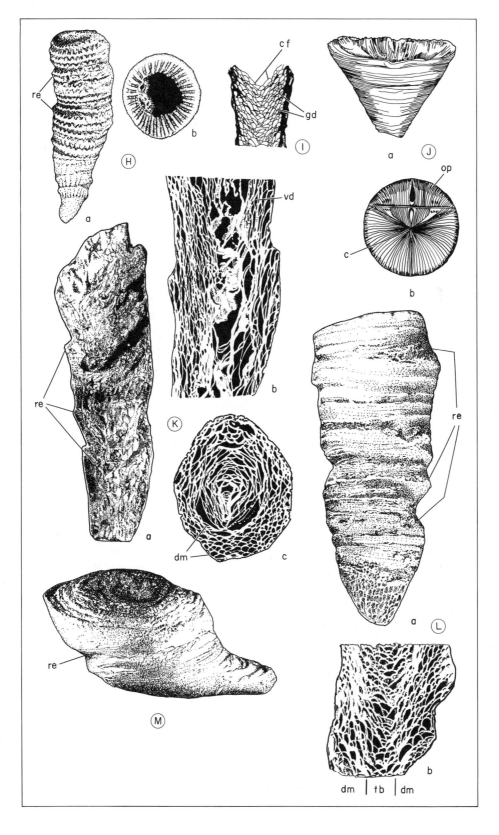

FIGURE 5.29 Suborders Columnariina and Cystiphyllina.

(*A*) *Favistella alveolatata*, M.–U. Ord., cosmopolitan, (*a*) Transverse section, note cerioid coralla and absence of marginarium; (*b*) longitudinal section, length 30.0 mm, note downturned edges of *tb*.

(*B*) *Spongiophyllum sedgwicki*, M. Dev. Europe. (*a*) Transverse section; (*b*) longitudinal section, length 8.0 mm.

(*C*) *Tabulophyllum traversense*, M. Dev., Michigan. (*a*) Side view, length 62.0 mm; (*b*) transverse section; (*c*) longitudinal section.

(*D*) (?) *Diversophyllum traversense*, M. Dev., Michigan. (*a*) Transverse section, diam. 8.0 mm; (*b*) longitudinal section.

(*E*) *Acanthophyllum* (?) *denckmanni*, (U–M. Dev., Germany. (*a*) Transverse section; (*b*) longitudinal section, length 30.0 mm.

(*F*) Lithostrotionella floriformis (Carb., North America), portion of upper surface (diam., of individual corallite, ca. 10.0 mm).

(*G*) *Lonsdaleia duplicata*, Lower Carb., Europe. Longitudinal section, length 35.0 mm.

(*H*) *Tryplasma radicula*, Sil., Michigan. (*a*) Side view showing repeated rejuvenescence (*re*), length 29.0 mm; (*b*) calyx, diam. ca. 12.0 mm. (For longitudinal section see Figure 5.20G.)

(*I*) *Cystiphyllum cylindricum*, M. Sil., Europe. (*a*) Transverse section; (*b*) longitudinal section, length 25.0 mm.

(*J*) *Calceola sandalina*, L.–M. Dev., Europe-Asia. (*a*) Convex side of corallite, length 49.0 mm, (*b*) septal arrangement in calice (*c*), diam. 30.0 mm, and interior of operculum (*op*) (diagrammatic).

(*K*) *Cayugaea subcylindrica*, M. Dev., Indiana. (*a*) Side view showing rejuvenescence; (*b*) longitudinal section, 44.0 mm; (*c*) transverse section.

(*L*) *Mesophyllum* (*Cystiphylloides*) *americanum*, M. Dev., S.W. Ontario. (*a*) Side view showing rejuvenescence, length 95 mm; (*b*) longitudinal section with *dm* and *tm* poorly differentiated; (*c*) *M* (*C.*) sp. (M. Dev., Michigan), growth form of typical corallum, length, 49.0 mm.

(*M*) *Skoliophyllum lamellosum* (L.–M. Dev., Germany) side view showing growth form with repeated rejuvenescence (*re*).

Illustrations for Figure 5.29 redrawn and adapted from Hill, 1956, and (*C*) Stumm, 1962; (*H*) Stumm, 1952; (*K*)–(*M*) Stumm, 1961–1962.

Symbols: *st*—stereozone, *tb*—tabulae, *tm*—tabularium, *ld*—dissepiments lonsdaleoid in whole or part (that is, large dissepiments with convex upper surface associated with vertical discontinuities in septa), *cd*—cystose dissepiments, *col*—columella, *as*—axial structure, *re*—repeated rejuvenescence (one mode of corallite increase, that is, corallite repeatedly becomes more constricted, leaving edge of older calice around constriction and then increases diameter; note that earlier growth stages are recapitulated to some extent), *dm*—dissepimentarium, *gd*—globose dissepiments, *cf*—calicular floor, *op*—operculum, *c*—calice, *vd*—vertically arranged dissepiments.

poor balance. Rolling, in turn, suggests active bottom currents.

There is no unanimity about interpretations even among specialists. Easton (1951) noted that curvature was a very ancient tendency in rugose corals and that it is relatively constant in a given species. The tendency is interpreted to have been an inclination toward the nutrient-oxygen source. Cuneate shape (that is, wedge shape) in Mississippian rugose corals was found to be unrelated to coralla curvature (environmentally determined) and, hence, genetic in origin. Oliver (1958) acknowledged the influence of environmental adaptations on coralla shape but stressed the variant forms described in the literature appeared genetically determined.

For operculate rugose corals, Richter (1929) showed by tank experiments that *form preceded function* in the cystiphyllinids such as *Calceola* (Figure 5.29, part *J*). Thus the slipper form of *Calceola* was secondary, that is, it was a consequence of the development of a moveable operculum. In the tank experiment the calicular end was oriented toward the strongest current, presumably equivalent to a nutrient-bearing current. The Termiers (1948) reached similar conclusions but doubted that the operculum opened as much as 90 degrees.

Without a doubt there is interplay here between a spectrum of genetically controlled form-potential and realization of any or all in variable environmental circumstances.

Measurements and Analysis

Useful ratios and measures have been employed by specialists in their study of the development of individual rugose corals within the same species or in comparing different species; some follow.

Septal Coefficient. Plotted as n/d, where n is the number of major or first cycle septa and d is the diameter in millimeters of the corallite at a given growth stage. One may plot n or n/d as ordinate and d as abscissa. (Voynovskiy-Kriger, 1954, in Chilingar, 1956; see also Oliver, 1960).

Tabularium Ratio. Here d^t/d, the tabularium diameter as a function of the total diameter, is plotted as ordinate and d as the abscissa (Oliver, 1960).

To distinguish form variants (called formae) within the same species, three factors may be used (Oliver, 1958; Wells, 1937):[4] *apical angle* (the pointed end of a solitary rugose coral is its apical end; measurement of the apical angle best represents the rate of diameter increase relative to length), *shape of corallum* (Figure 5.26), and *individual size* (length in millimeters).

Other Measurements

The distance between adjacent first-order septa a may be plotted against d, diameter in millimeters (Voynovskiy-Kriger).

Skeletal Thickening

In an earlier section it was observed that skeletal thickening (deposition of stereoplasm) may have paleoecological significance. Oliver (1960) saw that rugose corals associated with reef facies of Devonian age (Island of Gotland, Sweden, New York, Bohemia, England, and elsewhere) generally show excess stereoplasm. Such excess appears to have strengthened the skeleton and to have made it more wave-resistant.

However, non-reef Rugosa also have

(4Ref. Ch. 14.)

skeletons filled with stereoplasm. Some investigators suggest that skeletal thickening might be related to lithofacies; thin coralla with delicate structures occur in shales, which originally were clay muds deposited in quieter, deeper water; thicker, coarser types in limestones point to shallow, agitated waters where lime muds were deposited. Finally, an opposite interpretation of the Silurian Wenlock specimens of *Omphyma* from England has been given (Butler, 1937). There in shales, these forms had excess stereoplasm attributed to some depressive effect such as turbidity.

It seems generally agreed (Oliver, 1960) that deposition of excess stereoplasm was an adaptation to the environment.

Reef Facies

The larger compound Rugosa served as reef-framework structures. Solitary corals occur on reefs in association with reef-building tabulate corals, stromatoporoids, brachiopods, bryozoans, and so on. Figure 5.30 is an example of one such reefal facies of early Devonian age. The Coeymens reef limestone is composed of stromatoporoids, tabulate corals, and crinoidal debris. The reefs crop out in a series of knobs 20 ft high and 100 ft in diameter. Separation of these knobs reflects, to some extent (discounting erosion), the original interreef areas of variable size. Distribution of three rugose corals is shown. The apparent predominance of a given rugose coral in the eastern and central reef group (Figure 5.21A) may reflect the existence of interreef barriers to free passage of planula larvae. For corals generally distributed in all reef areas, such barriers apparently did not exist (Oliver, 1960).

Ross and Ross (1962, 1963) described a rugose coral fauna from the Pennslyvanian-Permian of the Glass Mountains of Texas (Figure 5.30B) in near-shore, shallow-water deposits. Locally, small patch reefs or large bioherms of shell debris (Captank formation) contained mollusks, rugose corals, and fusulinids (Ross and Oana, 1961).

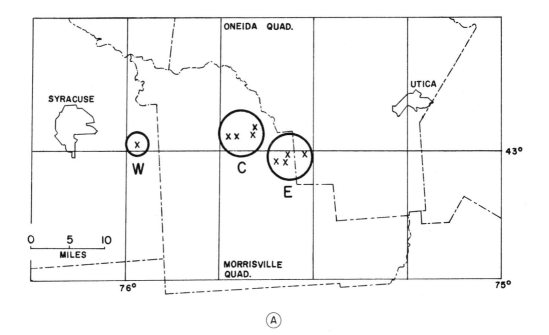

(A)

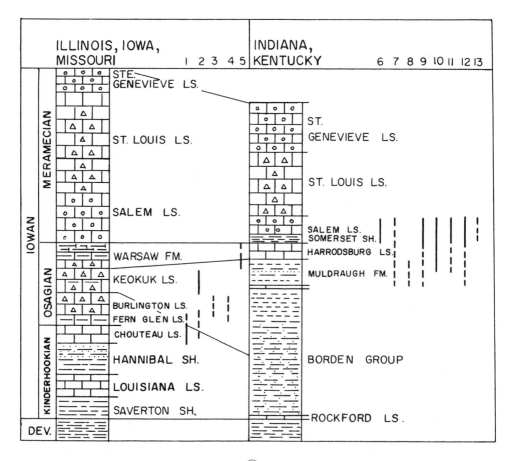

(C)

FIGURE 5.30 Geographic and stratigraphic data of some biohermal Rugosa.

(A) Rugose coral distribution and reef facies. Coeymans Limestone, N.Y. (after Oliver, 1960). Symbols: x—reefs, W—western reef group, C—central reef group, E—eastern reef group. *Briantelasma americanum* is prevalent in E (61 percent of all identified solitary corals). *Pseudoblothrophyllum helderbergium* is prevalent in C (74 percent of all identified solitary corals). *Siphonophrentis variabilis* is more generally distributed in W, C, and E.

(B) Rugose corals. Glass Mts., Texas (Penn.–Perm.).

(1) Generalized composite section. Note recurrent genera. (After Ross and Ross, 1962).

Coral species:

(1, 5, 15) *Neokoninckophyllum*: (1) *N. dunbari*; (5) *N. cooperi*; (15) *N. diciensis*.

(2, 3, 10, 19, 20, 23) *Dibunophyllum*: (2) *D. moorei*; (3) *D.* sp. A; (10) *D* sp.; (19) *D.* sp. B; (20, 23) *D. hessensis*.

(4, 12, 22; 11, 14, 16, 24) *Lophophyllidium*: (4, 12, 22) *L. solidium*; (11) *L. vallum*; (14) *L.*, compare *vidriensis*; (16) *L. skinneri*; (24) *L. vidriensis*.

(6) *Pseudozaphrentoides ordinatus*.

(7, 17, 21) *Amplexocarinia*: (7) *A.* sp; (17) *A.* sp. A; (21) *A?*, sp.

(8) *Lithostrotionella* sp.

(9) *Amplexizaphrentis* sp.

(13) *Leonardophyllum kingi*

(18) *Stereostylus tergidus*

(C) Composite columnar section (Mississippian) of five states, adapted from Easton, 1951. (1) *Neozaphrentis acuta*. *Triplophyllites* (T) species: (2–5), (8–12): (2,8) *ellipticus*, (3,9) *clinatus*, (4) *clinatus capuliformis*, (5) *reversus*, (10) *clinatus bicarinatus*, (11) *compressus*, (12) *compressus lanceolatus*, (13) new genus and species, (6) *Cyathaxonia venusta*, (7) *Hapsiphyllum ulrichi*.

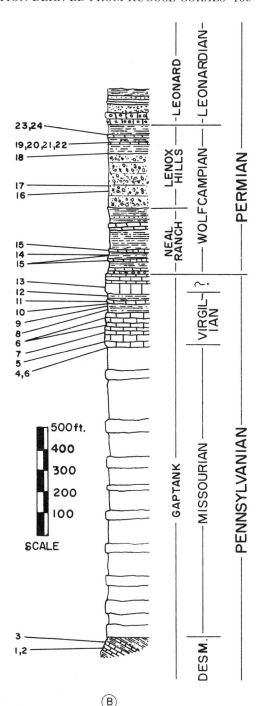

A brachiopod-coral fauna from the Lower Carboniferous (Lower Burindi Group) of New South Wales originally inhabited a topographical high now represented by interbedded siltstones and mudstones. Although this knoll is dominantly brachiopod, it contained a dense population of the rugose coral *Caninophyllum*, in a 1-ft-thick bed near the top of the section. This coralline population established itself on a sandy substrate. Isolated boulders contained species of *Lithostrotion*. Hill (1934) described a lithostrotionid fauna from overlying beds in this area and placed it as a correlate of the British D_2 zone (see Figure 5.34).

Biozones and Stratigraphic Correlation

A standard reference section for the Carboniferous of the British Isles has been a succession of coralline zones (*K* through *D*) with both brachiopod and coralline subzones (Figure 5.34). It was thus possible to correlate horizons elsewhere in the world. A Nova Scotian or Australian rugose fauna might then be referred to the British *D* zone (Hill, 1948). (Presently, other methods are used.)

Nelson (1960) established a sequence of lithostrotionid biozones in the Canadian Mississippian. As shown in Figure 5.33, he was able to correlate readily the Banff and the Highwood Pass areas—a distance somewhat less than 50 miles. Some of these biozones could be recognized over much greater distances. For example, the *Lithostrotionella jasperensis* zone could be picked out as far south as Arizona, and the *Lithostrotion mutabile* zone, well to the north in the Northwest Territory.

Ross and Ross (1962) established the vertical and horizontal distribution of a rugose lophophyllid-dibunophyllid fauna in

FIGURE 5.31 Principal areas in Western United States for Ordovician (chiefly U. Ord.) and Silurian coral faunas (after Helen Duncan).

the Pennsylvanian-Permian Glass Mountain region of Texas (Figure 5.30).

Langenheim and Tischler (1960) found that the coral fauna (mostly rugose) from the Tin Mountain limestone (Inyo County, California) included elements in common described from various Lower Mississippian beds: the Redwall limestone (Easton and Gutschick, 1953) and the Madison limestone of western North America.

The most abundant fossils in the Ordovician-Silurian carbonate rocks of the western states are corals, both rugose and tabulate (Figure 5.31). Duncan (1956) declared that at times these corals must be depended on exclusively to differentiate Upper Ordovician from Silurian strata and, on occasion, Ordovician-Silurian from certain coralline-rich Devonian dolomites.

TERMINOLOGY FOR RUGOSE CORALS (*Abbreviated from Moore, Hill, and Wells*)

If a term applies to any of the following orders, it is indicated by a corresponding letter: R — Rugosa, S — Scleratinia, T — Tabulata. A single letter denotes application of a given term only to the indicated order.

ALAR SEPTUM. One of two protosepta located about midway between cardinal and counter septa (R).

CALICE. Oral surface of corallite; generally bowl-shaped (R, S) (Figure 5.24).

CARDINAL SEPTUM. Protoseptum in the plane of bilateral symmetry of a corallite; insertion of newly formed metasepta occurs adjacent on both sides (R).

COLUMELLA. Solid or nonsolid calcareous axial structure; may project into calice as a calicular boss (R, S) (Figure 5.29, part F).

CORALLITE. Exoskeleton formed by an individual coral polyp (R, S, T).

CORALLUM. Exoskeleton of a coral colony or a solitary coral (R, S, T).

COUNTERLATERAL SEPTUM. One of two protosepta that adjoin counterseptum on either side (R).

DISSEPIMENT. Small domed plate formed by a cystlike enclosure in the peripheral region of a corallite (R, S, T).

DISSEPIMENTARIUM. Peripheral zone of a corallite interior occupied by dissepiments (R, S) (Figure 5.28, E).

EPITHECA. Sheath of skeletal tissue laterally surrounding a corallite (R, S, T).

FOSSULA. Interseptal space distinguished by its unusual shape and size (types: alar, cardinal, counter, closed, open) (R) (Figure 5.24).

HOLOCANTH. Trabecula consisting of a clear rod of calcite (R).

INCREASE. Addition of corallites to colonies by budding (see text for types) (R, T).

MAJOR SEPTUM. One of the protosepta or metasepta (R, S).

MARGINARIUM. Peripheral part of a corallite interior that consists of either dissepiment or dense deposits called the *stereozone* (R, T).

METASEPTUM. One of the main septa of a corallite other than the protosepta, generally extend axially beyond minor septa (R, S).

MINOR SEPTUM. One of the relatively short septa commonly inserted between adjacent major septa (R, S).

MONOCANTH. Simple trabeculae with fibers related to a single center of calcification (R).

OFFSET. New corallite in corallum formed by budding (R, S, T).

OPERCULUM. Lidlike covering of calice in some corallites (R).

PROTOSEPTUM. One of the first six septa of a corallite (R, S).

QUADRANT. Space in the interior of a corallite bounded by cardinal and alar septa (R).

RHABDOCANTH. Trabecula characterized by shifting centers of fibrous growth grouped around a main one (*R*).

SCLERENCHYME. Calcareous tissue of corallites; refers to skeletal parts markedly thickened (*R*).

SEPTAL GROOVE. An external longitudinal furrow corresponding in position to a septum on the inner side of the wall (*R*).

SEPTUM. Radially disposed longitudinal partition of corallite (*R, S, T*).

STEREOME. More or less dense calcareous skeletal deposit that covers and/or thickens various parts of corallite (*S*).

STEREOZONE. Area of dense skeletal deposits in a corallite; peripheral or subperipheral in position (*R, S, T*) (Figure 5.24*F, G*).

TABELLA. Small subhorizontally disposed plate in the central part of a corallite forming part of an incomplete tabula (*R, T*) (Figure 5.24*E*).

TABULA. Transverse partition of corallite, nearly plane or upwardly convex or concave, extending to outer walls or occupying only central part of corallite (*R, S, T*) (Figure 5.29*A*, part *b*).

TABULARIUM. Axial part of the interior of a corallite; tabulae are developed (*R, S, T*) (Figure 5.28*H*).

TRABECULA. Pillar of radiating calcareous fibers comprising skeletal elements in the structure of septum and related components (*R, S, T*, and the heterocorals). (See Figure 5.24*H–K* for types.)

Substrate and Depth

Hill (1948) recognized three major ecological groupings of Carboniferous corals (chiefly Rugosa):

1. Cold, deep, or murky seas; argillaceous-arenaceous limestones. (Small solitary rugose corals, nondissepimented—a "*Cyathaxonia*" fauna.)

2. Warm, shallow, clear seas; limestones, analogous to those where modern reef corals live. [Compound Rugosa and Chaetetida (see later section).]

3. Warm, shallow, clear seas, somewhat intermediate between (1) and (2). (Large, solitary Rugosa with dissepiments; caniniids and clisiophyllids chiefly.)

Evolution

Zaphrentites. Carruthers' study of *Zaphrentites* (that is, *Zaphrentis* of Carruthers, 1910) traced morphological changes vertically through some 4000 ft of the Scottish Lower Carboniferous. The rugose coral populations at different stratigraphic horizons formed a continuous series (from base of section to top) with several intergradations ranging from the geologically oldest *Z. delanouei* to *Z. parallella*, then to *Z. constricta*, and finally several horizons with *Z. disjuncta* (geologically youngest).

The chief variations were confined to two major structures: *cardinal fossula shape* (a tendency to narrow or constrict through time) and *length of major septa* (a tendency to shorten from the center toward periphery through time).

Most impressive to Carruthers and subsequent observers was the prevalence of *tachygenesis*. Tachygenesis designates the phenomenon in which descendant corals (higher up in the rock column) pass through growth stages in their youth that had characterized adults of the parental species (lower in the section). Thus, for example, *Z. parallella* went through a *Z. delanouei* stage in youth, and *Z. constricta* passed through both a *Z. delanouei* and a *Z. parallella* growth stage.

Longitudinal and transverse thin sections for petrographic study or cellulose peels of transverse sections are essential for detailed study of the internal structure of the Rugosa. With such data, evolutionary studies can be undertaken.

Evolutionary Origins and Linkages

The three rugose suborders are thought to have arisen in the following way. Streptelasmatina arose from the oldest rugose genus *Lambeophyllum* (Figure 5.27*A*); members of the Columnariina are possible descendants of *Columnaria alveolata*; those of the Cystiphyllina are possible derivatives of the Ordovician genus *Tryplasma* (Figure 5.29, part *H*).

The last survivors of the Rugosa in Permian time (Artinskian) were the Cyathaxoniicae, where the protosepta developed more strongly than the remaining septa. In this superfamily the family Polycoeliidae may have given rise directly to the earlier Scleratinia.

Hill (1956) italicized three "scleratinian" trends in the evolution of the Rugosa: predominance of protosepta over remaining septa; trend toward tertiary septal insertion between major and minor septa; and, finally, a trend to develop perforate septa (which are common in the Scleratinia).

Geochronometry[5]

Faul (1943) noted what he called "timed corals," that is, forms that in thin sections showed a periodic decrease in dissepiment size and an associated high development of carinae with bunching of tabulae. Since reef corals grow faster in summer and more slowly in winter, he reasoned that the denser parts of the coral skeleton (the constrictions) represented winter growth, and the unconstricted parts, summer growth. From these data he found the growth rate of *Hexagonaria* (formerly *Prismatophyllum*) to be about 1/2 to 1 cm per year. Some species of this genus from the Middle Devonian of Michigan were calculated to have lived more than 100 years.

UNRESOLVED PROBLEMS IN RUGOSA

The relationship between the geologically older order Tabulata (Chazyan, early Middle Ordovician) and the younger Rugosa (Black Riveran) is obscure. The two orders share some characteristics in common with scleratinians, such as skeletons of fibrous calcium carbonate, which are similar down to trabecula arrangement of fibers in the horizontal and vertical elements.

[5]See Appendix 1A and Figure 5.32.

Rugosa and Tabulata both display a comparable evolutionary trend in the development of the marginaria (Hill and Stumm, 1956). However, the Tabulata differ from the Rugosa in septation, presence of mural pores, and in other ways. It is in beds of Black Riveran time, in biofacies containing both tabulate and rugose corals, that one must search for new data on this question of affinities (Hill, 1956).

TABULATA

Introduction

Fossil corals assigned to the extinct order Tabulata, represented today in distinctive biofacies in the rock record, were once important as reef-builders in Paleozoic seas. However, to decipher data encoded in members of an order, it is necessary to establish first whether the order properly embraces related organisms. The lack of living representatives complicates the situation.

All of the following genera, once classed under the Tabulata, have since been reassigned: *Millepora* (Figure 5.4, VIII, IX hydrocorals), *Pocillopora*, *Seriatopora* (scleratinian corals), *Heliopora* (octocorals), *Fistulipora*, *Stenopora* (bryozoans), *Labechia* (stromatoporoids).

The matter is still more involved, as will be discussed later.

Taxonomy: Order Tabulata

Definition

1. *Corallum.* Compound (Figure 5.35*A–F*, for types).
2. *Corallites.* Slender, mostly small (Figure 5.35*A*).
3. *Longitudinal structures.* Septa, weak or absent (family Chaetetidae, aseptate); where present, short, equal, often twelve in number; each septa typically acanthine (spinose), that is, consists of a vertical series of thorns called trabeculae (Figure 5.35*H, I*).
4. *Intercorallite communication.* By means of mural pores on walls (Figure 5.35*I*), connecting tubes (Figure 5.35*E*), or irregular perforations.

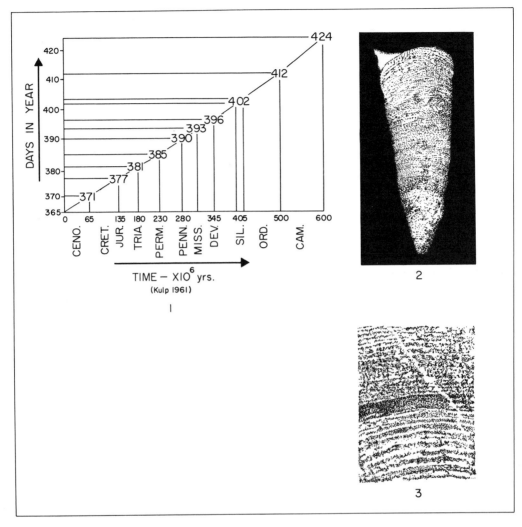

FIGURE 5.32 Corals and geochronology (after Wells, 1963).
(1) Decrease in numbers of days of the year as estimated from coral growth-line counts. (2) *Holophragma calceoloides* (L. Sil., Gotland), ca. 240 days old. Note epithecal growth lines. Enlarged. (3) Growth lines in Recent Coral (*Manicina areolata*, Dry Tortugas), enlarged. See Case Histories.

5. *Tabulae.* These are characteristic of the order; they are complete or, in some forms, funnel-shaped (Figure 5.35*A, H*).

6. *Transverse structures.* Extratabularial tissue is a coenenchyme in many genera (that is, a common marginarium composed of such transverse elements as dissepiments) (Figure 5.35*H*), small, domed plates or shallow, saucered plates (= *sola*) (Figure 5.35*C*).

7. *Geologic range.* Ordovician to Permian, (?) Triassic to (?) Eocene. Post-Paleozoic species assigned to the Chaetetidae may be coralline algae. These species are uncertain

(Hill and Stumm, 1956). Accordingly, the Tabulata may be regarded as essentially a Paleozoic order.

The order has been subdivided into six families in the Treatise, and that arrangement has been followed here.

Family 1. Chaetetidae. Massive coralla composed of extremely slender aseptate corallites with imperforate walls and complete tabulae. Middle Ordovician to Permian; (?)Triassic to (?)Eocene.

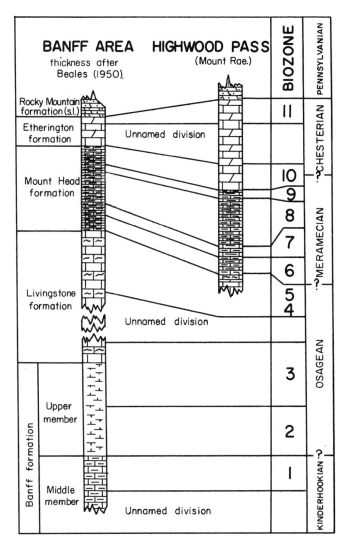

1. Lithostrotionella jasperensis
2. Lithostrotionella micra
3. Lithostrotion mutabile
4. Lithostrotionella bailliei x
5. Lithostrotion sinuosum
   Lithostrotionella bailliei, L. shimeri,
6. L. pennsylvanicum, Lithostrotion warreni,
   L. arizelum, L. sinuosum. ✳
   Lithostrotionella astroeiformis, L. shimeri,
7. L. pennsylvanicum, L. americana(?), L. banffense,
   Lithostrotion warreni.
8. Lithostrotion whitneyi.
   Lithostrotionella astroeiformis.
9. Lithostrotion arizelum
10. Lithostrotion genevievensis
11. Lithostrotionella stelcki;

✳ Species occurs in lower beds of zone

x Species occurs in upper beds of zone

Vertical Scale (feet)
400
300
200
100
0

FIGURE 5.33 Mississippian lithostrotionid zones (Canada) and stratigraphic correlation between two localities (after S. J. Nelson, 1960). The two localities are less than 50 miles apart.

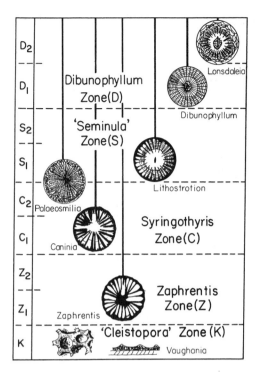

FIGURE 5.34 Carboniferous coral zones and brachiopod/coral subzones, British Isles (after L. D. Stamp, 1950. An Introduction to Stratigraphy British Isles. Thomas Murby & Co., London, 356 p., Neaverson, E., 1955, Stratigraphical Paleontology, Oxford Univ. Press. 739 p., indices, 18 pls., based on Lang et al.). Subzones; K_1—*Productus bassus*, K_2—*Spiriferina octoplicata*, Z_1—*Spirifer tornacensis*, Z_2—*Zaphrenthis konincki*, C_1—*Caninia cylindrica*, C_2—*Palaeosmilia* ϕ, S_1—*Caninia bristolensis*, S_2—*Productus corrugatohemisphericus*, D_1—*Dibunophyllum bourtonense*, D_2—*Lonsdaleia floriformis*.

The family subdivides into three subfamilies. These include subfamily Lichenariinae (see *Lichenaria*, Figure 5.36*A*, 2), subfamily Tetradiinae (see *Tetradium*, Figure 5.36*A*, *1*) and the subfamily Chaetetinae.

Family 2. Syringophyllidae. Massive; septa short, thick (but may be thin), spinose, and of equal number in a given species; mural pores (if present) interseptal, arranged in horizontal rows, each above a tabula; coenenchyme, if present, formed by intersection of septal and tabular extensions. Middle Ordovician to Lower Silurian (two subfamilies). See *Nyctopora*, Figure 5.36*B*.

Family 3. Heliolitidae. Massive coralla with slender tabularia separated by coenenchyme; each tabularium with twelve equal spinose septa and with complete tabulae. Middle Ordovician to Upper Devonian (four subfamilies). See *Heliolites*, Figure 5.36*C*; *Propora*, Figure 5.35*C*.

Family 4. Favositidae. Massive; typically without coenenchyme; corallite slender, with mural pores; septa short, equal, spinose, of variable number; tabula complete. Upper Ordovician to Permian; (?) Triassic (six subfamilies). Subfamily Favo-

FIGURE 5.35 Tabulate coral. Morphological features.
(*A*) Cerioid type: *Favosites* sp. (Helderbergian, M. Dev., N.Y.S.), Max. corallite length, ca. 35.0 mm.
(*B*) Meandroid type: *Desmidopora alveolaris* (Sil., Eng.), 8.0 mm wide.
(*C*) Coenenchymal type: *Propora tubulata* (Sil., Europe), ca. 7.5 mm wide.
(*D*) Cateniform type: *Halysites nexus* (M. Sil., Ky); fragment, 70.0 mm wide.
(*E*) Fasciculate type: *Syringopora ehlersi* Thunder Bay Ls., M. Dev., Mich. (a) Exterior section; (b) longitudinal section.
(*F*) Ramose type: *Thamnopora* sp.
(*G*) *Michelinia*. (a) *M. harkeri* (U. Penn?); (b) *M. tenuisepta* (L. Carb., Belgium).
(*H*) *Striatopora*. (a) / (b) *S. ornata:* (a) Surface section; (b) longitudinal section.
(*I*) *Alveolites labechei*, Sil., Eur. (a) Transverse section; (b) longitudinal section.
(*J*) *Heliolites porosa*, M. Dev. cosmop. (a) Transverse section; (b) longitudinal section.
Symbols: *t*—tabulae, *o*—offset, *pt*—prismatic tube, *mp*—mural pores, *coen*—coenenchyme, *s*—septum, *cor*—corallite, *it*—incomplete tabulae, *p*—palisade arrangement, *ct*—connecting tubule (also called transverse stolon), *ti*—infundibuliform tabulae, *c*—calice, *ma*—marginarium, *tr*—trabeculae, *tm*—tabularium, *ss*—spinose septa, *ct*—coenenchymal tubules, *ds*—dissepiment. Sources: (*A*) Tasch; (*B, C, G, H, J*) Hill and Stumm, 1959; (*D*) Buehler, 1955; (*E*) Watkins, 1959. See Figure 5.36 for Ref. (*G*) Nelson, 1962.

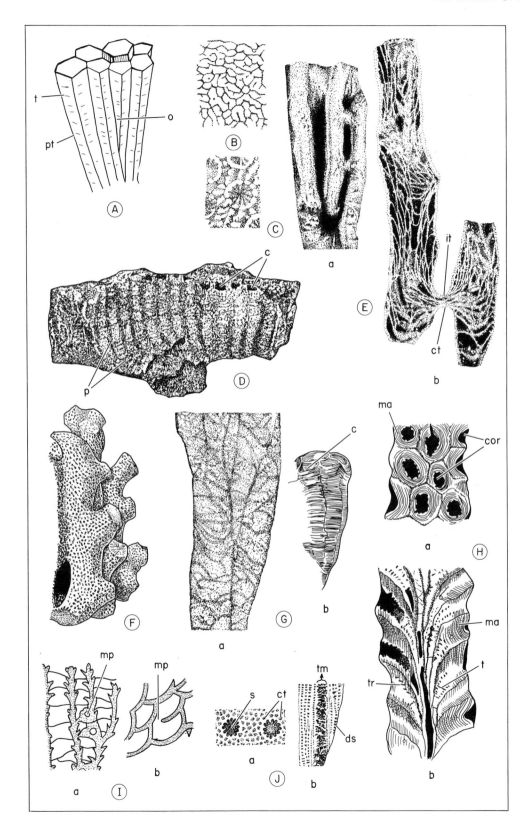

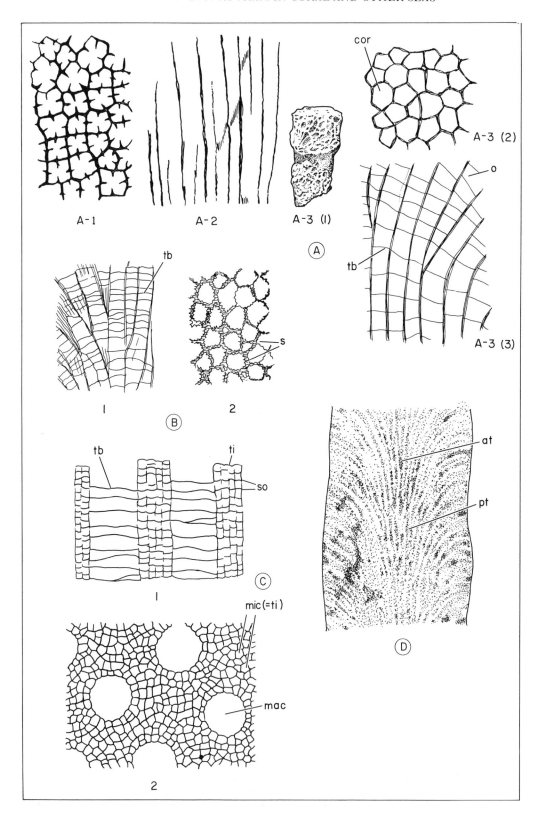

FIGURE 5.36

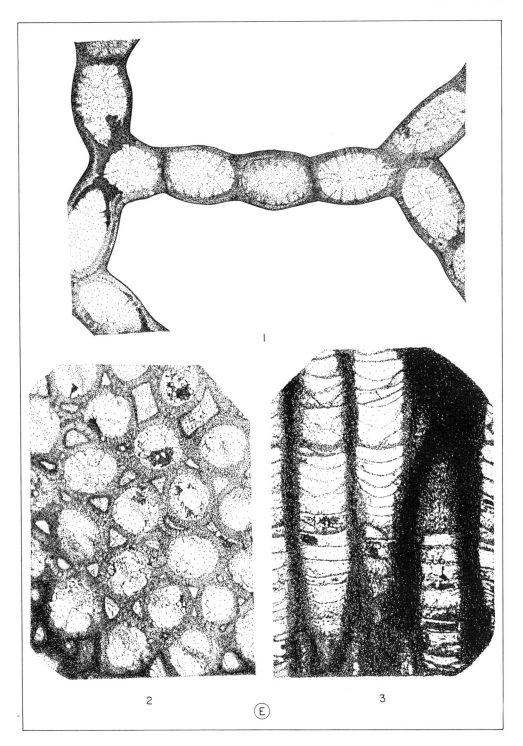

1

2

3

Ⓔ

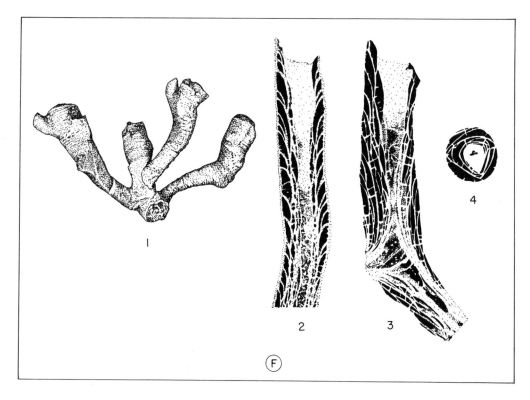

FIGURE 5.36 Representative genera of six tabulate coral families.
(*A*) Chaetetidae: *Tetradium fibratum,* M. Ord., N.A. (1) Transverse section showing quadripartite increase (4 offsets shown); (2) longitudinal section (length ca. 12.0 mm), showing erect corallites;
(3) *Lichenaria typa,* M. Ord., N.A. (1) Surface view of fragment (length 7.0 mm); (2) *Lichenaria* sp.
(Nevada), transverse section showing increase (compare size of corallites; smaller are offsets of
larger); (3) longitudinal section. (*B*) Syringophyllidae: *Nyctopora billingsi,* U. Ord., Texas. (1) Longitudinal section (length 8.0 mm); (2) transverse section showing short, spinose septa (note periodic
spacing of tabulae, *tb,* in ''(1)''.
(*C*) Heliolitidae: *Heliolites spongiosus,* Sil. (1) Longitudinal section showing the macrocorallites
(*mac*) and microcorallites (a reticulum of tubuli−*ti*), and sola (*so*), length, 7.0 mm; (2) transverse
section showing nature of tubuli reticulum surrounding macrocorallites.
(*D*) Favositidae: *Coenites juniperinus,* M. Sil., Gotland; longitudinal view with axial (*at*) and periaxial
tabulae (*pt*); X 4.
(*E*) Halysitidae: *Catenipora gracilis,* Sil., Canada. (1) Thin section showing fragile septal spines.
(2) *Halysites compactus,* Sil., Mich., horizontal section. (3) Vertical section, same specimen. All X 10.
(*F*) Auloporidae: (1) *Aulocystis jacksoni,* Wanakah shale (M. Dev., western N.Y.). Exterior view of
dendroid corolla with trumpet-shaped corallites, width ca. 6.0 mm; (2) *Aulocystis multicystosa* (M.
Dev., Mich.), longitudinal section showing cysts length, ca. 23 0 mm; (3), (4), *Aulocystis magnispina*
(M. Dev., Mich.), (3) longitudinal section showing well-developed spines on walls and cysts (these
spines represent septa); (4) transverse section.
Illustrations redrawn and adapted from: (*A*) Hill and Stumm, 1956; Duncan, 1956; (*B, C*) Duncan,
1956; (*D*) Stumm, 1961; (*E*) Buehler, 1955; (*F*) Watkins[1.]
Symbols: *cor*−corallite, *tb*−tabularia, *o*−offset, *s*−septa, *ti*−tubuli, *mac*−macrocorallite, *mic*−mi-
crocorallite, *so*−sola. [1]J. L., 1959. Jour. Paleontology, Vol. 33 (5), pp. 793−808, pls 108−111.

sitinae (*Favosites,* Figure 5.35*A*; *Desmidopora,* Figure 5.35*B*); subfamily Thecinae;
subfamily Pachyporinae (*Thamnopora,* Figure 5.35*F*; *Striatopora,* Figure 5.35*H*); subfamily Alveolitinae (*Alveolites,* Figures
5.35*I*, *Coenites,* Figure 5.36*D*); subfamily

Michelininae (*Michelinia,* Figure 5.35*G*);
sub-family Palaeacinae.

Family 5. Halysitidae. Phaceloid coralla (Figure 5.25, part *A*) composed of
cylindrical, oval, or subpolygonal coral-

lites united with one another along two or three sides, producing an anastomosing chainlike network. Corallite walls imperforate; microcorallites (= mesocorallites) are present between corallites in some forms (Figure 5.36E, *3*). Horizontal or arched tabulae are more closely set than the microcorallites. Some forms have twelve vertical rows of septal spines in the corallites. Ordovician to Silurian (*Halysites*, Figure 5.36E, *2–3*; *Catenipora*, Figure 5.36E, *1*).

Family 6. Auloporidae. Coralla compound, dendroid to phaceloid or reptant (that is, with creeping habit). Corallites cylindrical to trumpet-shaped; increase by lateral gemmation; transverse stolons (see *Syringopora*, Figure 5.35E) connect some forms; walls, solid; septa represented by spines, peripheral ridges — spines arranged in vertical rows; tabulae, if present, horizontal, convex or concave, closely or widely spaced, broken into tabellae in some forms; cysts present in a few genera (*Aulocystis*). (?) Silurian, Devonian to Permian. Two subfamilies Auloporinae (*Aulocystis*, Figure 5.36F, *1–4*), Syringoporinae (*Syringopora*, Figure 5.35E).

Skeleton

The tabulate skeleton is similar in crystalline structure to that of the Rugosa (Bryan and Hill, 1941).

INFORMATION DERIVED FROM TABULATE CORALS

Paleoecology

The three chief indicators of the essential conditions under which Paleozoic corals lived are corallum form, enclosing sediments, and composition of reef faunas. Hydrozoans of modern reefs are represented in Paleozoic reefs by stromatoporoids. When Vaughan (1911) wrote on this subject, the archeocyathids were considered to be simple corals; presently recognized tabulates (*Halysites* and *Heliolites*) were con-

sidered octocorals; and the alga, *Cryptozoon*, a stromatoporoid.

Vaughan (1911) and Wells (1957) reasoned that the following conclusions pertained to the Paleozoic corals as well as Recent.

1. Most Paleozoic corals (Rugosa and Tabulata) lived in ecological niches similar to those occupied by modern lagoonal reef corals.
2. *Depth.* Shallow; maximum depth about 50 meters, that is, within the influence of surface waves. To sustain this reasoning, Vaughan (1911) cited the evidence of coral breccia on Paleozoic reef flanks (see also Lowenstam, 1950; Lecompte, 1954).
3. *Light penetration.* Strong; corals lived in lighted (euphotic) zone.
4. *Temperature.* Annual minimum, 16°–21°C.
5. *Water.* Well oxygenated, agitated, and gently circulating.
6. *Substrate.* Bottom clean or relatively free of silt accumulation. Presence of silt would prevent planula larvae attachment.

Fossil Record

The prominent record of tabulate corals as reef-builders began in the Ordovician (Chazyan) of North America. The chaetetids, such as *Tetradium*, were very common. Tabulate-bearing reefs occurred throughout the Ordovician and are known from Alaska in the west, Baffin Island in the east, to Texas in the south.

Favosites and *Heliolites* were among the tabulates and other forms that built the Silurian (Niagaran) reefs of the Great Lakes area (Lowenstam, 1949). The Favositidae were dominant during the Silurian and most of the Devonian.

Although reef building by Carboniferous-Permian tabulates was insignificant, some tabulates were locally abundant. Examples are *Syringopora*, Mississippian, South Canadian Rockies (Figure 5.35E) and Pennsylvanian-Permian, midcontinental United States (Jeffords), and *Chaetetes*, midcontinental Pennsylvanian (Desmoinesian).

The triad of stromatoporoids, rugose corals, tabulates recurred in large numbers of reefs around the world (Lowenstam, 1950; Lecompte, 1954).

In a given formation and/or region, a particular coral may be very common. *Halysites gracilis* is prominent in the Red River formation (Selkirk member), Upper Ordovician of Manitoba (Leith, 1944); and *Acaciapora austini* is the dominant form in the Holdenville shale (Pennsylvanian of Labette County, Kansas) (West, 1964).

Stratigraphy

Tabulate and rugose corals often happen together in the same bed or zone, although one or the other may dominate. The four coral zonules of the fusulinid-bearing Nevadan Permian (Wolfcampian-Leonardian) are thought to be regional in extent, that is, characteristic of the Cordilleran geosyncline. Considered as a whole, this same fauna resembles that of the Uralian geosyncline and the Moscow Basin. Most of the fauna consists of massive rugosans, but the tabulate *Syringopora* is prominent in Zone 2 whereas *Cladochonus* (Family Auloporidae) occurs in Zone 1 (Wilson and Langenheim, 1962).

Figure 5.37 illustrates the utility of spe-

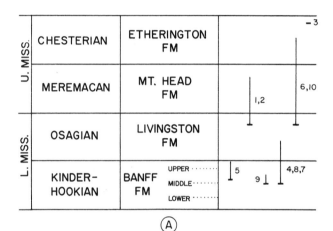

(A)

Syringopora sp.	Corallite Diameter (mm)	Number of Corallites per 100 sq mm	Other Features
1. S. bassai	3.6–4.3	1.5–4	it
2. S. dingmani	1.9–2.3	3–12	it
3. S. drummondi	2.4–2.8	7–12	"vt" to "it"
4. S. harveyi	1.3–2.0	6–15	it
5. S. harveyi forma agglomerata	1.6–2.0	17–22	it
6. S. magnussoni	2.6–3.3	3–6	"it" to "vt"
7. S. rudyi	3.0–3.5	2.5–4.0	it
8. S. surcularia	2.5–2.9	4–10	it
9. S. surcularia forma compacta	2.2–2.3	12–14	it
10. S. virginica	1.5–2.2	4–15	vt

(B)

FIGURE 5.37 (A) Tabulate coral species of *Syringopora* as stratigraphic indicators in the Mississippian Southern Canadian Rockies (data from S. J. Nelson, 1962a). (B) Numbers 1–10 corresponding to distinct *Syringopora* species). Chief characteristics used to distinguish *Syringopora* species and to determine their stratigraphic position. Note changes in corallite diameter and in frequency (spacing) of corallites: 17 to 22 per 100 sq mm, for example, denotes 17 to 22 individual corallites. (Compare (A).) Data from S. T. Nelson (1962a).
Symbols: *it* — infundibular tabulae, *vt* — vesicular tabulae.

cies of the tabulate coral *Syringopora* in the stratigraphic zonation of the Mississippian in the South Canadian Rockies (S. J. Nelson, 1962a). Other rarer tabulates, such as *Favosites* and *Michelinia*, are limited in stratigraphic occurrence in the Pennsylvanian-Permian of the North Yukon Territory; and the tabulate *Pleurodictyum* also occurs in the Banff formation in a limited vertical spread (S. J. Nelson, 1962b, text figure 1).

The chief characteristics that could be used to distinguish species of *Syringopora* are diameter of the coralla and number of corallites per 100 sq mm. Nelson (1962a) figured that species of this genus could be identified in the field by use of a caliper to measure corallite diameter and a card-

board or plastic cutout, 100 sq mm, against which to count the number of corallites occurring in that area in a given specimen. Use of the data in Figure 5.37 would then provide the required comparative information.

Reefs, Coral Seas, and Oil Accumulation

The reef area shown in Figure 5.40 illustrates what must have been a coral sea in Silurian time. These reefs vary in size both horizontally and vertically; some extend for a few feet, others for as much as 3.5 miles; some are a few feet high, others nearly 1000 ft. As seen in Figure 5.38, many reefs were discovered in the subsurface during the course of oil exploration. The famous reef at Thornton, just south of

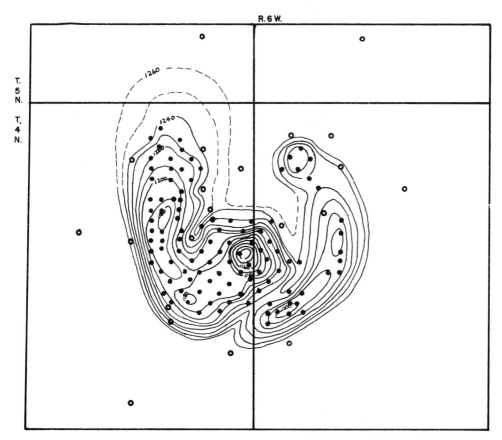

o DRY HOLE

• PRODUCING WELL

⌇⌇⁄¹²⁰⁰⌇⌇⌇CONTOUR ON TOP OF SILURIAN; INTERVAL 10 FT.; DATUM SEA-LEVEL

FIGURE 5.38 Buried Niagaran coral reef. Oil well production shown. Madison County, Illinois. (Reef ca. 6 square miles). After Lowenstam, 1950.

Chicago, Illinois, is a good illustration of a surface reefal occurrence. This reef has a diameter of greater than 1 mile.

One of the subsurface reefs shown in Figure 5.38 covers about 6 square miles and is about 500 ft thick. Most of the wells shown in the last-named figure obtain oil from the Niagaran reef body itself. Other wells retrieve oil from Devonian and Mississippian beds above the Niagaran reefs (Lowenstam, 1949). A core that penetrated one such Niagaran reef in Clinton County, Illinois, is shown in Figure 5.39. The specimen (*Favosites* sp.), part of a reef body, was encountered at a depth of 2242 ft below the surface.

The Leduc Oil Field, Alberta, Canada (Layer, 1949), had two main producing zones (D-2, D-3) in Upper Devonian dolomites that were from 500 to 900 ft and from 4850 to 5400 ft below the surface. The D-3 zone was a coral reef (bioherm) in Devonian time. Bryozoans and brachiopods were reef dwellers; the D-2 zone, also richly coralline, contained fragments of similar forms. However, the D-2 zone was bio-

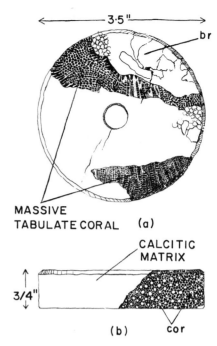

FIGURE 5.39 Silurian reef core, Clinton County, Illinois. (a) (b) *Favosites* sp. in brecciated carbonate matrix penetrated 19 ft below reef top that occurred 2223 ft below the surface. (Gulf No. 4, Warnicke, 6–2N-3W.) (a) Top view of core; (b) side view.

Symbols: *cor*—corallites, *br*—carbonate breccia (Tasch collection).

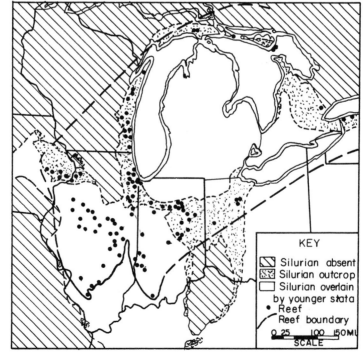

FIGURE 5.40 Silurian reefs, Great Lakes area (after Lowenstam, 1949).

stromal, that is, a blanket-type deposit. The Leduc Field, which embraced the two oil-producing horizons, covered an area of about 8100 acres. An estimated 100 million barrels constitutes the recoverable oil reserve contained in the two zones.

Deciphering reef trends from subsurface data is an important research tool in oil exploration, and the illustrations reviewed above (Silurian and Devonian reefs) explain why this is so. Corals may be a major or a minor component in such reefs, or may become dominant at a later stage of reef growth (for example, *Syringopora* is prominent in the initial stages of the Niagaran reefs) and then be succeeded by some other faunal elements. Subsequently corals may reappear in the same reef as minor components. Associated biotas include stromatoporoids, brachiopods, bryozoans, crinoids, algae.

Environmental Controls for Reef Trends

Numerous investigators have mapped reef trends. Hadding (1950), for example, noted that the Silurian reefs of Gotland occurred in SW-NE trending belts, and formed barrier reefs east and southeast of the coast. Lowenstam (1950) projected in some of the Niagaran reefs (Elmhurst Reef), a NE-SW alignment of both the elongated axis of individual colonies and of their spacing. Elsewhere reefs had an en-echelon arrangement.

Lowenstam (1950) proposed *current control* in the selection of a building site by planula larvae. Since relatively clean bottoms were necessary for planula attachment, the same investigator thought that in the Elmhurst reef, a transition from quiet to semiquiet water could have slightly increased circulation. In turn, such water movement would have allowed fewer land-derived clastics to settle on the bottom, thus permitting *Syringopora* larvae to attach.

Once wave-resistant reefs were formed, these structures themselves influenced both currents and wave motion.

A whole array of ecological factors— such as temperature, light penetration, general absence of turbidity, carbonate in solution—occurring in shallow-water zones established the essential "fence" conditions in the Paleozoic seas that delimited belts where planula larvae might attach.

Planula larvae seasonally must have reached enormous figures (see Scleratinia, this chapter). Those that wandered outside the ecologic fence did not survive; others, within the "fence," failed to find adequate attachment sites. Only those larvae that attached at sites favored by current control, lowered rate of sedimentation, and other factors, survived and gave rise to colonies that formed reef structures.

Reef trends may then be attributed to regional and local controls: ecologic fence, regional control; attachment site, local control.

Seasonal Growth

In tabulates, as in rugosans, there appears evidence of seasonal growth. Thus Duncan (1956) observed that in *Nyctopora* (Figure 5.36*B*) the tabulae were essentially horizontal and *periodically* more closely spaced. Rather than accept this feature as a diagnostic character, Duncan thought it more likely indicated seasonal growth.

Commensalism and Symbiosis

Several investigators have recorded examples of commensalism and/or symbiotic relationships in tabulate corals (Clarke, 1908; Yakovlev, 1926; Gerth, 1952; Watkins, 1959). Commensalism (*cum* = together; *mensa* = table) is a beneficial association between two or more individuals. A symbiotic relationship (*sym* = together; *bios* = life) exists when two organisms of different species interrelate in a beneficial way. The Devonian tabulate *Pleurodictyum* had a wormlike inclusion (a sipunculid worm, *Hicetes*; see Chapter 9), and this relationship was interpreted to be one of commensalism as in the case of Tertiary-to-Recent scleratinian corals. Yakovlev reported a symbiotic relationship of the tabulate *Aulopora* and several Devonian-Carboniferous brachiopods. *Aulopora* is frequently found attached to dorsal valves of brachiopods: *Spirifer*, *Cyrtia*, *Atrypa*, and others. Did the coral's nematocysts afford protection to these brachiopods? Did the coral feed from water currents generated by the brachiopods? Yakovlev answered in the affirmative to both of these questions and designated the relationship as "symbiosis." (Figure 5.41).

Dimorphism

In tabulate corals individuals are seen to be of the same size or unequal in size. From a study of *Favosites*, Jones (1936) suggested an explanation. External environmental influences could have affected the rate of upward growth and the rate of budding. Slow growth and rapid budding resulted in unequal corallites. Normal growth and moderate rate of budding yielded corallites of the same size.

LIVING SCLERATINIANS

Introduction

The oldest scleratinians are known from rocks of Middle Triassic age. Thus for approximately 210 to 220 million years up to the present day, almost all zooantharian corals, both solitary and colonial, have

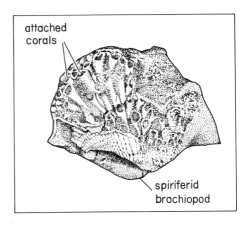

FIGURE 5.41 Commensalism or symbiosis? *Aulocystis commensalis* (family Auloporidae) attached to brachial valve of spiriferid brachiopod (M. Dev., Michigan). After Watkins, 1959. See Figure 5.36 for Ref.

been scleratinians. (The Rugosa and, effectively, almost all of the Tabulata were extinct by the close of the Paleozoic). As builders of coral reefs and islands and as dwellers in deep seas, they have engaged an army of researchers since the time of Charles Darwin.

Polyp Anatomy

The scleratinian polyp is cylindrical in form. The *oral end* consists of a disk bearing one or more marginal circlets of retractable hollow tentacles and a central mouth (Figure 5.42B, 1); the *aboral end* is defined by a *basal* or *pedal disk* by which solitary forms attach to the substrate (Figure 5.42C). The portion of the cylindrical

FIGURE 5.42 Scleratinian corals. Polyp and skeleton.
(A) *Flabellum curvatum*, Recent, Scotia Sea. (1) View of skeleton taken above; (2) detail of side of calice in (A) showing upper profile of denticulated septa.
(B) *Flabellum* sp. (1) Diagramatic: Sectioned to show relationship of polyp to skeleton, septa (*s*), mesenteries (*m*), and so on. (2) Schematic: Transverse action along line *x–x'* in (B, 1) to show position of calcareous septa relative to the fleshy platelike mesenteries. The mesenteries alternate in position with the septa (*s*) in life. Likewise, the polyp that rested on the skeleton (A, 1) was constructed like that shown in (B, 1). (C) *Caryophyllia profunda* (Rec., Norfolk Island, New Zealand). Corallum showing massive appearance of septa (*s*), basal plate (*bp*), and costa (*c*).
Symbols: *pe*—pedicel, *s*—septa (called sclerosepta or skeletal septa by some authors), *c*—calice, *p*—polyp, *m*—mesenteries (called septa by some authors, who then employ the term "sclerosepta" for calcareous radial partitions situated between mesenteries), *g*—gullet, *mf*—mesenterial filament, *epi*—epidermal wall, *c*—costae, *co*—ornamented costa (costa are septal extensions outside the wall).
Sources: (A) Squires, 1961; (B) Wells, 1956; (C) Squires, 1960.

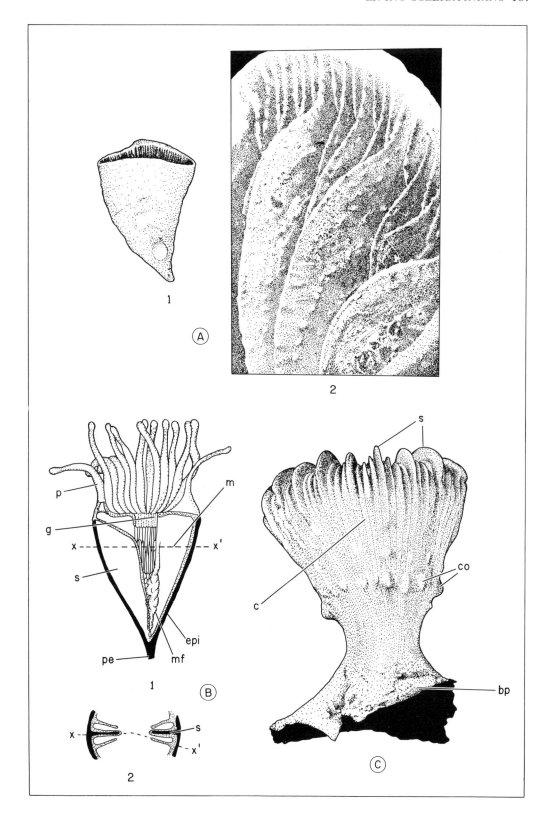

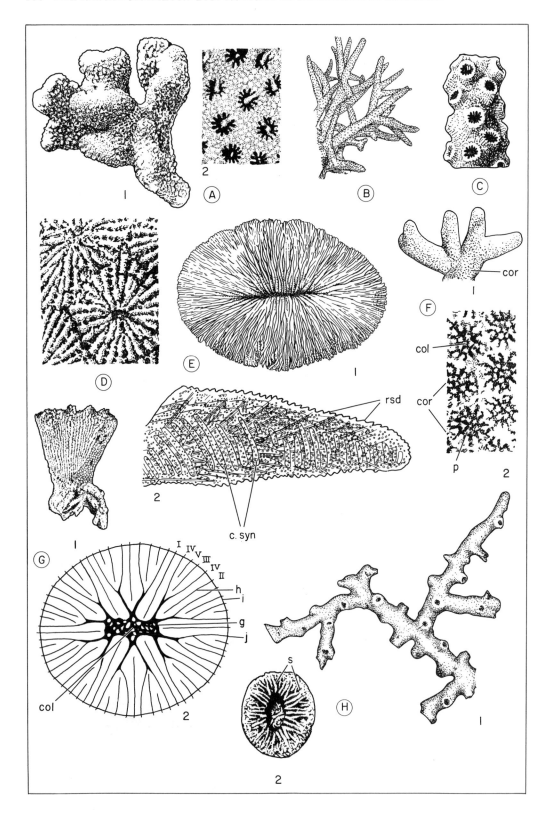

FIGURE 5.43 Genera of representative scleratinian families—I.

(*A*) Suborder Astrocoeniina. *Pocillopora, I. P. elegans,* (Rec., E. Indies), ramose colony bearing calices on short protuberances, X 0.5; 2. *P. egolouxi* (Rec., Torres Str., Australia) — note rudimentary septa; low boss in center of corallites is a columella.

(*B*) *Acropora formosa* (Rec., Formosa), ramose colony; this genus of a hermotypic (reef) coral embraces 40 percent of living scleratinians, or more than 200 species. The axial corallite is larger than the many corallites budded from it (that is, increase by extratentacular budding), X 0.2.

(*C*) *Dendracis mammillosa* (M. Oligo., Italy). A portion of coralla-like *Acropora* but without persistent axial corallite, X 2.

(*D*) Suborder Fungiina. *Siderastrea* (*Siderastrea*) *italica* (Mio., Poland), X 8. Note synapticular rings that define corallite walls; ceroid colonies of this genus formed by extratentacular budding. (The name in parenthesis is that of a subgenus—one of three subgenera in this genus.)

(*E*) The genus *Fungia*—range Miocene to Recent—contains six species groups of which only one is shown. (1) *F. suitari* (Rec., Hawaii), oral surface of solitary corallite showing numerous septa, X 0.5. (2) *Fungia* sp., showing characteristic compound synapticulae (*syn*) and rounded septal dentations (*rsd*); longitudinal section parallel to septal plane. The nature of septal dentation varies from species group to species group in this genus.

(*F*) *Porites* (*Porites*) *porites* (Rec., W. Indies). Ramose colony formed by extratentacular budding; has only two septal cycles; note single columellar trabecula (*col*) and presence of pali (*p*). One or more synapticular rings limit loosely united corallites. *Porites* is next to *Acropora* in importance as a living hermatypic coral.

(*G*) Suborder Dendrophylliina. *Balanophyllia: (1) B. tibusonensis* (Rec., Gulf of Calif.). Solitary, turbinate corallite with expanded base. Note costae and compare with (*G*) 2. (2) *B. elegans.* Diagramatic representation of the calice of an adult corallum. Roman numerals represent septal cycles. Other notations identified below.

(*J*) Note that costae (*j*) correspond to septa (*h*) and are their extrathecal continuation.

(*H*) *Dendrophyllia: (1) D. profunda* (Rec., off Florida), fragment of dendroid corallum, depth 686 meters, X 0.55. (2) *D. californica* (Rec., coast of Lower California), X 2.2. *Dendrophyllia* species are like those of *Balanophyllia* (*Balanophyllia*) but form dendroid colonies by extratentacular budding from the edge zone.

Sources: (*A*)–(*C*) Wells, 1956; (*D*) (*E*) (*F*) Wells, 1956; (*G*) Durham, 1942, 1949; (*H*) 1 Squires, 1959; (*H*)2, Durham, 1947.

Symbols: *coen*—coenosteum, *col*—columella (= g, in G2), *cor*—corallite, *pr*—protuberance, each bearing a calice, *p*—pali, *rsd*—rounded septal dentation, *c, syn*—compound synapticulae, *s*—septa, *syn. t*—synapticula theca, "i" in G2 (= a porous wall composed of one or more series of synapticulae), *cos*—costae.

polyp between the two disks is called the *column*.

The mouth leads to a hollow, interior gastrovascular cavity via a short *gullet* (Figure 5.42*B, 1*). The cavity, partitioned by vertical and radial fleshy mesenteries, alternates septal invaginations of the basal disk (Figure 5.42*B, 2*). Within the invaginations, up-foldings, or in-turnings of the basal disk, calcium carbonate is secreted to form the septa. Exclusive of the mesenteries, three layers of tissue compose the body layer of the polyp: ectoderm, jelly-like mesogloea, and endoderm. Zooxanthellae are common symbionts in the endodermal tissue of reef-dwelling corals. Stinging cells (nematocysts) are present in the outer tissue, the ectoderm. All these features also characterize the planular larval stage.

Much of the biological function of the polyp is conducted by the mesenteries and their filaments (Figure 5.42*B*): digestion, absorption, excretion, and gonad development.

In solitary scleratinian corals, the *edge zone* is defined as that part of the polyp that lies outside the wall; the gastrovascular cavity is continued in it (Figure 5.44 *A*); in colonial forms the edge zone continues between adjacent polyps, and it is then referred to as the coenosarc (Figure 5.44*B*). Structures or deposits by the coenosarc are called coenosteum and include costa, tabular or vesicular dissepiments (Figure 5.44*C, 1*), and sclerodermites (see discussion on skeletal morphology and Figure 5.44*C, 2*).

Skeletal Morphology

The fleshy polyp is enclosed in a hard, calcareous skeleton (*exoskeleton*). The complete skeleton (*corallum*) consists of several

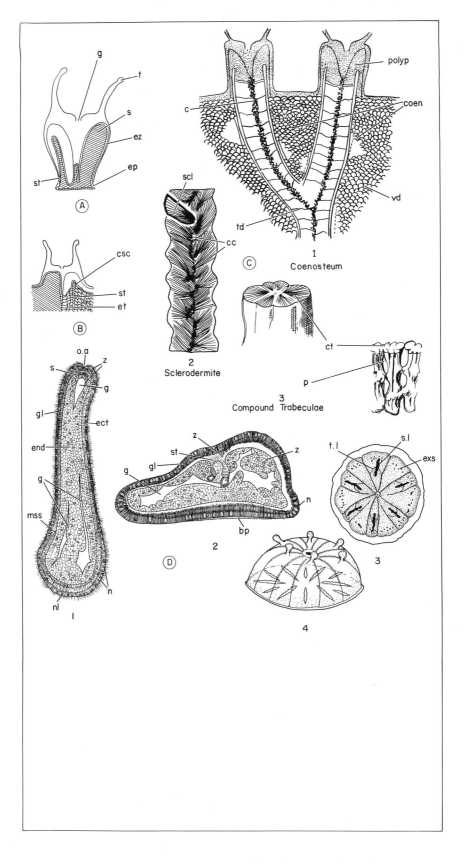

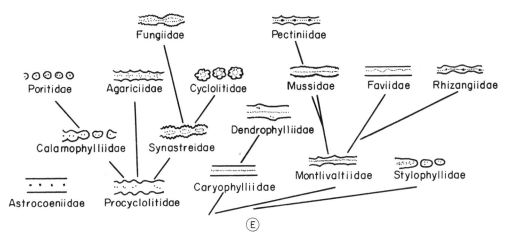

FIGURE 5.44 Septal structures. Major types.

(A) A solitary scleratinian, edge zone in a schematic vertical section.

(B) Colonial scleratinian, coenosarc in a schematic vertical section.

(C) Portion of corallum of *Galaxea:* (1) coenosteum, and so on; (2) sclerodermite; (3) fenestrate septum of *Cyclolites* showing compound trabeculae.

(D) (1–3) *Galaxea aspera*. (1) Planula larva, structures shown when stained, longitudinal section (ca. X 37); (2) larva, just settled, longitudinal section through gullet, 10μ thick (ca. X 52); (3) polyp development after the fourth day of attachment (ca. X 19). (4) *Siderastrea radians* polyp development a few days after fixation. Note the start of the first two cycles of entosepta and exosepta. (See "Terminology"; entosepta occur *within* entocoeles; exosepta in exocoeles.)

(E) Scleratinian septal structures, major types, and interrelationship between several families. The schematic figures represent horizontal sections of septa; groupings of trabecular centers shown by fine dots. To envision the longitudinal view of the same septa, imagine the given two-dimensional figure projected out of the page toward the reader.

Sources: (A)–(C) Wells; (C)2, Olgilvie, 1895; (C)3, Wells; (D)1–3, Atoda; (D)4, Duerden, 1904. (E) Vaughan and Wells, 1943.

Symbols: *t*–tentacles, *g*–gullet, *ez*–edge zone, *ep*–epitheca, *s*–septa, *st*–septotheca (thickened outer parts of septa forming corallite wall), *csc*–coenosarc, *et*–endotheca (dissepiment inside wall of corallum and corallite; compare Figure C, *td*–all outside corallite wall and hence are endothecal), *cor*–corallite, *c*–costa, *vd*–vesicular dissepiment, *td*–tabular dissepiment, *coen*–coenosteum, *col*–columella, *cc*–center of calcification, *scl*–sclerodermite (primary element in septa that consists of *cc* and surrounding fibrous clusters), *p*–perforation, *coel*–coelenteron, *ect*–ectoderm, mss(=mes)–mesogloea, *ect*–ectoderm, *z*–zooxanthellae, *oa*–oral aperture, *n*–nematocysts, *es*–endosepta, *exs*–beginning of exocepta represented by many small skeletal granules, *m*–mesenteries.

structures: the basal plate and its upward extension (*epitheca*) and septa normal to it, the *columella*, a central axial structure, and extrathecal structures (coenosteum) discussed above. Septal and thecal structures are frequently secondarily thickened by dense, calcareous deposits called *stereome*. The fleshy column wall of the polyp is enclosed by a skeletal *wall* or *theca*; this is secondary in origin. (The basal plate, the *initial* skeletal deposit, is followed by the septa, and the up-turned margin of this basal plate forms the epitheca.)

A cicular to prismatic crystals are characteristic for mineral aragonite. In the scleratinian skeleton (as well as in other corals

such as the Rugosa), it has been found that such aragonite fibers are elongate needles. Each such needle is a distinct crystal some 2 μ (microns) in diameter. They are arranged normal to what was originally a colloidal matrix secreted by the ectoderm (Bryan and Hill, 1940).

Atoda (1951*b*) found out that in *Galaxea aspera* on the third day after attachment of the planula larva (Figure 5.44D), septal traces first appeared as small skeletal granules in the endocoele. Such individual granules presumably consist of aragonite needles and represent one or more sclerodermites. This loose structure is lost on the fourth day, when septa develop an

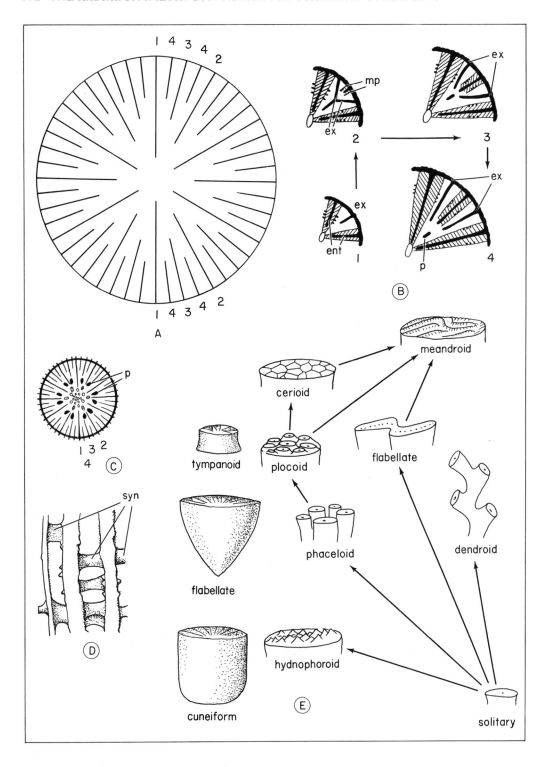

FIGURE 5.45 Septal cycles, substitution, and origin of pali in scleratinians.
(A) Normal order of septal cycles (number 1, first cycle; number 2, second cycle; and so on).
(B) A sequence (1)–(4) representing earlier to later sections of the same corallum and polyp at different levels. (1) Note entocoel with entosepta and exocoel with exosepta in earlier portion of coral. (2) Note bifurcation of the outer end of the exoseptum [compare to earlier developmental stage shown in (1)]; observe the insertion of a mesenterial pair (*mp*) between exoseptal branches, and beginning of an entoseptum. (3) A still later stage in which the internal portion of the exosepta [shown in (3)] remains as a remnant after substitution and formation of pali (*p*) in (4).1.
(C) *Trochocyathus rawsoni* Pourtalis, So. Africa (fam. Caryophyllidae), adult stage showing 12 pali. These arose by substitution of third-cycle entosepta for third-cycle exosepta; (see also (B)1–4. The exosepta have doubled [compare bifurcation in (B)2] forming the fourth septal cycle.
(D) *Microbacia*, portion of, showing septa and simple synapticulae connecting them.
(E) Colonial form of scleratinians depicting trend from solitary to complex colonial types. (A)–(E) modified after Wells, 1956.

irregular rodlike shape. (These are the six radially arranged endosepta.) Similar loose skeletal granules unify to form irregular rodlike septa in the exocoeles on the seventh or eighth day. These exosepta are situated transversely between each of the endosepta.

Ogilvie (1896) described and figured the fine structural element of the scleratinian septa. Clusters of aragonite fibers, noted above, formed fascicles or bundles that emanated from *centers of calcification*. Both centers and clusters together constitute the sclerodermites (Figure 5.44C, 2). Septa and derivative structures, such as the columella, show a pattern of sclerodermites.

A septum has been defined as a "palisade of trabeculae" (Wells, 1956). *Trabeculae*, as already discussed under Rugosa (see Figure 5.24H–K and text), are spines or rods developed by the sclerodermites. (See Figure 5.44E). In scleratinians they may be *simple* (that is, composed of single sclerodermites in series as in *Galaxea*, Figure 5.44C, 2) or *compound* (that is, composed of sclerodermites in bundles as in *Cyclolites*, Figure 5.44C, 3).

The characteristic mesenterial arrangement in scleratinians is shown in Figure 5.45A, B. A *hexameral primary plan* characterizes both mesenteries and septa. The first cycle consists of *protosepta*. These occur in the entocoels, as already noted (Figure 5.45B). The second cycle of six septa takes place in the spaces (exocoels) between the first (Figure 5.44D, 4). All cycles beyond the first are termed *metasepta* and appear one after the other. Thus the third

cycle introduces 12 septa, the fourth 24 septa (see Figure 5.45A), and so on. Eight cycles is generally the maximum.

Septal substitution (Figure 5.45B, 1–4) complicates the normal septal cycle shown in Figure 5.45A. It also gives rise to vertical pillars called *pali*, which are remnants of the substitution process. Pali particularly characterize the scleratinian family Caryophyllidae and occur in a circle or crown (Figure 5.45C). In some scleratinians (particularly the family Dendrophyllidae) exoseptal development in the fourth septal cycle exceeds entoseptal development. This arrangement is characteristic of what is called the *Pourtales plan*. Partial description of the family Dendrophyllidae reads: "Septa inserted following Pourtales plan at least in early stages." The description of the family Caryophyllidae reads in part: "Pali or paliform lobes common." These examples indicate the importance of specific morphological structures in identifying given families. To recall such structures simultaneously brings to mind the families they characterize.

Almost all scleratinians have *synapticulae*. These structures (rods or bars) perforate the mesenteries to connect adjacent septa. Members of the suborder Fungiina have fenestrate septa and are particularly so characterized (Figure 5.43D).

Reproduction

Both sexual and asexual reproduction occur in scleratinian polyps. Marshall and Stephenson (1933) studied the breeding of

Great Barrier Reef corals including forms like *Montipora*, *Acropora*, *Pocillopora*, *Porites*, *Favia*. Sections through the fertile region of *Favia doreyensis* showed that ova and testes occurred on the same mesentery (hermaphroditic condition). Hermaphroditism, also found in some species of other genera, is a rather common condition. Sections of the polyps of a colony of *Pavona cactus* showed that these were all males (unisexual). The gastrovascular cavity of a given polyp may contain dozens of larvae, which are freed from the mouth during the variable breeding season. Some forms breed in early summer (*Favia*); *Porites* breeds from summer into winter; others have a breeding period that seems to follow lunar periodicity.

Production of planulae (*Porites haddoni*) is illustrated. On March 21, 1929, Marshall and Stephenson (1933) filled three pails containing 3600 planulae of this species of *Porites*. On April 24, 14 colonies yielded only 83 planulae; on May 12, 12 colonies yielded 2500 to 3000 planulae. (In this record, collections extended intermittently from January 26, 1929, to July 22, 1929.)

Planulae are 1 to 3 mm long and are ciliated (Figure 5.44D, *1–2*). Atoda (1951*b*) noted that under laboratory conditions some 260 planulae (*Galaxea aspera*) of a population of 1172 settled within seven days; thereafter the number, settling after one month, steadily diminished. In the sea, also, other investigators report that most planulae will settle after a few days and attach by the aboral end. Polyp, mesenteries, and septa development, a few days after attachment, are illustrated in Figure 5.44D, *3–4*. If a number of planulae are crowded on the same bottom site, by skeletal fusion they can give rise to a colonial coralla.

One, or more than one, generation of planulae may be produced during the life span of a given polyp. In the former instance, after the production of a single planulae generation, the demise of the polyp or coral colony follows shortly thereafter (Gardiner, in Vaughan and Wells, 1943).

The mode of asexual reproduction or increase (intratentacular budding or fission, extratentacular budding, and transverse division) is an important determinant of colonial form in scleratinians (Figure 5.45E). Descriptions of family and genera contain entries on both modes of increase and the resulting morphology. Solitary polyps may come also from existing polyps by asexual increase (transverse division).

Types of asexual increase and the resulting colonial morphologies follow:

A. *Fission or Intratentacular Budding*. Hydnophoroid, phaceloid, plocoid, cerioid, meandroid, flabellate, dendroid.

B. *Extratentacular Budding*. Dendroid, reptoid, phaceloid, plocoid, cerioid (Figure 5.45E).

C. *Transverse Fission*. Solitary (shapes as in Rugosa with turbinate most common, and a few unlike rugosoid types, such as tympanoid, cuneiform, and flabellate.

Classification

Definition: **Order Scleratinia (after Wells, 1956).**

1. See definition of subclass Zoantharia and class Anthozoa.

2. Growth habit. Solitary or colonial.

3. Skeleton. Calcareous, external, secreted by ectoderm.

4. Skeletal organization:
(a) *Septa*. Radial partitions, located intermesenterially; formed within the basal upward infolds of polyp column wall.
(b) *Other structures*. Basal plate (Figure 5.42C); epitheca, dissepiments, synapticulae (Figure 5.43D), and mural (wall) structures.

5. Septal development and pattern. In ontogeny, septa follow the mesenterial pattern (Figure 5.44, D4). Addition of septa after six primaries occurs in successive cycles of 6, 12, 24, and so on; they are inserted in all six primary mesentery exocoeles in dorsoventral order (see Figure 5.22).

6. Reproduction. Sexual and asexual.

7. Geologic range. Middle Triassic to Recent.

There are over 2500 species of scleratinian corals. These belong to 32 families

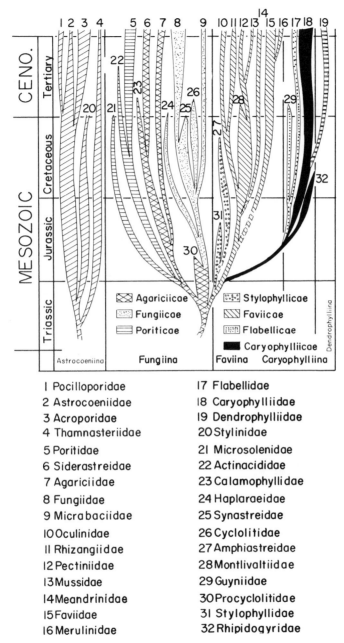

FIGURE 5.46 Derivative data: inferred pattern of evolution of scleratinian suborders and families (after Wells, 1956). Consult text discussion of this diagram.

1 Pocilloporidae
2 Astrocoeniidae
3 Acroporidae
4 Thamnasteriidae
5 Poritidae
6 Siderastreidae
7 Agariciidae
8 Fungiidae
9 Micrabaciidae
10 Oculinidae
11 Rhizangiidae
12 Pectiniidae
13 Mussidae
14 Meandrinidae
15 Faviidae
16 Merulinidae

17 Flabellidae
18 Caryophylliidae
19 Dendrophylliidae
20 Stylinidae
21 Microsolenidae
22 Actinacididae
23 Calamophyllidae
24 Haplaraeidae
25 Synastreidae
26 Cyclolitidae
27 Amphiastreidae
28 Montlivaltiidae
29 Guyniidae
30 Procyclolitidae
31 Stylophyllidae
32 Rhipidogyridae

embraced by five suborders, Figure 5.46. The suborders are as follows.

Suborder 1. Astrocoeniina. Chiefly colonial, corallites small (diameter, 1–3 mm); septa formed of a few simple trabeculae, appearing as simple spines to solid laminae inclined in series from walls; marginally dentate to smooth. Middle Triassic to Recent. Representative families: Pocilloporidae (Figure 5.43A), Acroporidae (Figure 5.43B).

Suborder 2. Fungiina. Solitary and colonial, corallites larger (2 mm diameter); septa formed by numerous trabeculae,

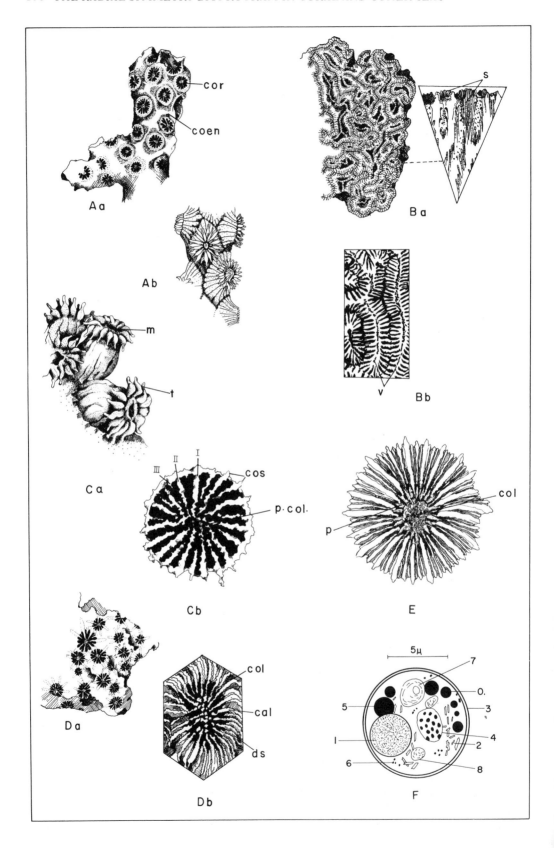

FIGURE 5.47 Genera of representative scleratinian families—II
(*A*) Suborder Faviina: *Oculina* sp. Rec. (*a*) Dendroid colony showing corallites united by dense coenosteum; septal pattern. (*b*) Polyps in various degrees of expansion. (Schematic). (*B*) Meandriina. (*a*) *Meandrina* sp., portion of brain coral displaying meandroid pattern; detail of septa, longitudinal cut. (*b*) *M. meandrites*, Rec., Bahamas, X1, detail showing continuous and discontinuous valleys (*v*), the series united by septotheca (= corallite wall of thickened outer parts of septa). Suborder Faviina. (*C*) *Astrangia danae*, Rec. Atlantic Coast: (*a*) polyps, variously expanded (schematic). (*b*) *Astrangia* sp., Rec. Texas, transverse section with 3 septal cycles shown, schematic. (*D–a*) *Montastrea annularis* Bahama Islands, B.W.I, portion of corallum, schematic, (*D–b*) *Montastrea cavernosa* Rec., Bahama Islands, B.W.I., corallite showing elevated calice with 4 complete septal cycles (first three reach columella). (*E*) Suborder Caryophylliina: *Caryophyllia profunda*. Rec. Norfolk Island, N. Z., attached to a cable, X 2.5, calice with 20 pali and stout, curled laths comprising columella. (Compare corallum, Figure 5.42. (*C*). (*F*) Typical zooxanthellae structure. Composite drawing. (1) Assimilation product; (2) calcium oxalate crystals, (3) cell wall; (4) nucleus; (5) oil; (6) pigment granules; (7) primary starch granules; (8) secondary starch grain.
Sources: (*Aa*) (*Ba*) Tasch collection; (*Ab*) Verrill; (*Bb*) Vaughn and Wells, 1943; (*Ca*) several authors; (*Cb*) Durham, 1949; (*D*) Squires, 1958; (*E*) Squires, 1960; (*F*) Yonge, 1957.
coen—coenosteum, *col*—columella, *p. col*—papillose columella, *ds*—dentate septa, *p*—pali, *m*—mouth, *t*—tentacle, *cos*—costae.

fenestrate, more or less perforate, united by simple or compound synapticulae (Figure 5.43*E*, *2*); margins beaded or dentate. Middle Triassic to Recent. Representative families: Siderastreidae (Figure 5.43*D*); Fungiidae (Figure 5.43*E*, *1*), Poritidae (Figure 5.43*F*).

Suborder 3. Dendrophylliina. Solitary and colonial, corallite size as in Suborder 2; septa formed by numerous trabeculae, basically laminar; generally secondarily thickened, irregularly perforate; with smooth or slightly dentate margins; synapticulae present. Upper Cretaceous to Recent. Embraces a single family: Dendrophylliidae. Genera illustrated include *Dendrophyllia* (Figure 5.43*H*) and *Balanophyllia* (Figure 5.43*G*).

Suborder 4. Faviina. Solitary and colonial, corallite size as in Suborders 2 and 3; synapticulae rarely present; septa composed of numerous trabeculae; septal margins dentate; dissepiments well developed. Middle Triassic to Recent. Representative families: Oculinidae (Figure 5.47*A*), Meandrinidae (Figure 5.47*B*), Rhizangiidae (Figure 5.47*C*), and Faviidae (Figure 5.47*D*).

Suborder 5. Caryophylliina. Like suborder Faviina, but with septal margins smooth. Jurassic to Recent. Representa-

tive families: Caryophylliidae (Figure 5.47 *E*), and Flabellidae (Figure 5.42*A*).

INFORMATION DERIVED FROM LIVING AND FOSSIL SCLERATINIANS

Evolutionary Trends

Septal structure (see suborder definition) is primary in deciphering evolutionary trends in scleratinians. The pattern shown in Figure 5.46 is essentially based on changes of such structures through time. The scleratinians arose some time during the Middle Triassic.

Wells (1956) distinguished three main evolutionary trends in the scleratinians. These pertain to the septa, primitive epithecal wall, and the colonial habit. These trends are as follows:

1. Increase or decrease in compactness of septa.
  (a) (Increase) Laminar septa. Faviina, Caryophylliina.
  (b) (Decrease) Porous septa. Dendrophylliina.
  (c) (Decrease) Fenestrate septa. Fungiina.
2. Edge zone development.
  (a) Septotheca, paratheca, synapticulotheca[6], and so on, develop (see Figures 5.44*A* and 5.43*D*).

---

[6]Edge zone, septotheca, and synapticulotheca previously defined. See text. Paratheca = corallite wall, formed by closely spaced dissepiments.

(b) Primitive epithecal wall suppressed (compare Figure 5.44*A, B*). Both (a) and (b) apply to Fungiina, Faviina, and Caryophylliina.

3. Development of colonial habit (hermatypic or reef-forming corals).

Ecology

Ecological Groups. Two large groupings of scleratinian corals are recognized. The groupings depend on the presence or absence of unicellular algal symbionts (zooanthellae, Figure 5.47). Their presence in abundance characterizes the first group—the hermatypic. The second group—the ahermatypic (deep-sea) corals—lack symbionts. Where present, such photosynthesizers apparently do not contribute oxygen corresponding to the coral's needs, as was once thought. Rather, they appear to function as "automatic agents of excretion" (Yonge, 1957) for the host animal. Thus these symbionts dispose of such coral wastes as carbon dioxide, ammonia, nitrates, and phosphates.

Bathymetry, Light Penetration, and Sedimentation. The close relationship between hermatypic corals and their algal symbionts which, living by photosynthesis, require radiant energy, appears to primarily determine the depths at which such corals can thrive. Depths of 20 meters support the greatest coral growth, although they range from 0 to 90.0 meters. At shallower depths of 4 to 5 meters, algal abundance in polyp tissue is maximized.

Marshall and Orr (1931) found for the Great Barrier Reef that corals could and did live in slightly turbid waters. Within limits (determined by their own ciliary movement and natural water circulation) they can rid themselves of sediment falling from above. However, during windy weather, when a change of bottom level occurs owing to the movement of substrate sediments, polyps become quickly covered and succumb in a day or two.

Restricted Areal Distribution. Modern coral reefs form a scatter pattern over an area of some 68 million square miles of the hydrosphere within the limits of the tropics and near-tropical regions.

This coralline belt has its own characteristics: water supersaturated in calcium carbonate (salinity, 30 to 40 parts per 1000); mean annual water temperature of 23° to 25°C; absence of turbidity and/or rapid sediment accumulation on the bottom requiring good water circulation.

The wide geographic distribution of such corals can be explained by the free-living planula larva stage (see Figure 5.44 *D*). Currents can transport the larvae to great distances.

Growth Rate and Growth Form. Essentially two factors govern growth rate: temperature and skeletal structure. Higher annual temperatures hasten coral metabolism and growth rate. Denser coralla grow more slowly than lighter, porous types. Annual increment in height may range from 5.0 to 80.0 mm. (Ma; 1966).

A reef 150 ft thick in the Florida-Bahama and West Indies area might accumulate at the rate of 0.083 ft/year to 0.019 ft/year; the rate depends on the particular coral species involved (Vaughan). Hoffmeister and Multer (1964) measured an elongated Pleistocene reef in the Florida Keys. They found its maximum thickness to be 170 ft and estimated a growth rate of 1 ft/70 years or 0.014 ft/year. However, the authors qualified their own estimate as probably inaccurate. The presence of unconformities (time breaks in deposition) in the reef itself pointed to a rise and fall of sea level during the Pleistocene. In turn, this indicated both sporadic and relatively slow growth. A *Montastrea* colony 7 ft in height was estimated to have grown at the rate of 1 ft/28.5 years. That would give a life age for the colony of 230 years. Taking into account degradation of the colony by organic and inorganic agents, reef sedimentary debris may represent another 230 years.

In the high-energy zone the major inorganic degradational forces (wave and current action) operate to break down the reef framework. It is here that coarser wave-resistant forms occur—for example, massive and stoutly ramose or branching. By contrast, in more protected, lower energy zones of the reef, the more delicate forms occur (Yonge, 1940).

Reef Ecosystem. Odum and Odum (1955) characterized the Eniwetok Atoll as an "open steady-state system"; here the algal-coelenterate complex allowed cyclic use and reuse of food and necessary nutrients to attain a near-balance in the gain and loss of organic matter.

Evaluation of the reef's *primary producers* (algae), *consumers* herbivores (coral polyps, echinoderms, mollusks, fish, annelid worms, and crustaceans), and carnivores (fish, mollusks, annelid worms, and so on) enabled these investigators to draw up an inventory of the ecosystem. They showed that the *biomass* (dry weight of protoplasm of the reef system) consisted of the contribution of each component noted above (Figure 5.5, parts *A, B*). (The biotic associates of reef corals, exclusive of decomposers like bacteria and foraminifera, are also shown in the pyramids compiled by Odum and Odum. In studying the various phyla, it is useful to remember the ways several of these are interrelated in the coral reef ecosystem.) Yonge (1940, p. 352) noted that all scleratinians are "specialized carnivores" adapted to capture zooplankton.

The Odum-Odum study determined that the total plant protoplasm (algal) exceeded annual animal biomass (coral polyps, and so on) by about 3 to 1.

Carbonate Deposition. At Eniwetok it was also found that calcareous deposition and organic deposition (that is, growth of tissue) were not necessarily in phase. The estimated carbonate deposition came to 1.6 cm/year; but it may have been eroded away at an equal rate. (See Chapter 15.)

In Jamaica on the fore-reef slope (see Figure 4.9) of reef communities, calcareous deposition by organisms is slow; there

Goreau and Hartman (1963) found, by Scuba diving, that large-scale activity by boring sponges (clionids; see Chapter 4) literally control the reefs. This is accomplished by erosion of the nonliving skeleton of potential reef-frame builders (massive corals).

Rock-boring organisms, according to both Gardiner, and Otter (1937) participate importantly in destroying coral reefs. These organisms attack and follow this order: first, boring algae, subsequently sponges, polychaete worms (Eunicidae; see Chapter 9), sipunculid worms and, finally, species of the lamellibranch pelecypod *Lithophaga*. In this way the coral rock is initially fragmented and then reduced to sand. Finally, sand-triturating organisms, such as holothurians (see Chapter 12), certain sipunculid, and polychaete worms pass the sand through their digestive systems and convert it into mud.

Speciation. It is apparent that two scleratinian genera abound in reef communities in both the Tropical Atlantic area and the Indo-Pacific area—*Acropora* and *Porites*. However, in terms of number of species, there are 50 times more species of *Acropora* and 10 times more of *Porites* in the Indo-Pacific area than in the Tropical Atlantic. In addition both areas have distinct coral genera that, nevertheless, cohabit similar ecologic niches on the reefs. Genera like *Pocillopora* that had thrived in both regions during the mid-Tertiary, as indicated by fossil faunas, are found now only in the extant Indo-Pacific reef faunas (Wells, 1957).

Studies of variation in species of the genus *Siderastrea* (Figure 5.43D), led Yonge (1935) to conclude that species were highly specialized for particular environments and moreover displayed flexibility to adapt to environmental pressures. For reef corals generally, he noted (1940) that species with a wide horizontal (geographic) spread embraced several *physiological* races.

An indication of the distribution of several species of the *same* scleratinian genus on a commonly shared limited reef tract (Great Barrier Reef) is illustrated in Figure 5.8.

Coral Atolls—Origin. The best evidence on the origin of coral atolls is the deep borings made on several of them (Figure 5.7, part *B*). The one at Eniwetok encountered olivine basalt at depths of 4222 and 4630 ft respectively (Ladd et al., 1953). In one deep drilling (Figure 5.7, part *B*), reef limestone of Eocene age that represented a Tertiary coral reef was found directly in contact with olivine basalt. Thus Darwin's surmise that atolls grew on submerged mountain tops was confirmed. The foundation of the Eniwetok coral atoll was clearly a basaltic volcano that rose an estimated 2 miles above the floor of the ocean and probably projected far above sea level originally. The fact that the planed upper portion of a Tertiary volcano has to now be reached by drilling more than 4/5 mile deep below present sea level indicates that considerable subsidence occurred from Eocene to the present in the area.

Continuous deposition of reef limestones occurred all during the Tertiary in atolls of the open Pacific (Figure 5.7, part *B*). This is indicative of hermatypic scleratinian activity in the area. The essentials of the reef ecosystem described by Odum and Odum, thus, must have persisted for some 58 million years at Eniwetok.

CRETACEOUS SCLERATINIAN CORAL REEF FACIES

Fossil occurrence of hermatypic corals in the rock column finds a good illustration in the Trinity Group (Lower Cretaceous) of central Texas (Wells, 1932). Three distinct coral reef horizons were found in the Glen Rose Formation. Although hermatypic corals occur in underlying beds of the Travis Peak formation, they did not build up any reefs. Possibly this may have been due to planulae attachment in a high energy zone.

All of these Cretaceous reefs were built on a foundation of caprinid pelecypod shells. The planulae of such scleratinians as *Isastrea, Orbicella, Astrocoenia* attached to a substrate of large masses or layers of molluscan shells. The pelecypods appear to have thrived as the corals grew upward.

The Glen Rose basal reef attained a height of 3 to 10 ft and spread over several square miles. The other two reefs are stratigraphically 100 ft and 200 ft, respectively, above the basal reef. This indicates recurrence of suitable reef-building conditions in the seas of the area during Glen Rose time. The middle and upper reef facies were both thicker (15 to 30 ft), had a more varied fauna, and a greater number of scleratinian corals than the basal reef. Some of the same reef species persisted in the area and occurred in the three reef horizons; several new species made their first appearance in the upper reef facies.

IDENTIFICATION OF A FOSSIL SCLERATINIAN TO THE SPECIES LEVEL

Data from J. W. Wells, 1945, G. S. A. Memoir 9, Part II. West Indian Eocene and Miocene Corals, p. 6, plate 2, figure 1.)

1. Locality. Eocene, Upper Scotland formation, Barbados, West Indies.

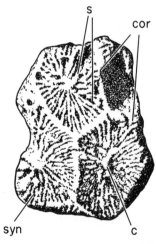

FIGURE 5.48 *Siderastrea scotica* Wells. Calices of holotype. Symbols: *s*—septa, *c*—columella, *syn*—synaptaculae. (Schematic, after photograph in Wells, 1945.)

2. Sample. $n = 3$, two mature and one immature specimen.

3. Description.

(a) Details of calicular surface obscure (Figure 5.48).

(b) Corallites, irregularly prismatic in outline; size, shape, and septal number variable: largest calice, 5.5×6.5 mm and 68 septa, smallest calice (immature) 2.5×3.0 mm and 34 septa.

(c) Thin septa united by synapticulae, especially near wall.

(d) Columella, a prominent plug, composed of one to three papillae.

4. Generic assignment. Characteristics listed under 3b, c, d above, place the specimens in the family Siderastreidae and genus *Siderastrea*.

5. Specific assignment. Only three previously known Eocene species of this genus by 1945. The first two are *S. funesta*, M. Eoc. S. Eur., 48 septa, and *S. parisensis*, Paris Basin, 24 septa. These two species were ruled out because of the fewer number of septa. A third known species, *S. hexagonalis*, L. Eoc., Alabama–Wisconsin, was close to the Barbados species in size of calice and number of septa. However, it differed in the weak development of the columella and in the more superficial calices. Accordingly, the Barbados specimens represented a new species—*Siderastrea scotica* Wells, 1941. (Figure 5.48).

QUESTIONS

Coelenterata (General)

1. (a) Outline steps in probable transition to the ancestral coelenterate (see Figure 5.1).

   (b) Compare the phylum definition with these steps; determine what phylum characteristics were already achieved by the end of this transition. What phylum characteristics came as a later evolutionary development?

2. What do the three classes of subphylum Cnidaria share in common?

3. Contrast the biological organization of a hydroid and scyphozoan; a hydroid and anthozoan.

Hydrozoa

1. How do the five living orders of Hydrozoa differ primarily?

2. Discuss millepore speciation, growth form, distribution, and contribution to coral reef structures.

3. With what biotic associates, in what lithological context did Nielsen's Danish fossil hydrocorals occur? Carozzi's Beaverhill Lake fauna?

4. Cite examples of stromatoporoids as stratigraphic markers, reef builders, and reef dwellers.

Scyphozoa

1. What plan characterizes the scyphozoan polyp and medusa?

2. Discuss the significance of paleosalinity and salt regulation in living medusae as derivative data bearing on the fossil record.

3. What did Kieslinger discover about the Solenhofen scyphomedusans?

4. Evaluate Sprigg's designation of the Pre-Cambrian as the age of jellyfishes.

Conulata

1. (a) What evidence did Kiderlen use to show that conulatids had quadrilateral radial symmetry?

   (b) What is the significance of this finding?

   (c) How can it be reconciled with three-sided conulatids?

2. Distinguish the following either by definition or sketch: periderm, lappet, ribs, septa, *schott.*

3. Explain why the pyramid corner in conulatids and the nature of the structures found there are considered by Sinclair to be "the most diagnostic features."

4. Discuss Figure 5.16.

Octocorallia

1. (a) What is the basic type of spicule structure in octocorals?
   (b) Discuss spicule structure.
   (c) What spicule characteristics are of taxonomic value?
2. (a) What are gorgonian windrows, and how do they accumulate?
   (b) How might such windrows appear in the rock column? Outline steps.
3. (a) What information may be obtained from growth form in octocorals?
   (b) Discuss octocorals as carbonate contributors to living reefs; during French Jurassic.
4. (a) What is "Bilobites"?
   (b) Why does its recurrence in Flysch deposits have value as derived information?

Zoantharia (General)

1. Sketch, label, and discuss the arrangement of mesenteries in the ancestral or stem zoantharian and derivative orders (Figure 5.22).
2. Discuss the basis for subdividing zoantharians into orders.

Heterocorallia

1. Compare the interpretation of Yabe and Hayasaka, and Schindewolf on septal insertion (Figure 5.23 and text).
2. How would you explain the narrow geologic range?

Rugosa

1. (a) Outline the ten points in the definition of the order.
   (b) Distinguish between the three suborders.
2. Sketch a rugose coral and label the marginarium, tabulae, major and minor septa.
3. Contrast amplexoid and lonsdaleoid septa.
4. Compare the genetic vs. the environmental influence in growth and form in Rugosa.
5. (a) Of what value are the following measurements and or ratios in studies of rugose corals: septal coefficient, tabularium ratio, apical angle?
   (b) What is the paleoecological significance of skeletal thickening?
6. (a) Discuss rugose corals in reef facies (Coeymans limestone, New York State); stratigraphic occurrence in the Glass Mountains (Texas); correlation in the Canadian and American Mississippian; occurrences in the Ordovician-Silurian of western United States.
   (b) What biotic associates occur in reef facies?
7. Discuss Hill's three major ecologic groupings of Carboniferous corals.

Tabulata

1. Review the definition of the order and contrast it with that of the Rugosa.
2. (a) Distinguish the five families.
   (b) Why are the Chaetetidae considered to be the most primitive tabulate corals? Sclerosponges?[1]
   (c) Why can the Tabulata be considered primarily Paleozoic corals?
3. Identify: mural pores, coenenchyme, trabeculae, infundibuliform tabulae, dissepiment, offset.
4. How did Nelson employ species of *Syringopora* as stratigraphic indicators in the Mississippian of the Canadian Rockies?
5. Outline Vaughan and Wells' paleoecological observations on rugose and tabulate corals.
6. (a) Based on Figure 5.39, reconstruct the subsurface Silurian reef showing drill penetration to the depth of the illustrated core.
   (b) Was the area in Figure 5.40 a Silurian coral sea? Explain.
   (c) Discuss the D-2 and D-3 zones in the Leduc Oil Field (Upper Devonian, Canada).
   (d) How does the environment control "reef trends"?
7. Cite an illustration of each of the following: seasonal growth, commensalism and symbiosis, dimorphism.

[1]*See J. Kazmicrczak, 1979 C N. Jb. Gool. Paläont.* Mh, 97-108

Scleratinia

1. (a) Discuss and illustrate, by a labeled sketch, the primary information in scleratinian polyp anatomy.

   (b) What is the order of skeletal deposit in the scleratinian skeleton?

2. Discuss each of the following: sclerodermite, trabeculae, endo- and exocoele, protosepta, Pourtales plan, synapticulae, pali, coenosteum.

3. (a) Cite the evidence on scleratinian planulae production of Great Barrier Reef corals and Atoda's laboratory observations.

   (b) How are those data related to reef distribution?

4. What is the relationship between the type of asexual increase and colonial coral morphology?

5. (a) Compare the definition of the order Scleratinia with that of the Rugosa and Tabulata.

   (b) Distinguish between the five scleratinian suborders and indicate their evolutionary relationship (*Hint.* See Figure 5.46.)

   (c) According to Wells what were the three main evolutionary trends in scleratinians?

6. (a) Distinguish between hermatypic and ahermatypic corals.

   (b) What is the role of zooxanthellae?

7. (a) What is the significance of unconformities in a fossil coral reef?

   (b) Discuss the findings of Odum and Odum on the reef ecosystem, carbonate deposition, reef zonation at Eniwetok (see Figure 5.5, part *A*, and Figure 5.5, part *B*).

   (c) Cite the order in which rock-boring organisms attack and break down coral reef structures?

8. (a) How did reef borings on Eniwetok shed light on the origin of coral reefs?

   (b) Study Figure 5.7, part *B*. What light does it afford on the deposition of reef limestones during the Tertiary?

REFERENCES

Coelenterata (General)

Hill, Dorothy, and J. W. Wells, 1956. Cnidaria: Treatise Invert. Paleont., Part F, Coelenterata, pp. F5, ff.

Hyman, L. H., 1940. The invertebrates: Protozoa through Ctenophora. McGraw-Hill. Chapters 5, 7, 8.

_____, 1956. Morphology of living coelenterates: Treatise Invert. Paleont., Part F, Coelenterata, pp. F10 ff.

Hydrozoa

Boschma, H., 1948. The species problem in Millepora: Zool. Ver. Mus. Leiden, No. 1, 115 pp., 15 plates, 13 figures.

_____, 1956. Milleporina and Stylasterina: Treat. Invert. Paleont., Part F, Coelenterata, pp. F90–F106.

Carozzi, A. V., 1961. Reef petrography in the Beaverhill Lake Formation, Upper Devonian Swan Hills Area, Alberta, Canada: J. Sed. Petrology, Vol. 31 (4), pp. 497–513, Figures 1–9.

Caster, K. E., 1945. A new jellyfish (*Kirklandia texana* Caster) from the Lower Cretaceous of Texas: Paleontogr. Americana, Vol. 3, pp. 168–220, plates 1–5, Figures 1–8.

*Fagerstrom, J. A., 1962. Middle Devonian stromatoporoids from South-eastern Michigan: J. Paleontology, Vol. 36 (3), pp. 424–430, plates 65–67.

Fischer, W. K., 1938. Hydrocorals of the North Pacific Ocean: Proc. U. S. Natl. Mus., Vol. 84, pp. 493–554, plates 34–76.

*Galloway, J. J., 1957. Structure and classification of the Stromatoporoidea: Bull. Amer. Paleontology, Vol. 37 (164), pp. 345–457, plates 31–37.

_____, 1960. Devonian stromatoporoids from the lower Mackenzie Valley of Canada: J. Paleontology, Vol. 34 (4), pp. 620–636, plates 71–77.

_____ and J. St. Jean, Jr., 1961. Ordovician Stromatoporoidea of North America:

*Stromatoporoids reassigned to Poritera. J. Kazmierczak (1976, *Nature*, 264: 49–51) concluded that stromatoporoids represented "permineralized colonies of coccoid cyanophytes" (i.e. were plants, not animals).

Bull. Amer. Paleontology, Vol. 43 (194), pp. 4–82, plates 1–13.

Hadding, Assar, 1950. Silurian reefs of Gotland: J. Geology, Vol. 58, pp. 402–409.

Harrington, H. J., and R. C. Moore, 1956. Trachylinida: Treat. Invert. Paleont., Part F, Coelenterate, p. F68.

Hudson, R. G. S., 1956. Tethyan Jurassic hydroids of the family Milleporidiidae: J. Paleontology, Vol. 30 (3), pp. 714–730, plates 75–77, 6 text figures. (Hydroids of uncertain position.)

Kühn, Othmar, 1939. Hydrozoa, in O. H. Schindewolf, Handbuch der Paläozoologie (Berlin), Vol. 2A, pp. A3–A68.

*Lecompte, Marius, 1956. Stromatoporoidea: Treat. Invert. Paleont., Part F, Coelenterata, F107.

Lenhoff, H. M., and W. H. Loomis, eds., 1961. The biology of Hydra and other coelenterates. Univ. of Miami Press, 448 pp.

Newell, N. D., 1954. Reefs and sedimentary processes of Rarioa: Atoll Research Bull., No. 36.

Nielson, K. B., 1919. Ein Hydrocoralfauna fra Faxe: Danm. geol. Undersøg. IV, Vol. 1, No. 10, 65 pp., 3 plates.

*Perkins, R. D., 1963. Petrology of the Jefferson limestone (Middle Devonian) of Southeastern Indiana: G.S.A. Bull., Vol. 74, pp. 1335–1356, 6 figures, 5 plates. (Stromatoporoid distribution.)

*Stearn, C. W., 1961. Devonian stromatoporoids from the Canadian Rocky Mountains: J. Paleontology, Vol. 35 (5), pp. 932–948, plates 105–107, 3 text figures.

Teichert, Curt, 1943. The Devonian of Western Australia: a preliminary review. Part I: Am. J. Sci., Vol. 241, pp. 69ff.

Vaughan, T. W., 1919. Fossil corals from Central America, Cuba, Porto Rico, etc.: U.S. Natl. Mus. Bull., Vol. 103, pp. 189–524, plates 68–152.

Werner, Bernhard, 1963. Effect of some environmental factors on differentiation and determination in marine Hydrozoa, with a note on their evolutionary significance: Ann. N.Y. Acad. Sci., Vol. 105, article 8, pp. 461–488.

Scyphozoa, Including Scyphomedusans and Conulatids

Bouček, Bedřich, 1939. Conularida, in O. H. Schindewolf, Handbuch der Paläozoologie:

Borntrager (Berlin), Vol. 2A, No. 5, pp. A113–A131, Figures 1–13.

Driscoll, E. G., 1963. Paraconularia newberryi (Winchell) and other Lower Mississippian conulariids from Michigan, Ohio, Indiana, and Iowa: Contrib. Mus. Paleontology, Univ. Michigan, Vol. 18 (3), pp. 33–46, 3 plates.

Finks, R. M., 1955. Conularia in a sponge from the West Texas Permian: J. Paleontology, Vol. 29 (5), pp. 831–836.

Kiderlen, Helmut, 1937. Die Conularien über Bau und Leben der ersten Scyphozoa: Neues Jahrb. f. Mineralogie: Beil-Band 77, Abt. B, pp. 113–169, Figures 1–47.

Kieslinger, Alois, 1939. Revision of Solenhofen Medusen: Paleontologische Zeitschr., Vol. 21, No. 4, pp. 287–296, 5 text figures.

Knight, J. B., 1937. Conchopeltis Walcott an Ordovician genus of the Conulariida: J. Paleontology, Vol. 11 (3), pp. 186–188, plate 29.

Maas, Otto, 1902. Ueber Medusan aus dem Solenhofer Schiefer und der Kreide der Karpathan: Paleontographica, Vol. 48, pp. 297–322, plates 22, 23, 9 text figures.

McKee, E. D., 1935. A Conularia from the Permian of Arizona: J. Paleontology, Vol. 9, pp. 427–429, plate 42.

Moore, R. C., and H. J. Harrington, 1956. Conulata, in Treat. Invert. Paleont., Part F, Coelenterata, pp. F54–F76 (see also pp. F29–F35).

Mayer, A. G., 1910. The Scyphomedusae, in Medusae of the world: Carn. Instit. Washington, Publ. 109, Vols. 1–3 (Vol. 3, pp. 499–835, plates 56–76, Figures 328–428).

Ross, R. J., Jr., 1957. Ordovician fossils from wells in the Williston Basin, Eastern Montana: Geol. Surv. Bull. 1021-M, pp. 454, 469.

Ruedemann, Rudolf, 1898. The discovery of a sessile Conularia: N.Y. State Geol. (Albany) Ann. Rpt. 15, pp. 701–720, plates 1–5.

Sinclair, C. W., 1952. A classification of the Conularida: Fieldiana, Geol. Chicago Nat. Hist. Mus., Vol. 10 (13), pp. 135–145.

————, 1944. Notes on the genera Archeoconularia and Eoconularia: Trans. Royal Soc. Canada, Series 3, Vol. 38, Section 4, pp. 87–95, plate 1.

————, 1943. The Chazy Conularida and their Congeners: Ann. Carn. Mus. (Pittsburgh), Vol. 29, article 10, pp. 219–240, 3 plates.

*[Stromatoporoids reassigned to Porifera]

Sprigg, R. C., 1949. Early Cambrian "jelly-fishes" of Ediacara, South Australia and Mt. John, Kimberly District, Western Australia: Trans. Royal Soc. S. Australia, Vol. 73, Part 1, pp. 72–98, plates 9–21, 10 text figures.

Sugijama, Toshio, 1942. Studies of Japanese Conularida: Geol. Soc. Japan Jour., Vol. 49, pp. 390–399, plate 15.

Termier, G., and H. Termier, 1950. Invertebrés de L'Ère Primaire. Herman et Cie. Vol. 2, pp. 108–117, plate 241.

Siphonophorida

Caster, K. E., 1942. Two new siphonophores from the Paleozoic: Paleontogr. Americana, Vol. 3, pp. 56–90, plates 1, 2, Figures 1–6.

Harrington, H. J., and R. C. Moore, 1956. Siphonophorida: Treat. Invert. Paleontology, Part F, Coelenterata, pp. F145 ff.

Anthozoans (Ceriantipatharia)

Wells, J. W., and Dorothy Hill, 1956. Ceriantipatharia: Treat. Invert. Paleont., Part F, Coelenterata, pp. F165–F166.

Anthozoans (Octocorallia)

Bayer, F. M., 1956. Octocorallia: Treat. Invert. Paleont. Part F, Coelenterata, pp. F166–F231.

———, 1955. Remarkably preserved fossil sea-pens and their recent counterparts: J. Washington Acad. Sci., Vol. 45, pp. 294–300, Figures 1, 2.

Cary, L. R., 1931. Studies on the coral reef at Tutuila American Samoa, with special reference to the Alcyonaria: Papers from the Tortugas Lab., Carn. Inst. Washington, Vol. 27 (3), pp. 53–98, plates 1–7, Figures 1–14, Tables 1–14, map.

———, 1918. The Gorgonaceae as a factor in the formation of coral reefs: Carn. Inst. Washington, Publ. 213, pp. 341–362, plates 100–105.

———, 1914. Observations upon the growth rate and oecology of the gorgonians: Papers from Tortugas Lab. (Carn. Inst. Washington), Vol. 5, pp. 81–89, plates 1–2.

Cayeaux, Lucien, 1939. Études de Gîtes minéraux de la France (Paris), Vol. 1, pp. 126, 129.

———, 1921. Rôle pétrographique des Alcyonaires fossiles, etc.: Comptes Rendus, Acad. Sci. Paris, Vol. 172, pp. 1189–1191.

Hyman, L. H., 1940. Reference cited under "Coelenterata (General)," pp. 538–566.

Lucas, Gabriel, 1950. Précisions sur les *Cancellophycus* du Jurassique: Comptes Rendus, Acad. Sci. Paris, Vol. 230 (13), pp. 1297–1299, Figure 1.

Montanaro Gallitelli, E., 1956a. Il Permiano del Sosio e i suoi Coralli: Palaeontographia Italica, Vol. 49 (n. ser. vol. 19), pp. 1–98.

———, 1956b. Trachypsammiacea, in F. M. Bayer, 1956, see above, pp. F190–F196.

Nielsen, K. E., 1925. Nogle nye Octocoraller fra Danienet: Meddelelser fra Dansk. Forening, Vol. 6, Part 28, pp. 1–6, Figures 1–3.

———, 1917. *Heliopora incrustans*, nov. sp., with a survey of the Octocorallia in the deposits of the Danian in Denmark: *idem*, Vol. 5, Part 8, pp. 1–13, Figures 1–17.

Wells, J. W., 1954. Recent corals of the Marshall Island, Bikini and nearby atolls. Oceanography (Biologic): U.S. Geol. Surv., Prof. Paper 260-I, pp. 385–486.

Wiens, H. J., 1962. Atoll environment and ecology: Yale Univ. Press. 504 pp., plates 1–88.

Anthozoans (Zoantharia – Heterocorallia)

Hill, Dorothy, 1956. Heterocorallia: Treat. Invert. Paleont., Part F, Coelenterata, pp. F324–F327.

Schindewolf, O. H., 1941. Zur Kenntnis der Heterophylliden, einer eigentumlichen palaeozoischen Korallengruppe: Paläont. Z., Vol. 22, pp. 213–306, plates 9–16.

Stuckenberg, A., 1895. Korallen und Bryozoen der Steinkohlen Ablagerungen des Ural und des Timan: Mém. Com. géol. St. Petersbourg, n. ser., Vol. 10, Part 3, 244 pp., 24 plates.

Yabe, H., and I. Hayasaka, 1915–1916. Paleozoic corals from Japan, Korea, and China: Geol. Surv. Tokyo, Vol. 22, pp. 55–70, 79–109, 127–142; Vol. 23, pp. 57–75.

Rugosa

Bryan, W. H., and Dorothy Hill, 1941. Spherulitic crystallization as a mechanism of skeletal growth in the hexacorals: Proc. Roy. Soc. Queensland, Vol. 52, pp. 78–91.

Butler, A. J., 1937. A new species of *Omphyma* and some remarks on the *Pycnactis-'Phaulactis* group of Silurian corals: Annals and Mag. Nat. Hist., Vol. 19, pp. 87–96.

Campbell, K. S. W., 1957. A Lower Carboniferous brachiopod–coral fauna from New South Wales: J. Paleontology, Vol. 31 (1), pp. 34–98, plates 11–17, 27 text figures.

Carruthers, R. G., 1910. On the evolution of *Zaphrentis delanouei* in Lower Carboniferous times: Geol. Soc. London Quart. J., Vol. 66, pp. 523–538, plates 36–37.

Chilingar, G. V., 1956. Review: About dynamics of septa development in tetracorals during ontogeny, by K. G. Voynovskiy-Kriger: J. Paleontology, Vol. 30, No. 2, pp. 406–411.

Duncan, Helen, 1956. Ordovician and Silurian coral faunas of Western United States: Geol. Surv. Bull., 1021-F, 231 pp., plates 21–27.

Easton, W. H., 1951. Mississippian cuneate corals: J. Paleontology, Vol. 25 (3), pp. 380–404, plates 59–61, 14 text figures.

———— and R. C. Gutschick, 1953. Corals from the Redwall limestone (Mississippian) of Arizona: Bull. So. Calif. Acad. Sci., Vol. 52, Part I, pp. 1–24, 3 plates.

Fagerstrom, J. A., 1961. The fauna of the Middle Devonian Formosa Reef limestone of Southwestern Ontario: J. Paleontology, Vol. 35 (1), pp. 1–45, plates 1–14, 1 text figure.

Faul, H., 1943. Growth rate of a Devonian reef-coral (*Prismatophyllum*): Am. J. Sci., Vol. 241, pp. 579–582 (a rugose coral reassigned to *Hexagonaria*).

Hill, Dorothy, 1934. The Lower Carboniferous corals of Australia: Proc. Roy. Soc. Queensland, Vol. 45, pp. 63–115, plates 7–11.

————, 1938. A monograph of the Carboniferous rugose corals of Scotland: Paleont. Soc. Mon., pp. 1–78, plates 1, 2.

————, 1948. The distribution and sequence of Carboniferous coral faunas: Geol. Mag., Vol. 85, pp. 121–148.

————, 1956. Rugosa: Treat. Invert. Paleont., Part F., Coelenterata, pp. F233–F323.

Jeffords, R. M., 1947. Pennsylvanian lophophyllid corals: Univ. Kansas Paleont. Contrib., Coelenterata, article 1, 84 pp., 28 plates.

Moore, R. C., and R. M. Jeffords, 1945. Descriptions of Lower Pennsylvanian corals from Texas and adjacent regions: Univ. Texas Publ. 4401, pp. 77–208, plate 14, Figures 1–214.

Nelson, S. J., 1960. Mississippian lithostrotionid zones of the southern Canadian Rocky Mountains: J. Paleontology, Vol. 34 (1), pp. 107–126, plates 21–25, 3 text figures.

Oliver, W. A., Jr., 1960. Rugose corals from reef limestones in the Lower Devonian of New York: J. Paleontology, Vol. 34 (1), pp. 59–100, plates 13–19, 3 text figures.

————, 1958. Significance of external form in some Onondagan rugose corals: J. Paleontology, Vol. 32 (5), pp. 815–837, plates 104–106, 3 text figures.

Pitrat, C. W., 1962. Devonian corals from the Cedar Valley limestone of Iowa: J. Paleontology, Vol. 36 (6), pp. 1155–1162, plates 158–159.

Richter, R., 1929. Das Verhaltnis von Funktion und Form bei den Deckelkorallen: Paleont. Z., Vol. 2, pp. 76–79, 2 figures.

Ross, J. P., and C. A. Ross, 1963. Late Paleozoic rugose corals, Glass Mountains, Texas: J. Paleontology, Vol. 37 (2), pp. 409–420, plates 48–50, 2 text figures.

————, 1962. Pennsylvanian, Permian rugose corals, Glass Mountains, Texas: J. Paleontology, Vol. 36 (6), pp. 1163–1188, plates 160–163, 11 text figures.

Ross, C. A., and S. Oana, 1961. Late Pennsylvanian and early Permian limestone petrology and carbon isotope distribution, Glass Mountains, Texas: J. Sed. Petrology, Vol. 31, pp. 231–244.

Stensaas, L. J., and R. L. Langenheim, Jr., 1960. Rugose corals from the Lower Mississippian Joana limestone of Nevada: J. Paleontology, Vol. 34 (1), pp. 179–188, 10 text figures.

Stumm, E. C., 1963a. Ordovician streptelasmid rugose corals from Michigan: Mus. Paleont. Contrib. Univ. of Michigan, Vol. 16 (2), pp. 225–246, 6 plates.

————, 1963b. Tortophyllina, Bethanyphyllum, Aulacophyllum and Hallia: *idem*, Vol. 18 (8), pp. 135–155, 10 plates.

————, 1962a. Part VII, The Digonophyllidae: *idem*, (31), Vol. 17 (9), pp. 215–231, 6 plates.

————, 1962b. Part X, Tabulophyllum: *idem*, Vol. 17 (14), pp. 291–297, 2 plates.

————, 1961. North American genera of the Devonian rugose coral family Digonophyllidae: *idem*, Vol. 16 (2), pp. 225–246, 6 plates.

————, 1952. Species of the Silurian rugose coral *Tryplasma* from North America; J. Paleontology, Vol. 26 (5), pp. 841–843, plate 125.

————, 1949. Revision of the families and genera of the Devonian tetracorals: G.S.A. Mem. 40, 92 pp., 25 plates.

———— and J. H. Tyler, 1962. Corals of the Traverse Group of Michigan: Part 9, *Heliophyllum:* Contrib. Univ. of Michigan, Vol. 17 (2), pp. 265–276, 3 plates.

———— and J. L. Watkins, 1961. The Metriophylloid coral genera *Stereolasma, Amplexiphyllum*, and *Stewartophyllum* from the Devonian Hamilton Group of New York: J. Paleontology, Vol. 35 (3), pp. 445–447, plate 58.

Termier, H., and G. Termier, 1948. Étude sur *Calceola sandalina* Linné: Rev. Sci. Ann. 86, pp. 208–218, 30 figures.

Wells, J. W., 1963. Coral growth and geochronometry: Nature, Vol. 197, No. 4871, pp. 948–950 (see Appendix I).

————, 1957. Corals: Treat. Mar. Ecol. and Paleoecol., G.S.A. Mem. 67, Vol. 2, pp. 773–782.

———— and Dorothy Hill, 1956. Zoantharia —general features: Treat. Invert. Paleontology, Part F, Coelenterata, F231.

————, 1956. Zoantharia, Corallimorpharia and Actiniaria: *idem*, p. F232.

Tabulata

Buehler, E. J., 1955. The morphology and taxonomy of the Halysitidae: Peabody Mus. Nat. Hist. Bull. 8, Connecticut, 72 pp., 12 plates, 2 text figures.

Bond, Geoffrey, 1950. The Lower Carboniferous reef limestones of Northern England: J. Geology, Vol. 58 (4), pp. 313–329, plates 1, 2, 6 tables, 3 text figures. (*Michelinia, Syringopora,* and so on).

Gerth, H., 1952. Die von Sipunculiden bewohnten lëbenden und jung tertiären Korallen und der wurmförmige Körper von *Pleurodictyum:* Paläont. Z., Vol. 25, pp. 119–126, plate 6, 2 figures.

Hill, Dorothy, and E. C. Stumm, 1956. Tabulata: Treat. Invert. Paleontology, Part F, Coelenterata, pp. F444–F475.

Jones, O. A., 1936. The contributing effect of environment upon the corallum in *Favosites,* with a revision of some massive species on this basis: Ann. Mag. Nat. Hist. 10, Vol. 17, pp. 1–24, plates 1–3, 12 figures.

Layer, D. B., et al., 1949. Leduc oil field, Alberta, a Devonian coral-reef discovery: A.A.P.G. Bull., Vol. 33 (4), pp. 572–602, 8 figures, 2 plates.

Leith, E. I., 1952. Schizocoralla from the Ordovician of Manitoba: J. Paleontology, Vol. 26 (5), pp. 789–796, plates 114–116.

Lowenstam, H. A., et al., 1956. Guidebook: The Niagara Reef at Thornton, Illinois: Ill. St. Geol. Surv., 19 pp., 5 figures.

————, 1950. Niagara reefs of the Great Lakes Area: J. Geology, Vol. 58 (4), pp. 430–486, 5 plates, 11 text figures.

————, 1949. Niagaran reefs in Illinois and their relationship to oil accumulation: State Geol. Surv. Rpt. 145, 36 pp., 9 text figures, plate 1.

Nelson, S. J., 1962a. Analysis of Mississippian *Syringopora* from the Southern Canadian Rocky Mountains: J. Paleontology, Vol. 36 (3), pp. 442–460, plates 71–75, 7 text figures.

————, 1962b. Permo-Carboniferous tabulate corals from Western Canada: J. Paleontology, Vol. 36 (5), pp. 953–964, plates 137–138, 4 text figures.

Stumm, E. C., 1961. Corals of the Traverse Group of Michigan: Part VI —*Cladophora, Striatopora,* and *Thamnopora:* Mus. Paleont. Univ. Mich., Vol. 16 (4), pp. 275–285, 2 plates.

Vaughan, T. W., 1911. Physical conditions under which Paleozoic coral reefs were formed: Bull. G.S.A., Vol. 22, pp. 238–252.

Wilson, E. C., and R. L. Langenheim, Jr., 1962. Rugose and tabulate corals from Permian rocks in the Ely Quadrangle, White Pine County, Nevada: J. Paleontology, Vol. 36 (3), pp. 495–520, plates 86–89, 4 text figures.

Yakovlev, N. N., 1926. The phenomena of parasitism, commensalism and symbiosis in the Paleozoic invertebrates: Ann. Soc. Paléont. Russia, Vol. 4, pp. 114–124, 4 figures.

Scleratinia

Atoda, Kenji, 1951a. The larva and postlarval development of the reef-building corals. III.

Acropora brüggemanni: J. Morphology, Vol. 89, No. 1, pp. 1–13, 9 text figures, 3 tables, plate 1.

————, 1951b. *Idem.* IV. *Galaxea aspera* Quelch: J. Morphology, Vol. 89, No. 1, pp. 17–30, 3 text figures, 3 tables, 3 plates.

Duerden, J. E., 1904. The coral *Siderastraea radians* and its postlarval development: Carn. Inst. Washington, Publ. No. 20, 130 pp., 18 plates, 13 figures.

Durham, J. W., 1949. Ontogenetic stages of some simple corals: Univ. Calif. Publ., Vol. 28 (6), pp. 137–176, plates 4, 5, 13 text figures.

————, 1947. Corals from the Gulf of California and the North Pacific West of America: G.S.A. Mem. 20, 46 pp., plates 1–14, Figures 1, 2, Tables 1–5.

————, 1942. Eocene and Oligocene coral faunas of Washington: J. Paleontology, Vol. 16 (1), pp. 84–104, plates 15–17, 1 text figure.

Goreau, T. F., and W. D. Hartman, 1963. Boring sponges as controlling factors in the formation and maintenance of coral reefs: R. F. Sognnaes, ed., Mechanisms of Hard Tissue Destruction: A.A.A.S. Publ. 75, Chapter 2, pp. 25–54.

Great Barrier Reef Expedition, 1928–1929. Scientific Rpt., Vol. 1. Includes, among other papers, C. M. Yonge on "The Physiology of Corals" (1930–1931); further studies on the physiology of corals by C. M. Yonge et al. (1931–1950); G. W. Otter "On rock-destroying organisms in relation to coral reefs" (1937); and "The biology of coral reefs" by C. M. Yonge (1940).

Hoffmeister, J. E., and H. G. Multer, 1964. Growth rate estimates of a Pleistocene coral reef of Florida: G.S.A. Bull., Vol. 75, pp. 353–358, 1 text figure, 2 plates.

Ladd, H. S., et al., 1953. Drilling on Eniwetok Atoll, Marshall Islands: A.A.P.G. Bull., Vol. 37, pp. 2257–2280, 2 plates, 5 figures.

Ma, T. 1966. Effect of water temperature on growth rate of reef corals. Oceanographica Sinica, Taipei, Formosa, 1959, Special Volume 10.

Manton, S. M., and T. A. Stephenson, 1935. Ecological survey of coral reefs: Brit. Mus. (Nat. Hist.) Gt. Barrier Reef Expedition, 1928–1929. Scientific Rpt., Vol. 3, No. 10, pp. 273–312, plates 1–16.

Marshall, S. M., and A. P. Orr, 1931. Sedimentation on Low Isles Reef and its relation to coral growth: Sci. Rpt. Gt. Barrier Reef Expedition, Vol. 1, pp. 93–133, plates 1–14, 7 text figures.

———— and T. A. Stephenson, 1933. The breeding of reef animals. Gt. Barrier Reef Expedition, 1928–1929, Part 1. The Corals, Vol. 3, No. 8, pp. 219–243, refs. plate 1, 6 text figures, 5 tables.

Odum, H. T. and E. P. Odum, 1955. Trophic structure and productivity of a windward coral reef community on Eniwetok Atoll: Ecol. Monographs, Vol. 25, pp. 291–320, 12 text figures, 17 tables.

Ogilvie, M. M., 1896. Microscopic and systematic study of Madreporian types of corals: Philos. Trans. Roy. Soc. (London), Series B., Vol. 187, pp. 83–345, 74 text figures, 1 table. (A classic monographic study on the fine skeletal structures.)

Squires, D. F., 1960. Scleratinia corals from the Norfolk Island Cable: Rec. Auck. Instit. Mus., Vol. 5 (3–4), pp. 195–201.

————, 1959. Deep sea corals collected by the Lamont Geol. Observatory. I. Atlantic corals: Amer. Mus. Novitates. No. 1965, pp. 1–42.

————, 1961. Pt. II. Scotia Sea corals: *idem*, No. 2046, pp. 1–48.

Vaughan, T. W. and J. W. Wells, 1943. Revision of the suborders, families, and genera of Scleratinia: G.S.A. Sp. Paper 44, XV + 363 pp., 51 plates, 39 figures, 3 tables. Bibliography with over 1000 entries.

Wells, J. W., 1957. Coral reefs: Treat. Mar. Ecol. and Paleoecol., Vol. 1. Ecology. G.S.A. Mem. 67, pp. 609–611, 2 figures, 9 plates.

————, 1956. Scleratinia: Treat. Invert. Paleontology, Part F, Coelenterata, pp. F328–F443.

———, 1945. West Indian Eocene and Miocene corals: G.S.A. Mem., Vol. 9, 25 pp., 3 plates.

———, 1942. Jurassic corals from the Smackover limestone, Arkansas: J. Paleontology, Vol. 16 (1), pp. 126–129, plate 21.

———, 1932. Corals of the Trinity Group of the Comanchean of central Texas: J. Paleontology, Vol. 6 (3), pp. 225–256, plates 3–39.

Yonge, C. M., 1957. Symbiosis: Treat. Mar. Ecol. and Paleoecol., G.S.A. Mem. 67, Vol. 1, pp. 429–442, 6 figures.

[Useful data available in: Knutson, D. W., R. W. Buddemeier and S. V. Smith, 1972. Coral chronometers: seasonal growth bands in reef corals. Science, Vol. 177, pp. 270–272. (Growth rate of seven different living coral species in six genera—0.46 to 2.9±cms/year.)].

chapter six

BOTTOM TENEMENTS: THE "MANY ANIMALS" OF THE SINGLE SKELETON (ECTOPROCTS/BRYOZOA)

INTRODUCTION

Encoded data derived from a living invertebrate may be long delayed. For most of human history, bryozoans went unnoticed although they have been around since Cambrian-Ordovician time. Not until the mid-nineteenth century were the anatomy of a bryozoan and that of an anthozoan polyp distinguished.

Focus will be on the subphylum Ectoprocta (*ektos* = outside; *proktos* = the anus) (Figure 6.3, part *A*). There are some 19,000 living and fossil species of ectoprocts, more than three-fourths are known only as fossils. There is no known fossil record of the skeletonless species of the other subphylum, Entoprocta (*ento* = inside). The name Ento- and Ectoprocta refers to the position of the anus, respectively inside or outside the lophophore (see Figure 6.3, parts *B* and *C*).

Entoprocts include slightly more than a handful of genera, but more than 1200 belong to the ectoprocts.

With one exception, the 155 ectoproct families are all marine. Living species, broadly distributed from polar regions to the tropics, range from the shallows to abyssal depths.

A student can simply collect living bryozoans in any fresh-water body or along marine beaches. In fresh water they will be attached to twigs or encrusted on rocks or shells, and on the coast they will commonly be attached to seaweed holdfasts cast on the sand.

LIVING BRYOZOANS

Classification

[Bassler (1953) recognized the phylum Bryozoa and two subphyla, the Ectoprocta and Entoprocta. Hyman (1959) and other biologists argue for suppression of the concept Bryozoa and recognition, instead, of two subphyla named as phyla. Only the Ectoprocta are defined and treated as a subphylum below.]

Subphylum Definition: Ectoprocta.

1. *Size.* Microscopic (individual zooids generally 1.0 mm or less (Figure 6.4, part *Ab*).
2. *Habitat.* Marine (with one exception — see class Phylactolaemata).
3. *Skeleton.* Zooecial walls may be gelatinous, chitinous, or calcified to a greater or lesser degree (Figure 6.3, part *D*). The exoskeletal case enclosing each zooid is a *zooecium*. Students of living forms (ectoproctologists) burn away the soft parts to study better the skeletal anatomy.
4. *Habit.* Sessile; exclusively colonial.
5. *Lophophore.* Circular or horseshoe in shape, surrounding mouth but not the anus; tentacle-bearing (Figure 6.3, part *A*).
6. *Digestive tract.* Recurved, placing anus near mouth.
7. *Organs and systems.* Lacks nephridia and a circulatory system.
8. *Reproduction.* Sexual generation produces free-swimming ciliated larvae that, after attachment, give rise, by asexual reproduction (budding), to a colony. The colony

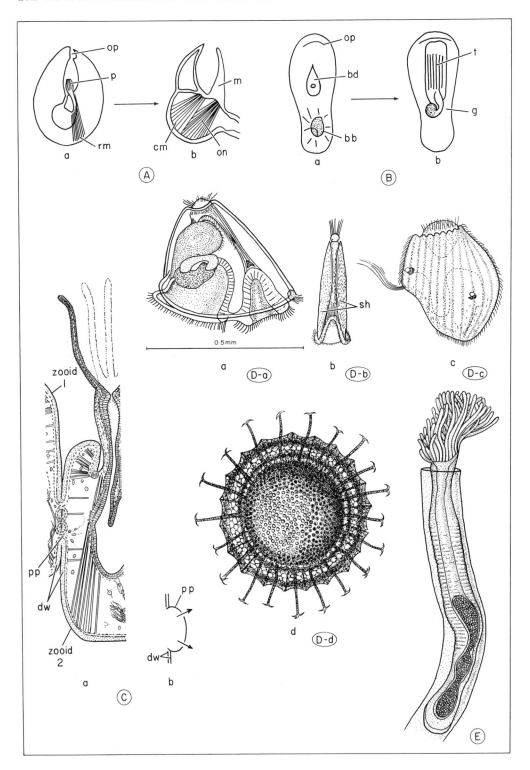

FIGURE 6.1 Ectoproct morphology and structures.

(A) Origin of the avicularia: (a) p—polyp, rm—retractor muscle, op—operculum, m—mandible (a modified operculum); (b) modified polypide, om—opening muscle, cm—closed muscle.

(B) Brown body, representing degenerate polypide: (a) bd—bud, op—operculum, bb—brown body; (b) regeneration of new polyp from brown body[1]; g—gut, t—tentacles.

(C) Cheilostomata: (a)(b) Contiguous zooids 1 and 2 with pore plate between. dw—double wall, pp—pore plate; arrows indicate direction of percolation of fluids from zooid 1 to zooid 2.

(D) Larvae: (a) Cyphonautes larvae (ctenostomes and cheilostomes) develop when eggs are shed into sea without previous brooding; (b) end view of cyphonautes larva, sh—shell, sensory bristles (uppermost on diagram); (c) Bugula (a cheilostome) larva, side view; (d) Cristatella mucedo, fresh-water bryozoan, statoblast with two circlets of hooked processes (schematic).

(E) Phoronis mülleri. A tubiculous worm, length, 3.0 mm; slime tube with some sand grains caking it houses body.

[1]Has not been confirmed by recent workers.

arises from a single progenitor called the *ancestrula* (Figure 6.4, part B1).

9. *Coelum.* Fluid-filled body cavity surrounding alimentary canal; contains reproductive organs (hermaphroditic) (Figure 6.3, part A, "*ov*" and "*te*").

10. *Geologic range.* (?) U. Cambrian; Ord.–Rec.

Affinities and Origin

Three phyla possess lophophores similarly constructed. These are the so-called lophophorate coelomates (that is, having both a lophophore and a coelum): the tubicolcous worm, phylum Phoronida (Figure 6.1, part E), subphylum Ectoprocta, considered in this chapter (Figure 6.4, part A), and the phylum Brachiopoda, discussed in Chapter 7.

The similarities among the three phyla point to a common ancestral type whence they evolved along separate descent lines. Such a type perhaps was wormlike (vermiform) with a distinct head, a lophophoral region and trunk (Hyman, 1959). The absence of a head in the three phyla is attributed to degeneration arising from the mode of life in postlarval stages, that is, attached to the bottom (brachiopods, ectoprocts) or relatively immobile (phoronids). [Compare J. D. Farmer, J. W. Valentine, and R. Cowen, 1973, Systematic Zool., Vol. 22(3), pp. 231–239.]

The Two Classes

All ectoprocts divide into two large classes based on the configuration of the lophophore, position of anus relative to the mouth, and presence or absence of a protective lip overhanging the mouth (called the *epistome*).

Class 1. Gymnolaemata. Circular ridge (lophophore) about mouth bearing tentacles; anus outside lophophore; epistome lacking. Generally, marine; with the exception of a few species of the order Ctenostomata found in fresh water. Three living orders (Ctenostomata, Cyclostomata, and Cheilostomata) and two extinct orders (Trepostomata and Cryptostomata) are embraced by this class. Range: (?) Cambrian; Ord.–Rec.

Class 2. Phylactolaemata. Horseshoe-shaped ridge (lophophore) bearing tentacles surrounds mouth and anus; epistome present. Exclusively fresh-water. Range: Cret.–Rec.

Reproduction and Larval Forms

Most living bryozoans (class Gymnolaemata) are bisexual. The same individual has both female and male reproductive organs, ovary and testes respectively (Figure 6.3, part A), and is therefore self-fertilizing (*hermaphrodites*). Separation of the sexes also occurs. It is common in the cyclostomes and known in several cheilostomes. Both conditions occur even in the same colony.

Although eggs may be shed directly into seawater (among many ctenostomes), *brooding* of fertilized eggs is the more common occurrence among living marine bryozoans. Studies indicate that some bryozoan sperm can survive about 10 minutes in seawater before loss of motility. A variety of sites for brooding are used (both internal and external), including brood chambers that are called *gonozooids*

(Figure 6.4, part *D*). The *coelum* (Figure 6.3 part *A*) is a frequent brooding place.

Living marine bryozoans breed seasonally, from two to six months of the year.

Ctenostomes and cheilostomes have two types of larvae: (1) Nonbrooded eggs shed into the sea, develop into *cyphonautes larvae* (Figure 6.1, part *D*) bearing a digestive tract and two valves that are cast off after attachment. These larvae can remain in the free-swimming stage from two to twelve months before attachment to the bottom. (2) Brooded eggs yield larvae lacking both valves and a digestive tract and so are unable to feed. In turn, this shortens the free-swimming stage. Larvae are round to oval shaped. Cyclostomes also yield nonfeeding larvae that are elongate to oval shaped and have a short, unattached stage.

Asexual development (budding) gives rise to a variety of colonial forms (Figure 6.4, part *Aa*). The process starts from an *ancestrula* (primary zooid). Borg (1926a) detailed the developmental stages for the living cyclostome *Crisia*, among others (Figure 6.4, part *B*). Envision a living polypide (Figure 6.3, *A*) inhabiting each zooecium shown in Figure 6.4, part *B*.

In fresh water, ectoprocts (Phylactolaemata) reproduce sexually or asexually. *Statoblasts* arise by asexual budding (Pennak, 1953, Chapter 12; Figure 6.1*A*, part Dd). A dual function enables the species to endure adverse environmental stress of colder temperature and contributes to a wider geographic distribution. In the spring, statoblasts with chitinous valves germinate and produce a new polypide.

Metamorphosis of Larvae and Calcification of the Ancestrula. Lutaud (1954) studied the metamorphosis of larvae and ancestrula calcification of the cheilostome *Escharoides coccinea*. The period of larva fixation to the bottom lasted from 20 to 25 minutes. Thereafter, over two to three hours, it underwent metamorphosis, that is, changing of larval organs into a mass of cells. Its shape changed from oval to subelliptical and subsequently to rectangular-round. A cuticular cover then developed; three to four hours after fixation, the first traces of very tiny polygonal calcite crystals (about 1 μ thick) were detected with a polarizing microscope. Subsequently patches of larger crystals developed. Twenty-four hours after fixation, all the frontal surface of the ancestrula was impregnated with calcite.

For further data, consult P. A. Sandberg, 1977. "Ultrastructure, mineralogy and development of bryozoan skeletons," *in* R. M. Woollacott and R. L. Zimmer, *Biology of Bryozoa*, Academic Press, pp. 93–128.

Body Case or Wall and Skeletal Structures. The calcareous layer of the cyclostome body wall relates to the outer cuticle and epidermis (Figure 6.3, part *D*). Such calcified layers explain the extensive fossil record of bryozoans.

Interzooidal pores and pore plates (Figure 6.1, part *C*) show that a *double wall* exists between contiguous zooids and that fluids pass between such zooids possibly by slow drip or percolation. Interzooidal pores also occur in ctenostomes and all cyclostomes.

Cumings and Galloway (1915) discussed and illustrated "communication pores" particularly in fossil trepostomes (Figure 6.2, part *A*).

Polymorphism. A complete bryozoan individual (a zooid) may be figured as having two parts, the soft parts called the *polypide*, and the skeleton the *zooecium* (Figure 6.3, part *A*). A colony composed of many individuals is called a *zoarium*. Borg (1926b) and other students of living marine bryozoans distinguish two large groupings of zooids: *Autozooids* (Figure 6.4, part *A*) and *heterozooids* (Figure 6.2, parts *D–F*). The existence of different kinds of zooids is denoted as "polymorphism." Freshwater forms have only autozooids.

Autozooids are designated as normal feeding individuals. Heterozooids embrace a variety of modified zooids differing in one way or another from the normal.

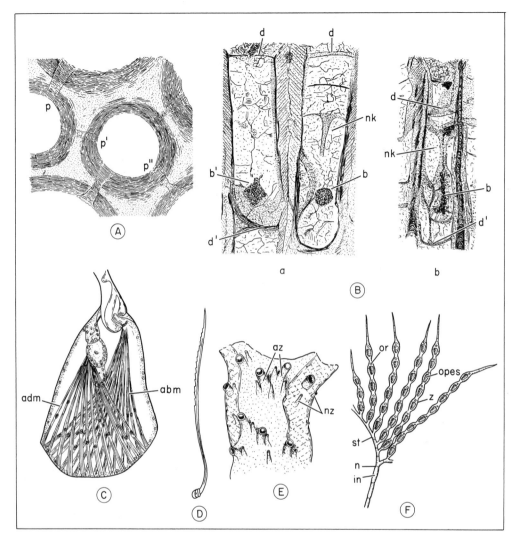

FIGURE 6.2 Ectoproct morphology and structures.

(A) *Heteropora tortilis* (Mio., Petersburg, Va.), tangential section, X 287. Communication pores (*p,p',p"*) between zooids (compare *C, a–b,* Fig. 6.1).

(B) (a) *Heterotrypa subramosa* Ulrich, longitudinal section, X 65; brown bodies; (b) *Batostoma variable,* longitudinal section, X 65, brown bodies in adjacent zooecium; *d* – diaphragm, *nk* – neck of cyst, *b* – brown body.

(C) Living cheilostome, showing the modified zooid that gave rise to the heterozooid vibraculum; *adm, abm* – muscles, upper left object – polypide remains, rotation of vibraculum (a modified operculum) occurs at apex by muscular action. It serves a sanitation function.

(D) Toothed vibracula bristle, schematic.

(E) *Diplosolen,* a cyclostome with modified autozooids: *az* – autozooid, *nz* – nannozooid.

(F) *Cothurnicella,* a stoloniferous cheilostome: *n* – node, *in* – internode, *o* – orifice, *z* – zooid, *st* – stolon. (Note: Stolons are kenozooids;) *ope* – opesium (= a large opening).

Sources: (Figure 6.1): (A)(B) – G. A. Kerkut, ed., *The Invertebrata,* 3rd ed., 1959, by Borradaile and Potts, by permission of Cambridge Univ. Press; (C) Marcus, E., 1926, Bryozoa, in G. Grimpe and E. Wagler (Eds), *Die Tierwelt der Nord-und Ostsee.* Part 7: (Da)(E) – Thorson, 1946. (Db)(Dc) – several authors; (Dd) – Pennak, 1953 (Figure 6.2): (A)(B) Cumings and Galloway; (C) Marcus, E., 1939 *Bryozoarios marinhas Brasileiros,* Vol. 3; (D) Harmer, S. F. 1926. The Polyzoa of the Siboga Expedition. Siboga Exped. Monogr., 4 parts (1945–1957); (E)(F) Calvet. Levinson and others.

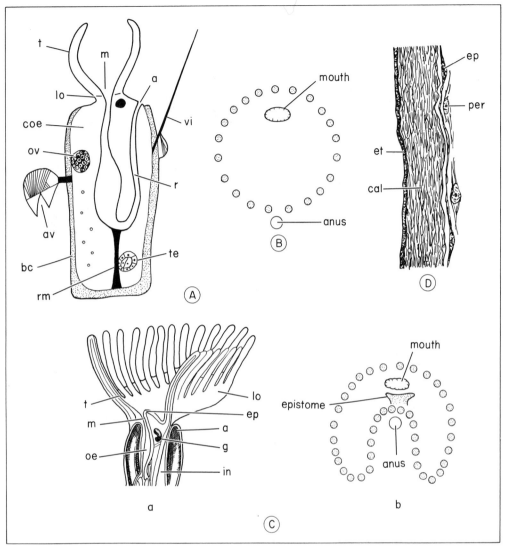

FIGURE 6.3 Ectoproct morphology.
(A) Schematic of an ectoproct; *bc*—body case, *rm*—retractor muscle, *m*—mouth, *a*—anus, *lo*—lophophore, *ov*—ovary, *te*—testes, *t*—tentacle, *coe*—coelum, *vi*—vibracula, *av*—avicularia, *r*—rectum.
(B) Gymnolaemata—diagram of apertural view of lophophore (compare "*C*").
(C) Phylactolaemata—two diagrammatic views of lophophore: (a) *Plumatella*, right half view of portion; (b) apertural view of lophophore; *ep*—epistome; *m, t, a, e*, as in (A) above.
(D) Schematic of body case or wall: *et* (= *ct*)—chitinous cuticle; *cal*—calcareous layer; *ep*—epidermis; *per*—peritoneum.
Sources: (A)–(C) G. A. Kerkut, ed., *The Invertebrata*, 3rd ed., 1959, by Borradaile and Potts, with permission of Cambridge Univ. Press; (D) Borg, 1926a.

All marine bryozoans display greater or lesser polymorphism. Emphasis will be given to the heterozooids.

In general, heterozooids result from a reduction of the polypide and consequent loss of nutritive and reproductive functions. The resultant zooid then serves another function. Two heterozooids

found only in the cheilostomes are the *vibracula* and the *avicularia*. The vibracula, although of restricted occurrence, nonetheless have intriguing structures. They are autozooids modified to form long, moving bristles that sweep the surface clean of debris and attached larvae (Figure 6.2, part *D*). Given an autozooid with polypide reduced, the operculum modifies into a long bristle as much as ten times the original length of the zooid; the resultant is a vibracula. By muscular action and proper articulation (Figure 6.2, part C). The bristle has great freedom of movement in any direction.

Given an autozooid with modification favoring the development of the operculum into a mandible (figuratively reminiscent of a bird's lower beak), and the related muscle apparatus into an enlarged muscle mass that fills coelom and controls the movement of the mandible—then the essential condition for an avicularia has been established (Figure 6.3, part *A*).

The avicularia also serves a sanitary function. When larvae of other animals attempt to encrust the outer surface of a cheilostome, the viselike grip of the avicularia immobilizes them.

Other types of heterozooids include *gonozooids* (autozooids modified for reproduction, Figure 6.4, part *D* 10) and *kenozooids* that are represented only by a body wall lacking the polypide, muscles, and so on. Kenozooids are common in some cyclostomes, and the stoloniferous ctenostomes; known in all orders, they occur less frequently. Examples of kenozooids are stolons, stalks, rhizoids (Figure 6.2, part *F*). Nannozooids (dwarf zooids) are also known in some cyclostomes and cheilostomes (Figure 6.2, part *E*). These are modified autozooids with a reduced and functionless digestive tract, a single tentacle, and so on. Hyman (1959) summed up the data on the interchangeability of various zooids. For example, kenozooids (stolons) transform into autozooids (buds).

How does polymorphism work genetically? Given *n* autozooids in a specified colony, plus numerous colonies, then natural selection must have operated on individual small variations existing in autozooids. Assume that the original genetic code specified a variable range of polypide size in autozooids in a particular species from large to small. Then, after modification on the molecular level through multi-

ple, multiple generations of autozooids, the new specification in selected instances could have been for polypide reduction, up to the total absence of any polypide at all: the kenozooids. However, since kenozooids (stolons) can transform into autozooids (buds), and a great interchangeability of zooid types exists also, it is more likely that "several alternate gene arrangements" (Mayr, 1963, pp. 150 ff.)[1] continue to coexist in a given population. Which arrangement(s) find expression at any given time also may be sponsored by selective factors in the environment.[1]

Brown Bodies. Many zooids possess brown bodies (Figure 6.1, part B). These are degenerated polypides. Multiple bodies of this kind may occur in the same zooid, and they indicate repeated polypide degeneration. Adverse oxygen, food, or temperature conditions, natural aging, or the onset of sexual reproduction—all these factors can account for degeneration (Hyman, 1959).

Such bodies are actually brown masses of disintegrated soft parts: tentacles, muscles, and so on. Ultimately they may be extruded via the anus in some cheilostomes and ctenostomes.

Brown bodies have been identified in fossil trepostomes (Figure 6.2, part *Ba*).

Colony Form, Size, Rate of Growth, and Substrates. There are no known solitary bryozoans. All are colonial. Most calcareous ectoprocts incrust; hence this dominates colonial forms. Other forms include variously branched (Figure 6.4, parts *A*, *D*) and stoloniferous (Figure 6.2, part *F*) colonies.

Substrates for incrustations of marine forms include almost all or any bottom debris or biogenic detritus [molluscan shells, other shell fragments, algal fronds, crustacean carapaces, and appendages (Osburn, 1957), even the bottom sand itself]. Bryozoan-algal relationships, as incrusting colony and substrata, respectively, also

[1] Ref. Ch. 14.

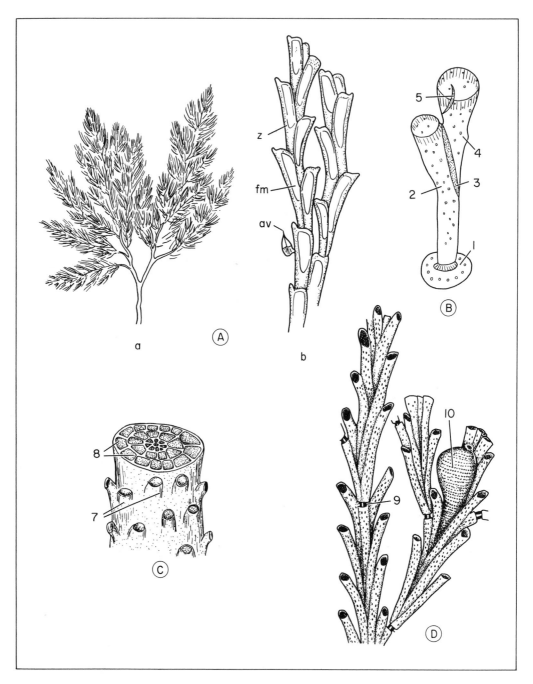

FIGURE 6.4 Ectoproct morphology.
(*A*) *Bugula*, a living cheilostome: (*a*) Colony of many branches; (*b*) detail of branch showing biserial zooids: (*z*) zooid; (*fm*) frontal membrane; (*av*) avicularia.
(*B*) *Crisia*, a living cyclostome: (1) Ancestrula; (2) Calcareous cylinder of first zooecium; (3) partition (septum) of second zooecium; (4) second zooecium; (5) partition (septum) of third zooecium.
(*C*) *Entalophora*, a living cyclostome: (7) Zooecia; (8) section of fused zooecia.
(*D*) *Crisia*: (9) Nearly separate tubular zooecia; (10) gonozooid (= brood chamber).
Sources: (*A*) Hyman, 1959 (from *The Invertebrates*, Vol. V, by Libbie H. Hyman. Copyright © 1959 by McGraw-Hill, Inc. with permission of McGraw-Hill Book Co.) and mounted laboratory specimen; (*B*) Borg, 1926*b*; (*C*) (*D*) after several authors.

occur in fossils (*Cambroporella*, Figure 6.5 part *A*). Paleozoic forms may reveal a symbiotic partnership (= bryozoan-algal *consortium*, Figure 6.5, part *B*). See Elias, 1954; Condra and Elias, 1944; Rigby, 1957.

Attachment to the substrate usually involves hooked, fibrous, or rhizoid types of stolons. Some forms, not incrusting, rather excavate burrows in shells. Silén (1956) reported a new shell-burrowing species (family Penetrantiidae, order Ctenostomata) in shells of the living marine pelecypod *Mytilus canaliculus* from New Zealand. He attributed this shell-burrowing capacity to the phosphoric acid secretion by the stolon ends.

Ectoproct colonies range in size from minute to gigantic as in some living cheilostomes. [Parallels to living occurrences may be seen in some stoloniferous ctenostomes incrusted on fossil brachiopod valves. Some fossil *Archimedes* (Figure 6.9, parts *Aa, Ad*) colonies attained considerable size.] Generally, however, smallness characterized colonies. Heights may attain 90 cm. *Bugula* (Figure 6.4, part *A*) attains heights ranging from as little as $\frac{1}{2}$ in. to as much as several inches.

Incrustations may spread over part, even the entire length of a suitable substrate. The giant fronds of the kelp *Laminaria* rise above their holdfasts 6 ft or more. Landsborough (cit. Hyman, 1959) reported a cheilostome colony of *Membranipora* estimated to consist of 2,300,000 zooids extended over an area 5 ft long by 8 in. wide on such a frond. This is a good example of niche preference where selective advantage results.

Colony growth *rate*, starting with the ancestrula, has been recorded by many investigators of cheilostomes. In the Adriatic, *Schizoporella sanguinea* colonies during the months of May through October displayed growth on suspended boards: June, 25 zooids; July, 400–500 zooids; October, 30,000 zooids. A Chesapeake Bay species of *Membranipora* had 10,000 zooids in the colony at the end of three weeks, while elsewhere another species of this genus yielded 20 zooids in two weeks. Clearly, rates of growth vary between and within genera (Friedl, Osburn, Matawari, O'Donoghue, cit. Hyman, 1959). (Large-size bryozoan fossils may represent relatively small periods of time reckoned by an inferred rate of growth.)

Distribution. As Osburn (1957) observed, Recent species are cosmopolitan and are found everywhere in salt water. They range from the Arctic to the Tropics, and they extend from shoreline deeper than 3000 fathoms. Many forms are circumpolar or circumtropical.

Deep-water genera are identical to shelf forms, but the species differ in the two habitats. Menzies and Imbrie (1958) reported that only geologically younger forms of bryozoans (that is, Tertiary–Recent) are found in the abyssal depths, and that Paleozoic types (Cyclostomata) are absent. This points to an invasion of deep sea by shelf cheilostomes, probably during the Tertiary.

Temperature and Salinity. Distribution of Recent bryozoans is essentially limited by temperature. Some species are limited to the Arctic or to Tropics, whereas others span these two regions.

Salinity decidedly affects development of bryozoan species. Normal seawater, or slightly lower than normal, presents optimal conditions. On the other hand, brackish estuarine waters (below 12 parts per 1000) support very few species. Osburn (1957) stresses, however, in these brackish water species, that population size and hence reproduction are not necessarily reduced.

Space Competition. The same living space on a piece of substrate—for example, a large pelecypod shell (*Pinna*)—may be the common site for attachment of as many as 40 bryozoan species.

FOSSIL ECTOPROCTS

Taxonomy

All ectoprocts, fossil and living, belong to five orders. (See subphylum definition.)

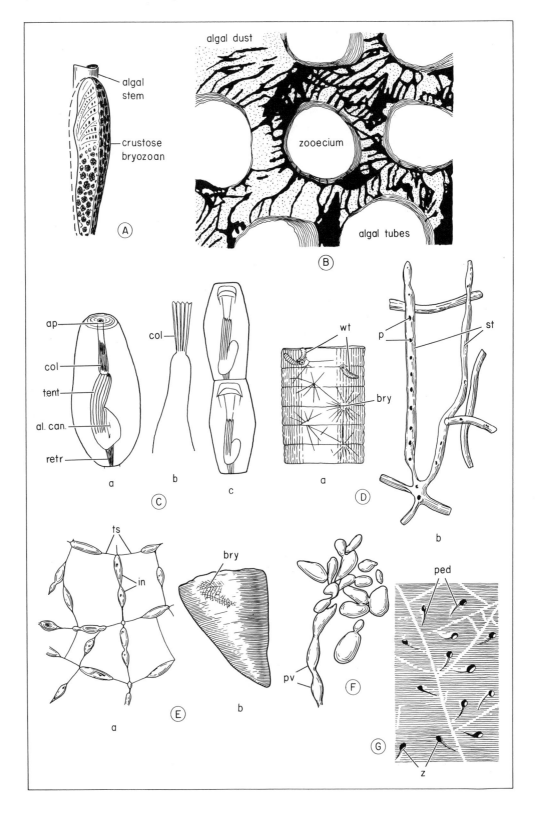

FIGURE 6.5 Bryozoa-algal relationships and living and fossil ctenostomes.
(A) *Cambroporella tuvensis* (L. Cam. Tuva); restoration of *Cambroporella*, a crustose bryozoa attached to stem of an alga.
(B) *Solenopora texana*, an alga, and *Acanthocladia guadalupensis*, a bryozoan-algal consortium, X 157; algal dust, algal tube (both of *S. texana*), z — zooecium of *Acanthocladia*.
(C) Living ctenostomes, *Valkeria*, Japan: (a) Schematic of zooid; (b) terminal end of zooid showing terminus of protracted collar (comblike, which explains the name "ctenostome"); (c) *Flustrella hispida*, schematic, 2 zooids with lips that close aperture; *ap* — aperture, *col* — collar, *tent* — tentacles, al. can. — alimentary canal, *retr* — retractor muscle.
(D)–(G) Representative fossil ctenostomes:
(D) *Vinella radiata* (U. Ord., Maysville, Ohio), X 1. (a) Encrustation on crinoid stem; note worm tubes (*wt*); (b) detail of *Vinella repens* (M. Ord., Minn.) similar to the encrustation on (a): *st* — stolon, *p* — pores (zooecia unknown). Note: stolons are kenozooids; see text.
(E) *Ropalonaria venosa* (U. Ord., Richmondian, Ohio), X 1. (a) detail of (b) encrusting rugose coral *Streptelasma* (see Chapter 5). (*in*) fusiform internodes or cells, *ts* — delicate tubular stolons partly embedded in host.
(F) *Ascodictyon fusiforme* (M. Dev., Mich.), X 10. A parasitic form showing little development of threadlike branching stolons, and clusters or isolates of pyriform vesicles (*pv*).
(G) *Spathipora sertum* (Mio., France), X 25, stolons perforate shell enamel; zooids (*z*) elongate, fusiform, and with long peduncle (*ped*) attached to buried stolons.

Each name order ends in the stem "-stomata" (= mouth). Alternate stem uses are "-stoma," "-stome."

Order 1. Ctenostomata (Greek ctenon = comb + mouth).

Zooecia membranous (chitinous or gelatinous). Simple mode of budding; zooids bud from tubular stolons that may be partially calcified. Terminal aperture (Figure 6.5, part *C, a*) closed by flexible body wall fold (= *collar*, Figure 6.5, parts *C, b-c*) bears comblike row of setae. Habit, strikingly commensural; in some shells, zoarium completely concealed (Figure 6.5, parts *D, E, G*). Range: Ord. – Rec. Representative genera include *Vinella* (Figure 6.5, part *D*), *Ropalonaria* (Figure 6.5, part *E*), *Ascodictyon* (6.5, *F*), *Spathipora* (6.5, *G*). Thin sections, generally not needed to identify Paleozoic ctenostomes.

Order 2. Cyclostomata (Greek cyclos = circle + mouth).

Calcareous tubular zooecia with circular apertures; walls thin, porous (*pseudopores*); polymorphism expressed in ovicells (= gonoecium) where reproduction occurs (Figure 6.6, part *K, b*); and in accessory tubes collectively called *kenozooecia*. The latter include short tubes (*nematopores*), pores with polygonal orifices (*tergopores*), vacuoles, and *mesopores*. Range: (?) U. Cambrian; Ord.–Rec. Hyman (1959) thought use of the "-pore" stem obscures the true polymorphic character of ecto-

proct heterozooids and favored "mesozooids" to "mesopores," and so on.

Cyclostome Features Observed in Thin Sections

Tangential Section. *Cross-section* of zooecia (circular to polygonal); *lunaria* [= crescent-moon shaped projection of peristome on proximal (posterior) side of zooecial tube, Figure 6.6, part *J, a*]; *vesicular interzooecial tissue*.

Longitudinal Section. Zooecial wall thickness and length; diaphragms (= calcareous platform or floor crossing zooecial tube or mesopore, Figure 6.6, part *H, a*); interzooecial vesicles; mesopores smaller than zooecia contain numerous diaphragms.

Representative Cyclostome Genera. *Fistulipora* (Figure 6.6, part *J, a–b*), *Ceramopora* (Figure 6.6, part *H, a–c*), *Meekopora* (6.6, part *I, a–c*), *Petalopora* (6.6, part *K, a–c*).

Order 3. Trepostomata (Greek trepos = change + mouth).

The name refers to a morphological change in the zooecial tube character during the zooid growth. Further details on immature and mature regions are below and in Figure 6.7, part *D, c*). Zoaria massive, lamellate, or stemlike. End walls called *diaphragms* when straight, and *cystiphragms* when curved, transversely partition long calcareous tubes (= zooecia). Each tube divides into an

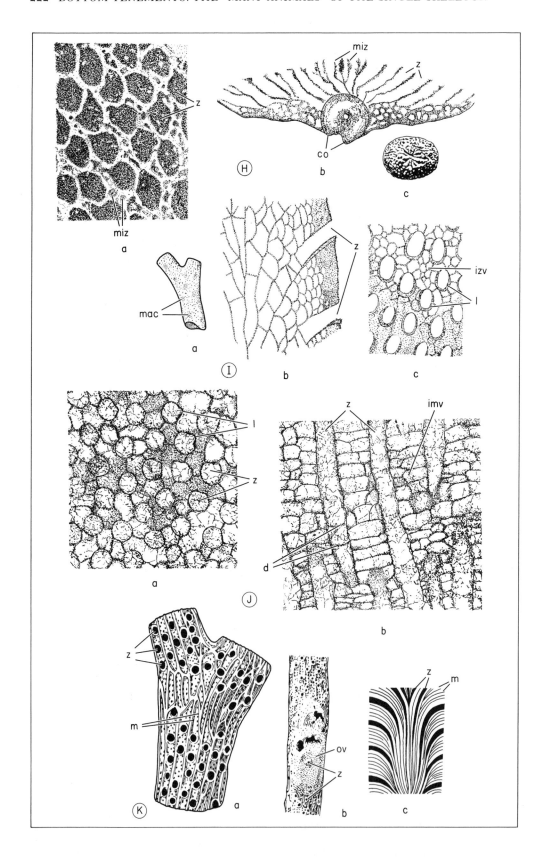

FIGURE 6.6 Representative cyclostomes.
(*H*) *Ceramopora imbricata:* (a) Tangential section showing zooecia (*z*), and interzooecial mesopores (*miz*) (Osgood Formation, Niagara, Indiana), X 20. (*b*) Longitudinal section of small zoarium attached to crinoid ossicle (*co*); (*z*) and (*miz*) as in part (*a*) (Silurian, Clintonian, N.Y., X 10). (*c*) Same as (*b*) but zoarium before sectioning, X 2.
(*I*) *Meekopora clausa:* (a) Zoarium X 1 (*mac* = maculae = clusters of kenozooecia). (*b*) Longitudinal section with interzooecial vesicles and zooecia (*z*), X 20. (*c*) Tangential section showing lunaria (*l*) and interzooecial vesicles (*izv*), X 20. Parts (a)–(c) from Miss. (Chesterian), Ky.
(*J*) *Fistulipora hemisphaerica,* Brownport Formation (Niagaran), Tenn.: (a) Tangential section near zoarial surface showing zooecia (*z*), lunaria (*l*), and interzooecial vesicles, X 20. (*b*) Longitudinal section showing zooecia (*z*), interzooecial vesicle in v. Note imbrications and interlocking columns; and diaphragms (*d*) in "z", X 20.
(*K*) *Petalopora costata,* Cret., France: (a) Surface depicting longitudinally arranged mesopores (*m*) and walls perforated by zooecial tubes (*z*), X 25. (*b*) Surface showing ovicell (*ov*) and zooecia (*z*), X 10; (*c*) Longitudinal section with walls perforated by zooecial tubes (*z*) and mesopores (*m*), X 25.
Sources [Figures 6.5 and 6.6: (*A*) Elias; (*B*) Rigby; (*C*) Silén; (*D*)–(*G*) (Hb) (*I*) (*K*) Bassler, 1953; (*Ha*) and (*J*) Perry and Hattin.

exozone and an *endozone*. The exozone, the *mature region*, occurs near the zoarial surface and is characterized by thickened walls, close spacing of diaphragms, and heterozooids positioned between zooecia (such as mesopores or acanthopores). The endozone, formerly called the *immature zone*, is situated in the axial part; characterized by thin walls, wide spacing of diaphragms, it contacts other zooecia on all sides (see Figure 6.7, part *D*, *c*, and Boardman, 1959). Zoarial surface commonly bears *monticules* (tubercles) and spotlike areas (*maculae*) at the level of, or below, the zoarial surface. Range: Ord. – Perm., (?) Triassic.

The order has two suborders: Suborder Amalgamata (with more or less complete coalescence of zooecial walls). Representative amalgamate genera: *Prasopora* (Figure 6.7, part *B*), *Leptotrypella* (Figure 6.7, part *C*), *Dekayella* (6.7, part *D*), *Eridotrypa* (6.7, part *E*), *Constellaria* (6.7, part *F*). The second suborder, the Integrata (individual zooecial wall boundaries distinct, noncoalesced), are represented by: *Trematopora* (Figure 6.7, part *G*), *Hallopora* (6.7, part *H*), *Anaphragma* (6.7, part *A*).

Boardman (1960*b*) showed that *Anaphragma mirabile*, in tangential section of young growth stage and in a thin-walled zoarium, had an Integrate appearance (Figure 6.7, part *A,a*). The section of an advanced growth stage, in a thick-walled zoarium, displayed a broadly Amalgamate

appearance. Sparling (1964) figured several tangential sections of one Amalgamata, genus *Prasopora simulatrix* (Figure 6.7, part *B*. These ranged from almost complete coalescence of walls (plate 161, Figure 2) to isolation of zooecia by numerous mesopores (plate 161, Figure 5) with wall boundaries commonly distinct. In *Anaphragma* and other forms, such variable relationships probably served a growth stage function.

Boardman (1960*a*), in a trepostome study from the Hamilton Group, proposed a new approach to placing such forms in suborders. Abandoning the unusable Amalgamata-Integrata suborders, he established two yet unnamed "main groups" (of possible subordinal value) based on longitudinal and tangential thin-section evidence.

Group 1 (= Suborder?). Genera characterized by laminae bending in a sweeping curve approaching the zooecial boundary (Figure 6.8, D); intersect the boundary at right angles (Figure 6.8, *A–D*).

Group 2 (= Suborder?). Genera with wall laminae remaining fairly straight approaching the zooecial boundary (Figure 6.8, *E*).

With two main groups, Boardman distinguished five types of wall structure: four types in Group 1 (Figure 6.8, *A–D*) and one type in Group 2 (Figure 6.8; *E*). These types were each named for a genus that characteristically displayed the structure.

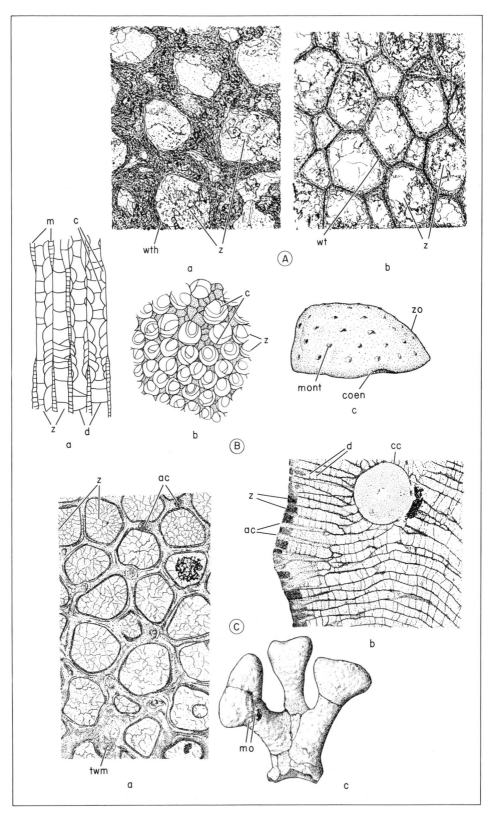

FIGURE 6.7

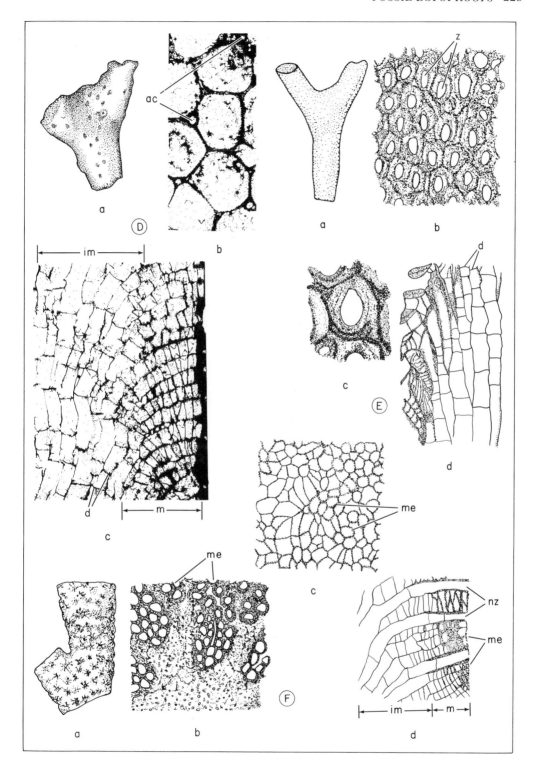

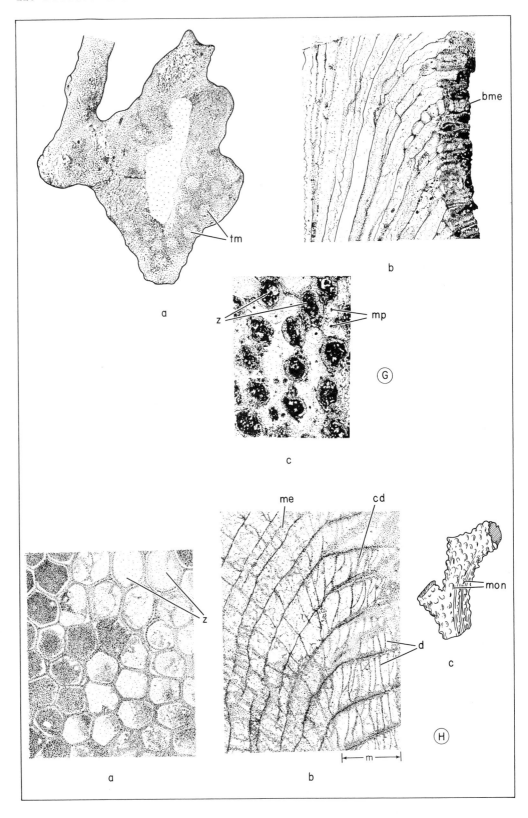

FIGURE 6.7 Trepostome genera – II.

(A) *Anaphragma mirabile*, Richmond Group, Illinois: (a) Tangential section, broadly *amalgamate* appearance (= advanced growth stage). X 50; (b) tangential section of another specimen illustrating the *integrate* appearance; z – zooecia, wth – thick wall; wt – thin wall, X 50.

(B) *Prasopora*: (a) *P. simulatrix* (M. Ord., Michigan), vertical section, m – mesopore, c – cystiphragms, z – zooecia, d – diaphragms, X 20; (b) tangential section, X 20; (c) *Prasopora sp.*, zo – zoarium, mont – monticule, coen – coenochyme, schematic.

(C) *Leptotrypella* (*P.*) *tuberata*, Hamilton Group, N.Y.S.; (a) Tangential section, z – zooecia, ac – acanthopores, twm – thickened wall of monticule, X 50; (b) longitudinal section, d – diaphragm, cc – crinoid columnal with zooecial growth around it, X 10; (c) zoarium, showing subramose growth and monticules, X 1/2.

(D) *Dekayella, D. praenuntia:* (a) Solid stony branch, X 2/3; (b) surface of same, X 6, *D. praenuntia echinata* (M. Ord., Wisconsin); (c) longitudinal section, X 60; zw – zooecial walls, m – mature zone, im – immature zone, ac – acanthopore.

(E) *Eridotrypa mutabilis*, Ord., Minn.: (a) Zoarium, X 1; (b) tangential section with thick walls, X 20; (c) same as (b) but detail, X 50; (d) longitudinal section, thick walls in mature region, and widely spread diaphragms (d) in immature, X 20.

(F) *Constellaria constellata*, Ord., Ohio: (a) zoarium, surface with stellate clusters, X1; (b) (c) tangential sections, clusters of mesopores (me) X 20; (d) longitudinal section, mature (m) and immature (im) region, spacing of diaphragms, clusters of mesopores, and normal zooids (nz), X 20.

(G) *Trematopora tuberculosa*, Sil. (Clinton), N.Y.: (a) External view of zoarium, tm – tuberculated monticules, X 1; (b) longitudinal section of (a), beaded mesopore chambers (bme) in inner region of exozone (= mature zone), X 20; (c) a different specimen, tangential view, smaller central pores (mp) in mesopore diaphragms of outer region of exozone, X 50.

(H) *Hallopora lydiana*, M. Ord., Cobourg beds, Ottawa, Canada: (a) Tangential section, mesopores practically absent, 6 to 7 zooecia/2mm usually, z – zooecia X 25; (b) same specimen, vertical section, note increase of diaphragms (d) in mature region (m), cd – tendency of some diaphragms to be cystose though most are horizontal, me – mesopore in immature region: (c) *H. ramosa*, zoarium with monticules (mon).

Sources: (A)(C), (G) – Boardman, 1959, 1960a, b; (B, a, b) Sparling, 1964; (Bc) – Tasch collection; (D, a, b) (E) (F) (H, c) Bassler, 1922. The Bryozoa or moss animals: Smithsonian Inst. Ann. Report 1920, pp. 339–380; 1953; (D, c–d) Perry, 1962; (H, a, b) Fritz, 1957).

Trepostome Features Observed in Thin Sections

Tangential Section. *Cross-section* of zooecia (generally polygonal with walls in contact, that is, amalgamate, or noncoalesced, that is, integrate); monticules and clusters of mesopores or mesopores between zooecia; acanthopores; cystiphragms (as in *Prasopora*). (Figure 6.10).

Longitudinal Section. Immature and mature zones; diaphragms; acanthopores; zooecial tubes and wall structures; mesopores; diaphragm pores; cystiphragms and cystose diaphragms (see Figure 6.7, cd).

Order 4. Cryptostomata (Greek cryptos = hidden + mouth).
Calcareous zoaria of fronds or stems; fronds delicate, reticulate; stems, slender; branchings of cylindrical or ribbon-like form. Similar to Trepostomata, except more abruptly differentiates between mature and immature region; relative shortness of zooecial tubes with each having an *orificial collar* (= "vestibule" of several authors). The collar extends from surface aperture to near the inner boundary of the mature zone often defined by a projecting shelflike *hemisepta* (= semidiaphragms). Interspaces between adjacent "orificial collars" filled by tissue (referred to as *vesiculose coenosteum* or *solid stereom*). Monticules and maculae occur. Polymorphism shows in two sizes of acanthopores (= acanthozooids) and in mesopores (= mesozooids). Range: Ord.– Perm. Representative genera include *Archimedes* (Figure 6.9, part *A*), *Fenestella* (Figure 6.9, part *B*), *Pachydictya* (6.9, part *C*), *Stictopora* (6.9, part *D*), *Rhombopora* (6.9, part *F*), and *Philodictya* (6.9, part *E*).

Order 5. Cheilostomata (Greek cheilos = lip + mouth).
Name refers to a hinged chitinous lip (operculum) closing aperture when the polypide retracts. Colonies branching, lamellate or incrusting. Zooecia short, boxlike (but generally rounded or angular chambers); arranged serially and side by side but with separate walls;

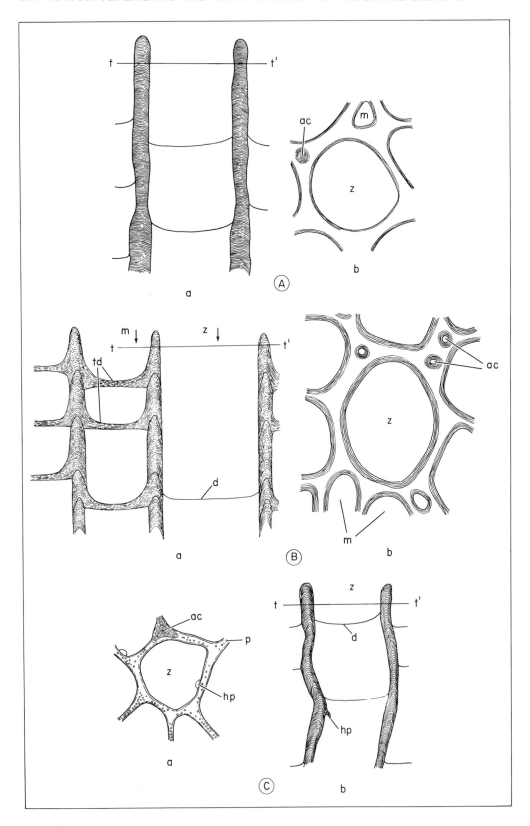

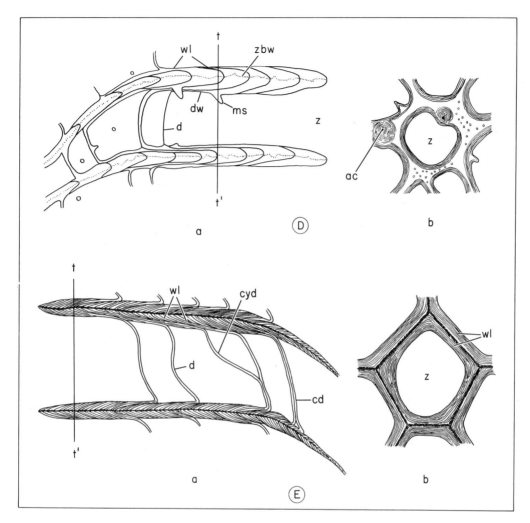

FIGURE 6.8 Five types of wall structures in M. Dev. Trepostomes (after Boardman, 1960a). Two groups (Suborders?): Group I, *A–D*; Group II, *E*.
Line *T–T'* in longitudinal view indicates position of trace of the plane of the tangential section; *z* – zooecia, *m* – mesopore (= mesozooid), *ac* – acanthopore (= acanthozooid), *hp* – heterophragm (= hooklike projection); *ms* – mural spine, *cyd* – cystoidal diaphragm, *cd* – compound diaphragm, *zbw* – zooecial wall boundary, *dw* – diaphragm wall, *d* – diaphragm, *wl* – wall laminae.
Entire classification is based on relative curvature of wall laminae as they approach *zbw*.

zooecial walls (frontal, basal, and four vertical) uniformly calcified except the frontal one (= frontal membrane, Figure 6.11, part *A*) which is chitinous (see also Lutaud, discussed earlier); the portion bearing the aperture, the "aperture field"; both calcified and chitinous, lateral and distal walls perforated by communication pores (Figure 6.11, part *C*); boxlike zooecia bear subterminal orifice provided with a hinged lip (operculum); polymor-

phism expressed in avicularia, vibracula (Figure 6.2, part *D*), and other heterozooids such as spines or dwarfed zooids (Silén, 1942) (Figure 6.2, part *E*, "nz"). Eggs matured in ooecia or ovicells called *brood chambers*. Range: (?)M. Jur.; Cret.– Rec.

Two suborders are recognized by whether zooids lack or have a compensatrix (= *ascus*, an internal sac opened to the exterior to regulate water pressure as

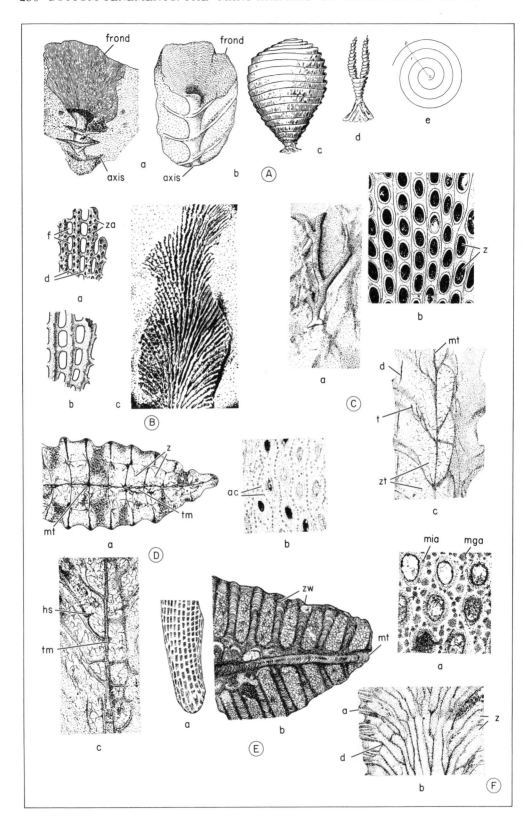

FIGURE 6.9 Representative cryptostome genera.

(A) *Archimedes:* (a) *A. wortheni* (Warsaw beds, Miss. Ill.), a portion of the zoarial network (*Fenestella*-like), and the screwlike axis of laminated tissue, schematic; (b) *A. stuckenbergi* (Kinderhookian, Missouri), portion of spirally twisted *Fenestella*-like frond, schematic; (c) (d) reconstructions of *A. wortheni* by M. Elias; (e) *Archimedes* spiral consists of successive and equal whorls, O — point of origin, r — straight line with extremity at O and revolving about it ($=$ radius vector), P — a point traveling along radius vector at a specified velocity. Equation of spiral: $r = OP = a\theta$ (where a = constant, θ = whole angle through which it has revolved). The increase here is an arithmetical progression; for a geometrical progression, $r = a^\theta$ where succeeding whorls increase in breadth — compare spiral axes of (A, a), (A, b), (A, d). After D'arcy Thompson, 1942, p. 752.

(B) *Fenestella alternata*. Chouteau group: (a) Palmate zoarium, X 1; (b) *Fenestella parallela*, obverse side, fenestrules (f), dissepiments (d), and arrangement of circular zooecial apertures (za), X 10; (c) reverse side, conical nodes, fenestrules, dissepiments, and note absence of za, X 10.

(C) *Pachydictya acuta acuta*, Champlainian, Ord., N.Y.S.: (a) Common aspect of bifurcate bifoliate branches of a colony, X 1; (b) tangential section showing arrangement of zooecial openings (z), X 20; (c) longitudinal section, mesotheca (mt) (when two series of zooecia grow back to back in bifoliate zoarium, a double laminae or mesotheca is formed); wide zooecial tubes (zt), nearly terminal diaphragms (d), and narrow tabulate interspaces between zooecia (t), X 20.

(D) *Stictopora blackensis*, Champlainian, N.Y.: (a) Transverse section showing rectangular shape of zooecial section (z) in mesothecal region (mt), median tubule (m), a small pore in mesotheca, and diaphragms, X 50; (b) tangential section showing zooecial openings and zooecial walls penetrated by acanthopores (= acanthozooids), X 50; (c) oblique longitudinal section, median tubule (tm), hemisepta (hs) (= shelflike platform in zooecial tube joined to wall on one side but extends only part way across tube), X 50.

(E) *Philodictya lanceolata*, Goldfuss: (a) Proximal (posterior) tip of specimen, X 5; (b) transverse section in lateral part of zooecium, lamellate, mesothecal plane (mt) and zooecial walls (zw), X 50.

(F) *Rhombopora lepidodendroides*, Beil Is., Perm., Kansas: (a) Tangential section, two sizes of acanthopores (= acanthozooids), megacanthopores (mga), micracanthopores (mia), and zooecial apertures (z), X 50; (b) longitudinal section, a — acanthopores, d — diaphragms; zooecial wall thickness increases in mature region, X 20.

Sources: (A) Condra and Elias, 1944; (B) Koenig, 1958; Jour. Paleontology, Vol. 32(1) pp. 126–143, pls 21, 22. (C) (D) J. P. Ross, 1964a; (E) J. P. Ross, 1960; (F) Perkins et al, 1962.

the polypides protruded or were withdrawn; a polypide can emerge from zooecium only if this extrusion is compensated by an equal volume of water, see Figure 6.11, part D).

Suborder 1. Anasca. Compensatrix absent; instead, the external hydrostatic system forms a cavity situated between cryptocysts (see Figure 6.11, part B, for definition) and the ectocyst. (The *ectocyst*, the chitinous outermost layer, is differentiated from the *endocyst*, the very fine epithelial membrane that constitutes the living part of the bryozoan.) A calcareous, chitinous secretion forms the zoarial skeleton occurring between the ectocyst and endocyst (Canu and Bassler, 1920, p. 46).

Suborder 2. Ascophora. Compensatrix present (Figure 6.11, part D). Representative genera include suborder Anasca:

Bugula (Figure 6.4, part *A*), *Reginella* (Figure 6.11, part *E*), *Membranipora* (Figure 6.11, part *A, a–b*), *Callopora* (Figure 6.11, part *B*). Suborder Ascophora; *Psilosecos* (Figure 6.11, part *G, a–c*).

INFORMATION DERIVED FROM FOSSIL ECTOPROCTS

Stratigraphy

J. P. Ross (1964a), in a study of New York State Black Riveran and Trentonian (U. Ord.) cryptostome bryozoans, found that certain species had a restricted vertical range useful in stratigraphic zonation. Thus two subspecies of *Stictopora labyrinthica* did not extend above the Black Riveran (Chaumont Formation, Figure 6.9, part *D*); a few subspecies of *Pachydictya acuta* were confined to Trentonian Shore-

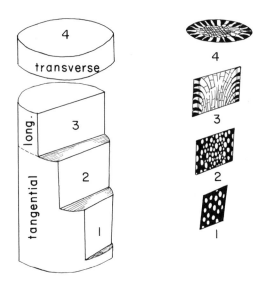

FIGURE 6.10 Thin sections of a trepostome (Paleozoic). Left figure, schematic representation indicating orientations for respective cuts on a given specimen. Right figures show changing aspects of internal structures from one type of thin section to another. Often for species identification several different types of sections are needed. 1, 2-Tangential section; 1, shallow; 2, deep; 3-Longitudinal section; 4-Transverse section. Numbers correspond in left and right figures. (Redrawn, and adapted from Ross and Ross, 1965).

ham Formation and basal Denmark Formation; another species spanned the Upper Denmark Formation and the Lower Coburg Formation (*Championdictya*).

Still other species with a more extensive vertical spread cut across several formations. Even so, using *overlapping stratigraphic ranges* characterizes formations lacking distinctive index species. The Kirkfield Formation, for example (Figure 6.12, part *A*), would contain a suite including *Escharopora recta, Pachydictya acuta acuta, P. acuta tabulata, P. acuta tristolone, Diploclema trentonense,* and *Stictopora blackensis*. The formation below it would include all of these except *Diploclema*; the formations above it would have the same suite plus the *P. acuta* subspecies and varieties indicated in Figure 6.12, part *A*. Adequate differences distinguish a particular formation.

The Ottawa formation (M. Ord., Ottawa–St. Lawrence Lowland) contained chiefly a trepostome fauna. Fritz (1957) saw a striking similarity of fauna to Loeb-

lich's Bromide fauna from Oklahoma (1942), Troedsson's M.–U. Ord. Northern Greenland fauna (Cape Calhoun Formation), as well as faunas of equivalent age from Minnesota and the Baltic. Perry (1962), in turn, recognized that various elements of the M. Ord. Spechts Ferry bryozoan fauna of Illinois, Wisconsin, and Iowa shared commonly with equivalent beds in Minnesota and the Baltic and with Loeblich's and Fritz' faunas.

Sparling (1964) showed species and subspecies of a single trepostome genus (*Prasopora*, Figure 6.12, part *B*) helped to zone a M. Ord. core from Michigan and thus permitted correlations with equivalent age beds in Minnesota.

Boardman (1960a) recognized morphologically distinct trepostome species in the Hamilton Group (M. Dev., N. Y.), and named *zonules* (locally recognizable biostratigraphic units) after them. Remember that such bryozoan zonules with a wide lateral (geographic) spread are confined within individual Hamilton Group members, *Leptotrypella* (L.) *furcata*, for example, found throughout the Windom Member (Moscow shale), spreads laterally for some 100 miles (Boardman, 1960, Figure 6 and Register of Localities).

Archimedes (Figure 6.9, part *A*) and fenestellid bryozoans are common in the Mississippian. In the Chesterian, of Illinois, and western Kentucky, this genus is the most prolific bryozoan (McFarlan, 1942). In the underlying Meramacean Group (Warsaw limestone), bryozoans in marly beds and shales abound on almost every slab. Williams (1957) found that *Archimedes* and fenestellid types were dominant forms in the Warsaw seas. Fenestellids and branching bryozoans are common too in the overlying Spergen limestone. In the younger St. Louis–Ste. Genevieve limestones, bryozoans are more prominent at some exposures than corals or other forms.

The older Osage Group in the Upper Mississippi Valley contain common fenestellid and ramose-type bryozoans. These occur in large faunules (Keokuk limestone); in the Burlington limestone, bryozoans and other fossils are released from weathered chert nodules. The still older Kinderhook Group displays many bryozoan individuals but few species (Louisiana limestone), much fenestellid biodetritus (Hamburg oolite and Hannibal shale), and a wide lateral bryozoan spread (Chouteau limestone). The genus is known also from the Upper Carboniferous and Lower Permian of the U.S.S.R.

Bryozoans are ubiquitous in the marine Pennsylvanian and Permian deposits. In the uppermost marine phase of Pennsylvanian cyclothems, they are especially numerous, although common elsewhere in cyclical deposits. Among more prominent forms in such context are *Batostomella*, *Rhombopora*, fistuliporoids, fenestrate types, and *Meekopora* (Moore and Dudley, 1944).

In Australia, particular Permo-Carboniferous bryozoan species proved useful in local, regional, and intercontinental correlations. Thus beds within New South Wales, Tasmania, and Queensland yield the same bryozoan species, *Fenestella fossula* and *Polypora woodsi*. Both the eastern and western provinces of Australia share several Permian species, for example, *Fenestella horologia*. In turn, western Australian Permian bryozoans show affinities to those described from Timor, the Middle and Upper brachiopod *Productus* limestone fauna of the Salt Range, equivalents in midcontinental North America (Crockford, 1951).

By early Mesozoic, the bryozoan world faunas took on a different aspect. Several orders apparently became extinct (cryptostomes, trepostomes—although trepostomes possibly extended into the Triassic); others remained sparse (ctenostomes). Speciation was active among the cyclostomes. Thus, by Upper Cretaceous, there were hundreds of new species. The cheilostomes did not arise until Middle Jurassic time in Northwestern France. That left a gap from the Permian close to Middle Jurassic time—an estimated 80 million years—which allowed the cyclostomes to radiate. Cheilostomes expanded in post-Mesozoic, Tertiary-Pleistocene, and Recent seas (Bassler, 1953; Canu and Bassler, 1923).

Mesozoic bryozoan information, exclusive of Europe, is scarce (Bassler, 1953; Buge, 1952; Voigt, 1930; Cayeux, 1897). British Mesozoic fossils (1962) show bryozoans absent in Triassic (Keuper and Rhaetic) and Lower Lias (Lower Jurassic). Single species *Berenicea archiaci* occurs in the M. Lias, is not again reported until the Inferior oolite beds (M. Jurassic), but thereafter recurs through the Oxford Clay. Subsequently, no bryozoans are reported until the Upper Chalk (Senonian and Maestrichtian, U. Cret.). In the Upper Chalk, two bryozoan genera were represented by one species each.

Abundant cyclostomes and cheilostomes occur in the Upper Cretaceous chalks of Europe (Denmark, Holland, Belgium, France, and Germany) (Cayeux, 1897).

Tertiary and Quaternary bryozoans, having a wide lateral spread with numer-

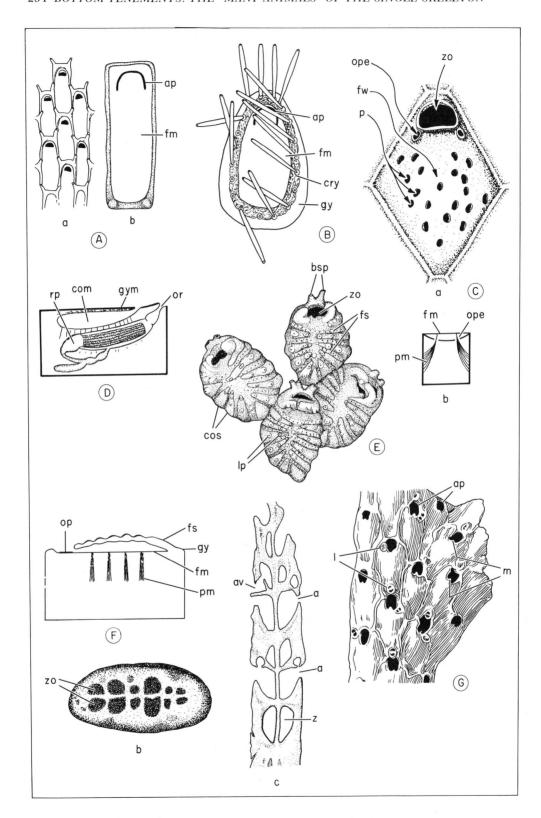

FIGURE 6.11 Aspects of Cheilostome morphology and representative types.

(A) Anasca: *Membranipora membranacea*, Rec., Atlantic: (a) Surface view of a colony, (b) zooid enlarged, *fm* — frontal membrane; *ap* — aperture.

(B) *Callopora whiteavesi*, zooid, *fm* — frontal membrane, *cry* — cryptocyst (shelflike, calcareous laminae beneath frontal membrane), *gy* — gymnocyst (peripheral calcified portion of frontal membrane, fm).

(C) (a) Tangential section in plane of colony; surface features of a Tertiary cheilostome; *zo* — zooecial opening, *ope* — opesiule (one of small grooves in cryptocyst for passage of depressor muscle attached to frontal membrane), *p* — pores; (b) cross-section, relationship of muscles to opesiule, *pm* — parietal muscle, *fm* and *ope* as above, *fw* — frontal wall.

(D) Ascophoran, showing compensatrix (*com*) opesiule, orifice (*or*), retracted polypide (*rp*), and gymnocyst (*gym*).

(E) *Reginella floridana* (Rec., Gulf of Mexico, mudlumps), portion of encrusting zoarium, X 50, *bsp* — bifid spines, *zo* — zooecial opening, *fs* — frontal or (costal) shield [formed into a rigid shield by fusion of spines and subsequent calcification of frontal wall of spines; *fs* forms a ceiling of a room, the floor of which is the frontal membrane (Silén, 1942, pp. 47 ff)], *cos* — costule, *lp* — lumen pores, two to a costule.

(F) Formation of compensatrix by fusion of spines to form frontal shield (*fs*), thus forming ceiling of room, and so on, as described in *E*; *fm* and *gy* as above; *pm* — parietal muscles (compare *D*, "com").

(G) *Psilosecos moralis*, Vincentown sand, Eoc., N. Jersey. Right figure (= a) Portion of bilaminar zoarium; *mucro, m* (rounded or spinelike projection at posterior edge of rim surrounding orifice), and *lyrula, l* (median, anvil-shaped tooth located below the mucro), and zooecial aperture, (*ap*), X 25 (b) transverse action showing zooecia exactly oppostite one another X 20. (c) longitudinal section, avicularia, *av* — zooecium, *z*, and orifice or aperture, *a*, X 20.

Sources: (A,a) (G,b) Bassler, 1953; (B) (F) Silén, 1942; (C,a) Ross and Ross, 1965; (D) several authors; (E) Cheetham and Sandberg, 1964; (G,b,c) Canu and Bassler, 1933.

ous species, have been carefully studied by many workers (Canu and Bassler, 1913, 1920, 1933; Martin, 1943; Stach, 1936, 1937; Cheetham and Sandberg, 1964, and Cheetham, 1963; Lagaaij and Gautier, 1965, among others).

The Vincentown limesand of New Jersey, regarded as Upper Cretaceous, has been accepted as Eocene in age. This formation has some layers composed almost solely of fragmentary cyclostomes and cheilostomes. The bryozoan fauna is readily retrieved by sieving the unconsolidated sediments. Canu and Bassler (1933) found similarities between the Vincentown bryozoan fauna and the European Maestrichtian (U. Cret.) and Danian (Paleocene), even to the identity of some species. The species occurrence found in the Eocene of Maryland and the Gulf coastal states was also reported in European deposits. In such instances, one cannot rely solely on a single faunal element, such as bryozoans, to provide a decisive age determination.

Two lithofacies occur in alternate layers in the Vincentown — calcareous and glauconitic. The calcareous facies is rich in foraminifera and bryozoa, echinoids, corals and worms. The glauconitic facies is rich in foraminifera (Richards). Clearly, the complete facies fauna must be studied to establish the formation's geologic age.

Sedimentary Environments

Considering that biodetritus and other bottom debris are favored by bryozoan larval attachment, apparently bryozoan fossils should occur commonly in the "impure calcareous clastic facies" (Duncan, 1957).

Boardman (1960a, pp. 9–10) noted the distribution of trepostomes in the Hamilton Group of New York State: *dark shale facies* — no trepostomes; *sandstone facies* — no trepostomes; *siltstone facies* (transitional between the optimum environment of calcareous shale and inhospitable sandstones) — few trepostomes; restricted occurrence is attributed to absence of suitable substrate for zoaria attachment; *coral-bearing limestone facies* — while encrusting fistuliporoids are common, trepostomes are rare, but increase in numbers where a

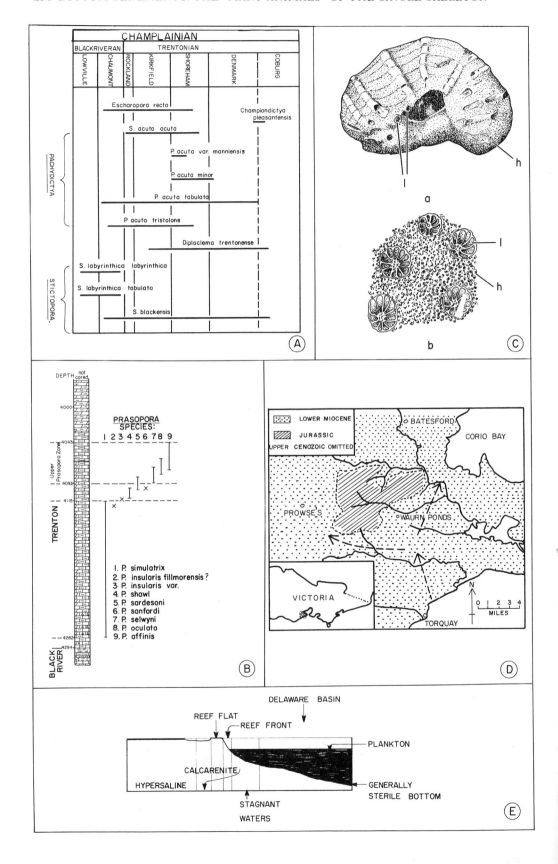

FIGURE 6.12 Examples of derivative data from fossil bryozoans.
(*A*) Restricted vertical ranges of some cryptostome species (U. Ord., N.Y.), J. P. Ross, 1964, J. Paleont., Vol. 38: (1), text figure 2. *P. acuta acuta* not *S. acuta acuta*.
(*B*) Zonation of a core by *Prasopora* species (after Sparling, 1964, J. Paleont., Vol. 38 (6), text figure 1).
(*C*) Bryozoa-Coelenterate association (adapted from Buge, 1952), a scleratinian coral (*l*) (*Lithodendron parasiticum*); *h*—the cheilostome bryozoan, *Heloporella palmata*, Miocene (Helvetian), Touraine.
(*D*) Inferred current trends (Early Miocene, Geelong District, Victoria, Australia, adapted from L. W. Stach, 1936).
(*E*) Chief Biotopes (= ecological zones) during Guadalupian time, Permian, Texas: Black areas—heavy water poorly aerated at depth. White area—aerated waters. Bryozoans on reef flat and reef front (see text). Modified from Newell, *et al.*, 1953.

lithofacies change to a calcareous shale or mudstone occurs (this facies indicates a shallow, rough water environment); *calcareous mudstone facies*—this was the optimum environment for trepostome development; *in situ* ramose trepostomes occur.

Duncan (1957) suggested that siltstone and sandstone occurrences may be explained by prior anchorage in a more suitable substrate and subsequent burial in silt and/or muds. Perry (1962) recorded bryozoan presence in the Spechts Ferry formation (M. Ord.), especially in argillaceous limestones and calcareous shales.

In situ bryozoans could serve as sediment traps for calcium carbonate. Depending on the persistence and/or varying energy level of the water, *in situ* bryozoans would be fragmented and their pieces redeposited. Bryozoan detritus is abundant in the reef talus of the Permian Guadalupian Mountain reefs. Plotting percentage abundance in siliceous residues of five arbitrarily chosen bryozoan growth forms (fenestrate, ramose, massive encrusting, pinnate, and small conical) against distance from reef source, Newell et al. (1953, pp. 113, 147–150 and Figure 67) found an apparent anomaly. They found that the percentage of bryozoan fragments residues increased basinward even though the percentage of general organic remains decreased in the same traverse. Selective winnowing and transport destruction were hazarded as possible explanations. Perhaps the grade size of bryozoan fragments was closer to transporting current competence, and other faunal elements may have been destroyed more readily.

Calcarenites were the common bryozoan-bearing rocks in the area.

The interreef sediments of the Chazy formation (Ord., N.Y. and Vermont) also contain considerable bryozoan detritus. Deposition in high wave-energy environments was posed by the lithological context in which Chazyan cryptostomes occurred, namely, biosparites, intrasparrudites, intrasparites, and dolomites. Intraclasts often contained foliaceous bryozoans transported from a calm water environment (J. P. Ross, 1963, pp. 584–586; 1964a, pp. 945–946). Other formations with much bryozoan detritus include the Spergen limestone (admixed with oolites and small shells) and the Vincentown limesands.

Moore and Dudley (1944) summarized the lithologies where Pennsylvanian and Permian bryozoans are common: limestones including fine, medium, coarse-grained, pure-to-impure, earthy, siliceous, ferruginous, carbonaceous, and oolitic. They are also abundant in calcareous shales and in beds described as massive, flaggy, thin-wavy-bedded, or having a shaley character.

In the eastern United States, Upper Devonian bryozoans most frequently occur as impressions in shales and sandstones, or recrystallized specimens embedded in limey sandstones (McNair).

Bryozoan Bioherms (?)

Bryozoan accumulations have been called *bioherms* and *reefs* (Grabau, 1924, and many others), *reef knolls* and *tangled mattes* (Newell, 1953), *lenslike accumulations* and *biostromes* (Duncan, 1957, after many authors), *prairies* (Termier and Termier); *thickets* (J. P. Ross, 1964*b*), *nests* (Condra, and Schoewe et al., in Moore and Dudley, 1944).

One views bryozoan prairies and thickets as precursors of both biostromes and bioherms (Duncan, 1957; J. P. Ross, 1964*b*). The midcontinental "nests" of *Meekopora* and a few other bryozoans occur as "circular patches, 2 to 3 feet in diameter, made up of closely-packed bryozoan zoaria surrounded by an upraised rim of barren limestone" (Moore and Dudley, 1944). Likely, "nests" and "prairies" are equivalent terms.

There are, however, significant considerations necessary to evaluate a given accumulation as biohermal, biostromal, or some other type.

1. Given a reef, to what extent were bryozoans the essential frame builders? (Examination of reports on the Silurian reefs of Gotland, Niagaran reefs of the Great Lakes, the Permian reef complex of Texas, and others indicates that bryozoans were not important reef builders although they contributed importantly to bioclastic reefrock or, as Hadding put it, to the "matrix" but not the "reef stock." Newell et al. (1953) suggested that fenestrate forms found in growth position probably did serve as frame builders on reef knolls (Permian of Texas) below wave base, but only in early growth stages.

2. Some investigators have reported extensive destruction by recrystallization of organic structures in reefs leaving fenestellid or other bryozoans as relicts (Zekkel, 1941, Kazanian, Russian Plain; Trechmann, 1914, Permian, Durham Magnesium limestone). In such situations, it appears difficult to evaluate the relative importance of bryozoans compared to other faunal elements as frame builders. Furthermore, the process outlined in item 3 applied in some instances.

3. Item 1 above requires further clarification for all reefs where the so-called *Stromatactis* is an important component.

Such reefs include the Niagaran (Silurian) reefs of Illinois and Indiana, Frasnian reefs of Belgium (see also Chapter 5), and the Mississippian knoll reefs of England.

Thus Textoris and Carozzi (1964) found that *Stromatactis* in the Indiana reefs "consists mainly of *Fistulipora* colonies" These became partly or wholly dissolved and subsequently filled with calcite spar. Solution and recrystallization have often obscured all organic structures (Frasnian reefs, Niagaran reefs of Illinois, studied by Lowenstam and others).

The four stages in the development of an ideal Niagaran reef, according to Textoris and Carozzi, are as follows. Stage 1: Crinoid accumulation forms a mound of fossiliferous calcisiltite below wave base. Stage 2: With the addition of *Stromatactis* (essentially dissolved bryozoan colonies subsequently in-filled), the mounds of Stage 1 become a massive corelike body that enlarges. Stage 3: The mound is built into the wave-base zone. Well-bedded *Stromatactis* calcisiltite alternates with deposits of fossiliferous calcisiltite and mud-supported crinoidal calcarenite. Stage 4: Mound is now mostly above wave base and is enlarged by a bioconstructed core built by stromatoporoids. On the core flanks, well-bedded crinoidal calcarenite accumulates. *Stromatactis* (and hence, bryozoans) become rare.

The low-energy calcisiltite core (Stage 2) is not thought to be due to the bryozoan fistuliporoids as the organic constructors or frame builders. Both the limited size of fistuliporoids and insufficient population density argue to support this interpretation. Rather the core is attributed to physical-chemical recrystallization that led to a complex network of interconnected *Stromatactis* cavities. Subsequent steps: an internal mechanical deposition of bioclastics and a physico-chemical developed calcite spar in-fill. (See Figure 6.13 for stages.)

Some of the preceding citations support Duncan's view (1957) that bryozoan contribution made to reef rock was sporadic and that many bryozoan accumulations

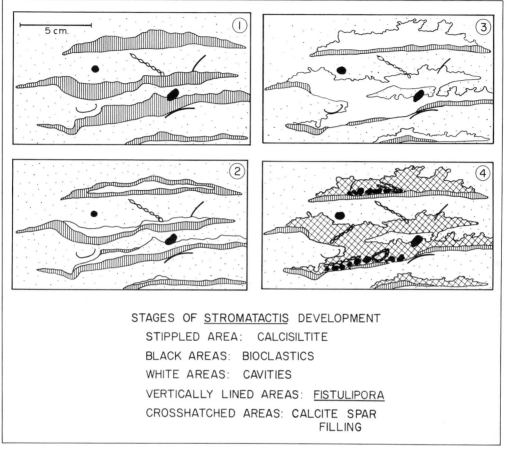

STAGES OF <u>STROMATACTIS</u> DEVELOPMENT

STIPPLED AREA: CALCISILTITE

BLACK AREAS: BIOCLASTICS

WHITE AREAS: CAVITIES

VERTICALLY LINED AREAS: <u>FISTULIPORA</u>

CROSSHATCHED AREAS: CALCITE SPAR
FILLING

FIGURE 6.13 Stages in *Stromatactis* development (after Textoris and Carozzi, 1964, A.A.P.G. Bull., Vol 48(4), p. 412). Development is thought to proceed from 1 to 4.

previously called bioherms are biostromes. However, Pray (1958) suggested that fenestrate bryozoa may have contributed more in bioherm genesis than Duncan allowed.

J. P. Ross (1964) defined four general kinds of sedimentary deposits in the Chazyan, two of which were *biohermal*: (*a*) small bioherms with thickets of ectoprocts and other fossils in growth position, (*b*) large bioherms composed of lenticular beds essentially bioclastic with local pockets of calcareous fossil debris grading from clay-to-pebble size.

Biostromal bryozoan accumulations are widespread in Devonian and Carboniferous rocks. Pray (1958) reported fenestrate bryozoans to be the major faunal component in the core facies of crinoidal bioherms (Osagian, Mississippian, New Mexico). The "core facies" was chiefly of aphanitic calcite and comminuted fenestrate bryozoans. Bryozoans, either in growth position or in biodetritus, are thought significant in the genesis of these mound- and ridge-shaped structures. These New Mexican reefs are similar to other Mississippian reefs.

In many parts of the world (Germany, Great Britain, U.S.S.R), Permian bryozoan accumulations (bioherms) are important. The Zechstein reefs are a good example (Grabau, 1924; Mägdefrau, 1933). Little is known about bryozoan reefs during the Mesozoic. There were several occurrences during the Tertiary—among others, Miocene reefs (Nalivkin, 1960, geologic map) of the Kerch peninsula that partly separates the Black Sea and the Sea of Azov, and also Pliocene deposits of England (Coralline Crag).

The Kerch knolls, hillocks, or stacks of unstratified lenticular limestone are largely composed of one bryozoan species, *Membranipora lapidosa*, and are referred to as "Membraniporakalk." Other bryozoans were found by Martin (1943), namely fenestrate and ramose types. Reef masses attain heights up to 100 ft (Grabau, 1924).

The Coralline Crag rocks appear to have been built up on submarine banks of organic debris (Bryozoa and shell fragments of sand size) in shallow water. Current-bedded sand indicates strong current action (Harmer, 1898).

Biotic Associates

Articulate brachiopods and various mollusks are common associates of bryozoans. Other associates include echinoderms— echinoids in the Vincentown limesand, or crinoids in numerous Mississippian bioherms; corals—small-horn, tabulate and, more rarely, hermatypic; foraminifers— fusilinaceans in the Texas reefs and foraminifers in the Vincentown limesand, in the Eastern European reef facies; serpulid worms in Miocene reefs of the Russian-Roumanian area, and the Cretaceous (Upper Maestrichtian) of Western Europe and elsewhere.

Massive bryozoans were associated (Guadalupian time) with cemented brachiopods and various algae (reef flat facies) in the Texas Permian reefs. Ramose, pinnulate, and fenestrate bryozoans were deposited with pedunculate brachiopods among other forms (reef-front facies) (Figure 6.12, part *E*).

Fossil ctenostomes adhere to, or penetrate, shells of myalinid pelecypods and many brachiopod forms such as the abundant *Composita* and *Chonetes* (Condra and Elias, 1944).

In the Sappington sandstone of Southwestern Montana, two species of *Fenestella* were obtained from insoluble residues when the enclosing algal nodule (= oncolite) was dissolved in hydrochloric acid (Malone and Perry, 1965). Other algal

and bryozoan relationships have been discussed (Figure 6.5, part *B*).

Several workers (Elias, Newell, Lowenstam, among others) have reported scarce-to-absent bryozoans in biotopes where sponges thrived and large burrowing pelecypods, brackish water inarticulate brachiopods, stromatoporoids, hydrocorallines (Duncan, 1957).

Commensalism and Symbiosis

Some bryozoan incrustations on fossil mollusks have aroused considerable interest. Buge (1952) illustrated a "symbiotic" association of the bryozoan species *Callopora* incrusting an unidentified gastropod containing hermit crab fragments. This suggested that the hermit crab had used the shell. Dunbar (1928, pp. 164–165) reported a rare instance of a cheilostome bryozoan imprisoned between successive volutions of a deeply involute shell of a Cretaceous ammonite, *Sphenodiscus*. What made this occurrence rare was not the bryozoan and cephalopod association— surficial incrustations on species are common—but rather the unusual attachment site denoted a detrimental commensal life association for the cephalopod.

In his study of early Paleozoic bryozoans of the Baltic, Bassler (1911, pp. 106, 200–201, 208–210) recorded incrustation examples of various mollusks. *Spatiopora* commonly incrusts shells of the orthocone cephalopods in the Makoqueta shale of Iowa and elsewhere. Of greater interest are the reports of bryozoan incrustations on the mollusk *Hyolithes*. The bryozoans involved are *Mesotrypa expressa* and *Leptotrypa hexagonalis*. The latter species usually occurs as incrustations on *Hyolithes baconi* (see Chapter 8).

Voigt (1930) described a serpulid worm invading a Cretaceous bryozoan zooecia as suggestive of parasitism.

A life association between a Miocene scleratinian coral and a cheilostome bryozoan (Figure 6.12, part *C*) was interpreted as the reason that the bryozoan *Heloporella* developed a more massive globose colony. The association did not seem to have harmed the bryozoan. In fact, Buge (1952) speculated about the growth directions of coralline calices and their coincidence with the bryozoan zoaria (Figure 6.12, part *C*). Evidence led him to dismiss the "symbiotic" relationship and to favor *"association biologique."*

Population Density

Four to six specimens of a single cryptostome species or of different species were commonly represented in the more than 600 thin sections studied by J. P. Ross from the Champlainian of New York (1964*b*).

Growth Habit, Bathymetry, and Energy Zones

Stach (1936, 1937), from zoarial form study in existing cheilostomes, found that *a definite relationship exists between the various zoarial types and their habitat*. He further indicated that application to fossil bryozoans might be made and illustrated his point referring to cheilostomes of the Pliocene deposits of Victoria, Australia. Subsequent workers applied this approach to the Eocene cheilostomes (Cheetham, 1963), to living cheilostomes, to cyclostomes in the Rhône Delta (Lagaaij and Gautier, 1965) and to fossil trepostomes (Boardman, 1960).

Stach recognized nine zoaria groups. [The first part of the group name below indicates a particular family (Membraniporidae) or a particular genus (*Vincularia*).]

I. *Membraniporiform.* Unilaminate zoaria with entire basal lamina attached to a substratum. Essentially adapted for littoral and sublittoral zone. Lagaaij and Gautier (1965) defined group types: (*a*) zoaria usually, but not necessarily, unilamellar, incrusting a solid substratum; dorsal wall of zooecia entirely calcified; (*b*) unilaminate zoaria incrusting a generally flat, flexible substratum; calcification of dorsal wall of zooecia absent or weakly expressed.

II. *Petraliform.* Zoarium unilamellar; basal lamina attached by chitinous rootlets to loose and irregular substrate such as coralline algae and sponges. Littoral zone type. Specialized types with fenestrate zoaria (*Petralia undata*) produce rootlets that form an attachment filament, fastening zoarium to sandy bottom. Hence it can thrive in fairly deep water affected by currents since the fenestrate zoarium offers reduced resistance to current action.

III. *Eschariform.* Zoarium bilaminate, foliaceous; strongly calcified; attached to substrate by rootlets (= radules) or direct adherence. Sublittoral zone to depths of at least 10 fathoms.

IV. *Reteporiform.* Zoarium attached, rigid; strongly calcified; fenestrate. High-energy zone (strong wave and current action).

V. *Vinculariform.* Zoarium attached, consisting of erect, rigid subcylindrical branches. Low-energy zone (deep or quiet waters).

VI. *Cellariform.* Zoarium reticulated with subcylindrical internodes consisting of numerous zooecia. Essentially adapted for littoral zone where algae usually form the base of attachment. May extend to greater depths.

VII. *Catenicelliform.* Zoarium of articulated internodes of few zooecia. Essentially adapted for littoral zone where wave action is strong.

VIII. *Flustriform.* Flexible, chitinous zoarium, lacks calcareous supporting skeleton. Littoral zone.

IX. *Lunulitiform.* Zoarium free, hollow-conical; the zooecia opens on the outer face; not attached to substrate in adult life. Sandy bottom, with strong current action; upper depth limit, 15 fathoms.

Other forms fit special bryozoan faunas as in the Rhône Delta. Cheetham (1963) included Petraliform types under Membraniporiform, and grouped IV and V under Eschariform. Varied flexibility is necessary in applying Stach's groups.

An excellent illustration derives bathymetric and paleo-oceanographic data from the use of nine groups as given by

Stach (1936). In Figure 6.12, part D, Lower Miocene limestones are distributed on all sides of a Jurassic outlier (Barrabool Hills).

Stach sampled the Miocene and found groupings of cheilostomes: Waurn Ponds, eschariform with fewer cellariform and lunulitiform types; Prowe's Quarry, eschariform, reteporiform, and vinculariform types, about equal proportions; Batesford Quarry, predominantly vinculariform with minor cellariform and eschariform elements. From bryozoan fossils, Stach inferred the whole oceanographic system of the area (see Figure 6.12, part D). He was able to define the deflection: a generally northerly current by the Jurassic buttress or barrier that stood as an island in a Miocene sea. Because of this very deflection alone, at Prowe's and Batesford Quarries, quiet water conditions had prevailed, whereas at Waurn Ponds there were high-energy turbid conditions.

Boardman (1960a) applied Stach's findings to the trepostomes of the Hamilton Group of New York State. In the coral-bearing limestone facies, there were predominantly incrusting colonies and a bioclastic matrix. This suggested agitated water (high-energy zone); in the calcareous mudstone facies, complete zoaria, predominantly ramose colonies, coexisted in quiet conditions (low-energy zone).

Growth

Nonuniform growth was postulated by Perry and Hattin (1958, p. 1049) for the Niagaran fistuliporoid bryozoan colonies. They distinguished zones of thin vesicles less than 0.75 mm thick; larger vesicles flanked these zones anteriorly and posteriorly. The thin zones, interpretedly, reflected times of retarded growth under adverse environmental conditions. Excess or severely limited salinity often has been attributed as a cause in growth retardation.

Study of nearly complete ramose trepostome zoaria (New York Hamilton Group) led Boardman (1960a) to formulate a growth hypothesis founded on both

growth and resorption. Possible steps would be (1) lateral gemination of thin-walled tubes in neanic zone introduced new zooecia; (2) branches lengthened by distal (forward) growth of zooecia; (3) around growing tip (see Figure 6.14, part B), a thick-walled ephebic zone formed; (4) polypides resorbed the skeleton (in Figure 6.14, part B, compare the "remnant zone" with those on either side of it; before resorption, all zooecia appeared complete with aperture, diaphragms, and thicker anterior walls; after resorption, only the "remnant zone" of the growing tip remained); (5) another cycle followed as in (1) above. [There was a direct connection (that is, continuous tubes) between new neanic thin-walled tubes and thick-walled remnants of previous cycles.]

Variability of calcite resorption from one cycle to another may have been wholly or partially controlled by recurrent temperature-dependent or other environmental conditions.

It appears that in these ramose colonies, *resorption* of calcitic skeletal material kept the growth process going. As colony growth continued, zooecia at the growing tip (Figure 6.14, part B) deviated from the branch center, crowded out by new neanic tubes formed by gemination.

Boardman figured that large colonies died from base upward and that normal growth continued only in the forward region (Figure 6.14, part B). The dead zoaria parts bearing overgrowths, served to support growing parts, keeping them upright, preventing their burial by a constant sediment influx on the bottom.

In Figure 6.14, part B, at the growing tip, each living chamber contained a polypide (see Figure 6.3, part A). The most anterior (distal) diaphragms probably served as a "floor" to support the growing polypide. Successively posterior (proximal) diaphragms in the same zooecia represented "older floors" that served as supports during early individual growth stages (see also Chapter 8, Cephalopoda, phragmacone and septa).

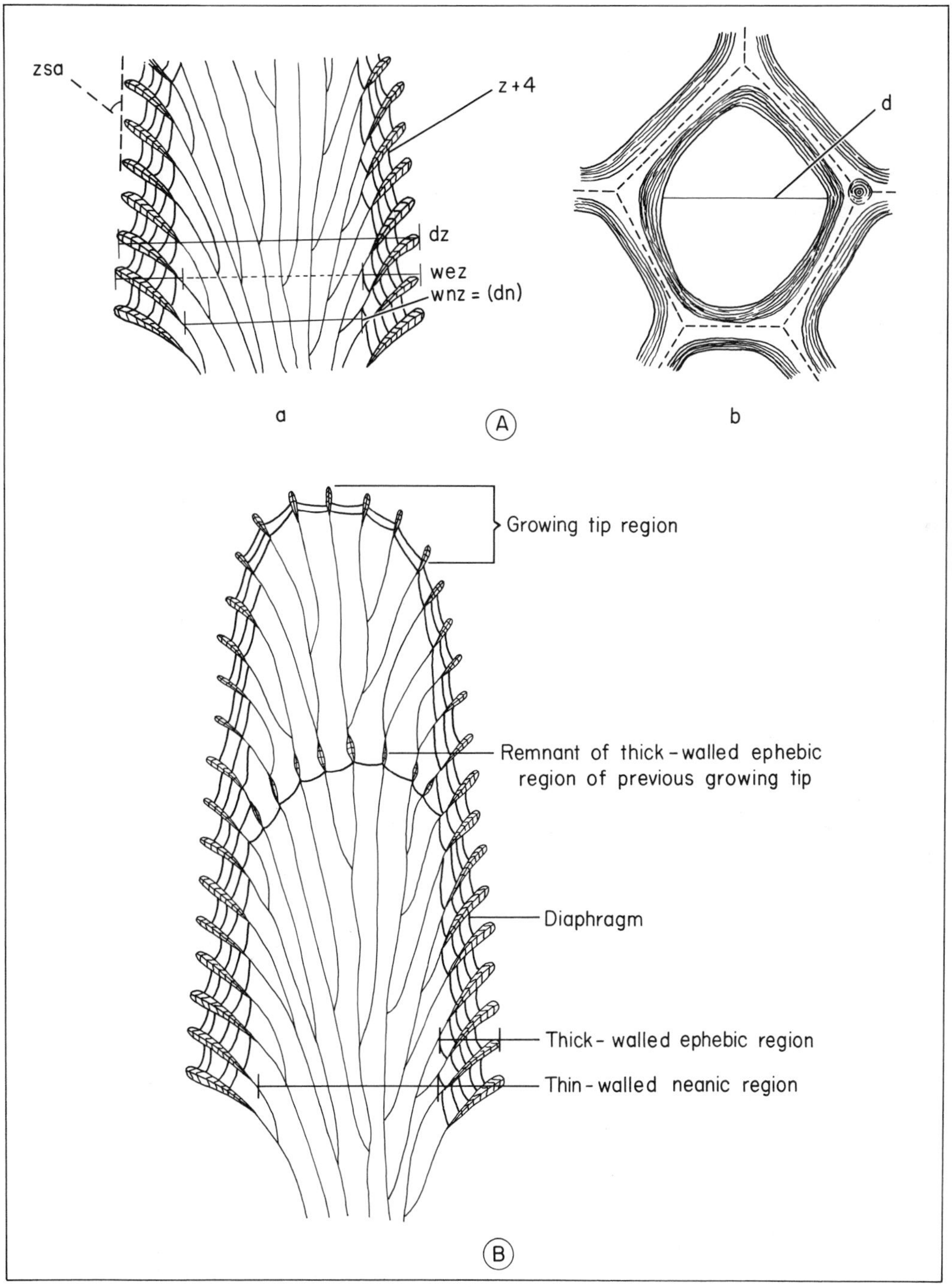

FIGURE 6.14 Standard measurements (after Boardman, 1960a).
(*A–a*) Ramose zoarium, longitudinal section; schematic diagram, *zsa* — zooecial surface angle, *dz* — diameter of zoarium, *wez* — width (diameter) of ephebic zone (= mature region), *wnz* — width (diameter) of neanic zone (= immature region), *z* + 4 = zooecium with four diaphragms. (*b*) Simple zooecium, tangential section, *d* — diameter of zooecium. (*B*) Schematic diagram, longitudinal view indicating essential growth regions and resulting measurable areas.

At death, a given polypide decayed and an abandoned zooecia remained.

Hydrostatic System and Bathymetric Tolerance

Canu and Bassler (1920) recorded numerous variations in the zoarial hydrostatic system (external) in genera of the cheilostome suborder Anasca. The genus *Lunulites* can "control" its volume; it can increase or decrease it by parietal muscular action or by development of tuberosites on the noncelluliferous face.

Judging from hydrostatic system variability in living cheilostomes, apparently selective pressure favored modifications that enhanced volume control and, thereafter, greater bathymetric tolerance. One assumes that Mesozoic and Tertiary cheilostomes also, for these same reasons, inhabited greater depths than other orders (see Duncan, 1957).

Measurements

Some of the standard measurements on bryozoans are shown in Figure 6.14, part A, B. Others include the following.

I. Axial ratio $= d_n/d_t$ (d_n = neanic diameter; d_t = total diameter, see Figure 6.14, part B).

II. A measure of growth stage: axial ratio (*AR*) is plotted against number of zooecial diaphragms.

Paleoclimates

Darteville (1933), Elias, Crockford (1951), among others, have attempted paleoclimatological inferences from study of Tertiary (Eocene) and Upper Paleozoic bryozoan faunas, respectively. Among study data used, are inferred temperature tolerances of different bryozoan species; impoverished faunas (that is, presence of few species); absence of a given species or a group of species in a traverse across a region or country, in beds of equivalent age; distinction between so-called cold-water and warm-water faunas; distribution—cosmopolitan or local; ecological inferences, such as depth, temperature, and related data on other faunal elements inhabiting the same basin as bryozoans—fossil fish in the Paris and Belgian basins; inferred migrations of entire bryozoan faunas from one region to another.

Buge (1952) found as a general rule that one can distinguish three large groupings of living bryozoans: (*a*) truly cosmopolitan species largely indifferent to temperature; (*b*) tropical species; (*c*) species of temperate and cold zones. In the European Tertiary (Neogene), certain genera that had thrived in the temperate zone are represented today in tropical and equatorial zones (*Entomaria, Steganoporella*, and so on). This, in turn, implies migration. Other living genera did not migrate from the zone in which they had thrived during Tertiary time. Crockford (1951) inferred a northward migration of eastern Australian Permian bryozoan faunas "as a whole" —possibly influenced by cold-water currents along the coastline of the time.

Speciation and Subspeciation

During Leonardian through Guadalupian time in the Texas Permian, Newell et al. (1953, Figure 83; compare Figure 81) reported there were 23 bryozoan species. One species (*Septopora* sp.) did not persist through Guadalupian time; two of the four species of *Domopora* made their apparent first appearance in Guadalupian time; and the same holds for two of the three species of *Fenestrellina* and one of the possible two species of *Lioclema*. In this area the only new genus to make its apparent first appearance in Guadalupian time was *Phyllopora*. It follows that about six species made their apparent first appearance in Guadalupian time.

The bryozoans during the preceding Leonardian time were shallow-basin inhabitants. However, by Guadalupian time they had become reef inhabitants (Figure 6.12, part *E*).

In the area the essential bryozoan complex (that is, 17 species) persisted from Leonardian through Guadalupian time. This occurred despite a change from (*a*) shallow-basin conditions (Early Brush Canyon time) to (b) growth of small patch reefs during the Leonardian (Getaway time) to (c) barrier reef conditions where differentiation gradually occurred between the reef front and the reef flat behind it (Guadalupian time, Figure 6.12, part E).

A clue to this tough persistence can be gleaned from the data on new species that appeared in item (b) above but not in (a). Thus, of the total of 23 species, ten first appeared in (b) and species of (a) occurred in (b) also—a total of 14 species. Accordingly, bryozoan species continuance may be attributed to the favorable environment created first by patch reefs and continued by the barrier reef.

The ranges of four subspecies and one variety of the cryptostome *Pachydictya acuta* overlap in New York State's Trentonian beds (Shoreham formation, see Figure 6.12, part A). Of these, three subspecies (*P. acuta*, subspecies *acuta*, subspecies *minor*, subspecies *tristolone*) occurred in the same algal-bryozoan biopelmicrite at Canajoharie Creek (J. P. Ross, 1964a, table 1, text figures 3 and 7) and were apparently contemporaneous.

Competition for the same substrate is intimated: *P. acuta acuta* and *P. acuta tristolone* ranged from Rockford and/or Chaumont, respectively, to Upper Shoreham time but did not survive into Denmark time. *P. acuta minor* first appeared in basal Shoreham time and continued briefly into basal Denmark time. The subspecies *P. acuta tabulata*—the longest-ranging subspecies—the sole survivor of *Pachydictya acuta* stock, endured all through Denmark time (see Figure 6.12, part A).

Ross (1964a, p. 12) commented that *Pachydictya acuta tabulata* apparently migrated from the northwest (Black Riveran time) to the southeast toward Canajoharie Creek (Trentonian time).

How can the comparative *tabulata* subpopulation vigor be explained? Possibly it migrated responding to adverse sedimentary conditions; this implies natural selective factors at work. Those that did not migrate did not survive. Furthermore, absence of *tabulata* subpopulation from the Canajoharie Creek area spared it from selective pressures working there. Several subspecies competed for the same substrate; reasonably, this led to or contributed to their several extinctions.

IDENTIFICATION OF A BRYOZOAN TO SPECIES LEVEL

[Data from M. A. Fritz, 1957, Bull. Geol. Soc. Canada, Vol. 42, p. 18, plate X, figures 1.4 (see Figure 6.7, part H, this chapter.]

Source of Material. Ottawa formation (Upper phase, Cobourg beds), M. Ord., Ottawa, Canada.

Given. An undetermined fossil assemblage labeled "Bryozoans" from the above-named formation.

Problem. To identify to the species level certain specimens collected from the Cobourg beds at two localities in Ottawa.

Procedure

1. Group specimens according to external features. [Objects are ramose, the cylindrical branches 5 to 10 mm in diameter; surface smooth or bearing a few tubercles (= monticules?) and numerous minute openings (= zooecia?), cover visible surfaces (Figure 6.7, part H, c).]

   Are these objects ectoprocts? See phylum definition. If the openings are zooecia, objects may be zoaria—an assemblage of numerous bryozoan colony zooids. If correct, thin-section study should reveal typical bryozoan internal structures.

2. Prepare tangential and vertical (longitudinal) thin sections of apparent zoaria (Figure 6.10). When examined under transmitted light, specimens will look like those in Figure 6.7, part A, a–b.

Examination of Thin Section

Tangential Section (Figure 6.7, part H, a)

i. Numerous zooecia (polygons, hexa-gons, and pentagons) are present; they usually contact each other on all sides. This type of structure proves that we are dealing with ectoprocts. Monti-cules are seen to be composed of simi-lar but fewer zooecia than the sur-rounds, that is, five to six in 2 mm compared to six to seven in 2 mm.

ii. Zooecial walls thick.

iii. Well-marked black line of demarcation between zooecia.

iv. Mesopores, almost absent in larger specimens; sparse in smaller.

Vertical (Longitudinal) Section (Figure 6.7, part H).

i. Axial region (immature region). Wall, very thin where diaphragms develop-ed regularly. Diaphragms are spaced about one zooecial tube diameter apart in this region.

ii. Mature region. (*a*) Comparatively long; walls distinctly thickened; diaphragms increase in number — six or more in a distance equal to one zooecial tube diameter; (*b*) diaphragms often hori-zontal, some cytose (approximating vesicular tissue in places), see Figure 6.7, part H, cd; (*c*) in submature zone (at contact of immature and mature re-gion), mesopores may or may not be present; if present, they are closely tabulated.

3. To what bryozoan order do these specimens belong? The presence of a mature and immature region *without* abrupt change and the lack of hemi-septa, acanthopores, and so on, elimi-nate Cryptostomata and indicate as-signment to order Trepostomata.

4. Although of admitted limited utility, the suborder Integrata is determined by noncoalesced zooecial walls separa-ted by a dark, divisional line.

5. In this suborder, which family is rep-resented? Of the four families in the suborder (see also Treatise G, Bryozoa, pp. G-107–G-119), elimination (desig-nated by "X") can be made as follows:

   Family Amplexoporidae. Lacks meso-pores and has abundant acanthopores (X).

   Family Trematoporidae. Mesopores closed and acanthopores present (X).

   Family Phylloporinidae. Zoaria anas-tomising or reticulating slender bran-ches; acanthopores present (X).

   The remaining family, Halloporidae, essentially lacks acanthopores, has ramose zoaria, tabulate mesopores, and so on. This fits the sample.

6. What genus within the family Hallo-poridae meets the requirements of the studied sample? Of five genera in the family (Treatise G, Bryozoa, pp. G-112–G-113), eliminate (X) as follows.

   Genus Sonninopora. Has minute acanthopores in mature region (X).

   Genus Panderpora. Zooecia with wide-spaced and curved diaphragms; hemispherical zoaria (X).

   Genus Halloporina. Zooecia with strongly crenulate (crumpled or fol-ded) wall, lacking diaphragms (X).

   Genus Calloporina. Zoarium discoid and convex (X).

   The only remaining known genus is *Hallopora*, which fits the problem fos-sil.

7. What species in the genus *Hallopora* is represented? Of the described species, those with closest affinities to the un-named at hand are *H. multapulata* (also found at the same localities as the questionable sample), *H. ampla*, and *H. splendens.*

   The studied sample differed from these species in (*a*) large zooecia, (*b*) thick walls, (*c*) lack of mesopores, and (*d*) cystose diaphragms. Taken together these four features distinguish the un-known from the named-group species.

8. Accordingly, a new species was erected: *Hallopora lydiana* Fritz, 1957.

QUESTIONS

1. How would you distinguish a branching ectoproct from an anthozoan, a hydrozoan? Cite the chief differences.

2. (a) Discuss polymorphism in ectoprocts following Borg's subdivisions. Stress the various types of zooids and how they probably originated.
   (b) Outline the stages of development after larval attachment (refer to Figure 6.4, part *B, Crisia*); draw a complete zooid at each stage.
   (c) How do zooids of a colony communicate in life?

3. (a) What are "statoblasts"?
   (b) Define "brown bodies."

4. What substrates may be inferred from the fossil record of bryozoans?

5. Distinguish between the five ectoproct orders in terms of zooecia, shape and composition, polymorphism, reproduction, geologic range and structures seen in fossil thin sections.

6. (a) Discuss bryozoan stratigraphic utility in the following: (1) Ordovician, Devonian, Carboniferous, Permian, Cretaceous, Tertiary-Quaternary.
   (2) Local and regional correlations.
   (3) Intercontinenal correlations.
   (b) Relate specific bryozoans as reef or bioherm components.
   (c) Identify and state the significance of *Stromatactis, Membranipora,* fenestellids.

7. (a) How are energy level and zoarial form related?
   (b) What information did each of the following investigators obtain: Boardman Crockford, Ross, Newell et al., Rigby, Canu and Bassler, Sparling, Williams?

REFERENCES

Ectoproct/Bryozoa

Bassler, R. S., 1953 Bryozoa. Treat. Invert. Paleont. Part G, 236 p., refs.

————, 1911. The early Paleozoic bryozoa of the Baltic provinces: U.S.N.M. Bull. 77, 348 pp., 12 pls, text figures.

Boardman, R. S., 1960*a*. Trepostomatous bryozoa of the Hamilton Group, New York State: Geol. Survey, Prof. Paper 340, 83 pp., plates 1–22, 27 text figures.

————, 1960*b*. A revision of the Ordovician bryozoan genera Batostoma, Anaphragma, and Amplexopora: Smithsonian Misc. Coll., Vol. 140 (5), pp. 1–23, plates 1–7.

————, 1959. A revision of the Silurian bryozoan genus Trematopora: *idem*, Vol. 139 (6), 13 pp., plates 1, 2; text figure 1.

Borg, F., 1926*a*. Studies on recent cyclostomatous bryozoa: Zool. Bidrag, Vol. 10, pp. 182–498, 14 plates, 107 text figures.

————, 1926*b*. Body wall in bryozoa: Quart. J. Micros. Soc., Vol. 70, No. 280, N.S., pp. 583–597.

British Museum (Nat. Hist.), 1962. British Mesozoic fossils: pp. 1–45, 62 plates, biblio.

Buge, Émile, 1952. Classe des Bryozoaires, in Jean Piveteau, ed., Traité de Paléontologie, Vol. 1, pp. 687–749.

Canu, F., and R. S. Bassler, 1933. The bryozoan fauna of the Vincentown limesand: U.S.N.M. Bull., Vol. 165, 91 pp., plates 1–21.

————, 1920. North American early Tertiary bryozoa: U.S.N.M. Bull., Vol. 106, 878 pp., 112 plates.

————, 1913. North American and later Tertiary and Quaternary bryozoa: U.S.N.M. Bull., Vol. 125, 302 pp., 47 plates.

Cayeux, L., 1897. Contribution a l'étude micrographiques de terrains sédimentaires, . Mém. Soc. Géol. Nord. Vol. 4, part 2.

Cheetham, A. H., 1963. Late Eocene zoogeography of the eastern Gulf Coast region, Mem. G.S.A., 113 p.

Cheetham, A. H., and P. A. Sandberg, 1964. Quaternary bryozoa from Louisiana mudlumps: J. Paleontology, Vol. 38 (6), pp. 1013–1046, text figures 1–59.

Condra, G. E., and M. K. Elias, 1944. Study and revision of *Archimedes* (Hall): G.S.A. Sp. Paper No. 53, 1946, refs., 41 plates.

Crockford, Joan, 1951. The development of bryozoa faunas in the Upper Paleozoic of Australia: Proc. Linnean Soc. New South Wales, Vol. 76, pp. 105–122.

Cumings, E. R., and J. J. Galloway, 1915. Studies on the morphology and histology of the Trepostomata or monticuliporoids: G.S.A. Bull., Vol. 26, pp. 349–374, plates 10–15.

Darteville, Edmond, 1933. Contribution à l'étude des Bryozoaires fossiles de L'Eocène de la Belgique: Soc. royale zool. Belgique Annales, Vol. 63 (1943), pp. 55–116.

Dunbar, C. O., 1925. On an ammonite shell investing commensal bryozoa. Am. J. Sci., 5th Ser., Vol. 16, pp. 164–165.

Duncan, Helen, 1957. Bryozoans. Treat. Ecol. and Paleoecol., G.S.A., Mem. 67, Vol. 2 (Paleoecology), pp. 783–800.

Elias, M. K., 1954. *Cambroporella* and *Coeloclema* Lower Cambrian and Ordovician bryozoans: J. Paleontology, Vol. 28 (1), pp. 52–58, plates 9, 10.

Fritz, M. A., 1965. Bryozoan fauna from the Middle Ordovician of Mendoza, Argentina: J. Paleontology, Vol. 39 (1), pp. 141–142, plate 19.

Grabau, A. W., 1924. Principles of stratigraphy (reissued, 1960, Dover. Publ., Vol. I, II), Vol. I, pp. 433–444; Bryozoan "reefs," and so on.

Hadding, Assar, 1941. The pre-Quaternary sedimentary rocks of Sweden, VII, Reef limestones: Lunds Univ. Arsskrift, n.f., Avd. 2, Bd. 37, No. 10, 137 pp.

Harmer, F. W., 1898. 1. The Pliocene deposits of the east of England: The Lenham beds and the Coralline Crag: Geol. Soc. London Quart. Jour., Vol. 54, pp. 308–354.

Hyman, L. H., 1959. The Invertebrates: McGraw Hill. Vol. 5, Chapters 19–20.

Korn, Hermann, 1930. Dei cryptostomen Bryozoen des deutschen Perms: Leopoldina, Vol. VI, pp. 341–375, plates I–IV.

Lagaaij, R., and Y. V. Gautier, 1965. Bryozoan assemblages from marine sediments of the Rhône delta, France: Micropaleontology, Vol. 11(1), pp. 39–58, 34 text figures.

Lutaud, Geneviève, 1954. Progression de la calcification chez *Escharoides*: Arch. Zool. Exp. Gen., Vol. 9, Notes et Revue 1, pp. 36–50, 5 text figures.

Mägdefrau, K., 1933. Zur Entstehung der mittel deutschen Zechstein-Riffe. Centralbl. Mineralogie, 1933, Abt B, pp. 621–624.

Malone, P. G., and T. G. Perry, 1965. Fenestellid bryozoans from oncolites in the Sappington sandstone of southwestern Montana: J. Paleontology, Vol. 39 (1), pp. 41–44, plate 14.

Martin, G. P. R., 1943. Zur Kenntnis der tertiaren Bryozoenriffe (Sarmat) auf der Halbinsel Kertsch: Deutsche Geol. Gesell. Zeitschr., Vol. 95, pp. 133–137.

McFarlan, A. C., 1942. Chester bryozoa of Illinois and western Kentucky. J. Paleontology, Vol. 16, pp. 437–458, pl. 65–68.

Moore, R. C., and R. M. Dudley, 1944. Cheilotrypid bryozoans from Pennsylvanian and Permian rocks of the midcontinent region: Univ. Kansas Publ., K.G.S. Bull. 52, Part 6, pp. 229–408, plates 1–48.

Nalivkin, D. V., 1960. The geology of the U.S.S.R. Pergamon Press, 153 p, maps, index.

Osburn, R. C., 1957. Marine bryozoa: Treat. Marine Ecol. and Paleoecology, G.S.A. Mem. 67, Vol. 1, pp. 1109–1112.

Parkinson, D., 1957. Lower Carboniferous reefs of northern England: A.A.P.G. Bull., Vol. 41, pp. 511–537.

Pennak, R. H., 1953. Fresh water invertebrates of the United States. Ronald Press, N.Y., 726 pp., appendices, index.

Perry, T. G., 1962. Spechts Ferry (Middle Ordovician) bryozoan fauna from Illinois, Wisconsin, and Iowa: Ill. State Geol. Surv., Circular 326, 36 pp., 7 plates.

————, and D. E. Hattin, 1958. Astogenetic studies of fistuliporoid bryozoa: J. Paleontology, Vol. 32 (6), pp. 1039–1050, plates 129–131, text figure 1.

Pray, L., 1958. Fenestrate bryozoan core facies, Mississippian bioherms, Southwestern United States: J. Sed. Petrology, Vol. 28 (3), pp. 261–273, figures 1–4.

Rigby, J. K., 1958. Relationship between *Acanthocladia guadalupensis* and *Solenopora texana* and the bryozoan algal consortium hypothesis: J. Paleontology, Vol. 31 (3), pp. 603–606, 2 text figures.

Ross, J. P., 1960. Type species of *Ptilodictya — Ptilodictya lanceolata* (Goldfuss): J. Paleontology, Vol. 34 (3), pp. 440–445, plates 61–62, text figure 1, table 1.

————, 1961*a*. Ordovician, Silurian and Devonian bryozoa of Australia: Bur. Min. Res., Geol. and Geophysics Bull. 50, 111 pp., 28 plates.

————, 1961*b*. Larger cryptostome bryozoa of the Ordovician and Silurian, Anticosti Island, Canada, Part II: J. Paleontology, Vol. 3 (2), pp. 311–344, plates 41–45.

————, 1963. Ordovician cryptostome bryozoa, standard Chazyan Series, New York and Vermont: G.S.A. Bull., Vol. 74, pp. 577–608, 10 plates.

————, 1964*a*. Champlainian cryptostome bryozoa from New York State: J. Paleontology, Vol. 38 (1), pp. 1–32, plates 1–8, 11 text figures, table 1.

————, 1964*b*. Morphology and phylogeny of early Ectoprocta (Bryozoa): G.S.A. Bull, Vol. 75 (10), pp. 927–948, 10 text figures, 2 tables.

————, and C. A. Ross, 1965. Bryozoans, in B. Kummel and D. Raup, eds., Handbook of Paleontological Techniques. Freeman and Co., pp. 40–44.

Silén, L., 1942. Origin and development of the cheilo-ctenostomatous stem of Bryozoa: Zool. Bidrag. Uppsala, Vol. 22, pp. 1–59, 64 text figures.

————, 1954. Developmental biology of Phoronidea of the Gullmar Fiord area (West Coast of Sweden): Acta Zoologica Stockholm, Vol. 35 (3), pp. 215–257, 24 text figures.

————, 1956. On shell-burrowing Bryozoa from New Zealand: Trans. Roy. Sco. New Zealand, Vol. 84, Part 1, pp. 93–96, 1 text figure.

Sparling, D. R., 1964. *Prasopora* in a core from the Northville area, Michigan: J. Paleontology, Vol. 38 (6), pp. 1072–1081, plates 161–162, 3 text figures.

Stach, L. W., 1936. Correlation of zoarial form with habitat: J. Geology, Vol. 44, pp. 60–65.

————, 1937. The application of the Bryozoa in Cainozoic stratigraphy: Australian and New Zealand Assoc. Adv. Sci., Rpt. 23rd Meeting, Auckland, pp. 80–83.

Textoris, D. A., and A. V. Carozzi, 1964. Petrography and evolution of Niagaran (Silurian) reefs, Indiana: A.A.P.G. Bull., Vol. 48 (4), pp. 397–426.

Trechmann, C. T., 1914. On the lithology and composition of Durham Magnesium limestones: Geol. Soc. London, Quart. Jour., Vol. 70, pp. 232–256.

Voight, Ehrhard, 1930. Morphologische und stratigraphische Untersuchungen uber die Bryozoenfauna der oberan Kreide: Leopoldina VI, pp. 379–563, plates 1–39 (U. Cretaceous cheilostomes in northwest Germany, the Baltic, and Holland).

Williams, J. S., 1957. Paleoecology of the Mississippian of the Upper Mississippi Valley region, in H. S. Ladd, ed., Treat. Mar. Ecol. and Paleoecology, Vol. 2, Chapter 12.

Zekkel, J. D., 1941. Importance of reefs in the Kazanian stage: Akad. Nauk. SSSR Doklady, Vol. 23, pp. 569–571.

Two publications show newer trends in ectoproct classification, that is, revision of genera: R. S. Boardman and J. Utgaard, 1966, J. Paleont., Vol. 40 (5), pp. 133–142; J. Utgaard, 1968. J. Paleont., Vol. 42 (6), pp. 1444–1455. Chemical composition of ectoprocts related to ecology provide useful data. (T. J. M. Schopf and F. T. Manheim, 1967, J. Paleont., Vol. 41, pp. 1197ff).

Boardman, R. 1971. Smithsonian Contrib. Paleobiology. No. 8, 28 p; 11 pls (functional morphology details). (Newell et al., 1953 See refs. Ch. 5.)

Larwood, G. P.; ed. 1973. Living and fossil Bryozoa (Academic Press); for many excellent papers on modern researches. McLeod, J. D., 1978. The oldest Bryozoans: new evidence from the early Ordovician: Science, vol. 200, pp. 771-773, figure 1. An abundant Arkansas-Missouri fossil, based on well-preserved specimens discovered in the matrix, was shown to be a new bryozoan species (genus *Dianulites*, Order Trepostomata).

chapter seven
MOORED TO THE SEA BOTTOM: PUMP-FILTER MODULES IN SPACE AND TIME (THE BRACHIOPODA)

INTRODUCTION

Although known since the Middle Ages brachiopod affinities have been debated by leading investigators—embryologists, zoologists, paleontologists, and others— for several hundred years. Lingulids were so plentiful on exposed Japanese mud flats that they were collected by the peck and eaten whole or in part and the shell discarded (Yatsu, 1902).

The brachiopod shell has two valves. Since there are 260+ living species of brachiopods, one would think that detailed anatomical studies could readily

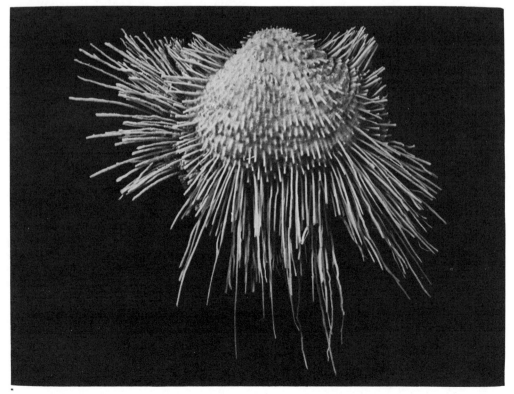

A Permian productid brachiopod from Khisor Range, West Pakistan. The corona of dense slender spines anchored the adult ventral valve to the substrate. (See Figure 7.12A.) (courtesy of Dr. R. E. Grant.)

resolve the affinity problem. However, assignment proved elusive.

Classification makes strange bedfellows. Brachiopods have been shifted back and forth from one phylum to another: arthropods (barnacles), mollusks (class, family, order of — associated with bivalves including rudists), echinoderms (larval starfish), annelids (segmented worms), among others (Muir-Wood, 1955).

Brachiopod relationships to bryozoans and phoronid worms were first glimpsed in the last quarter of the nineteenth century. Today they are grouped, with reasonable confidence, in the lophophorate coelomates, each component of which represents a distinct phylum.

Ager et al. (1965) found the following speculation attractive: lophophorates were independently derived from Protozoa. This is quite a contrast to the vermiform ancestral type that Hyman favored.

Fossil and living brachiopod species, in some 1700 genera, constitute a negligible fraction, slightly more than 1 to 3 percent, of all known animal species, vertebrate and invertebrate. About 30,000 species are extinct. The bulk of these lived in Paleozoic seas (G. A. Cooper). The phylum, however, has a fossil record extending through Mesozoic, Tertiary-Quaternary time; today some 260 species survive.

The Brachiopoda have long since been replaced as principals among benthonic marine forms (see Chapter 8). Living brachiopods appear to be lingering survivors of a declining phylum (Williams, 1965). Nevertheless, brachiopods are of extraordinary value in biostratigraphic problems.

LIVING BRACHIOPODS

Phylum Definition

(After Williams, 1956; Borradaile et al., 1958[1]; Ager et al., 1965; and others.)

[1](See ref. Ch. 8).

Phylum Brachiopoda (arm + foot). The long since discarded generic name *Lampus* (= lamp), *Lampus terebratula* denoted the resemblance of terebratulid brachiopod shell form to that of a Grecian lamp (Muir-Wood, 1955).

1. Coelomate, unsegmented invertebrates.
2. Skeletal system:
(a) Bivalved. *Dorsal* or *brachial* valve contains the lophophore; *ventral* or *pedicle* valve, the larger of the two which has at the posterior end, a muscular pedicle. Valves bilaterally symmetrical; hinged (articulates), unhinged (inarticulates) (Figure 7.1).
(b) Shell composition. Calcareous (articulates) and in the inarticulate Craniidae. Chitinophosphatic (inarticulates).
(c) Spicules. Make no apparent contribution to the solid skeleton but do occur in connective tissue of terebratuleans, and so on.
3. Attachment. Attached to substrate by pedicle; secondarily cemented, or free. (In some fossil forms attachment is aided by anchoring spines.)
4. Mantle extensions. Epithelial tissues of body wall line valves. Outer epithelium secretes the calcareous shell; inner epithelium forms an oblique wall between two cavities, mantle and body (Figure 7.1B).
5. Cavities. The *body* or *coelomic cavity* is in the posterior third of the shell; *mantle* or *brachial cavity* occupies the remainder. The main body occupies the body cavity; mantle cavity is largely occupied by the lophophore suspended between the mantles. Body cavity extends into mantles through *mantle canals* that serve a respiratory function (Figure 7.1D, a).
6. Lophophore. A complex, ciliated, circumoral feeding organ; variable in shape (Figure 7.2); general structure — lobulations margined by cirri and filaments (tentacles) (Figure 7.1A, B). For mode of functioning, see below.
7. Systems:
(a) *Digestive*. Intestine in articulates has a blind terminus; fecal pellets voided through mouth; inarticulates have an anus.
(b) *Circulatory*. An "open" system: longitudinal vessel in dorsal mesentery — contractile parts of this vessel constitute the "heart." (In *Crania* there are several contractile vesicles.) The circulatory system has connec-

tions to mouth, mantle, and generative organs, all of which end blindly; that is, circuit is completed in mesentery tissue spaces. Fluid that circulates is colorless, coagulable, noncellular.

(c) *Nervous system.* Two major ganglia and connectives.

(d) *Reproductive.* Sexes separate; rarely hermaphroditic as in living *Argyrotheca*.

(e) *Excretory.* One or two pairs of nephridia.

(f) *Muscular.* Three sets of muscles (articulates): adductors to close valves; diductors to open valves; peduncular – muscle of the pedicle; adjustors – can move shell sidewise and up or down. In inarticulates, muscles serve to hold valves together in absence of hingement; also to open and close them and to allow rotation and sliding; one or two pairs of adductors; three pairs of oblique muscles (median, internal, external). Muscle attachment sites on interior valve surfaces remains as *muscle marks* (inserted, raised, platformed) and are important in brachiopod taxonomy.

(g) *Respiratory.* Mantle canals discussed above (see Figure 7.1*D*); a secondary respiratory function is also attributed to the lophophore.

Habit and Habitat

Exclusively marine; benthonic; epifaunal; sessile. (Living lingulids can endure temporary, fresh to brackish water conditions but only in a state of suspension of normal metabolism. Generally they are immersed in salt water of normal salinity (Rudwick, 1959). Geologic range: Cambrian to Recent.

The phylum is subdivided into two classes: Inarticulata and Articulata. (These are defined in the section on fossils.) In addition, there is one class of uncertain affinities.

Pump and Filter System

An ideal maximum-efficiency-model pump-filtration system within an enclosed space is diagrammed in Figure 7.2*B* (compare Mollusca). Important in such a mechanical system is (*a*) creation of an inhalent and exhalent aperture and chamber, (*b*) which in turn insures an increase

of pressure in the system as filtered water fills the exhalent chamber and passes out of the mantle cavity via the exhalent aperture.

The ciliated filaments of the bilaterally symmetrical lophophore meet the first requirement by dividing the mantle cavity into inhalent chamber and aperture. Thereafter, ciliary movement creates currents for water circulation while nondiscriminating filtration occurs between the filaments—that is, extraction of food and other particles such as silt (Figure 7.2*A*). A pressure increase follows from the division of the mantle cavity into chambers. Important in such a filter-feeding systems' efficiency is the nonmixing of filtered and unfiltered water. Efficient operation of such a system requires severe restriction of possible current arrangements (Rudwick, 1960, 1965).

Application of the model to living and fossil forms, based on the structure of the lophophore, showed that two major types, and only two types, of current systems are possible in spire-bearing forms (spirolophes; exhalent type (*Spirifer*), where water is drawn inward into the interior of the spirals that contain filtered water, and the inhalent type (*Atrypa*), where water is drawn into the mantle cavity laterally and antero-laterally and then into the spiral interior that contains unfiltered water (Figure 7.2*C, D*).

Factors affecting feeding efficiency are actual configuration of the *spiralia* (which determines conical surface area covered by lophophore filaments), and the *free space outside the spiralia* for circulation of filtered water (*Atrypa*) or unfiltered water (*Spirifer*). The last is achieved by close molding of the *brachidium* to the shell (Rudwick, 1960).

Ontogenetic Variation in Filtration

Beecher originally—and subsequently numerous other workers (Atkins, Williams, Rudwick, among others)—recognized increasing complexity of the lophophore during ontogenetic development.

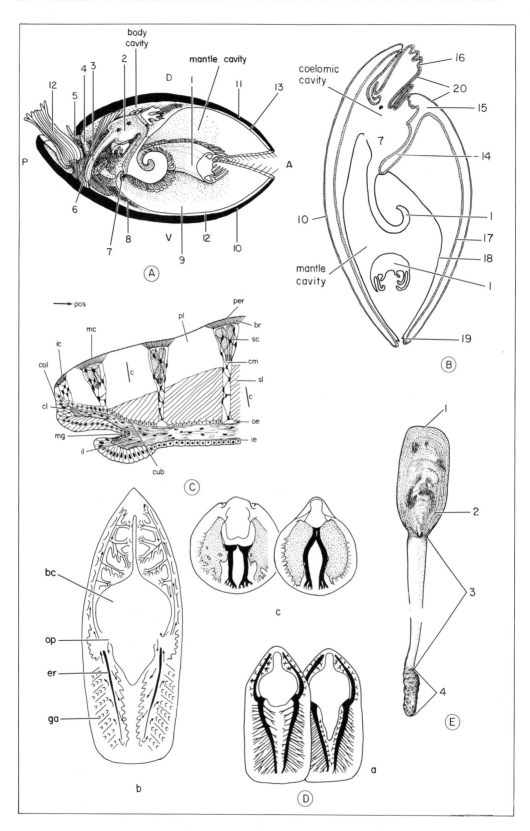

FIGURE 7.1 Gross anatomy and orientation of the Brachiopoda.
(A) *Terebratulina*, an articulate, schematic. (1) *Lophophore*, with spiral part to left (= spirolophe) and cross-section to the right (that is, the so-called left side arms). *Digestive Tract:* (2) stomach; (3) blind intestine (only inarticulates have an anus); (7) mouth. (4–6) *Muscles:* (4) dorsal adjustor; (5) didactor; (6) adjustors. (12) muscular stalk (= pedicle). *Reproduction:* (8) gonads. *Valves:* (10) pedicle, or ventral valve (V); (11) brachial (or dorsal) valve (D). *Mantle:* (12) (= 12') ventral mantle; (13) dorsal mantle. P = posterior; A = anterior. (B) Terebratulid, submedian longitudinal section, showing the two cavities (mantle and body) and epithelial tissue distribution, schematic; (1) (7) (10) as in (A); (14) lophophore support; (15) shell (16) pediculate epithelium; (17) outer epithelium; (18) inner epithelium; (19) mantle edge (= generative zone), (20) periostracum. (C) Generalized section through anterior mantle of an articulate brachiopod; two mantle lobes: *il* – inner lobe, (outer lobe, above "mg"), *mg* – mantle groove, *ie* – inner epithelium, *oe* – outer epithelium, *cm* – caeca and structures of same, *sc* – secretory cell, *br* – brush; shell layers: *per* – periostracum (generated by cells of *mg*); *pl* – primary layer (secreted by columnar cells, *col,* of outer boundary of outer lobe; secondary layer (*sl*) (secreted by cuboidal cells, *cub,* of outer lobe; diagonal lines with *c* on the right represent *C*-axes of calcite fibers of each shell layer (Note their parallel orientation). (D, a–c). Mantle Canal Systems: (a)(b) Inarticulates: (a) *Lingula*, right pedicle valve, left; brachial valve, right; *vascula lateralia* – black; (b) Dorsal valve of *Glottidia*, body cavity extensions into the mantle (= canals); arrows show directions of circulation; *bc* – body cavity, *op* – opening of mantle cavity into *bc, er* – epithelial ridge, *ga* – gill ampullae (= tubular extension into mantle cavity that magnifies respiratory area). (c) Articulates: *Terebratulina*, the terminology of canal segments; *vascula media* – black; *vascula genitalia* and gonadal sacs – stippled; pedicle valve, right; brachial valve, left. (Note canal system differs in each valve.) (E) Preserved specimen of living inarticulate *Lingula* sp. (Philippine Islands), X 1: (1) anterior; (2) dorsal (brachial) valve; (3) pedicle; (4) sand tube.
Sources: (A)–(C)(Dc) A. Williams, 1956. Cambridge Philos. Soc. Biol. Rev. Vol. 31., pp. 247–287, text figs 1–7; Da, Williams and Rowell, 1965; (Db)E. Morse, 1902. Boston Soc. Natural Hist. Mem. Vol. 5, pp. 313–386, pls 39–61; (E) Tasch after J. M. Lammons, unpublished report, 1960.

As the animal grew in size, it would require more food. Being a suspension feeder, this would necessitate, for survival, a complexing of the current system and hence the structures of the lophophore. If species size was small, a simple lophophore type would characterize the adult—the same type through which larger species passed in early ontogenetic stages (see Figure 7.2F).

The stages generally recognized all stem from the *trocholophe*, or trocholophous stage (Figure 7.2F, *a*). Given a trocholophe, which is a complete ring around the oral disk that generally bears a single row of filaments, and any of the following developments could yield a more complex type: lengthening, twisting, or coiling of the brachial axis. Deformation of the circularity of the trocholophe by a median indentation yields a *schizolophe* (Figure 7.2 F, *b*)—each brachia of a pair having a row of paired filaments. If the original circle were distorted into a pair of spirally coiled brachia, that would be the *spirolophe* (Figures 7.1B and 7.2F, *b*). If brachia

were folded into one or more lobes in addition to a median indentation, the type would be *ptycholophe*.

Of all types, the spirolophe is most widespread today among both inarticulates and articulates. Rudwick (1965) conjectured that this may have been true also in the geologic past. Both observations indicate a high adaptive value for the spirolophe-type lophophore.

Other types are:

1. *Zygolophe.* Each brachium's straight or crescent side arm bearing two rows of filaments.

2. *Plectolophe.* Characterizes terabratuloids; terminal portion of brachia coil in median plane.

3. *Deuterolophe.* Spirally coiled part of lophophore bearing double brachial fold and double row of paired filaments (Figure 7.2F, *c*).

Type of lophophore can be and has been inferred for many fossil forms (Figure 7.2). Internal structures of the brachial valve, and comparative studies of living brachiopods, permit such inferences.

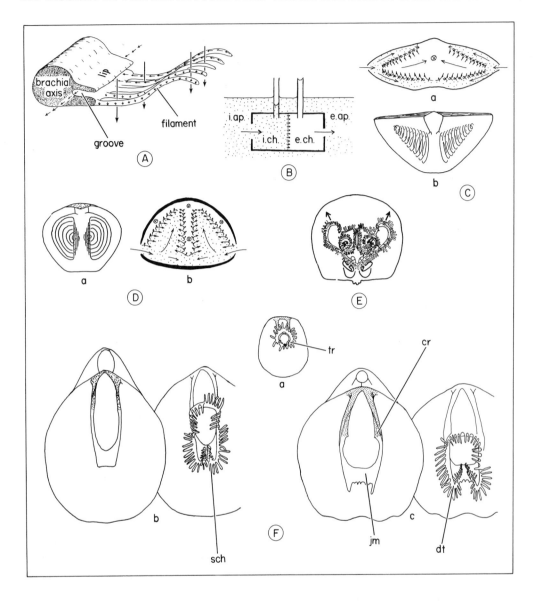

FIGURE 7.2 Lophophore structure and some types.

(A) Diagram of portion of left brachium of *Tegulorhynchia;* basic structure and action of lophophore are indicated. Small, broad arrows show direction of lateral cilia action; long arrows indicate the water currents produced by cilia action; horizontal arrows, action of frontal cilia; arrows in food groove (overhung by brachial lip), ciliary currents leading toward the mouth.

(B) Model filter-feeding system of maximum efficiency (schematic): *i.ap. e.ap*—inhalent and exhalent apertures; *ich, ech*—inhalent and exhalent chambers, manometer shows pressure changes across system (Note rise of meniscus in *ech*); stippled area is unfiltered water; unstippled, filtered water.

(C) Filter feeding inferred for fossils. *Spirifer striatus:* (a) Current system, symbols as in *B;* (b) spiral brachidia (ventral side).

(D) *Atrypa reticularis:* (a) Spiral brachidia, ventral side; (b) current system, symbols as in *B.*

(E)–(F) A few stages of lophophore development (see text). (E) Fossil articulate, *Productus s.s.,* schizolophe. (F)(a–c) The Ordovician spiriferid *Protozyga elongata,* developmental stages of brachidial apparatus (left) and inferred ontogeny of lophophore development (right): (a) trocholophe, *tr,* (b) schizolophe, *sch; cr*—crus; *jm*—jugum; (c) deuterolophe (*dt*).

Sources: (A)–(D) Rudwick, 1960; (E) Stehli, 1956a; (E) Williams, 1956; Williams and Rowell, 1965.

Food

The particles filtered by living brachiopods encompass diatoms and dinoflagellates – the number-one and number-two primary producers in the modern seas. What of the diet of Paleozoic and Mesozoic forms? (See page 297) Paine (1963) reported naked flagellates (*Monochrysis*, and so on), the diatom *Coscinodiscus*, sand grains – 100 μ diameter, small gastropod veligers, and even nauplids (125 μ diameter) in the gut of larval *Glottidia pyramidata*. Atkins (1959) reported dinoflagellates (peridineans) in the nutrient intake of a Recent brachiopod.

Spawning and Larval Stages

With a few exceptions, sexes are separate. Generally fertilization and development are external except for a few forms that carry their young in brood pouches (Figure 7.3*G*, *H*). Any of the following may serve as a brachiopod brood pouch: a portion of the lophophore arms (tentacles temporarily form a makeshift brood pouch), the mantle cavity, or enlarged nephridia.

Eggs and sperm are discharged into the sea during breeding season, which may be a restricted time of the year or intermittently during an entire year. *Terebratella inconspicua* (Figure 7.3*A–D*) is known to spawn between April and May – the austral winter in New Zealand waters; *Terebratulina septentrionalis*, at various times from April to August in Maine waters. In Red Sea waters, lingulids breed from June to October and in the Indian Ocean, in October only. Spawnings may extend from March through a part of September in the southern Red Sea. *Lingula* breeds from July to August in Japanese waters, December and February in Burmese waters, and Chuang (1959) reported that larvae appeared all year round in Singapore waters.

Several interesting observations were made by Chuang: (*a*) The smallest spawning *Lingula* in his study had a ventral valve length of 22.6 mm and the largest, 50.0 mm. This indicated that sexual maturity started early and "spawning power" is retained in old age. (*b*) All ova of a given individual were not extruded at the same time. (*c*) Spawning in a female occurred in bursts separated by rest intervals, the first burst being the most intensive. Paine (1963) recorded 20 minutes for males and 60 minutes for females for duration of ejection of sexual products in *Glottidia*. (*d*) Spawning went on both day and night. (*e*) Several thousand ova were extruded per female per day. (*f*) Neither the presence of males nor crowding was necessary for females to spawn.

Allowing for variations, other reports on articulates and inarticulates are similar to those already given (Percival, Yatsu, Morse, Ashworth, Blochmann, among others). The exact mode of coordination in the shedding of sex cells in the sea, in contrast to the same event in a laboratory Petri dish culture, is unclear. Percival (1944) found an approximate 1 : 1 ratio of males to females in his New Zealand collection. Paine also found a similar ratio in *Glottidia pyramidata* and observed that their close association on bottom stones in their natural habitat facilitated fertilization. He further commented that in the observed population spawning occurred "just before or at high tide" and explained that this probably facilitated seaward movement to cleaner oceanic water.

The inarticulate *Glottidia* released an estimated 47,000 eggs in one burst, and during an entire spawning season of 3 to 4 months the maximum discharge for a female was 150,000 ova (Paine, 1963). This may be contrasted with Chuang's study of *Lingula* – in an equivalent period, something less than one-third the *Glottidia* maximum was obtained. Such large numbers of eggs, which compared to the sea urchin are not large numbers at all, tend to insure species survival despite egg wastage. *Lingula* eggs were spherical in shape, enclosed in a vitelline membrane, and bore a visible nucleus. Their mean size was between 90 and 100 μ (1 $\mu = 10^{-3}$ mm). Chuang observed that unfertilized ova

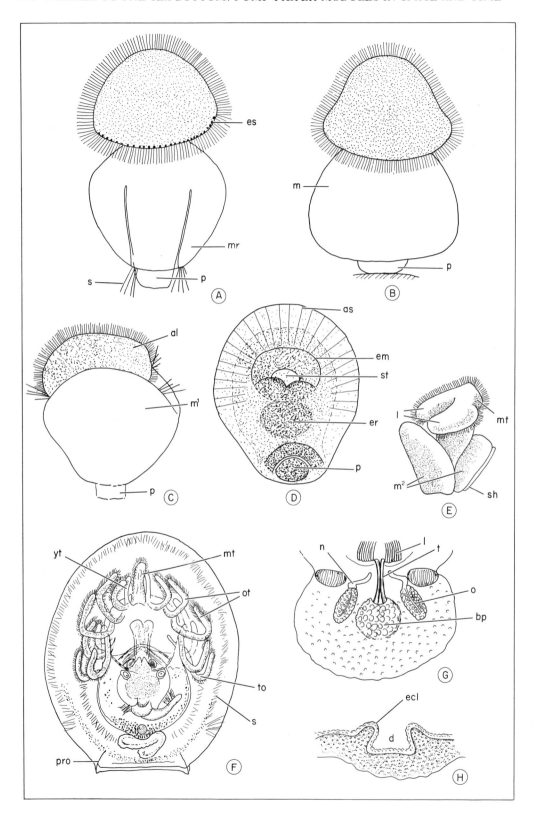

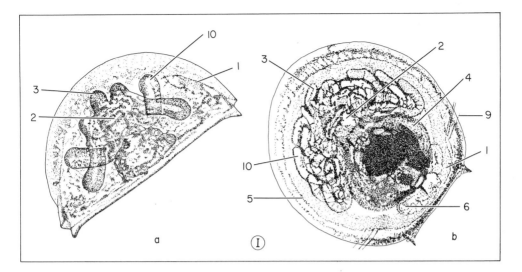

FIGURE 7.3 Larval stages and related data.

(*A*)–(*D*) *Terebratella inconspicua,* Rec., N.Z. (*A*) Motile (free-swimming) stage after emergence from parental mantle cavity. (*B*) Two hours after attachment. (*C*) Mantle reversal begins to envelope partially the anterior lobes. (*D*) Rudiments of lophophoral cirri appear.

(*E*) (*F*) *Lingula anatina,* Rec., Japan. (*E*) Body projecting from mantle lobe—note tentacle development on lophophore and shell secreted by mantle lobe, (*F*) Late larval stage, eight pairs of tentacles; *es*—pigment spots, *mr*—mantle ridge, *p*—peduncle, *s*—seta, *m*—mantle before reversal, *al*—anterior lobe, *m'*—mantle after reversal, *st*—stomodeum (invagination of ectoderm), lobes below *st*—rudiments, 1st pair lophophoral cirri, *er*—rudiment, adult enteron, *yt*—youngest to oldest tentacles, *mt*—median tentacle, *ot*—other tentacles, *pro*—protegulum (the first formed shell of organic material simultaneously secreted by both mantles), *l*—lophophore, *m²*—mantle lobe, *sh*—shell.

(*G*) (*H*) *Lacazella mediterranea,* Rec., Medit., schematic. (*G*) Both valves opened, upper (brachial) showing only a portion of the lophophore (*l*) and tentacles (*t*) bent into brood pound (*bp*); *o*—ovaries, *n*—nephridium. (*H*) Median section of *G* showing external calcareous lamella (*ecl*) of ventral valve that in females is indented to form a depression for the brood pouch.

(*I*) Planktonic, living inarticulate *Glottidia pyramidata:* (*a*) At the 3-pairs of lophophoral cirri stage (age 6 to 7 days); 1—protegulum; 2—mouth; 3—median tentacle of lophophore; 10—lophophoral cirrus. (*b*) Motile larva just prior to settling; 4—food in gut; 5—larval shell; 6—pedicle coiled within valves.

Sources: (*A*)–(*D*) Percival, 1944; (*E*)–(*H*) Yatsu, 1902; Lacaze-Duthiers, 1861; (*I*) Paine, 1963.

disintegrated after one to two days in seawater.

Fertilized eggs generally undergo cell division, passing through blastula and gastrula stages, and become motile larvae (Figure 7.3*A*). Brachiopod larvae are not infrequent in planktonic hauls in modern seas. This holds even though the motile existence of articulates may persist a few hours, days, or weeks, and that of inarticulates, from three weeks to less than 90 days (*Glottidia* and *Lingula,* respectively). The most vulnerable period of existence for a brachiopod is the motile larval stage. The high larval mortality explains the high egg production per individual (Figure 7.4*B*, 1). This last insures sufficient survi-

val to maintain the species. Rudwick (1965) and Paine (1963) have observed that after spatfall (that is, attachment to substrate), the individual's chance of survival to maturity are greatly enhanced.

After the motile stage the larva settles to the bottom. The pelagic stage is now replaced by a benthonic one. Many forms attach to a substrate by a peduncle, some cement, others bury in sand or occupy burrows (Figure 7.3*B*). Once settled on the substrate, the larva passes through a progressive series of changes, of which *mantle reversal* is of importance in articulates. (Inarticulates do *not* undergo mantle reversal.) During mantle reversal, toward the end of the enclosure of the anterior

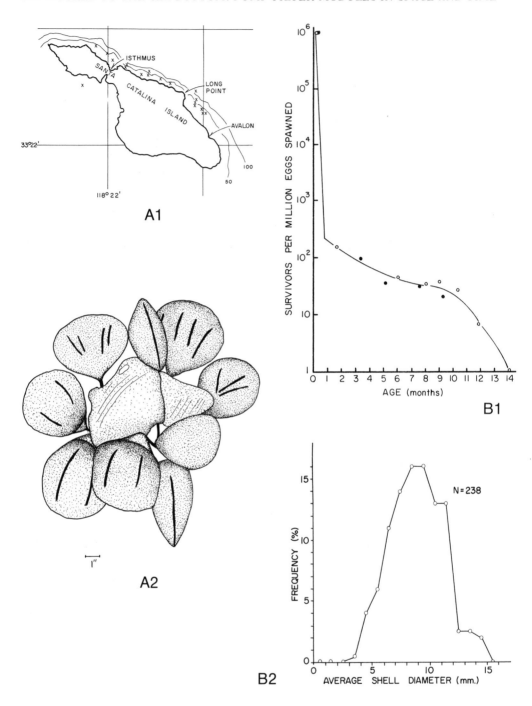

FIGURE 7.4 Information derived from living brachiopods.
(*A*) (*1*) Distribution of brachiopods around Santa Catalina Island off the Californian coast. Contour lines (50, 100) represent depth in fathoms; *x* = brachiopod collecting sites. (*2*) *Laqueus californicus* individuals attached in grapelike clusters to two shells of *Terebratalia occidentalis* (*n* = 15). Collected around Santa Catalina Island. Chiefly occur at depths of 30 to 60 fathoms. Shells translucent. Long, flexible pedicles provide leeway in attachment according to Stehli. Data from N. T. Mattox, 1955.
(*B*) (*1*) Survival curve for living *Glottidia pyramidata*. Florida. Curve of densities of two sub-populations (represented by **O** and ●) based on age. After R. T. Paine, 1963. (2) Living *Discinisca strigata*. Size distribution within a population according to frequency of average shell diameter. Northern Gulf of California. After Paine, 1962.

lobe by the mantle lobe, Percival (1944) saw that the outer surface of the mantle flaps became "glistening, white, and smooth"—that is, the *protegulum* or larval valve had been secreted (Percival, 1960).

During metamorphosis, which represents those anatomical changes that transform larva to adult, the interior lobe changes in form and arrangement, to give rise to rudiments of mouth, lophophoral cirri, gut, coelom, and so on. Subsequently these structures attain more mature expression. In *Terebratella inconspicua*, increasing number of lophophoral filaments accompanies increasing size of the animal. Early development of the lophophore proceeds without any calcified loop structural supports—these develop later.

Individual Life Spans

Life spans for individual brachiopods vary between species and may go as high as eight years (Rudwick, 1965). Percival (1944) found four year-classes (multimodal curve) in the New Zealand population of *Terebratella inconspicua*. Rudwick (1962) obtained a different distribution in the same area and thought that the largest individuals were probably more than four years old.

In another study, Paine (1962, Figure 1) found, in collections of the inarticulate *Discinisca strigata* from Baja, California, that only a single year-class was represented (unimodal curve, $n = 238$).

Paine (1963) linked variation in age structure (year-classes) in subpopulations of *Glottidia pyramidata* to (*a*) degree of permanency of the bottom, (*b*) availability of new recruits—that is, spat. A stable sand bottom would harbor several year-classes and obtain new recruits yearly.

There are, of course, still other factors involved. Storms and erosional processes can disrupt a given substrate such as a sand bar with a population of *Glottidia pyramidata* or a mudflat with a population of *Lingula*. Postlarval movements thus occur, and adult animals can be washed or transported to a different area. Exist-ence of an extensive bottom vegetation can act as a sieve to screen out young forms and prevent their establishment in otherwise suitable areas (Paine, 1963).

Population Density

In Lyttelton Harbor, New Zealand, Percival (1944) collected 711 specimens of *Terebratella inconspicua* from four adjacent stones covering an area of about 2 to 3 square decimeters. Paine (1962) performed a transect of the intertidal zone (northern Gulf of California) starting from the high-water mark to the low-water mark. At 10-ft intervals, he overturned six flat rocks each about 1 ft square and recorded the number of individuals of *Discinisca strigata*—$n = 122$ (Figure 7.5). The species did not occur in great clumps but as single individuals.

Mattox (1955) recorded 43 individuals of the inarticulate *Glottidia albida* (from a bottom sampler collection) at depths of 45 fathoms, taken from a 2-ft area. He also found that the articulate *Laqueus californicus* occurred in grapelike clusters—15 individuals of this species attached by their pedicles to shells of two *Terebratalia occidentalis* (Figure 7.4*A*, 2). Another such cluster consisted of 31 individuals of *Laqueus* attached to the shell of a larger and older *Laqueus*.

A density of 50 individuals per sq ft was recorded for *Glottidia pyramidata* on a stable

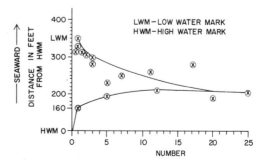

FIGURE 7.5 Number of *Discinisca strigata* recorded in transit of intertidal zone. Northern Gulf of California. Sampling interval every 10 ft from high-water mark (HWM) to low-water mark (LWM). Maximum depth, 20 ft. Sample: *n* = 122. Data from Paine, 1962.

grass-crowned sandbar, Florida (Figure 7.4*B*, 1).

Other reports indicate that single dredge hauls in Norwegian waters and elsewhere yielded 150 inarticulate craniids (*Crania anomala*) and 100 *Hemithyris psittacea*, respectively. The latter were attached to each other or to pebbles (Hyman, 1959). Other species of this last-named genus are found in Alaskan and Japanese waters today. It is apparent that whether living brachiopods occur singly or in clusters, their distribution may be described as *patchy*.

Biotic Associates and Predators

In 1955 Mattox deliberately recorded conspicuous animal associates of the brachiopods in the Catalina Island area (Figure 7.4*A*) to enhance ecological understanding. He found a community of 115 different kinds of animals and nearly 500 individuals living together in a 6-sq-ft area at a depth of 36 fathoms on a shell-sand bottom.

Animals included three individuals of *Laqueus californicus* and two of *Terebratalia occidentalis*. The polychaetous annelids were most abundant and diverse in this community, mollusks and then echinoderms being the next largest groups. Brachiopods were negligible components— 1/100 of the total sample.

Floridian specimens of *Glottidia* were incrusted by a hydroid or small pelecypods, or served as attachment sites for gastropod eggs and algae. As many as five pelecypods would attach to one brachiopod valve (Paine, 1963). Besides the big three associates that Mattox had found — polychaetes, mollusks, and echinoderms — Paine noted a phoronid worm, some parasites, ciliates, and trematode worms. Brachiopods have also been found in the stomachs of fish. Carnivorous snails are known to prey on sessile brachiopods.

Bathymetry, Salinity, Substrate

Figures 7.4*A* and 7.5 show the depth at which characteristic Californian and Flor-

idian articulates and/or inarticulates were taken. Off the Catalina Island, brachiopods were chiefly retrieved from depths of 30 to 80 fathoms with a few specimens at 120 fathoms, and a few others at 861 fathoms. There are striking changes in size and color with changing depth. Up to 50 fathoms, *Laqueus californicus* consisted of large forms having a pink coloration; beyond this depth, smaller forms occurred in which the pink had faded to white.

The Floridian inarticulates are distributed in greater numbers in the shallows — maximum depth 20 ft, intertidal zone.

According to Schuchert (1911) bathymetric distribution of living brachiopods ($n = 158$ at the time he wrote, but now increased to $n = 260$ species) was as follows: low tide—3 percent; strand to 90 ft—28 percent; 90 to 600 ft—40 percent. Thus 71 percent occur from the strand to a depth of 600 ft. The bulk of inarticulates range from low-water mark to 90 ft; some do extend to 600 ft, where numerous articulates first occur (Figure 7.4*A*). (Articulates, of course, range from the strand line, with many living in shallow waters.)

Some forms have a very broad range extending from shallows to abyssal depths — *Macandrevia cranium* ranging to 4000 meters; *Terebratulina retusa*, to 3600 meters (Hyman, 1959). Abyssal species that are permanent inhabitants at such depths include *Pelagodiscus atlanticus*, to 5000 meters, and *Abyssothyris wyvillei*, to 5500 meters (but apparently absent from abyssal trenches).

Thomson (1927) found that modern terebratelloids were distributed with reference to the 600-ft (100-fathom) line as follows: above it, 32 percent; below it, 29 percent; across it, 39 percent. Lacaze Duthiers (1861) obtained his Mediterranean thecideids from 60 to 80 fathoms.

Brachiopods are predominantly shallow, continental shelf forms (Rudwick, 1965), although they cover a very broad range of depths. Today they occur in Arctic and Antarctic waters and all seas in between. They thrive in abundance in polar waters, the Mediterranean, waters around

New Zealand, Australia, and Japan (Cooper, 1954). Some 31 living brachiopod species are spread along the Pacific coast of North America from Bering Strait to Panama (Hertlein and Grant, 1944).

Salinity. Generally 35 ppt (parts per thousand) is normal for living brachiopods (Rudwick, 1965). There are several reports of lower or higher tolerance nevertheless (Yatsu, 1902; Hertlein and Grant, 1944; Thayer, 1974).

Thayer's experimental data on articulates indicated a much broader salinity tolerance potential. Examples: *Hemithyris psittacea*, 14–39 ppt; *Terebratalia transversa*, 17–45 ppt; *Terebratulina unquicula*, 16–38 ppt; *Laqueus vancouverensis*, 17–45 ppt. These forms lived in depths of nearly constant salinity (31 ± 1 ppt).

Earlier reports (Cori, 1933, see Hyman, 1959, p. 594) that brachiopod salinity tolerance cannot go below 30 ppt, are thus no longer valid.

Substrate. Living brachiopods prefer hard substrates: sand, rock, pebble, gravel, shell, shell debris. Paine's Floridian inarticulates preferred flat cobbles. Mattox's Catalina brachiopod fauna came from rocky, pebbly, arenaceous clay and sand bottoms or were attached to shells. Cooper's (1954) Bikini Atoll brachiopods were attached to coralline slabs. Yatsu's lingulids were burrowed on a clay mud-flat. Mediterranean thecideids fixed to corals and some to bryozoans. Certain existing brachiopods—*Terebratulina*, and others—favor soft organic substrates, horny worm tubes or algal stems (Rudwick, 1965).

Shell Secretion, Structure, and Composition

From the paleobiological view, the brachiopod shell is of extraordinary interest.

Articulate Shell. A brief resumé of brachiopod shell structure is given in Chapter 15. Here some of the same aspects are

explored in more detail, and additional data are reviewed. Figure 7.1*B–C* outlines the essentials in articulate valve secretion.

1. The larval stages (Figure 7.3). During mantle reversal in articulates Percival observed that the outer surface of the mantle flaps (lobes) became white and smooth, indicating initial shell formation (protegulum).

2. (a) To appreciate shell formation, it is necessary to visualize the mantle edge (the generative zone) and its outer and inner epithelium, as well as the mantle groove—an important epithelial generative zone (Figure 7.1*B–C*).
   (b) Immersed in a salt-water medium of normal salinity, all parts of the brachiopod shell must be derived from the dissolved salts and organic constituents. (Review Chapter 15.)
   (c) The honeycombed *periostracum*—an organic (proteinaceous) external skin of the shell—is spun out of the closed end of the mantle groove and adheres to the outer lobes' inner surface up to the tip. Beyond this point, it separates more and more from the outer epithelium as the biocrystals of calcite that it secretes intercalate (Figure 7.1*B* and 7.6*F*).
   (d) Williams (1956, 1965) suggested that a physiological change must occur at the tip of the outer mantle lobe within the columnar epithelial cells. These cells initially secreted the proteinaceous periostracum, and this is followed by a new regimen—the secretion of biocrystals of calcite. These form a two-layered shell. In lingulids, alternation of organic and phosphatic layers can be accounted for by an equivalent physiological alternation in organic and inorganic secretions of the outer epithelium.

The Carbonate Layers:

1. Primary layer (*P*). Cryptocrystalline outer layer, secreted as noted in 2(d) above and forms a continuous sheet. The primary layer's thickness is determined by number of secreting cells. It is finely fibrous, and the long axis of fibers is parallel to the crystallographic *C*-axis of calcite and normal to shell surface.

2. Secondary layer (S_1). As seen in Figure 7.1*C*, posterior to the peripheral zone of growth (which controls general enlargement of shell), the columnar cells become

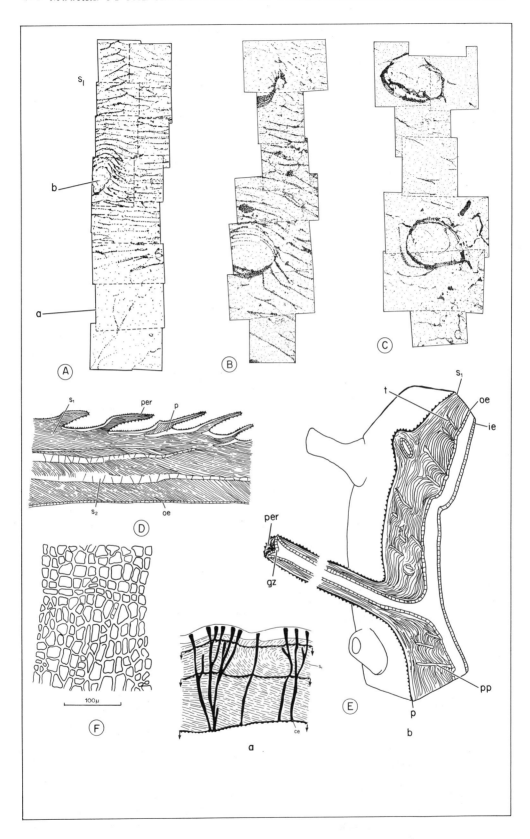

FIGURE 7.6 A fourth shell layer (adventitious calcite) in Recent brachiopods.
(A) *Gryphus stearnsi*, Recent. Transverse section of brachial valve; composite electron micrograph; s_1—uppermost portion of inner carbonate layer (note prismatic crystals of Alexander's s_2 layer), *b*—punctae, *a*—adventitious calcite, large crystals. (Observe curvature of crystals around punctae.)
(B) (C) *Pictothyris picta*, long. sec., brachial valve. Electron micrographs. (B) Portion of inner carbonate layer (S_1); (C) the adventitious layer, which occurs below base of B (see A). Oval areas = punctae. (Note lack of curvature of calcite around punctae in C, and compare to A.)
(D) *Atrypa* sp., M. Dev. United States. Periostracum (per), primary shell (P), the two secondary layers (S_1—fibrous; S_2—prismatic) in relationship to outer epitheca (oe). Relationship is inferred.
(E) Punctation. (a) *Rhipidomella* sp., M. Dev., U.S.A., cross-section, branching caeci, endopunctate condition. (b) *Productella* sp., U. Dev., U.S.A. (reconstruction). Pseudopunctate. P—primary shell, S_1—secondary shell, ce—caecum, oe, ie—outer and inner epithelia, pp—pseudopunctae (= shell penetrated by taleolae, t—calcite rods), per—periostracal, gz—generative zone.
(F) *Hemithyris psittacea*. Cellulose acetate peel of periostracum displaying honeycombed structure.
Sources: (A)–(C) Sass, Monroe, Gerace, 1965; (D)–(F) Williams, 1965.

cuboidal. Each cuboidal cell secretes a single calcite fiber (intracellular deposition) which, as it grows, stretches the membranous cell wall. This accounts for the complete enclosure of each fiber in a delicate cytoplasmic sheath. The C-axis of the fibrous calcite is oblique to the long axis of the fibers but parallel to the primary layer.

3. Prismatic layer (S_2). Alexander (1948) described a third relatively thick prismatic layer of coarse calcite prisms. The latter were in continuity with earlier-formed oblique fibers. Prisms are aligned at right-angles to substrate in which they were formed. Williams (1956) recognized Alexander's prismatic layer as a modification of the secondary layer described above in spiriferaceids, rostrospiraceids, and pentameraceids.

4. Adventitious layer (A). Sass et al. (1965) discovered still another layer—the fourth layer—composed of adventitious calcite. They used the electron microscope (Figure 7.6). It consists of relatively large, irregularly shaped crystals that do not curve around punctae as calcite crystals do in the overlying S_1 layer. It is added to the shell of mature individuals. In light of Williams' studies discussed above, it would seem that after the S_1–S_2 layers are deposited, a fourth secretion occurs.

The large irregular biocrystals of the adventitious layer are unlikely to have been secreted as were fibers of the S_1 layer—that is, by stretching the wall membrane of individual cuboidal cells. Crystal size and irregular configuration seem to preclude this.

Although details of internal shell structure, cardinalia, and so on, will be discussed subsequently, here it can be noted that (*a*) they are composed of secondary shell biocrystals and (*b*) they are attributed to accelerated growth of the outer epithelium (Williams, 1956, 1965).

Spiculation. Calcite crystalline bodies of irregular shape, granulated and variously arranged, are dense in connective tissues of some articulates. Inarticulates appear to lack spicules. Spiculation is found in lophophore and its filaments, body wall, mantle over the gonads (Atkins, 1959) and nephridial processes (Heller, 1931, see Hyman, 1959). Spicules apparently serve to stiffen the connective tissue. In *Platidia davidsoni* (Recent) the heavily spiculated lophophore was thought to compensate for slight development of brachial supports (Atkins, 1959a). Stehli (1956a) reached a similar conclusion for other forms.

Although of limited development, brachiopod spiculation is reminiscent of an internal skeleton insofar as it gives support within tissues (Thomson, 1927). Each spicule, according to Thomson, is a single crystal of transparent calcite, 2.5 mm or less in diameter. It yields a uniaxial figure under polarizing microscope. The flat calcite plates are often perforated and have processes that give a stellate-like configuration to the entire crystal. Upper and lower surfaces of the plates bear spicules.

Best available evidence indicates that spicules make no contribution to the brachiopod solid skeleton and are rare in the fossil record (Williams, 1956, 1965).

Some brachial supports or loops (discussed subsequently) develop calcareous spines. According to Elliott (1953b), this suggests "some power of secretion by the cirri" (cirri = lophophore filaments or tentacles). Such filaments are densely spiculated in terebratelloids. Some filamentous spicules may thus be resorbed and secreted to form loop spines. A check on Elliott's suggestion would be the following determination: Is loop-spine calcite in optical continuity with loop calcite? An alternate origin of loop spines could be *selective resorption.*

Calcite Resorption. Although loop growth and enlargement are discussed elsewhere in this chapter, here it may be remarked that this internal calcareous skeletal structure, composed of thin, rigid ribands of shelly matter, grows and enlarges by resorption and secretion of calcite (Elliott, 1953b).

Biochemistry. In addition to the chemical and other data on brachiopods given in Chapter 15, Jope's work (1965) on the organic content of the brachiopod shell provides previously unavailable analyses (Table 7.1).

The periostracum of articulates yields about 0.4 to 0.5 percent hexosamine [derived from an unidentified mucopolysac-

TABLE 7.1. Amino Acids in Brachiopod Shells (In per cent of total amino acid residues present)

| | Inarticulates | | | | | Articulates | | | | | |
| | Recent | | | | | Recent | | | Fossil | | |
	1[a]	2	2a	3	4	5	6	7	8	9	10
Aspartic acid	+[b]	+	15.6	0		+	14.9	23.5	+	+	
Glutamic acid	+	+	9.1	0		+	7.5	5.8	+	+	+
Cystine	+	+					0.3	0.3			
Serine	+	+	8.9	0			9.7	7.0			
Threonine	0	0	4.5	0			3.2	3.7			
Glycine	++	++	7.2	++	+	+	31.5	32.4	+	++	+
Alanine	++	+	13.2	+			1.6	4.1	+	+	+
Proline	++	0	4.5	0			3.3	1.1	+	+	
Valine	+	+	5.4	0			1.6	2.3	+	+	+
Methionine	0			0							
Leucine/Isoleucine	+	0	8.6	0			5.5	1.9	+	+	+
Phenylalanine	++	+	2.4	+	+	+	4.6	0.7		++	
Tryptophane	+	+		+	+		0.9				
Tyrosine	+	+	3.1	+	+		7.8	1.2			
Lysine	+	+	2.7	0	+		0.8	0.7		0	
Histidine	+	+	5.5	+	+		5.5	11.3		++	
Arginine	+	+	7.3	+			1.2	3.9			
Hydroxyproline	+	0	2.1	0							
μM amino acids per gram/fossil or matrix										0.5	0.5

[a]1–3, Whole shells, 1. *Discinisca lamellosa,* 2. *Lingula* sp., 2a. Periostracum only (*Lingula* sp.), 3. *Crania anomala.* 4–5, Whole shells, 4. *Notosaria nigricans,* 5. *Macandrevia cranium.* 6–7, Periostracum only, 6. *Laqueus californicus,* 7. *Terebratalia transversa.* 8–10, Fossil, 8. *Spirifer* sp. (Carboniferous), 9. *Linoproductus* (Carboniferous), 10. *Plaesiomys (Dinorthis) subquadrata* (Ordovician).
[b]+ = Amino acids present in small or moderate amounts; ++ = Amino acids present in high proportions; 0 = No amino acids detected; μM = micromoles, denotes measure proportional to number of molecules per unit mass.
Source. H. M. Jope, 1965.

charide (?) of the caeca]. The protein content is variable: 21.8 percent in *Laqueus*; 13.3 percent, *Terebratalia*. Table 7.1 shows the amino acid breakdown of these.

The periostracum is spun out of proliferating epithelial cells. Cell nuclei contain the DNA molecules, which serve as the template specifying the kind of protein that will comprise the periostracum and the mode for realizing this specification —transfer and messenger RNA. Honeycomb structure (Figure 7.6*F*) of the periostracum may reflect spatial arrangements of molecules.

The Inarticulate Shell. The essential aspects of inarticulate shell composition and chemistry are covered in Chapter 15. Here additional data are considered.

Larval and postlarval development of the shell of *Lingula anatina* can best be related to stages denoted by number of pairs of lophophore filaments that have appeared (Yatsu, 1902; Williams, 1965):

1. **Single Pair of Filaments.** Still within egg; mantle flaps—the outer epithelial tissue— secrete chitinous circular cuticle (= larval shell or protegulum) (Figure 7.3*E, F*). Ventral and dorsal valves joined at hinge line, which is a fold of the circular cuticle. Hingeline width is 280 μ and remains so. Subsequently, embryo breaks out of egg membrane and leads a free-swimming existence for less than 90 days, after which it attachs to the substrate.

2. **Seventh-Eight Pairs of Filaments.** Circular outline; break of cuticle at hinge line near brachial valve, resulting in overlap of pedicle valve on brachial valve.

3. **Tenth-Fifteenth Pairs of Filaments.** Attachment. Shell growth from circular to elliptical configuration.

The periostracum is secreted by epithelial tissue. Table 7.1 shows the amino acid composition of this tissue. Both chitin (10.6 percent) and protein (67.9 percent) are present in the pigmented periostracum of *Lingula*. Pigmentation is due to iron hydroxide and is distributed chiefly in growth lines of the embryonic shell (Jope, 1965).

For understanding of the chitin-protein linkage, see discussion in Chapter 15. *Crania* with a calcareous shell appears to be transitional between articulates and inarticulates. It lacks chitin. Jope (1965) thinks that chitin formation in brachiopods is related to the presence of phosphate—all inarticulates except *Crania* being calcareous-phosphatic.

Shell Composition. Alternating phosphatic layers [carbonate fluorapatite = francolite (McConnell, 1963)] and organic layers [chitin, protein (?)]. In addition, the outer skin (periostracum) consists of chitin-protein. Together, these components comprise the lingulid shell.

The layers are of variable thickness throughout, and in Figure 7.7*A* the organic layer is several times thicker than the phosphatic (see also Figure 15.5*C* and Chapman, 1914).

Shell Repair. Although little is known about shell repair in living brachiopods, it is apparent that if the mantle edges (that is, the epithelial cells) are either damaged or malfunctioning, normal secretion of shell will be interrupted. The accomodations that will then be made, such as sealing off of the injured site by uninjured mantle cells and resumption of normal calcite secretion, will be reflected by a visible repair site. (See "Paleopathology".)

Caecal Invaginations and Punctation. Accompanying shell layer secretion and growth are caecal invaginations of the mantle (Figure 7.1*C*)—the endopunctate condition (Williams, 1956, 1965). Invaginations are represented by holes or openings (punctae) in the shell structure.

In Figure 7.1*C*, "*ic*," an initial appearance of a caecum is shown. As the shell grows, the caeca migrate posteriorly (Figure 7.1*C*, see arrow), and two or three elongate mucus cells appear inside the lumen. Calcite of the primary layer encases the entire secretory portion to the brush tip.

What of the junction between primary

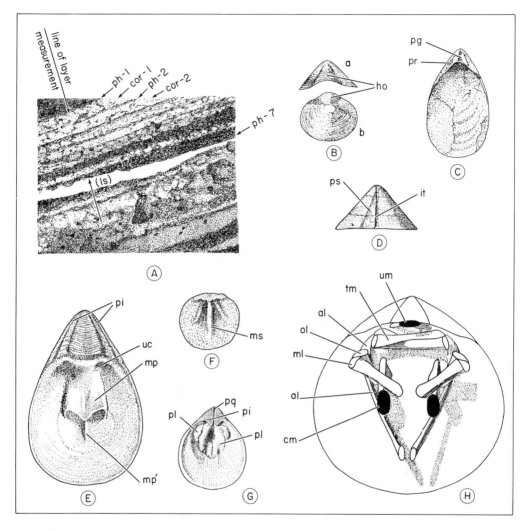

FIGURE 7.7 Inarticulates: shell structure and morphology.

(A) Portion of transverse section of *Lingula* sp. (Rec., Philippine Islands), dorsal valve; dark layer, highly phosphatic; light bands, corneous (= chitin), × 154. *ls* — layer separation due to processing.

(B) *Paterina*, pedicle valve, *ho* — homeodeltidium, *pro* — propares.

(C) *Lingulella*, pedicle valve, *pg* — pedicle groove.

(D) *Acrotreta*, posterior view; *it* — intertrough; *ps* — pseudointerarea.

(E) *Trimerella*, *uc* — umbonal cavity, *mp* — mouth platform, *mp'* — median partition, *pi* (= ps).

(F) *Spondylotreta*, brachial valve interior; *ms* — median septum.

(G) *Bicia gemma* pedicle valve, *pl* — platform, *pl* — posterior lateral muscle scars, *pi* (= ps); *pg* — pedicle groove.

(H) *Obolus*, muscle system; dorsal view after removal of dorsal valve; *um* — umbonal muscle, *tm* — trimedian muscles, *al* — anterior laterals, *ol* — outside laterals, *cm* — central muscles.

Sources: (A) After photograph by J. M. Lammons, unpublished report, 1960; (B)–(D) G. A. Cooper, 1944; (E)–(G) adapted from Ager et al., 1965, and H. Bulman, 1939. Geol. Mag. Vol. 76, pp. 434–444, 4 text figs.

and secondary layers relative to the caecum? A stalk forms that connects the caecum to the outer epithelium. This cylindrical stalk, in turn, is accommodated by secondary layer fibers (calcite crystals). According to Sass et al. (1965 and Figure 7.6), these crystals wrap around and parallel the walls of the punctae, at least in some terebratelloids.

All articulates do not have punctae — indicating a lack of caecal invaginations — this being the impunctate condition.

Where rods of granular calcite (taleolae) penetrate the secondary layer (S_1), the volume occupied by such rods, if completely cleared of mineral content, would appear as true punctae. Accordingly this is called the pseudopunctate condition. Pseudopunctae are conical deflections of the S_1 layer and may or may not have taleolae. They apparently represent continuous centers of growth that secrete it at the same rate as the S_1 layer.

Of the three conditions of punctation— *endopunctate*, *impunctate*, and *pseudopunctate*—the impunctate is apparently the most primitive. Furthermore, endopunctate forms in the fossil record closely resemble living terebratuloids in punctation patterns (surficial view) or branching patterns (cross-sectional view). This permits extrapolation to fossil endopunctate shells (Figure 7.6E).

Punctation is also present in inarticulates. An endopunctate condition is known in living and, by extrapolation, fossil craniids. Apparently other calcareous inarticulates are impunctate. Valves of forms such as *Lingula* and *Discinisca* show very fine perforations. These punctae may be attributed to cytoplasmic strands ensuing from the mantle rather than from the caeca (see Williams, 1965; Chapman, 1914.)

[Punctation, particularly in articulates, is an important taxonomic character of fossil valves.]

FOSSIL BRACHIOPODS

Introduction

Assignment of scientific names to fossil requires intimate knowledge of shell characters.

Important shell structures relate to various biological systems and functions (Table 7.2). The muscle system in fossils is represented by muscle marks and fields. Associated with musculature are varied structures. Shell form is determined by growth rates (Rudwick, 1959) and reflects the mode of life (Figure 7.9B, part G, e).

Surficial ornamentation, such as spines, fulfilled either an attachment or sensory function, whereas branching costellae in some forms apparently reflect a setae grille that in life sieved out large particles.

Mantle canals relate to the circulatory system.

Musculature and Related Features

The musculature of articulates consists of two pairs of diductors to open the valves, two adductors to close valves, and two pairs of adjustors (ventral and dorsal) to move shell relative to pedicle. The musculature of inarticulates is more complex (Figure 7.7H) and includes muscles to open, close, rotate, and slide the valves. How are these muscles reflected in shell features?

Muscle Scars or Marks. Articulate muscle scars (Figure 7.8B, part J), where preserved in fossils, are of taxonomic value. For example, genus *Strophomena* (M. Ord.– U. Ord.) was characterized by "a ventral muscle scar, subcircular to suboval, with strong lateral boundary ridges."

Inarticulate musculature, where living equivalents are available (*Lingula*, and so on), can sometimes be extrapolated to fossil material. Musculature of *Obolus* (Figure 7.7H) reconstructed by reference to *Lingula*, in turn, serves as a model for some other inarticulates. Thus *Lingulella*, *Dicellomus*, and so on, have musculature in whole or part similar to *Obolus*.

Structures. Pedicle and brachial valves of articulates contain various structures related to musculature (Table 7.2). Because of their importance in taxonomic descriptions, some of these major structures and their variants will be briefly reviewed. Certain inarticulates also have muscle platforms, such as the trimerellaceans (Figure 7.7E) and the Obollelidae (Figure 7.7G).

Spondylium (Pedicle Valve). Divergent dental plates unite to form this structure

TABLE 7.2. Some Articulate Shell Structures and their Biological Relationship and Function

Shell Structure	Valve	Biological or Other Relationship	Function	Figure
Dental plates	P^a	Musculature	Attachment for ventral adjustor muscles	7.9B, part I
Spondylium	P	Musculature	Elevates complete ventral muscle field	7.8B, parts L, M
Tichorhinium	P	Musculature	Accommodates base of pedicle (?)	7.9B, part I
Cardinal process	B^b	Musculature	Separation/attachment of diductor muscles	7.8A, part D
Cruralium (counterpart to spondylium)	B	Musculature		7.8B, part N
Tooth (ridges)	P	Valve	Articulation	7.9B, part H
Denticules	P	Valve	Articulation	
Sockets	B	Valve	Articulation	7.8B, part H
Cardinalia (includes cardinal process, socket ridge, crural bases and their accessory plates.)	B	Defines the sockets	Articulation, muscle attachment, partial support of lophophore	7.8B, part H
Brachidial apparatus (Connected to cardinalia and includes spiralia, loop, jugum.)	B	Filter/pump system	Lophophore support	7.8B, part H, I
Delthyrium	P	Attachment to substrate	Pedicle opening	7.8A, part G
Deltidial plates ⎫ Deltidium ⎭	P	Attachment to substrate	Restricts pedicle opening during growth	7.8B, part F
Pseudodeltidium	P		Covers delthyrium	7.8A, part C
Notothyrium (dorsal counterpart of delthyrium)	B			7.8A, part B
Chilidium			Covers notothyrium	7.8A, part C

^aP = pedicle valve.
^bB = brachial valve.

that elevates the entire ventral muscle field. Two main types are *spondylium simplex* (Figure 7.9B, part H), *spondylium duplex* (Figure 7.9B, part K). A third type, *spondylium triplex*, resembles the other two but has further support of two lateral septa (example: *Antigonambonites planus*).

This shell feature is found in many forms but especially denotes clitambonitoids (*spondylium simplex* or *triplex*), pentameroids [*spondylium duplex* (see "Classification")]. Its counterpart in the brachial valve is the *cruralium* (Figure 7.8B, part N). especially denotive of pentameraceans.

Cardinal Process (Brachial Valve). Part of the "Cardinalia," this structure serves for separation and attachment of diductor muscles (Figure 7.8A, part D). It consists of a swollen head (myophore) on a shaft. The head can be unilobed, bilobed (*Dalmanella*), trilobed (*Resserella*), or quadrilobed (subfamily Chonetanae, for example, may have either a bi- or a quadrilobed process).

The cardinal process of rhynchonellids, terebratuloids, and spiriferoids has a comblike appearance (seen in thin section).

Shell Structures Related to Articulation

The chief elements in the articulation of valves are paired hinge teeth (pedicle valve) and opposing sockets to receive them (brachial valve). The circular/triangular teeth are underlain by dental plates that serve as a support (Table 7.2).

The Cardinalia. Besides the cardinal process, the Cardinalia include socket ridges, crural bases, and their accessory plate. A dual role is served: articulation (definition of sockets) and partial support for the lophophore (Figure 7.8B, part *H*).

Shell Structures Related to Lophophore

Calcareous processes attached to the dorsal valve (cardinalia and/or septum) support the lophophore of many brachiopods. The brachidial apparatus (*brachidium*) is connected to the cardinalia through its anterior projections—the *crura*. The *loop* proper is attached to (is, in fact, an outgrowth of) the crura. The crura have two pointed processes projecting obliquely inwards and ventrally—the *crural processes*—which are the proximal ends of the loop proper. Instead of loops, calcareous ribbons (*spiralia*) may grow out of the crura. Another structure, the *jugum*, is a medially placed connection between the two primary lamellae of the spiralia (Figure 7.2F) (Thomson, 1927).

The simplest type of brachidium consists of two crura. The Rhynchonellacea are characterized by crura.

As discussed earlier, the lophophore performs a critical filter/pump function. Loop and spiralia are both variable in configuration and orientation within the valves.

Loops may be long or short (Figure 7.10B, C) and consist of ascending or descending branches; transverse and connecting bands, and may or may not be attached to the median septum (Figure 7.10A). Loops in many terebratuloids undergo ontogenetic changes (Stehli, 1956a).

Shell Structures Related to Pedicle Emergence Area

Two posterior notches in certain articulates form a diamond-shaped aperture (orthaceans, enteletaceans, and pentameroids). On the pedicle valve, the notch is called *delthyrium* through which the pedicle protrudes; in the brachial valve, the notch is called *notothyrium* (Figure 7.8A, part B). Both notches are restricted during growth by development of lateral plates. Those restricting the delthyrial aperture are *deltidial plates* (Figure 7.8B, part I). Those that restrict the notothyrial opening are called *chilidial plates*. Plates may remain discrete or, if growth continues, they ultimately meet to form a single structure (*deltidium* in pedicle valve; *chilidium* in brachial valve) (Figure 7.8A, part C).

The actual opening of the shell from which the pedicle emerges is the *foramen* (Figure 7.8A, part Aa). Its position relative to the deltidium and ventral beak is variable. In the order Strophomenida the cover to the delthyrium is called the *pseudodeltidium* (Figure 7.8A, part C). This cover always dorsally encloses the apical foramen. Its counterpart, the chilidium, is lost in the course of strophomenid evolution (Muir-Wood, 1955; Williams, 1965).

Convex external plates that cover in part or whole the delthyrium and notothyrium in the inarticulate *Paterina* and related forms, are called homeodeltidium and homeochilidium (Figure 7.7B).

External Features

Shells may be smooth (the primitive condition) or ornamented. Such ornament consists of lines, striae, ribs, costae, plicae (folds), granules, tubercles, or spines and is either radial or concentric or both. If the latter, growth lines will cross the radial ornament and, at points of intersection, lead to granulate, imbricate, tuberculate, or spinose outgrowths (Thomson, 1927).

Ribbing. Costa (coarse) and costellae (fine) are prominent radial ridges on

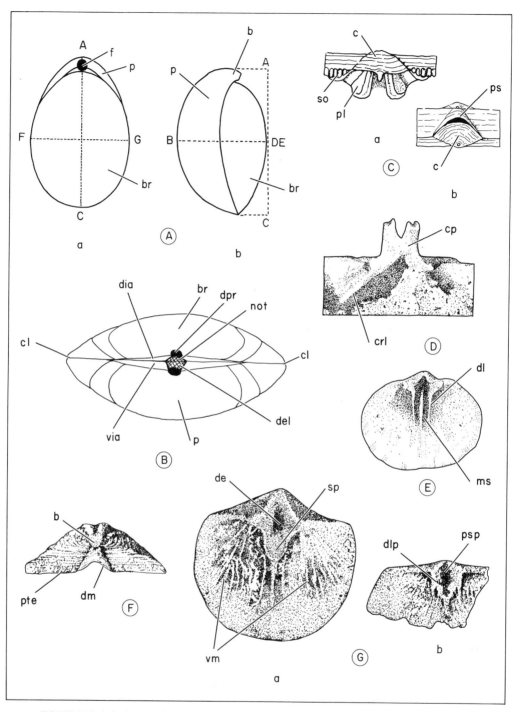

FIGURE 7.8A Articulates—I. Shell morphology and structures.

(A) Terebratuloid-type brachiopod; *p*—pedicle valve, *br*—brachial valve, *b*—beak, *f*—pedicle foramen, *c*—commissure. Measurements: *AC*—length, *FG*—midwidth, *BDE*—height (thickness) (Muir-Wood and Cooper, 1960).

(B) Generalized articulate; pedicle emergence area and characteristic structures; *not*—notothyrium, *del*—delthyrium, *dia, via*—dorsal and ventral interarea, *pl*—protegulum, *cl*—cardinal extremity (Williams and Rowell, 1965).

(C) *Douvillina (Mesodouvillina) neocomensis*. (*a*) Ventral view, dorsal (brachial) valve, X 5; (*b*) posterior view, median portion of interarea (pedicle valve); *c*—chilidium, *so*—socket plate, *pl*—cardinal process loop, *ps*—pseudodeltidium, X 6 (Williams, 1953).

272

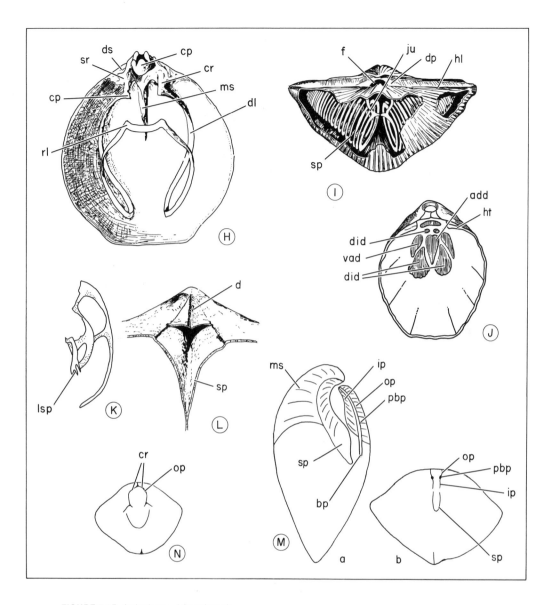

FIGURE 7.8B Articulates – I (continued).

(*D*) *Derbya cymbula,* dorsal valve, anterior view of bifid cardinal process (*cp*) and crural lamellae (*crl*) that diverge from it, X 1.5.

(*E*) *Enteletes damblei,* dorsal view, ventral valve with median septum (*ms*) and dental lamellae (*dl*).

(*F*) *Kozlowskiellina velata,* posterior view, pedicle valve with palintrope (*pte*) and deltidium (*dm*); *b* – beak (pedicle foramen below arch of *dm;* Amsden and Boucot, 1958).

(*G*) (*a*) *Ladogiella* sp., pedicle valve, interior view; spondylium (*sp*), delthyrium (*de*), vascula media (*vm*), (*b*) *Apheoorthis* sp., pseudospondylium (*psp*) and dental plates (*dlp*) (after G. A. Cooper, 1944).

(*H*) *Neothyris lenticularis,* brachial valve, interior view of brachidium: *cp* – cardinal process, *cr* – crura, *ms* – median septum, *dl* – descending portion of loop, *rl* – recurved portion of loop, *cp* – crural process, *sr* – socket ridge, *ds* – dental sockets (after Thomson, 1927).

(*I*) Spiriferacean brachiopod (Carboniferous), part of brachial valve removed to show spiralia (*sp*); jugum (*ju*); other features shown are *f* – foramen, *dp* – deltidial plates, *hl* – hinge line.

(*J*) *Magellania* sp., ventral valve, muscle scar field; *did* – didactor muscle scars, *vad* – ventral adjustor muscle scars, *did'* – accessory didactor muscle scars, *add* – adductor scars, *ht* – hinge teeth.

(*K*) *Macandrevia cranium,* lateral view of loop, with loop spines (*lsp*). (See text discussion.)

(*L*) *Pentamerus* sp. (Sil., N.Y.), silicified specimens; *d* – deltidium, *sp* – spondylium, X 3.

(*M*) Discrete outer brachial plates showing relationship to spondylium (*sp*): (*a*) Longitudinal section, pedicle valve, (*b*) transverse section at arrow. N. *Clorindella areyi* (Sil., N.Y.) outer brachial plates (*op*) united to form cruralium (*cr*), schematic; *a, b* – as in *M* above; *ip* – inner plate; *pbp* – posterior extension of brachial process, *ms* – median septum. (Illustrations: (*I*)–(*N*) after several authors.)

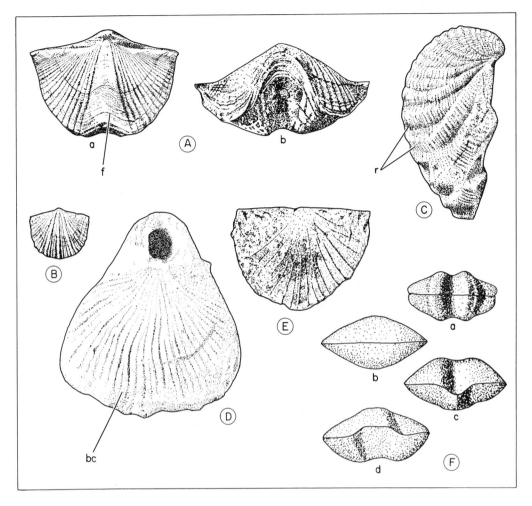

FIGURE 7.9A Articulates—II. Shell morphology and structures.
(A) *Spinocyrtia granulosa* (M. Dev., N.Y.). (1) Brachial valve with fold (*f*); (2) pedicle valve showing sulcus, reduced.
(B) *Fasciculina fasciculata* (L. Ord., Okla.), enlarged to illustrate coarsely fasciculate ornamentation; costa (coarse); costellae (fine).
(C) *Fluctuaria undata* (L. Carb., Belgium), lateral view; *r*—rugae.
(D) *Phragmothyris cubensis* (Eoc., Cuba), enlarged; *bc*—bifurcate costellae (= multicostellate condition).
(E) *Actomena orta* (M. Ord., Esthonia), brachial valve, exterior displaying the unequally parvicostellate condition, X 1.5.
(F) Types of folding. Dorsal anterior view: (*a*) bilobate; (*b*) rectimarginate; (*c*) sulcate; (*d*) uniplicate.

articulate shells. These form characteristic patterns: fasciculate, multicostellate, parvicostellate, and so on (Figure 7.9*A*, parts *D, E*). In some forms (Orthida, Strophomenida), patterns of branching costellae may reflect setae grilles that in life covered the apertures and sieved out oversized particles (Rudwick, 1965).

Percival (1960) pointed out another function of corrugated or ridged surfaces of articulate valves. In living rhynchonellid shells the corrugated surfaces collect fine mud. In turn, the interridge mud serves as a substrate for a variety of small animals such as foraminifera, sponges, hydroids, polychaetes, bryozoans, young mollusks, various arthropods—ostracods, copepods, and so on—and others. In addition, attachment sites for brachiopod larvae are also provided.

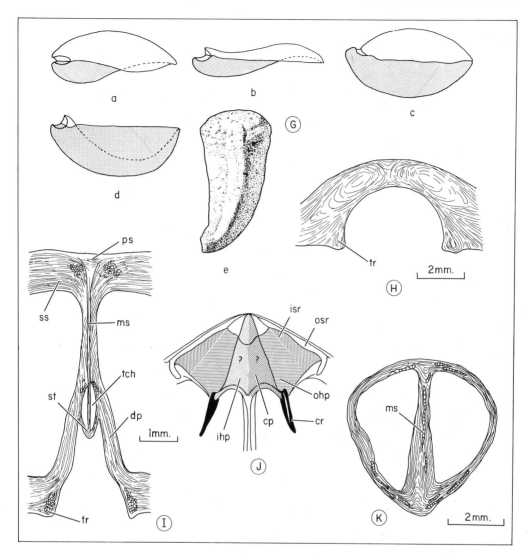

FIGURE 7.9B Articulates—II (continued).
(*G*) Profiles of brachiopod configurations: (*a*) convexo-concave (genus *Carinferella*); (*b*) resupinate (subfamily Strophomeniae); (*c*) biconvex (subfamily Terebratulinae); (*d*) concavo-convex (suborder Chonetidina exclusive of one genus); (*e*) conical (*Gemellaroia marii*); *bv*—brachial valve, *pd*—pseudodeltidium. Schematic; gray—pedicle valve; white—brachial valve. (*H*) *Skenidoides oraigensis* (M. Ord.), transverse section, spondylium simplex; *tr*—tooth ridge. (*I*) *Cystina* sp., *dp*—dental plates, *tch*—tichorhinium, *st*—strut, *tr*—tooth ridge, *ms*—median septum, "*ps*" and "*ss*"—primary and secondary shell. (*J*) *Laqueus* sp. (Rec.), *osr*—outer socket ridge, *isr*—inner socket ridge, *ihp*—inner hinge plate, *ohp*—outer hinge plate, *cr*—brachiophore process or crus (pl. crura), *cp*—crural plate. (*K*) *Gypidula dudleyensis* (U. Sil., England) spondylium duplex; *ms*—median septum.
Sources: Figures for 7.9A and 7.9B: (*H*)–(*K*), Williams and Rowell, 1965; others, after several authors.)

A greater density of biotic associates can be expected more frequently in costellate fossil brachiopods than in smooth forms.

Types of Folding. Articulate shells may show either *opposite* or *alternate* folds (Figure 7.9*A*, part *F*). Folds are major elevations of the valve surface, appearing convex in transverse view. The reverse of a fold is the *sulcus*, a major depression of the valve surface, concave in transverse view. Both fold and sulcus are radial from the umbo.

In the *opposite* type of folding, fold opposes fold on each valve and sulcus, sulcus with the commissure rectimarginate (Figure 7.9*A*, part *F*); in the *alternate* type, fold opposes sulcus and vice versa, with commissure being waved.

Rugae. Concentric or oblique wrinkles of the shell characterize some forms (Leptaenidae, Linoproductidae—Figure 7.9*A*, part *C*).

Alation and Mucronation. The cardinal extremities (Figure 7.8*B*, part *I*) terminate the posterior margin. These are the lateral extremities of the hinge. They may be angular, obtuse, or rounded. Where produced as winglike extensions or ears (*alae*), the condition is known as "alate", where produced into points, the condition is known as "mucronate" (*Microspirifer*, Figure 7.10*F*).

Lobation. In some forms, folds and sulcus may develop separate lateral lobes. This condition is exemplified in the family Pygopidae (L. Jur.–M. Cret.). In this family there are a dorsal median sulcus and a posterior ventral fold. In the young forms these develop two lateral lobes. In adults the lobes are in contact and fused through still enclosing a median perforation (Figure 7.10*D*).

Shell Configuration

The biconvex configuration characterizes unspecialized adults. Associated with specialization are a variety of modifications of this archetypal shape (Figure 7-9*B*, part *G*): convexo-concave, resupinate, concavo-convex, plano-convex among others. A tendency for one or both valves to assume conical form is more marked in the inarticulates and generally affects one valve or the other; for example, both valves of the inarticulate Orbiculoidea are subconical, to cite one variation of this tendency. The coralloid shell form occurs in several articulate families (Figure 7.9*B*, part *G*, *e*).

Slopes of Interarea

Either the ventral and dorsal interareas of articulates (Enteletidae, Figure 7.8*A*, part *B*) or the pseudointerarea of the pedicle valve of inarticulates (Acrotretidae, Figure 7.7*D*) may have various slopes relative to the plane of commissure (a plane passed through the line of junction between valve margins).

Classification

Williams and Rowell (1965) observed that since the phylum Brachiopoda is founded mainly on paleontological data, a prerequisite for classification is a continuing search for a genetic basis that actually reflects brachiopod evolution. Such a classification is not possible in the present state of knowledge. Among reasons: *prevalence of homeomorphy* throughout the history of the phylum (for example, equivalent shell shapes recur over and over again in unrelated stocks); *convergent evolution*—shell characters arise independently in unrelated stocks.

The classification used in the Treatise is built from the genus up to higher categories. As of 1965, 1700+ genera were recognized. Two factors were primary considerations: (*a*) The taxonomic importance of a given shell character such as cardinalia, musculature, and so on, was determined on its redundancy—the number of species or genera in which it appears. (*b*) The vertical (geologic) spread of a given genus.

Brachiopod Classes

[See phylum definition for generalities.]

Class Inarticulata. Valves, chitino-phosphatic or calcareous, with outer skin, periostracum respectively chitino-proteinaceous or proteinaceous (Table 7.1); punctate or impunctate; articulation absent to rare (*Dicellomus*); muscles and body wall hold valves together; muscle system includes muscles to open, close, rotate, and slide valves (Figure 7.7*H*); calcareous

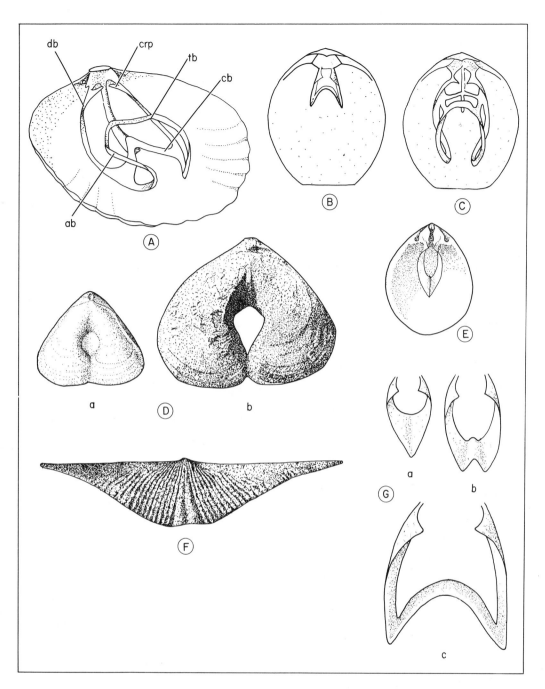

FIGURE 7.10 Loops and other features of articulate valves.

(A) Terebratellid loop, brachial valve indicating various parts; *crp* — crural process, *ab* — ascending branch, *tb* — transverse band, *cb* — connecting band, *db* — descending branch.

(B)–(C) Recent terebratuloids. (B) Short loop, characteristic of terebratulaceans. (C) Long loop, characteristic of terebratellaceans.

(D) Lobation. (a) *Pygope deltoidea,* (France), brachial valve, with imperfectly fused lobes, X 0.6; (b) *Pygites diphoides* (France), brachial valve, reduced, with large median perforation.

(E) *Centronella* (suborder Centronellidina) (L. Dev.–M. Dev.), most primitive terebratuloid loop — a calcareous riband suspended between the crura. (This type of loop is common to almost all Paleozoic terebratuloids.)

(F) *Mucronation. Microspirifer mucronatus* (M. Dev., N.Y.), with mucronate cardinal extremities.

(G) Three stages (a–c) in loop ontogeny of the Middle Paleozoic terebratulacean. *Cranaena.*

Illustrations are schematic; sources: (A) Williams and Rowell, 1965; (B)(C)(G) Stehli, 1956a; (D)(E) Muir-Wood et al., 1965. Treat. Invert. Paleont. Pt H. Brachiopoda. Vol. 2, pp. H 728–H 864.

lophophore supports represented by shelly outgrowths of the posterior margin of the brachial valve lacking, but some of the Acrotretida bear a modified median septum or other plate structures that apparently provided some degree of support. [Lophophore in modern articulates goes from initial trocholophous to schizolophous and culminates in the spiral spirolophe stage (see Figure 7.1*A*).]

Attachment by pedicle, which develops from ventral mantle (for example, in lingulids), or cementation of pedicle valve to substrate (craniaceans) or no attachment at all (trimerellids). Geologic range: L. Cambrian–Recent.

The class is divided into four orders.

Order 1. Lingulida. Biconvex, beak genally terminal in both valves; pedicle where present, emerging posteriorly between valves. L. Cambrian–Recent. Two superfamilies: Superfamily Lingulacea, *Lingula* (Figure 7.11*A*), *Obolus* (Figure 7.7*H*), *Lingulella* (Figure 7.11*B*). Superfamily Trimerellacea, *Trimerella* (Figure 7.7*E*).

Order 2. Acrotretida. Outline usually circular-subcircular; shell phosphatic or calcareous (the latter punctate); pedicle

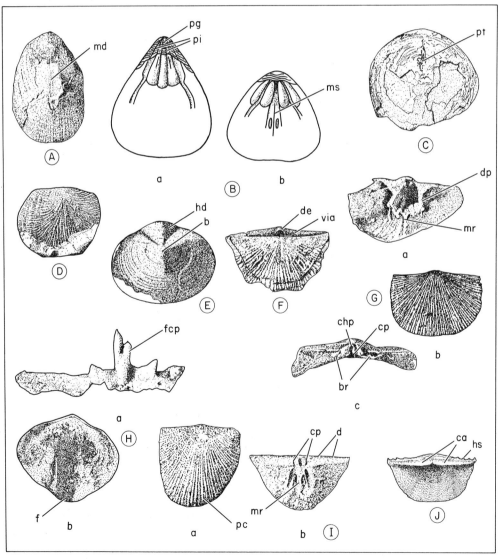

FIGURE 7.11

FIGURE 7.11 Representative inarticulate and articulate fossil brachiopods.
(*A*) *Lingula iowensis*, Maquoketa shale (U. Ord., Iowa), dorsal view, partly exfoliated. Note median depression (*md*) and five marginal striae, X 1.
(*B*) *Lingulella wanniecki*, Cambrian, Salt Range, Pakistan. (*a*) Ventral valve, *pg* – pedicle groove, *pi* – pseudointerarea, and muscle impressions; (*b*) dorsal valve, *ms* – median septum, and muscle impressions, X 16.
(*C*) *Orbiculoides falklandensis* (Dev. Falkland Islands), exfoliated pedicle valve, X 1.5; *pt* – pedicle slit.
(*D*) *Orthiscrania divaricata* (Ord., Europe), convex shell, pedicle valve showing costellae and marginal beak, X 1.5.
(*E*) *Micromitra williardi* (L. Cam., Alabama), top view of ventral valve; *b* – beak, *hd* – homeodeltidium; there is no visible foramen.
Articulata: (*F*) *Xenorthis* sp. (U. Cam., Maryland), complete individual, dorsal view, X 2; *d* – deltidium, *via* – ventral interarea.
(*G*) *Tritoechia typica*, L. Ord., Oklahoma. (*a*) tilted ventral valve; *dp* – dental plates, *mr* – median ridge, X 2; (*b*) dorsal valve exterior, X 2; (*c*) same, posterior view; *chp* – chilidial plates, *cp* – simple cardinal process, *br* – brachiophores (= thin rods at margins of socket), X 3. (*H*) (*a*) *Triplesia insularis* (U. Ord., Ireland), ventral view of forked cardinal process (*fcp*); (*b*) *Triplesia extans* (M. Ord., N.Y.). Brachial valve with fold (*f*); shell roughly trilobate, sulcus on pedicle valve, X 5.
(*I*) *Strophodonta* (*Brachyprion*), sp., L. Dev., Oklahoma. (*a*) Ventral valve, exterior; *pc* – parvicostellate, X 1.5; (*b*) dorsal valve, interior, X 1.5; *cp* – cardinal process, *mr* – median ridge and two lateral ridges divide adductor muscle scars, *d* – denticulate hinge line.
(*J*) *Chonetes granulifer*, dorsal view of young specimen; *hs* – hinge spines, *ca* – wide cardinal area, reduced.
(Sources: (*A*) Wang, 1949. G. S. A. Mem. 42; (*B*) Schindewolf and Seilacher, 1955. Akad. Wiss. Literatur. Mainz. Abhand. Math. Natur. Klasse. No. 10, pp. 303–314; (*C*)(*D*) Rowell, 1965a. Antarctic Res. Ser., Vol. 6; 1965b, Treatise, Vol. *H* (1); (*E*) Walcott, 1908; (*F*)(*G*) Ulrich and Cooper, 1938; (*H*) Williams and Wright, 1965; (*I*) Williams, 1953; (*J*) Dunbar and Condra, 1932.)

confined to pedicle valve, if present; beak marginal–subcentral. L. Cambrian–Recent. Two suborders: Suborder Acrotretidina with three superfamilies: (*a*) Acrotretacea, *Acrotreta* (Figure 7.7*D*); (*b*) Discinacea, *Orbiculoides* (Figure 7.11*C*); (*c*) Siphonotretacea. Suborder Craniidina with a single superfamily, Craniacea, *Orthiscrania* (Figure 7.11*D*).

Order 3. Obolellida. Calcareous shell, biconvex, outline subcircular-elongate oval; well-defined pseudointerarea; pedicle emergent site variable; marginal beak (brachial valve). L. Cambrian–M. Cambrian. One superfamily: Obolellacea, *Bicia* (Figure 7.7*G*).

Order 4. Paterinida. Phosphatic shell; outline rounded or elliptical; pedicle valve convex to hemiconical; triangular delthyrium divides pseudointerarea; the former is variably closed by the homeodeltidium. Homeochilidium partially closes notothyrium. L. Cambrian–M. Ordovician. One superfamily, Paterinacea, *Paterina* (Figure 7.7*B*); *Micromitra* (Figure 7.11*E*).

(See Class Uncertain, after Articulata.)

Class Articulata. Valves calcareous, composed of an outer proteinaceous skin (periostracum) and two to four layers, cryptocrystalline primary and fibrous/prismatic secondary, and occasionally an additional layer, the adventitious; punctate, pseudopunctate, or impunctate; smooth or ornamented; articulation by hinged teeth (composed of secondary shell) and sockets; *cardinalia* (cardinal process, socket, ridges, and so on) is associated with articulation, lophophore support, and muscle attachment. Modified socket ridges (*crura*) may give rise to loops or spires (*brachidium*); muscle system, diductors and adductors for opening and closing valves and a set of adjustors to move valves relative to pedicle. Mantle reversal after larvae settle is known in living species but may not have occurred in some extinct articulates. Lophophore (see text discussion); anus absent. Geologic range: L. Cambrian–Recent.

The class is subdivided into six orders and two others that are "uncertain."

Order 1. Orthida. Unequally biconvex, strophic, generally impunctate shells;

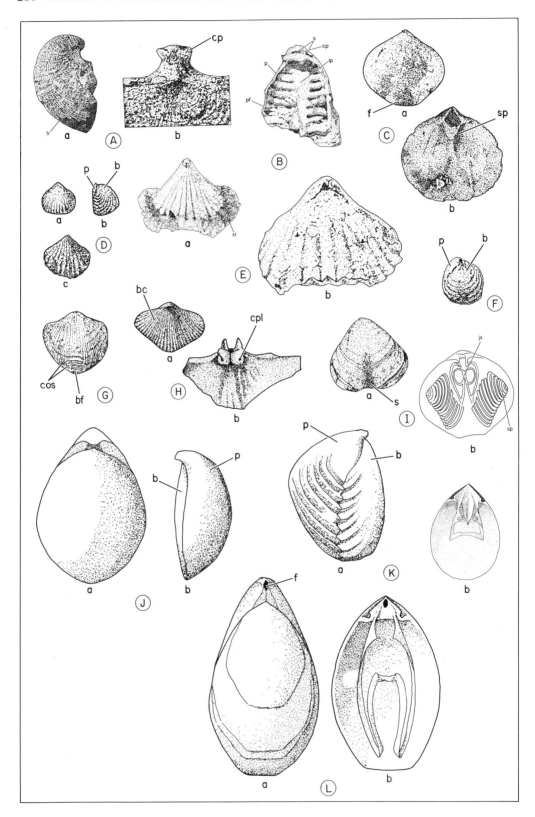

FIGURE 7.12 Representative articulate fossil brachiopods.

(A) *Dictyoclostus semireticulatus*, L. Carb., England. (a) Convex pedicle valve with curved trail (*tr*), X 1; (b) portion of brachial valve, showing massive cardinal process (*cp*), X 2.

(B) *Leptodus americanus* (M. Perm., Texas), brachial view, with portion of internal plate and upper vestigial true brachial valve (upper triangular portion) with sockets (*s*), cardinal process (*cp*), and pedicle valve (*p*) showing everted posterior flap (*pf*), X 1.5.

(C) *Diaphelasma complanatum*, L. Ord., Manitou Formation, Colorado. (a) Dorsal view, *f* — fold (sulcus in pedicle valve), X 2; (b) ventral valve, *sp* — spondylium, X 3.

(D)(a–c). *Sphaerirhynchia glomerosa* (L. Dev., Oklahoma), (a) brachial valve; (b) pedicle valve; (c) lateral view, *p* — pedicle valve; *b* — brachial.

(E)(a) *Stenoscisma venustum* (Leonard, Glass Mts. Texas), dorsal view with well-developed stolidium (= narrow broad frill projecting at distinct angle to shell's main contour), reduced; (b) *Stenoscisma* sp., schematic, *a*, transverse section through both valves reveals a spondylium (pedicle valve) opposite the brachial camarophorium (= a spoon-shaped trough).

(F) *Rhynchopora geinitziana*, U. Perm., U.S.S.R. lateral view, *p* — pedicle valve, *b* — brachial valve with fold, X 2.

(G) *Atrypa oklahomensis* (L. Dev., Oklahoma), brachial view; note brachial fold (*bf*) and costellate surface, X 1.

(H) *Trematospira multistriata*, L. Dev., N.Y. (a) Brachial valve with fold; note bifurcating costae (*bc*), X 1; (b) brachial valve anterior with quadrilobate cardinal plate (*cpl*), X 3.

(I) *Meristella atoka*, Sil., Oklahoma. (a) Pedicle valve with sulcus (*s*), X 1; (b) *Merista typa* (U. Sil., Maryland); brachidium showing spiralia (*sp*) and stem jugum (*js*) that bifurcates posteriorly. X 1.

(J) *Centronella glansfagea*, L. and M. Dev., North America. (a) Brachial view; (b) lateral view of concavo-convex shell; *p* — pedicle valve, *b* — brachial valve, X 2.

(K)(a) *Dielasmina plicata*, (Perm., Pakistan Salt Range); *p* — pedicle valve, *b* — brachial valve, enlarged; (b) brachial valve interior with terebratuliform loop; enlarged.

(L)(a) *Cryptonella planirostra*, displaying cryptonelliform loop, enlarged, *f* — foramen.

(Sources: (A) Muir-Wood and Cooper, 1960; (B) Williams, 1965; (C) Ulrich and Cooper, 1938, G.S.A. Special Paper 13; (D)(G)(Ia) Boucot and Amsden, 1958; (E) R. E. Grant, 1965a. Smithson. Misc. Coll. Vol. 148 No. 2; 1965b. Treatise, Vol. H (1–2); (F) McLaren 1965, Treatise, Vol. H (2), p. H632; (H)(Ib) Boucot et al., 1965; (J)–(L) Muir-Wood, et al., 1965, see Fig. 7.10 for Ref.)

hinge line and interareas are well developed; delthyrium/notothyrium generally open but may be closed by pseudodeltidium and chilidium; cardinal process rarely absent or forked. (Strophic = shell with true hinge line parallel to hinge axis.) Lower Cambrian–Upper Pennsylvanian. Three suborders: Suborder Orthidina: *Xenorthis* (Figure 7.11F). Suborder Clitambonitidina: *Tritoechia* (Figure 7.11G). Suborder Triplesiidina: *Triplesia* (Figure 7.11H).

Order 2. Strophomenida. Shell generally pseudopunctate, plano-to-planoconvex and biconvex, resupinate (see Figure 7.9B, part G) or geniculate (= angular head in lateral profile); pseudodeltidium and cardinal process rarely absent; socket ridge and interareas vestigial or absent, but in some forms may be strongly developed; spines may occur on both valves. L. Ordovician–L. Jurassic. Four suborders: Suborder Strophomenidina: *Strophodonta* (subgenus, *Brachyprion*) (Figure 7.11I).

Suborder Chonetidina: *Chonetes* (Figure 7.11J). Suborder Productina: *Dictyoclostus* (Figure 7.12, A). Suborder Oldhaminidina: *Leptodus* (Figure 7.12B). (A complete specimen of *Waagenoconcha* with corona of anchoring spines is illustrated in Figure 7.12A.)

Order 3. Pentamerida. Shell impunctate, generally biconvex with pedicle spondylium; delthyrium open or partly closed by deltidium; brachial processes braced posteriorly by supporting plates that generally terminate blindly. M. Cambrian–U. Devonian. Suborders: Suborder Syntrophiidina: *Diaphelasma* (Figure 7.12C). Suborder Pentameridina: *Pentamerus* (Figure 7.8B, parts L, M); *Clorindella* (Figure 7.8B, part N).

Order 4. Rhynchonellida. Shell usually rostrate, impunctate; partly closed delthyrium; median septum (brachial valve) often supports septalium/hinge plates; dental plates usually present; spondylia present in some families but normally

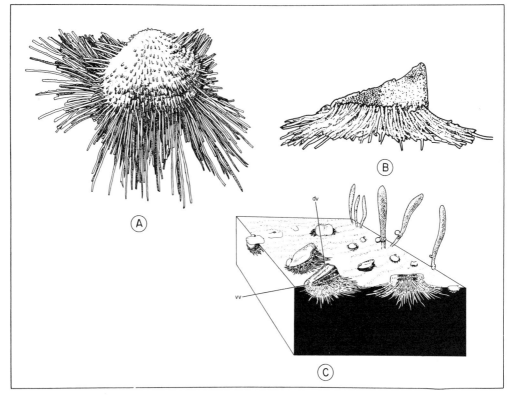

FIGURE 7.12A *Waagenoconcha abichi,* Permian, Productus limestone, Pakistan. (*A*) Silicified specimen, ventral valve having dense corona of long slender spines used to anchor to substrate; (*B*) side view with anchoring spines correctly oriented; (*C*) relationship to substrate shown (after Grant, 1966).

lacking. M. Ordovician–Recent. Subdivided into three superfamilies. [Rostrate = pedicle valve beak projects over narrow cardinal margin. Septalium = troughlike structure of brachial valve between hinge plates (or homologues) characterizes many Mesozoic rhynchonellids.]

Superfamily Rhynchonellacea: *Sphaerirhynchia* (Figure 7.12*D*). Superfamily Stenoscismatacea: *Stenoscisma* (Figure 7.12*E*). Superfamily Rhynchoporacea: *Rhynchopora* (Figure 7.12*F*).

Order 5. Spiriferida. Shell generally biconvex (plano-convex rare) with relatively large body cavity; punctate or impunctate; spiral brachidium characteristic (except for one family); jugum and circular foramen may be present or absent; delthyrium open or closed. M. Ordovician–Jurassic. Four suborders: Suborder Atrypidina; *Atrypa oklahomensis* (Figure 7.12*G*). Suborder

Retziidina: *Trematospira* (Figure 7.12*H*). Suborder Athyridina: *Meristella* (Figure 7.12*I,a*), *Merista* (Figure 7.12*I,b*). *I*); *Kozlowskiellina* (Figure 7.8*B*, part *F*); *Cystina* (Figure 7.9*B*, part *I*).

Order 6. Terebratulida. Shell punctate with functional pedicle; delthyrium more or less closed; adult loop and lophophore variable; loop arises from cardinalia/medium septum (in part); dental plates present or absent; internal calcareous spicules in some families. L. Devonian–Recent. Three suborders: Suborder Centronellida: *Centronella* (Figures 7.10*E*, 7.12*J*). Suborder Terebratulidina: *Dielasmina* (Figure 7.12*K*). Suborder Terebratellidina: *Cryptonella* (Figure 7.12*L*).

In addition, there are two suborders that cannot be placed at present among any of the six defined articulate orders. Suborder Thecideidina (Triassic–Recent):

Thecidea (Figure 7.13*A*). Suborder Dictyonellidina (M. Ordovician–Permian): *Isogramma* (Figure 7.13*B*, *a*); *Megapleuronia* (Figure 7.13*B*, *b*).

The Thecideidina may be highly modified terebratuloids. An extraordinary feature without parallel in the other suborders is the umbonal plate of the Dictyonellidina.

Class Uncertain. The order Kutorginida[2] [L. Cambrian – (?)M. Cambrian] is tentatively assigned here. It may have derived independently of both inarticulates and articulates (Williams and Rowell, 1965): *Kutorgina* (Figure 7.13*C*).

INFORMATION DERIVED FROM FOSSIL BRACHIOPODS

Archetypal Brachiopod

The problem of the origin of the two classes (Inarticulata, Articulata) – possibly three classes – is shrouded in difficulties. A brief citation of the "given" will clarify the nature of the problem confronting specialists.

1. Representatives of the two classes and a possible third are all present in the Lower Cambrian. Included are the four inarticulate orders, Acrotretida, Obolellida, Paterinida, Lingulida, and the articulate order Orthida, as well as Kutorginida.

2. Primitive features of inarticulates are considered to be (1) chitino-phosphatic shell, (2) impunctate shell, (3) absence of mantle reversal on settling of the larvae, (4) pedicle arising as outgrowth of inner epithelium of ventral mantle (Figure 7.14 *B*) (by contrast, in living articulates, the pedicle arises in the embryo from a pedicle rudiment distinct from a mantle rudiment; both rudiments occur together whereas, in inarticulates, the pedicle arises only after evagination of the ventral mantle), (5) paired lophophore filaments.

3. Morphological studies and anatomical considerations have led to two observations: (*a*) Among articulates, mantle re-

²Rudwick (1970) assigns this order to the Inarticulata.

versal was introduced somewhere during the evolution of the Rhynchonellida (and subsequently came to characterize the Spiriferida, Terebratulida, and Thecidellina). The oldest Rhynchonellida are Ordovician in age (Figure 7.14*B*). All other brachiopods, articulates and inarticulates, it is inferred, did not undergo mantle reversal. (*b*) The pedicle arises from larval rudiments in the following articulates: Billingsellacea, Clitambonitidina, Triplesiacea, and Strophomenida (that is, these articulates retained primitive inarticulate features).

In general terms, with the above-listed three items, it is possible to envision brachiopod evolution in its beginnings as follows.

(a) Given one or more large interbreeding populations living in upper Pre-Cambian seas, the archetypal brachiopod stock. These populations would be close to modern chitino-phosphatic inarticulates but probably more generalized in organization.

(b) In this ancestral stock, the genetic code prescribes the "primitive features" listed above. If the loci on the DNA molecule are unaffected, the primitive features will persist. If mutational or other changes occurred at one or more of these loci, primitive features, some or all, would be replaced by new shell and body characters. Natural selection, of course, would mediate.

(c) That such "changes" did occur is apparent from our review of the "given." What specifically occurred? An undifferentiated calcareous shell had to appear, as well as a multilayered calcareous shell (Figure 7.14*B*). Both, presumably, derived from a chitino-phosphatic shell of the archetypal stock. Since in all cases the shell is secreted by the mantle epithelial cells, loci on the DNA molecule that control such secretory events in the mantle had to undergo change. Jope (1965, Table 1-2) showed clearly that in all shell types (articulates and inarticulates), the periostracum is organic with protein present in

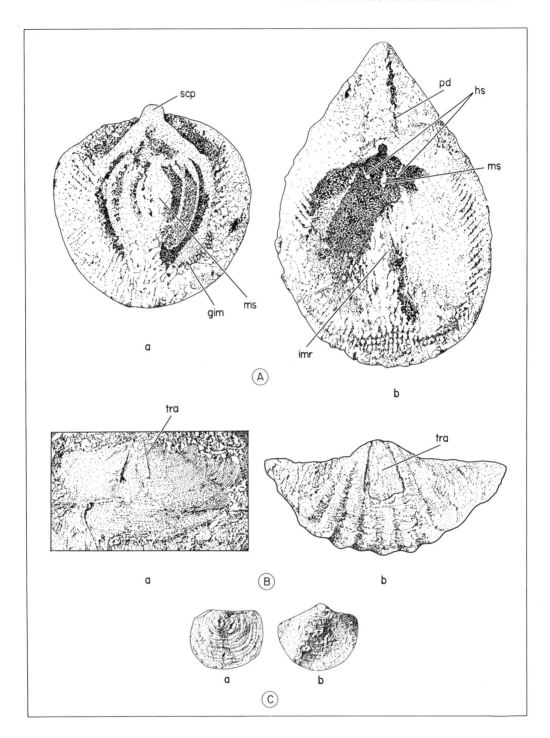

FIGURE 7.13 Brachiopods of uncertain order or class assignment.
(A) *Thecidea papillata*, U. Cret., France. (a) Brachial valve interior; *gim* – wide granular internal margin, *scp* – square cardinal process, *ms* – branched median septum, X 6, (b) Pedicle valve interior; *pd* – pseudodeltidium, *imr* – internal median ridge, *ms* – median septum, *hs* – hemispondylium (= two small plates bearing adductor muscles; characterizes the thecideacean ventral umbo).
(B)(a) *Isogramma texanum* (Penn., Texas), rubber replica of pedicle valve exterior showing concentric lines and triangular area (*tra*) marked medially by a narrow, low ridge (the *tra* demarks the triangular umbonal plate without parallel in any other brachiopod suborder), X 1; (b) *Megapleuronia grecoi* (Permian, Sicily), posterior view of pedicle valve showing *tra* and external ornamentation, X 1.
(C) *Kutorgina cingulata*, L. Cam., Vermont. (a) Pedicle valve exterior; (b) brachial valve exterior, X 1.5.
Sources: (A) Elliott, 1965 Treatise, vol H (2); (B) Cooper, 1952; (C) Rowell, 1965 Treatise, vol H (1).

both; that in Recent chitino-phosphatic inarticulates calcium is present in various forms as the phosphate, carbonate, sulphate, and fluoride; that minute-to-trace amounts of phosphate were present in living articulates. Hyman (1958)[3] found that the pedicle in living articulates is chitinous.

Thus genetic loci governing chitin secretion were obviously modified to eliminate all chitin from the articulate shell while permitting the pedicle epithelium to secrete it in the articulate pedicle. This last event was a retention of a primitive feature from the ancestral stock, whereas the former was a genetic innovation. Other loci affected by genetic changes must have been those controlling calcium metabolism. The normal content of calcium in body fluids of inarticulates was enhanced at the secretory level by epithelial cells, and the end result was crystal formation. At the same time, high phosphate of the inarticulate shell was reduced but retained as a trace in the body fluid. This was probably concomitant with loss of chitin secretion (Jope, 1965).

Regarding the protein of the articulate periostracum compared to that of the inarticulate's, the loci on the DNA-RNA molecules for most amino acids must have remained unaltered although quantitative differences do exist (Table 7.1). Thus the same suite of amino acids occurs in both with few exceptions. Hydroxyproline is present in inarticulates but not articulates. Cystine is present in the articulate

periostracum but not in the inarticulates (though present in the whole shell of inarticulates). These few differences in amino acids must reflect mutations specifying addition and/or subtraction of one or another.

So viewed, the transition from a chitino-phosphatic primitive ancestral shell came about by a few mutations affecting very specific genetic loci: loci for chitin secretion, loci for calcium metabolism, loci for amino acid composition of periostracum protein. Several of these could have overlapped or been triggered by the same mutations and/or other genetic changes (recombinations).

(d) Such genetic modification did not occur in a static original ancestral population. Rather, fragmentation of this parental population must have occurred and subpopulations formed. In turn, this led to geographic, and then reproductive, isolation. All during this time subpopulations were interbreeding, and whatever genetic changes occurred were subjected to natural selection. Those that conferred a selective advantage were favored. [It is clear from the persistence of primitive types in modern seas (the chitino-phosphatic inarticulate), that all primitive features were not inadaptive. Side by side with this development, rise and radiation of the articulates indicate that the genetic modifications affecting shell composition were of great adaptive value.]

(e) Some idea of the enormous interchange of genetic material associated with the rise of the two classes, and possibly a third from a common ancestral stock, can

[3] See ref Ch. 15.

be gained by the following theoretical model: Assume that the ancestral stock thrived in Pre-Cambrian seas some 100 million years before the Cambrian. Allow, as known data suggest for the life span of individual brachiopods, a lower limit of four, and an upper limit of eight years for average life span of individuals in this population. Then, in the 100-million-year span, assuming continuous reproduction, there would be 12.5 to 25.0 million brachiopod generations. Associated with natural mutational events, high larval mortality

would also act as a selective factor. Further selection would occur after larvae had settled on the substrate. Such selection from generation to generation would gradually favor those changes in the genetic code of greatest adaptive value.

(f) Accompanying the evolution of the calcareous and layered calcareous shell were a variety of other notable changes. Some of these concerned only shell structure: those associated with *articulation* (teeth and sockets), *lophophore supports* in the brachial valve, *musculature*. Others re-

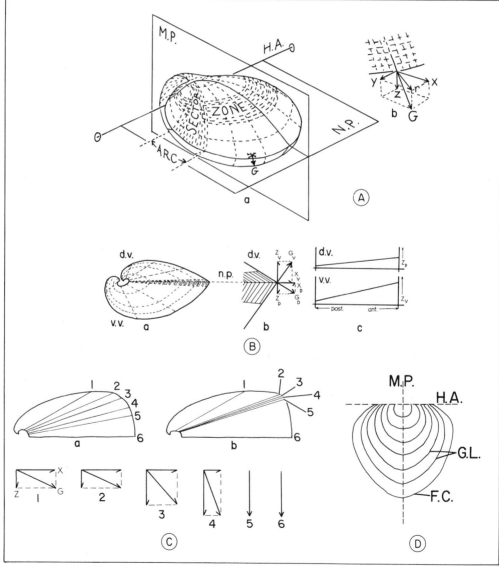

FIGURE 7.14A

FIGURE 7.14A Information derived from fossil brachiopods—I. Growing edge of dorsal valve of rectimarginate shell and its vectoral components: G—rate of growth at one point on dorsal valve edge, NP—normal plane, MP—median plane, HA—hinge axis (= line in space about which valves open and shut).

(A) (a) Perspective view; (b) vectoral components of G.

(B)(a) Same as A,a, lateral view (note logarithmic spiral form of shell); a pair of contemporaneous final zones are shaded; (b) growth rates operative during formation of final zones are shown in enlarged valve edge (anterior side) and resolved into components. Components are Gd—dorsal, Gv—ventral growth rates; the anterior component is resolved into x and z (x_p, x_v); vertical component—z_p, z_v; (c) anterior-posterior gradients (successive points traced around commissure from hinge axis (posterior side to anterior side) and each point analyzed as in b). Vertical scale arbitrary.

(C) Change of growth pattern during ontogeny; a, b, lateral views: 1–6, successive stages at which growth rates were as shown in vector diagrams. In both cases, zone between stages 5 and 6 is a vertical zone.

(D) Commissure plans and corresponding growth stages. (Start from the final commissure (FC) that is, the edge of a given valve or the line of junction of the margins of two valves of a given fossil.) Next place this commissure (and subsequent ones to be described) with reference to the median plane (MP) and the hinge axis (HA). (See part Aa for three-dimensional relationship of the latter.) Now select growth lines (GL) of successively earlier growth stages, and superimpose on the same reference coordinates, as shown. The history of changing commissure plans for the given individual can thus be graphically seen.

(Illustrations redrawn after Rudwick, 1959.)

lated to soft-part anatomy: median shift of musculature in articulates with loss of inarticulate's oblique muscles, suppression of one of the two rows of paired lophophore filaments, redistribution of organs in body cavity—for example, shift of inarticulate gonads confined to the mesenteries well within the body cavity to the articulate condition, migration of gonads out of body cavity to postero-lateral mantle pouches (Williams and Rowell, 1965).

One may envision limited mutations at very few gene loci that were involved in such modifications, since many of these features are related. For example, the distributional shift in muscle attached bases in articulates could have been triggered by the same genetic modification that specified teeth and sockets for articulation and filling the mantle cavity with crural-type lophophore supports.

Allowing 10 to 20 + million generations (a much greater number may have been involved) and the continuous operation of natural selection, then it is less difficult to understand the appearance of the two, possibly three, brachiopod classes at the start of the Cambrian. Similarly, the existence of all four inarticulate orders at the start of the Cambrian implies a long Upper Pre-Cambrian speciation history.

Shell Growth and Form and Its Significance

Increase. The size of brachiopod valves is a function of marginal increments secreted by mantle epithelial cells. These additions are permanent (resorption is unknown except at the pedicle foramen). Shell increase by marginal (peripheral) increments can proceed in three different ways after secretion of the initial valve (protegulum).

Holoperipheral Increase. Increase in valve size all around margins, in anterior, posterior, and lateral directions (example: *Orbiculoidea*).

Mixoperipheral Increase. Posterior sector of valve increases in growth anteriorly and toward other valve; otherwise, like holoperipheral increase, in an anterior and lateral direction (example: *Dinorthis*).

Hemiperipheral Increase. Shell increase in all directions except posterior (example: articulate productids and inarticulate lingulids).

This method of describing shell increase is merely to facilitate understanding and does not enter into taxonomic descriptions. On the other hand, a more insightful and useful method was devised by Rudwick (1959). Figure 7.14A, part A, a, shows

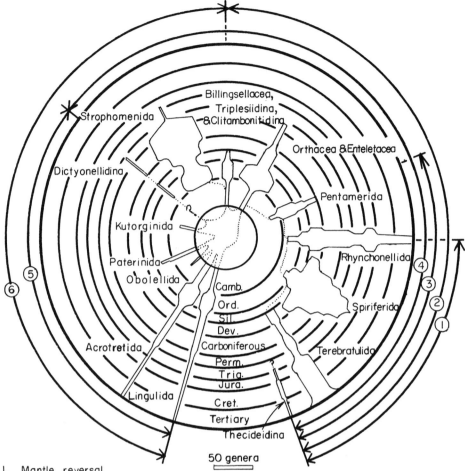

1. Mantle reversal
2. Pedicle from rudiment
3. Persistent non-fibrous primary layer
4. Fibrous secondary layer
5. Chitino-phosphatic or undifferentiated calcareous
6. Pedicle from ventral body wall

FIGURE 7.14B Brachiopod phylogeny ("-ida" endings = orders; "-ina" = suborders; "-acea" = superfamilies). The variable thickness of each group plotted corresponds to increase and decrease of number of genera through geologic time. Scale width = 50 genera. This diagram shows assumed affinities and relationships between groups (dotted lines) and known first appearance in the fossil record. Redrawn after Williams, 1965.

in perspective view, a rectimarginate shell (see also Figure 7.9*A*, part *F*, *b*), that is, growing edge of valve lies in a single plane (= normal plane).

In addition to the two planes, median and normal, and the hinge-line axis, arc, sector, and zone are defined.

Zone, Arc, Sector

Zone. All the shell material deposited during a certain period of growth defines a zone. It can be a strip of valve surface between any two growth lines or between one growth line and the final valve edge. (Concentric ornamentation forms distinctive zones on a given valve.)

Arc. Any section of the growing edge is defined as an arc.

Sector. Shell material deposited by an *arc* during the whole growth of the valve defines a sector. A tract of valve surface radiating from the umbo is well defined by radial ornamentation and forms distinctive sectors.

Vectoral Growth Components. At any given point on the valve edge, a certain *rate* of growth (G) will prevail; G can be resolved into component growth rates, that is, rates of growth in different directions (growth can be unequal in different directions). The component growth rates (Figure 7.14*A*, part *A*, *b*) are as follows.

Anterior component (x). It lies in normal plane and in a plane parallel to median plane of symmetry and acts *anteriorly*.

Lateral component (y). It lies in normal plane and acts *laterally* in a direction perpendicular to the median plane (that is, parallel to hinge axis).

Vertical component (z). It lies in a plane parallel to the median plane and acts toward the opposite in a direction perpendicular to the normal plane.

Radial component (r). It is the resultant of x and y and acts *radially* in the normal plane.

Of these, x, y, and z are major components.

At this point, take another look at the definitions of the various kinds of peripheral increase discussed earlier. Components x and y apply to *anterior* and *lateral* directions referred to in the former definitions. Being more exactly defined relative to reference planes (normal and median), they provide a more precise and meaningful account of growth in the anterior and lateral directions. Furthermore, definition of peripheral increase lacks the vertical component (z) that adds another dimension to valve increase, one that corresponds to the actual growth of valves.

Logarithmic Spiral Form and Antero-Posterior Gradient. · The same valve illustrated in Figure 7.14*A*, part *A*, *a* is shown in lateral view in Figure 7.14*A*, part *B*, *a*. Thus the lateral component (y) is eliminated and one can study the anter-

ior and vertical components (x and z, respectively). The valves have a logarithmic spiral form; that is, the ratio between vertical (z) and anterior (x) components at each point on the commissure remains constant during growth. (Commissure is the line of junction between margins or edges of valves).

Successive growth stages on each valve are identical in form but differ in absolute size. By contrast, suppose that the ratio of $z:x$ changes during ontogeny; the logarithmic spiral form would then be lost, and successive growth stages will differ not only in size, but in form as well.

Rudwick (1959, p. 5) emphasized the importance of the antero-posterior gradient (Figure 7.14*A*, part *B*, *c*). In the logarithmic spiral valve, all of the following are dependent on this gradient ($z:x$): relative height and orientation of interareas (Figure 7.11*F*); degree of incurvature; prominence of beak (see also Figure 7.14*A*, part *B*, *a*) and the convexity of the valve as a whole. However, a *directional change* in growth pattern occurs during the ontogeny of numerous brachiopods. A directional change in form results.

If, for example, x progressively decreases relative to z (Figure 7.14*A*, part *C*), relative convexity of the valve will progressively increase. Taking the two valves together (Figure 7.14*A*, part *B*) with valve edges in contact, then only the vertical component (z) can vary independently on each valve—the other components, lateral and anterior, being dependent. Valves of unequal convexity can be accounted for by the independent variations of z.

Commissure Plans. Suppose that one wanted to eliminate the z component as is done in the peripheral increase definitions above; then the x and y components will determine, without regard to valve convexity, the form of the commissure. Figure 7.14*A*, part *D*, illustrates a plot of successive commissure plans on a given valve. The median plane and hinge axis can be used as reference lines on which to superimpose not only the final commissure (*FC*), but selected growth lines. Commissure plans can be useful in comparing shells of similar or variable convexity.

Nonlinear Growth and Deflections and Deformations. Thus far, only linear growth patterns have been considered. In such patterns, *z* increases uniformly along successive points of the commissures (antero-posterior gradient). In nonlinear growth patterns, the *z* component is curved and the valve edge in lateral view will be flexed instead of rectimarginate. There is no reversal in these growth patterns during ontogeny. Rectimarginate shells remain rectimarginate. Laterally flexed shells remain so. Flexures may be strong or weak, dorsal or ventral, and may or may not influence the form of the valve surface.

Given a strong lateral flexure in moderately convex valves, an effect is noticeable: a dorsally flexed commissure results in a *sulcate* dorsal valve and a *keeled* ventral valve; or a ventrally flexed commissure results in sulcus and keel appearing in ventral and dorsal valve, respectively. (Weak flexures can only be observed in a lateral view of the shell.)

Given a rectimarginate commissure and deflect only the median arc out of the normal plane (Figure 7.14*A*, part *A*, *a*), the result is a median fold in the brachial valve and a sulcus in the pedicle valve. These are common occurrences in fossil brachiopods.

Rudwick observed that most deflections of the commissure are due to (*a*) a "localized anomaly," (*b*) the dual effect of the radial (*r*) and vertical (*z*) components of growth. Because deflections tend to be dominated by either the *r* or the *z* components, two main types of deflection can be identified: *r*-dominant (radial deflection) and *z*-dominant (vertical deflection). A third type, composite deflection, occurs where *r* and *z* are both responsible for deflection. The three types are illustrated in Figure 7.15.

Generally, when fossil brachiopods are described, the shell deformations (fold, sulcus, plications) are noted. Thus, for example, the genus *Platyspirifer* has "fold and sulcus weak." However, deformations are produced by the types of deflections discussed above (Figure 7.15) and cannot be understood except in terms of growth components (vertical, radial, or both). Furthermore, study of deformations alone can be misleading since (*a*) identical deflections can yield identical deformations, (*b*) similar deformations can arise from different deflections (Rudwick, 1959). Deflections, insofar as they affect form of valve surfaces, are influenced by general patterns of growth.

Some of the effects of serial radial deflections (Figure 7.15*A*, *b*) given by Rudwick are (*a*) on biconvex shells, a series of radial costae on each valve; (*b*) on perfectly plane valves, no costae; (*c*) on uniquely biconvex shells, costae are more prominent on more convex valve. Serial vertical deflection also produces costae on both valves, but costa on one valve corresponds to furrows between two costae on the other.

Application of some of the concepts and observations reviewed can readily be seen in the Jurassic species *Globorhynchia subobsoleta* (Figure 7.15*B*, 3). Compared to the size of each shell, the median deflection is quite variable. However, the "wave length" (commissure trace) is markedly invariable. (Invariable shell characters are most useful in systematic discrimination within and between species populations.) If one merely notes the deformation as Buckman did in describing the genus "with arcuate uniplication, and low dorsal fold," the constancy of the commissure trace is bypassed. The latter is an even more reliable shell feature and represents deflections causing the deformation.

In nonrectimarginate shells (nonlinear growth pattern) the deflections and shell distortions that they cause can be superimposed. Extremely complex shell form results.

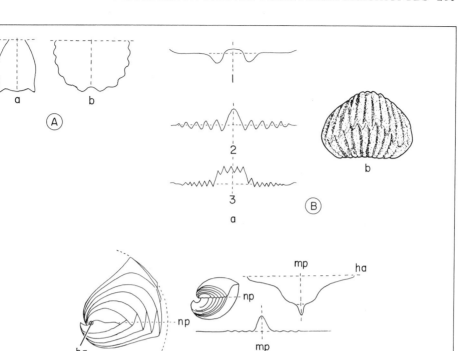

FIGURE 7.15 Information derived from fossil brachiopods—II. Types of deflections of the commissure.

(*A*) Radial deflections: (Horizontal dashed line = median plane; vertical dashed line = hinge axis. (1) *Digonella digona,* paired radial; (2) *Argyrotheca cuneata,* serial radial.

(*B*) Vertical deflections (horizontal dashed line = normal plane (see Figure 7.14*A,* part *Aa*); vertical dashed line = median plane). These commissure traces are obtained by wheeling the shell over a flat surface (plasticine) and drawing trace by camera lucida. (1) *Composita ambigua;* (2) *Spiriferina münsteri;* (3) (Ba3 and Bb) *Globirhynchia subobsoleta,* adult. In (1)–(3) the dorsal side is above.

(*C*) Composite deflections (form governed by anomalous vertical and radial components): *Rhynchonelloidea subangulata,* left figure, lateral view; *HA*—hinge axis, *NP*—normal plane, *MP*—median plane, dashed arc-deflected commissure. Three right figures: *Spirifer triangularis,* lateral view shows shell in which both radial and vertical median deflections are superimposed; upper right, commissure plan with vertical deflection eliminated; lower left, commissure trace with radial deflection eliminated.

(Illustrations redrawn after Rudwick, 1959, except *Bb,* after Ager et al., 1965.)

Genetic Control of Growth Patterns. What controls govern rates and direction of growth at point *G* (Figure 7.14*A,* part *A, a*)? Similarly, for local anomalies due to *r, z,* or both components that account for most commissure deflections, one can ask the same question.

Since the overall effect of shell growth pattern, biconvex, sulcate, and so on, characterizes brachiopod families, subfamilies, genera, and species, the ultimate control must be genetic. The major growth vectors at a given point on the growing edge of dorsal or ventral mantle undoubtedly, as in all vectoral forces, react on and counteract each other in a physical sense.

The genetic specification blueprints the general growth pattern (biconvex, and so on). Of *n* possible variations in commissure, costation, and deformation due to deflections and superimposed on this general pattern, natural selection will weed out the inadaptive variations. Those variations that actually appear and come to characterize a given species population must be regarded as having adaptive value. Local anomalies mentioned earlier can be attributed to mutational events affecting restricted gene loci.

Paleoecology of Shell Shape. Many strophomenids lost the pedicle during ontogeny. The pedicle atrophied and the pedicle foramen was sealed up (Williams, 1953). The same genetic modifications associated with this loss may also have been related to gene sites governing shell shape (concavo-convex). The strongly concavo-convex configuration was an apparent adaptation for an unattached adult exis-

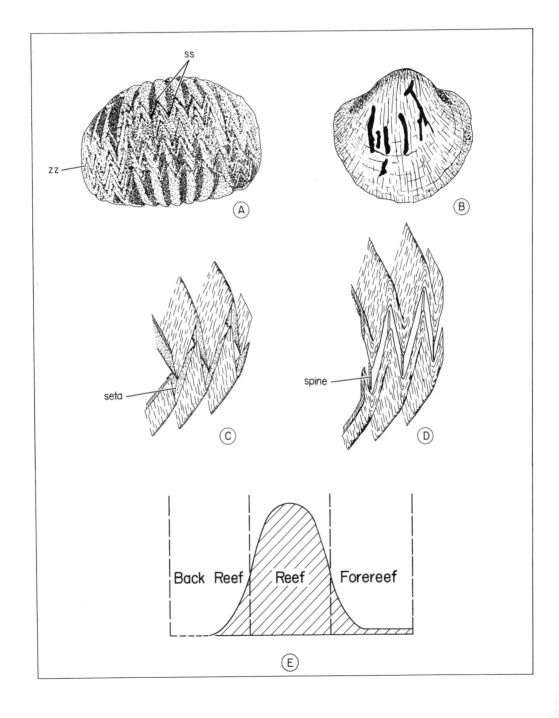

tence on a clay mud or lime mud bottom. Generally it is inferred that strophomenids lived with convex valves on the sea floor and hinge line imbedded in the mud.

During ontogeny several strophomenids went from a moderately to a strongly concavo-convex shell (evidenced by geniculation of the shell). Rudwick (1965) inferred that the gently concavo-convex shell might have permitted occasional swimming whereas the strongly concavo-convex shell may have represented a reversion to a sessile immobile existence.

The habit of convex valves resting on the mud substrates, with commissure held above and clear of it, must have characterized all brachiopods with strongly concavo-convex shells, according to Rudwick. Shell configuration is thus seen to be highly adaptive.

Zigzag Serial Deflections at the Commissure. Although zigzag deflections are most varied and abundant in the Rhynchonellacea (Figure 7.15*B*, part 3), they occur in nearly every brachiopod superfamily. This condition persisted from Middle Ordovician to Lower Cretaceous time. It disappears from Tertiary–Recent articulates (with an exception in the Eocene).

Rudwick (1964) examined the geometry of such zigzag deflections and concluded that they must have served a protective function. Herta Schmidt first demonstrated this function for various rhynchonellids such as *Uncinulus*.

The zigzag deflections proper and accessory diaphragms, setae, and spines in them served as a sieving device, helping to exclude harmful oversized particles (Figure 7.16*A*, parts *A–D*) and could also have performed as a warning device to snap valves shut. (In living articulates, the mantle edge responds to tactile and chemical stimuli and, when stimulated, immediately snaps shut. Zigzag deflections and their accessories could have been a further elaboration of this system.)

The fact that zigzag deflections are obsolescent in Tertiary–Recent articulates probably reflects a more highly involved warning system in such forms—greater sensitivity of mantle edges, for example. One can expect many evolutionary experiments, some of which prove adaptive, in the history of any group of animals. Persistence of zigzag deflection over 300 to 400 million years testifies to its efficacy. Nevertheless, mechanisms that were once adaptive for however long, can become inadaptive.

Zigzag deflections are apparent adaptations to the animal's internal organization rather than a response to any definite environmental pressure. Such pressures might include hypersaline conditions in Permian seas, during which time zigzag forms abounded.

FIGURE 7.16A and B Information derived from fossil brachiopods—III.
(*A*)–(*D*) Zigzag deflections and some accessory protective devices. (*A*) *Uncinunellina jabiensis* (U. Perm., Timor); *zz*—zigzag deflection, anterior view; *ss*—setal foramena at crest of zigzags; X 3. (*B*) See Fig. 7.16*B*, legend. (*C*)(*D*) Protected zigzag slits: (*C*) With crestal protection by setae raised on internal platform. (*D*) With crestal protection by internal spines. (*C, D* schematic, interior view.) (*E*) Distribution of Permian brachiopods in Capitan reef, back-reef and fore-reef facies. Area under curve denotes population density (schematic) on reef proper compared to fore-and back areas. (data from Newell et al., 1953; see Ch. 4 for Ref.) (*F*) *Atrypa waterlooensis* (Dev., Iowa), shell covered with injuries of boring sponges and annelid worms that penetrated near shell margin; (1-one such site on shell) (Fenton and Fenton, 1932. Amer. Midland Naturalist, Vol. 13, pl. 10.) (*G*) Shell repair in *Rafinesquina nasuta* (U. Ord., U.S.A.); *dme*—damaged mantle edge, *ss*—secondary shell deposited to cover injured site, *cc*—converging costellae, (schematic, after Williams, 1965). (*H*) Epifauna of a Devonian spiriferid, *Spinocyrtia iowensis* (U. Dev. Iowa) (after Ager, 1961): (1) tabulate coral, *Aulopora*, (2) encrusting ectoproct, *Paleschara*, (3) branching ectoproct, *Hederella*, (4) serpulid worm, *Spirorbis*. (*I*) Permian *Composita* (Ft Riley Limestone, Kansas) with numerous attached *Spirorbis* on each valve (*sp*) (Tasch Collections), (*J*) Ring of attached spines, *Megousia auriculata* (Word Limestone, Perm., Texas), X 4 (Muir-Wood and Cooper, 1960); *rs*—attachment ring of spines in pedicle valve umbo. Figures (*A*)–(*D*) after Rudwick, 1964. [Continued on p. 296].

(Endospines covering the interior of productid valves served a dual function—protective against predators and as a strainer excluding various objects. Obviously, these were as efficacious as zigzag deflection and are indicative of different solutions to the same types of problems.)

Paleoecology (General)

Depth. Bathymetric data on fossil brachiopods are scattered through a vast literature (see discussion on bathymetry of living forms).

There are several approaches to estimating depths: substrates and habitats, biotic associates, depth range of living equivalents, geologic history of such groups, mode of attachment, color markings, and others.

During the Paleozoic, many articulate brachiopod faunas were associated with reef facies (the Permian Delaware Basin, Texas, is a good example, Figure 7.16*A*,

low depths. Tertiary *Lingula* of New Zealand lived in depths of 0 to 60 ft (Allan, 1936). Ferguson (1963), in a paleoecological analysis of *Lingula squamiformis* (Upper Mississippian, Scotland), found what he interpreted to be successive stages in a marine transgression of the area (Table 7.3).

Various inarticulates fossilized with graptolites in shales were interpreted as pelagic—attached to floating seaweeds (Ruedemann, 1934, see Ch. 13). Included were *Acrotreta, Lingulella, Lingula, Orbiculoides.* Since brachiopods, living and fossil, require well-oxygenated water, their occurrence in black shale facies suggests that they must have lived above the fouled substrate—as epiplankton attached to vegetation (Rudwick, 1965).

The Permian genus *Oldhamina* was partly buried in mud deposited at a depth estimated to be not less than 100 meters (Noetling). Menzies and Imbrie (1958) compiled an age-depth data chart for

TABLE 7.3 Bathymetry of *Lingula Squamiformis*, **Upper Mississippian, Scotland**

Lithology	Topozone	Depth	Population Density
Marine Limestone			
	IV	Deeper than 300 ft	Absent
	III	140 to 300 ft	Declines as depth increases
Shale	II	Mean sea level to 140 ft	Abundant
	I	Upper part of tidal zone	Sparse
Fire clay			

Source. Ferguson, 1963.

part *E*). Most of these brachiopods were reef dwellers—*Prorichthofenia, Leptodus,* among others. Some lived in the shallow basin, for example, productids; few on the shelf (Newell et al., 1953, see ref. Ch. 6). Rarely did any forms live in deeper water (*Leiorhynchus bisulcatum*). Devonian reef facies of Europe are characterized by species of *Stringocephalus,* among others (Gignoux, 1955). Some Jurassic to Cretaceous brachiopods also have reefal associations.

Many studies of fossil inarticulates, particularly *Lingula,* indicate relatively shal-

extant brachiopods with a fossil record (Table 7.4).

Data in Table 7.4 show (*a*) that no brachiopod living at greater than 200 meters originated in the Paleozoic (an expected result, owing to post-Permian extinction of most Paleozoic forms), and (*b*) that present-day inhabitants of deeper waters are chiefly brachiopods that arose in Mesozoic-Tertiary time [compare the thin-shelled fossil *Pygope* that occurred in the European Jurassic-Cretaceous in the deep-water facies of the Alpine region (Gignoux 1955)].

TABLE 7.4. Geologic Age – Present Depth of Living Genera of Brachiopods

Present Depth (Meters) Data	0–10	10–200	200–2000	2000+
Geologic Age				
Paleozoic	12%[a]	3%	0%	0%
Mesozoic	25	23	21	10
Tertiary	63	74	79	90

[a]Percent of modern brachiopods that originated in Paleozoic, Mesozoic, or Tertiary time and inhabit a given depth at present.

Source: Menzies and Imbrie, 1958.

Brachiopod migration to deeper water appears to be a Mesozoic–post Mesozoic event.

Cambro-Ordovician brachiopods (articulates and inarticulates) are often found in sandy and shelly facies indicative of the littoral zone. In the European Jurassic, terebratulids and rhynchonellids abound in a near littoral facies. These last forms also occur in Cretaceous chalk beds (Gignoux, 1955).

Predation, Symbiosis, and Paleopathology

Predation. A glimpse of predation in Paleozoic seas is obtained from evidence of boring snails, annelid worms, and sponges on fossil brachiopods (Fenton and Fenton, 1932, plate X). (Figure 7.16B).

Cooper (1957) remarked that the great abundance of Middle Devonian brachiopods as well as their sessile habit indicated that they may have been among the great sources of food in such seas for various predators—gastropods, cephalopods, fishes. (See "Paleopathology" for fish-reptile predation on Mesozoic brachiopods.)

Another interpretation of brachiopod abundance (in the Saverton shale, Kinderhookian, Upper Mississippi Valley) suggests that predominance could denote lack of effective enemies (J. S. Williams, 1957).

Cooper's suggestion might similarly apply to other portions of the Paleozoic as well. So viewed, the dense populations of Paleozoic brachiopods appear as one of the links in the food chain in such seas. This area of investigation has been neglected.

Symbiosis. A host of attached forms also occur on shells of the same Cedar Valley atrypas. These include *Spirorbis*, *Hedonella*, and young corals. *Spirorbis* is commonly attached to compositas in the Permian Ft. Riley limestone of Kansas. It is to be expected that some sessile brachiopod shells would serve as substrates for larval attachment in Paleozoic seas (Figure 7.16A, part H).

Yakovlev interpreted a coral-brachiopod association as a symbiotic relationship: The coral *Aulopora* benefited from the water currents created by the pump action of the brachiopod lophophore whereas the nematocysts of the coral served to protect the brachiopod. Clarke noted a relationship in the Lower Devonian (Germany) where the coral *Pleurodictyum problematicum* attached to the brachiopod *Chonetes sarcinulates*. Ager (1961 and this text, Figure 7.16A, part H) described the epifauna of a Devonian spiriferid.

Both Recent and Paleozoic brachiopod larva attach(ed) to brachiopod shells. Trechmann cited the example of a Permian *productoid* with ventral and dorsal valves bearing long hinge spines. To these, two species of brachiopods were attached (see further discussion on "Attachment Loops").

Paleopathology. Fossilized shells of *Atrypa reticularis* from the Devonian of Iowa show evidence of numerous minute injuries and subsequent repair. Equivalent damaged and repaired, or permanently distorted, brachiopod shells are recorded from many parts of the rock column. Several brachiopod shells in the European Mesozoic bear fish or reptilian tooth bites or markings (*Pygope*, *Menzelia*). Others from the Paleozoic–Mesozoic (*Rafinesquina*, *Rynchonella*) are deformed, fractured, or otherwise distorted (Tasnádi-Kubackska, 1962, see refs. Ch. 1) (Figure 7.16A parts F, G; 7.16B).

Brachiopods are gregarious, and this

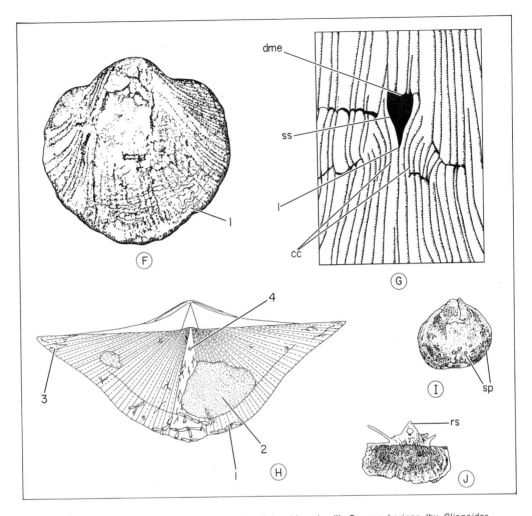

FIGURE 7.16B Information derived from fossil brachiopods—III. Sponge borings (by *Clionoides thomasi*) (in black), on brachial valve of *Atrypa waterlooensis* (Dev. Iowa) (Fenton and Fenton, 1932. Amer. Midland Naturalist Vol. 13, pl. 7). (See Figure 7.16*A*, part B.) (*F–J* legends, p. 293).

results in crowding. Living so close together has resulted in malformation and deformation of their growing shells (Cooper, 1957). Disease has been discounted as a causative factor in most deformed brachiopod shells, crowding being the preferred explanation.

The evidence reviewed above shows that some brachiopods known as fossils were subjected to *crowding, genetic defects at birth,* and *accidents.* Horn corals, for example, in the Cedar Valley sea floor toppled over and crushed the delicate *Atrypa* valves. Hungry predators of all kinds attacked the sessile faunas. Brachiopod shells also had the capacity for shell re-

pair (Williams, 1965). *Rafinesquina nasuta,* Figure 7.16*A,* that lived in Upper Ordovician seas, shows evidence of a damaged mantle edge, secondary shell deposits at the injured site, sealing off of the injured site, and convergence of costellae.

Attachment. Fossil brachiopods had various types of attachment to the substrate: attached by pedicles, originally attached by pedicles but pedicle later reduced or atrophied leaving shell free on the substrate (*Strophomenida*), attached by cementation to hard substrates, originally cemented but later free on substrate (common with concavo-convex shells such as

Oldhamina), attached or anchored by spines (Rudwick, 1965).

In the Productoidea, Muir-Wood and Cooper (1960) recognized three types of living habit dependent on type of spine characteristic of each genus. Some productoids were attached by anchor spines to solid objects throughout life (Aulostegidae, Richthofeniidae, among others). Some, like *Scacchinella*, formed banks or patch reefs, living in clusters on crinoidal coarse gravel. Other productoids broke free from attachment in later stages (*Megousia*) and lived free. Since Glass Mountain productoids apparently had their spines tangled by bryozoan branches or seaweeds, freedom from attachment could have been incomplete. The third group were never attached and always lived free on the sea floor (Echinoconchidae, Dictyoclostiidae, Productellidae, Leioproductidae). This group developed long spines that allowed suspension between vertical coralline or vegetal growths, or conferred resistance to being flipped over by current action.

Unklesbay and Niewoelner (1959) found infant productoid brachiopods (*Productella*) in the Louisiana limestone of Missouri that bore *attachment loops*. They were attached to spines of a mature spinous brachiopod. Muir-Wood and Cooper (1960) found that spat of several genera of the Costispiniferinae of the Permian Glass Mt. fauna were attached by a ring of clasping spines, as was the linoproductoid *Megousia* (Figure 7.16*B*, part *J*).

An epiplanktonic mode of life, attachment to floating seaweeds, has already been noted for black shale brachiopods.

Feeding. Experiments based on the working-model (paradigm) approach led to clarification of the feeding mechanism of a Permian richthofenid brachiopod (Figure 7.17*A*, *a–c*). Essentially, rapid movements of the dorsal valve (a lid) created powerful currents and eddies from which mantle surfaces collected food particles in suspension (Rudwick, 1961).[4]

[4]R. E. Grant does not accept this interpretation.

Similarly, it is possible to reconstruct the lophophore for many fossil brachiopods (Figure 7.2*E–F*) and from this to infer the nature of currents and eddies as well as inhalant and exhalant apertures. The internal spine mesh served as a strainer (Figure 7.17*A*).

The problem of what types of food particles were extracted by these mechanisms will have different answers for different geologic periods. Diatoms and dinoflagellates are common food particles of living brachiopods. In Permian seas, diatoms were absent, not yet having evolved. The primary producers of Paleozoic seas were probably unicellular algae (acritarchs) and the dinoflag·llates (Tasch, 1967).

Biostratigraphy and Zonation. The generalities of brachiopod stratigraphic distribution can be quickly gleaned from Figure 7.18. In terms of number of genera in the several orders, three maxima are clearly shown for Ordovician, Devonian, and Permian. The Ordovician was a time of maximum marine transgression in North America. New ecologic niches were open, allowing rapid radiation and evolution of orthid, strophomenid, and pentamerid brachiopods (see Cooper, 1956). Rise in inarticulate genera also occurred at this time but thereafter declined.

Devonian time may be viewed as the heyday of brachiopod evolution in terms of greatest generic diversity.

Can the drop in number of genera (U. Carb.) be attributed to marine transgression and consequent relict nonmarine coastal swamps? The latter were to give rise to the European-North American Coal Measures. Williams (1965) implied that this might be so. Periodic marine invasions over such coastal swamps characterize the cyclothem deposits. Under such conditions brachiopods, once again, had newly opened niches. Carboniferous seas were referred to in the older historical geology texts as the "*Productus*" seas, so abundant were spinose brachiopods worldwide.

The Permian maxima were followed by the extinction of most Paleozoic genera. Three superfamilies had dominated most

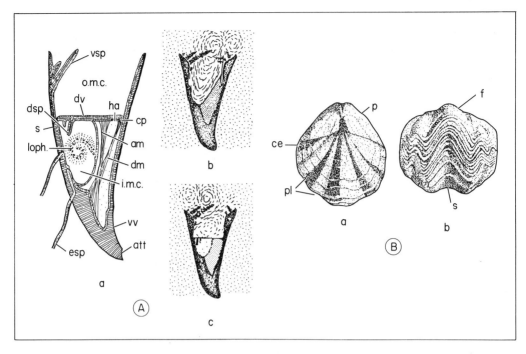

FIGURE 7.17 Information derived from fossil brachiopods—IV.

(A) (a) Reconstructed anatomy of a richthofenid brachiopod showing lophophore position; *dv*—dorsal valve (= lid); *vv*—ventral valve; *imc, omc*—inner and outer mantle cavities; *vsp, dsp, esp*—ventral, dorsal, and external spines; *att*—attachment area; *S*—shelf on which lid rests; *cp*—cardinal process to which divaricator muscle (*dm*) was attached, and adductor muscle (*am*). (b)–(c) *Prorichthofenia permiana*, reconstructed feeding mechanism; (b) dorsal valve fully open; (c) dorsal valve fully closed. Suspended particles indicated by irregular stippling; direction and relative velocity of particles shown by "tails" behind dots; visceral cavity shown by close regular stippling.

(B) A common Mesozoic brachiopod, *Spiriferina* (Trias.—L. Jur.), S. *walcotti*, (Lias., British Isles): (a) Brachial valve, X 1; *p*—pedicle valve, *pl*—plicate folds, *ce*—rounded cardinal extremities; (b) anterior view, X 1; *f*—fold, *s*—sulcus.

Sources: (A) Rudwick, 1961; (B) Boucot *et al.*, 1965.

of the Paleozoic: orthids, strophomenids, and spiriferids.

Late Mesozoic–Recent distribution (Figure 7.18) corresponds to the phylum's declining phases).

These generalities can be supplemented by some selected references. Cambrian faunas are dominated by trilobites and inarticulate brachiopods. Schindewolf (1954) described Lower Cambrian Salt Range inarticulates—*Neobolus, Lingulella, Botsfordia*. These particularly characterized sandstones of the *Neobolus* shale but some forms, such as *Botsfordia*, also occurred in the overlying Magnesium sandstones.

Lochman's investigation of the Cambrian of Montana and Wyoming (1957),

showed both main classes of brachiopods present. In the M. Cambrian, calcareous-phosphatic brachiopods are regularly found with trilobite molts in shales and limestones. The U. Cambrian (Franconian stage) articulates such as *Eoorthis* and *Billingsella* formed widespread reefs or meadows. *Eoorthis* larvae attached to hard, calcareous substrates of algal meadows composed of *Collenia magna*. Their shells accumulated in a bed a few inches thick. Still higher up in this zone, *Billingsella* larva settled on hard substrates and also formed meadows. Somewhat thicker beds were composed of such *Billingsella* shells.

Bell (1941), who described the brachiopods of the above study, saw that *Eoorthis* and *Billingsella* are almost always found to-

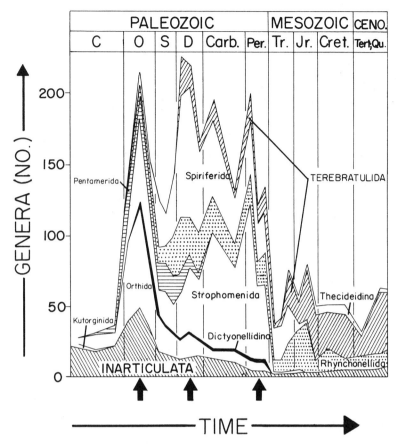

FIGURE 7.18 Maxima (arrows at base) and minima in number of brachiopod genera through geologic time (after Williams, 1965).

gether. The seas of the time must have been favorable since he described six new species of *Billingsella* and one new variety from an "essentially contemporaneous population" in Montana.

Brachiopods attained a maxima in number of genera during the Ordovician marine transgression of North America. Both local and regional correlations are possible with Ordovician brachiopods—for example, the *Sowerbyella* and *Dalmanella* zones in the Galena Group of Illinois-Missouri.

A relative sparsity of graptolites, compared to relative abundance and variety of brachiopods in the limestone facies (L. Ordovician, Oklahoma, West Texas, and Great Basin region) was noted by Cloud and Barnes (1957).

The Niagaran Reefs of the Great Lakes area have brachiopods second only to crin-

oids in abundance. During the rough-water or wave-resistant stage of reef development a great variety of species were represented. Large populations of heavy-shelled inarticulates (Trimerellacea) formed great communities in some places; pentamerids were especially prominent.

Lowenstam (1957) inferred that an increase in population of heavy-shelled forms like *Trimerella* denoted functional adaptation. Presumably this was in response to the movement of skeletal sands in shoal water areas. He also observed that the reef-dwelling *Conchidium* lacked a sinus and was coarsely plicate, whereas inter-reef and open-water shelf-dwelling *Rhipidium* had a sinus and the Trimerellacea and *Pentamerus* species were smooth shelled. Could these features be attributed to selection pressure? Lowenstam suggested that forms lacking a sinus might have had a selective advantage. Similarly, selection might also have favored smooth shelled forms living in protected (unagitated) reef areas, over plicate types.

During the M. Devonian, brachiopods were plentiful and radiated into all marine environments. Accordingly, they occurred in all facies, although some were excluded from fouled environments (black shale facies). Brachiopods are important zone fossils for all parts of the Devonian. A good example is *Stringocephalus* (Cooper *et al*). Its presence in rock exposures of a given locality denotes the upper half of the Middle Devonian. *Paraspirifer* is another important index fossil. In Germany it marks the base of the Middle Devonian, whereas *Uncites* species characterizes reefal facies. The European Upper Devonian is marked by *Spirifer verneuili.*

Paraspirifer is abundant in the United States — at the top of the Onondaga, Columbus, and Jefferson limestones. At particular localities, given horizons can be marked by almost any brachiopod that is adequately representative and distinctive.

An instructive example of the spectrum and speciation of Devonian brachiopods can be found in a study of the Traverse Group of Michigan (Imbrie, 1959) (Figure 7.19). Maximum strophodont speciation coincided with the Alpena limestone reefal facies. Could this coincidence perhaps explain such local speciation (24 species and 11 subspecies)? Imbrie thought it could. Fractionation of a large and variable strophodont population into subpopulations (geographic isolation) would have been facilitated by diversity of ecologic niches created by biohermal structures growing from the sea floor.

The Mississippian of the Upper Mississippi Valley contains a characteristic brachiopod fauna. Shallow-water articulates in the Upper Saverton shale (Lower Kinderhookian) includes such forms as *Schuchertella, Cyrtina, Productella,* and *Syringothyris.* In the overlying Louisiana limestone, some of these genera and still others occur — all very small forms. In

the next sequence of beds, Hamburg Oolite and Hannibal shale, several of the same genera recur along with rhynchonellids and *Stenocisma,* as well as *Camarotoechia.* The Upper Kinderhookian also has its distinctive brachiopod fauna (J. S. Williams, 1957).

Among the abundant Pennsylvanian brachiopods are such common genera as *Composita, Derbya, Mesolobus, Linoproductus,* and *Lingula* as well as *Orbiculoidea* for the inarticulates. Permian Glass Mountain limestones have yielded beautifully preserved brachiopod faunas in siliceous residues, such as *Prorichthofenia, Leptodus, Meekella, Chonetes.*

Despite the fact that mollusks, particularly cephalopods, dominate Mesozoic faunas, there are numerous occurrences of brachiopods. Lingulids are frequently encountered in certain shale and siltstone facies of Triassic and Jurassic beds. The articulate *Spiriferina* was widespread from Lower to Upper Triassic time: Portneuf limestone of southeastern Idaho (Kummel, 1957), Canadian Arctic (Logan, 1965), Muschelkalk, Germany (Gignoux, 1955). Rhynchonellids and terebratulids are common in chalky and sandy limestones (upper Middle Jurassic) and in sandy beds (lower Upper Jurassic) (Imlay, 1957).

Pygope characterized the deep-water facies of the Alpine region (Jurassic) and spiriferinids persisted from Triassic to end of the Lias (Corroy, 1927; Astre, 1938).

Lower Cretaceous chalks contain many of the same forms found in the Jurassic — rhynchonellids and terebratulids (Fage, 1934). *Pygope* continued in the deep-water facies. The European Upper Cretaceous contained varied thecidean brachiopods (*Bifolium, Eolacazella*) (Elliott, 1953a). Eleven species of *Argyrotheca,* extant in modern seas, are known from the Upper Cretaceous of Europe and America (Elliott, 1951). *Platidia,* found in modern seas, is also represented in the European Upper Cretaceous.

The terebratuloids are found throughout the Tertiary worldwide. *Argyrotheca* species are known from the Paleocene

	MICHIGAN																	SURROUNDING AREAS										
	Alpena and Presque Isle Cos.									Cheboygan Co.					Emmet & Charlevoix Co.			Mo.	Ill.	N.Y.				Ont.			Ind.	Ohio
	Bell	Rockport Quarry	Ferron Point	Genshaw	Newton Creek	Alpena	Four Mile Dam	Norway Point	Potter Farm	Ferron Point	Genshaw	Koehler	Gravel Point	Bebe School	Gravel Point	Middle Petoskey	Upper Petoskey	St. Laurent	Rapid	Centerfield	Ludlowville	Wanakah	Enfield	Arkona	Hungry Hollow	Widder	Logansport	Silica
Chonetes ensicosta									X					X	X													
C. mediolatus	X	X	X			X				X								X						X			X	
Douvillina distans						X																		X			X	
Longispina emmetensis					X	X									X													
L. lissohybus		X																										X
L. subcalva	X									X																		
Megastrophia concava					X															X								
M. gibbosa						X																		X			X	
Oligorhachis oligorhachis						X									X													
Pentamerella aftonensis													X		X													
P. alpenensis					X	X									X													
P. lingua		X								X																		
P. pericosta						X	X						X											X	X		X	
P. petoskeyensis							X									X												
P. tumida				X	X								X															
Pholidostrophia geniculata						X									X									X	X			
P. ovata							X																		X			
Helaspis luma		X	X									X																
Protoleptostrophia lirella					X	X																X						
Rhipidomella trigona																	X						X					
Schizophoria ferronensis	X	X	X							X																		X
S. traversensis			X							X																		
Schuchertella anomala						X									X													
S. crassa	X	X	X							X														X				
Sieberella romingeri			X							X																		
Sphenophragmus nanus					X										X									X	X		X	
Strophodonta crassa						X										X												
S. erratica					X	X																						
S. extenuata	X	X	X	X																				X				
S. extenuata extenuata	X									X														X				
S. extenuata ferronensis	X																							X				
S. fissicosta				X											X													
S. nanus				X	X										X													
S. pentagonia																	X		X	X								
S. proteus					X										X													
S. titan titan					X										X													

FIGURE 7.19 Stratigraphic ranges of selected brachiopod species of the Traverse Group (M.-U. Dev., Michigan), and surrounding areas. (Vertical entries at head of columns are formation names. Traverse Group is exposed in the Thunder Bay region and ranges from M. Dev.: Bell Formation to top of Norway Point Formation; U. Dev.: from base of Potter Farm.) Greatest speciation in genus *Strophodonta* occurred during Alpena limestone time (see text). Note *Strophodonta extenuata* ranging through time represented by four formations. This is interpreted to have been a continually evolving unit. The two restricted subspecies of this species indicate some fractionation of the parental population (geographic isolation) in the basin during Ferron Point time. (After Imbrie, 1959.)

through Miocene of Florida, Mexico, Texas, Alabama, and New Jersey (Vincentown Formation). The Australian Tertiary (Crespin and Chapman, in Thomson, 1927; Allan, 1939) contains such forms as *Crania*, *Thecidea*, *Tegulorhynchia*, and various terebratuloids.

There are several Cretaceous and Tertiary brachiopods in western North America. Among these, Tertiary genera are also found in Pleistocene beds and thrive in modern seas: *Glottidia, Hemithyris, Laqueus, Morrisia,* and *Terebratalia* (Hertlein and Grant, 1944). The first appearance of *Laqueus* was in Miocene time, *Glottidia*, Eocene.

Brachiopod Zoogeography

Boucot et al. (1969) attempted to reconstruct early Devonian brachiopod zoogeography. Cosmopolitan Silurian brachiopod faunas were followed by early Devonian faunas restricted to definite marine zoogeographical provinces. These

distinct brachiopod assemblages characterized each of the following Devonian provinces at the stage indicated.

Lower, Lower Devonian (Gedinnian)
Appalachian Province — Gaspé to North Mexico
Old World Province — Eurasia, S.E. Australia, Nevada

The above two-province boundaries were relatively stable during the next stage (Lower Coblenzian).

Upper, Lower Devonian (Upper Coblenzian)
Malvinokaffric Province — southern South America, southwest Africa, Antarctica
New Zealand Province
Cordilleran Province — mixture of Appalachian and Old World Province forms

During the rest of the Lower Devonian–Middle Devonian, more cosmopolitan distribution of brachiopods was typical. However, the Late Appalachian Province maintained its identity into late Devonian.

A recent study of some Antarctic Lower Devonian brachiopods can illustrate the value of this approach. Among the articulates were three genera having Malvinokaffric affinities: *Pleurothyrella, Australospirifer, Cryptonella* (?)(Boucot and Johnson, 1965). Lacking more precise stratigraphic correlations between continents, the zoogeographical approach permits at least some evaluation of the given fauna relative to other contemporary ones by placing an areal restriction on distribution (occurring only within the Province).

Speciation

Brachiopod speciation studies are reviewed in Chapter 14, where the findings on selected populations are analyzed in terms of modern evolutionary theory.

Loop and Lophophore Evolution in Order Terebratulida

For confident identification, Mesozoic–Cenozoic terebratuloids require a knowledge of internal structures, particularly loop development, because of homeomorphy (Muir-Wood, 1965). External shell

features of Paleozoic terebratuloids require, for taxonomic purposes, data on the brachidium and cardinalia (Stehli, 1965). Parallelism and/or convergence in shell development enhances this need.

Knowledge of loop development and evolution permits assessment of "the nature and disposition of the lophophore in wholly extinct groups" (Williams and Wright, 1961). Thus a portion of the critical soft-part anatomy of brachiopods known only as fossils can be reliably reconstructed from serial sections of fossil valves. These sections reveal stages of development of the brachidial apparatus (Figure 7.2*F*).

A simple approach to the subject is to refer loop development and type to the three suborders.

Suborder 1. Centronellidina. L. Dev.–Perm. Archaic terebratuloids belong to this suborder. The most primitive forms bear a centronelliform loop (Figure 7.10*E*). Owing to metamorphosis during ontogeny, loops attain different stages of development. These stages are named after genera bearing equivalent types of adult loops. Thus the centronelliform stage refers to the adult loop in the genus *Centronella*. Other loop stages are discussed in the text. More advanced forms have more complex loops.

During Lower Devonian time, a marked

adaptive radiation occurred among the Centronellidina. One of the families that appeared at this time, the Mutationellidae, embraced genera that displayed notable diversity in loop development (see Figures 7.20*J* and 7.20*A*). From this family, which was genetically variable for loop type, arose in Lower Devonian time the two other suborders.

Suborder 2. Terebratulidina. L. Dev.– Rec. This suborder is characterized by a short loop that supported the lophophore only at its base. Accordingly, calcareous spicules in lophophore tissues aid in support. A primitive form, *Cranaena*, probably derived from the Mutationellidae. [For loop development, see Figure 7.10*G* and note that the earliest loop development stage — centronelliform — is skipped in the ontogeny of modern terebratuloids (Stehli, 1956*a*).] The same superfamily that includes *Cranaena* has specialized descendants that can have highly complex loops.

Suborder 3. Terebratellidina. L. Dev.– Rec. The suborder is characterized by long loops that provide more complete lophophore support. [Nevertheless, in Recent species of *Platidia*, spiculation is strong and the main support of the lophophore is by spicules (Atkins, 1959).]

Paleozoic forms have a *cryptonelliform* loop (see Figure 7.20*A*, where a terebratulacean also attains a cryptonelliform-like loop, and Figure 7.20*B*, representing the type for the cryptonelliform loop in the terebratellacean, *Cryptonella*).

Elliott (1953*b*) observed that the terebratelloid loop always begins with the appearance of a median septum, often in the form of a septal pillar. He considered this structure fundamental to the Mesozoic–Cenozoic superfamily Terebratellacea.

Terebratelloid loops can be grouped into three series that correspond to their development in the families Dallinidae, Terebratellidae, and Kraussinidae (Figure 7.20*C–E*). The loop stages are frenuliform (genus *Frenulina*), magadiniform

to magelliform (genera *Magadina* and *Magellania*), and kraussiform (genus *Kraussina*). In the course of ontogeny, the loop passes through several different stages that will differ depending on evolutionary affinities of the given terebratelloid. Elliott (1953*b*) suggested chromosome studies of living forms as a check on the validity of the modern practice of grouping species with the *same adult loop pattern* in the same genus.

The three series and the stages that loops pass through in ontogenetic development are as follows.

Series 1. Dallinidae. Campagiform (Figure 7.21*a*), frenuliform (Figure 7.21*b*), terebrataliiform (Figure 7.21*c*), dalliniform — the stage at which the adult loop is completely separated from the median septum (Figure 7.21*d*).

Series 2. Terebratellidae. Premagadiniform, magadiniform, magelliform, terebratelliform, magellaniform (Figure 7.20*D*).

Series 3. Kraussinidae. Kraussiniform, mergerliniform, megerliform (Figure 7.20*E*).

In the sequence of genetically programmed steps, numerous variations appear in different genera and species of the same family and even for different species of the same genus. Modern *Platidia* is a good example of this (compare the variant brachial apparatus in three Recent species, Figure 7.20*G–I*). While loops evolve in response to the need for greater lophophore support, their variation (in the adult stage) bears no obvious relationship to the generally plectolophous lophophore structure (Elliott, 1953*b*).

As Stehli (1956*a*) put it, the plectolophous condition of lophophore development occurs well before the adult loop stage that will support it. As a matter of fact, in modern *Platidia*, the main support of the lophophore is by spicules (Atkins, 1959). Even though in *P. davidsoni* (Figure 7.20*H*) the brachial support is much reduced compared to *P. anomioides* (Figure 7.20*I*), the adult plectolophe condition of the lophophore in each requires spicule support.

Elliott (1953*b*) surmised that in the crow-

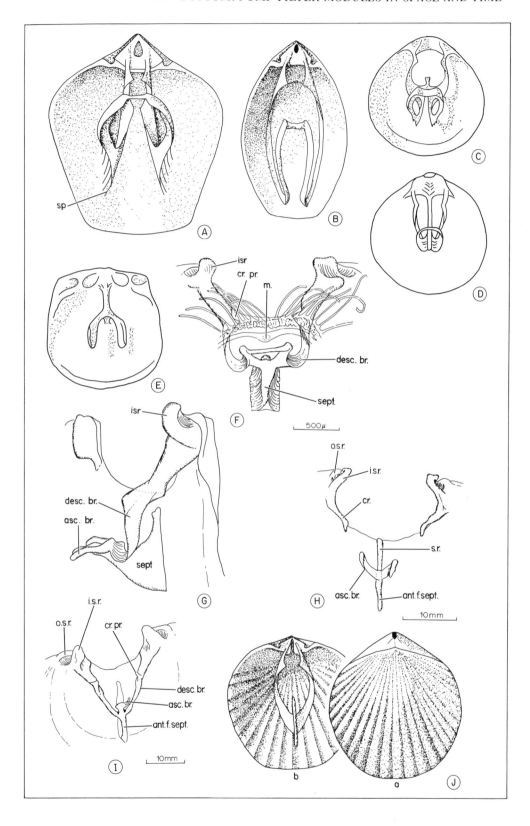

FIGURE 7.20 Loops—stages, metamorphosis, evolution.
(*A*) Terebratulaceans. Family Mutationellidae (genus *Cryptacanthia,* M. Penn.—L. Perm.), schematic representation of a cryptonelliform-like loop; *sp*—spinose extremities.
(*B*) Terebratellacean (Paleozoic): *Cryptonella* (L. Dev.—Perm.), cryptonelliform loop.
(*C*)–(*E*) Terebratellaceans, Mesozoic-Cenozoic: (*C*) *Macandrevia,* frenuliniform loop; (*D*) *Terebratella,* with a magdiniform to magelliform loop; (*E*) *Kraussina,* with a kraussiniform loop.
(*F*)–(*G*) *Platidia annulata* (Recent): (*F*) Brachial support relative to lophophore position and mouth; (*G*) side view of (*F*); *m*—mouth, *desc. br.*—descending branch, *sept*—septum, *cr. pr.*—crural process, *isr*—inner socket ridge, *s.r*—septal ridge, *ant. f. sept.*—anterior face of septal pillar (= median septum), *asc, br.*—ascending branch.
(*H*)–(*I*) Brachial supports: (*H*) *Platidia davidsoni;* (*I*) *Platidia anomioides* (shell length of *H* = 6.6 mm; of *I* = 4.0 mm).
(*J*) *Mutationella podolica,* (Czortków Stage, L. Dev., Poland): (*a*) Brachial valve; (*b*) interior of "*a*" with primitive centronelliform loop, enlarged.
(Sources: (*A*)(*B*) Stehli, 1965; (*C*)–(*E*) Elliott, 1953; (*F*)–(*I*) Atkins, 1959 *a, b;* (*J*) Stehli, 1965; Kozlowski, 1929.)

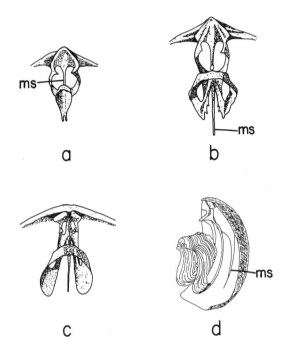

FIGURE 7.21 Terebratellacean loop—metamorphosis during ontogeny in *Dallina septigera,* (?) Eoc.—Rec. Progressive stages: (*a*) campagiform stage; (*b*) frenuliniform stage; (*c*) terebrataliform stage; (*d*) dalliniform stage (see text). Upper portions of figures are posterior, lower portions, anterior. (*a*)–(*d*): Steps in loop development described in text and related to stages above.
Illustrations redrawn after: Stehli, 1956*a;* Frile, 1877; Fischer and Oehlert, 1891, pl. IV; Elliott, 1953*b;* Atkins, 1960*a.*

ded living conditions of brachiopod colonies despite a sessile mode of life on the sea floor, passive competition for food must have prevailed. In turn, this would have sponsored a tendency toward more efficient lophophores that bring food particles to the brachiopod mouth.

Evolution of lophophores proceeded in the direction of greater complexity, that is, toward the plectolophous. The primitive centronelliform loop is assumed to have been associated with the less complex schizolophe stage (Stehli, 1956*a*). Accordingly, brachial supports (loop patterns) would presumably have been influenced by the same or related gene complex controlling the trend to more complex lophophores.

Translated into the actual physical mechanism, what is involved in different loop patterns and their ontogenies, according to Elliott, is "differing degree and time of onset of calcification during development of the same type of lophophore" (modern terebratelloids are plectolophous in adult stages). It follows that the *precise sites* of calcite deposition in the epithelial sheaths, or of resorption at subsequent stages, are the actual loci where the genetic code was (and is) translated into this or that loop pattern.

Origin of the Order Terebratulida

The loop of the earliest known spiriferid *Protozyga elongata*, Ord., Okla. (Figure 7.2*F*), is homologous with the centronelliform type (Figure 7.10*E*). This suggests a spiriferoid parental stock for the order. It is thought that the terebratuloid loop could have derived from a spirebearing population by suppression of the calcareous spires (that is, through prerequisite gene modifications). Another possibility—since loop precedes spires ontogenetically, perhaps the original isolates from the parental spiriferid population were quite indistinguishable from loopbearing spiriferids in the earlier stages, only subsequently differentiating. The oldest known centronellids (L. Dev., Poland) include *Mutationella*, among others, and presumably it was from such stock that centronellid radiation and differentiation took place in early Devonian time (Koslowski, 1929; Williams and Wright, 1961; Stehli, 1961*b*, 1965).

Identification of an Articulate Brachiopod to Species Level

Data from T. W. Amsden, in Amsden and Boucot, 1958.

Given. A sample of articulate brachiopods from the Haragan Formation of Oklahoma (L. Dev.)

Population. $n = 200+$.

***Characteristics* (External).**
   1. *Shape, and so on.* Subequally biconvex shells, pedicle beak overlaps dorsal valve; shell outline elongate oval (length greater than width); both valves with shallow sulcus (better developed on pedicle valve; obscure on some shells).

   2. *Ornamentation, and so on.* Rounded costellae separated by narrow interspace (ribs coarsen with growth of shell and may become obsolete in the sulci). Concentric growth lamellae also present. Shell substance punctate.

   3. *Internal Hinge Plate.* Supported by stout median septum that has two flanges posteriorly directed (Figure 7.22*D*).

Class. Are these articulate or inarticulate shells? These are articulates. Why? Because of the articulation (cardinalia), the calcareous biconvex shell, overlap of pedicle beak on dorsal valve.

Family. The punctate biconvex shells with the relatively large body cavity suggest one of the spiriferids. Why? The indicated valves would have accommodated

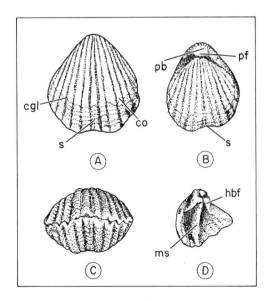

FIGURE 7.22 *Rhynchospirina maxwelli* Amsden, n. sp., 1958. (*A*) Pedicle view; (*B*) brachial view; (*C*) anterior view; (*D*) brachial cardinalia; *co*—costellae (note interspaces between), *s*—pedicle sulcus, *pb*—pedicle beak, *pf*—pedicle foramen, *hpf*—hinge plate flanges, *ms*—median septum, *cgl*—concentric growth lamellae.

a spiral brachidium. Although spires and jugum have not been observed in any of the sample available, their presence is inferred.

The combination of costate, sulcate, and punctate characteristics eliminates all spiriferid families that are smooth, with bifurcating costae, strong folds, or are impunctate. (If details of the brachidium were known, these would be important aids in reaching a decision.) The family that best appears to fit the sample is Rhynchospirinidae. This family has "medial costae finer than on the flanks." Our study sample can be accommodated to this description since in some specimens the costellae tend to become obsolete on the sulci side.

Measurements. Length/thickness ratio, 1.2–1.6; length/width ratio, 0.9–1.12; costellae (anterior end), small individuals 8–12 per 5 mm, larger individuals 4–8 per 5 mm.

Genus. To what genus in the family can the sample be assigned? There are only two genera in this family: *Rhynchospirina* and *Homoeospira*. Externally, these two are alike. Internally, both cardinal plate and median septum differ. In *Rhynchospirina*, the cardinal plate projects posteriorly into the umbonal cavity of the pedicle valve; in *Homoeospira*, it does not. In the former genus, the median septum reaches to the anterior half of the shell (Figure 7.22*D*) but does not in *Homoeospira*.

Accordingly, on the basis of these internal structural differences, the sample is assigned to the genus *Rhynchospirina*.

Species. The final step is species assignment. Measurements made earlier help to establish that the sample constituted a single species.

There are two species each of *Homoeospira* and three of *Rhynchospirina*. The reason that species of the former need still to be considered is because of past assignments to this genus that are now questionable.

Homoeospira subgibbosa. Similar to sample in length/width and length/thickness ratios; but has obscure pedicle sulci and coarser costellae; internally, the species has an abbreviated median septum. *Does not fit the sample.*

Homoeospira foerstei. Small size compared to sample; coarse costellae and well-developed sulci. *Does not fit the sample.*

Rhynchospirina electra. Smaller shell than sample with weaker convexity and deeper sulcus. *Does not fit the sample.*

Rhynchospirina formosa (Hall). Less coarse costellation than study sample; brachial sulcus almost nonexistent; pedicle sulcus weak; similar brachial hinge plate but median septum "low" compared to "high" in sample. *Does not fit the sample.*

Rhynchospirina globosa (Hall). Thicker, more robust shell, lacking sulci traces; costellation similar to sample. *Does not fit the sample.*

Since no known or closely related species matches the size and characteristics of sulcation, costellation, as well as internal structural features observed in the study sample, it is identified as a new species: *Rhynchospirina maxwelli* Amsden, 1958 (Figure 7.22).

QUESTIONS

1. Does the phylum Brachiopoda have any unique characteristics that distinguish it from (a) all other phyla, (b) other lophophorate coelomates?

2. Account for the post-Paleozoic decline of the brachiopods.

3. (a) What is the significance of inhalant and exhalant chambers?

   (b) How are these brought into being?

4. (a) Relate each of the following internal shell structures to their biological functions or associations: cardinalia, brachidium, articulation structures.

(b) Identify each of the following in the same terms as (a): spondylium, cruralium, delthyrium, notothyrium, chilidium.

(c) From a theoretical model, outline possible steps leading to the origin of the two main classes of brachiopods from a pre-Cambrian population.

5. Identify and cite a species in which each of the following occurs (refer to the Treatise): jugum, crura, bifid cardinal process, foramen, pseudodeltidium, centronelliform loop, median septum, fasciculate, setae grills, rectimarginate, sulcus, commissure, mucronate condition, rugae, plica, rostrate shell, endospine.

6. (a) Cite the most significant information concerning brachiopods obtained by Newell, et al., Agar, Fenton and Fenton, Williams, Muir-Wood and Cooper, Menzies and Imbrie, Ferguson, Lowenstam, Boucot et al.

(b) Cite examples of the utility of brachiopods in zonation.

(For useful ecological data on living articulates see H. C. McCammon, 1973. J. Paleont., v. 47 (2):266–278; 1969, Same, v. 43:976–985. J. B. C. Jackson, T. F. Goreau, and W. D. Hartman, 1971. Recent brachiopod-coralline sponge communities and their paleoecological significance: Science, vol. 173, pp. 623–624.

REFERENCES

Ager, D. V., 1959. The classification of the Mesozoic Rhynchonelloidea: J. Paleontology, Vol. 33 (2), pp. 324–332, plate 49, 4 text figures.

————, 1961. The epifauna of a Devonian spiriferid: Quart. J. Geol. Soc. London, Vol. 117, pp. 1–8; "Discussion," pp. 8–10.

———— et al., 1965. Systematic Descriptions, in Treat. Invert. Paleontology (*H* — Brachiopoda), Vol. I, pp. H-256 ff.; Vol. 2, pp. H-523 ff. (F. G. Stehli, Vol. 2, pp. H 730 ff).

Allan, R. S., 1939. Studies of the Recent and Tertiary Brachiopoda of Australia and New Zealand: Canterbury College, Christchurch, N.Z. Canterbury Mus. Records: Vol. 4 (5), pp. 231–248, plates 29–31.

Alexander, F. E. S. 1948. A revision of the genus *Pentamerus* (James Sowerby, 1813) (abbrev. title). Geol. Soc. London Quart. J., Vol. 103, pp. 143–161, 1 plate.

Amsden, T. W., and A. J. Boucot, 1958. Stratigraphy and paleontology of the Hunton group in the Arbuckle Mountain Region: Oklahoma Geol. Surv. Bull. 78, 170 pp., 14 plates, 42 text figures.

Astre, G., 1938. *Persistance de Spiriférines dans les mers aaleniennes*: Soc. Hist. nat. Toulouse Bull., Vol. 72.

Atkins, D., 1956. Ciliary feeding mechanisms of brachiopods: Nature, Vol. 177, pp. 706–707.

————, 1959a. The growth stages of the lophophore of the brachiopods *Platidia davidsoni* (Eudes Deslongchamps) and *P. anomioides* (Scacchi and Philippi) with notes on the feeding mechanisms: J. mar. biol. Assn. U.K., Vol. 38, pp. 103–132.

————, 1959b. A new species of *Platidia* (Brachiopoda) from the La Chapelle Bank Region, *idem*, pp. 133–142.

————, 1960a. A note on *Dallina septigera* (Lovén) (Brachiopoda, Dallinidae): *idem*, Vol. 39, pp. 91–99.

————, 1960b. The ciliary feeding mechanism of the Megathyridae (Brachiopoda) and the growth stages of the lophophore: *idem*, Vol. 39, pp. 459–479.

————, 1961. The growth stages and the adult structure of the lophophore of the brachiopods *Megerlia truncata* (L.), and *M. echinata* (Fischer and Oehlert): *idem*, Vol. 41, pp. 95–111.

Bell, W. C., 1941. Cambrian Brachiopoda from Montana: J. Paleontology, Vol. 15 (3), pp. 193–255, plates 28–37.

Boucot, A. J., and E. D. Gill, 1956. *Australocoelia*, a new Lower Devonian brachiopod from South Africa, South America, and Australia: J. Paleontology, Vol. 5, pp. 1173–1178, plate 126, 1 text figure.

————, and J. G. Johnson., 1965. Articulate Brachiopoda. A.G.U. Antarctic Res. Series. Vol. 6, pp. 255–261, pls. 3–8.

————, J. G. Johnson, and J. A. Talent, 1969. Early Devonian brachiopod zoogeography: G.S.A. Special Paper No. 119, 113 pp.

Chapman, F., 1914. Notes on shell-structure in the genus *Lingula*, Recent and fossil: Roy. Micros. Soc. London, Vol. 34, pp. 28–31, plate 5.

Chuang, S. H., 1959. The breeding season of the Brachiopod *Lingula unguis* (L.): Biol. Bull., Vol. 117 (2), pp. 202–207.

Cloud, P. E., 1942. Terebratuloid Brachiopoda of the Silurian and Devonian: Geol. Soc. Amer. Sp. Paper 38, 138 pp., plates 1–26.

Cooper, G. A., 1944. Phylum Brachiopoda, H. W. Shimer and R. R. Shrock, Index Fossils of North America, pp. 277–365, plates 105–143.

————, 1952. Unusual specimens of the brachiopod family Isogrammidae: J. Paleontology, Vol. 26 (1), pp. 113–119, plates 21–23.

————, 1954. Recent brachiopods: Bikini and nearby atolls, Marshall Islands: U. S. Geol. Surv., Prof. Paper 260-G, pp. 315–318, plates 80–81 (*Argyrotheca, Crania, Thecidellina*).

————, 1955. New genera of Middle Paleozoic Brachiopoda: J. Paleontology, Vol. 29 (1), pp. 45–63, plates 11–14, 1 text figure.

————, 1956. Chazyan and related brachiopods: Smithsonian Misc. Coll., Vol. 127, Part 1, pp. 1–1024 (text), Part 2, pp. 1025–1245, plates 1–269.

———— and A. Williams, 1952. Significance of the stratigraphic distribution of brachiopods: J. Paleontology, Vol. 26, pp. 326–337.

Corroy, G., 1927. Les Spiriféridés du Lias europeén, etc.: Ann. de Paléontologie, Vol. 16.

Dunbar, C. O., and G. E. Condra, 1932. Brachiopoda of the Pennsylvanian system in Nebraska: Nebraska Geo. Surv. Bull. 5, 373 pp., plates 1–44.

Elliott, G. F., 1951. On the geographical distribution of terebratelloid brachiopods: Ann. & Mag. Nat. History, series 12, Vol. 4, pp. 305–334.

————, 1953*a*. The classification of the thecidean brachiopods: *idem*, series 12, Vol. 6, pp. 693–701, plate 18.

————, 1953*b*. Brachial development and evolution in terebratelloid brachiopods: Biol. Reviews, Vol. 28, pp. 261–279.

————, 1955. Shell structure of thecidean brachiopods: Nature, Vol. 175, pp. 1124.

Fage, G., 1934. Les Rhynchonelles du Crétacé inférieur des Charentes: Bull. soc. géol. France, Vol. 4.

Ferguson, Laing, 1963. The paleoecology of *Lingula squamiformis* Phillips during a Scotish Mississippian marine transgression: J. Paleontology, Vol. 37 (3), pp. 669–681, 8 text figures.

Fischer, P., and D. P. Oehlert, 1891. Brachiopodes: Exped. Sci. Travailleur et du Talisman (1880–1883), 139 pp., 8 plates (Paris).

Frile, Herman, 1877. The development of the skeleton of the genus *Waldheimia:* Archiv. Math. Naturvidens, pp. 380–386, 6 plates.

Gignoux, M., 1955. Stratigraphic geology. W. H. Freeman & Co., 654 pp.

Grant, R. E., 1966. A Permian productid brachiopod: life history: Science, Vol. 152 (No. 3722), pp. 660–662, Figures 1, 2.

Hertlein, L., and U.S. Grant IV, 1944. The Cenozoic Brachiopoda of western North America: Univ. Calif. (Los Angeles) Publ. Math. Phys. Sci., Vol. 3, pp. 1–172, 21 plates.

Hyman, L. H., 1959. The Invertebrates. McGraw Hill, Vol. 5, 766 pp. (Brachiopoda, Ch. 21).

Imbrie, John, 1959. Brachiopods of the Traverse Group (Devonian) of Michigan: Am. Mus. Nat. Hist. Bull. 116, Art. 4, pp. 349–409, plates 48–67.

Jope, H. M., 1965. Composition of the brachiopod shell: Treat. Invert. Paleont. (*H –* Brachiopoda), Vol. 1, pp. H-156 ff.

Kozlowski, R., 1929. Les brachiopodes gothlandiens de la Podolie polonaise: Paleont. polonica, Vol. 1, 254 pp., 12 plates, 95 text figures (Warsaw).

Lacaze-Duthiers, F. J. H. de, 1861. Histoire naturelle des brachiopodes vivants de la Méditerranée: Première Monographie: Historie de la Thécidie (*Thecidium mediterraneum*): Ann. Sci. Nat., series 4 (Zool.), Vol. 15, pp. 259–330, plates 1–5.

Logan, A., 1965. Middle and Upper Triassic spiriferinid brachiopods from the Canadian Arctic. G.S.A. Program (Kansas City), Abstracts, p. 97.

Mattox, N. T., 1955. Observations on the brachiopod communities near Santa Catalina Island, in Essays in the natural sciences in honor of Capt. Hancock. U. Calif. Press (Los Angeles), pp. 73–83, Figures 1–5.

McConnell, Duncan, 1963. See ref. Chapter 15.

Menzies, R. J., and John Imbrie, 1958. On the antiquity of the deep sea bottom fauna: Oikos, Vol. 9 (2), pp. 192–210.

Muir-Wood, H. M., 1955. A history of the classification of the phylum Brachiopoda:

Brit. Mus. (Nat. Hist.) London, pp. 1–93, Figures 1–12.

———— and G. A. Cooper, 1960. Morphology, classification, and life habits of the Productoidea (Brachiopoda): G.S.A. Mem. 81, 447 pp.

Paine, R. T., 1962. Filter-feeding pattern and local distribution of the brachiopod, *Discinisca strigata:* Bio. Bull., Vol. 123, pp. 597–604.

————, 1963. Ecology of the brachiopod *Glottidia pyramidata:* Ecological Monographs, Vol. 33, pp. 187–213, 17 text figures.

Percival, E., 1944. A contribution to the life history of the brachiopod *Terebratella inconspicua* Sowerby: Roy. Soc. N.Z. Trans., Vol. 74, (1), pp. 1–24, plates 1–7, text figures 1–4.

————, 1960. A contribution to the life history of the brachiopod *Tegulorhynchia nigricans:* Micros. Soc. Quart. J., Vol. 101, pp. 439–457, text figures 1–6.

Rudwick, M. J. S., 1959. The growth and form of brachiopod shells: Geol. Mag., Vol. 96, pp. 1–24.

————, 1960. The feeding mechanisms of spire-bearing fossil brachiopods: Geol. Mag., Vol. 97, pp. 369–383.

————, 1961. The feeding mechanisms of the Permian brachiopod *Prorichthofenia:* Paleontology, Vol. 3, Part 4, pp. 450–471, plates 72–74.

————, 1962. Filter-feeding mechanisms in some brachiopods from New Zealand: J. Linn. Soc. Zool., Vol. 44, pp. 592–615.

————, 1964. The function of zigzag deflection in the commissure of fossil brachiopods: Paleontology, Vol. 7 (1), pp. 135–171, plates 21–29.

————, 1965. Ecology and Paleoecology: Treat. Invert. Paleont. (*H*–Brachiopoda), Vol. 1, pp. H199–H214.

Sass, D. B., E. A. Monroe, and D. T. Gerace, 1965. Shell structure of Recent articulate Brachiopoda: Science, Vol. 149 (3680), pp. 181–182, figures 1, 2.

Schindewolf, O. H., 1954. Über einige kambrische Gattungen inartikulater Brachiopoden: Neues Jahr. f. Geol. Paleont., Vol. 12, pp. 538–557.

Schuchert, C., 1911. Paleogeographic and geologic significance of Recent Brachiopoda. Bull. G.S.A., vol. 22, 258–275.

Stehli, F. G., 1956a. Evolution of the loop and lophophore in terebratuloid Brachiopoda:

Evolution, Vol. 10, pp. 187–200, 9 text figures.

————, 1956b. Notes on oldhaminid brachiopods: J. Paleontology, Vol. 30, No. 2, pp. 305–313, plates 41–42, 1 text figure.

————, 1956c. Shell mineralogy of Paleozoic invertebrates: Science, Vol. 123, pp. 1031–1032.

————, 1961a. New terebratuloid genera from Australia: J. Paleontology, Vol. 35, pp. 451–456.

————, 1961b. New genera of upper Paleozoic terebratuloids: *idem,* Vol. 35, pp. 457–466.

Stenzel, H. B., 1939. New Eocene brachiopods from the Gulf and Atlantic Coastal Plains: Texas Univ. Bur. Eco. Geol. Contrib. to Geology, pp. 717–730.

Tasch, P., 1967. The problem of the primary food chain in the sea through geologic time. Proc. Second Intern. Conf. Palynology (Utrecht, Holland, 1966), pp. 283–290.

Thomson, J. A., 1927. Brachiopod morphology and genera (Recent and Tertiary): New Zealand Board Sci. & Art. Manual, No. 7, 338 pp., 2 plates, 103 text figures.

Treatise on Marine Ecology and Paleoecology, H. S. Ladd, ed., 1957. G.S.A. Mem. 67, Vol. 2. (Papers by C. Lochman, p. 117; P. E. Cloud and V. E. Barnes, p. 163; H. Lowenstam, p. 215; G. A. Cooper, p. 249; J. S. Williams, p. 279; B. Kummel, p. 437; R. W. Imlay, p. 469; J. B. Reeside, p. 505.)

Unkelsbay, A. G., and W. B. Niewoehner, 1959. Attachment loops of infant brachiopods from the Louisiana limestone of Missouri: J. Paleontology, Vol. 33 (4), pp. 547–549, plate 74.

Walcott, C. D., 1908. Cambrian Geology and Paleontology: Cambrian Brachiopoda No. 3. Smithsonian Misc. Coll., Vol. 53, pp. 53–137; ————, 1924. *Idem,* Vol. 67, pp. 477–533.

————, 1912. Cambrian Brachiopoda: U. S. Geol. Survey, Mon. 51, Part 1, 872 pp., 76 text figures; Part 2, 363 pp., 104 plates.

Williams, Alwyn, 1953. North American and European strophedontids: their morphology and systematics: Geol. Soc. America, Mem. 56, pp. 1–67, plates 1–13.

————, 1957. Evolutionary rates of brachiopods: Geol. Mag., Vol. 94 (3), pp. 201–211.

————, 1965. "Stratigraphic distribution" (see next reference).

———— and A. T. Rowell, 1965. "Brachiopod

anatomy," "Evolution and phylogeny," "Classification," in Treat. Invert. Paleont. (*H*–Brachiopoda), Vol. 1, pp. H-6, H-57, H-164, H-199, H-237.

———— and A. D. Wright, 1961. The origin of the loop in articulate brachiopods: Paleontology, Vol. 4 (2), pp. 142–176.

Yatsu, Naohidé, 1902. On the development of *Lingula anatina:* J. Coll. Sci. Imperial Univ. Tokyo, Japan, Vol. 17, article 4, pp. 1–112, plates 1–8.

(A useful paperback for students — M. J. S. Rudwick, 1970. Living and fossil brachiopods. Hutchinson Univ. Library, 184 pages, 99 figures.) C. W. Thayer, 1974. Salinity tolerances of articulate brachiopods. G.S.A. Abstracts with Program (NE Section, Baltimore, Md., pp. 80–81).

chapter eight
THE MOLLIS CLANS: WORLD OF THE GEOMETRICAL SHELL

INTRODUCTION

Fossil and now living mollusks comprise about 11 percent of all animal species that have ever evolved during the whole earth history. The animal species, invertebrate and vertebrate, number over 2 million. Mollusks exceed the total number of species of Protozoa, Porifera, Coelenterata Worms, Bryozoa, and Brachiopoda — all taken together (9.8 percent).

Among all invertebrates, mollusks make the greatest contribution to our food consumption. Every year the United States harvests oysters, clams, scallops, and mussels (class Pelecypoda), exceeding 130 million pounds. Even in dim human history beginnings, primitive man valued the mollusk food source as kitchen middens (prehistoric shell heaps) attest (Buchsbaum; Light; Storer and Usinger; Easton).

Mollusks occur in all sorts of marine, fresh-water, and land habitats at all latitudes. The Galathea Deep Sea Expedition (1950–1952) recovered a bivalve (*Glomus*) from a depth of 10,190 meters in the Phillipine Trench, and a marine snail (*Odostomia*) from 8210 meters in the Kermadec Trench. Moreover mollusks are known from mountainous areas some three miles above sea level. One or another molluscan species has constantly inhabited our planet for the past 625 million years (Cambrian to Recent).

Some mollusks can live in trees (*Liguus*

fasciatus roseatus) and feed on barks and leaves (Buchsbaum, 1948) — these are the vegetarians; others are carnivorous predators. Still others are parasites: The snail *Odostomia plicata* draws its nutriment from punctures in the skin of the peacock worm's tentacles (Jägersten, 1960). Mollusks bore into shells, mud, wood, or rock; some form bioherms (oyster reefs); some move by jet propulsion in the sea. Larval stages of pelecypods (some gastropods and cephalopods) contribute important components to plankton hauls (Thorson, 1946).

Furthermore, the cephalopod brain has an experimentally proven learning capacity (Wells, 1962)!

Beauty distinguishes molluscan shells — their color, form, and sculpture. Some mollusks (order Nudibranchia) having lost their shells, only the larvae have whorled shells. Among others, certain cephalopods have reduced shells (*Spirula*). Living mollusk range embraces miniscule snails (1.0 mm) to giant squids 50 ft long, weighing an estimated 2 tons, as well as giant clams (*Tridacna*) 5 ft long, weighing 500 pounds. [Equivalent size spreads are known from the fossil record: Giant orthocone cephalopods in the Ordovician attained 9 ft in length; giant Mesozoic ammonites, the size of a wagon wheel; minute snails, less than 1/4 in. in diameter (basal Maquoketa shale of Iowa).]

The fossil record of mollusks — chiefly cephalopods, pelecypods, and gastropods

—is very extensive and furthermore of first-rate importance in paleontology, biostratigraphy, paleoecology, and other studies.

LIVING MOLLUSKS

Admitting the diversity in external appearance and habits, all mollusks do share a common structural plan (Figure 8.2), and this alone suffices to unite them in the same phylum.

Phylum Definition

(*L. mollis* = soft. The name refers to the soft internal body in contrast to the generally hard-shelled exterior.)

1. *Segmentation.* Unsegmented, except for a recently discovered *Neopilina* (and related fossil monoplacophorans), which is segmented (Figure 8.8*C*, *G*).

2. *Symmetry.* Bilateral symmetry (lost in the Gastropoda).

3. *Four Body Regions* (see Figure 8.1):

(a) *Head.* With tentacles and eyes (lost in bivalves.
(b) *Foot.* Ventral and muscular.
(c) *Visceral Mass* (or Hump). Coiled in gastropods; internal organs concentrated in this mass.
(d) *Mantle* (or Pallium). A soft skin or sheet of tissue overgrowing visceral mass; dorsal; secretes calcareous shell with an organic matrix (see Chapter 15); produced into free flap(s) partially enclosing *mantle cavity*.
4. *Mantle Cavity.* This space, at posterior end of visceral mass, contains paired gills and is the "respiratory chamber." Anus, excretory and reproductive systems open into the mantle cavity; thereafter exhalent currents remove their products.

5. *Gills* (= ctenidia). Cilia on the gills create, by their movement, inhalent water currents ventrally and an exhalent current

that leaves the mantle cavity dorsally (Figure 8.1*B*). Originally used for breathing, but in the Class Pelecypoda, greatly enlarged gills sieve to obtain minute food particles. (See Figure 8.24*A* for types of gills.)

6. *Digestive Tract.* Complete with mouth, commonly with jaws, buccal cavity contains radula (horny ribbon covered with many rows of minute, hard, recurved teeth—a rasping organ); salivary glands; and always a stomach into which hepatopancreas ("liver") opens. (Pelecypoda lack jaws and radula.)

7. *Organs and Organ Systems. Circulatory.* Generally a three-chambered heart (ventricle and two auricles); blood may contain the respiratory pigment, haemocyanin; arterial and venous systems, haemocoels; and in cephalopods only, capillaries. *Respiratory.* Gills (generally one pair but in the class Amphineura, up to 80 pairs); lung in mantle cavity (class Gastropoda, order Pulmonata); respiration also by mantle or epidermis. *Excretory.* Kidneys or pericardial glands with such function. *Reproductive.* Sexes usually separate; hermaphroditism widespread; fertilization, commonly external; gonads paired, with ducts; development via trochophore larva followed by veliger larva, or secondarily, direct (that is, no larval stages) (Figure 8.3).

8. *Shell.* An external calcareous skeleton characterizes mollusks. There are a few exceptions.

9. *Geologic Range.* Cambrian to Recent.

Affinities

Discovery and study of living segmented mollusks (Lemche, 1957)—*Neopilina*—have confirmed the long-suspected affinities of annelid worms, arthropods, and mollusks. Furthermore, it is now beyond doubt that the Mollusca separated from

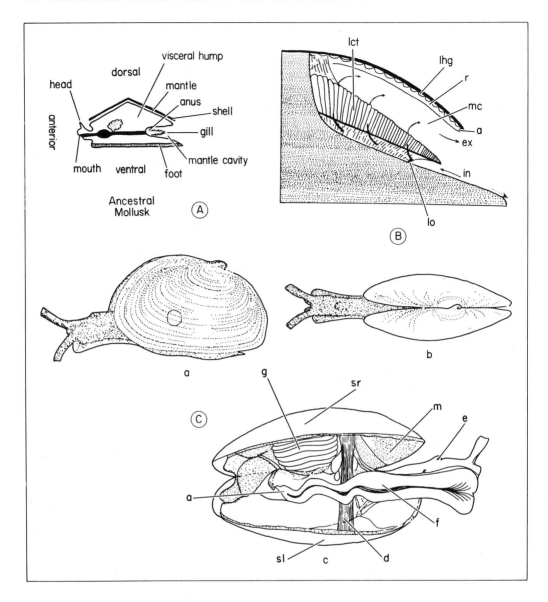

FIGURE 8.1 Ancestral mollusk.

(A) Compare this figure to four-body regions described in phylum description; black area – digestive tract; schematic.

(B) Respiratory chamber (= mantle cavity, see A) of ancestral mollusk. Pallial organs (*lo, lhg*) and currents (*Ex, In*) were presumably associated with cleaning out sediments and metabolic wastes; *lhg* – a mucous gland, consolidates sediments before removal by currents: *a* – anus, surface of foot with backward-directed ciliary current shown by backward directed arrow labelled "in"; *lct* – left ctenidium; *In, Ex* – inhalent and exhalent currents generated by ciliary action. (A) after Lancaster; (B) after C. M. Yonge.

(C) *Tamanovalva limax* Kawaguti and Baba, a bivalved opisthobranch gastropod (order Sacoglossa, see classification, this chapter). Recent, sublittoral zone, below Sea of Japan (at Bisan Seto). (*a*) Lateral view from left; (*b*) dorsal view, note protoconch (left valve only) sinistrally coiled (helicoidal); probably the original valve, X 0.66; (*c*) ventral view, X 8, ligament joins two valves, no hinge teeth. (The bivalve condition in a normal univalve group – Gastropoda – lends support to derivation of Bivalvia + Gastropoda, and so on, from an ancestral-type figured in A above.) After Cox, 1960*b; f* – foot, *e* – eye, *m* – mantle edge, *sr, sl* – right and left valve respectively, *g* – gill, *a* – anus, *d* – adductor muscle. A hypseloconid mollusk (See Monoplacophora), the possible direct ancestor of Cephalopoda (Yochelson, Flower, Webers, 1973. Lethaia, v. 6:274–309). [B. Runnegar and J. Pojeta, Jr., 1974. Science, vol. 186, pp. 311–317, suggest a dual origin for molluscan classes. Monoplacophora gave rise to Gastropoda, Cephalopoda, Rostroconchia, and possibly Placophora. Rostroconchia (Pojeta et al., 1972, vol. 177, p. 264) gave rise to Scaphopoda and Pelecypoda.]

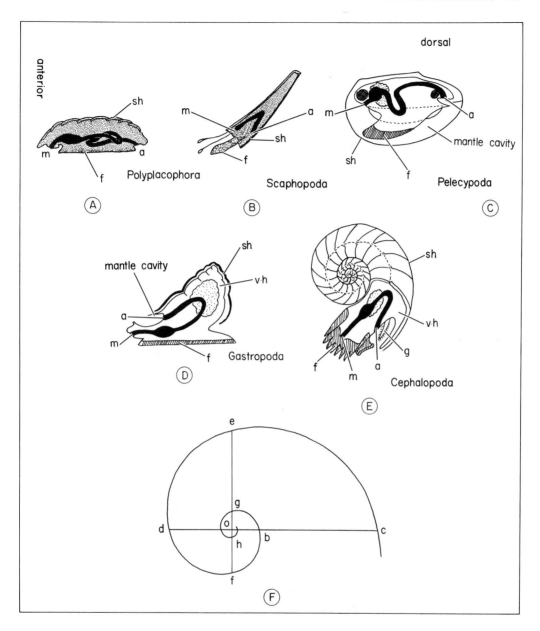

FIGURE 8.2 Common bioprogram in five classes of mollusks. (*A*) Polyplacophora. (*B*) Scaphopoda. (*C*) Pelecypoda. (*D*) Gastropoda. (*E*) Cephalopoda. Compare relationship of mouth and anus, and position of foot in each; *m*—mouth, *a*—anus, *sh*—shell, *f*—foot, *g*—gill *v.h*—visceral mass, *m*—mantle. (*F*) The equiangular spiral (after D'Arcy Thompson).

the Annelida *after* the appearance of segmentation (metamerism) (C. M. Yonge, 1960).

Shell Structure

Details of shell structures and illustrations are given in Chapter 15. Under each molluscan class, further discussion of shell structure will be detailed. Here certain general observations may be reviewed.

Bøggild (1930) recognized five types of shell structures in mollusks: homogeneous, prismatic, foliated, nacreous, and crossed laminar. Mackay (1952), in a

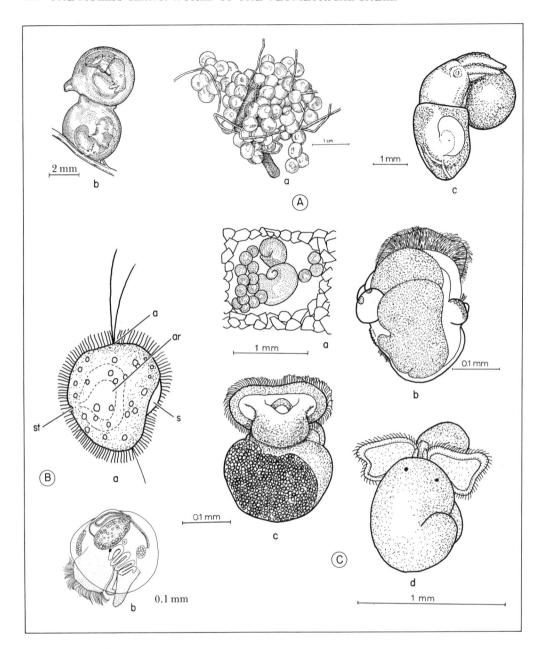

FIGURE 8.3 Direct development and larval stages in Mollusca.
(A) *Sepietta oweniana* (Orb.), Cephalopoda. (a) Egg mass *in situ* on piece of seaweed, N. Kattegat, depth 8 meters, X 2; (b) two eggs with embryos visible through shell; (c) embryo dissected out of egg, right side; observe yolk sac. Note that embryos have all the characteristics of the adult (other than yolk sac) in contrast to other mollusks where a series of larval stages precede metamorphosis into the adult stage (compare *B* and *C*).
(B) (a) The pelecypod, *Mytilus edulis* L., trochophore stage at 24 hours; a—apical tuft; ar—arch-enteron, st—stomadaeum, s—shell; (b) later larval stage, veliger, 400µ length.
(C) *Natica catena*, snail, developmental stages. (a) Transverse section of egg mass showing egg space with embryos; (b) young veliger. (Note: Trochophore larval stage passed, generally, in egg capsules; hatched forms emerge as in *N. catena*, as veligers.) At this stage, velum, foot, larval heart, excretory organ, and first traces of a shell are visible; (c) (d) still later veligers. (A), (B, b) (C) after Thorson, 1946; (B, a) after Raven, 1958.

petrographic study of shell structure in modern mollusks, found the prismatic and the crossed laminar to be more common.

What causes molluscan shell structure? Mackay reasoned that often the mineral composition, that is, aragonite or calcite, caused the observed structures. Thus he found that nacreous and crossed laminar structures were usually produced by aragonite, whereas mineral calcite composed the foliated structure. The other structures noted above might be either calcite or aragonite, but no dual occurrences of both polymorphs in the *same* layer occurred.

Prismatic Structure. According to Bøggild, there are three types of prismatic structure: normal, complex, and composite. Mackay found normal prismatic to be most common. Branched or straight prisms were noted. Mackay thought that complex prismatic structure might be due to twinning of calcite or aragonite. Bøggild's composite prismatic structure (1930, p. 48) had a layer of large prisms — first order — composed of second order (that is, smaller and finer prisms), which made an angle with the first.

Foliated Structure. This structure is common in calcite. It resembles cross-bedded sandstone in appearance. Where present, it occurs in inner layers, in a major portion of the shell.

Nacreous Structure. Structure consists of equidistant, equally thick leaves of aragonite (ca. 0.001 mm) separated by thin leaves of organic substance — conchiolin (see Grégoire, Chapter 15, Bøggild, 1930, p. 250).

Crossed Laminar Structure. (See Figure 15.6*E*.) First-order lamels are rectangular in form with longer axis parallel to shell's surface, and with shorter axis generally in a vertical position. Each such lamel shows uniform extinction under

crossed nicols and hence is a single crystal.

Second-order lamels (thickness a little less than 0.001 mm) are the building units of first-order lamels. They are oriented normal to the face of first-order lamels. Their edge forms an angle of 41° with the margin of first-order lamels. Given any two adjoining first-order lamels, the second-order lamels will be inclined in opposite directions (they cross at angles of 82° or 98°). It is this arrangement that confers the crossed laminar structure and thus strengthens the shell.

Bøggild found that in the several classes of mollusks, first-order lamels are markedly different.

Homogeneous Structure. Under plain light, no structure can be located. However, under crossed nicols, individual crystals can be seen. Mackay noted that this structure is often "transitional" to the other structures, particularly in the prismatic layers.

The direction in which a shell is sectioned can transform a so-called prismatic layer on one slide into a crossed-laminar structure on another. Thus a section cut parallel to the first-order lamels shows crossed laminar structure, whereas a section cut at right angles to the above direction exhibits prismatic structure (Mackay, 1952, pp. 9, 27).

Shape and Form of Shell

Coiling, indeed various degrees of uncoiling, are generously represented in, and expressed with remarkable diversity by, molluscan shells. D'Arcy Thompson (1942) brought to a synthesis the extensive literature on the geometry of the molluscan shell. Effectively he showed that among mollusks differences in the form of the shell were a matter of degree. The almost infinite variety of shell shapes could be reduced to "the variant properties of a simple curve" — the equiangular spiral (Figure 8.2). In an equiangular spiral, the whorls continuously increase in breadth at a steady, unchanging ratio.

Thus the configuration of the spiral pearly *Nautilus*, a conical gastropod, and a tusk-shaped scaphopod are products of the same simple curve. The *Nautilus* shell is a cone rolled up. The conical gastropod is only a limiting cone of the spiral shell. The shell of the scaphopod, *Dentalium*, represents a small portion of a single whorl of the spiral (Thompson, 1942, pp. 759, 792–793).

More recently, Raup (1961, 1962)[1] studied the geometry of coiling in gastropods. He noted weaknesses in Thompson's use of certain basic parameters and attempted a fresh approach. By supplying data on four shell parameters into a digital computer, Raup obtained graphical reconstructions of hypothetical or "computer" snails.

Although further discussion and details are deferred at this time, the significance of this work may be considered briefly at this point.

What does an array of nonexistent, never existing ("hypothetical") snail configurations denote? D'Arcy Thompson (1942, p. 849) expressed doubt that natural selection played a role in the production of the diverse configurations of molluscan shells. Rather, he attributed them to the mathematical properties of a simple curve. However, Raup's computer snails make clear that, of *n* possible configurations of the gastropod shell (obtained by changing numerical data on each of four parameters), only very specific ones were *selected* in the actual course of evolution. Why? Apparently, potential, unused, unrealized configurations were either inadaptive or less adaptive than those that were realized.

Not the mere geometry of the shell, as such, but the properties—the advantages one configuration confers over another in the biological life of a given snail—determine the ones natural selection prefers.

Gastropod torsion (that is, twisting of

the visceral mass) and spiral coiling, are unrelated in time (C. M. Yonge, 1947; 1960, pp. 1–12). It is unknown "to what extent coiling preceded torsion." Furthermore, the two are "not connected." A greater and also even a lesser degree of coiling is known to have *preceded* torsion (see also Eales, 1950).

Some explanation for coiling is necessary. Yonge (1960, pp. 1–12) envisioned coiling as a solution to the problem presented by increasing height of the gastropod shell. The result provided more compact arrangement of the lengthening visceral mass.

Thus coiling of gastropod shells leading to diverse configurations was itself a response, an adjustment to pressures created by the animal's growth. In this sense, successful snail configurations (contrasted to "hypothetical" configurations) may be identified as having adequately met such pressures.

In gastropods the environment may also influence shell shape (C. M. Yonge, 1963). The limpet *Patella vulgata* (H. B. Moore, 1934) illustrates this response (Figure 8.4*A*). The uncoiled conical shell of limpets is the very simplest of gastropod shells. The shell's elliptical edge fits a given patch of rock to which it attaches. How can the variations in shape in Figure 8.4*A* be explained? C. M. Yonge believes it relates to the way limpets solve the desiccation problem.

Thus Type I (high-water) shells (Figure 8.4*A*, a) allow for least evaporation around the edges because of the narrow base. Associated with muscular contraction to pull the shell close to the rock is a tendency to inwardly pull the mantle edge. That diminishes the circumference on which lime for the shell is deposited and thereby leads to a narrow base. Type III (low-water) shells (Figure 8.4*A*, c) were deposited while the limpet was under water; muscular contraction was negligible. Hence the mantle spread outward, and lime deposition for the shell occupied a wider rim than in Type I shells.

A clear example of shell variation in

[1]See Appendix 1, Part 2, for Refs.

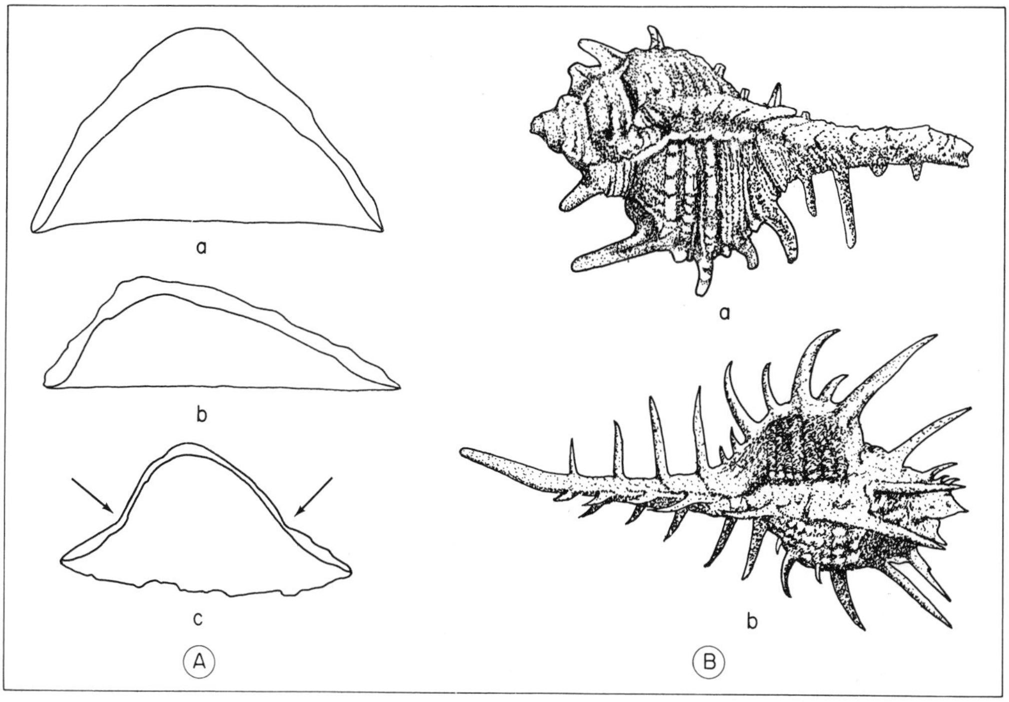

FIGURE 8.4 Variable shell shape and ecology.
(A) Limpet, *Patella vulgata*. (a) High-water level; shell high in proportion to base; (b) low-water level; shell flatter and not so steeply conical; (c) animals from (a) transferred to pools; subsequent growth denoted by arrows, called "ledging," that is, tendency to flatten out.
(B) *Murex tribulus*. Great Bitter Lake, near Fayed, Egypt; note variations in spines ranging from tubercles to up to 12.0 mm in length and the angles (22°–53°) they make with surface of shell. *a*—migrant subpopulation; *b*—Red Sea population.
(A) after H. B. Moore, 1934; see also C. M. Yonge, 1963, p. 141; (B) M. Taylor, 1954.

gastropod, *Murex tribulus*, in two environments was given by Taylor (1954, and see Figure 8.4B). The Great Bitter Lakes came into being with the opening of the Suez Canal. Among immigrant organisms entering this new habitat from the Red Sea came *Murex tribulus*. The Red Sea population of this species (Figure 8.4B) lives between the coast and a coral reef. Its sharply pointed spines may have served a protective function against predators. By contrast the migrant subpopulation of the species (Great Bitter Lakes) in some three-quarters of a century, since isolation from the Red Sea population, have developed more modified spines. That is, spines, tubercle or blunt-ended, rise from the surface at subdued angles.

This last-named tendency is attributed to the probable absence of Red Sea predators in the placid Great Bitter Lakes environment. If this trend continues, the Great Bitter Lakes subpopulation, already a geographic isolate from the parent population, may become a genetic isolate as well. If so, it will have become a true new species of *Murex*.

Coiling in cephalopods, Dunbar (1924) considered to be a response to the development of buoyant air chambers.

Food and Predation

Gastropods are chiefly herbivores. For the maintenance of the limpet *Patella vulgaris*, it has been estimated that about 75 cm of incrusting algae are needed to support 1 cu. cm of limpets—during the first year of life. Carnivorous snails also occur, as evidenced by boreholes (1.0 mm diameter) in many bivalve shells. These snails bore into and eat other mollusks

(*Natica* is a carnivorous sand-burrowing snail; the whelk, *Buccinum*, attacks scallops).

Marine clams take in their food along with water from the sea and strain it out along the edge of the gills. Ciliary action of the gills carries particles to the mouth. Siphonal development matches the type of bottom feeding. Thus short siphons are known in forms that are suspension feeders (*Cardium*). Elongate siphons characterise forms that are deposit feeders (*Tellina*); so they consume bottom detritus and organic debris (C. M. Yonge, 1963).

All cephalopods are predaceous. The octopus, for example, seizes crabs in its sucker arms and breaks open the shell with its horny jaws and radula. The squid preys on small fish.

Man is one of the predators on mollusks since they are articles of human diet. Some mollusks, likewise, prey on other mollusks. A good example is the slipper limpet *Crepidula*; it smothers oysters by settling on them. Still other predators range from starfish, boring sponges, polychaete worms to certain whales. Starfish (*Asterias*) swallow their prey whole, later disgorging the empty shell. Boring sponges and polychaete worms are oyster predators. The chief food of certain whales is squid.

Oyster Reefs and Reef Patches

The distribution of modern oyster reefs in Texas Bays and a portion of the Gulf of Mexico is shown in Figures 8.5 and 8.6. Ladd (1957) described the characteristics of existing oyster reefs of some bays on the central Texas coast. In cross-section, these reefs are low mounds with dead shells in the higher center; living oysters and biotic associates range on the slopes. The reef's actual shape is oval-to-spindle, or it may extend from the shore as a narrow bar.

The dominant form found is the oyster *Crassostrea virginica* and the mussel (*Brachiodontes*). In addition, cemented or clinging to the larger shells are various smaller

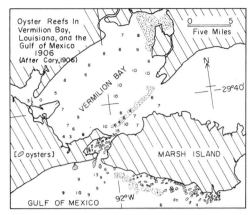

FIGURE 8.5 Oyster reefs in Gulf of Mexico (Hedgpeth, 1953).

barnacles, serpulid worms. Some reefs have numerous small gastropods.

Ladd notes that in the upper reaches of the bays, mussels are ubiquitous, literally competing with and covering the oysters. Puffer and Emerson (1953) found that the oyster reef biotope consisted of eleven living mollusks, the chiton, *Ischnochiton*, three pelecypods, and seven gastropods. Material washed from shells was found to contain foraminiferal tests and ostracode valves (Ladd, 1957, p. 627).

In Copano Bay (northeast end), Ladd reported, one shell ridge 20 ft wide remained dry at high tide. It consists essentially of worms and pieces of oyster shells, many riddled by boring forms.

Norris (1953) commented on one reef (Panther Reef, Lower San Antonio Bay) that was five miles long and seldom more than 150 yards wide. He cited that reefs were closely spaced and had vertical thicknesses of 12 to 20 ft. Many reefs stood as low islands during low water.

The oyster biocoenosis of the Texas Bays and Gulf is characterized by a mean temperature range of 16° to 25°C (within this range, *Crassostrea virginica*); regular influxes of fresh water result in salinities of 10 to 30 parts per 1000; relatively shallow depths but not tidal exposure; and a suitable substrate (mud or sandy mud). Natural freshening of the water can result in living relicts of no longer flourishing

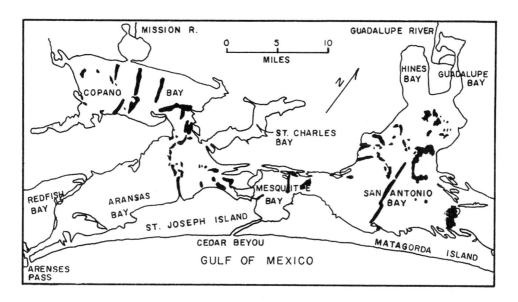

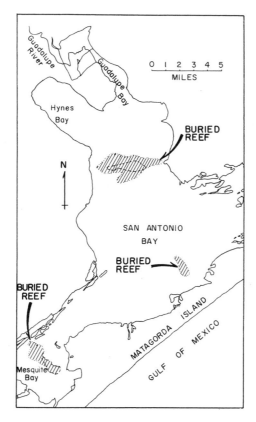

FIGURE 8.6 Shell reefs in some Texas bays. (*A*) Modern reef, blacked areas. (*B*) Buried reefs. (After Norris, 1953.)

oyster reefs. These relicts occur as clumps of oysters called towheads (Hedgpeth, 1953).

Obviously, a biostromal deposit, such as made by oysters, serves as a sediment trap.

Bottom Communities

Thorson (1957) demonstrated (*a*) the uniformity of the level sea bottom with its fauna, plus (*b*) the existence of "parallel communities" in world seas. These communities have common characteristics. They live on the same kinds of substrate, at similar depths; the same genera are found in widely separated communities (but different species depending on geographical regions).

Among these level bottom communities, mollusks play a large role (Figure 1.3). With particular emphasis on mollusks, compositions of these communities are reviewed below.

1. Macoma Communities. Estuarine, from tidal zone to depths of 10 to 60 meters or deeper. Characterizing genera [lamellibranchs—*Macoma*, *Mya*, *Cardium* (in some cases also *Scrobicularia*); worms— the polychaete *Arenicola*]. When silt/mud increases, deposit-feeding *Macoma* and *Arenicola* thrive; where sand increases, *Cardium*, a suspension-feeder, dominates. Localities: East Greenland fjords, Barents Sea, White Sea, and so on, inner Danish waters, the Sea of Bothnia, Aland and the Baltic, San Juan Archipelago in the Vancouver-Seattle region, and Japan.

2. Tellina Communities. Shallow water; inhabit exposed beaches from tidal zone to 5–10 meters deep. Characterizing genera: lamellibranchs—*Tellina*, *Donax*, and *Dosinia*; the echinoderm *Astropecten*; and, in tropical regions, the prosobranchs, *Terebra* and *Ancilla*. Substrate: pure bottoms of generally hard sand. The standing crop comprises several generations. Localities: Scottish coast, Danish and German North Sea coast, northern part of the Adriatic Sea, Black Sea, coastal

zone off Roumania, New England coast, in-shore part of Texas coast, New Zealand, Senegal, French West Africa.

3. Venus Communities. Open sea; on sandy bottoms from 7–10 meters to 30–40 meters in depth. Characterizing genera: lamellibranchs—*Venus*, *Spisula* (or *Mactra*), *Tellina*, and *Thracia*; the prosobranch *Natica* (= *Polynices*); starfish, *Astropecten*, and the polychaete *Ophelia*, plus or minus the irregular sea urchins, *Echinocardium* or *Spatangus*.

In the *Venus* community, changes in the sedimentary substrate will bring one or another form into predominance. Thus, if sand is loose, *Spisula* predominates; if sands harden, *Tellina*; if sands are coarse and mixed with shell gravel, species of *Spisula* and *Tellina* will be larger and their shells will be more solid. Moreover, a characteristic microfauna of small nudibranchs, *Polygordius caecum*, will appear.

Where *Spisula* (or *Mactra*) dominate, productivity reaches among the greatest known for the sea bottom. Localities: circumpolar; inner Danish waters, German Bays, and East Frisian Islands, Dogger Bank, southern North Sea, south of the Isle of Man, off Faro, Portugal, S.W. Italy, middle and inner part of Adriatic Sea, Black Sea, coast of Roumania, Scottish area of North Sea, Plymouth, England, S.W. Ireland, the Faroes, N.W. coast of Sweden, Persian Gulf, coast of Madras, New England coast.

The cited examples indicate the cosmopolitan distribution of molluscan-rich level bottom communities. Other communities characteristic of estuarine or sheltered areas are *Syndomya* community, the *Turritella* and *Cerithium* community (either on sandy mud or the mud of a sand or mud bottom) in fjords from 10 to 38 meters (*Turritella*) or 4 to 17 meters (*Cerithium*). There are communities dominated by foraminifera, amphipods (*Haloops tubicola*), crabs (exclusively from warm seas). There are also level bottom communities known exclusively in Arctic and sub-Arctic waters, and several occur without known parallels elsewhere. One such community of brittle stars and scallops (*Amphilepsis norvegica* and *Pecten vitreus*) is found in the deepest part of the Skagerrak and Oslo Fjord, Norway.

From these examples, it is clear that a *small number of genera* do dominate the infauna of recent seas and that it is possible, as Thorson emphasized (1957, p. 521), to designate at present those few genera that dominate modern level bottom communities as the "index fossils" for the present part of Recent time. Thus an examination of the rock column a few million years from now may be expected to yield fossilized remains of the forms named above. The densest populations represented will correspond to those that, in modern seas, are predominant in their given communities. The faunal associates will be those that occur and/or come to dominate living communities depending on substrate and other ecological factors.

The basis for regional and intercontinental stratigraphic correlations can be better understood as one reviews the localities of the listed mollusk-dominant level bottom communities. Placed in the rock column, the equivalent communities at each named locality, regardless of how remote from any other, would constitute contemporaneous populations. The beds in which they occur would represent time horizons. (The same applies, of course, to the nonmolluscan level bottom communities. Thus a crab or amphipod community might be contemporaneous with a mollusk-dominated community. These, in the rock column, would represent distinct biofacies while being the time equivalents of mollusk beds around the world.)

Sedimentary Environment and Climatic Indicators

Parker (1964) has shown how living macroinvertebrate assemblages (dominantly molluscan but including other such faunal elements as crustaceans and echinoderms) characterize various environments in the Gulf of California and the continental slope of Western Mexico. The assemblages serve as climatic indicators and point to sea level changes. For example, Environment Type VI (outer shelf, 66 to 120 meters, clay bottom, Southern Gulf of California) — of the five common species found in this environment, four were mollusks and one a polychaete worm.

Adaptive Modification

The catalogue of adaptive modifications in the Mollusca is long. The molluscan mantle cavity, a respiratory chamber, still serves that function while being modified into a feeding chamber in bivalves and some gastropods, and into an organ of locomotion in cephalopods and some bivalves. In terrestrial gastropods it serves as a "lung" (Yonge).

In the Gastropoda many adaptations are associated with pressure created by the problem of "sanitation," that is, how to rid the mantle cavity of feces and sediment (Figure 8.1*B*).

In burrowing bivalves the foot and siphons have variously modified (Figure 8.7*A*). Adaptive value favors special attachment mechanisms (cementation, byssus threads, Figure 8.7*C*); protection or nutrient a borehole provides — in rock (Figure 8.7*F*), shell (8.7*D*), or wood (8.7*B*). A wide spectrum of modification covers such items as loss of gills in nudibranch gastropods, protective coloration in nudibranchs and cephalopods (Wells, 1962), and shell pigmentation and banding, a means of waste disposal, or to serve protective, or other functions (Comfort, 1950). Furthermore, shell configurations have altered to solve problems of desiccation (Figure 8.4); changes have arisen from symbiosis (Bowsher, 1956) or other causes. Still other modifications include bypassing of larval stages, uncoiling of shell, bioluminescence in squids, modified mode of reproduction or locomotion (the scallop *Pecten* and living cephalopods move by jet propulsion).

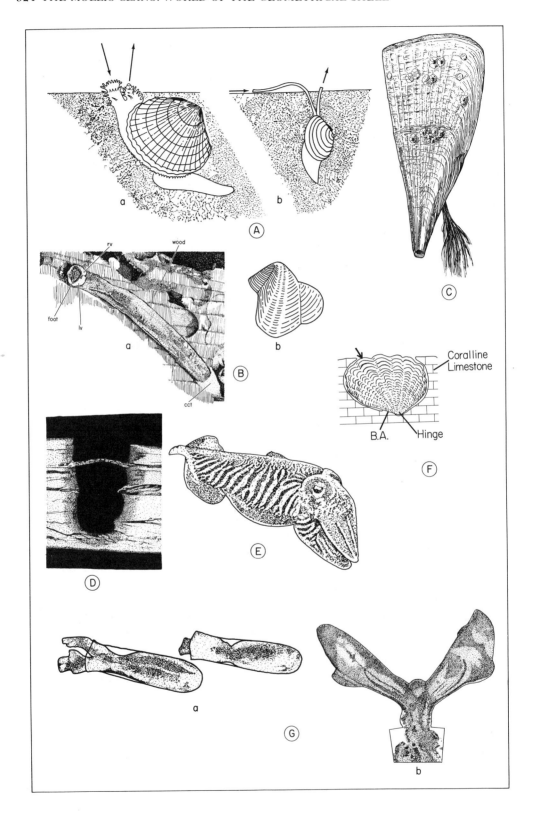

a

b

A

rv   wood

foot

lv

a

B

cct

b

C

b

Coralline
Limestone

B.A.   Hinge

F

D

E

a

G

b

FIGURE 8.7 Adaptive modifications in selected Mollusca.

(A) Sand burrowing bivalves: (a) *Cardium edule,* suspension feeder; arrows indicate in-current and ex-current passage of water through siphons; compare to (b); (b) Tellina tenuis, deposit feeders; compare siphons to (a), X 1.

(B) (a) Wood-boring bivalve, shipworm *Teredo navalis;* cct—secreted calcium carbonate tube encasing entire elongate body, rv—right valve, lv—left valve; siphons exposed to water, in life, X 3.
(b) Left valve of *Teredo* showing ridges with sharp teeth that scrape wood, forming burrows; these valves are contracted by powerful adductor and smaller interior adductor muscles, X 1.5.

(C) *Pinna fragilis* (reduced); long silklike byssus thread for bottom attachment; barnacles attached to valve (after Yonge).

(D) Hole in *Crassostrea virginica* made by the oyster drill *Urosalpinx cinerea follyensis;* microradiograph; white streaks-zones of dense conchiolin; width of holes cut at right angles to shell surface = 1.0 mm.

(E) Cuttlefish, *Sepia*—showing protective coloration.

(F) Boring in coral rock by *Tridacna crocea;* schematic, left side view. Anterior, to left, posterior; to right; B.A.—massive byssus (covered by shell) attachment site. I observed many such borings on the reef flat, Heron Island Great Barrier Reef. Arrow points to surface where mantle is visible through narrow opening between valves which shuts tight when disturbed after expelling a jet of excurrent water.

(G) Marine gastropods (pteropods) with foot modified for swimming; (a) note thin vase-like shell and (b) swimming foot extended.

Illustrations drawn, modified and adapted from following sources: (A) (Bb), Yonge, 1963a (see references under "General"); (Ba) photograph, Buchsbaum, 1948; (D) Carriker et al., 1963. A.A.A.S. Publ. 75; (E) photograph, Raoul Barba cit. Buchsbaum, 1948; (F) Yonge, 1963b, A.A.A.S. Publ. 75; (G) (a), photograph, William Beebe, cit Buchsbaum 1948; (G) (b) Kornicker, 1959.

Brain and Behavior

Wells (1962) studied brain and behavior in the cephalopods *Sepia* and *Octopus*; they are capable of receiving the same sorts of sensory information as we do, but they cannot do much with information. *Octopus* can be taught visual geometry, to discriminate sphere from cube.

FOSSIL MOLLUSKS

Eight Molluscan Classes

As presently sorted, the phylum Mollusca consists of eight classes. Three of these classes are the placophorans (plate-bearers): Monoplacophora (*mono =* single), Polyplacophora (*poly =* many), and Aplacophora (*a =* without). Only the aplacophora has no fossil record. The other five classes include Scaphopoda, Hyolitha, Gastropoda, Pelecypoda, and Cephalopoda. Major attention goes to the last three classes because of their importance in the fossil record.

Class Monoplacophora

Mollusks with cap, spoon-shaped or arched single shell; some groups bilaterally symmetrical; others, longitudinally curved and deviating from symmetry. [Internal metamerism (segmentation) is characterized by pairing of ctenidia, muscles, and other structures.] Geologic range: Lower Cambrian to Recent.

Representatives of the three orders include Order Tryblidioidea: (?) *Scenella* (Figure 8.8A), *Pilina* (Figure 8.8B), *Neopilina* (8.8C); Order Archinacelloidea: *Archinacella* (8.8D), *Hypseloconus* (8.8E); (?)Order Cambridioidea: *Cambridium* (8.8F).

Derived Information from Fossil Monoplacophora

Discovery of **Neopilina** *and its Bearing on the Fossil Record.* C. M. Yonge (1957) characterized Lemche's discovery (1957) of *Neopilina* (see also Clarke and Menzies, 1959) as "a zoological event of the first order." Why? There has always been some doubt that mollusks were related to annelids and arthropods, both of which show metamerism. *Neopilina* is segmented (that is, has internal metamerism) in a way similar to that of annelid worms and arthropods. Thus the doubt about affinities has been removed. As a

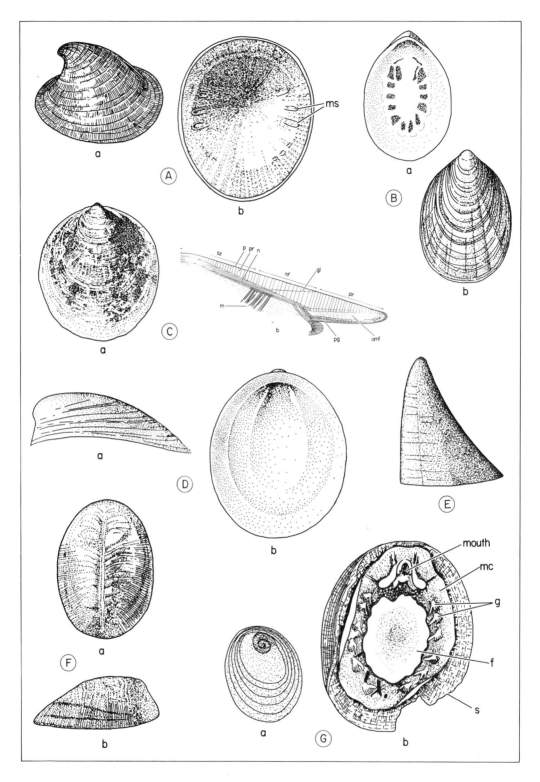

FIGURE 8.8 Class Monoplacophora.

(A) (a) (?) *Scenella reticulata*, L. Cam., Newfoundland, X ca. 3.0; (b) *Scenella* sp., M. Cam., Canada (B.C.); *ms*—muscle scars; X 3.5.

(B) *Pilina unguis*, M. Sil. Gotland; (a) interior; paired muscle scars, compare with (G); (b) dorsal view, X 0.7.

(C) *Neopilina (Neopilina) galathae*, Rec., West Coast, Central America; (a) Dorsal view of shell, X 3.5. (b) N. (N). *galathae*, shell formation and structure, enlarged; *sz*—sterile zone, *p*—periostracum, *pr*—prismatic layer, *nl*—nacreous layer, *nf*—nacre forming zone, *gl*—growth line, *omf*—outer marginal fold, *m*—muscles, *pg*—periostracum gland.

(D) *Archinacella powersi*, M. Ord., Wisc., (a) left side; (b) apertural view, X 1.3.

(E) *Hypseloconus elongatus*, U. Cam., Wisc., left side, X 0.7.

(F) *Cambridium nikiforovae*, L. Cam., N. Asia; (a) dorsal side view of steinkern; (b) right side view; observe internal markings (edges and furrows normal to edges of aperture), X 2.

(G) (a) N. (N). *galathae*, apical part of larval shell (protoconch), asymmetrical coiling, X 3.5; (b) *Neopilina (Vema)* sp. Rec., West Coast, Peru; ventral view with paired gills (g); like *Neopilina (Neopilina)*, but has six instead of five pairs of gills and lacks coiled protoconch, X 5.3.

Sources: (A) (B), (D–F) Knight and Yochelson, 1960; (C) (a) Lemche, 1957; (C)(b) Lemche and Wingstrand, 1959; (G)(a, b) Clark and Menzies, 1959.

result, mollusks, annelids (see Chapter 9), and arthropods (see Chapter 10) are traceable to common segmented ancestors.

Neopilina is a Rosetta stone or a key to inferred but never proven evolutionary relationships between three phyla. Even further, it helps to indicate and clarify intermolluscan affinities. Thus it is a link between the chitons (class Polyplacophora) and the nautiloids (class Cephalopoda); it has a radula positioned like that in the Aplacophora (wormlike Mollusca) and appears to be similar to class Gastropoda only in the dextral coiling of its larval shell (Figure 8.8G, a; Lemche, 1957).

However, a few general features sharply discriminate mollusks from annelid worms and arthropods: restriction of an ectodermal cuticular skeleton to the dorsal side; the muscular foot; and the pallial (mantle) groove with gills (Lemche and Wingstrand, 1959).

Before *Neopilina* was discovered (note that *neo* = new; *Pilina* is the name of a Silurian genus), the monoplacophorans were thought to have become extinct by Middle Devonian time. Furthermore, the presence of an asymmetrically coiled shell (protoconch) was unsuspected (Yochelson, 1958; see Figure 8.8G, a). The discovery has extended the range from the older Paleozoic to the Recent, with a large gap remaining in between; it has given clear evidence of the existence of a radula and of paired gills (muscle scars

but not gill impressions are known in fossil monoplacophorans). It also provides evidence on other internal anatomical structures that must also have characterized Paleozoic fossil forms in life.

Distribution. Fossil monoplacophorans (Figure 8.8) are known from Europe, Northeast Asia, North America, and Australia. Lemche's species of *Neopilina* was dredged off the Mexican coast (see also Menzies and Robinson, 1961); the species of Clarke and Menzies was dredged from the Peru-Chile Trench off Northern Peru.

Substrate. In the North American Upper Cambrian, monoplacophorans are locally abundant in sandstone; other parts of the Cambrian and Ordovician yield forms from limestones. In beds of M. Ordovician through the Devonian, fossils occur in both limestones and calcareous shales (J. B. Knight). *Neopilina (N.) galatheae* lives in dark, clay mud at abyssal depths; it is distributed patchwise in dense clusters. Rasetti (1954) collected *Scenella* (Figure 8.8A) from both Middle Cambrian limestones of the Upper Mt. Whyte formation and in the Lower Cambrian Peyto limestone member of the St. Piran sandstone in British Columbia.

Yochelson (1958) described some Lower Ordovician monoplacophorans from

Missouri. These occurred as steinkerns preserved in chert in Gasconade dolomite.

Bathymetry. Associated fossils found with monoplacophorans are characteristic of Early and Middle Paleozoic epicontinental seas. This contrasts with living *Neopilina*, which inhabit deep waters — 18,000 ft (Vema) and about 11,000 ft (Galathea). There was thus a post-Devonian, and from the evidence (Menzies and Imbrie), a post-Paleozoic migration of some monoplacophorans to deep waters. In turn, that could explain their "sudden" disappearance from the fossil record (M. Devonian) dominated by shallow continental shelf deposits.

Nutrition. *Neopilina* (*N.*) *galatheae* feeds on radiolarians and diatoms in the abyssal depths; crowding of radiolarian scleracoma and centric diatoms in their intestines mix with much undefined detrital matter. Both Lemche and Wingstrand (1959) and Yonge (1957) agree that *Neopilina* is a deposit feeder. There is good likelihood that Paleozoic monoplacophorans similarly organized were also deposit feeders.

Shell Mineralogy. A difference has been noted between the shell mineralogy of Silurian monoplacophorans (*Tryblidium* and *Pilina*) and living *Neopilina*. Lowenstam (1963) found that the fossil forms from Gotland had an outer calcitic and an inner aragonitic layer. All three collections of *Neopilina* species (Schmidt, 1959; Lowenstam, 1963) were found to be composed entirely of aragonite. The difference is attributed to evolution.

Biotic Associates. A common associate of monoplacophorans (such as *Scenella*) in Cambrian beds is trilobites (*Plagiura-Kochaspis* Zone, Mt. Whyte formation; *Bonnia-Olenellus*, St. Piran sandstones). *Pilina* and *Tryblidium* occur in an algal-coral-stromatoporoid reef facies (Silurian of Gotland). With migration to abyssal depths, an array of new faunal associates occur. In the Galathea hauls (Lemche and Wingstrand, 1959, p. 63) along with *Neopilina*, the following were recovered: bathypelagic and bottom-dwelling crustaceans; echinoderms (asteroids, ophiuroids, crinoids, holothuroids); fish; and so on.

Class Polyplacophora

A. G. Smith, 1960, recognizes class Amphineura (old name: Loricata) and two subclasses — the chitins, Polyplacophora, and the vermiform, Aplacophora, which lacks a fossil record. However, C. M. Yonge (1960) proposed that the two subclasses be treated as classes — a proposal followed here. Shore-living mollusks, characterized by an encircling girdle (nude or ornamented with spicules, spines, bristles, or chitinous protuberances, see Figure 8.9*A*, *d*); in the girdle, partly or wholly embedded, are a dorsal series of eight calcareous articulating plates with variable overlap. The plates may be viewed as separate shells. Head differentiated; with flat foot and ventral sole adapted for creeping; gills numerous, positioned in ventral groove between foot and girdle; radula present. Sexes separate. Geologic range: U. Cambrian–Recent.

The class subdivides into two orders: Paleoloricata (primitive chitons with valve structure lacking articulamentum; Figure 8.9*A*, *b* and Neoloricata (with articulamentum).

Order Paleoloricata, in turn, has two suborders, but the Neoloricata has four. Genera are in each order shown in Figure 8.9. [Representative genera include Paleoloricata, suborder Chelodina, genus *Gotlandochiton* (Figure 8.9*B*, *a*); suborder Septemchitonina, genus *Septemchiton* (Figure 8.9*B*, *c*); order Neoloricata, suborder Lepidopleurina, genus *Lepidopleurus* (Figure 8.9*B*, *f*); suborder Ischnochitonina, genus *Chiton* (Figure 8.9*B*, *h*); suborder Acanthochitonina (Figure 8.9*B*, *j*); suborder Afossochitonina, genus *Afossochiton* (Figure 8.9*B*, *k*).

Mineralogy and Valve Structure. The polyplacophoran shell is composed of aragonite (Lowenstam, 1962). The individual "valves" (plates or shells) have four (Knorre, 1925, p. 573) or five layers (Bergenhayn, 1930, pp. 4–13) (Figure 8.9*A*, *b*).

Size. Average size is from 1 to 3 in.; both minute (0.25) and giant species (13 to 14 in.); *Cryptochiton*, from west coast of North America, is such a giant chiton.

Distribution, Habitat, Associates. All marine; sluggish bottom crawlers. Chitons live on or under rocks and some can tolerate silt and soft bottoms. Paleozoic forms seem to have generally preferred soft substrates. They occur in association with other mollusks and are found in reef facies (Permian of West Texas and Silurian of Gotland). Herbivores feed on marine algae. Omnivores may have a varied diet ranging from young barnacles to hydroids and bryozoans. Chitons, characteristic of the littoral zone, can range to 13,800 ft.

Radula. A bilaterally symmetrical lingual ribbon set with chitonous denticles (= radula) is present in most, but not all, chitons. It is used for feeding. Algae are removed from rocky surfaces by the rasping action of the radula. Lowenstam (1962) found that the denticles of chiton radulae are capped with magnitite. It is the very first indication that magnitite or maghemite can be precipitated by a biological agent.

Chitons graze at night and return to a particular position ("home") on a given rock or substrate (homing "instinct"). Lowenstam (1962) suggested that, conceivably, chitons magnetically guide themselves home.

Information Derived from Fossil Chitons Fossil Record and Explosive Evolution. There are about 73 Paleozoic species (22 percent of all known fossil species), 31 Mesozoic species (9 percent of all known

fossil species), and 248 Cenozoic species (Tertiary–Pleistocene) (69 percent of all known fossil species). Over 500 species and subspecies have been described from the Recent. A. G. Smith (1960) refers to this as "explosive evolution" during Recent time.

Expansion of this magnitude, in an evolutionary sense, is sponsored by modern forms having a selective advantage over more primitive Paleozoic types. Particularly this would apply to Paleozoic chitons with weak valve imbrication (overlap or articulation). The valves completely lack projecting structures (insertion plates and so on, see Figure 8.9*B*, *d*). Considering the greater facility afforded by a well-articulated series of valves in remaining attached to rock substrates, and in better protection afforded when "curl-up" was necessary owing to wave action, these valve characteristics could have bestowed a selective advantage on Mesozoic and Cenozoic chitons.

Class Scaphopoda

Bilaterally symmetrical, marine, benthonic mollusks. Body elongate, completely wrapped around by mantle and tusklike shell; head and appendages project from wider anterior opening (= aperture), other end of shell (= apex, the posterior end) extends above the surface of the sea floor; both exhalent and inhalent currents enter the mantle cavity at this end; apex opening may be simple, or variously slitted or notched; head with many prehensile processes (= captacula; Figure 8.10*B*) collects food particles; no eyes or tentacles; foot reduced for burrowing; ctenidia absent; respiration via mantle. Radula short, well developed; separate cerebral and pleural ganglia; sexes separate, reproduction without copulation; trochophore larva. Shell secreted by mantle, calcareous, three-layered, composed of aragonite. Geologic range: (?)Ord.–Dev.–Rec.

Two families are recognized in this class: Dentaliidae (sculptured shell) and

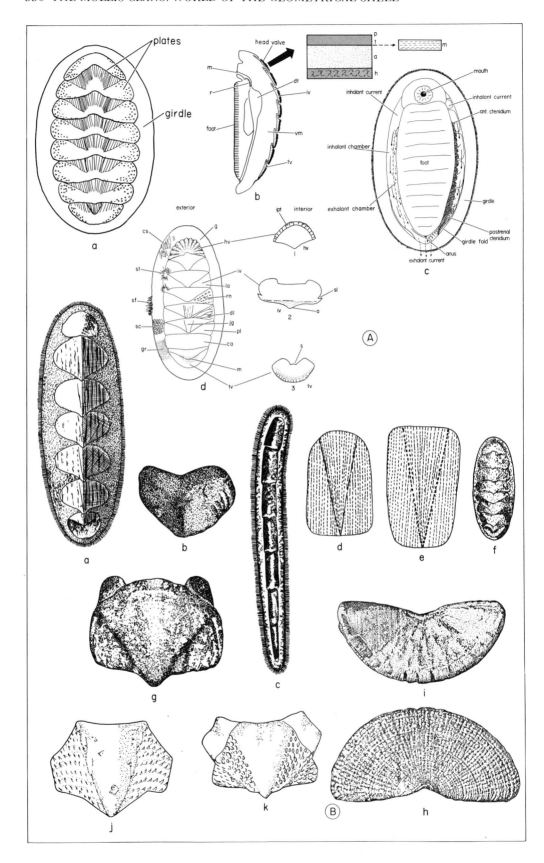

FIGURE 8.9 Polyplacophora.

(A) Anatomical and Shell Features: (a) *Chaetopleura spiculata*, dorsal aspects; *g*—girdle, *pl*—plate; (b) median longitudinal section, valve articulation; *tv*—tail valve; *hv*—head valve; *dt*—digestive tract; *r*—radula; arrow points to section through one plate; *p*—periostracum, *t*—tegmentum, *a*—articulamentum, *h*—hypostracum; *m*—mesostracum; observe tilelike projection by which posterior plate underlies next anterior; *vm*—visceral mass, *f*—foot, *m*—mouth; (c) *Lepidochitona cinereus*, ventral aspect, gills and currents shown; (d) exterior view of skeletal morphology; *cs*—calcareous spines, *sf*—spiculose fringe, *sc*—scales, *gr*—granules, (d3) *tv*—tail valve, head valve radially ridged, (d2)—intermediate valve, *rn*—radially nodulose, *dl*—divaricating lines, *jg*—jugum, *pl*—pleural area, *l*—lateral area, *ca*—central area, *m*—mucro; interior view of valves: (d1) head valve, *ipt*—insertion plate teeth; (d) *iv*—intermediate valve, *sl*—slit, *a*—apex.

(B) Selected fossil genera. (a)(b) *Gotlandochiton* (suborder Chelodina). (a) *G. interplicatsis*, Sil., Gotland, reconstruction, X 1. (b) *G. troedssoni*, intermediate valve, X 3.4. (c)(d)(e) *Septemchiton* (suborder Septemchitonina); *S. vermiformis*, U. Ord., S. Scotland, (c) reconstruction, X 1.6; (d)(e) head and intermediate valves, sculpture indicated a unique wormlike form, possibly living in sand X 8. (f)(g) *Lepidopleurus latodepressus*, Carb., Scotland: (f) reconstruction, X 0.5; (g) dorsal view, intermediate valve, valve sculpture of granules, X 1.5. (h)(i) *Chiton cretaceus* (suborder Ischnochitonina), U. Cret., Tenn.: (h) tegmentum, surface granulose, X 2.6; (i) inside surface; insertion plate teeth, X 2.6. (J) *Acanthochitona casus* (suborder Acanthochitonina), L. Mio., Victoria, So. Australia, intermediate valve, 1.5 mm long, 2.0 mm wide; note ornamentation, (k) *Afossochiton* (T). *magnicostatus* (suborder Afossochitonia), M. Mio., Victoria, So. Australia, intermediate valve, 5.0 mm long, 6.5 mm wide; note carina, beak, and raise diagonal ribs.

Sources: (Aa) M. E. Pierce, 1950; (Ab, d) A. G. Smith, 1960; (Ab-insets) Tasch 1972; (Ac) Yonge, 1939. Quart. J. Micros. Soc. (London), Vol. 81, pp. 367–390; 1960; (Ba–c) Smith, 1960; (Bh, i) Berry, 1940; (Bj, k) Ashby and Cotton, 1939.

Siphonodentaliidae (shell generally smooth and glassy). The Dentaliidae embrace three genera, two of which are extinct: *Dentalium* (Figure 8.10*C*), *Plagioglypta* (Figure 8.10*E*), and *Prodentalium* (Figure 8.10*F*).

[The genus *Dentalium* is subdivided into 16 subgenera based on the apical character of the shell (Henderson, 1920). Figure 8.10*C* is one of these in the subgenus *Antalis*.]

The family Siphonodentaliidae embraces three genera; all are extant: *Cadulus*, *Entalina*, and *Siphonodentalium*. One of the six subgenera of *Cadulus* is illustrated—*Cadulus* (subgenus *Polyschides*) (Figure 8.10*D*).

Information Derived from Living and Fossil Scaphopods

Distribution, Habitat, Associates. Scaphopods are benthonic animals; rarely found in the littoral zone, scaphopods are equally distributed in the neritic and bathyal realms (numerous specimens of *Dentalium* were recovered with the monoplacophoran *Neopilina*). Scaphopods that partly embed in bottom mud or in sand are deposit feeders (C. M. Yonge, 1960). The captacula capture foraminifers and other small organisms. Paleozoic species are reported from limestone, sandstone, and shale associated with other mollusks.

Fossil Record and Recent Expansion. Forms are known from the Russian Ordovician (*Plagioglypta*). There is then a gap in the Silurian followed by a good Devonian representation (*Plagioglypta*, *Prodentalium*). There are several American reports from the Permo-Carboniferous. Among others, the Pennsylvanian Magdalena group, New Mexico (Young, 1942); Millsap Lake formation, northwest Texas (A. K. Miller, 1949); other reports from Colorado (Girty, 1903) and elsewhere (J. B. Knight, 1933, cit. Yochelson, 1957, p. 820) and from the Permian Kaibab formation of Arizona (Nicol, 1944). Mesozoic forms (Gardner, 1878; Richardson, 1906) and Tertiary forms (Julia Gardner, 1947) are widespread.

Modern forms did not appear until early Cretaceous. During the Tertiary, the number of genera and subgenera expanded. However, only in Recent time has this expansion been striking. Lud-

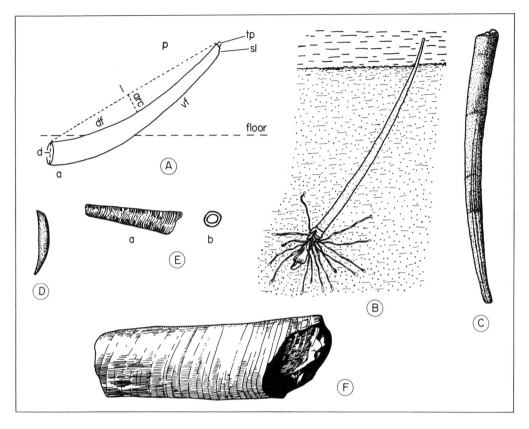

FIGURE 8.10 Scaphopoda: Morphology and fossil types.

(A) Morphological features: *d* – diameter, *a* – anterior, *df* – dorsal face, *l* – length, *tp* – terminal pipe, *sl* – slit or notch, *vf* – ventral face.

(B) *Dentalium* in act of feeding; buried in aquarium silt. Note head with several prehensile processes (= *captacula*) that collect food particles. Exhalent and incurrent currents pass through the apical opening exposed in the water.

Fossil types: (C) *Dentalium* (*Antalis*) *chipolanius*, M. Mio., Fla. Side view, length 40 mm.

(D) *Cadulus* (*Polyschides*) *lobion*, max diam., 0.9 mm, side view, M. Mio., Fla.

(E) *Plagioglypta undulata*, Trias., Aus. (Tyrol), X 1. (*a*) Side view, (*b*) transverse section.

(F) *Prodentalium raymondi*, U. Carb., N. Mex., X 0.5, fragment, side view.

Illustrations redrawn and adapted from: Ludbrook, Smith, Julia Gardner.

brook (1960, Figure 29) noticed that the number of living species exceeds the Tertiary total (see also Recent expansion of polyplacophorans).

The picture of scaphopod development through time with expanding numbers of genera, subgenera, and species indicates great genetic fluidity in the class. The apical character of the shell (presence of slit(s) or notch – Henderson, 1920) distinguishes the subgenera within the two families but is a genetically variable feature.

In turn, variability of apical characters relates to exhalent and inhalent currents which enter and leave at the apical openings. These serve sanitation, respiratory and, in part, nutritional functions. *Those apical characters of the shell that confer a selective advantage in these functions are being favored by natural selection.*

The rock record in future ages may very well show that scaphopods have become a much larger class, more important paleontologically.

Class Gastropoda

Four classes of mollusks have the stem "poda" (Greek *podos* = foot). Gastropoda

(Gr. *gastro* = stomach), Pelecypoda (Gr. *pelekus* = ax); Scaphopoda (Gr. *scaphe* = bowl); Cephalopoda (Gr. *cephalo* = head — this is probably a misnomer). The significance of these names can be seen in Figure 8.2. Pelecypoda are now referred to as Bivalvia.

The class definition covers the following items (after Cox, 1960, and others).

1. Head. Distinct, more or less fused with the foot (Figure 8.11*A*); with eyes and tentacles in unspecialized forms (eyes are absent in the deep-sea group Coccolinacea).
2. Foot. Solelike and adapted for creeping; much modified in pelagic and some other forms (see pteropods, Figure 8.7*G*).
3. Radula. Normally present.
4. Nervous system. Cerebral and pleural ganglia are distinct.
5. Torsion. In some forms this is an early ontogenetic event; in others, condensed ontogeny omits torsion but it is inferred to have occurred in ancestral forms. Result of torsion is a reorientation of organs of the pallial complex (Figure 8.11*D*) compared to their positions in hypothetical ancestral mollusks (compare Figure 8.1*A* and Figure 8.11*D*).
6. Symmetry. Varying degrees of bilateral asymmetry are present in all living representatives. Extinct Bellerophontacea may have had a high degree of bilateral symmetry in early ontogeny. Symmetry may be displayed in the shell alone (*Patella*), in the adult shell alone (*Fissurella*), and asymmetry may be expressed only in the various systems (nervous, circulatory, reproductive, and/or digestive and excretory).
7. Shell. Where present, the shell is single (= univalve), calcareous, closed apically, endogastric when spiral, and lacking regular chambers. (Figure 8.12, 8.13).

    Endogastric is the normal condition in most adult gastropods. It is a mode of coiling in which the shell extends backward from the aperture over the extruded head-foot mass.

    Exogastric is the condition before torsion. It is a mode of coiling in which the shell extends forward from the aperture.

    Irregular septa occasionally seal off the earliest formed parts of the shell. It is this

that distinguishes gastropod shells from those of most cephalopods.

8. Geologic range: Lower Cambrian–Recent.

Classification. The class is divided into three subclasses: (1) Prosobranchia, (2) Opisthobranchia, (3) Pulmonata. These subclasses are differentiated as follows.

1. Gill position. *Forward* in Prosobranchia; *rear* in Opisthobranchia where present. Where gills are absent, the mantle cavity has been modified as a lung (Pulmonata), *branchia* = gills; *proso* = forward; *opistho* = backward; *pulmo* = lung.
2. Torsion. Torsion occurs in subclasses 1 and 3. Detorsion occurs in subclass 2. Each of these subclasses will be considered in turn.

Subclass 1. Prosobranchia. Gastropods that exhibit torsion have varishaped shells; foot commonly bears corneous or calcareous operculum to close the aperture (Figure 8.11*A*); visceral loop twisted into figure eight (which is called the *streptoneurous* condition); mantle cavity opens to front; ctenidium (or ctenidia) in front of the heart; head with single pair of tentacles; sexes distinct except in a few genera. Habitat — fresh water, terrestrial, or marine. Geologic range: Lower Cambrian–Recent.

The subclass Prosobranchia embraces two orders: Archaeogastropoda and Caenogastropoda. Soft-part anatomy forms the basic distinction: nature of the ctenidia (*Aspidobranch*, with two rows of gill leaflets primitively, one in advanced forms, or *Pectinibranch* with one pair of gill leaflets); heart with one or two auricles; and so on. Since fossils show none of these features, how can they be classified? Both orders include extant representatives. Accordingly, comparative study of such shells with fossil equivalents showing similar structures permits classification. Soft-part anatomy is inferred. (See Figure 8.4 for derived information from shell structures.)

Order 1. Archaeogastropoda. Ctenidia "aspidobranch"; lack siphons or proboscis; male without prostate and penis; heart

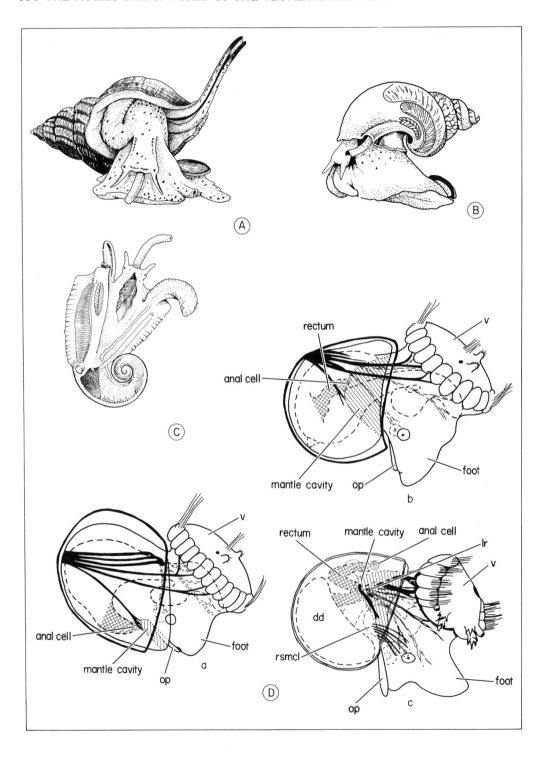

FIGURE 8.11 Morphology of soft part anatomy of the Caenogastropod *Buccinum undatum.*
(A) Note operculum on foot (lower right); head bearing eyes and tentacles; central tube (proboscis) ends in a mouth; inhalent siphon projects on right. (B) Note mantle cavity (that chiefly serves a respiratory function) contains the ctenidium (the second leaf-like structure above proboscis in diagram); see "C" for further view of ctenidium (on left of figure).
(C) Genitalia: penis to right of head; testis occur in outer whorl (upper portion). Heart: auricle and ventricle shown as two adjacent small bodies directly below ctenidium.
(D)(a–c). Stages of torsion in veliger larva of *Patella vulgata.* (a) pretorsion, 70 hours old; (b) 76 hours, 90° torsion (observe position of anal cell and mantle cavity); (c) final position of rectum, mantle cavity and anal cell; (complete rotation of 180° by end of 3½–4 days. Crofts attributed initiation of torsion to action of the asymmetrical retractor muscle). Compare to Da; v – velum, op – operculum, dd – digestive diverticulum, rsmcl – right shell muscle (columellar); lr – larval retractor.
Illustrations redrawn and adapted from Cox, 1960; Crofts, 1955, Zool. Soc. London, Proc. Vol. 125, pp. 711–750, figs 12, 13, 17; Yonge, 1960.

with two auricles; inner layers of shell often, but not always, nacreous. Geologic range: Lower Cambrian–Recent.

The order further subdivides into eight suborders, with one uncertain. [One or more representative genus in each suborder is illustrated in Figures 8.14 and 8.15. Suborder Bellerophontina: *Cyrtolites* (Figure 8.14A), *Bucania* (Figure 8.14 B), *Coreospira* (Figure 8.14C). Suborder Macluritina: *Maclurites* (8.14D). Suborder Pleurotomariina: *Ceratopea* (8.14E), *Trepospira* (8.14F), *Pleurotomaria* (8.14G), *Emarginula (Entomella)* (8.14H). Suborder Patellina: *Scurriopsis* (8.15A). Suborder Trochina: *Holopea* (8.15B), *Platyceras* (8.15C). Suborder Neritopsina: *Naticopsis* (8.15D). Suborder Murchisoniina (*Hormatoma*) (8.15E). Suborder Uncertain: *Bucanospira* (8.15F).]

Order 2. Caenogastropoda. Varishaped, porcelaneous, asymmetrical shell; right ctenidium absent, left, monopectinate (= single set of flattened filaments along axis) and absent in some families; siphon or proboscis often present; usually male forms with penis; heart with one auricle. Habitat: marine, fresh water, terrestrial. Geologic range: Ordovician to Recent.

The order embraces five superfamilies. A few representative genera are illustrated in Figure 8.16. [Superfamily Loxonematacea: *Loxonema* (Figure 8.16A). Superfamily Cerithiacea: *Turritella* (Figure 8.16B). Superfamily Cyclophoracea: *Anthracopupa* (Figure 8.16C). Superfamily Rissoacea: *Rissoa* (8.16D); superfamily

Subulitacea: *Subulites* (8.16E, F).]

From Bilateral Symmetry to Asymmetry. In the Lower Cambrian the oldest known true gastropods first appeared. Family Coreospiridae members (Figure 8.14C) from the Lower Cambrian possessed both bilateral symmetry and coiled shells. Class Gastropoda, it is generally thought, arose from ancestors with complete bilateral symmetry not undergoing torsion. In addition, instead of a shell, the mantle may have contained calcareous spicules (see also Figure 8.1). In the Coreospiridae, another genus, *Latouchella*, the shell spire curves through some 90 degrees. Knight (1952) postulated that perhaps it was the "first" bellerophont and first prosobranch. Yonge (1960), however, surmised that "greater coiling than this" preceded torsion.

At any rate, the first asymmetrical gastropod shells appeared in the Upper Cambrian (Pleurotomariacea and Macluritacea, see Figure 8.14).

How the Orders Relate. Although the Caenogastropoda order clearly derived from the Archaeogastropoda, the precise lines of descent are still hazy. The former appears to have polyphyletic origins [that is, they arose along different descent lines among archaeogastropods instead of one (= monophyletic)]. One descent line widely supported among specialists follows: Pleurotomariacea – Murchisoniacea – Loxonematacea – Cerithiacea (compare Figures 8.14, 8.15, and 8.16 for representative genera in these superfamilies).

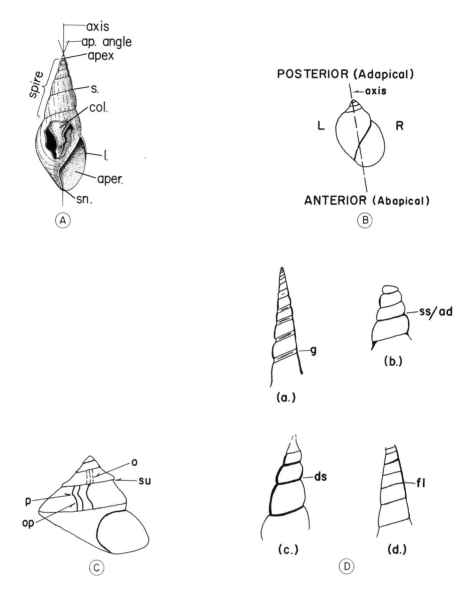

FIGURE 8.12 Morphology, orientation, and other features of gastropod shells. (*A*) Gross morphology: apical angle; apex; *s* — suture; *col* — columella (exposed by break in shell); *l* — lip; *aper* — aperture, *sn* — siphonal notch. (Rotate shell 90° to right for attitude in life, with head-foot projecting from aperture) (compare 8,11*B*). (*B*) Directional map of gastropod shell: L — left, R — right side of axis; any part of high spired shell towards axis-adaxial; any part away from axis-abaxial; approach to posterior is the adapical direction, to anterior the abapical direction on shell. (*C*) *Streptotrochus rugulosus* (U. Sil., Czechoslovakia). Three types of growth lines are shown on the same shell: *o* — orthocline (= perpendicular to suture — *su*), *p* — prosocline (in direction of growth), *op* — opisthocline (opposite to prosocline, away from direction of growth). Two other terms are: proscyrt (lines arch forward; compare prosocline), opisthocyrt (lines arch backwards; compare to opisthocline). Description of growth line orientation enters into taxonomy. Terms may apply to alignment of ridges, tubercles, nodes and so on, as well as lines. (*D*) Some gastropod shell suture types. (See "*s*" in *A* above: (*a*) grooved suture (*g*): (*b*) shallow suture (*ss*): (*c*) deep suture (*ds*); (*d*) flush suture (*fl*); side of whorl flat; suture follows basal helicone angle. (In some forms shallow suture is adpressed that is whorl's outer surface overlaps so as to ultimately converge.)
Illustrations redrawn and adapted from Cox 1960 and others.

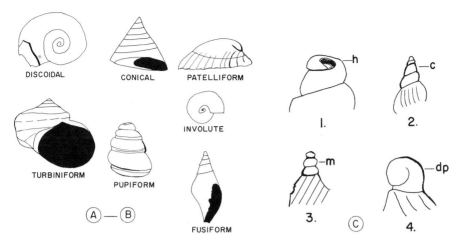

FIGURE 8.13

(A)–(B) Various gastropod shell configurations-Names. (For other configurations: trochiform, see Figure 8.14G turriculate, Figure 8.12D). (Several sources, including Tasch collections).
(C) Some coiling types in gastropod apical whorls (= protoconch). (1) Heterostrophic (h) – numerous types. (2) Conical (c) – can be steep or broadly conical. (3) Mammillated (m). (4) Paucispiral (= few whorls), deviated type (dp).
Data from Cox, 1960.

Subclass 2. Opisthobranchia. Gastropods exhibit detorsion (that is, straightening out of the nervous system – the *euthyneurous* condition); ctenidium, if present, are displaced to the right side and rear of body; shell tends to be reduced or absent; operculum commonly absent in adults (except in the primitive genus *Acteon*); mantle cavity becomes shallow, migrates to right and is lost with shell; hermaphroditic; with retractile penis; exclusively marine; free-swimming veligers show torsion and have operculum (Figure 8.3C). Geologic range: (?)Devonian; Mississippian to Recent.

Where a shell is present, it is either *aciculate* (slender, tapered to sharp point) as in the family Pyramidellidae or, often, low-spired, involute or convolute. (*Involute:* last whorl envelops earlier ones; *convolute:* last whorl completely embraces and conceals earlier ones.) Apical whorls or nuclear whorls of the embryonic shell (*protoconch*) coiled in opposite direction to rest of shell. This condition is referred to as *heterostrophic*.

This subclass, although prominent in present seas, has a relatively poor representation in the fossil record. Following Knight (1944) and others, it is subdivided into three orders with a fossil record: Pleurocoela, Pteropoda, and Nudibranchia. An additional order Uncertain (following Cox, 1960a), is also included.

Order Pleurocoela. (This order is often called the Tectibranchia.) Shell commonly dextral (right-handed), involute or convolute; both shell and mantle cavity tend to become obsolete. Geologic range: Mississippian to Recent. Superfamily Acteonacea: *Acteonina* (Figure 8.17A), *Acteonella* (Figure 8.17B).

Order Uncertain. This order embraces the superfamily Pyramidellacea. Living species of Pyramidellacea have been shown to be opisthobranchs; some are highly specialized ectoparasites lacking a radula (Cox, 1960a). Fossil forms under this superfamily show no evidence of lack of a radula or of parasitism. Geologic range: (?)Devonian; Mississippian – Recent.

[Superfamily Pyramidellacea: *Donaldina* (Figure 8.17C); *Streptacis* (Figure 8.17D).]

Order Pteropoda. Pelagic mollusks, naked or covered with thin, generally transparent, varishaped shell; often operculate. (Living pteropod shells – see Fig-

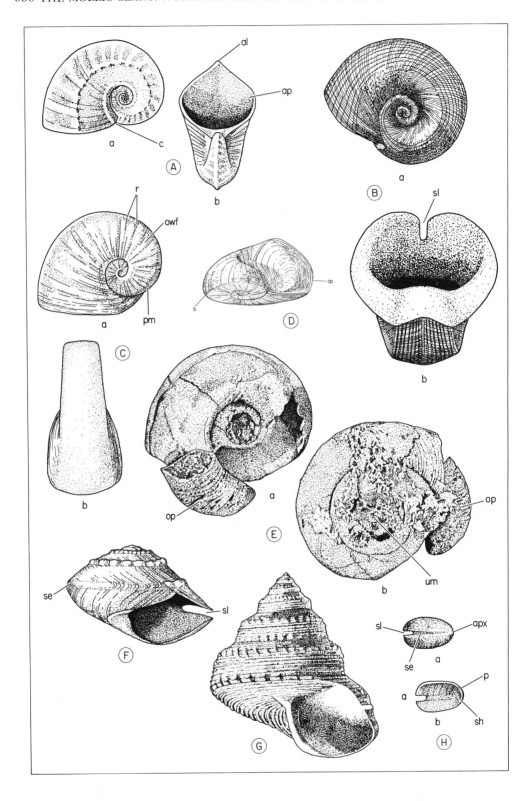

FIGURE 8.14 Representative fossil genera of order Archeogastropoda.
Suborder Bellerophontina: (*A*) *Cyrtolites ornatus*, U. Ord., N.Y., X 2; (*a*) left side, (*b*) apertural side; *al*—anterior lip with sharp angular sinus, *ap*—aperture, *c*—sharp median carina. (*B*) *Bucania sulcatina*, M. Ord., N.Y., X 1.3; (*a*) left side, (*b*) apertural side; *sl*—slit. (*C*) *Coreospira rugosa*. L. Cam., Korea; (*a*) left side, (*b*) anterior, X 5; *r*—rugae, *owf*—flattened outer whorl face, *pm*—protruding margin; note bilateral symmetry and coiling.
Suborder Macluritina: (*D*) *Maclurites logani*, M. Ord., Canada, X 0.3, apertural view; *op*—operculum in place, *b*—flat base.
Suborder Pleurotomariina: (*E*) *Ceratopea unguis*, L. Ord., Arkansas, max. diam., 24.0 mm; (*a*) top view showing operculum (*op*) glued into presumed natural position, (*b*) umbilical view, *um*—umbilicus. (*F*) *Trepospira sphaerulata*, U. Penn., Ill., X 2; *se*—selenizone (= spiral band of crescentic growth lines or threads generated by narrow notch or slit), *sl*—slit. (*G*) *Pleurotomaria anglica*, L. Cret., France, X 0.3; note trochiform configuration and slit that generates selenizone near midwhorl. (*H*) *Emarginula* (subgenus *Entomella*) *clypeata*, Eoc., France, X 6; (*a*)–(*b*) two views; *sl*—slit, *se*—selenizone, *apx*—apex, *a*—anterior, *p*—posterior, *sh*—projecting shelf.
Illustrations redrawn and adapted from following sources: (*A*)–(*D*), (*F*), (*G*) Knight, Batten, Yochelson, 1960; (*E*) Yochelson and Bridge, 1957.

ure 8.17*E*—accumulate in quantities to form "pteropod ooze.") Geologic range: Mesozoic–Recent.

Representative genera: *Vaginella* (Figure 8.17*E*); *Clio* Figure 8.17*F*.

Order Nudibranchia. Ctenidium lost, usually replaced by adaptive gills or dorsal processes; shell absent except in rare *Sacoglossa* (Figure 8.1*C*). Known only from the Recent.

Subclass 3. Pulmonata. Mantle cavity acts as a lung; without ctenidium; nervous system in adults becomes symmetrical following torsion (that is, the central nervous system has straightened out—the euthyneurous condition); usually with well-developed shell; radula with teeth reduced; mostly terrestrial and freshwater; hermaphroditic. Geologic range: Mesozoic to Recent.

Specialists on Paleozoic gastropods do not recognize any pulmonates in the pre-Mesozoic record. Cox (1960*a*, I, p. 146) established that unquestionable Basommatophora (such genera as *Lymnaea*, *Physa*, *Planorbis*, and *Ellobium*) first appear in abundance in late Jurassic fresh-water deposits. Earliest reliable reports of land pulmonates (Stylommatophora) are from the Upper Cretaceous deposits in Southern France.

Order 1. Basommatophora. Pulmonates with shell (spiral, cap, or bowl-shaped) and single pair of noninvaginable tentacles with eyes at base. Genital orifices

separated. Habitat—intertidal marine, coastal and inland terrestrial, and freshwater. Representative genera include *Ferissia* (family Ancylidae) (Figure 8.18*D*); *Planorbis* (Figure 8.18*E*); *Physa* (Figure 8.18*F*).

Order 2. Stylommatophora. Pulmonates with helical shell (see *Helix*, Figure 8.18*A*) may be reduced to calcareous granules (*Arion*); two pairs of invaginable tentacles with eyes borne on tips of posterior pair. Genital orifices open into a common vestibule. The order is entirely terrestrial, and numerous species show great adaptive radiation (that is, exploitation of varied land habitats). Representative genera include *Helix* (Figure 8.18*A*); *Lymnaea* (8.18*B*); *Succinea* (8.18*C*).

Origin of the Pulmonata. C. M. Yonge (1960) and J. E. Morton (1955) attribute the greatest probability to pulmonate origin through the primitive family, the Ellobidae (see Figure 8.18*G–I* for recent representative genera of this family).

INFORMATION DERIVED FROM FOSSIL GASTROPODS

Temperature

Extrapolation of temperature ranges of living marine gastropods to Tertiary equivalents sometimes can be achieved on the basis of comparative shell morphology (Figure 8.19*B*, *C*). The littoral genus

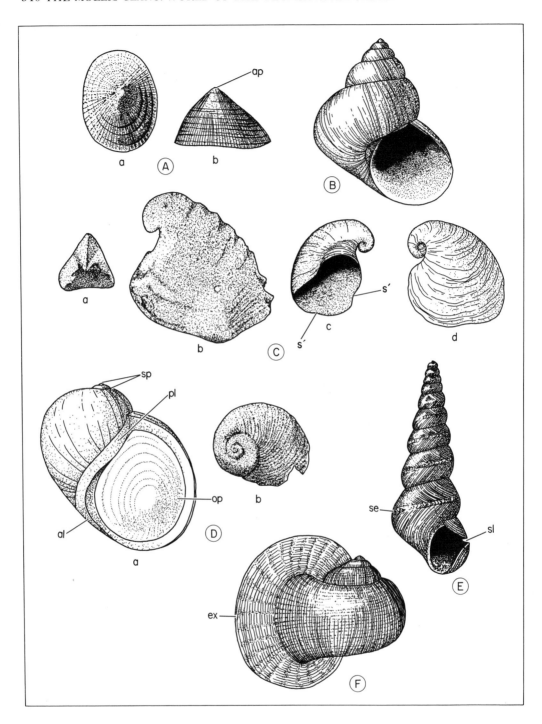

FIGURE 8.15 Archeogastropoda.
Suborder Patellina: (*A*) *Scurriopsis* (subgenus *Scurriopsis*) *neumavri*, L. Liassic, Sicily, X 1; (*a*) from above, (*b*) left side, note broadly elliptical shell and position of apex (slightly anterior to median).
Suborder Trochina: (*B*) *Holopea symmetrica*, M. Ord., N.Y., X 2.7; note turbiniform configuration, rounded whorls. (*C*) *Platyceras* (subgenus *Euthyrachis*) *indianense*, Four Mile Dam Is., M. Dev., Michigan, X 1; (*a*) anterior view showing sharp carina and untorted beak, (*b*) left side of mature individual showing three spine bases on apertural extremity; (*c, d*) *P.* (subgenus *Platyceras*) *vetustum*, Miss., Ireland, X 0.7: (*c*) oblique apertural view, *s's'* — sinuate apertural margin, (*d*) apical view. Note coiled early whorls and capuliform configuration (= a simple depressed cone with eccentric apex and near-apical part of shell slightly coiled).
Suborder Neritopsina: (*D*) *Naticopsis*; (*a*) *N.* (subgenus *Naticopsis*) *phillipsi*, L. Carb., Ireland; *op* — operculum in place, *sp* — spire, *pl* — parietal lip (= part of inner lip); *al* — part of the inner lip; terminal part of columella (= solid or hollow pillar surrounding axis of coiled shell); (*b*) *N. kaibabensis*, silicified specimen: top view, X 3.
Suborder Murchisoniina: (*E*) *Murchisonia* (subgenus Hormatoma) *gracilis*, M. Ord. Canada, X 1; *sl* — slit, *se* — selenizone (see Figure 8.14F for definition).
Suborder "Uncertain". (*F*) (?) *Bucanospira expansa*, M. Sil., Texas, X 1; *ex* — explanate (= outspread and flattened apertural lip).
Illustrations redrawn and adapted from following sources: (*C, a, b*) Tyler, 1965; (*D, b*) Chronic, 1952; others, Knight, Batten, Yochelson, 1960.

Ceratostoma, C. delorae (Figure 8.19*A*, *a*) from the Middle Miocene in California bears closest morphological resemblance to Recent species of the genus that range northward to Alaska (Figure 8.19*A*, *b*).

Spoor, Color Markings

Spoor. Living gastropods are known to leave characteristic trails or tracks in crawling over firm, wet substrates such as sand flats exposed at low tide. Locomotion in gastropods is controlled by muscular contraction of the sole of the foot. Contractions occur in a series of waves — either a single series of waves (*monotaxic*), a series in each half of the sole (*ditaxic*), or bipartite waves in each half of the sole (*tetrataxic*) (Cox, 1960*a*).

Figure 8.19*D* shows a ditaxic track with median groove made by modern *Nucella lapillus*. Figure 8.19*E* shows a portion of a fossilized ditaxic trail in a Pennsylvanian sandstone. The fossil trail matched that made by living *Littorina* (Powers, 1922).

Several workers have described trails and burrows ascribed to fossil gastropods. Among others, Fenton and Fenton (1931*a*, 1931*b*, 1937) described and figured such presumed Paleozoic spoor.

Bucher (1938) uncovered a brachiopod layer in the Richmondian (Upper Ordovician) of Ohio; numbers of *Dalmanella meeki* valves had a round gastropod bore hole attributed to *Lophospira*. Mechanical

action of the radula can account for these holes. The common American oyster drill (*Urosalpinx cinerea*) is a modern predator on oysters; it kills 30–100 oysters in a single season. (Figure 8.19*F*).

The Tertiary also contains numerous records of such predation on pelecypods. Also selected nonmolluscan shells in the Paleozoic seas were attacked by gastropod predators.[2] Such data help to establish some food chain relationships in ancient seas.

Color Markings and Their Significance. Foerste (1930), Girty (1912), and J. B. Knight (1933), among others, have reported retention of color bands in late Paleozoic gastropods. Color retention has been recorded in less than 300 fossil gastropod species. Fewer than 50 of Paleozoic age (chiefly Carboniferous): *Holopea harpa* (Ordovician) is one of the oldest examples.

Color bands are, of course, due to pigments. Comfort's (1950) study of the biochemistry of molluscan shell pigments indicated that the more primitive molluscs (Archeogastropoda, among others) include porphyrins, indigos, and unidentified acid-soluble pigments. Also he noted in the archeogastropods, as in the lower

[2] May be polychaete worm burrows (Richards and Shabica, 1969, J. Paleo., 43(3):838).

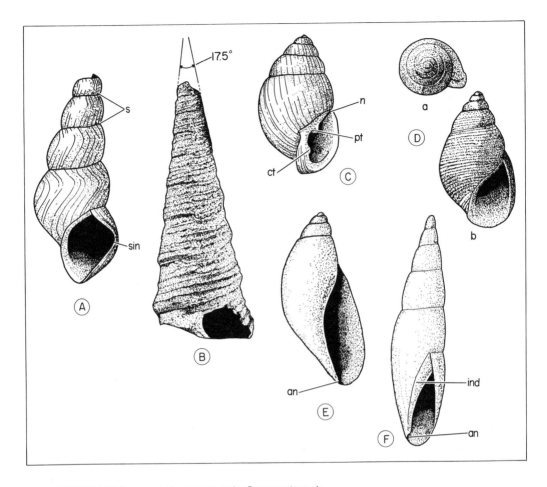

FIGURE 8.16 Representative genera, order Caenogastropoda.

(A) *Loxonema sinosum* (M. Sil., Eng.), X 1; s—suture, sin—sinus.

(B) *Turritella morteni*, top layer (Ratcliff Wharf, Va., Eoc.), greater diam. 18.0 mm, apical angle = 17.5°.

(C) *Anthracopupa ohioensis* (L. Perm., Ohio), schematic; n—notch on inner side of labrum (outer lip), pt—parietal tooth, ct—columnar tooth; note pupiform configuration, that is, like insect pupa.

(D) *Rissoa litiopaopsis*, Alum Bluff Group, Fla., M. Mio. (a) Apical view, (b) apertural view, ht. 3.4 mm, diam. 1.9 mm.

(E)–(F) *Subulites*. (E) *Subulites* (subgenus *Cyrtospina*) *tortilis* (M. Ord., Tenn.), X 1; note curved axis of shell and that last whorl occupies more than one-half total shell height; an—anterior notch. (F) *Subulites* (subgenus *Subulites*) *subelongatus* (M. Ord., N.Y.), X 0.3; ind—inductura (smooth shelly layer extending from inner side of aperture over parietal region, and so on), an—anterior notch.

Illustrations redrawn and adapted from following sources: (B) Palmer, 1937 (See "Scaphopoda" for reference); (D) Gardner, 1947; others, Knight, et al., 1960a.

bivalves, that shell pigments deposited in the shell are a means of waste disposal. Shell pigmentation functions as protective coloration. The presence of particular pigments in some gastropod and other molluscan shells may result from diet. The appearance time of particular shell pigments is known to be genetically determined in some forms. Relative albinism,

occurring in some individuals of the snail *Helix aspersa*, is attributed to tyrosinase deficiency in the secreting tissue.

Symbiosis and Adaptive Modification

A symbiotic relationship between Paleozoic platycerid gastropods (including *Cyclonema*, *Naticonema*, and *Platyceras*—

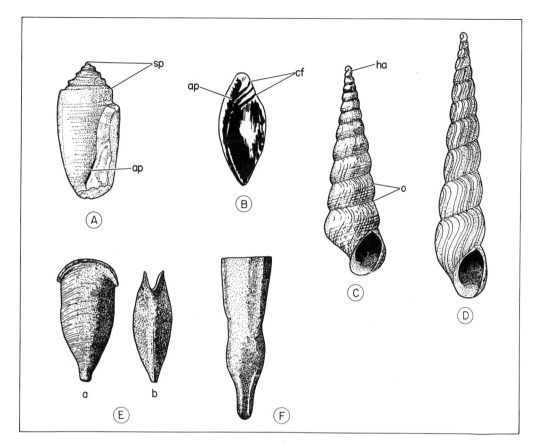

FIGURE 8.17 Representative genera of subclass Opisthobranchia.
Order Pleurocoela: (*A*) *Acteonina carbonaria* (L. Carb., Belgium), X 4; *sp* — short graduate spire, *ap* —
high narrow aperture. (*B*) *Acteonella ghazirensis*, Cenomanian-Turonian, schematic.
Order Uncertain: (*C*) *Donaldina grantonensis* (L. Carb., Scotland), X 10, apertural view; *ha* — hetero-
strophic apex, *o* — ornament confined to lower part of whorl. (*D*) *Streptacis whitfieldi* (M. Penn., Illi-
nois), X 10, apertural view, compare (*C*).
Order Pteropoda: (*E*) *Vaginella chipolana* (Mio., Fla.), X 6; *a* — ventral, *b* — lateral, aperture constricted; apex blunt, closed by septum, (*F*) *Clio* (subgenus *Creseis*) *hastata* (Eoc. — Oligo., Ala.,
Miss.), X 50; small inflation characteristic of immature shell, (See Figure 8.7*G,a–b*).
Illustrations redrawn and adapted from several authors.

Figure 8.20*A*), crinoids, and cystoids is well documented (Bowsher, 1955, 1956, and see Treatise, Mollusca I, pp. I-239).

Platycerid fossils commonly occur on the calyxes over the anal vent (Figure 8.19*G*). In life, the mantle cavity would have covered the anal opening of the crinoid host. Also, the platycerid aperture closely conforms to the shape of the calyx. These observations point to a coprophagous habit. The attached parasitic platycerids apparently lived on a fecal pellet diet emitted by the crinoid host.

Great variation in platycerid shell shape and coiling arose from this symbiotic relationship. The shape and ornamentation of the crinoid host importantly influenced both type and number of reentrants and salients of the platycerid aperture. The initial regularity and coiling of the shell (close helicoidal) (Figure 8.20*B*) was gradually lost through time; the result, irregular uncoiled shells (Figure 8.20*C*).

The coprophagous habit among platycerids probably arose in Ordovician time coincident with the appearance of the crinoid host. Populations of camerate crinoids (see Chapter 12) having a well-developed *tegmen* readily available for

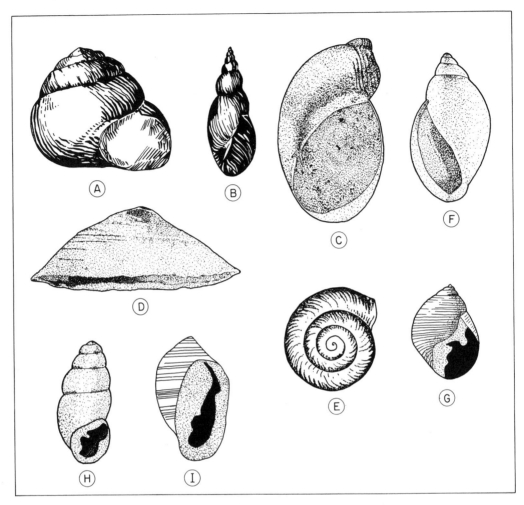

FIGURE 8.18 Representative genera of subclass Pulmonata.
Order Stylommatophora: (A) *Helix globosa* (Oligo., Isle of Wight), X 0.66. (B) *Limnaea longiscata* (Oligo., Isle of Wight), X 0.66. (C) *Succinea grosvenori* (Pleist., Peorian loess, Kansas), X 4.
Order Basommatophora: (D) *Ferissia parallela,* patelliform shell (Pleist., New England), X 7.
(E) *Planorbis euomphalus* (Oligo., Isle of Wight), X 0.66. (F) *Physa anatina,* Laverne fm. (L. Plio.), X 4.
Living genera of three subfamilies of the primitive family Ellobidae (see text): (G) *Pedipes afer* (subfam. Pedipedinae). Recent, enlarged (H) *Carychium minimum* (subfam. Carychiinae). Recent, enlarged. (I) *Cassidula mustelina* (subfam. Ellobiinae), Recent.
Illustrations redrawn and adapted from Neaverson, Leonard, and Thiele.

FIGURE 8.19 Information derived from living and fossil gastropods—I.
(A) (a) *Ceratostoma delorae,* M. Mio., Calif., X 1. (b) *Ceratostoma foliatum.* Rec., Alaska, X 1.
(B)(C). (B). Geologic range of Ceratostoma species (Miocene-to-Recent).
(C) Temperature range (degrees Centigrade) of Recent *Ceratostoma* species. (Extrapolate temperature shown back through Pleistocene and Tertiary).
(D) *Nucella lapillus,* track.
(E) Portion of fossilized ditaxic trail, Pennsylvanian.
(F) *Dalmanella meeki* with gastropod boring.
(G)(a) *Platyceras*—of coprophagus habit attached to the crinoid *Platycrinites,* (b) attached to *Ulocrinus* (arm of crinoid removed in drawing).
Illustrations redrawn and adapted from following sources: (A)–(C) Hall, 1959; (D) Cox, 1960a; (E) Powers, 1922; (F) Bucher, 1938; (G) Bowsher, 1956.

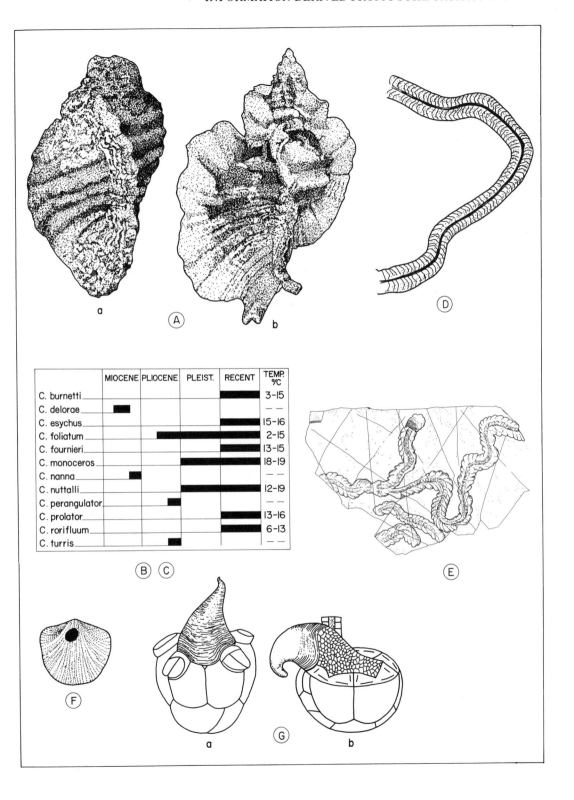

	MIOCENE	PLIOCENE	PLEIST.	RECENT	TEMP. ⁰C
C. burnetti				▬	3-15
C. delorae	▬				— —
C. esychus				▬	15-16
C. foliatum		▬		▬	2-15
C. fournieri				▬	13-15
C. monoceros			▬	▬	18-19
C. nanna	▬				— —
C. nuttalli			▬	▬	12-19
C. perangulator			▬		— —
C. prolator				▬	13-16
C. rorifluum				▬	6-13
C. turris			▬		— —

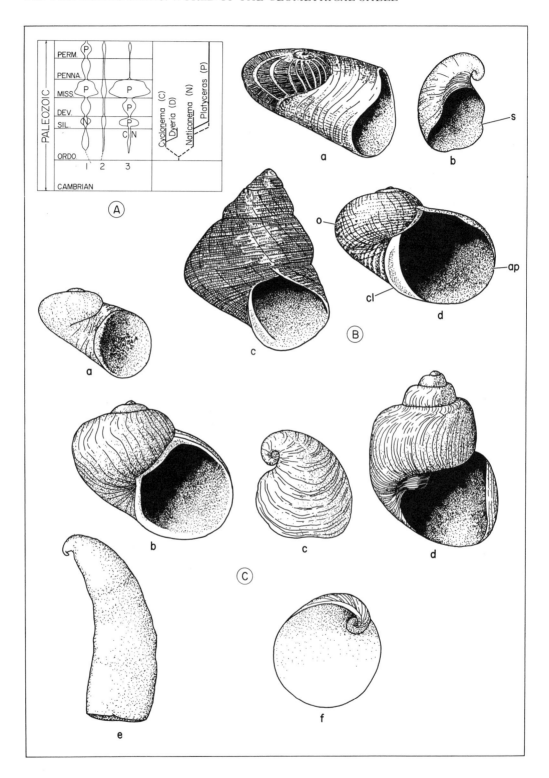

FIGURE 8.20 Information derived from fossil platyceratids—II.
(*A*) Geologic range and phylogenetic relationship of the platyceratid genera and comparative range and abundance of crinoid hosts: 1. Inadunata. 2. Flexibilia. 3. Camerata.
(*B*) Related platyceratid genera: (*a*) *Cyclonema* (subgenus *Dyeria*), X 1. (A third subgenus of *Cyclonema* not shown is *Ploconema*.) (*b*) *Platyceras* (subgenus *Platyceras*) oblique apertural view, X 0.7. (*c*) *Cyclonema* (subgenus *Cyclonema*) X 2. (*d*) *Naticonema*, X 2.7.
(*C*) Subgenera of *Platyceras*: (*a*) *P.* (*Tubomphalus*), X 1.3. (*b*) *P.* (*Platyostoma*). (*c*) *P.* (*Platyceras*) (same as *B, b*); apical view. (*d*) *P.* (*Visitator*) X 0.7. (*e*) *P.* (*Orthonychia*), a steinkern. (*f*) *P.* (*Praenatica*), X 0.7, view of aperture; *s*—apertural margin deeply sinuate, *cl*—columnar lip, *o*—ornamentation.
Illustrations redrawn and adapted from Bowsher, 1955, Knight, Batten, Yochelson, 1960.

platycerid attachment had diminished considerably by the close of Mississippian time. This compelled the coprophagous Pennsylvanian-Permian platycerids to attach to the inadunate crinoids where spaces between the arms limited available space for anal attachment. Corresponding to this change in host, there was increased variability of aperture and shell shape in late Paleozoic forms compared to the Devonian-Mississippian platycerids.

One would expect that crinoid hosts and coprophagous symbiont platyceroids should become extinct together. That apparently did occur at the close of Permian time. Coprophagy in marine gastropods living as symbionts on Mesozoic or Tertiary crinoids is unknown, although certain Recent noncoprophagous gastropods in the family Melanellidae (Thiele, 1931, I, pp. 226 ff.) live either a free existence or as true parasites in association with echinoderms.

Platyceras shells must have become detached from their hosts when the crinoid plates separated and scattered upon the demise of the host. This would account for isolated shells occurring as fossils. These detached shells found as fossils, according to Bowsher, might allow identification of the type of crinoid host (camerate or inadunate) to which, in life, the shell had been attached. Study of apertural irregularities would be the critical factor in arriving at such determinations.

Figure 8.20*C* illustrates the six subgenera of the genus *Platyceras*. Extreme variability is attributed by specialists to the stationary habit, that is, attached to crinoid calyxes (Knight et al., 1960, I, pp. 239 ff.). One wonders about the dynamics of this variability in shell configuration and apertural characteristics even with evidence on change in crinoid hosts (camerate to inadunate) in post-Mississippian time. How was such variability established?

Given *n* platycerid larvae in the sea in which a crinoid "garden" grows, there is some evidence that a suitable place on the calyx for larval attachment may have been a selective factor, although generally there does not seem to have been marked selectivity of crinoid hosts on the generic or specific level. Furthermore, it is likely that the coprophagous habit was also a random acquisition That is to say, *n* larvae, having affixed to crinoid calyxes over the anal vent, were able to survive on a diet of crinoid waste. Repetition of this attachment and diet from generation to generation established coprophagy.

From one generation to another within a given *Platyceras* species, selective pressures apparently operated in a way that sinuate apertural margins and uncoiling of shell, among other features, were favored. Thus individual platycerids would be successful or unsuccessful in the coprophagous habit to the extent that irregularities of their apertural margins conformed to those of their crinoid hosts. Natural selection would then weed out those individuals whose genetic message was inadaptive in this regard.

Siliceous Residues

Yochelson (1956) based his Permian gastropod study in Southwestern United States largely on siliceous residues remaining after acid digestion of 45–50 tons of limestone. New evidence on formation

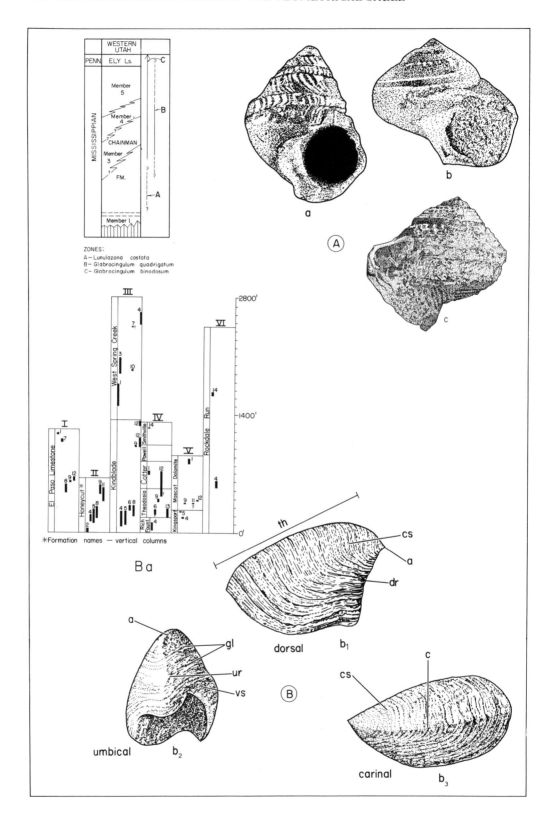

WESTERN UTAH

PENN. ELY Ls.

MISSISSIPPIAN

Member 5

Member 4

CHAINMAN FM.

Member 3

Member I

ZONES:
A — Lunulazona costata
B — Glabrocingulum quadrigatum
C — Glabrocingulum binodosum

a

b

c

A

2800'

1400'

0'

III West Spring Creek

VI Rockdale Run

I El Paso Limestone

II Honeycut*

Kindblade

IV Cotter Powell Smithville

Rich Theodosia Fount.

V Mascot Dolomite

Kingsport

*Formation names — vertical columns

B a

th

cs

a

dr

dorsal

b₁

a

gl

ur

vs

umbical

b₂

cs

c

carinal

b₃

B

FIGURE 8.21 Information derived from gastropods—III.
(A) Gastropod biozones (Chainman fm. W. Utah, Mississippian) Generalized columnar section.
(a)–(c) Biozone fossils: (a) *Lunalazona costata*, (b) *Glabrocingulum quadrigatum*, (c) *Glabrocingulum binodosum*, X 2.
(B) *Ceratopea*: (a) stratigraphic occurrence, Localities: I. West Texas, II. Llano Uplift. III. Arbuckle Mts. IV. Ozark Uplift V. Eastern Tennessee. VI. Western Maryland. Species: 1. *C. ankylosa*. 2. *C. aretina*. 3. *C. buttsi*. 4. *C. capuliformis*. 5. *C. corniformis*. 6. *C. germana*. 7. *C. hami*. 8. *C. incurvata*. 9. *C. keithi*. 10. *C. knighti*. 11. *C. subconica*. 12. *C. tennesseensis*. 13. *C. torta*. 14. *C. unguis*. (b)(1–3) morphology of *Ceratopea* opercula, (1) dorsal view, (2) umbilical view, (3) carinal view, schematic; *th* — thickness, *cs* — carinal surface, *ur* — umbilical ridge, *a* — apex, *dr* — dorsal ridge, *c* — carina, *gl* — growth lines, *vs* — ventral surface.
Illustrations redrawn and adapted from (A) Sadlick and Nielsen, 1963; (B) Yochelson and Bridge, 1957.

age plus a tremendous increase in excellent specimens was obtained by Haas (1953) in his study of acid-etched late Triassic gastropods from Central Peru.

Biofacies, Ecological Units, Stratigraphy

Numerous studies on faunas of the Pennsylvanian-Permian provide ecological data on gastropods (Yochelson, 1956; McCrone, 1963; Johnson, 1962; Chronic, 1952, among others). Yochelson (1956) found that the majority of the Permian West Texas gastropods lived in algal patches or algal reefs in shallow water. This study suggested possible distinction between limestone vs. shale gastropods. McCrone (1963) studied the Red Eagle Cyclothem (Lower Permian of Kansas), reported occurrence of gastropods in algal osagites, and thought greater abundance of gastropods in limestones denoted carbonate preference.

Johnson (1962) attempted statistical recognition of Pennsylvanian ecological units using fossil data found in cyclothem collections (63 species in 152 collections). Through punch-card manipulation, he was able to discriminate objectively 19 groups of mutually associated species. For example, Group II contained five gastropod species (*Cymatospira, Euphemites, Glabrocingulum, Pharkidonotus,* and *Strobeus*), two pelecypod species (*Astartella, Nuculopsis*), one cephalopod species (*Pseudorthoceras*), and an articulate brachiopod species (*Crurithyris*).

Subsequent analysis of these groups led to three inferred large ecological associations (or units); two were brachiopod and one, gastropod. The "gastropod association" represented a grouping of near-shore inhabitants that lived on a substrate firmer than that of the brachiopod *Orbiculoides*.

The molluscan fauna (pelecypods and gastropods) of the Kaibab formation (Permian, Arizona) is divisible into two groups: (1) smooth, thick-shelled, large forms, and (2) finely ornamented thin-shelled forms. The dual nature of mollusks suggested that each group inhabited a different bottom environment. Generally, Kaibab mollusks lived in warm, shallow water, less than 100 ft in depth.

A study of the ontogenetic variation of some pleurotomarian gastropods from western United States indicated three gastropod biozones in the Mississippian Chainman formation (Figure 8.21A). These biozones correlate and overlap previously established goniatite and brachiopod zones (Sadlick and Nielsen, 1963).

Yochelson (1964a) described a new gastropod genus, *Modestospira*, from the "Orthoceraskalk," Island of Bornholm, Denmark. Although he had 12 fossil collections from this region, *similarity of matrix and preservation* suggested their restriction to one and the same stratum. This inference, in turn, indicated correlations with Swedish and English beds.

The massive calcareous opercula of *Ceratopea* (Figure 8.21B), often silicified, have proved useful in regional correlations in Eastern and southwestern United States, and may prove equally valuable for intercontinental correlations: Greenland, Spitzbergen, Northern Scotland, and Newfoundland. Yochelson (1964*b*) concluded from a study of East Greenland forms of this genus that morphologic development of given *Ceratopea* opercula can permit "prediction of the general level at which these forms may occur." Thus the Greenland specimens (Narwhale Sound Formation) seem to have occurred in a zone that is an age-correlate of the middle third of the Kindblade Formation of Oklahoma (see Figure 8.21*Ba*) (Yochelson and Bridge, 1957).

The Upper Cretaceous of North America, Europe (Germany, France, Holland, Belgium), Southern India, Southern Africa—the desert, for example—contains extensive molluscan faunas including many gastropod species (Wade, 1926). One of the largest representations of Upper Cretaceous gastropod species (174 species) occurs in the Coon Creek fauna (Ripley Formation, Tennessee). Wade (1926, p. 21), based on the complete faunal evidence, concluded that the Coon Creek fauna thrived in agitated waters (near-shore or intertidal) near a coast of a low-lying land mass.

The common Cretaceous gastropods of the Atlantic Coastal Plain are *Turritella*, *Gyrodes*, *Pyropsis*, *Anchura*, *Volutomorpha* (Richards, 1953, p. 287). Recognition of such common forms in the field is an important aid in stratigraphic studies.

Stanton (1947) monographed the pelecypods and gastropods of the Comanchean beds (Lower Cretaceous to Lower Upper Cretaceous) of Texas, Arkansas, Kansas, New Mexico, and Arizona. Twenty-six families and 42 genera and subgenera were represented. In general, while a given gastropod species might have been common within Comanchean outcrops in a given state (intrastate), for example, Texas or Kansas, infrequently they were found to be present in equivalent outcrops in other states. By contrast, given Comanchean pelecypod species (*Gryphaea*, *Exogyra*) more often occurred

interstate (Stanton, 1947, pocket insert). One may infer that the gene pools of Comanchean gastropods were more restricted. Thus species of the genus *Nerinea* were found only in Texas localities.

Tertiary-Pleistocene molluscan faunas are also very widespread. In the Miocene of New Jersey, Maryland, Virginia, the Carolinas, and Florida, gastropods and pelecypods occur in equal numbers. The bulk of all Miocene invertebrate fossils in these areas are mollusks—about 75 percent of the total (Richards, 1953, p. 337). Palmer (1937) described the extensive Claiborne fauna (Eocene of Texas, Louisiana, Mississippi, Alabama, and South Carolina) of mollusks, including gastropods. Generally, preservation is excellent: in one genus, *Turritella*, 12 species and six varieties (subspecies?) could be distinguished.

Since Tertiary-Pleistocene molluscan faunas have numerous living representatives, it is possible to relate present habitat to the fossil record. Generally, but not all, the Pleistocene molluscan species found in the Palos Verdes Hills terrace deposits are known to have lived at depths of less than 10 to 25 fathoms. During the Pleistocene, these hills formed an island. A series of 12 terrace deposits is represented. Of these, eight are fossiliferous. The Pleistocene gastropods lived in rock-cliff and tide pool environments, judging from living equivalents. Today these facies can be found on the twelfth terrace 1215 ft above sea level, as well as on lower terraces (Woodring et al., 1946).

Pleistocene mollusks (gastropods), often abundant, have left shells well preserved in silt deposits. They occur in lentils of sand, sandy silts of Nebraskan tills or Kansan age, silts interstratified with volcanic ash, for example, or, in loess banks in the glaciated portion of the Mississippi Valley. Leonard (1952) recorded densities as high as 5000 shells per cu ft of matrix at some places.

Frye and Leonard (1952) reported that distinctive gastropod assemblages characterized each major Pleistocene cycle in Kansas; furthermore, molluscan faunas permitted zonal subdivision of massive silt deposits. In turn, such zones in Kansas could be correlated with others throughout the glaciated portions of the Mississippi Valley. Interstate correlations between Kansas and Nebraska, for example, for Wisconsonian subages have been questioned (Frankel, 1957).

Despite obvious restrictions, reconstructions of Pleistocene ecological conditions have been importantly advanced by molluscan faunal data.

Four Tertiary and Quaternary species of minute endodont land snails (*Ptychodon*) were recovered from drill holes in Bikini, Eniwetok, and Funafuti atolls. They occurred in lagoonal deposits (at depths of 170–1800 ft below sea level). Important derivative data were retrieved from this evidence: (*a*) Within a long

history of subsidence, coral atolls had several periods of emergence as land snail fossils proved; (*b*) one mode for the distribution of life in the Pacific during Tertiary-Quaternary time, could have been the "high island stepping stones," that is, during periods of emergence (Ladd, 1958).

CLASS BIVALVIA

(Other names widely used for this class are Pelecypoda and Lamellibranchiata. Bivalvia is the oldest name applied to the class.)

1. Body. Bilaterally symmetrical, much compressed laterally. (For evolution of the mantle and shell from the extended condition see Figure 8.22*A*, *B*) Downgrowth of mantle on each side has completely enveloped "head," foot, and visceral mass. Head reduced to rudimentary condition

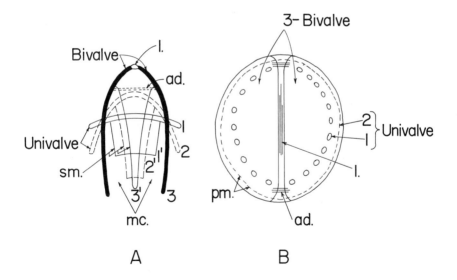

FIGURE 8.22 Postulated stages in evolution of Bivalvia.
(*A*)(*B*). (On left, transverse sections through shell superimposed; on right, dorsal views superimposed. Numbers *1, 2, 3* correspond to the same stage of bivalve evolution in *A* and *B*; *1', 2', 3'* denote shell muscles (*sm*), and foot (base of respective figures). (*A*)(*B*)−*1*, start with a slightly arched univalved shell. Compare Figure 8.1; circles in *B* denote shell muscles. *2*, increased concavity of univalve; further pallial muscle development (*pm*) represented by inner circular dashed line. *3*, bivalved stage comes about by further lateral compression, secretion of a ligament (*l*), formation of adductor muscles (*ad*) from portion of pallial muscle that unite. Each mantle lobe now has a center of calcification. Note progressive deepening of mantle cavity (*mc*), and lengthening of foot in the evolutionary process. Try to envision the genetic sequence leading from *1* to *3*.
Illustrations adapted from Yonge, 1953, Figure 1.

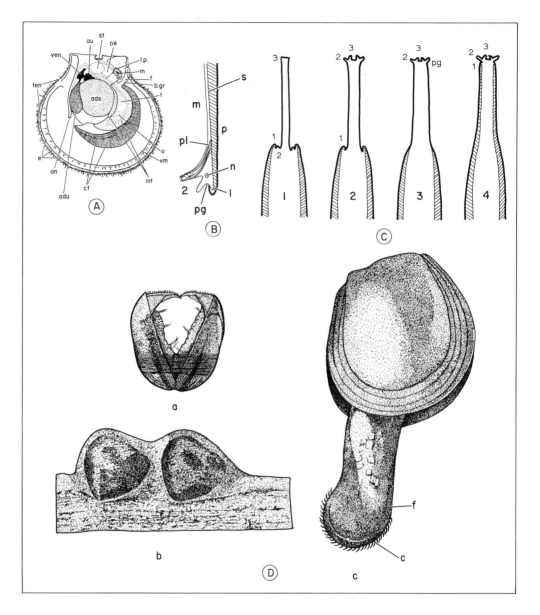

FIGURE 8.23A Anatomical and other features of Bivalvia.

(A) The scallop, *Pecten maximus*, right valve and ctenidium cut away to show general anatomy. Trace digestive system from mouth (*m*) through stomach (*st*) to anus (*an*); note position of ovary (*o*) and testis (*t*), mantle tentacles (*ten*), and eyes (*e*); adductor muscle (monomyarian) (*ads*); *ct* – lamella of left ctenidium, *int* – intestine; *ven*, *au* – ventricle and auricle (= the heart) schematic.

(B) Section through mantle edge of a bivalve: *m* – mantle; *n* – nerve; *p* – periostracum; *pg* – periostracal groove; *pl* – pallial muscle; *s* – shell; 1, 2, 3 – outer (secretory), middle (sensory), inner (muscular) lobes. (Number 3 not shown, above "2" and "pl").

(C) Formation of siphons: (1) where inner mantle lobe alone is fused (example: *Mytilus*); (2) where middle lobes are also fused (examples: *Venus, Cardium*); (3) where siphons are covered with periostracum (example: *Aloidis*); (4) where all lobes are completely fused (example: family Cuspidariidae). [1, 2, 3 as in (B) above.]

(D) Reproduction in fresh-water mussels: (a) hooked glochidium of *Symphynota*, anterior end view; (b) parasitic phase – glochidia of *A, cataracta* upon fin of carp; (c) young mussel (*Lampsilis ligamentina*) one week after close of parasitic period. Note calcified bivalved shell; *f* – foot, *c* – cilia (23 X 20 mm). Illustrations redrawn and adapted from: (A) Dakin, cit. Borradaile and Potts; (B) C. M. Yonge, 1948. Nature, Vol. 161, p. 198; (C) Lefevre and Curtis, 1910.

(that is, site of mouth opening, lacking eyes, tentacles, and radula).

A series of stalked eyes are found around the mantle in *Pecten*. They serve a sensory function and respond to a decrease in light. There are also sensory tentacles (Figure 8.23*A*) extending from the margin of the mantle. Thus the enclosing mantle (the middle fold) has taken over the sensory function that the head originally had (Figure 8.23*A*) (C. M. Yonge, 1948; Borradaile and Potts, 1938; M. A. Pierce, 1950).

2. Labial palps. A pair of *labial palps* that serve a particle-sorting function are situated on either side of the mouth (Figures 8.23*A* and 8.24*C*).

3. Mantle. The mantle edge consists of three lobes: one of these is adjacent to the shell that it secretes; a middle lobe, which has a sensory function (see above comment), and the innermost lobe, which is muscular (Figure 8.23*A*, *B*).

4. Dorsal hingement. The two lateral valves are joined dorsally by ligament and hinge (Figure 8.25*A*). [Hinge teeth and ligaments are important valve characteristics in fossil forms.]

5. Valve closure. Valves are closed ventrally by muscular contraction (one or two transverse adductor muscles). (Muscles scars on fossil valves are an important characteristic.)

6. Foot. Generally the foot is wedge-shaped (laterally compressed), and commonly used for slow, ploughing movements through bottom sediments. Wide modification of the foot occurs — for burrowing and/or in connection with various forms of attachment to the substrate. Cementation, as in Ostreidae, for example, may lead to loss of the foot.

7. Ctenidia (gills). Two ctenidia occur in the mantle cavity and are frequently enlarged; with diverse ciliary patterns (cilia create currents and ctenidia serve as food-

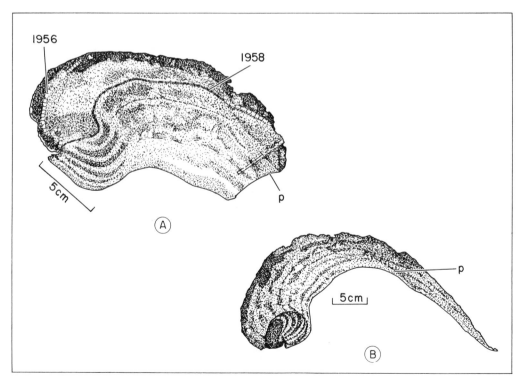

FIGURE 8.23B Growth of the giant clam *Tridacna gigas*, Bikini Atoll, Recent. (Dates indicate radioactive shell layers.) The growing shell incorporated strontium-90 from fallout of nuclear explosions at the atoll in the years 1956, 1958. The regular bands (interpreted to represent one year's growth) could thus be counted, giving an age of 9 years for the complete shell (length, 52 cm). (*A*) transverse section, *p*—pallial mark; (*B*) entire section, same clam. Data from Kelshaw Bonham, 1965. *Science*, Vol. 149 (3681) p. 300.

collecting or sieving devices for small food particles). Bivalves are referred to as "ciliary feeders." (The structure of the ctenidia is an important characteristic in the classification of living bivalves but has little utility in the classification of fossils.)

Stasek (1963 and see Figure 8.24*A*) recognized three categories of ctenidium and palp types:

Category 1. Ventral tips of at least the first few, or usually of many anterior filaments of the inner demibranch (Figure 8.24*A*) *inserted unfused* into a distal oral groove. (Nuculacea, Mytilacea, Unionacea, Astartidae, possibly Trigonacea).

Category 2. The same as Category 1 but the anteriormost filaments *inserted and fused* to a distal oral groove. [Bivalves included here are: Carditacea, Isocardiacea, Cyprinacea, Chamacea, Cardacea, Veneracea (in part), and so on.]

Category 3. The same as Category 1 but *no insertion*; fusion to inner palp lamellae of antero-ventral margin of inner demi-

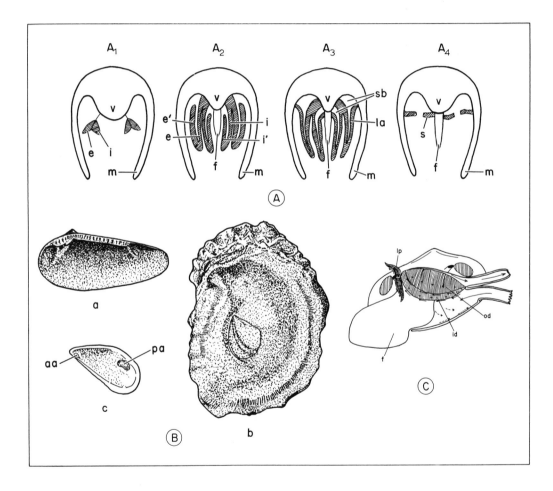

FIGURE 8.24 Bivalve structures.
(*A*) Gill types: *A1*—Protobranch. *A2*—Filibranch. *A3*—Eulamellibranch. *A4*—Septibranch. *e, i*—external and internal row of filaments respectively; *e', i'*—the same turned back; *sb*—suprabranchial chamber; *s*—septum; *f*—foot; *v*—visceral mass. (Adapted from M. E. Pierce, 1950.)
(*B*) Muscle scar types: (*a*) Isomyarian (= 2 adductor muscles coequal in size) (*Ctenodonta*); (*b*) Monomyarian (= 1 posterior adductor muscle) (*Exogyra*); (*c*) Anisomyarian (= anterior adductor muscle, reduced or absent) (*Mytilus*), *aa, pa*—anterior and posterior adductor (note reduced anterior muscle). (after several authors).
(*C*) *Amphidesma*, a representative eulamellibranch bivalve; ciliary currents shown; *id, od*—inner and outer demibranch of ctenidia; *lp*—labial palps; *f*—foot; inhalent and exhalent siphons on right. (See text for Stasek categories of ctenidia (gill) and palp types.) (Adapted from Morton and Yonge, 1964.)

branch may occur. (Among others, Pteriacea, Pectinacea, Ostreacea.)

8. Reproduction. Of the 10,000 described bivalve species known in 1943, the sexual condition in more than half was unknown. Most of the remainder are known and in those cases, the sexes are separate (dioecious condition). Less than 400 species were found to be deviates (hermaphrodites, bisexual or ambisexual—monoecious condition) (Coe, 1943).

Marine bivalves go through a trochosphere and veliger larval stage (Figure 8.3*B*). Bivalved larvae (glochidia) of freshwater mussels (Unionidae) are hatched in a modified portion of the gills known as brood pouches or marsupium; they are incapable of locomotion upon release from the egg membrane and die in water unless they contact a suitable host fish, in which the post-embryonic is passed as a parasite (Figure 8.23*A*, part D) (Lefevre and Curtis, 1919).

9. Siphons. The mantle cavity is partitioned into an upper (dorsal) and lower (ventral) cavity. Each cavity or chamber opens at the respective siphon—exhalent (dorsal); inhalent (ventral). A given active bivalve must maintain a continuous flow of water. Water enters through the inhalent siphon, passing through the gill laminae where the food particles are trapped or strained out, and exits through the exhalent siphon. (For mode of formation of siphons by fusion of mantle edges, see Figure 8.23*A*, part C.)

One individual mussel about 5 in. long will pass 4.5 pints per hour through its gill chambers during summer months and a reduced quantity during the winter (C. M. Yonge, 1963). Given a mussel bed with a population of some 300,000 individuals, then, using these figures, the inhalent water should run about $170,000 \pm$ gallons per hour.

10. Other systems and organs. Olfactory organs (osphradia), auditory and equilibrating organs (otocysts), tactile papillae, and a nervous system; a closed circulatory system (hemolymph, heart with single or paired cardiac ventricle and two auricles); digestive system (stomach, intestinal canal, proteinaceous crystalline style in a small pocket of the intestine that continuously forms and dissolves and aids carbohydrate digestion, oral and anal ends at opposite poles of the body—see Figure 8.23*A*); excretory system consists of paired nephridia discharging independently of rectum (Dall, 1895); respiration is achieved by the exchange of gases through medium of mantle and the inhalent current and *not* through the ctenidia (Borradaile and Potts, 1958).

Purchon (1960) recognized five stomach types in Bivalvia based on various structures and their relationships.

11. Shell. Bivalved shell of calcium carbonate (calcite, aragonite, or both). Shell layers variable, from two to three (periostracum, prismatic layer, and a foliated or nacreous layer). Trueman (1942) reported three layers in addition to the periostracum in *Tellina tenuis*, the outer radial prismatic layer, middle layer of crossed lamellae, and an inner complex layer. He observed that some species of this genus lacked the outer prismatic layer whereas others possessed shell structures comparable to that of *T. tenuis*.

Modern *Tridacna gigas* shells from Bikini atoll, studied by radioautography (Strontium 90), revealed in transverse sections the presence of a series of regular or annular bands. One of these bands preceded the 1956 layer of radioactive accumulation from an atoll nuclear explosion, two bands intervened in 1958, and six bands followed to the time of collection. The age of this clam was determined by counting these bands; nine years old, length of 52 cm (see Figure 8.23*B*).

12. Geologic range: Ordovician to Recent.

FOSSIL BIVALVES

While biologists favor bivalve classification predicated on gill structure (for types, see Figure 8.24*A*; Franc, 1960), this is hardly practical for the study of fossil valves because such information is lost.

One can scarcely find two students of bivalves who use the same classification. This can be confusing to the nonspecialist. The difficulties in classifying bivalves may be attributed to two factors: (*a*) highly specialized forms, and (*b*) adaptive variations with parallel development from one family to another (Morton, 1963).

Dall's system (1913) was founded on hinge dentition; Thiele's (1934–1935) was a composite (dentition, gill structure, musculature—see Figure 8.24 and 8.25*B* for types). Franc (1960), as mentioned, favored gill structure. Newell (1965) supported by Cox's deep insight (1960) tried to reconcile the varied data into a new synthesis. He recognizes six subclasses and 14 orders. The chief criteria employed are dentition, ligament type, shell microstructure (after Bøggild), and

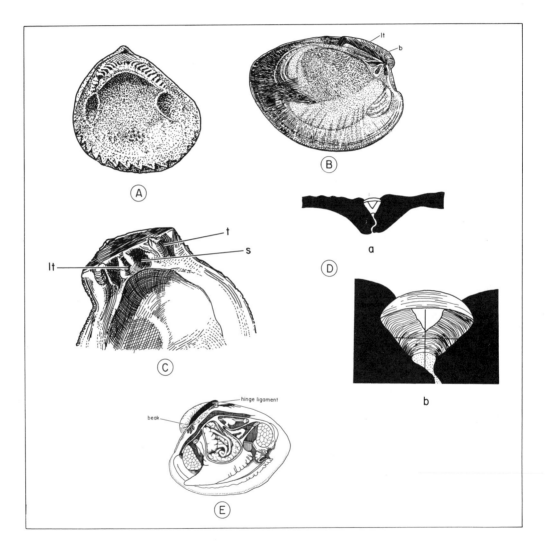

FIGURE 8.25A Ligament positions and types in bivalves.
Positions: (A) *Amphidetic*, type of ligament that extends on either side of the beak. (B) *Opisthodetic*, ligament (*lt*) withdrawn wholly behind beak (*b*). After Neumayr.
Types: (C) *Alivincular*, ligament (*lt*) more or less a flattened cord extending from one umbo to another, with long axis transverse to plane of valve margin (= plane of commissure) and axis of motion; usually associated with amphidetic area; *t*—tooth, *s*—socket, (D) *Multivincular*, ligament that consists of a reduplication of alivincular type at intervals upon amphidetic area or upon posterior limb of cardinal margin; (*a*) single elastic lamellar ligament band above; (*b*) massive fibrous ligament below. (E) *Parivincular*, external ligament comparable to cylinder split on one side, attached by severed edges, one edge to each valve; long axis corresponds to vertical plane between valves; always opisthodetic. After Dall.
Illustrations redrawn and adapted from: (A) Gardner, 1926; (B)(C) Deschaseaux, 1952; (D) Newell, 1937; (E) Brown, 1950.
A – *Glycymeris, B* – *Venus, C* – *Spondylus, D* – *Arca, E* – *Venus.*

habit. Keyed into Newell's system at the superfamily level are data on gill structure, gill cilia, stomach type, labial palp type, and shell microstructure.

The six subclasses of Newell are Paleotaxodonta, Cryptodonta, Pteriomorpha, Palaeoheterodonta, Heterodonta, Anomalodesmata. This classification appears in the Treatise on Invertebrate Paleontology and is followed here. Descriptions include subclasses, their respective orders, and illustrative fossil examples embraced by each. [Shrock and Twenhofel (1953) followed Thiele's system, whereas Moore, et al. (1952) preferred Dall's. Newell's system, as mentioned, is preferred in this book because it is based on a serious reevaluation of all available evidence both biological and paleontological.]

Classification

Subclass 1. Paleotaxodonta. Primitive taxodont hinge (that is, teeth all alike, Figure 8.25*B*, *Ab*); shell microstructure, nacreous and cross-lamellar; equivalved. Ctenidia, protobranch (see Figure 8.24*A*), function chiefly in respiration; soft bottom detritus feeder. Geologic range: Ordovician to Recent.

[Order Nuculoida, genus *Nucula* (Figure 8.26*A*, part *A*).]

Subclass 2. Cryptodonta (= Paleoconcha of several authors). Thin-shelled forms without lateral teeth or well-developed hinge, Ctenidia, protobranch, used to feed and to breathe; generally equivalved; shell microstructure, homogeneous (see earlier discussion of Bøggild). Geologic range: Ordovician to Recent.

[***Order Solemyoida.*** Homogeneous aragonite ostracum; siphonate burrowing protobranch. Genus *Solemya* (Figure 8.26 *A*, part *B*).]

[***Order Praecardioda.*** Paleozoic cryptodonts. Genus *Praecardium* (Figure 8.26*A*, part *C*).]

[***(?)Order Conocardioda.*** Paleozoic cryptodonts; anteriorly gaping; cellular cell structure. Genus *Conocardium* (Figure 8.26*A*, part *D*).]

Subclass 3. Pteriomorphia. Shell structure, ligaments, gills, and stomach all are variable; commonly byssate (that is, attached by byssus to substrate) in adults; generally members of the epifauna but a few adapt to a boring habit in firm substrate. Fossil record suggests phyletic unity within the subclass. Geologic range: Ordovician to Recent.

[***Order Arcoida.*** Isomyarian (that is, with two adductor muscles approximately equal in size, see Figure 8.24*C*), filibranchs with crossed-lamellar shells; generally equivalved. Representative genera: *Arca* (Figure 8.25*B*, *Aa*). *Glycymeris* (Figure 8.25*B*, *Ab*), *Parallelodon* (Figure 8.26*A*, part *E*).]

[***Order Mytiloida.*** Anisomyarian (that is, with adductor muscle scars markedly unequal); equivalved, filibranchs, and eulamellibranchs (see Figure 8.26*A*, part *F*) with prismato-nacreous shells; mainly byssate in adults. Representative genera: *Mytilus* (Figure 8.26*A*, part *F*); *Pinna* (8.26*A*, *G*).]

[***Order Pterioida.*** Anisomyarian and monomyarian; byssate in adults; nacreous, crossed-lamellar, or foliate internally. Two suborders: Suborder 1. Pteriina. Representative genera: *Pteria* (Figure 8.26*A*, part *H*), *Pecten* (Figure 8.23*A*, part *A*), *Pseudomonotis* (Figure 8.26*A*, part *I*). Suborder 2. Ostreina. Representative genera: *Ostrea* (8.27*G*), *Exogyra* (8.27*H*), *Gryphaea* (8.27*I*).](Nonmarine types, Figure 8.29).

Subclass 4. Palaeoheterodonta. Possesses free or incompletely fused mantle margins, an opisthodetic, parivincular ligament (see Figure 8.25*A*, *E*), and prismato-nacreous shell (*Opisthodetic* = location of external ligament behind beak; *parivincular* — long axis of ligament parallel to hinge line.) Geologic range: Middle Cambrian to Recent.

[***Order Actinodontoida.*** Early Paleozoic precursors of many bivalve orders; teeth absent or radial, poorly differentiated, originating at the beaks, equivalved. Genus *Modiomorpha* (Figure 8.27*A*).]

[***Unionoida.*** Variable Upper Paleozoic

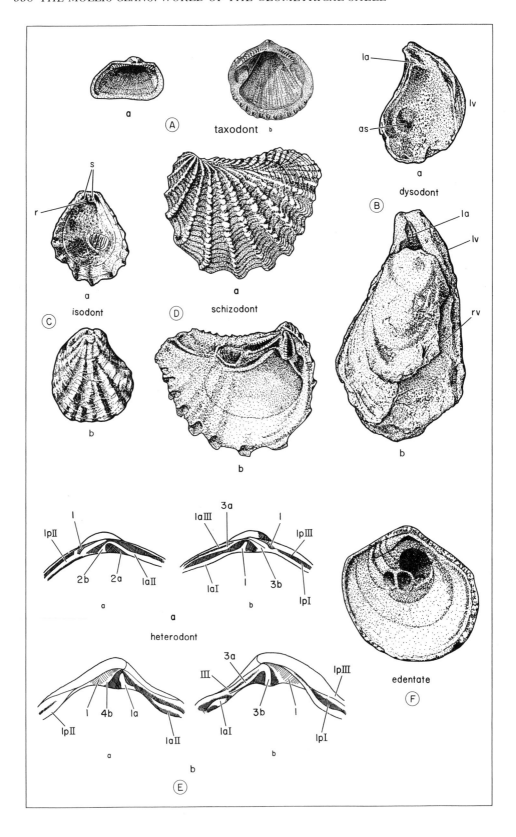

FIGURE 8.25B Dentition in bivalves.

(A) Taxodont. *Arca* (*Barbatia*) *adamsi*, M. Mio., Fla., schematic. (a) Interior of left valve; (b) *Glycymeris suborta*, M. Mio., Fla., interior of left valve.

(B) Dysodont. *Ostrea franklini* (Comanchean) Ark; (a) left valve showing adductor scar and ligament area; (b) right and left valve in contact, *lv*—left valve or attachment valve, *rv*—right valve, *la*—ligament area.

(C) Isodont. *Plicatula densata*, M. Mio., Fla. (a) Interior of right valve showing resilifer (R) and sockets (s) shown in black on either side of teeth, shown in white; adductor scar = hatched area; (b) exterior of left valve.

(D) Schizodont. *Trigonia thoracica*, U. Cret., Coon Creek, Tenn. (a) Left valve, exterior view; (b) left valve, interior view, and striated tooth; sockets and adductor muscle scar. Heterodont.

(E) (a) *Cyrena gravesii*, schematic; a left and b right valves; teeth numbered after Bernard and Munier-Chalmas (see text); (b) *Lucina neglecta*, Mio., a left and b right valves.

(F) Edentate, *Anomia tellinoides*, U. Cret., Coon Creek, Tenn., right valve, interior view showing lack of dentition and byssal attachment and adductor scar.

Illustrations redrawn and adapted from Gardner, Stanton, Bernard, and others.

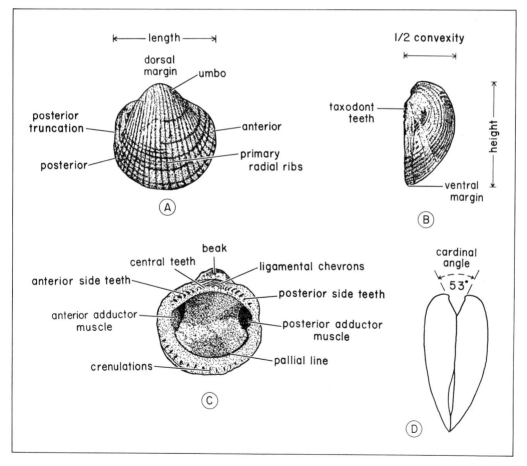

FIGURE 8.25C Bivalve shell morphology and orientation.
Glycymeris (*Glycymerita*) *veatchii*. U. Cret., Calif. (A) Exterior view. (B) Posterior view. (C) Interior view. (Duplivincular ligament type.) (D) *Barbatia velata*, Rec., Galapagos Islands, anterior view, showing small cardinal angle, schematic. After Nicol, 1950.

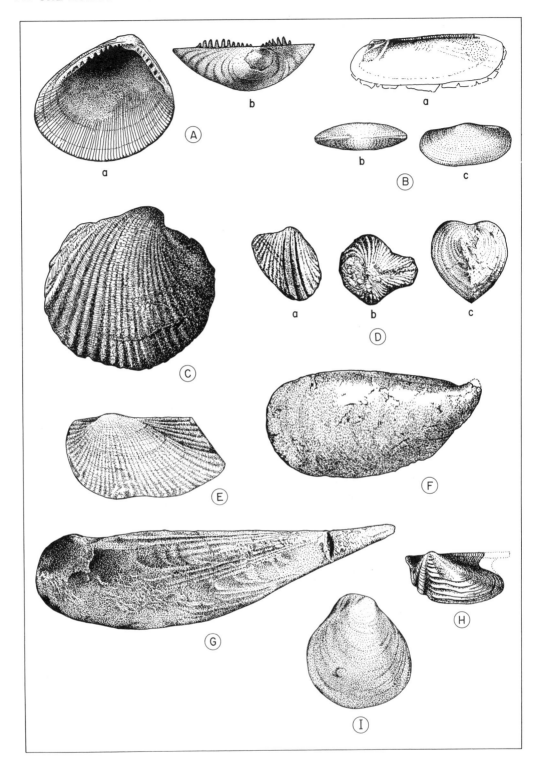

FIGURE 8.26A Representative fossil bivalves—I. (*A*) *Nucula.* (*B*)(*a–c*) *Solemya.* (*C*) *Praecardium.* (*D*)(*a–c*) *Conocardium.* (*E*) *Parallelodon.* (*F*) *Mytilus.* (*G*) *Pinna.* (*H*) *Pteria.* (*I*) *Pseudomonotis.* All figures schematic. Illustrations redrawn and adapted from several authors.

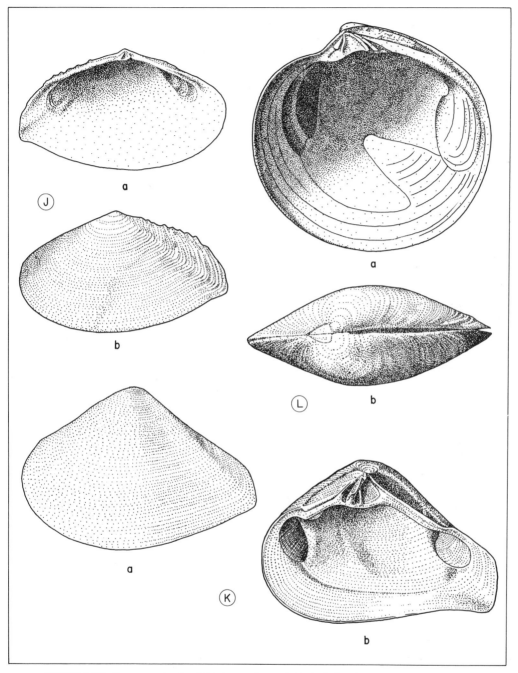

FIGURE 8.26B Representative fossil bivalves—I (cont.) (*J*)(*a, b*) *Tellina*. (*K*)(*a, b*) *Crassatellites*. (*L*)(*a, b*) *Dosinia*. All figures schematic. Illustrations redrawn and adapted from Gardner, 1926.

and post-Paleozoic nonmarine forms, probably derived from the Actinodontoida; eulamellibranchs. This order is probably polyphyletic. Genus *Unio* (Figure 8.27*C*) *Carbonicola*, Figure 8.29*A*.

[**Order Trigonoida.** Trigonal marine shells; lateral teeth lacking; filibranchs;

shell microstructure, homogeneous. Representative genera: *Trigonia* (Figure 8.25*B, D*), *Myophoria* (8.27*B*).]

Subclass 5. Heterodonta. Heterodonts (that is, lateral teeth occur on either side of cardinal teeth either situated below the

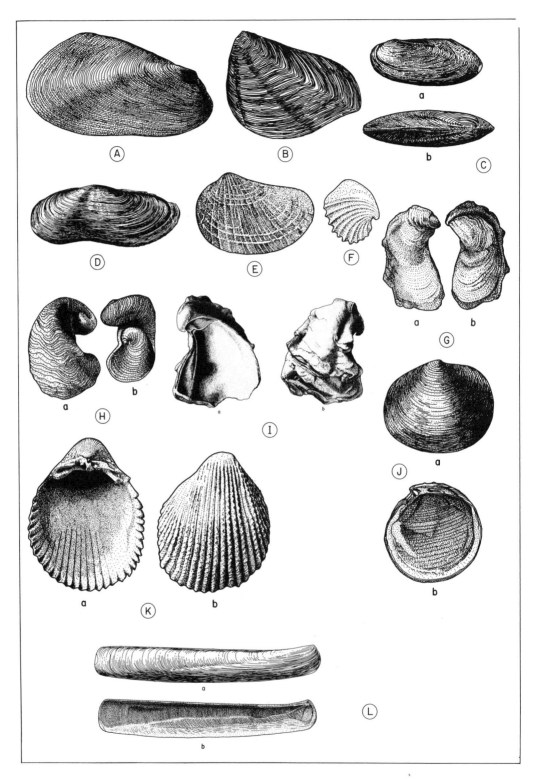

FIGURE 8.27 Representative fossil bivalves—II. (A) *Modiomorpha.* (B) *Myophoria.* (C)(a, b) *Unio.* (D) *Mya.* (E) *Pholadomya.* (F) *Verticordia.* (G)(a, b) *Ostrea.* (H)(a, b) *Exogyra.* (I) *Gryphaea.* (J)(a, b) *Lucina.* (K)(a, b) *Cardium.* (L)(a, b) *Ensis.* All figures schematic. Illustrations redrawn and adapted from several authors.

beak or on both sides); complex, crossed-lamellar eulamellibranchs. Geologic range: Silurian to Recent.

[**Order Hippuritoida.** Coral-like rudistids with thick shells and thickened teeth (pachydonts); chiefly attached extinct forms. Representative genera: *Chama* (Figure 8.28*A*), *Megalodon* (Figure 8.28*B*), *Hippurites* (8.28*C*), *Caprina* (8.28*D*), *Diceras* (8.28*E*), *Monopleura* (8.28*F*), *Durania* (8.28*G*).]

[**Order Veneroida.** Active heterodonts. This order subdivides into two suborders: suborder 1. Lucinina, embracing ten superfamilies; suborder 2. Arcticina, embracing five superfamilies. Suborder 1. Representative genera: *Lucina* (Figure 8.27 *J*), *Cardium* (Figure 8.27*K*), *Crassetellites* (8.26*B*, K), *Ensis* (8.27*L*), *Tellina* (8.26*B*, J). Suborder 2. Representative genera: *Dosinia* (8.26*B*, L), *Venus* (8.25*A*, *E*).]

[**Order Myoida.** Asthenodonts (that is, teeth obsolete) with degenerate hinge; generally with siphons and united mantle margins; shell complex, crossed-lamellar; this order probably is phyletic. Two suborders: suborder 1. Myiina, genus *Mya* (Figure 8.27*D*); suborder 2. Pholodina, genus *Teredo*. Figure 8.7*B*].

Subclass 6. Anomalodesmata. Usually fossorial (that is, of burrowing habit), without well-developed hinge teeth; often with siphons; mantle margins united; ligament associated in all but the most primitive forms with an internal resilium and lithodesma. (*Internal resilium* = a triangular ligament structure composed of fibrous conchiolin both elastic under compression and situated in a central pit, resifer, or chondrophore along the inner margin of each valve, that is, below the valve margin. *Lithodesma* = calcareous pieces that support the internal resilium.) This subclass has internally nacreous shells. Geologic range: Ordovician to Recent.

[**Order Pholadomyina.** Burrowers with primitive hinge; eulamellibranchs. Genus *Pholadomya*, (Figure 8.27*E*).]

[**Order Poromyoida.** Septibranchs. Genus *Verticordia* (Figure 8.27*F*).]

Bernard Munier-Chalmas Notation for Heterodont Teeth

To facilitate detailed comparisons of dentition in the hinge teeth (which are projections from a hinge plate) of heterodonts, Bernard and Munier-Chalmas devised a widely employed notational system (1895, pp. 116–119). This system permits a written dental formula for the right and left valve (numerator and denominator, respectively, of a fraction).

This notation may best be explained by reference to Figure 8.30. On the left is the left valve, on the right, the right valve. Reading directly from the *anterior* side of the left valve, the formula is *LA*, II, IV, *CA*, 2*a*, 2*b*, 4*b*, 6*b*, L (= ligamental groove —an item that can be omitted from the formula) and *LP* II, IV, VI. (In the above formula, *LA* = lateral anterior teeth, *CA* = cardinal teeth, and *LP* = lateral posterior teeth.) Reading from the anterior side of the right valve, the following sequence occurs: I, III, V, *L*, 5*b*, 3*b*, 1, 3*a*, I, III, V. (In the last-named sequence, the first set of Roman numerals represent lateral anterior teeth, whereas the last set, the lateral posterior teeth and the Arabic numerals, or numerals with letters, represent the cardinal teeth.)

The Bernard Munier-Chalmas notation can readily be followed; the designation, anterior and posterior, eliminated by using an arrow that indicates the anterior and *L* or *R*, the left or right valve. Thus an abbreviated *Cyrena* formula could be expressed:

$$\frac{R - \text{I, III}, 3a, 1, 3b, \text{I, III}}{L - \text{II}, , 2a, 2b, 4b, \text{II}}$$

There is no need to represent corresponding socket for a given hinge tooth since it is understood to be present. For lateral anterior I of the right valve, in the above formula, there is a corresponding unnumbered socket; for *LA* II of the left valve, the corresponding socket is on the

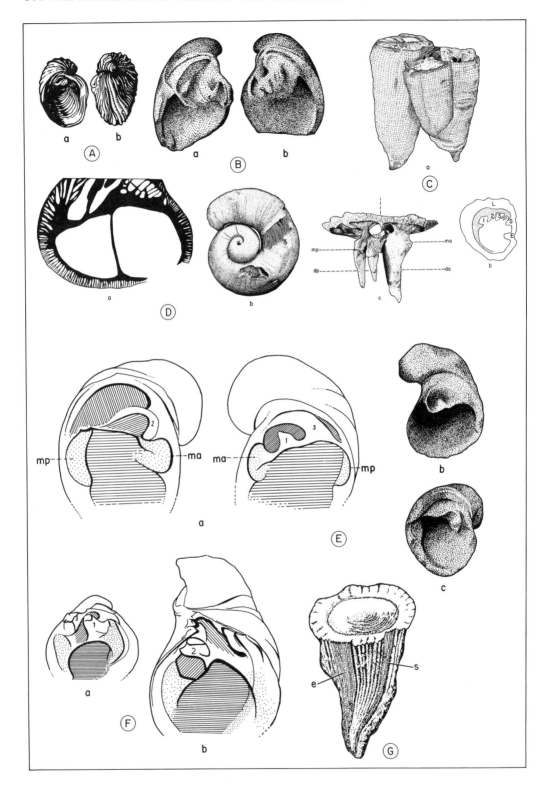

FIGURE 8.28 Representative fossil bivalves—coral-like rudistids. Schematic. (A) *Chama.* (B) *Megalodon.* (C) *Hippurites:* (a) *H. radiosus;* (b) *H.* sp. right valve, transverse section; (c) *H. radiosus.* (D) *Caprina.* (a) Transverse section, left valve. (b) side view. (E)(a–c) *Diceras.* (F) *Monopleura.* (G) *Durania,* both valves, side view. E and S = siphonal bands.
Left and right valve to left and right when paired; *ma, mp* = anterior and posterior muscle insertion; 1–3, teeth; *da, dp* — anterior and posterior teeth; *L* — ligamental ridge.
Illustrations redrawn and adapted from: (A) Neaverson; (B, C, D, F, H) Jean Piveteau, ed., Traité de Paléontologie, 1952, Vol. II; Colette Deschaseaux "Classe de Lamellibranches," in Traité, Vol. II; (G) Toucas, cit. Deschaseaux and Coogan.

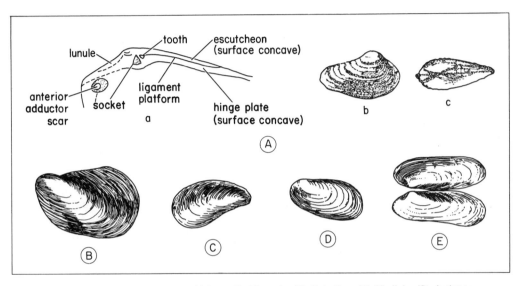

FIGURE 8.29 Nonmarine bivalves. (A)(a–c) *Carbinocola.* (B) *Naiadites.* (C) *Modiola.* (D) *Anthraconauta.* (E) *Anthraconaia.* All figures schematic. Illustrations redrawn and adapted from M. J. Rogers, 1965.

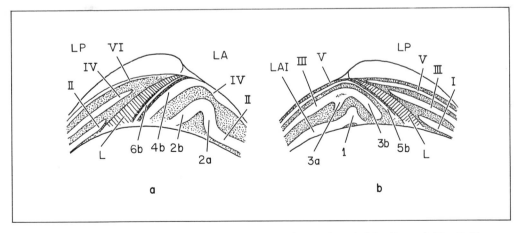

FIGURE 8.30 Theoretical case to illustrate application of dental formula (after Bernard). [See D. W. Boyd, cit. N. D. Newell, 1969, Treat. Invert. Paleont., Pt. N., Bivalvia, Addendum, p. N908, for discussion of limitation of this notation.] (See text for details.) (a) Left valve. (b) Right valve.

right valve, and so on; the same holds true for the cardinal teeth. (Figure 8.31).

Cox (1960) figured that while the above dental notation has been a valuable tool to indicate tooth homologies among the heterodonts, it could be misleading if applied to other groups, such as the rudistids (Figure 8.28). Further considerations about dentition show (*a*) that the absence of teeth (*edentulous* condition) may not be primitive since it occurs in genera belonging to families with well-developed teeth, and (*b*) that comparable tooth systems may arise in different lineages by convergence.

INFORMATION DERIVED FROM FOSSIL BIVALVES

Hingement

The whole matter of bivalve hingement may be viewed as a strictly mechanical problem involving stress-strain relationships. Suppose we are given two valves, any configuration, the following specification: (1) they meet at one end where the ligament serves as a true hinge with otherwise no permanent points of contact, and (2) appropriate musculature allows for valve opening and closure. In this model several mechanical problems must be solved: (*a*) Lateral slippage along the hinge axis [a condition experimentally recorded by Anthony (1905, cit. Newell, 1953)]; (*b*) imperfect closure of valve margins.

Clearly, unrestricted lateral slippage will act to prevent articulated valve closure. In turn, a poor marginal valve fit would pose a threat to individual survival because of the vulnerability of the exposed soft-part anatomy (in nonburrowers).

The diversity of dental patterns and ligament types (Figure 8.25*A, B*) apparently arose as adaptive responses to the problem raised in (*a*) and (*b*) above. Trueman (1953) showed a close correlation between ligament type and mode of life. Teeth and sockets (located below

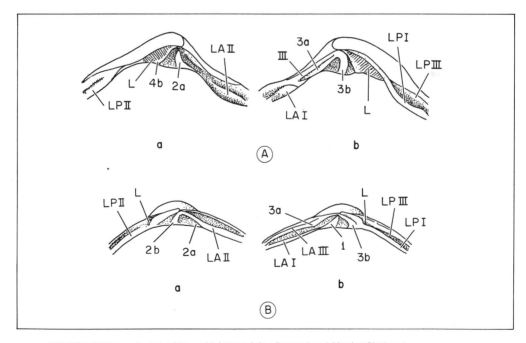

FIGURE 8.31 Heterodont dentition: chief types (after Bernard and Munier-Chalmas).
(*A*) Cyrenoid type of dentition: *Cyrena gravesii*, (*a*) left valve, (*b*) right valve, schematic. Cyrenoid-Corbiculoid.
(*B*) Lucinoid type of dentition: *Lucina neglecta*, (*a*) left valve, (*b*) right valve, schematic.

hinge axis) serve as a guidance system to insure snug valve closure. (This occurs in reasonably small tolerances of "go, no-go"). Dentition thus functions as a supplement to the ligament (Newell, 1953). Where comparable dentition occurs in different lineages, equivalent solutions to similar mechanical stresses are indicated.

Because natural selection favors "snug marginal fit" of valves during closure in nonburrowers, this condition may have a high selective value (see Newell, 1953). Dentition that ensures this condition and its evolution thus relate to a critical feature in bivalve survival. Ligament type likewise has adaptive value.

Radiation of Paleozoic Bivalves

Vogel's study (1962) of the oldest known toothed bivalves from the Middle Cambrian of Murero, Spain, brought forth new data on the most primitive hinge dentition and on the radiation of Paleozoic bivalves (Figure 8.32). He found that the most primitive hinge had teeth running parallel to the hinge line (either one tooth socket, as in Figure 8.32A, or a more advanced form with two teeth sockets, as in Figure 8.32B); and erected a new genus to embrace these forms (genus *Lamellodonta*, order Palaeoheterodonta).

As Newell (1965) commented, although nuculoids (Figure 8.26A) are considered the most primitive living bivalves, "they are not known to be the most ancient." Vogel's new genus containing the oldest tooth forms do not have a nuculoid hinge. The revised estimate of phylogenetic relationships is shown in Figure 8.32G. It will be seen that the Lamellodontidae appear to be ancestral to lineages that gave rise to the nuculoid bivalve.

Havlíček and Kríž (J. Paleontology, 1978, vol. 52, p. 972) would remove the middle Cambrian bivalve, *Lamellodonta*, from Bivalvia and place it under the inarticulate brachiopods.

The probable origins of heterodonts (Figure 8.32M), mytilaceans (Figure 8.32 C), pteriaceans (8.32K), and ariaceans are

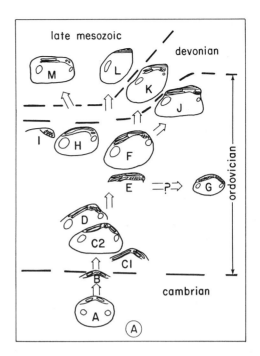

FIGURE 8.32 Inferred phylogeny of Paleozoic bivalves. (Modified from Vogel, 1962, Figure 17, as interpreted by Newell, 1965.) (A), (B) Lamellidontidae, Cambrian. (E), (F) Cyrtodontidae, Ord. (G) Nuculoida, Ord. (H), (I) Modiomorphidae, Ord. J. Parallelodontidae, Ord. (K) Pteriniidae, Dev. (L) Myalinidae, Late Paleozoic. (M) Permomorphidae, Late Paleozoic (not Late Mesozoic). (A–D, H, I) subclass Palaeoheterodonta; (M) subclass Heterodonta; (E, F, J–L) subclass Pteriomorphia.

also illustrated in Vogel's reconstruction.

There is a group of primitive Ordovician bivalves such as *Babinka* (Figure 8.34) with a pattern of four muscle scars below the beak reminiscent of triblidiid gastropods (see also Monoplacophora and Figure 8.8) (Vokes, 1954). Vokes thought this pattern reflected the ancestral stock from which bivalves derive. Cox (1960) also concluded that *Babinka* "approximated to the theoretical concept of the newly-evolved bivalve" disregarding the small, obscure, radially disposed hinge teeth said to be present. In this respect, *Babinka* was preceded in time by Vogel's Spanish Cambrian lamellodonts, which lacked this muscular pattern.

The role of *Babinka* in bivalve phylogeny is only vaguely known at present (Cox, 1960). Newell assigned the family to which it belongs (Babinkidae Horný) to the subclass Heterodonta, apparently derivative through several evolutionary stages of the Lamellodonta.[1] (Figure 8.34).

Evolution of Oysters and Scallops

Newell (1954), with particular attention to the fossil record and the stratigraphic sequence of fossil bivalves, discriminated between various trends: parallel evolution (similar morphological stages in separate but related lines) and convergent evolution (more distantly related groups came to resemble each other in specific features such as dentition, through time) (Figure 8.33).

Recent arcaceans have taxodont dentition and, hence, have been placed close to the nuculoids in the evolutionary scale (see Figure 8.32G). However, Mesozoic and Paleozoic ancestors of arcaceans have a distinctly different dentition from the nuculoids, and are more reminiscent of primitive pteriaceans, mytilaceans, and so on. The taxodont condition of Recent arcaceans is therefore an example of convergent evolution.

Similar findings show parallel evolution in ligament types (Newell, 1954). Living Pteridae and living oysters have a similar ligament type. Yet fossil evidence shows such ligament developed along two lines from different ancestors; both possessed a duplivincular ligament type (that is, a ligament responsive to both tensional and compressive forces via lamellar and fibrous tissue components).

[1] If reassigned, this derivation would be invalid.

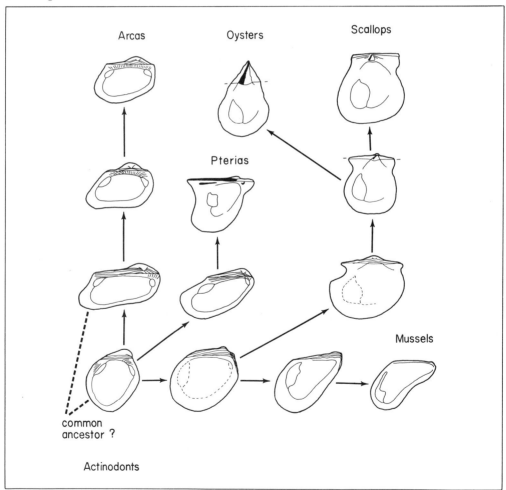

FIGURE 8.33 Inferred origins of mussels, oysters, scallops, and other bivalves. After Newell, 1954, with modifications suggested by Cox, 1960.

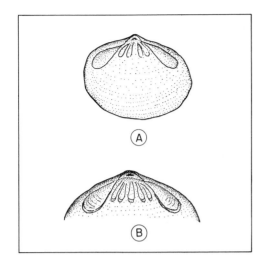

FIGURE 8.34 (A), (B) *Babinka*—a primitive Ordovician bivalve. Illustrations redrawn and adapted from Vogel, 1962.

Figure 8.33 summarizes the evolutionary lines leading to oysters, and so forth. The bulk of living bivalve families and superfamilies appear to have originated in pre-Jurassic seas (Newell, 1954).

Rudistid Evolution, Banks ("Reefs"), Extinction

Morphology and Derivation. A group of bivalves, known only as fossils, the rudistids (Figure 8.28), are remarkable in several ways: (*a*) They have a relatively brief geologic range from Upper Jurassic to Upper Cretaceous, (*b*) their geographic spread is broad (Figure 8.35), and (*c*) they developed several valve characteristics that were unusual for bivalves, including such structures as pillars and pseudopillars, canals, oscules, cellular-prismatic structures, accessory cavities, as well as elongation of the lower valve, either right or left, and reduction of the upper valve, either left or right, to a lid or a cover.

The Devonian–Upper Triassic bivalve *Megalodon* (Figure 8.28*B*) apparently gave rise to the Upper Jurassic form *Diceras* (Figure 8.28*E*). *Diceras*, existing in great profusion, was thick-shelled and cemented to the bottom by the right valve. It appears to have given rise after Kimmeridgian time, Upper Jurassic, to two developmental lines: rudistids with the *left* valve fixed or cemented to the bottom (genera such as *Requienia* and *Toucasia*), and those with right valve fixed or cemented [genera such as *Hippurites* (Figure 8.28*C*), *Radiolites*, *Caprina* (Figure 8.28*D*), and *Monopleura* (Figure 8.28*F*)].

Generally, the fixed valve had a tooth and two sockets, whereas the free valve had two teeth and one socket.

Rudist Banks. A Turonien carbonate facies (Lower Cretaceous of Europe) might consist of several superimposed rudistid banks, sometimes called Hippurites Reefs. The upper bank would show large numbers of the genus *Radiolites* in transverse section, a few *Hippurites*, some algal debris, and foraminifera. Below this, a bioclastic carbonate would be composed of *Radiolites*, echinoderm, and bryozoan debris. Below this bank another one

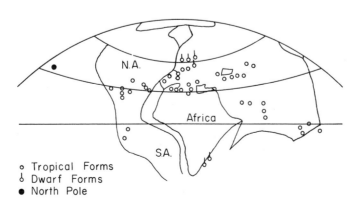

o Tropical Forms
♂ Dwarf Forms
● North Pole

FIGURE 8.35 Worldwide rudistid distribution (Cretaceous, etc.). After Dacqué.

would occur containing a smaller population of erect and crowded rudistids with the unattached valve uppermost, with very few microforms, and almost no other fauna. The basal bed in this section would again be a bioclastic composed of the debris of larger rudistids. Thus it is possible to follow the successive development through time of rudistid reefs or banks (Deschaseaux, 1952).

However, elsewhere isolated rudistids occur, as solitary forms. Although rudistids are warm-water forms, both gigantic and dwarfed individuals are known (Dacqué, 1926, and see Figure 8.35). Giant forms include *Titanosarcolites*, which attained a length of 6 ft (Neaverson, 1955) and occurs in the Upper Cretaceous, Kemp Clay of Texas as well as Jamaica, Cuba, and other West Indian Islands (Stephenson, 1938). This form apparently had a recumbent habit since length and weight would render it unstable in an erect position on the sea floor.

An example of a rudist facies and its relationship to other molluscan facies (cephalopods and other bivalves) is shown in Figure 8.36. The rudistids include *Toucasia*, *Requienia*, and *Monopleura*. Rudistids are also reported from the Niobrara of Colorado (Griffith, 1949) and Kansas (genus *Durania*). They range as far north as Montana [Upper Pierre shale — *Titanosarcolites* (?)].

Growth. A study of a caprotinid rudist growth series from the Middle Albian of Texas indicated that the two valves are actually almost equivalved up to a valve length of 7.5 cm and a total shell volume of about 4.5 cc. At this stage, the attached conical lower valve's linear growth rate decreases (that is, slows down), whereas the free coiled upper valve continues to increase until valve length attains $17 \pm$ cm and a shell volume of 20.0 cc. Thereafter, the linear growth rate of the upper valve decreases. The result is that the younger, smaller shells differ in form from the top-heavy adult valves (Perkins, 1965).

The seven families of rudists include family Diceratidae, Requieniidae, Monopleuridae, the family referred to in the above-named growth series studies, Caprotinidae, the Caprinidae, Hippuritidae, Radiolitidae. Only the Requieniidae are fixed by the left valve. All the other families contain individuals fixed by the right valve. The Diceratidae are fixed by either valve. A family name, such as Diceratidae, is based on the name of the genus, *Diceras* (see Deschaseaux, 1960, p. 2162).

Shell Structure. Cox (1934), in a study of the structure of the Persian rudist *Trechmannella persica* (family Caprinidae), carefully described the shell.

The lower, or right valve, attached to the substrate and considerably elongated, was composed of two layers: outer layer remarkably thick (up to 15.0 mm), inner layer rather delicate, being cellular. The

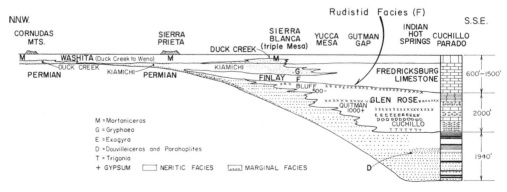

FIGURE 8.36 Rudistid facies (*F*) in relation to other molluscan biofacies in the Upper Cretaceous of Texas. Adapted from Sellards et al., 1932.

outer layer is built of layered prismatic calcite, sloping at 45 degrees to the main axis of the valve—a cone-in-cone arrangement—and forms a beveled rim into which the upper (left) valve fits. Two longitudinal swellings along the ventral side of the lower valve are thickenings of the outer shell layer: *siphonal bands*.

The smaller, upper (left) valve is also composed of two layers. The outer layer, while compact in structure, is comparatively thin. The inner layer consists of a series of narrow, radial cells separated by thin partitions.

Extinction. The last rudistid survivors are known from the rock column of the Catalonian Pyrenees in beds indicative of a brackish environment. Deschaseaux (1948) found no support for the concept that environmental changes suffice to account for their extinction in Danien time. The end of the Cretaceous, the so-called time of the "great dying," saw the extinction of numerous marine invertebrates; and therefore a more likely general explanation, such as Bramlette's (1965), may have applicability. The dying

also may be related to changes in the primary food chain in the sea at that time (Tasch, 1966).

Stratigraphic Zonation. Four successive rudist zones are recognized in Southern Europe. They have great stratigraphic utility for tracing and zoning the Upper Cretaceous (France, Italy, the Balkans) (Gignoux, 1955).

Paleoecology

Reefs, Biostromes. Laughbaum (1960) traced an oyster biostrome in the Denton Formation (Lower Cretaceous) from south to north, a distance of about 70 miles, through three counties of Texas. The essential results are summarized in Figure 8.37.

The investigator recognized four successive biofacies (*A–D*). The paleoecological inferences for each biofacies were: (*A*) *temperature* (tropical to semitropical), *depth* (shallow, neritic), *salinity* (normal marine); (*B*) constant thickness of limestone over a distance of 23 miles suggested quiet water or low energy conditions;

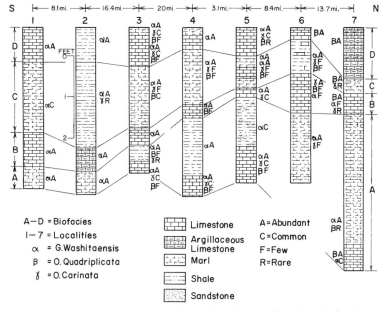

FIGURE 8.37 Oyster biostrome (Denton formation, L. Cret., Texas). (The three dominant oyster species involved are: *Gryphaea washitaensis* (α), *Ostrea carinata* (β), and *Ostrea quadriplicata* (γ). Distance apart between sections in miles. Adapted from Laughbaum, 1960.

salinity as in (*A*); warm, shallow water conditions (locality 7 — the shallowest locality of biofacies *B*); (*C*) water slightly turbid, and shallower water to the north; (*D*) temperature, tropical to semitropical; depth, shallower to the north; deposition in the upper part of the neritic zone.

Dwarfing/Commensalism in Nonmarine Bivalves.

Broadhurst (1959) described a new nonmarine bivalve species, *Anthraconaia pulchella* (see Figure 8.29), from the Coal Measures of Manchester, England. In considering the paleoecology, he related "dwarfing" of this species to failure of food supply, foul bottom conditions, or both (1959, Figure 6). Elsewhere in the Coal Measures (Eagar, 1952) stunting of individuals was attributed to silting.

Trueman (1942) reported an example of supposed commensalism between *Spirobis* worm tubes (marine polychaetes) and various lower Carboniferous nonmarine bivalves (*Naiadites, Carbonicola, Anthroconauta*). These tubes, in sizable numbers, were attached to particular parts of the bivalve shell. From these data the investigator determined the position and type of mantle border. Thus *Naiadites* appeared to have had a mantle border with an incurrent region that may be compared to that of modern *Mytilus*.

Size in Marine Bivalves.

Maximum attainable size in marine bivalves is thought to be controlled by temperature. The gigantic forms among living and fossil marine bivalves (living *Tridacna gigas*, fossil *Inoceramus*, and Upper Cretaceous rudists) are all from a warm-water environment. The largest Paleozoic bivalves occurred in the Permian of New South Wales and attained lengths to 200 mm. Living bivalves of the Arctic and Antarctic waters never grow to large size. Small species, less than 10.0 mm, are abundant in the Antarctic (Nicol, 1964*b*).

Bathymetric and Climatic Indicators.

Tertiary to Recent molluscan faunas (including bivalves) from the west coast of the United States have long been recognized as valuable bathymetric and climatic indicators (see Raup and Lawrence, 1963; and critical discussion of this paper by McAlester et al., 1964, for east-coast Pleistocene molluscs). Thus the bivalve *Venericardia planicosta* occurs with fossil palms in the Eocene of Puget Sound (a cool-temperate biota); a typical subtropical fauna of the Upper Miocene of Santa Margarita — San Pablo includes *Ostrea titan, Cardium quadrigenarium*, and *Pecten estrellanus*; an Upper Pliocene fauna at San Diego includes *Arca, Pecten*, and *Ostrea* species, among others (this is a warm-temperate fauna).

Rate of Bivalve Evolution.

In a concise instructive paper, Nicol (1953) evaluated the average duration of 20 late Cenozoic bivalve species from the Atlantic Coastal Plain. The average duration came to 6.5 million years; that is, after this lapse of time, the species either became extinct, or gave rise, by genetic modification, to a new species. One species from the Middle Miocene of Virginia, *Euloxa latisulcata*, survived an estimated one million years. Species of the same genus (*Glycymeris*, for example) will have variable durations in time. A few species will be short-lived, that is, 1 to 2 million years, while numerous others (18 species of *Glycymeris*) will endure from 2 to 15 million years.

Associated with longer duration in time, a bivalve species population will be more numerous in individuals, will have a broader geographic spread, and display greater variability. The variable duration of species within a given genus can help to clarify the reason why certain of the longest enduring species remain as relicts or last survivors in a given region. A noteworthy example of this last point are the last rudists in Catalonia that preceded the final extinction of the group on a worldwide scale.

A possible follow-up can permit estimates of the *probable number of generations* within a given Tertiary species whose duration has been inferred. In turn, that would place in focus the genetic variability (expressed in shell morphology) involved in the production of the observed older-younger members of the studied species population. (See E. G, Kauffman, 1977, pp. 112–116, ref. p. 439.)

Availability of Ecological Niches in Paleozoic Seas? The rarity of shell-fixation to the bottom by Paleozoic bivalves was utilized as a point of departure to explore available ecologic niches in Paleozoic seas. One rare bivalve example having this habit in Paleozoic time was *Pseudomonotis* (Pennsylvanian-Permian).

Following Dacqué's explanation, Nicol (1944) also attributed the rarity to: (*a*) bivalve occupancy of the same environment with dominant Paleozoic brachiopods; (*b*) many brachiopods attached to hard substrates by shell fixation (see Chapter 7); (*c*) the fact that numerous brachiopod genera disappeared at the end

of the Permian and continued this numerical decrease throughout the Mesozoic; (*d*) this left open ecologic niches where attached brachiopods had previously been; (*e*) since the greatest adaptive radiation of bivalves occurred from Mesozoic time on, shell fixation to bottom substrates naturally became more widespread; an example, among the oysters (Ostreidae), which do not appear in the Paleozoic.

Amino Acid Content of Fossil Bivalve Shells. See Table 2.2, Chapter 2, for amino acid content of Tertiary mollusks (compare Hare, 1963).

Paleotemperatures. Dodd (1964) defined nine structural shell types in living *Mytilus californianus*. These are depicted in Figure 8.38*A*. When the known mean annual temperature of the collecting locality was plotted against structural type (Figure 8.38*B*), the structural-type numbers 1–9 increased with decreasing tem-

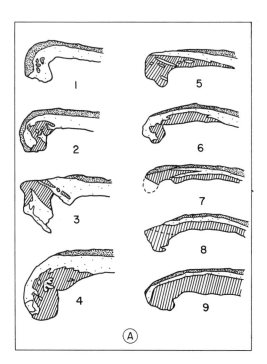

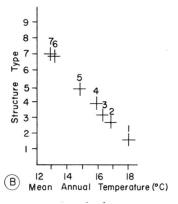

FIGURE 8.38 Structural shell types in living *Mytilus californianus* and inferred paleotemperatures from fossil bivalves. After Dodd, 1964.
(A) Tracings of photograph showing various structural types [numbers 1–9; compare (B)], X 1. Lined area—beak calcite and inner prismatic layer; clear area—beak aragonite, blocky aragonite area, plus nacreous layer; stippled area—outer prismatic layer. (B) Structural-type variation with mean annual temperature (°C).

perature. The investigator thought that shell microstructure of unworn or slightly unworn fossil bivalves could yield data on paleotemperatures.

CLASS CEPHALOPODA

[The class name (meaning head + foot) may be a misnomer, according to Morton and Yonge (1964), because the circlet of tentacular arms could be of head (cephalic) rather than foot (pedal) origin. See Figure 8.39*B*.] Included in this class are the living pearly *Nautilus*, octopus, squid, argonaut, cuttlefish, as well as fossil ammonoids, nautiloids, bactritids, and coleoids. (For evolution of body form and mode of life, see Figure 8.39*A*.)

Body. Bilaterally symmetrical.

Head. Well developed, with mouth circled by tentacles and containing a buccal mass (with radula and upper and lower jaws or mandibles). In the dibranchiates (*Loligo*), food, after poisoning in the buccal gland, is actually *bitten* by the jaws and swallowed by aid of the radula (Yonge, 1960, pp. *I*-33–*I*-34). Cephalopods are predaceous carnivores. Compare Figure 8.39*B*, part C. There are large eyes. The head anterior in position is surrounded by a circle or crown of mobile and prehensile tentacles (arms).

Foot. Highly modified; in part represented by a funnel or siphon called the muscular *hyponome*, through which the mantle cavity ejects a concentrated exhalent water current. This current serves for jet propulsion. (Signals initiated by

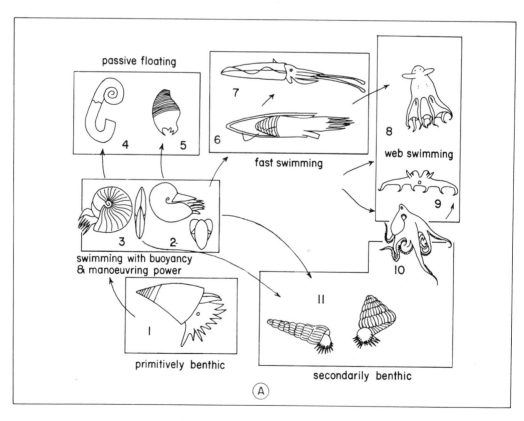

FIGURE 8.39A The evolution of body form and mode of life in the Cephalopoda. (1) Hypothetical primitive benthic cephalopod. (2) Nautiloid. (3) Ammonoid. (4) *Macroscaphites.* (5) *Morphoceras.* (6) Belemnoid. (7) A modern teuthoid. (8) *Vampyroteuthis.* (9) *Opisthoteuthis.* (10) *Octopus.* (11) *Turrilites.* (12) *Trochoceras,* to right of (11). After John Morton, 1963.

stimulation of the eye are transmitted rapidly via giant muscle fibers to retract the muscles of the hyponome, and so on. This causes a sudden jet expulsion of water and darting movements of the animal in an opposite direction.)

Ctenidia. Single pair of gills in dibranchiates such as *Octopus*; two pairs of gills in tetrabranchiates such as *Nautilus.*

Systems. Circulatory (heart, with ventricle and one or two pairs of auricles); *respiratory* (blood is pumped by "branchial hearts," that is, muscular dilatations to, and oxygenated in, ctenidial capillaries); *nervous* [greatly concentrated and highly organized; giant nerve fibers control movements of mantle and funnel (hyponome) in the three divisional central nervous system]; *Muscular*; *Digestive*; *Reproductive.*

Reproduction. Sexes separate, marked sexual dimorphism in some cephalopods (*Argonauta*); development direct, that is, embryo-to-adult without trochophore or veliger larval state (see Figure 8.3); fertilization internal; [copulation by seasonally modified arms called the *hectocotylus* in dibranchiates, and permanently modified arms (= *spadix*) in *Nautilus*].

Skeleton and Shell. Head is built around internal cartilagenous skeleton that serves as a support or enclosure for the nervous system and provides attachment sites for many main muscles and hyponome.

Typically, with a univalved coiled or straight, chambered shell. Septa separate the chambers that connect by a fleshy siphuncular cord that is a prolongation of the mantle. The undivided body chamber holds the living animal (Figure 8.39*B*, part D). The divided portion of the conch is called the *phragmacone*. In living and fossil dibranchiates, the shell is reduced, internal, or completely absent (Figure 8.39*A*). Buoyancy is adjusted by: moderating gas in the chambers or pumping water

into cuttle bone (*Sepia*) (Figure 8.39*B*, part *E*, see Chapter 15).

Habit and Habitat. Predaceous carnivores, exclusively marine; active swimmers.

FOSSIL CEPHALOPODS:

Subclasses

Neobiologists recognize three subclasses of the class Cephalopoda: Nautiloidea, Ammonoidea, Coleoidea (Morton and Yonge, 1964, pp. 5–6). In addition to these three subclasses, paleobiologists require three additional subclasses to accommodate fossil forms: Endoceratoidea, Actinoceratoidea, Bactritoidea (Teichert and Moore, 1964, pp. K2–K4; Sweet, 1964, p. K12). The essential code (taxonomy) for each subclass follows.

Species

Although the class Cephalopoda has more than 10,000 fossil species (nautiloids, bactritids, ammonoids, coleoids), merely 650 species are extant in modern seas. These include only coleoids (octopus, squid, cuttlefish, and argonaut) and *Nautilus.*

Orientation and Skeletal Morphology

Orientation. The standard orientation of coiled and straight cephalopods is shown in Figure 8.40.

Morphology. Foremost among subclasses, exclusive of the Ammonoidea and Coleoidea, are the siphuncular structures, position, and so on.

Siphuncular Structures. The general morphology of an orthoceratid cephalopod is shown in Figure 8.41*A*. Two divisions of the siphuncle are recognized in fossil forms: *ectosiphuncle* (septal necks and connecting rings) and *endosiphuncle* (all space and calcareous tissues embraced by the ectosiphuncle).

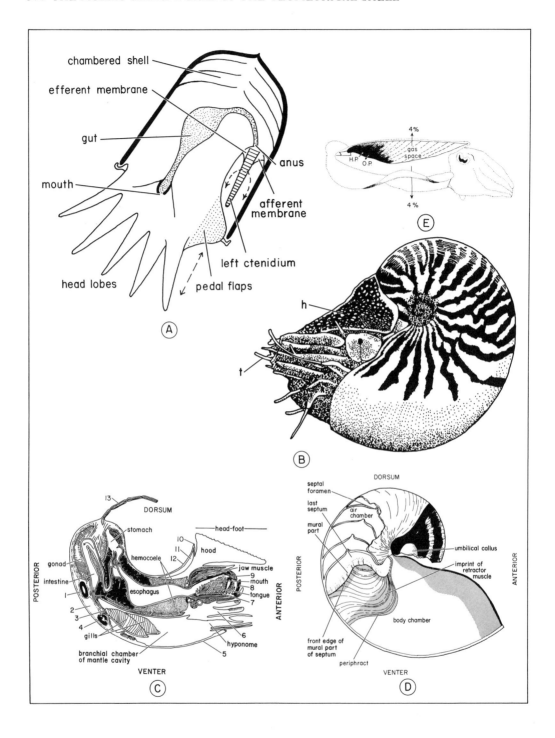

FIGURE 8.39B Probable origin, shell, and other features of cephalopods (schematic).

(A) Postulated early stage in evolution of cephalopod from ancestral mollusk (compare Figure 8.1B). Note chambering of shell by septa, in domelike shell of primitive ancestor; development of tentacles, head lobes, modification of foot to form pedal flaps at entrance to mantle cavity; migration of anus to ventral position.

(B) *Nautilus macromphalus*, Rec., New Caledonia. Living forms seen in action in Catala's film taken at Noumea, New Caledonia, Carnival Under the Sea (Wichita State Univ. Library).

(C) Fleshy parts of *Nautilus;* note orientation, anterior-posterior, dorsum, venter. 1. heart. 2. pericardial chamber. 3. anus. 4. kidney. 5, 10–12 mantle: 5, 12-fold; 10 – edge of; 13 – siphuncular cord.

(D) Mature *Nautilus pompilius* shell, cut along plane of bilateral symmetry; note body chamber (phragmacone, septa, septal foramen through which siphuncle passes).

(E) *Sepia*, showing internal shell (cuttle bone); oldest and most posterior chambers almost full of liquid (in black); newest, ten plus complete chambers are completely filled with gas. These chambers confer buoyancy and animal can retain normal posture. *HP* – hydrostatic pressure of sea water is balanced by osmotic pressure (*OP*) difference between cuttle bone liquid and the blood. The divergent arrows marked 4% denote upward lift conferred by the cuttlebone (= 4% of animal's weight in air), and downward sinking force due to animal's excess weight that it overcomes.

Illustrations redrawn and adapted from Yonge, 1960; Stenzel, 1964; Denton, 1964.

TABLE 8.1. Types of Septal Necks (see Figure 8.41A part B)

1. Achoanitic. Rudimentary septal necks; slight inflection of septa around septal foramen (see Figure 8.41A) (restricted to the Ellesmerocerida).

2. Retrochoanitic. Septal necks directed backward (nautiloids).

3. Loxochoanitic. Like (2) above, but short, straight, pointing obliquely toward interior of siphuncle (Ellesmerocerida, some Endocerida, and Orthocerida).

4. Orthochoanitic. Like (2) above, but straight, cylindrical, and extending only a short way to the preceding septum [Ellesmerocerida, Orthocerida (superfamily Orthocerataceae), and most coiled forms (Tarphycerida, Barrandeocerida, Nautilida)].

5. Suborthochoanitic. Like (2) above, but short, straight, and with slightly outwardly inclined tips but with no measurable brim.

6. Cyrtochoanitic. Like (2) above, but curved so as to be concave outward (Actinocerida, Oncocerida, Discosorida).

7. Hemichoanitic. Like (2) above, but extending one-half to three-fourths of the distance to the preceding septum. [Endocerida (Proterocameroceratidae) and rarely in Orthocerida.]

8. Subholochoanitic. Like (2) above, but approximately equal in length to the distance between two septa and deflected inward at their tips leaving an appreciable gap between two successive septal necks (occurrence as in (3) above).

9. Holochoanitic. Like (2) above, but extended backward through the length of one camera (Endocerida, but also in some genera of Nautilida, such as *Aturia*).

10. Macrochoanitic. Like (2) above, but reaching backward beyond the previous septum and invaginated into preceding septal neck (occurs only in Endocerida).

Source. After Teichert and Glenister, 1954, and Teichert, 1964.

Ectosiphuncle. Septal neck types are illustrated in Figure 8.41A, part B, and defined in Table 8.1. These can enter into both family and generic descriptions. Thus family Thylacoceratidae Teichert and Glenister, from the Australian Ordovician (1954) embraces genera that have subholochoanitic to macrochoanitic septal necks, whereas a given genus such as *Thylacoceras* has subholochoanitic necks.

Connecting Rings. These are tubular membranes (composed of calcite and conchiolin) that consist of septal necks of *elliptoechoanitic* conchs (that is, all types of septal necks except holo-, subholo-, and macrochoanitic). They may be thin or thick (Figure 8.41A, part C).

Endosiphuncle. Types of endosiphuncular structures include *diaphragms* (Figure 8.41B, part A), *longitudinal partitions*

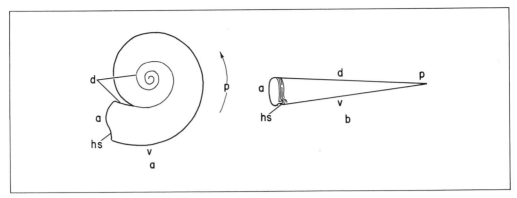

FIGURE 8.40 Standard orientation of straight and coiled cephalopods; *hs* — hyponomic sinus, corresponds to location of the flexible hyponome; *a* — anterior; *p* — posterior; *v* — ventral (note that *hs* is always ventral); *d* — dorsal. Adapted from Teichert.

(Figure 8.41*B*, part *C*), *endocones* (8.41*B*, part *B*), and endosiphuncular lining or deposits (8.41*B*, part *D*). Still other types are parietal deposits and a central cylindrical tube.

Cameral Deposits. The particular position distinguishes three basic types of calcitic cameral deposits (Figure 8.41*B*, part *F*). Thus, if located on the anterior or concave surface of the septum, the deposit is *episeptal* (Type 1); where episeptal deposits cover the wall of the septa, they are called *mural deposits* (Type 2); and deposits on the posterior or convex surface of a septum are named *hyposeptal* (Type 3).

Flower (1955) realized that (*a*) after septal development, there is a considerable lag before cameral deposits are laid down, (*b*) when the latter appeared, they did so at the apical end, and (*c*) thereafter, they grew forward very rapidly. Furthermore, within a given species, the interval of vacant camerae is a constant, for example, five in Figure 8.41*C*, part *d*. Accordingly, youngest deposits (*Y*) will occur directly behind the vacant camera, and oldest (*O*) in the apical direction (Figure 8.41*C*, part *d*). Such considerations are important in fossil fragment's study.

Cameral deposits are weighted on the ventral side (Figures 8.40; 8.41*C*, parts *a–d*). Presumably, they were secreted by tissue. There has been dispute as to

whether deposits were formed during the life of the animal or in the post-mortem state.

Excellent preservation of cameral deposits in the Pennsylvanian *Pseudorthoceras knoxense* from asphalt-impregnated limestones of Oklahoma have permitted electron microscopic and biochemical studies (Grégoire and Teichert, 1965). Hare (1963) found eight amino acids in shell and cameral deposits: lysine, histidine, arginine, aspartic acid, tyrosine, serine, glutamic acid, glycine, and alanine. This finding, taken together with electron micrographs that show abundant fragments of nacreous conchiolin and aragonite crystal imprints in cameral deposits, has convinced many specialists of the *organic* origin of such deposits.

(For the significance of siphuncular and cameral deposits, see Figures 8.41 *A–C* and 8.42.)

Shell Truncation

An interesting morphological feature known as shell truncation or *decollation* occurs in two nautiloid orders — Orthocerida — at least two families, and all of the order Ascocerida (Figure 8.43*A*, part *H*; 8.43*B*, parts *a–c*). Periodic truncation (severing) of the apical portion of the phragmacone (= *deciduous* conch) apparently occurred once or over and over again up to a maximum of some 24 times

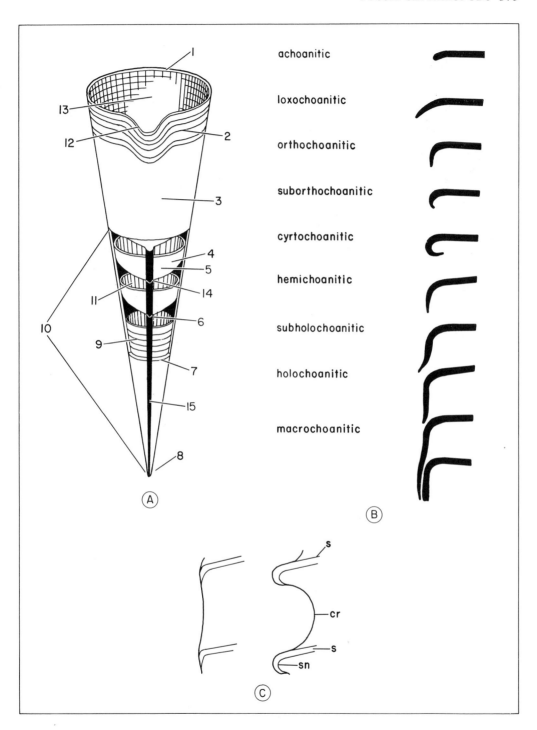

FIGURE 8.41A Skeletal morphology-1: Siphuncular structures.

(A) Schematic representation of orthocone conch (wide spacing of anterior septa of phragmacone exaggerated to indicate actual shape. (1) peristome, (2) growth lines, (3) body chamber. (4) mural part of septum, (5) septum; (6) septal neck, (7) shell wall, (8) initial camera, (9) suture, (10) phragmacone, (11) camera, (12) hyponomic sinus, (13) aperture, (14) septal foramen, (15) portion of siphuncle.

(B) Types of septal neck (see "definitions" in text).

(C) Connecting rings (compare left and right figures); s—septum, sn—septal neck, cr—connecting ring.

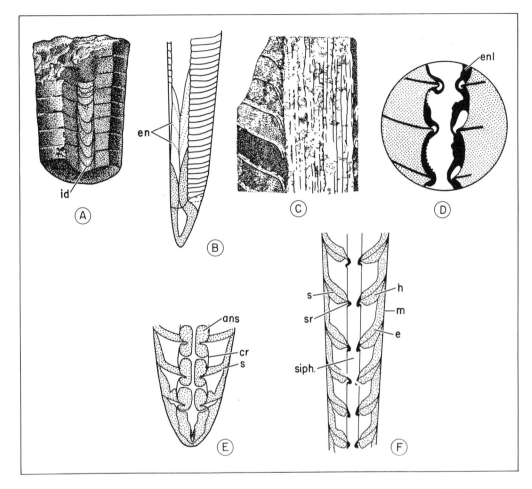

FIGURE 8.41B Endosiphuncular structures (schematic): (A) *Robsonoceras robsonense,* L. Ord., Brit. Col., *id* — irregular diaphragm; (B) *Chazyoceras valcourense,* M. Ord., New York, restored longitudinal section, *en* — endocones. (C) *Evenoceras rozhcovense,* L-M. Ord., Siberia, longitudinal section of siphuncle and part of phragmacone; note longitudinal partitions. (D) *Cayutoceras casteri,* U. Dev., New York, longitudinal section of siphuncle, *enl* — endosiphuncular lining: (E) *Carbactinoceras torlevi,* L. Carb., Europe, lateral longitudinal section, *ans* — annulosiphonate deposits, *cr* — connecting ring, *s* — septum.
(F) Cameral deposits: *Geisonoceras teicherti,* reconstructed longitudinal section showing types of cameral deposits, *h* — hyposeptal, *m* — mural, *e* — episeptal, *s* — septum, *siph* — siphuncle, *sn* — septal neck.

during early stages of growth. The anterior part of the phragmacone, and the body chamber (= mature *breviconic* conch) remaining in the mature stages of growth, bore at its base two structures. These were the *septum of truncation* and the *siphuncular displacement canal* (Figure 8.43*A*, part H*b*). The latter indicates a small ventral displacement in the siphuncle as it crossed the septum of truncation.

The mature (ascoceroid) conch de-

veloped saddle-like dorsal camerae which were gas-filled (Figure 8.43*A*, part H*a*). These, sometimes followed by adventitious, or geronto-pathologic septa (see Figure 8.43*A*, part H*b*), were like those of the severed orthoconic phragmacone. ("Geronto-pathologic" refers to a septa secreted during old age.)

It is thought that truncation may have resulted from a resorptive process (Tasnádi-Kubacska, 1962) by way of the si-

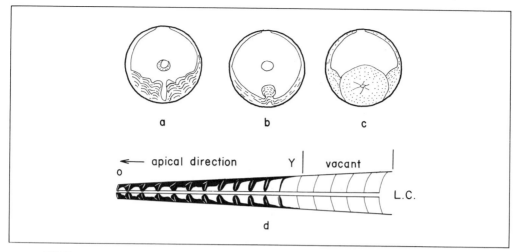

FIGURE 8.41C Skeletal morphology—I: Siphuncular structures (cont.)
Bilateral symmetry of cameral deposits. (Marked concentration of weight on ventral side of shell.)
(a) *Pseudorthoceras*, (b) *Michelinoceras*, (c) *Actinoceras*, (d) *Geisonoceras* sp. showing vacant camerae behind living chamber (*L.C.*); youngest (*Y*) cameral deposits, and oldest (*O*).
Illustrations for Figure 8.41A, B, C redrawn and adapted from Teichert, 1964; Teichert and Glenister,
1954; Flower, 1957.

phuncle and cameral mantle. So viewed, the deciduous conch, representing early growth stages, served as a calcium carbonate storage site.

Deciduous and mature conchs of the same orthoceratid genus have been mistakenly assigned to separate genera (*Brachycycloceras*, for example).

Taxonomy

Subclass Nautiloidea. Small to large conchs. Configuration variable (straight, cyrtoconic, coiled-involute); cameral deposits common in uncoiled forms. Siphuncle—position and diameter variable with long or short segments. Septal necks —commonly orthochoanitic (see Figure 8.41*A*, part B, and definitions) but may be cryptochoanitic; connecting rings of variable thickness with or without siphuncular deposits. [Differentiated from Endoceratoidea by lack of endocones (Figure 8.41*B*) and from Actinoceratoidea by lack of complex annular deposits.] Geologic range: Upper Cambrian to Recent. [Embraces about 700 genera (Kummel, 1954).]

The subclass is divided into eight orders for each of which one or more genera are figured. [Order Ellesmerocerida, the ancestral nautiloid stock, embraces all Cambrian and Lower Canadian nautiloids: *Burenoceras* (Figure 8.43*A*, part B*b*), *Protocycloceras* (Figure 8.43*A*, part A*a*), *Shideleroceras* (8.43*A*, A*b*). Order Orthocerida: *Cayutoceras* (8.41*B*, D). Order Oncocerida: *Tetrameroceras* (8.43*A*, G), *Oncoceras* (8.44*F*). Order Discosorida; *Phragmoceras* (8.44*A*). Order Tarphycerida: *Lituites* (8.44*B*). Order Barrandeocerida: *Peismoceras* (8.43*A*, D); *Uranoceras* (8.44*C*). Order Nautilida: *Indonautilus* (8.43*A*, E), *Millkoninckoceras* (8.43*A*, F), *Nautilus* (8.44*D*); *Aturia* (8.44*E*).]

Subclass Endoceratoidea. Medium to very large conchs (a reconstructed Harvard specimen is about 5800 mm long— Teichert and Kummel, 1960). Generally straight, longiconic (but some breviconic and slightly curved forms). Siphuncle: medium-sized to large, cylindrical, ventral, or subventral generally. Septal rings: achoanitic to macrochoanitic. Connecting rings: simple or complex. Siphuncular deposits: close-packed endocones or radially arranged longitudinal lamellae fill posterior part of siphuncle (Figure 8.41*B*,

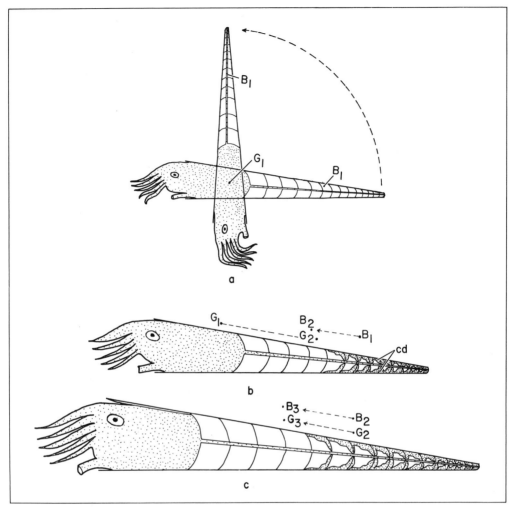

FIGURE 8.42 Cameral deposits and horizontal mode of life (G_1, G_2, G_3 = center of gravity; B_1, B_2, B_3 = center of buoyancy).
(a) G_1 lies on living chamber, B_1 on phragmacone, resulting in vertical position of shell.
(b) Cameral deposits (cd) in apical part of phragmacone causes shift from G_1 to G_2 and B_1 to B_2.
(c) As animal grew, further cameral deposits caused shift of G_2 and B_2 and B_3 permitting continuance of horizontal mode of life.
Illustrations redrawn and adapted from Flower, 1957.

part *D*). Cameral deposits are absent. Geologic range: Lower Ordovician to Upper Ordovician; (?)Middle Silurian.

Two orders are embraced by this subclass: Endocerida and Intejocerida. [Order Endocerida: *Chazyoceras* (Figure 8.41 *B*, part *B*); *Manchuroceras* (Figure 8.43*A*, part *B*); *Emmonsoceras* (8.45*B*). Order Intejocerida: *Evenoceras* (8.41*B*, part *C*).]

Subclass Actinoceratoidea. Medium to very large conchs that are generally straight. Siphuncle: with septal necks, cryptochoanitic to recumbent, segments tend to be wider than long although the opposite also occurs; two sizes of siphuncles — *large*, with a broad contact of segments and ventral wall, and *small*, that is subventral to subcentral in position; within siphuncle, annular deposits fill much of the space; cameral deposits: episeptal and hyposeptal (Figure 8.41*B*, part *F*, and 8.45*A* and *C,b*). Other genera have only deposits on the concave surface of the cryptochoanitic neck (*Stützring* of Teichert). Geologic range: Middle Ordovician to Pennsylvanian. [Embraces a single order, Actinocerida, *Carbactinoceras* (Figure 8.41*B*, part *E*), *Armenoceras* (8.45*C*)].

Subclass Bactritoidea. Orthoconic to cyrtoconic conchs, longiconic or breviconic. Apical angle, small or large; siphuncle, in contact with ventral wall, narrow. Suture, with V-shaped ventral lobe. Protoconch, globular to egg-shaped. Geologic range: Ordovician to Permian. [Contains a single order, Bactritida, which in turn embraces two families: Family Bactritidae, *Pseudobactrites* (Figure 8.45*D*), family Parabactritidae, *Parabactrites*.]

Relationship between Cephalopod Groups

All cephalopod evolution is traceable to the Cambro-Ordovician nautiloids, the Ellesmerocerida (Figure 8.43*A*. From this ancestral stock in Ordovician time, arose all the nautiloid orders: Orthocerida, Discosorida, Tarphycerida. The Orthocerida gave rise to the Ascocerida; the Tarphycerida gave rise to Oncocerida and Barrandeocerida (all in Lower Ordovician time); from the Oncocerida in Middle Silurian time arose the Nautilida.

The subclasses Endoceroidea and Actinoceratoidea are traceable (the links are unclear) to the Ellesmerocerida. The subclass Bactritoidea appears to have evolved from an orthoceratid stock in Upper Silurian time. According to Erben (1964), the family Bactritidae of the last-named subclass gave rise to the Protobelemnoidea and the family Parabactritidae to the Belemnoidea *s.s.* [Note: *s.s.* (*sensu stricto*) — in a restricted sense, that is, the subclass has been sharply delimited; *s.l.* (*sensu lato*) — in the broad sense.]

Other investigators (Spath, 1933, Table 2) attribute the rise of the belemnoids to orthoceratid ancestors.

In turn, from the belemnoids, the modern dibranchiates evolved in Triassic time although the links are nazy. [Further discussion comes under subclass Coleoidea (Order Belemnitida).]

The intermesh of phylogenetic linkages between cephalopod groups completes itself with the memorandum that the nautiloids apparently became the ancestral stock of the ammonoids.

The Ammonoids

Origins. There have been differences of opinion on the derivation of ammonoids. (Details below.) Spath (1933) favored derivation from coiled Silurian nautiloids (*Barrandeoceras* and equivalent types; see Figure 8.44*C*). This derivation was considered impossible by Schindewolf (1954) because characters found in *Barrandeoceras* types differed profoundly from coiled ammonoids: siphuncle position, cup-shaped initial chamber, rapidly expanding first whorls, and mode of involution. Instead, both he and Erben (1964) favored derivation from a line starting with *Bactrites*.

Two postulated derivations of coiled ammonoids are as follows.

Erben (1964): *Bactrites → Lobobactrites → Cyrtobactrites → Anetoceras* (A typical most primitively coiled ammonoid from the Lower Devonian in Germany.)

Schindewolf (1954): *Bactrites → Lobobactrites → Gyroceratites → Mimagoniatites.*

Emphasis falls on the first two forms regarded as primitive ammonoids by Schindewolf; in turn these derived from orthocone nautiloids. (*Anetoceras* and *Gyroceratites* belong to the same family and *Mimagoniatites* to another family. However, both families are embraced in the same ammonoid superfamily, Anarcestaceae.) See Figure 8.45*E* for comparative suture lines illustrating Schindewolf's postulated line of descent.

Morphology. Whether ammonoids evolved directly or indirectly from nautiloid cephalopods through the Bactritidae, salient morphological features need review before defining the subclass.

Suture Line. Ammonoid sutures act as the single most useful (and most used)

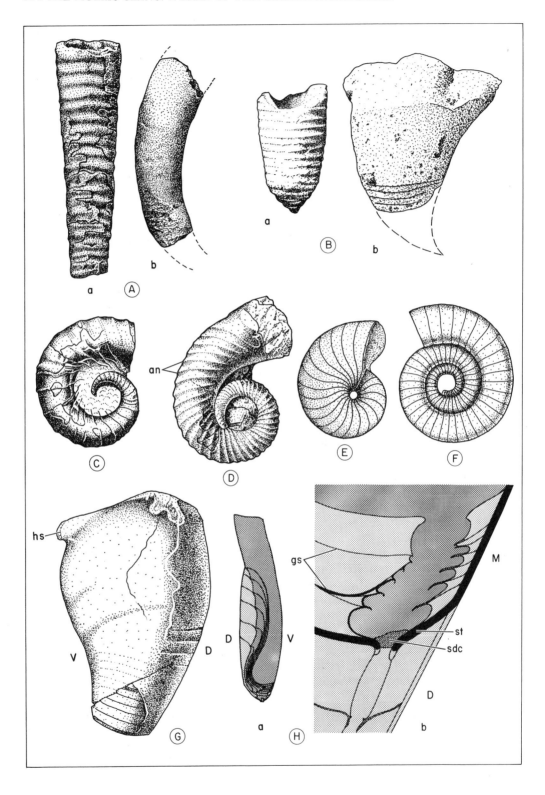

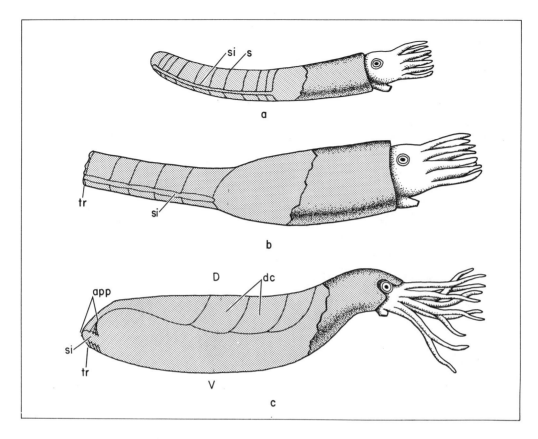

FIGURE 8.43A Shape, truncation, and curvature in nautiloids.

(*A*) Longicones; (*a*) *Protocycloceras lamarcki,* L. Ord. Canada, annulate orthocone (ortho = straight); (*b*) *Shideleroceras sinuatum,* cyrtoconic (that is, curved conch in which less than one whorl is completed), lateral view, U. Ord., Ohio. [A; Bb — Order Ellesmerocerida].

(*B*) Brevicones: (*a*) *Manchuroceras excavatum.* L. Ord., Tasmania; endogastrically curved [see (*G*)]; (*b*) *Burenoceras barnesi,* lateral view, L. Ord., Texas.

(*C*) Gyrocone: *Goldringia cyclops,* M. Dev., New York.

(*D*) Torticone (= coiled in heliocoidal spire): *Peismoceras optatum,* M. Sil. – U. Sil., Czech; *an* — annulations.

(*E*) Involute coiling: *Indonautilus kraffti,* M. Trias. – U. Trias., Himalayas.

(*F*) Evolute coiling: *Millkoninckoceras konincki,* L. Carb., Europe.

(*G*) Endogastric condition [= venter (*v*) on or near concave (inner) side], *Tetrameroceras picinctum,* M. Sil., Europe, compare (*H*).

(*H*)(*a*) Exogastric condition [= venter (*v*) on or near convex (outer) side], *Ascoceras manubrium,* U. Sil., Sweden; (*b*) *Ascoceras* sp. showing septum of truncation (*st*) and siphuncular displacement canal (*sdc*), gerontic septa (*gs*), mature conch (*M*), and deciduous conch (*D*).

Glossoceras lindstroemi, U. Sil., Sweden, three growth stages restored in part (*a–c*) to illustrate truncation and orientation of the conch; *si* – siphuncle, *s* – septa, *tr* – truncated [truncated part of phragmacone = deciduous portion; body chamber and attached anterior part of phragmacone (*app*) = mature portion], *D* – dorsal, *V* – ventral, *dc* – gas-filled dorsal camera.

Illustrations redrawn and adapted from Teichert, Unklesbay, Teichert and Glenister.

feature in Paleozoic form classification. For Mesozoic forms, these suture lines are supplemented by whorl configuration, ornamentation, mode of coiling, apertural type, and so on.

How was the suture formed? Repeatedly, the animal grew, it vacated the pre-vious body chamber, secreted a floor (septum) behind it, it moved forward and occupied a new undivided body chamber. *The suture line is the wall reflection or trace of the juncture line between septum and the inner wall of the whorl.* (See text page 401 for probable significance of suture lines in the life of the animal.) [Movement forward, less than a full cameral length in a given whorl, can result in overlapping of septal sutures during late ontogenetic stages (Arkell, 1957).]

Three categories of ammonoid sutures are shown in Figure 8.46. *Goniatitic* sutures have all, or almost all, lobes plain, that is, undivided and not denticulate; *Ceratitic* sutures have serrate or denticulate lobes with entire saddles; *Ammonitic* sutures have both lobes and saddles denticulate.

Suture names can be misleading. (1) Goniatitic sutures occur beyond the Paleozoic, into Triassic and into Cretaceous time. They occur in forms unrelated to

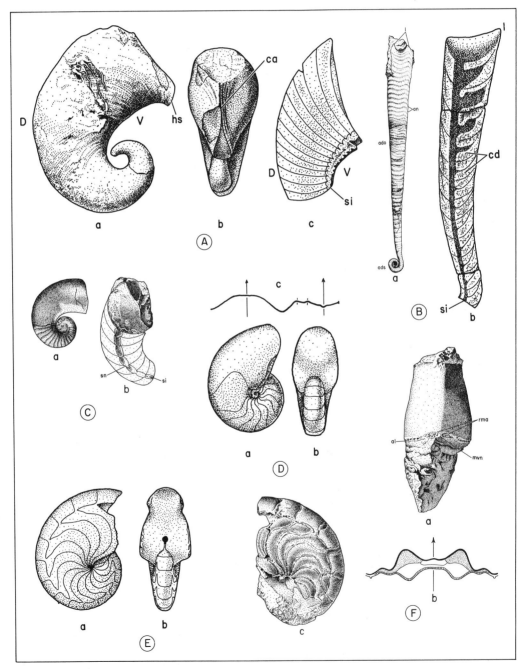

FIGURE 8.44

FIGURE 8.44 Selected nautiloid genera.

(A) Order Discosorida. *Phragmoceras lamellosum*, M. Sil., Sweden, X 0.45; (a) Lateral view, *D*—dorsal, *V*—ventral, *hs*—hyponomic sinus; (b) ventral view, *ca*—constricted aperture; (c) *P. farcimen*, Sweden, X 0.7, dorso-ventral section showing siphuncle (*si*) with internal deposits, *D*—dorsal, *V*—ventral, *ca*—constricted aperture.

(B) Order Tarphycerida. *Lituites lituus*, M. Ord., Germany; (a) lateral view, fully mature specimen, X 0.5, *ads*—adapical spiral, *ado*—adoral slightly sigmoid orthocone, *an*—annulation, *l*—lappets; (b) cameral deposits (*cd*), siphuncle (*si*), M. Ord., Sweden, X 1.

(C) Order Barrandeocerida. *Uranoceras*: (a) *U. hercules*, Sil., Wisc., Ill., X 0.3; note tendency to uncoil; (b) *U. uranium*, M. Sil., Czech., X 0.7; *si*—siphuncle, *sn*—septal neck.

(D) Order Nautilida. *Nautilus pompilius*, Rec., S.W. Pacific, X 0.5.

(E) *Aturia*: (a) *A. curvilineata*, Mio., Venezuela, X 0.3; (b),(c) *A. augustata*, (b) Mio., Washington, (c) internal mold, Oligo., Washington; (b),(c) enlarged.

(F)(a) *Oncoceras* sp., Upper Ord., Canadian Arctic, X 1.5, ventral view; *mvn*—median ventral notch, *rma*—retractor muscle scars, *ai*—annular elevation; (b) schematic representation of *Nautilus* showing annular elevation, retractor muscles attached to stippled areas, longitudinal mantle muscles attached to heavy black line, compare (F)(a).

Illustrations redrawn and adapted from Teichert, 1964; Miller and Downs, 1950; Sweet, 1959; Mutvei, 1957.

true Paleozoic goniatites. (2) The same extension is true of ceratitic sutures that are known from Upper Mississippian and occur in Jurassic and in Cretaceous forms. These are referred to as "pseudoceratites." (3) Characteristic ammonitic sutures occur in the early Permian (Miller, Furnish, Schindewolf, 1957; Arkell, 1957). (4) Any category of sutures may occur in the same family, the Tropitidae, for example, Upper Triassic; goniatitic and ceratitic sutures may characterize species of the same genera (*Sandlingites*, for example); ceratitic sutures can display a tendency to become subammonitic and complex within a given suture line of an individual ammonoid and within the same species (*Nicomedites*, for example).

Sutures may be grouped in still a different way. Schindewolf (1954) recognized three grades of suture development: trilobate, quadrilobate, and quinquelobate.

Trilobate Grade. This represents the most elementary Paleozoic goniatite suture. It consists, as the name indicates, of three lobes (protolobes): ventral or external lobe, lateral lobe, and dorsal or internal lobe (see Figure 8.46). These protolobes are found in all goniatites, clymenids, ceratites, and ammonites. Subsequent evolution leads to the development of secondary lobes (metalobes) and, in more advanced goniatites, to subdivisions of the saddles to form new lobes.

Quadrilobate Grade. This was the pri-

mary suture in Triassic ceratites. To the three protolobes, a first umbilical lobe is added (see Figure 8.46B).

Quinquelobate Grade. Since Jurassic and Cretaceous ammonites arose from a Triassic ceratite line, they possess similar lobe formations, excepting one detail. They are quinquelobate and have two umbilical lobes in addition to the three protolobes.

Because of the sutures being important in classification, additional terminology is used (Figures 8.46 and 8.47A). The external (visible) portion of the umbilical lobe is called *suspensive lobe*. Subdivisions of the suspensive lobe are called *auxiliaries* (auxiliary lobe). During development minor *accessory* lobes are formed between the primaries. When these lobes are as large as the primary lobe (and often indistinguishable from them), they are designated "adventitious" (Figure 8.47A, b). In complex ammonitic sutures, minor frills on lobes are *lobules*; leaf-shaped frills on saddles are *folioles*.

Arkell (1957) cautioned that (a) considerable variation in sutural details within the same species and (b) bilateral asymmetry can result in distinctive differences in suture line on opposite sides of a whorl. *Dactylioceras commune* displays such asymmetry in a marked way (Swinnerton and Trueman, 1917, Figures 9–10, and see Figure 8.47A, c, herein).

There are a variety of ways to draw (reproduce) a suture: camera lucida, or

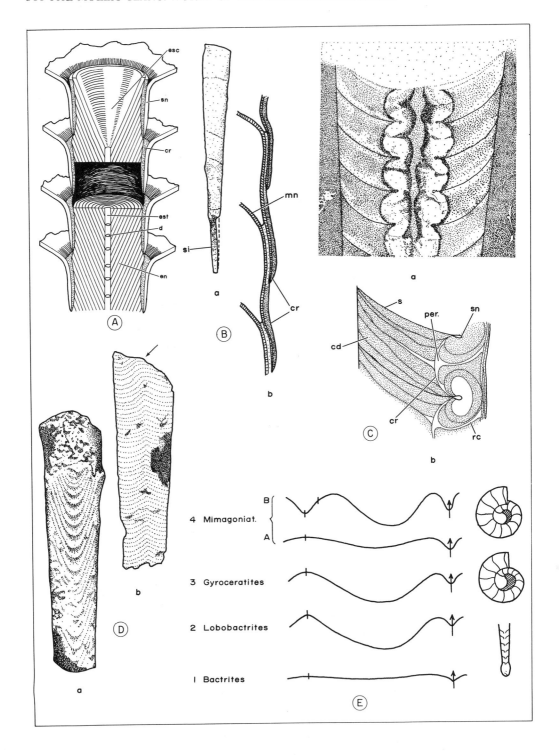

FIGURE 8.45 Selected endoceratids, actinoceratids, bactritids, and others.
(A) Siphuncular structures in endoceratids: *en* – endocones, *d* – diaphragm, *est* – endosiphuncular tube, *cr* – connecting ring, *sn* – septal neck, *esc* – endosiphocone.
(B) Endoceratoidea, *Emmonsoceras aristos,* Lower M. Ord., New York; (a) part of phragmacone and protruding siphuncle (*si*) length, 38 cm; (b) Diagram of siphuncle wall, *mm* – macrochoanitic neck, *cr* – connecting ring.
(C) Actinocerida, *Armenoceras centrale,* schematic; (a) longitudinal section, Cambro-Ord., E. Jehol, enlarged; (b) diagrammatic sketch to show neck rings (*sn*), septa (*s*), radial canals (*rc*), connecting rings (*cr*), perispatium (space between annulosiphonate deposits (*per*) and free part of connecting ring), endosiphuncular linings and cameral deposits (*cd*).
(D) Bactridoidea, *Pseudobactrites peneaui,* L. Dev., France, X 1.5; (a) Ventral view showing linguiform ventral sinus, (b) ventro-dorsal view depicting prominent saddle (arrow).
(E) Comparative suture lines in *Bactrites* through *Mimagoniatites* illustrating Schindewolf's (1954) thesis of ammonoid descent. The ammonoid *Gyroceratites* with a loss of coiled shell has a suture the same as *Lobobactrites,* from whom it descended. *Mimagoniatites* starts off with a *Gyroceratites* suture line (4A) but the adult part of the shell shows further evolution resulting in a more complex suture (4B). For details of interpreting sutures, see next section, "Ammonoids."
Illustrations redrawn and adapted from: (A) Teichert, 1964; (B) Flower, 1955; (C) Kobayashi and Matumoto, 1942; (D) Erben, 1957; (E) Schindewolf, 1954.

directly inking on a plastic strip applied to the shell surface, and so on (see Furnish and Unklesbay, 1940, for details).

In measuring any suture, it is desirable to indicate shell diameter at the place where it occurs, because suture lines change during ontogenetic development (see Figure 8.48).

Phylogeny and Ontogeny. One traces progressive growth stages from juvenile (neanic) to old age (gerontic) in a study of successive suture lines becoming more complex during the individual's life. In turn, such ontogenetic sequences reflect the line of evolutionary descent.

A classic example is seen in Figure 8.48. Progressive growth stages (represented by suture lines *A'–D*) of the genus *Uddenoceras* indicate the sutural changes through the lifetime of a given individual (= ontogeny). Some mature sutures of other Pennsylvanian genera, southwest United States, are found to correspond to one or another of the growth stages of *Uddenoceras.* Thus *Uddenoceras* passed through *Pronorites, Prouddenites,* and *Uddenites* stages of development. This indicates its phylogenetic line of descent.

Lines of phylogenetic descent have also been traced in ammonoid collections from a vertical sequence of strata; changes occur in ornament and lappets (*Kosmoceras,* Middle Jurassic of England) or other shell features including configuration.

D. Collins' (G. A. Canada, M. A. Canada, and G.S.A. Joint Program, Toronto,

Canada, 1978, Abstracts with Programs, p. 382) study of a new collection of 2,100 complete specimens of *Kosmoceras* indicated much greater diversity through time (3 meters of section, samples collected every centimeter) than did Brinckmann because, in his celebrated study of this evolving ammonoid, Brinckmann lumped several vertical layers together in a single sample.

In Figure 8.47*B* the protoconch, the first chamber, should be envisioned as a secretion by the ammonoid larva. This is thought to have been a free-swimming, possible a drifting, form. A free-living larval stage can explain cosmopolitan distributions. Thus the extensive correlations of Mesozoic strata in different parts of the world by ammonoid faunas (Arkell, 1956) can be attributed to the very fortuitous worldwide spread of the free-swimming ammonoid larvae.

Shell Structures and Accessories (Aptychi). Objects associated with ammonites that served the function of closure of the aperture and hence were opercula are collectively called *aptychi.* [One pair of calcareous or corneous plates that completely close the aperture is an *aptychus*; a single chitinous disk that partly closes the aperture is an *anaptychus*, a univalve type of aptychus (Lower Devonian to Cretaceous); fused bivalved calcareous aptychus plates are called *synaptychus* (Upper Cretaceous) (Figure 8.49).]

Classification of aptychi relies heavily on

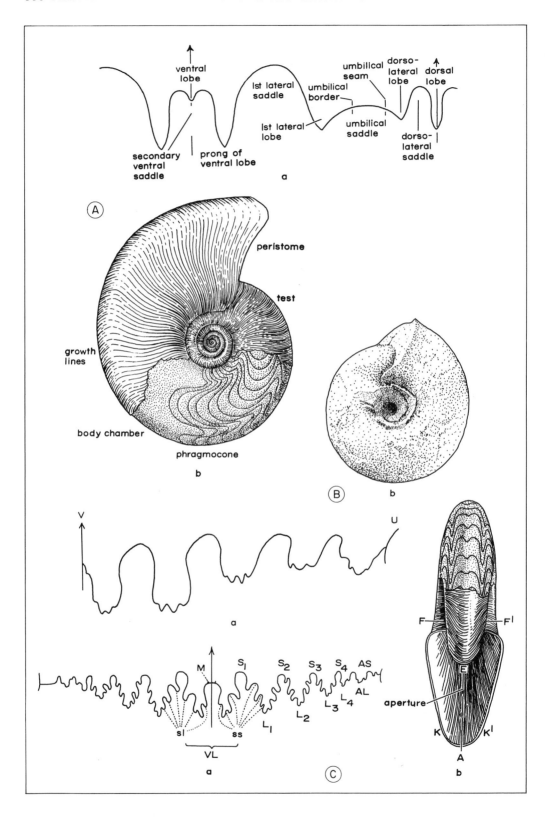

ventral
lobe

1st lateral
saddle

umbilical
seam

dorso-
lateral
lobe

dorsal
lobe

umbilical
border

1st lateral
lobe

umbilical
saddle

secondary
ventral
saddle

prong of
ventral lobe

dorso-
lateral
saddle

a

Ⓐ

peristome

test

growth
lines

body chamber

phragmocone

b

Ⓑ   b

V   U

a

M   S₁   S₂   S₃   S₄ AS

AL

sl   ss   L₁   L₂   L₃ L₄

VL

a   Ⓒ

F   F¹

E

aperture

K   K¹

A

b

FIGURE 8.46 Ammonoid sutures—types.

(A) Goniatitic. *Manticoceras* sp., U. Dev.; (a) characteristic goniatitic suture, *v*—ventral lobe, *d*—dorsal lobe, *u*—umbilicus; arrows point in apertural direction (see text); (b) lateral view showing test exterior and internal mold with sutures; growth lines, *p*—periostome (aperture), *in*—internal mold, test, external, *ph*—phragmacone; external = external suture, portion from mid-ventral lobe to umbilical seam; internal = internal suture, portion from umbilical seam to mid-dorsal lobe. [Note that in tracing suture from mid-dorsal to mid-ventral lobe across the umbilicus only one-half of the complete suture (internal and external) is revealed. (For details, see text.)]

(B) Ceratitic. *Owenites koeneni.* L. Trias., Calif.; (a) typical ceratitic suture showing one-half of the external suture which, to be complete, would show a mirror image (symmetry) on the left side of the arrow; (b) mature specimen, 44 mm diam., *Meekoceras gracilitatus,* Nevada, lateral view.

(C) Ammonitic: (a) *Perrinites cummingsi,* Carb., Texas, showing complete external suture; *M*—median or ventral saddle in ventral lobe, *sl*—secondary lobes, *ss*—secondary saddles, S_1–S_4—saddles (= forward curve of suture line in apertural direction) L_1–L_4—lobes (= backward curve of suture line toward posterior end of whorl) *As* and *Al*—auxiliary saddles and lobes (= saddles and lobes beyond the second lateral saddle, that is, S_3 to *As*, L_3 to *A*1. (b) *KAK'* = outer part of whorl = venter, *FEF'* = inner part of whorl =dorsum.

Illustrations redrawn and adapted from: (A) Miller et al., 1957; (B) Kummel and Steele, 1962; (C) Plummer and Scott, 1937.

surficial ornamentation (punctae, granulae, ribs, furrows, and so on) and secondarily, on configuration and thickness. The structure of these plates is shown in Figure 8.49*F, b.*

Bivalved forms have been mistaken for brachiopods (*Orbiculoides*), pelecypods, phyllopods, and phyllocarids. Improper determinations are corrected readily by studying the cross-section of a given plate. Aptychi may occur in beds where ammonites are rare. Aptychi are common to abundant in particular Mesozoic beds [bituminous shales and lithographic limestones (Solenhofen) in the Upper Jurassic and Lower Cretaceous] (Arkell, 1957; Basse, 1952; Trauth, 1937). Less frequent is the occurrence of aptychi and anaptychi in place in a given ammonite (Figure 8.49*C, a*). They do occur in Paleozoic beds, but rarely. (Carboniferous Cleveland shales, Ohio; Upper Devonian of Germany, Trauth, 1935, plates 2, 3.)

Locally, isolated aptychi may serve as index fossils where they form "aptychus beds." Calcitic aptychi are also more resistant to dissolution and diagenesis than are aragonite conchs (Fisher and Fay, 1953). The Kansan Niobrara chalk is a good example with more aptychi than conchs.

When aptychi sealed the apertures (Figure 8.49*F,* retracted portion), both ammonite body and tentacles must have been completely withdrawn into the body chamber. Thus the buoyant shell would react as a passive drifter in the sea, its resis-

tance to movement totally a function of its shape. Contact with the environment could have been maintained through aptychi pores or through a gap left from incomplete aperture closure (Figure 8.49*C*).

Classification

Subclass Ammonoidea. Tetrabranchiate cephalopods, characteristically coiled tightly in a plane (equilateral spiral, see "Derivative Data") and symmetric with a bulbous, calcareous protoconch (Figure 8.47*B*); septa that form angular sutural flexures readily visible on an internal mold, and a small marginal siphuncle. Geologic range: Lower Devonian to Upper Cretaceous.

[The subclass is divided into eight orders, each having one or more genera: Order Anarcestida (the ancestral stock of the ammonoids): *Gyroceratites* (Figure 8.50*A*). *Manticoceras* (Figure 8.49*A, a*). *Agoniatites* (Figure·8.50*B*). Order Clymeniida (dorsally situated marginal siphuncle): *Soliclymenia* (Figure 8.50*C*), *Platyclymenia* (8.50*D*). Order Goniatitida (basic structure of eight lobes): *Tornoceras* (8.50*E, a*), *Imitoceras* (8.50*E, b*), *Goniatites* (8.50*E, d*), *Gonioloboceras* (8.50*F, a*), *Stacheoceras* (8.50*F, b, c*), *Gastrioceras* (Figure 8.51*A*), *Schistoceras* (8.51 *B, b*), *Texoceras* (8.51*C*). Order Prolecantitida (extends to the Upper Triassic); *Prolecanites* (8.51*D*), *Pronorites* 8.51*E*), *Medlicottia* (8.51*F*).]

Order Ceratitida (characterized by greater complexity of suture and development of ornamentation) embraces most Triassic ammonoids; originated from

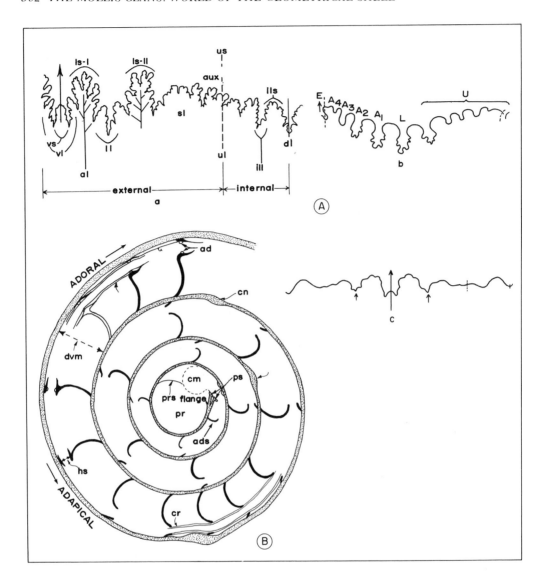

FIGURE 8.47 Further details of ammonoid sutures.

(A) (a) Ammonoid suture. Note umbilical lobe. Umbilical seam (broken line) separates internal from external sutures. Umbilical lobe (ul) is centered on umbilical seam (us) and occurs in both internal and external sutures. Visible portion of ul is a suspensive lobe, that is, subdivided into auxiliary lobes. Arrow points toward aperture and represents median line of venter (see Figure 8.46KAK'): median line of dorsum shown by unbroken line on far right; vs—ventral saddle, vl—ventral lobe, dl—dorsal lobe, ls-I, ls-II—first and second lateral saddle, al—accessory lobe, ll—lateral lobe, sl—suspensive lobe, ill—internal lateral lobe. (b) *Engoceras* sp., L., Cret., Texas: E—external lobe, l—lateral lobe, U—umbilical lobe, A_1–A_4—adventitious lobes. (C) Asymmetrical suture line in *Dactylioceras commune;* arrows indicate first lateral lobe to right and left of ventral lobe (large arrow): to left, lateral lobe is trifid, to right, bifid.

(B) Schematic enlargement of dorso-ventral section of typical ammonoid emphasizing most adapical sutures. Peripheral arrows indicate forward (adoral) and backward (adapical) directions: ad—auxilliary deposits. cm—caecum (closed pouch at apex of siphuncle). prs—prosiphon. pr—protoconch (first chamber of shell closed by proseptum—ps). cn—constriction. cr—connecting rings. ads—adapical septum. hs—height of siphuncle, dvm—dorsoventral measurement.

Illustrations redrawn and adapted from (A, a, b) Arkell, 1957: (A, c) Swinnerton and Trueman, 1917: (B) Miller et al., 1957.

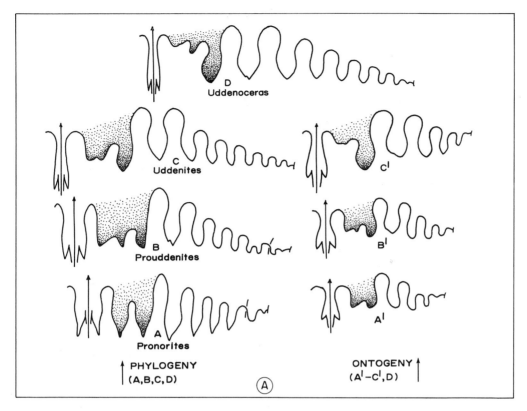

FIGURE 8.48 A classic example of an ontogenetic sequence that corresponds to a phylogenetic line of descent: *A'–C', D: Uddenoceras*, progressive ontogenetic stages. *A–C:* Phylogenetic line of descent leading to *D:* (A) *Pronorites,* (B) *Prouddenites,* (C) *Uddenites. A–D:* all mature sutures. All specimens of Pennsylvanian age from southwestern United States. Development was in direction of greater complexity of suture line through time. After A. K. Miller et al., 1957.

the Carboniferous–Permian Daraelitidae which had a ceratitic suture (Figure 8.52*A*, *a*); subdivided into eight superfamilies. One representative genus is illustrated for each: *Dieneroceras* (Figure 8.52*A*, *b*), *Meekoceras* (Figure 8.52*B*), *Groenlandites* (8.52*C*), *Nevadites* (8.52*D*), *Tardeceras* (8.52*E*), *Lobites* (8.52*F*), *Arcestes* (8.52*G*), *Ptychites* (8.52*H*), *Pompeckjites* (8.52*I*).

Order Phylloceratida (leaf-shaped saddle endings; ancestral to all post-Triassic stocks) itself derives from the superfamily Noritaceae, which includes *Meekoceras* (see Figure 8.52*B*). There is one superfamily; a single genus illustrates it: *Calliphylloceras* (Figure 8.53*A*, part *A*).

Order Lytoceratida (sutures of a few complex elements with endings, mosslike and not phylloid; single valve aptychi in *Lytoceras*). Five superfamilies, one genus illustrates each: *Lytoceras* (Figure 8.53*A*,

part *B*), *Spiroceras* (Figure 8.53*A*, part *Bb*), *Ancyloceras* (8.53*A*, *Ca*), *Baculites* (8.53*A*, *Cb*), *Scaphites* (8.53*B*, part *D*).

Order Ammonitina (characterized by thick test and strong ornament; coiling normal; embraces all post-Triassic stocks except Phylloceratida, Lytoceratida and immediate branches plus derivatives; families of the Ammonitina are traceable to one or another of the last-named orders). Subdivided into nine superfamilies, one genus is illustrated: superfamily Psilocerataceae, *Psiloceras* (Figure 8.53*B*, part *E*); superfamily Eoderocerataceae, *Eoderoceras* (Figure 8.53*B*, part *Eb*); superfamily Hildocerataceae, *Bouleiceras* (8.53*B*, *Fa*); superfamily Haplocerataceae, *Oxycerites* (8.53*B*, *Fb*); superfamily Stephanocerataceae, *Cardioceras* (8.53*B*, *G*); superfamily Perisphinctaceae, *Perisphinctes* (8.53 *B*, *Ha*); superfamily Desmocerataceae,

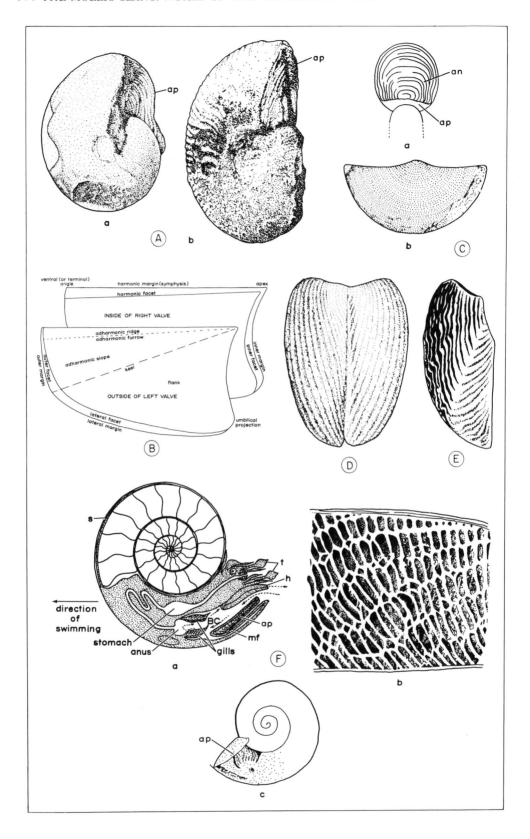

FIGURE 8.49 Ammonoid Opercula.

(*A*) (*a*) *Manticoceras* sp., U. Dev., Germany, X 1; *ap* — aptychus in place; (*b*) aptychus (*Lamellaptychus*) of (*D*) closing the aperture of *Oppelia*, M. Jur., Eng.

(*B*) Morphological features of aptychi.

(*C*) Anaptychi; (*a*) schematic of anapychi closing aperture of the ammonite genus *Lytoceras;* (*b*) *Sidetes*, Cret., X 1.

(*D*) *Lamellaptychus*, M. Jur. — L. Cret.

(*E*) *Rugaptychus*, U. Cret. Generic names applied to aptychi and anaptychi are given in parentheses.

(*F*) (*a*) Hypothetical reconstruction of living ammonite showing position of retracted aptychi (*ap*) on ventral side of body chamber. (Median section of shell.) After M. Schmidt and F. Trauth, cit. Arkell, 1957, *t* — tentacle, *h* — hyponome, *mf* — mantle fold, *s* — siphuncle. (*b*) *Laevaptychus*, x-section, showing lamellar layer, cellular layer. (*c*) Restored conch with head withdrawn and aptychi in place. After Fischer and Fay, 1953.

Illustrations redrawn and adapted from Arkell, and others.

Desmoceras (8.53*B*, *Hb*); superfamily Hoplitaceae, *Placenticeras* (8.53*B*, *Ia*); superfamily Acanthocerataceae, *Adkinsia* (8.53*B*, *Ib*).

INFORMATION DERIVED FROM FOSSIL NAUTILOID AND AMMONOID SHELLS

Nautiloids (Shell Composition and Structure)

In Chapter 15, the evidence on this theme is fully discussed. Best recent evidence appears to weight the argument in favor of biocrystallization of cameral deposits (Grégoire and Teichert, 1965).

Significance of Cameral and Siphuncular Deposits

If a nautiloid were heavier than water, and if there were either no buoyant mechanism at all or an inadequate one to counteract the combined body and shell weight, the animal would sink (negative buoyancy). Conversely, if it were lighter than or equal to the density of seawater, it would enjoy positive buoyancy. In actuality, living *Nautilus* has about the same density as seawater.

The problem of maintaining horizontal locomotion is further complicated by the presence of both gas and fluid in the chambers of living *Nautilus* (and by extrapolation, many fossil nautiloids). Still other factors are septal spacing (close or wide), which would alter the given chambers volume available for the gas or fluid;

presence or absence of cameral and siphuncular deposits; and degree of coiling.

The gas presumably was left behind when the cameral fluid was withdrawn from younger camerae. In the chambers of living *Nautilus,* the argon/nitrogen ratio exceeds that to be found in atmospheric air by a factor of three (Bidder, 1962). It is at a pressure of about 1.0 atmosphere. Bidder (1962) measured the greatest fluid volume in the younger camerae of *Nautilus* and the least as well as the most viscous in the most apical ones. This combined fluid ballast and buoyant gas closely compares to that of *Sepia* (see Figure 8.39*B*, part E) and presumably was present in many nautiloids now known as fossils. The liquid in *Sepia* is controlled by a siphuncular membrane that closes the chambers; however, in *Nautilus*, control is through the siphuncular epithelium (possibly ancestral to the siphuncular membrane of *Sepia*).

As shown in Figure 8.42, the tendency of cameral gas would be to confer positive buoyancy in the apical sector of the cone, thus orienting the animal's head and hyponome vertically downward. (The hyponome is the means by which the animal jet-propelled itself.) Since this was the least suitable position, some compensation had to be made to counteract the positive buoyancy of the apical end of the cone. (Horizontal locomotion was the preferred mode in light of the animal's anatomy.)

Suppose now that discrete weights of

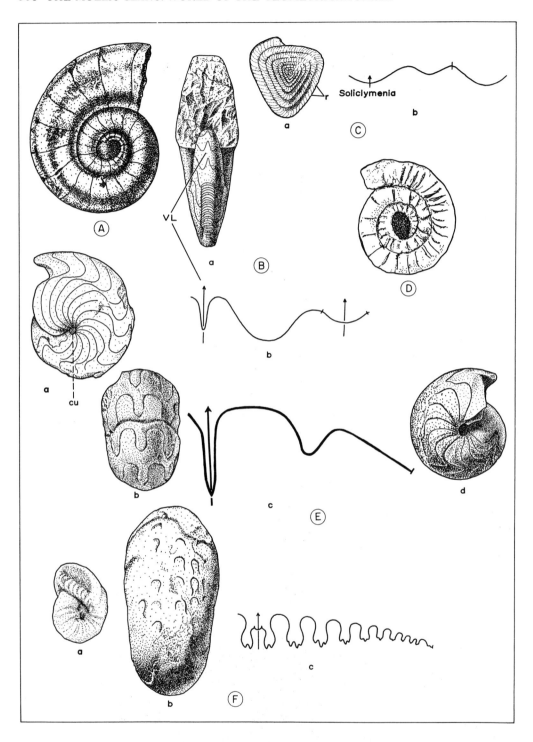

FIGURE 8.50 Paleozoic ammonoid genera—I.
Order Anarcestaceae. (A) *Gyroceratites gracilis*, M. Dev., Ger. Lateral view of discoidal conch showing whorls in contact and large umbilical perforation; Goniatitic suture, enlarged.
(B) *Agoniatites vanuxemi*, M. Dev., N.Y. (a) Ventral view of discoidal conch, shell turned about to living position to correspond to orientation of *VL* (ventral lobe) of suture. (b) Suture showing small ventral, broad lateral, and dorsal lobes, enlarged.
Order Clymeniida. (C) *Soliclymenia paradoxa*, U. Dev., Ger. (a) Lateral view, triangular coiling and densely crowded ribs (r), X 4. (b) Mature suture with broadly rounded ventral lobe enlarged.
(D) *Platyclymenia* (*Pleuroclymenia*) *americana*, Dev., Montana, X 1. Suture simple, with lateral and dorsal lobes only, enlarged.
Order Goniatitida. (E) (a) *Tornoceras* (*Tornoceras*) *uniangulare*, U. Dev., Michigan, lateral view showing suture, closed umbilicus (cu). (b) *Imitoceras brevilobatum*, L. Miss., Missouri, portion of third volution showing characteristic external suture. (c) *Imitoceras grahamense*(?), Penn., Kansas, diam. = 3.0 mm for illustrated suture. (d) *Goniatites choctawensis*, Miss., Western Utah, immature specimen: lateral view (diam. 5.75 mm) of globular conch.
(F) (a) *Gonioloboceras goniolobum*, Penn., Kansas, ventro-lateral view showing high spired gastropod imbedded in aperture, X 4.5 (b) (c) *Stacheoceras gordoni*, Perm., N. Calif. (b) internal mold. (c) External suture at diam. 23.0 mm, prominent bifid ventral lobe, nine lateral and auxilliary lobes, and an umbilical lobe, X 4.
Illustrations redrawn and adapted from: (A, Bb, Cb) Miller, et al., 1957; (Ba) Miller, 1938; (Ca) Plummer and Scott, 1937; (D) House, 1962; (Ea) Miller, 1938; (Eb) Miller and Collinson, 1951; (Ec, Fa) Tasch, 1953; (Ed) Miller et al., 1952; (Fb, c). Miller, Furnish, Clark, 1957. J. Paleontology, Vol. 31 (6) pp. 1057–1068.

ballast in the form of calcareous cameral deposits were added to the cameral walls, weighting the ventral side of the shell (Figure 8.42). Furthermore, let there be similar calcareous siphuncular deposits (Figure 8.41), both types of deposits showing bilateral symmetry. Then, in this situation, the buoyant tendency of the cameral gas would be counterbalanced, and both center of buoyancy and gravity would be positioned one above the other. A horizontal equilibrium position in the water column would thus pertain.

Teichert (1933) estimated weight/volume relationships in actinocerids (Figure 8.45C). Among his very interesting findings were as follows: (a) Cameral and siphuncular deposits are concentrated apically. (b) Cameral deposits are far more significant in hydrostatic function than in siphuncular. (The hydrostatic function involves compensation for buoyant effect of the cameral gas to permit a final horizontal positioning.) (c) This is true because cameral deposits equal to one-fifth the radius of the camera would counteract the buoyant effect of the cameral gas, whereas siphuncular deposits equal to one-third the diameter would achieve the same effect. (d) Forms with excess cameral and siphuncular deposits

were, if anything, overcompensated and hence too heavy to swim; a benthic existence is indicated.

In attempting to account for the entire range of forms among fossil nautiloids, other complications arise. Closely spaced septa in the ancestral ellesmerocerids and among other brevicones may have been insufficiently buoyant (filled with gas) to counteract the gravitational drag on the body weight of the animal. It is also possible that the weight of closely spaced septa may have served as concentrated ballast (Flower, 1955). In turn, these observations suggest that such forms lived on the bottom as a motile epifauna and crawled by use of tentacles (Furnish and Glenister, 1964). Other benthic crawlers included *Turrilites* with a turreted spiral conch (Lower Cretaceous), possibly *Cyrtoceras* (of the late Silurian), a small brevicone, *Nipponites*, a sessile Upper Cretaceous form from Japan, and also possibly *Gonioceras*, a likely benthonic crawler (Berry, 1928, plate 2).

Flower (1955) posed that in the endocerids, where the endocones (Figure 8.41B, part B) fill the large ventral siphuncle, the hydrostatic role of the cameral deposits may have been served by them. Still other hydrostatic adaptations evolved per-

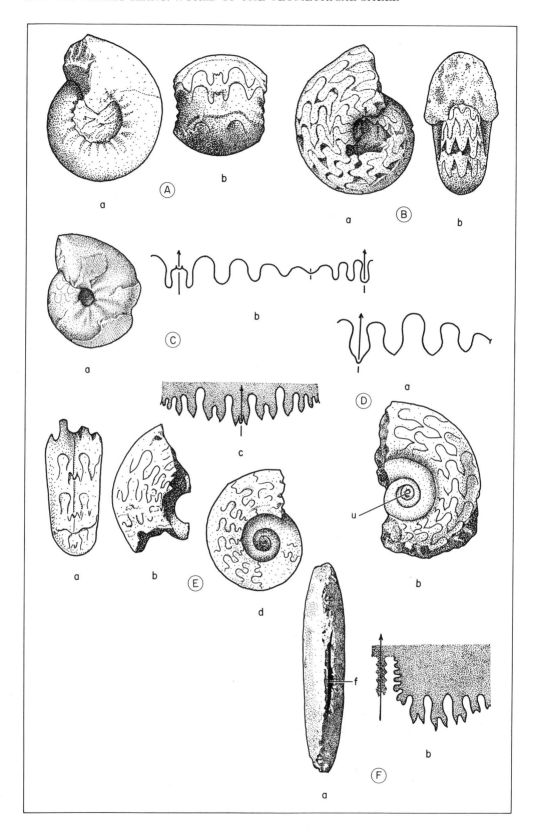

FIGURE 8.51 Paleozoic ammonoid genera—II.
Order Goniatitida. (A) (a) *G. listeri*, lateral view, Penn., Illinois, X 2. (b) *Gastrioceras montgomeryense*, Penn., Texas, ventral view with characteristic suture, diam. = 9.2 mm.
(B) *Schistoceras hyatti*, U. Carb., Texas, diam. = 43 mm: (a) lateral view; (b) ventral view—characteristic suture of adult conch; note chemical erosion of mold along suture pattern.
(C) *Texoceras texanum*. M. Perm. Texas: (a) lateral view showing portion of reticulate surface and suture, X 1.3; (b) suture showing subequal rounded lobes, X 3.
Order Prolecantitina. (D) (a) *Prolecanites hesteri*, L. Carb., Eng., goniatitic suture, enlarged; (b) *P. americanus*, U. Miss., Indiana, lateral view of fragmented discoidal conch, note large umbilicus (u).
(E) (a–c) *Pronorites arkansasensis*, Penn., Texas: (a) ventral view of fragment, X 2; (b) lateral view of same specimen, X 2; (c) external suture with trifid ventral lobe. (d) *P. grafordensis*, lateral view of immature specimen, Penn., Texas, X 4.
(F) *Medlicottia copei*, Perm. Texas: (a) thin discoidal conch, ventral view with characteristic deep furrow (f), X 1; (b) external suture, first ventral saddle, high and digitate, X 1.
Illustrations redrawn and adapted from: (A) Plummer and Scott, 1937; (B) Smith 1903, (C) Miller et al., 1957, after Girty; (Da) Miller et al., 1957; (Db) Miller and Garner, 1954; (E) (F) Plummer and Scott, 1937.

mitting diverse solutions to the problem of horizontal locomotion despite actual buoyancy. These included shell truncation, streamlined configurations, modification of septa, and relocation of septa, as well as coiling.

Transition from a nekto-benthonic existence in juvenile stages to a purely nektonic habit in maturity can be inferred for the ascocerids (Figure 8.43*A*) and also certain orthoceratids. To solve the problem of hydrostatic equilibrium, these forms underwent successive shell truncations, apparently, to diminish apical buoyancy. The calcareous deposits of the shed stages were probably resorbed. In maturity, two major changes occurred: the development of a fusiform configuration, excellently streamlined for horizontal locomotion, and the development of saddle-like camerae presumably gas-filled in a dorsal position (that is, above the body chamber). Thus the problem of horizontal equilibrium was solved by bringing the two factors of buoyancy and gravity one above the other.

Dunbar (1924) attributed exogastric curvature (Figure 8.43*A*, part *H*) to a tendency to counteract the buoyant effect of apical gas in orthocones. This was achieved by (*a*) adding larger new camerae throughout growth, (*b*) with the consequent elevation of the apex and shortening of the shell's horizontal length; (*c*) the end-result *coiling* placed the camerae above the body chamber. Coiling in nau-

tiloids is therefore interpreted to be one solution to the buoyant camerae problem (Nautilida, Tarphycerida, Barrandeocerida).

Furnish and Glenister (1964) emphasized the lack of primary cameral and siphuncular deposits in the last-named orders. This observation led to the conclusion that such deposits were unnecessary to maintain horizontal equilibrium since coiling had brought the two centers of buoyancy and gravity close to each other. Only one tarphycerid family (Lituidae) had considerable calcareous deposits (Figure 8.44*B*), which may account for its unusual type of coiling. Berry (1928) thought that the adult *Lituites* must have drifted head downward passively sinking toward the food source and fed on the bottom.

Flower (1954) reviewed nautiloid coiling evidence and concluded: (1) primitive cephalopods were dominantly endogastric cyrtocones (Figure 8.43*A*, *G*); (2) their descendants had accommodated to larger shells and hence to that greater buoyant effect of cameral gas; (3) a change from a cyrtoconic (less than 1/2 volution) to a coiled shell (5/4 + volutions) apparently occurred four times in nautiloid evolution: during the Canadian, Middle Silurian, and twice in the Devonian; (4) the recurrent change from cyrtocone to gyrocone or torticone was *saltational* (by sudden jumps) and not gradational; it could have been under ecological rather than under

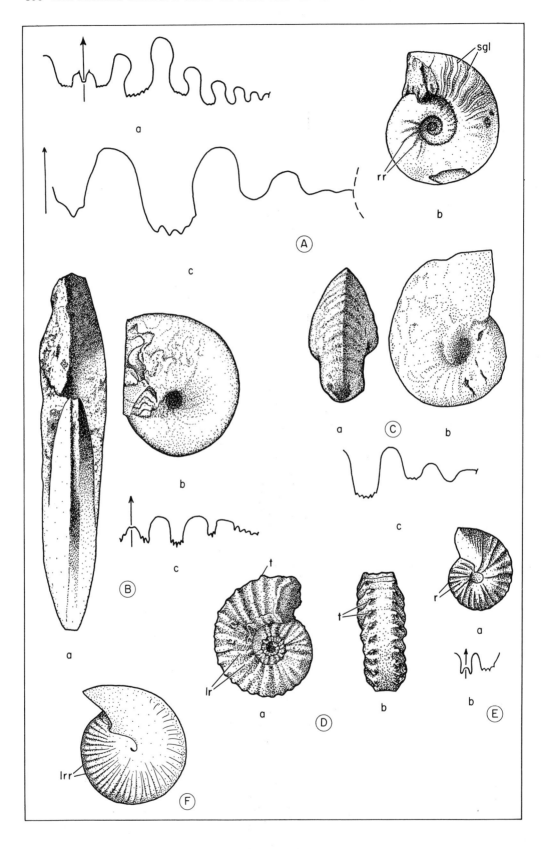

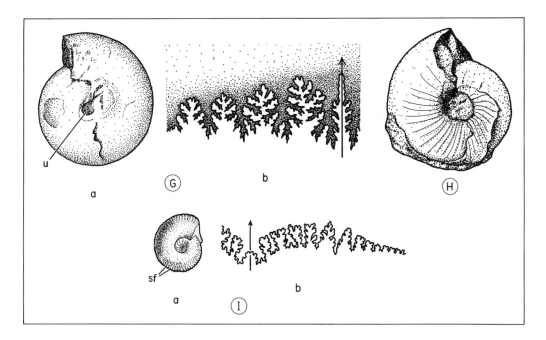

FIGURE 8.52 Mesozoic ammonoid genera—I (Triassic).

(A) (a) Suture of *Daraelites* (Daraelitidae), Perm., Sicily to show type of suture ancestral to those of Order Ceratitina (range, Perm.-Trias.), enlarged. (b) (c) Order Ceratitina, *Dieneroceras spathi*, L. Trias., Nev.: (b) lateral view of evolute and compressed conch; note radial ribs (*rr*) and sinuous growth lines (*sgl*), X 1; (c) suture, enlarged.

(B) *Meekoceras gracilitatus*, L. Trias., Nevada: (a) ventral view, showing furrow, reduced; (b) lateral view with partially reduced suture, X 1; (c) suture, X 1.

(C) *Groenlandites nielseni*, M. Trias., Peary Land: (a) Sharp angular venter X 2; (b) lateral view of mature crushed involute shell with small deep umbilicus, X 2; (c) incomplete ceratitic suture where diam. = 17.0 mm, X 6.5.

(D) *Nevadites merriami*, M. Trias., Nevada, reduced: (a) lateral view of evolute conch, *t*—tubercules, *r*—ribs; (b) same specimen, flattened venter and tubercles at end of ribs (suture, not shown, is ceratitic).

(E) *Tardeceras parvum*, U. Trias., Calif.: (a) lateral view showing ribs (*r*), X 2; (b) ceratitic suture, X 4.

(F) *Lobites ellipticus*, U. Trias., Alps, X 1 (suture, not shown, is goniatitic), lateral view of involute shell with excentric last volution; note low radial ribs (*lrr*).

(G) *Arcestes* (*Proarcestes*) *meeki*, M. Trias., Nev.: (a) lateral view of early mature involute conch, showing narrow and deep umbilicus (*u*), X 1; (b) ammonitic suture, X 3.

(H) Ptychites meeki, M. Trias., Nev., lateral view of subovoid, discoidal conch, X 2.

(I) *Pompeckjites layeri* (family Pinacoceratiidae), U. Trias., Alps, Balkan Timor, Calif.: (a) lateral view showing sinuous folds (*sf*) on ventral portion of whorl sides, reduced; (b) ammonitic suture with complex adventitious and auxilliary elements (see suture definition, Figure 8.47A.) Ammonite suture specialization reaches its highest complexity in this family.

Illustrations redrawn and adapted from: (A, B) Kummel and Steele, 1952; (C) Kummel, 1953; (D–F, G–I) Smith, 1914.

genetic control. (Item 4 may be clarified by noting that ecological controls also operate, though indirectly, through natural selection influence on the genetic code.)

The Role of Septation

As the shell grew, septa permitted the animal's body mass to remain close to the aperture. A second function of the septa was a depository for metabolic calcium. Where closely spaced, as in the ellesmerocerids, septa could serve as ballast. Septation, it is generally acknowledged, likely preceded the occurrence of gas-filled camerae (Flower, 1955). Similarly, in light of Bidders' study of living *Nautilus* and Denton's of *Sepia*, it is reasonably assumed that liquid ballast preceded the accumulation of gas in the camerae.

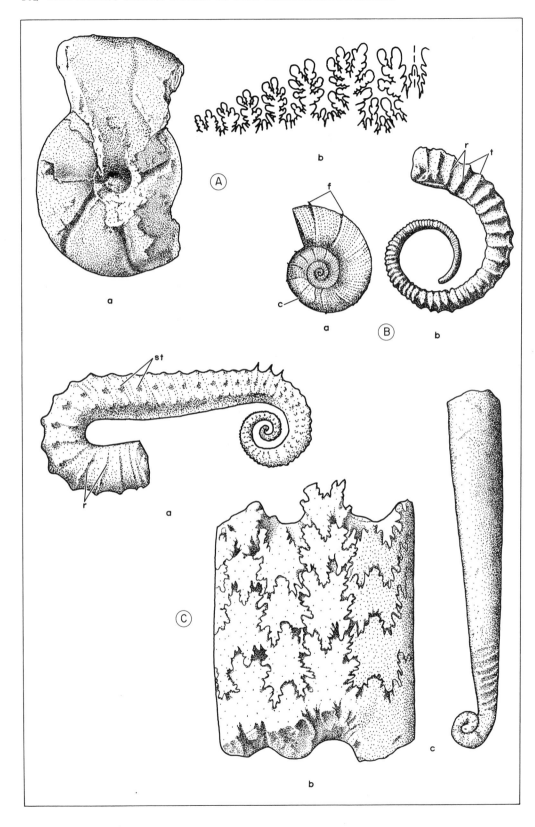

Depth and Biotic Associates

Modern *Nautilus* occurs at depths of 3 to 300 fathoms and nocturnally migrates to inshore shallow waters. Fossil ortho-cones presumably inhabited clear, shallow water. In algal and coral bioherms, they occurred in great abundance. Other associates of fossil nautiloids include: graptolites (Ruedemann's "Sargasso Sea"), other mollusks such as pelecypods, gastropods, and some ammonoids, trilobites (*Isotelus, Calymene*), brachiopods, crinoids, and fusulinaceans. Fish remains and eurypterids are found sometimes with nautiloids as are incrusting forms such as bryozoans.

Feeding

Modern *Nautilus* is a carnivore and a scavenger. It feeds largely on small decapod crustaceans but is also known to eat fish as well. There is some evidence that certain fossil orthocones also were carnivores (Flower, 1955). Dunbar (1924) thought that some nautiloids with restricted apertures may have fed on microscopic food. Morton and Yonge (1964, p. 49) also inferred a microphagous plankton-feeding mode of life for some Ordovician brevicones.

Function of the Tentacles

Bidder (1962) found marked differences in the function of the three types of tentacles in living *Nautilus*: Type I — preoral and postoptic, "alert function"; Type II — nineteen pairs of digital tentacles within Type I — "search function". Type III — a close set circle of numerous buccal tentacles (= labials) surrounding the powerful beak — "feeding function" (see Figure 8.39B, parts A, B). When the animal was alerted, Type I and part of Type II tentacles were extended; responding to the presence of fish or crab, more of Type II tentacles were thrust out to form a "cone of search." These last remained extended while the animal circled in various directions to locate food.

In fossil nautiloids generally, there may have been a parallel division of labor in the tentacular apparatus to permit alert, search, and feeding functions.

Lithology and Geologic Record

Lithology. Nautiloids occurred in, or are associated with, the following types of limestones, among others: massive, dense, fine-grained, pure and impure, algal and coralline, black or light, shaley and sandy, sublithographic and fragmental. Internal molds may be cherty, phosphatic, pyritic, dolomitic, and calcitic. They are found in the black shale facies and in other shale interbeds of limestone. Various sandstones have yielded nautiloid fossils: the Upper Canadian marly sandstones of Tasmania and the Tertiary sandstones in many places (Sari sandstone, Japan, Lomitos sandstone, northwest Peru). Nautiloids more infrequently accumulated into coquinas owing to wave action (Maquoketa shale). In some beds, nautiloid concentration is so great that these fossils constitute a distinct cephalopod facies such as the *Orthoceras* limestone of Sweden. In other beds nautiloids may be a sparse component of the biota.

Geologic Record. The first nautiloids appeared in the Cambrian (Flower, 1954).

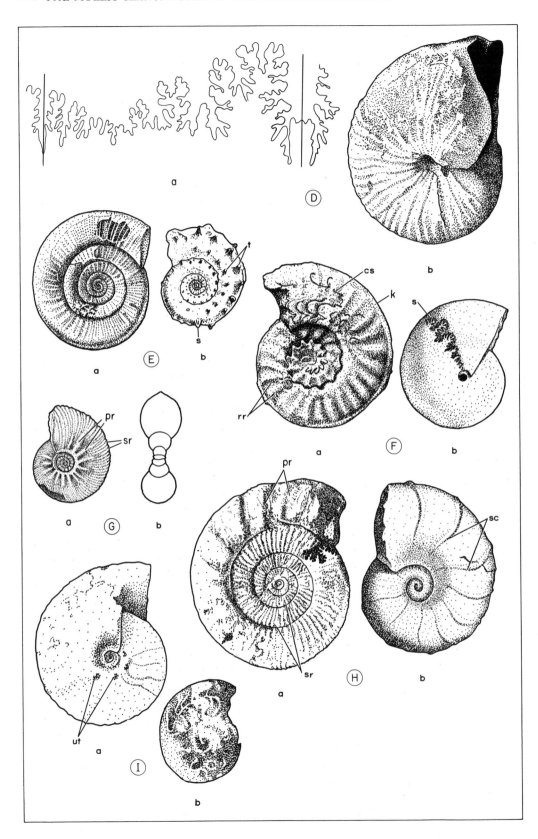

FIGURE 8.53B Mesozoic Ammonoid Genera – II (Triassic – Cretaceous) (cont.)

(*D*) (*a*) *Scaphites plesus*, Cret., Kansas, suture, X 3; (*b*) lateral view.

Order Ammonitina. (*E*) (*a*) *Psiloceras planorbis*, L. Jur., Eng., lateral view showing smooth, compressed, evolute conch, X 1; (*b*) *Eoderoceras bispinigerum*, L. Jur., Eng., X 0.5; *t* – tubercle, *s* – spines.

(*F*) (*a*) *Bouleiceras nitescens*, L. Jur., Madagascar, X 1; *cs* – ceratitic suture, *k* – keel, *rr* – rursiradiate ribbing (= inclined backward, proceeding from umbilical area towards venter); (*b*) *Oxycerites aspidoides*, M. Jur., Switzerland, X 0.3, *s* – complex suture; note finely frilled endings on lobes and saddles.

(*G*) *Cardioceras americanum*, U. Jur., Wyo.: (*a*) lateral view, max. diam. = 50 mm; *pr* – primary ribs, *sr* – secondary ribs; (*b*) cross-section of inner whorls, X 6.

(*H*) (*a*) *Perisphinctes* (*P.*) *variocostatus*, U. Jur., Eng., reduced, lateral view; note change in sharp ribbing (*sr*) of inner and middle whorls to coarse primary ribs of outer whorl (*pr*). (*b*) *Desmoceras* (*D.*) *latidorsatum*, L. Cret., France, enlarged, *sc* – sigmoid constrictions form strong rounded ribs.

(*I*)(*a*) *Placenticeras planum*, U. Cret., N. Mex., internal case, diam. = c. 100 mm, *ut* – umbilical tubercles; (*b*) *Adkinsia tuberculata*, U. Cret., Texas, X 2, tubercle-like thickenings on umbilical border are not seen; only small pyritic specimens are known in the family Flickiidae that embraces this genus.

Illustrations redrawn and adapted from: (*A*) Imlay, 1953; (*Ba*) Basse, 1952; (*Bb, Ca, E, F, Hb*) Arkell et al., 1957; (*Cb, Ia*) Reeside, 1927; (*D*) Elias, 1953; (*G*) Reeside, 1919; (*Ib*) Bose, 1927.

Ordovician time ushered in their rapid evolution that climaxed in the Silurian. Cephalopod facies recur in beds of Chazyan to Mohawkian age and include Michelinoceratidae, Stereoplasmoceratidae, Oncoceratidae, Valcouroceratidae, Barrandeoceratidae, Actinoceratidae, Endoceratidae. Both Lower and Upper Silurian, as well as Lower Devonian beds, yield sparse nautiloid faunas. However, the Upper Devonian carbonate facies contains a sizeable fauna.

Some highlights of the Paleozoic record of nautiloids are as follows: (1) Upper Canadian time – the appearance of the first coiled cephalopods (Tarphyceratida); (2) Devonian time – nautiloids give rise to ammonoids; (3) Mississippian time – coiled nautiloids come to dominate over other types, (4) Pennsylvanian-Permian time – coiled types from shallow-water facies competed with ammonoids – the competition continued through the Mesozoic (A. K. Miller, 1949). According to Flower (1957), associations rich in nautiloids produce few ammonoids, but notice that ammonoid-rich facies are sparse to lacking in nautiloids.

In the Mesozoic, and the Cenozoic soon after the close of the Cretaceous, the nautiloids showed their greatest impetus. That period was a time of "great dying" in the seas of the world (Bramlette, 1965); also it marked the extinction of the am-

monoids. After this demise of a competitor, the nautiloids experienced their "last surge" (eruptive evolution). From the ancestral *Eutrephoceras*, which persisted into the Cenozoic, *Climonia* and its apparent descendant *Hercoglossa* gave rise to four diverse genera (Figure 8.54*A*, *B*). Of these, only two survived into the Tertiary – *Aturia* and *Nautilus* – and only the last-named is still extant (A. K. Miller, 1949).

Stenzel (1957) said that relative to other mollusks in the same beds, Tertiary nautiloids occurred in the following ratio: 1 nautiloid per 1000 to 10,000 other mollusks.

Ammonoids (Geometry and Physics of Ammonoid Shells)

Generally, ammonoids developed tight coils about the protoconch (Figure 8.47*B*). The geometry of this coiling conforms to the equiangular spiral (*see* Figure 8.2). Before this aspect is examined further, note the evolute/involute types of coiling. An external depression, the *umbilicus*, occurs on each side of the shell; the center in the axis of coiling is embraced by the last whorl. Coiling that involves little or no whorl overlap yields a wide umbilicus (= the evolute type of coiling). Conversely, considerable whorl overlap produces a narrow umbilicus (= the involute type of coiling).

The wide range of coiling type (or

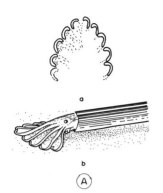

Upper Cretaceous	Paleocene	Eocene	Oligocene	Miocene	Pliocene	Pleistocene (and Recent)
		Aturia				
		Aturoidea				
		Woodringia				
		Hercoglossa				
		Deltodanautilus				
		Cinomia				
		Nautilus				
	Eutrephoceras					

(A)

(B)

Mud flat

AUSTIN

Epineritic

Engonoceras   Oxytropidoceras

Infraneritic

Mortoniceras   Turrilites

Hamites   Douvilleiceras

Epibathyal   Infrabathyal

Desmoceras   Lytoceras   Phylloceras

high tide | 5-7fathoms 20 | 80-100 fathoms
Low tide |

LITTORAL | MUD FLAT | EPINERITIC | INFRANERITIC | EPIBATHYAL | INFRAB. | ABYSSAL

Adapted from Scott, 1940

(C)

FIGURE 8.54A Information derived from fossil nautiloids and ammonoids.
(*A*) (*a*) Impression (spoor) presumably left by the orthocone *Orthonybyoceras.* Upper Ord., Ohio; (*b*) interpretation of how they were made.
(*B*) Eruptive evolution of Tertiary Nautiloids.
(*C*) Bathymetric distribution of some Texas Cretaceous ammonoids.

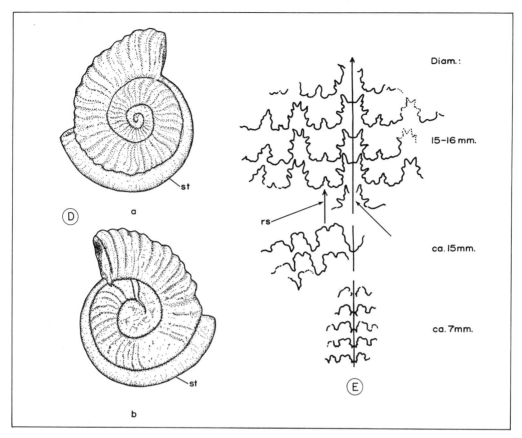

FIGURE 8.54B Information derived from fossil nautiloids and ammonoids (cont.)
(*D*) Symbiosis in the Liassic ammonoid *Schlotheimia, st — Serpula* tube: (a) right side, (b) left side, X 1.5.
(*E*) *Hysteroceras varicosum*, last five sutures (diam. = 15–16 mm) turned upside down; *rs* + arrows = reverse sutures. Westermann, 1972, Lethaia, pp. 165–167 restudied this specimen and found the alleged suture to be an artifact of inverse reassembly of fragments.
Illustrations redrawn and adapted from: (*A*) Flower, 1955; (*B*) Miller, 1949; (*C*) Hedgpeth, 1953, after Gayle Scott; (*D*) Schindewolf, 1934; (*E*) Haas, 1941.

uncoiling) in ammonoids (Figure 8.55) relates to certain arithmetical constants characteristic of the equiangular spiral. When the constant angle of the spiral (Figure 8.55*F,b, G*) is between 0° and 20°, the shell will appear as a straight cone; increase to 50° or greater will yield a gently curved shell; further increase to 80° ± will yield a closely coiled shell as in the ammonoids and in *Nautilus*; at a still further increase toward 90°, coil length rapidly increases, and at 90° the spiral vanishes. These and related observations are fully explained in D'Arcy Thompson, 1942, Chapter XI. Also consult Currie (1942, 1943).

The spiral angle (alpha) can vary in individual development, that is, increase or decrease. Currie (1943) saw a change in growth gradient in species of *Promicroceras* by plotting a median whorl height against the shell's diameter, and the logarithm of median whorl height against number of half-whorls. Among changes in gradient (slope of the curve), one happened at about the eighth to tenth half-whorl, another at a half-whorl to one whorl later. These gradient changes corresponded to variation in the spiral angle. The first (eighth to tenth half-whorl) change might be attributed to spiral angle increase, and the second to a spiral angle decrease (Figure 8.56).

Trueman (1941) found that in many

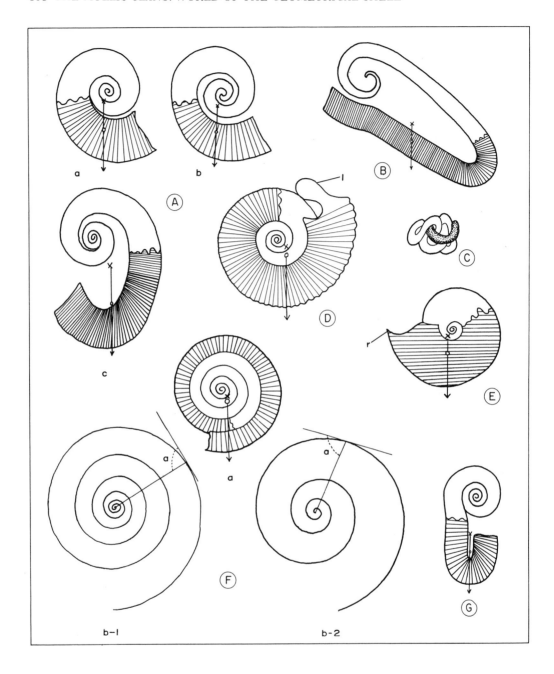

FIGURE 8.55 Natural rest and/or floating positions in coiled and uncoiled ammonites (*x* – approximate position of center of buoyancy; small circles – center of gravity; lined area – body chamber).
(*A*) Three species of *Crioceratites* showing change in the two centers with progressive uncoiling: (*a*) *C*. sp. 1; (*b*) *C*. sp. II: (*c*) *C*. sp. III.
(*B*) *Lytocrioceras jauberti* (note extent of uncoiling and location of the two centers).
(*C*) *Nipponites mirabilis*, coiled in succession of *U*'s, forming a tangle in three-dimensional view.
(*D*) *Normannites*; *l* – lappet.
(*E*) *Ludwigia* sp., *r* – rostrum.
(*F*) (*a*) *Caloceras* sp (*bl*) *Dactylioceras* sp., spiral angle alpha (*α*) = 85°30′; (*b*2) *Ludwigia* sp., alpha (*α*) = ca. 83.
(*G*) *Macroscaphites yvam*.
Illustrations redrawn and adapted from: (*A, B, D–G*) Trueman, 1941; (*C*) Morton, 1960.

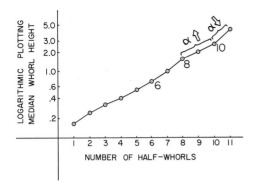

FIGURE 8.56 Change in growth gradients (slope of curve) and corresponding changes in variation of spiral angle in *Promicroceras capricornoides.* Logarithmic plotting of median whorl height at intervals of half-whorl, $\alpha \uparrow$ = increase of spiral angle; $\alpha \downarrow$ = decrease in spiral angle. Curve shows changes in slope between the second and third half-whorl at the sixth, eighth to tenth, and beyond tenth half-whorl. Modified from Currie, 1943.

ammonites the length of the body chamber showed a definite relationship to spiral angle (see Figure 8.55). He saw that short body chamber corresponded to a low spiral angle and one occupying more than a whorl, to a spiral angle of 86°. A ratio of air or gas chamber (phragmacone) volume to that of body chamber volume was found commonly; to be about 1:2 or 1:3; more rarely, ratios of 1:1 or 1:8 occurred. During individual development, these ratios remained constant.

In the same study, Trueman established the centers of buoyancy and of gravity (gravity above buoyancy) in several ammonites (Figure 8.55). This indicated probably the rest and floating position an ammonite assumed in life.

Flume experiments on various coiled nautiloids and ammonoids yielded quantitative data relative to a streamlining test. Coated plaster casts of 21 species were used (Kummel and Lloyd, 1955). The measure of streamlining was found by determining the drag coefficient ($C_{D'}$) on specimens submerged in a moving fluid. In general, evolute forms tended to be less streamlined than involute forms; although if compressed, evolute forms also showed favorable streamlining. However coarsely ornamented, widely umbilicate forms with depressed whorl sections displayed less favorable streamlining. (Figure 8.57).

Depth, Mode of Life, Biotic Associates

Depth. There are various indirect clues to the depths at which ammonoids lived. They were able both to descend to bathyal oceanic depths and ascend toward the surface. How? The simple man-made submersibles add ballast for descent and transfer ballast for ascent. Based on Denton's work on *Sepia* (Figure 8.39*B*, part E) and Bidders' on *Nautilus*, it is possible that ammonite fluid ballast was added for descent and cameral gas for

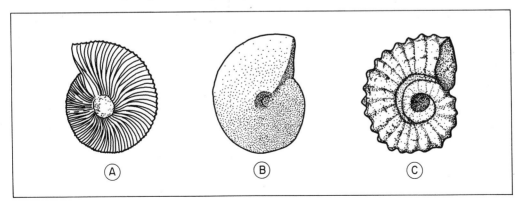

FIGURE 8.57 Flume experiments in comparative streamlining of coiled cephalopods; $C_{D,}$ = drag coefficient (see text for details). (*A*) *Macrocephalites*, involute, compressed, $C_{D'}$ = 13.6; (*B*) *Perrinites*, same type as "*A*", $C_{D'}$ = 15.4: (*C*) *Douvilleiceras*, evolute, coarsely ribbed, loosely coiled, $C_{D'}$ = 16.4. After Kummel and Lloyd, 1955. (See J. C. Chamberlain, Jr. and Gerd E. G. Westermann, 1976. Paleobiology, Vol. 2, pp. 316–331. Maintenance of maximum efficiency through all developmental stages will be favored by progressive roughening of the shell for small individuals or species, and increasing smoothness for large shells.)

ascent. This appears to satisfy simple physical requirements of the theory.

The attendant buoyant effect of passively generated gas entering the partial vacuum of the camerae does compensate for external pressure at depth, yet removes the animal to higher levels. If, however, the fluid-gas arrangement did exist, flooding camera with fluid would submerge the test, withdrawal of fluid would leave the camera filled with gas to give it lift. A possibility exists that fluid ballast and buoyant gas, nicely adjusted at depth, (*a*) prevented a tendency for excessive buoyancy and (*b*) compensated for external pressure. More complete evacuation of fluid from the camerae would then permit ascent to the surface.

The ascent continued precarious since the gas-filled camerae exerted a greater and greater pressure as seawater pressure diminished. In that event, the shell should have burst outward if the decreased external pressure were uncompensated. Arkell (1949) proposed a structural specification that provided resistance to bursting upon ascent from depth. Suppose the areal surface of septa attachment to shell wall were enlarged, then this would substantially strengthen the shell. This could explain a puzzling development in ammonoid evolution: the complexing of sutures. Arkell's interpretation signifies that sutural elaboration was adaptive and reflected a structural compensation for ascent.

Arkell (1957) thought that, generally, complex sutures probably reflected deepwater forms while simplified sutures denoted a shallower habit or sluggish benthic existence. Irregularly coiled forms — *Nipponites* and turbinate shells (*Turrilites*) presumedly were respective sessile forms and vagrant benthos. These interpretations, explaining where such forms may have lived on the sea floor, do not clarify the depths. Gayle Scott (1940) assigned depths of 20 to 100 fathoms for *Turrilites* and *Hamites* in the Cretaceous in Texas.

Figure 8.54*A*, part *C*, indicates several postulated depths at which Texas Creta-ceous ammonoids lived. The whole ammonoid assemblage can pass as mobile benthonic or nekto-benthonic. Shell morphology, sculpture, and coiling were the chief criteria used to group these ammonoids. Generally, sculptured types lived at depths of 20 to 100 fathoms (infraneritic zone); smooth, ovate forms, *Desmoceras*, lived at or below 100 fathoms (epibathyal zone); and smooth, obese forms lived in deep water (infrabathyal). Deep-water inhabitants included *Phylloceras* and *Lytoceras*, among others.

Spoor. The best example of ammonoid spoor was given by Rothpletz (1909). He found impressions of a perisphinctid venter beside the fossilized ammonoid in a Jurassic slate. His interpretation placed the living ammonoids in shallow water and, as the tide receded, indicated that its venter touched bottom mud first. [Compare Figure 8.54*A*, *A*].

Substrate and Biotic Associates. Paleozoic goniatites chiefly inhabited shallow water. Commonly they are recovered from mudstones, carbonaceous and calcareous shales, and shaley limestones (Miller, 1957). Some occur in marcasite concretions in the Devonian of New York State, for example. Others are found in breccias, coal balls, ironstones, and calcareous nodules of the British Coal Measures. They have also been recorded in the tuffaceous marls of Timor, associated with stromatoporoid and *Receptaculites* bioherms, occasionally with coarse conglomerates (Devonian of Western Australia). Additional occurrences include estuarine, shallow bay-accumulated black limestones, as well as shales with calcareous hard bands (Yorkshire, England).

Generally, cephalopods are sparse in the Devonian (U.S.A.) and Carboniferous (England) limestones. Teichert (1943) concluded from the goniatite facies in the West Australian Devonian that goniatites lived in the outer reaches of a lagoon (10 to 50 miles off the coast of that time) margined by a barrier reef.

Rarely do ammonoids live in coral reef facies or in current-bedded rocks (Arkell, 1956). Many ammonoids are thought to have led a life among marine vegetation arising from loose clay mud substrates. If true, that helps explain the many Mesozoic pyritized ammonoid faunas found in shales (Tasch, 1953; Vogel, 1959) or marls (Spath, 1928). In Great Britain some Jurassic ammonoids are associated with shell banks in noncoralline areas. Times of shell bank formation presumably were times when the sea shallowed. Brinkmann (1929) noticed that among ammonoids, at such times (Middle Jurassic, Oxford clay), "evolutionary leaps" occurred. In Alaska Jurassic ammonoids are picked up in concretions in sandstones and siltstones.

Among chief biotic associates of goniatites and ammonites were the same ever-present mollusks. Other forms include echinoids, serpulid worms, brachiopods. In the Middle Permian of Southern Mexico, ammonoids occur associated with sponges, worms, crinoids, bryozoans, mollusks, brachiopods, and foraminifera (Miller et al., 1941). The Callovian (Jurassic) beds of the Alaska peninsula contained, besides ammonoids, belemnoids, brachiopods, crustaceans, worm tubes, and floral remains (Imlay, 1953).

Estimated Growth Rate of Ammonoids

Schindewolf (1934) studied a collection of Liassic ammonoids in which serpulid worm tubes grew in the venter (Figure 8.54B, part D). He inferred that for a tube to encircle a single whorl would have taken from 0.5 to 3.0 years. It follows that since the serpulid worm grew along with the ammonoid, one could assign this growth rate to a single whorl. Septal counts of the last whorl estimate one week to one month as the time required to build one single ammonoid camera (that is, the time between secretion of two successive septa).

Larval, Juvenile, and Adult Stages

Free-swimming, ammonoid larval stages

could have been carried by ocean currents to the far corners of the earth. Cosmopolitan distribution of many Mesozoic genera and species would thus be readily explained.

Measurement of a shell population on a bedding plane might reveal multiple size classes; Trueman (1941) demonstrated this. Could these correspond to successive *seasonal* broods? Currie's work on Jurassic ammonites (1944) provided information. She identified growth stages in the ontogenetic development of specific ammonites. She estimated that specific shell forms attaining to the twelfth to thirteenth half-whorl represented the shell growth of four seasons or years from the beginning of postembryonic growth. The number of half-whorls cited are equivalent to six whorls. Accordingly, following Currie's estimate, 1 1/2 whorls would represent 1.0 year of growth time.

Relative small size and uncrowded simple sutures distinguish juvenile from adult specimens in the same population. Crowded septa generally denote late adult (gerontic) growth stages. Multiple whorls, relatively large size, and maximum suture complexity in a given shell are thought to correspond to adult stages.

Gigantism and Dwarfism

Both tendencies—to excessive or to curtailed growth—are recorded in ammonites (Miller and Youngquist, 1946; Kummel, 1953; Tasch, 1953; Vogel, 1959). These are not restricted occurrences. The Triassic, for example, abounds in so-called dwarf faunas; then Cretaceous (Senonian) beds and many Jurassic deposits as well, all over the world, provide multiple giant ammonite examples.

Predators, Body Injuries, and Cannibalism

Decapod crustaceans—which were prominent predators on some ammonites—bit out pieces of the body chamber. Body chamber injuries repaired often left distortions and monstrosities (Tasnádi-Kubacska, 1962, pp. 61–68). Spath

(1945) cautioned that irregularities in ornament often do not permit one to distinguish normal from abnormal ammonites. Furthermore, apparent abnormalities found in limited samples (triangular coiling in *Clymenia*, for example) often turned out to be normal when more material became available.

Figure 8.54*B*, part *E*, illustrates an injured shell that has the last few suture lines reversed (Haas, 1941)[3]. Schwarzbach (1936) evaluated the contents found in the body chamber of a Jurassic ammonoid (*Oppelia*): broken, tiny aptychi and several segmented siphuncles. Did these represent a brood of young oppelids cannibalized by the adult? If true, this occurrence probably denotes an inadequate food supply for the adult population. Other examples of probable malnutrition in ammonoids are known (Tasnádi-Kubascska, 1962).

Geologic Record and Stratigraphy

Of the 1800 or so genera of known ammonoids, less than 10 percent are Paleozoic; 90 percent are Mesozoic. Kummel (1954) emphasized three critical periods in the evolutionary history of ammonoids: near extinction in Late Permian, again in Late Triassic, and then after a period of extended radiation from root stocks, final extinction in the Late Cretaceous. (The root stock from which hosts of Jurassic and Cretaceous ammonoids derived were the Lytoceratina and the Phylloceratina; see Figure 8.58*A*.)

The origin of ammonoids, interpreted by Schindewolf, has been discussed (Figure 8.45*E*).

The index fossil value of ammonoids during the Paleozoic can be gleaned from Table 8.2 (A. K. Miller, 1957). Twenty-four Paleozoic ammonoid zones are recognizable. Each genus indicated represents a faunal zone regionally recognized; *Mimosphinctes*, in Germany; intercontinental correlations with *Goniatites*. Species of the given genus found in North America, Europe, or North Africa designate beds

³See notation. Figure 8.54*B*, part *E*.

of Upper Mississippian age. Table 8.2 shows that many Paleozoic ammonoids were distributed worldwide.

It has been long remarked that ammonoids were "the leading actor(s) in the drama of Mesozoic stratigraphy." The following summary indicates this plausible evaluation. There are 29 recognizable ammonite zones in the Triassic of the world (Scythian through Rhaetian stages) (Kummel, 1957, Table 1). An even finer subdivision, possible in the world Jurassic (Hettanian through Purbeckian stages), has 60-plus zones. Jurassic zonal index ammonites are figured in Arkell (1954, pp. 760–790).

Arkell (1949, p. 40) estimated that the known 55 ammonite zones of the Jurassic represented about 500,000 years per zone. Two essential revisions are needed. Kulp's time scale (1961) almost doubled the time span for the Jurassic (46 ± 10^6 years) and 60-plus ammonite zones are now recognized. Dividing 60 into 46 million years yields about 760,000 years per zone.

Buckman designated the *hemera* (plural, *hemerae*) as a time unit that represented a span during which any one index ammonite was dominant. Some 62 hemerae comprised 11 zones in the Lower Lias. Using the older time scale, Arkell (1949) estimated an average of about 88,000 years per hemera. This figure would be modified upward by the newer scale. A magnitude of $100,000 \pm$ years crudely approximates the needed correction. Detailed studies, such as Brinkmann's (1929) on *Kosmoceras*, correlated with these geochronological estimates, could possibly yield evaluation of the year value for certain morphologic changes in ammonoid shells.

The stages of the Cretaceous of many areas of Western Europe (Berriasian through Maastrichtian) divide into 36 ammonite zones (Wright, 1957, Table 4). Additional zones are recognizable in restricted areas.

Thus some 125 traceable ammonite zones subdivide the Mesozoic.

Although no estimate is available on ammonite population densities in the

TABLE 8.2. Paleozoic Ammonoid Zones

Number of Zones	System	Series	Zone	Type Area	Other Areas
5	Permian	Upper	*Cyclolobus*	SaltR.	Himal.-Armenia-Madag.-Greenl.
		Middle	*Timorites*	Timor	USA(Tex.)-Mex.(Coah.)
			Waagenoceras	Sicily	Tunisia-Pamirs-China-Timor-USA(Tex.)-Mex.(Coah.)
			Perrinites	USA(Tex.)	Mex.-Colom.-USA-Can.-Urals-Crimea-AsiaM.-Timor.
		Lower	*Properrinites*	USA(Tex.)	USA(Kans.)-Timor-Urals-Crimea
7	Pennsylvanian	Upper	*Uddenites*	SW.USA	Urals
			Prouddenites	SW.USA	USA(Ohio-Pa.)-Urals
		Middle	*Eothalassoceras*	SW.USA	Turkestan
			Wellerites	SW.USA	USA(Ohio)-Urals
			Owenoceras	SW.USA	?Argentina
		Lower	*Paralegoceras*	USA(Iowa)	SW.USA-C.USA-Peru-N.Afr.-S.China
			Gastrioceras	Eng.	
4	Mississippian	Upper	*Eumorphoceras*	USA(Okla.)	USA-W.Can.-Alaska-Eu.-N.Afr.
			Goniatites	Eng.	USA-W.Can.-Alaska-Eu.-N.Afr.
		Lower	*Beyrichoceras*	Eng.	Eu.-N.Afr.-C.USA
			Protocanites	USA(Ind.)	USA-Eu.-N.Afr.-Austral.(N.S.W.)
8	Devonian	Upper	*Wocklumeria*	C.Eu.	N.Afr.
			Clymenia	C.Eu.	N.Afr.
			Platyclymenia	C.Eu.	N.Afr.-W.Austral.-USA (Mont.)
			Cheiloceras	C.Eu.	N.Afr.-W.Austral.-USA (Pa.)
			Manticoceras	USA(N.Y.)	USA-W.Can.-China-Austral.-N.Afr.
		Middle	*Maenioceras*	Ger.	USA-E.Can.-N.Afr.-Austral.
			Anarcestes	Ger.	USA-E.Can.-N.Afr.-Austral.
		Lower	*Mimosphinctes*	Ger.	

Source. After A. K. Miller et al.

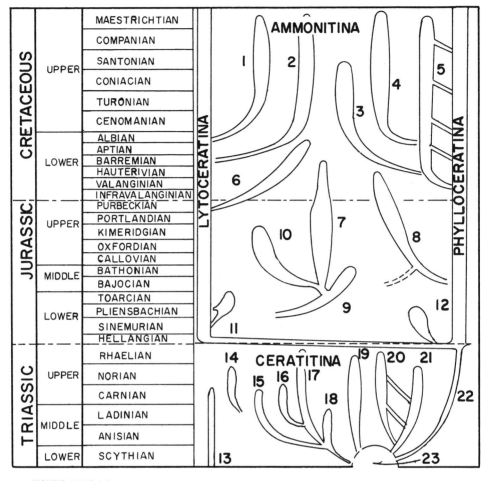

FIGURE 8.58A Inferred phylogeny and geologic spread of Mesozoic ammonoids (after Kummel, 1954, based on Arkell, Wright, and Kummel).
(1) Scaphitaceae. (2) Hamitaceae. (3) Hoplitaceae. (4) Acanthocerataceae. (5) Desmocerataceae. (6) Criocerataceae. (7) Perisphinctaceae. (8) Oppellaceae. (9) Harpocerataceae. (10) Stephanocerataceae. (11) Eoderocerataceae. (12) Arietitaceae. (13) Pronoritaceae. (14) Tropitaceae. (15) Ptychitaceae. (16) Lobitaceae. (17) Arcestaceae. (18) Meekocerataceae. (19) Pinacocerataceae. (20) Trachycerataceae. (21) Ceratitaceae. (22) Phyllocerataceae. (23) Xenodiscaceae.

rocks of these zones, it is clear that enormous populations of multiple species thrived in the seas of the time.

Continental-Drift Theory

Arkell (1949) figured the worldwide distribution of certain genera was puzzling until one envisioned "a more compact arrangement of Jurassic land masses." He referred to the genera *Dactylioceras, Stephanoceras,* and *Macrocephalites.* Figure 8.58B represents a plot of two Bajocian ammonites on DuToit's map of Gondwana.

Geosynclinal Theory

Alpine, Caucasian, and Himalayan geosynclinal belts show a sedimentary record, (*a*) discontinuous and (*b*) changeable (cyclic) as in ordinary epicontinental and shelf seas elsewhere during Jurassic time (Kummel, 1961, Figures 9-4, 9-5, 9-15; Brinkmann, 1960, pp. 70–114). The Alpine geosyncline region had shifting shallows, islands, troughs, and deeps. Prior to the study of ammonite faunas that helped unravel biostratigraphy, such geosynclinal accumulations had not yielded adequate data on bathymetry,

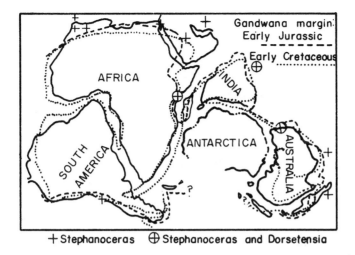

+ Stephanoceras   ⊕ Stephanoceras and Dorsetensia

FIGURE 8.58B Du Toit's map of Gondwana showing distribution of Bajocian ammonites in Europe and western Cordillera of South America; + − *Stephanoceras*, o − *Stephanoceras* and *Dorsetensia*. Slight modification by Arkell, 1949, of Gondwana margins on the east coast of India. (Arkell, 1949.)

ecology, or dynamic variable environment (Arkell, 1949, pp. 413–414).

Near-Extinctions and Extinctions

The times of near-extinction of ammonites (end of Trias; during Barremian) and then final disaster (late Cretaceous — Maestrichtian (Figure 8.58*A*) are attributed to (*a*) regressive epicontinental seas shrinking the area for favorable niches, and (*b*) possible predation by giant cuttlefish, sharks, or rays (Arkell, 1949).

Coleoids

Introduction. All living cephalopods (*Nautilus* excepted) belong to this subclass. These include squids, cuttlefishes, sepioles, and octopods. In addition, extinct belemnoids also belong with Coleoids. Belemnoids have an impressive fossil record in contrast to the other coleoids.

Considering the abundance, the diversity of cephalopods in the geologic past, one might hastily conclude coleoids were a dead end — a vestigial dying remnant of a once populous class. Instead of the end, however, living coleoids are the beginning — a new burst, an expansion in cephalopod evolution (Morton, 1960).

Up to now the chambered shell was our dominant consideration. Solutions of buoyancy problems became critical in evolutionary development. However, by increasing their locomotion speed, the coleoids, except *Sepia*, freed themselves to solve problems created by gravity drag. This was achieved by shell reduction (internal when present) or by shell loss, *Octopus*. (*Sepia*, confined to the continental shelf, remained dependent on buoyant mechanisms.) Note that in the modern seas some squids are the fastest marine invertebrates (A. M. Bidder, 1964).

Furthermore, new ecologic niches could be exploited after the external shell load loss freed coleoids from their benthic or nekto-benthic existence. Thus most living coleoids are both pelagic and bathypelagic. Only a few forms favor littoral or sublittoral habitat. Accompanying these new freedoms, came the remarkable upgrading and development of sensory and cerebral powers (Morton and Yonge, 1964). The brain in *Octopus*, with more than 168 million neurons, and in *Sepia*, apparently largely concerns itself with learning and remembering (M. J. Wells, 1962).

Subclass Coleoidea (formerly Dibranchiata). Living forms characterized by tentacles (arms) eight to ten in number; these

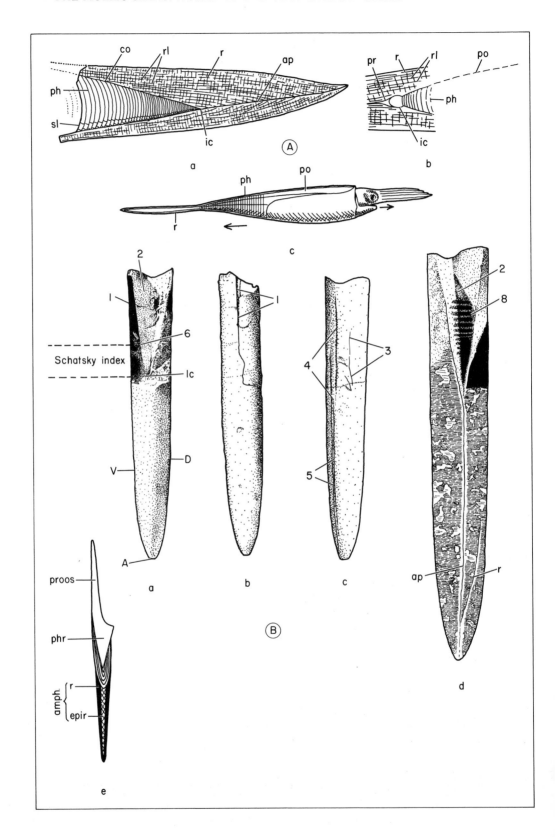

FIGURE 8.59 Belemnite structures.

(A) (a) *Homolotheuthis spinata.* Jur., Ger., schematic, median view; (b) section through end of phragmacone *Hibolithes hastatus,* Jur., France, X 5; (c) Reconstruction of young rostrum of *Passaloteuthis* sp., Jur., France, schematic; *ph* – septate phragmacone, *si* – siphuncle, *co* – conotheca, *rl* – rostral lamellae, *r* – rostrum, *ap* – apical line, *ic* – initial chamber or protoconch ("embryonic bulb"), *pr* – primordial rostrum, *po* – pro-ostracum, right arrow – direction of jet, left arrow – direction of movement.

(B) *Belemnitella praecursor micronatiformis,* U. Chalk (Cret., Eng.).

(a)–(c) Guard, internal and external aspects of same specimen; (a) *A* – apical end, *D* – dorsal side, *V* – ventral side, *ic* – embryonic bulb; (1) ventral fissure, (2) alveolus, (6) beginning of ventral fissure, (b) Guard with (l) ventral fissure, (c) Guard showing. (3) lateral single furrow, (4) dorso-lateral depression, (5) dorso-lateral double furrows. Schatsky index = measurement made from protoconch to (6). (d) Guard, *B..praecursor media* (L. Campanian, U.S.S.R.), X 1/2, 2 – alveolus bearing septal markings of chambered phragmacone, *ap* – apical line, *r* – rostrum. (e) Schematized belemnite shell showing relationship of epirostrum (*epir*) to rostrum (*r*) (*r* + *epir* = amphitheca).

Illustrations redrawn and adapted from: (A) Naef, 1922, (B) Jeletsky, 1955, Müller-Stoll, 1936.

may specialize for seizing prey or for reproduction (copulatory arm-hectocotylus); single pair of ctenidia and kidneys; ink sac; complex eyes; shell, either internal or external; internal shell is contained in a naked sac-forming mantle, may be in various reduction stages or completely absent. Geologic range: Misissippian to Recent. [The subclass was divided into two orders by Naef (1922): Decapoda, Octopoda (Figure 8.62).]

***Order* Decapoda.** Ten arms; internal shell. (There are three suborders: Belemnoidea, Teuthoidea, Sepoidea.)

Suborder Belemnoidea. Extinct decapods with internal calcareous shells; shells composed of three parts, a massive *rostrum* served as ballast; a septate *phragmacone*, buoyant in life; and a tonguelike projection of the phragmacone, the pro-ostracum (= vestige of a living chamber) (Figure 8.59).

Commonly fossilized as a so-called "guard" or "cigar," that lacks the pro-ostracum but contains the other skeletal units. Septate phragmacone may be visible when eroded or can be made so by either thin-sectioning or radiography of the fossil guard. Geologic range: Mississippian to Eocene. [The suborder embraces five families: Aulacoceratidae, Chitinoteuthidae, Belemnitidae, Neobelemnitidae, Belemnoteuthidae. One genus in each family is illustrated: Family Aulacoceratidae, *Ausseites* (Figure 8.60*A*); *Choanoteuthis* (Figure 8.60*B*); onychites, *Dictyoconites* (8.60*F*);

family Belemnitidae, *Homoloteuthis* (8.59*A*, a), *Hibolithes* (8.59*A*, b), *Passaloteuthis* (8.59 *A*, c), *Belemnitella* (8.59*B*); family Neobelemnitidae, *Belemnosella* (8.60*D*); family Chitinoteuthidae *Chitinoteuthis* (8.60*C*); family Belemnoteuthidae, *Xiphoteuthis* (8.60*E*).]

Suborder Sepoidea. Decapods; the internal calcareous shell has a rostrum and pro-ostracum reduced and a phragmacone tending to curve. (Curvature finds maximum expression in *Spirula*.) Geologic range: Jurassic to Miocene. [*Sepia* (Figure 8.61*C*, c), *Spirulirostra* (Figure 8.61*A*), *Voltzia* (8.61*B* and *C*, a), *Belosepia* (8.61*C*, b), *Spirula* (8.61*D*).]

Suborder Teuthoidea. Decapods, with a horny, internal shell, much reduced; nonseptate phragmacone (= conus) may or may not be present; rostrum, either lacking or very small. For taxonomy, both morphology and orientation of the pro-ostracum (= gladius or pen) are used. Geologic range: Jurassic to Recent. [*Leptoteuthis* (Figure 8.61*E*), *Trachyteuthis* (Figure 8.61*F*).]

***Order* Octopoda.** Eight arms, all longer than the body; internal shell lacking in living form, but in fossils, a reduced shell is known (Figure 8.61*G*). *Argonauta* females secrete an external shell via the dorsal arms (known also in fossils). Geologic range: Cretaceous to Recent. [Three suborders are embraced by this order; for two, there is a sparse fossil record, and one, Cirroteuthidea, lacks a fossil record.

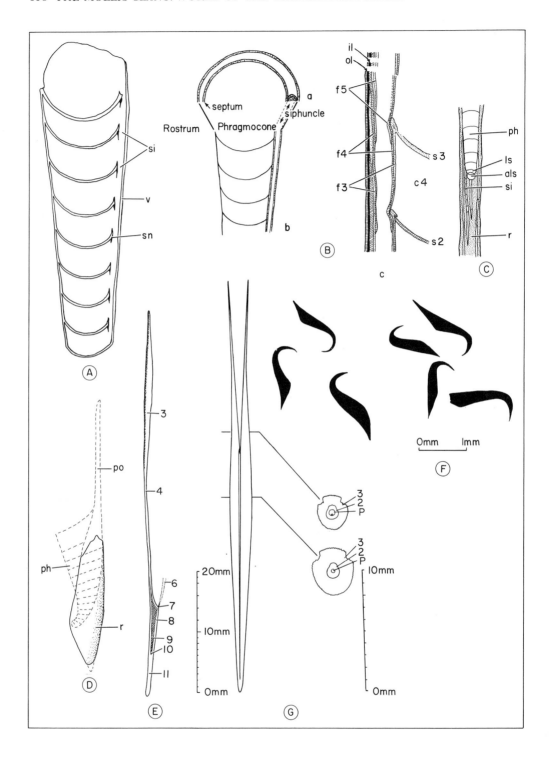

FIGURE 8.60 Representative belemnoids.

Family Aulacoceratidae. (A) *Ausseites ausseanus*, Trias., Austrian Alps, reduced vertical section of large phragmacone showing siphuncle (*si*) close to venter (*v*); *sn* – septal necks.

(B) *Choanoteuthis mulleri*. Trias., Nevada: (a) Transverse and (b) longitudinal thin sections, X 2; (c) diagram of portion of unique holochoanitic siphuncle (see "Nautiloids" for definition) composed of successive invaginated funnels (septal funnels (*f3, 4, 5*)) extending through more than two chambers. [Note that funnels succeed and appear to strengthen inner layer (*il*) and outer layer (*ol*) of the shell wall and bend to decrease width of siphuncle as well as to reinforce shell wall. X ca. 8.]

(C) Family Chitinoteuthidae. *Chitinoteuthis* sp., schematic; *ph* – phragmacone, *r* – chitinous rostrum, *ls* – terminal septum, *als* – appendix of *ls* – note that resorption of early chamber and protoconch is indicated by presence of the siphuncle (*si*) beyond *als*; other structure parallel to *si* are ridges or carina.

(D) Family Neobelemnitidae. *Belemnosella americana*, Eoc., Miss., schematic reconstruction, internal view; *po* – proostracum, *r* – rostrum, *ph* – phragmacone.

(E) Family Belemnoteuthidae. *Xiphoteuthis elongata.*, L. Lias, Eng., lateral view, 3 – proostracum, 4 – thinner portion same as 6; dashed lines: muscle inserted at 7 – margin of shell, 8 – conotheca, 9 – phragmacone, 10 – protoconch, 11 – rostrum, reduced.

(F) Family Aulacoceratidae. *Dictyoconites groenlandicus*, Perm., Greenland, arm hooks (onychites) from two localities. (Reassigned to *Permoteuthis groenlandicus* by Jeletsky, 1966.)

(G) Reconstructed ephebic rostra of (F) showing (a) longitudinal section at left, cross-section at right (taken at height of 32.0 mm): lower section, ontogenetic stages from youngest to oldest are P(=1) – nepionic deposits, 2 – neanic deposits, 3 – ephebic (a fourth stage, not shown, may be called ephebic-gerontic or gerontic, upper section, the phragmacone (P) is shown as it appears in cross-section in relation to the other concentric bands of the guard.

Illustrations redrawn and adapted from Fischer, 1947, 1951; Flower, 1944; Naef, 1922; and others.

Suborder Paleoctopoda, *Paleoctopus* (Figure 8.61G); suborder Polypoidea, *Argonauta* (Figure 8.61H).]

Information Derived from Fossil Coleoids

Origin and Evolution. How were coleoids differentiated from their nautiloid orthocone ancestor (Figure 8.62)? The long, rounded phragmacone in *Aulacoceras* (Triassic) bears many resemblances to that of *Orthoceras*: the septa are very widely spaced and the rostrum is relatively small. In addition, the siphuncle, long, slender, partly inflated between septa, is located near the ventral face; the conotheca is smooth (Flower, 1944; Roger, 1952; Swinnerton, 1946).

Another differentiation from the ancestral stock was the development of a chitinous rostrum (Chitinoteuthidae, Figure 8.60C) with a phragmacone reminiscent of the Aulacoceratidae and hence, of *Orthoceras*. In an aulacoceratid, from Greenland Permian (genus *Dictyoconites*), Fisher (1947) detected impressions of possible cameral deposits. In a Mississippian aulacoceratid belemnite (genus *Eobelemnites*), Flower figured (1945, Figure 1) septal

neck connecting rings and cameral deposits (see earlier description of nautiloids). Such features persisted in higher forms.

An ancestral external orthoconic shell became, through time, a reduced internal shell. Ultimate results of this tendency led to reducing the living chamber to a thin conchiolin plate (as in living *Loligo*) and to reducing the phragmacone. In Paleozoic orthocones, the phragmacone contributed important buoyancy; it reduced to a small, hollow, nonstructured cone (living *Omnastreptis*, Indian Ocean) (Swinnerton, 1946); also there was complete loss of the rostrum; and in *Octopus*, the entire shell.

Some noteworthy events in belemnoid history were differentiation from an ancestral stock in the older Paleozoic (Figure 8.62); radiation leading to four families by Lower Jurassic time – two were extinct by Middle Jurassic time (Aulacoceratidae and Chitinoteuthidae); rise of a new subfamily of the Eubelemnitidae (= Belemnitidae), the Duvaliinae in Upper Jurassic; and extinction during Upper Cretaceous of the Eubelemnitidae. The Eubelemnitidae formed the bulk of world belem-

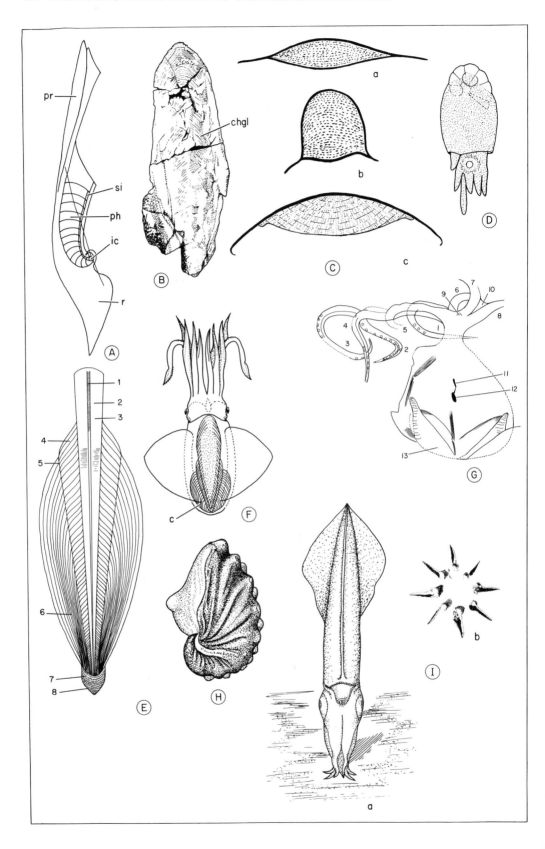

FIGURE 8.61 Representative fossil coleoids, Decapods (Sepoidea, Teuthoidea), Octopoda.
Sepoidea. (A) *Spirulirostra americana*. Mio., Mexico, X 1, lateral view; *pr*—pro-ostracum flattened, and with two lateral wings, seen from below; *si*—siphon; *ph*—phragmacone; *ic*—protoconch; *r*—rostrum (compare *D*).
(B) *Voltzia palmeri*. U. Jur., Cuba, reduced, entire specimen, dorsal view of shield; *chgl*—central field with chevron-shaped growth lines; [compare *Ca*].
(C) Comparison of mid-sections of *Voltzia* (a), *Belosepia* (b), and modern *Sepia* (c).
(D) *Spirula*. Rec., young stage. Note loss of rostrum and pro-ostracum (compare *A*, its prototype).
Teuthoidea. (E) *Leptoteuthis syriaca*. U. Cret., Lebanon, reduced; 1—median rib, 2—median plate, 3—median field, 4—lateral field, inner asymptote, 6—lateral plate, outer asymptote, 8—conus or alveolus.
(F) *Trachyteuthis hastiformis*, U. Jur., Ger., schematic, reduced, to show relationship of internal shell and the conus (c) to the rest of the animal; median plate (dotted) is partly granulose; the various asymptotes and striations of the shell are also shown.
Octopoda. (G) *Paleoctopus newboldi*, U. Cret., Lebanon, reduced; 1–8—arms, 9—mandible, 10—velum, 11—canal, 12—ink sac, 13—U-shaped shell (overlying fish skeleton, and crustacean remains found fossilized with the octopod—these are not shown to set off the octopod).
(H) *Argonauta sismondae*, Pliocene, Italy, lateral view, X 1.
(I) (a). Living *Loligo pealii*, female, bouncing over bottom on tentacle tips—a possible way in which *Ib* was formed on the sandy substrate; schematic. A recent film by Cousteau showed hundreds of squid contacting the substrate in this way. (b) Fossil spoor of an apparent squid-like dibranchiate cephalopod, *Asterichnites octoradiatus*, sandstone beds, Mowry shale, U. Cret., Wyo; holotype, reduced.
Illustrations redrawn and adapted from Berry, 1928; Schevill, 1950; Naef, 1922; Roger, 1946; Brown and Vokes, 1944.

noid population from at least Lower Jurassic time, a span of some 100 million years. Another important event was the rise of a new Tertiary family, the Neobelemnitidae. This family (embracing genera with dual characteristics of reduced rostrum and tendency to curve) foreshadowed the Sepoidea. Before Miocene time, the last of the belemnoids, the neobelemnites, were extinct.

Stratigraphy (Belemnite Zones). Only the belemnoids play an important role in stratigraphy. Other coleoids, while of interest paleobiologically, are too sporadically encountered to serve in zonation and in correlation (Berry, 1926; Roger, 1946; Schevill, 1950).

Paleozoic and Mesozoic beds can be dated and correlated on the basis of abundant faunas exclusive of belemnoids. Thus, even though some genera of the

Aulacoceratidae are widespread in the Triassic (*Dictyoconites*, for example, Miller, 1961), the age of the beds in which they occur can be determined independently. Contrasting Jurassic ammonites and belemnoids, Gignoux (1955) decided that ammonites not only were abundant, but that some were facies-related unlike belemnites.

Both in Northwestern Europe, including Great Britain and Poland, and in Middle through Southern Russia, excluding the Crimea and the Caucausus, belemnite zones are common in the Upper Cretaceous (Jeletsky, 1950, 1955):

Coniacian Stage	*Actinocamax*
Santonian Stage	*Gonioteuthis* and *Belemnitella*
Lower Campanian Stage	*Gonioteuthis* and *Belemnitella*
Upper Campanian Stage	*Belemnitella*
Maastrichtian Stage	*Belemnella*

Several species, so-called flat belemnites (*Duvalia*), make up the deep Mediterranean facies of the Lower Cretaceous (Tithonian to Hauterivian) (Gignoux, 1955; Brinkmann, 1960).

Belemnoids are widely distributed in the North American Mesozoic. Some forms,

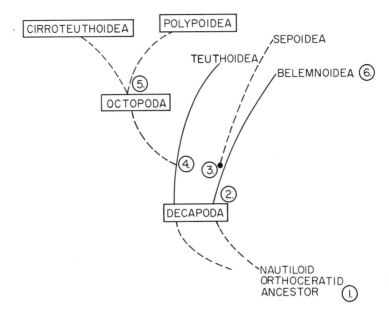

FIGURE 8.62 Origin and relationship between coleoid orders and suborders.
(1) The Belemnoidea are thought to have originated from an U. Silurian ancestor of the type of *Orthoceras pleuronotum* or *Bactrites*.
(2) Major expansion of Belemnoidea began in the Lias but differentiation of this coleoid branch began in the Paleozoic (Aulacoceratidae, Figure 8.60*A*, Belemnoteuthidae, Figure 8.60*E*).
(3) Neogene.
(4) U. Jurassic.
(5) U. Cretaceous.
(6) Extinction in the Neogene (the Neobelemnitidae were the Tertiary line, see Figure 8.60*D*). Illustrations redrawn and adapted after Roger, 1946, 1952; Flower, 1944, 1945; Berry, 1928. Compare Jeletsky, 1966, "Comparative Morphology, Phylogeny and Classification of Fossil Coleoidea", Univ. Kansas, Paleont. Contrib., Article 7. Jeletsky's reconstruction tied Sepoidea (Sepiida) to the line of 4 and 5, and had all three (octopods, teuthids, and sepids) diverge from a common late Triassic line, which extended back to Late Carboniferous time. Starting in the U. Dev. there were two main offshoots of a common ancestor, Bactritida and Aulacocerida as one set of descendents, and Phragmoteuthida (a new order) and Belemnitida. A considerable portion of Jeletsky's reconstruction is based on assumed time ranges.

Aulacoteuthis, occur in the Lower Cretaceous from Alaska into California. Some of the Aulacoceratidae (*Choanoteuthis, Dictyoconites, Metabelemnites*, and so on) occur in the Upper Triassic of California, Nevada, and Mexico. *Pachyteuthis* is found in the Upper Jurassic of South Dakota, Wyoming, and Utah. Several species of *Belemnitella* are known in the chalks of the Upper Cretaceous—Great Plains (Niobrara Formation, *B. praecursor*), the Atlantic into Gulf Coastal Plains (*B. mucronota*). United States Gulf plains repeat the chalks of England, Northern France, and Germany (see "Paleotemperatures," below). *Actinocamax* comes from the Upper Cretaceous of Manitoba; the Eocene of

Alabama has yielded one Neobelemnitidae, *Belemnosella*.

Everywhere the Paleozoic belemnoid record is sparse. The very oldest known representatives occur in a boulder in the Mississippian Caney shale of Oklahoma.

Paleotemperatures. Urey and co-workers, almost two decades ago, (1951, and so on) established a scale, related the O^{16}/O^{18} ratio in calcium carbonate shells of living forms and seawater temperature where they lived, and applied it to the fossil record. Jurassic and Cretaceous belemnite guards were found to be very fine paleothermometers. Reason: unaltered rostra consist of stable megascopic crystals of calcite radially arranged around the apical line.

Among the findings: mean seawater temperature in which Jurassic belemnites lived was 17.6°C (± 6.0° maximal seasonal variation) contrasted to 15° to 16°C for Upper Cretaceous belemnites. By counting concentric bands on the guard (four ontogenetic stages are recognizable in belemnite guards — see Figure 8.60G), it was possible to infer that warmer waters during youth stages turned colder during adult stages. Seasons (three summers and four winters) after the nepionic stage indicated age — four years. Ladd (1957) thought this variation could be explained by reference to modern squid habits — a living relative. Thus warmer waters in youth might correspond to life in shallow coastal bay waters, and colder to migration in adulthood to the cooler open sea.

Subsequent paleotemperature studies, chiefly those of Bowen (1961a–d, 1962, 1963), who studied belemnite rostra from all over the world, have provided interesting inferences on Jurassic and Cretaceous climates. His data permitted checks on the location of the poles determined by paleomagnetic studies, plus many analytical evaluations: aspects of Drift theory; occurrence or nonoccurrence of glaciation in Gondwana area; seasons of time during the Mesozoic; oxygen isotope variations in the Mesozoic seas.

Figure 8.63 plots paleotemperatures determined from oxygen isotope ratios of belemnite rostra for Cretaceous time. Two temperature maxima are shown: Albian and Coniacian, and Santonian. Temperature diminution in post-Albian time corresponds to the early Chalk seas period.

Furthermore (1961c, 1963), Bowen inferred during the Jurassic a more widespread equatorial (tropical) and warm-arid (semitropical) climatic belt than at present. The temperature gradually rose throughout the Mesozoic (although not unidirectionally because Figure 8.63 shows a rise and fall of temperatures during the Cretaceous).

Ammonoid distribution around the Gondwanaland periphery (Figure 8.58B) should be evaluated in light of inferred paleoclimates for Jurassic time.

Paleopathology. Numerous investigators have figured injured and regenerated belemnite rostra (Tasnádi-Kubacska, 1962; Müller Stoll, 1963). Some forms involve species of *Duvalia, Hibolithes, Chitonoteuthis*. After injury, calcite deposition resumed. This led to either shell thickening (indicative of a littoral habit), offset of the newly grown portion, or some other repair compensation for the injured part.

Coprolites may contain belemnite guards. When present, they provide a clue to marine reptile diet. The so-called *Bel-*

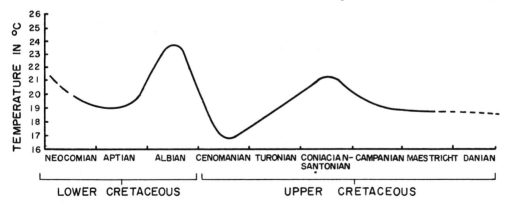

FIGURE 8.63 Paleotemperatures of the Cretaceous as determined by oxygen isotope analysis of belemnite rostra. Note two maxima, Albian and Coniacian-Santonian. Diminution of temperature corresponds to the early period of the Chalk Seas. After Bowen, 1961d. (See W. Stahl and R. Jordan, 1969. Earth and planetary science letters, Vol. 6, pp. 173–178, on post-mortem changes in isotope composition and ways to increase reliability of samples.)

emnitenschlactfeld (belemnite slaughterhouse) has been attributed to such predation: It consists of numerous belemnite guards or guard fragments on the same bedding plane. Another possible explanation could be contemporaneous demise of a large population in an area (Naef, 1922, Figure 81).

Fischer (1947) examined a limestone slab with a $50\pm$ sq cm area from the Permian, Greenland. He recorded 38 rostra discriminated as follows: nepionic, earliest stage, 1; neanic, 1; ephebic, 24; gerontic, old age, 6. Some Fischer evidence suggests the belemnites accumulated in a stagnant lagoon (black pyritic shales and limestones). Further, marine water washed planktonic open sea belemnoids into the lagoon. This last evidence came from deposits of radiolarians, fish, plus belemnoids. Death must have followed quickly after the belemnoids entered the stagnant lagoon.

Fossil Spoor. Associated with fish scales and an ammonite fragment, a curious set of stellate imprints (8-rayed) occur in sandstone of the Mowry shale (Upper Cretaceous, Wyoming) (Figure 8.61*I, b*). Vokes (1944) reviewed pertinent evidence concerning a fossil squid from lithographic limestone of Bavaria. This creature also left a stellate 8-rays (tentacles) imprint. Reconstruction of living *Loligo pealii* bouncing over the bottom on tentacle tips bent outward was available in literature (Figure 8.61*I, a*). The last item suggested a way Mowry imprints may have been made. Conclusion: Mowry imprints were made by an Upper Cretaceous dibranchiate.

Hyolithids and Other Forms

There is a group of small, conoidal shells whose affinities have been debated. Included are coleolids, cornulatids, tentaculatids (Bouček, 1964), hyolithellids, hyolithids, and a miscellaneous assortment grouped with these forms or with the pteropods (Fisher, 1962, pp. W-98 ff.); see Chapter 5.

Taxonomic confusion has been clarified by Yochelson (1961*a, b*, 1963), Marek and Yochelson (1964), Paulsen (1963). In addition, noted in Chapter 5, conulatids, once included with hyolithids, were reassigned to the scyphozoans, a result of work of Knight (1937) and Kiderlen (1937). Cambrian *Hyolithellus* has been reassigned as a close relative of the hemichordates in the class Pogonophora (Paulsen, 1963). *Tentaculites*, of questionable affinity, is thought to be nonmolluscan (Yochelson, 1963, 1965), possibly with annelid worm affinities (Bouček, 1964). These forms possess shell walls pierced by radial canals. A new molluscan class has been proposed for hyolithids and other forms discussed below.

K. M. Towe (Science, 1978, p. 626) found the shell mineralogy and structure of *Tentaculites* reminiscent of an articulate brachiopod.

Before considering the new class Hyolitha, it will prove worthwhile, when attempting to classify small, conoidal fossil shells, to review in condensed form Yochelson's nine criteria (1961*b*) for distinguishing molluscan shells in assemblages: (1) Predominantly $CaCO_3$ (calcite and/or aragonite). (2) Shell layered, not pierced by holes. (3) Prominent growth lines. (4) Logarithmic growth pattern. (5) Basically univalved (modifiable to bivalved condition). (6) Basic bilateral symmetry (modifiable to asymmetrical condition). (7) Lacks trace of apical attachment disk or foramen. (8) Presence of longitudinal or transverse septa. (9) Opercula may be associated with the shell. Shells having most of these features are acceptable molluscan.

Class Hyolitha. Operculate, shell elongate, tapering; apical portion commonly septate. Geologic range: Lower Cambrian–Middle Permian. Marek and Yochelson (1964) recognized three families within this class, defined and figured below:

Family Hyolithidae. (Embraces the bulk of forms in this class.) Semicircular expansion of ventral edge of aperture (= lip);

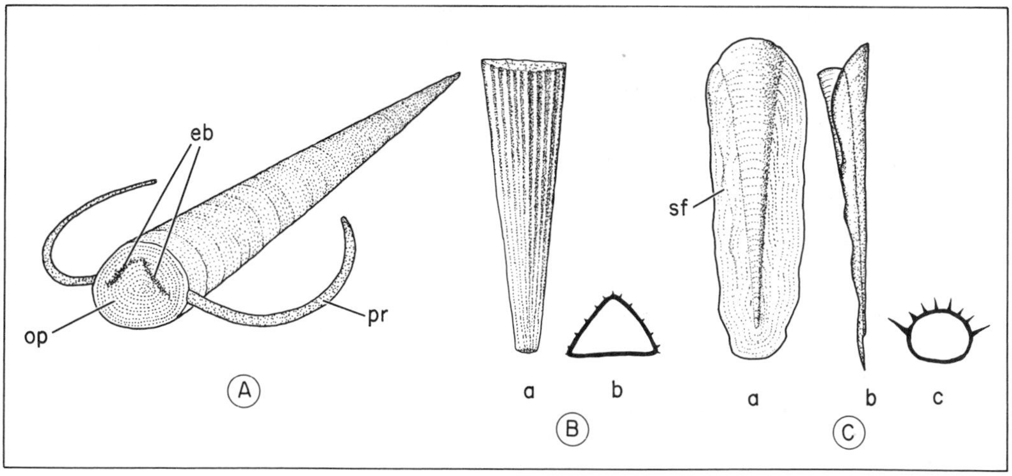

FIGURE 8.64 Class Hyolitha.
(A) Family Hyolithida. *Hyolithes striatulus*, Late Ord., Czech., X 1; *op*—operculum, *pr*—"prop" (elongate paired structures between *op* and aperture), *eb*—exterior reflections of interior paired depressions where probe tip attached.
(B) (*a*–*b*) Family Orthothecidae. *Orthotheca intermedia*. Dev., Czech., X 3.
(C) (*a*–*c*) Family Pterygotheciidae, *Pterygotheca barrendei,* Dev., Czech, schematic; *sf*—swimming frill.
Illustrations redrawn and adapted from Marek and Yochelson, 1964; Fisher, 1962.

presence of elongate paired structures (= props) between operculum and aperture. L. Cambrian–M. Permian. *Hyolithes* (Figure 8.64*A*) (Yochelson, 1961*a*).

Family Orthothecidae. Orthoconic shells lack noticeable lips or shelves; aperture almost at right angles to shell axis. L. Cambrian–M. Devonian. *Orthotheca* (Figure 8.64*B*).

Family Pterygothecidae. Aperture with ventral lips bear small notches along edges at lip base. L. Ordovician–Devonian. *Pterygotheca* (Figure 8.64*C*).

Information Derived from Hyolithids

Distribution and Density. The bulk of the class Hyolitha is composed of the genus *Hyolithes*. During Cambrian time, this genus had worldwide distribution and occurred in abundance. There are upward of 300 known Cambrian species. An unusual example of single species occurrence in a formation is the Middle Cambrian Burgess Shale of British Columbia. Some 200 specimens of *Hyolithes carinatus* with remarkable preservation are known from this formation (Yochelson, 1961*b*). Hyolithids diminish in the Ordovician and

even more so thereafter until the end of the Paleozoic; present evidence lists them extinct.

Paleoecology. Since *Hyolithes* has been regarded as similar to living pteropods (see Chapter 5), the animal was thought to have led a planktonic existence.

Yochelson, originally (1961*a*), and Marek and Yochelson (1964) reinterpreted the *props*. In this new view these structures were a third hard part (the operculum and tubelike shell were the other two).

Marek, in 1963, had suggested that the props could have served to "pole" the shell across the bottom substrate. This explanation, in light of *all* available evidence, seemed plausible to Marek and Yochelson (1964).

Hyolithes thus appears to have been *vagrant benthos* instead of a component of plankton.

The ecology of *Pterygotheca, Orthotheca,* and related forms seems similar to that of *Hyolithes* (although no *props* are present). Some Orthothecidae specimens, however, cannot be considered benthonic with any assurance. The "swimming frill" (see Figure 8.64*C, b,* "*sf*") of *Pterygotheca* may be an overgrown bryozoan colony. If so, this single structure that makes it appear so different from *Hyolithes* would be eliminated (Marek and Yochelson, 1964).

IDENTIFICATION OF AN AMMONOID TO THE SPECIES LEVEL

[Data from Miller and Furnish, 1958, *J. Paleont.*, Vol. 32 (2), pp. 269 ff. See references for complete title.]

Given. Several specimens and fragments (internal molds) of cephalopods from the Burlington limestone (L. Mississippian), Missouri. Some of these had originally been assigned to *Goniatites* species.

Preservation. All specimens occur in white, chalky chert.

Shell Characters. Surficial ornamentation: Internal mold surface bears prominent growth line traces; four to six rounded constrictions per volution (Figure 8.65*B*).

Measurements. Figure 8.65*A* indicates that in life conchs attained a diameter of 75+ mm. Where the phragmacone is 50 mm diameter, the whorls are about 25 mm wide, and 23 mm high. Umbilical diameter is about 1/5 that of the complete conch. On one specimen the umbilical diameter is about 11 mm.

Suture. What is the characteristic suture pattern? Goniatitic (see text discussion of suture types). Characteristics: Sides of ventral lobe are nearly parallel; lateral lobe, V-shaped. Internal sutures form three long, slender, pointed lobes reminiscent of other species of genus *Muensteroceras*.

Genus. Genus determination is primarily based on the proportions of the ventral lobe of sutures. At the time of this particular study, four genera were considered possibilities:

1. *Beyrichoceras* **(Foord, 1903).** Ventral lobe relatively wide, prongs diverge.
2. *Muensteroceras* **(Hyatt, 1884).** Ventral lobe of suture narrow, with parallel sides.
3. *Beyrichoceratoides* **(Bisat, 1924).** Ventral lobe indistinguishable from *Muensteroceras*.
4. *Bollandoceras* **(Bisat, 1952).** Ventral lobe indistinguishable from *Muensteroceras*.

[Genera 3 and 4 have been placed, as suggested by authors, in the synonymy of *Muensteroceras* (see Miller *et al*, 1957, p. L. 57).]

Comparison of the ventral lobes of sutures of the studied specimens (Figure

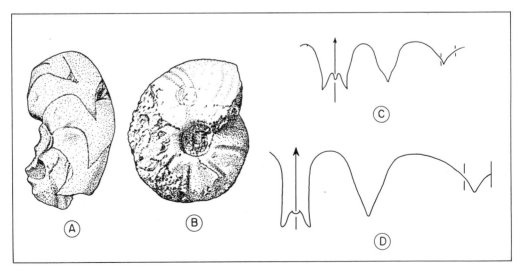

FIGURE 8.65 The cephalopod to be identified. (*A*) Portion of phragmacone, X 1: (*B*) nearly complete specimen, X 1; (*C*) external suture reconstructed from (*A*), at estimated diameter of complete shell of 45–50 mm, X 2; (*D*) *Beyrichoceras hornerae*, compare this suture's ventral lobe with that of (*C*). Illustrations redrawn and adapted from Furnish and Miller, 1957.

8.65*C, D*) denoted the closest relationship was to *Muensteroceras*. Accordingly, Burlington specimens clearly belong to this genus.

Other Generic Characters. Conch, discoidal to globular. Umbilici, small to moderate.

Species. Five closely related Mississippian species of *Muensteroceras* were considered: (*a*) *M. parallelum* (the type species of the genus) — umbilicus, relatively small: smaller than study specimens. (*b*) *M. oweni* (found associated with *a*) — larger umbilici than in study specimens. (*c*) *M.(?) mitchelli* — close in proportions to study specimens but the ventral lobes of sutures were not so nearly parallel. (*d*) *M. pfefferae* — a crushed specimen; ventral profiles apparently were considerably narrower than those of study specimens. (*e*) *M. eshbaughi* — a crushed specimen; ventral profiles like those of (*d*).

Thus three different criteria were employed to eliminate species (*a*)–(*e*): *diameter of umbilicus* (too small, too large); *minor morphology of ventral lobe*; and *comparable ventral profile* (*d* and *e* were much narrower than study specimens). That signifies that each of these shell features were thought to be genetically determined and unlikely to occur in any members of the populations of species (*a*) through (*e*). Accordingly, the authors designated their study specimens as a new species, *Muensteroceras rowleyi*.

QUESTIONS

General

1. Outline basic information in the phylum.
2. Explain significance of bivalved gastropods.
3. (a) Sketch characteristic molluscan shell structure, after Bøggild.
   (b) Explain (a) via Mackay (1952).
4. Compare living and fossil oyster reefs.
5. Relate Thorson's work on level bottom communities to the biostratigraphic problems of regional and intercontinental correlations; use fossil mollusks.

Monoplacophora

1. What is the primary information in a monoplacophoran?
2. Why has *Neopilina* been called the "Rosetta stone"?
3. How can you account for the "sudden" disappearance of monoplacophorans from the rock record in the post-Paleozoic?

Polyplacophora

1. Discuss the structure of the polyplacophoran shell.
2. What is the preferred living zone for a polyplacophoran?

Scaphopoda

1. Explain the paleobiological significance of the subdivision of the Dentaliidae into 16 subgenera.
2. Where have fossil scaphopods been reported in the American Paleozoic? Mesozoic? Tertiary?

Gastropoda

1. Itemize the primary data in class Gastropoda and the three subclasses.
2. (a) Draw and label the hypothetical ancestral mollusk. What changes were needed to derive a primitive gastropod? (Consult Runnegar and Pojeta Jr., 1974.)
   (b) Identify and/or sketch the following: involute, protoconch, convolute, heterotrophic.
3. Sketch and label five representative archeogastropods, caenogastropods.
4. Identify: selenizone, umbilicus, aperture, carina, columnar lip, rugae, parietal lip, whorl, columnella, opercula, sutures, inductura, keel. (*Hint.* See illustrations.)
5. Evaluate gastropod fossil studies: Yochelson, 1959; McCrone, 1963; Johnson, 1962; Sadlick and Nielson, 1963; Yochelson, 1964*a*. Stress, in your argument, biofacies, ecological units, and stratigraphic correlations.

6. On what evidence did Ladd (1958) postulate the distribution of Tertiary-Quaternary life in the Pacific via "high island stepping stones"?

Bivalvia

1. What are essential bits of information in the primary code of the Bivalvia?

2. (a) Study Figure 8.22, then sketch postulated stages leading from primitive mollusk to the bivalve stage.
   (b) Delimit critical genetic specifications for changes noted in 2a.

3. Define each term of these related items:
   (a) Isomyarian, monomyarian, anisomyarian.
   (b) Protobranch, filibranch, eulamellibranch, septibranch.
   (c) Amphidetic, opisthodetic, alivincular, multivincular, parivincular.
   (d) Taxodont, dysodont, isodont, schizodont, heterodont, edentate.

4. (a) Compare Newell's classification (1965) to that of Stasek, Franc, Dall, Purchon, Thiele.
   (b) Can you cite the critical bits of information distinguishing each of Newell's six subclasses?

5. How does dentition relate to the ligament? (Evaluate selective advantage of each.)

6. (a) What new light has been shed on Vogel's (1962) M. Cambrian *Lamellodonta* and its taxonomic placement?
   (b) Discuss probable origin of mytilaceans, pteraceans, arcaceans, and heterodonts (see Figure 8.32).
   (c) Do the same as in 6b for oysters and scallops (Figure 8.33).

7. (a) Cite examples of parallel and convergent evolution from Newell's researches.
   (b) Describe the development of a rudistid bank.
   (c) Give highlights on extinction of rudistids, biostratigraphic zonation in the Upper Cretaceous by rudist zones, shell structure in *Trechmanella*, shell fixation.

8. Identify: cardinal angle, resilium, costa, lithodesma, resilifer, byssus, dorsal, ventral, posterior, anterior, prodissoconch, chondrophore, myophore, byssal notch.

9. In beds of what age would you hope to find *Pinna, Pseudomonotis, Pecten, Trigonia, Pholadomya, Nucula, Chama, Anthraconauta*?

10. Study Figure 8.38*A* and explain how Dodd (1964) related shell structural type in a bivalve to paleotemperatures.

Cephalopoda — Exclusive of the Coleoids

1. State the primary information in cephalopods.

2. Review steps from the origin of cephalopods beginning with the ancestral mollusk. (*Hint.* Figure 8.39*B*.) (Consult Yochelson et al., 1973).

3. (a) Identify: hyponome sinus, phragmacone, living chamber, siphuncle, septal foramen, septa.
   (b) Draw a coiled and a straight cephalopod; label, dorsal, ventral, anterior, and posterior sides.

4. (a) How do endo- and exogastric shells differ? Evolute and involute?
   (b) Relate shell truncation to the concept of carbonate resorption.

5. (a) Cite the critical bits of information distinguishing the six subclasses.
   (b) Which of these bits are missing from the fossil record?

6. Illustrate and explain Schindewolf's thesis on ammonoid descent. (*Hint.* Figure 8.45*E*.)

7. Draw, then fully label, the three ammonoid types of sutures.

8. How do all cephalopods relate phylogenetically? (Show the transition, the postulated links between all subclasses.)

9. (a) State suture characteristics of: ceratitic, goniatitic, ammonitic.
   (b) What three grades of suture development did Schindewolf (1954) define? Of what value is this distinction?
   (c) Account for asymmetrical sutures.
   (d) Define: suspensive, auxiliary, accessory, adventitious lobes, lobules, and folioles.

10. (a) Study Figure 8.48 and outline how ontogeny reflects phylogenetic descent.
    (b) Evaluate the utility of such derivative data for stratigraphy and correlations.

11. (a) Distinguish between internal and external sutures.

(b) What is the very minimum fragment of one ammonoid that can allow specific and generic identification?

12. What geologic age could you assign to each of the following: *Manticoceras, Actinoceras, Owenites, Bactrites, Lituites, Nautilus, Emmonsoceras, Oppelia, Goniatites, Meekoceras, Lytoceras.*

13. Discuss the work of Dunbar, Berry, Teichert, and Flower pertaining to the buoyancy problem in nautiloids.

14. (a) Compare the utility of ammonoids and nautiloids in biostratigraphic zonation and interregional/intercontinental correlations in the Paleozoic, the Mesozoic, and the Tertiary.

(b) What is the spiral angle function? (See Currie, 1943.)

15. (a) Cite estimated growth rate of ammonoids. How is it determined?

(b) Cite ammonoid spoor value in paleoecology.

(c) Can seasonal broods in fossil ammonoid populations be detected? Show how.

Cephalopoda — Coleoids

1. (a) Account for the new expansion in cephalopod evolution represented by extant coleoids.

(b) Cite the primary code of the subclass Coleoidea.

(c) How could you distinguish a coleoid phragmacone from that of an orthoconic nautiloid?

2. (a) What specific features distinguish the two coleoid orders?

(b) How would you proceed to identify the internal structures of an assemblage of belemnite guards?

(c) Identify: epirostrum, embryonic bulb, apical line, pro-ostracum, rostrum, septate phragmacone, nepionic, neanic, and ephebic-gerontic deposits, onychites, *Belemnitenschlactfeld.*

3. Compare belemnoids of the Paleozoic, Mesozoic, and Tertiary (Eocene).

4. Discuss major events in belemnoid history.

5. How did Bowen utilize data on oxygen isotope ratios obtained from belemnite rostra to determine paleotemperatures, Cretaceous paleoclimates, continental-

drift theory, Gondwana glaciation, and the seasonal growth of belemnoids?

Hyolitha

(It has been proposed that this class be removed from the Mollusca.)

1. Review Yochelson's (1961) nine criteria for distinguishing which small conoidal fossils are molluscan.

2. Why was this class established? Give its primary code. Explain why Runnegar et al. (1974) reject it as belonging to the Mollusca.

REFERENCES

General

Buchsbaum, Ralph, 1948. Animals without backbones. Univ. of Chicago Press, Chapters 17, 18.

Comfort, A., 1950. Biochemistry of molluscan shell pigments: Proc. Malacological Soc. London, Vol. 28, pp. 79–85.

Jägersten, Gösta, 1960. Life in the sea. Basic Books, pp. 123–147.

Ladd, H. S., 1951. Brackish-water and marine assemblages of the Texas coast, with special reference to mollusks. Texas Univ. Instit. Mar. Sci. Publ. Vol. 2 (2), pp. 125–163.

Light, S. F., 1954. Intertidal invertebrates of the central California coast. Univ. of Calif. Press (rev. R. I. Smith et al.), pp. 211–270.

Morton, J. E., 1960. Molluscs: An introduction to their form and function. Harper and Bros., 220 pp. (useful bibliography).

Naef, Adolf, 1926. Studien zur generellen Morphologie der Mollusken: Part 3. Ergebnisse und Forschritte der Zoologie, Vol. 5, Jena, pp. 28–107.

Nicol, J. A. Colin, 1960. The biology of marine animals. Interscience Publishers, 674 pp. (reprinted 1964).

Raven, Chr. P., 1958. Morphogenesis: the analysis of molluscan development. Pergamon, 268 pp.

Tasnádi-Kubacska, A., 1962. Paläopathologie, Gustav Fischer, Jena, 240 pp.

Thiele, Johannes, 1931. Handbuch der systematischen Weichtierkunde, Vol. 1. Loricata/Gastropoda, 778 pp. Vol. 2. Scaphopoda/Bivalvia/Cephalopoda. 375 pp (reissued A. Ascher and Co., Amsterdam, 1953).

Thompson, D'Arcy W., 1942. Growth and Form. Macmillan, 1097 pp.

Thorson, Gunnar, 1946. Reproduction and larval development of Danish marine bottom invertebrates (with section on lamellibranch larvae by C. Barker Jørgensen). Meddelser fra Komm. For Danmarks Fiskeri-Og Havundersøgelser, Copenhagen, 484 pp.

_____, 1957. Bottom Communities (Sublittoral or Shallow Shelf) G.S.A. Mem. 67, Vol. 1, Ch. 17.

Wilbur, K. M., and C. M. Yonge, 1964. Physiology of mollusca. Academic Press, N.Y.

Yonge, C. M., 1960. General character of mollusca. Treat. Invert. Paleont., Part 1, Mollusca I, pp. 113–136.

_____, 1963a. The Seashore. Atheneum Press, N.Y., 326 pp.

[Runnegar B. and J. Pojeta, Jr., (1974) ref. p. 439]

Monoplacophora

Clarke, A. H., Jr., and R. L. Menzies, 1959. *Neopilina (Vema) ewingi*, a second living species of the Paleozoic class Monoplacophora: Science, Vol. 129, pp. 1026–1027, Figure 1, Table 1.

Horny, R. J., 1961. *Archeopraga*, a new problematic genus of monoplacophoran molluscs from the Silurian of Bohemia: J. Paleontology, Vol. 37 (5), pp. 1071–1073.

Knight, J. B., 1959. Occurrence and classification of fossil Monoplacophora (Mollusca). XVth Interntl. Congress Zool. London, Section IV, pp. 1–3.

_____ and E. L. Yochelson, 1958. A reconsideration of the relationships of the Monoplacophora and the primitive gastropod. Proc. Malacological Soc. London, Vol. 33, pp. 37–48.

_____ and E. L. Yochelson, 1960. Monoplacophora. Treat. Invert. Paleont., Part 1, Mollusca I, pp. 177–189.

Lemche, Henning, 1957. A new living deepsea mollusc of the Cambro-Devonian class Monoplacophora: Nature, Vol. 179, pp. 413–416, 4 text figures.

_____ and K. G. Wingstrand, 1959. The anatomy of *Neopilina galatheae*. Galathea Report. Danish Science Press Ltd., Copenhagen, Vol. 3, pp. 9–57.

Lowenstam, Heinz, 1963. Biologic problems relating to the composition and diagenesis of sediment, in T. W. Donnelly, ed., The

Earth Sciences: problems and progress in current research. 195 pp. (p. 179).

Menzies, R. J., and D. J. Robinson, 1961. Recovery of the living fossil mollusk, *Neopilina,* from the slope of the Cedros Trench, Mexico: Science, Vol. 134, pp. 338–339.

Rasetti, Franco, 1954. Internal shell structure in the Middle Cambrian gastropod *Scenella* and the problematic *Stenothecoides:* J. Paleontology, Vol. 28, pp. 59–66, plates 11–12.

Schmidt, W. J., 1959. Bemerkungen zur Schalenstruktur von *Neopilina galatheae,* in Galathea Report, Vol. 3 (see above for complete reference), pp. 73–77, plates 1–2.

Yochelson, E. L., 1958. Some Lower Ordovician monoplacophoran mollusks from Missouri: J. Washington Acad. Sci., Vol. 48 (1), pp. 8–14, Figure 1–13.

Yonge, C. M., 1957. Reflexions on the monoplacophoran *Neopilina galatheae* Lemche: Nature, Vol. 179, pp. 672–673.

Polyplacophora (Chitons)

Bergenhayn, J. R. M., 1930. Kurze Bermerkungen zur Kenntnis der Schalenstruktur und Systematik der Loricaten. Kungl. Svenska Vetenskaps. Handl. Tredje Seren. Vol. 9 (3), pp. 3–51, 10 plates text figures.

Berry, C. T., 1940. Some fossil Amphineura from the Atlantic Coastal Plain: Acad. Nat. Sci. Philadelphia, Proc., Vol. 91, pp. 207–216, plates 9–12.

Cotton, B. C., and F. K. Godfrey, 1940. The Molluscs of South Australia, Part 2. So. Australian Branch, Brit. Sci. Guild Adelaide (Fossil Chitons, pp. 569–577).

Koch, H. J., 1951. A new chiton from South Africa: Proc. Malacological Soc. London, Vol. 28 (6), pp. 211–212, plates 23–25.

Knorre, Heinrich von, 1925. Die Schale und die Rückensinnesorgane von *Trachydermon (Chiton) cinereus* L. und ceylonischen Chitonen der Sammlung Plate: Jenaische Zeitschr. Naturwiss, Vol. 61 (new series Vol. 54), pp. 469–632, plates 18–35, text figures 1–17. Jena.

Lowenstam, Heinz, 1962. Magnetite in denticle capping in Recent chitons (Polyplacophora): Geol. Soc. America Bull., Vol. 73, pp. 435–438, 1 plate.

Pierce, Madeline E., 1950. Mollusca: *Chaetopleura spiculata,* in F. A. Brown, Jr., ed., Selected invertebrate types. John Wiley, pp. 318–319.

Smith, A. G., 1960. Amphineura, in Treat. Invert. Paleont. Part 1, Mollusca I, pp. 141–174.

Scaphopoda

Deschaseaux, Colette, 1952. Classe des Amphineures, in Jean Piveteau, ed., Traité de Paléontologie, Vol. II, pp. 210–215; Classe des Scaphopoda, pp. 216–219.

Gardner, J. S., 1878. On the Cretaceous Dentaliidae: Geol. Soc. London Quart. Jour., Vol. 34, pp. 56–65.

Gardner, Julia, 1947. The molluscan fauna of the Alum Bluff of Florida. Part 8, U.S.G.S. Prof. Paper 142-H, pp. 493–638 (Scaphopod plate LXII).

Henderson, J. B., 1920. A monograph of the East American scaphopod mollusks: U.S. Nat. Mus. Bull. 111, 150 pp., plates 1–20.

Ludbrook, N. H., 1960. Scaphopoda, in Treat. Invert. Paleont. Part 1, Mollusca I, pp. 137–141.

Palmer, K. Van Winkle, 1937. The Claibornian Scaphopoda, Gastropoda, and dibranchiate Cephalopoda of the Southern United States: Amer. Paleont. Bull., Vol. 7 (Parts 1 and 2), pp. 1–548, plates 1–90.

Richardson, L., 1906. Liassic Dentaliidae: Quart. J., Geol. Soc. London. Vol. 62, pp. 573–596.

Yochelson, E. L., 1957. Scaphopods and chitons of the Paleozoic, in Treat. Mar. Ecol. Paleocol.: G.S.A. Mem. 67, Vol. 2, pp. 819–820.

Young, J. A., 1942. Pennsylvanian Scaphopoda and Cephalopoda from New Mexico: J. Paleontology, Vol. 16 (1), pp. 120–125, plate 20, 2 text figures.

Gastropoda

Avnimelech, M., 1945. Revision of fossil pteropods from Southern Anatolia, Syria, and Palestine: J. Paleontology, Vol. 19 (6), pp. 637–647.

Boycott, A. E., 1928. Conchometry. Proc. Malacological Soc. London, Vol. 18, pp. 8–31.

Bowsher, A. L., 1955. Origin and adaptation of platyceratid gastropods: Univ. Kansas Paleont. Contr., Mollusca, art. 5, pp. 1–11, plates 1–2, Figure 1.

————, 1956. The effect of the crinoid host on the variability of Permian platyceratids, in E. L. Yochelson, Permian Gastropoda of the Southwestern United States: Amer.

Mus. Nat. Hist. Bull., Vol. 110, art 3, pp. 261–263.

Bucher, W. A., 1938. A shell boring gastropod in a *Dalmanella* bed of Upper Cincinnatian age: Am. J. Sci., 5th Ser., Vol. 36 (211), pp. 1–7.

Chronic, Halka, 1952. Molluscan fauna from Permian Kaibab Formation, Walnut Canyon, Ariz.: Geol. Soc. Amer. Bull., Vol. 63, pp. 95–166, 15 figures, 11 plates.

Collins, R. L., 1934. A monograph of the American Tertiary pteropod mollusks, in Johns Hopkins Univ. Studies in Geol. No. 11, pp. 137–161.

Cox, L. R., 1960a. Gastropoda: general characteristics of gastropods, in Treat. Invert. Paleont., Part 1, Mollusca I, pp. 184–1169.

————, 1960b. A bivalve gastropod: Nature, Vol. 185 (4715), pp. 749–751, Figures 1–2.

Eales, N. B., 1950. Torsion in Gastropoda. Proc. Malacological Soc. London, pp. 53–61, plates 4–5.

Fenton, C. L., and M. A. Fenton, 1931a. Apparent gastropod traces in Lower Cambrian: Amer. Midland Nat., Vol. 12, pp. 401–405.

————, 1931b. Some snail borings of Paleozoic age: *idem*, Vol. 12, pp. 521–528.

————, 1937. Burrows and trails from the Pennsylvanian rocks of Texas: *idem*, Vol. 18, pp. 1079–1084.

Fisher, D. W., 1962. Small conoidal shells of uncertain affinities, in Treat. Invert. Paleont., Part W, Miscellanea, pp. W98–W143.

Frankel, Larry, 1957. The value of Pleistocene mollusks as index fossils of Wisconsin sub-ages in Nebraska: J. Paleontology, Vol. 31 (3), pp. 641–647.

Frye, J. C., and A. B. Leonard, 1952. Pleistocene geology of Kansas: State Geol. Surv. Kansas Bull. 99, 210 pp. (Mollusca, pp. 144–180).

Gardner, Julia, 1947. The molluscan fauna of the Alum Bluff of Florida. Part 8, U.S.G.S. Prof. Paper 142-H, 638 pp., plates LII–LXII.

Girty, G. H., 1912. Notice of a Mississippian gastropod retaining coloration: Am. J. Sci. Vol. 34 (4th Ser), pp. 338–340.

Haas, Otto, 1953. Mesozoic invertebrate faunas of Peru. Part 1, General Introduction. Part 2, Late Triassic Gastropods from Central Peru: Amer. Mus. Nat. Hist. Bull., Vol. 101, 310 pp., 17 plates.

Hall, C. A., Jr., 1959. The gastropod genus

Ceratostoma: J. Paleontology, Vol. 33 (3), pp. 428–434, plates 61–63.

Johnson, R. G., 1962. Interspecific associations in Pennsylvanian fossil assemblages: J. Geology, Vol. 70, pp. 32–55.

Kornicker, L. S., 1959. Observations on the behavior of the pteropod *Creseis acicula* Rang. Mar. Sci. of the Gulf and Caribbean Bull., Vol. 9 (3), pp. 331–336.

Knight, J. B., 1930–1934. The gastropods of the St. Louis, Missouri Pennsylvanian outlier. J. Paleont., **4** (1930), 1–88; **5** (1931), 1–15; **5** (1931), 173–229; **6** (1932), 189–202; **7** (1933), 30–58; **7** (1933), 359–392; **8** (1934), 139–166; **8** (1934), 433–447.

————, 1939. Handbuch der Paläozoologie, O. H. Schindewolf, ed. Vol. 6, 1: Gastropoda by W. Wenz. Berlin (Gebruder Borntraeger, 1931, pp. 230–232).

————, 1941. Paleozoic gastropod genotypes: Geol. Soc. America Sp. Paper 32, 510 pp., 96 plates, 32 figures.

————, et al., 1944. Paleozoic gastropods, in H. W. Shimer and R. R. Shrock, Index Fossils of North America, pp. 437–479.

————, 1952. Primitive fossil gastropods and their bearing on gastropod classification: Smithson. Misc. Coll. Vol. 117, no. 13, 56 pp., 2 pls.

————, et al., 1960a. Systematic descriptions, in Treat. Invert. Paleont., Part 1, Mollusca I, pp. 1169 ff.

————, R. L. Batten, E. L. Yochelson, and L. R. Cox, 1960b. Supplement—Paleozoic and some Mesozoic Caenogastropoda and Opisthobranchia, pp. 1310 ff. (See 1960a).

Ladd, H. S., 1958. Fossil land snails from Western Pacific atolls: J. Paleontology, Vol. 32 (1), pp. 183–198, plate 30, 5 text figures.

Lane, C. E., 1959. Some aspects of the general biology of *Teredo*, in Marine boring and fouling organisms. Seattle Univ. Washington Press, pp. 137–144, Figures 1–23.

McCrone, A. W., 1963. Paleoecology and biostratigraphy of the Red Eagle cyclothem (Lower Permian) in Kansas: State Geol. Surv. Kansas Bull., 164, 75 pp.

Moore, H. B., 1934. The relation of shell growth to environment in *Patella vulgata*. Proc. Malacological Soc. London, Vol. 21, pp. 217–222, plates 23–25.

Morton, J. E., 1955. The evolution of the Ellobiidae with a discussion on the origin of the Pulmonata: Zool. Soc. London Proc., Vol. 125, pp. 127–168.

———— and C. M. Yonge, 1964. Classification and structure of the Mollusca, in K. M. Wilbur and C. M. Yonge, Physiology of Mollusca, Chapter 1 (see, under "General," Wilbur and Yonge, 1964).

Naef, Adolf, 1913. Studien zur generellen Morphologie der Mollusken. Part 1: Über Torsion und Asymmetrie der Gastropoden. Ergebnisse und Fortschritte der Zoologie, Vol. 3, pp. 74–160 (Jena).

Powers, S., 1922. Gastropod trails in Pennsylvanian sandstones in Texas: Am. J. Sci., 5th ser., Vol. 53, pp. 101–107, text figures 1–3.

Rodda, P. V., and W. L. Fisher, 1964. Evolutionary features of *Athleta* (Eocene, Gastropoda) from the Gulf coastal plain: Evolution, Vol. 18 (2), pp. 235–244.

Sadlick, W., and M. F. Nielsen, 1963. Ontogenetic variation of some Middle Carboniferous pleurotomarian gastropods: J. Paleontology, Vol. 37 (5), pp. 1083–1103, plates 148–150, 8 text figures.

Stanton, T. W., 1947. Studies of some Comanche pelecypods and gastropods: U.S. Geol. Surv. Prof. Paper No. 211, 116 pp.

Taylor, Monica, 1954. Variation of the shells of *Murex tribulus* from the Great Bitter Lake, Egypt: Nature, Vol. 174 (4441), pp. 1111–1112.

Termier, Geneviève, and Henri Termier, 1952. Classe des Gasteropodes, in Jean Piveteau, ed., Traité de Paleóntologie, Vol. 2, pp. 365–446.

Tyler, J. H., 1965. Gastropods from the Middle Devonian Four Mile Dam Limestone (Hamilton) of Michigan: J. Paleontology, Vol. 39, pp. 341–349, plates 47–48.

Wade, Bruce, 1926. The fauna of the Ripley Formation on Coon Creek, Tennessee: U.S. Geol. Surv. Prof. Paper No. 137, 192 pp.

Wenz, W., 1938. Gastropoda, in O. H. Schindewolf, ed., Handbuch der Paläozoologie (see Knight, 1939, in this section).

Woodring, W. P., et al., 1946. Geology and paleontology of Palos Verdes Hills, California: U.S. Geol. Surv. Prof. Papers No. 207, 123 pp.

Yochelson, E. L., 1956. Permian Gastropoda of the Southwestern United States: American Mus. Nat. Hist. Bull., Vol. 110, art. 3, pp. 173–276.

————, 1957. Notes on the gastropod *Palliseria robusta* Wilson: J. Paleontology, Vol. 31 (3), pp. 648–650.

————, 1964a. *Modestospira*, a new Ordovician gastropod: J. Paleontology, Vol. 38 (5), pp. 891 ff.

————, 1964b. The early Ordovician gastropod *Ceratopea* from East Greenland: Meddelelser om Grønland, Vol. 164, No. 7, 11 pp. (Copenhagen).

———— and R. W. Kopf, 1956. A new Devonian spinose platyceratid gastropod: J. Paleontology, Vol. 30 (5), pp. 1170–1172, plate 125.

———— and Josiah Bridge, 1957. The Lower Ordovician gastropod *Ceratopea*: U.S. Geol. Surv. Prof. Paper No. 294-H, pp. 281–302.

Bivalvia

Bernard, F., 1895–1897. Sur le développement et la morphologie de la coquille chez les lamellibranches. Première note (1895), Vol. 23, pp. 104–154; Deuxième note (1896a), Vol. 24, pp. 54–82; Troisième note (1896b), Vol. 24, pp. 416–449; Quatrième et Dernière note (1897), Vol. 24, pp. 559–566. Soc. Géol. de France Bull., 3d ser.

Bøggild, O. B., 1930. The shell structure of the mollusks: Mém. Acad. Roy. Sci. Lettres Danemark, Copenhagen, ser. 9, Vol. 2, pp. 230–326, plates 1–15, 10 text figures.

Bonham, Kelshaw, 1965. Growth rate of giant clam *Tridacna gigas* at Bikini atoll as revealed by radioautography: Science, Vol. 149 (3681), pp. 300–302.

Bramlette, M. N., 1965. Massive extinction in biota at the end of Mesozoic time: Science, Vol. 148, pp. 1696–1699.

Branson, C. C., 1957. Pelecypoda of the Paleozoic: Treat. Mar. Ecol. and Paleocol., G.S.A. Mem. 67, Vol. 2, pp. 817–818.

Broadhurst, F. M., 1959. *Anthraconaia pulchella* sp. nov. and a study of paleoecology in the Coal Measures of the Oldham area of Lancashire: Quart. J. Geol. Soc. London, Vol. 114, pp. 523–545.

Coe, W. R., 1943. Sexual differentiation in mollusks. 1. Pelecypods: Quart. Rev. Biology, Vol. 18 (2), pp. 154–164.

Cox, L. R., 1933. The evolutionary history of the rudists: Geol. Assoc. Proc., Vol. XLIV, part 4, pp. 379–388, text figures 43–46.

————, 1934. On the structure of the Persian rudist genus *Trechmanella* (formerly *Polyptchus*) with the description of a new species: Proc. Malacological Soc. London, Vol. XXI, pp. 42–46.

————, 1960a. Moulds of intestines in fossil *Nuculana* (Lamellibranchia) from the Lias of England: Paleontology, Vol. 2, Part 2, pp. 262–269.

————, 1960b. Thoughts on the classification of the Bivalvia: Proc. Malacological Soc. London, Vol. 34, pp. 60–85.

Dall, W. H., 1895. Tertiary mollusks of Florida. Part 3. A new classification of the Pelecypoda: Wagner Free Inst. Sci. Philadelphia Trans., Vol. 3, pp. 483–565.

Davies, A. M., 1933. The bases of classification of the Lamellibranchia: Proc. Malacological Soc. London, Vol. 20, pp. 322–326, text figures 1, 2.

Davies, Tudor T., 1965. Effect of environmentally induced growth rate changes in *Mytilus edulis* shell: G.S.A. Program (Kansas City), abstracts, p. 41.

Degens, E. T., and R. H. Parker, 1965. Significance of shell protein variation to environment and molluscan phylogeny: G.S.A. Program (Kansas City), abstracts, p. 43.

Deschaseaux, Colette, 1948. Le problème de l'extinction des groupes étudié chez les Rudistes: La Revue scientifique, Vol. 86 (2), pp. 83–86.

————, 1952. Classe des Lamellibranches, in Jean Piveteau, ed., Traité de Paléontologie, Vol. 2, pp. 220–364 (Rudistids, pp. 323–362).

————, 1960. Les premiers bivalves, in P. P. Grasse, ed., Traité de Zoologie, Vol. 5, pp. 2134 ff.

Dodd, J. R., 1964. Environmentally controlled variation in the shell structure of a pelycypod species: J. Paleontology, Vol. 38 (6), pp. 1065–1071, plate 160, 6 text figures.

Franc, André, 1960. Classe des Bivalves, in P. P. Grasse, ed., Traité de Zoologie, Vol. 5, pp. 1858 ff.

Gardner, Julia, 1926. The molluscan fauna of the Alum Bluff Group of Florida: U.S. Geol. Surv. Prof. Paper No. 142-A, 64 pp.

Hedgepeth, J. W., 1953. An introduction to the zoogeography of the Northwestern Gulf of Mexico with reference to the Invertebrate Fauna: Inst. Mar. Science, Vol. 111 (1), pp. 111–211.

Kauffman, E. G., 1973. Cretaceous bivalves, in A. Hallam, Ed., Atlas of paleobiogeography. Elsevier, pp. 353–383.

Laughbaum, L. R., 1960. A paleoecologic study of the Upper Denton Formation, Tarrant, Denton and Cooke Counties, Texas: J. Paleontology, Vol. 34 (6), pp. 1183–1197.

Lefevre, George, and W. C. Curtis, 1910. Studies on the reproduction and artificial propagation of fresh water mussels: Bur. of Fisheries Bull., Vol. 30, pp. 109–197, plates 6–17.

Mackay, I. H., 1952. The shell structure of the modern molluscs, in J. H. Johnson, Studies of organic limestones and limestone-building organisms: Colorado School of Mines Quart., Vol. 47, No. 2, pp. 1–27.

McAlester, A. L., 1960. Pelecypod associations and ecology in the New York Upper Devonian: G.S.A. Program (Denver), abstracts, p. 156.

————, 1965. Life habits of the "living fossil" bivalve *Neotrigonia*: G.S.A. Program (Kansas City) abstracts, p. 102.

————, I. G. Speden, and M. A. Buzas, 1964. Ecology of Pleistocene mollusks from Martha's Vineyard—a reconsideration: J. Paleontology, Vol. 38 (5), pp. 985–991.

Morton, John, 1963. The molluscan pattern: evolutionary trends in modern classification: Linnean Soc. London Proc., Vol. 174, Part 1, pp. 53–72.

Newell, Norman, 1937. Late Paleozoic pelecypods: Pectinacea: State Geol. Surv. Kansas, Vol. 10, 116 pp.

————, 1942. Late Paleozoic pelecypods: Mytilacea: State Geol. Surv. Kansas, Vol. 10, Part 2, 78 pp.

————, 1954. V. Mollusca, in B. Kummel, ed., Status of Invertebrate Paleontology, 1953. Mus. Comp. Zool. (Harvard) Bull., Vol. 112 (3), pp. 161–172.

————, 1955. Permian pelecypods of East Greenland: Medellelser om Grønland, Vol. 110, No. 4, 34 pp.

————, 1965. Classification of Bivalvia: Amer. Mus. Novitates, No. 2206, Jan. 29, 1965, pp. 1–25, 3 text figures, 1 table.

————, et al., 1953. The Permian reef complex of the Guadalupe Mountain Region, Texas and New Mexico. A study of paleoecology. W. H. Freeman and Co. 209 pp.

Nicol, David, 1944. Observations on *Pseudomonotis*, a Late Paleozoic pelecypod: The Nautilus, Vol. 57 (3), pp. 90–93.

————, 1945. Genera and subgenera of the pelecypod family Glycymeridae: J. Paleontology, Vol. 19 (6), pp. 616–621, 2 text figures.

————, 1950. Origin of the pelecypod family Glycymeridae: J. Paleontology, Vol. 24 (1), pp. 89–98, plates 20–22, 2 text figures.

————, 1953. Period of existence of some late Cenozoic pelecypods: J. Paleontology, Vol. 27 (5), pp. 706–707.

————, 1963. Further comments on biotic associations and extinction: Systematic Zoology, Vol. 12 (1), pp. 38–39.

————, 1964a. Inferences derived from general analysis of Recent and fossil marine pelecypod faunas: J. Paleontology, Vol. 38 (5), pp. 975–983.

————, 1964b. An essay on size in marine pelecypods: J. Paleontology, Vol. 38 (5), pp. 968–974.

Owen, G., 1953. The shell of the Lamellibranchia: Quart. J. Micros. Sci., Vol. 94, pp. 57–70.

Parker, R. H., 1956. Macro-invertebrate assemblages as indicators of sedimentary environments in east Mississippi Delta region: A.A.P.G. Bull., Vol. 40 (2), pp. 295–376, 8 plates.

————, 1964. Zoogeography and ecology of macro-invertebrates of Gulf of California and continental slope of Western Mexico, in Marine Geology of the Gulf of California —a Symposium: A.A.P.G. Mem. No. 3, pp. 331–368.

Packard, E. L., and D. L. Jones, 1965. Cretaceous pelecypods of the genus *Pinna* from the west coast of North America: J. Paleontology, Vol. 39 (5), pp. 910–915, plates 107–108.

Perkins, B. F., 1965. Analysis of a caprotinid rudist growth series: G.S.A. Program (Kansas City) abstracts, p. 123.

Pierce, Madelene, F., 1950. *Venus mercenaria*, in F. A. Brown, Jr., ed., Selected invertebrate types. John Wiley, pp. 324–334.

Purchon, R. D., 1960. Phylogeny in the Lamellibranchia: Univ. of Malay Press, Singapore (Proc. Cent. and Bicent, Congress Biol. Singapore, Dec. 2–9, 1958).

Raup, D. M., and D. P. Lawrence, 1963. Paleoecology of Pleistocene mollusks from Martha's Vineyard, Massachusetts: J. Paleontology, Vol. 37, pp. 472–485, Vol. 38 (5), pp. 991–993.

Rogers, Margaret J., 1965. A revision of the species of nonmarine Bivalvia from the Upper Carboniferous of eastern North America: J. Paleontology, Vol. 39 (2), pp. 663–668, plates 83–85, 1 text figure.

Schmidt, W. J., 1924–1925. Bau und Bildung der Prismen in den Muschelschalen: Mikrokosmos, Vol. 12, pp. 49–73.

Stanley, J. M., 1970. Relation of shell form to life habits in the Bivalvia (Mollusca): G.S.A. Mem. 125.

Stasek, C. R., 1963. Synopsis and discussion of the association of ctenidia and labial palps in bivalved Mollusca: The Veliger, Vol. 6 (2), pp. 91–97, 5 text figures.

Stephenson, L. W., 1922. Some Upper Cretaceous shells of the rudistid group from Tamaulipas, Mexico: U.S. Nat. Mus. Proc., Vol. 61, pp. 1–13.

————, 1938. A new Upper Cretaceous rudistid from the Kemp Clay of Texas: U.S. Geol. Surv. Prof. Paper No. 193-A, 7 pp. plates 1–5.

Trueman, E. R., 1942. The structure and deposition of the shell of *Tellina tenius*: J. Roy. Micros. Soc., Vol. 62, pp. 69–92, plate 1, 13 text figures.

Vermunt, L. W. J., 1937. Cretaceous rudistids of Pinar del Rio Province, Cuba: J. Paleontology, Vol. 11 (4), pp. 261–275.

Vogel, Klaus, 1962. Muscheln mit Schlosszähnen aus dem Spanischen Kambrium und ihre Bedeutung für die Evolution der Lamellibranchiaten: Abhandl. Math.–Nat. Kl. Acad. Wiss und Lit. Mainz, 1962, No. 4, pp. 192–244, 19 figuren, plates 1–5.

Vokes, H. E., 1954. Some primitive fossil pelecypods and their possible significance: J. Washington Acad. Sci., Vol. 44 (8), pp. 233–236.

Yonge, C. M., 1926–1927. Structure and physiology of the organs of feeding and digestion in *Ostrea edulis*: J. Mar. Biol. Assn. U. K., Vol. XIV (n.s), pp. 295–382.

————, 1953. The monomyarian condition in the Lamellibranchia: Roy. Soc. Edin. Trans., Vol. LXII, Part 2, No. 12, pp. 443–476.

————, 1954. Form and habit in *Pinna carnea* Gmelin: Roy. Soc. London Philos. Trans. Series B, Vol. 237 (Biol. Sciences 1952–1954), pp. 335–372.

Cephalopoda

Abel, O., 1916. Paläobiologie der Cephalopoden aus der Gruppe der Dibranchiaten. Fischer, Jena.

Arkell, W. J., 1949. Jurassic ammonites in 1949: Sci. Progress, No. 147, pp. 401–417, plate 1.

————, 1953. Seven new genera of Jurassic ammonites: Geological Mag., Vol. 90, pp. 36–40, plate 1.

————, 1956. Jurassic geology of the world. Oliver and Boyd (Edinburgh), 804 pp., 46 plates.

————, B. Kummell, and C. W. Wright, 1957. Mesozoic Ammonoidea: Treat. Invert. Paleont., Part L, Mollusca 4, pp. L80 ff.

Basse, E., 1952. Céphalopodes, Nautiloidea, Ammonoidea, in Jean Piveteau, ed. Traité de Paléontologie, Masson (Paris), Vol. 2, pp. 522–688, plates 1–24, Figures 1–60.

Berry, E. W., 1922. An American *Spirulirostra*: Amer. J. Sci., Vol. 3, pp. 327–334, text figures 1–5.

————, 1928. Cephalopod adaptations—the record and its interpretations: Quart. Rev. Biol. (Baltimore), Vol. 3, pp. 92–108, Figures 1–6.

Bidder, Anna M., 1962. Use of the tentacles, swimming and buoyancy control in the Pearly Nautilus: Nature, Vol. 196 (4853), pp. 451–454.

————, 1964. The Coleoidea, in K. M. Wilbur and C. M. Yonge, eds. Physiology of Mollusca. Academic Press, pp. 54–56.

Böse, Emil, 1927. Cretaceous ammonites from Texas and Northern Mexico. U. Texas Bull. No. 2748, pp. 143–306, plates 1–18.

Bowen, Robert, 1961a. Paleotemperature analyses of Mesozoic Belemnoidea from Germany and Poland: J. Geology, Vol. 69 (1), pp. 75–83.

————, 1961b. Paleotemperature analyses of Mesozoic Belemnoidea from Australia and New Guinea: G.S.A. Bull., Vol. 72, pp. 769–774.

————, 1961c. Paleotemperature analyses of Belemnoidea and Jurassic paleoclimatology: J. Geology, Vol. 69 (3), pp. 309–320.

————, 1961d. Oxygen isotope paleotemperature measurements on Cretaceous Belemnoidea from Europe, India, and Japan: J. Paleontology, Vol. 35 (5), pp. 1077–1084.

————, 1963. O^{18}/O^{16} paleotemperature measurement in Mesozoic Belemnoidea from Neuquén and Santa Cruz Provinces, Argentina: J. Paleontology, Vol. 37 (3), pp. 714–718.

Brinkmann, Roland, 1929. Statistisch-Biostratigraphie Untersuchungen an mitteljurassischen Ammoniten über Artbegriff und Stammesentwicklung: Abhandl. Gesell. Wiss. Göttingen, math.-phys. K1., Neue Folge, Vol. 13, Part 3, pp. 1–249, plates 1–5, 56 text figures, 129 tables (See Arkell et al., 1957, page L 101, in this section, for critical comment on this and related studies.)

————, 1960. Geological evolution of Europe (J. E. Sanders, trans.). Hafner Publ. Co. (New York), 156 pp., 46 figures, 19 plates, 18 correlation charts.

Brown, B., and H. E. Vokes, 1944. Fossil imprints of unknown origin: Amer. J. Sci., Vol. 242 (12), pp. 656–672.

Bruun, Anton Fr., 1955. New light on the biology of *Spirula*, a mesopelagic cephalopod, in Essays in the Natural Sciences in honor of Captain Allan Hancock. U. So. Calif. Press (Los Angeles), pp. 61–69, plates 1–2.

Currie, Ethel D., 1943. Growth stages in some species of *Promicroceras*: Geological Mag., Vol. LXXX, pp. 15–22.

————, 1944. Growth stages in some Jurassic ammonites: Roy. Soc. Edinburgh Trans., Vol. 61, pp. 171–198.

Denton, E. J., 1964. The buoyancy of marine molluscs, in Physiology of Mollusca. (See, under "General," Wilbur and Yonge, Chapter 13.)

Dunbar, C. O., 1924. Phases of cephalopod adaptation, *in* Organic adaptation to environment. Yale Univ. Press.

Elias, M. K., 1933. Cephalopods of the Pierre Formation of Wallace County, Kansas, and adjacent area: Univ. Kansas Sci. Bull., Vol. 21 (9), pp. 289–333, plates 28–42.

Erben, H. B., 1964. Bactritoidea: Treat. Invert. Paleont., Part K, Mollusca 3, pp. K491–K505.

Fischer, A. G., 1947. A belemnoid from the Late Permian of Greenland: Meddelelser om Grønland, Vol. 133, No. 5, 22 pp.

————, 1951. A new belemnoid from the Triassic of Nevada: Am. J. Sci., Vol. 249, pp. 385–393, plates 1–2.

————— and R. O. Fay, 1953. A spiny aptychus from the Cretaceous of Kansas: State Geol. Surv. Kansas Bull. 102, Part 2, pp. 77–92, plates 1–2, text figure 1.

Flower, R. H., 1944. *Atractites* and related coleoid cephalopods: Amer. Midland Naturalist, Vol. 32 (3), pp. 756–770.

————, 1945. A belemnite from a Mississippian boulder of the Caney shale: J. Paleontology, Vol. 19 (5), pp. 490–503, plate 65.

————, 1954. Cambrian cephalopods: New Mexico Inst. Mining and Tech., State Bur. Mines and Min. Res. Bull. 40, 44 pp., plates 1–3.

————, 1955a. Saltations in nautiloid coiling: Evolution, Vol. IX(3), pp. 244–260, text figures 1–3.

————, 1955b. Trails and tentacular impressions of orthoconic cephalopods: J. Paleontology, Vol. 29 (5), pp. 857–867, 4 text figures.

————, 1955c. Cameral deposits in orthoconic nautiloids. Geological Mag., XCII(2), pp. 89–103.

————, 1955d. New Chazyan orthocones: J. Paleontology, Vol. 29 (5), pp. 806–830, plates 77–81, 1 text figure.

————, 1957. Nautiloids of the Paleozoic, in Treat. Mar. Ecol. Paleoecol., Vol. 2, G.S.A. Mem. 67, pp. 829–852, text figures 1–6.

————— and B. Kummel, Jr., 1950. A classification of the Nautiloidea: J. Paleontology, Vol. 24 (5), pp. 606–616, 1 text figure.

Foerste, A. F., 1930. The color patterns of fossil cephalopods and brachiopods, with notes on gasteropods and pelecypods: Univ. Michigan Mus. Paleont. Contrib., Vol. 3 (6), pp. 109–150, plates 1–5.

Furnish, W. M., and A. G. Unklesbay, 1940. Diagrammatic representation of ammonoid sutures: J. Paleontology, Vol. 14 (6), pp. 598–602, 3 text figures.

————— and B. F. Glenister, 1964. Paleoecology, Nautiloidea-Ellesmerocerida: Treat. Invert. Paleont., Part K, Mollusca 3, pp. K114–K124, K129–K159.

Glenister, B. F., and W. M. Furnish, 1961. The Permian ammonoids of Australia: J. Paleontology, Vol. 35 (4), pp. 673–736, plates 78–86, 17 text figures.

Grégoire, Charles, and Curt Teichert, 1965. Conchiolin membranes in shell and cameral deposits of Pennsylvanian cephalopods, Oklahoma: Okla. Geol. Notes, Okla. Geol. Surv., Vol. 25 (7), pp. 175–201, plates 1–11.

Haas, Otto, 1941. A case of inversion of

suture lines in *Hysteroceras varicosum* (Sow.): Am. J. Sci., Vol. 239, pp. 661–664, text figures 1–2, plate 1.

Hare, P. E., 1963. Amino acids in the proteins from aragonite and calcite in the shells of *Mytilus californianus*: Science, Vol. 139, pp. 216–217, Table 1.

Imlay, R. W., 1948. Characteristic marine Jurassic fossils from the Western Interior of the United States: U.S. Geol. Surv. Prof. Paper 214-B, pp. 13–23, plates 5–9.

————, 1955. Characteristic Jurassic mollusks from Northern Alaska: U.S. Geol. Surv. Prof. Paper 274-D, pp. 69–93, plates 8–13.

Jeletsky, J. A., 1955. Evolution of Santonian and Campanian *Belemnitella* and paleontological systematics exemplified by *Belemnitella praecursor* Stolley: J. Paleontology, Vol. 29 (3), pp. 478–509, plates 56–58, 1 text figure.

————, 1950. *Actinocamax* from the Upper Cretaceous of Manitoba: Geol. Surv. Canada Bull. 15, pp. 1–27, 3 plates, 2 text figures.

Kobayashi, Teiichi, and Tomomasa Matumoto, 1942. Three new Toufangian nautiloids from Eastern Jehol: Japanese J. Geol. Geogr., Vol. 18 (4), pp. 313–317, plates 30–31.

Kummel, B., 1953. Middle Triassic ammonoids from Peary Land: Meddelelser om Grønland, Vol. 127, No. 1, pp. 1–19, 4 text figures, 1 plate.

————, 1954. V. Mollusca. Cephalopoda, in Status of Invert. Paleontology, 1953: M.C.Z. Bull. (Harvard), Vol. 112 (3), pp. 181–192.

————, 1961*a*. The Spitzbergen arctoceratids: M.C.Z. Bull. (Harvard), Vol. 123 (9), pp. 499–529, 6 figures, plates 1–9.

————, 1961*b*. History of the Earth. W. H. Freeman and Co., 544 pp.

————, 1964. Nautiloidea–Nautilida, in Treat. Invert. Paleont., Part K, Mollusca 3, pp. K383–K457 (with systematic descriptions as indicated by W. M. Furnish and B. F. Glenister).

———— and R. M. Lloyd, 1955. Experiments in relative streamlining of coiled cephalopod shells: J. Paleontology, Vol. 29 (1), pp. 159–170, 5 text figures.

———— and Grant Steele, 1962. Ammonites from the *Meekoceras gracilitatis* zone at Crittenden Spring, Elko County, Nevada:

J. Paleontology, Vol. 36 (4), pp. 638–703, plates 99–104, 20 text figures.

Miller, A. K., 1938. Devonian ammonoids of America: G.S.A. Sp. Paper 14, 262 pp., 30 plates.

————, 1947. Tertiary nautiloids of the Americas: G.S.A. Mem. 23, pp. 1–234, plates 1–100.

————, 1949. The last surge of the nautiloid cephalopods: Evolution, Vol. 3 (3), pp. 231–238.

———— and W. Youngquist, 1946. A giant ammonite from the Cretaceous of Montana: J. Paleontology, Vol. 20 (5), pp. 479–484, plates 73–75.

———— and H. R. Downs, 1950. Tertiary nautiloids of the Americas: supplement: J. Paleontology, Vol. 24 (1), pp. 1–17, plates 1–10.

———— and C. Collinson, 1951. Lower Mississippian ammonoids of Missouri: J. Paleontology, Vol. 25, pp. 454–487, plates 68–71.

————, W. H. Furnish, and O. H. Schindewolf, 1957. Paleozoic Ammonoidea: Treat. Invert. Paleont., Part L, Mollusca 4, pp. L11–L79.

———— and W. H. Furnish, 1958. Middle Pennsylvanian Schistoceratidae (Ammonoidea): J. Paleontology, Vol. 32 (2), pp. 253–268, plates 33–34, 9 figures; Goniatites of the Burlington Limestone in Missouri: J. Paleontology, pp. 268–279, plate 35, 5 text figures.

Miller, Halsey, 1961. Belemnoides del Triasico Superior del Estado de Sonora, in Paleontologia del Triasico Superior de Sonora, Part IV. Paleont. Mexicana, No. 11, pp. 3–11, plate 1.

Müller-Stoll, Hanns, 1936. Beiträge zur Anatomie der Belemnoidea: Nova Acta Leopoldina, new series, Vol. 4, No. 20, pp. 160–217, 5 text figures, 14 plates.

Mutvei, Harry, 1957. A preliminary report on the structure of the siphonal tube and in the precipitation of lime in the shells of fossil nautiloids: Arkiv. För Mineralogi och Geologi, Vol. 2, No. 8, pp. 179–188, 4 text figures, 1 plate.

Naef, Adolf, 1922. Die fossilen Tintenfische: Carl Fischer, Jena, 322 pp., 101 figures.

Reeside, J. B., 1919. Some American Jurassic ammonites of the genus *Quenstedtoceras, Cardioceras*, and *Amoeboceras*, family Cardioceratidae: U.S. Geol. Surv. Prof. Paper 118, 64 pp., 24 plates, 1 figure.

_____, 1927. The cephalopods of the Eagle sandstone and related formations in the western interior of the United States: U.S. Geol. Surv. Prof. Paper 151, 40 pp., 45 plates.

Roger, Jean, 1945. Les invertébrés des couches a poissons du Cretácé superieur du Liban: Mém. Soc. Géol. France, new series, Vol. XXIII, Mém. No. 51, pp. 1–92, plates I–X.

_____, 1948. Découverte d'une coquille de *Sepia*: Soc. Géol. France Bull. S.R., Vol. 17, pp. 225–232, 2 text figures (abbreviated title).

_____, 1952. Sous classe des Dibranchiata Owen 1836, in Jean Piveteau, ed., Traité, de Paléont., Vol. II, pp. 689–755.

Roll, Artur, 1935. Über Frasspuren, in Ammonitenschalen: Zentralbl. Min. Geol., Part B, pp. 120–124, Figures 1–11.

Rothpletz, August, 1909. Über die Einbettung von Ammoniten in die Solnhofener Schichten: Abhl. k. bayer. Akad. Wiss. München (Kl.2), Vol. 24, pp. 313–337, plates 1–2.

Scott, Gayle, 1940. Paleoecological factors controlling the distribution and mode of life of Cretaceous ammonoids in the Texas area: J. Paleontology, Vol. 14 (4), pp. 299–323, 9 text figures.

Schevill, W. E., 1950. An Upper Jurassic sepoid from Cuba: J. Paleontology, Vol. 24 (1), pp. 99–101, plate 23.

Schindewolf, O. H., 1934. Über Epöken auf Cephalopoden Gehaüsen: Paläont. Zeitschr., Vol. 16, pp. 258–283, plates 19–22, Figures 1–3.

_____, 1954. On development, evolution, and terminology of ammonoid suture line, in B. Kummel, ed. Status of Invert. Paleont.: M.C.Z. Bull., (Harvard), Vol. 112 (3), pp. 217–237.

Schwartzbach, Martin, 1936. Zur lebenweise der Ammoniten: Natur und Volk, Vol. 66, pp. 8–11, 3 figures.

Spath, L. F., 1928. The belemnite marls of Charmouth, in Lang, W. D. et al., Quart. J. Geol. Soc. (London), Vol. 84, pp. 222–232, plates 16–17.

_____, 1933. The evolution of the Cephalopoda: Biol. Rev., Vol. 8, pp. 418–462, Figures 1–13.

_____, 1939. On a new belemnoid (*Conoteuthis renniei*) from the Aptian of the

Colony of Moçambique: Colónia de Moçambique Serv. Indus., Minas e Geol., Bol. 2, Lourenço Marques, pp. 13–15, Figures 1–2.

_____, 1945. Problems of ammonite nomenclature X. The naming of pathological specimens: Geological Mag., Vol. 82, pp. 251–255.

_____, 1950. The study of ammonites in thin, median sections: Geological Mag., Vol. 87, pp. 77–84.

Stenzel, H. B., 1941. The Eocene dibranchiate cephalopod genus *Belemnosella* Naef 1922: J. Paleontology, Vol. 15 (1), p. 90 (abbreviated title).

_____, 1964. Living *Nautilus*: Treat. Invert. Paleont., Part K, Mollusca 3, pp. K59–K93.

Sweet, W. C. 1964. Nautiloidea-Orthocerida, Oncocerida, Barrandeocerida: Treat. Invert. Paleont., Part K, Mollusca 3, pp. K216–K261, K277–K319, K368–K382.

Swinnerton, H. H., 1950. Outlines of Paleontology. 3rd ed., Edward Arnold & Co., 370 pp.

_____ and A. E. Trueman, 1917. The morphology and development of the ammonite septum: Quart. J. Geol. Soc. (London), Vol. 73, Part 1, pp. 26–58, plates 2–4, 13 text figures.

Teichert, Curt, 1933. Der Bau der actinoceroiden Cephalopoden: Palaeontographica, Part A, Vol. 78, pp. 11–230, plates 8–15, Figures 1–50.

_____, 1943. The Devonian of Western Australia, a preliminary review: Am. J. Sci., Vol. 241, pp. 69–94, 167–184.

_____, 1964. Morphology of Hard Parts, Biostratonomy, Endoceratoidea, Actinoceratoidea, Discosorida: Treat. Invert. Paleont., Part K, Mollusca 3, pp. K13–K153, K124–K127, K160–K189, K190–K216, K320–K342.

_____ and B. F. Glenister, 1954. Early Ordovician cephalopod faunas from Northwestern Australia: Amer. Paleontology Bull., Vol. 35, No. 150, 110 pp., 10 plates.

_____ and B. Kummel, 1960. Size of endoceroid cephalopods: Breviora, M.C.Z. (Harvard), No. 128, Dec. 20, 1960, pp. 1–7, Figure 1.

_____ and R. C. Moore, 1964. Introduction, Classification and stratigraphic distribution, Introductory discussion: Treat. Invert. Paleont., Part K, Mollusca 3, pp.

K2–K4, K94–K106, K127–K129.

Trauth, Friedrich, 1934. Die aptychen der Paläozoikums: Preussiche Geol. Landesanstalt, Berlin Jahrbuch, Vol. 55, pp. 44–83, plates 2–3 (printed 1935).

Trueman, A. E., 1941. The ammonite body chambers with special reference to the buoyancy and mode of life of the living ammonite: Quart. J. Geol. Soc. (London), Vol. 96, pp. 339–383, Figures 1–17.

Urey, H. C., et al., 1951. Measurement of paleotemperature of the Upper Cretaceous of England, Denmark, and southeastern United States: G.S.A. Bull., Vol. 62, pp. 399–416.

Wells, M. J., 1962. Brain and behavior in cephalopods. Stanford Univ. Press, Calif., 154 pp.

Hyolitha

Fisher, D. W., 1962. (See under "Gastropoda," D. W. Fisher, 1962.)

Marek, Ladislav, and E. L. Yochelson, 1964. Paleozoic mollusk, *Hyolithes:* Science, Vol. 146 (3652), pp. 1674–1675, 3 text figures.

Paulsen, Valdemar, 1963. Notes on *Hyolithellus* Billings 1871, class Pogonophora Johannson, 1937: Biol. Medd. Dan. Vid. Selsk., Vol. 23, No. 12, 14 pp.

Yochelson, E. L., 1961. The operculum and mode of life of *Hyolithes:* J. Paleontology, Vol. 35 (1), pp. 152–161, plates 33–34, 1 text figure.

————, 1963. Problems of the early history of the Mollusca: XVI Interntl. Congr. Zool. Proc. (Washington, D. C.), p. 187.

[See Runnegar and Pojeta, Jr., 1974. Science, vol. 186, p. 311. Hyolitha considered to be a probable extinct phylum related to the Sipunculoidea, Annelida, and Mollusca through a common pre-Cambrian ancestral stock. E. G. Kauffman, 1977. Evolutionary rates and biostratigraphy. In E. G. Kauffman and J. E. Hazel, Eds. Concepts and methods in biostratigraphy. Dowden, Hutchinson, and Ross, Inc., pp. 109–141 utilized various rapidly evolving Cretaceous molluscan lineages (bivalves, gastropods, ammonites) that are closely related in the evolutionary process to test the "predictability in evolutionary rates" and "adaptive traits."]

chapter nine

WORM'S EYE VIEW: THE WORLD OF POLYCHAETES AND OTHER WORMS

INTRODUCTION

Worms have been on our planet from at least the Upper pre-Cambrian and possibly even longer. In numbers and moreover in singular variety, worms defy imagination. Any old sample of garden soil is likely to contain millions of nematode worms. Millions of earthworms (*Lumbricus*) will be driven from soil burrows during every rain fall. A half-billion parasitic *Trichinella* may simultaneously bore through the body of one infected man. Astronomical numbers of one species—the marine polychaete palolo worm (*Eunice viridens*)—discharge eggs or sperm ("swarm") during certain phases of the lunar cycle. Enough palolo worms so that the event turns the sea around Samoa and Fiji into a milky soup—a rare food delicacy.

Worms can range in size from microscopic objects 1/10 mm long, to giants from 10 to 11 ft long (Australian earthworm *Megascolecides australis*; the polychaete *Eunice gigantea*) to 80 ft long in ribbon worms. At all depths and in all seas marine worms are found throughout the world.

Worms have radiated into almost every known habitat: fresh, brackish, marine waters; thermal springs; soils; the intestines, blood and tissues of man, as well as into other animals. Some live anywhere—in the interstitial spaces between sand grains (arachianellids—*Prodrilus rubropharyngeus*), under stones, attached to algal fronds (*Spirorbis*), in hydroid forests, in calcareous tubes (tubiculous worms like *Serpula*), burrowed in soft substrates or in molluscan shells (*Polydora*), or they inhabit special sites, the gill cavity of lobsters (*Histrobdellia homari*).

The parasitic existence of our many worms (blood flukes, tapeworms, nemerteans, such as hookworms, trichina worms, filaria worms) gets the blame for painful and deadly human diseases plus animal diseases.

CLASSIFICATION

Any animal with an elongate body, much longer than wide, is vermiform (wormlike). Among other creatures having vermiform bodies are insect larvae (arthropods), ctenophores (comb jellies), phoronid worms, and the Onchyophora. To this list can be added some mollusks (Chapter 8), echinoderms (Chapter 11), the so-called "blindworms" (legless lizards), and the wormlike hemichordates. Among proper worms, one can distinguish unsegmented types (flat, ribbon, round, spiny-headed, horsehair, and others) and segmented types (Annelida).

Should all of these quite unlike creatures sharing the condition of being vermiform be brought under a single phylum? Gradually, investigators, even Linnaeus in 1788, began to pick away at the clumsy Vermes phylum, removing this or that group. Phylum Annelida, segmented worms, was established in 1809; flatworms, Platyhelminthes, were first grouped together in 1816; protozoans, rotifers, nematodes, and others were removed at a later time, and so it went (Hyman, 1951a).

There are as many species of living and extinct worms—segmented and unsegmented—as all the known species of Porifera, Coelenterata, Bryozoa, and Echinodermata taken together. This amounts to 3-plus percent of the animal kingdom. The three largest groups, in terms of numbers of living species, are Nematoda,

10,000 species; Platyhelminthes, 7000 species; and the Annelida, 6000 species (Mayr et al., 1953, Table 1. See References, Chapter 1).

This chapter places major emphasis on the annelid worms (polychaetes) that are abundant Paleozoic fossils (polychaete worm jaws-scolecodonts) and/or important reef builders (serpulids). The sparse fossil record of other worm phyla necessitates less study. Fossil spoor (burrows, tracks, trails) of various annelids are also commonly encountered in field work.

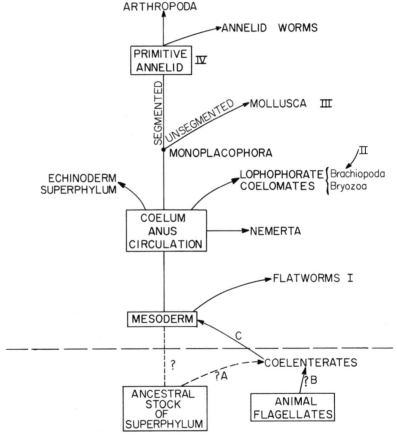

FIGURE 9.1 Divergent branches of Annelid superphylum, showing relationship of annelid worms and arthropods. Data from Simpson et al., 1957 and others. (See also Chapter 8, discussion of Monoplacophora, and Chapter 6, on Bryozoa.)

I–IV: the main branches traceable to a remote ancestral stock. Some minor acoelomates possess an anus and circulatory system (vascular cavities) (phylum Nemertea) and must have branched off somewhat later than the flatworms. Horizontal dashed lines indicate rise of Annelid Superphylum. (A), (B), (C) Three alternate interpretations of the rise of coelenterates and flatworms. (A), Arose from ancestral primitive metazoan stock of Annelid superphylum; (B), arose from animal flagellates; (C), coelenterates gave rise to flatworms by development of mesoderm and bilateral symmetry, the traditional view now held by many zoologists. (Nemerta = Nemertea.)

C. R. Stasek, 1972, Chem. Zool., vol. 7, pp. 1–41, figs. 1A–D and 14, indicated that the molluscan framework could have been established by stages from free-living flatworms. Thus, the mucoid coat is thought to be the precursor of a cuticle and later molluscan shell development. Flatworms branched off prior to the Paleozoic rise of the several molluscan classes.

AFFINITIES

Many zoologists envision two superphyla (each a grouping of several phyla), one the echinoderm superphylum (Chapter 11), the other the annelid superphylum (Simpson et al., 1957). This annelid superphylum accounts for the Mollusca (Chapter 8), Arthropoda (Chapter 12), as well as Bryozoa, Brachiopoda, Platyhelminthes, and Annelida.

Platyhelminthes start as probably the most primitive of the worms. They lack anus and coelom, and are thought to have diverged from the ancestral stock before these structures evolved (Figure 9.1[1]). Once evolved, other divergent branches include lophophorate coelomates (brachiopods and bryozoans) plus unsegmented Mollusca (but showing affinities through the segmented Monoplacophora—Chapter 8). Final divergence possibly occurred from the primitive annelid which divided into two branches—annelid worms and arthropods (Chapter 10).

LIVING AND FOSSIL WORM PHYLA

Acoelomata (Lacking a Coelom)

The phyla grouped under Acoelomata include Platyhelminthes, (flatworms, turbelarians, trematode and cestode parasitic worms), with no known fossil record as of present writing; the Nemertea, Nematoda, and Nematomorpha—discussed later— have a fossil record, and additional minor phyla with no known fossils.

Shared Characters

Acoelomate worms have the common characters: (1) all unsegmented, (2) all possess a mesoderm, (3) where a primary body cavity exists, it lacks a clearcut epithelial boundary, hence, is not considered a true coelom—may contain fluid, vacuolated cells, or mesenchymal cells. (Borradaile, et al., 1958, note that in Nemertea, the body cavity divides into a series of canals for circulation. This development

[1]See Stasek, 1972 Figure 9.1].

is defined as the *first* appearance of a vascular system in the animal kingdom. Figure 9.1 indicates that this form must have diverged from the line that gave rise to the coelomates after the branching off of flatworms. The oldest fossil Nemertea is Upper Jurassic in age.)

Phylum Nemertea (Proboscis Worms)

Worms with elongate, cylindrical or cordlike body; proboscis—threadlike, eversible (can be extended to grasp prey or withdrawn); anus; vascular system; marine forms have pilidium larvae; sexes separate; free-living or parasitic; mostly marine, but some fresh-water. Geologic range: Jurassic to Recent. [*Amphiporus* (Rec.) (Figure 9.2*A*, *a*), *Legnodesmus* (U. Jur.) (Figure 9.2*A*, *b*).]

Phylum Nematoida (also called Nematoda) (Thread or Round Worms)

Thread-shaped body pointed at both ends; secretes elastic proteinaceous cuticle that is molted several times during ontogeny; free-living and parasitic; habitats: soil, fresh and marine waters. Geologic range: L. Carb.–Rec. [Two classes based on presence or absence of phasmids (= caudal sensory organs). Only the second class Phasmidia (with phasmids) has a fossil record.] [*Rhabditis* (Rec.) (Figure 9.2*B*, *d*), *Scorpiophagus* (L. Carb.), (Figure 9.2*B*, *e*), *Oligoplectus* (Oligo.) (Figure 9.2*C*, *g*). Nematodes are also known from Quaternary mammalian (including human) remains (Dollus, 1950).]

Phylum Nematomorpha (Horsehair Worms)

Unsegmented hair-shaped body; sexes separate: genital ducts open into hind gut to form cloaca; marine and fresh-water; parasitic—larva have boring organ to enter body insect cavity leaving host at maturity. Geologic range: Eocene–Recent. [Nematomorph tied in a Gordian knot (Rec.) schematic (Figure 9.2*C*, *h*), *Gordius* (Eocene) (Figure 9.2*C*, *i*).]

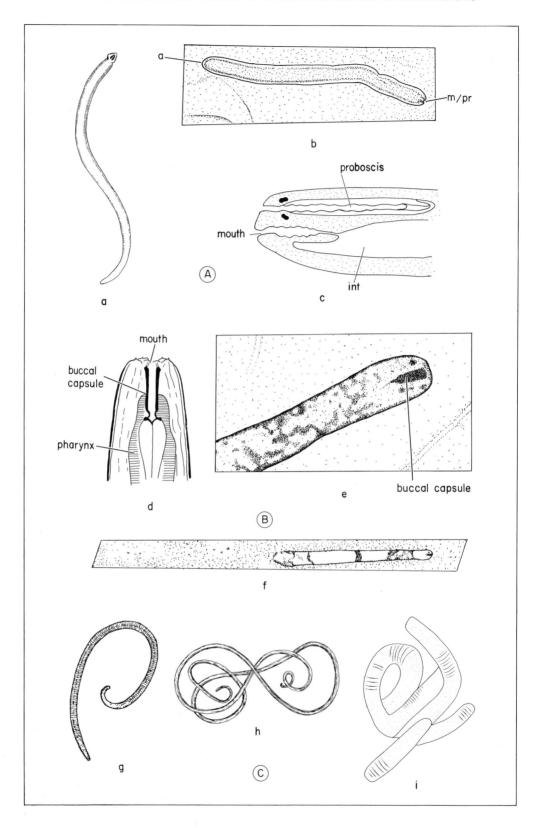

FIGURE 9.2 Representative Acoelomate worms—living and fossil (A) Phylum NEMERTEA: (a) *Amphiporus binoculatus*, Rec., Puget Sound, schematic. (b) *Legnodesmus ehlersi*, U. Jur., Solenhofen, Germany, X 1; *a*—anal end, *m/pr*—mouth, proboscis end of Ac. (c) Schematic longitudinal section of nemertean: proboscis, oral and intestinal cavities (= *int*), X 1. (B) and (C) Phylum NEMATOIDA (= NEMATODA). (d) Living nematode, *Rhabditis filiformis*, detail, schematic. (e) Fossil nematode, *Scorpiophagus baculitormis* in tibia of the scorpion *Gigantoscorpio willsi* (L. Carb., Scotland); compare buccal cavity with (B, d). (f) Complete individual of (B, e), schematic.
(C) (g) *Oligoplectus succini*, Oligo., Germany. (h) Phylum *Nematomorpha*: Typical nematomorph in a gordian knot, schematic. (i) *Gordius tenuifibrosis*, Eoc. (Braunkohle), Germany, body filiform, enlarged.
Illustrations redrawn and adapted from: (Aa, Bd). The Invertebrates, Vol. III, by Libbie Hyman. Copyright 1951 by McGraw-Hill, Inc. With permission of McGraw-Hill Book Co. (Ac) G. A. Kerkut, ed., The Invertebrata, 3rd ed., 1959, by Borradaile and Potts. By permission of Cambridge Univ. Press. (Be–f) Størmer, 1963; (Ab, Cg, Ci) Howell, 1962, after several authors.

Other Worm Phyla

There are two other quite distinct worm phyla with both living and fossil representatives, defined and illustrated here. (They are not grouped with acoelomates or with the annelids.)

Phylum Chaetognatha (Arrow Worms). Swimming, torpedo-shaped marine worms; unsegmented coelomates; elongate, have three regions: head, trunk, tail bearing lateral and caudal fins; head with pair of eyes and two groups of chitinous teeth and jaws; lacks localized vascular, excretory systems. Geologic range: Cambrian; Recent. [*Sagitta* (Rec.) (Figure 9.3*A*, *a*), *Amiskwia* (M. Cambrian, Burgess shale) (Figure 9.3*A*, *b*).]

Phylum Sipunculoidea (Peanut Worms): Unsegmented (or weakly segmented) coelomate, marine or estuarine worms; body cylindrical or subcylindrical with retractile introvert at anterior end maybe armed with chitinous hooks. Geologic range: Cambrian–Recent. [A Recent sipunculoid worm (Figure 9.3*B*, *a*), *Lecathylus* (Sil.) (Figure 9.3*B*, *b*).] Possibly related to Hyolitha.

Phylum Annelida

(1) Segmented worms with coelom. (2) A single preoral segment (prostomium) (Figure 9.4*B*). (3) Muscular body wall—external, circular layer of muscles; internal, longitudinal layer. (4) Other systems: *nervous*—a central nervous system with paired gan-

glia in each segment; *excretory*; *reproductive*—fertilization, external or internal, with or without trochophore larval stage (Oligochaetes bypass the larval stage). (5) Cuticle—thin, nonchitinous. In the Arthropoda (Chapter 10), the cuticle is markedly thickened. The annelid cuticle has chitinous bristles (chaetae) secreted by, and imbedded into ectodermal pits in each segment. (*Chaetae* are important in classification; thus Polychaeta, Poly+chaeta=many bristles; Oligochaeta, Oligo+chaeta=few bristles. Geologic range: Proterozoic–Recent. Note that arthropods lack chitinous chaetae but some forms do possess hollow outgrowths of the cuticle.)

INFORMATION DERIVED FROM LIVING POLYCHAETES AND OTHER WORMS

Rise of the Annelids

A glance at Figure 9.1 indicates the critical part played by the development of a mesoderm. (This tissue is intermediate between the outer ectoderm and the inner endoderm.) One must envision a long, still dimly perceived, evolutionary sequence. Mutations at specific gene loci, the operation of natural selection favoring some gene combinations (genotypes) over others, must have worked before the intermediate tissue, the mesoderm, could evolve from common ancestral planula-like gastrula (Figure 9.4*A*) of both coelenterates and annelids (Figure 9.1).

Once the mesoderm had become part

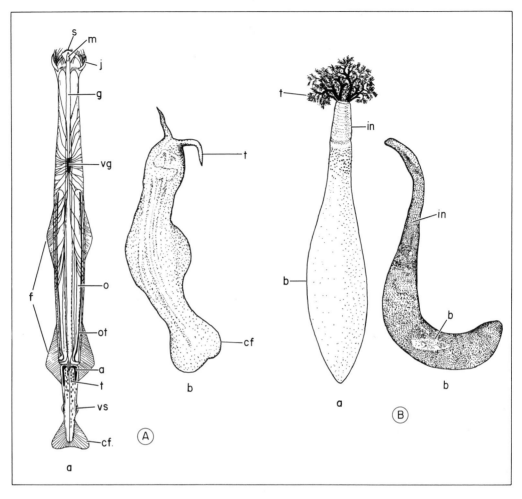

FIGURE 9.3 Chaetognaths and sipunculoids—living and fossil.
Phylum Chaetognatha: (*A*)(*a*) *Sagitta* sp. (arrow worms), X 2: *s*—spines, *m*—mouth, *j*—jaws, *vg*—ventral ganglion, *o*—ovary, *ot*—oviduct, *a*—anus, *t*—testes, *f*—fin, *cf*—caudal fin. (*b*) *Amiskwia saggitiformis*, M. Cambrian, Burgess shale, Canada (Brit. Col.), X 3: *t*—tentacles (paired), *cf*—caudal fin, compare to (*a*).
Phylum Sipunculoidea: (*B*)(*a*) *Dendrostomum pyroides*, a sipunculoid worm: *t*—tentacles, *in*—introvert, *b*—body, (schematic). (*b*) *Lecathylus gregarius*, Racine Dol., Sil., Illinois, X 1, labels as in (*B, a*).
Illustrations redrawn after Borradaile et al.;* Walcott; Fisher, 1952.
(*From G. A. Kerkut, ed., The Invertebrata, 3rd ed., 1959 by Borradaile and Potts. By permission of Cambridge Univ. Press.)

of the genetic code, further changes in this code could ensue. The selective process promoted developments. At first, discrete spaces appeared in the mesoderm for the accumulation of waste products. Later a more continuous body cavity grew. This was the coelom.

The coelom walls in the course of time became sites for significant anatomical modifications. These expressed further mutations, recombinations, and selections: a new muscular system, a more efficient excretory system (for wastes previously discharged into discrete mesoderm spaces), a circulatory system, specialized functions for various tissues, along with enhancement of the nervous system (Snodgrass, 1938).

A brief time before the coelom evolved, the body became segmented (ectoderm

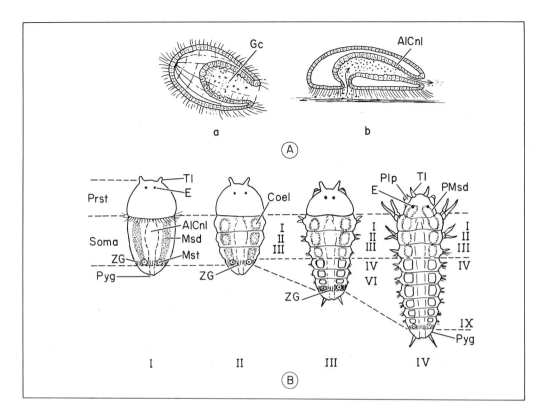

FIGURE 9.4 Rise of the annelids.
(A) (a) Theoretical planula-like gastrula—postulated common ancestor of Coelenterata and Annelida: Gc—gastrocoele. (b) Hypothetical annelid ancestor: Al Cnl—alimentary canal.
(B) Direct primary segmentation of body of larval polychaete worm; growth of worm (I–IV) by successive addition of somites in subterminal zone of growth; coel—coelomic cavity; e—eye; I–III, primary larval somites, IV–IX, secondary somites; Msd—mesoderm; Mst—mesodermal teleoblast (post larval somites arise here); soma-body region between prostomium (Prst) and pygidium (Pyg), in which somites are formed; Tl—tentacle; Pmsd—prostomial mesoderm; Zg—zone of growth at end of soma.
Reprinted from R. E. Snodgrass: A Textbook of Arthropod Anatomy. Copyright 1952 by Cornell University. Used by permission of Cornell University Press.

and mesoderm). This phenotypic expression has been attributed to need for greater body movement efficiency (Figure 9.4B) (Snodgrass, 1938). Segmentation may very well have resolved this need and counterselection pressures. Thus a triploblastic (ectoderm, mesoderm, endoderm) segmented, coelomate animal could evolve soon after the mesoderm appeared.

Relationship of Coelenterates and Annelids

Several alternate proposals explain apparent affinities of coelenterates, flatworms, and annelids.

1. Coelenterates arose from the primitive Metazoan stock that gave rise to the Annelid superphylum (Figure 9.1A).
2. Coelenterates arose from animal flagellates (Chapter 5).
3. Coelenterates gave rise to flatworms by development of a mesoderm and bilateral symmetry. Flatworms, in turn, gave rise to higher forms by modification. (Stasek, 1972.)
4. Both coelenterates and annelids had a common planula-like gastrula ancestor (Snodgrass, 1938).

Polychaete Larva and Bottom Faunas

Polychaetes of Danish waters (Øresund) (Thorson, 1946) furnish a well-studied example. Among four most frequent

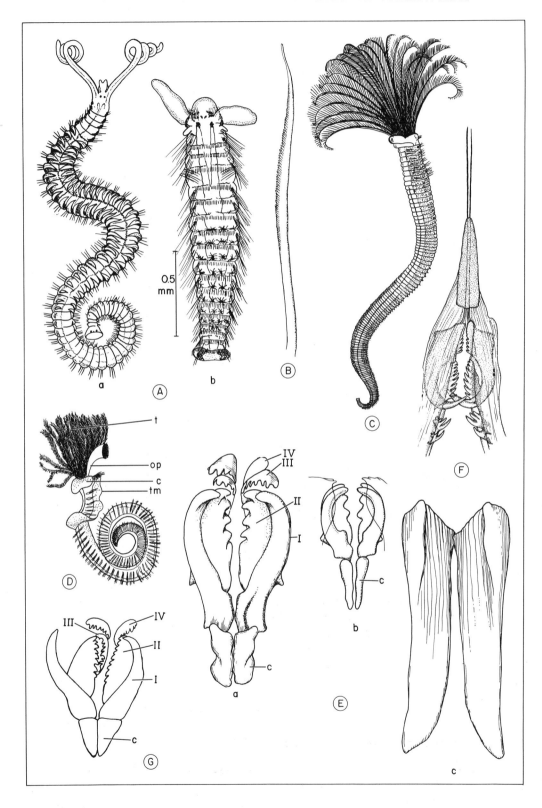

FIGURE 9.5A Living Polychaetes—with or without jaw apparatus and/or tubes.
(A) Order Sedentarida: Living *Polydora ciliata,* (a) adult, (b) larval form, Danish waters, schematic.
(B) *Dodecaceria laddi,* Eniwetok Atoll. One type of modified chitinous setae that, in this species and *Polydora,* may serve the same rasping function, boring into coralline rock, as the jaw apparatus of the Errantida.
(C) The sabellid worm, *Megalomma vesiculatum;* tentacles, tube shown. Rec., Calif., schematic.
(D) Serpulid worm, *Mercierella enigmatica,* removed from its tube and viewed from left side (schematic); note c—collar, t—tentacles (branchiae), op—operculum, tm—thoracic membranes.
(E) Pharyngeal structures ("scolecodonts"). (a) *Eunice afra,* North Marshall Islands, Rec., enlarged: maxillae (I–IV), carriers (c). (b) *Palola siciliensis,* from *Heliopora,* No. Marshall Islands, Rec., dorsal, X 45. (c) mandibles, ventral view.
(F) A littoral species *Drilonereis cylindrica,* Rec., Gulf of Mexico, four maxillae, ventral view, enlarged.
(G) *Onuphis* sp., Rec., Philippines: carrier (c) maxilla I–IV; note unpaired piece, III, X 17.
Illustrations redrawn from (A, a) Smith et al. (A, b) Thorson, 1946; (C)–(D) Hartman, 1954a; (E) Hartman, 1954b; (F) Hartman, 1951; (G) Treadwell, 1931.

components of plankton hauls from these waters were polychaete larvae dominated by *Polydora ciliata* (Order, Sedentarida). Adult *Polydora* forms inhabit ("infect") shells of the living mollusk *Littorina littoria* (Figure 9.5A). In these waters, about 142 polychaete species represent larval and adult forms. Adults are fairly high numerically in the bottom fauna).

Examination of 73 polychaete species indicated 55 percent had a long larval life. There were, of course, polychaetes with nonpelagic development (*Lumbriconereis,* among others); still others with a short pelagic life (several species of *Spirorbis* offer brood protection so young are released close to the metamorphosis time into adult stages), also a few with both pelagic and nonpelagic development (*Nereis pelagica*).

A single *Polydora ciliata* lays about 400 eggs per batch. Hatching to metamorphism in pelagic forms may span from two to four weeks. There are several maxima for polychaete larva; although some species occur during every month of the year without regard to seasons.

About 15.5 percent of Øresund polychaete plankton are errant polychaeta, that is, forms with jaw apparatus (= scolecodonts). The much greater remaining percentage consists of sedentary polychaetes, that is, forms without jaws. Infaunal polychaetes (below the intertidal zone—burrowed) are almost 66 percent of all polychaete larvae, but the epifaunal polychaetes (living on rocks and shells, etc.) contribute 34-plus percent.

Nutrition and Predation

In Øresund some polychaete larvae feed on plentiful larval lamellibranchs as well as on phytoplankton and zooplankton. Where the former diet holds, the entire stock of lamellilbranch larva can be decimated. In turn, the food chain (fish food) may be considerably influenced.

Very few animals find much food value in polychaete larva. Nevertheless, they are eaten by young herrings, several ctenophores, medusae, and arthropods (Thorson, 1946).

In coral atolls of the Pacific, many annelids derive nutrient from coral clumps or microorganisms within the clump (Hartman, 1954a. Reference, Chapter 1).

Bathymetry

Polychaete worms are found in all oceans; they range from fresh and brackish waters (*Nereis japonica*) to littoral (Hartman, 1951), and shallow, marine waters, including tidal pools. In the central Arctic, they range from shoal waters to bathyal depths. In the Caspian and Black Sea basins, they show a wide salinity tolerance (as elsewhere), and have a vertical range of 650 meters (Hartman, 1957).

Many polychaete faunas associate with coral reefs (Great Barrier Reef, Australia; North Marshall Island atolls; and so on).

Fauvel (1914, cit. Hartman, 1957) found that littoral species at the poles occur even in the abyssal regions in the tropics.

Temperature and Distribution

Temperature exerts a dominant control in speciation. There are cold- and warm-water polychaete faunas. A distribution factor is larval dispersal by currents; for example, New Zealand and Antarctica have about 62 polychaete species in common. West Indies and Brazil have many common species. Many polychaete faunas show affinities; the Marshall Island polychaetes show affinities to faunas from other parts of the Indo-Pacific area (Hartman, 1954*a*).

Biochemistry and Mineralogy

Water and Carbon. All marine worm analyses (several phyla) show an average 80 percent water. Fifty percent of the dry marine worm residue is Carbon.

Nitrogen. In polychaetes, nitrogen is 9.0 to 15.0 percent of the dry matter. This is protein nitrogen.

Calcareous and Phosphatic Tubes. The tubes of some forms—Serpulidae and Onuphidae—may be organic (chitin) with incorporated mineral grains or broken shell fragments, or may be composed of magnesium and calcium compounds with a 2.5 percent organic content (Tables 9.1, 9.2). Such forms occur in masses or in reefs in several areas of the sea—for example, the entrance to Baffin Bay, Texas.

The calcareous content, as expected, ranges from 89.0 to 99.0 percent, but the magnesium content ranges to 9 percent. Table 9.2 indicates the phosphatic content found in polychaete tubes—generally with more phosphorus than calcium.

Blood Pigments. A blood pigment series from various worms contains iron (erythrocruoin, chlorocruoin, haemerythrin). Copper and other metals have not been reported from such pigments. (See "Worm Jaws" below for copper.)

Heavy Metals. In tubes of *Serpula* species, Vinogradov found 0.09 percent manganese. If copper or zinc is present, they occur in traces. All tubiculous polychaetes do not contain manganese (Sabellidae).

TABLE 9.1. Composition of Calcareous Polychaete Tubes (In Percent of Ash Residue)

Organism	$CaCO_3$	$MgCO_3$	$C_3P_2O_8$	$CaSO_4$	Region
Serpula triquetra	—	4.455	—	—	North Sea
Filograna complexa					
(*Serpula complexa*)	99.01	—	0.99	Trace	England
Hydroides dianthus	89.66	9.72	0.62	Trace	Massachusetts

Source. After Vinogradov, 1953. (See Chapter 15, for reference).

TABLE 9.2. Composition of Phosphatic Polychaete Tubes (In Percent of Dry Residue)

Organism	CaO	MgO	P_2O_5	F_2O_3	SiO_2	SO_3	Region
Onuphis conohylega	22.70	—	7.50	—	—	—	USSR (Gulf of Kola)
Leodice polybranchia	5.12	4.40	6.43	2.33	28.05	6.06	
Hyalinaecea artifex	5.30	8.57	20.32	0.01	4.33	—	

Source. After Vinogradov, 1953. (See Chapter 15, for reference).

Nickel and cobalt are present but in very small amounts (three to four decimal places). Silver and lead are also irregularly present in some forms in barely detectable quantities. The figures in percent of dry matter are cited for one polychaete having a jaw apparatus: *Nereis japonica* — 2.43 Ca, 1.00 Mg, 0.408 P, 0.019 Fe, 0.550 S, 0.550 Al, no arsenic, and a trace of Zn (Yamamura, 1934, cit. Vinogradov).

Halogens. All marine worms contain iodine. In tubiculous forms, the organic tube iodine content may be as much as 0.6 percent of dry matter. The tubes referred to are horny tubes composed of keratin, protein, and so on. Chitinous and calcareous tubes are iodine-poor, according to Vinogradov. However, Cameron (cit. Swan, 1950, p. 302) found this held true only until he brought the tubes into complete dissolution. Thereafter he found the iodine content rose. Cameron noticed no connection between calcium-secreting sites and iodine concentration. Bromine constitutes 0.1 percent of the dry tube matter of the Eunicidae (with jaw apparatus). The Br/I ratio varies according to organic content.

Worm Jaws. Schwab (1966) found, by spectrographic analysis, that the jaws of living *Nereis virens* yielded a high calcium concentration plus trace copper amounts. (For further discussion, see Fossil Worm Jaws.) (See also Figure 9.5B.)

Mineralogy of Worm Tubes

Lowenstam (1954) presented the first evidence for (*a*) the polymorphic composition in the polychaete worm tube of *Eupomatus gracilis* (Recent, Bermuda), that is, the presence of aragonite and calcite, also (*b*) the seasonal variation in aragonite/calcite ratios in such tubes (Figure 9.6).

Calcium Metabolism in Serpulid Polychaetes

Calcareous tubes of serpulid worms grow larger by the addition of anterior increments. But what structures are respon-

sible? That question was put by E. F. Swan (1950).

The test animals he used included a brackish-water tubiculous polychaete that abounds in Lake Merritt, Oakland, California, *Mercieriella enigmatica* (Figure 9.5*A*, part *D*), and two cosmopolitan marine species, *Serpula vermicularis* and *Sabella media*.

Swan employed three techniques: histochemical, direct observation, and tracing by use of radioactive strontium.

Intent on locating the precise structures in the polychaete responsible for absorption, transportation, and secretion of calcium in the tube, Swan found histochemical studies the most illuminating.

The calcium carbonate of the tube followed the route indicated below, at least in part:

1. Calcium absorbed from brackish or marine waters.
2. It is then stored by spherule-containing cells of the stomach, and by the anterior part of the intestines. (These were observed.)
3. The route from the cells of the digestive tract to the major subcollar glands is unclear, but probably via the circulatory system.
4. At the major subcollar glands, there is presumably, a supply of CO_2. Both calcium transport to this site and immersion of tissues in seawater — are the other necessary conditions for a biomineralization event.

   An equilibrium is established between the bicarbonate (formed from dissociated water molecules and CO_2) — HCO_3^- and $CO_3^=$. This equilibrium is upset in the direction of more $CO_3^=$ due to the difference in CO_2 in the surrounding seawater and the subcollar glands proper. The result, Swan noted, leads to calcium precipitation into glands and likely some into the immediate seawater layer contacting gland tissues (Figure 9.5*A*, part D).
5. All the calcium for the tube may not have passed through the animal's body. Some, as radioactive strontium experiments showed, could have come by direct exchange with seawater or lakewater.

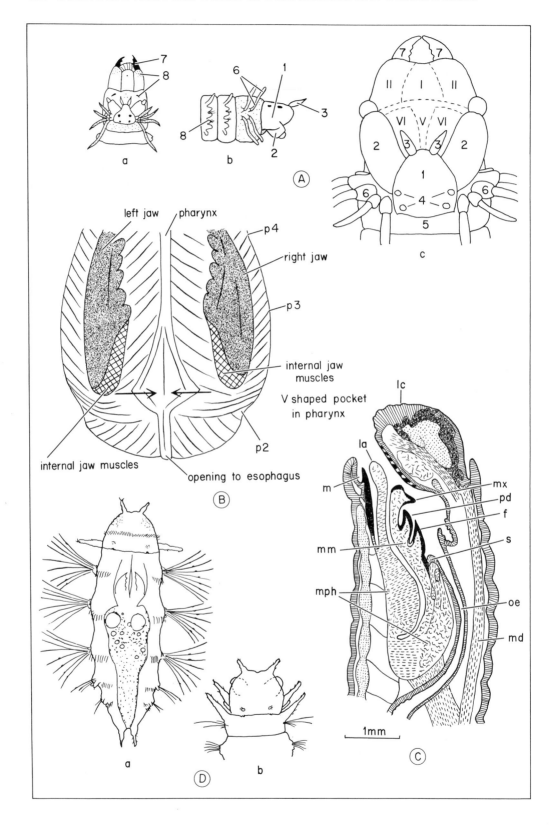

FIGURE 9.5B Polychaete jaws — in living forms in context of related anatomy.

(A) *Nereis* sp; (a) Dorsal view of head showing everted proboscis; note jaws (7 — see A, b) and position of paragnaths (8); (b) side view, labels as in "c" (c) dorsal view of head; observe jaws (7) in relation to areas of proboscis (I, II, V, VI). Not seen: Areas I–IV on maxillary ring; areas V–VIII on oral ring. 1 — prostomium; 2 — palps; 3 — prostomial tentacles; 4 — eye; 5 — peristomium; 6 — peristomial cirri; 7 — jaw ("scolecodont"); 8 — chaetae on parapodia.

(B) *Nereis* sp., dorsal half of retracted proboscis, ventral view. Jaws in natural or resting position in relation to various muscles that influence and control jaw movements: P_2 — posterior jaw muscle; P_3 — proximal jaw muscle; P_4 — anterior jaw muscle, X 12.5. V-Shaped pocket in posterior portion of pharynx could permit inward movement as indicated by arrows by action of P_2, P_3. Such movement would cause anterior ends of jaws shown in (A, a) to spread out.

(C) *Eunice punctata*. Medial cut through anterior portion of living animal showing position of jaw apparatus and related musculature: *mx* — maxilla; *pd* — dental plate; *f* — forceps; *s* — supports; note position of mandible (*m*) in relation to maxillae; relevant musculature for jaw muscles: *mm* — upper jaw; *md* — retractor of pharyngeal pouch (by this muscle, armored pharynx is withdrawn; opposite movement is eversion (Figure Aa, Ac); *mph* — musculature of pharyngeal pouch that controls rasping movement of mandible (lower jaw). Other structures shown: *la* — anterior lobe of pharynx; *lc* — cephalic lobe; *oe* — oesophagus.

(D) *Nereis vexillosa*, Friday Harbor, Washington, schematic. (a) 5-Segmented larva; maxilla retracted. (b) During process of cephalization 1st setigerous segment modifies to form the peristomium (See Ac, 5). Head of 8-segmented larva shows peristomial tentacles.

Illustrations adapted and redrawn from: (A, a–c) Borradaile et al., 1958,* (A, b) Hartman, 1954b; (B) based on original dissection and drawing by Jerry Stude, 1961; (C) simplified from Heider, 1925, Figure 7, p. 69; (D) M. W. Johnson, 1943.

*From G. A. Kerkut, ed., The Invertebrata, 3rd ed., 1959, Borradaile and Potts. By permission of Cambridge University Press.

There are further qualifications. Jean Hanson (1948) studied the tube of the serpulid *Pomatoceras triqueter* and also saw it was secreted by glands identified by Swan as the major subcollar gland. The investigator found the enzyme, alkaline phosphatase, in the epithelia of the digestive tract, in the subcollar glands, and at other

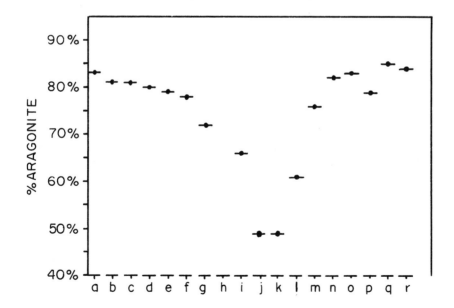

FIGURE 9.6 Tube of the living polychaete worm *Eupomatus gracilis*, Bermuda: variation of aragonite with consecutive growth increments (a–r). Two maxima and one minima of the aragonite/calcite ratio are plotted. Oscillations (if one connects points on the graph) suggest seasonal temperature control of this ratio (after Lowenstam, 1954).

sites. This enzyme is likely to have served an important function in calcium transfer and secretion of the tube. (See Chapter 15: work of Wilbur et al.)

Another puzzle is *shell repair*. When the posterior end of the calcareous tube of a living polychaete is broken, the break is repaired by calcium-plate formation. Since the shell apparently grows from the anterior end, some explanation is needed. Swan speculated that bits of unconsolidated carbonate secreted by the major subcollar glands might be carried backward to the injured site by ciliary action and cemented in place as the bicarbonate content decreased (being greater anteriorly).

Polychaetes, Sipunculoids, and Other Worms as Rock and Shell Borers

Gardiner (1903) and subsequent investigators have found polychaete worms to be the most effective agent in the breakdown of reef rock built by corals and calcareous algae. (It should be remembered that Goreau found sponges to act a similar part in some instances; see Chapter 4.) More recently, Hartman (1954a) found that most annelids of the northern Marshall Islands had a destructive effect on reef structures. Some were forms with hard pharyngeal structures (jaws, scolecodonts) such as *Eunice afra, Palola siciliensis* (Figure 9.5A, part E). Their destructive activity (Eunicea) is achieved by the rasping action of the ventral mandibular plate (Figure 9.5A, part Ec) while the dorsal maxillary pieces (Figure 9.5A, Ea) probably serve for grasping (Hartman, 1954a).

In the absence of pharyngeal hard structures, modified setae may perform similar functions (*Polydora*, for example, Figure 9.5A, and see Yonge, 1963, p. 6). Still a third mode of rock penetration for forms lacking hard structures is chemical action (*Hypsicomus*, family Sabellidae, Sedentaria).

Thus the three ways in which annelid worms can destroy corals and other rocks involve jaws, modified setae, and chemical action (Yonge, 1963). [Errant polychaetes (that is, with jaws) do destroy reefs, but sedentary polychaetes (serpulids) contribute to reef growth via massed worm tubes as in *Salmacina* and *Eupomatus* (Hartman, 1954a).]

Some sipunculoids are common coral rock borers (*Aspidosiphon* sp.). Sand burrowing forms use small cuticular plates while coral rock borer plates fuse to form a massive shield—the boring tool. Yonge (1963) remarked that the primitive condition was sand-burrowing—sand being swallowed along with organic material. The transition to rock boring was apparently one of accommodation to a new geologic niche rather than development of a unique mode of life. The coral rock borers have a boring tool that seems to have been favored by natural selection since organic nutrient for the worm is obtained from the same rocks into which they bore. The weeding-out process of sand borers in this novel niche can be envisioned.

A flatworm, the turbellarian *Pseudostylachus ostreophagus*, native to Japanese waters, was inadvertently introduced into Puget Sound along with Japanese oysters. This worm bores a keyhole-shaped opening in the oyster *Crassostrea gigas* (Yonge, 1963).

FOSSIL POLYCHAETES

Taxonomy

Phylum Annelida (see definition, p. 447). The phylum embraces two classes: Polychaetia and Myzostomia.

Class Polychaetia (Sandworms, Tube-Secreting Worms, and Others). Segmented; somites with lateral parapodia (Figure 9.5B, part A) bear chaetae (= bundles of bristles); sexes, usually separate; fertilization commonly external; trochophore larval stage; budding in some species (as in colonial *Salmacina incrustans*, Jean Hanson, 1948, Figure 1). Dominantly marine, but earthworms and leeches are nonmarine. Geologic range: Cambrian to Recent.

The class is divided into three orders, discussed and illustrated after necessary preliminaries: Order 1, Errantida (with jaws); Order 2, Sedentaria (burrows or tubes); Order 3, Miskoiida (extinct order — no jaws, no tubes).

Errant Polychaetes

The jaw apparatus (mandibles, maxillae, carriers, paragnaths) occurs in living forms as a complete assemblage (Figure 9.5*B*; for terminology, see Figure 9.7*B*; for relationships of mandible, maxillae, and so on, see Figures 9.5*A* and 9.5*B*, part *C* — *Eunice punctata*). However, in the fossil record such assemblages generally, but not always, are found disassociated. Accordingly, they most often occur as isolated pieces — maxilla, carrier, mandible. As a result,

the taxonomy is confused. Scolecodont natural assemblages have generally been named (*Paulinites, Xanioprion, Polychaetaspis*, and so on, Figure 9.8). However, each known fossil assemblage contains components that have been found as isolated pieces in the fossil record; and they were given separate generic names, *Arabellites, Eunicites*, and so on. (Such components consist of forceps, dental plates, paragnaths, and so on.)

Two names have come into existence: one for the natural assemblage and one for component pieces of the same assemblage. The International Code for Zoological Nomenclature (1961) came into existence to insure worldwide uniformity in the naming of animals. Specific rules have been established to deal with whole,

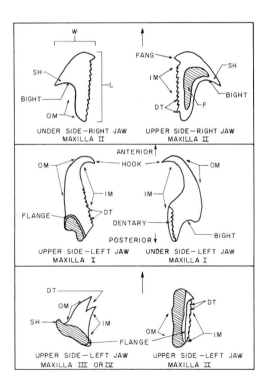

FIGURE 9.7A Errant polychaete jaw apparatus.
(*A*) Buccal armature, dorsal, upper side; *c* — carrier; *F* — forceps or maxilla I; *DP* — dental plate, maxilla II; *UP* — unpaired piece, maxilla III; *P* — paragnath, maxilla IV.
(*B*) Mandibles, upper side, ventral to buccal armature (= a rasping tool to bore into rock while maxillae are for grasping).
(*C*) Schematic representation of buccal armature, upper side. After Treadwell, 1921, modified by R. K. Sylvester, 1959.
[Note: Figure 9.7A, part A — the pulp cavity shown on forceps dorsal side should be dashed to indicate presence not on dorsal, but on ventral side. See legend, Figure 9.7B.] Arrow indicates anterior direction.

FIGURE 9.7B Terminology for Scolecodont Maxilla. All of the pieces are part of the upper jaw. [Kielan Jaworowska, (1961, p. 242) notes that Sylvester gave a wrong definition of fossa. In all Recent Eunicidae pulp cavities and their openings occur in the lower (ventral) side of the jaws. Usage of "upper side, or underside of the jaw" is incorrect.] Dashed area — fossa; *W* — width; *L* — length; *IM, OM* — inner margin, outer margin, respectively; *sh* — shank, *DT* — denticles.

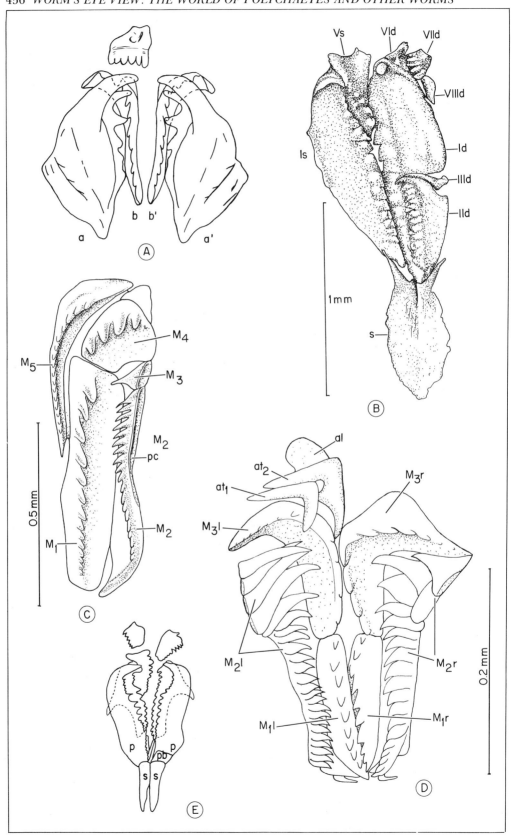

FIGURE 9.8 Some natural fossil Scolecodont assemblages.
Eunicites. (A) *Eunicites* sp. undet., Ft. Riley Is., Permian, Kansas (Components: *aa*[1] − maxilla I, *Arabellites comis*; *bb*[1] − *Arabellites falciformis*; *Eunicites* sp., the unpaired piece).
(B) *Polychaetapis wyszogrodensis* Kozlowski: *s* − supports; *ls*, *ld* − left and right forceps; *lld* − basal piece; *llld* − intercalary tooth; *Vs* − unpaired piece; *Vld* − right posterior maxilla; *Vlld* − right anterior maxilla; *Vllld* − right lateral tooth (= paragnath).
(C) *Vistulella kozlowski* Kielan-Jaworowska, Ord., Poland; dorsal view; M_1–M_5 maxillae; *pc* − pulp cavity (= fossa).
(D) *Xanioprion borealis* K.−J., Ord., Poland; schematic; M_1l, M_1r − left (*l*) and right (*r*) maxilla; at_1, at_2 − anterior teeth; *al* − attachment lamella.
(E) *Paulinites paranaensis* Lange, Dev., Brazil, Dorsal view; *s* − carrier; *pb* − basal plate of right forceps; *pp* − left and right forceps; other components: 2 asymmetrical subtriangular dental plates; *ar* unpaired piece, and 2 oblong asymmetrical paragnaths (see "B" for definition).
Illustrations redrawn from Tasch and Stude; Kozlowski, Kielan-Jaworowska, Lange.

as opposed to isolated, units of the same animal: (*a*) Given a single jaw piece, say a Maxilla I, that was found fossilized and named (genus and species); (*b*) at a later time a complete jaw assemblage is recovered from the fossil record and one of the component pieces of this assemblage is that named in (*a*) above; (*c*) then it would have followed that the whole assemblage take the name of the isolated part as in (*a*). This practice has not been followed. The result is a dual nomenclature.

In addition to *Paulinites* Lange, other more or less complete fossil scolecodont assemblages include *Kettnerites* and *Pernerites* Šnajdr; *Polychaetaspis* and *Polychaetura* Kozlowski; *Vistulella*, *Atraktoprion*, *Kalloprion*, *Ramphoprion*, and *Xanioprion* − all named by one of the leading workers in this area, Kielen-Jaworowska (Figure 9.8).

As shown in Table 9.3, nine form genera are represented by component pieces − mandibles, carriers, right and left forceps, right and left dental plates, an unpaired piece, right and left paragnaths. In fossil collections, such isolated pieces will predominate; therefore it is essential to recognize these as components of complete jaw assemblage(s).

Polychaete Taxonomy

The three orders of polychaetes are the Errantida, Sedentaria, and Miskoiida.

Order 1. **Errantida.** Except for head/anal regions, somites are all alike, as are parapodia (Figure 9.5*B*, part *A*, *b*8); pro-

trusible armored pharynx (Figure 9.5*B*, part *A*, *a*) consists of upper jaw components (carriers or supports, dental plates, intercalary teeth, paragnaths plus several paired or unpaired maxilla − called forceps). Polychaete jaws are used for grasping. The lower jaw components (paired mandibles) serve for rasping solid rock (Figure 9.5*B*, part *C*). All jaw components (= scolecodonts) are chitinous (see Chapter 15, "Chitin"). Both upper and lower jaw components together, constitute the natural scolecodont assemblage. Geologic range: Ord.−Recent.

[Representative form genera, exclusive of natural assemblages shown in Figure 9.8, include *Arabellites* (Figure 9.9*A*), *Staurocephalites* (Figure 9.9*B*), *Eunicites* (9.9*C*), *Lumbriconereites* (9.9*D*), *Leodicites* (9.9*E*).]

One family, the Sprigginidae (Figure 9.9*F, a*), tentatively placed under this order, is known only from the Cambrian of Southern Australia (Glaessner, 1959). The one living segmented worm (Errantida) resembling it is *Tomopteris longisetis* (Figure 9.9*F, b*).

Tubiculous and Burrowing Polychaetes

Order 2. **Sedentaria.** Operculate and nonoperculate annelids; generally unarmored, nonprotrusible pharynx (that is, without jaws); live in burrows or secrete organic (chitinous) calcareous or phosphatic tubes. Tubes may incorporate sand grains, shell fragments, fish scales, and other foreign particles. Tubiculous annelids, not organically attached to their tubes, can readily be removed from them;

TABLE 9.3. The Natural Scolecodont Assemblage *Paulinites* **Lange and its Included Component Form Genera**
(Named from Isolated Jaw Pieces Found Outside of the Brazilian Devonian)

Component Pieces	Form Genera	Age and Region
Mandibles	*Diopatraites fustis*	Ord., Ontario
Carriers	*Marphysaites optus*	Ord., Ontario
Right forceps	*Nereidavus ontarioensis*	Dev., Ontario
	Nereidavus harbisonae	Dev., Ontario
	Oenonites asperus	Sil., Scandinavia
Left forceps	*Arabellites hamatus*	Sil., Gotland
	Nereidavus harbisonae	Dev., New York
	Nereidavus planus	Dev., Ontario
	Oenonites asperus	Sil., Scandinavia
Right dental plate	*Leodicites raemanni*	Dev., New York
	Leodicites variedentatus	Sil., New York
	Arabellites cultriformis	Dev., Ontario
	Arabellites falciformis	Dev., Ontario
	Lumbriconereites webbi	Ord., Minnesota
Left dental plate	*Ildraites howelli*	Dev., New York
	Arabellites prosseri	Dev., Canada
	Arabellites dauphinensis	Dev., Canada
Unpaired piece	*Oenonites impardentatus*	Ord., Ontario
	Arabellites acutidentatus	Ord., Minnesota
	Leodicites streetvillensis	Ord., Ontario
Right paragnath	*Arabellites (?) obliquus*	Sil., Canada
	Eunicites nanus	Dev., Canada
	Eunicites placidus	Dev., New York
Left paragnath	*Eunicites seamani*	Dev., New York

Source. Data from Lange, 1949; see Figure 9.7 for terminology.

tubes are attached or free; conical to coiled. Habitat: brackish or marine. Geologic range: Cambrian to Recent.

[*Serpula* (Figure 9.10*A*), *Rotularia* (9.10*C*). *Spirorbis* (9.10*D*, *G*), *Glomerula* (9.10*E*), *Hicetes* (9.10*F*), *Pectinaria* (9.10*H*).]

***Order 3*. Miskoiida.** Marine polychaetes with elongate bodies; somites — numerous and uniform; row of setae around mouth; abundant branched parapodia; retractile elongate proboscis. Middle Cambrian, Burgess shale, British Columbia, Canada.

Worms of Uncertain Affinities

There are numerous worms whose taxonomic position is truly unclear. Two of these are illustrated: *Palaeoscolex piscatorium* (L. Ord., England, × 3) (Figure 9.11*A*) and *Tullimonstrum gregarium* (Penn., Illin-

ois) occur in concretions found in an old strip mine dump heap (Figure 9.11*B*).

In addition, numbers of fossil worm spoor exist (burrows, castings, and so on); the ubiquitous *Scolithus* burrows are a good example. Häntzschel reviewed and figured the whole array of such trace fossils and problematica (1962). Fossil spoor can be used, as any other fossil, to distinguish a given bed; example: *Scolithus* sandstone.

INFORMATION DERIVED FROM ANNELID WORMS

Reworking of Substrates

Darwin found that the common earthworm, *Lumbricus* (class Oligochaeta), can turn over, during burrowing, as much as 18 tons per acre per year. Such worms are recognized in the fossil record ever since

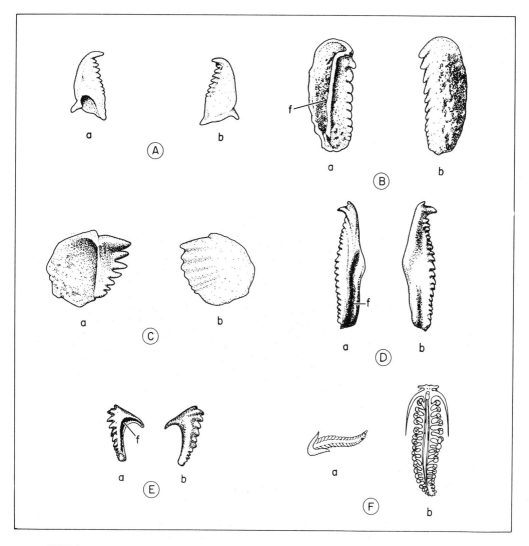

FIGURE 9.9 Representative Scolecodont form genera (individual components—maxilla—of a complete assemblage). All figures enlarged.

(*A*) *Arabellites comis.* Snyder Creek Formation. U. Dev., Central Missouri; (*a*), (*b*) jaw—underside (ventral) and upperside (dorsal).

(*B*) *Staurocephalites aequilateralis* L. Miss., Missouri. Left jaw, (*a*) underside (ventral); (*b*) upperside (dorsal); *f*—fossa.

(*C*) *Eunicites paranaensis,* Maxilla IV, left jaw, (*a*) underside (ventral); (*b*) upperside (dorsal); X 30, same formation as (*A*) above.

(*D*) *Lumbriconereites bicornis.* Watertown, N.Y., Maxilla I, right jaw, (*a*) underside (ventral); (*b*) upperside (dorsal).

(*E*) *Leodicites exilis,* Maxilla II, right jaw; (*a*), (*b*) lateral sides.

(*F*) Family Sprigginidae: (*a*) *Spriggina floundersi,* pre-Cambrian, Australia, schematic; (*b*) living *Tomopteris longisetis,* compare to (*F, a*), schematic.
Illustrations after several authors.

Carboniferous time. Most likely, many fossil soils of Paleozoic, Mesozoic, Tertiary to Cenozoic age, were reworked by oligochaetes (see Figure 9.10*I*). Hazen's Paleocene fossil (1937) was actually a sandfilling of a burrow made and occupied by an earthworm. The impression of the somites on the burrow walls is attributed to peristaltic movements of the worm crawling upward in its burrow (see Häntzchel, 1957, 1962).

Similar reworking of sandy substrates

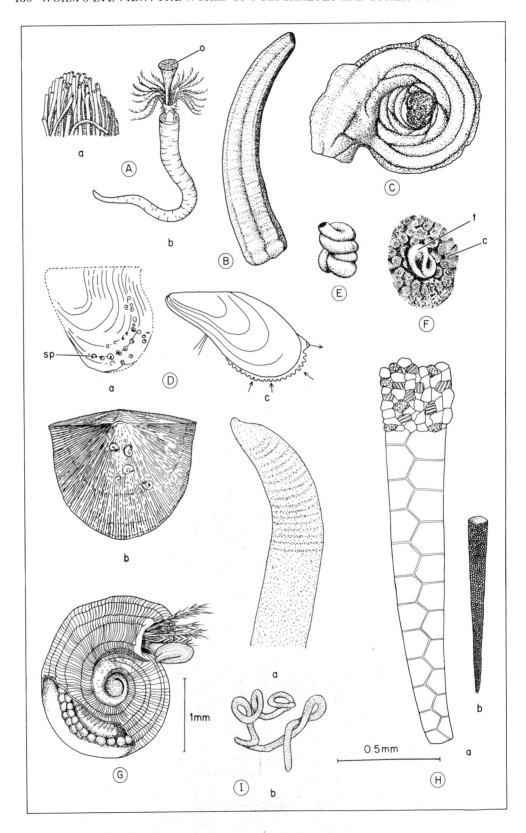

FIGURE 9.10 Tubiculous polychaetes and burrowing oligochaetes—living and fossil.

(A) *Serpula*, (a) *S. socialis*, M. Jurassic, Germany, enlarged; (b) *S. contortiplicata*, Recent: o—operculum, X 10.

(B) *Hamulus onyx*, Cook Creek, U. Cret., Texas, side view, X 4.

(C) *Rotularia cortezi*, Eocene, Texas, attached surface, slightly enlarged.

(D) (a) *Spirorbis* sp. attached to nonmarine lamellibranch *Naiadites* aff. *triangularis*, a commensal relationship (see text), slightly enlarged. Arrow points to a single growth line bearing a probable contemporaneous population of spirobids (sp). (b) *Spirorbis* sp. Dev., Europe, attached to articulate brachiopod shell, schematic. (c) Incurrent and excurrent apparatus in *Mytilus* (see Chapter 8). Compare to (D, a).

(E) *Glomerula gordialis*, U. Cret., Eur., X 4.

(F) *Hicetes innexus*, U. Dev. (Hamilton) N.Y., slightly enlarged; tube (t) built within the coral (c) *Pleurodictyum* (see Chapter 5).

(G) Living *Spirobis granulatus* adult, cut to show chain of eggs close to maternal body. Taken from Hittarps Reef, Øresund, depth 10–16 meters, attached to shell of small living *Littorina littorea*; scale shown.

(H) *Pectinaria*. (a) *P. koreni*, larval tube with rhomboid pattern and sand grain structure characteristic of adult tube, Øresund, enlarged. (b) *P. belgica*, Miocene-Recent, U.S.A., note sand grains, compare to (H, a).

(I) Class Oligochaeta: (a) Fossil earthworm (?), Paleocene, Wyoming, X 3 +; (b) *Lumbricopsis permicus* Permian, Czechoslovakia, enlarged.

Illustrations redrawn and adapted from: (Aa, Db) Zittel; (Ab, Ba) Wade, 1922; (C) Julia Gardner, 1939; (Da) Trueman, 1942; (E, F, Hb) Howell, 1962, after several authors; (G, H) Thorson; (I) Hazen, 1937.

in modern shallow seas was reported by Davidson (1891). He figured that a high tonnage of bottom sand/acre off the coast of Great Britain passed through the lobworm alimentary canal. Equivalent activity is thus suggested in shallow marine substrates in the geologic past onward from the Proterozoic time. Shrock (1935) commented that the number of worm castings found in the Indiana Salem limestone showed that marine worms of the time had reworked much substrate.

Reefs

Various tubiculous and other polychaetes formed reef structures in the geologic past. Some reefs are formed by still extant types (*Serpula*—Gulf Coast; *Sabellaria*—North Sea). Hedgpeth (1957) reported on a Gulf Coast serpulid reef. At the mouth of Baffin Bay, such reeflike serpuloid masses occur. These rocklike clumps of tubes, several feet in diameter, rise 2 to 4 ft above sandy lagoons. No living serpulids have been found on this reeflike mass in recent years. Close to the harbor at Vera Cruz, Mexico, the most proximate serpulid reefs occur in normal marine waters.

Fossil reef structures are also inferred from the great density and packing of individual *Scolithus* (L. Cambrian, Sweden—Hadding, 1929); *Spirorbis* (Carboniferous of Europe—Haack, 1923). Dense serpulid worm tubes occur in the serpulite limestones; they measure up to 50 meters thick, in the German Jurassic (Gignoux, 1955).[2] Parsch (1956) described a varied serpulid fauna from these limestones. He related his collections to laminated facies or to sponge facies and placed the serpulids in stratigraphic context. Figure 9.10*A*' indicates the frequency of serpulids in respective subgenera through the Jurassic stages. Certain species were found restricted to one or another subdivision in the Lias. Most species were long-ranging in the other stages (Parsch 1956, Table 1-3).

Götz (1931) reviewed the whole subject of fossil serpulid worms. They lived in sand, shell, glauconitic, or clay mud substrates. Depths were not restricted to shallows or estuarine environment. Some forms extended as deep as 3000 ft. Among the chief predators on serpulids were echinoids, crabs, snails, and fishes. H. W. Ball (1960)[3] reported on the serpulid *Rotularia* (Figure 9.10*C*) from the Upper Cretaceous of James Ross Island, Graham

[2]Reference, see Figure 4.5*C*, Ch. 4.

[3]Falkland Islands Dependencies Surv. Sci. Rpt. No. 24.

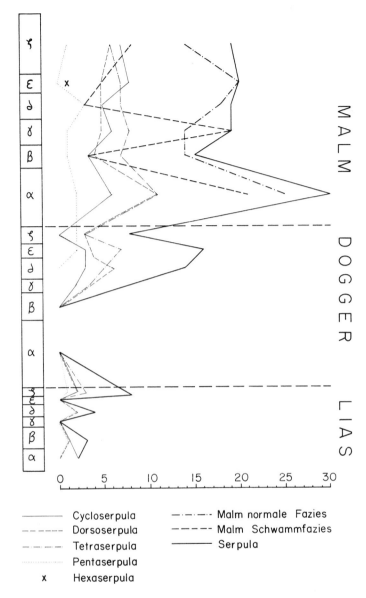

FIGURE 9.10A Frequency diagram of *Serpula* fauna in S. W. German Jura Mountains. (Lias, Dogger, Malm are European stages of the Jurassic, Schwammfazies = sponge facies; normale Fazies = laminated facies.) Lines indicate number of serpulids of respective subgenera at given horizons; for example, at middle *alpha* (Malm), 2–3 specimens of *Pentaserpula*, about 7 of *Cycloserpula*, and 30 of *Serpula* were found. After K. O. A. Parsch, 1956.

Land. *Rotularia* occurred in postdepositional nodules of fine-grained, silty, glauconitic, calcareous sandstones along with decapods (see Chapter 11); the surface of these nodules were extensively marked with trace fossils. That suggested a highly organic original sediment riddled by dense populations of worms.

Composition and Fine Structure of Fossil Scolecodonts

Schwab (1966) studied two fossil scolecodonts from the Upper Ordovician, Ohio, and from the Independence shale, Upper Devonian, Iowa. Spectrographic analysis revealed that both specimens have large

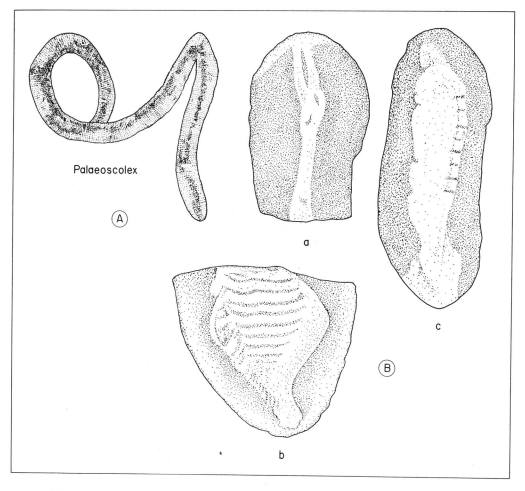

FIGURE 9.11 Worms of uncertain taxonomic position.
(A) *Paleoscolex piscatorum*, L. Ord., England, X 3.
(B) *Tullimonstrum gregarium*, Penn., Illinois; (a) distal end of proboscis in concretion, 5.8 cm long;
(b) tail in concretion, 6.7 cm; (c) nearly complete individual proboscis is turned back over head; in
concretion, 12 cm long;
Illustrations redrawn after (A) Whittard, 1953; (B) E. S. Richardson, Jr., 1966.

amounts of calcium, copper, silica, and magnesium. The scolecodonts were found to consist of dense lamellar, dark-brown to black outer cuticle – the organic layer (of rhombic/cubic cellularity – Figure 9.12B, a) – and a lighter colored inner lining, presumably with numerous minute "tubules" (Figure 9.12B, b; see also Tasch and Stude, 1966). The mineral composition was fluorapatite ($3Ca_2P_2O_8CaF_2$).

Other halogens, iodine and bromine, occur in living jawed polychaetes (Eunicidae). The ratio of iodine to bromine will vary with the organic content.

Commensalism and Parasitism

There are several records of worm fossil associations with or within corals (Figure 9.10F), various mollusks (oysters, cephalopods, fresh-water lamellibranchs), brachiopods (Figure 9.10D), insects, crinoids, and so on (Tasnádi-Kubacska, 1962).

A carefully studied occurrence is that of *Spirorbis* on shells of the nonmarine lamellibranch *Naiadites* in the British Carboniferous (Trueman, 1942). Trueman noted that this apparent commensalism was particularly common at certain times.

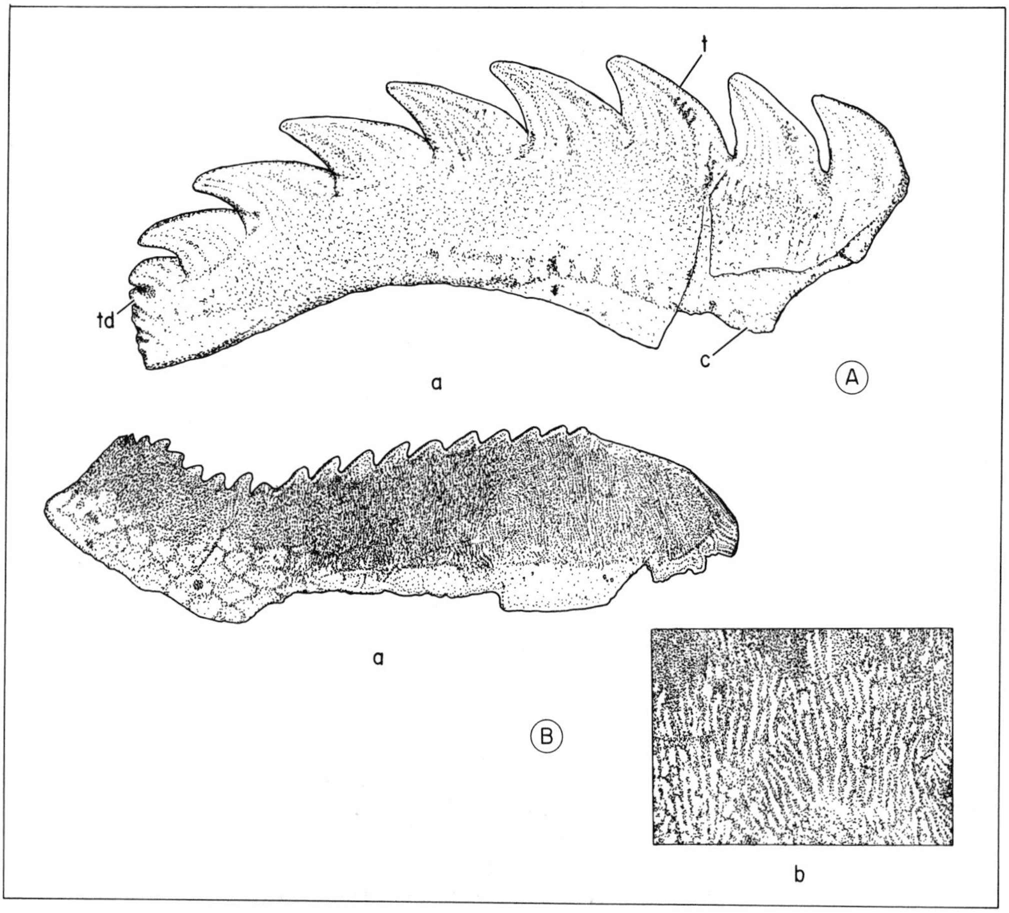

FIGURE 9.12 Scolecodont fine structure.
(*A*) *Arabellites falciformis* (?), Ft. Riley Limestone. Perm. (Wolfcamp), Kansas; right jaw, transmitted light; *c*—crenulated margin probably composed of cubic cells; *t*—fibers (= "tubules" of Schwab); *td*—reversely curved tusklike denticles, greatly enlarged.
(*B*) (*a*) Unidentified Scolecodont. Independence Shale, U. Dev., Iowa; lateral view showing "growth cells" that presumably gave rise to tubules [compare (*c*) and (*t*) of (*A*)], X 500.
(*b*) Same; enlargement of "growth cells" to indicate tubules, X 1000.
Illustrations after (*A*) Tasch and Shaffer, 1961; (*B*) Schwab, 1966.

Geological age		Graptolite succession	Sweden	Estonia		Atraktoprion cornutus	Atraktoprion sp. a	Kalloprion ovalis	Ramphoprion elongatus	Xanioprion borealis
Upper Ordovician		Pleurograptus linearis	Slandrom Limestone	E	Rakvere Stage	+	—	+	—	+
Middle Ordovician	Caradocian	Dicranograptus clingani	Macrourus Limestone	DIII	Oandu Stage	—	—	—	—	—
				DII	Keila Stage	—	—	—	+	—
		Diplograptus multidens	Ludibundus Limestone sensu lato	DI	Johvi Stage	—	—	+	—	—
				CIII	Idavere Stage	—	—	+	—	—
		Nemagraptus gracilis		CII	Kukruse Stage	—	+	+	—	+

FIGURE 9.12A'. Stratigraphic range and position of scolecodont and natural assemblages from Polish boulders. Scolecodont genera present (+), absent (—). After Kielan-Jaworowska.

Several spirorbids attached along a *single* growth line of the pelecypod (Figure 9.10D, a). What could it mean? Trueman thought it represented contemporary members of a colony. Sites of spirorbid attachment were always in particular parts of the shell. Position on a given shell varied somewhat between different genera. Such selective attachment might relate to the nature of the host's mantle border because that marked the limits of in-current streams of water (see Figure 9.10D, c). For the lamellibranch *Naiadites*, Trueman compared the in-current streams to those in living *Mytilus*.

Evolutionary History of Scolecodonts

Kozlowski (1956) and others concluded that individual elements in the polychaete worm jaw apparatus have been slightly modified forward from Ordovician, possibly even Cambrian or pre-Cambrian time. Lange (1947) noticed that the Devonian scolecodonts in Brazil seemed but slightly modified from older forms.

Hartman's (1954a), atoll annelid study, provided a valuable clue for paleobiologists treating scolecodont assemblages. The most destructive species (*Palola siciliensis*), destructive in the sense of boring into coral-algal rock, had a mandible much increased in size compared to less destructive species. Since the mandible (Figure 9.7A, part B) is a rasping tool for rock-boring while the maxilla are merely grasping elements, it is reasonable to expect that greater variability would be displayed by the raspers. Thus, in specific instances, depending on the nature of the fossil material at hand, it may be desirable to emphasize the size, configuration, and so on, of the mandible as well as the maxilla. Presumably, a thicker, larger mandible afforded a selective advantage in the transfer from less consolidated substrates to coral-algal rock.

Burgess Shale Annelids

The Middle Cambrian Burgess shale annelids consisted of 11 genera assignable to widely separated families. This led Walcott (1911) to decide that for all classes the fundamental character had developed prior to Middle Cambrian time. Since Proterozoic sedimentaries abound in spoor of marine worms, it is reasonable to infer a still older pre-Cambrian origin for the primitive annelid. Another indication of early radiation and evolution from the ancestral stock is the occurrence of tubiculous (*Sabellidites*) and burrowing worms (*Scolithus*) in the Lower Cambrian of the Leningrad region, Sweden, Scotland, United States, and so on (Howell, 1957).

The abundance of scolecodonts (errant polychaetes) in the Paleozoic ever since Ordovician time also indicates separation of errant and sedentary polychaetes *before* Ordovician time.

Biostratigraphy

Brazil. Lange (1949) pointed out the twofold stratigraphic value of scolecodonts: They are characteristically Paleozoic (although, of course, known from the Mesozoic—Solenhofen, Jurassic; the Lebanon Cretaceous—Roger, 1946); they prove useful when other microfossils are sparse or absent.

The scolecodont collections from the Ponta Grossa shale (Lower Devonian, State of Parana, Brazil) were made over several years. Scolecodonts were "more or less accidentally encountered" in general fossil collections in the shale. A deliberate biostratigraphic study designed to collect scolecodonts very likely would reveal occurrence on specific horizons of certain species and transgression of successive horizons by others. (The writer had the opportunity of collecting from the Ponta Grossa shale in company with Dr. Lange and others during 1968 and made this point in the field; see Fort Riley study below.)

Poland. Remarkably abundant and well-preserved scolecodont assemblages from the Polish Ordovician were reported by Kielan-Jaworowska (1961–1962, and so on) and earlier by Kozlowski (1956). Earlier in this chapter some details of this fauna were discussed. The Polish scolecodonts occurred in erratic boulders composed of fine to coarse-grained limestone. Biotic associates included Foraminifera, Chitinozoa, graptolites, dasycladacean(?) algae, hydroids, brachiopods, trilobites, crinoids, and conodonts, among others.

The stratigraphic position of several of these scolecodont assemblages is shown in Figure 9.12*A'*. It is seen that a single species distinguishes certain stages (*Atraktoprion* sp., *Ramphoprion* sp.). Other species are longer ranging, for example, *Kalloprion* species.

United States. An extensive biostratigraphic study within a single formation is that of the Ft. Riley limestone, Permian (Wolfcampian) of Kansas (Tasch and Stude, 1966). Localities and distances between quarries are shown in Figure 9.13.

Two datum planes were used in the five

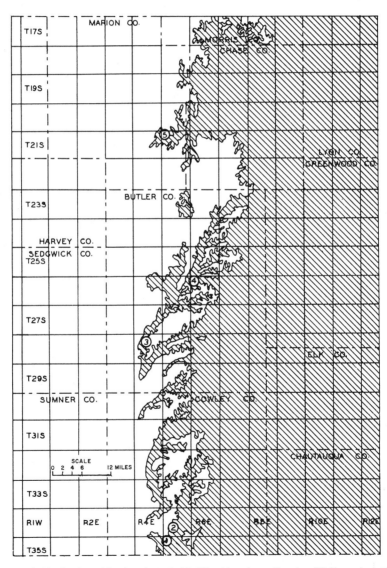

FIGURE 9.13 Distribution of Scolecodonts in Ft. Riley Limestone, Permian (Wolfcampian) of Kansas. Localities: (1) Camp Horizon, Old Quarry along bluff of Arkansas River, SW, sec. 6, T35S, R5E, Cowley County. (2) Silverdale Quarry, Cowley County. (3) Augusta Quarry, Butler County. (4) El Dorado Quarry, Butler County. (5) Florence Quarry, Marion County.

quarries studied: the calciphosphatic in-articulate brachiopod *Orbiculoides* zone, and the top of the Florence flint. All scole-codont-bearing beds were located at a given section by Stude and Tasch and Stude in terms of inches above or below the *Orbiculoides* zone – which is 18 ± ft above the top of the Florence flint.

Geographic Spread

How did the Ft. Riley scolecodont spe-cies distribute themselves on the Wolfcam-pian sea floor over a distance of some 86 miles? (See Figure 9.13.) Four species went into all localities: *Arabellites comis*, *A. falciformis*, *Staurocephalites fortrileyensis*, and *Nereidavus wolfcampis*. These species stayed at four or five localities: *Arabellites plesiocomus*, *Nereidavus orbiculoides*, *Ildraites anatinus*. Some species were apparently restricted to the sea floor at two localities. Still other species were found only at a single locality.

This geographic spread indicates a gen-erally restricted and local distribution of most scolecodont species represented in the Ft. Riley collection.

Also noted was a constant association of scolecodonts now assigned to different species and/or genera at several localities. This probably denotes that many isolated maxilla belonged to the same natural as-semblage but were disarticulated after the animal's demise or possibly during pro-cessing.

Vertical Spread and Lithofacies

The Ft. Riley was divided into seven lithofacies listed from oldest to youngest: (*A*) intrabiosparite, (*B*) algal intrabiospar-ite (*Osagia*), (*C*) clayey intrabiosparite, (*D*) clayey fossiliferous dismicrite, (*E*) clayey fossiliferous biomicrite, (*F*) mixed, and (*G*) foraminiferal biosparite (a buff-colored oxidized zone). How were the scolecodont species distributed according to these lithofacies?

1. Lithofacies G – an oxidized zone, was bar-ren of scolecodonts at all localities.

2. The same forms that had the greatest geo-graphic spread also had the most complete vertical distribution occurring in all litho-facies except G.

3. Some species were vertically restricted to a single lithofacies, for example, *Ildraites flor-encis* and *Ildraites bifurcatus*. These forms were restricted to lithofacies D, a dismicrite. *Ildraites pseudobifurcatus* and two other *Il-draites* species were restricted to either litho-facies A, an intrabiosparite, or B, an algal intrabiosparite.

In general, *Ildraites* species were restrict-ed to dismicrite–biomicrite, whereas most *Lumbriconereites* species were confined to intrabiosparites. By contrast, species of *Ara-bellites*, *Staurocephalites*, and *Nereidavus* ranged through all lithofacies.

4. Every species in the collection occurred in the algal *Osagia* intrabiosparite lithofacies. How can this association be explained? Hart-man's observations of living polychaetes of the coral atolls (1954a, p. 621) reaffirmed ol-der reports of jawed annelid efficiency in rock boring through coral and coralline al-gae. Extrapolating to the Wolfcampian sea floor on which *Osagia* grew, it is plausible to assume a similar relationship between then-living polychaetes (now represented by scolecodont fossils) and the hard algal sub-strate into which they bored.

Anaerobic Conditions

Experiments with living polychaete worms and amphipods (Chapter 10) in closed jars led Dean et al. (1964) to con-clude that fossil polychaetes found com-pactly coiled indicate anaerobic conditions at the time of demise.

Geologic Occurrence

Fossil scolecodonts have been found in Ordovician to Permian rocks in many places of the world. They are less com-monly reported from the Mesozoic. Ex-amples of reports in the literature are Or-dovician (Stauffer, 1933), Silurian (Got-land) (Martinsson, 1960), Devonian (Eller, 1938), Mississippian (Lower Carbonifer-ous) (Sylvester, 1959), Permian (Zechstein) (Seidel, 1959). As yet undescribed scole-codonts from the Pennsylvanian Leavenworth Formation of Kansas are in the writer's collections. *Eunicites* species have also been reported from the Jurassic lithographic limestones, Solenhofen, Germany (Ehlers, 1869, Palaeontographica, Vol. 17), and the Upper Cretaceous fish beds of Lebanon (Roger, 1946). More recently, many reports have been forthcoming, for example, from the Polish Permian and Triassic.

In the midcontinental United States there is a reasonably complete representation of scolecodonts (in Kansas, Oklahoma, Missouri) ranging from Mississippian to Guadalupian (Tasch and Stude, 1966). Paleozoic scolecodonts, taking account of the Zechstein occurrence, range from Ordovician to Permian (Lower Ochoan).

[J. Jansonius and H. Craig (Canadian Petrol. Geol. Bull., 1971, Vol. 19 (1), pp. 251–302) published a useful reference work; a dual classification of disjunct vs. assemblages of scolecodonts is proposed and illustrated for disjunct forms. An excellent bibliography adds to the value of this paper.] [For five requirements to be met in the reconstruction of a fossil polychaete worm jaw apparatus from disjunct elements, see Szaniawski and Gazdicht, 1978, Acta Palaeontologia Polonica, 23 (1), pp. 6–7. Further data on the Italian and American Cretaceous has been forthcoming (Charletta and Boyer, 1974).]

QUESTIONS

1. According to B. F. Perkins (1966), rock borings, as distinct from burrows in carbonate sequences, "imply stratigraphic breaks with histories of (1) emergence, (2) lithification, and (3) submergence."
   (a) How would you distinguish a worm rock boring from a burrow in the rock column?
   (b) How would you tell a worm boring from that of a sponge or mollusk?
   (c) Discuss polychaete rock boring in modern coral atolls.
   (d) Discuss different modes of penetration of rock by various types of worms.

2. What subsequent developments in evolution are previewed in the history of the annelid superphylum?

3. (a) Draw and label the complete armor (jaw apparatus) of a living errant polychaete. Do the same for a fossil assemblage.
   (b) What function does the mandible perform? The maxilla?
   (c) What is the mineralogical and chemical composition of scolecodonts?

4. Where are existing serpulid reefs located? Describe one.

5. Discuss the biostratigraphic utility of scolecodonts.

6. (a) Review the paleobiological significance of the following researches:
      (1) Lowenstam on serpulid worm tubes.
      (2) Occurrence of living polychaete species in plankton hauls.
      (3) Worm castings and burrows relative to reworked substrates.

7. (a) Cite varied habitat of living worms and relate to the fossil record.
   (b) What is the significance of so many distinct phyla of worms?

8. Richardson found a new problematical fossil worm, *Tullimonstrum*, in nodules located in an abandoned coal mine pit. The creature is bilaterally symmetrical; head region *unsegmented*; trunk and tail region *segmented*; anterior proboscis with *internal teeth* in slot. Tail region has lateral triangular *fins*. Consider each underlined word of the abbreviated description; cite arguments for or against assigning it to one or another of the worm phyla discussed in this chapter.

REFERENCES

Charletta, A. C., and P. C. Boyer, 1974. Scolecodonts from the New Jersey Coastal Plain: Micropaleontology, vol. 20 (3), pp. 354–366, pls. 1–4.

Davidson, C., 1891. On the amount of sand brought up by lobworms to the surface: Geological Mag., n. ser., Vol. 8, pp. 489–493.

Dean, David, J. S., Rankin, Jr., and E. Hoffman, 1964. A note on the survival of polychaetes and amphipods in stored jars of sediments: J. Paleontology, Vol. 38 (3), pp. 608–609.

Dollus, R. Ph., 1950. Liste de Némathelminthes connus à l'etat fossile: Soc. Géol. France (Jan. 9, 1950), pp. 82–85.

Eller, E. R., 1938. Scolecodonts from the Potter Farm Formation of the Devonian of Michigan: Carnegie Mus. Annals, Vol. 27, pp. 275–286.

Fisher, W. K., 1952. The sipunculid worms of California and Baja, California: U.S.N.M. Proc., Vol. 102, No. 3306, pp. 371–440.

Gardiner, J. S., 1903. The fauna and geography of the Maldive and Laccadive Archipelagoes. Cambridge University Press, Vol. 1, pp. 1–400.

Gardner, Julia, 1939. Notes on fossils from the Eocene of the Gulf Province: U.S.G.S. Prof. Papers 193-B, 37 pp., plates 6–8.

Glaessner, M. F., and B. Daily, 1959. The geology and late pre-Cambrian fauna of the Ediacara fossil reserve: Records So. Austral., Mus., Vol. 13 (3), pp. 369–401, plates 41–47, text figures 1–2 [*Spriggina.*]

Götz, G., 1931. Bau und Biologie fossiler Serpuliden: Neues Jahrb. für Mineral. Geol. und Paläont., Part B, Vol. 66, pp. 385–438.

Haack, W., 1923. Zur Stratigraphie und Fossilfürhrung des Mittleren Bundsandsteins in Norddeutschland: Jahrb. der Preuss. Geol. Landesanstalt, Vol. 42, pp. 560–594.

Hadding, A., 1929. The pre-Quaternary sedimentary rocks of Sweden III. Lunds Universitats Arrskrift N.F. Avd. 2, Vol. 25, No. 3.

Hanson, Jean, 1948. Formation and breakdown of serpulid tubes: Nature, Vol. 161 (4094), pp. 610–611, Figure 1.

Häntzschel, W., 1957. Quer-Gliederung bei rezenten und fossilen Wurmröhen: Senckenbergiana, Vol. 20, pp. 145–154.

————, 1962. Trace fossils and Problematica: Treat. Invert. Paleont., Part W. Miscellanea, pp. W177–W245.

Hartman, Olga, 1947. Polychaetous annelids, Part V, Eunicea, Pts. VI–VIII, 512 pp., 63 plates.

————, 1951. The littoral marine annelids of the Gulf of Mexico: Inst. Mar. Sci., Univ. of Texas, pp. 7–124, 27 plates.

————, 1954a. Marine annelids from the northern Marshall Islands: Geol. Surv. Prof. Paper 260-Q, pp. 619–642, text figures 169–178.

————, 1954b. Key to the families of Polychaeta, in R. I. Smith et al., eds., Intertidal invertebrates of the central California coast. U. Calif. Press, pp. 70–107.

————, 1957. Marine worms: Treat. Marine Ecol. and Paleoecol. G.S.A. Mem. 67, Vol. I (Ecology), pp. 1117–1128.

Hazen, B. M., 1937. A fossil earthworm (?) from the Paleocene of Wyoming: J. Paleontology, Vol. 11, p. 250.

Hedgpeth, J. W., 1957. Biological Aspects, in K. O. Emery and R. E. Stevenson: Estuaries and Lagoons: Treat. Marine Ecol. and Paleoecol. G.S. Mem. 67, Vol. I (Ecology), Chap. 23 (p. 716).

Heider, K., 1925. Über. Eunice: Zeitschrift. Wiss. Zoology, Vol. 125, pp. 55–90 (Leipsig).

Howell, B. F., 1962. Worms: Treat. Invert. Paleont., Part W. Miscellanea, pp. W144–W177.

Hyman, L. H., 1951a. The Invertebrates: Platyhelminthes and Rhynchocoela. The acoelomate Bilateria, Vol. II. Acanthocephala, Aschelminthes and Entoprocta. The pseudocoelomate Bilateria, Vol. III. Smaller coelomate groups, Vol. V. McGraw-Hill.

Johnson, M. W., 1943. Studies on the life history of the marine annelid *Nereis vexillosa*: Biol. Bull., Vol. 84, pp. 106–114.

Kielan-Jaworowska, Zofia, 1961. On two polychaete jaw apparatuses: Acta Palaeontologica Polonica, Vol. 6 (3), pp. 237–254, plates 1–7.

————, 1962. New Ordovician genera of polychaete jaw apparatus: *idem*, Vol. 7 (3–4), pp. 291–332, plates 1–13.

Kozlowski, Roman, 1956. Sur quelques appareils masticateurs des annélides polychètes ordoviciens: Acta Palaeontologica Polonica, Vol. 1 (3), pp. 165–204, text figures 1–20.

Lange, F. W., 1949. Polychaete annelids from the Devonian of Paraná, Brazil: Am. Paleontology Bull., Vol. 33 (104), pp. 1–102, plates 1–6.

Martinsson, A., 1960. Two assemblages of polychaete jaws from the Silurian of Gotland: Bull. Geol. Inst. Univ. Uppsala, Vol. 39, pp. 1–7.

Parsch, K. O. A., 1956. Die Serpuliden-Fauna Des Südwestdeutschen Jura: Palaeontographica, Vol. 7, Part A, pp. 211–240, plates 19–20, Tables 1–3.

Richardson, E. S., Jr., 1965. Wormlike fossils from the Pennsylvanian of Illinois: Science, Vol. 151, pp. 75–76.

Richter, R., 1927. "Sandkorallen" — Riffe in der Nordsee: Natur und Museum, pp. 49–62.

Roger, Jean, 1946. Les Invertébrés des Couches a Poissons du Cretacé Supérieur du Liban: Mem. Sur. Géol. France, p. 69, plate 10, Figure 9.

Seidel, Sieglinde, 1959. Scolecodonten aus den Zechstein Thuringens: Freiberger Forscheingshefte, Vol. 76, pp. 1–32, plates 1–4.

Shrock, R. R., 1935. Probable worm castings ("coprolites") in the Salem limestone of Indiana: Indiana Acad. Sci. Proc., Vol. 44, pp. 174, 175.

Schwab, K. W., 1966. Microstructure of some fossil and Recent scolecodonts: J. Paleontology, Vol. 40 (2), pp. 416–423, plates 53, 54, 3 text figures.

Šnajdr, Milan, 1951. On errant polychaeta from the lower Paleozoic of Bohemia: Sborník Paleont. Geol. Surv. Czechoslovakia, Vol. 18, pp. 241–291, plates 1–10 (Ord.–Dev.).

Snodgrass, R. E., 1938. Evolution of the Annelida, Onchyophora, and Arthropoda: Smithsonian Misc. Coll., Vol. 97 (6), pp. 1–149.

Stauffer, C. R., 1933. Middle Ordovician Polychaeta from Minnesota: G.S.A. Bull., Vol. 44, pp. 1173–1218.

Størmer, Leif, 1963. Gigantoscorpio willsi, a new scorpion from the lower Carboniferous of Scotland and its associated preying microorganisms. Oslo Univ. Press, 124 pp., 22 plates, 45 text figures.

Swan, E. F., 1950. The calcareous tube-secreting glands of the serpulid polychaetes: J. Morphology, Vol. 86, pp. 285–315.

Sylvester, R. K., 1959. Scolecodonts from central Missouri: J. Paleontology, Vol. 33 (1), pp. 33–49, plates 5, 6, 3 text figures.

Tasch, Paul, and J. R. Stude, 1966. Permian scolecodonts from the Ft. Riley limestone of southeastern Kansas: Wichita State Univ. Bull., Univ. Studies No. 68, Vol. 42 (3), pp. 3–35.

Taylor, A. L., 1935. A review of the fossil nematodes: Helminthol. Soc. of Washington Proc., Vol. 2, pp. 47–49, Figure 6.

Treadwell, A. L., 1921. Leodicidae of the West Indian Region: Carnegie Inst. Washington (Dept. of Mar. Biology), publ. 293, Vol. 15, 127 pp., 9 plates.

———, 1922. Leodicidae from Fiji and Samoa: idem. Vol. 18 (68 text figures, plates 1–8).

Trueman, A. E., 1942. Supposed commensalism of Carboniferous spirorbids and certain non-marine lamellibranchs: Geological Mag., Vol. 79, pp. 312–320, 3 text figures.

Voigt, Erhard, 1938. Ein fossiler Saitenwurm (Gordius tenebrosus n. sp.) aus der eozänen Braunkohle des Geiseltales: Nova Acta Leopoldina, n.f., Vol. 5, pp. 351–360.

Wade, Bruce, 1922. The fossil annelid genus Hamulus Morton, an operculate serpula: U.S.N.M. Proc., Vol. 59, pp. 41–46, plates 9, 10.

Walcott, C. D. 1911. Middle Cambrian annelids: Smithsonian Misc. Coll. Vol. 57, No. 5, pp. 109–142, pls 18–23.

Wells, G. P., 1945. The mode of life of Arenicola marina L.: J. Mar. Biol. Assn., Vol. 26, pp. 107–207, text figures 1–10.

Whittard, W. F., 1953. Palaeoscolex piscatorum gen. et. sp. nov., a worm from the Tremadocian of Shropshire: Geol. Soc. London Quart. J., Vol. 109 (2), pp. 125–132, plates 4–5.

Wrigley, Arthur, 1951. Some Eocene serpulids: Proc. Geol. Assn. London, Vol. 62, Part 3, pp. 177–202.

Yonge, C. M., 1963. Rock-boring organisms, in R. F. Sognnaes, Mechanisms of hard tissue destruction. A.A.A.S. Publ. 75, pp. 1–24.

chapter ten
PHYLUM BEYOND CENSUS: THE JOINTED LEG ANIMALS (ARTHROPODS – I)

INTRODUCTION

More arthropods exist on our planet than any other form of animal life. Insects outnumber all others, with an estimated 1 million species. That continuous battle against insect pests in home, garden, farm, and orchard testifies to the unrelenting vigilance necessary to prevent an irreversible insect invasion of niches man so far considers his own. The Coleoptera (beetles and weevils) with 350,000-plus species holds top number among insects. Species of the Chelicerata (spiders, and so on) exceed living protozoans; more crustacean species are living than fish species.

What does this species abundance signify? The arthropod body plan (Figure 10.1), plus multiple adaptations it permitted, led to a highly successful evolution.

As expected, the fossil record of arthropods (trilobites, ostracodes, and insects) is extensive, often abundant. Particular forms, such as copepods, may be very sparsely represented as fossils; occurrence may be limited as with the merostomes (eurypterids, xiphosurans). Yet for perhaps a billion or more years one or another of the arthropods has been an occupant of the earth.

Viewing modern arthropod distribution, it is clear there are almost no ecologic niches lacking representatives of this ubiquitous phylum: in marine, brackish, fresh water, thermal, saline waters, polar seas (euphausids called "whale krill" in polar seas – McWhinnie, 1964), at all alti-

tudes, at all latitudes (tropics to the poles), in soils, forests, deserts, in temporary patches of water (branchiopod crustaceans, ostracodes, insect larvae), in oceanic deeps (crabs and crustaceans), as ectoparasites on other creatures (mites, order Acarina), and so on.

Certain arthropod crustaceans (lobsters, crabs, shrimps) are fished and served as food to man. Specially prepared insects (grasshoppers, termites, and others) are consumed, valued, sold as delicacies; bee's honey has for centuries been marketed worldwide. Insects can facilitate plant pollination while others destroy (consume) every vestige of vegetation. Some insects are disease vectors.

There are numerous singular characteristics of arthropods. Some are unique (flight, ecdysis); other features (bioluminescence) are shared by 40-plus orders of invertebrates. *Flight* – the only flying invertebrates are insects (more in later section) that have evolved mechanisms for sustained aerial flight. *Social organization/ division of labor* – the beehive, termite, wasp, and ant communities highly organize themselves. Populations are divided into castes (workers, soldiers, and so on). *Architecture* – nest-building by termites and web-spinning by spiders follow distinctive geometric and structural patterns that can be used for taxonomic purposes (termites) or for pharmaceutical drug tests (the orb-weaver spider, *Araneus* – Curtis, 1965). *Guidance Systems* – bees can orient

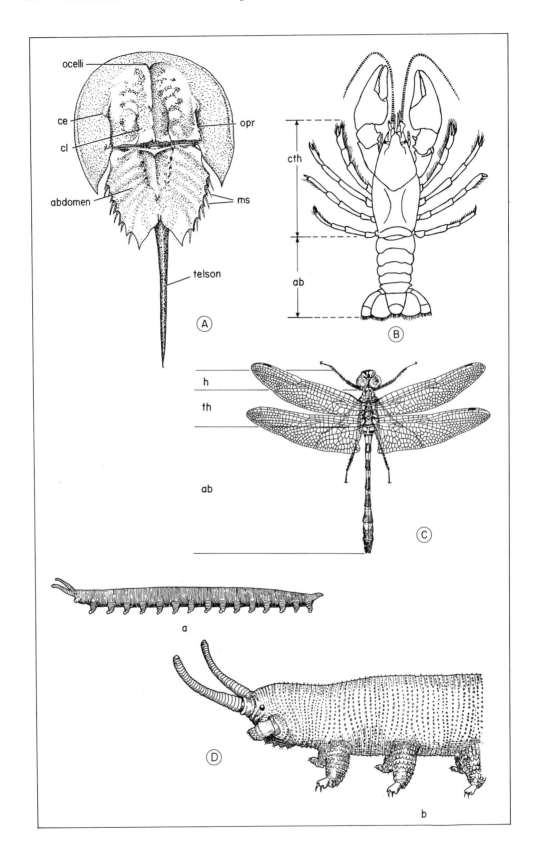

FIGURE 10.1 Some representative living Arthropods (schematic).
(*A*) Living merostome (subphylum Chelicerata), the horseshoe crab, *Limulus polyphemus;* prosoma
(= cephalothorax), abdomen, telson; *cl*—cardiac lobe, *opr*—opthalmic ridge, *ce*—compound eye,
ms—moveable spine, ocelli.
(*B*) Living crustacean (subphylum Mandibulata), the crayfish, *Cambarus bartoni; ab*—abdomen,
cth—cephalothorax.
(*C*) Living dragonfly (Mandibulata, Insecta), *Macromia magnifica; h*—head, *th*—thorax, *ab*—abdomen.
(*D*) Living onchyophoran (superphylum Protarthropoda), *Peripatus novae-zealandiae;* (*a*) complete
individual; (*b*) lateral view of anterior part of body; note antenna, oral papilla, legs, eye.
Illustrations redrawn and modified from (*A*) Stormer, (*B*) Ross, (*C*) Borror and DeLong, (*D*) Snodgrass.

by polarized light (von Fritsch, 1950) and some crustaceans, *Daphnia*, for example, can detect it (Waterman, 1961).

Bees may employ as many as three types of signals (olfactory, optical, and mechanical) to communicate with their hive. Their "language" is so highly developed that Lindauer (1961) investigated 'comparative linguistics" among the taxonomic relatives of *Apis mellifera. Light Production*—among crustaceans three classes (ostracodes, copepods, and malacostracans) contain self-luminous organs (that is, glands manufacture chemicals that support luminescence). *Ecdysis*—Arthropods shed successive "skins" or exoskeletons as they grow.

SHARED BASIC FEATURES AND DIFFERENCES (ANNELIDA-ONCHYOPHORA-ARTHROPODA)

All three groups mentioned display metamerism. This involves segmental repetition of organs derived from body ectoderm and mesoderm. Furthermore, there are homologies (not identities) in musculature, nervous, and circulatory systems.

(1) Arthropods reached a high development by Lower Cambrian time (trilobites, for example). (2) Extant Onchyophora (Figure 10.1*D*) and Arthropoda diverge markedly in anatomical characters. (3) The two known onchyophoran fossils (see below) suggest, moreover, that by Cambrian time, vermiform and segmented primitive onchyophorans had developed

appendages for locomotion. This is to say that they had diverged from creeping ancestral annelids to become walking worms. Items (1) to (3) imply an earlier divergence of arthropods from the onchyophoran-arthropod line of descent.

In evaluating arthropod evolution, consider as a possibility that "arthropodization" could have occurred more than once. Arthropodization embraced several specializations—exoskeletal growth by molting, jointed appendages, striated muscles, among others (Tiegs and Manton, 1958, pp. 255–264, 324). If so, as now seems likely, then arthropods would not be monophyletic—descended from a single line. Rather, they must be polyphyletic—with two or more lines of descent. In each, arthropodization would have occurred independently—a condition known as *convergence.* (cf. Manton, 1973, ref. p. 605.)

The concept that crustaceans in particular had a polyphyletic origin is doubted because of compelling evidence on the primitive living fossil cephalocarid *Hutchinsonella* (see Chapter 11). Sanders (1955, 1963*a*, 1963*b*) has shown how the branchiopod, ostracode, copepod, and primitive malacostracan limb could have been derived from that of the crustacean cephalocarid.

Annelids have a soft, nonchitinous cuticle (see Chapter 9). Both arthropods and onchyophorans have in common a chitinous ectodermal cuticula, although the onchyophoran cuticle is unjointed. Other common features are segmental ambula-

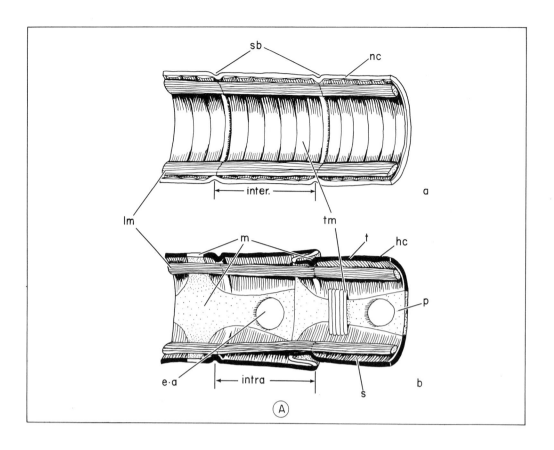

FIGURE 10.2 Somites, Tagma, cephalization in arthropods contrasted with annelids.
(A) Body segments (somites) of annelid (a) and arthropod (b); nc—nonchitinous cuticle, hc—hard
exoskeleton, sb—segment borders, lm—longitudinal muscle. Note intersegmental (inter) position of
lm in (a) contrasted to intrasegmental (intra) position in (b)—see text; tm—transverse muscle of (a),
(b); sclerite (= hardened portion of exoskeleton) divisible into a dorsal tergite (t), ventral sternite (s),
and intermediate membraneous lateral parts (pleurite—p); m—membrane.
(B) Cephalization in polychaete worms and arthropods (on the assumption that first antennae are
prosomal appendages). (a) Adult polychaete, note I–II united; (b) insect embryo with acron in an ar-
chicephalon; may bear two pairs of appendages; (c) theoretical protomandibulate arthropod head in
a protocephalon composed of acron and one somite; (d) chelicerate arthropod in which acron extends laterally and dorsally over several somites united in a prosoma (all figures schematic).
I–IV—somites; Acr (= P)—acron, head, or prostomium; t—tentacles; m—mouth; plp—pedipalp; 1st
Ant.—first antenna, an appendage of the acron; 2 ant.—second antenna, an appendage of first somite; md—mandible; 1 Mx, 2 Mx—first and second maxillae; prnt—pre-antenna; e—lateral eye; lm—
labrum; prtc—protocephalon; chl—chelicera (equivalent to second antenna).
(C) An alternate interpretation of (B), (a) Cephalic segmentation in an annelid-like arthropod prototype; (b) same is generalized typical arthropod (schematic). Note that first antenna is not an appendage of acron.
(D)–(H) Malacostracan Anaspides tasmaniae, Rec. (D), Entire animal, left side, with appendages.
(E) Third left pereiopod and adjoining part of tergum, lateral view. These walking limbs are thought
to have deviated least from original limb structure. Subsequent diversified limbs in (F)–(H) below,
should be viewed as modification of limb figured in (E). (F) Head and anterior part of body, with appendages. (G) Left maxilliped, anterior. (H) (a, b) Posterior view: (a) Left 2nd maxilla, (b) left 1st
maxilla; (c) mandibles and their musculature. II–XVIII—segments; 1 ant., 2 ant.—first and second
antenna; mxpd—maxillipeds, prpds—pereiopods (ambulatory legs), plpds—pleopods, urpd—uropod,
eppds—epipodites; VII, t—tergum, plate VII; ltg (pl)—lateral tergite, pleuron; cxpd—coxopodite,
bspd—basipodite, expd—exopodite, iscpd—ischiopodite, mrpd—meropodite, crppd—carpodite,
propd—propodite, dactpd—dactylopodite, endpd—endopodite; md—mandible; 1 mx, 2 mx—1st and
2nd maxillae; prtc—protocephalon; r—rostrum, pgn—paragnath; endt(s)—endite(s); mdB—base of
mandible; mol—molar process area of mandible; gnl—gnathal lobe of mandible; plp—palp; inc—incisor process of mandible; a—pleural articulation of coxa; lg—intergnathal ligament; da, dp—dorsal
anterior and posterior muscle respectively. Illustrations redrawn and modified after (A) Weber, 1952;
Størmer, 1959; (B) Snodgrass, 1938; (C) Weber; Hupé; Størmer et al., 1959; (D)–(H) Reprinted from
R. E. Snodgrass. A Textbook of Arthropod Anatomy. Copyright 1952 by Cornell University. Used by
permission of Cornell University Press.

474

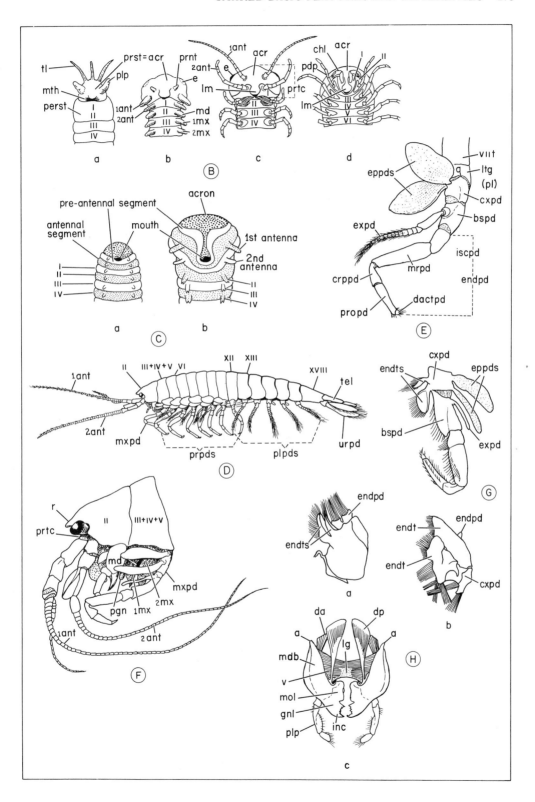

tory appendages (outgrowths of body wall) and segmental excretory organs.

Compared to annelids, arthropods, aside from a distinctive cuticular development discussed later, lack chaetae (but do have bristles), show marked diversification in structure and function of limbs. Limbs are modified into jaws, and further for sensory and reproduction functions, and so on. They also have stout jointed legs. Furthermore, there is a tendency toward cephalization and caudalization in some forms (that is, a tendency for fusion of head and posterior segments respectively). Musculature and nervous system in arthropods are more complex than in annelids.

ARTHROPOD CUTICLE AND ITS SIGNIFICANCE

Of the three groups discussed, arthropods alone underwent a unique cuticular development. (See Chapter 15, Figure 15.9B, for structure of arthropod cuticle.) Hard plates develop in the cuticular layer, and these are separated by areas of flexibility. Once a thickened, hard yet flexible cuticle (exoskeleton) developed, new possibilities for variation opened up for the arthropod. What possibilities?

1. Given: A model with a jointed cuticle, that is, consists of separate hard pieces joined by flexible membranous areas (Figure 10.2A, b). Further, given a continous muscular layer of the body wall (the ancestral annelid condition – Figure 10.2A, a). Problem: How to transform the primitive musculature so that it can move the several hard pieces of the exoskeleton? The solution slowly attained by genetic modification over countless generations, as well as operating natural selection, is simple to relate. Continuous musculature of the ancestral annelid became discontinuous musculature. Longitudinal muscle action, for example, changed from intersegmental (Figure 10.2A, a), as in annelids, to intrasegmental in arthropods (Figure 10.2A, b). When contracted, the longitudinal muscle caused telescoping of body segments (Snodgrass, 1938, p. 79).

2. The successful arthropod invasion

onto dry land may be attributed, at least partly, to antidesiccation and skeletal supportive function of the thickened arthropod exoskeleton. The tendency toward loss of water by evaporation from the body surface is counteracted by the hard cuticle (Borradaile et al., 1958).

In annelids, the postoral segments tend to be effectively monotonous. Some exceptions in polychaetes show a degree of fusion of somites (Figure 10.2B, a). By contrast, in arthropods, grouped somites (= tagmata) (Tiegs and Manton, 1958, p. 267) are a common condition. These may or may not be fused. Fusion of head segments (cephalization) formed the cephalon – one tagma; fusion of posterior segments (caudalization) formed the trilobite pygidium – another tagma. Other fused somites may form a cephalothorax in crustaceans; fused similar somites may characterize a given body portion such as the abdomen in insects. Each grouping formed individual tagma (Figure 10.1).

Two alternate views of cephalization in arthropods (compared to polychaetes) are shown in Figure 10.2B, C. Specialists differ in their interpretations. Snodgrass and others favored the interpretation that the first antenna was a preoral appendage (Figure 10.2B). Tiegs and Weber, and Størmer, among others, favored the view that it was postoral (Figure 10.2C). Hupé's work on Lower Cambrian trilobites led him to favor postoral.[1]

Cephalization in Arthropoda was not a uniform process. Accordingly, the number of post-oral segments that fused to form a distinct head varies: 4–6, Trilobita, Myriopoda, Insecta (with one exception); 6, Chelicerata (Xiphosura, 7); 1, 3 or 4, Crustacea (Størmer).

ARTHROPOD APPENDAGES

In the early arthropod evolution line all legs functioned as ambulatory appendages (Snodgrass, 1952). Subsequently, genetic modification and selection pressures favored diversification of appendages.

We may take the malacostracan crusta-

---

[1]Cisne (1974) supported the "pre-oral" interpretation based on new evidence.

cean *Anaspides tasmaniae* as an example (Figure 10.2*D–H*). The ambulatory limbs (pereiopods, Figure 10.2*E*) are thought to be closest to the unmodified primitive ancestral limb. Essentially these limbs compose a segmented shaft with seven segments. (Names of segments are given in the legend accompanying Figure 10.2.) The final segment is the clawlike dactylopodite. The joint between two segments, the meropodite and carpopodite, marks the *knee bend*.

What modifications of the archetypal walking limb gave rise to the maxillipeds (Figure 10.2*G*) that participate in the feeding function? (1) No change in the number of segments. (2) Reduction of the two epipodites of the coxopodite (see *Eppds* in Figure 10.2*E, G*). (3) Marked reduction of the exopodite (see *Expd* in Figure 10.2*E, G*). (4) Two new endites fringed with long hairs are borne by the maxilliped coxa (not present in the archetypal limb) (see *Endts*, Figure 10.2*G*). (5) The maxillipeds (in *Anaspides*) occupy the position of what should have been the first pair of ambulatory legs. However, along with changes itemized in (2)–(4), the maxillipeds reoriented (turned forward) so as to lie at the sides of the mouth parts (mandible, first and second maxilla) in front of them (Figure 10.2*F*).

What modifications led to the transformation of archetypal walking limbs into mandibles? (See Figure 10.2*H*.) (1) Snodgrass (1952, p. 137) remarked that the three-segmented palpus (*Plp*: Figure 10.2*H*) is visible evidence of reducing the distal portion of a walking limb (see Figure 10.2*E*). (2) Body muscles attached to the mandible base (*mdB* in Figure 10.2*H, c*), responsible for mandibular movement, correspond to coxal muscles of an ordinary leg. In turn, the mandibular base corresponds to the coxopodite (Figure 10.2*E* and *H, c*). (3) The free gnathal lobe of the mandible (*gnL* in Figure 10.2*H, c*) is interpreted to be a *new* coxal endite (which can be seen to be a toothed incisor process— vaguely reminiscent of polychaete worm jaws—an effective biting jaw).

Finally, relative to the maxillae, what necessary modifications changed an archetypal walking limb into such feeding structures? In Figure 10.2*F* it will be seen that the maxillae, positioned in front of the maxillipeds, are suspended from somites III, IV, V (fused into a single plate). In *Anaspides*, but not all crustaceans, maxillae which pass food to the mouth have been simplified structurally, so that they bear no resemblance to a walking limb. (*a*) In the first maxilla (I Max) a reduced coxopodite—*Cxpd*—is present. (*b*) A small lobe (*Endpd* in Figure 10.2*H, b*) may represent a reduced endopodite (see Figure 10.2*E*). (*c*) In the second maxilla (II Max) the movable bilobed segment bearing long hair brushes (*Endpd*) may likewise represent a reduced endopodite (Snodgrass, 1952, pp. 136–137).

Even this limited review of modified anterior walking limbs now serving as parts of the "feeding" apparatus (maxilliped, maxilla, mandible) indicates that *two major tendencies* characterize limb development in the crustacean *Anaspides*: (1) *reduction* and *modification* of walking limb segments; (2) appearance of *new endites* in association with such reduced limb segments.

How can this diversification trend of appendages in arthropods be viewed genetically; how acknowledge the many variants for different classes? First, the ancestral stock originally shared a common gene pool. Subsequently, this became fractionated by migration, and so on. Subgene pools came into existence. Within these, hundreds of thousands, possibly millions, of generations came and went. Mutations, recombination of genetic material (see Chapter 14), and/or invasion by migrants, brought outside genetic material, contributed to the available gene variability at any given time. Contrary selective pressures, of course, operated all along. The modified limbs, reviewed are all head parts, all related to feeding. Clearly, the kinds of available foods and the existing population's capacity to benefit (by foods) exerted selection pressure in preferred directions. Too many appendages specialized for locomotion could have acted at any one stage of development as a different contrary selection pressure.

The crustacean considered has as its principal food the algal slime and organic detritus distributed on rocks and plants, which are its habitat. It will also eat small creatures of its own species (cannibalism), worms, dead insect larvae, and tadpoles. The toothed incisor process of the mandible would be a valuable biting tool for the last-named items on the diet. S. M. Manton (1963, Figure 50) noticed that it is generally agreed that the ability to tackle comparatively large, even hard food "is an advance" on soft-food feeding. Elsewhere (1963, p. 20) Manton stressed that fine-food feeders need a mandible chiefly for rubbing and squeezing but not for biting.

It seems to follow that the need to accommodate to a mixed diet of fine, small, and soft food, along with coarser, larger, hard foods, created strong selection pressures in *Anaspides* precursors. In this regard, various living crustaceans continue more than one mode of feeding (detritus, filter, and in some instances, cannibalism). [Lockheed, Sanders, Manton, see Manton, 1963, "discussion"]. Manton stressed the idea that, at the primitive level, "many functions are done by *every* limb including the mandible: feeding, locomotion, respiration, etc."

The appearance of new endites in modified limbs reflect modifications of the genetic code already favored by natural selection. It was at some of the multiple loci (see Chapter 14) of the twin DNA strands that changes occurred on the molecular level. These were externalized in the loss, fusion, or reduction of walking limb segments, or in the gain of new substructures. The spread of this modified genetic code through successive arthropod populations through time may or may not have been successful in an evolutionary sense; in the final analysis, natural selection, rather more precisely, the resolution of selective pressures decided the choice.

Snodgrass (1952) came to view arthropods as mobile tool chests. The "tools" were their appendages diversified to feed, to reproduce, to swim, to walk, and so on. In this regard, he held, arthropods were unmatched among other animals. Accompanying these evolving built-in "tools" was a high developing nervous system necessary for their use.

Attempts to establish homologies between the jaw (mandible) in all arthropod classes, or between crustaceans and hexapods (Snodgrass, 1952), ended based on untenable premises (Manton, 1963, pp. 112, 136; Figure 61). The jaw mechanism of the Chelicerata, Crustacea, Hexapoda, and Myriopoda (the last two classes probably of common origin) are thought to have evolved independently.

MOLTING CYCLE

As understood by specialists molting is not the mere shedding of the old integument (= ecdysis), but it embraces all steps from preparation for shedding the old integument, the actual shedding, postecdysis linear size increase of the animal, and subsequent tissue growth. Molting too often is construed in the narrower simply exuviation sense (= ecdysis).

In decapod brachyurans, whose molting cycle is best known, Drach (1939) and Passano (1960) recognized four intermolt stages (*A–D*, Table 10.1) preceding ecdysis or shedding of the old skeleton. (Study Table 10.1 with the steps outlined by Travis et al. in Chapter 15 and in Figure 15.9*B*.)

Activity, Feeding, Water Absorption

How active is the given crustacean during the molting cycle? Does it feed continuously? What of water absorption during the cycle?

Activity Level. Stage A—newly molted; activity slight; increases to "full" through Stage D (D$_2$) when exocuticle secretion begins, and reduces to vanishing from D$_2$ to Stage E.

TABLE 10.1. Molting Cycle: Intermolt Stages of a Decapod Crustacean (Brachyura)

Stage	Name	Characteristics	Duration (per cent)	
A	— Newly molted			
A_1		Continued water absorption and initial mineralization	0.5	
A_2	— Soft	Exocuticle mineralization	1.5	2.0
B	— Paper shell			
B_1		Endocuticle secretion begins	3.0	
B_2		Active endocuticle formation, chelae hard; tissue growth begins	5.0	8.0
C	— Hard			
C_1		Main tissue growth	8.0	
C_2		Tissue growth continues	13.0	
C_3		Completion of exoskeleton; membranous layer formed	15.0	
C_4		"Intermolt"; major accumulation of organic reserves	30.0^+	66.0^+
or				
$C_4{}^T$	— Permanent anecdysis	Terminal stage in certain species; no further growth	Permanent	
D	— Proecdysis			
D_0		Epidermal and hepatopancreas activation.	$10.0^+?$	
D_1		Epicuticle formed and spine formation begins	5.0	
D_2	— Peeler	Exocuticle secretion begins	5.0	
D_3		Major portion of skeletal resorption	3.0	
D_4	— About to molt	Ecdysial sutures open	1.0	24.0^+
E	— Molt	Rapid water uptake and exuviation	0.5	0.5

Source. Slightly modified from Drach, 1939; Passano, 1960.

Feeding. Stage A–B (B_1), no feeding; feeding starts at B_2 and continues to Stage D (D_1), thereafter reducing to zero up to Stage E.

Water. Water absorption steadily declines from 86 percent during Stage A (A_2) to 60 percent during Stage D (D_0) and then begins to rise at Stage D_4; thereafter it rises rapidly at Stage E.

What is the comparative duration of the intermolt stages? Seen in Table 10.1, Stage C endures longest (66.0 percent). This is the time of main tissue growth and accumulation of organic reserves. Stage D is next in length of duration (24 percent). (In less heavily calcified forms such as the prawn *Palaemon serratus*, these intervals become reversed: Stage D covers 60-plus percent and Stage C, 21.0 percent of the time.) Substage C_4T sometimes occurs (see Table 10.1)—in which event the growth cycle terminates without further ecdysis (= anecdysis).

Molt-Control Mechanisms

Figure 10.3 schematically shows comparable molt-control mechanisms or pathways for crustaceans and insects. A few preliminaries show why it is thought that the molt-control mechanisms for these two groups are fundamentally similar, differing only in detail (Passano, 1960).

1. *Y*-organ (ovoid paired glands, 1/2 mm in diameter, situated ventrally, lateral to the eye socket in decapod brachyurans). Experiments such as gland extirpation, transplantation, and so on, proved that this gland (*Y*-organ) controls molting in malacostracans.

2. In further demonstrations the *Y*-

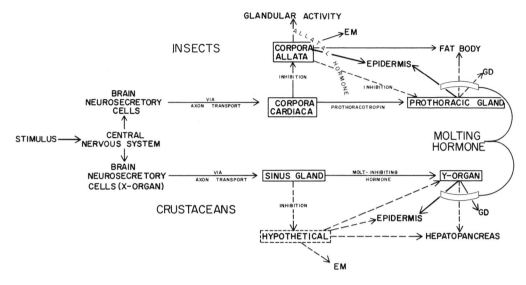

FIGURE 10.3 Organ and secretory controls of arthropod molting (Crustacea; Insecta). Dashed lines = inferred connections; *GD*—gonadal development; *EM*—egg maturation. Adapted from Passano.

organ closely resembled the thoracic and prothoracic glands in certain insects (hemimetabolous types, that is, in which metamorphosis is gradual or incomplete).

3. Both the crustacean *Y*-organ and the insect prothoracic gland cyclically exude neurosecretions (hormones) that correlate with given stages of the molting cycle.

In both insects and crustaceans, molt-control mechanisms are brought into action by some specific stimulus to the central nervous system (CNS).

Insect. (1) Neurosecretory cells of the brain yield a substance that migrates to and remains in the *corpora cardiaca* (= paired organs). (2) Under propitious circumstances, such as light, temperature, or very specific stimuli, this stored substance, a hormone, prothoracotropin, is released and activates the prothoracic gland. (3) Thus activated, these glands release a molting hormone (insect alpha and beta ecdysones). (4) This hormone triggers a sequence of biochemical events including growth of the epidermal tissue. (5) Shortly after event (3), paired endocrine glands (*corpora allata*) are activated and secrete another hormone. This is the allatal hormone, also referred to as the larval/

juvenile hormone. (6) The allatal hormone, if produced in sufficient quantity, checks the tendency of the epidermis once stimulated by the molting hormone in event (4) to grow indefinitely to adulthood. A normal nymphal molt cycle thus occurs. [Preadult growth stages (= instars) in insects are referred to as *nymphs*. Adult stages = *imago*. Imbalance in allatal vs. molting hormones can lead to either precocious miniature imagos or retention of juvenile characters (Passano, 1960).]

(7) The prothoracic glands often, but not always, regress and disappear at the end of the final molt. They may, however, remain partly or fully functional for a lifetime (Apterygota).

Crustacea. (1) Neurosecretory cells of the brain (= *x*-organ) (Carlisle and Passano, 1953) yield material that migrates to and is stored in the Sinus gland. (2) Under propitious stimuli of the CNS, the *x*-gland secretions are released into the blood. (3) This substance, a molt-inhibiting hormone, (Figure 10.3) chiefly functions to inhibit the *Y*-organ defined earlier. (4) The *Y*-organ activated from a hypothetical source releases a molting hormone that is chemically close to insect ecdysomes. (5) In turn, event (4) triggers a sequence that includes gonadal development and epidermal tissue growth. (The molt-inhibiting hormone presumably serves a similar function as the allatal hormone in insects.)

Passano (1960) and other specialists, encouraged by advances in the study of molt control depicted in Figure 10.3 and discussed above, inferred the likelihood of "a single generalized arthropod mechanism."

From the point of view of paleobiology, it is important to have in mind the possibility of imbalance in inhibitor: sponsor molting hormones to account for some anomalies found in fossilized arthropods. One might find juvenile characteristics in adult stages, for example, or a reduced size among some adults, and so forth. In addition, the oldest Cambrian record of trilobites (*Olenellus* zone) provides evidence of regular molting cycles. Presumably, if there is and has been a generalized mechanism, then some variation of hormonal control must have mediated trilobite molting as well. That would place the evolution of the precursors of these biochemical events somewhere in the pre-Cambrian.

BLOOD, EXCRETA, RESPIRATORY ORGANS

Although relevant anatomy details come under the discussion of the several arthropod classes, some striking differences may be mentioned now (Borradaile et al., 1958; Waterman, 1960). Some lower crustaceans and a few insects have hemoglobin, whereas the horseshoe crab, *Limulus*, scorpions, and some spiders, have hemocyanin. Nitrogen excreta consists chiefly of amines and ammonia compounds (Crustacea), urates (Insecta), guanine (Arachnida). Aquatic arthropods breathe through the body integument and/or gills (see epipodites, *eppds* in Figure 10.2E). Among terrestrial forms there are two modes of respiration (lung books and/or trachae).

PALEOECOLOGY

As illustrative types of derivative data available on living arthropods, three selected examples represent the enormous literature. These concern the decapod crustacean, *Cancer* (a true crab), the merostome, *Limulus*, and the role of crustaceans in the marine food chain.

Temperature-Controlled Distribution

MacKay (1943) plotted the worldwide distribution of 19 species of the crab, *Cancer*, in relation to belts of equal surface temperature (Figure 10.9A). His thesis stated that temperature controlled present-day distribution. If so, he reasoned, finds of fossil crabs assignable to the marine genus *Cancer* might permit paleo-temperature determinations by reference to Figure 10.9A). (*Cancer*, apparently, first appeared in Eocene time.)

Limulus

1. Ionic Concentration of Sera/Seawater. Cole (1940) determined the ionic concentrations of Na, K, Ca, Cl, Mg, SO$_4$, in the blood serum of *Limulus polyphemus* compared to the seawater of its habitat (Figure 10.9B). The graph shows how, as the animal moved from marine water to brackish water, each ion in its blood serum varied independently of the other (for example, K).

2. Salinity and Temperature. L. *polyphemus* is found in the estuarine facies from Maine to Yucatan (Shuster, 1957). The salinity tolerance ranges from 11.0 to 30.6 parts per 1000, while the temperature range is 12.5°C to 28.5°C. Depths range up to 12.5 fathoms (75.0 ft). Forms may spend winters buried in bottom mud or sand.

3. Ectocommensals. The large encrusting biota (= ectocommensals) of *Limulus* are absent from all immature and most adult limuli, according to Shuster (1957).

CRUSTACEAN ROLE IN MARINE FOOD CHAINS

Crustaceans play a critical role in the food chain in the sea. In particular, in many seas of the world, copepods (*Calanus finmarchius*) and also euphausiacean crustaceans (Barents Sea, for example) are basic components of fish food. *Calanus* is eaten by herrings, haddock, and various fish fry (Zenkevitch, 1963). Manteufel (1941, cit. Zenkevitch, 1963, p. 104) estimated that in western Murman Peninsula, in 1934, 200,000 tons of herring entered an area. In one year, they consumed an estimated four million tons of *Calanus* plankton. To feed all of Barents Sea herring, according to Zenkevitch, the annually required zooplankton exceeds thousands of millions of tons.] From the gigantic blue whale that feeds on euphausiaceans ("whale krill") to the tiniest fish fry, most aquatic animals, fish and mammal, depend for food on small crustaceans, usually planktonic, sometimes benthonic (McWhinnie, 1964; Schmitt, 1957) (Figure 10.9*C*).

INFORMATION DERIVED FROM FOSSIL ARTHROPODS

Classification: Phylum Arthropoda (arthros = joint + poda = foot)

1. *Metazoans. Segmented, bilaterally symmetrical,* and possess *jointed legs.*
2. *Body cuticle chitinous.* With membranous connections between articulating segments. (Exoskeleton, comparatively hard.) Shed during growth (see "Molting Cycles," above).
3. Somites have paired appendages, variously specialized or reduced.
4. Systems. *Nervous*—brain: anterior, connected to paired ventral nerve cords and ganglia in each somite. *Sensory Organs*—ocelli or compound eye; organs for smell, taste, and touch. *Excretory*—lack true nephridia. *Respiratory*—gill, trachae, book lung, or body surface are means of respiration. *Circulatory*—heart, dorsal with arteries but no veins.
5. Sexes. Generally separate; fertilization chiefly internal.
6. Geologic range. Pre-Cambrian (?); Lower Cambrian–Recent.

The phylum, for convenience, may be subdivided into two supersubphyla: The small Protoarthropoda and the enormous, dense, varied Euarthropoda that contains the bulk of all arthropods, fossil and living.

Protoarthropoda

Vermiform metazoans, lacking rigid exoskeleton and jointed appendages; pair of modified appendages (= mandibles) for feeding. (?)Pre-Cambrian, Cambrian–Recent. [The main forms include *Asheaia* (Figure 10.4*A*), *Xenusion* (Figure 10.4*B*), *Peripatus* (Figure 10.1*D*).] [*Xenusion* is more closely related to the Arthropoda than to the Onchyophora.]

Euarthropoda

Arthropods with hardened, chitinous exoskeleton; segmented body; jointed appendages; body divisible into head, thorax, and abdomen (cephalothorax and so on also occurs due to fusion); larval stages precede adult forms; growth by molting. Systems include circulatory, nervous, respiratory. Sense organs well developed. (?)Pre-Cambrian, Cambrian–Recent.

The supersubphylum, Euarthropoda, embraces four subphyla: Trilobitomorpha (trilobites, and so on), Chelicerata (spiders, and so on), Pycnogonida (sea spiders), Mandibulata (crustaceans, myriopods, insects, and so on). Of these groups three classes have outstanding importance—Trilobita, Crustacea, Hexapoda (subclass Insecta). (Extensive treatment in Chapter 11.) Here, components of the Trilobitoidea, merostomes, arachnids, and sea spiders come under review.

Subphylum Trilobitomorpha. Aquatic arthropods; antennae preoral; trilobite or modified trilobite biramus appendages (Figure 10.4*C*). Lower Cambrian–Middle Permian. Two classes: Trilobita (Chapter 11); Trilobitoidea: *Marrella* (Figure 10.4*D*), *Leanchoilia* (10.4*E*), *Waptia* (10.4*F*), *Cheloniellon* (10.4*G*).

Subphylum Chelicerata. Arthropods of terrestrial, as well as aquatic habit. First head appendages are the chelicerae or pincers (2 to 4–jointed), then the pedipalps (originally ambulatory but secondarily modified for grasping, sensing, or chewing functions). Body plan consists of prosoma (cephalothorax of six postoral segments plus preoral portion) and opisthosoma (= abdomen) of primarily some 12 segments, segment VII is commonly reduced, segment VIII bears the genital opening (Figure 10.4H; Figure 10.1A). Cambrian–Recent.

The chelicerates are divided into two subclasses: Class I, Merostomata; Class II, Arachnida. Representative merostomes include subclass Xiphosura [*Euproops* (Figure 10.4I, a), *Pringlia* (Figure 10.4I, b), *Paleolimulus* (10.8E–F), *Neostrabops* (10.8D), *Mesolimulus* (10.8G)], subclass Eurypterida [*Eurypterus* (10.5A), *Pterygotus* (10.5B), *Megalograptus* (10.5C), *Dolichopterus* (10.5D), *Hughmilleria* (10.5E)]. Representative arachnids include *Palaeopisthacanthus* (Figure 10.6C, compare 10.6A), *Eodiplurina* (Figure 10.6G), *Palaeocharinus* (10.6H).

Subphylum Pycnogonida. Exclusively marine arthropods superficially resemble Chelicerata but differ from them because the third pair of appendages modify as *ovigers* (in the male carries eggs during incubation); anterior part of the body produced as a proboscis; abdomen markedly reduced. Sexes separate but hermaphroditic forms occur. Four pairs of walking legs generally, and four corresponding trunk somites (however, a twelve-legged form is known). [*Nymphon*, Rec. (Figure 10.7A); *Palaeopantopus*, Dev. (Figure 10.7B).] Arthropods of uncertain position: *Palaeoisopus*, Dev. (Figure 10.7C).

Information Derived from Trilobitomorphs (Exclusive of Trilobites)

Preliminary Remarks (Trilobitoidea). Almost all forms that Størmer assigned to the Trilobitoidea came from the same formation (Middle Cambrian Burgess Shale) and the same locality (locality 35k near

Field, British Columbia–see Walcott with note by Resser, 1931). Preserved in thin, carbonized films in shale, the fossils did not represent molts but, rather, complete individuals showed remarkable details of soft-part anatomy. A selected few of the 35,000 fossils from the Burgess Shale are shown in Figure 10.4C–F. The Canadian Geological Survey has made new collections.

Are the Trilobitoidea a Homogeneous Grouping? It is preferable to view this class as erected for convenience to embrace all nontrilobite forms with trilobite-type appendages. However, all nontrilobites with trilobitan limbs are not included. Sanders (1957, p. 118) pointed out that the crustacean cephalocarid had a trilobite-like limb. (Reference, Chapter 11).

Morphology and Habit. Benthonic, pelagic, and planktonic forms are included in the Trilobitoidea. This has interest because all these types together were fossilized in the Burgess Shale. Flat forms greater than 10 cm long (*Cheloniellon*, Figure 10.4G) apparently were bottom scavengers. Shrimplike forms (*Waptia*, Figure 10.4F) were pelagic and apparently used both abdominal limbs and broad, terminal rami for propulsion (Walcott, 1931). *Marrella* (Figure 10.4D) seems to have adaptively modified its exoskeleton for a planktonic existence, that is, the cephalic shield extended as flat horns or floating organs that would retard sinking when the limbs were at rest.

Gills and Oxygen Content of the Burgess Shale Sea. In most trilobitoideans gills developed strongly (see Figure 10.4D–G). Størmer (1959) speculated that this might indicate a low oxygen content of the Burgess Shale Sea. Thus even a form like *Marrella* living in the uppermost (usually oxygen-rich) zone had marked gill development. (See Whittington, 1971.)

Larval Stages. No complete sequence of growth stages is known for any Burgess Shale trilobitoidean. However, Burgess

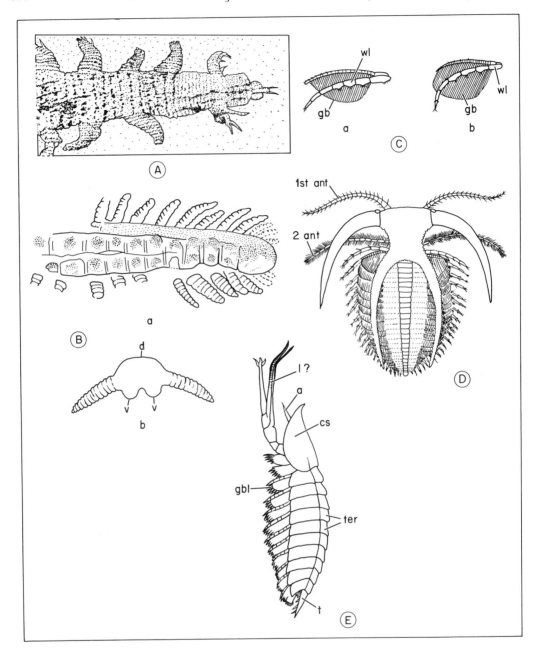

FIGURE 10.4 Onchyophoran, Trilobitoideans and some Chelicerates.

(A) Asheaia pedunculata, Burgess sh, W. Canada, X 4. Note paired appendages and unsegmented body.

(B) (a, b) Xenusion auerwaldae, erratic in glacial drift, Sweden. (a) Reconstruction of animal seen in relief; note paired appendages and rounded prominences in each segment; (b) cross-section, *d* – dorsal, *v* – ventral. One possible interpretation.

(C) Trilobitomorph appendages, diagrammatic: (a) trilobite, *Triarthus;* (b) Marellomorph. *Marrella splendens,* M. Cam. Burgess shale; *wl* – walking leg, *gb* – gill branch.

(D) Dorsal side of (*C, b*), X 3; 1st ant; 2 ant – first and second antenna.

(E) Leanchoilia superlata. M. Cam., Burgess shale. X 0.7; dorsal view, *l?* – first pre-oral limb, *cs* – cephalic shield, *ter* – tergite, *gbl* – gill branches, *t* – telson, *a* – pre-oral antennule.

(F) Waptia feldensis. M. Cam. Burgess Shale, X 1.5; dorsal side, reconstruction.

(G) (a) Cheloniellon calmani, L. Dev., Ger., X 0.4. Note: pre-oral 1st Ant., trilobite-like appendages (that is, walking legs with gill branches); *f* – furca. (b) *Mollisonia symmetrica.* M. Cam., Burgess sh.,

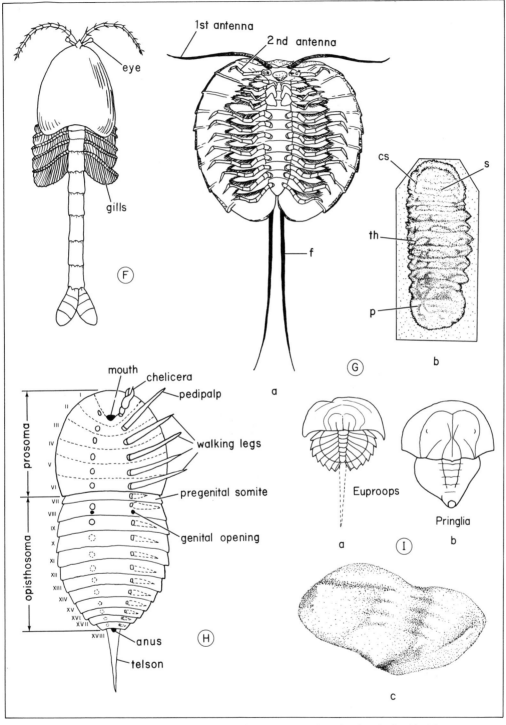

dorsal side; note trilobation and unfused segments (transverse lobes) (s) of cephalic shield (cs); p — pygidium, *th* — thorax, X 0.7.
(*H*) Schematic representation of a chelicerate.
(*I*) (*a, b*) Xiphosurans, (*a*) *Euproops rotundatus,* U. Carb., Eng. X 2.5; (*b*) *Pringlia birtwelli.* U. Carb., Eng., X 2; (*c*) *Pringlia leonardensis,* Perm., Kansas, X 6; note bulbous posterior boss.
Illustrations redrawn after Walcott, Heymons, Størmer, Tasch.

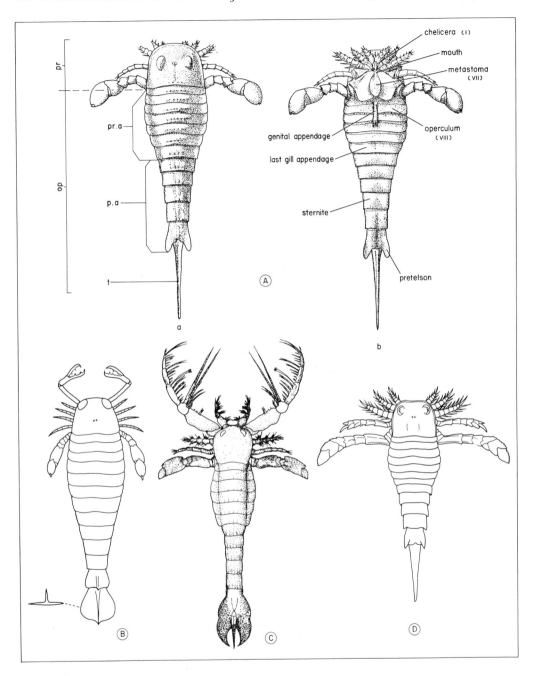

FIGURE 10.5 Eurypterids.
(A) (a, b) *Eurypterus fishcheri.* M. Sil., Baltic. (a) Dorsal view, (b) ventral view, schematic; *pr*—prosoma, *op*—opisthosoma, *pra*—preabdomen, *pa*—postabdomen, *t*—telson.
(B) *Pterygotus (Pterygotus) Rhenaniae,* L. Dev., Ger., X 0.45, dorsal view.
(C) *Megalograptus ohioensis,* U. Ord., Ohio, X 0.12.
(D) *Dolichopterus machocheirus,* Sil., N.Y., X 0.3, dorsal view.

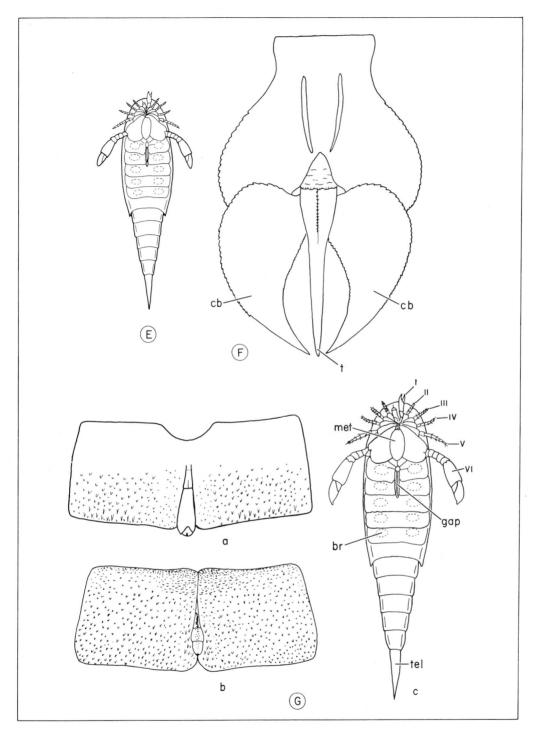

FIGURE 10.5 continued

(*E*) *Hughmilleria norvegica*, L. Dev., Norway, X 0.5, ventral side.

(*F*) Detail of (*C*); *t*—telson; *cb*—cercal blades (these served as protective or copulatory function).

(*G*) (*a–b*) *Megalograptus ohioensis* (Ord., Ohio) (*a*) Male opercula, median appendange (= *ma*) genitalia; (*b*) female opercula compare "*ma*" in (*a*) and (*b*); (*c*) same as "*E*" ventral view, genitalia (*gap*); *br*—gills.

Illustrations redrawn and adapted from: (*A*)–(*F*) Størmer, 1955, after several authors, (*C*) Størmer, 1944, (*G*) Caster and Kjelleswig-Waering, 1946, (*E*).

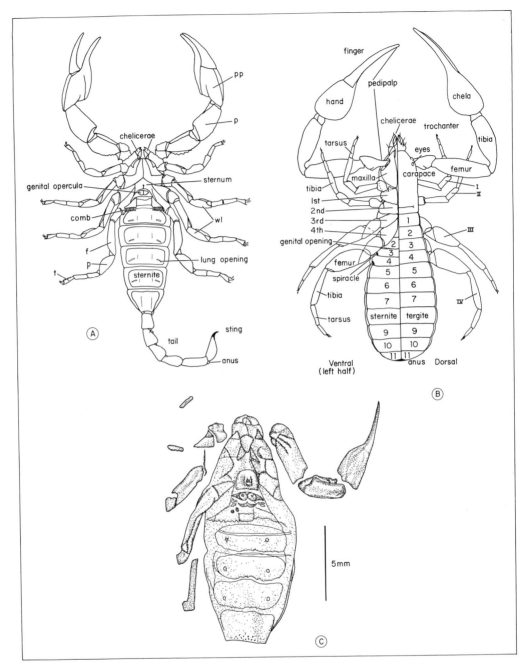

FIGURE 10.6 Arachnids.

(A) Scorpions. *Opisthacanthus leptarus* (Panama, Rec.), ventral side; *pp*—pedipalp, *p*—patella, *t*—tarsus, *f*—femur, *wl*—walking leg.

(B) Pseudoscorpion. *Hesperolpium sleveni* (Rec.), schematic, I–IV—legs; 1st to 4th—coxa.

(C) Fossil scorpion. *Palaeopisthacanthus schucherti* (Mazon Creek, Ill., Carb.), ventral view [compare to (A) above].

(D) Crab Spider. *Misumenopis sp.* (family Thomisidae, Rec.), *pdp*—pedipalp, *l*—legs, *e*—eyes.

(E)–(F) Structural features of spiders (Order Araneida), (E) Ventral view (generalized, schematic); *ch*—chelicera, *pdp*—pedipalp, *ex*—coxa, *fn*—fang of chelicera, *lbm*—labium, *stn*—sternum, *l*—legs, *epg*—epigynum, *spr*—book lung spiracle, *spr₂*—tracheal spiracle; *spn*—spinnerets, *ef*—epigastric furrow. (F) Leg; *cx*—coxa, *tr*—trochanter, *fm*—femur, *ptl*—patella, *tb*—tibia, *mts*—metatarsus, *ts*—tarsus, *tcl*—tarsal claws.

(G) Fossil spider. *Eodiplurina cockerelli* (Oligo., Colo.), X 6.5.

(H) Arachnids (subclass Soluta, Order Trigonotarbida), *Palaeocharinus rhyniensis* (Old Red ss. Scotland). (a) Dorsal side, X 18, (b) sternum, X 60 (distinguished from spiders by marginal plates on abdominal tergites).

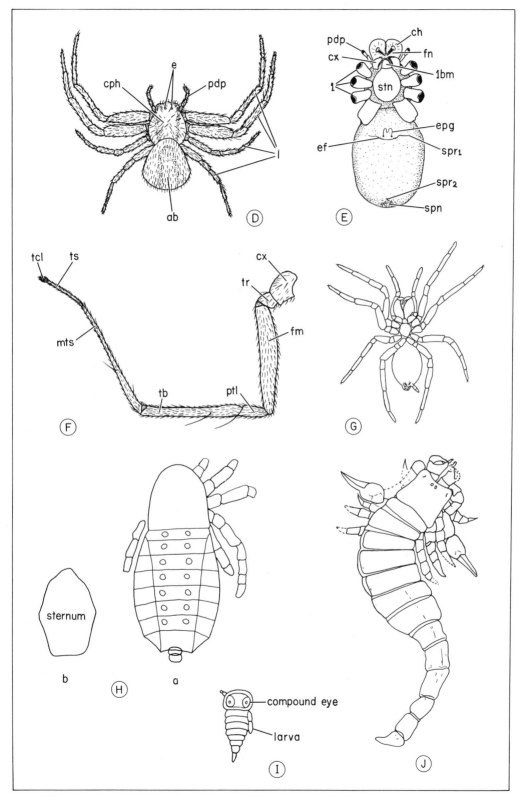

(*I*) Youngest known larval stage of a eurypterid. *Stylonurus myops* (Sil., New York), X 11.
(*J*) Fossil scorpion, *Palaeophonus caledonicus* (Sil., Scotland), schematic.
Illustrations redrawn after Petrunkevitch, Størmer, and others.

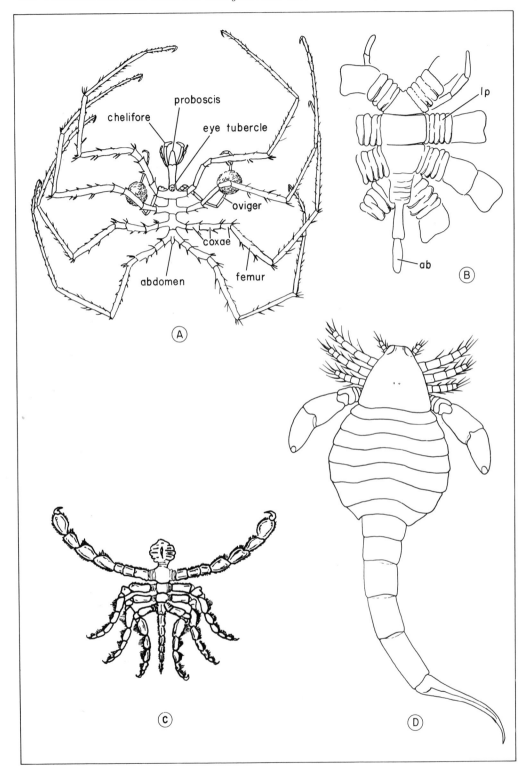

FIGURE 10.7 Living and fossil Pycnogonids, and others.
(A) Living sea spider (subphylum Pycnogonida). *Nymphon rubrum* (diagrammatic, enlarged) (Sars, modified by Hedgpeth).
(B) *Palaeopantopus maucheri* (L. Dev., Hunsrück shale, Ger.), dorsal side of body, proximal side part of appendages, enlarged, *ab* — abdomen, *lp* — lateral processes (Petrunkevitch cit. Hedgpeth).
(C) Of uncertain position (*Incertae sedis*); *Palaeoisopus problematicus* Broili (L. Dev., Hunsrück shale. (Ger.), schematic; Incorrectly assembled by Broili and corrected by Dubinin, 1957, Doklady Akad. Nauk. U.S.S.R., vol. 117, pp. 881–884. (Dr. Wilhelm Stürmer sent two photographs clarifying the position of the claws).
(D) Eurypterid, *Carcinosoma scorpionis* (Sil., New York), dorsal view of male, reconstructed. Compare Figure 10.6 *J* (Clarke and Ruedemann, cit Størmer).

Shale forms like *Mollisonia* (Figure 10.4*G*, *b*), with five unfused transverse cephalic lobes, may represent an early larval stage (see Trilobite Ontogeny, Chapter 11).

Primordial Arthropod Fauna. Tiegs and Manton (1958) stressed that fossils like *Marrella* and *Cheloniellon* bore witness to a primordial arthropod fauna probably the ancestral pre-Cambrian stock(s) of Crustacea and Trilobita. [Sanders (1955) decided that the trilobitan limb of the crustacean cephalocarid *Hutchinsonella* did not support the independent origin view of crustaceans and trilobites.] This primordial fauna was characterized by serially uniform trilobite-like limbs, lack of jaws, presence of large labrum — features associated with particle and detritus feeding.

It is interesting that forms like *Burgessia*, *Cheloniellon*, and *Waptia* combine crustacean and noncrustacean features, whereas other Burgess Shale fauna are trilobitelike (*Leanchoilia*, and so on). Despite these differences, *all* forms show either trilobitan or modified trilobitan appendages.

The concept of "mosaic evolution" was invoked to explain primitive and advanced characters present in the same trilobitoideans (Tiegs and Manton, 1958). After all, some forms seem to give a preview of trilobites while others, of crustaceans or merostomes. How can this be explained? Mosaic evolution refers to the unequal evolution of the components of a type [which is a mosaic of primitive/advanced, and general/specialized features (Mayr, 1963)[2]]. In a shift to a new adaptive zone, varied selection pressures may favor

a given structure — the trilobitan limb, for example. Accordingly, this and related characteristics may be acquired repeatedly and independently in stock after stock.

MEROSTOMES

Exoskeleton

The available information on merostome exoskeleton composition is limited. Lockheed (1955) spoke of the xiphosuran exoskeleton as uncalcified and made leathery by a high content of scleroprotein. The thinness of the xiphosuran test has been mentioned (Raymond, 1944). Cayeux (1916, cit. Vinogradov, 1953, p. 45) found a lack of calcium carbonate in the *Limulus* cephalothorax and saw that, largely chitinous externally, it contained calcium phosphate internally. Other investigators have shown the presence of P, S, and the halogens, Cl, Br, I. Vinogradov (1953) found 0.05 percent ash in the carapace dry matter of *Limulus polyphemus* a very small amount compared to the decapod crustaceans, for example). This ash contained 0.1 to 0.2 percent CaO but little phosphate (X-ray analysis failed to record apatite or any crystalline phosphate).

From scattered data, it is likely that the exoskeleton of xiphosurans, both living and fossil, is composed of a chitin-protein complex (see Chapter 15) with minor salt impregnations and trace halogen amounts.

The Upper Cambrian aglaspids by qualitative chemical tests were found to possess a phosphatic exoskeleton (Raasch, 1939). The eurypterid exoskeleton (or the exuviae of successive molts) tested chitinous (Caster and Kjellesvig-Waering, 1964; Størmer, 1955).

[2]Reference, Chapter 14.

Evolutionary Trends

In xiphosurans there was a marked increase in size through time. Smaller forms occur in the Paleozoic (3 to 5 cm), somewhat larger forms in the Mesozoic, and the largest in the Cenozoic (30 to 60 cm). Modern immature limulids range in size from 45 to 75 mm. As in other arthropods, xiphosurans from Paleozoic to Mesozoic time experienced abdomen shortening and a trend toward gradual abdominal segment fusion.

Eurypterid size increased in time although large forms occurred all during the Ordovician-Permian. Some Devonian forms attained the size of 180 cm (*Pterygotus*—which Størmer labeled the largest known arthropod). Limb modification from walking legs to swimming legs and grasping organs, as well as elaboration of ornamentation (scales, tubercles, knobs, spines), showed other evolutionary tendencies. Clarke and Ruedemann (1912 cit. Størmer, 1944) thought that the strong development of scales, both coarse and pointed, in *Lepidoderma* (= *Anthroconectes*) denoted racial senescence.

In some eurypterids, tail development was marked—an apparent adaptation for greater abdomen flexibility in swimming (Tiegs and Manton, 1958). *Pterygotus*, for example, had its telson modified into a tail fin and rudder to facilitate swimming. Other curious specializations evolved, as in the caudal armature of *Megalograptus*. This armature consisted of scimitar-like cercal blades capable of considerable muscle-controlled vertical movement (Figure 10.5*F*). The structure was apparently a protective or a copulatory device (Caster and Kjellesvig-Waering, 1964).

Habitats

Xiphosurans migrated from pure marine to brackish/fresh-water environments during Devonian time. By Permo-Carboniferous time, many species were freshwater forms. However, marine, brackish, and fresh-water forms are known during the Mesozoic. Today limulids are exclusively marine although one genus tolerates estuarine and river water.

Caster and Kjellesvig-Waering (1964) view the eurypterids and all merostomes as "primordially and typically marine." Kjellesvig-Waering (1961) recognized two marine and one brackish water ecologic setting in which Silurian eurypterids lived. Størmer (1955) noted that eurypterid remains are sparse in marine deposits and generally occurred in fresh/brackish water deposits. Sparse eurypterids and a few xiphosurans (*Paleolimulus*, *Anacontium*, and so on) occur in fresh/brackish water facies in the Wellington Formation of Kansas (Dunbar, 1923; Raymond, 1944; Tasch, 1961, 1963).

Raasch (1939) made an exhaustive study of aglaspids found with typically marine trilobite/brachiopod faunas. He arrived at the conclusion that *Aglaspis* had similar habits and habitats as modern *Limulus*.[3]

Eurypterid Origin?

The genus *Paleomerus* (Lower Cambrian, Sweden) (Figure 10.8*H*) combines xiphosuran/eurypterid characters. The loose articulation of tergites and compound eyes are aglaspid characters, while the abdominal segments is an eurypterid feature. According to Størmer (1955), this Lower Cambrian form represents a transitional stock linking the Cambrian aglaspid-xiphosuran to the Ordovician eurypterid. A primitive eurypterid such as *Brachyopterus stubblefieldi* (Ord., Wales) (Figure 10.8*I*) with an aglaspid-like body and short walking legs might have been the transition-type between a *Paleomerus* aglaspid and the more specialized eurypterids with diversified limbs and a more pronounced tail (Tiegs and Manton, 1958, p. 312).

Larval Stages

Larval stages of fossil xiphosurans are inferred principally from the embryo-

[3]Briggs, Bruton and Whittington, (1979, Paleontology v. 22 (1) (167–180) restudied appendages of *Aglaspis* and concluded it is neither a merostome, any other chelicerate, nor related to trilobites.

logical researches of Iwanoff (1932–1933) on living *Limulus*. The earliest larval stage of a living limulid is four-segmented. The first free larval stage has often been called "the trilobite larva." It is reminiscent of a mature *Euproops* in configuration (Figure 10.8*B*, compare 10.4*I, a*). Attempts to relate the limulid larva to that of the trilobites have not succeeded. Limulid larva differ from trilobite larva in their unsegmented abdomen (that is, segments are fused), limulid limbs, and head structure (Tiegs and Manton, 1958, p. 301).

Little is known about the ontogeny of eurypterids. However, size distributions are known in fossil collections, that indicate younger and older growth stages. The youngest larva known (2.0 to 3.0 mm long) is that of *Stylonurus myops* (Silurian of New York) (Figure 10.6*I*). It has fewer abdominal segments than the adult. Its compound eyes were apparently seated on ovate nodes. The growth from larval to early adult stages can be appreciated by comparison of a postlarval stage with the larval stage. In *Stylonurus dolichopteroides* (L. Dev., Norway) the length of the prosoma is more than five times that of a larval form (Figure 10.6*I*).

Substrates, Biotas

Eurypterids are found in shales, limestones, dolomites, sandstones, and siltstones. *Hughmilleria* and *Pterygotus* were found in dark grey limestone of the Wills Creek Formation (Silurian, West Virginia), associated with ostracodes and lingulid brachiopods. Typical marine forms are fossilized with eurypterids of the Mocktree Shale, of the Welsh Borderland (Kjellesvig-Waering, 1961). These include trilobites, starfish, bryozoans, articulate and inarticulate brachiopods.

Three ecological groupings have been suggested for the eurypterids found in the Welsh Borderland: (1) A completely marine association—rich marine biota plus the eurypterids of the Carcinosomatidae and Pterygotidae. (The Upper Ordovician eurypterid *Megalograptus* from Ohio belongs to this normal marine phase.)

(2) The intermediate group between normal marine and brackish. The Eurypteridae belong to this facies. Shales, silts, limestones, biogenic/organoclastic shelly facies—characterize some lithologies. Eurypterid faunas of New York State, Northern Ohio, and the Baltic area lived in such contexts. (3) Fresh—brackish to marine group—this environment was characterized by clastics. Hughmilleridae and Stylonuridae lived in such a sedimentary setting. (However, *Hughmilleria* also occurred in limestones, and hence one must also envision both clay and lime mud substrates.)

Xiphosurans are known from sandstones (Upper Devonian, Pennsylvania; Old Red Sandstone), fresh-brackish argillites, carbonaceous shales, Tertiary lignites, and so on. Some forms are common in the Coal Measures (*Pringlia, Euproops*) together with fossilized plants. In the fresh-brackish water faunas of the Kansas Wellington Formation, insects and xiphosurans are biotic associates.

Raasch (1939) found that the merostome aglaspids came from the predominantly arenaceous St. Croixan deposits of the Upper Mississippi Valley (an outcrop belt with a 400-mile maximum diameter). Common biotic associates were trilobites, brachiopods and, less frequently, crinoids represented by fragments and graptolites, among others.

Limulid Spoor

Even in the absence of actual fossilized remains of a limulid xiphosuran, it is possible to infer their presence from tracks left in the sediments. The Jurassic Solenhofen Plattenkalk bears the well-known trails of *Limulus walchi*. Caster (1938) restudied tracks in the shale band of a flagstone (Upper Devonian of Pennsylvania) attributed to an amphibian. He showed that they were made by a merostome, probably a close relative of the Devonian xiphosuran *Protolimulus*.

Predation/Cannibalism

Caster and Kjellesvig-Waering (1964) found what they interpreted to be the

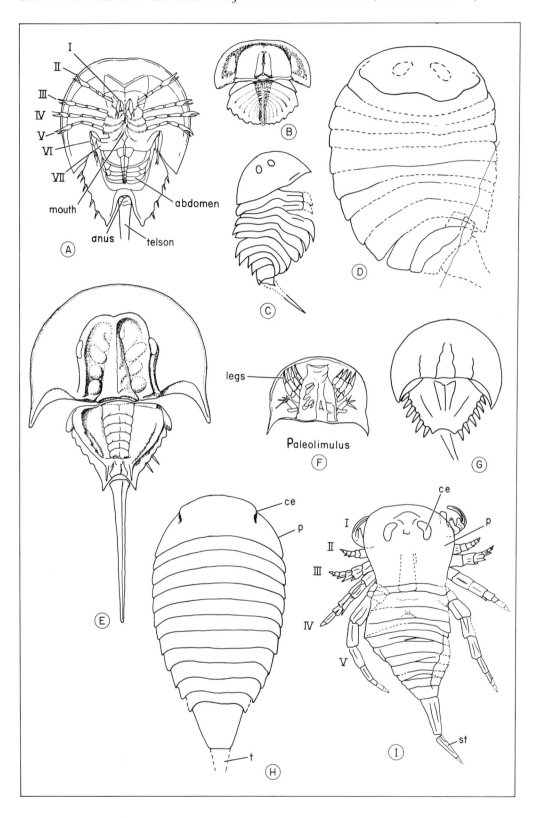

FIGURE 10.8 Anatomical features of merostomes.
(A) *Limulus polyphemus,* ventral view, appendages, and so on (compare Figure 10.1A), schematic. I, chelicera; II–VI, walking legs (modified after Størmer).
(B) *Tachypleus gigas,* first free larval stage (= *Euproops* or "trilobite" larva), compare Figure 10.4I, a (after Iwanoff).
(C)–(D) Family Aglaspidae. (C) *Aglaspis simplex* (U. Cam., Wisc.), dorsal view, X 0.7 (after Raasch).
(D) *Neostrabops martini* (U. Ord., Ohio), X 0.75 (after Caster and Macke, cit. Størmer).
(E)–(F) *Paleolimulus avitus* (Perm. Kans.) (E) Dorsal view, reconstruction, schematic (after Dunbar).
(F) Ventral view, chelicerae and chelate legs, X 1.
(G) *Mesolimulus walchi* (Jurassic, Ger.), X 0.4. (Kirchner cit. Størmer).
(H) *Paleomerus hamiltoni* (L. Cam., Sweden), dorsal view, X 0.8; *t* – telson, *ce* – compound eyes, *p* – prosoma (after Størmer).
(I) *Brachyopterus stubblefieldi* (Ord., Wales), dorsal view, X 1.5. Note aglaspid-like body (compare with *H*) – a primitive condition. (Thus, from a possible transitional type to more specialized eurypterids via *Paleomerus*-type ancestor.) I–V – appendages; I–III, walking legs; I – modified into clasping organs; *ce* – compound eye; *p* – prosoma; *st* – styliform telson (after Størmer).

coprolites of a eurypterid *Megalograptus* associated with the latter. These objects contained undigested eurypterid fragments along with the *Megalograptus.* The investigators, realizing the prevalence of postmating cannibalism among living scorpions and spiders, suggested a similar explanation for these unusual coprolites.

Sex Determination in Eurypterids

The genital appendages and opercula of eurypterids differ sufficiently to permit sex determination [(*Eurypterus fischeri, Pterygotus* (*P.*) *rhenoniae* (Størmer, 1944, Figure 9), *Megalograptus ohioensis* (Caster and Kjellesvig-Waering, 1964, and this chapter, Figure 10.5C, G)].

ARACHNIDS

Main Evolutionary Trends

As in other arthropods, two major trends characterize arachnid evolution. Segments were (1) fused, and/or (2) lost. Among the scorpions, the five posterior segments were not lost; rather, they were retained as a tail in an attenuated condition (Figure 10.6A). Among the spiders (order Araneida) primitive families had abdomens composed of 12 distinct segments. They were reduced to 5 or 6 in other families, and completely fused in adults. Among mites and ticks (Order Acarida) abdominal segmentation was lost and the body showed no external segmentation.

Fossil Record

Some of the oldest Paleozoic arachnids include the scorpion *Palaeophonus* (Silurian of Gotland) (Figure 10.6J), the acarid *Protocarus crani* from the Rhynie Chert of Scotland (Devonian, Old Red Sandstone), and Carboniferous spiders with segmented abdomens and presumably four pairs of spinnerets – from which webs' silk was spun – such as *Arthrolycosa* and *Protolycosa.*

Petrunkevitch (1953) concluded that by Silurian time all orders of Arachnida were either already extant or in the process of evolving.

Post-Paleozoic beds yield a negligible arachnid record. This persisted until Tertiary-Quaternary time. The Paleozoic surpasses the Mesozoic in fossil arachnids (Petrunkevitch, 1955). Famous arachnid-bearing deposits and sites include Baltic Tertiary amber faunas, Aix-en-Provence deposits, Colorado's Florissant insect beds.

Lithologies

Besides preservation in amber – where the entire arachnid is preserved intact – remarkable preservations are found in lithographic limestones (Jurassic, Germany), European and American Coal Measures (Mazon Creek, for example). Some arachnids have been recovered from Carboniferous ironstones, Tertiary onyx marble, Devonian chert – among other lithologies.

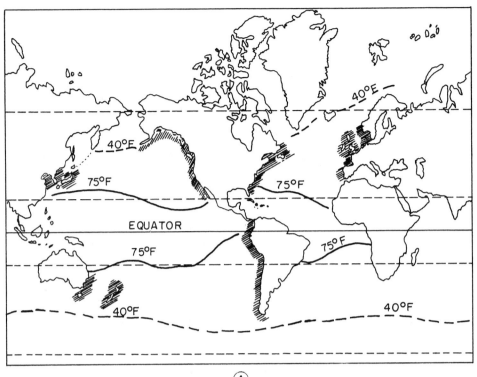

(A)

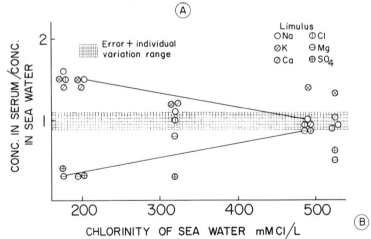

(B)

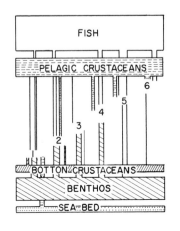

(C)

FIGURE 10.9 Derivative data from living arthropods.
(A) Modern distribution of decapod crustacean *Cancer* (shaded area). Mean surface water isotherm shown. After McKay, 1943.
(B) Composite salinities of blood serum of *Limulus polyphemus* and seawater of their habitat. After Cole (1940) cit. Shuster, 1957.
(C) Role of crustaceans in diet of chief commensal fish of Barents Seas. Left to right: 1, haddock; 2, *Anarchishos*, 3, sand dab; 4, ray; 5, cod; 6, sea bass. [1–3, benthos feeders; 6, plankton eater; 5, about 15 + percent crustaceans in diet but chiefly pelagic fish]. After Zenkevitch, 1963.

Skeleton

A solid, chitinous exoskeleton encloses the arachnid body. Exoskeleton in some arachnids is tri-layered (former order Pedipalpi assigned to three orders): outer — thin pigmented layer; basal — hypodermic layer; internal — stratified layer. Scorpions have the pigmented layer below a lighter outer one. In acarids the laminated layer is pierced by fine canals vertical to the surface (Størmer, 1944).

Scorpions and Eurypterids

Silurian scorpions such as *Palaeophonus* closely resemble eurypterids like *Carcinosoma scorpionis* (Petrunkevitch, 1955; Størmer, 1944). In the past, scorpions were considered descendants of eurypterids. Doubtless there is a superficial resemblance between scorpions and certain eurypterids. Størmer (1945, p. 63) pointed to the eurypterid structure of the walking legs of particular early scorpions. He also compared the joint occurrence of Silurian scorpions and eurypterids. However, such morphological homologies are actually an example of convergent evolution of separate lines of descent, rather than representing a common descent line. Petrunkevitch (1955, p. 44) pointed to many fundamental differences between scorpions and eurypterids—among others, the position of the mouth.

SEA SPIDERS AND PROBLEMATICA

Pycnogonids vs. Chelicerates

Sea spiders (subphylum *Pycnogonida*) (Figure 10.7A) differ from Chelicerates (subphylum Chelicerata) in striking ways: among others, ovigers, abdomen, proboscis. The third pair of appendages are modified as ovigers; the abdomen is reduced, almost vestigial; a proboscis develops in the anterior portion of the body (Hedgpeth, 1955). There are, nevertheless, resemblances between pycnogonids and arachnids, for example, chelicerae, legs, eyes, feeding, and digestion (Snodgrass, 1952).

Fossil Record

Despite 500 extant species, 100 in a single genus, *Nymphon*, there remains only one fossil species, *Palaeopantopus maucheri* (Figure 10.7B). This fossil is represented by two specimens from the Hunsrück shale (L. Dev., West Germany). Note that this species has a body with six somites, no visible proboscis, and a small segmented abdomen. *Palaeoisopus*, also from the Hunsrück shale, is a problematical arthropod of uncertain affinities.

QUESTIONS

1. Discuss the arthropod body plan (see Figure 10.1); indicate what factors have made it successful.

2. How are annelid worms and arthropods related?

3. What is arthropodization? How does it bear on such questions as mono- vs. polyphyletic origin and convergent evolution?

4. Discuss the structure and significance of the arthropod cuticle.

5. Study Table 10.1, Figure 10.3, and Figure 15.9B (Chapter 15) and outline the major molting cycle events of a living decapod crustacean and an insect.

6. What bearing does your answer to question 5 have on the fossil record of arthropods?

7. Indicate the kinds of recoverable data available on living merostomes, crustaceans in the food chain, and oxygenation in the Burgess Shale sea.

8. How do eurypterids differ from xiphosurans and arachnids?

REFERENCES

Borradaile, L. A., and F. A. Potts et al., 1958. The Invertebrata, 3rd Ed., rev. G. A. Kerkut, Cambridge Univ. Press, pp. 389–396.

Bruun, A. F., 1956. Animal life of the deep sea bottom, in A. F. Bruun, ed., The Galathea Deep Sea Expedition. MacMillan, pp. 149–195.

Carlisle, D. B. and L. M. Passano, 1953. The X-organ of Crustacea: Nature. Vol. 171, pp. 1070–1071.

Carthy, J. D., 1965. The behavior of arthropods. Univ. Rev. Biol. W. H. Freeman, 134 pp.

Caster, K. E., 1938. A restudy of tracks of *Paramphibius*: J. Paleontology, Vol. 12 (1), pp. 3–60.

———— and E. N. Kjellesvig-Waering, 1964. No. 32. Upper Ordovician eurypterids of Ohio: Palaeontographica Americana, pp. 301–342.

Curtis, Helen, 1965. Spirals, spiders, and spinnerets: Amer. Scientist, Vol. 53 (1), pp. 52–58.

Drach, P. 1939. Mue et cycle d'intermue chez Crustacés Décapodes: Ann. instit. océanogr. (Paris) (n.s.), Vol. 19, pp. 103–391.

Dunbar, C. O., 1923. Kansas Permian insects: Part 2. *Palaeolimulus*, a new genus of Paleozoic Xiphosura, with notes on other genera: Amer. J. Sci., 5th Ser., Vol. 5, pp. 443–454.

Fritsch, K. von, 1950. Bees: their vision, chemical sense, and language. Cornell Univ. Press, New York.

Heymons, R., 1928. Über morphologie und Verwandtschaftliche Beziehungen des *Xenusion auerwaldae* Pomp. aus dem Algonkium: Z.f. Morphol. u. Ökol. d. Tiere, Vol. 10, pp. 307–329.

Iwanoff, P. P., 1932–1933. Die embryonale Entwichlung von *Limulus moluccanus*: Zool. Jahrb., Vol. 56, pp. 164–343.

Kjellesvig-Waering, E. N., 1961. The Silurian eurypterids of the Welsh Borderland: J. Paleontology, Vol. 35, pp. 781–835, 5 plates, 45 text figures.

Lindauer, M., 1961. Communication among social bees. Harvard Univ. Press, 131 pp.

Manton, S. M., 1963. Jaw mechanisms of Arthropoda with particular reference to the evolution of Crustacea, in Phylogeny and evolution of Crustacea. Sp. Publ., Mus. Comp. Zool. (Harvard), pp. 111–140 (see discussion, pp. 141–144).

MacKay, D. C. G., 1943. Temperature and world distribution of crabs of the genus *Cancer*: Ecology, Vol. 24 (1), pp. 113–115.

McWhinnie, M. A., 1964. Temperature responses and tissue respiration in Antarctic Crustacea with particular reference to the krill *Euphausia superba*, in Biology of Antarctic Seas: Antarctic Res. Ser., Vol. 1, pp. 63–72.

Passano, L. M., 1960. Molting and its control, in T. H. Waterman, ed., The physiology of Crustacea. Academic Press, Vol. 1, pp. 473–536.

Petrunkevitch, A., 1955. Arachnida: Treat. Invert. Paleontology (P) Arthropoda 2, pp. P42–P162.

————, 1953. Paleozoic and Mesozoic Arachnida of Europe. G.S.A. Mem. 53, 128 pp., 58 plates.

————, 1949. A study of Paleozoic Arachnida: Trans. Conn. Acad. Arts and Sci., 1949, Vol. 37, pp. 69–315.

Raasch, G. O., 1939. Cambrian Merostomata: G.S.A. Spec. Paper 19, 116 pp.

Raymond, P. E., 1944. Late Paleozoic xiphosurans: Mus Comp. Zool. (Harvard) Bull., Vol. 94 (10), pp. 475–508.

Sanders, H. L., 1955. The Cephalocarida, a new subclass of Crustacea from Long Island Sound: Nat. Acad. Sci. Proc., Vol. 41, pp. 61–69.

Schmitt, W. L., 1966. Crustaceans. Univ. Michigan Press, Ann Arbor, 194 pp.

————, 1957. Marine Crustacea (except ostracodes and copepods): Treat. Ecol. and Paleoecol. G.S.A. Mem. 67, Vol. I, pp. 1151–1159.

Shuster, Jr., C. N., 1957. Xiphosura (with especial reference to *Limulus polyphemus*. Treat. Ecol. and Paleoecol., G.S.A. Mem. 67, Vol. I, pp. 1171–1173.

Snodgrass, R. E., 1952. A textbook of arthropod anatomy (Facsimile issued by Hafner Publ. Co., 1965, 339 pp.)

————, 1938. Evolution of the Annelida, Onchyophora, and Arthropoda: Smithsonian Misc. Coll., Vol. 97 (6), 149 pp.

Størmer, L., 1958. "Chelicerata," "Merostomata:" Treat. Invert. Paleontology. (P) Arthropoda 2: P1; P4.

————, 1952. Phylogeny and taxonomy of fossil horseshoe crabs: J. Paleontology, Vol. 26 (4), pp. 630–639, 3 text figures.

————, 1944. On the relationships and phylogeny of fossil and Recent Arachnomorpha: Skrifter Det. Norske Videnskaps-Akademi, Oslo, Vol. 1 (5), 158 pp.

Tasch, P., 1963. Paleolimnology, Part 3. Marion and Dickinson Counties, Kansas, with additional sections in Harvey and Sedgwick Counties; Stratigraphy and Biota: J. Paleontology, Vol. 37 (6), pp. 1233–1251, plates 172–174.

————, 1961. Paleolimnology, Part 2. Harvey and Sedgwick Counties, Kansas; Stratigraphy and Biota: J. Paleontology, Vol. 35 (4), pp. 836–865, plates 97–98.

Tiegs, O. W., and S. M. Manton, 1958. The evolution of the Arthropoda: Biol. Bull., Vol. 33, pp. 255–337, 18 figures.

Walcott, C. D., 1931. Addenda to description of Burgess Shale fossils: Smithsonian Misc. Coll., Publ. 3117, Vol. 83 (3), pp. 1–44, plates 1–23, figures 1–11 (ed. C. E. Resser).

Waterman, T. H., 1961. Light sensitivity and vision, in The physiology of Crustacea. Academic Press, Vol. 2, pp. 1–64.

Whittington, H. B., and W. D. I. Rolfe (eds.), 1963. Phylogeny and evolution of Crustacea: Spec. Publ., Mus. Comp. Zool. (Harvard): 192 pp.

Zenkevitch, L., 1963. Biology of the seas of the U.S.S.R. Interscience, N.Y., 841 pp.

(Whittington H. B. 1971. Redescription of *Marrella splendens* (Trilobitoidea) (abbrev. title). Geol. Surv. Canada, Bull. 209. (H. B. W. concluded that this Burgess shale species was benthic in habit (possibly planktonic in early larval stages) and a particle/detritus feeder). (Cisne, J. L., 1974. Trilobites and the origin of arthropods: Science, Vol. 186, pp. 3–18. When greater "energetic efficiency" became an important selection factor, arthropods developed fossilizable hard parts through a shift from body support by the body cavity to the cuticle. The cuticle was later strengthened to become the exoskeleton.)

chapter eleven
THE SKELETON SHEDDERS (ARTHROPODS – II)

TRILOBITA

Introduction

Trilobita, Crustacea, and Hexapoda (Insecta) are, geologically speaking, the most important arthropods for biostratigraphy and interregional correlations.

Trilobites

Trilobites were described in the seventeenth century, although soft-part anatomy remained unknown until the last part of the nineteenth century. They were classed as merostomes, crustaceans, or even mollusks.

Class Trilobita has several unusual factors: (1) It has been extinct since close of Paleozoic. (2) It arose and evolved from Cambrian to Permian time – a span of some 300 to 400 million years. (3) During the entire Paleozoic era, 10,000 species evolved (classed in about 1500 genera). They are now placed in seven orders (plus one or two uncertain orders). (4) These species, all marine, inhabited Paleozoic seas of every continent including Australia, and Antarctica (Saul, 1965).

Trilobite Affinities

Størmer (1944)[1] found four major characteristics common to Trilobita, Arachnida, Xiphosura, and Cambro-Devonian arthropods: (1) trilobation of dorsal shield (also indicated in eurypterids); presence of well-defined head shield, and tendency to develop styliform telson; according to Iwanoff (1933), the telson of *Limulus* is homologous with median dorsal spines in

[1]Ref. Ch. 10.

primitive trilobites (see Figure 11.1*A*, *B*); (2) presence of four postoral, larval, or primary somites (see Chapter 10); (3) appendages of postoral somites are either trilobitan limbs or modifications of this type (see Chapter 10); (4) the intestinal diverticulae (pouches or glandular ramification of intestines – see Figure 11.1*C–E*) are very strongly developed.

These shared characteristics led to the conclusion that these groups all belonged to a common Arthropoda branch – the superphylum Euarthropoda.

General Morphology; Trilobation

Figure 11.2*A* – dorsal view – indicates the longitudinal trilobation that gave trilobites their name; the central segmented lobe is the axial lobe (*ax*) and on either side of it are the pleural lobes (*pl*). At right angles to the trilobite organization are transverse divisions: anterior (cephalon), median (thorax), posterior (pygidium).

Cephalon. Cephalon or head region is a rigid tagma (see also Chapter 10) formed by fusion of several anterior somites. The cranidium is the rest of cephalon minus free cheeks (*fc*) that separate during molting (Figure 11.2*A*, dorsal view). It includes glabella and area on either side bounded by sutures (*fs*).

Ventral side of the cephalon has three skeletal structures: rostral plate, hypostoma (Figure 11.2*D*), and metastoma (Figure 11.2*A*, *m*). A single rostral-hypostomal plate can be formed by fusion, or the rostral plate and hypostoma can be attached by a hypostomal suture (Figure 11.3*D*), stalk, or peduncle (*Paedeumias*). Free hy-

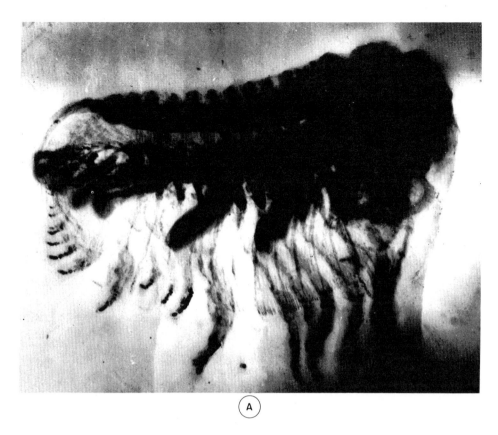

(A)

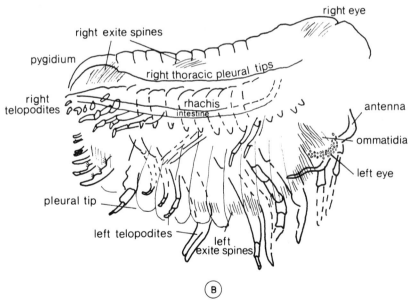

(B)

FIGURE 11.0 *Phacops* sp. (L. Devonian, Hunsrück shale). (a) X-ray photograph, courtesy of Prof. Wilhelm Stürmer. (b) Labeled sketch of (a) after Dr. Jan Bergström. (For details, see, W. Stürmer and J. Bergström, 1973, Paläont. Z., Vol. 47, pp. 104–141.)

postoma also occur, unattached, and membrane-supported. Isolated hypostoma and/or rostral hypostomal plates are known from somewhat less than one-third of all known trilobite species, and are important fossil objects.

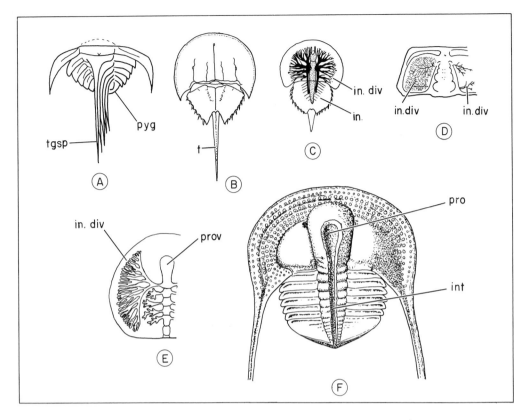

FIGURE 11.1 Some homologous features in trilobites and related forms (schematic).
(A) Trilobite, *Elliptocephalus asaphoides,* showing telsonic structures; *tgsp* – dorsal tergal spines, compare to (B); *pyg* – pygidium.
(B) The living xiphosuran, *Limulus polyphemus; t* – styliform telson, compare to (A).
(C) *Limulus polyphemus; in div* – intestinal diverticule, *in* – intestine.
(D) Trilobite, *Elyx laticeps* (M. Cam., Sweden), compare *in div* with (C) and (E).
(E) *Natavia compacta* (M. Cam., Burgess shale arthropod), see Chapter 10; compare *in div* with (C) and (D).
(F) *Omnia ornata* (U. Ord., Bohemia), X 2; pro-proventriculum (= stomach) located beneath glabella; *int* – intestine (on ventral side); the mouth probably led to stomach through a short esophagus (Harrington, 1959) and anus was located posteriorly on last pygidial segment, compare Figure 11.2A.
Illustrations redrawn after: (A)–(E) Størmer, 1945; (F) Beecher cit. Harrington, 1959.

Mouth occurs between hypostoma and metastoma (Figure 11.2*D*).

Sutures
Characteristics. Cephalic sutures were a characteristic trilobite feature associated with molting (Figure 11.2*A, fs*), completely lacking in agnostids and eodiscids, nonfunctional in others. These sutures are represented by narrow lines of weakness that correspond to soft, uncalcified portions of the exoskeleton. Such lines cut across the mineralized integument. Along sutures cephalon separates into isolated pieces – free cheeks, for example, detach.

Sutures are symmetrical and occur in pairs. They may be dorsal-intramarginal, marginal, or ventral-intramarginal. The first and last meet only the dorsal and ventral sides, respectively.

Types. There are three major types of sutures. Accordingly, one may ask relative to the genal angle, "Is the suture inside, in front of, or, does it bisect that angle?" (Figure 11.3*E*). If *inside*, it is *opisthoparion*; if *in front of*, it is *proparion*; if it *bisects* genal angle, it is *gonatoparion* (Figure 11.3*A–C*).

Some authors give distinct names to different types of opisthoparion sutures. These names are based on given genera.

Thus a cedariform-type of opisthoparion suture is named for *Cedaria*.

On the ventral side of the cephalon, the following types of sutures may or may not occur; rostral, hypostomal, and others, such as connective (Figure 11.3*D*). During the life of a given trilobite, breakage occurred along these lines of weakness on the ventral side during molting. This accompanied equivalent separation on the dorsal side. The end result was several shed or detached pieces and a remaining intact skeletal exuvia or molt.

The hypostoma apparently served a supportive function for the trilobite antennule. Following Öpik (1937), who thought that the antennary muscles were attached to ventral bosses on the tips of the hypostome's anterior wings (Figure 11.3*F*, *a*), Whittington and Evitt (1953) made a further suggestion. They proposed that in cheirurid trilobites, probably some others also, lateral notches of the hypostome (Figure 11.3*F*, *n*) served as passageways for the forwardly directed antennule.

Nonfunctional Facial Sutures. Some facial sutures appear to have been nonfunctional; they did not demark lines of weakness, because partial or complete symphysis strongly bound opposite sides of sutural markings. In such cases, as with phacopid trilobites (Devonian *Phacops*, and so forth) molting apparently occurred at the cephalon-thorax juncture.

In some forms lacking facial sutures, such as the Harpidae, marginal sutures occur (along the outer edge of the cephalon separating dorsal exoskeleton plate from ventral doublure). Harpid-type sutures are thought to have been evolved from the opisthoparion type found in genus *Entomaspis* (Rasetti, 1952); see Figure 11.4*B*.

Growth could only occur in trilobites by shedding the old skeleton. An exit or escape from the old skeleton had to be preceded by a break along the facial sutures, marginal sutures, or cephalic-thoracic juncture, or along ventral sutures, or some combination of these.

The eyeless agnostids are thought to descend from eye-bearing trilobites having proparion facial sutures and the ptychoparid sutural pattern (that is, with facial, rostral, connective, and hypostomal sutures all functional during ecdysis). (This pattern is primitive and basic for all trilobites. Among facial sutures, the opisthoparion is considered primary. Every other variant from the ptychoparid sutural pattern is thought to be due to migration and regression of the dorsal/ventral sutures.)

Migration and Disappearance of Sutures. Tendency toward reduction and/or loss of eyes in some trilobites (Phacopidae, Conocoryphidae, Raphophoridae) is often accompanied by equivalent migration, reduction, and disappearance of facial sutures (Figure 11.4*A*). In other instances, sutures became tightly knit or joined (ankylosed) and hence did not break during molting, being nonfunctional.

Glabella. A raised sector along the axis of the cephalon is called the *glabella*. It may be smooth or retain, as it often does, traces of the original cephalic somites (segments that existed precephalization and in the ancestral stock). These traces appear as raised, lateral, glabellar furrows (Figure 11.2*A*).

Every aspect of the glabella is variable — size, shape, convexity. Glabella shape has value for discriminating one genus from another (Figure 11.5*A–C*) although, generally, any single character is inadequate alone.

Thorax. The sector of the trilobite skeleton that is composed of movable, articulating segments (tergites) is the thoracic region. These tergites have a median, axial part and paired lateral outgrowths (pleurae) (Figure 11.2*A*). The freely movable thorax allowed the living animal to enroll in order to protect the soft parts on its ventral side.

Agnostids with two thoracic segments and eodiscids with three such segments (Figure 11.5*E*) when enrolled, resemble the bivalved crustaceans (Swinnerton, 1950).

The number of thoracic segments is

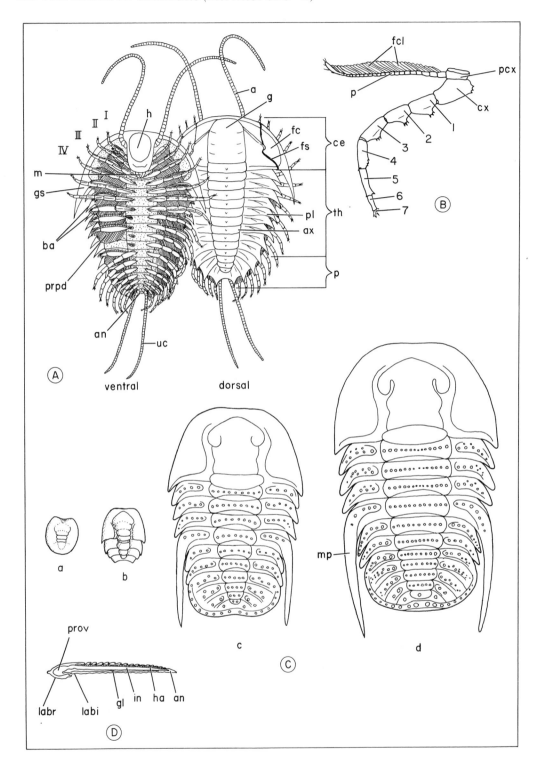

FIGURE 11.2 General morphology of trilobites.

(A) *Olenoides serratus* (M. Cam., B. C., Canada). Reconstruction, ventral and dorsal view, X 1; *ce* — cephalon, *a* — antenna, *g* — glabella, *fc* — free cheek, *fs* — facial suture, *ax* — axis, *p* — pygidium, *th* — thorax, *pl* — pleura, *uc* — uniramous (single branch) cerci, *an* — anus, *prpd* — preepipodite, *ba* — biramous (double branch) appendages, I–IV — postoral somites, *gs* — genal spine, *h* — hypostoma or labrum, *m* — metastoma.

(B) Trilobite biramous appendage, general plan (schematic); *pcx* — precoxa; *cx* — coxa; *p* — prpd, as in A; *fcl* — filaments; 1, trochanter; 2, prefemur; 3, femur; 4, patella; 5, tibia; 6, tarsus; 7, pretarsus; 1–7, telopodite.

(C) *Shumardia pusilla* (Sars), L. Ord., Tremadoc, Eng. (a) Protaspis, X 40; (b) meraspis, degree zero X 40; (c) meraspis, degree 5, X 30; (d) holaspis, X 30; *mp* — macropleura.

(D) *Triarthrus eatoni,* median section, inferred internal, structures: *labr* — labrum (or hypostoma); *labi* — labium; *gl* — ganglion of ventral nerve cord, *in* — intestine, *prov* — proventriculum, *ha* — heart; *an* — anus; *m* — mouth (opening before *labi*).

Illustrations after several authors.

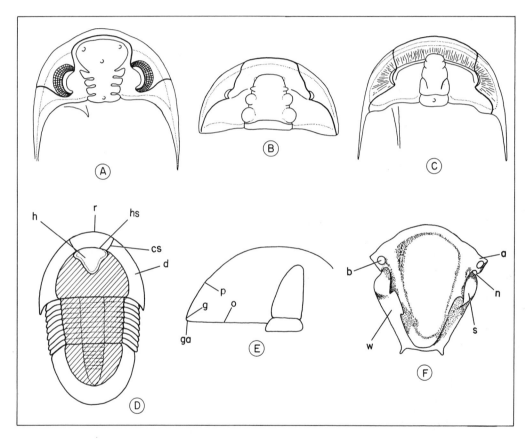

FIGURE 11.3 Trilobite facial sutures, and so on.

(A) Proparian. *Odontochile hausmanni,* (Dev., Bohemia), X 1.

(B) Gonatoparian. *Flexicalymene senaria* (M. Ord., Bohemia), X 3.5.

(C) Opisthoparian. *Ptychoparia striata* (Bohemia), X 4.0.

(D) Ventral sutures of cephalic area, *Stygina latifrons* (U. Ord., Ireland), enlarged: *h* — hypostoma, *hs* — hypostomal sutures, *r* — rostral plate, *cs* — connective sutures separate *r* from *d* (doublure) — ventral reflection of cephalic exoskeleton).

(E) Position of sutures relative to genal angle (*ga*). *o* — opisthoparian, inside genal angle; *g* — gonatoparian, bisects genal angle; *p* — proparian, in front of genal angle (Tasch).

(F) Structure of a hypostome (after Whittington and Evitt, 1953), *Ceraurinella typa, schematic, a* — anterior wing; *b* — wing process, *n* — lateral notch, *s* — shoulder of lateral border; *w* — posterior wing.

(A)–(D) redrawn and adapted from Harrington.

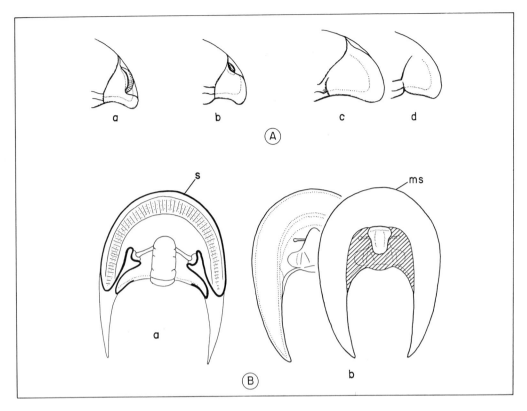

FIGURE 11.4 Facial sutures—derivation and lateral migration.
(A) Eye loss in Phacopidae and accompanying lateral migration of facial sutures: (a) *Phacops cir-cumspectans,* (b) *Cryphops* (?) *ensae,* (c) *Dianops limbata,* (d) *Ductina ductifrons;* (a)–(d) U. Dev., Germany.
(B) (a) *Entomaspis radiatus* (U. Cam., Missouri), X 5.6; s—suture, ms—marginal sutures; (b) *Para-harpes hornei* U. Ord., Scotland), X 2 [(B,a) suture derived from (B,b)].
Illustrations redrawn after (A) Richter and Richter, 1920; (B,a) Rasetti, 1952; (B,b) Whittington, 1950.

often variable within given families and genera. In *Elrathia kingii,* population of $n = 339$, thoracic segment number varied from 10 to 13 in adult specimens (Bright, 1959). In some forms, however, it is a fixed number as in the agnostids, with two, and the Asaphidae, with eight.

Considerable interest surrounds the origin of the trilobite thoracic region. Although the trilobite cephalon is clearly composed of primary somites, the thorax and pygidium are formed of secondary somites. Størmer (1942, p. 124) advocated, and more recently Hessler (1962) illustrated, secondary segmentation of the trilobite thorax. The precise mode of origin of a thoracic segment was discussed by Stubblefield (1928) from a study of the larval stages in *Shumardia pusilla* (Figure 11.2C).

He found that (*a*) new segments formed at the pygidial posterior, (*b*) a new segment formed behind these while those formed in (*a*) advanced forward, and (*c*) ultimately these advancing segments became secondarily three thoracic segments (that is, they were released from the anterior end of the pygidium).

Størmer observed that each thoracic pleura contained portions of two succeeding somites coalesced along the pleural furrows. Musculature needed to move tergites of the thorax, and position of internal anatomy relative to axial portion of the thorax, is shown in Figure 11.5D. Internal anatomy consists of digestive, circulatory, nervous, and other systems.

Pygidium. Tendency toward caudalization in trilobites has already been noted

(*Bumastus*, Figure 11.5*H*). Number of segments that comprise a given pygidium is quite variable: from a single tergite to more than 30. In a North African genus *Daquinaspis*, for example, pygidium is small, short, and consists of a single axial ring, and rounded axial termination (Figure 11.5*G*). In *Dalmanites*, by contrast, pygidium is long and consists of 11 to 16 rings, and a pleural field of 6 to 7 ribs.

Commonly the size of pygidia is compared to that of the cephalon and a corresponding terminology is used. If the pygidium is smaller than the cephalon, it is *micropygous*; if subequal, *isopygous*; and if larger, *macropygous*. More important taxonomically is shape of pygidium and number of axial rings, pleural ribs, and pleural spines.

The anus was apparently borne on the terminal segment of the pygidium (Figure 11.2*A*).

Generally, pygidia are semielliptical in shape; pygidial shapes are extremely varied. They often attain bizarre arrangements owing to spination (Figure 11.5*J*). Axial sectors of the pygidium are usually composed of successive rings. However, where caudalization has advanced, these will be subdued or missing; pleural ribs and furrows, likewise.

Exoskeleton Structure and Composition. The trilobite exoskeleton was chitinous. Thin section studies of various trilobites (Størmer, 1930) demonstrated that its dorsal integument corresponds to that of living arthropods. It is trilayered. These three layers are the equivalent of the endocuticle of living arthropods (see Chapter 15). Conceivably, the very thin layer 1 in *Tretaspis* (Figure 11.5*K*, 1) is equivalent to the thin epicuticle in modern arthropods (Chapter 15).

Some of the original pigment of the Number 1 outer pigmented layer is occasionally preserved in fossil pygidia, and more rarely, thoraces. Williams (1930) found two rows of chocolate-brown spots on the pygidial areas of *Phillipsia(?) tenuituberculata*. Teichert (1944) also reported

many small, round, scattered spots on pleural and axial surfaces of the pygidium *Ditomopyge meridionalis*. A fan-shaped arrangement of transverse light (dark-gray) and dark (black) bands marked the axis and pleurae of *Anomocare victata* (Raymond 1922). Such reports indicate a whole spectrum of protective coloration for camouflage and possibly sexual display.

Bøggild (1930) found mineral calcite in trilobite carapaces. Numerous investigators have reported a high phosphate content — 17 percent in *Paradoxides* and 19 percent in *Olenellus*. Elsewhere, in *Paradoxides* species, a range of 0.5 to 1.5 percent phosphate (P_2O_5) has been reported (Vinogradov, 1953).

Bright's study of *Elrathia kingii* from the Middle Cambrian of Utah (1959) gave insight into postmortem mineralization. Carapaces were replaced by cryptocrystalline calcite, iron oxide, and silica. Secondarily silicified trilobite faunas are not unusual (Whittington and Evitt, 1953).

The calcite/phosphate ratio in trilobite carapaces probably differed from that in modern crustaceans (see Chapter 15). Presumably phosphate was much higher. If so, trilobites would then have been phosphorus concentrators. However, secondary enrichment by replacement from enclosing sediments must be taken into account.

Inferred Biology of Trilobites

Biology of trilobites must be inferred from fossils alone. See Bergström, 1973.

Appendages and Limb Function. There are three kinds of known appendages in trilobites: paired, multijointed (15 to 50 joints), uniramous (single branched) antennae of varying lengths; nondifferentiated biramous appendages (four of which are cephalic) and in a single species, multijointed, uniramous cerci arising from the last pygidial segment. Probably modified appendages, these are reminiscent of the cephalic antennae (Størmer, 1944).

Exclusive of antennae and cerci, lack of differentiation of trilobites' biramous ap-

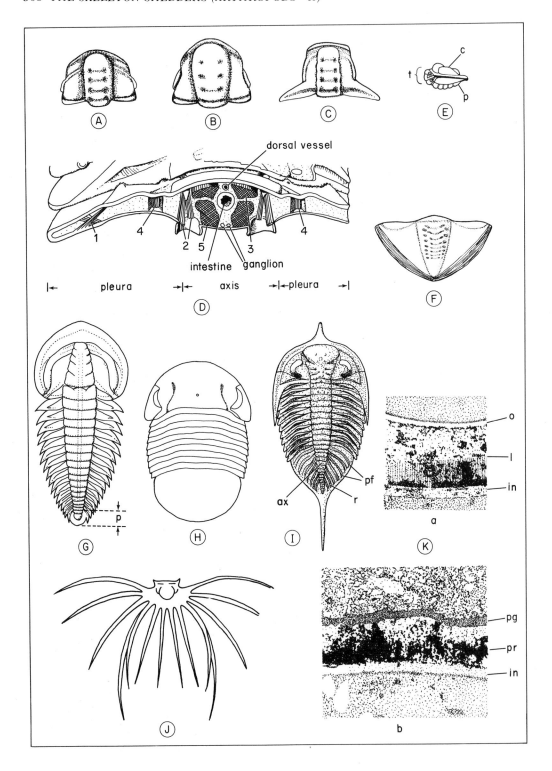

FIGURE 11.5 Glabella, thorax, pygidia, and others.

(A)–(C) Variable glabellae in genera of given family; Oryctocephalidae: (A) *Oryctocephalus*, M. Cam., X 2; (B) *Oryctocephalites*, M. Cam., X 2; (C) *Oryctocara*, M. Cam., X 5.

(D) Trilobite thoracic region, with internal anatomy and musculature of two thoracic segments (modified after Hupé, 1953); (1)–(5) muscles. Schematic reconstruction.

(E) *Serrodiscus helena* (L. Cam., Newfoundland). Enrolled specimen of an eodiscid trilobite appears as bivalved crustacean; c – cephalon, p – pygidium, t – three thoracic segments. (After Kobayashi, 1944, plate 2, Figure 9.)

(F) *Ampyx nasutus*, pygidium, with two paired rows of muscle scars on axis; (after Whittington, 1950).

(G) *Daquinaspis ambroggii* (L. Cam., Morocco), X 1.5. Note size of pygidium (p).

(H) *Bumastus* (B.) *barriensis* (U. Ord., England). Smooth pygidium lacking defined axis or pleural furrows, X 0.5.

(I) *Dalmanites caudatus* (Sil., England), axis (ax) with 16 rings; pleural field (pf) with 6–7 pairs of ribs (r), X 0.7.

(J) *Ancyropyge romingeri* (M. Dev., Michigan), X 1.3.

(K) Trilobite integument seen in thin section. *Tretaspis seticornis:* (a) Glabellar region, cross-section, X 110; o – outer layer, l – lamination of integument, in – inner layer; (b) axial furrow, transverse section; pg – pigmented layer, pr – principal layer, in – inner layer. Adapted from Størmer, 1930.

Illustrations redrawn and modified from: (A–C) (G–J) Harrington, et al., after several authors.

pendages constitutes a primitive feature (see Chapter 10, "Arthropod Appendages"). However, as Snodgrass stressed (1952), trilobites are "in no sense" a primitive arthropod.

The general plan of a trilobite biramous limb is shown in Figure 11.2B. (Data from 19 trilobite species.) There are two branches: the so-called walking leg composed of articulating podites (telopodite), and the gill-bearing branch (pre-epipodite). Through the precoxa (pcx), these limbs attached to the ventral integument.

While biramous appendages have uniform structures and configurations, they differ in size and position. Size increases from I to III or V on the cephalon, then generally decreases toward the pygidium (see Figure 11.2A).

Functionally, the biramous limb appears to have had a threefold purpose: *ambulatory* (walking, the telopodites), *natatory* (swimming, the pre-epipodites and/or telopodites, according to some authorities), and *respiratory*. Every limb can perform many functions at the primitive level (Manton, 1963).

Trilobite Spoor. (See Figure 1.7.) *Cruziana* indicates that the animal crawled along the soft-bottom substrate and burrowed into surficial mud layers.

Sensory Apparatus. The sensory apparatus of trilobites may be divided into three groupings: appendages, eyes, and miscellaneous (hypostomal maculae, median sensory organ, integumentary sensory organ). The paired, uniramous, antennal, and posterior cerci are inferred to have been sensory organs. According to Raymond (1920) the sensory caudal rami of *Neolenus* suggest that the animal could move backward as well as forward.

Lateral paired eyes are a prominent skeletal feature in numerous fossil trilobites. They range from quite small and separated (Harpidae) to enormous and fused, as in *Symphysops* (Figure 11.6A). The visual surface is borne on the free cheeks and is shed during molting. Two types of eyes are distinguished: the compound eye (holochroal), and the aggregate eye (schizochroal). The compound eye consists of closely packed hexagonal lenses (from 100 to 15,000) in direct contact with one another. The aggregate eye has fewer lenses (from 2 to 400), which are larger and separated by sclerotic tissue. They are not in contact. The phacopid and some cheirurid trilobites are characterized by aggregate eyes (Whittington and Evitt, 1953), whereas forms like *Peltura* had compound eyes.

A tendency toward eye reduction and/or loss (that is, blindness) occurred in many trilobites: Conocoryphidae, Eodiscidae,

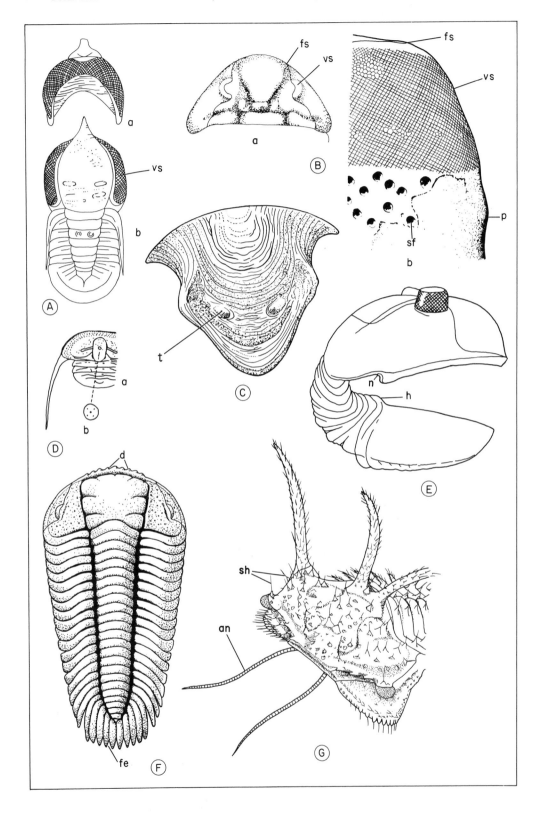

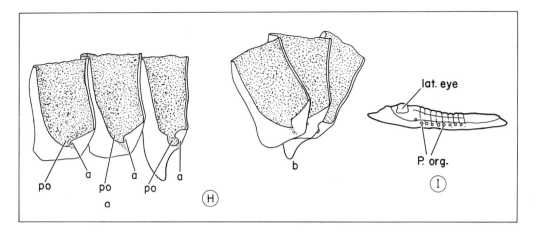

FIGURE 11.6 Visual systems and other trilobite structures.
(*A*) Large confluent eyes, *Symphysops subarmatus* (U. Ord., Great Britain), X 1.2. This eye arrangement presented a continuous visual surface (*vs*); (*a*) ventral view of *vs*, (*b*) dorsal view. After Hupé, 1953.
(*B*) (*a*, *b*). *Asaphus*. (*a*) Cephalon with, facial suture (*fs*), *vs*—visual surface (compare to *B*, *b*); (*b*) *Asaphus cornutus* (Ord., Russia), portion of eye with ocular peduncle (*p*); *vs*, *fs* as in (*B*,*a*); *sf*—sensorial fossettes. Modified from Hupé, 1953. [See Towe, 1973, Science, v. 179, p. 1007].
(*C*) Hypostome, *Scutellum polyactin*, with macula bearing faceted and reticulate tubercles. Adapted from Lindstrom, cit., Harrington, 1955.
(*D*) Median glabellar tubercle, *Tretaspis seticornus*, schematic. (*a*) Meraspid larva with median tubercle, (*b*) details of median glabellar tubercle. After Størmer, 1944.
(*E*, *F*) Enrollment structures. (*E*) *Asaphus expansus* (Ord., Esthonia): hook (*h*)—the vincular hook fits into notch (*n*), the vincular notch, upon completion of enrollment. (*F*) *Pliomera fischeri* (M. Ord., Esthonia); slightly enlarged. Observe anterior denticle row along cephalic border that, on enrollment, meshes with free extremity of pygidium (*fe*). After Harrington, 1955.
(*G*) *Ceratocephala lacinata* (M. Ord.), reconstruction; sensory hairs (*sh*) occupying perforations in spines and tubercles, and antennules (*an*). After Whittington and Evitt, 1953.
(*H*) Panderian openings ("pores"?) in *Dimeropyge spinifera*—three adjacent pleurae extended: (*a*) The raised interior edge (*a*) of Panderian opening (*po*); (*b*) (*a*) acts to limit closure of the three pleurae enrolled, X 30. After Whittington and Evitt, 1953.
(*I*) Same as (*H*), *Isotelus gigas*, *po* (= P. org) may be pores leading to the nephridia of each metamere; *lat. eye* = lateral eye.

Agnostidae, Olenidae, Shumardiidae, and others. Blindness was secondary. Retention in some blind forms of ocular structures exclusive of the visual surface (that is, ocular protuberances, palpebral lobes, and so on) is evidence. Blindness was apparently unrelated to environment since blind forms are known among pelagic, epiplanktonic types and vagrant benthos (Harrington, 1959; Hupé, 1953).

The "miscellaneous" grouping includes hypostomal maculae (Figure 11.6*C*). Was a visual function served by faceted and reticulate tubercles on the ventral skeletal piece (hypostome)? If so, what of the smooth macula on other hypostomes? Ventral eyes connected to the brain by nerve fibers are known in other arthropods (*Limulus*, for example). Lindstrom favored the view that tubercles were visual lenses. An absence of tubercles on any hypostomal macula would then indicate degenerate ventral eyes, that is, loss of lenses. A comparable tendency occurred in the dorsal optical apparatus of many blind species.

All workers are not agreed on this interpretation. Whittington and Evitt could not confirm this view in their samples of silicified cheirurids. They preferred maculae as places of muscle attachment. Harrington (1959) argued against the muscle-attachment interpretation and for a visual function.

Another of the "miscellaneous" grouping is the median glabellar tubercles, once held to be structures of a median eye (Figure 11.6*D*). Størmer and others held that

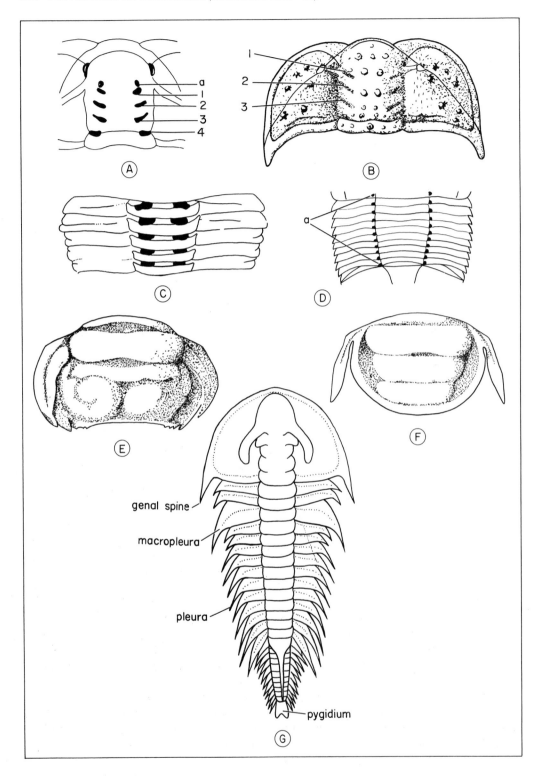

FIGURE 11.7 Muscle attachment sites; dimorphism, in trilobites.

(A) *Cybele grewingski*, schematic (after Öpik), (1)–(3) attachment sites for muscles that moved the three cephalic paired appendages; (a) scars left by muscles that moved uniramous antennae; (4) possible attachment site for dorsal and ventral muscles.

(B) *Cybele bellatula* (L. Ord., Norway), X 3. Three pairs of glabellar furrows (1–3).

(C) *Lonchodomas rostratus* (L. Ord., Norway). Black areas indicate attachments for thoracic appendage muscles.

(D) *Ectillaenus katzeri* (M. Ord., Bohemia), axial furrow apodemes—probable attachment sites for dorsal and ventral muscles (see Figure 11.5D).

(E, F): (E) *Redlichia forresti* (M. Cam., W. Australia), female pygidium, enlarged. (F) Same, male pygidium, enlarged.

(G) *Olenellus vermontanus* (L. Cam., Vermont), X 1. Observe macropleura thought to be genital segment-bearing gonopores.

Illustrations redrawn and adapted from: (A, E–F) Öpik, (C–D) Hupé, (G) several authors.

this structure was a "sensorial complex." It did play some undefined role in receiving and/or transmitting sensory data.

Several different structures of the integument may have served a sensory function—chiefly tactile. Wherever preserved, ventral appendages carried stiff setae/bristles; sensorial fossettes on the eye platform (Figure 11.6B, b), perforated tubercles, and spines in forms like *Ceratocephalia* carried sensory hairs (Figure 11.6G) (Hupé, 1953; Whittington and Evitt, 1953). Were these perforations external openings of cutaneous/subcutaneous glands (Harrington, 1959)?

Information from these several sensory sources had to be sorted and a proper response transmitted over nervous system (Figure 11.5D) to a local muscle complex. Speaking on another theme (function of limbs), Harrington (1959) expressed doubts that an arthropod as primitive as a trilobite had a nervous system so highly developed that it could deliberately coordinate movements for walking and for swimming.

Muscular System. Suppose that there is a lack of evidence on places of trilobite muscle attachment; still it is possible to infer muscle system from multiple specimens with preserved ventral appendages (Figures 11.2A, 11.5D).

There is, however, excellent evidence on muscle site attachment both dorsal and ventral. A dorsal pit (in some forms, glabellar furrows—Figure 11.7A–B) corresponds to invaginations of dorsal skeleton (apodemes). Similar muscle attachment scars occur on pygidium (Figure 11.5F) and thorax (Figure 11.7C–D).

Such musculature provided the living trilobite with capabilities for moving antennae, versatile movements of appendages, capacity to enroll for protection (Figure 11.6E–F), to crawl out of the old skeleton during the molting cycle, and to perform burrowing and reproductive functions.

Digestive System. Structure of the trilobite digestive system is poorly known. Figure 11.1F shows a fortuitous example of preservation that reveals hollow interior of proventriculum (or stomach) and intestine. On ventral side (Figure 11.2D), mouth led through brief esophagus to stomach, and anus opened in terminal segment of pygidium. There is a structure, "the glandular diverticule" of the digestive tube (= genal caeca). Fossilized preservation of such structures in certain trilobite cephala is shown in Figure 11.1D. Similar structures are common in either trilobitomorphs or chelicerates—arachnids (*Limulus*, Figure 11.1C) and several Burgess Shale arthropods (Figure 11.1E).

What food did trilobites consume? Størmer (1944) observed that lack of jaws[2] suggested that most species were mud feeders (exclusive of epiplanktonic forms, of course). Snodgrass (1952) doubted that trilobites with long filamentous antennae were usually or commonly mud/sand burrowers.

Raymond (1920) thought that the spiny coxal lobes (see Figure 11.2B for *coxa*) primarily functioned in food gathering

[2]Bergström, 1973.

and preparation, bringing food forward from one spiny lobe to the next until it reached the mouth. Since these spines did not meet in a midline, how could they serve for food passage, particularly of minute particles (Størmer, 1944; Manton, 1963, p. 21)? However, Sanders (1963) inferred trilobites could have fed, like cephalocarids, utilizing large masses of detritus passed forward from limb to limb to mouth. If true, then there was no necessity for limbs to meet along midline. Sanders envisioned the trilobite mode of feeding as a filter-feeding current system.[3] He also considered that trilobites may have been predaceous like modern *Limulus*.

Snodgrass (1952) observed that worms and other soft-bodied creatures abounded in bottom muds and sand that were habitat for Cambrian trilobites. Presumably, such forms were part of the trilobite diet. Other worms (see "Paleopathology" below) found their diet in the trilobite carapace.

Reproduction, Dimorphism, and Excretion. The macropleura (Figure 11.7*G*) found in some trilobites have been interpreted to be the genital segments bearing gonopores (Hupé, 1953), but this is highly speculative.

On the other hand, *Olenelloides armatus* Peach (5.0 mm long) with a macropleurid pair every third segment (Figure 11.7*A*, part A) may represent an asexual budding sequence (three buds); compare *Olenellus* or *Paedeumias* (see Sharov, 1966, p. 38). Lemche (1957) was the first to propose that the so-called abortive segments in *Olenellus vermontanus* (Figure 11.7*G*), *O. thompsoni*, and other olenellids might be evidence of the type of asexual budding that occurs in certain primitive polychaete worms (Figure 11.7*A*, part *C*). These purported abortive segments include two to six extra segments behind the spine-bearing one (Figure 11.7*A*, part B).

Lemche's interpretation is provocative and conceivable. Lemche cautioned that this type of asexual budding applied only to primitive trilobites lacking a true pygidium.

Another type of evidence is *dimorphism* [that is, two distinct variants in the same skeletal part, for example, the pygidium of a given species (see Table 11.1; Figure 11.7*E*, *F*)]. Öpik (1958) recognized male and female characteristics in specimens of *Redlichia forresti*.

The Panderian organs are really two different kinds of structures: Panderian protuberances (Figure 11.6*H*) and Panderian openings or pores (Figure 11.6*I*). Protuberances appear to limit closure (that is, prevent overgliding) during enrollment; pores may serve an excretory function — being external openings of metameric nephridia. Trilobites presumably had nephridia in each segment. Hupé argued that Panderian organs were an artifact. Whittington's finding of a circular opening on the doublure of a silicified *Isotelus* free cheek apparently conflicts with Hupé's view (Whittington, 1941, plate 75, Figure 47).

Respiratory System. Respiration in trilobites was external; Bergström (1973) noted that x-rays did not reveal any respiratory organ, as such. If gills were present, they would be in larger trilobites only, while all others probably respired through the integument on the ventral side and the basal segment of the exites. (Figure 11.0*a* and *b*).

Larval Stages. Larval sequences are known for numerous trilobites (Whittington, 1959). During ontogeny, trilobite length increased as much as 50 to 400 times. Three larval stages have been arbitrarily defined: (1) protaspid, (2) meraspid, (3) holaspid. The *protaspid* covers all stages of development after fertilization of the egg and before appearance of a definite transverse suture subdividing the dorsal shield into head and tail region. Protaspids are trilobed with axial segment usually bearing four to five rings. They are subcircular

---

[3]Bergström (1969) also concluded that some trilobites were filter feeders.

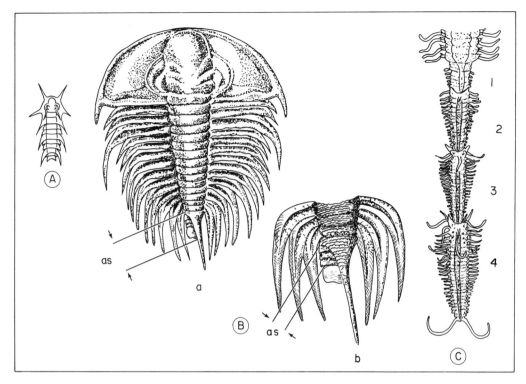

FIGURE 11.7A Asexual budding (?) in olenellid trilobites. After Sharov, 1966; Malaquin, 1893 cit. Lemche, 1957; and others.

(A) *Olenelloides armatus* Peach, schematic. Every third body segment is a macropleura, compare to (C).

(B) *Olenellus thompsoni* (L. Cam., Vermont), X 0.6; (a) dorsal exoskeleton, *as* — abortive segments; (b) detail of *as* — asexual budding (?).

(C) The polychaete *Autolytus edwardsi*, posterior part of body with four (1–4) rudiments of new animals of different ages budding off; schematic.

shields, 0.25 mm to 1.0 mm in length. The smallest known larva in agnostid trilobites is a meraspid, degree zero, 1.0 to 2.0 mm long.

The *meraspid* corresponds to the sequence leading to formation of thoracic segments. Raw (1925) divided the meraspid period into several steps (= degrees). One segment in the thorax equaled degree 1, two segments, degree 2, and the condition prior to any thoracic segment development between cephalon and pygidium was designated degree zero. Figure 11.2C shows how meraspid degree zero appears in comparison to degree 5 with a well-defined pygidium and thoracic segments intervening between it and cephalon. During the meraspid period — a time of rapid development — the skeleton increased in average size over the protaspid stage, some six to 12 times.

What other changes occur during the meraspid period? Form and proportion of glabella (widening anteriorly in many, but not all forms), loss of some furrows (median) and deepening of others (lateral), changes in convexity, increase in eye facet number, development of optical system, and reduction in hypostomal size.

The holaspid *period* covers time of growth after completion of the thorax. Holaspids of different genera and species vary in length from 1.8 mm (*Shumardia*, Figure 11.2C) to 400+ mm (even larger forms are known, up to 70 cm).

What sort of changes occur during the holaspid stage? In a large sample of *Dipleura dekayi* (Devonian) (Cooper, 1955) that ranged in length from 8.0 to 19.0 mm, there was a loss of glabellar furrows, widening of thoracic axial lobe, and pygidial fusion of segments into a smooth axis.

TABLE 11.1. Dimorphism in *Redlichia forresti*

Taxonomic Character	Female	Male
Number of free segments in thorax	17	17
Semiattached segments	1	1
Dorsal spine in fourth segment	+	+
Dorsal spine in twelfth segment	+	+
Intergenal spines	+	+
Rostral shield shorter than frontal limb	+	+
First axial ring enlarged	+	+

Sex Character	Female	Male
Axial spines stout	−	+
Axial spines slender	+	−
Pleural spines long	−	+
Pleural spines short	+	−
Size large, average 100 mm	+	−
Size small, average 15 mm	−	+
Posterior pleura extended beyond pygidium	−	+
Posterior pleura not extended beyond pygidium	+	−
Occurrence frequent	+	−
Occurrence rare	−	+

Source. After Öpik, 1958.

Other changes included varying outline and proportions of cephalon and axial lobes, and variable spinosity, convexity, and size.

To what major events in the animal's life did the several larval stages correspond? Clearly, the circular-to-ovate protaspid, sometimes with spinous cephalon and also pygidial spines, appears to have been planktonic. Swinnerton (1950) noted that in protaspids with eyes, the eyes were functional, close to head shield margin. This condition characterized floating and free-swimming types. In forms like *Triarthrus*, original marginal eyes apparently migrated away from the margin. Such migration corresponded to change from pelagic to benthonic habit.

As the animal grew (meraspid stage), intestines lengthened, new walking appendages regularly developed along the thoracic sector (meraspid degree), and greater respiratory surface became available through external gills of pre-epipodites. If the pre-epipodite was a usable swimming appendage, then the animal's swimming capacity would have been enhanced at each degree of this stage.

Macropleural development is known in the meraspid stage of many trilobites [(?) *Paradoxides, Paedeumias yorkense, Shumardia pusilla* – Figure 11.2*C*]. If Hupé's speculation is correct, then growth of macropleura corresponds to growth of paired genital segments bearing gonopores. Similarly, Swinnerton's speculation, if correct, would denote that many meraspids with nonmarginal eyes lived a benthonic existence (that is, transferred from planktonic to benthonic habit).

During the holaspid period, cephalization (loss of furrows) and caudalization (fusion and/or smoothing of pygidial axis) were advanced. Such rigid tagma would permit powerful movements of the pygidium during enrollment, or incomplete enrollment-type movements to aid escape. The animal could thrust its pygidium far under the ventral side, and automatic withdrawal of dorsal cephalon from a predator would result—equal and opposite reaction.

Molting Cycle, Exuvia, and Population Density. Details of the trilobite molting cycle must be inferred. All that is given is a series of trilobite exuvia—cranidia, pygidia, free cheeks, hypostomata. It is reasonable to assume that a cycle equivalent to, but not necessarily identical with, that of living arthropods applied to all known trilobites (Chapter 10; see also Chapter 15 for further details).

Separation of the exoskeleton along facial and/or other sutures triggered movement by the trilobite to escape its rigid enclosure. Raw (1927) estimated that in the trilobite, *Leptoplastoides salteri*, at least 29 separate molts (represented by shed exuvia) spanned the growth cycle from earliest protaspid to late holaspid stages. Shaw (1955, p. 138) estimated the number of molts from *Plicatolina kindlei*. He allowed five molts to the protaspid stage and one molt for each of the known 16

thoracic segments. This brought the total to 16. To this were added 10 molts for missing segments not represented in fossilized exuvia. For this species, Shaw estimated 30± molts, the estimate being reduced to 20 to 25 molts when thoraces were fewer. The actual number used in Table 11.2 was 20, to keep calculations simple.

Generally, the true relationship can be established between number of individual trilobite exoskeletal fragments (exuviae) and numbers of individuals (that is, population density or n) by use of a simple equation:

$$n = \frac{\text{maximum number of exuviae} \atop \text{(cranidia or pygidia)}}{\text{the estimated number of} \atop \text{molt stages}}$$

In Table 11.2 there are 56 cranidia among the fossil exuvia representing the species *Parabolinella triarthroides*. Thus $n = 56/20 = 3$. Two of these individuals had completed their growth cycle, whereas one

TABLE 11.2. Trilobite Exuvia, Number of Molt Stages, and Corrected Population Density. Upper Gorge Fauna, Upper Cambrian, N.W. Vermont

Species	Est. No. of Molt Stages	Fossilized Exuvia				n	Percent of Total Fauna
		A^a	B	C	D		
1. *"Terranovella" gelasinata*	22[b]	36	0	4	0	2	1.8
2. *Parabolinella triarthroides*	22	56	15	0	0	3	2.16
3. *Plicatolina kindlei*	21	21	0	0	8	1	0.9
4. *Missisquoia typicalis*	20	61	23	0	0	3	2.6
5. *Symphysurina minima*	16	42	32	8	5	3	2.6
6. *Hardyoides glabrus*	15	21	2	1	0	1	0.9
7. All others (agnostids):	7						
(a) *Geragnostus bisectus*	—	0	8	0	0	4	3.5
(b) *Geragnostus bisectus typica*	—	15	0	0	0	2	1.8
(c) *Geragnostus bisectus brevis*	—	5[c]	0	0	0	1	0.9
(d) *Pseudagnostus araneavelato*	—	19	12	0	0	3	2.6
(e) *Pseudagnostus bilobus*	—	15	11	0	0	2	1.8
(f) *Homagnostus* sp.	—	5	1	0	0	1	0.9
(g) *Litagnostus raymondi*	—	Ten shields		0	0	1	0.9

[a] A—Cranidia; B—Pygidia; C—Free Cheeks; D—Hypostome.
[b] Species numbers 2, 3, 4, and 6 were assumed to have had 20 molts each; species 5 and 6 were taken at 15 molts each for calculation of number of individuals ($=n$).
[c] Assuming seven molt stages for agnostids, the five cranidia in 7c and 7f represent incomplete growth series of a single individual.
Source. Adapted from A. B. Shaw, 1953.

had four additional holaspid stages to complete.

Shaw (1953) demonstrated that when a number of individual trilobites are tallied as above (rather than a number of exuviae shed by a single trilobite), the trilobite component of a fossil fauna (brachiopods and miscellaneous forms) was only 21 percent. By actual count, trilobite exuviae came to 454 out of 724 total fossils collected, or 65 percent.

Enrollment. Ability to enroll was common to many, but not all, trilobites, that is, to roll up the pygidium until it touched, interlocked with, or contacted the ventral anterior side of the cephalon. Lower Cambrian micropygous forms like the Olenellidae presumably could not enroll. They must have sought protection of the bottom substrate to guard their vulnerable ventral side (see Figure 11.11*G* for a remopleurid trilobite that developed a long, forked, hypostome and could not enroll).

For many trilobites there must have been a selective advantage in enrolling so securely that all ventral soft parts were protected from predators, being sealed in a so-called protective exoskeletal box. Thus special structures developed for interlocking or close fit of the enrolled pygidium and cephalon (Figure 11.6*E–F*). These included a *vincular hook* that fit into a vincular notch upon completion of enrollment; anterior denticles along the cephalon border that enmeshed with the free extremities of the pygidium during enrollment; and posterior protuberances that acted to limit closure (that is, overgliding) of pleura during enrollment (Figure 11.6*H*).

Completely enrolled, a given trilobite was effectively invulnerable to most predators. However, during Devonian–Mississippian time, trilobites became likely prey to sharks whose pavement teeth were quite capable of crushing them whether enrolled or not (Dunbar, 1960, Figure 180). Carboniferous trilobites and species of *Calymene* are commonly found enrolled.

Classification. Class Trilobita

1. Arthropods of marine habit only.
2. Exoskeleton, mineralized, chitinous, bent inward ventrally forming doublure.
3. Dorsal exoskeleton, subelliptical arched or flat, and longitudinally trilobed (hence the name "trilobite"), that is, three distinct lobes.
4. At right angles to the trilobation, the body is divided transversely into *cephalon* (head shield or tagma), articulated with freemoving thoracic segments (= *thorax*), in turn, articulated with *pygidium* (tail shield or tagma).
5. Cephalon with raised axial portion (= *glabella*) carrying furrows that denote original somites preceding cephalization.
6. Complex eyes are present in many forms, but blindness is secondarily acquired.
7. Sutures—lines of weakness—of cephalic shield along which breaks occur during ecdysis, include many types: facial, marginal and, on ventral side, rostral, hypostomal, median, or lateral. Sutures may be obsolete or functionless.
8. Appendages include paired uniramous antennae, paired biramous walking limbs and, rarely, antenniform cerci.
9. Respiration through integument and exites.
10. Sexes separate; possibly expressed by dimorphism (see Table 11.1); reproduction by copulation; larval stages; protaspid, meraspid, holaspid; larva planktonic; adults generally benthonic, although some epiplanktonic adults and swimmers among spinous forms.
11. Molting cycles occurred approximately 20 to 30 times during the life of an individual; each molt is represented by exuviae that include cranidia, free cheeks, pygidia, hypostome and, occasionally, rostral plate.
12. Geologic range: Lower Cambrian to Middle Permian.

Present Status of Classification

Great emphasis has been placed on facial sutures as the basis for classification. However, almost every skeletal feature has at one time or another been used: eyes, thoracic segments, pygidia, and so forth (Harrington, 1959).

Whittington (1954), a little over a decade ago, could find no acceptable classification system for trilobites but tentatively accepted Hupé's superfamilies (1953). He noted that at least four orders might be indicated: one to include the eodiscoids and agnostiids; a second to include olenelloids and redlichioids; a third for the corynexochoids; and a fourth for the ptychoparioids and allied superfamilies. Cheiruroids might require a separate order. Lichids and odontopleurids were not considered in these groupings.

The Treatise volume on trilobites (1959) recognized four of the proposed orders that Whittington indicated. The Cheiruridae were included in the order Phacopida, and the Lichidae and Odontopleuridae were each placed in separate orders. The Treatise classification is primarily founded on the following three features: (1) cephalic axial characters (glabella, shape, furrows, and so on), (2) pattern of sutures, (3) caudalization of pygidial segments. In this sense it is more comprehensive than previous ones.

Class Trilobita now covers seven orders, 13 suborders, and 30 superfamilies.

Order 1. **Agnostida.** Small trilobites with subequal cephalon/pygidium; two to three thoracic segments. Range: L. Cam.–U. Ord. [Suborder Agnostina; *Pleuroctenium* (Figure 11.8*A*); suborder Eodiscina, *Pagetia* (Figure 11.8*B*), *Serrodiscus* (Figure 11.5*E*).]

Order 2. **Redlichiida.** Cephalon, large, semicircular, usually with well-developed spines; thorax with numerous segments; pygidium, diminuitive, rudimentary; opisthoparion or ankylosed facial sutures; glabellar furrows prominent; eyes, elongate and crescentic. Range: L. Cam.–M. Cam. [Suborder Olenellina, *Daquinaspis* (Figure 11.5*G*), *Olenellus* (Figure 11.7*G*), *Elliptocephalus* (Figure 11.1*A*); suborder Redlichiina, *Redlichia* (Figure 11.7*E–F*); suborder Bathynotina, *Bathynotus* (Figure 11.8*C*).]

Order 3. **Order Corynexochida.** Elongate; subelliptical, exoskeleton, chiefly micropygous; semicircular cephalon with well-developed genal spine; glabella, long, subparallel side, in some, general expanding anteriorly; eyes ridged and elongate; opisthoparion facial suture; rostral plate either rudimentary or fused to hypostoma; thorax with 5 to 11 segments, spinous terminations; pygidium with marginal spines but may have smooth border. Range: L. Cam.–U. Cam. [*Olenoides* (Figure 11.2*A*), family Oryctocephalidae (Figure 11.5*A–C*).]

Order 4. **Ptychopariida.** (This is a large order with five suborders and 26 superfamilies.) Trilobites with more than three thoracic segments; chiefly opisthoparian sutures but some proparion and/or marginal; glabella tapers forward, glabellar furrows, where present, are subparallel linear depressions; preglabellar field present but often reduced; hypostoma separated from cephalon by suture or uncalcified membrane; rostral plate may or may not be present. (This order is thought to be the close relative of the Redlichiida and ancestral to most or all post-Cambrian trilobites exclusive of the agnostids.) [Suborder Ptychopariina, *Shumardia* (Figure 11.2*C*), *Ptychoparia* (Figure 11.3*C*), *Bumastus* (Figure 11.5*H*), *Asaphus* (11.6*B*), *Entomaspis* (11.4*B, a*), *Paraharpes* (11.4*B, b*), *Symphysops* (11.6*A*), *Lonchodomas* (11.7*C*), *Pemphigaspis* (11.8*D*), *Ameura* (11.8*E*).]

Order 5. **Phacopida.** (Post-Cambrian trilobites only; derived from Order 4.) Typically with proparion sutures but may be gonatoparion or opisthoparian, or lacking; varishaped, forward expanding (Phacopina) or tapering forward glabella (Calymenina); varied development of glabellar furrows that are not always present; preglabellar field short or lacking; rostral plate may or may not be present; thorax with 8 to 19 segments; median to large pygidium (occasionally small). Range: L. Ord.–U. Dev. [Suborder Cheirurina, *Ceraurinella* (Figure 11.3*F*), *Pliomera* (Figure 11.6*F*), *Cybele*

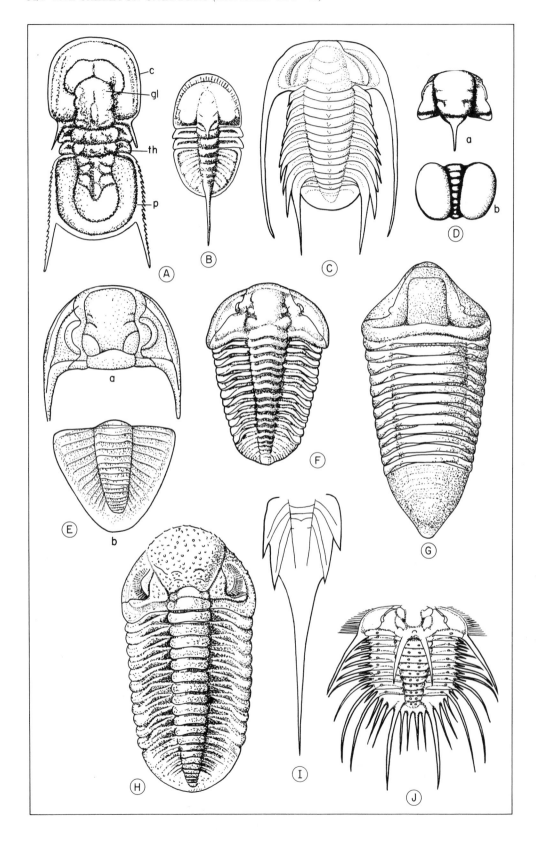

FIGURE 11.8 Representative trilobites.
(A) Suborder Agnostina. *Pleuroctenium granulatum,* X 4.5 (M. Carb., Bohemia); *c* – cephalon, *gl* – glabella, *th* – thorax, *p* – pygidia.
(B) Suborder Eodiscina. *Pagetia significans* (M. Cam., Australia).
(C) Suborder Bathynotina. *Bathynotus holopyga* (Upper L. Cam., Vermont), X 1.
(D, E): (D) Suborder Ptychopariina *Pemphigaspis bullata* (U. Cam., Minn.), X 4: (a) cranidia, (b) pygidia. (E) *Ameura sangamonensis* (Penn. Ill.), X 2; (a) cephalon, (b) pygidium.
(F, G): (F) Suborder Calymenina, Calymene blumenbachii (M. Sil., England), dorsal view, reconstruction, X 0.7. (G) *Trimerus* (*Dipleura*) *dekayi,* reconstructed exoskeleton; note fused pygidia and absence of furrows in glabella, schematic.
(H) Suborder Phacopina, *Phacops fecundus* (Dev.), reduced.
(I) Order Lichida. *Uralichas avus* (M. Ord., Bohemia), X 1.3; pygidia.
(J) Order Odontopleurida. *Odontopleura ovata* (M. Sil, Bohemia), reconstructed exoskeleton, X 1.
Illustrations redrawn and modified after Harrington, et al., after several authors. (B) Kobayashi, (D) Palmer, (H) Hupé, (J) Whittington.

(11.7*A*, *B*); suborder Calymenina, *Calymene* (11.8*F*), *Flexicalymene* (11.3*B*); suborder Phacopina, *Phacops* (11.8*H* and 11.0), *Dalmanites* (11.5*I*).]

Order 6. **Lichida.** Distinctive cephalic and pygidial features; glabella, broad, extends to anterior border; pair of lateral glabellar furrows longitudinally elongated; lateral glabellar and occipital lobes tend to fuse with one another and other parts of the cranidium; opisthoparion sutures; large pygidia; median to large cephalon; pygidial pleural segments flattened, three pairs of leaflike, spinose pleura; tuberculate dorsal surface. Range: L. Ord.–U. Dev. [*Uralichas* (Figure 11.8*I*).]

Order 7. **Odontopleurida.** Convex cephalon; glabella expanded at occipital ring; tapers forward; median or paired spines on lobes, may arise from occipital ring; two to three pairs of lateral glabellar lobes; convex genal regions; eye lobes prominent; free cheeks margined by spines; thorax with eight to ten segments; pleural spines; pygidia short, subtriangular, axis with two to three rings; external surface tuberculate to spinose. Range: M. Cam.–U. Dev. [*Odontopleura* (Figure 11.8*J*).]

Information Derived from Trilobites

Biozones, Biostratigraphy, and Correlation. Trilobites are particularly useful zonal fossils in Cambrian beds on a worldwide scale. The Lower Cambrian *Olenellus*

zone is encountered in outcrops of Newfoundland, Vermont, New York, Pennsylvania, as well as Scotland and Greenland (see Figure 11.9). Equivalents also occur in the Californian Cambrian and the Pacific province. This zone, or its equivalents, is often a critical reference point in trying to date underlying beds. For example, a problematical fossil molluscan (*Wyattia reedensis*) from Inyo County, California, was found 900 meters below strata bearing the earliest known olenellid trilobites in the section (*Fallotaspis* sp.). Accordingly, it was designated pre-Cambrian (Taylor, 1966).

A Cambrian biofacies with the same components as those in the Bonaterre dolomite of Missouri can be collected in the Warrior Formation of Pennsylvania. Such equivalences are common. Palmer (1954) correlated the trilobite fauna of the Riley Formation of Central Texas with similar faunas in Minnesota, Montana, and Wisconsin. V. Paulsen (1964) showed how the trilobite fauna of Northwest Greenland could be correlated with Middle Cambrian trilobite zones of the world (Figure 11.9*A*, *B*).

Among others, Middle Cambrian forms include species of *Paradoxides*, *Olenoides*, *Ptychoparia*, (Figures 11.2*A*, 11.3*C*). Upper Cambrian forms include species of such common forms as *Dikelocephalus* and *Olenus*.

Trilobite faunas are also quite useful in post-Cambrian stratigraphy: Ordovician (Harrington and Leanza, 1958, Figure 3; B. N. Cooper, 1953; Hintze, 1952; Ross,

CORRELATION OF MIDDLE CAMBRIAN FAUNIZONES, STAGES, AND FORMATIONS

PACIFIC REALM ZONES	ATLANTIC REALM STAGES	ATLANTIC REALM ZONES	NW. GREENLAND FORMATION, MEMBERS	ALBERTA BR COLUMBIA FORMATIONS	WESTERN NEWFOUNDLAND FORMATIONS	AUSTRALIA (QUEENSLAND, NORTHERN TERRITORY) STAGES AND ZONES	CENTRAL SIBERIA ZONES	HWANGHO BASIN (MANCHURIA, NORTH KOREA) STAGES
BOLASPIDELLA	PARADOXIDES FORCHHAMMERI	LEJOPYGE LAEVIGATA		?	MARCH POINT	PARADOXIDES FORCHHAMMERI	SOLENOPARIA	KUSHANIAN
		JINCELLA BRACHYMETOPA	BLOMSTERBAEK LIMESTONE M.	ELDON			CENTROPLEURA	
BATHYURISCUS–ELRATHINA		TRIPLAGNOSTUS–GONIAGNOSTUS		STEPHEN	TREYTOWN POND		CICERAGNOSTUS	TAITZUAN
	PARADOXIDES PARADOXISSIMUS		CAPE WOOD FORMATION (CAPE RUSSELL M.)	?	CLOUD RAPIDS	UPPER PARADOXIDES PARADOXISSIMUS		
GLOSSOPLEURA				CATHEDRAL		XYSTRIDURA–DINESUS	ORYCTOCEPHALUS	
ALBERTELLA	ECCAPARADOXIDES OELANDICUS		?			REDLICHIA	TOLLASPIS	TANGSHIHAN
PLAGIURA–POLIELLA	?			MT. WHYTE			ERBIA	SHIHCHIAOAN
							?	MISAKIAN
								?

A

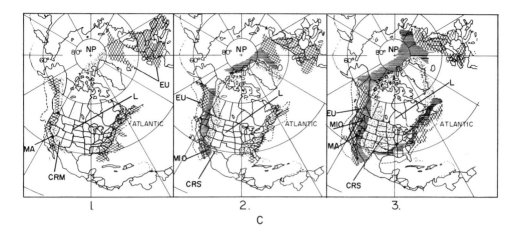

C

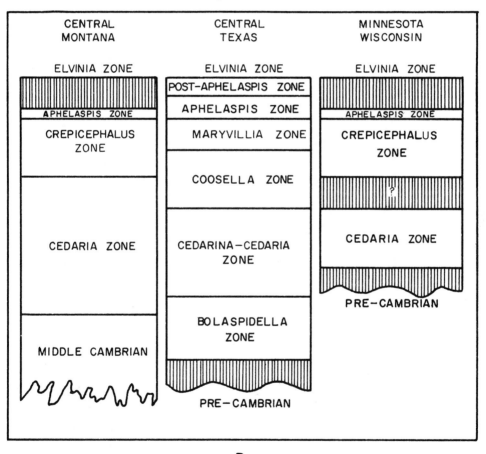

B

FIGURE 11.9 Information derived from trilobites. I. Correlation and paleogeography.
(*A*) Worldwide correlation of M. Cambrian trilobite zones. After V. Poulsen, 1964.
(*B*) Correlation of trilobite zones of Riley Formation, central Texas, with typical Dresbachian (Lower U. Cam., Minnesota, and so on). After A. R. Palmer, 1954.
(*C*) Paleogeography of North America and N. W. Europe based on distribution of trilobite faunas. Modified from Lochman and Wilson, 1958. (1) Lower Cambrian; (2) Middle Cambrian; (3) Upper Cambrian; *MN* — mixed assemblage (that is, shelf and trough fauna); *Eu* — eugeosyncline, *Mio* — miogeosyncline, *CRM* — craton (shelf) margin, *CRS* — cratonic shelves. *L* — land.

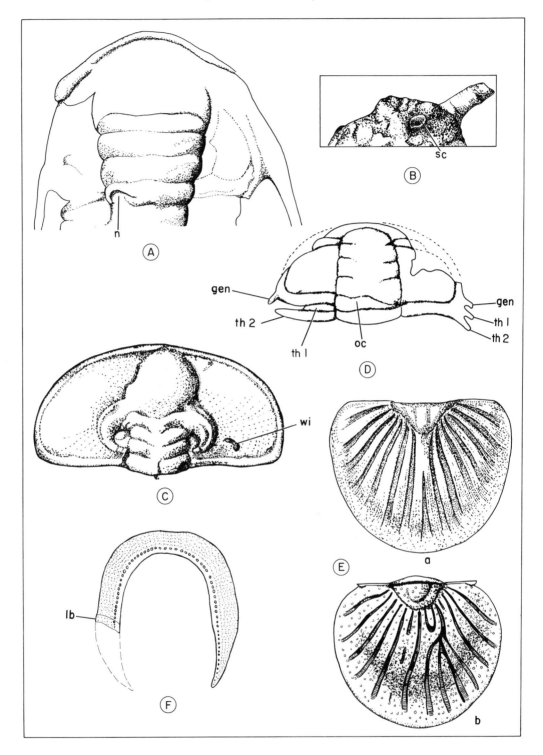

FIGURE 11.10 Information derived from Trilobites. II. Paleopathology.
(A) *Centropleura phoenix* (M. Cam., Queensland), *n* — notch, on left margin of occipital lobe subsequently healed (wound was inflicted before secretion of test). After Öpik.
(B) *Tricrepicephalus paracomus* (U. Cam., Missouri), X 3; *sc* — scar at base of left pygidial spine (after Lochman).
(C) *Olenellus getzi* (L. Cam., Pennsylvania), *wi* — worm impression (boring) on trilobite right cheek, X 1. After Ruedemann and Howell.
(D) *Rossaspis superciliosa* (L. Ord., Utah), X 1; *gen* — genal spines; *oc* — occipital lobe; *th1* — left abortive pleura; *th* 1, *th2* — on right, fused (Ross after Hupé).
(E) *Scutellum*. (a) *S. paliforum* (L. Dev., Bohemia), normal pygidium, (b) *S. flabelliforum* (M. Dev., Germany), pathological pygidium (Richter and Richter cit. Hupé).
(F) *Harpes* cf. *ottawaensis* (L. Trenton, Quebec), X 1.5; *lb* — ventral side of left brim; arrow bent to traverse break in brim, repaired by solid skin. After Sinclair.

1949; A. E. Wilson, 1947, Table 1); Silurian (calymenid trilobites, and so forth); Devonian (phacopids, and so on); (Delo, 1940; Richter and Richter, 1926, 1955).

Paleogeography, Bathymetry, Substrates.
The older concept that certain trilobites belong to the Pacific or Atlantic provinces (Gignoux, 1955) was dispelled by restudy of expanded available data on trilobite distribution (Lochman and Wilson, 1958). Trilobites from both provinces occurred in mixed faunas (that is, shelf/trough). Three tectonic divisions and corresponding environments were recognized: *Craton*: stable/unstable shelf area — formerly the Pacific realm; extracratonic — including troughs and (*a*) *miogeosyncline*, (*b*) *eugeosyncline* — equivalent to the Atlantic province. (More recent work has further modified this interpretation.)

Environmental characteristics of each of these three tectonic units are also those of the Cambrian trilobites that inhabited their waters. *Craton* — salinity, variable; depth, 0 to 50 ft, eastern North America; shelf assemblages found in sandstone, dark brown, dark gray, calcareous, sandy shale, and pure to sandy limestones and dolomites. Cordilleran area: Late Cambrian beds represent transgressive sands and shales. Craton on shelf area characterized by continental land mass sedimentation. *Miogeosyncline* — shallow water (from shoals to moderate depths, but deeper water along outer half of trough); varied salinity (but no fresh or brackish components). In Cordilleran areas carbonates are prominent.

Eugeosyncline — control by open ocean; normal oceanic salinity; shallow around islands but depth of 400 to 600 ft beyond these was inhabited by benthos. Poorly vented basins represented the euxinic environment. Olenid trilobites especially adapted to this environment (Henningsmoen, 1957). Characteristic lithologies include sand, sandy shale, gray, green, purple, and black siliceous shales; siltstones and claystones; thin, dark-gray, argillaceous limestones — beds, nodules, and lenses; graywackes. Clastics dominate the euxinic biofacies; carbonates subordinate, impure, and areally limited; limestones sometimes associated with volcanics and dark cherts.

Lochman has remarked that most trilobite genera in any Cambrian assemblage were *vagrant benthos*, although the Agnostida may have been sessile benthos buried in bottom ooze. Generally, Cambrian trilibites avoided surf and strong wave zones and sought out quieter habitats such as shallows of protected bays or moderate depths off open coasts (100 to 200 ft deep).

Paleopathology and Regeneration.
Compare the occipital lobe in Figure 11.10*A* and *D*. In these quite unrelated genera, one individual experienced a wound (notch in the occipital lobe, that is, in the fleshy soft part prior to secretion of the test), which subsequently healed; the other (*Rossaspis*) shows a malformed occipital lobe and malformation or fusion of thoracic segments. Clearly, these three sites of abnormal growth were triggered by a chemical or mechanical operation in some segment of the genetic code of the given animal.

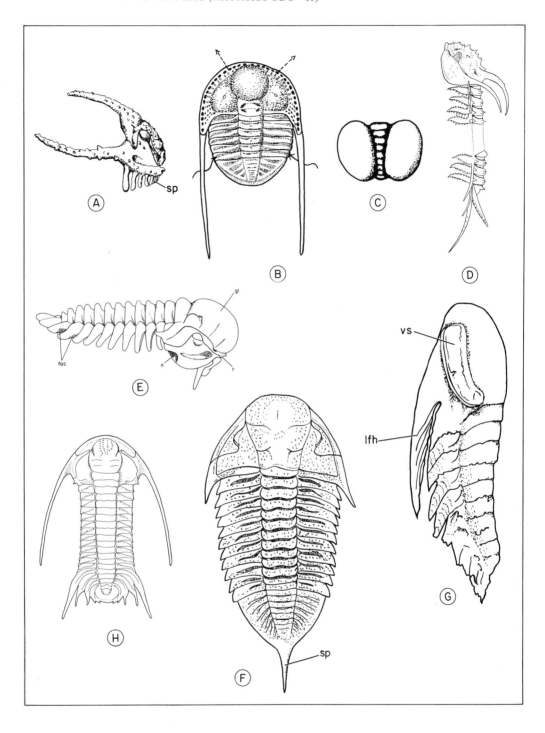

FIGURE 11.11 Information derived from trilobites. III. Some adaptive modifications.

(A) *Acidaspis brightii* (M. Sil., England), X 2, *sp* — spines facilitate feeding and/or respiration.

(B) *Tretaspis seticornis* (U. Ord., British Isles), arrow indicates water intake by flexion of thorax and expulsion via pits of perforate flange.

(C) *Pemphigaspis bullata*, swollen pygidium suggests planktonic existence.

(D) *Ceratocephala lacinata*, X 2.3 (see Figure 11.6G); spinosity facilitated flotation.

(E) *Sphaerexochus pulcher*, multiple adaptation: strongly curved glabella (*gl*) for buoyancy; *n* — notch in border of free cheek that received tips of thoracic pleura during enrollment; *fac* — facet; *r* — rostrum.

(F) *Dalmanitina* (D.) *socialis* (Ord.-Sil., Czechoslovakia), *sp* — pygidial spine thought to have served a mooring function as tail spine in *Limulus*.

(G) *Hypodicranotus striatulus* (Trenton Is., New York), X 3; lateral view; *vs* — visual surface of left eye; *lfh* — long fork of hypostome that reached pygidia and prevented enrollment.

(H) *Centropleura phoenix* (M. Cam., Queensland). Cephalon and pygidial adaptations interpreted to have favored a free floating and pelagic (swimming) existence. Reduced pleura length diminished effectiveness of protective coiling. Reconstruction after Öpik, 1961.

Illustrations after Whittington, Palmer, Whittington and Evitt, and others.

Other abnormalities or so-called monstrosities (= teratologies) are found in sizable segments of a population. For example, two out of every five individuals or some 40± percent of 2000 specimens in *Trinucleus pongerardi* (Oehlert, 1895) displayed such teratology as bifurcation of the posterior end of each, or at least of one of the genal spines. Hupé (1953) speculated whether hybridization might not have occurred by mating of normal individuals (that is, without bifurcate genal spines) and a mutant individual with bifurcate spines.

Considering the large representation of this teratology, one can assume that it was due to a mutation. Such teratologies are infrequent and incidental in the evolution of the family Trinucleidae. This indicates that selective pressures favored the normal condition (nonbifurcate genal spine).

Numerous pathological pygidia have been described. Lochman described one of *Tricrepicephalus paracomus* in which the left pygidial spine, a hollow structure, had been snapped off, leaving a wound in the side of the pleural lobe (Figure 11.10B). The wound healed, covered by new epidermis. However, the spine did not regenerate at the molt represented by the exuvia being studied.

Comparison of Figure 11.10E, a and b, indicates the nature of the pathology in the latter. Deformation in *Scutellum* may be due to a parasitic infection or to repair of a tear that occurred at time of molting.

Sinclair (1947) described the repair of an injury to the left brim of *Harpes* cf. *ottawaensis* (Figure 11.10F). Another of Sinclair's specimens, a species of *Ceraurus*, showed partial replacement of a genal spine tip after breakage and loss.

Parasitic worms may have imbedded in trilobite carapaces in order to get at the digestive glands or imbedded in other parts of the skeleton (compare Figures 11.1D and 11.10C). Portlock, Peach, and others (cit. Ruedemann and Howell, 1944) have reported such worm borings in trilobite tests.

Adaptive Modifications. Trilobites display a variety of modifications of apparent adaptive value (and, hence, favored by natural selection). Among the Odontopleuridae (*Acidaspis*, and so on), the interior lateral cephalic border bears a row of vertically directed spines of graduated lengths. The cephalon could have rested on tips of spines (Figure 11.11A). This appears to have been an adaptation for resting close to the sea floor, slightly raised above it to facilitate feeding (Whittington, 1956).

Begg (1944) concluded from a study of a sample (*n* = 65) of one of the Trinucleidae (genus *Tretaspis*) that the animal could execute upward and downward flexor motions of the thorax while resting immobile on the bottom. Two possible explanations were forthcoming: (1) Upward flexion could free the external gills (preepipodites) from bottom mud and hence facilitate respiration; (2) upward flexion may have been a feeding mechanism that caused in-current water-bearing food particles to rush under the thorax and freed most nutrient particles when downward flexion expelled water through pits of perforate fringe (Figure 11.11*B*).

Adaptations for planktonic existence can be seen in *Pemphigaspis* with a swollen pygidium and expanded glabella (Figure 11.11*C*). Swimming and floating adaptations are clearly evident in spinose forms like *Ceratocephala* (Figure 11.11*D*) (Whittington and Evitt, 1953). In *Centropleura*, Öpik (1961) thought reduced length of the anterior 13 pleurae and flatness of pleural lobes of the thorax were swimming adaptations. He regarded species of this genus as pelagic hunters.

The same trilobite can, of course, display multiple modifications, for example, for buoyancy (expanded glabella that may have contained fat), for enrollment (notch to receive thoracic pleurae) (Figure 11.11*E*).

Many trilobites (olenellids, dalmanitaceans, and so on) have prominent pygidial spines. These are thought to have served a mooring function as in *Limulus*. The latter, a mud-burrower and/or plougher, forces its tail spine into the substrate as a pivot from which it can thrust its body forward (Figure 11.11*F*).

Other trilobites had enormous visual surfaces (some were confluent anteriorly): *Cyclopyge*, *Symphysops*, and the remopleurids (*Hypodicranotus*) (Whittington, 1952, and see Figure 11.11*G*). It is thought (Swinnerton, 1947) that trilobites with enlarged visual surfaces lived in an open water habitat that was illuminated from below and above—implying a level high up in the euphotic zone.

Whittington (1952) redescribed a known remopleurid trilobite that had restricted body movements as a result of a unique structure—long, forked hypostome, the forks of which reached backward to anterior of the pygidium. This same species, *Hypodicranotus striatalus*, also had a very large visual surface. Forks of the hypostome provided protection of soft ana-

tomy on the ventral side; hence the creature's inability to enroll for protection was compensated. Large eyes and general ovate configuration suggest a floating or swimming existence. The fossil occurred with shallow-water marine benthos. However, floaters or swimmers would fall to the bottom upon demise and become fossilized with benthos if any were present.

Finch (1904, pp. 179–181, and cit. Raymond, 1920, pp. 102) found a slab (2 to 3 ft × 1 ft) of Maquoketa shale with 15 complete trilobites (*Vogdesia vigilens*). Each of these had the cephalon extended horizontally near the surface, while thorax and pygidium projected downward. Might these forms have been in the habit of burying themselves posterior end first, up to the eyes, waiting for prey? Finch thought so. Raymond (1920) concurred and further noted that forms like *Bumastus* (Figure 11.5*H*), *Illaenus*, *Asaphus* (Figure 11.6*E*), among others, had shapes similar to that of *Vogdesia*. Besides, all of these forms possessed incurved pygidia and large or tall eyes. Accordingly, Raymond suggested that all such forms had adapted to the habit of digging into the substrate from the posterior end.

If we follow Raymond's line of reasoning, the indicated characteristics might be viewed as having selective value. A contrary selective pressure would be the danger of a large influx of mud sufficient to smother the mud-packed individuals of, say, the *Vogdesia* population.

It is likely that the pygidia of many trilobites, viewed as solid tagma, could have been quite versatile. Such versatility could have expressed itself as aiding propulsion, digging, a protective function for soft parts during enrollment and, in rare instances, mooring through a pygidial spine.

Geologic History and Extinction

Study of Figure 11.12 can prove instructive.

1. *Origination of Orders.* Of the seven trilobite orders, five arose in the Cambrian, two in the Ordovician. No new or-

FIGURE 11.12 Stratigraphic distribution of trilobite orders and Late Paleozoic families. Adapted and modified from Harrington, 1959.

Order	Cambrian LMU[a]	Ordovician LMU	Silurian LMU	Devonian LMU	Mississippian LU	Pennsylvanian LMU	Permian LMU
Agnostida	xxxxxxxxxxxxxxxx						
Redlichiida	xxx						
Corynexochida	xxxxx						
Ptychopariida	xxxxxxxxxxxxxxxxxxxxxxxxxxxxxxxxxxxxxxxxxxxxxxxxxxxxxxxxxxxxxxxxxxxxxxxxxxxx						
Phacopida	xxxxxxxxxxxxxxxxxxxxxxxxxxxxxxxxxxxxxxxxxxxx						
Lichida	xxxxxxxxxxxxxxxxxxxxxxxxxxxxxxxxxxxxxxxxxxx						

Late Upper Paleozoic Families							
Proetidae			xxxxxxxxxxxxxxxxxxxxxxxxxxxxxxxxxxxxxxxx				
Phillipsiidae						xxxxxxxxxxxxxxxxxxxxxxxxxxxxxx	
Otarionidae			xxxxxxxxxxxxxxxxxxxxxxxxxxxxxxxxxxxxxxxxxxxxxxxxxx				
Brachymetopidae				xxxxxxxxxxxxxxxxxxxxxxxxxxxxxxxxxxx			

[a]L – Lower; M – Middle; U – Upper.

ders appeared between Ordovician to Permian time. Cambrian to Ordovician history thus holds the key to the essential range of diversity distinguishing trilobite orders.

2. *Extinction of Orders*. By the end of Cambro-Ordovician time, three orders were extinct (Redlichiida, Corynexochida, and the Agnostida, in that sequence). Another two orders were extinct by the end of the Devonian. Only the Ptychopariida spanned the entire Paleozoic and, within that order, only certain families of the superfamily Proetacea persisted to Middle Permian time (see Figure 11.12).

Thus about three-sevenths of trilobite history was completed by the end of the Cambrian and another two sevenths by the end of the Devonian. To think of extinction of trilobites as a sudden event in Middle Permian time is misleading.

The class, as seen through serial extinctions of the several orders, was retrograde or stagnant in an evolutionary sense from post-Devonian to Middle Permian time. In terms of origination of new families, a sign of evolutionary vigor, this evaluation can be extended back to the Silurian. It does not mean that no further speciation occurred. What it does imply is that only two new families appeared — one in the Devonian, Brachymetopidae, and another in the Mississippian, Phillipsiidae. (The

Silurian also saw the rise of a single family — the Dalmanitidae.)

3. *Comparative Million-Year Span of Orders*. In terms of millions of years ($\times 10^6$), the span of time that each of the seven orders survived follows: Ptychopariida, 340; Odontopleurida, 205; Agnostida, 175; Phacopida and Lichida, 155; Corynexochida, 100; and Redlichiida, 70.

4. *Ventral Appendages*. Fossilized appendages of trilobites from Lower Cambrian (*Olenoides*, and so on) through Lower Devonian (*Phacops*, and so on) (Raymond, 1920; Harrington, 1959) reveal that the basic plan of trilobite biramous appendages (all uniform and nondifferentiated) did not change by genetic modification through time.

5. *Exoskeleton*. As noted earlier, two major tendencies in trilobite exoskeletal evolution were *cephalization* (that is, complete fusion of glabellar furrows) and *caudalization* (that is, complete fusion of pygidial segments).

Permo-Carboniferous Speciation

Since an acceptable genetic classification is presently unavailable, the difficulty in deciphering trends and lineages is apparent. There are exceptions in isolated

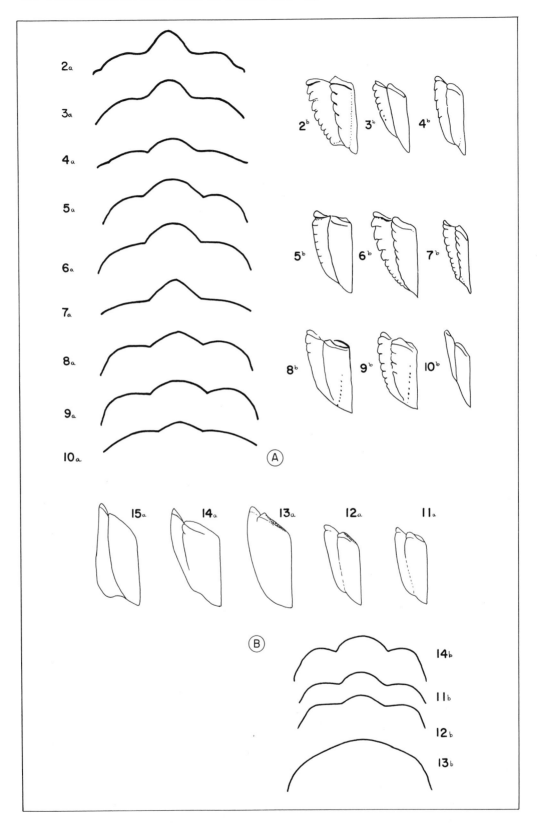

FIGURE 11.13 Pygidial variation in species of Lower Carboniferous trilobites. Data from Richter and Richter, 1949, 1951, 1959. Schematic.
(A) Genus *Cyrtosymbole*, Subgenus *Waribole:* (2a, b) C. (W) *abruptirhachis*, subgenus *Macribole:* (3a, b) C. (W) *drewerensis drewerensis;* (4a, b) C. (W) *duodecimae;* (5a, b) C. (M) *blax;* (6a, b) C. (M) *hercules;* (7a, b) C. (M?) *ogilivis.*
Genus *Liobolina:* (8a, b) L. *nebulosa;* (9a, b) L. *submonstrans;* genus *Diacoryphe;* (10a, b) D. *pfiefferi.* (B) Genus *Liobole:* (11a, b) L. *glabroides;* (12a, b) L. *glabra glabra;* (13a, b) L. *coalescens;* (14a, b) L. *glabra hiemalis;* genus *Cystispina;* (15) C. *nasifrons.*

cases, as in Lochman's study of some Upper Cambrian and Lower Ordovician trilobite families (1956).

Despite limitations, it is valuable to consider Permo-Carboniferous speciation in view of Middle Permian extinction of trilobites.

Phillipsia is thought to be the ancestral type, or close to this type, from which most Carboniferous trilobites have been derived (Weller, 1937) (Figure 11.13*A*). Genera of the family Phillipsiidae known from the Lower Permian include *Ditomopyge*, *Delaria*, *Paladia*, and *Ameura* (Figure 11.14*I*). Phillipsid genera of the last trilobite radiation (Middle Permian) include *Permoproetus*, *Paraphillipsia*, *Neoproetus*, *Vidria*, *Pseudophillipsia*, *Anisopyge*, *Neogrif-*

fithides (Figure 11.14*A–H*), and *Ditomopyge artinskiensis.*

What changes in primitive character occurred from the *Phillipsia*–ancestral type through different phillipsid lineages (Figure 11.14)? (1) Glabella shape and structure changed, (2) size of eyes changed, (3) nature of the anterior cephalic border changed, (4) pygidial form changed. How? The glabella expanded in front. The visual surface increased. Anterior to the glabella, the flat marginal border was lost; a preoccipital lobe developed, and the basal glabellar furrows were straightened. The number of pygidial axial segments increased, but not in step with the number of pleural segments. The flange of the pygidia became sharply differentiated; pygidial axis elevated, and in one lineage, axial crest flattened and pleural lobes became geniculate (that is, bent under).

Types of pygidial changes in Lower Carboniferous trilobites of genus *Cyrtosymbole* (subgenus *Macrobole*) and genus *Liobole* are shown in Figure 11.13, based on studies by the Richters.

Richters' (1959) diagnosis of superfamily Proetacea recognized numerous subgenera; *Proetus*, with 2 subgenera; *Cornuproetus*, 3; *Deckenella*, 4; *Cyrtosymbole*, 6 (see Figure 11.13, *A*); *Drevermannia*, 4; *Brachymetopus*, 2, among others. Also, former subgenera of genus *Phillibole* (Richter and Richter, 1949), namely *Liobole* and *Cystispina*, are now recognized as genera. This trend towards greater subdivision of Permo-Carboniferous trilobites is also found in Hessler's studies of Lower Mississippian Proetidae of the United States (1962, 1963). Reason for this trend is the marked variation observed in such populations.

In Hessler's studies, a new subgenus of *Proetus*, *Padoproetus*, was erected and a

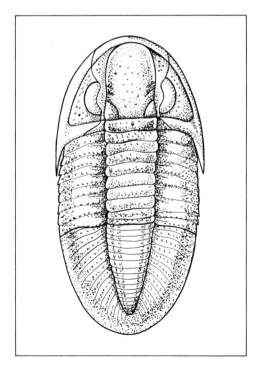

FIGURE 11.13A *Phillipsia.* A primitive genus ancestral (?) to most Carboniferous trilobites (Schematic).

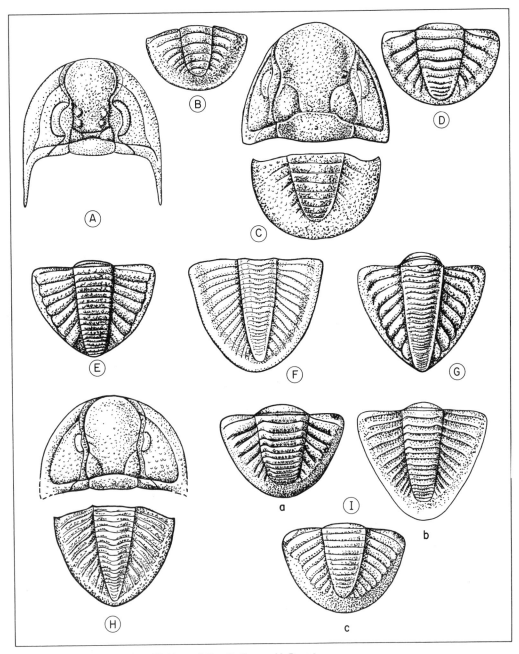

FIGURE 11.14 The last trilobite radiation (L. Perm. – M. Perm.)

(*A*) *Delaria antiqua* (M. Perm., Texas), X 3.5.
(*B*) *Permoproetus teschi* (M. Perm., Crimea), X 1.5.
(*C*) *Paraphillipsia karpinskyi* (M. Perm., Crimea), X 3.
(*D*) *Neoproetus indicus* (M. Perm., Timor), X 2.5.
(*E*) *Vidria vespa* (M. Perm., Texas), X 2.
(*F*) *Pseudophillipsia sumatrensis* (M. Perm., Sumatra), X 1.
(*G*) *Anisopyge perannulata* (M. Perm., Texas), X 2.6.
(*H*) *Neogriffithides gemmellaroi* (M. Perm., Crimea), X 1.3.
(*I*) (*a*) *Ditomopyge scitula* (Penn., Illinois), X 2. (*b*) *Ameura sangamonensis* (Penn., Illinois) X 2.
(*c*) *Paladin morrowensis* (L. Penn., Arkansas), X 2. Note, (*a*)–(*c*) extended into Lower Permian.
Illustrations redrawn and adapted from Harrington; J. M. Weller; and others.

specimen, formerly recognized as *Phillipsia samsoni*, redefined as the genotype of a new genus, *Breviphillipsia*. Furthermore, the species *Proetus ellipticus* was taken as the genotype of a new genus, *Elliptocephala*. A comparison of the skeletal morphology of some Lower Carboniferous genera is given in Table 11.3.

As clearly shown in Figure 11.13, *A* and *B*, marked variation existed in Permo-Carboniferous gene pools. Obviously, considerable similarities are also evident and suggest that equivalent gene combinations were available (Table 11.3). Aside from modification of genotypes for the pygidial genetic specifications, other changes noted above reflect similar modification of the genetic code. Such changes involved glabella, visual surface, anterior glabella border, and so on. Again, natural selection must have been the arbiter, deciding which modifications were to be favored, that is, not to be selected out.

Last survivors among the trilobites, Middle Permian genera of the family Phillipsiidae, all shared certain characteristics: pygidium and cephalon approximately same size (isopygous); glabella large and cephalon lacking preglabellar field; pygidial border tending to steepen; all individuals small and exoskeletal carapace markedly ornamented with granules, tubercles, or very short spines. However, despite these shared characteristics, amount of variation between species and genera was marked.

Any visible pathological or aberrant tendencies cannot be attributed to these Middle Permian trilobites destined to be the last of their class. The comparatively large glabella was a likely juvenile character, whereas marked axial segmentation of the pygidia appears to have been a primitive feature. On the other hand, their ornamentation may be considered relatively insignificant when compared with the exotic spinose development of Lower and Middle Paleozoic forms (*Tetraspis, Odontopleura*, and so on).

There are many explanations to account for Middle Permian extinction of trilo-bites. One of the most persistent of these is the concept of racial old age or senility (*phylogerontismus*). The marked diversity between forms in the Permo-Carboniferous, the spectrum of variation, and extensive geographic spread all argue against a loss of racial vigor—whatever that means!

The geographic spread for Lower and Middle Permian phillipsid genera is impressive: Crimea, Sicily, United States, Central Himalaya, East Indies, Southeast Asia, England, Europe, southern Manchuria.

These observations suggest actively interbreeding regional populations, passage of mutant and/or recombined genes through local gene pools of the time, and introduction of new genotypes into such local pools through migrant populations. Concurrent with all of this must have been the normal operation of natural selection that preferred a given genotype to some other.

Racial senility? This concept is absurd.

Each of the following explanations have been put forth: Trilobites became extinct because of rise of the pavement-toothed shark and other predatory fish (particularly Devonian and post-Devonian); tendencies in the Lichidae, and so on, toward excessive ornamentation (spinosity) represented inefficient utilization of the creature's biological processes; climatic zonal changes that opened or closed trilobite migration routes. (The cosmopolitan distribution of Permo-Carboniferous forms lends little support to this last hypothesis.) Additional explanations include such items as change in ocean currents arising from successive orogenies (Dunbar, 1960; Hupé, 1953; Richter and Richter, 1951). Bergström (1969) suggested that the "little evolutionary flexibility in trilobite appendages" could explain their total extinction with the rise of more advanced and faster moving competitors.

The most plausible hypothesis is that which allows for both predators (fish, for example) and competitors (other inverte-

TABLE 11.3. Comparative Skeletal Morphology of Five Lower Carboniferous Trilobite Genera

Structure	Proetus	Phillipsia	Breviphillipsia	Elliptophillipsia	Phillibole
Outline of carapace	Semicircular	Semioval	Subtriangular	Parabolic	Parabolic
Vaulting of carapace	High	High	Medium	Low	Medium
Cephalic border	Thick to thin, evenly rounded	Medium-sized, with flat outer surface	Medium-sized, usually with flat outer surface	Thin, with flat outer surface	Thin, evenly rounded
Lateral border furrow	Sharp	Sharp	Sharp	Sharp	Broad
Anterior border	Present	Present	Present	Present	Present, with narrow p:eglabellar field
Genal angle	With genal spines	With genal spines	Blunt	With genal spines	With genal spines
Glabella	0.94–1.13 times as long as wide; sides straight, converging anteriorly; widest at basal lobes	1.15–1.49 times as long as wide; sides nearly straight, usually converging anteriorly; widest at basal lobes	1.06–1.64 times as long as wide; sides nearly straight, but slightly constricted to 2p furrow; converging anteriorly; widest at basal lobes	1.38–1.45 times as long as wide; sides nearly straight, converging anteriorly or subparallel; widest at basal lobes	1.15–1.35 times as long as wide; sides convex; parallel forward to level of δ, then converging anteriorly; widest at basal lobes
1p furrows	Curved	Curved	Curved	Curved	Curved
Basal lobes	Present in Padoproetus, convex	Convex	Convex	Convex	Rather flat
2p-4p furrows	Defined in Padoproetus, separated	Well-separated	Well-separated	Well-separated	Obscure in P. conkini
Eyes	Medium-large	Small	Medium	Large	Medium
Thoracic segments	Ten	Nine	Nine	Nine	Nine
Pygidial ratio	1.58–2.18 times as wide as long	1.11–1.32 times as wide as long	1.21–1.52 times as wide as long	1.54–1.56 times as wide as long	1.54–1.88 times as wide as long

	6–11 rings / 5–8 ribs	15–19 rings / 12–16 ribs	9–13 rings / 9–12 ribs	9–10 rings / 9 ribs	11–13 rings / 5(+?)–8 ribs
Pygidial segmentation					
Pygidial border	Present	Absent	Usually absent	Absent	Absent
Posterior half-ribs	Normally developed	Suppressed	Usually suppressed, or forming anterior slope of pleura	Normally developed	Normally developed
Ornamentation	Granules	Spines	Smooth, granules, spines, tubercles	Granules	Granules

Source. After Hessler.

brates such as the abundant bottom feeders and scavengers of the Paleozoic).

Identification of a Trilobite to the Species Level

[Date from L. A. Hintze, 1952, Lower Ordovician trilobites from western Utah and eastern Nevada: Utah Geol. and Min. Sur. Bull. 48, p. 243, 8 plates; also consult Ross (1951, pp. 97–100) and Treatise volume on trilobites.]

Given. Four trilobite pygidia, 2 cranidia, 1 incomplete and 1 complete fossil with attached incomplete cephalon, thorax, and attached pygidia, and an isolated hypostome. (For terminology, consult this chapter.)

Description of Fossils. The pygidia display a flat, rather broad border; pleural fields are smooth or show faint ribs. Another feature is the median notch (see below), which gives a bilobed appearance to adult pygidia (but is absent in immature pygidia less than 5.00 mm in length).

Thoracic segments, 8; facial sutures are intramarginal (that is, along border of cephalon close to margin), and palpebral lobes are large. Posterior edge of free cheek and genal spine meet at almost a 90-degree angle. Anterior wings of hypostome are quadrangular.

Familal and Generic Level. To what trilobite genus can these fragments be assigned? To what family?

Tendency toward loss of apparent segmentation of the cephalon and pygidium, rounded margin of the pygidium, as well as the 8 thoracic segments, all suggest family Asaphacea. This family is divided into several subfamilies. One of these, the subfamily Symphysurininae, has a hypostome with quadrilateral anterior wings. (This feature also occurs in the subfamily Asaphinae but, in addition, the hypostome has a deep notch in the posterior margin – lacking in the specimen being studied here.) The subfamily embraces five genera.

Four of these genera can be eliminated as follows:

Symphysurina. Facial sutures are marginal (the unknown specimens have intramarginal sutures); differs from the unknown in shape of cranidium, genal spines, and terminal spines on posterior end of pygidium in the type species.

Kobayashia. Has larger frontal area (*fa*, see Figure 11.14*A*) than the unknown; a narrower glabella in front of the eyes; lateral glabella processes are more strongly expressed; few but more strongly expressed ribs on inner part of pygidial pleural field (in contrast to relatively smooth pleural field in the unknown).

Parabellefontia. Unlike the unknown in its almost complete obsolescence of the cephalic and pygidial axes; in other words, the cranidium and pygidia are smooth, and the glabella and pygidial axis are undefined.

Varvia. Facial suture marginal; axis of pygidium very short with faint depressed border. These features are unlike those of the unknown specimens.

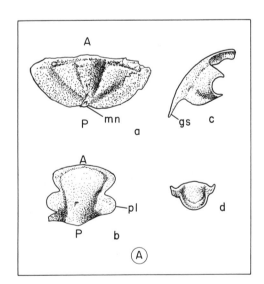

FIGURE 11.14A Unknown species (L. Ord., Western Utah). (a) Pygidia, (b) cranidia, (c) free cheek, (d) hypostome. Drawn from photographs in Hintze, 1952, plate IV.
Symbols: *pl* – palpebral lobe; *A* – anterior; *P* – posterior; *mn* – median notch; *gs* – genal spine; (*fa*) = frontal area not marked on figure, located directly below letter *A* in (*b*).

The only other genus in this subfamily is *Bellefontia*. It has intramarginal facial sutures, a well-distinguished pygidium with broad, flattened border, and relatively smooth pleural fields. These features correspond to those in the unknown. Accordingly it is assigned to *Bellefontia*.

Since there are two subgenera, to which one does the unknown material belong?

Bellefontia (Subgenus Bellefontia). It has a narrow frontal area and posterior end of the pygidium is rounded—as in the unknown specimens. The other subgenus (*Xenostegium*) has a longer frontal area than subgenus *Bellefontia* and bears a terminal spine on the pygidium. Obviously, the unknown must be assigned to *Bellefontia (Bellefontia)* species.

Species Level. To what species shall these specimens be assigned? Hintze (1952) used three sets of criteria, placing greatest reliance on the pygidial feature "median notch."

Cranidium. There are slightly larger palpebral lobes (*pl* in Figure 11.14*A*) in the unknown species than in any of the following known species: *B.(B.) colliena, B. (B.) nonius,* and *B. (B.) chamberlaini.*

Pygidium. Median notch in pygidial rim that gives a bilobed appearance is a unique feature and does not occur in other species of the genus. (The concave border of many genera in the family Asaphidae should not be confused with "median notch" since it is not an indentation, but rather, an upward, narrow flexure of the medial sector of the pygidial border.)

Free Cheek. The angle alpha (α) formed by the posterior edge of the free cheek and the genal spine is closer to 90° in the unknown species than in other species of the genus. For example, in *B.(B.) chamberlaini*, alpha is greater than 90°.

Hintze grouped his fossil specimens into a new species and gave them the name *Bellefontia ibexensis*. [Only subsequently did Ross (1956) propose two subgenera, making the more complete name at present for Hintze's species *Bellefontia (Bellefontia) ibexensis*.]

QUESTIONS

Arthropods — II
Trilobita

1. Why are trilobites such excellent index fossils in the Lower Paleozoic? Cite the various factors involved.

2. When we speak of "trilobites and *related* forms," what is the anatomical basis for the term "related"? Cite examples and class characteristics.

3. How are the following pairs associated: facial sutures and molting; population density and trilobite exuviae; major and nonfunctional sutures; enrollment and selective advantage?

4. (a) What are the three major features on which the Treatise classification of trilobites rests?

   (b) Distinguish between order Agnostida and all other trilobite orders; ptychoparids and phacopids; lichids and odontopleurids.

5. What was the relationship between Cambrian trilobites and the tectonic units: craton, miogeosyncline, eugeosyncline?

6. Review the last trilobite radiation and various explanations for trilobite extinction.

7. (a) State geological age represented if your hammer loosened a piece of rock containing: *Phacops, Isotelus, Centropleura, Tricrepicephalus, Cruziana, Ameura?*

   (b) How would you interpret a find of an olenellid in pre-Cambrian schist?

8. Which of the following trilobite fragments would permit you to identify the given fossil down to the genus level: free cheek, left half of thorax, pygidial axis, hypostome, genal spine, fragment of a cranidium, meraspid stage, perforate fringe? Give reasons.

9. A group of marine biologists have announced that they will undertake a deliberate search on the continental shelves in the hope of finding extant trilobites. Discuss the rationale for such a search in light of the data on trilobite evolution and extinction.

REFERENCES

Trilobita

Begg, J. L., 1944. On the fringe of *Tretaspis:* Geological Mag., Vol. 81 (3), pp. 113–117, plate 5.

Bright, R. P., 1959. Paleocologic and biometric study of the Middle Cambrian trilobite *Elrathia kingii* (Meek): J. Paleontology, Vol. 33 (1), pp. 83–98.

Cooper, B. N., 1953. Trilobites from the Lower Champlainian Formation of the Appalachian Valley: G.S.A. Mem., Vol. 55, pp. 1–69.

Cooper, G. A., 1935. Young stages of the Devonian trilobite *Dipleura dekayi Green:* J. Paleontology, Vol. 9 (1), pp. 3–5, plate 1.

Delo, D. M., 1940. Phacopid trilobites of North America: G.S.A. Sp. Paper 29, 135 pp., 13 plates.

Finch, G. E., 1904. Notes on the position of the individuals in a group of *Nileus vigilans* found at Elgin, Iowa: Proc. Iowa Acad. Sci., Vol. 11, pp. 179–181, plates 1–4.

Harrington, H. J., et al., 1959. Trilobita (including general description, morphology, ontogeny, classification, systematic descriptions): Treat. Invert. Paleontology, (0) Arthropoda, pp. 0–40ff.

_____ and A. F. Leanza, 1957. Ordovician trilobites of Argentina: Univ. Kansas Press, Lawrence, Kansas, 251 pp.

Henningsmoen, G., 1957. The trilobite family Olenidae: *Norske Videnskapsakademien* (Oslo) Mat.-Naturv. Kl. Skr. No. 1; 303 pp., 31 plates, 19 figs.

_____, 1951. Remarks on the classification of trilobites: Norsk. geol. tideskrift, Vol. 29, pp. 174–217.

Hessler, R. R., 1963. Lower Mississippian trilobites of the family Proetidae in the United States, Part 1, J. Paleontology, Vol. 37 (3), pp. 543–563.

_____, 1962a. The Lower Mississippian genus Proetidae (Tril.): J. Paleontology, Vol. 36 (4), pp. 811–816, plate 119.

_____, 1962b. Secondary segmentation in the thorax of trilobites: J. Paleontology, Vol. 36 (6), pp. 1305–1312, plate 176.

Hintze, L. H., 1952. Lower Ordovician trilobites from Western Utah and Eastern Nevada: Utah Geol. and Min. Surv. Bull. 48, 243 pp.

Hupé, P., 1953. Classe des Trilobites, in Jean Piveteau, ed. Traité de Paleontologie, Vol. 3, pp. 44–246.

Kielan, Z., 1954. Les trilobites mesodevoniens des Monts de Sainte-Croix: Palaeontologia Polonica, No. 6, pp. 1–47, plates 1–7.

Lemche, H., 1957. Supposed asexual reproduction in trilobites: Nature, Vol. 180, No. 4590, pp. 801–802.

Lochman (Lochman-Balk), C., 1956. The evolution of some Upper Cambrian and Lower Ordovician trilobite families: J. Paleontology, Vol. 30 (3), pp. 445–462, plate 47, 7 text figures.

_____, 1941. A pathologic pygidium from the Upper Cambrian of Missouri: J. Paleontology, Vol. 15 (3), pp. 324–325, 3 text figures.

_____ and J. L. Wilson, 1958. Cambrian biostratigraphy in North America: J. Paleontology, Vol. 32 (2), pp. 312–350.

Oehlert, D. P., 1895. Sur les *Trinucleus* de l'ouest de la France: Bull. Soc. Géol. de France: Troisieme Série, Vol. 23, pp. 299–336, plates 1, 2.

Öpik, A. A., 1961. The geology and paleontology of the headwaters of the Burke River, Queensland: Bur. Min. Res., Geol. & Geophys. Bull. 53, 196 pp.

_____, 1958. The Cambrian trilobite Redlichia: organization and generic concept: Bur. Min. Res., Geol. & Geophys. Bull. 42, 36 pp.

Palmer, A. R., 1954. The fauna of the Riley Formation in Central Texas: J. Paleontology, Vol. 28, pp. 709–786, plates 76–92, 6 figures.

————, 1951. *Pemphigaspis*, a unique Upper Cambrian Trilobite: J. Paleontology, Vol. 25, pp. 762–764, plate 105.

Poulsen, V., 1964. Contribution to the Lower and Middle Cambrian paleontology and stratigraphy of Northwest Greenland: Meddel. om Grønland (Copenhagen), Vol. 164, No. 6, 82 pp. (English).

Rasetti, F., 1952*a*. Revision of the North American trilobites of the family Eodiscidae: J. Paleontology, Vol. 26, pp. 434–451, plates 51–54.

————, 1952*b*. Ventral cephalic sutures in Cambrian trilobites: Am. J. Sci., Vol. 250, pp. 885–898.

————, 1948. Cephalic sutures in *Loganopeltoides* and the origin of "hypoparian" trilobites: J. Paleontology, Vol. 22, pp. 25–29, plate 7.

Raymond, P. E., 1922. A trilobite retaining color markings: Am. J. Sci., Vol. 4, ser. 5, pp. 461–464, Figure 1.

————, 1920. The appendages, anatomy, and relationships of trilobites: Conn. Acad. Arts & Sci. (New Haven), Mem. Vol. 7, pp. 1–169, plates 1–11, Figures 1–46.

Raw, F., 1927. The ontogenies of trilobites and their significance: Am. J. Sci., Vol. 14, ser. 5, pp. 7–35, 131–149, 240, 27 figures.

————, 1925. The development of *Leptoplastus salteri* and other trilobites: Quart. J. Geol. Soc. London, Vol. 81 (2), pp. 223–324, plates 15–18.

Richter, R., and E. Richter, 1955. Oberdevonische Trilobiten, Nachträge, 2, Phylogenie der oberdevonischen Phacopidae: Senckenberg, leathaea (Frankfurt A.M.), Vol. 36, pp. 56–72, Figures 1–2.

————, 1951. Der Beginn des Karbons im Wechsel der Trilobiten: Senckenbergiana, Vol. 52, pp. 219–266, plates 1–5.

————, 1949. Der trilobiten der Erdbach-Zone (Kulm) ein Rheinischen Schiefergebirge un deim Harz: Senckenbergiana, Vol. 30, pp. 63–94, plates 1–5.

————, 1926. Die Triloliten des Oberdevons, Belträge zur Kenntnis devonischer Triloliten, 4. Abhandl. preuss. geol. Lande-sanst. (Berlin), Vol. 99, pp. 1–314, plates 1–12, Figures 1–18.

Ross, R. J., Jr., 1951. Stratigraphy of the Garden City Formation in Northeastern Utah and its Trilobite fauna: Peabody Mus. Nat. Hist., Yale Univ. (New Haven) Bull. 6, pp. 1–161, plates 1–36.

————, 1949. Stratigraphy and trilobite faunal zones of the Garden City Formation, Northeastern Utah: Am. J. Sci., Vol. 247, pp. 472–479, 2 figures.

Ruedemann, R., and B. F. Howell, 1944. Impression of a worm in the test of a Cambrian trilobite: J. Paleontology, Vol. 18 (1), p. 96, plate 19.

Saul, J., 1965. Trilobita in J. B. Hadley, ed., Geol. and Paleont. of the Antarctic, Antarctic Res. Ser., Vol. 6, AGU, pp. 269–271.

Shaw, A. B., 1953. Paleontology of Northwestern Vermont, III. Miscellaneous Cambrian fossils: J. Paleontology, Vol. 27 (1), pp. 133–146.

————, 1952, Part 2, *idem*, Vol. 26 (3), pp. 458–483.

————, 1951. Part 1, *idem*, Vol. 25 (1), pp. 97–114.

Sinclair, G. W., 1947. Two examples of injury in Ordovician trilobites: Am. J. Sci., Vol. 245 (4), pp. 250–257, plate 1.

Størmer, L., 1951. Studies in trilobite morphology, Part III. Norsk. Geol. Tidskr. (Oslo), Vol. 29, pp. 108–158.

————, 1942. *Idem*, Part II, *idem*, Vol. 21, pp. 49–164.

————, 1939. *Idem*, Part I, *idem*, Vol. 19, pp. 143–213.

Swinnerton, H. H., 1950. Outlines of Paleontology. E. Arnold & Co., 3rd ed., 370 pp.

Taylor, M. E., 1966. Pre-Cambrian Molluscan-like fossils from Inyo County, California: Science, Vol. 153 (3732), pp. 198–201, Figures 1–4.

Ulrich, E. O., and C. E. Resser, 1933. The Cambrian of the Upper Mississippi Valley, Part 2, Trilobita. Bull. Publ. Mus. City of Milwaukee, Vol. 12 (1), pp. 123–306.

————, 1930. *Idem*, Part 1, *idem*, Vol. 12 (1), pp. 1–122.

Weller, M., 1937. Evolutionary tendencies in American Carboniferous trilobites: J. Paleontology, Vol. 11 (4), pp. 337–346.

Whittington, H. B., 1956. Type and other species of Odontopleuridae (Trilobita): J. Paleontology, Vol. 30 (3), pp. 504–520.

————, 1951. Some Middle Ordovician trilobites of the families Cheiruridae, Harpidae, and Lichidae: J. Paleontology, Vol. 25 (5), pp. 587–616.

————, 1950. Sixteen Ordovician genotype trilobites: J. Paleontology, Vol. 24 (5), pp. 531–565.

Whittington, H. B., and R. W. Evitt II, 1953. Silicified Middle Ordovician trilobites: Geol. Soc. A. Mem. 59, 95 pp.

Williams, J. S., 1930. A color pattern on a new Mississippian trilobite: Am. J. Sci., Ser. 5, Vol. 20, pp. 61–64.

Wilson, A. E., 1947. Trilobita of the Ottawa Formation of the Ottawa-St. Lawrence Lowland: Can Dept. Mines and Resources, Geol. Surv. Bull. 9, 62 pp.

Bergström, J., 1973. Organization, life and systematics of trilobites. In Fossils and Strata, No. 2, Universitetsforlaget, Oslo. 1969. Remarks on the appendages of trilobites: Lethaia, Vol. 2, pp. 395–414.

CRUSTACEA – I

Introduction

Crabs, lobsters, ostracodes, barnacles, conchostracans, crayfish, and others, are embraced by superclass Crustacea. The name, which refers to the crustaceous or *hard shell* or shield, is not always applicable since, in some crustaceans, the carapace is either poorly developed or absent (Anostraca, Mystacocarida, Cephalocarida).

Two features characterize different lines of crustaceans: (1) naupliar stage of development, (2) possession of two pairs of antennae — adult, preoral antennules (first antenna) and the antennae (second antenna) (Figure 10.2). [There are exceptions to item (1). Among others, branchiopod conchostracan *Cyclestheria hislopi* bypasses the naupliar stage (Sars, 1887). Item 2 above distinguishes crustaceans from the Insecta (which have first pair of antennae only) and Myriopoda (which have no antennae).]

Jointed feet and mandibles also designate the class. However, these are respectively common to the phylum and subphylum Mandibulata.

Living Crustaceans

Modified Appendages. Walking limbs were reviewed and illustrated in Chapter 10. These related to specialized head appendages such as mandible, maxillipeds, and maxilla. The high degree of specialization of crustacean appendages includes modification, not only for *sensory* purposes (I and II antennae) and *feeding* (mandible, maxillae, and maxillipeds, Figure 10.2), but also for *reproduction* (egg-carrying or sperm transfer). Other modifications are for *grasping* (offense and defense) — the chelipeds — and partially for *respiration* (abdominal swimmerettes in crayfish, for example). Finally, uropods (Figure 10.2) served a *swimming* function.

Dahl (1963) indicated the probability that the primitive crustacean limb was concerned with both locomotion and feeding, that is, before modification, as remarked above and in Chapter 10. It fed by transporting large particles to the mouth. In the primitive living crustacean *Hutchinsonella* (Sanders, 1957), all limbs have two functions — first and second antennae have a sensory plus a swimming and crawling function, whereas remaining limbs serve for locomotion as well as feeding. Specialization of appendages, in arthropods generally, began anteriorly and worked its way backwards, according to Calman (1926). Thus posterior limbs are regarded as more primitive. This would place greatest selective pressure on the ancestral crustacean for modification of the most anterior walking legs for sensory-feeding functions.

Primitive and Ancestral Crustacean (Urcrustacean). In the past few decades two new primitive crustaceans were discovered. *Derocheilocaris typica* (Figure 11.15*D* and, for larval stages, Figure 11.16*G*) of the order or subclass Mystacocarida. This is a very small form, 0.5 mm in length, inhabiting interstitial capillary water between sand grains on intertidal sandy beaches of Massachusetts and elsewhere (Europe, Africa, Chile) (Pennak and Zinn, 1943). These forms are not known as fossils. In 1955 a few specimens of *Hutchinsonella macracantha* (Figure 11.15*A–C*) were collected from Long Island Sound. Subsequently, a larger sample (*n* = 320) was retrieved from Buzzards Bay, Massachu-

setts, in subtidal mud (Sanders, 1955, 1957). This species is assigned to subclass Cephalocarida.

Of the two primitive crustaceans noted, the cephalocarids are most primitive, and closest to the basal crustacean stock. It follows that the ancestor of all other crustaceans, including the mystacocarida, was cephalocarid-like. [*Hutchinsonella* is blind. Blindness, being a secondary acquisition, is not a primitive condition — a point stressed by Erik Dahl (1963, Chapter 14). Thus the ancestral crustacean had eyes.]

No generally accepted fossil cephalocarids are known. H. K. Brooks (1955, p. 853, fn) tentatively assigned a Pennsylvanian form *Tesnusocaris goldichi* from Texas to the cephalocarids (Figure 11.15*E*). Is this primitive crustacean a cephalocarid? There are several notable obscurities and differences that make this assignment questionable.

By contrast, lipostracan *Lepidocaris* (see Branchiopoda below) from the Rhynie Chert of Scotland is thought to be closest to the living cephalocarid. [Living *Hutchinsonella* is, however, more primitive than the Devonian *Lepidocaris* (both are blind and about 3.0 mm long) because (1) its first maxilla is only slightly modified, (2) its second maxilla is wholly unmodified, (3) its appendages are trilobite-like (Sanders, 1957).]

Sanders (1957, 1963) found that limb patterns in different crustacean subclasses were either directly comparable to cephalocarid counterparts (Figure 11.17*A* first maxilla, for example), or could be derived from them by reduction and simplication. He used the generalized limb patterns of living cephalocarids as a Rosetta stone to unravel the derivation of appendages in other crustacean subclasses.

Suppose the following are given: a cephalocarid first maxilla which, in larval form, consists of a flattened leaflike exopod (*Exo*) bearing setae laterally and distally; a multisegmented endopod (*End*) with each joint bearing two setae on its

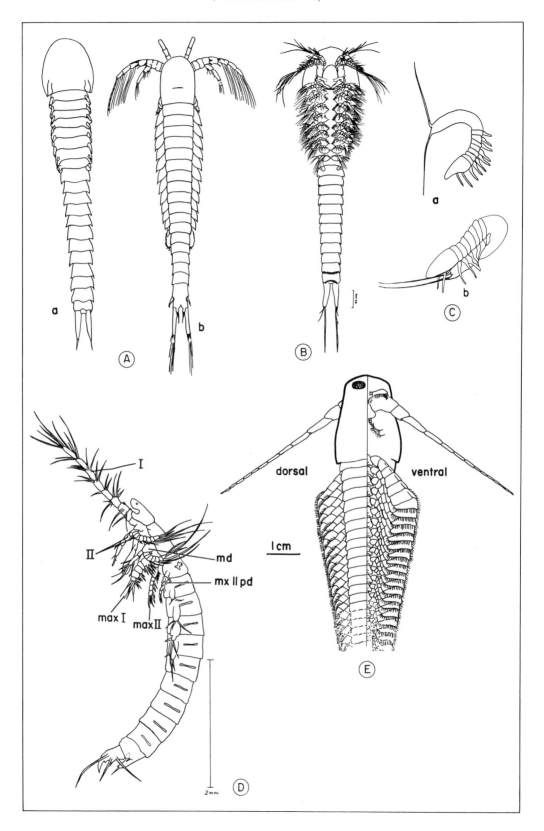

FIGURE 11.15 Primitive crustaceans: living and fossil.
(A) (a) A cephalocarid, *Hutchinsonella macracantha*, Rec., Buzzards Bay, Mass. (b) The fossil lipostracan, *Lepidocaris rhyniensis*, M. Dev., Rhynie Chert, Scotland. Compare (A,b) and (A,a) for closeness of configuration and appendages.
(B) Ventral view of adult of (A,a), actual length 0.2 mm.
(C) Same as (A,a), showing versatility in body movements. (a) Note flexibility of cercopods. (b) Note how posterior end is drawn through ventral appendages. Schematic.
(D) A mysticocarid, *Derocheilocaris typica*, Rec., Massachusetts, lateral view; observe vermiform body; cephalic thorax bearing I, II antennae; Max I, II – maxilliped (*Mxllpd*), and 4th thoracic limb; 0.5 mm long.
(E) *Tesnusocaris goldichi*, Penn., Texas, length 77 mm. An assumed cephalocarid that is unacceptable as such, according to Hessler.
Illustrations redrawn and adapted from Scourfield, Sanders, Pennak and Zinn and H. K. Brooks.

medial surface and carrying three or more spines distally; and a protopod (*Pro*) with two or more endites bearing gnathid spines or setae. Then, to derive the first maxilla of each of the following subclasses through evolution, the indicated reduction and simplification of the cephalocarid-like ancestral limb was necessary (Figure 11.17*A*).

Class Malacostraca. (Protozoeal larval stage.) Decrease in number of endopodal segments; reduced size of exopod.

Classes Copepoda, Ostracoda. Endopodal segments fused; all other components persist.

Class Mystacocarida. Complete loss of the exopod; otherwise identical with the primitive first maxilla of the cephalocarid larvae.

Class Branchiopoda. First maxilla highly reduced but morphologically and functionally similar to that of the adult cephalocarid (Sanders, 1963).

Serial homology between cephalocarid limbs and those of other crustacean subclasses demonstrated for the first maxilla can be extended to other appendages: cephalic, thoracic, and trunk (Sanders, 1963*b*, Figures 33–38; 1957, Figure 6). The first and second antennae and mandibles of still another subclass — the Cirripedia — also are traceable to the equivalent cephalocarid limb.

Thus, as the most generalized primitive form known among living or fossil crustaceans, extant cephalocarids are reminiscent of, and close to, the ancestral crustacean. The ancestral crustacean must now be regarded as cephalocarid-like. It probably arose somewhere during Upper pre-Cambrian time. In the oldest Paleozoic, ostracodes were the only crustaceans known from the Lower Cambrian. All other crustaceans were apparently Devonian or post-Devonian in origin.

However, since the cephalocarid-like ancestral crustacean had trilobitan limbs, as do Lower Cambrian trilobitomorphs, it must have diverged from a common ancestral stock with trilobites and other trilobitoidea in late pre-Cambrian time. That time of origination would have allowed the ancestral stock to give rise to modified limbs of Cambrian ostracodes, Devonian lipostracans, and other branchiopods.

Structure and Chemistry of the Cuticle (Exoskeleton). Exoskeletal mineralogy, skeletogenesis, and cuticle structure of a decapod crustacean are discussed and figured in Chapter 15 (see Figure 15.9). Carapaces or shells of other crustaceans are discussed under the several orders.

Molting Cycles. Critical stages in the best-known molting cycle, that of a decapod crustacean, were reviewed in Chapter 10 (molt control mechanisms) and in Chapter 15 (chemical aspects). Table 10.1 merits careful study. The molt cycle of ostracodes will be more fully explained under discussion of that subclass.

Food Chain. Crustaceans generally constitute the marine zooplankton herbivorous segment grazing on phytoplankton (diatoms, dinoflagellates, and so on).

In turn, crustaceans are an important element in the fish food chain. Copepods,

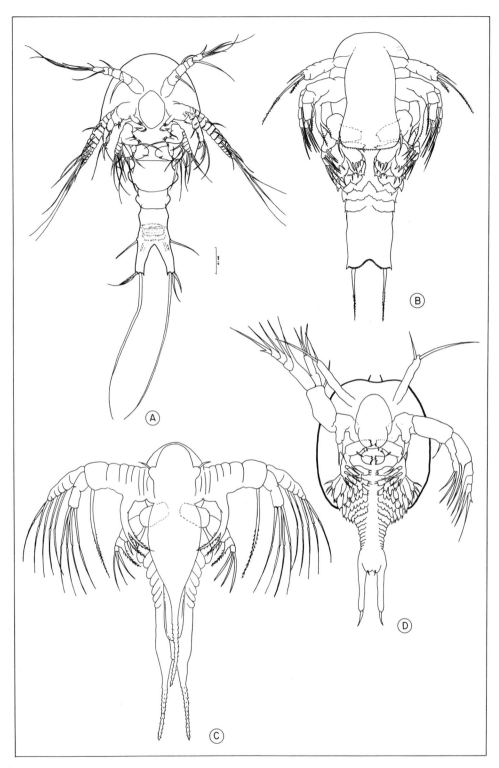

FIGURE 11.16 Crustacean larval stages.
(*A*) *Hutchinsonella macracantha* stage 1, naupliid; ventral view.
(*B*)–(*E*) Branchiopod naupliids. (*B*) lipostracan, *Lepidocaris;* (*C*) conchostracan. *Limnadia lenticularis:* (*D*) notostracan. *Triops caneriformis;* (*E*) anostracan. *Artemia salina.*
(*F*)–(*I*) Nonbranchiopod naupliids: (*F*) copepod, *Temora longicornis;* (*G*) mysticocarid, *Derocheilocaris ramanei;* (*H*) cirriped, *Balanus balanoides;* (*I*) Decapod malacostracan, *Penaeus duorarum.* All schematic, after Sanders, 1963.

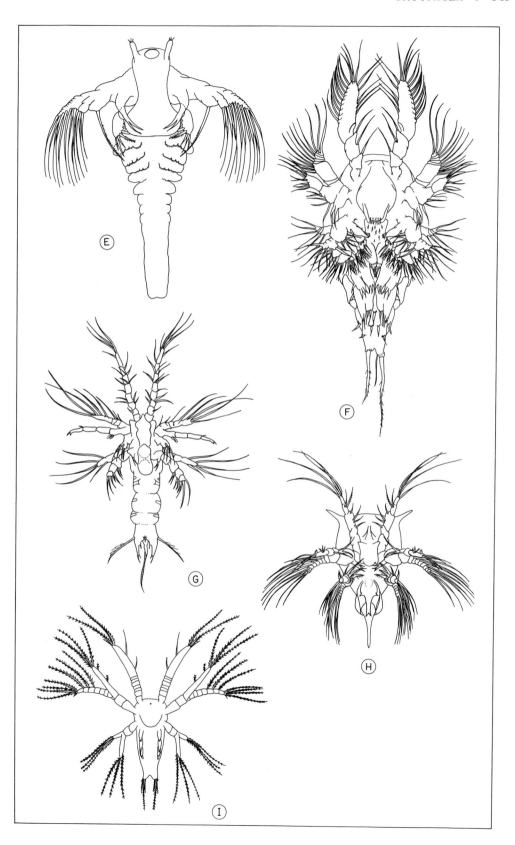

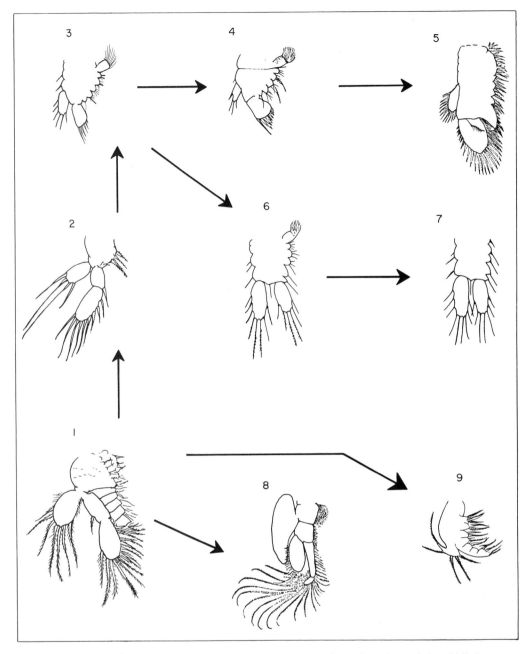

FIGURE 11.17 Inferred evolution of different crustacean appendages from the cephalocarid limb. (After Sanders, 1955, 1963). (1) Cephalocarid *Hutchinsonella,* 4th thoracic appendage. (2) Same as (1), 8th thoracic appendage. (3) Lipostracan *Lepidocaris,* 2nd or 3d thoracic limb. (4) Same as (3), 1st thoracic limb. (5) Anostracan *Branchinecta paludosa,* foliaceous thoracic appendage. (6) Cephalocarid *Lepidocaris,* 4th to 6th (?) trunk limb. (7) Sames as (6), biramous copepodan 7th (?) to 11th trunk limb. (8) Malacostracan, *Nebalia bipes,* thoracic appendage. (9) Malocostracan (Decapoda), *Penaeus* larva, 2d maxilla. (1–9, endopodite on the right).

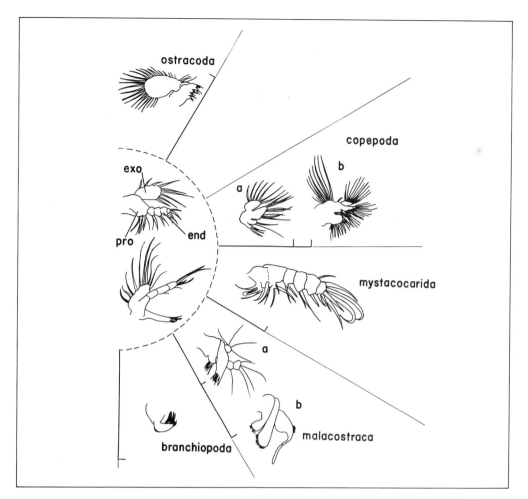

FIGURE 11.17A Limb derivation (first maxilla) in crustacean subclasses. Derivation from cephaloca-rid-like crustacean prototype. *Pro*−protopodite that bears both the expodite (*exo*) and endopodite (*end*). Length of heavier portion of radiating lines is a measure of similarity of limb. Subclass Ceph-alocarida, inside circle; upper labeled figure, naupliar; lower, adult. Subclass Copepoda: (*a*) nau-pliar, (*b*) adult. Subclass Malacostraca: (*a*) decapod protozoeal larval stage, (*b*) adult. See text for explanation. After Sanders, 1963.

for example, are chief food for larval teleost fish. Mysids, amphipods, euphaus-ids, marine cladocerans, and various crus-tacean larvae−such as decapod, ostra-code, and barnacle−are significant com-ponents of the herring food chain.

[Read Figure 11.18 as follows: Phyto-plankton (diatoms, and so on) with lesser contributions by protozoans, radiolarians, foraminifers, and tintinnids, are con-sumed by zooplankton, chiefly crustacean copepods. Most copepods are consumed directly by young and adult herrings, but others supply the herring diet indirectly through intermediaries, such as larval mollusks.]

Autotomy and Regeneration. Some crus-taceans, particularly lobsters and crabs, can shed (separate) an injured or unin-jured appendage at a preformed break-age site. This capacity is known as autot-omy. Separation may be sponsored by the animal itself or a predator (W. L. Schmitt, 1966, pp. 174–175). Regeneration of a new limb subsequently occurs at site of a cast-off appendage.

Regeneration of lost appendages, such as antennae, caudal filaments, thoracic appendages, and so on, is also known in other crustaceans−cladocerans, isopods, amphipods, cirripeds, and copepods (Bliss, 1960). Injured carapaces or valves

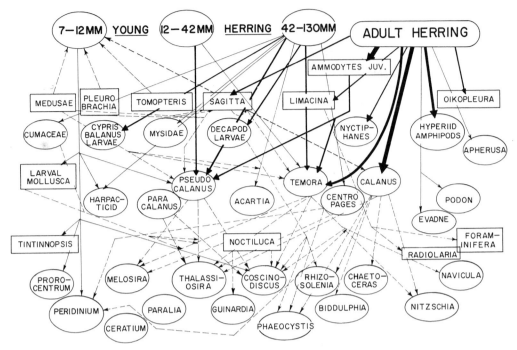

FIGURE 11.18 Food chain in the sea. Crustacea and other food of the herring. After Hardy, 1924. Crustaceans represented include: amphipods, mysids, euphausids (*Nyctiphanes*), decapod, ostracode, and barnacle larvae; marine cladocerans (*Poda, Evadne*) and most abundant, copepods (*Pseudocalanus, Paracalanus, Calanus, Temora*, and so on).

that were repaired have been recorded in both fossil and living branchiopod crustaceans (Tasch, 1961).

Crustacean Spoor, Shelters, and Predation. Living crustaceans excavate burrows in soil, marine substrates, beach sand, tidal muds (shrimp, crabs), wood, rock (chalk, claystone, sandstone, shale) — and isopods are involved in this type of burrow. Other crustaceans construct enclosing tubular houses of algal mats or agglutinate fragments of algae, hydroids, and so on, by glandular secretions (shrimp and amphipods, for example).

In some instances, as in the Malayan rice fields situated in tidal areas, a mound-topography is created by subway-type excavation of shrimplike *Thalassia anomala*. These mounds may attain heights of two feet or more. Various branchiopod crustaceans leave distinctive tracks or trails in bottom sediments — the conchostracan *Cyzicus*, notostracan *Triops*, and others.

Besides self-built subterranean bur-

rows, crustacean shelters and hideouts include trees, the tree-climbing mangrove crab *Aratus posonii*, for example, or *Birgus latro*, the high-climbing cocoanut-eating crab. Other shelters include molluscan shells and sponges — crabs enter sponge canals or sponges may encrust on the crustacean carapace; seaweeds and rock anemones; preexisting serpulid worm and other burrows; encrusting bryozoans; water in tropical pitcher plants; certain Algerian thermal waters (ostracodes); transported hollow plant stems (isopods).

In addition, parasitic copepods and branchiurans infect certain fish (copepod genus *Pennella* and branchiuran genus *Argulus*) and whales; barnacles are parasitic on sea turtles and whales, among others; boring forms of barnacles excavate in corals, echinoderm and molluscan shells, and other barnacles (Schmitt, 1966; Schöne, 1961).

Although generally not so regarded, crustaceans are important destroyers of economic crops and predators on mollusks. Crops include rice (notostracan *Triops* and land crab *Cardisoma*), cotton (crayfish), and other crops. Crabs chiefly are dangerous pests — in oyster farms, for example, where they destroy soft shell clams (Schmitt, 1966).

Reyment (1966) reported gastropod predation on Recent mollusks and larval shells of ostracodes in the Niger Delta. Ratio of drilled (that is, with drillhole in shell) to undrilled ostracode valves was 1:65.

Additional spoor are *Krebsaugen* (Chapter 1, Figure 1.7, "11") and successive castoff exoskeletons (molts) in crabs, branchiopods, ostracodes.

Complex Behavior and Learning. Crustaceans display a variety of complex behaviors and adaptations for life on land, variations in temperature and salinity (Kinne, 1963). These behaviors are associated with such activities as burrowing, swimming, feeding (orientation), courtship and copulation, protective camouflage and avoidance of painful stimuli, defensive and attack situations, among others. Some of these behaviors are genetically determined and inflexible, for example, larval locomotor movements; others, such as feeding, are plastic and modifiable by experience.

Crustacean Bioluminescence and Light Sensitivity. Many ostracodes (*Cypridina* and so on), copepods, such as *Cyclops*, and certain malacostracans (*Mysis, Euphausia,* and so on) are self-luminous. They are bioluminescent, that is, manufacturing their own light in luminous glands. Others, such as certain isopods and amphipods, are generally luminous secondarily — infected by luminous bacteria (Harvey, 1961).

Bioluminescence appears to serve varied "signal" functions: sex recognition and attraction, attraction of prey, warning, defense, and so on. Various crustaceans in the geologic past, Mesozoic malacostracans and Paleozoic ostracodes, are likely to have been primarily or secondarily luminescent. If so, phosphorescences in modern seas due to luminescent crustaceans are likely to have existed in Paleozoic/Mesozoic seas as well.

The cladoceran *Daphnia* can analyze polarized light. Several other crustaceans as well as daphnids discriminate colors (Waterman, 1961).

Classification.

Superclass Crustacea [*crustaceous* carapace]. This category embraces cephalocarids, shrimp, ostracodes, copepods, crabs, lobsters, barnacles, and so on (Calman, 1909; Borradaile et al., 1955; Snodgrass, 1952).

Head. The crustacean head is composed of several fused cephalic somites; two are preoral and bear antennules and antennae, respectively, and three are postoral and bear mandibles, maxillae, and maxilla, respectively.

Thorax. This unit is composed of a variable number of postcephalic somites in different states of fusion — from two somites in ostracodes, to over 40 in some branchiopods. [Head and thorax may be fused to form a cephalothorax as in the crayfish (Figure 10.1*B*).]

Abdomen. Abdominal somites and limbs may or may not be present between the last genital somite and telson; they are lacking in barnacles, ostracodes, copepods, and cladocerans.

Appendages. Modified to serve varied functions as discussed earlier.

Carapace. Crustaceans generally have shell, shield, or carapace. In lieu of these, some forms, such as anostracans, have a tough endoskeleton that serves a supportive function.

Respiration. Aquatic crustaceans generally respire through gills or body surface; land crustaceans often have pseudolungs or pseudotrachea (modified gill chambers) (Borradaile et al., 1958).

Excretion Generally excretion occurs through two pairs of glands.

Sexes. In crustaceans, sexes are often separate but hermaphroditism occurs in sessile barnacles and parasitic isopods. Parthenogenesis occurs in many branchiopods and ostracodes. Some branchiopod notostracans have ovotestes.

Larval Stages. Crustaceans commonly hatch as naupliar larvae and develop to adulthood by successive moltings (Figure 11.16). Naupliar stage may be bypassed; the crustacean can hatch out in a more advanced stage, metanauplius or zoea (Figure 11.17*A*).

Geologic range: Cambrian to Recent.

QUESTIONS

Crustacea—I

1. (a) What anatomical features distinguish crustaceans from other arthropods?
   (b) Discuss range of adaptive modification of appendages.
2. (a) What have researches of Sanders et al. contributed to our understanding of crustacean evolution;
   (b) the Urcrustacean?
3. (a) Discuss the role of crustaceans in marine food chain.
   (b) Attempt to explain how this role might have evolved through geologic time.

REFERENCES

Crustacea—I

Bliss, D. E., 1960. Autotomy and regeneration, in T. H. Waterman, ed., The Physiology of Crustacea. Academic Press. Vol. I, pp. 561–589.

Borradaile, L. A., and F. A. Potts, 1958. See references, Chapter 10.

Brooks, H. K., 1955. A crustacean from the Tesnus Formation (Pennsylvanian) of Texas: J. Paleontology, Vol. 29 (5), pp. 852–856.

Calman, W. T., 1926. The Rhynie crustaceans: Nature, Vol. 118, pp. 89–90.

_____, 1909. Crustacea, in Ray Lankester, ed., A Treatise on Zoology. Adam and Chas. Black, London, Part 7, 382 pp. (Reprinted, A. Asher & Co., Amsterdam, 1964.)

Dahl, E., 1963. Main evolutionary lines among Recent Crustacea, in H. B. Whittington and W. D. I. Rolfe (eds.), Phylogeny and Evolution of Crustacea: Mus. Comp. Zool. (Harvard) Sp. Publ., 1963, Chapter 1.

Harvey, E. N., 1961. Light production, in T. H. Waterman, ed., The Physiology of Crustacea. Academic Press, Vol. II, pp. 171–190.

Hessler, R. R., 1964. The Cephalocarida, Comparative Skeletomusculature. Connecticut Academy of Arts and Sciences Mem., Vol. 16, 59 pp., 47 figures.

Jones, M. L., 1961. *Lightiella serendipita*, gen. nov., sp. nov., a cephalocarid from San Francisco Bay, California: Crustaceana, Vol. 3 (1), pp. 31–46.

Kinne, O., 1963. Adaptation, a primary mechanism of evolution, in Phylogeny and Evolution of Crustacea: Mus. Comp. Zool. (Harvard), Sp. Publ. 1963, Chapter 3.

Pennak, R. W., and D. J. Zinn, 1943. Mystacocarida, a new order of Crustacea from intertidal beaches in Massachusetts and Connecticut: Smithsonian Misc. Coll., Vol. 103 (9), pp. 1–11, plates 1, 2.

Raymont, J. E. G., 1963. Plankton and Productivity in the Oceans; Macmillan, N.Y., 614 pp.

Reymont, R., 1967. Paleoethology and fossil predators: Kansas Acad. Sci. Trans., Vol. 70 (1), pp. 33–50.

Sanders, H. L., 1963a. Significance of the Cephalocarida, in Phylogeny and Evolution of Crustacea: Mus. Comp. Zool. (Harvard), Sp. Publ., 1963, Chapter 12.

—————, 1963b. The Cephalocarida: functional morphology, larval development, comparative external anatomy. Conn. Acad. Arts and Sci. Mem., Vol. 15, 42 pp., Figures 1–38.

—————, 1957. The Cephalocarida and crustacean phylogeny: Systematic Zoology, Vol. 6 (3), pp. 112–128.

Sars, G. O., 1887. On Cyclestheria hislopi (Baird): Forhandl. Videnskabs.—Selskabet., Vol. 1, pp. 3–60, plates 1–8.

Schöne, H., 1961. Complex behavior, in T. H. Waterman, ed., The Physiology of Crustacea. Academic Press, Vol. II, pp. 465–520.

Tasch, P., 1961. Valve injury and repair in living and fossil conchostracans: Kansas Acad. Sci. Trans., Vol. 64 (2), pp. 144–149.

CRUSTACEA—II (COPEPODS, BARNACLES, CONCHOSTRACANS, AND CHIEFLY OSTRACODES)

The superclass Crustacea embraces eight classes—Cephalocarida, Branchiopoda, Ostracoda, Copepoda, Branchiura, Mystacocarida, Cirripedia, and Malacostraca. Of these, cephalocarids, mystacocarids, and branchiurans have no known fossil record at the present writing. Copepods, with few exceptions, are all Recent (Harding, 1956; Palmer, 1960). Accordingly, more extended treatment will be given to the Branchiopoda, a long-neglected group, the ubiquitous fossil Ostracoda, and the Malacostraca.

Class Cephalocarida

Small, primitive crustaceans with generalized (undifferentiated) appendages. Recent only. Hutchinsonella (Figure 11.15A–C).

Class Mystacocarida

Microscopic vermiform, primitive crustaceans inhabiting interstitial waters of intertidal beaches. Recent only. Derocheilocaris (Figure 11.15D).

Class Branchiura

Small crustaceans modified for temporary ectoparasitic existence on fresh and salt water fishes; so-called "fish lice." Recent only. Argulus (Figure 11.19A, B).

Class Copepoda

Free or parasitic crustaceans, lacking carapace and compound eyes; with six pairs of trunk limbs that are mostly biramous except for I and VI. Holocene; Miocene; Pleistocene–Recent. Calanus (Figure 11.19A, part A).

[The Copepoda embrace eight orders. All but two of these are known only from the Recent. Order Harpacticoida and Cyclopoida are represented by fossil species from the Miocene of the Mojave Desert, California (Palmer, 1960). The species assigned to the first of these belongs to an existing genus, Cletocamptus (Figure 11.19 A, part Ab). The species belonging to the second order is undetermined. These and some Pleistocene (Holocene) estuarine forms (Harding, 1956) are the only copepods known as fossils at present. Both have a modern aspect. Tertiary forms were retrieved from insoluble residues of small, calcareous nodules about one inch in diameter. These nodules weathered out of laminated shales of the Barstow Formation. Copepods are critical components of the present-day marine food chain (Figure 11.18). Conceivably, they played some role in the food chain of the Tertiary seas.]

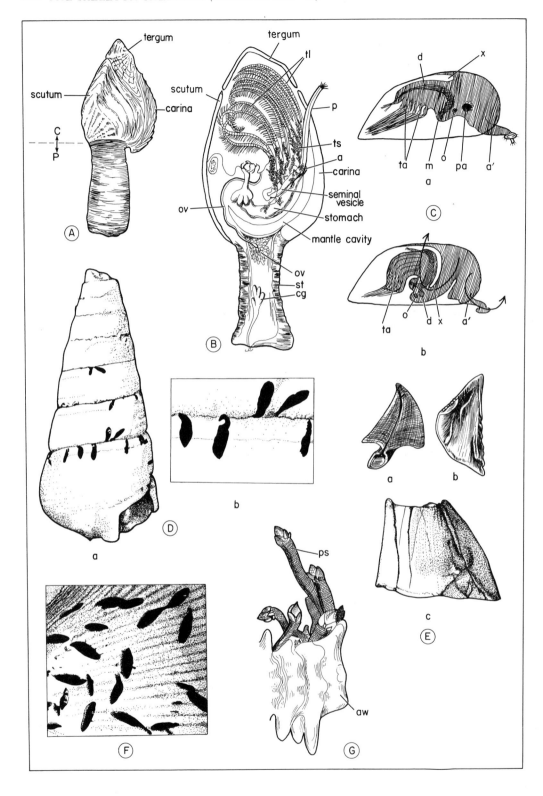

FIGURE 11.18A Living and fossil crustaceans—I. Class Cirripeda.
(A)–(C) *Lepas anatifera*. (A) Complete individual, Recent, showing capitulum (body above peduncle) = *C*; stalk or peduncle = *P*; fleshy mantle or carapace entirely encloses body and consists of 5 calcareous plates: carina (median dorsal) and on each side of it, two plates, scutum and tergum. (B) Longitudinal cut through (A) showing adult anatomy; *tl*—thoracic limbs, *p*—penis, *ts*—testes, *a*—anus, *mc*—mantle cavity, *ov*—ovary, *st*—stalk, *cg*—cement gland. (C), Successive larval stages. (C) (*a, b*) *Cypris* stage, a bivalved shell that attaches to substrate by the cement gland secretion through antennule (*a'*); *d*—gut, *x*—dorsal fold, *pa*—compound eye, *o*—naupliar eye, *ta*—thoracic appendages I–IV, *m*—mouth. (D) (a) Host shell of gastropod *Ceritella protcori* contains burrows of the acrothoracian barnacle *Rogerella cragini*, Cret., Texas; (b) enlargement of burrows that have a maximum size of 1.9 mm length X 0.7 mm width.
(E) *Balanus concavus*, Mio., Maryland. (a) Exterior of left tergum, X 1; (b) exterior of left scuta, X 1; (c) *Balanus concavus eseptatus*, Mio., Port-au-Prince, Haiti, lateral view, X 1.5.
(F) Acrothoracian burrows in brachiopod *Linoproductus*, Perm., Red Eagle Is., Oklahoma, X 6.
(G) *Pollicipes concinnus*. Cret., Oxford Clay, England. A group of specimens found adhering to an ammonite whorl (*aw*), X 1; *ps*—peduncle side. After Charles Darwin.
Illustrations after Calman and several other authors.

Class Cirripedia

Crustaceans, fixed in the adult stage and free-swimming in larval stages (nauplius, cypris with bivalve shell—Figure 11.18 A, *Ca, b*); carapace, absent only in problematical order Apoda, forms mantle enclosing body and limbs, strengthened by multiple calcareous plates (Table 15.6, Chapter 15; see Figure 11.19*B* for plate names). Attached to substrate by antennules from which cement gland opens; these are vestigial in adult and antennae also disappears; trunk limbs, typically six pairs, biramous and cirriform; usually hermaphroditic, but separate sexes are known as well as degenerate complemental males living in or on edge of mantle cavity. Geologic range: Ordovician(?), Silurian–Recent (Figure 11.19, *A*).

The class embraces five orders: Thoracica, Acrothoracica, Apoda, Rhizocephala, Ascothoracica. Of these, three orders lack a fossil record. The Apoda, *Proteolepas*, is a maggot-like form never found or reported since Charles Darwin's original description. Its position among the cirripeds is dubious (W. A. Newman, 1962). The Rhizocephala (*Sacculina*, for example) are parasites on decapod crustaceans, while the Ascothoracica imbed in tissues of certain coelenterates (Antipatharians) and echinoderms (Calman, 1909).

Both of the other two orders have a fossil record. The Thoracica contains the common pedunculate "goose" barnacle,

Lepas, and the unstalked "acorn" barnacle, *Balanus*, among others.

Order **Thoracica.** Alimentary canal present; six pairs of biramous thoracic limbs; adult permanently attached by preoral region. Four suborders: Two suborders not illustrated here are Verrucomorpha and Brachylepadomorpha. Suborder Lepadomorpha, *Lepas*, Recent (Figure 11.18*A*, part *A*); *Pollicipes*, Cret. (Figure 11.18*A*, part *G*); *Euscalpellum*(?), Eocene–Oligocene; suborder Balanomorpha, *Balanus*, Miocene (Figure 11.18*A*, part *E*).

Order **Acrothoracica.** Sexes separate; trunk limbs reduced in number; live in burrows. Paleozoic–Recent. *Rogerella*, Cret. (Figure 11.18*A*, part D); unidentified species, Perm. (Figure 11.18*A*, part *F*).

Life Cycles

Barnacles hatch from eggs as free-swimming naupliids with few exceptions. This accounts for their worldwide distribution. Along one kilometer of British shore line, estimated density of intertidal barnacle spat (*Balanus balanoides*) among nannoplankton was 20 million to one trillion (H. B. Moore, 1935). In a two-week cycle of naupliar development, it passes through six stages.

No limbs or segments are added during the first five stages. At terminal naupliar

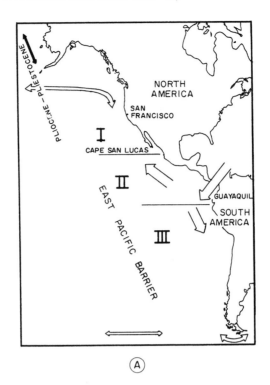

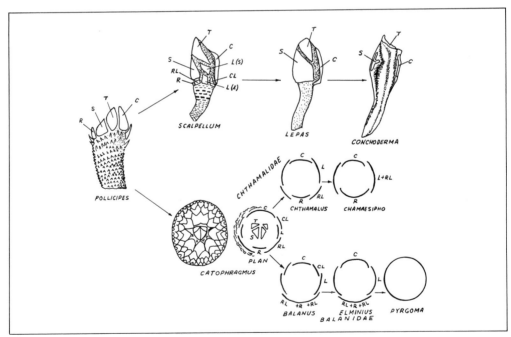

FIGURE 11.19 Information derived from living and fossil cirripeds.

(A) Major migration routes and modern distribution provinces of the acorn barnacle—the Balano-morpha—chiefly *Balanus*. Arrows show routes; wider arrows denote more important routes taken by free-swimming barnacle larvae. Oligocene-Miocene: closing of seaways (end of Miocene) across Central America and northern South America, led to shutting off communication with Atlantic-Caribbean tropical fauna. Pliocene-Pleistocene-Tertiary: Bering land barrier fragmented (Pliocene) allowing establishment of communication with northern Atlantic fauna (see arrows, Provinces I and II). Provinces: I, temperate N.E. Pacific; II, tropical E. Pacific; III, temperate S.E. Pacific.

(B) Evolution of plate systems in some barnacles, starting from a *Pollicipes*-like progenitor in the Jurassic. Plates: C—carina, CL—carino-lateral, L—lateral, L(I)—inferior lateral, L(S)—superior lateral, T—tergum, R—rostrum, RL—rostro-lateral, S—scutum. Scales omitted from right side of plan view of *Catophragmus*. Scuta and Terga plates omitted from the plan view of Chthalamidae and Balanidae. Note reduction of scales on stalk on upper series (side view); lower series (plan view) shows loss of stalk leading to acorn barnacles (Balanamorpha) and also loss or fusion of some plates.

Illustrations redrawn and adapted from (A) Zullo, 1966; (B) Green, 1961.

stage, it adds the first maxilla and a segment. When it next molts, the first bivalved cypris stage occurs (Figure 11.18*A*, part C) with a large number of added segments and limbs (Sanders, 1963, p. 164, Figure 70). Bottom settling and attachment to some substrate generally follows by a few hours metamorphosis into a cyprid larva (W. A. Newman, 1962, unpublished remarks).

Affinities and Ancestry

Calman (1909) remarked that great structural differences separated barnacles and other crustaceans. He thought this indicated early divergence of cirripeds from the known line of crustacean descent. W. A. Newman (1962) envisioned the ancestral type of cirripeds as an epibenthonic, possibly burrowing form with a rather large, broad, shieldlike carapace. He suggested that cyprid larva bivalves are reminiscent of the ancestral type. Bivalves were flattened out as a dorsal shield as occurs in living Ascothoracica cyprids. According to Newman, despite lack of a fossil record, this last-named order appears to be primitive. Professor M. F. Glaessner surmises that some of the pre-Cambrian shieldlike carapaces might have been such a cirriped ancestral type.

Evolution

Figure 11.19*B* illustrates two postulated lines of evolution starting with a *Pollicipes*-like progenitor. The upper line proceeded

to *Scalpellum* and then to *Lepas* with reduction in scales of the peduncle and mantle plates becoming more regular or lost completely. A second line proceeded through *Catophragmus* by reduction of stalk. From *Catophragmus* with eight plates (see plan view, Figure 11.19*B*), the Chthamalidae and Balanidae were derived by loss and fusion of plates. [Compare each of the following with the plan view of *Catophragmus—Pyrgoma* and *Chamaesipho* (Green, 1961, pp. 11–12.]

Skeleton, Mineralogy, and Chemistry

Skeleton. Fleshy mantle or carapace is bolstered by a variable number of protective plates. Of these, the *carina* is median-dorsal, flanked by the *tergum* (two plates) and *scutum* (two plates). Darwin referred to the scutum as the most important and persistent valve in pedunculate forms. Margins of scuta and terga open and shut for exsertion during feeding and retraction of *cirri* during times of disturbance (Figure 11.18*A*, part A). Other valves are "laterals" (lateral, inferior lateral, superior lateral, rostrolateral, carino-lateral) and rostrum.

Mineralogy and Chemistry. Plates of living barnacles are entirely mineral calcite (Chave, 1954). However, fossil forms may be all chitin. Chave found that hard supporting elements of the fleshy barnacle carapace contained less than 5 percent $MgCO_3$. Quantity of $MgCO_3$ seemed

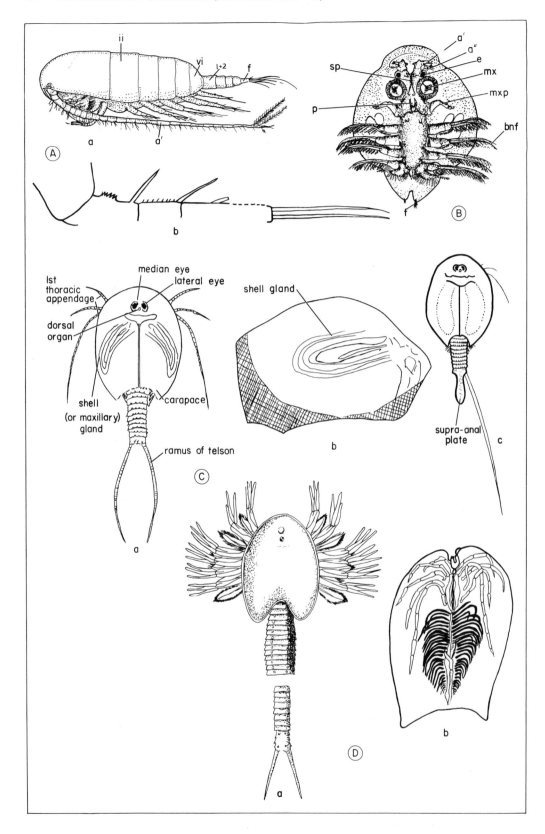

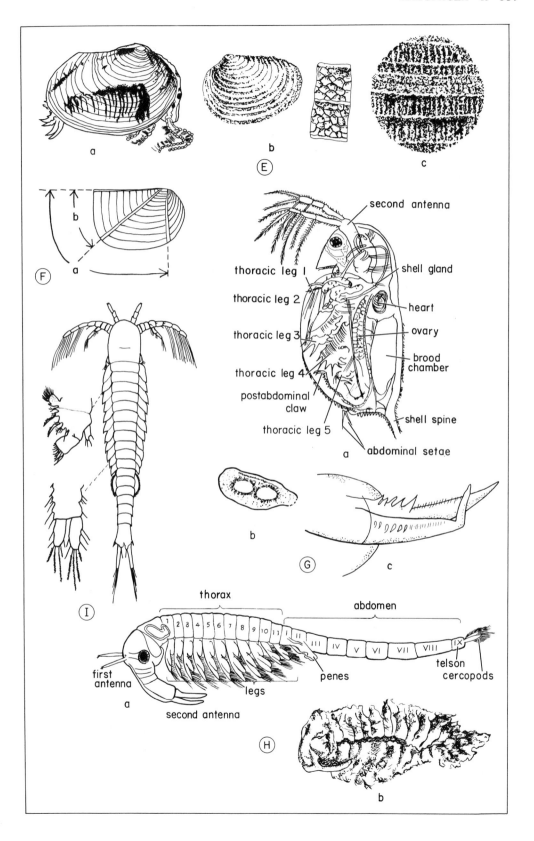

second antenna

thoracic leg 1

thoracic leg 2

thoracic leg 3

thoracic leg 4

postabdominal claw

thoracic leg 5

shell gland

heart

ovary

brood chamber

shell spine

abdominal setae

a

b

c

E

F

G

H

I

thorax

abdomen

first antenna

second antenna

legs

penes

telson

cercopods

a

b

FIGURE 11.19A Living and fossil crustaceans—II.

(A) Copepoda: (a) *Calanus finmarchius*, Rec., female, X 16, side view; *a'* —antennule; II, VI—thoracic somites; 1 + 2—first and second abdominal somites coalesced; *f* —caudal furca. After Sars. (b) *Cletocamptus* sp., female, leg 4, exopod. Mio., Mojave Desert, Calif., schematic.

(B) Branchiura: *Argulus americanus*, *a'* —antennule, *a"* —antenna, *e* —paired eyes, *mx* —sucker formed from maxilla, *mxp* —2nd maxilliped, the 1st thoracic appendage, *bnf* —biramous natatory feet, *f* —furca, *sp* —poison spine, *p* —suctorial proboscis. Rec., female, seen from below, X 5.

(C) Branchiopoda. Order Notostraca, (a, b) *Triops*: (a) *Triops cancriformis*, Carb.—Rec., schematic; (b) *Triops beedei*, shell gland, internal cast. Penn., Oklahoma, enlarged. (c) *Lepidurus: L. apus*, dorsal view, Rec., schematic, X 1.5.

(D) (a) Order Kazacharthra: *Jeanrogerium sornoyi*, L. Jur., U.S.S.R., X 2, complete individual reconstructed; (b) order Acerostraca, *Vachonisia rogeri*, L. Dev., Germany, X 9; ventral side shown by X-ray.

(E) (a) *Cyzicus* (*Cyzicus*) *cycladoides*, Rec., France; (b) *Cyzicus* (*Euestheria*) *minuta*, U. Trias., polygonal ornamentation; (c) *Cyzicus* (*Lioestheria*) sp., longitudinal striae as hachure-type ornamentation.

(F) *Leaia tricarinata*, Coal Measures (Westphalian). No. France, left valve, X 10. After Pruvost, 1919. Alpha (α) = angle alpha, extends from dorsal margin to anterior rib; beta (β) = angle beta, extends from dorsal margin to posterior rib.

(G) Order Cladocera: (a) *Daphnia pulex*, Rec., male, schematic; (b) fossil *Daphnia* sp., ephippium that originally bore two ephippial eggs (0.2–0.4 mm diameter), Oligo. (Braunkohle), W. Ger.; (c) *Bosmina longispina*, claw, subfossil from Connecticut Lake, enlarged.

(H) Order Anostraca: (a) *Branchinecta paludosa*, Rec., male, schematic; (b) Anostracan, (?) M. Mio., lacustrine deposits, Mojave Desert, Calif., enlarged.

(I) Order Lipostraca. *Lepidocaris rhyniensis*, reconstruction, dorsal view, adult male, showing appendages, M. Dev., X 25.

Illustrations redrawn and adapted from Calman, Palmer, Novojilov, Lehmann, Deevey, Tasch, and others.

to be temperature-controlled. Lowenstam (1963, Figure 9) showed that cirriped plates had a $SrCO_3$ content (in mol %) of 1.0 + (for further analyses, see Table 15.6 and discussion on "Arthropods," Chapter 15).

Age of Barnacles

Since Balanomorpha show greatest expansion in the world's intertidal regions, Pilsbry designated this period "Age of Barnacles." In the high littoral zone, barnacles are presently undergoing *explosive evolution* (W. A. Newman, 1962). This involves rapid expansion to fill out all available niches within the high littoral zone. It also involves geographic isolation leading to reproductive isolation and ultimately rise of new species. Barnacles may very well be more abundant and common fossils in shallow, inshore deposits of the future rock column. (Figure 11.19*A*).

Ecology and Paleoecology

Attachment and Burrows. *Balanus balanoides*, the common barnacle, forms dense populations along rocky coasts at or near the water line, on the west coast of Sweden (Jägersten, 1960). They also attach to ships, piers, and so on, and constitute a

menace to man's enterprises. Both the modern acorn barnacle, *Coronula*, and the stalked barnacle, *Conchoderma*, are common parasites on humpback and sperm whales. They encrust on the skin of dolphins as well as whales. *Conchoderma* attaches itself to incrusted acorn barnacles (Slijper, 1965). One humpback whale bore an estimated 1000 pounds of barnacles as parasites (Zenkevitch, cit. Slijper, 1965).

Modern acrothoracicans live in burrows in molluscan shells (for example, columella of living whelks). In the Paleozoic–Mesozoic they burrowed in shells of mollusks, plates of echinoderms (Schmidt and Young, 1960; Branson, 1964), brachiopod shells, rugose corals, and ectoproct bryozoans (Rodda and Fisher, 1962).

Nutrition. Diet of modern balanids consists of plankton and organic detritus. Having six pairs of thoracic limbs bearing cirri, these sweep surrounding waters for food content (Jägersten, 1960).

Habit. There are no fresh-water barnacles. The entire class occurs in a marine environment.

Bathymetry. While abundant in the shallows, as is evident by their common attachment to shells of shallow water invertebrates, some barnacles occur at great depths—to 2000 fathoms.

Fossil Record. The fossil record rules out older speculations that class Cirripedia probably originated in Upper Mesozoic time. A clearcut record of fossil acrothoracic barnacles, represented by distinctive burrows in shells and other skeletal fragments, can be traced to the Devonian (Codez and Saint-Seine, 1958). Upper Paleozoic burrows are not uncommon (Rodda and Fisher, 1962). Fossil fragments ascribed to thoracic barnacles of Cambro-Ordovician age (*Lepidocoleus*, *Turrilepas*, *Strobilepas*, among others) are inadequate to establish such affinities. However, two Paleozoic thoracic barnacles are accepted as authentic in the Treatise: one from the Upper Silurian of Esthonia, *Cyprilepas holmi*, and one from the Middle Carboniferous of the U.S.S.R., *Praelepas jaworski* (Newman, Zullo, and Withers, 1969).

The Mesozoic record of fossil barnacles exceeds that of the Paleozoic. Darwin remarked on the large number of pedunculate species in the English Cretaceous chalks in contrast to the comparatively small number of individuals.

The Tertiary-Recent record of thoracic barnacles is most abundant (for example, see Cheetham, 1963). *Balanus* made an apparent first appearance in the Eocene and commonly numerous individuals are crowded together on a molluscan shell substrate or other Tertiary skeletal material. Balanids are the most abundant barnacle fossil in the South American Tertiary, Chile, Peru, and Ecuador (Camacho, 1966, pp. 468–469).

Biometrics. Codez and Saint-Seine (1958) made various measurements in order to compare burrows made by different acrothoracic barnacle species in the geologic past. Measurements were expressed as ratios so as to yield an index number (ratio × 100). Thus l/L, p/L, and L/λ [where L, l, p, and λ are respectively length (L), width (l), maximum depth (p),

and length of aperture (λ) of a given burrow on a given shell]. When l or p, for example, is graphed as a function of L or λ, one can distinguish distinct genera, although overlap does occur.

Class Branchiopoda

Crustacea with dorsal shield or shell separated into two distinct parts; also univalved, or carapace may be completely absent; number of trunk somites variable; posterior part of trunk without limbs; trunk limbs generally foliaceous and lobed, Lower Devonian–Recent.

As presently understood (Tasch, 1969), the Branchiopoda are divisible into three subclasses each of which embraces two or more orders. Among the branchiopods, conchostracans are most abundantly represented as fossils.

Subclass Calmanostraca. [Order Notostraca, *Triops* (Figure 11.19*A*, part *Ca*), *Lepidurus* (Figure 11.19*A*, part *Cc*); order Kazacharthra, *Jeanrogerium* (Figure 11.19*A*, part *Da*); order Acercostraca, *Vachonisia* (Figure 11.19*A*, part *Db*).] [Stürmer and Bergström (Paläont. Zeitschr., 1976, Vol. 50, p. 78) based on new x-ray analysis of fossils of *Vachonisia*, question Lehmann's assignment of it to the branchiopod crustaceans.]

Subclass Diplostraca. [Order Conchostraca, *Cyzicus* (Figure 11.19*A*, part *E*), *Leaia* (Figure 11.19*A*, part *F*); order Cladocera, *Daphnia* (11.19*A*, *G*, *a–b*), *Bosmina* (11.19*A*, *Gc*).]

Subclass Sarsostraca. [Order Anostraca, *Branchinecta* (11.19*A*, *Ha*), *Genus* (11.19*A*, *Hb*); order Lipostraca, *Lepidocaris* (11.19*A*, *I*).]

Information Derived from Fossil Branchiopods

Biostratigraphy. Fossil conchostracans are among the most useful index fossils in the restricted facies in which they occur. They are extremely valuable in nonmarine

Permo-Carboniferous deposits of the world, Asian Mesozoic, and Devonian of many regions (U.S.S.R., Canada, U.S.A.). Figure 11.20*A* illustrates the utility of conchostracan genus *Rhabdostichus* in delimiting a portion of the Williston Basin Upper Devonian (Wilson, 1956).

Microstratigraphy and Season Determination. Conchostracan beds are often suitable for microstratigraphic studies as in the Permian Wellington Formation of Kansas. In such instances, seasons of time (Figure 11.20*B*) and total age of a given thickness of conchostracan-bearing sedimentary rocks can be evaluated (Figure 11.20*C*).

Glacial and Postglacial Lakes. Parts of the cladoceran body (detached abdomen, mandibles, claws, and so on) as well as the univalve itself (ephippia, and so on) have been found in borings of postglacial lake bottoms (Swain, 1956). Similar kinds of fossils also occur in samples of glacial lake bottoms (Frey, 1958). Obviously, fossil cladocerans can be a valuable aid in biostratigraphic studies of Pleistocene and post-Pleistocene lakes.

Evaporites and Paleoclimatology. Borrate, halite, and gypsiferous deposits may contain fossil branchiopods or their spoor, such as fecal pellets (Palmer 1957; Eardley, et al., 1957; Tasch, 1963*a*).

Branchiopod habitation of marginal marine sites, marginal pools of evaporating basins, and lacustrine environments can yield paleoclimatic indications. Successive generations of conchostracans in a thin slab of calcareous argillite denote seasonal wetting and drying cycles.

Ecdysis and Life Span. Along with other crustaceans, branchiopods, with rare exceptions, repeatedly shed their skeletons (molt) in the course of growth. Conchostracans molt approximately every three days. (There are exceptions to this in given regions such as Tunisia.) Accordingly, by counting growth bands on a given fossil

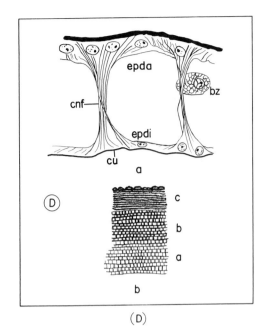

(D)

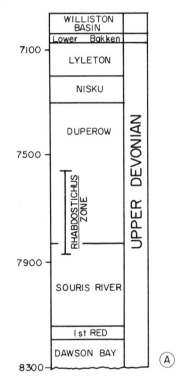

FIGURE 11.20 Information derived from fossil Branchiopoda: biostratigraphy, geochronology, and shell features.

(*A*) Conchostracan *Rhabdostichus* zone in the U. Dev. of the Williston Basin. After J. L. Wilson, 1956.

(*B*) A thin fissile shale (Permian) 3.2 in. thick contained 25 successive conchostracan generations (Wellington shale, Noble Co., Oklahoma). These are evaluated on a season-by-season basis (Tasch, 1964).

Conchostracan generation number | Depth below top of bed (mm.) | Time—value of sediment interval between any two successive conchostracan—bearing beds (in years) (1 season = 1 year)

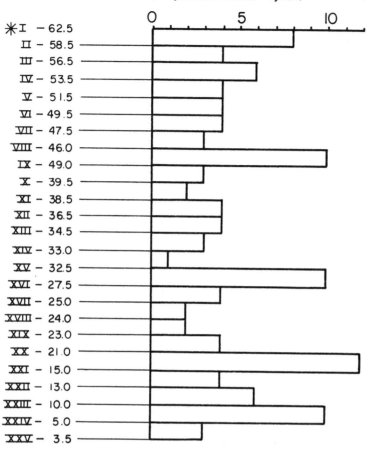

✳ = 1 season (yr.)

XIV – XV = same season

(B)

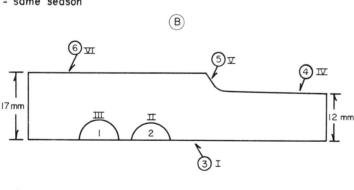

◯ = Specimen numbers

I – VI = Conchostracan generation number

(C)

(C) Geochronology: A banded argillite 17.0-mm maximum thickness from the Ohio Range, Antarctica, contained six distinct generations of fossil conchostracans (Roman numerals). Dividing slab thickness by smallest sediment interval separating any two successive generations, gives a value of 30 seasons = 30 ± years (Doumani and Tasch, 1965).

(D) (a) Slice, a few microns thick, through shell of living conchostracan *Limnadia lenticularis* L.; *ca*—outer cuticle, on top; *cu*—inner cuticle, *ep*—epidermis, *bz*—blood cells, *cnf*—connective fibers. After Nowikoff, 1905. (b) Outer, intermediate, and inner chitinous structures of the shell. Note canals in intermediate layers.

561

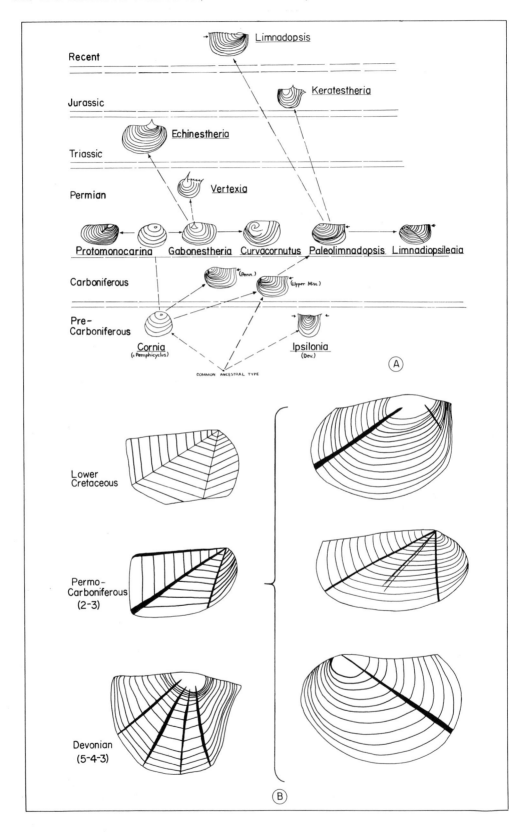

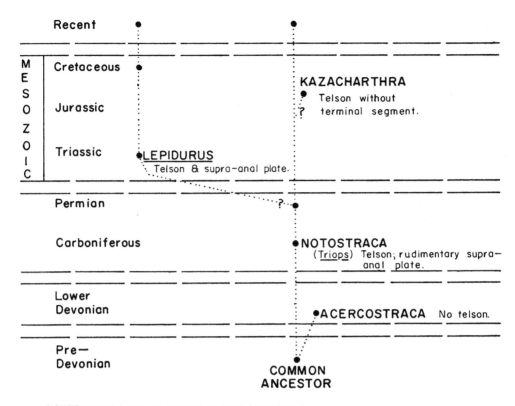

FIGURE 11.21 Information derived from fossil Branchiopoda.
(*A*) Evolution of conchostracans with spines, serrate dorsal margins and posterior recurvature.
(*B*) Evolution of ribbed conchostracans. Left column, lower figure—*Pteroleaia* (U. Dev., Can. Arctic), middle figure—*Leaia* (Permo-Carb.), right column, other Permo-Carb. ribbed forms: lower figure, *Monoleaia;* middle figure, *Paraleaia;* upper figure, *Massagetes.* Left column, upper figure, *Japanoleaia*—flat, diagonal ridges where ribs had been in older forms.
(*C*) Inferred relationships between three branchiopod orders.
Illustrations after Tasch, 1963*b*.

valve, total life span of the individual can be determined.

If life span of the *oldest* individual represented on a given bedding plane is determined, that will approximate actual duration of the given water body (pond, pool, puddle) at a particular time in the Paleozoic or Mesozoic.

Paleoecology
Fresh-Water Facies. Branchiopod fossils of any kind are generally indicators of fresh to brackish water environments. Such biofacies may interfinger with marine facies or pinch out after limited lateral spread.

Biotic Associates. Among others, several branchiopod orders may occur together, the notostracan *Triops,* anostracan *Streptocephalus,* conchostracan *Cyzicus.* Var-

ious insects are frequent associates (Tasch, 1964, Figure 2). Occasionally, marine forms, such as trilobites, cephalopods, and so on, are found fossilized with conchostracans.

Evolution. The only branchiopod order not illustrated in Figure 11.21*A* is the Cladocera. This order was thought to have been derived from some conchostracan during Cretaceous–pre-Oligocene time. More recently, a Russian investigator (Smirnov) published data, yet to be independently evaluated, on Permian(!) cladocerans from Kazakhstan.

Shell Composition and Structure. A shell slice through living conchostracan *Limnadia lenticularis* (Figure 11.20*D, a*) shows an outer and inner cuticle secreted by the respective epithelial tissues. Hollow spaces ("blood chambers") in the chitinous shell may be impregnated with calcite and, in life, contain blood cells. The shell serves a respiratory function (Tasch, 1967b). When sectioned, it reveals multiple superimposed chitinous layers (Figure 11.20D, b).

There are internal and external shell structures in conchostracan valves. Internal structures include canals, pores, punctae. External structures include costae, growth bands (a new band being added for each molt about every three days), ribs. Alpha and beta angles are measured on ribbed forms. (For angles, see Figure 11.19A, part F.)

Leaiid conchostracans may have from one to five ribs; estheriellid conchostracans are characterized by numerous costae. Ornamentation may be used to distinguish subgenera of the common genus Cyzicus, which is still extant. Other valve characteristics include spines on growth bands, causing a serrate condition of the dorsal margin, or arising from beak, or spines formed by posterior prolongation of the dorsal margin. Recurvature of posterior margin is also an important valve character in some genera.

Continental Drift. Follow-up of the research on Antarctic conchostracans referred to in Chapter 1 (Tasch, 1967c, 1967d) has added new data pertinent to Drift theory.

The presence of fossil leaiids and cyziciid conchostracans in the Permian of the Ohio Range, Antarctica, is puzzling if the continents were at present distances apart. Fresh to brackish water migratory routes for conchostracans that could account for their presence, all require greater closeness of southern hemisphere continents at the time.

Once the Gondwanaland concept is accepted and a variety of independent geophysical and oceanographic evidence supports it—a greater closeness of southern hemisphere continents and northern hemisphere continents follows. The puzzle is then quickly resolved. Thus, if South Africa and South America were in contact or proximity along their continental shelves, Carboniferous ribbed conchostracans (genus Leaia) from Africa (Morocco) could have migrated to southern Africa and west to Brazil. That would account for Brazilian and South African leaiids.

African leaiids probably reached western Australia in Carboniferous time, and these forms migrated by Permian time to New South Wales and eastern Australia. Restoring Antarctica and Australia to their positions before Drift places New South Wales very close to Antarctica. Clearly, Permian leaiids of New South Wales could have migrated readily to Antarctica across a swampland (indicated by considerable fossilized vegetation—the Glossopteris flora).

Leaiids from the Moroccan Carboniferous probably were derived from European stock (Spanish), and this links Europe, Africa, South America, Australia, and Antarctica faunally through the ribbed conchostracan leaiids. In North America also, Carboniferous leaiids occur. Recent geophysical reconstructions bring Ireland and Newfoundland closer together. Data on Leaia tricarinata independently would make this type of reconstruction necessary. Thus it is an index fossil in the Coal Measures of Europe and occurs in those of Nova Scotia. This distribution denotes that leaiids migrated back and forth before Drift. It follows that American leaiids are actually tied in with the Permian Antarctic faunas along the circuitous routes already discussed.

Class Ostracoda

Laterally compressed crustaceans; head undifferentiated, enclosed within bivalved carapace that is more or less calcified. (Occasionally, as in a new suborder, carapaces are phosphatic, composed of mineral apatite. The family Halocypridae contains mainly species with uncalcified chitinous valves.) The outer chitinous coat of carapace valves is preserved infrequently in fossils; valves are hinged along dorsal

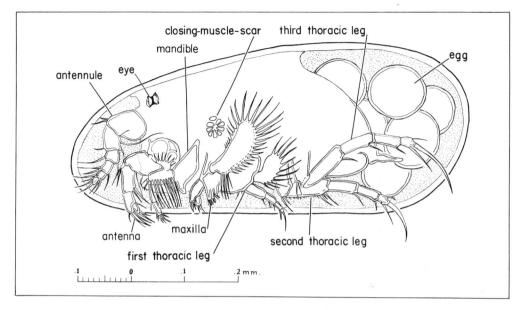

FIGURE 11.22A Morphology of a representative podocopid ostracode. *Darwinula stevensoni,* Rec. (female, left valve removed, compare definition Class Podocopida). Illustration redrawn and modified from Kesling, 1961.

margin. Appendages: four cephalic, one to three thoracic, and a pair of furcal rami; no abdominal appendages. Habitat: chiefly marine and fresh-water, but terrestrial forms are known (Figure 11.23*I*). Range: Lower Cambrian–Recent (Figure 11.22*B*).

Ostracodes form a class that is probably second in abundance and importance for stratigraphy only to foraminifera and conodonts. It embraces five orders and 25 superfamilies.

***Order* Archaeocopida.** Cambro-Ordovician forms with slightly calcified or entirely phosphatic shell; long, straight hinge line; eye tubercle; probably precursors of post-Cambrian ostracodes. [*Bradoria,* (Figure 11.23*A*).] Muller's new suborder Phosphatocopina (1965) was assigned to Bradorina Raymond, 1935, which is a synonym of Archeocopida Sylvester-Bradley (1961).

***Order* Leperditicopida.** Ordovician-Devonian forms with a long, straight hinge and thick, well-calcified shell; surface usually smooth with valves unequal to subequal. (?) Upper Cambrian–Lower Ordovician–Upper Devonian [*Leperditia* (Figure 11.23*B*)].

***Order* Palaeocopida.** Dorsal margin long and straight. Surface smooth-to-ornamented; dimorphic structures common (frills, brood pouches, lobes, and so on). Lower Ordovician–Middle Permian, (?)Recent. [Suborder Beyrichicopina, *Beyrichia* (Figure 11.22*B,* part C); suborder Kloedenellocopina, *Kloedenella* (Figure 11.23*D, a, b*).]

***Order* Podocopida.** Dorsal margin generally curved; where straight, it is shorter than the total length; three different types of adductor muscle scar patterns; circular aggregate of many scars (suborder Metacopina); reduction of individual scar number, aggregate or biseral (suborder Platycopina); or discrete, and variously arranged (suborder Podocopina); hinge margin differentiated chiefly in one suborder only, Podocopina. Lower Ordovician to Recent. [Suborder Podocopina, *Bairdia* (Figure 11.23*E, a, b*), suborder Metacopina, *Healdia* (Figure 11.23*F, a, b*); suborder Platycopina, *Cytherella* (*Cytherella*) (Figure 11.23*G*).] (Also, Figure 11.22*A*).

***Order* Myodocopida.** Most planktonic ostracodes are included in this order:

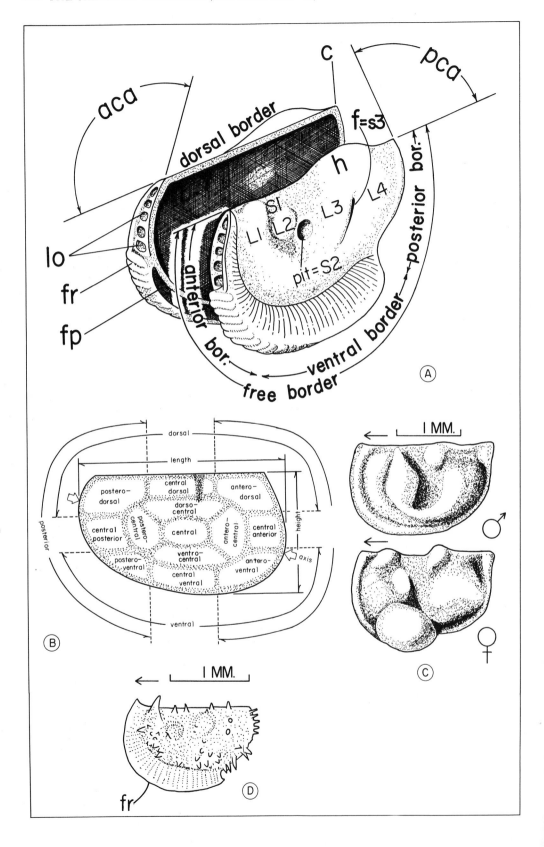

FIGURE 11.22B Ostracod morphology.
(*A*) Surface features, angles, borders of a palaeocopid ostracode (compare order definition): *pca* — posterior cardinal angle, *aca* — anterior cardinal angle, *c* — corner, *lo* — loculi, *fr* — frill, *fp* — false pouch, L_{1-4} — lobes (L_1 — anterior, L_4 — posterior); S_{1-3} — sulcus, pits, or fissures.
(*B*) Geography of ostracode valve.
(*C*) Dimorphs: *Beyrichia kirki;* male, female; compare lobes in each, and note brood pouch.
(*D*) Ornamentation: *Hollina spiculosa*, spinose surface of frill (*fr*).
Illustrations redrawn after Kesling, 1951a.

valves, subequal, smooth or ornamented; anterior rostra and notch may or may not be developed; second antenna modified for swimming; dimorphic. Ordovician to Recent. [Embraces two suborders: Suborder Myodocopina, subfamily Cypridinacea, *Cypridina* (Figure 11.23*H*).]

The Ostracode Carapace

Size, Shape, Structure. The ostracode bivalved carapace is generally a small object less than 1.0 mm long, although in some forms it reaches a length of 5.0 mm. Shape is variable from ovate to quadrate. The two valves, right and left, are either unequal or subequal in size. Inequality in size necessarily leads to the larger valve overlapping or enclosing the smaller.

Structurally, seen in section, the ostracode valve consists of a hard (calcitic) and soft (epidermal) layer. A slice through such a valve reveals the following layering from surface to inner face — a thickness of 22 to 45 μ (Figure 11.22*C*, part *A*): (1) Surficial chitin coat, (2) calcium carbonate layer, (3) chitin coat of outer epidermis, (4) outer epidermal cell layer, (5) space containing varied cells, (6) inner epidermal cell layer, (7) antero-ventral part of (6) above expands to form calcitic duplicature of skeleton.

The hard layers above are composed of microcrystalline calcite, (the long axis of these crystals generally are disposed at right angles to valve surface (Kesling, 1951*b*)). [K. T. Müller (1965) described the first recorded phosphatic ostracode carapaces from Carbonaceous limestones of the Swedish Upper Cambrian. Outer lamella of the carapace in the new suborder, Phosphatocopina, has an inner phosphatic and outer chitinous layer.]

Duplicature is "welded" to (2) above by a chitinous layer. Various pores penetrate layers 1–3; some bear sensory hairs.

Valve Hingement. Mode of dorsal articulation of ostracode valves is genetically variable. Such variation finds expression in different kinds of hingement. In broad terms, dorsal contact (hingement) between valves is either smooth, without any interlocking devices, or achieved only through such devices: modified hinge bars, grooves, cardinal teeth, sockets, and so on.

Major hingement types include (see Sylvester-Bradley, 1956) the following.

Adont. Simple bar and groove (Figure 11.22*C*, part *D*, 1).

Prionodont. Like "adont" but crenulated along ridge and groove (Figure 11.22 *C*, part *D*, 2).

Lophodont. One cardinal tooth at each extremity of one valve, with corresponding sockets on other valve (Figure 11.22*C*, part *D*, 3).

Merodont. Bar with crenulated teeth at each extremity articulating with corresponding groove and socket of other valve (Figure 11.22*C*, part *D*, 4).

Entomodont. Consists of four elements: median bar and groove with anterior coarsely crenulate element and smooth or crenulate remaining portion, and anterior/posterior crenulate teeth and sockets (Figure 11.22*C*, part *D*, 5).

Amphidont. Four-element hingement reminiscent of "entomodont" type; one valve with well-defined toothlike projection and extremities, separated by median furrow with deep smooth socket at anterior end; opposite valve with projections and depressions reversed (Figure 11.22*C*, part *D*, 6).

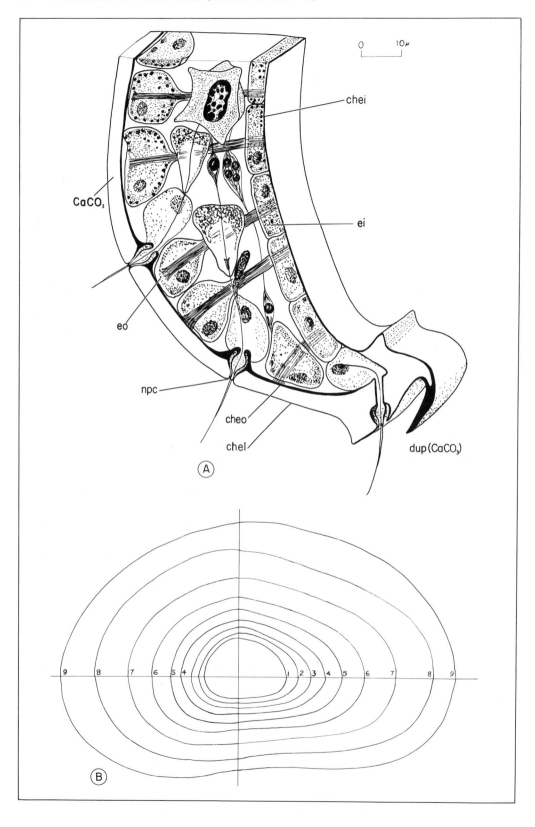

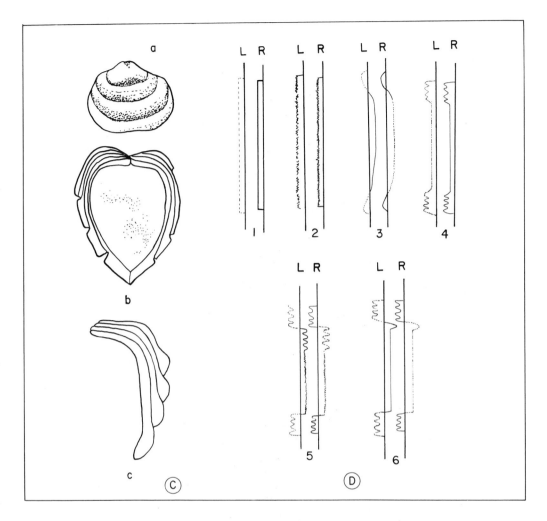

FIGURE 11.22C Ostracode carapace structure.
(A) Antero-ventral part of right valve of living *Cypridopsis;* CaCO₃ – microcrystalline calcite, chiefly at right, to valve surface; epidermal layers: *eo* – outer layer, *ei* – inner layer; chitin coatings of outer epidermal layer (*cheo*), chitin coatings of calcareous layer (*chel*), chitin coatings of inner epidermal layer (*chei*); duplicature (*dup*) composed of CaCO₃; *npc* – normal pore canal. Unlabeled cells and structures include nerve, subdermal and supporting fibers.
(B) Successive instars (molts) from smallest #1 to largest (= adult) #9, of *Cypridopsis vidua*, Rec.
(C) *Cryptophyllus sulcatus*, U. Ord., Indiana; (a) lateral view of presumed right valve, X 40; (b) transverse section showing shell layers of unshed molts (instars), X 80, compare (c).
(D) Hinge types, dorsal view, diagrammatic (see text for definitions). (1) Adont, (2) Prionodont, (3) Lophodont, (4) Merodont, (5) Entomodont, (6) Amphidont. *L, R* – left and right valve, respectively. Illustrations redrawn and adapted from Kesling, 1951*b*, and Sylvester-Bradley, 1956.

Type of hingement is extensively used in Mesozoic and Cenozoic ostracode taxonomy. (Example: family Pectocytheridae, Lower Cretaceous to Recent, hinge line, a modified merodont type.)

[Instars of post-Paleozoic ostracodes with an *amphidont* hinge in maturity often show development from adont hinge (youngest instar) to merodont and amphidont type (after the last molt) (Van Morkhoven, 1961).]

Surface Features; Cycles. Surficial features of ostracode valves include pore canals, sulci, pits, and lobes (Figure 11.22 B, part A), marginal frills (Figure 11.22B,

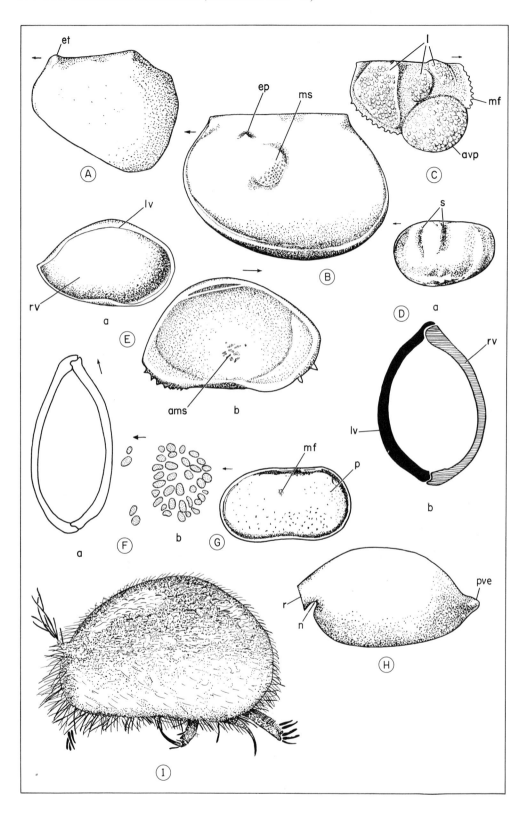

FIGURE 11.23 Representative fossil and recent ostracodes.
(A) Order Archaeocopida. *Bradoria scrutator,* L. Cam., Nova Scotia, X 4. Left valve showing eye tubercle (*et*) in antero-dorsal corner (arrow points to anterior end).
(B) Order Leperditicopida. *Leperditia* sp., U. Sil., Sweden. Left valve, X 2, *ep* — eye protuberance, *ms* — adductor muscle scar.
(C) Order Palaeocopida. *Beyrichia fittsi.* L. Dev., Oklahoma, female, right valve, X 24; *mf* — marginal frill, *avp* — antero-ventral pouch, *L* — trilobate.
(D) (a) *Kloedenella parvisulcata,* male, left valve, U. Sil., X 30; *s* = 2 sulci in anterodorsal half. (b) *Kloedenella nitida,* transverse section showing left valve (*lv*) overlapping right valve (*rv*), enlarged.
(E) Order Podocopida. (a) *Bairdia oklahomensis,* Perm., Illinois, X 35. Right valve (*RV*) overlapped by left valve (*LV*); (b) *B. formosa,* Rec., Medit., left valve, X 25; note discrete muscle scar pattern (*ams*).
(F) (a) *Healdia cara,* Penn., Ill., longitudinal section showing *LV* overlapping *RV,* X 50; (b) *Healdia* sp., Penn., Texas, circular muscle scar with many aggregate spots from *RV* interior, X 200.
(G) *Cytherella* (*Cytherella*) sp., Neogene, Brit. Isles, female, left valve, X 38; *mf* — muscle field, *p* — pits.
(H) *Cypridina inermis,* Rec., Malaya, female, left valve, X 30; *r* — rostrum, *n* — notch, *pve* — posterior ventral extremity.
(I) *Mesocypris terrestris* (family Cypridae), a terrestrial ostracode collected from forest humus, Knysa Forest, Natal, So. Africa, right valve showing muscle scars and "hairy" covering of shell. After Harding, 1955.
Illustrations redrawn and adapted from Sylvester-Bradley et al.

part *D*), varied spines and denticles, reticulations, granules, and striae. A broad band of variation is observed in surface ornamentation, marginal structures, and kind of frill, where present.

Geography of the Ostracode Valve. In order to locate any feature of the ostracode valve with some precision, the valve is divided into sectors relative to dorsal, ventral, anterior, and posterior borders (Figure 11.22*B*, part *B*). In Figure 11.22*B*, part *A*, suppose you wish to locate *L2* or *S3*. The lobe, *L2*, is close to dorsal and anterior border and hence, it is antero-dorsal in position. *S3*, a fissure, is closest to posterior border and central sector of the valve and, hence, is postero-central in position.

Muscle Scars. Muscles attached to the ostracode valve include flexor and extensor muscles for appendages, mandible-adjustor, valve-closing adductor muscles. Evidence of adductors is most frequently preserved in fossils (Figure 11.23*B*). Generally, the point of adductor muscle attachment to the shell's calcareous part is represented in fossils as a *boss* or *muscle scar.* The scar may be composed of multiple secondary scars that form a characteristic field or pattern.

Scar fields can be discerned by various means: stains, fluoridization of valve, coating valve surface with oil or water — the last-named method being quickest and simplest. The little that is known of muscle scar patterns suggests a general trend toward reduction in number of secondary scar elements. This is seen when geologically older valves (Ordovician, and so on) are compared to geologically younger ones from late and post-Paleozoic (Scott, 1961).

Muscle scars are particularly useful in classification of Mesozoic and Cenozoic ostracodes. However, other valve characters are used as well.

Orientation. Muscle scars or total muscle scar pattern — that is, adductor plus mandibular, antennal, and other scars — plays an important role in orienting the ostracode valve. Generally, muscle scar pattern is anterior in position. In living forms, adductor scars are situated in front of (anterior to) the midpoint of the length (see Howe and Laurencich, 1958, pp. 26–33; Van Morkhoven, 1962, Vol. 1, pp. 47–58).

Other criteria (Scott, 1961) that may be applied to orient fossil ostracode valves are as follows: (1) In valves bearing sulci, the anterior portion of valve is deter-

mined by position of S2. Where it occurs marks anterior portion of adductor muscle scars (Figure 11.22B, part B). (2) Often the widest end of valve is posterior in dimorphic forms.
(3) The more complex elements mark anterior end in a given hinge structure, whereas hinge channel tends to be widest posteriorly. (4) Posterior end, in outline and instar stages, is usually more tapered than anterior. (5) Major spines tend to be directed backward (that is, posteriorly). (There are several exceptions to some of these criteria, of course.)

Dimorphism, Reproduction, and Eggs. Valves of male and female ostracodes of the same species may display in structure and size striking, negligible, or subdued differences. Female valve may bear brood pouch (Figure 11.22B, part C) (Henningsmoen, 1953). In hollenellids (Figure 11.22 B, part D), differences in form of frill will distingush male from female. Dimorphic structures are important for classification of members of the order Palaeocopida (compare diagnosis of orders).

Kesling (1951a) observed that sex determination and mode of reproduction in Paleozoic ostracodes are speculative subjects. Some guidance can be offered by extrapolation from, and comparison with, living forms. Living ostracodes (superfamily Cypridacea) have two types of reproduction: (a) copulation of male and female (called *syngamy*), and (b) female fertile eggs yielding females (parthenogenesis). Type (b) characterizes freshwater ostracodes. These two modes of reproduction are rarely interchangeable in a given species. *Cyprinotus incongruens* is an exception.

So-called nuptial swarming occurs (*Philomedes globosa*). Benthonic females ascend from depths (600± ft) to mate with planktonic males (Kesling, 1961).

Ostracode eggs have a high tolerance for cold. Certain individuals are born cold-tolerant and survive even freezing to solid ice stage (Kesling, 1951b). Apparently ice crystals do not damage the egg's soft

substance. These eggs can also withstand variable periods of complete desiccation. Some forms remain viable after 30 years in such a state (Sars, 1894). Dried pond muds from Texas, Kansas, and elsewhere have yielded cultures of ostracodes in the writer's laboratory. This capacity to withstand prolonged desiccation is shared with many branchiopods. Muds that were cultured for conchostracans often have yielded ostracodes and vice versa.

The ostracode egg has an outer cover composed of two concentric, subspherical, chitinous walls that may be secondarily calcareous. A fluid of water consistency is held between these walls. Capacity to resist desiccation has been attributed to presence of this fluid (Van Morkhoven, 1962, Vol. I, pp. 139).

Ecology

Diet. Living ostracodes observed in varied natural environments and aquarium cultures, have a mixed nutrient intake (Van Morkhoven, 1962, Vol. I, pp. 140). They feed on diatoms, bacteria, protozoans, organic detritus of higher plants and animals, growing pond weed stalks (lacustrine environment), eelgrass, or coralline algae (marine environment) (Swain, 1955).

Ostracode Biofacies. Many investigators have examined ostracode suites from diverse environments: marine, brackish, fresh, and subenvironments, such as, marginal or prodeltaic. Among others, studies briefly considered here include Swain (1955), San Antonio Bay, Texas; Benson (1959, see also 1961), Todos Santos Bay Region, Baja, California; Kornicker (1961), Bimini Area, Great Bahama Bank.

In the San Antonio Bay Region, Swain (1955) distinguished three ostracode biofacies: (1) river and prodeltaic facies, (2) bay facies divisible into midbay, marginal and lower bay facies, (3) open gulf facies. Ostracode populations in the bay exceeded those in the gulf by a factor of five.

Benson (1959, see also 1961) distinguished

several ostracode biofacies in Todos Santos Bay Region: estuarine, rocky tide pool, salt-water lagoon and marsh, shallow shelf (coarse sand of deeper water and fine sand and silt environments) (Figure 11.24).

The ecological setting of Recent Bairdinae in the Bimini area led to the following findings: *temperature* exerted major control over number of *Bairdia* species in a given area; ostracodes were rare to absent on oolitic substrates; *salinity* ranged from 30 to 40 parts per thousand. Subtidal and shelf were chief sediment types dense with *Bairdia* specimens; ostracode collections were greatest in depths of less than six meters (Kornicker, 1961).

Ontogenetic Stages. After hatching from the egg, ostracodes, like other crustaceans, grow by molting. Body appendages and structures are serially expressed during the growth cycle; at molts 1 to 3,

for example, in several species, no thoracic legs are present as yet. These appear leg at a time in successive molts.

Ecdysal cycle in an ostracode will be as follows (with some modifications for a given species):

I. Hatching from egg.

II. Instar I. The first set of valves replace egg shell; first and second antennae and mandibles are the body structures.

III. Instars II–VII. Valves are shed and replaced repeatedly by secretion (that is, each instar number refers to shed pair of valves); after each instar, additional body structures previously present as beginnings, are defined and fully developed. Thus the maxilla first appears at Instar II and develops by Instar III; thoracic legs appear serially, that is, Th_1, Th_2, Th_3, from Instar IV through VI, and so on.

IV. Instar VIII. All appendages are in final form; beginning of genitalia; mouth shifts position forward.

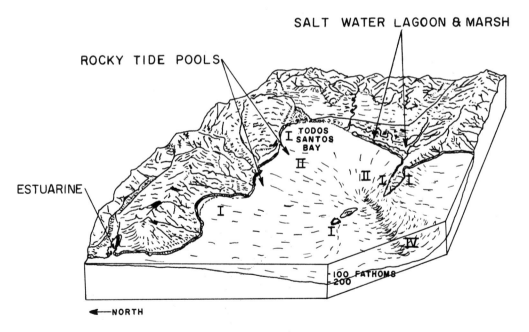

FIGURE 11.24 Recovery of information from living ostracodes. Ostracode biofacies, Todos Santos Bay Region, Baja, California, Mexico. Redrawn and modified from R. H. Benson, 1959.
Biofacies I (shallow shelf, coarse sands): Species of *Hemicytherura, Brachycythere, Cytherura, Bradleya, Quadracythere,* and so on.
Biofacies II (shallow shelf, fine sands and silt): *Brachycythere, Palmenella, Cytherura, Hemicythere,* and so on.
Biofacies IV (deeper water): *Cytheropteron, Bythocypris.*
Salt-Water Lagoon and Marsh: *Puriana, Xestoleberis, Loxoconcha, Cyprideis.*
Rockey Tide Pools: *Haplocytheridea, Loxoconcha, Xestoleberis, Brachycythere, Candites.*
Estuarine: *Cyprideis, Cypridopsis.*

V. Adult. Genitalia developed; valves with adult characteristics (hinge, duplicature, and so on) in final form (Kesling, 1961, p. Q20, Table; Van Morkhoven, 1962, Table 8).

The result of this cycle is a series of shed carapaces or *instars*, graded in size and morphology, covering ontogeny of a single individual from larval stages to adulthood. (See derivative data from fossil ostracodes.)

Fowler's Equation. A simple mathematical formula was devised by Fowler (1909; compare Kesling, 1951*b*, pp. 101–102) to permit computation of length an ostracode valve should attain at a given instar stage. There have been marked differences in interpretation of this equation's utility (Kesling, 1951*b*; Spjeldnaes, 1951; Shaver, 1953).

Salinity. Genus *Cyprideis* generally embraces brackish water forms—salinity to 10 parts per mille (see Figure 11.24, salt water, lagoon, and estuarine species). However, species of this genus, such as *C. edentata*, were found in highly saline coastal waters on Bonaire, Netherlands Antilles—salinity to 80 parts per mille (Klie, 1939). Some species of the genus contain both nodose and smooth-valve forms. If you follow one along a salinity gradient ranging from fresh to marine water, ratio of nodose/smooth forms will increase so that nodose valves exceed smooth ones.

Some *Cyprideis* species, along with those of other genera such as *Loxoconcha*, while not halophiles, are certainly halo-tolerant. Thus, where Florida Bay waters exceed 55 parts per 1000, species of these genera are represented by high population densities (Benson, 1961).

From studies of ostracodes in the Skaggarak-Baltic Sea, Elofson (1941, p. 464) summarized findings on variations encountered in number of species along a salinity gradient—a traverse from fresh to brackish to marine environments. [In what follows, salinity is given in parts per 1000 – ($^0/_{00}$) followed by number of spec-

cies.] The results: $33^0/_{00} - 78$; $32^0/_{00} - 66$; $30^0/_{00} - 61$; $30^0/_{00}$ to $17^0/_{00} - 52$; $17^0/_{00}$ to $10^0/_{00} - 42$; $10^0/_{00}$ to $2^0/_{00} - 18$.

Wagner (1957) applied Elofson's and other researches to Quaternary ostracode faunas of the Low Countries. A typical locality was the Northeast Polder—polders being reclaimed areas of sea floor in the Netherlands. He was able to show that a given fossil ostracode fauna could be treated as if it were a living fauna. In turn, that permitted extrapolation of ecological data derived from studies like that of Elofson's to the Quaternary fossil record. This applied to such items as salinity, depth, substrate, population density.

[One of the strengths of Wagner's study was use of comparatively large and equinumerous samples of ostracode valves (23,000± specimens) for determination of number and density of different species represented at each horizon. Because each sample effectively had the same number of specimens, meaningful comparisons between different horizons could be made.]

In general, allowing for variants and exceptions, the lower the salinity, smoother the ostracode valve. Thus fresh-water forms tend to be smooth and thin, whereas marine forms are heavily ornate and thick-valved.

Kornicker and Wise (1960) experimentally established environmental boundaries of the marine ostracode *Hemicythere conradi*. After an acclimation period (temperature, 20° to 26°C), in salinity of $33^0/_{00}$, individuals of this species survived salinity ranges from 6 to $65^0/_{00}$.

Substrates, Depths, Habit. Living ostracodes abound in calcareous sands and muds, and in shallow waters of the continental shelf. However, some species range to bathyal or abyssal depths. Other species are burrowers or crawlers over the bottom, and some are pelagic (Elofson, 1941). A few species of ostracodes are terrestrial (Harding, 1953, 1955).

Shape, Form, Structure as Environmental Indicators. The valves of living ostracodes are often excellent indicators of environments. One can distinguish among swimmers, crawlers, and burrowers by reference to the valve alone (Table 11.4). Natural selection favors those valve features that prevent burial in soft substrates, permit location of interstices on coarse substrates, strengthen valves against sediment impact, facilitate burrowing.

Information Derived from Fossil Ostracodes

Instars and Successive Instars. Spjeldnaes' study (1951) of the growth of *Beyrichia jonesi* (Silurian of Gotland) may be taken as illustrative. Successive instars (Figure 11.25A) can be detected in any fossil valve population by plotting length (that is, greatest length parallel to hinge line) against frequency of occurrence in the population. Sharp peaks generally denote separate instars. If certain instars are missing, the curve will be discontinuous where they should have been represented.

As a check, one could apply Fowler's equation, $L_n = L_1 k^{n-1}$ (where k = the growth factor, L_1 = length of first stage (instar), L_n = length of the nth stage). Fowler (1909,

TABLE 11.4. Ostracode Valves as Environmental Indicators

Habit	Valve Features
Swimmers	Smooth, lightweight, thin valves, high relative to length; simple hinge.
Crawlers	(*A*) Soft substrates. Flat venter that may be produced as wings (alae), frills or keels for support on soft muds.
	(*B*) Coarse substrates. Spinose; spines encase sensory setae, or ribbed to strengthen valves against shifting sand.
Burrowers	(*A*) Soft substrates. Smooth, elongate valve.
	(*B*) Interstitial between sand grains. Small, short, robust valves.

Source. Data from R. H. Benson, 1961.

p. 224) found that in early growth, "each stage increases at each molt by a fixed percentage of its length" (= growth factor k) "which is approximately constant for the species and sex." Spjeldnaes' study indicated that the mean growth factor was 1.28 ± 0.02. (See earlier discussion of Fowler's equation).

In Figure 11.25A, population of valves is $n = 972$. Of these, 130 individual valves represented the fourth instar, all less than 0.5 mm in length. Stage 11 has less than 10 valves greater than 2.0 mm in length. Fossil ostracode collections, whether dominant or minor elements of the total fossil population, will contain such growth series. Fowler's equation permits one to infer growth stages represented.

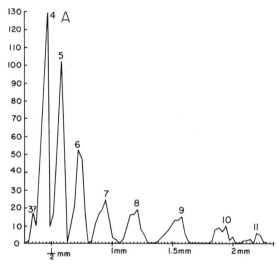

FIGURE 11.25 Information derived from fossil ostracodes. (A) Growth stages. *Beyrichia jonesi*, Gotland, $n = 972$ measured valves; numbers 3–11 are growth stages (instars); length in mm plotted against frequency in number of individual valves.

Biostratigraphy. Paleozoic ostracodes have greatest utility in biostratigraphy and regional correlations in the pre-Mississippian (Harris, 1957). They are also important in beds of Mississippian (Benson and Collinson, 1958), Pennsylvanian (C. L. Cooper, 1946), and Permian age (Sohn, 1954). Illustrative Pennsylvanian and Permian occurrences include Illinois cyclo-

Species	Depth of water (ft)[a]	Alum Bluff			Choctawhatchee				Duplin
		Oak Grove	Shoal River	Chipola	Yoldia	Arca	Ecphora	Carnuncelloria	Marl N–C°
Bairdoppilata triangulata	19–239 (> 60)						x	x	x
Haplocytheridea proboscidiala	20–95 (< 35)						x	x	x
Orionina bermudae	27			x		x	x	x	x
Puriana rugipunctata	19–239 (< 50)		x		x	x	x	x	x
Echinocythereis garretti	60					x	x	x	
Pterygocythereis americana	19–239 (> 75)		x	x		x	x	x	
Aurila amygdala	19–76	x	x	x					
Aurila conradi	19–131					x	x	x	x
Cytheretta sahnii	20–63 (< 35)	x		x			x	x	
Hulingsina ashermani	32–239 (> 60)	x		x		x	x	x	x
Pellucistoma magniventra	24–63 (< 35)							x	x
Campylocythere lacvissima	19–95 (< 25)								x
Cytherura johnsoni	19–131								x

[a]Figures in parentheses indicate depths at which Recent species are most common.

FIGURE 11.25 (cont.) (*B*) Bathymetry. Eastern Gulf of Mexico ostracodes in Miocene and Recent sediments; depths at which Recent ostracodes occur are extrapolated to Tertiary forms.

Species	Lower Decorah	Middle Decorah	Upper Decorah	Lower Galena
Aparchitidae				
Aparchites fimbriatus (Ulrich)	rrrrrrrrrrrr	rrrrrrrrrrrr	rrrrrrrrrrrr	
Aparchites ellipticus Ulrich	rrrrrrrrrrrr	rrrrrrrrrrr?		
Aparchites macrus (Ulrich)	rrrrrrrrrrrr	rrrrrrrrrrrr	rrrrrrrrrrrr	
Aparchites paratumida Swain & Cornell, n. sp.	rrrrrrrrrrrr	rrrrrrrrrrrr	rrrrrrrrrrrr	
Bullatella granilabiata (Ulrich)	rrrrrrrrrrrr	rrrrrrrrrrrr	rrrrrrrrrrrr	
Macronotella scofieldi Ulrich			rrrrrrrrrrrr	rrrrrrrrrrrr
Saccelatia arrecta (Ulrich)	aaaaaaaaaa	aaaaaaaaaa	aaaaaaaaaa	
Saccelatia angularis (Ulrich)		rrrrrrrrrrrr		
Saccelatia arcuamuralis Kay	rrrrrrrrrrrr	rrrrrrrrrrrr	rrrrrrrrrrrr	
Saccelatia bullata Kay	rrrrrrrrrrrr	rrrrrrrrrrrr		
Saccelatia cletifera Kay		rrrcccccrrrr	rrrrrrrrrrrr	
Aechmina ionensis Kay		rrrrrrrrrrrr	rrrrrrrrrrrr	
Leperditellidae				
Schmidtella umbonata Ulrich	vvvvvvvvv	vvvvvvvaa	aaaaaaaaaa	
Schmidtella affinis Ulrich	aaaaaaaaaa	aaaaaaaaaa	aaaaaaaaaa	
Schmidtella brevis Ulrich		rrrrrrrrrrrr		
Schmidtella incompta Ulrich	aaaaaaaaaa	rrrrrrrrrrrr	aaaaaaaaaa	aaaaaaaaaa
Pedomphalella intermedia Swain & Cornell, n. sp.		rrrrrrrrrrrr		
Pedomphalella subovata Swain & Cornell, n. sp.	rrraaaaaaaa	accccccccca	aaaaaarrr	rrrrrrrr?
Cryptophyllus oboloides (Ulrich & Bassler)	rrr	aaaaaaaaaa	aaaaaaaaaa	aaaa?
Byrsolopsina planilateralis (Kay)	a	aaaaaaaaaa	aaaaaa	
Byrsolopsina centipunctata (Kay)	rrrr	rrrrrrrrrrrr	rrrrrrrrrrrr	
Byrsolopsina ovata (Kay)	rrrrrrrrrrrr	rrrrrrrrrrrr	rccccccccc	cccc
Leperditellidae				
Byrsolopsina normella Swain & Hansen, n. sp.		cccccccccc		
Parenthatia punctata (Ulrich)		vvvvvvvvv	vvvvvvvvv	
Parenthatia camerata Kay			vvvvvvvvv	
Eurychilinidae				
Nodambichilina symmetrica (Ulrich)		rrrrraaaaa	aaaaaaaaaa	rrr
Tsitrella simplex Swain & Hansen, n. sp.		aaaaaa	rrrrrrrrrrrr	
Punctaparchitidae				
Punctaparchites rugosus (Jones)	aaaaaaaaaa	aaaaaaaaaa	aaaaaaaaaa	
Punctaparchites multipunctata (Kay)				
Macrocyproides trentonensis (Ulrich)	aaaaaaaaaa	aaaaaaaaaa	aaaaaaaaaa	aaaaaaaaaa

r, rare (1–5); *c*, common (6–10); *a*, abundant (11–25); *v*, very abundant (26 +).

FIGURE 11.25 (cont.) (*C*) Biostratigraphy. Stratigraphic distribution of some Decorah ostracodes (Ord., Minnesota); abundance of species on slide of approximately 100 specimens: *r*, rare (1–5); *c*, common (6–10), *a*, abundant (11–25); *v*, very abundant (26 +).

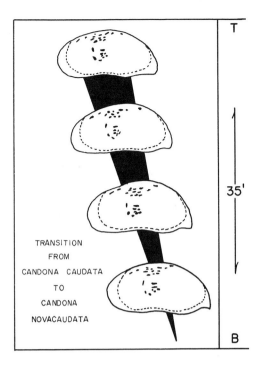

TRANSITION
FROM
CANDONA CAUDATA
TO
CANDONA
NOVACAUDATA

FIGURE 11.25 (cont.) (*D*) Evolution. Postglacial (Holocene) core from Lake Erie; change in shape of carapace from bottom (*B*) to top (*T*) of a core 35 ft long—from *Candona caudata* to *Candona novacaudata*. Illustrations adapted and redrawn from: (*A*) Spjeldnaes, 1951; (*B*) Benson and Coleman, II, 1963; (*C*) Swain et al., 1961; (*D*) Benson and MacDonald, 1963.

them marine phase; Ohio Pottsville Series (Marple, 1952); and Permian Glass Mts. of Texas, insoluble residues.

Representative studies of pre-Mississippian ostracodes include Simpson group of Oklahoma (Harris, 1957), Decorah shale of Minnesota (Swain et al., 1961, and see also Figure 11.25C), Silurian of New Jersey and so on (Swartz and Whitmore, 1956), and the Devonian of central Pennsylvania (Swartz and Swain, 1941).

Ostracode utility in post-Paleozoic stratigraphy is so important that Van Morkhoven devoted a whole chapter to it (1962, Vol. I, Chapter 6); and Howe and Laurencich (1958) wrote an entire book on Cretaceous ostracodes alone. Generalities that emerge from these works, numerous papers and monographs are the following: (*a*) Ostracodes inhabited marine, brackish, and fresh waters, and these facies may overlap or interfinger—explaining one aspect of their biostratigraphic and ecologic value. (*b*) Fragments of valves may permit identification, which is an asset of great value when one studies well cuttings of Mesozoic-Tertiary beds. (*c*) Restricted vertical ranges are often encountered in post-Paleozoic ostracode genera and species. (*d*) From Jurassic to Recent, several ostracode genera have restricted vertical ranges and wide lateral distribution, which makes them excellent index fossils.

A good example is restricted range of different species of *Camptocythere* in the marine Jurassic of northwestern Europe. These species, which are characterized by a conspicuous right valve, help zone and define the Lower Dogger (M. Jurassic (Table 11.5).

Despite overlap of several species, extended vertical ranges beyond horizons shared by both species suffice to denote distinct Dogger subdivisions.

Ostracode species of such genera as *Protocythere* and *Cythereis* are stratigraphically important in Lower Cretaceous. Ostracode facies markedly change in the Upper Cretaceous: merodont-hinged *Protocythere* species are replaced by species of same genus with amphidont hinges and by species of genus *Veenia*. More typical Cenozoic genera include *Paracyprideis*, Upper Cretaceous (Cenomanian to Recent), *Brachycythere* (Cenomanian to Oligocene). Many Cretaceous genera continued into the Tertiary (Reyment, 1960; Van Morkhoven, 1962).

Bathymetry. Benson and Coleman (1963) studied 30 species of Recent marine ostracodes from eastern Gulf of Mexico. Of these, about 15 species are known as fossils in Miocene and Pliocene of southeastern United States and Cuba. Known maximum depths of Recent forms could be extrapolated backward to Miocene equivalents. Thus populations of *Bairdoppilata triangulata* occur in water depths of 19 to 239 ft, but most commonly at greater than 60 ft. This species, known as fos-

TABLE 11.5. Ostracode Biostratigraphy of the Middle Jurassic of Germany. (Based on Six Species of the Genus *Camptocythere*)

		Middle Jurassic (Dogger)		Lower Jurassic (Lias)
Series	*Bajocian*	*Aalenian*		*Toarcian*
Species	*Gamma (γ)*	*Beta (β)*	*Alpha (α)*	*Zeta (ζ)*
1. *Praecox*			xxxxxxxxxxxxxxxxxxx[a]	
2. *Modesta*		xxx		
3. *Taveolata*		xxx		
4. *Obtusa*		xx		
5. *Wedia*	xxxxx			
6. *Pusilla*	xxxxxx			

[a]The "x" indicates range; note overlap in ranges of species 2 and 3, and species 5 and 6. There are no species in the Bajocian.
Source. Data from Triebel, cit. Van Morkhoven, 1962.

sil in Miocene deposits of Florida and North Carolina, is assumed to have lived at equivalent depths as Recent forms.

Evolution. Bettenstaedt (1958) studied 22 successive populations of *Cytherelloidea ovata* in the Lower Cretaceous (Hauterian) of northwest Germany. Many of these population ranges overlapped. Sixteen variants were found that provided a continuous vertical spectrum of gradual ornamentation change. Besides being of biostratigraphic value (each variant designated a particular stratum), variation in surficial ornamentation reflected important genetic changes on the molecular level.

Actual transition from older to younger horizons in Hauterian beds was from strong ribbed forms to almost smooth forms. Ornamentation and habitat of ostracodes are known to be related. Ribbing could have served to strengthen valves of individuals inhabiting high energy, shallow marine zones. By contrast, loss of ribbing would denote either deepening water or more brackish water. The latter seems more likely since Recent *Cytherelloidea* species are known in both shallow marine and occasionally brackish water environments (Van Morkhoven, 1963, Vol. II, pp. 21–23).

At any rate, the selective factor appears to have been either *energy level* (presence or absence of wave action) or *variable salinity*, with either a decrease in energy level and salinity, or both, serving to sponsor smoother shell form through time.

Benson and MacDonald (1963) described an interesting gradual transition from *Candona caudata* to *Candona novacaudata*. The transition was observed in a post-glacial (Holocene) 35-ft long core from Lake Erie (Figure 11.25D). The time represented is stated to be 6000 to 8000 years. In the core's basal portion, the ratio of *C. caudata* to *C. novacaudata* was 4:1, and *C. caudata* had a pointed, hooked posterior (female, left valve) and a relatively shorter and higher carapace. Also females exceeded males. By the time represented at the core's top, ratio of the two species was reversed with *C. novacaudata* population dominant and an increase in males. As shown in Figure 11.25D, the pointed, hooked posterior grades upward into a blunted, noselike posterior (see *C. novacaudata* female, left valve).

Although the Lake Erie study does not explore probable selective factor(s) accounting for observed vertical displacement of *C. caudata* by *C. novacaudata*, it is apparent that the tendency toward a population with a blunted noselike posterior (female left valve) must have been favored by natural selection. In turn, this could denote some important environmental change through the time represented in the core.

The main evolutionary trends that Reyment observed (1960, Vol. I, pp. 46–49) in Cretaceous-Paleocene ostracodes of the Nigerian delta were in the direction of increase or decrease in size (*Cythereis deltaensis*, for example, decreased in size). One species of *Veenia* displayed "no changes whatsoever" in the usual growth parameters from Upper Maastrictian to Lower Paleocene—a span of eight to ten million years.

Predation. Reyment's study of Recent and Paleocene ostracodes of the Niger Delta showed that predaceous gastropods drilled holes in Paleocene ostracode valves more frequently than in their Recent equivalents.

Density of Recent ostracodes in the Niger Delta is about 620 ostracodes per gram. The predation discussed does not act in any significant sense to limit or thin out such population density. Rather the chief cause of premature death in these Recent ostracodes was fortuitous burial beneath sediments (and presumably the same applied in the geologic past. Schäfer (1962, cit. Reyment, 1966b) inferred that a given marine ostracod burrowing or crawling on bottom substrate may be in jeopardy of being buried alive somewhat less than a dozen times in its lifetime.

Transition to the Terrestrial Environment. The terrestrial environment as used here refers specifically to niches on dry land as distinct from those in streams, lakes, and so on.

Even though the problem has not yet been encountered in the fossil record of ostracodes, it is of interest that living African fresh-water ostracodes have migrated to the terrestrial environment and successfully occupied some niches there. Harding (1953, 1955) observed that wet or slightly damp moss was the route to dry land taken by fresh-water ostracodes (*Mesocypris*, Figure 11.23*I*) as well as by several species of harpacticoid copepods.

Transition for the ostracode *Mesocypris* was accomplished with very few adaptations. The latter included a hairy shell, possibly to help resist desiccation by trapping drops of moisture on individual hairs; tight-fitting valves; powerful second antennae to aid locomotion: "swimming" setae reduced in length and vestigial; tips of second antennae and the asymmetrical furcal rami armed with exceptionally long claws; a food-directing mechanism to guide morsels to the mouth, consisting of sharp spines on end of mandibular palp, setae of mandible, maxilla, and first leg.

One or another of these adaptations occur in some other nonterrestrial genera. From these, selective pressures exerted in the given environment led to preservation of characters or combinations of characters of greatest adaptive advantage.

Harding (1955) inferred that evolution of a terrestrial mode of life in *Mesocypris* probably occurred "quite recently."

QUESTIONS

Crustacea-II

General, Copepods, Barnacles, Branchiopods

1. (a) Briefly distinguish each of the eight classes of superclass Crustacea from one another.

   (b) What do you conclude from these statements: "Copepods have a sparse fossil record." "Known fossil copepods have a modern aspect."

   (c) Is it possible that Tertiary copepods may have played an important role in the food chain? Explain your answer.

2. (a) Sketch and label the plate system of a pedunculate barnacle.

   (b) How can a fossil acrothoracic barnacle be detected and identified?

   (c) Discuss the work of Codez and Saint-Seine (1958) on barnacle biometrics.

3. Why is ours called the age of barnacles?

4. (a) What essential features link all Branchiopoda?

   (b) Identify by order: *Triops, Daphnia, Cyzicus, Branchinecta.*

   (c) Sketch and label a ribbed and un-ribbed conchostracan.

5. (a) How can fossil cladocerans aid biostratigraphic studies in glacial and postglacial lakes?

   (b) Discuss seasonal intervals and geochronology in conchostracan-bearing beds.

Crustacea – II

Ostracoda

1. How would you establish the fact that Klaus Müller's phosphatic genus of ostracodes has valves of primary and not secondary phosphate?

2. (a) Compare the five orders of ostracodes. How are they alike and how do they differ?

   (b) What is the relative taxonomic value of each of the following features of the ostracode valve: shape and size; muscle scar pattern; ornamentation; hingeline?

3. Draw an ostracode valve and let your pencil run over it stopping at random. Then locate yourself precisely at five different such stops.

4. Cite the main findings of the following studies: Klie, Elofson, Wagner.

5. What criteria can aid in orienting fossil ostracode valves? (Try this with several fossil valves.) Why is orientation important?

6. Given a collection of fossil ostracode valves, how can you tell which belong to a single individual?

7. If you found two distinct suites of fossil ostracodes that contained the following genera, what age could you tentatively assign to the beds containing them?

   Suite A: *Cytherella, Paracypris, Loxoconcha,* and *Brachycythere.*

   Suite B: *Primitella, Bollia, Healdia, Amphissites.*

REFERENCES

Crustacea — II

Cirripedia

Branson, C. C., 1964. Barnacle burrows and shells of Oklahoma fossils: Okla. Geol. Notes, Vol. 24 (4), pp. 98–99.

Calman, W. T., 1909. "Cirripedia" in Ray Lancaster, ed., A Treatise on Zoology; Crustacea. Chapter 5, pp. 106–136.

Cheetham, A. H., 1963. Gooseneck barnacles in the Gulf Coast Tertiary: J. Paleontology, 37 (2), pp. 393–400.

Codez, J., and R. de Saint-Seine, 1958. Révision des Cirripédes acrothoraciques fóssiles: Soc. Géol. de France Bull., 7 (6), pp. 699–719, plates 37–39, text figures, tables.

Darwin, C., 1851. A monograph on the fossil Lepadidae or the pedunculated cirripeds of Great Britain: Palaeontographical Soc., 80 pp.

Green, J., 1961. A Biology of Crustacea: Quadrangle Books, Chicago, pp. 10–12.

Jägersten, G., 1960. Life in the Sea. Basic Books, N.Y., pp. 102–108.

Kolosváry, G., 1958. Ein neuer operculaten Cirripedies aus der Kreide: Paläont. Z., 32, pp. 30–39.

Novojilov, N. I., 1955. Restes de Cirripèdes particuliers du Devonien moyen de la Sibérie du Sud. C. R. Acad. Sci. URSS, Vol. 100 (6), pp. 1161–1162.

Pilsbry, H. A., 1925. Miocene and Pleistocene Cirripedia from Haiti: Proc. U. S. N. M. 65 (Art. 2), pp. 1–3, plate 1.

Schlaudt, C. M., and K. Young, 1960. Acrothoracic barnacles from the Texas Permian and Cretaceous: J. Paleontology, Vol. 34 (5), pp. 903–907.

Withers, T. H., 1953. Catalogue of fossil Cirripedia, Vol. 3, Tertiary. Brit. Mus. (Nat. Hist.), 396 pp., 64 plates. *Idem*, Vol. 2. Cretaceous, *idem*, 433 pp., 50 plates *idem*, Vol. 1, Triassic and Jurassic, *idem*, 131 pp., 12 plates.

————, 1915. Some Paleozoic fossils referred to the Cirripedia: Geol. Mag., Vol. 6, pp. 112–123.

Zullo, V. A., 1966. Zoogeographic affinities of the Balanomorpha (Cirripedia: Thoracica) of the Eastern Pacific, in R. I. Bowman, ed., The Galápagos. Univ. of Calif. Press, pp. 139–144.

Copepoda

Harding, J. P., 1950. A rare estuarine copepod crustacean, *Enhydrosoma gariensis*, found in the Holocene of Kent: Nature, Vol. 178, pp. 1127–1128.

Johnson, M. W., 1934. The life history of the copepod *Tortanus descaudatus* (Thompson and Scott): Biol. Bull., Vol. 67 (1), pp. 182–200.

Palmer, A. R., 1960. Miocene copepods from the Mojave Desert, California: J. Paleontology, Vol. 34 (3), pp. 447–452, plate 63, 1 text figure.

Branchiopoda

Copeland, M. J., 1962. Canadian fossil Ostracoda, Conchostraca, Eurypterida, and Phyllocarida: Canadian Geol. Surv. Bull. 91, pp. 12–16, plates 3–4.

Frey, D. G., 1958. The late glacial cladoceran fauna of a small lake: Arch. Hydrobiol., Vol. 54, pp. 14–270, plates 35–41, 113 text figures.

Kobayashi, T., 1954. Fossil estherians and allied fossils: Tokyo Univ. Jour. Fac. Sci., Vol. 9, sec. 2, Part 1, pp. 1–192, 30 text figures.

Lehmann, W. M., 1955. *Vachonia rogeri*, n. gen., n. sp., ein Branchiopod aus dem unterdevonischen Hunsrückschiefer: Paläont. Zeitschr. Vol. 29, pp. 126–130, plates 11, 12, text figures 1, 2.

Novojilov, N., 1959. Position systematique des Kazachartha (Arthropodes) d'après de nouveaux materieux des monts Ketmen et Sajkan (Kazakhstan), S. E. et N. E.: Soc. Géol. France Bull. 7ᵉ ser., Vol. 1, pp. 265–269, plates 8, 9.

————, 1958. Recueil d'articles sur les phyllopodes conchostracés: Ann. Mines. Bur. Rech. Géol. Géophys. & Min. Serv. Inf. Géol. (Jan. 1958), No. 26, 117 plates.

Novikoff, M., 1905. Untersuchungen über den Bau der *Limnadia lenticularis* L., Zeitschr. Wissenschaft. Zool. Leipzig, pp. 561–616, Tables 19–22.

Pruvost, P., 1919. Introduction à l'étude du Terrain Houiller du Nord et du Pasde-Calais. La Fauna Continentale du Terrain Houiller du Nord de la France: Minist. Travaux Publ., Mém.

Swain, F. M., 1956. Stratigraphy of lake deposits in central and northern Minnesota: AAPG Bull., Vol. 40, pp. 600–653, 29 text figures.

Tasch, P., 1969. Branchiopoda. Treat. Invert. Paleontology. Part R. Arthropoda, Vol. 4, pp. R-218–R-291.

————, 1967*d*. Antarctic leaiid zone: seasonal events: Gondwana correlations, Gondwana Stratigraphy, U.N.E.S.C.O. (1969).

————, 1964. Periodicity in the Wellington Formation of Kansas and Oklahoma, in D. F. Merriam, ed., Symposium on Cyclic Sedimentation: Bull. 169, Vol. 2, pp. 481–496.

————, 1963*a*. Fossil content of salt and association (cor. associated) evaporites, in Symposium on Salt, No. Ohio Geol. Soc., pp. 96–102.

————, 1963*b*. Evolution of the Branchiopoda, in H. B. Whittington and W. D. I. Rolfe, eds., Phylogeny and Evolution of Crustacea. Mus. Comp. Zool. (Harvard), Chapter 11, pp. 145–157, Figures 63–67, Table 5.

Wilson, J. L., 1956. Stratigraphic position of the Upper Devonian branchiopod *Rhabdostichus* in the Williston Basin: J. Paleontology, Vol. 30, pp. 959–980.

Ostracoda

Bartenstein, H., 1959. Feinstratigraphisch wichtige Ostracoden aus dem nordwestdeutschen Valendis: Paläont. Z., Vol. 33 (4), pp. 224–242.

Bettenstaedt, F., 1958. Phylogenetische Beobachtungen in der Mikropaläontologie: Paläont. Z., Vol. 32 (3, 4), pp. 115–140.

Benson, R. H., 1961. Ecology of ostracode assemblages: Treat. Invert. Paleont., Pt. Q. Arthropoda 3, pp. Q-56–Q-63.

————, 1959. Ecology of Recent ostracodes of the Todos Santos Bay Region, Baja, California, Mexico: Univ. Kansas Paleont. Contrib., Arthropoda, Art. 1, pp. 1–80.

Benson, R. H., 1955. Ostracodes from the type section of the Fern Glen Formation: J. Paleontology, Vol. 29 (6), pp. 1030–1039.

———— and G. L. Coleman, II, 1963. Recent marine ostracodes from the Eastern Gulf of Mexico: Univ. Kans. Paleont. Contribution, Art. 2, pp. 1–52.

———— and H. C. MacDonald, 1963. Postglacial (Holocene) ostrcodes from Lake Erie: Univ. Kans. Paleont. Contribution, Art. 4, pp. 1–26.

————, et al., 1961. Systematic Descriptions, Treat. Invert. Paleontology, Part Q. Arthropoda 3, pp. Q99–Q414.

———— and C. Collinson, 1958. Three ostracode faunas from the Lower and Middle Mississippian strata in southern Illinois: III. State Geol. Surv. Circ. 255, pp. 1–25, plates 1–4, 4 text figures.

Elofson, O., 1941. Zur kenntnis den marinen Ostracoden Schwedens mit besonderer Berücksichtigung des Skaggeraks: Zoologiska Bidrag. f. Uppsala, Vol. 19, pp. 217–534, 52 text figures, 42 plates (Ecological Studies Part 10, pp. 406ff).

Fowler, G. M., 1909. Biscayan plankton collected during a cruise of H.M.S. Research, 1900: Linnean Soc. London, Trans. Zool., 2nd ser., Vol. 10 (9), pp. 219–238.

Gutentag, E. D., and R. H. Benson, 1962. Neogene (Plio-Pleistocene) freshwater ostracodes from the Central High Plains: State Geol. Surv. Kansas, Bull. 157 (4), 1–60, plates 1–2, Figures 1–15.

Harding, J. P., 1953. The first known example of a terrestrial ostracode, *Mesocypris cypris terrestris* n. sp.: Ann. Natal. Mus. Vol. 12 (3), pp. 359–365.

Harris, R. W., 1957. Ostracoda of the Simpson Group: Okla. Geol. Surv. Bull. 75, 283 pp.

Henningsmoen, G., 1954*a*. Lower Ordovician ostracodes from the Oslo Region, Norway:

Norsk Geol. Tidssk., Vol. 33 (1–2), pp. 41–68.

———, 1954*b*. Upper Ordovidian ostracodes from the Oslo region, Norway: *Idem*, Vol. 33 (1–2), pp. 69–108.

———, 1954*c*. Silurian ostracodes from the Oslo Region, Norway: *Idem*, Vol. 34 (1), pp. 15–71, plates 1–8.

———, 1953. Classification of Paleozoic straight-hinged ostracodes: Norsk Geol. Tidssk., Vol. 31, pp. 185–288.

Howe, H. V., R. V. Kesling, and H. W. Scott, 1961. Morphology of living Ostracoda: Treat. Invert. Paleontology: Part Q. Arthropoda 3, pp. Q3–Q17.

——— and L. Laurencich, 1958. Introduction to the study of Cretaceous ostracodes: Louisiana State Univ. Press, 518pp.

Kesling, R. V., 1961. "Reproduction of Ostracoda," "Ontogeny of Ostracoda"; Treat. Invert. Paleontology: Part Q, Arthropoda 3, pp. Q17–Q19, Q19–Q20.

———, 1951*a*. Terminology of ostracode carapaces: Contrib. Mus. Paelont. Univ. Michigan, Vol. 9 (4), pp. 93–171.

———, 1951*b*. The morphology of ostracode molt stages: III. Biol. Monogr., Vol. 21 (1–3); Univ. Ill. Press, Urbana, 116pp.

Klie, W., 1939. Ostracoden aus den marinen Salinen von Bonaire, Curaçao und Aruba: Zool. Ergebnisse einer Reise nach Bonaire, Curaçao und Aruba im Jahre 1930: Capita Zoologica, Vol. 8 (25), pp. 1–19.

Kornicker, L. S., 1961. Ecology and taxonomy of Recent Bairdiinae (Ostracoda): Micropaleontology, Vol. 7 (1), pp. 55–70, plate 1.

——— and C. D. Wise, 1960. Some environmental boundaries of a marine ostracode: Micropaleontology, Vol. 6 (4), pp. 393–398.

Marple, M. F., 1952. Ostracodes from the Pottsville series in Ohio: J. Paleontology, Vol. 26 (6), pp. 924–939.

Müller, K. J., 1965. Ostracoda (Bradorina) mit phosphatischen Gehäusen aus dem Oberkambrium von Schweden: N. Jb. Geol. Paläont., appendix, Vol. 121, pp. 1–42, Tables 1–29.

Reyment, R. A., 1960–1966. Studies on Nigerian Upper Cretaceous and Lower Tertiary Ostracoda: Part 1, Senonian and Maastrichtian Ostracoda: Acta Univ. Stockholmiensis. Contrib. in Geology, Vol. 7, 232pp.

———, 1963. *Idem*, Part 2, Danien, Paleocene, and Eocene Ostracoda: *idem*, Vol. 10, 284pp.

———, 1966. *Idem*, Part 3, *idem*, Vol. 14, 135pp.

Scott, H. W., 1961. "Shell morphology of the ostracode valve", "Orientation of ostracode valves": Treat. Invert. Paleontology, Part Q, Arthropoda 3, pp. Q21–Q37, Q44–Q47.

Sohn, I. G., 1954. Ostracoda from the Permian Glass Mountains, Texas: U.S.G.S. Prof. Paper 264-A, 24pp., 5 plates.

———, 1950. Growth series of ostracodes from the Permian of Texas: U.S.G.S. Prof. Paper 221-G, pp. 33–43, plates 7, 8.

Spjeldnaes, N., 1951. Ontogeny of *Beyrichia jonesi* Boll.: J. Paleontology, Vol. 25, pp. 745–755.

Swain, F. M., 1955. Ostracoda of San Antonio Bay, Texas: J. Paleontology, Vol. 29 (4), pp. 561–646.

Swain, F. M., J. R. Cornell, and D. L. Hansen, 1961. Ostracoda of the families Aparchitidae, Aechminidae, Leperditellidae, Drepanellidae, Eurychilinidae, and Punctaparchitidae from the Decorah shale of Minnesota: J. Paleontology, Vol. 35 (2), pp. 345–372, plates 46–50, 2 text figures.

——— and J. A. Peterson, 1951. Ostracoda from the Upper Jurassic Redwater shale member of the Sundance Formation at the type locality in South Dakota: J. Paleontology, Vol. 25 (6), pp. 796–807.

Sylvester-Bradley, P. C., 1956. The structure, evolution, and nomenclature of the ostracod hinge: Brit. Mus. (Nat. hist.) Bull. Geology, Vol. 3 (1), pp. 1–21, plates 1–4.

Tressler, W. L., 1959. Ostracoda, in H. B. Ward and G. C. Whipple, Fresh Water Biology, W. T. Edmonson, ed., 2nd Ed., pp. 657–734.

Van Morkhoven, F. P. C. M., 1962–1963. Post-Paleozoic Ostracoda: Vol. 1 (General), 196pp.; Vol. 2 (Generic Descriptions), 473pp. Elsevier, Amsterdam.

Wagner, C. W., 1957. Sur les ostracodes du Quaternaire Récent des Pays-Bas et leur utilisation dans l'étude géologique des dépots holocenes. Dissertation, Univ. of Paris, Printed by Mouten, The Hague, 260pp.

CRUSTACEA—III (MALACOSTRACANS)

Fossil and Living Malacostracans

Class Malacostraca. [Name refers to soft (*malacos*) shell (*ostracon*) of several class members. The class embraces, among others, decapod crustaceans—lobsters, shrimp, crayfish, and so on, and such forms as the sandhopper (amphipod) and pill bug (isopod).]

Diagnosis. Crustacea characterized by the following features are included in the class Malocostraca:

1. *Eyes.* Compound, typically stalked.

2. *Carapace.* Variously developed, bivalved in Leptostraca, vestigial in others, lacking in superorder Syncarida, and in decapods a single shield. Generally covering thorax (see Chapter 15 for decapod carapace composition and skeletogenesis, and text for sections on composition and structure of phyllocarid cuticle).

3. *Somites.* Thoracic somites, eight; abdominal somites, six to seven—appendage bearers.

4. *Telson.* Rarely with caudal furca.

5. *Genital Apertures.* Female, always located on sixth thoracic somite; male, on the eighth thoracic somite (see Figure 11.26).

6. *First Antenna.* Biramous.

7. *Development.* Often direct; development in brood pouch in order Peracarida; young rarely hatched in naupliar stage, but there are exceptions among some decapod malacostracans. Where larval stages are present (Figure 11.28*A*), a zoeal stage and, in some forms, a metazoeal stage or mysis stage are distinct.

8. *Geologic Range.* Cambrian to Recent.

Treatise authors recognize three subclasses: (1) Phyllocarida, (2) Uncertain (embraces all known phyllocarid forms assigned to Discinocarina), and (3) the subclass containing the bulk of malacostracans, Eumalacostraca.

Ancestral Malacostracan and the Caridoid Facies.   Assume the following set of characteristics (see Figure 11.26*A* and *B*): (1) carapace enveloping thorax, (2) stalked eyes, (3) biramous antennules and first antenna, (4) scale-like exopodite on the second antenna, (5) natatory (that is, used in swimming) exopodites on thoracic limbs, (6) an elongated and ventrally flexed abdomen, (7) tail fan spread on either side of the telson (Calman, 1909, p. 144; Siewing, 1963, pp. 100–102). Taken altogether, these characters constitute a type of organization known as the "caridoid facies."

Furthermore, assume a given group of certain crustaceans embracing several distinct orders—such as mysids, euphausids, decapods—share the caridoid facies in common.

From the two assumptions above, it follows that the caridoid facies is a highly generalized kind of organization, occurring as it does in several crustacean orders. Observation of such commonality in a suite of morphological characters was the basis for erecting the class Malacostraca. In addition, it can be inferred that an ancestral malacostracan originally possessed the characteristics of this facies and that it was genetically transmitted to successive descendants (Figure 11.26*A* and *B*).

Could different orders now grouped in class Malacostraca have arisen along distinct lines of descent from different ancestral stocks (= polyphyletic descent)? The answer is twofold: (1) Recurrences and constants—in addition to the caridoid facies that recurs throughout the class, there are still other constants. The genital openings, for example, are always in a constant situation, and there are a constant number of segments in head, thorax, and abdomen. These, Siewing (1963) held, show that orders of the class do form a *natural* group suggesting monophyletic descent. (2) Genetic code—is it possible that such recurrences and constants throughout the class are transmitted without equivalent unchanging configura-

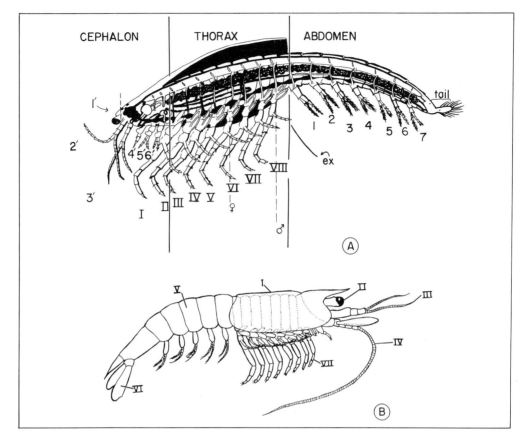

FIGURE 11.26 Malacostraca – general anatomy.
(A) Schematic diagram showing anatomy of the hypothetical ancestral malacostracan. Adapted from Siewing, 1963. Cephalic appendages, 1' – embryonic preantennulary, 2' – antennule, 3' – antenna, 4' – mandible, 5' – maxillula, 6' – maxilla (for a criticism, see Sharov, 1965, p. 54). Thoracic appendages I–VIII, female genital opening always on appendage VI, male genitalia, appendage VIII; abdominal appendages exclusive of tail, 1–7.
(B) Lateral view of a generalized malacostracan showing that combination of characters called the "caridoid facies" (adapted from Calman, 1909). This facies consists of I, carapace covering thoracic region; II, movable stalked eyes; III, biramous antennules; IV, a scalelike exopodite on the antenna; V, ventrally flexed abdomen; VI, a tail fan consisting of lamellar rami and a telson; and VII, natatory exopodites on thoracic limbs (see *ex* in Figure A).

tions in portions of the genetic code? No.

Sanders (1963) found that the caridoid facies appeared late in larval development. He arrived at this conclusion from an analysis of decapod *Penaeus* larval stages. The precaridoid morphology was shown to be "remarkably similar" to the Cephalocarida. These data referred back to the probable malacostracan ancestor, that is, a cephalocarid-like crustacean.

All of these factors indicate a monophyletic origin for the class.

The Subclasses. Of the three subclasses, only phyllocarids and eumalacostracans will be treated here.

Subclass 1. Phyllocarida. Malacostraca with large carapace of two valves covering cephalothorax only. Along dorsal margin, a valve hinge line may or may not be

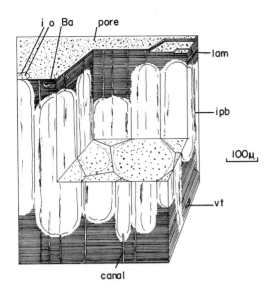

i o Ba   pore

lam

ipb

100µ

vt

canal

FIGURE 11.26B Reconstruction of cuticle of the phyllocarid *Ceratiocaris papilio*. Replacement calcite is shown in white. Close ruled lines represent liminae in the collophane. *Ba* — Balkenlagen (presumed traces of oriented micells or chitin crystallites) (see Chapter 15), common in beetle cuticles; *o* — outer wall; *i* — inner pillar of calcite replacing a prism; *lam* — laminae; *ipb* — interprismatic boundary; *vt* — vermiform tubule. Ghost laminae and canals are seen as fainter lines through replacement calcite. (After Rolfe, 1962.)

present. At front of the carapace, there's a distinctive feature, a movable, articulated, lanceolate, rostral plate (see Rolfe, 1967, and Figure 11.27*A*). Three orders: Order 1. Leptostraca, *Nebalia* (Figure 11.27*A*, part *A*); Order 2. Hymenostraca, *Hymenocaris* (Figure 11.27*A*, part *C*); Order 3. Archeostraca, *Nahecaris* (Figure 11.27*A*, part *B*); *Ceratiocaris*.

Subclass 2. Eumalacostraca. Shrimplike crustaceans, generally with carapace; furca and median spine on telson reduced and functionally replaced by the caudal fan, composed of uropods and body of telson (H. K. Brooks, 1962, also Figure 11.27*B*, part *D*). Five superorders.

Superorder 1. Eocarida. Embraces three extinct orders in which carapace is not fused with thoracic somites. *Eocaris* (Figure 11.27*B*), *Palaeopalaemon* (Figure 11.27 *A*, part *D*).

Superorder 2. Syncarida. Malacostracans without carapace. *Squillites* (Figure 11.27*B*, part *B*).

Superorder 3. Pericarida. If carapace is present, it will not be fused with more than four thoracic somites. *Anthracocaris* (Figure 11.27*B*, part *C*), *Kilianicaris* (Figure 11.27*B*, part *D*). (This superorder includes mysids, amphipods, isopods, and so on.)

Superorder 4. Eucarida. Carapace fused with all thoracic somites. Order Decapoda embraces eumalacostracans that possess ten legs: Suborder 1, Natantia, includes *swimming* forms such as true shrimps and prawns. Suborder 2, Reptantia, includes *crawling* forms such as crayfish, lobster, and crab. *Cancer* (Figure 11.27*B*, parts *E–H*).

Superorder 5. Hoplocarida. Carapace is shallow and fused with three of the seven thoracic somites. *Pseudosculda* (Figure 11.27*B*, part *I*). (This superorder embraces stomatopods, see Figure 11.28*B*.)

Information Derived from Fossil and Living Malacostracans

(Phyllocarids are considered separately after Malacostracans.)

Biotic Associates. Fossils found with eumalacostracans are varied arthropods (ostracods, branchiopod conchostracans, insects), also arachnids, merostomes, and myriapods. In addition, other forms found as associates include mollusks, graptolites, diverse plant remains, reptiles, and amphibians.

Living amphipods, crabs (coral gall crabs), shrimp are found associated with various coelenterates; other crabs live in or on echinoderms, such as the sea cucumber and sea urchin (Green, 1961). In the Upper Jurassic, some decapods (Galatheidae, for example) were associated with algal and sponge reefs (Roger, 1953), as are some living eumalacostracans.

Substrate and Lithofacies. Well-preserved fossil malacostracans have been retrieved from clay ironstone concretions and sublithographic/lithographic or bioclastic limestone facies; thin-bedded,

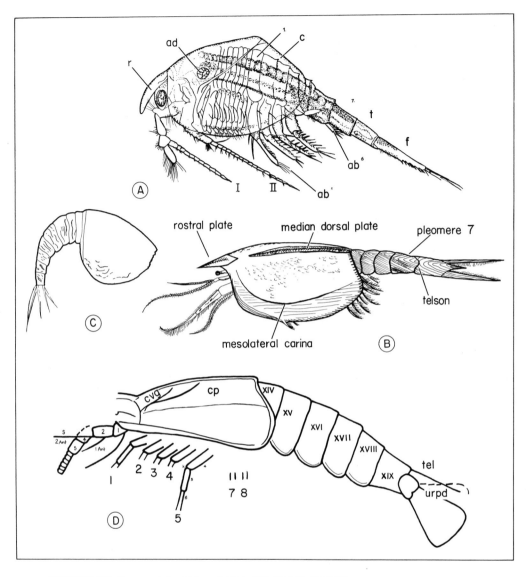

FIGURE 11.27A Representative Malacostracans. Subclass Phyllocarida.

(*A*) Order Leptostraca. *Nebalia bipes,* female, Rec., lateral view. After Calman, 1909. I, II—first and second antenna; *ab*[1], *ab*[6]—first and sixth abdominal appendages; *ad*—adductor muscle of carapace; 1–7—first and seventh abdominal somites; *c*—bivalved carapace; *t*—telson; *f*—caudal furca.

(*B*) Order Archeostraca, suborder Rhinocarina: *Nahecaris stuertzi,* L. Dev., Hunsrück. Carapace structure after Broili, revised by Rolfe, 1963.

(*C*) Suborder Hymenostraca, *Hymenocaris vermicauda,* U. Cam., Wales.

(*D*) *Paldeopalaemon newberryi,* U. Dev., Ohio, X 2; *Urpd*—uropod; *tel*—telson; abdominal segments—Roman numerals; appendages—Arabic numerals.

Illustrations redrawn and modified from: (*A*) Calman, 1909; (*B*) Rolfe, 1963; (*C*) Zittel, 1900, after Salter; (*D*) Brooks, 1962.

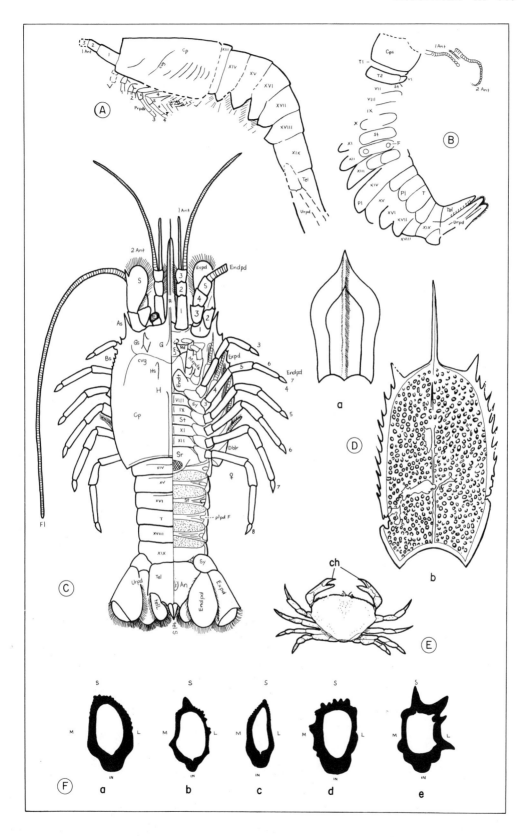

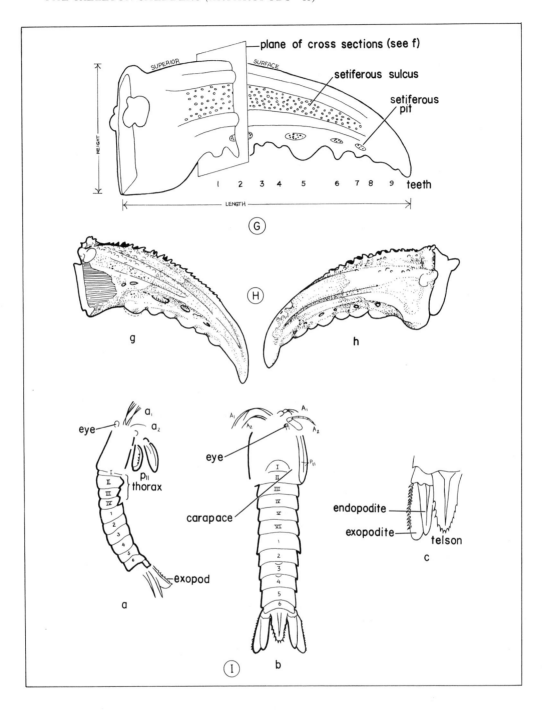

FIGURE 11.27B Representative fossil genera of the five superorders of subclass Eumalacostraca. Superorder 1. Eocarida: (*A*) *Eocaris oervigi,* U. Middle Dev., W. Ger., X 3. Superorder 2. Syncarida. (*B*) *Squillites spinosus,* Heath sh., Miss., Montana, X 8. Superorder 3. Peracarida. (*C*) *Anthracocaris gracilis,* Mazon Creek, Penn., Illinois, schematic. (*D*) (a) *Kilianicaris lerichi,* U. Jur., reduced; (b) *Anthrapalaemon grossarti,* 2nd *Leaia* horizon, Carb. (Westphalian), No. France, X 1.5. Superorder 4. Eucarida. Order Decapoda. (*E*) *Cancer* sp., the edible crab, schematic; *ch* — chelipeds. (*F*) Five species of *Cancer* differentiated by cross-section of cheliped dactyl (compare *G*): (a) *C. productus,* (b) *C. magister,* (c) *C. gracilis,* (d) *C. jordani,* (e) *C. branneri* — all Pleistocene, Los Angeles area. California. *S* — superior surface, *IN* — inferior surface, *M* — medial surface, *L* — lateral surface. (*G*) Morphology of dactyl, showing plane of cross-section and terminology, schematic. (*H*) *Cancer jordani,* dactyl, medial surface, drawn from the fossil. Superorder 5. Hoplocarida. (*I*)(a–c) *Pseudosculda laevis,* U. Cret., Lebanon.

Illustrations redrawn and adapted from: (*A–C*) H. K. Brooks, 1962; (*D, a*), Van Straelen in Piveteau, 1963; (*D, b*), Dalinval, 1947; (*F–H*) Menzies, 1951; (*I*) Roger, 1945.

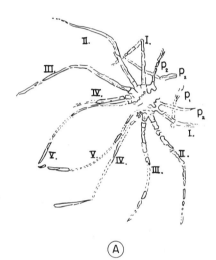

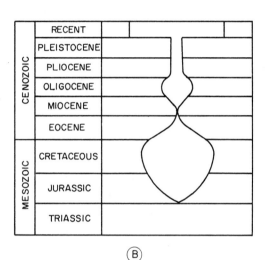

(A)   (B)

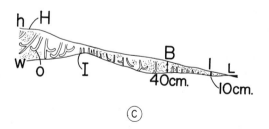

(C)

FIGURE 11.28 Information derived from fossil malacostracans.
(*A*) Probable larva of *Palinurina* revealed on slab of Solenhofen Is. (U. Jur.), under ultraviolet radiation. Fluorescent photograph by Leon, 1933, slightly enlarged. P_1, P_2 — palps; I–IV — walking legs.
(*B*) Geological history of order Stomatopoda. After Berry, 1939.
(*C*) Schematic beach profile showing burrows of various crabs; (*B*) zone of *Ocypoda ceratophthalma;* *BL* — zone of *Ilyoplax formosensis;* *O* — burrow of *Ocypoda;* *I* — burrow of *Ilyoplax.* After Hayasaki, 1935.

lagoonal dolomites; phosphatic concretions; subgraywackes; black, carbonaceous/graptolitic shales; argillaceous sandstones. Most decapod fossils have been retrieved from calcareous concretions occurring in marine sandstones, greensands, and shales (H. K. Brooks, 1962).

Habitat, Salinity, and Ionic Regulation.

Living eumalacostracans have invaded most habitats. Originating in marine environments, they have migrated to freshwater and terrestrial habitats (Pearse, 1929; Edney, 1960; Green, 1961). Blood salts in marine decapods and other eumalacostracans are maintained by ionic regulation. This involves maintaining higher values of the ions Na^+, K^+, Ca^{++}, and lower values of Mg^{++} and SO_4^{--} than seawater (Robertson, 1960).

Such regulation occurs by various means and is affected by temperature. Controls for ionic regulation are gills, antennal glands, and gut. Loss of water and dissolved salts occur via antennal gland secretions of Mg^{++} and SO_4^{--}. Further loss occurs by excretions such as urine. Both such losses are balanced by uptake of water and its contained ions through the gills.

Woodlice (Isopoda, family Oniscoidea) are completely free of water and live wholly on land. Apparently this transition occurred across the littoral zone, possibly before Middle Mesozoic time (Edney, 1960). Other malacostracans have migrated to fresh-water environments and maintained higher salt concentrations in their blood than prevails in the fluid media. The fossil syncarid *Paleocaris* from the Belgian Coal Measures lived in a marine habitat, whereas all other known syncarids are fresh-water residents (H. K. Brooks, 1957). In this example, transition to fresh water must have occurred in post-Carboniferous time.

Roger (1953) postulated three times of transition from marine to fresh-water environments for malacostracan decapods: Upper Jurassic, Upper Cretaceous, and Upper Eocene. Multiple transitions

were determined independently for branchiopod conchostracans (Tasch, 1963).

According to Pearse (1929), Edney (1960), and others, a critical factor in transition to terrestrial environments is ability to conserve water. A hard exoskeleton, such as the malacostracan carapace, could and did function in this manner. Bliss (1963) showed that function of the pericardial sacs in terrestrial crabs was related to water conservation.

The fossil record of malacostracans shows that estuarine (Woodward, 1903), swamp, and lagoonal environments were often favored, [G. S. Carter (1931) found that most of the swamp faunas in the Paraguayan Chaco swamps, including crustaceans, were similarly adapted for living in an oxygen-poor medium. Pennsylvanian coal swamps were also probably oxygen-poor.]

Modern forms range from the littoral zone (Pearse, 1929) to abyssal depths. Orders Cumacea, Tanaidacea. Isopoda, Amphipoda, and Decapoda occur in the abyss. Decapod migration to the deeps must have been post-Triassic; isopod migration, sometime in the Jurassic, and amphipod migration was likely a Tertiary event (see Menzies and Imbrie, 1958).

Some malacostracans were found in trawls of the deeps during the Galathea Expedition. The Phillipine, Sunda, and other trenches at depths of 1+ to 2 miles below sea level yielded malacostracans among other fauna (Bruun, 1956). Other modern forms inhabit rivers and lakes. Thus 300 living species of the amphipod *Gammarus* inhabit Lake Baikal (J. L. Brooks, 1950), and others inhabit terrestrial niches (Green, 1961).

Data on ionic regulation in modern forms extrapolated to the fossil record could add useful paleoecologic information. Geochemistry of enclosing sediments could be supplemented. Certain chemical data, for example, may be inferred from known regulation of blood salts in living forms. Given a sizable population of decapods, their magnesium uptake or release into basinal waters could be a factor in the magnesium concentration prevailing in local and restricted areas of the sea.

Adaptive Modification, Habit. Several modifications of skeletal anatomy are found in malacostracans. Those forms that are no longer characterized by the caridoid facies have abandoned swimming and crawl or burrow instead. The latter are no longer filter feeders as are swimming malacostracans. Natural selection favored those changes in the thoracic endopodites that made a new mode of feeding possible. The common shore crab *Carcinus* is an example. Its great chelae can seize food to pass between the mandibles (Borradaile *et al.*, 1958).

Pearse (1929) found a reduction in number of gills and a decrease in gill volume relative to body volume in modern crabs that had migrated landward, or had adjusted to modern living. Edney (1960) noted that some terrestrial crabs can live in air for months after gill removal whereas others show increased vascularization of the gill chamber wall (that is, increased blood circulation to these walls). Bliss (1963) found that pericardial sacs in terrestrial crabs were enlarged over those of marine forms.

Fossil isopod species (genus *Palaega*) from the Solenhofen limestone had well-developed eyes and mouth parts. This suggested that they were relatively shallow-water predators, that is, water of depths of 40 to 50 ft. Modern deep-sea forms, by contrast, have reduced eyes or are blind. Legs of fossil forms also indicate that on the Liassic sea floor they lived as vagrant benthos (Reiff, cit. H. K. Brooks, 1957).

A shrimplike body (that is, a large abdomen relative to length of cephalothorax) was thought to be an adaptation to a nektonic or swimming habit. Some Paleozoic eumalacostracans had such bodies — *Eocaris* (Figure 11.27B, part A), *Devoncaris*, and some others (H. K. Brooks, 1962).

Fossil Record and Evolution

Fossil Record. Species of *Eocaris* and *Devonocaris* belong to the apparently oldest eumalacostracans found in Germany and the United States (Moscow Formation, New York). *Palaeopalaemon*, another eocarid, ranges from the Upper Devonian to the Lower Mississippian and occurs in Ohio, Kentucky, and Iowa (Figure 11.27A, part D) (Brooks, 1962).

Compared to the Devonian, the Scottish Mississippian, in particular, (but also the American), has yielded many more fossil eumalacostracans. Mississippian genera include the pericarid *Anthracocaris* and syncarids like *Squillites* (Figure 11.27B, part B).

Pennsylvanian beds, particularly ironstone concretions and black shales, frequently contain abundant eumalacostracan fossils. The classic fauna is that of the Illinois Mazon Creek area (Richardson, 1953; Brooks, 1962). Other classic faunas in which abundant and varied fossil forms occur cover those of the Jurassic Solenhofen limestone, in which numerous decapods are excellently preserved, and the Cretaceous limestone fauna of Lebanon with some 30 eumalacostracan species (Roger, 1946).

Eumalacostracans are found in many parts of the Paleozoic throughout the world: Permian of Brazil (Beurlen, 1931), Permian of South Africa (Broom, 1931), and Carboniferous of Canadian Maritime Provinces (Copeland, 1957).

Tertiary faunas include terrestrial isopods (Oniscoidea) ranging from Eocene to Oligocene, and amphipods such as *Palaeogammarus* (Miocene) — both preserved in Baltic ambers. Mesozoic decapods are quite widespread in Europe and America. Cretaceous decapods have been found in marls, sands, chalks, and clays in many localities: New York, New Jersey, Delaware, Maryland, North and South Carolina, the Gulf states, Wyoming, South Dakota, and Alberta and British Columbia, Canada. Tertiary-Pleistocene decapods are worldwide: Panama, Australia,

Antilles, Hungary, Fiji Islands, the United States, Europe, Mexico (Roger, 1946).

Evolution. Of the three evolutionary reconstructions figured (Figures 11.29*A* and *B*; 11.30), Siewing's is unrelated to the rock column; that of Brooks concentrates on the Paleozoic, but follows Siewing's interpretation of Recent orders; Glaessner's embraces Mesozoic and Cenozoic time as well. Since this last paper considers trends, it is discussed first.

In what directions did evolution go after initial differentiation of malacostracan body plan into carapace and abdomen? (See Figure 11.26.) Evolution proceeded in two divergent directions: (1) toward consolidation of original differentiation into carapace and abdomen as in decapods, and (2) a return to uniform segmentation including loss or reduction of carapace as in the isopods and amphipods.

Glaessner (1957) delineated three trends in decapod evolution that occurred since their apparent first appearance in the Triassic (Figure 11.30): (1) appearance of swimming forms (Natantia), and (2) retention of cephalo-thorax-abdomen division in benthonic forms (Reptantia, tribe Astacura), (3) reduction of abdomen and elimination of its locomotor function. [Decapod suborder Reptantia is divisible into four tribes: Astacura, (true lobsters and crayfish), Palinura (spiny lobsters and related forms), Anomura (hermit crabs, and so on), and Brachyura (true crabs).]

During the course of evolution of benthonic forms, Glaessner suggested that there must have been strong selective pressures for abdomen reduction. Three different tendencies toward this end can be cited: *reduction in calcification*, which led to burrowing or concealment in molluscan shells as with hermit crabs; *reduction in size* with attendant loss of important thoracic skeletal structures as in some anomurids; and the only tendency that proved to be successful in an evolutionary sense—*a coupling* of reduced abdominal size with internal skeletal recon-

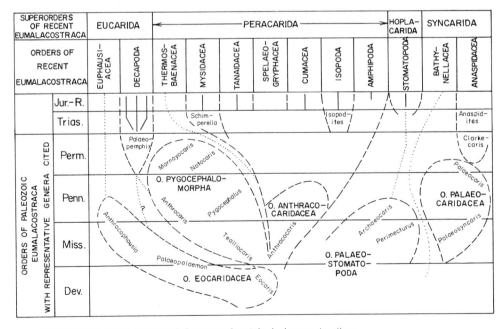

FIGURE 11.29 Malacostracan phylogeny: paleontological reconstructions.
(A) Fossil and Recent Eumalacostraca—phylogeny and classification. After H. K. Brooks, 1962, 1963. *Thermosbaenacea* assigned to the Peracarida in the Treatise.

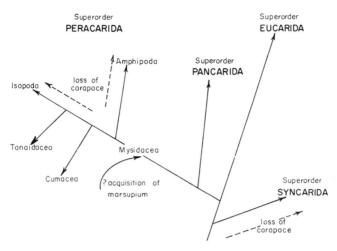

FIGURE 11.29 (cont.) (*B*) Zoological reconstruction. After Siewing, 1963. Marsupium brood pouch is rarely seen in fossils; compare *B* and *A*.

struction of the thorax. The last occurred in the Brachyura. Two advantages were conferred on brachyurans by this coupling: improved locomotion and ecologic range extended from marine littoral zone to terrestrial habitats (for example, tree-climbing crab, *Aratus pisonii*—Schmitt, 1966).

Long-tailed Reptantia (abdomen not

reduced) have declined since the Mesozoic, whereas short-tailed Reptantia, such as Brachyura in which the abdomen is reduced, are ascendant. The latter are both numerous, markedly differentiated, and broadly distributed (Glaessner, 1957).

Siewing (1963) observed that the Syncarida clearly branched off the main stem of descent before Pericarida and Eucarida.

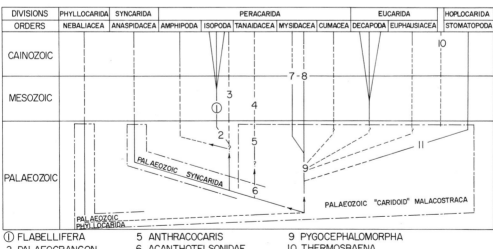

① FLABELLIFERA
2 PALAEOCRANGON
3 PHREATOICIDEA
4 PALAEOTANAIS
5 ANTHRACOCARIS
6 ACANTHOTELSONIDAE
7 MYSIDAE
8 LOPHOGASTRIDAE
9 PYGOCEPHALOMORPHA
10 THERMOSBAENA
11 PERIMECTURIDAE

FIGURE 11.30 Malacostracan phylogeny. After Glaessner, 1957, 1963. Compare 11.29A and *B*. Note that Paleozoic phyllocarids by this reconstruction are related to, or give rise to, the Paleozoic ancestral malacostracans (see Figure 11.26).

The fossil record supports this view (Brooks, 1962).

Manton (1963) would trace pericardian characters observed in some living forms, to an earlier stage than that represented by mysid-like animals that had progressively lost a carapace. Siewing also held that the Pericarida are connected to the mainstem of the Malacostraca through the Mysidacea (Figure 11.29B). He also rejected the paleobiological approach shown in Figures 11.29A and 11.30 in which the pericarid groups are independently derived from one another. There was no known proof, Siewing argued, that Pericarida were directly derived from Syncarida as figured by Glaessner.

As for Decapoda, they may have had a polyphyletic origin as Brooks indicates (Figure 11.29A). Glaessner, however, thinks that they were derived from eocarid ancestors. On such questions more and better fossil evidence is needed before decisions can be reached.

Information Derived from Phyllocarids

Carapace and Cuticle. Carapace length in phyllocarids ranges from 8.00 mm in an Upper Silurian form from Oklahoma to 4.00 cm in a Recent form. The carapace is variably sclerotized. A restricted zone remains unsclerotized to allow valve hingement and flexibility. (This could have been a selective factor.)

Giant specimens of *Ceratiocaris papilio* from the Middle Silurian of Scotland permitted a detailed analysis of cuticle microstructure (Figure 11.26B, and see Chapter 15 for decapod cuticle). Cuticle thickness varied in different parts of the exoskeleton, from 0.03 mm for the carapace to 0.60 mm for the style, and in the mandible tip of one specimen, a thickness of 2.50 mm was determined. No major subdivisions of the cuticle were observed, that is, epi-, exo-, and endocuticle.

Composition of the cuticle was found to be collophane (cryptocrystalline apatite).

Structures observed in the cuticle included pores and canals, laminae, Balkenlagen, prisms, polygons, and secondary structures such as vermiform tubules, holes, and bumps, as well as cracks (see Figure 11.26B).

Fossil Record of Phyllocarids. Despite the sparse record of phyllocarids, fossils are locally abundant in laminated siltstones of Middle Silurian, southern Scotland, Upper Devonian of Cologne, Germany and western Australia, the Middle Pennsylvanian Mecca shale of Indiana (Rolfe, 1967), and elsewhere. With one exception, phyllocarids are marine and fossilized with primitive vertebrates and eurypterids. They are also known from graptolitic shales and have been found with orthocone cephalopods.

There have been several finds of fossil phyllocarids with appendages preserved: Middle Cambrian of British Columbia, Middle Silurian of Scotland, Devonian of Germany and United States, Lower Carboniferous of France, and Lower Permian of South Korea.

Fossil and Living Myriopods

Diagnosis. Superclass Myriopoda (the Greek *myrioi* = 10,000+ footed) is a group of arthropods that are many-legged.

1. Land-living arthropods.
2. Air breathing through trachea as in insects.
3. Wormlike, with multiple segments bearing legs.
4. Body divided into distinct head and undifferentiated trunk.
5. Head bears a single pair of antennae; mandible without palp (toothed for chewing); one or more pairs of maxillae.
6. Sexes separate.
7. Skeleton chitinous; may be strengthened by carbonates as in decapods; regularly molted.
8. Geologic range: Upper Silurian to Recent.

Six myriapod classes are recognized. One contains only extinct forms, the class Archipolypoda; one class is uncertain. Two classes either lack a fossil record or have a sparse one — Pauropoda and Symphyla (Cenozoic) (Laurentiaux, 1953, p. 386). The remaining two classes are extant and prominent for their fossil record. Of these several classes, three are discussed and illustrated.

Class Archipolypoda. Body segments with two dorsal and two ventral plates, each with two pairs of legs. *Euphoberia* (Figure 11.31*E, a, b*).

Class Chilopoda. Common centipedes (generally less than 100 legs as in *Scutigera* (Figure 11.31*A, a*); one pair of walking legs on each segment except the first (which bears poison claws), and the last two (which are legless). *Cermatia* (Figure 11.31*B*).

Class Diplopoda. Essentially vegetarian millipede arthropods with chitinous skeleton impregnated with calcareous deposits; multiple segments of two types; anterior four segments bear one pair of legs; posterior (abdominal) segments consist of many double segments, each bearing two pairs of legs (*Julus*, Figure 11.31*C, a*); individuals may have up to 200 individual legs, or 100 pairs, *Pleurojulus* (Figure 11.31*D*).

Information Derived from Living and Fossil Myriopods

Ecology and Paleoecology. (*Feeding Habits, Habitat*). Chilopods are carnivorous in habit using poison claws to capture and kill living prey. Some centipedes live under stones and logs and even beneath loose bark. Others frequent more open places (Snodgrass, 1952). Fossil chilopods, such as *Cermatia*, structurally similar to living forms, probably had a like habit and habitat. Its preservation in amber suggests proximity to outer bark of then living trees.

Living diplopods feed primarily on decaying vegetation — selected dead leaves and shrubs. A fossil diplopod (*Xylobius*

sigillariae, Carboniferous, Nova Scotia) occurs as a cast in a Coal Measures *Sigillaria* tree stump, indicating a similar habit in the geologic past (H. K. Brooks, 1957).

Biotic Associates. Dawson (1860, cit. H. K. Brooks, 1957) found a Carboniferous diplopod fossilized with a mollusk (land snail), worms (*Spirorbis* tubes), reptilian and plant remains, and other arthropods.

Fossil Record. The bulk of known myriapod fossils occur in Permo-Carboniferous beds in America (Mazon Creek, for example), Europe, and England. Tertiary amber faunas from the Baltic region are also known. Compared to other arthropod faunas found as fossils, myriapods have been infrequently encountered. Perhaps fortuitous conditions necessary for fossilization were very rare in their terrestrial niches. Enclosure in amber, clay ironstones, or as imprints in carbonaceous shales was among the several modes of preservation.

Myriopoda and Hexapoda. There has long been a "Symphyla" theory (class Symphyla of the superorder Myriopoda) of insect origin with many adherents (Tiegs and Manton, 1958, pp. 273–279). The Symphyla (Figure 11.32*A*) are closer to the diplopods (Figure 11.31*C*) in such features as mandible structure, position of gonads (ventral), and genital opening (anterior on third body segment) (Snodgrass, 1952). However, they not only have an essentially insect-type head, but also share with the hexapods, such as *Campodea*, (*a*) the beaded antennae (see Figure 11.32*B*), (*b*) abdominal styles, moisture-absorbing exsertile vesicles (Figure 11.32*C*), and (*c*) terminal cerci.

Besides cited homologies between Symphyla and insects, there are still others that have been derived from embryological data (Tiegs and Manton, 1958). These are thought to signify that the

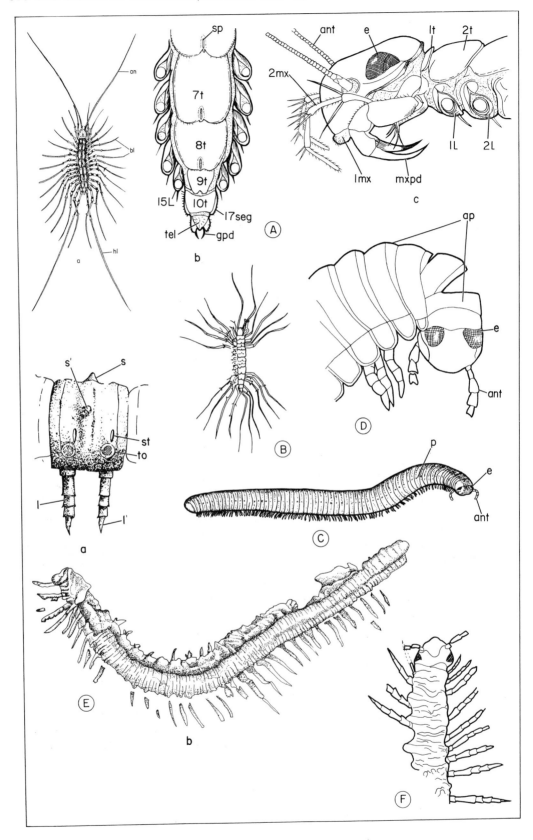

FIGURE 11.31 Living and fossil Myriopoda: class Chilopoda:
(A) *Scutigera coleoptrata,* Rec., the common house centipede, schematic. (a) Dorsal view of complete individual; *an* — antennae, *bl* — body legs, *hl* — hind legs; (b) dorsal view of posterior part of body, female; (c) head and anterior body segment, left side; *Sp* — spiracle (= tracheal tube opening on body), 7, 8, and so on = tergum, *tel* — telson, *gpd* — gonopod, 15 *L*, and so on = leg, *ant* — antenna, *E* — eye, *1mx, 2mx* — first and second maxilla, *mxpd* — maxilliped, a poison claw; *L* = first leg.
(B) *Cermatia illigeri,* Oligo., amber, Baltic.
Class Diplopoda: (C) *Julus terrestris,* a common milliped with about 70 body segments, Rec. *P* — pores for excretion, *E* — eyes, *ant* — antenna, X 3.5.
(D) *Pleurojulus levis.* Perm., Bohemia, anterior, schematic; *e* — eye, *ant* — antenna, *ap* — apode, legless segments.
Class Archipolypoda: (E)(a): *Euphoberia granosa.* Penn., Illinois; *L* — legs, *s* — spines, X 2. (b) *Euphoberia ilarenai,* Penn. (L. Stephanian), Spain; trunk segment showing stigmata (*st*); *to* — opening of body wall of trachea; *s* — *s'* — spines; 1, 1' — legs, enlarged.
(F) *Kampecaris forfarensis,* L. Dev. (Old Red *ss*), Scotland, a protochilopod, according to Sharov (1966).
Illustrations redrawn and adapted from (A) Snodgrass, 1952; (B) Laurentiaux, 1963; (C) Borradaile et al., 1958, after Koch; (D, F) Sharov, 1966, after Peach and Fritsch; (E, a) Shimer and Shrock, 1944; (b) Melendez, 1948.

Symphyla grade of organization foreshadowed that of insects. Furthermore, it was reasoned that from a many-legged (polypodous) Protosymphyla ancestor, development of longer legs, reduction of legs, and development of a six-footed (hexapodous) gait led to the several insect lines of descent (Imms, 1936; Sharov, 1966).

However, Manton found such basic differences between symphylans and hexapods that she concluded, "A suggested archi-Symphyla ancestor of insects must be abandoned" despite probable common origin of Myriopoda and Hexapoda. These differences include distinctive mandibular mechanisms, hexapods, for example, have unjointed rolling jaws, while myriapods have jointed transversely biting jaws (Manton, 1963, Figure 61; cf. Manton, 1973).

New Arthropod Phylum — Uniramia. Manton (1973) argued that comparative functional evidence on how animals fed, how muscles and limbs operated, and so on, eliminated the possibility of "an ancestral insect." Since all existing theories of insect origin postulate "functionally impossible ancestral stages," they are therefore invalid.

In this interpretation, there were "five separate groups of multilegged, soft-bodied, terrestrial arthropods" sharing several anatomical features which independently evolved hexapodous limbs.

Among the entomologists who Manton had in mind was Sharov, whose 1966 book was found to be full of "abundant phantasies and errors of fact" by her and other investigators. One of those referred to Sharov's view of a monophyletic arthropod origin (single line of descent) rather than polyphyletic origin (separate lines of descent) that is favored by modern evidence on comparative functional morphology.

Manton established a new arthropod phylum, Uniramia, to include three subphyla: Onchyophora (see p. 473), Myriopoda, Hexapoda. Crustacea and Chelicerata are the other two arthropod phyla, while Trilobita are closest to Chelicerata since both have comparable biramous trunk limbs.

Differentiation of Myriopod Classes. Manton (1963, p. 138) has found that the marked differences in external trunk characteristics of several myriopod classes correlate with life habit. Fast-moving, predatory chilopods, for example, contrast with burrowing vegetarian diplopods (compare Figure 11.31*A* with 11.31*C*).

In chilopods, evolution of the entognathous condition (mouth parts sunk below head surface) was probably stimulated by its predatory, crevice-living habit. Extreme head-flattening also, according to Man-

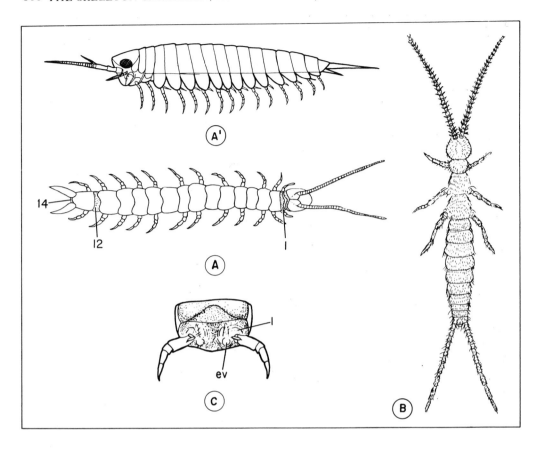

FIGURE 11.32 Information derived from living and fossil Myriopoda.
A. Class Symphyla, *Scutigerella immaculata.* Rec., adult, schematic.
B. Campodea, wingless insect (Order Diplura). (See *A.*)
C. Same as *A; l*—lobe of body supporting leg, *ev*—eversible vesicle.
Illustrations redrawn and adapted from Snodgrass; Ross.

ton, provided a selective advantage in this respect.

Fossil Arthropleurids

Class Arthropleurida. The extinct arthropleurids are a unique, restricted group of arthropods whose affinities have long been in dispute (Figure 11.33A). Waterlot, who named the class, thought that they had closest affinities to trilobites (that is, with biramous appendages). Størmer disputed this interpretation, as did, more recently, Rolfe. Limbs were found to be uniramous. The most recent taxonomic assignment places the class under the Myriapoda.

Chief Diagnostic Features

Size. Arthropleurids are gigantic in size. They ranged up to 6 ft in length and undoubtedly are the largest terrestrial arthropod ever to have inhabited the planet. During Coal Measures time (Carboniferous — Westphalian to Stephanian), which totally spans their existence, gigantism also characterized some insects.

Body Segments and Appendages. The number of body segments is unknown but, as determined from almost complete specimens, they ran from 23 to 27 (postcephalic segments). There is one pair of limbs per segment and hence, the creature was uniramous (Figure 11.33A, B).

Similarities and Differences with Myriapods. Arthropleurids, Rolfe (1967) further observed, differed from other myriapods particularly in larger number of leg segments — 8 to 10 — and in the presence of a rosette plate (Figure 11.33B).

Closest affinities are thought to be with class Diplopoda. A further distinction was arthropleurid tendency to gigantism.

Paleoecology

Lithology. Specimens occur in nonmarine shales and Carboniferous clay ironstone concretions. Where they have been found in marine situations, they probably were washed in from a nonmarine habitat.

Habitat. Association with plants is common, which suggests that arthropleurids lived in a coal swamp environment (limnic and paralic facies). The test of one fossil, *A. armata,* preserved fragments of a lycopod (carbonized pieces of epidermal tissue and tracheid cells with scalariform pitting). According to Rolfe, this indicated an herbivorous or an omnivorous habit.

Biotic Associates. In addition to coal swamp vegetation, some arthropleurids have been found fossilized with insects, chelicerates, some nonmarine ostracodes, branchiopods, lamellibranchs, fish, *Spirorbis* species, and myriopods.

Predation and Large Size. It appears that one ecological benefit due to gigantic size was uninterrupted growth in an environment largely free of predators or competitors. Chief predators for arthropleurids were probably coal swamp amphibians.

Extinction. The restricted range of arthropleurids generally corresponds, as Rolfe observed, in space and time with the European-American floral belt, that is, the Coal Measures swamp environment. With disappearance of such widespread swamplands, the special arthro-

TABLE 11.6. Species of Genus Arthropleura and Their Ranges

European Coal Basin Zones (Continental)	Arthropleura species
Stephanian	*fayoli*
Upper Westphalian	D — C — *robusta, britannica, pruvosti* $\vert$ *armata*[a] B — *mammata* $\vert$ *mailleuxi* A —
Lower Westphalian (Namurian)	Questionable Arthropleurids

[a]Species outside of bracket span more than one subdivision, C and D in the case of *A. armata*, and A and B in the case of *A. mailleuxi*.

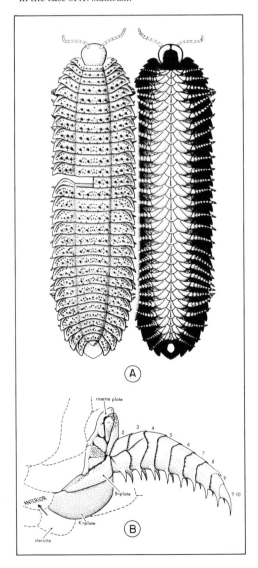

(A)

rosette plate

3 4 5 6 7 8 9 ? 10

2

B-plate

ANTERIOR

K-plate

sternite

(B)

pleurid ecological niche would have been eliminated. Thereafter, reduction in numbers would ultimately lead to extinction.

At least seven species are known in genus *Arthropleura* (see Table 11.6).

Arthropleurid speciation peak occurred during Upper Westphalian time in Zones B and C when six of the seven species coexisted. In zone C time, four species made their apparent first appearance. Appearance of these new species might have been associated with gene pool fractionation or extension when coal swamp pools shrank in volume or overflowed.

A. fayoli, last of the arthropleurids was not a species that survived through time but rather, a *new* species. Obviously it arose from a geographic isolate and died out, probably through failure to yield adequate numbers of viable offspring.

FIGURE 11.33 *Arthropleura* — a probable myriapod. (A) *Arthropleura armata*, U. Carb., dorsal (left), ventral (right) aspects; anterior = top of figure, posterior = bottom of figure. Body segment number conjectural; half of one tergite removed to show anterior border of underlying tergite. Note uniramous limbs, one pair to a segment. No limbs are shown on first segment of trunk. Dashed lines show restored parts of body. After Rolfe, 1967. (B) Reconstruction of left leg of a median size adult *Arthropleura* sp., antero-ventral aspect oblique to axis of limbs. Limb appears foreshortened. Detail of one of the limbs seen in (A), showing various plates.

QUESTIONS

Crustacea — III
Malacostracans

1. Distinguish a malacostracan from a trilobite, a eurypterid, and a branchiopod. Cite specific differences.

2. (a) What is the caridoid facies?

   (b) Explain the genetic significance (molecular level) of this facies.

   (c) How does it bear on the malacostracan ancestor?

3. Briefly review the fossil record of eumalacostracans in Paleozoic, Mesozoic, and Cenozoic beds.

4. (a) Compare malacostracan transition to the terrestrial environment to that of certain ostracodes.

   (b) What did Carter find about swamp faunas that might be applied to the Pennsylvanian swamp age?

   (c) Did the modern marine distribution of malacostracans hold during Paleozoic? Mesozoic? Explain your answer.

5. (a) What three evolutionary tendencies in malacostracan evolution did Glaessner describe?

   (b) Discuss adaptive modification in malacostracans.

   (c) In what lithologies would you look for fossil malacostracans?

Myriopods

1. Distinguish myriopods from all other arthropods.

2. (a) What role have myriopods filled in biological models of insect evolution?

   (b) How do fossil arthropleurids resemble and differ from myriopods?

   (c) Discuss gigantism and extinction in arthropleurids.

REFERENCES

Crustacea — III
Malacostracans

Bachmayer, F., 1948. Pathogene Wucherungen bei jurassischen Dekapoden: Sitzuns. sber. Oesterr. Akad. Wiss-math. nat. Kl. Part 1, Vol. 157, No. 6–10.

Beurlen, K., 1931. Crustacean reste aus den Mesosaurierschichten (Unterperm) von Brasilien (São Paulo): Palaeont. Zeitschr., Vol. 13, pp. 35–50.

Berry, C. T., 1939. A summary of fossil Crustacea of the Order Stomatopoda, and a description of a new species from Angola: The Amer. Midland Nat., Vol. 21 (2), pp. 461–471.

Bliss, D. E., 1963. Pericardial sacs of terrestrial Brachyura, in Phylogeny and Evolution of Crustacea: Mus. Comp. Zool. (Harvard), Sp. Publ. 1963, pp. 59–78.

Brooks, H. K., 1962. The Paleozoic Eumalacostraca of North America: Bull. Amer. Paleontology, Vol. 44 (202), pp. 164–274.

_____, 1957. Chelicerata, Trilobitomorpha, Crustacea (Exclusive of Ostracada) and Myriopoda: Treat. Mar. Ecol. and Paleoecol., G.S.A. Mem. 67, Vol. 2, pp. 895–929.

_____ and K. E. Caster, 1956. *Pseudoarctolepis sharpi* n. gen., n. sp. (Phyllocarida) from the Wheeler Shale (Middle Cambrian) of Utah: J. Paleontology, Vol. 30 (1), pp. 9–14.

Brooks, J. L., 1950. Speciation in ancient lakes: The Quart. Rev. Biol., Vol. 25 (1), pp. 30–60; Vol. 25 (2), pp. 131–176.

Broom, R., 1931. On the Pygocephalus-like Crustacea of the South African Dwyka: Zool. Soc. London, Proc. Part 1, pp. 571–573, plate 1.

Carter, G. S., 1931. The fauna of the swamps of the Paraguayan Chaco in relation to its environment: J. Linnean Soc. London, Zool., Vol. 37, pp. 327–368.

Copeland, M. J., 1957. The arthropod fauna of the Upper Carboniferous rocks of the Maritime Provinces: Canad. Dept. Mines and Tech. Surv., G. S. Canada Mem. 286, 60 pp.

Dalinval, M. A., 1947. Découverte d' *Anthrapalaemon* et d'un second horizon à *Leaia* dans le Terrain houiller du Nord: Soc. Géol. du Nord Ann., Vol. 17, pp. 27–33.

Glaessner, M. F., 1960. The fossil decapod Crustacea of New Zealand and the evolution of the Order Decapoda: N.Z. Geol. Surv. Paleont. Bull. 31, 63 pp., 7 plates.

_____, 1957. Evolution trends in Crustacea (Malacostraca): Evolution, Vol. 11 (2), pp. 178–184.

Green, J., 1961. A Biology of Crustacea. Quadrangle Books, Chicago, 165pp.

Hayasaki, I., 1935. The burrowing activities of certain crabs and their geological significance: Amer. Midland Nat., Vol. 16.

Léon, D. R., 1933. Ultraviolettes Licht entdeckt Versteinerungen auf "leeren" Platten: Ein Pantopod im Jura-Kalk: Natur und Museum, Vol. 63, pp. 361–364, text Figures 1–2.

MacKay, D. C. G., 1942. Temperature and the world distribution of crabs of the genus *Cancer*: Ecology, Vol. 24 (1), pp. 113–115.

Menzies, R. J., 1951. Pleistocene Brachyura from the Los Angeles area, Cancridae: J. Paleontology, Vol. 25 (2), pp. 165–170, 13 text figures.

Pearse, A. S., 1929. Observations on certain littoral and terrestrial animals at Tortugas, Florida, with special reference to migrations from marine to terrestrial habitats: Papers from the Tortugas Lab., Carnegie Inst. Washington, Vol. 26, pp. 265–272.

Rathburn, M. T., 1935. Fossil Crustacea of the Atlantic and Gulf Coastal Plain: G. S. A. Spec. Paper No. 2, 119 pp.

Rhodes, F. H. T., and A. A. Wilson, 1957. The arthropod species *Anthrapalaemon dubius* and related forms: J. Paleontology, Vol. 31, pp. 1159–1166.

Richardson, E. S., 1956. Pennsylvanian invertebrates of the Mazon Creek area, Illinois: Fieldiana, Geology, Vol. 12, pp. 1–76.

Roger, J., 1953. Sous-Classe des Malacostracés, in Jean Piveteau, ed., Traité de Paléontologie, Maison et Cie, Vol. 3, pp. 309–378.

Rolfe, W. D. I., 1969. "Phyllocarida," "Arthropleurida," Treat. Invert. Paleontology, Part R, Arthropoda 4, Vol. 1, pp. R29 ff; Vol. 2, pp. R607 ff.

_____, 1962a. A syncarid crustacean from the Keele beds (Stephanian) of Warwickshire: Paleontology, Vol. 4 (2), pp. 546–551, plate 68.

_____, 1962b. The cuticle of some Middle Silurian ceratiocarid Crustacea from Scotland: Paleontology, Vol. 5 (1), pp. 30–51, plates 7–8.

Scott, H. W., 1938. A stomatopod from the Mississippian of central Montana: J. Paleontology, Vol. 12 (5), pp. 508–510, 2 text figures.

Sharov, A. G., 1966. See reference, "Insecta," this chapter.

Siewing, R., 1963. Studies in malacostracan morphology: results and problems, in Phylogeny and Evolution of Crustacea. Mus. Comp. Zool. (Harvard), pp. 85–103.

Waterman, T. H. (ed.), 1960. The Physiology of Crustacea. Academic Press, New York, Vol. 1. (See especially contributions by J. D. Robertson, Chapter 9, "Osmotic and Ionic Regulation," pp. 317 ff., Gwyneth Parry, Chapter 10, "Excretion," pp. 341 ff.; E. B. Edney, Chapter 11, "Terrestrial Adaptation," pp. 367 ff.

Whittington, H. B., and W. D. I. Rolfe (eds.), 1963. Phylogeny and Evolution of Crustacea. Mus. Comp. Zool. (Harvard). (Particularly papers by R. Siewing, S. M. Manton, H. L. Sanders, D. E. Bliss, and Chapters 11, 12, and 15 for remarks by H. K. Brooks, M. Glaessner, and P. Tasch.)

Myriopoda

Brade-Birks, S. G., 1928. An important specimen of *Euphoberia ferox* from the Middle Coal Measures of Crawcrook: Geol. Mag., Vol. 65, pp. 400–406, plate 16, text figures 1–3.

Brooks, H. K., 1957. "Myriapoda," in H. S. Ladd, ed., Treat. Mar. Ecol. and Paleoecol., Vol. 2, pp. 928–929.

Chamberlain, R. V., 1949. A new fossil chilopod from the Late Cenozoic: Trans. San

Diego Soc. Nat. Hist., Vol. 11 (7), pp. 117–120, plate 7.

Imms, A. D., 1936. The ancestry of insects: Soc. Brit. Ent. Trans., Vol. 3, pp. 1–32.

Laurentiaux, D., 1953. Classe des Myriapodes, in Jean Piveteau, ed., Traité de Paléontologie: Masson et Cie, Vol. 3, pp. 385–396.

Meléndez, B., 1948. Un Miriapodo fósil en el Estefaniense de Llombera (Léon). Bol. R. Soc. española Hist. Nat., Vol. 46 (Madrid).

Peach, B. N., 1889. On some myriapods from the Paleozoic rocks of Scotland: Physical Soc. Edinburgh Proc., Vol. 14, pp. 113–126.

Peach, B. N., 1882. On some myriapods from the Lower Old Red Sandstone: Physical Soc. Edinburgh Proc., Vol. 7, pp. 177–188.

Snodgrass, R. E., 1951. Comparative studies of the head of mandibulate arthropods. Comstock Publ. Co., Ithaca, N.Y., 116 pp.

Waterlot, G., 1953. Proarthropodes d'affinities incertaines. Order des Arthropleurida, in Jean Piveteau, ed., Traité de Paléontologie: Masson et Cie, Vol. 3, pp. 247–254.

ARTHROPODA – III (INSECTA)

Living and Fossil Hexapoda

Introduction. There are more species of land-living insects on our planet than species of any other living animal (with or without backbone). Exclusive of insects, only about 33 percent of all other animal species remain. This was not always the case. True insects (class Insecta) are unknown before the Devonian (and abundantly from Upper Carboniferous time on). Related springtails (class Collembola) also can be traced back to the Devonian.

The name Hexapoda (referring to six-legged arthropods) has been used as a synonym for the class Insecta (Snodgrass, 1952; Brues, Melander, Carpenter, 1954). The term is frequently encountered in both modern and older literature in combinations such as "the Onchyophora-Myriopoda-Hexapoda stem" (Manton, 1963). These are now constituents of one of the three arthropod subphyla of Uniramia (Manton, 1973). One cannot prove that all forms having three pairs of legs (hexapodous condition) had a common ancestry, since the condition separately evolved several times. Manton pointed this out more than a decade ago.

Adaptive Mechanisms (Insecta). Morphologic structures, ecologic factors, and behavior in insects have a broad spectrum of variability. These are even more striking because both insect individuals and whole species have highly prescribed existences with few degrees of freedom (Ross, 1948). However, the latter condition arises from specialization as in honey bee or gall wasp. Specialization itself is a product of natural selection, which picks and chooses from the existing range of potential variability.

Examples of adaptive modification include the following.

Mouth parts. Modified for chewing (toothed mandible) or sucking (beak or proboscis).

Respiration. Trachea, tracheoles bearing fluids that circulate; air sacs (grasshoppers) to let air in/out of tracheal system; larval gills of some aquatic insects (that can obtain oxygen from water — may be blood gills, rectal or tracheal gills).

Legs. Modified for jumping, burrowing, springing, striding, feeding, cleaning, grasping, and so on.

Antennae. Variation in size, form, and range of functions, which may be olfactory, auditory, tactile, and so on (Borror and De-Long, 1954.

Still other examples of adaptation occur in the following.

Metamorphosis. There are various kinds of cycles, from simple (ametabolous) — embryo, juvenile, adult — to highly complex (holometabolous) — embryo, larval stages, pupa, adult (or imago) (Guthrie and Anderson, 1957).

Sound and light-producing mechanisms.

Social castes. Among bees, ants, and termites.

Winged and wingless condition.

Protective coloration. As in moths and beetles, among others (G. G. Simpson et al., 1957).

Integument and Ecdysis

Integument. The insect cuticle serves a threefold function (Locke, 1964). It is a skin covering the body. In a more rigid state (skeleton), it forms wings and appendages; yet joints are flexible, and there is an elastic component. Mouth parts, too are formed of hardened cuticle. The inner cuticle layer serves as a food reserve during molting periods and times of starvation.

Some insect cuticles are comparatively quite hard (Moh's scale — about 3). Mandibles, for example, can penetrate various metals (silver, copper, tin, zinc) and scratch calcite (Bailey, cit. Locke, 1964).

Organization of entire integument consists of cuticle, underlying epidermis that secretes it, and basement membrane (Figure 11.34*A*). Details of structure, including cement and wax layers and stages in molt-intermolt sequence, are shown in Figure 11.34*B*.

Two subdivisions of the cuticle are outer epicuticle and inner procuticle. Procuticle includes exocuticle and endocuticle (the bulk of the integument). Epicuticle is a nonchitinous, lipid/protein layer and contains cement, wax, and a cuticuline layer. Procuticle is chitin-protein in composition. The insect can draw food reserves in time of emergency or stress situations from this layer.

Insects shed the old, rigid skeleton to allow body growth, as with arthropods generally. Plasticity does persist in some insect integuments, but it is a brief, transient phase at most. Steps in the molt-intermolt sequence are given in Figure 11.34*B*.

Ecdysis. Briefly, the ecdysial program includes epidermal cell secretion of a new epicuticle under the old endocuticle; digestion and resorption of old endocuticle by an enzymatic fluid, and redeposition of these products as a new endocuticle below the new epicuticle; shedding of old cuticle (about 85 percent of endocuticle being resorbed); hardening of outer region of endocuticle and secretion of a wax layer on outer surface of the epicuticle (Locke, 1964).

For insects generally, there are five to six molts in a lifetime. In particular instances, this number may be expanded by a factor of ×5 (Ross, 1965).

Discarded insect cuticles, exclusive of wings, are sometimes found as fossils. By definition, *instar* denotes an insect

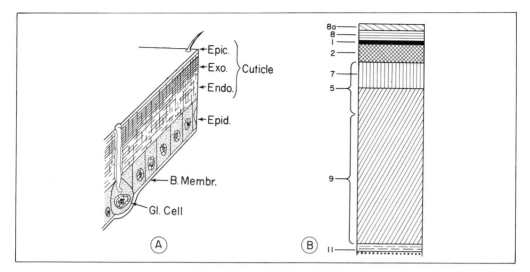

FIGURE 11.34 Insect integument.

(A) Slice of insect integument, schematic, cuticle, epidermis (*epid*), and basement membrane (*B, membr*); *epic*—epicuticle = nonchitinous lipid-protein; *exo*—exocuticle and *endo*—endocuticle = procuticle, chitin-protein.

(B) Structure of insect integument. Numbers below show order in which components were deposited, and define steps of the molting process: (1) Cuticulin, (2) inner epicuticle formation. Not shown are (3) and (4)—(3) activation of molting fluid, (4) cuticle resorption. (5) Endocuticle deposition. (6) Not shown—molting and plasticity. (7) Hardening and darkening of endocuticle (= stabilized endocuticle). (8) Wax secretion (molt stage), (8a) cement secretion by dermal glands (other layers secreted by epidermis). (9) endocuticle deposition (after molting). (10) Not shown—intermolt wax secretion. (11) ecdysial membrane formation. Note that (5) and (9) endocuticle deposition and (8) and (10), wax secretion, occur in both molt and intermolt periods.

Illustrations redrawn and adapted from (A) Richards, 1953; (B) Locke, 1964; and others.

growth stage during the molting process, and *stadium*, time between two molts.

Development and Origin of Insect Wings

Development. Our knowledge of insect history is derived largely from fossilized insect wings. In life such wings were body wall expansions, formed of flattened double layers. Between the double layers are nerve and tracheal extensions of body nerve and tracheal systems; blood circulates in it, and it also contains body fluids (Snodgrass, 1935). Wing cuticula composition is the same as that of body wall. It is the persistent, hard-to-break-down chitin component that accounts for numerous fossilized insect wings — or at least impressions of these.

Origin. The interpretation of a paranotal or paratergal outgrowth, exemplified in *Lemmatophora typica* from the Kansas Permian (Figure 11.35D), as a probable insect-wing archetype (Snodgrass, 1931, Sharov, 1966) is disputed by Kukalova-Peck [1978. Origin and evolution of insect wings and their metamorphosis as documented by the fossil record. Jour. Morphology, v. 156 (1): 53–126] based on a detailed study of the fossil evidence. Thus it is the "articulated wings of Paleozoic nymphs" that are primitive, while the solid paranota are a subsequent development associated with protection of "tender wing tissue" and "to streamline the body." According to Kukalova-Peck, it was during the juvenile and not the adult age that "all major evolutionary steps" in wing development "happened."

True flight could not develop in insects until a suitable hingement of wings

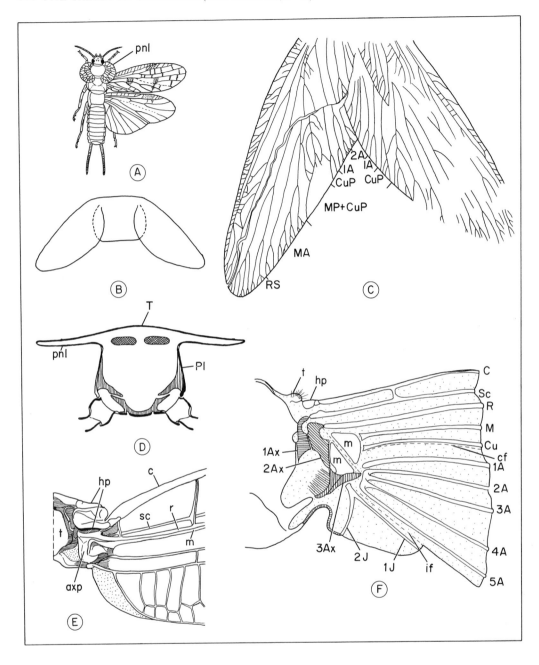

FIGURE 11.35 Origin and evolution of insect wings.

(A) *Lemmatophora typica*, Wellington sh., Perm., Kansas (Tillyard, 1928).

(B) *Eopterum devonicum*, U. Dev., U.S.S.R. (cit. Sharov, 1966). This fossil represents crustacean uropods (not wings as Sharov suggested) and is reassigned to the Eumalacostraca (Rohdendorf, 1972).

(C) Family Paoliidae, the most ancient Carboniferous pterygotes; *Sustaia impar*, showing folded wings. Note developed venation. (Sharov, 1966).

(D) Cross-section (schematic) of thoracic segment with paranotal extension of tergum (*T*); compare (A) above; *pnl* – paranotal, *Pl* – pleural.

(E), (F) Wing connection to insect thorax. (E) Articulation on insects that do not flex wings; wing base of dragonfly. *Axp* – axillary plate, *hp* – humeral plate, *t* – thorax. After Snodgrass, 1935. (F) Membranous hinge composed of axillary sclerites (1–3 Ax), *hp* – humeral plate, *m* – median plate, *t* – tegula (Ross, after Snodgrass). For wing venation (*C, Sc, M, R*), see Figure 11.38A.

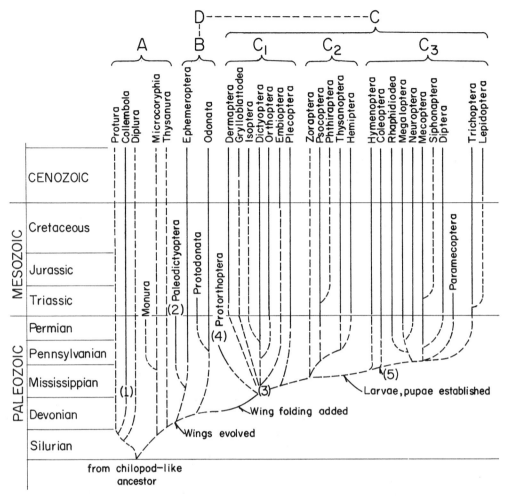

FIGURES 11.36–11.37 Evolution of insects. Adapted from H. H. Ross, 1965. Broken lines = inferred dating; solid line = data on fossil and extant insect orders. (1) *Rhyniella praecursor* fits here (see Fig. 11.39C). (2) Included in this branch are Protohemiptera, Megasecoptera, and possibly other extinct orders. (3) *Eopterium devonicum* deleted; reassigned to Crustacea (see p. **608**). (4) Included in this branch are Protoelytroptera, Caloneurodea Protoperlaria, and possibly other extinct orders. (5) Elytra or wing covers develop in beetles (Coleoptera). These are the first pair of wings, hard and shell-like that fold back; the second membranous pair of wings that are used in flight, at rest, fold up under the elytra. Upper letters (A)–(D): (A) Apterygota or wingless insect, (D) Pterygota or winged insects, include (B) and (C). (B) Paleoptera, (C) Neoptera; (C₁) orthopteroid orders, (C₂) hemipteroid orders, (C₃) neuropteroid orders. The order Diplura is replaced by Entotrophi—bristle tails; orders Protura and Collembola are designated classes on a par with class Insecta by Carpenter in the forthcoming Treatise.

to body evolved. This was achieved by the evolution of a membranous hinge composed of many small sclerites (= axillary sclerites) and others (Figure 11.35F). Of course, body musculature likewise had to be modified. Five paired sets of muscles in each wing are needed for flight movements (upstroke, downstroke, forward and backward movements, and pivotal rotation of each wing on its long axis)

(Snodgrass, 1935).

Insects that did not go beyond hinge-ment stage of wing evolution fall into two large groupings (Ross, 1965, and Figure 11.36–11.37): the Paleoptera (Figure 11.35E), those that do not flex wings in flight (dragonflies, among other Odonata, and so on), and the Neoptera, those that develop a folding mechanism for positioning wings when not in flight (Snodgrass, 1935; Ross, 1965. (The majority of insects are in the last group.) The latter development of these tendencies has been attributed to selective pressures arising from a habitat among overgrown vegetation.

Was the Palaeoptera condition due to a loss of a former capacity to flex wings in flight, as has been speculated (Sharov, 1966)? Complex musculature and sclerite movement as well as possible plaiting and folding of wings, as in the grasshopper, are involved in wing flexion (Snodgrass, 1935). If so, modified genetic combination governed or affected those capacities that disappeared from the Carboniferous Palaeoptera gene pools. However, this speculation was based on misidentification of a crustacean as an insect (Figure 11.35*B*) and is unacceptable.

Wing Venation. Examination of wings of most living insects and those found fossilized (Figure 11.35*C*) reveals a characteristic pattern of convex and concave lines (veins) that may branch, and other shorter lines that cross between some of these (= cross-veins).

Such characteristic patterns (= venation) are shown through detailed anatomical studies to be relics of the ramifying insect respiratory system. That system is composed of a network of air tubes and tubules (trachea and tracheoles) that oxygenated almost every cell of the insect body and extended to wing buds and wings as they grew. Subsequently, after growth ceased and blood circulation could perform the aeration function as in the cockroach, tracheae hardened and became detached at the base where they joined body tracheae. Thereafter they served only as wing structural supports.

Venation pattern (Figure 11.38*A*) is the single most valuable taxonomic characteristic of fossil insect wings. Numerous categories from order to genera are identified by telltale wing venation. (It may be useful to recall that what is actually

being classified is a hardened, relict, tubular respiratory network.)

Venation patterns shown changes through geologic time. Sharov (1966) reiterated the once-rejected concept that geologically older insects had richer and more ramified venation. Or one can say that the venation tended toward a reduction in number of cross-veins and vein branches. This was also achieved by vein union and realignment—which led to enhanced wing structural support after aeration function was completed and trachaea had hardened (Ross, 1965).

Did this need for greater structural support within the wing serve as a selective factor? Were particular types and realignment of veins (that is, those with greater support potential when air tubes hardened) favored over other possible arrangements? The fact that all insect venation patterns, however diverse, are traceable to an original archetypal pattern (Figure 11.38*B*) hints at the not insignificant adaptive advantages conferred by modification of the archetypal venation.

Terminology of Veins and Cross-Veins. Main veins and principal named cross-veins of the insect wing are figured and illustrated in Figure 11.38*A* and defined below. (Modified slightly from Zimmerman in Tasch and Zimmerman, 1962.) It should be borne in mind that the venation illustrated is not definite but merely illustrative of most possibilities and does not characterize a given insect. Furthermore, venation may be different than shown by being reduced, fused, and so on, or more complex as in odonates with triangle, nodus, antenodal, and so on.

Costa (*C*) is the heavy, unbranched vein forming anterior margin of the wing.

Subcosta (*Sc*) is the vein immediately posterior to *C* and is always concave (in a depression).

Radius (*R*) is usually heaviest vein in the wing and always convex (along a ridge). Distinction between obverse and reverse of the wing is determined readily by finding whether *Sc* and *R* are convex or con-

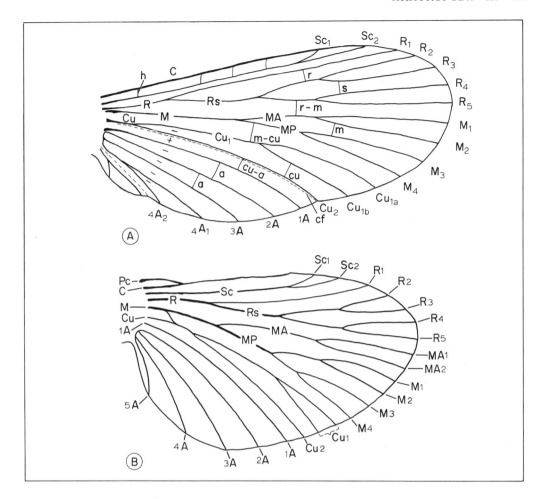

FIGURE 11.38 Venation.
(A) Identification of essential wing venation. After Zimmerman, in Tasch and Zimmerman, 1962, and others. See text for cross-vein names and *(B)* below for main vein names; *cf* — cubital furrow, a crease along any fold separating cubital and anal veins.
(B) Archetypal venation. Names given to veins according to the Comstock-Needham system. *Pc* — precosta at base of wing; *c* — costa, marginal in modern insects; *Sc* — subcosta, 2-branched; *R* — radius, 5-branched; *M* — media, 6-branched; *Cu* — cubitus, 3-branched; *A* — anal, varied members of (*MP, MA* — media posterior, media anterior; *Rs* — radial sector). After Snodgrass, 1935 and others.

cave. If *R* is convex, then the *obverse* half of the wing is being examined; if concave, then the *reverse* side is in view. The obverse half represents the wing as though the dorsal view of the actual were being observed, *R* branches into *R₁* and into the *radial sector (Rs)*, which is again divided into four branches, *R₁* is convex and *Rs* is concave.

(+, −): The positive sign is used to indicate convex veins; the negative sign, concave veins.

Media (M) is the next branch. (In Per-

mian insects there are usually two main branches, Media anterior (*MA*) and Media posterior (*MP*). The former is convex, the latter concave. Both frequently branch with the typical condition showing four branches of *M* reaching margin of the wing.)

Cubitus (Cu) has two branches; *Cu1* is convex and frequently branched, *Cu2* is concave and unbranched.

Anal veins (*1A, 2A, 3A*, and so on) are frequently set apart from the other veins (that is, cubitus) by a crease, the *cubital*

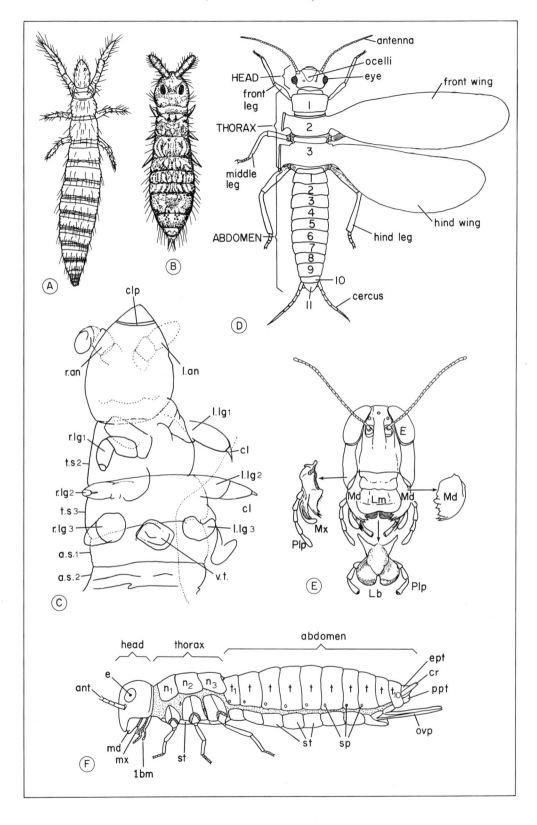

FIGURE 11.39 The three classes of hexapods.
Class Protura: (A) *Acerentulus barberi,* Rec. Class Collembola; (B) *Achorutes armatus,* Rec., compare (C). (C) *Rhyniella praecursor,* Dev., Rhynie chert, Scotland; *r* and *l* – right and left; *lg* 1–3 – legs; *ts₂.₃* – thoracic segments; *an* – antennae; *cl* – claw; *clp* – clypeus; *as* – abdominal segment. Class Insecta: (D) Winged insect, adult, schematic. (E) Grasshopper head, anterior view; *e* – eye, *md* – mandibles, *Mx* – maxilla, *Plp* – palpus. (F) Lateral view of an insect showing head, thorax, and abdominal structures; *n* – nota of thorax, *t₁.₁₀* – tergites, *st* – sternite, *sp* – spiracles, *ovp* – ovipositor, *cr* – cercus, *ept* and *ppt* – epiproct and periproct.
Illustrations redrawn and adapted from: (A, B, E) Ross, 1965; (C) Scourfield, 1960; (D) Scourfield, cit. Ross, 1965; (F) Borror and De Long, 1954.

furrow [(*cf*) – which occurs directly below and parallel to *Cu*2] along which the wing folds (Ross, 1965). These veins may form a fan (especially in the hind wings of the Protorthoptera and Protoperlaria) and in some forms, such as the Corrodentia, may be strongly convex.

Jugal veins (1*J,* 2*J*) – these are not shown. Slightly beyond 4*A₂* a crease may occur (the *jugal furrow*) that separates the anal from the jugal fold. When present, short jugal veins occur in the jugal fold (Ross, 1965).

Cross-Veins. These are named on the basis of the veins they connect or on their wing location (Borror and DeLong, 1954). They are written in small letters. (Exceptions to the rule are indicated by an asterisk below.)

humeral (*h*) – cross-vein in between *C* and *Sc*

*costal (*c*) – connects costa to subcosta or *R₁* (Ross, 1965)

radial (*r*) – connects branches of the radius

*sectoral (*s*) – connects *R₃* and *R₄*

radio-medial (*r-m*) – connects radius to media

medio-cubital (*m-cu*) – connects media to cubitus

cubital (*cu*) – branches of cubitus

cubito-anal (*cu-a*) – connects cubitus to anal (not shown are cross-veins between *CU₂* and 1*A*

anal (*a*) – various anals

Classification. Three classes are discussed: Protura and Collembola [considered as orders by some authors, but as classes by Mayr et al. (1955), Sharov (1966), Carpenter (1967)] as well as Insecta. Greatest emphasis is given to Insecta because of the considerable fossil record.

Diagnosis. *Class Protura.* (Telsontails).

Minute, wingless, blind, terrestrial forms with 11 abdominal segments and lacking antennae. Range: Recent only. [*Acerentulus,* Rec. (Figure 11.39*A*).]

Class Collembola (Springtails). Soft-bodied, wingless, abdomen of six or fewer segments; springing or jumping apparatus (= furcula); conspicuous near end of abdomen. Range: Devonian to Recent. [Representative genera: *Rhyniella,* Dev. (Figure 11.39*C*); *Protentombrya,* Cret. (Figure 11.40*A*).]

Class Insecta. Generally small to minute arthropods (consult Figure 11.39*D*).

Diagnosis of Insecta

1. *Body.* Divided into three distinct segments: head, thorax, abdomen.

2. *Head.* Bearing one pair of antennae, compound eyes, ocelli (three pairs), and three pairs of mouth parts for chewing, sucking, lapping, and so on [mandible or toothed jaws, and two pairs of maxillae each bearing a jointed palpus (Figure 11.39*E*).] [Specialized head structures are modified walking limbs in insects as in other arthropods (see Chapter 10). During evolution, trunk gnathal segments united with the protocephalon to form a definitive head (Snodgrass, 1935).]

3. *Thorax.* Three thoracic somites with pair of jointed legs on each. Legs generally have clawlike ending and are eight- or nine-jointed.

4. *Wings.* Three types of conditions occur: Where wings are present, they are usually two pair arising from the second and third thoracic somites; a pair may be lacking on third somite or on both second and third somites (wingless condition). Wings are outgrowths of body walls and usually bear network of veins and crossveins.

5. *Abdomen.* Usually composed of 11 segments or fewer (6–8) with last segment re-

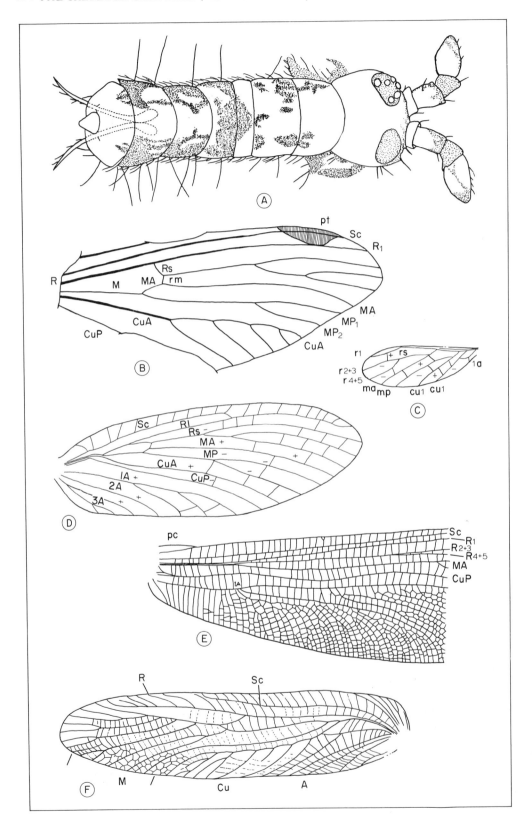

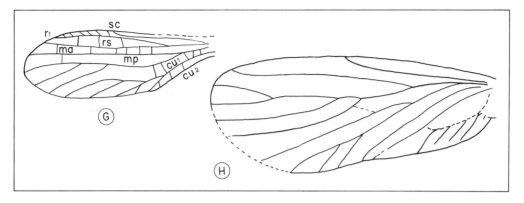

FIGURE 11.40 Representative fossil hexapods—I (schematic).
(A) Class Collembola: *Protentombrya walkeri,* amber fauna, Cret., Cedar Lake, Canada, schematic.
Class Insecta:
(B) Order Paleodictyoptera, *Kansana pulchra,* L. Perm., Kansas.
(C) Order Megasecoptera, *Asthenohymen minutus,* L. Perm., Okla.-Kans.
(D) Order Diaphanopterodea, *Parelmoa revelata,* L. Perm. Okla.
(E) Order Protodonata, *Typus gracilis,* L. Perm., Okla.
(F) Order Protorthoptera, *Metoedischia magnifica,* U.S.S.R.
(G) *Probnis speciosa,* Perm., Okla.
(H) Order Miomoptera, *Miomatoneura frigida,* Perm., U.S.S.R.
Illustrations redrawn and adapted from Carpenter, 1937, 1947; Tillyard, 1937; Martynov, 1928; Tasch
and Zimmerman, 1962, J. Paleontology, Vol. 36 (6) pp. 1319–1333, 22 text figures.

duced. Reduction of segments can occur by either fusion or telescoping. The apical portion may bear appendages (Figure 11.39*D*) and/or male-female genitalia, for example, the ovipositor (Figure 11.39*F*).

6. *Legs.* Eight- or nine-jointed (may be less), generally terminating in claws.

7. *Respiration.* Occurs through ramifying network of tracheae (air tubes) that open at paired spiracles, occurring on several segments (Figure 11.39*F, sp*).

8. *Sex and Development.* Sexes separate; eggs and spermatazoa; males unknown in some species; parthenogenesis common and hermaphroditism not unknown. Development is direct or occurs with metamorphosis (see "adaptive mechanisms," above).

9. *Immature Stages.* Transformation from larva to adult (metamorphosis) involves body and functional changes; wings and genitalia develop; size increases. Some changes are of the Jekyll-Hyde type, that is, from crawling, vermiform, grublike body to winged, flying insect. Larvae may have special appendages that adults lack. (Some wingless insects show negligible change from larva to adult and effectively do not experience metamorphosis as described above.)

In forms with gradual or incomplete metamorphosis (hemimetabolous), the *nymph* is a preadult instar (cockroach, grasshopper). Forms with complete metamorphosis (holometabolous) go through a larval stage (feeding), and a resting stage (nonfeeding, quiescent), that is, *pupa* stage preadult (Coleoptera, Diptera).

10. *Habitat.* Land-living primarily; aquatic habitats also occupied.

11. *Geologic Range.* Devonian; Upper Carboniferous to Recent.

Insect Orders. The 36 orders of insects recognized by F. M. Carpenter in the forthcoming Treatise on Invertebrate Paleontology insect volumes are tabulated in Table 11.7. Of these, nine are extinct, and of the 27 remaining orders that are extant, two lack a fossil record. Both in terms of first appearance and disappearance of extinct forms and the first origination of living forms, the Permo-Carboniferous record is highly significant. Well-known insects formed dense populations during those ages: silverfish, grasshoppers, crickets, mayflies, beetles, roaches, and dragonflies, among others.

TABLE 11.7. Tabulation of the 36 Orders of Insects (Class Insecta – Living and Fossil)

Extant Order	*Popular Name*	*Time of First Appearance*
1. Entotrophi	Bristletails	Devonian(?)
2. Thysanura	Silverfish	Upper Carboniferous
3. Ephemeroptera	Mayflies	Upper Carboniferous
4. Orthoptera	Grasshoppers, crickets	Upper Carboniferous
5. Blattodea	Roaches	Upper Carboniferous
6. Odonata	Dragonflies	Early Permian
7. Psocoptera (formerly Corrodentia)	Book lice	Early Permian
8. Hemiptera	Bugs	Early Permian
9. Mecoptera	Scorpion flies	Early Permian
10. Neuroptera	Ant lions, Dobson flies	Early Permian
11. Perlaria	Stone flies	Late Permian
12. Thysanoptera	Thrips	Late Permian
13. Coleoptera	Beetles	Late Permian
14. Phasmatòdea	Walking sticks, leaf insects	Triassic
15. Dermáptera	Earwigs	Jurassic
16. Trichoptera	Caddis flies	Jurassic
17. Diptera	Flies, mosquitos	Jurassic
18. Hymenoptera	Bees, wasps, ants	Jurassic
19. Isoptera	Termites	Early Tertiary
20. Embioptera	Embiids, web spinners	Early Tertiary
21. Siphonaptera	Fleas	Early Tertiary
22. Lepidoptera	Butterflies, moths	Early Tertiary
23. Strepsiptera	Twisted wing flies	Early Tertiary
24. Manteodea[a]	Praying mantids	Oligocene
25. Anoplura	Sucking lice	Pleistocene
26. Grylloblattodea[a]		Recent
27. Mallophaga	Bird lice	Recent

Extinct Order		*Known Geologic Range*
28. Paleodictyoptera		Upper Carboniferous–Permian
29. Megasecoptera		Upper Carboniferous–Permian
30. Diaphanopterodea		Upper Carboniferous–Permian
31. Protodonata		Upper Carboniferous–Permian
32. Miomoptera		Upper Carboniferous–Permian
33. Caloneurodea		Upper Carboniferous–Permian
34. Glosselytròdea		Permian
35. Protelytróptera		Permian
36. Protorthoptera		Upper Carboniferous–Triassic

[a]Formerly suborders of Orthoptera.

Source. Data from F. M. Carpenter, 1953 and 1967, in press.

Bees, ants, and wasps (Hymenoptera), and ordinary flies and mosquitos (Diptera) did not appear until mid-Mesozoic–Jurassic time. Some other orders apparently originated during this time also — earwigs and caddis flies. Not until early Tertiary did the air swarm with butterflies and moths, and were other habitats infested with termites. One order of sucking lice (Anoplura) did not evolve until Pleistocene.

***Order* Palaedictyoptera.** Palaeopterous insects with membranous wings, hind wing similar to fore wing, or with larger anal lobe; all main veins present in both wings; *archedictyon* or weak cross-veins present. (Archedictyon is a fine network of cuticular ridges occurring on some insect wings.) Upper Carboniferous–Upper Permian. (*Kansana*, Figure 11.40*B*.) [Palaeopterous = insects with wings not folded over abdomen, outstretched; Neopterous = insects with wings folded over abdomen.]

Order **Megasecoptera.** Palaeopterous insects; wings equal or subequal with nearly identical venation; all main longitudinal veins present forming an alternation of convexities and concavities; wings with distinct cross-veins not irregular or reticulate. Upper Carboniferous–Upper Permian. (*Asthenohymen*, Figure 11.40*C*.)

Order **Diaphanopterodea.** (Formerly included under Megasecoptera.) Neopterous insects. Venation: radius with bend near base; cross-wing distribution irregular, do not form rows. Radial sector and anterior media either independent or with some anastomosis; posterior media (MP) branched or unbranched. Upper Carboniferous–Permian. (*Parelmoa* (Figure 11.40*D*.)

Order **Protodonata.** Large to very large predaceous insects; wing span may be as great as 2 plus ft. Fore and hind wings similar in venation except for broader anal area in hind wing; posterior media (*MP*) and anterior cubitus .(*CuA*) absent or obsolescent; other cross-veins known in living Odonata lacking (nodus, arculus). Upper Carboniferous–Upper Permian. (*Typus*, Figure 11.40*E*.)

Order **Protorthoptera.** Small to large insects (relatives include grasshoppers and roaches); pronotal lobes usually expanded to form pronotal shield. Forewing, tegminous (= toughened); venation normal; hind wing has expanded anal area, that is, more than six veins; or the latter is smaller than the *remigium* (= area of wing exclusive of jugal and anal areas). Upper Carboniferous–Upper Permian. [*Metoedischia* (Figure 11.40*F*); *Probnis*, (Figure 11.40*G*).]

Order **Miomoptera.** (Formerly assigned to orders Protoperlaria and Protorthoptera.) Small insects; primitive net and wings absent, only oblique cross-veins persist (but may be rare or absent). Hairs in rows on basal part of longitudinal vein; *SC* usually much shortened; *R* considerably longer than the *SC* and usually forming a row of oblique veins toward costal margin. Posterior wings tend to be dilated in anal region; venation as in anterior wing (Martynov, 1928). Upper Carboniferous to Permian. (*Miomatoneura*, Figure 11.40 *H*.)

Order **Caloneurodea.** Small to large insects (relatives include Protorthoptera and Orthoptera); fore and hind wings similar in shape and venation; hind wing lacks enlarged anal area (see Protorthoptera). Order is based on a family (Caloneuridae) that has four anal veins. Other families of the order have less than four. Numerous and strongly developed cross-veins. Upper Carboniferous–Upper Permian. (*Apsidoneura*, Figure 11.41*A*.)

Order **Protelytróptera.** Small insects (relatives include earwigs and roaches); forewing modified to form convex *elytra* (thickened leathery or horny front wing), with venation reduced and weak; hind wings longer and broader than forewings; anal area expanded, much larger than the *remigium* (see Protorthoptera for definition), with longitudinal and transverse fold. Lower and Upper Permian. (*Protelytron*, Figure 11.41*B*.)

Order **Glosselytródea.** Small insects (relatives of Caloneurodea). Forewing tegminous; conspicuous precostal area and submarginal vein (subcosta, *Sc*, and posterior cubitus, *CuP*) following the distal half of wing margin—that is, half farthest from body. Permian. [This order has been delimited since its range was originally given as Permian to Upper Jurassic (Brues et al., 1954).]

Information Derived from Fossil Insects

Entrapment and Modes of Preservation. The fossil insect record was accumulated by chance burial and/or entrapment. Modes of preservation were diverse. One

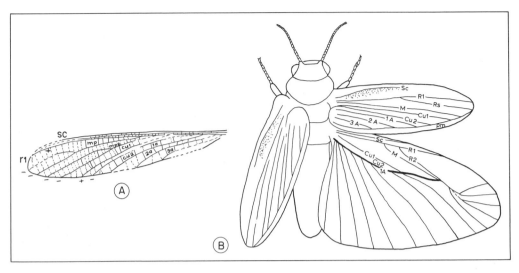

FIGURE 11.41 Representative fossil hexapods—II (schematic).
(*A*) Order Caloneurodea: *Apsidoneura reducta,* Perm., Okla.
(*B*) Order Protelytroptera: *Protelytron permianium,* Perm. Kansas, reconstruction.
Illustrations after Tasch and Zimmerman, 1962; Carpenter, 1933.

must distinguish between three-dimensional preservation of a complete cadaver (as in amber faunas), and burial, subsequent organic decay, and sedimentary cover that results in the formation of casts and molds of insect wings, and so on.

Methods of entrapment include hardened tree gums (ambers), asphalts and asphalt seeps, volcanic ash falls, and mineralized waters (forming onyx marble, for example—Pierce, 1957). The Miocene of the Mojave Desert yielded insects that had perfect three-dimensional preservation, being silcified or calcified or replaced by an evaporite (Palmer, 1957). As shown subsequently (see "Spoor" below), insects are preserved in petrified woods, seeds, coprolites of lizards, birds, and so on. Most widespread are impressions of wings in varied sediments—indurated clay muds, silts, peats, lignite, carbonaceous shales, among others. Often ironstones or other concretions contain insect fossils.

Representative Fossil Insect Faunas. The Illinoisian Mazon Creek fauna (Middle Pennsylvanian) contains a representative fossilized insect assemblage. Of 200 animal species, insects constituted more than one-half of the total (Richardson, 1956). These fossils were retrieved from ferruginous concretions in coal strip mines. Originally the environment was a low-relief coastal plain (delta-like) that supported a rich biota of plants, vertebrates (fishes chiefly), and invertebrates.

A Permian fauna from the Wellington Formation of Kansas and Oklahoma has been much studied (Tillyard, Carpenter, Tasch, and Zimmerman). Thirteen insect orders are represented in the Tasch collection. Insect wing impressions were preserved in hard, argillaceous limestone in association with clam shrimps and floral debris. The coastal swamp environment supported intermittent fresh-water situations. Publications of Tillyard and Carpenter, in particular, have made this one of the world's best-known faunas.

The Solenhofen lithographic limestone (Jurassic of Bavaria, Germany) is equally famous for its remarkably preserved vertebrates as it is for its invertebrates (Handlirsch, 1925; Carpenter, 1932). Ten insect orders were fossilized. In these, Carpenter has shown only larger insects were preserved while all others had been decomposed or consumed by fish. Among the orders, the Odonata are most abundant, followed by Hymenoptera, Coleoptera, Blattaria, and so on.

Although "amber" insect fauna commonly denotes the Baltic fauna, others are known, such as the Canadian amber fauna of Cretaceous age (Carpenter et al., 1937). The insects occurred in transported amber that had been deposited along a mile of beach in association with wood debris. The amber stretched in a band 30 ft wide and about 2 ft deep, and had a total estimated weight of something less than 750 tons. Insect orders represented included Homoptera, Hymenoptera, Diptera, and Coleoptera – the last-named specimens being inadequate for taxonomic purposes (see Langenheim, 1960, for an Alaskan Cretaceous amber insect fauna).

Thousands of insect specimens have been retrieved from paper-thin, unusually light to the touch, lacustrine, brown shales incorporating volcanic ash. These comprise the well-known Florissant deposits of Colorado (Miocene age). Preservation is sometimes so excellent that it reveals objects at the microscopic hair level.

Spoor. Fossil spoor of various insects are often encountered from Late Paleozoic to Pleistocene. Obviously, Late Mesozoic, Tertiary, and Pleistocene examples are more extensive.

Calcitic to argillaceous-calcitic molds of *larval chambers* of mining bees are known from fresh-water deposits – Bridger Formation, Upper Eocene of Wyoming (Figure 11.42*A*); see Brown, 1934. Mold of a comb of a *wasp's nest* was found when an ironstone concretion was split. Age was Upper Cretaceous (Utah) (Brown, 1941).

The Green River Formation of Wyoming yielded an unusual type of fossil – *caddis fly cases*, constructed almost entirely of ostracode valves laid down in a compact mosaic (Bradley, 1924). This fossil formed a conspicuous layer that alternated with ostracode-bearing marl.

Excreta of termites from the Califor-nian Pliocene deposits were preserved in opalized wood as hexagonal pellets called "frass" (Rogers, 1938) (Figure 11.42*B*).

Channels, tunnels, and burrows attributed to the engraver or bark beetles, abound in the tree species *Araucarioxylon arizonicum* in the Petrified Forest of Arizona (Chinle, Triassic). These were preserved in agatized wood from this forest. Some channels were 5 mm wide (Walker, 1940).

Eggs and egg masses are not uncommon in Tertiary beds. Such egg masses may hold upward of 20,000 eggs imbedded in albuminous matter (Figure 11.42*D*) (Scudder, 1890).

Many different species of insect trails from the Permian of Germany have been reported. Insect-cut leaves are known from several sites including the Lower Eocene of Kentucky (Berry, 1931). The round cutouts were attributed to either bees or caddis grubs. An oak leaf (*Quercus cognatus*) from the Miocene shale of Washington bore *insect galls* reminiscent of those made by living cynipids (gall wasps) (Hoffman, 1932). Fossilized leaves and petrified woods are excellent potential sources of insect spoor (see Brues, 1936).

Paleopathology. Several disease-carrying vectors in the form of familiar insects are found in the Tertiary record: flea, *Palaeopsylla klebsiana* (Eocene, Baltic amber), and the tse-tse fly, *Glossina oligocena* (Eocene, Brown Coal, Germany, and Oligocene of Colorado, Figure 11.42*E, F*).

The role of certain insects as disease vectors that can infect and wipe out whole populations of modern mammals was excellently reviewed by Osborn (1929) in his monograph on titanotheres of Wyoming, Dakota, and Nebraska. Osborn suggested, as have others (Tasnádi-Kubacska, 1962), that such pathogenic agents as the tse-tse fly could have been factors in titanothere extermination. This suggestion is bolstered by the further observation that the environment in Wyoming and surrounds during parts of the Tertiary was not unlike that of the modern African

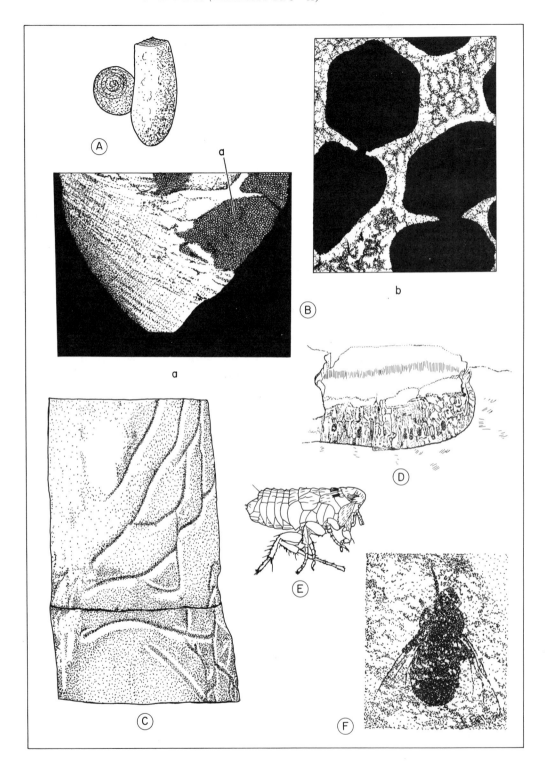

A

a

a

B

b

C

D

E

F

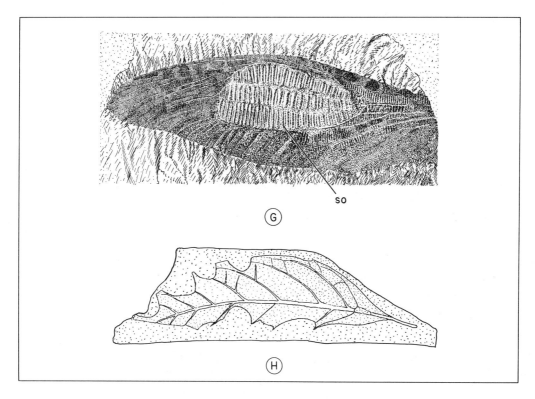

FIGURE 11.42 Information derived from fossil insects—I.
Spoor:
(A) *Celliforma spirifer,* fossil mold of larval chamber of mining bee (Hymenoptera), U. Eoc., Bridger Basin, Wyoming, X 1. Note characteristic spiral seal.
(B) Fossilized termite (Isoptera) pellets in opalized wood, Plio., Santa Barbara, Calif., schematic. (a) Polished specimen of wood showing areas with termite pellets, X 1; (b) detail from thin section of (a), showing opalized pellet of hexagonal cross-section (0.3–0.5 mm), interstices with opal and chalcedonic opal, enlarged.
(C) *Paleoscolytus divergus,* channel made by larvae of an engraver or bark beetle (Coleoptera), Chinle, Trias., Petrified Forest, Natl. Monument, Ariz.; channel width, 5.0 mm, bored in tree species *Araucarioxylon arizonicum,* entire fossil—wood and channel agatized.
(D) *Corydalites fecundum* (Neuroptera), part of egg mass in lignitized beds, Crow Creek, Colo., Tert.; egg mass 5 cm long X 2 cm wide X 1 cm high; individual egg length, 2.6 mm; total eggs = 2000±.
Paleopathology:
(E) *Palaeopsylla klebsiana,* fossil flea with specialized mouth parts, amber fauna, Eoc., Baltic.
(F) *Glossina oligocena,* tse-tse fly, Oligo., Colorado, disease vector.
Specialized Structures:
(G) Stridulating organ (so), *Clatrotitan andersoni* (Orthoptera), Trias., New South Wales. Austral. (H) *Spoor:* fossil leaf of *Icacorea prepaniculata,* L. Eoc., Kentucky, cut by a leaf-cutting bee (Neuroptera). Note scallop-shaped bites, 6.0–9.0 mm diameter.
Illustrations redrawn and adapted from (A) R. W. Brown, 1934; (B) A. F. Rogers, 1938; (C) Walker, 1940; (D) Scudder, 1890; (E, F) Tasnadi-Kubacska, 1962; (G) Carpenter, 1953; (H) E. W. Berry, 1931.

habitat of the tse-tse fly and other disease-transmitting insects.

Clearly, even where they were not vectors in disease transmission, insect eggs were attached to rodent hairs (Oligocene) and larval forms developed in decaying mammalian bones, carcasses, and excrement during Tertiary and Pleistocene time. In a fossil bone fragment of a bird, a blowfly pupuria, was found; and in crocodile feces, insect coproporphyrins occurred. Certain dung beetles (*Palaeocopris,* Pleistocene, California) that bred in bison, horse, and camel dung, disappeared from the coastal plain area when these herbivores did (Pierce, 1957).

Biostratigraphy, Microstratigraphy of Insect Beds. New insect beds were discovered in the Kansas-Oklahoma Permian Wellington Formation in conjunction with a program to map fossil clam shrimp-bearing beds. The biostratigraphic sequence was found to recur in the formation at approximately every 100 ft vertically. Thus six successive insect beds are now known in a total thickness of 700 ft (Tasch, 1962, Table 1, and Table 11.8 below).

Microstratigraphic study of these insect beds permitted reconstruction of the biotic spectra (Figure 11.43). Associated with the insects were evaporites (hopper crystals, salt casts), clam shrimps, mollusks, plants, carbonized wood, and so on.

Since several of these insect beds, the *Asthenohymen-Delopterum* bed, for example, are traceable from Oklahoma into Kansas individually, they serve very well as a *datum plane*. By reference to such planes, it was possible to determine the precise stratigraphic position and regional correlatives of a given clam shrimp facies that were usually quite restricted (Tasch and Zimmerman, 1962).

In a section some seven meters thick representing the Florissant Lake basin southern portion, a fossiliferous sequence of alternating shales and sandstones was found (Scudder, 1890, pp. 21–22). Eight insect-bearing beds were separated by nonfossiliferous sandstones or nonlaminated shales. Melander (1949) thought that this rhythmicity in insect occurrence had seasonal connotations. Successive ash falls that account for the insect beds, apparently corresponded with the spring season of the year when dipterids and other insects were abundant.

Pierce (1944, 1957) attempted to zone strata in the La Brea asphalt pit (Pleistocene) using entrapped insects as zonal fossils. Zones A–C were defined below 2 ft of commercial asphalt grading from black to brown. Only extinct insects were recovered in Zone B, which was 48 in. below the top. In the lower part of Zone A, almost 20 in. below the top, insects found were distinct from those found in Zone B and showed identity with Recent forms.

Evolution. The essentials of insect order evolution are delineated in Figure

TABLE 11.8. Stratigraphic Position of Wellington Formation Insect-Bearing Beds (Permian, Leonardian, Kansas-Oklahoma)

Identification Number	Name of Bed	Locality	Elevation above Top of Marine Herington Limestone (ft)	100-ft Units above Herington Limestone
6	YIB	Kay Co., Okla.	630 to 700	6th
5	A-D	Noble and Kay Counties, Okla.; Sumner Co., Kan.	558 to 559	5th
4	Midco[a]	Noble and Kay Counties, Okla.	550±	
3	Carlton[b]	Dickinson Co., Kan.	250 to 300	3rd to 2nd
2	TFB	Marion, Harvey, and Sedgwick counties, Kan.	220±	
1	OIB	Sumner Co., Kan.	50 to 70	1st

[a]Midco, three beds in 4.7 ft.
[b]Carlton, three beds in 5.5 ft.
Datum. Top of Herington Limestone. Identification Numbers: (6) YIB, youngest insect-bearing bed, Wellington XIX (Tasch, 1962); (5) A-D, *Asthenohymen-Delopterum* bed (Tasch and Zimmerman, 1959, 1962); (4) Midco (Raasch, 1946); (3) Carlton (Dunbar, 1924); (2) TFB, 10-ft bed – designates the elevation above the top of the Annelly gypsum of this insect-bearing bed (Tasch, 1961); (1) OIB, oldest insect-bearing bed, Wellington XVIII (Tasch, 1962).

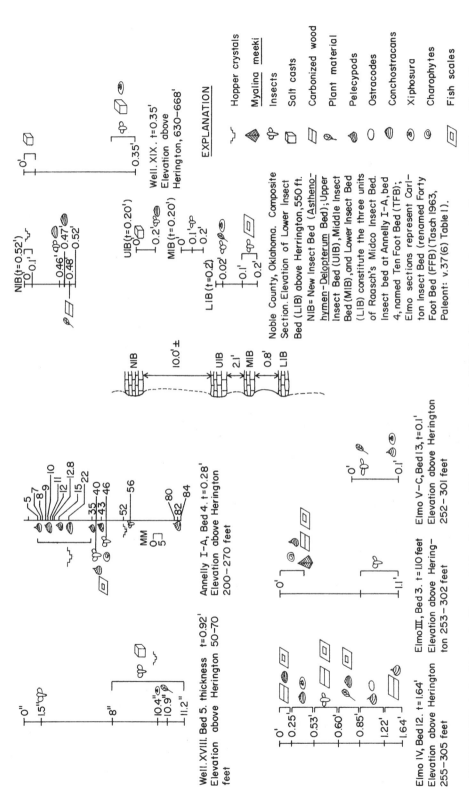

FIGURE 11.43 Derivative data from fossil insects—II. Microstratigraphy. Biotic spectra of insect beds. Wellington Formation, Perm., Leonardian, Kansas and Oklahoma. After Tasch, 1964. [Paleont. = J. Paleont. 1963].

11.36. Four major stages are recognized (Carpenter, 1953; Sharov, 1966).

Stage 1. Wingless (Thysanura — silverfish, and so on).

Stage 2. Wings evolve (Paleoptera — earliest winged insects).

Stage 3. Wing folding evolves (Neoptera.)

Stage 4. More complex metamorphism evolves. (A change occurs from the series of nymphal stages in neopterous insects as a prelude to adult stages, to larval stages and pupa that little resemble the adult.)

There are, of course, a considerable number of smaller evolutionary events such as adaptive modification of the first pair of wings in beetles (Coleoptera) as elytra or hardened wing covers. There was a tendency toward large size in Coal Measure insects. This, in part, has been attributed to fierce competition between insects whose nymphs preyed on larval amphibians, and amphibians that consumed insects. There were more than 1500 fossil insect species of Pennsylvanian age. Other evolutionary events included development of claws, enlargement and fusion of internal head supports, modification of certain sclerites, suppression of juvenile ocelli, loss of ovipositor structures, and so on (Ross, 1965, Figure 159).

Accompanying rise of flowering plants (angiosperms) in the Cretaceous (probably earlier, in the Jurassic), were insects that adapted to pollen and nectar feeding.

Bees, moths, flies, and beetles are among modern insects that pollinate fruits and legumes. They constitute part of the 20 percent of modern insects that depend on flowers for nutriment at one or another life cycle stage. Clearly, wind-pollination has been significantly supplemented or supplanted, since some 65 percent of all angiosperms are pollinated by insects (Ross, 1965).

Rise of the angiosperm should be viewed as having provided and extended new ecologic niches and food sources for numerous insects.

Three of the major evolutionary stages — development of wings, wing-folding, and complete metamorphosis — had survival value and hence provided a selective advantage. Wings allowed escape from a host of Late Paleozoic insect predators (not only amphibians or reptiles, but scorpions, spiders, and so on), while wing flexure when at rest, permitted insects to blend with foliage or ground pattern and so escape detection by predators. Effectively, 100 percent of all existing insects are winged types (99.9 percent) — witness of this feature's evolutionary success. Neopterous insects with complete metamorphosis had immature stages unlike that of the adult. This characteristic allowed such stages to invade different habitats, such as other animals' tissues. It also permitted them to feed on other kinds of food. Such diversity and flexibility apparently had survival value (Carpenter, 1953).

QUESTIONS

Insecta

1. (a) How does insect ecdysis resemble and differ from the crustacean cycle?

   (b) Compare organization of the insect integument and that of decapod crustaceans; trilobitomorphs.

2. (a) What special significance attaches to each of the following: *Lemmatophora typica*, archetypal venation, *Rhyniella praecursor*.

   (b) Draw a wing and trace on it each of the following: media, costa, radius, anal and jugal veins, and cross-veins — *Cu-A, m-cu.*

3. For each of the following insect faunas, indicate typical lithology and kinds of insects that occur: Solenhofen, Florissant, Canadian Cretaceous.

4. Study Figure 11.43 and indicate biotic associates of Permian insects in the mid-continental United States.

5. What are the four major insect evolution stages that paleoentomologists recognize? Explain the selective advantage of each.

6. (a) Discuss the variety of fossil insect spoor and its context in the rock column.

   (b) In beds of what age would you expect to find fossil ants, roaches, termites, bees, butterflies?

   (c) Discuss Tertiary insects as disease vectors.

REFERENCES

Insecta

Berry, E. W., 1931. An insect cut leaf from the Lower Eocene: Am. J. Sci., Ser. 5, Vol. 21, pp. 301–303, Figure 1.

Bradley, W. H., 1924. Fossil caddice fly cases from the Green River Formation: Am. J. Sci., Ser. 5, Vol. 7 (43), pp. 310–312.

Brown, R. W., 1941. The comb of a wasp nest from the Upper Cretaceous of Utah: Am. J. Sci., Vol. 239 (1), pp. 54–56, 1 plate.

———, 1934. *Celliforma spirifer*, the fossil larval chambers of mining bees: J. Washington Acad. Sci., Vol. 24 (12), pp. 532–539, 1 figure.

Brues, C. T., 1936. Evidences of insect activity in fossil wood: J. Paleontology, Vol. 10 (7), pp. 633–643, 3 figures.

———, A. L. Melander, and F. M. Carpenter, 1954. Classification of Insects: Mus. Comp. Zool. (Harvard) Bull., Vol. 108, 827 pp.

Carpenter, F. M., 1953. The evolution of insects: Am. Scientist, Vol. 41 (2), pp. 256–270.

———, 1947. Lower Permian insects of Oklahoma, Part I. Introduction and the orders Megasecoptera, Protodonata and Odonata: Am. Acad. Arts and Sci. Proc., Vol. 76, pp. 25–53.

———, 1933. Part 6. The Lower Permian insects of Kansas: Am. Acad. Arts and Sci. Proc., Vol. 68, pp. 411–450.

———, 1932. Part VIII. Jurassic insects from Solenhofen in the Carnegie Museum and the Museum of Comp. Zool.: Ann. Carn. Int. Mus., Vol. 21, pp. 97–129.

———, et al., 1937. Insects and arachnids from Canadian amber: Toronto Univ. Geol. Ser. No. 40, Contrib. to Canadian Min., pp. 7–62.

Guthrie M. J., and J. M. Anderson, 1957. Zoology. Wiley, 655 pp. (pp. 454–457: Insect life cycles and metamorphosis).

Hackman, R. H., 1964. Chemistry of the insect cuticle, in M. Rockstein, ed., Physiology of Insecta. Academic Press, pp. 471–506.

Handlirsch, A., 1925. Handbuch der Entomologie. Jena, Gustav Fischer, Vol. 3 (Paleontology, and so on).

Hoffman, A. D., 1932. Miocene insect gall impressions: Botanical Gazette, Vol. 93 (3), pp. 341–342, 1 figure.

Imms, A. D., 1936. The ancestry of insects: Soc. Brit. Entomology Trans., Vol. 3, pp. 1–31.

Langenheim, Jr., R. L., C. L. Smiley, and Jane Grey, 1960. Cretaceous amber from the arctic Coastal Plain of Alaska: G.S.A. Bull., Vol. 71, pp. 1345–1356, 2 figures.

Laurentiaux, D., 1953. Classes Insectés, in Jean Piveteau, ed. Traité de Paléontologie, Vol. III, pp. 397–527.

Locke, M., 1964. The structure and formation of the integument in insects, in M. Rockstein, ed., The Physiology of Insects, Academic Press, Vol. 3, pp. 379–470.

Manton, S. M., 1973. Arthropod phylogeny — a modern synthesis: J. Zool. London, Vol. 171, pp. 111–130.

Martynov, A., 1928. Permian fossil insects of northeast Europe. Travaux du Musée Géologique: Académie des Sci., de l'URSS, Vol. IV: 118 pp.

Melander, A. L., 1949. A report on some Miocene Diptera from Florissant, Colorado: Am. Mus. Novitiates, No. 1407.

Osborn, H. F., 1929. The Titanotheres of ancient Wyoming, Dakota, and Nebraska: U.S.G.S. Monograph, 1955, Vol. 2, pp. 869–874.

Osnovy paleontologia, 1962. [A Russian volume with extensive illustrations and descriptions of fossil insect orders by specialists.] Vol. 9, 534 pp. (Moscow).

Palmer, A. R., et al., 1957. Miocene arthropods from the Mohave Desert, California: U.S.G.S. Prof. Paper 294-G: 237 ff., plates 30–32.

Pierce, W. D., 1957. Insects, in Treat. Mar. Ecol. and Paleoecol. G.S.A. Mem. 67, Vol. 2, pp. 943–951.

————, 1951. Insect fossils in onyx marble and modern entrapment in calcite waters: G.S.A. Bull., Vol. 62 (12), pp. 15–23.

Richards, A. G., 1953. Structure and development of the integument, in W. D. Roeder, ed., Insect Physiology. pp. 1–21. 1953. Also, Chemical properties of cuticle, in same volume, pp. 23–41.

Richardson, Jr., E. S., 1956. Pennsylvanian invertebrates of the Mazon Creek Area. Introduction: Fieldiana: Geology, Vol. 12 (1), pp. 3–12.

Rogers, A. F., 1938. Fossil termite pellets in opalized wood from Santa Maria, California: Am. J. Sci., Vol. 36 (215), pp. 381–392.

Ross, H. H., 1948. A Textbook of Entomology, 1st ed. 1965, 3rd ed., Wiley, 517 pp.

Scourfield, D. J., 1940. The oldest known fossil insect: Nature, Vol. 145 (3682), pp. 799–801, 4 text figures.

Scudder, S. H., 1890. The Tertiary insects of North America: U.S.G.S. of the Territories, Vol. 13, pp. 11–621, plates 1–28.

Sharov, A. G., 1966. Basic arthropodan stock with special reference to insects: Pergamon Press, 239 pp.

Simpson, G. G., C. S. Pittendrigh, L. H. Tiffany., 1957. Life. An Introduction to Biology. Harcourt, Brace and Co., Inc.

Snodgrass, R. E., 1935. Principles of Insect Morphology, McGraw-Hill.

Tasch, P., 1964. Periodicity in the Wellington Formation of Kansas and Oklahoma, in D. Merriam, ed., Symposium on Cyclic Sedimentation: K.G.S. Bull., 169, Vol. 1, pp. 481–494.

Tasch, P., 1962. Vertical extension of mid-continent Leonardian insect occurrences: Science, Vol. 135 (3501), pp. 378–379, Table 1.

Tillyard, R. J., 1937. Kansas Permian insects, Part 19: The order Protoperlaria (*cont.*): The family Probnisidae: Am. J. Sci., Vol. 33, pp. 401–425.

————, 1930. The evolution of the class Insecta: Roy. Soc. Tasmania, Papers and Proc., 1930, pp. 1–89.

————, 1928. Kansas Permian insects, Part 10: Am. J. Sci., 5th ser., Vol. 16, pp. 185–220.

Walker, M. V., 1938. Evidences of Triassic insects in the Petrified Forest National Monument, Arizona, in Proc. U.S.N.M., Vol. 85 (3033), pp. 137–141, plates 1–4 (issued 1945).

chapter twelve
PATTERNS OF FIVE: CREATURES WITH TUBE FEET (ECHINODERMATA)

INTRODUCTION

A group of spiny skinned invertebrates of modern and ancient seas are so distinctive that they lack any near relatives among other phyla (Fell and Pawson, 1966). Included in this phylum are a broad array of living forms that superficially appear to differ markedly from each other: feather stars (crinoids), sea urchins, brittle stars, sea cucumbers, starfish (Figure 12.1A). To these must be added the many extinct forms such as blastoids, cystoids, edrioasteroids, and so on. The common features shared by all members of the phylum are the following: (1) Body structures in adults occur in patterns of five (pentamery), and (2) presence of tube feet.

FIGURE 12.1 Density of a natural crinoid comulatid community. Depth 650 meters off Northern Spain. Original photograph by A. S. Laughton, cit. Marr, 1963.

Originally the Greek name of the phylum [*echino* + *dermata*, or spiny (prickled) + skin] was applied only to the class that embraced sea urchins—the marine echinoids, and the terrestrial hedgehog.

A smooth body surface is a rarity in echinoderms; spines may be absent, however, and tubercles or bosses may constitute the surficial irregularities. In sea cucumbers, warts occupy the surface in lieu of calcareous projections (Hyman, 1955).

LIVING ECHINODERMS

Physiology

Echinoderms are brightly colored, chiefly because of pigment (Vevers, 1966). Some are luminescent (Ophiuroidea) (Millott, 1966a). Many display complex behaviors, such as burrowing, migratory, and reproductive—the last-named being reminiscent of the beginnings of social behavior (Reese, 1966). The capacity to cast off appendages (autotomy), to regenerate new individuals from body pieces as in asteroids and ophiuroids and so on, and to replace shed appendages is well known (Swan, 1966). Some echinoderms emit toxic secretions; others release poisons into the tissues of animals through the *pedicellariae* (echinoids) (Alender and Russell, 1966).

Some echinoderms are eaten as foods by man: dried cucumbers (holothurians) called *trepang* or *bêche de mer* are sold as a

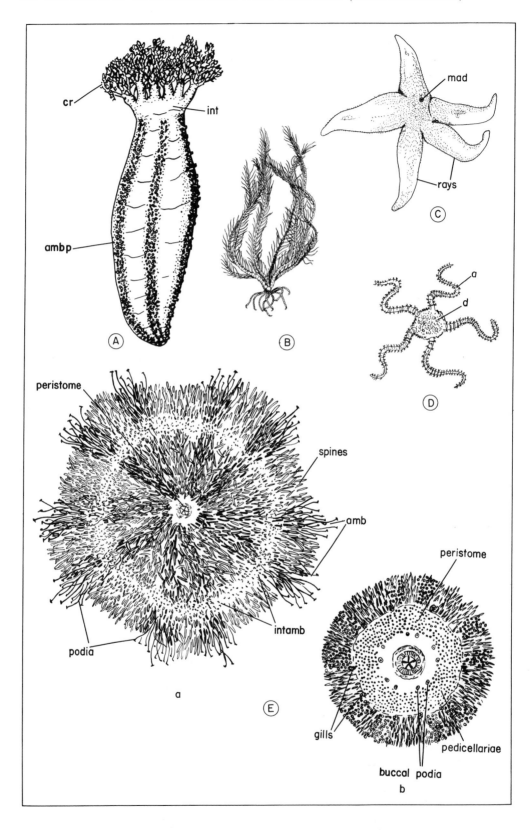

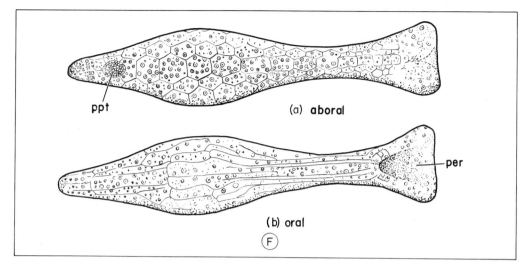

ppt (a) aboral

per

(b) oral

(F)

FIGURE 12.1A Echinodermata types—living classes (schematic).

(*A*) Class Holothuroidea (sea cucumbers): *Cucumaria frondosa,* New England; *cr*—crown, *int*—introvert, *Amb. p.*—ambulacra and podia for locomotion.

(*B*) Class Crinoidea (feather star): *Antedon bifida,* a common extant comatulid crinoid; arms, bearing food grooves (not seen), and cirri at base.

(*C*)(*D*) Class Stelleroidea, subclass Asteroidea (starfish): (*C*) Preserved material, asteroid, Recent; (*D*) subclass Ophiuroidea (brittle star or serpent star), Recent, *mad*—madreporite; *d*—dish; *a*—arm.

(*E*)(*a*–*b*) Class Echinoidea: *Tripneustes ventricosus,* Rec., Bahamas; *amb*—ambulacrum, *int*—interambulacrum.

(*F*)(*a*–*b*) *Echinosigra paradoxa* (sea urchin); deep-sea form, Indian Ocean; note position of peristome (*per*) at end of neck-shaped projection and body plates; periproct on aboral side.

Illustrations redrawn and adapted from: (*A*) Pawson, 1965; (*B*) Mounted specimen; (*C*) Gregory, 1900; (*D*) Hertwig; (*E*) from The Invertebrates, Vol. IV, by Libbie Hyman (Copyright 1955 by McGraw-Hill, Inc. With permission of McGraw-Hill Book Co.); (*F*) after several authors.

food delicacy to the Chinese. Echinoid gonads, raw or roasted, are valued as a nutritious food in many parts of the world (Hyman, 1955).

Population Density, Habitat

In numerical terms, echinoderms are one of the most abundant of living forms; at abyssal depths they account for 90 percent of the biomass. The Galathea Expedition (Bruun et al., 1956), for example, found echinoderms dominant in their deep sea trawls; greatest in number were the sea cucumbers, followed by the brittle stars (ophiurans). Although not abundant, remarkably distended deep sea urchins were also found in these trawls (Figure 12.1*A*, part *F*). To glimpse the density at shallower depths, see Table 12.1.

TABLE 12.1. Comparative Population Densities of Selected Living Echinoderms

Species	Population Density $(x/m^2)^a$
Crinoids:	$(\text{to } 54)^b$
Antedon vulgaris	16
Echinoids:	(to 800)
Echinus esculentus	4+
Lytechinus anamesus	"vast herds"
Ophiuroids:	(to 710)
Ophiothrix fragilis	100^c
Amphiura community	400–500
(*filiformis* and/or *chiajei*)	

[a] Number of individuals (*x*) per square meter of sea floor (m^2).

[b] See Figure 12.1, 65 per square meter.

[c] This density extended over 6–10 miles of sea floor.

Source. Data from Reese, 1966.

Echinoderms are ubiquitous. They occur in all seas, as for example, crinoid comulatids in Antarctic waters. They are distributed in all latitudes and range from intertidal to abyssal depths. Although some brackish water forms are known, there are no fresh-water echinoderms. The phylum is exclusively marine. Ophiuroids are apparently the most successful living echinoderms in terms of their horizontal spread and judging from their cosmopolitan distribution in the littoral zone, as well as in deeper waters (Hyman, 1955).

An abundance of echinoderms are represented in the plankton as larval stages although they play a negligible role in the food chain. Northern echinoderms, for example, spend three to four weeks as part of the plankton before metamorphosis into a young urchin (Thorson, 1946).

Species, Living and Fossil

There are somewhat fewer than 20,000 known echinoderm species (living and fossil) and, of these, only some 5000+ are living. Taking all known animal species, both fossil and living, into account, echinoderms constitute less than 2 percent of the total.

The enormity of the number of echinoderm species brought to extinction through geologic time, can best be exemplified by the crinoids and echinoids: The former has six times the number of extinct species to living; the latter, eight times. Only in one echinoderm group, the sea stars (subphylum Asterozoa, class Stelleroidea), is the trend reversed. In this group it is the living species that far exceed the number of extinct species, by a factor of X8.

Clarification of Affinities

The precise placement of the spiny-skinned phylum among the invertebrates were, in one degree or another, taxonomic problems for over two centuries. Holothurians, for example, went unrecognized for quite a while. Ultimately, problems were resolved belatedly (mid-nineteenth century). Resolution came about serially; echinoderms were originally assigned to the mollusks by Linnaeus and later to the coelenterates by Cuvier, and so on.

Embryological researches helped to eliminate some strange bedfellows in echinoderm classification. It was shown that, unlike coelenterates, they possessed a distinct body cavity (= coelom) — a condition known as *coelomate*. Furthermore, they were unlike mollusks, annelid worms, and arthropods since the coelom derived from the enteron, or gut cavity — *enterocoely*. In the three groups named, the coelom was derived by splitting of the mesoderm — *schizocoely* (Nichols, 1962).

Echinoderm Embryology and the "Echinoderm Superphylum"

Because of some morphological and biochemical characteristics that are shared in common by echinoderms, protochordates, and vertebrates, an echinoderm superphylum has been envisioned by some workers (Simpson et al., 1957). The following four items are the ones that are cited to sustain this view.

1. There is the same basic plan in hemichordate larvae (*tornaria*) and echinoderm larvae (*dipleurula*) (Nichols, 1962).

2. Echinoderms and chordates both develop from undifferentiated cells following fertilization of the egg and cell cleavage.

3. In some chordates and in echinoderms, the mesoderm and coelom develop in the same way (that is, enterocoely).

4. Echinoderms and chordates have a common biological signature, that is, the phosphagen creatine, as contrasted with the Annelid superphylum that was characterized by arginine. [Fell and Pawson (1966) found two types of phosphagen in echinoderms: creatine phosphate in ophiuroids and both types of sea urchins, and arginine phosphate in crinoids, asteroids, and holothurians.].

Fell (1948) argued strongly against Item 1. He pointed out that embryo and larva of echinoderms are variable and independent of the adult stage. Furthermore, on larval similarities, ophiuroids are closer to echinoids than to asteroids; asteroids would be nearer to holothurians. However, paleontological and morphological evidence deny this (Fig. 12.2).

Since within the phylum Echinodermata larval similarities are not indicators of taxonomic affinities, how can they be beyond and outside the phylum with hemichordates?

Apparently, neither on biochemical nor on embryological grounds can the original argument be sustained.

When one excludes so-called larval resemblances among echinoderms, the critical evolutionary role ascribed to the dipleurula deflates (Nichols, 1962, Figure 26). According to theory, modification of the dipleurula ultimately gave rise to a hemichordate (ascidian?), attached and free-living echinoderms respectively.

Despite the above disclaimers, other morphological aspects suggest that there is a remote relationship between primi-

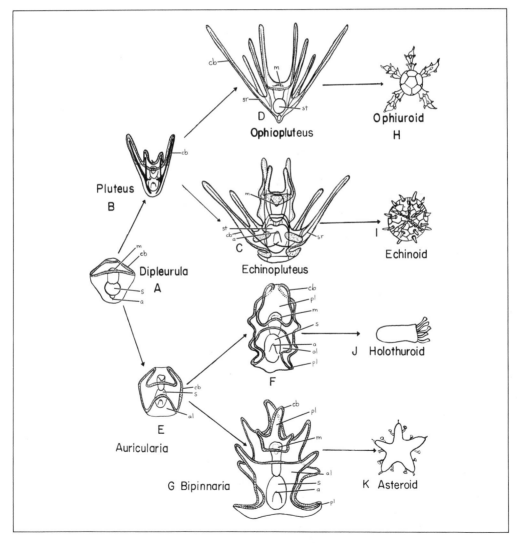

FIGURE 12.2 Echinoderm embryology (after Fell, 1948). Symbols: *m* – mouth, *cb* – ciliated band, *pol* – preoral lobe, *pll* – postlateral lobe, *sr* – skeletal rod, *st* – stomach, *a* – anus, *al* – anal lobe. F, G – *pl*, upper = *pol*; *pl*, lower = *pll*.

tive vertebrates and primitive echinoderms. Thus Romer (1967), Simpson et al. (1957), among others, envision a common ancestral stock for both pterobranchs (hemichordates) and primitive echinoderms. The common ancestor was a primitive sessile arm-feeder (Romer, 1967, Figure 1; compare Barrington, 1965, p. 9, Figure 7).

LIVING AND FOSSIL ECHINODERMS

Classification

[After Hyman, 1955; Fell and Pawson, 1966.]

Phylum — Echinodermata. (Includes holothurians or sea cucumbers, echinoids or sea urchins, asteroids or sea stars, crinoids or sea lilies, and a variety of extinct forms such as blastoids and cystoids.)

1. Noncolonial coelomates.
2. Characterized by pentaradiate — radial or meridional symmetry, which is secondary. That is, this symmetry is superimposed on a bilateral symmetry (found in some fossils and in larval forms of living echinoderms) (Figure 12.3*C*).
3. Definite head or brain lacking.

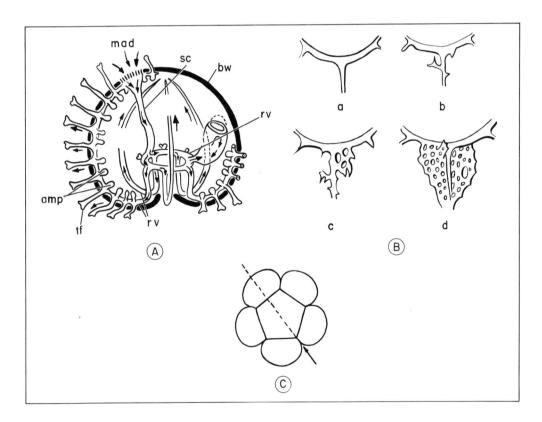

FIGURE 12.3 Echinoderms—diagnostic characteristics.
(A) Water vascular system; arrow shows direction of water flow taken in from seawater at madreporite (*mad; bw* — body wall; *sc* — stone canal, *amp* — ampulla, *tf* — tube feet, *rv* — radial vessel, *rv'* — ring vessel (basal "rv", = *rv'*, schematic). After Fell and Pawson, 1966.
(B)(a–d) Endoskeleton. Fenestrate, two-dimensional framework of calcite in developing genital plates of echinoid, *Anthocidaris crassispina* (after Onoda 1931; compare Raup, 1967).
(C) Pentamerous symmetry of enchinoderm skeleton; aboral view, schematic (after Nichols, 1962); with central plate (bearing anal opening usually) and ring of five plates (crinoid basal plates or genital plate in asteroids, ophiuroids, and echinoids. Note lack of any opposing sutures—a condition of least weakness for the skeleton. [Onoda reference p. 699].

4. Water vascular system. Echinoderms have an internal closed system of reservoirs and ducts that bear watery fluids. Part of the coelom (that is, the hydrocoel) becomes the water-bearing system of vessels. Projections of these vessels (= podia or tube feet) occur in a pentamerous arrangement where radial symmetry prevails, that is, a fivefold equispaced arrangement of ambs about the mouth. The internal system communicates with the external seawater environment through pores or poreclusters (= madreporite) at one or another stage.

[Tube feet serve food-gathering, locomotor, and sensory functions, as well as respiratory and burrow-building. Nichols (1962) remarked that their original function was probably respiratory. Protraction and retraction of tube feet is achieved by modification of the hydraulic pressure of the fluids they contain. Such control is essentially muscular (Nichols, 1966).]

5. Endoskeleton. This structure is composed of separate plates, spicules (holothurians), or pieces of calcium carbonate (Figure 12.3*B*). Each plate responds optically as a single calcite crystal.

[Several investigators (Donnay, 1956; Garrido and Blanco, 1947; Nissen, 1963, cit. Raup, 1967) indicate that *polycrystallinity* is involved; each skeletal element may be composed of numerous submicroscopic fibers (crystallites) occurring as a bundle, and all with nearly identical crystallographic orientations (Raup, 1967). Skeletal elements are generally spongy or fenestrate — with reticulations of multiple holes, and often developed from a single, triradiate crystal (see Chapter 15; Raup, 1967; Gordon, 1926, 1927, 1929).

[The skeletal tissue is embedded in the middle layer or dermis of the body wall.]

External spines or protuberances (warts or bosses), characterize many echinoderms.

6. Differentiation. The echinoderm body has a distinct oral and aboral surface. (Other animals with radiate symmetry have bodies similarly differentiated.)

7. Organ Systems. *Nervous* (primitive; a radiate pattern of gangliated nerve cords). *Excretory* and *respiratory* (generally wanting in any distinct form, although such functions are obviously performed). *Circulatory* (a blood lacunar system; no definite vessels). *Reproductive* [sexes separate but not distinguishable externally; sex cells often discharge into, or fertilized in, seawater; hermaphroditism known (holothurians, ophiuroids); occasional brooding of young; larva are bilaterally symmetrical].

8. Habitat. Exclusively marine.

9. Geologic range: Cambrian to Recent.

Echinoderm Embryology

Knowledge of larval development, larval types, and so on, among living echinoderms can shed some light on certain aspects of the rock record of this phylum.

Larval Development. There are four different modes of development among modern echinoderm larvae: internal (ophiuroids); external (asteroids, echinoids, and holothurians). Of the latter, echinoids are pelagic plankton feeders, and holothurians and asteroids, also pelagic, feed on their own yolks (Thorson, 1946).

Commonly, sperm are shed into seawater by the males and subsequently, as a result, females are induced to spawn. Fertilization, larval development (over a period of three to ten weeks), and metamorphosis all occur in seawater for such species (Hyman, 1955; Nichols, 1962; Thorson, 1946). [Eggs shed in a single spawning may number from a few million to tens or hundreds of millions in echinoids and asteroids (Thorson, 1946). Of course, smaller quantities are also known in modern ophiuroids.]

Role of Modern Echinoderm Larvae in the Plankton. In Danish waters (Øresund), Thorson (1946) found that all echinoderm larvae constituted only 1.9 percent of the plankton. Ophiuroid larvae dominated. The echinoderm larva thus play a negligible role in the food chain that begins in the plankton. An enormous waste of pelagic larvae occurs; great numbers were killed by brackish waters to which they had been carried from higher-salinity areas.

Larval Types. The basic echinoderm larval type, called the *dipleurula* (Figure 12.2), has a ciliated band forming a closed loop about the mouth. The main types of echinoderm larvae (*pluteus auricularia,* and *bipinnaria*—Figure 12.2) were long ago shown to arise from a dipleurula antecedent (Fell, 1948).

Given the essential anatomy of a dipleurula (Figure 12.2), a pluteus larval type will evolve *if* the ciliated band of the former can be extended along paired arms or processes. If the arms continue to develop and are strengthened (calcareous spicules develop internally), then dipleurula transforms into an echinopluteus or an ophiopluteus. [The spicules are subsequently discarded (Nichols, 1962) or partially resorbed (Gordon, 1926).]

The dipleurula may transform into a barrel-shaped configuration (Figure 12.2E). A looping of the ciliated band would follow giving rise to the *auricularia* larva of the holothurian, and the *bipinnaria* larva of the asteroid. (For larvae of crinoids, see Crinozoa.)

Free-living vs. Attached Tendencies. Up to 1965, it was common to recognize two subphyla in the phylum Echinodermata—the Eleutherozoa (sea stars, sea urchins, sea cucumbers, and so on), free-living forms, and the Pelmatozoa (sea lilies or feather star crinoids, and various extinct forms such as blastoids), forms that were more-or-less permanently attached to the bottom.

For two sound reasons, this approach to classification was found to be untenable (Fell, 1965, and Fell and Pawson, 1967): (1) The tendencies "free-living" or

"attached" arose independently and at different times in varied echinoderm groups. (2) Grouping under these subphyla united strange bedfellows (unrelated stocks). The fact that sea stars, sea urchins, and so on, lived unattached to a substrate merely signifies convergent evolution (that is, the same tendency arose in alien populations).

Major Determinants of Echinoderm Evolution. According to Fell (1967), there were two major pressures that determined the course of echinoderm evolution. These selective pressures were independent of each other even though they were exerted at the same time: (1) patterns of echinoderm body symmetry— that is, an internal requirement to conform to one or another of such patterns; (2) the advantage conferred by *modification, loss or gain of given body organs* (locomotor, feeding) under pressures imposed by a given habitat.

Patterns. During evolution of the echinoderm water vascular system, an arrangement of vessels, ducts, and so on (= *hydrocoel*), developed in certain ways. There were two major trends: hydrocoel growth along an elongate axis—*meridional* or longitudinal gradient (Figure 12.4, Upper Panel) or *radial* growth (Figure 12.4, Lower Panel). As seen in Figure 12.4, the meridional gradient characterizes echinoids and holothurians (class Echinozoa), whereas radially divergent growth characterizes crinoids (class Crinozoa) and asteroids (class Asterozoa).

These hydrocoel growth gradients exerted an internal pressure to assume one or another type of body symmetry, because other organs followed similar gradients. In the crinoids, not only does the water vascular system continue into the pinnulated arms (Figure 12.4, lower panel), but so do other organ systems— nervous and reproductive. In addition, the ambulacral grooves also extend into the arm. Thus the organs follow the radial gradient established by the hydrocoel.

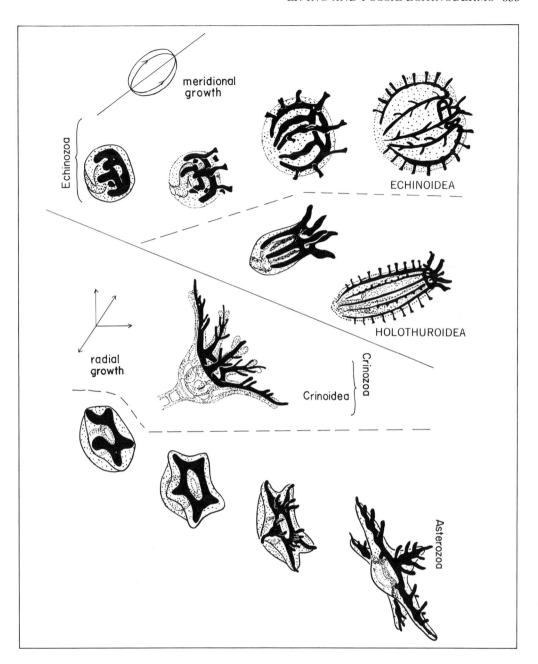

FIGURE 12.4 Hydrocoel growth gradients. Class Echinozoa (echinoids, holothuroids), upper panel. Left-hand corner indicates *kind* of gradient—growth along meridional (longitudinal) lines. Lower panel: class Crinozoa and class Asterozoa. Left-hand corner indicates kind of growth gradient—growth along radially divergent lines. Black portion of figure = hydrocoel or water vascular system; to left, early postmetamorphic stages; to right, subsequent growth stages. (See text discussion for role in echinoderm evolution.)
Illustrations redrawn after Fell, 1963.

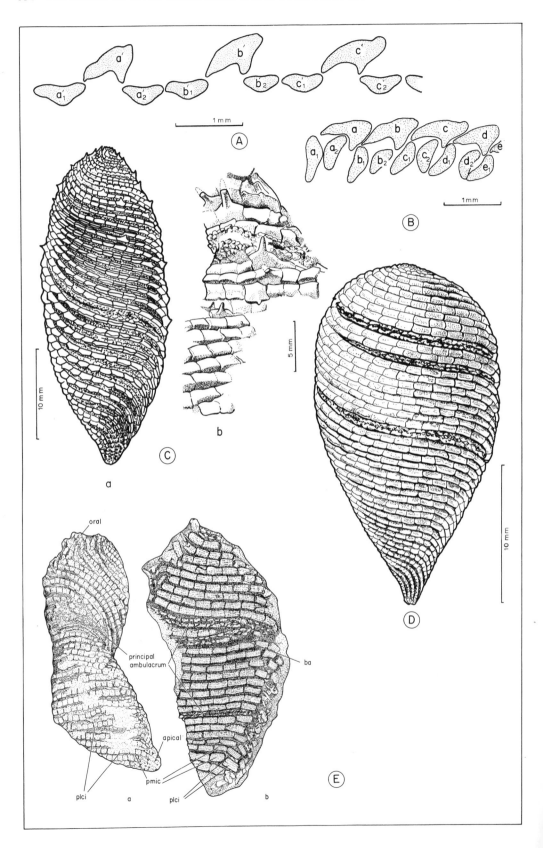

FIGURE 12.5 Helicoplacoids.

(A, B, D, E_b) *Helicoplacus gilberti* (type species), L. Cambrian, Poleta Formation, Calif.: (A) Expanded plates of test; section parallel to oral-aboral axis, and normal to surface; (B) the same, but with retracted plates; (D) restored complete specimen in retracted state; E_b — incomplete flattened specimen, enlarged. Note fine longitudinal ribs (see text) in plates of medial interambulacral, ba — branch ambulacra; pmic and plci — plates of medial interambulacral columns and plates of lateral columns of interambulacrals.

(C)(E) *Helicoplacus curtisi:* (C_a) Restored in semiexpanded state (fusiform) in (C) and (D) oral pole is up; aboral end points downward; (C_b) incomplete specimen (holotype) with ambulacral groove and spines rising from adjacent plates. E_a — *Helicoplacus* sp., flattened nearly complete specimen and primary ambulacra, as well as opposite poles (oral; apical).

Illustrations redrawn after Durham and Caster, 1963, 1966a.

Body Organs. Specialization of the tube feet for burrowing, feeding, etc. that were originally respiratory in function, denotes selection at work (Nichols, 1962, 1966).

Symmetry, Gradients, Classification

Skeletal material of both living and fossil echinoderms reflect a given pattern of body symmetry; these point to preferred growth gradients followed by the hydrocoel — that is, meridional or radial.[1] Fell accordingly proposed four new subphyla: Echinozoa, Homalozoa, Crinozoa, Asterozoa. A fifth subphylum, Blastozoa, was erected by Sprinkle (1973) for four classes formerly assigned to Crinozoa.

Subphylum 1. Echinozoa. Armless, essentially globoid echinoderms. Hydrocoel growth gradient, meridional (Figure 12.4 upper panel). Embraces six classes, only two of which are extant (Echinoidea, Holothuroidea) (Figure 12.1A, part E).

Subphylum 2. Homalozoa. Cambrian to Devonian echinoderms. This extinct group lacked radial symmetry.

Subphylum 3. Crinozoa. Globoid, placoid echinoderms with arms and having partial meridional symmetry. In the more evolved classes (Crinoidea), ambulacra develop radially divergent systems. Two major classes belong to this subphylum and, of these, only one is extant — the Crinoidea (Figure 12.1A, part B). The other is the Paracrinoidea (Sprinkle, 1973).

[1]These, were thought, to be "a more reliable guide to affinity."

Subphylum 4. Asterozoa. Echinoderms that are free-living, radially symmetrical, with star-shaped body lacking a stem. One class, the Stelleroidea, is divided into two extant subclasses: Somasteroidea and Asteroidea (Figure 12.1A, parts C, D).

Subphylum 5. Blastozoa (Sprinkle, 1973). Echinoderms with characteristic biserial brachioles, commonly mounted on recumbent ambulacra; pore or fold-like respiratory structures; globular multiplated calyx, stem or holdfast. Geologic range: Lower Cambrian–Upper Permian. (Includes Eocrinoidea, Parablastoidea, Rhombifera, Blastoidea).

FOSSIL ECHINODERMS

Subphylum Echinozoa

The subphylum embraces the following extinct classes: Helicoplacoidea, Edrioasteroidea, Ophiocistoidea, and Cyclocystoidea. In addition to the living Holothuroidea and Echinoidea, the fossil record of these classes extends from the Paleozoic.

The Helicoplacoids. Class Helicoplacoidea [*helix* = spiral + *plakos* = flat plate]. Free-living helicoplacoid echinoderms with spirally pleated, expansible, and flexible test (Figure 12.5, A), and primary ambulacrum having a single branch (Figure 12.5E, a). Lower Cambrian (*Olenellus* zone). [*Helicoplacus* (Figure 12.5), *Waucobella* (Figure 12.5A).]

Recoverable Information

Skeletal Morphology. The individual calcitic rectangular and rhombic plates comprising the test are arranged in a spiral fashion (Figure 12.5E).

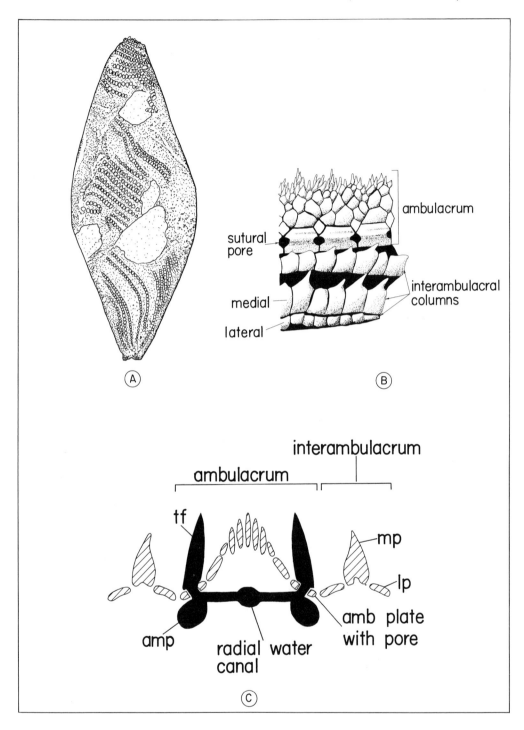

FIGURE 12.5A Helioplacoids. *Waucobella nelsoni*, Poleta Formation, L. Cambrian, Calif., (*a*)–(*c*): (*a*) *Holotype X 1.6*; (*b*) ambulacrum, lateral view, enlarged, foreground, (*c*) cross-section schematic, expanded state, enlarged; *tf* — respiratory type foot; *amp* — inferred ampulla of water vascular system; *mp* — plate of medial interambulacral column; *Lp* — plate of lateral interambulacral column with sutural pore.
Illustrations redrawn after Durham, 1967.

Two major types of plate systems are present: those defining the primary ambulacral groove (composed of four or more rows of platelets), the secondary ambulacra as well, and the interambulacral areas (where the platelets are larger than those of the ambulacrum). [Added data on the plates were published by Durham (1967), for example, Subclass Polyplacida —a mosaic of small plates (interambulacral?) organized into expansible pleats.]

The primary food groove (ambulacrum) runs from the oral end (mouth) for several complete spiral turns (Figure 12.5E), and sometimes bifurcates (that is, branching ambulacra appear— see Figure 12.5E, a). The interambulacral areas generally display a series of pole-to-pole platelets. That is to say, there are three columns of such plates spread from the apical to the oral end (Figure 12.5E, a). It was the movement of the external central column relative to the internal column (of the interambulacral area) that retracted or expanded the test (Figure 12.5, A, B). Durham and Caster (1963) remarked that this mode of expansion was unique among known echinoderms.

The mouth, while not seen, is inferred at the oral end. It is estimated to have been minute, about 1 mm in diameter. An anal orifice is also inferred to have terminated the tapered apical end.

Inferred Biology. When completely expanded (Figure 12.5C), the helicoplacoid body increased its volume by a factor of 2. This is thought to have been the *usual* state.

A muscle system of some type appears to be indicated by the capacity to vary body configuration (Durham and Caster, 1963).

Tube Feet. Fell (1966) interpreted the ambulacrum as being in a primitive state. From this it followed that if tube feet were developed, they could have served a restricted function—respiratory or sensory, for example. Since then, Durham (1967) described a new species (Figure 12.5A, part C) and reconstructed its tube feet. Their presence is clearly

indicated by sutural pores (Figure 12.5A, part B).

Nutrition. A lack of any special food-gathering organs suggests a cilia-feeding system that could direct mouthward organic particles along the spiral intergrooves of the interambulacral areas. The comparatively small mouth size indicates small particle feeders.

Fell (1966) surmised that helicoplacoids were detritus feeders, swallowing large quantities of mud from which organic particles were retrieved. The mode of locomotion discussed below led to this suggestion.

Locomotion. An accordion-like progression over or through the bottom muds is indicated by the plate arrangement (Figure 12.5).

Skeletal Growth. Plates originated at the apical pole and secretion of biocrystals making up subsequent plates brought the older ones into a more oral position.

If, in the reconstruction shown in Figure 12.5, one were to separate tapered apical end from the rest of the animal, this would represent the immediate postlarval plates that were possibly secreted by an attached larva.

Free-living vs. Attached States. Conceivably, in the early postlarval stages, the creature was attached. However, when maturity was attained and it was able to move accordion fashion and, in an expanded state, to increase its volume, it became free-living. It is also possible that the retracted state corresponded to a stationary position on the sea floor, while an expanded state reflected movement through the water (Durham and Caster, 1966). Furthermore, it may have been attached by the apical end even in adulthood at certain times.

Symmetry. A twofold symmetry occurs in fossil helicoplacoids. There is an original radial symmetry of the interambulacrum that is subsequently made spiral by torsion. In addition, there is the bilateral symmetry of the single primary ambulacra that is "imposed" on the original radial pattern of the interambulacrum (Durham and Caster, 1966).

Speciation. Two species of *Helicoplacus* have been recognized: *H. gilberti* and *H. curtisi*. Exclusive of size differences, the two species are similar yet differ chiefly in two features: (1) *longitudinal ribs* on plates of the medial interambulacral column — present in *H. gilberti* (Figure 12.5E, b), lacking in *H. curtisi*; (2) *medial spines* — present in *H. curtisi* (Figure 12.5C), lacking in *H. gilberti* (Figure 12.5E).

The 31 specimens and multiple dissociated plates on which the two species were based occurred on the same bedding plane. In life, therefore, it can be assumed that the two species coinhabited the same living space on the clay mud sea floor. Obviously, speciation must have been preceded by geographical isolation — all within the one-square-mile area or near to, yet outside of, it. Subsequent genetic isolation (isolated gene pools) would have followed.

The immediate parent population had all the characteristics shared by both species to which it gave rise, including structures and mechanisms to control variations in skeletal volume. The code governing such features was transmitted unchanged. However, the gene loci specifying interambulacral medial plate structure, for example, were modified. This modification, in turn, accounts for the presence of the fine longitudinal ribs, and so on, in one species population and not the other.

The matter of spinosity is of interest. Presumably, it had some adaptive value. The positional relationship of spines on the test is one of the main differences when one compares the two species. Besides medial spines, *H. curtisi* has spines which are expressed near the apical pole as incipient spines. Spines are also present around the oral pole (Figure 12.5C, b). By contrast, lacking medial spines, *H. gilberti* displays spines of every fourth to ninth plate approaching the oral pole.

In mud ingestors adoral spinosity could

have served a sanitary function. Thus it could have prevented cluttering of food grooves with mud. It may also have served a mud-loosening function during the concertina-like locomotion through the unconsolidated substrate. Such functions and others not discussed would have given the possessors of adoral spinosity a selective advantage.

The fine longitudinal ribs on medial plates of *H. gilberti* may have been related to feeding. A ciliated epidermis presumably directed food particles to the ambulacral grooves *over* the plates and grooves of the interambulacral area. Transfer could have been facilitated by the fine ribs.

Another observation that can be made on helicoplacoid speciation concerns biocrystals. Figure 12.5C, b shows how the spines are attenuated extensions of the medial and other plates. The genetic code specification for such spinosity was thus regional — that is, additional secretion of calcite biocrystals at restricted plate sites and in a controlled direction. Spines were thus projections of the plate. It is likely that equivalent but slightly different genetic specifications could account for spinosity of both species. In the examples cited, spine position and size, as well as irregular plate configurations, must have been latent in the precursor population.

The above discussion can be amplified by newer data provided in Durham's 1967 paper. In it he described two new helicoplacoid species and two new helicoplacoid genera — *Waucobella* and *Polyplacus*. The two new helicoplacoid species are *H. everndeni* and *H. firbyi*. In the first of these, the ambulacra is as in *H. gilberti* but differs from *H. curtisi* by its broad, low spine (on every second to fifth medial interambulacral plate) and characteristics such as length of its medial interambulacral plate. *H. firbyi* is distinguished by its greater number of medial interambulacral plates per half-volution (35 compared to 15 for *H. gilberti*). The last-named species is also distinguished from *Waucobella nelsoni* by its more inflated shape. *H. firbyi* also lacks spines.

Waucobella nelsoni has a well-developed spine on every other medial interambulacral plate (Figure 12.5*A*, part *B*).

The new subclass of which *Polyplacus kilmeri* is the only representative, is characterized by helical organization and a mosaic of small, interambulacral (?) plates arranged in pleats that could expand. The material in this instance is incomplete and the ambulacra is unknown. These forms are so unlike the helicoplacoids, while retaining a broadly similar organization, that their position is as yet undetermined.

These new data help somewhat to clarify the genetic interpretation. All four species of *Helicoplacus* and the new genus *Waucobella* shared a common cluster of class characters (flexible, helical organization, plate systems, and so on). When variation occurred, it was not only in spinosity (size, position, shape, as well as presence or absence of spines) but in the plates (number, dimensions, and so on) and skeletal configuration as well.

The lack of spines in *H. firbyi* suggest that the biocrystallization that went to form spines under different genetic coding was switched to produce extra medial plates due to some modification of one or more gene loci. The new data show that *H. gilberti* constitutes one-third of all the fossil helicoplacoids in number of specimens and that *H. curtisi* (as redefined by Durham, 1967) is comparatively rare. Second in abundance is *H. everndeni* at the same California locality as *H. gilberti*. Conceivably, fractionation of the *gilberti* population gave rise to the other species (*H. firbyi*) that occurs in the same interval and area as *H. gilberti*.

Paleoecology

Biotic Associates. Helicoplacoids have been found fossilized in California localities with various trilobites (*Nevadella gracile*, for example), some archeocyathids, an eocrinoid, *Eocystites*, and in

the Nevada section, an edrioasteroid like *Stromatocystites* (see subsequent description, this section), as well as some invertebrate brachiopods (Durham and Caster, 1963).

Substrate and Biostratigraphy. The California forms occurred in a shale with graded bedding, which was intercalated in a thick sequence of archeocyathid-bearing limestones. (See Chapter 3 on archeocyathid paleoecology.)

Horizontal distribution of the helicoplacoids in the White-Inyo Mountain region covered an area of one square mile—now represented at three different localities. Vertically, they occur in the *Olenellus* zone of California in beds about 2500 ft above the base of a thick section (6700 ft, total thickness).

Evolutionary Relationship.
The ancestral stock that gave rise to the helicoplacoids is thought to have existed in pre-Cambrian seas, to have been free-living (at least partially so) and to have had a skeleton lacking torsion.

Furthermore, it is thought that the pre-Cambrian, pretorsion ancestor of the helicoplacoids was so generalized that it differentiated (by small genetic steps, of course) to give rise to edrioasteroids, holothurians, and possibly some early echinoids such as *Aulechinus* (Fell, in Fell and Pawson, 1966, Figure 1–17*b*; Durham and Caster, 1966). That is, this ancestor presumably gave rise to all Echinozoan classes. Closer inspection of these affinities will be undertaken in the discussion of the several classes.

There are no known direct descendants of the helicoplacoids—that is, subsequent forms that could control skeletal volume and locomotion by expansion and contraction. Presumably the genetic code for this unique feature came into existence with the class.

Originally, the particular set of genetic specifications involved probably conferred a selective advantage; in the long run, it was inadaptive.

REFERENCES

General[2]

Alender, C. B., and F. E. Russell, 1966. Pharmacology, in R. A. Boolootian, ed. Physiology of Echinodermata (= PE). pp. 529–540.

Barrington, E. T. W., 1965. The Biology of the Hemichordata and Protochordata: Univ. Reviews of Biology. W. H. Freeman, 167 pp.

Bruun, A. F., 1956. Animal Life of the Deep Sea Bottom, in The Galathea Deep Sea Expedition, 1950–1952. pp. 149–195.

Fell, H. B., 1948. Echinoderm embryology and the origin of chordates: Biol. Rev., Vol. 23, pp. 81–107.

————, 1963. The evolution of the echinoderms: Ann. Rpt. Smithsonian Instit., 1962, pp. 457–490.

————, 1967. Echinoderm ontogeny, in R. C. Moore, ed. Treat. Invert. Paleont., Part S, Echinodermata 1, Vol. 1, pp. S60–S85.

———— and D. L. Pawson, 1966. General Biology of Echinoderms. PE, pp. 1–48.

Hyman, L. H., 1955. The Invertebrates: Echinodermata, The Coelomate Bilateria. McGraw-Hill. Vol. 4, 705 pp.

Jägersten, Gösta, 1960. Life in the Sea. Basic Books, N.Y. 184 pp.

Marr, J. W. S., 1963. Unstalked crinoids of the Antarctic Continental Shelf. Notes on their natural history and distribution: Philos. Trans. (B), Vol. 246, pp. 327–379.

Millott, Norman, 1966. Coordination of spine movements in echinoids, PE, pp. 465–485.

————, 1966a. Light production, PE, pp. 487–501.

Nichols, David, 1962. Echinoderms. Hutchinson Univ. Library, London, 180 pp.

————, 1966. Functional morphology of the water-vascular system, PE, pp. 219–244.

Raup, D. M., 1966. The endoskeleton, PE, pp. 379–395.

————, 1968. Theoretical morphology of echinoid growth: J. Paleontology, Vol. 42, Part II of II. Paleont. Soc. Mem. 2, pp. 50–63.

Reese, E. S., 1966. The complex behaviour of echinoderms, PE, pp. 157–218.

Romer, A. S., 1967. Major steps in vertebrate evolution: Science, Vol. 158 (3809), pp. 1629–1637.

Swan, E. F., 1966. Growth, autotomy, and regeneration, PE, pp. 397–434.

Thorson, Gunner, 1946. Reproduction and larval development of Danish marine bottom invertebrates, etc. Meddelelser fra Komm. for Danmarks Fisheri-og-Havundersøgelser. Serie: Plankton, Vol. 4 (1), 484 pp.

Vevers, H. G., 1966. Pigmentation, PE, pp. 267–275.

Helicoplacoids

Durham, J. W., 1967. Notes on the Helicoplacoidea and early echinoderms: J. Paleontology, Vol. 41 (1), pp. 97–102.

———— and K. E. Caster, 1963. Helicoplacoidea: A new class of echinoderms: Science, Vol. 140 (3568), pp. 820–822.

————, 1966. Helicoplacoids, in R. C. Moore, ed. Treat. Invert. Paleont. (= TIP), U, Echinodermata 3, pp. U131–U136.

Fell, H. B., and R. C. Moore, 1966. General features and relationships of echinozoans, TIP, pp. U108–U118.

———— and D. L. Pawson, PE, pp. 1–48.

Raup, D. M., 1966. PE, pp. 379–395.

---

[2]Sprinkle, J., 1973. Morphology and evolution of Blastozoan and Echinoderms. M.C.Z. (Harvard).

EDRIOASTEROIDS

Class Edrioasteroidea

(After Regnéll, 1966; Fell and Pawson, 1966, Bell, 1975; Bell and Sprinkle, 1978). Extinct echinozoans composed of many plates, with ambulacral system confined to the upper hemisphere of the test, and composed of five ambulacra (= pentamerous symmetry). Ambulacra radiate from the mouth and follow partial meridional courses on the test (Figure 12.4, Echinozoa) but may curve clockwise and/or counterclockwise thereafter. They achieve length and area increase by spiral torsion on a given plane (Figure 12.6C).

Both mouth and anus are on the upper surface of the test. Arms, brachioles, and thecal pores lacking; hydropores are present as openings or intake of the water-vascular system. Geologic range: Lower Cambrian–Lower Carboniferous (Mississippian) (Kesling, 1960; Ehlers and Kesling, 1958).

Two orders, two suborders, and five families are recognized by Bell (1975) and one new one by Bell and Sprinkle (1978): Order Isophorida [family Lebetodiscina; family Carneyellidae (*Anglidiscus*, Figure 12.6D; *Timeischytes*, Figure 12.7C; family Isorophoridae; family Agelacrinitidae (*Lepidodiscus*, Figure 12.7D)]. Order Edrioasteroidea. [Family Edrioasteridae (*Edrioaster*, Figure 12.6A–C)]; family Totiglobidae (*Totiglobus*) with a fully plated aboral surface – a unique and probably "a primitive trait."

INFORMATION DERIVED FROM EDRIOASTEROIDS

Inferred Biology–Organs and Other Anatomical Systems

Digestive. Tube feet movement in seawater captured organic particles on mucoid surfaces. These were subsequently swept along the ambulacral grooves by ciliary currents to the mouth (Fell and Pawson, 1966). (There is a fivefold ambulacral system that radiates from the mouth to the peristomal region.) The food particles enter the gullet that leads to the stomach. [Internally, there are a series of plates called "periradials" and "interradials" – five each that constitute the internal peristomal skeleton (compare Figure 12.6A). These plates formed a mouth frame through which the gullet passed into the stomach. The whole frame is called the "substomal chamber" (Figure 12.6F) (Regnéll, 1966; Foerste, 1914).]

Excretory–Anus. Periproct, often covered by a pyramid of plates that served a valve function.

Water Vascular System and Related Apparatus. In ambulacral grooves, each amb (plate) held along its midline an external water vessel (= tube foot). When inflated, tube feet extended from the interambulacral podial pores (Figures 12.6B, 12.6D). (See also Figure 12.3A.)

Hydropore (gonopore?). Often referred to ambiguously as the "third aperture," this is an external pore or slit (Figure 12.6A), presumably for water intake (Kesling and Ehlers, 1958). Apparently it was connected to the internal water vascular system. It may have also served as an exit for germ cells. Accordingly, the name "gonopore?" is added above.

Respiratory. Some degree of anal respiration is thought to be indicated by the anal pyramid. The latter presumably was a valvular control that was related to opening and closing of the anal or cloacal opening (Regnéll, 1966; Williams, 1918, pp. 76–78). [In living sea cucumbers (*Holothuria tubulosa*,) when the posterior end was capped, respiration was reduced more than 50 percent of normal (Farmanfarmaian, 1966).]

Muscular. If a cloacal pump existed, as discussed above, then an oval sphincter muscle and cloacal wall muscles governing relaxation and contraction can be postulated. Similarly, associated with a flexible theca, in some degree was some mode of muscular control.

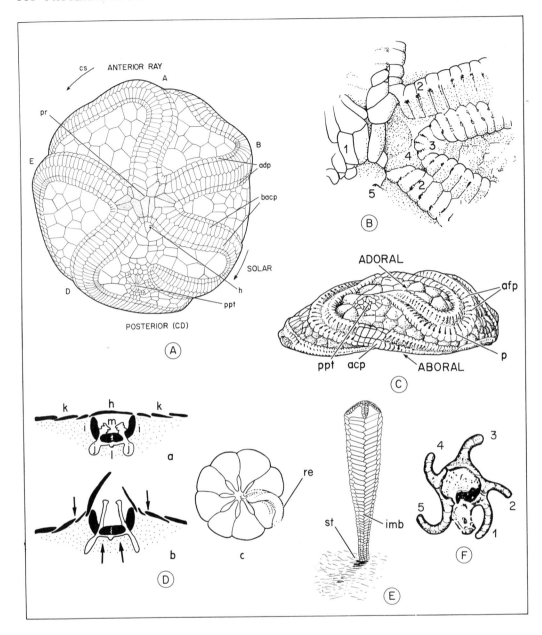

FIGURE 12.6 Edrioasteroid morphology.

(A) *Edrioaster bigsbyi* (M. Ord.). Adoral surface. Letters *A–E* (Carpenter's system of labeling): *A* = *A*-ray or anterior ray defined as being located opposite *CD* — interray (posterior). Note: periproct occurs between *C* and *D*. Periproct (*ppt*) designates limits of anus; peristomal region (*pr*) of mouth. Curvature of ambulacra either solar (clockwise) or *CS* (counterclockwise); *adp* — adambulacral plates or floor plates (= *afp* in C).

Space between ambulacra *A* and *B*, *B* and *C*, and so on, covered with interambulacral plates. Hydropore (*h*) often called "third aperture" (see text); *bacp* — biserial ambulacral cover plates.

(B) Same species as (A), with portion of floor plates, and so on, of *B*- and *C*-ambulacra on the right. On the left, (1) ambulacral cover plates, compare (A): (2) ambulacral floor plates underlie ambulacral cover plates which have been removed; (3)(4) peristomal elements, compare (A): (5) hydropore, see (A). Note pores in (2). These floor plate pores enter into the family description of the Edrioasteridae. Pores (vertically, pore canals) were podial pores; that is, a tube foot protruded from it.

(C) Same species (A), posterior view, with ambulacra cover plates (*acp*), ambulacra floor plates (*afp*), as well as pores (*p*); adoral (arrow) and aboral surfaces. [Figures (A)–(C) schematic.]

(D) *Anglidiscus fistulosus* (L. Carb., England), hypothetical transverse section of ambulacral groove; *h* — ambulacral cover plate, *i* — ambulacral floor plate, *j* — ampulla, *k* — interambulacral plate, *l* — radial canal (water vascular system), *m* — tube foot. (a) Animal deflated, groove closed. (b) Inflated, groove open. Arrows indicate pressure exerted on interambulacral plate (*k*) by inflation of saclike test, thus allowing swelling of tube foot; (*c*) Inner side of anal pyramid. *re* – possible rectum, impression; enlarged.

(*E*) *Pyrgocystis* (Ord.-Sil.), schematic; stemmed type; *st* – stem, *imb* – imbricating plates of stem (see Rievers, 1961, plate 2, Figure 1).

(*F*) *Streptaster septembrachiatus.* Central view of oral face seen from below; substomial cavity (chamber).

Illustrations redrawn and adapted from: (*A*)–(*C*) Bather, 1915; (*D*)–(*F*) Anderson, 1939; (*E*) Nichols, 1962.

Biomineralogy. Some older members of the class (family Stromatocystitidae) had a flexible test. This is attributed to (*a*) weak calcification of the skeleton, that is, thin or minute calcitic plates, and (*b*) strands of mesenchymal tissue attached to nonimbricate plates (Figure 12.7*A*, *b*) that allowed limited expansion of the test.

Skeletal Plates. The various plates of the edrioasteroid test fall into groups: ambulacral plates (floor and cover or roof plates, Figure 12.6*A*–*C*); interambulacral plates; peripheral plates (Figure 12.7*D*); peristomal plates; anal pyramid plates and aboral plates.

Some interambulacral plates retain tubercles with which, in life spines were articulated (Nichols, 1962). Among other possibilities, edrioasteroid skeletal plates could be minute or large, imbricate or nonimbricate (Figure 12.7*D*), fused or irregular polygonal or scalelike, biserial or uniserial. (If biserial, there were two rows of interlocking cover plates that met in a line; if uniserial, these were floor plates.)

Certain plates delineated the oral or mouth region (= peristomal plates). The anal opening was covered by plates of the anal pyramid. The plates of the ambulacra do not override or rest on, but are located adjacent to and between, the interambulacral thecal plates.

Population Density. In a portion of a 6-in. Carboniferous core, Anderson found 15 almost complete and 18 fragmentary tests of *Lepidodiscus fistulosis* (1939, plate 5). These forms were obviously attached directly to the clay mud substrate.

Size and Shape. The diameter of the adult theca varies between genera and species. Diameters range from 5.0 mm and less to 40 mm and 60 mm.

The original spheroidal test is commonly flattened or compressed into discoidal form. Forms with flexible tests tend to elongate along the vertical axis, that is, to be subconical. If they adhere to hard substrates aborally, they are turret-shaped (Figure 12.7*B*). Some forms had a nonpermanent attachment: A tough membrane acted as a disk and was presumably stretched over the concave aboral surface.

Biotic Associates, Substrates. A limited number of sessile forms attached directly to mud bottoms. In such cases, according to Foerste, aboral plates became obsolete and attachment was achieved by a fleshy surface. Mostly hard substrates were favored – bryozoans (Figure 12.7*B*, *b*), corals, sponges, and so on.

Specialists surmise that the presence of podial pairs (Figure 12.6*B*) imply tube feet projection during the inflation of the test and, further, that this signified a capacity for limited positional shifts while attached (Regnéll, 1966). In other cases, movement along the bottom may have been possible for some attached forms.

Generally, preferred substrates were in the littoral zone and the lithologies represented in the fossil record indicate calcareous, arenaceous, and micaceous mud bottoms.

Symbiotic Relationships. In some cases, attachment was to whatever exoskeleton of dead individuals was avail-

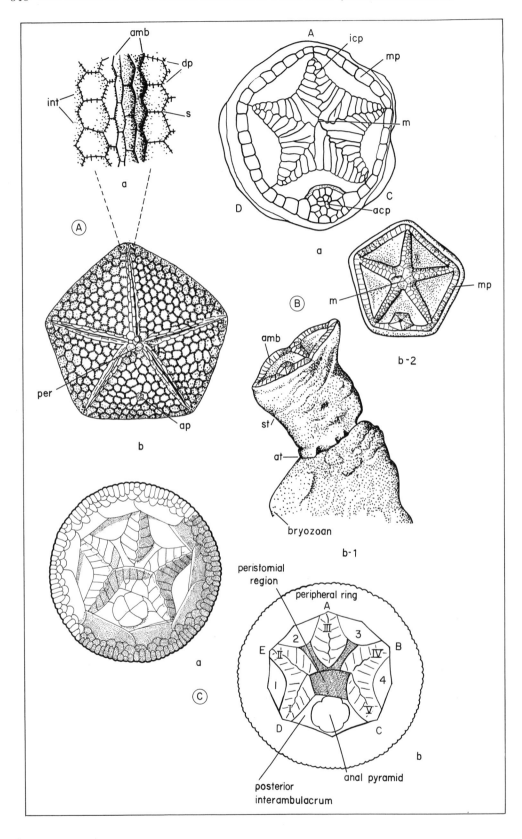

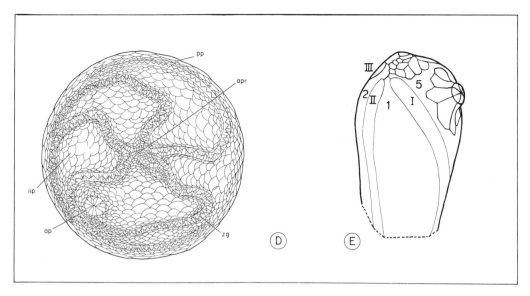

FIGURE 12.7 Representative genera of edrioasteroids.

*(A) *Stromatocystites pentangularis*, M. Cam., Czechoslovakia. (a) Detail, median section of theca with sutures (s) between interambulacral plates (*int*) and markings (*dp*) left by mesenchymal strands that connected such plates, thus conferring flexibility to test; *amb*—median ambulacral. (b) adoral view, reconstruction, thecal diameter, 35 mm.

*(B) *Cyathocystis*. (a) *C. oklahomae*, Lower Bromide, Oklahoma (Ord.). (A)–(C(D): Rays, labeled; *icp*—interlocking cover plates of ambulacra; *mp*—marginal plates; *m*—mouth, under five large plates; *acp*—anal covering plates; thecal width, 7.0 mm, height, 5.5 mm; (b) (1–2) *C. plautinae*, M. Ord., Esthonia: (1) Lateral view of theca attached to bryozoan (ca. 15.0 mm long); *amb*—ambulacra, *st*—saclike theca, *at*—attachment end (aboral), (2) adoral view, *m*—mouth; *mp*—marginal plates; schematic.

Family Carneyellidae (C) *Timischytes megapinacotus*, Four Mile Dam limestone, M. Dev. (a)–(b) Adoral view: (a) reconstruction, schematic; (b) labeled parts of theca; numbers I–V for ambulacra (Jaekels' system) shown with equivalent letters (see Figure 12.6A). Note the extended peristomal region and peripheral ring, both characteristic.

Family Agelacrinitidae. (D) *Lepidodiscus squamosus*, L. Miss. (Keokuk), Indiana, adoral view, reconstruction, maximum diameter of test, 1.70 in.; *iip*—imbricating interambulacral plates, *ap*—anal pyramid, *pp*—peripheral plates, *apr*—asymmetrical peristomal region, *zg*—zigzag (or serrate) central line where ambulacral plates meet. (For Family Edrioasteridae, see Figure 12.6A–C.)

*(E) *Lispidecodus plinthotus*. I–V, ambulacral area; 1–5, interambulacral area (reconstruction after R. V. Kesling, schematic). Locate peristomal plates and periproct plates. Note thecal shape.

Illustrations redrawn from Kesling et al., several papers; Strimple, 1955. *= families not recognized by Bell (1975).

able. That is, attachment was by chance. In other cases, when edrioasteroids attached themselves to living individuals, a form of commensalism may have existed. Gekker (1917) described such a relationship for *Cyathocystis plautinae*. Regnéll (1966) expressed the opinion that the general rule had not been to attach to dead animals.

Major Tendencies in Edrioasteroid Evolution

Beginning in the Lower Cambrian, edrioasteroids had flexible tests, dorsally attached to substrates, and simple ambulacra—uniserial floor plates. Later development led to forms with a stronger expression of torsion, as well as longer and more curved ambulacra, and buried floor plates in each ray (Fell and Pawson, 1966). Regnéll (1950, 1966) is of a different opinion. He envisioned progression from biserial to uniserial.

General ambulacral extension and enlargment appear to be related to the efficiency of gathering of food particles. An increased span of the ciliated food grooves would confer a selective advantage, since more such particles would be retrieved—along the radial plus the curved course of the ambulacra. [Bather (1915*b*), following Foerste, attributed the curvature of the ambulacral course to the pull of gravity on individuals resting on bottom slopes.]

Derivation of Edrioasteroids

Fell (1965, 1966) derived the edrioasteroids from lines of descent that radiated from a preancestral *Helicoplacus* stock. He also observed that tube feet of edrioasteroids were functionally similar to those of the Crinoidea.

REFERENCES

Edrioasteroids

Anderson, F. W., 1939. *Lepidodiscus fistulosis* sp. nov. from Lower Carboniferous rocks, Northumberland, Great Britain. Geol. Surv. Bull. 1, pp. 67–81, plate 5, text figures 1–10.

Bassler, R. S., 1935. The classification of the Edrioasteroidae. Smithsonian Misc. Coll., Vol. 93, Part 8, pp. 1–11, plate 1.

————, 1936. New species of American Edrioasteroidae. Same, Vol. 95, Part 6, pp. 1–33, plates 1–7.

Bather, F. A., 1915. Studies in Edrioasteroidea, I–IX. Reprinted with additions from the Geological Magazine. Published by the author at "Fabo" Marryat Road, Wimbledon, England.

Ehlers, G. M., and R. V. Kesling, 1958. *Timeischytes*, a new genus of hemicystitid edrioasteroid from the Middle Devonian Four Mile Dam limestone of Michigan: J. Paleontology, Vol. 32 (5), pp. 933–936, plate 121, 1 text figure.

Fisher, D. W., 1951. A new edrioasteroid from the Middle Ordovician of New York: J. Paleontology, Vol. 25 (5), pp. 691–699, 8 text figures.

Harker, P., 1953. A new edrioasteroid from the Carboniferous of Alberta: J. Paleontology, Vol. 27 (2), pp. 289–295, 3 text figures.

Kesling, R. V., and G. M. Ehlers, 1958. The edrioasteroid *Lepidodiscus squamosus* (Meek and Worthen): J. Paleontology, Vol. 32 (5), pp. 923–932, 1 text figure.

Kesling, R. V., 1967. Edrioasteroids with a unique shape from Mississippian strata of Alberta: J. Paleontology, Vol. 41, pp. 197–202, 2 text figures.

Regnéll, G., 1966. Edrioasteroids, in R. C. Moore, ed. Treat. Invert. Paleont., Part U, Echinodermata 3, Vol. 1, pp. U136–U173.

Rievers, J., 1961. Eine neue *Pyrgocystis* (Echinod. Edrioasteroidea) aus den Bundenbacher Dach schiefern (Devon.): Munich. Bayer. Staatssammlung Paleont. Hist. Geol. Mitteil., Vol. 1, pp. 9ff. and plate 2, figures 1–4.

Strimple, H. L., and D. A. Griffham, 1955. A new species of *Cyathoeystis*: J. Wash. Acad. Sci., Vol. 45 (11), pp. 353–355.

Williams, S. R., 1918. Concerning the structure of *Agelacrinites* and *Streptaster*, Edrioasteroidea of the Richmond and Maysville Divisions of the Ordovocian: Ohio J. Sci., Vol. 19, pp. 59–86, plates 1–9. (Consult Regnéll, 1966, for reassignments.)

Complete revision of edrioasteroid taxonomy in Bell, B. M., 1974. Study of North American Edrioasteroidea, New York State Mus. Memoir 21. Bell and Sprinkle, J., 1978. J. Paleontology, Vol. 52 (2), pp. 243–266.

OPHIOCISTIOIDS
[*Ophis* = snake; *kiste* = box]

Class Ophiocistioidea

(After Ubaghs, 1966; Fell and Pawson, 1966; Fedotov, 1926.) Free-living, extinct, pentaradiate echinozoans. Test with calcareous plates. Aboral surface with periproct (Figure 12.8*C*). Adoral side flat, with central mouth and buccal apparatus consisting of five interradial jaws. The apparatus is margined by a flexible peristomal membrane Figure 12.8*A*). Five triserial ambulacra spread from mouth, that is, each with three columns consisting, of medial periradials and two lateral adradials. Such ambulacrum supported multiple pairs of very large scale-covered tube feet (Figure 12.8*D*). Sieve plate (madroporite) and gonopores (Figure 12.8*B*) on the same interray. Geologic range: Lower Ordovician to Middle Silurian; (?) Middle Devonian.

Four families are recognized by Ubaghs (1966): Family Eucladiidae (*Eucladia*, Figure 12.8*B*), Family Sollasinidae (*Sollasinia*, Figure 12.8*A*), Family Volchoviidae (*Volchovia*, Figure 12.8*C*), and (?) Family Rhinosquamidae (not illustrated).

Information Derived from Fossil Ophiocistioids

Inferred Biology: Organs and Other Anatomical Systems

Digestive. Some of the tube feet may have served to grasp and push prey toward the mouth (Ubaghs, 1966). The buccal apparatus had five long plates (Figure 12.8*A*) that may have acted as rasping teeth. The entire apparatus possibly served a masticatory function. In *Volchovia* it resembles the Aristotle's lantern in sea urchins.

Excretory. Periproct—site of anus; in fossils, the anal vent is often preserved as a series of small plates arrayed in pyramidal fashion; its location on the aboral side helped to establish adequate sanitation.

Function of Tube Feet. Marked variation in the structure of the preserved podia in ophiocistioids denotes functional differences (Nichols, 1962; Ubaghs, 1966). Tube feet were covered by a series of scalelike imbricating, calcareous plates. Some of these were comparatively enormous, suggesting greater strength. Smaller tube feet are thought to have served a sensory function. Larger tube feet served a locomotor function as well as feeding accessories—directing food toward mouth (Fell and Pawson, 1966).

Respiratory. Nichols (1962) remarked that the calcareous armor on the tube feet was poorly suited for gaseous exchange, in contrast to the naked tube feet of other echinoderms (see Nichols, 1966). How then did these creatures respire? One fossil cast (*Sollasinia woodwardi*) indicated the existence of probable pores between imbricating plates of the tube feet. Nichols inferred that papillae probably extended along the length of the tube feet as in living crinoids and ophiuroids. If so, they would mark the precise site of uptake of oxygen from the water.

Skeletal. The skeleton of varisized, calcareous, polygonal plates was rigid. Adjacent plates were effectively soldered by calcareous cement (Fell and Pawson, 1966). Some fossils have plates on one side only, with a tough organic integument that is slightly calcified, on the other side (Ubaghs (1966). Others were entirely encased in a plated skeleton. In some ophiocistioid fossils, minute spines are found. These represented aboral tubercles on the skeleton.

Attachment Surface. Disklike forms similar to *Volchovia* apparently attached to hard substrates on the sea floor by their flexible ventral surfaces. The entire body, in such cases, acted as a sucking disk as in certain modern ophiuroids.

Water-Vascular System. Little is known of this system beyond the existence of a sieve plate for water intake (= madroporite) and the presence of tube feet, which had to be expanded or retracted by operation of this system.

Muscular System. Nothing is known of this system. Musculature may have been needed by the buccal apparatus as with

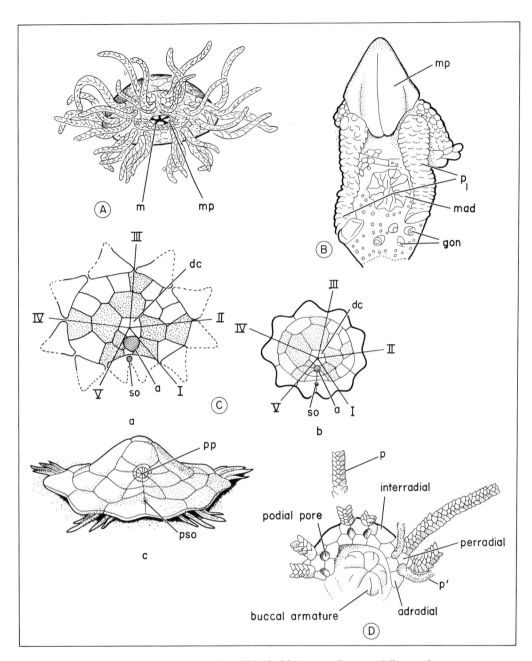

FIGURE 12.8 Ophiocistioidea: general morphological features and representative specimens.

(*A*) *Sollasina woodwardi,* U. Sil., England: *m* — mouth, *mp* — mouth plates (= dental plates of buccal apparatus). Note longitudinal rows of podia. Reconstruction after Nichols, 1962.

(*B*) *Eucladia johnsoni,* U. Sil., England, Part of *B*-ray and *BC* — interray; P_1 — first pair of podia; *mad* — madreporite; *gon* — gonopore. For letters of rays, see Edrioasteroids.

(*C*)(*a*) *Volchovia norwegica.* Ord., Oslo, Norway. Dorsal shield. I–IV, rays; *dc* — dorsocentral plate; *a* — anal opening (periproct); *so* — second opening [(gonopore/hydropore(?))]. Dashed lines, restored marginal plates; rest of plates shown as fossilized. (Long transverse axis along IV–II = 70 mm.)

(*b*) *Volchovia mobilis.* L. Ord., Leningrad region. U.S.S.R.; reconstructed after Gekker, 1938, Dorsal shield.

(*c*) Same as *C*(*b*). Reconstruction after Gekker; *pp* — periproct, *pso* = *so* = second opening.

(*D*) *Euthemon igerna* Sollas. U. Sil., England, X 3.5; *p* — podia, *p'* — buccal podia, buccal armature pentaradiate (see *A* above).

the more complex Aristotle's lantern of echinoids and in the movement of skeletal spines—if they were capable of movement in life.

Reproductive. Presence of probable genital pores has been inferred from the appearance in *Eucladia* of four prominent hollow papillae and of a perforated tubercle in *Sollasinia* (Fedotov, 1926; Ubaghs, 1966).

Shape and Size. Size varied from 7.0 mm to 90.0 mm in diameter. Shapes generally conformed to low domal configurations that became distorted by flattening and crushing during the fossilization process.

Distribution. Some fossil ophiocistioids are widely distributed. *Volchovia*, for example, occurs in the Ordovician shales of Norway (Oslo region) and limestones of the Leningrad region as well as in the Upper Ordovician of Ohio (Gekker, 1938; Regnell, 1949; Ubaghs, 1966). Others are known from incomplete material at a single locality, such as *Rhinosquama* from the Middle Devonian of Germany (Richter, 1930). Several fossils are known only from the English Silurian (*Eucladia*, and so on).

Affinities and Derivation. Structures and organization of ophiocistioids suggest homologies with ophiuroids (Fedotov, 1926; Regnéll, 1948; Nichols, 1962; Ubaghs, 1966). Ophiocistioids, however, are distinct in lack of arms, possession of an anal orifice and plates covering it, a single unpaired gonad with a gonopore on the same interradial as the madreporite (Figure 12.8*B*), and a distinctive buccal apparatus. One can thus rule out ophiuroids as being close, in an evolutionary sense, to the ophiocistioids. (In terms of gradients, Echinozoa, such as the ophiocistioids, lacked a radial gradient, possessed by the Asterozoa—a sufficient indication of evolutionary diversion.)

Fedotov (1926), Nichols (1962), among others envisioned a common ancestor for echinoids and ophiocistioids because both groups had several similar features (theca of plates; teeth or plates in buccal apparatus; and a peristome strengthened by spicules or small plates).

REFERENCES

Ophiocistioids

Fedotov, D. M., 1926. The plan and structure and systematic status of the Ophiocistia (Echinodermata): Zool. Soc. London Proc. pp. 1145–1157.

Regnéll, G., 1948. Echinoderms (Hydrophoridea, Ophiocistia) from the Ordovician Upper Skiddavian, etc.: Norsk Geol. Tidsskrift, Vol. 27, pp. 14–58, plates 1–2.

Richter, R., 1930. Schuppenröhren als Anzeiger von zwei im deutschen Devon neuen Echinodermen-Cruppen (Edrioasteroidea Billings und Ophiocistia Sollas?): Senckenbergiana, Vol. 2, pp. 279–304.

Ubaghs, G., 1953. Classe des Ophiocistioides (Ophiocistioidea), in Jean Piveteau, ed., Traité de Paleontologie, Vol. 3, pp. 843–856, text figures 1–7, Masson et Cie (Paris).

————, 1966. Ophiocistioids, in R. C. Moore, ed. Treat. Invert. Paleont., Part U—Echinodermata 3, Vol. 1, pp. U174–U188.

[Nichols, 1962, 1964, reference, p. 646].

CYCLOCYSTOIDS

Class Cyclocystoidea

(After Kesling, 1963, 1966; Fell, in Fell and Moore, 1966; Regnéll, 1945, 1960.) Small, discoid echinoderms that display the echinozoan pattern. [Fell (Fell and Moore, 1966) defined an echinozoan pattern as one with distinct radial meridional symmetry, and entirely lacking outspread extensions such as arms or brachioles (Crinozoa) or rays (Asterozoa).] Calcareous plated theca composed of oral and aboral disks both of which are attached to a framework of thecal plates (= submarginal ring). A flexible marginal ring consists of minute imbricating plates. Pentaradiate ambulacra bifurcate and develop a dendritic pattern away from the mouth, and join with perforations (ducts) through each proximal plate of the submarginal ring (Figure 12.9). The distal part of each such plate bore facets with which ducts were aligned. Geologic range: Middle Ordovician to Middle Devonian.

Only a single family and genus is recognized — Cyclocystoididae (*Cyclocystoides*, Figure 12.9).

INFORMATION DERIVED FROM FOSSIL CYCLOCYSTOIDS

(Inferred Biology: Organs and Other Anatomical Systems

Digestive and Excretory. All that is known of the digestive system is the occurrence of a presumed aperture (= mouth?) at the center of the oral disk and the presumed circumanal area of plates. Despite this limited evidence, the presence of an ambulacral system that becomes pentaradiate around the presumed mouth area, indicates the route taken by food. Brachioles have been suggested (indicated by facets presumed to be for their attachment) as a food-gathering apparatus.

Respiratory. There is no clear evidence pertaining to a respiratory function. Possibly some of the ornamentation of the oral surface of the submarginal plates (tubercles, papillae, punctae, and so on), particularly papillae so served.

Skeletal. The ring of submarginal plates is a most striking feature. Associated with this ring are various structures — oral and aboral disk, perforation of the plate (ducts), indication of the ambulacral system, and a relationship to such ducts, and the margining ring of imbricating plates.

All plates were calcitic. Some were thicker, some longer than others. Plates were apparently embedded in a tough organic integument (oral disk) that gave greater flexibility to the oral surface relative to the aboral.

Muscular System and Attachment. It is thought that the aboral surface attached temporarily to substrates by suction created by muscular contraction of the marginal ring (Kesling, 1966), and may have been free-living.

Reproductive. There is no indication of the presence of a gonopore.

Water Vascular System. Other than the concavities (beveled in the oral surface of the submarginal plates) that join together to form a continuous channel, and the ambulacral system already described (Figure 12.9B), there is no evidence of a connection between these two systems by means of a hydropore.

Nervous and Sensory. There are no data on these systems.

Stratigraphy, Fossil Record. The oldest cyclocystoid (*Cyclocystoides billingsi* Wilson) known is from the Middle Ordovician of North America (Ottawa, Canada). It shows the characteristic submarginal ring, and so on. In turn, this specialization in form (discoid) and structure (flexible plated theca), points to a considerable departure from a more generalized ancestral stock probably of pre-Cambrian age (see Fell, in Fell and Moore, 1966, Figure 96).

Cyclocystoids persisted in North America through the Lower Silurian. Already in Upper Ordovician times some forms migrated to England and subsequently to Scotland and Gotland. When they became apparently extinct in North America, most of Europe, and the British Isles, some forms persisted through Lower and

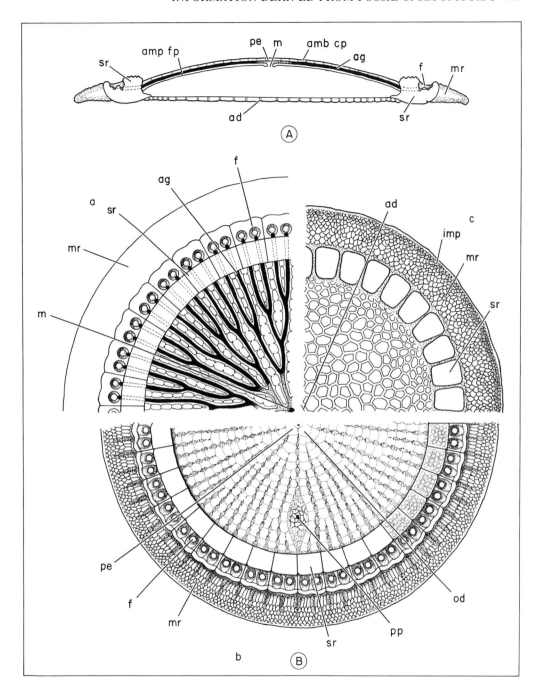

FIGURE 12.9 Actual and inferred morphology of the cyclocystoids. Hypothetical reconstruction adapted from Kesling, 1966.
(A)–(B) *Cyclocystoides halli*, M. Ord., (Trentonian), Canada: (A) Cross-section; body of animal contained in black area between disks. (B) Lower hemisphere, one-half of oral side; upper right quadrant, one quarter of aboral side; upper left quadrant, ambulacral cover plates (*amb cp*) removed to show ambulacral grooves (*ag*) in black; *od* — oral disk, *ad* — aboral disk. *mr* — marginal ring, *sr* — submarginal ring, *pp* — periproct, *m* — mouth, *pe* — peristomal plates, *imp* — imbricating plates, *f* — first ambulacral floor plates (*amb fp*).

Middle Devonian time in Belgium (Ardennes) and Germany (Westphalia) (Regnéll, 1948; Kesling, 1966, Figure 145).

This relict cyclocystoid species, *C. devonicus* Sieverts-Doreck, represented by molds and casts of the oral and aboral side, displays many of the detailed structural features shown in Figure 12.9. These include ambulacral system, submarginal ring, facets, ducts. Thus these and similar features apparently persisted from the oldest to the youngest known specimens— a span of some 45 million years, representative of a comparatively stable genetic configuration.

Origin and Affinities. These are different viewpoints on the probable origin of the cyclocystoids. Kesling (1966, Figure 144, text figures U201–203) favors derivation from a diplopore-bearing cystoid ancestor. He based his argument on the "exceptional similarities" to the cystoid *Tholocystis*. Resemblances included five-rayed branched ambulacra that merge into brachioles, brachiole facets borne on plates thicker than others in the test, and a comparatively large flexible base (see Begg, 1934, who also favored cystoid affinities).

Kesling further postulated a common Cambrian ancestor for the cystoid *Tholocystis* and *Cyclocystoides*, both of which made an apparent first appearance in the Middle Ordovician.

There are difficulties in Kesling's interpretation. No plates are known with diplopores; no objective evidence on the presence of brachioles (and hence none on brachiole facets); the remarkable stability in all anatomical structures, commented on in the previous section.

Kesling and many other investigators (Bather, Foerste, Piveteau) saw strong resemblances to the edrioasteroids, particularly in the marginal ring of imbricating plates (see Figure 12.9*B*). Since this factor played a role in the flexibility of the test and its capacity to fix to the substrate by suction, it is difficult to see why it should be less significant in evolutionary considerations than the other postulated structures. [Regnéll (1960) rejected classification under edrioasteroids and properly so, just as classification under the cystoids would not take into account all presently known factors.]

The approach of Fell (Fell and Moore, 1966) seems more suitable for the existing evidence, that is, both edrioasteroids and cyclocystoids derived from a common pre-Cambrian echinozoan ancestor.

REFERENCES

Cyclocystoids

Begg, J. L., 1934. On the genus *Cyclocystoides:* Geological Mag., Vol. 71 (839), pp. 220–224, plate 11, Figures 1–9.

Foerste, A. F., 1920. Racine and Cedarville cystids and blastoids with notes on other echinoderms: Ohio J. Sci., Vol. 21 (2), pp. 33–78, 4 plates, 4 text figures.

Kesling, R. V., 1963. Morphology and relationship of *Cyclocystoides:* Univ. Michigan, Mus. Paleontology Contrib., Vol. 18 (9), pp. 157–186, 5 text figures.

————, 1966. Cyclocystoids, in R. C. Moore, Ed., Treat. Invert. Paleont., (U) Echinodermata 3 (1), pp. U188–U210.

Regnéll, G., 1960. Intermediate forms in early Paleozoic echinoderms: Interntl. Geol. Congress, Rpt. 21st Session, Part 22, pp. 71–80.

Sieverts (Sieverts-Doreck), Hertha, 1951. Uber *Cyclocystoides* Salter und Billings und eine neue Art aus den belgischen und rheinischen Devon: Senckenbergiana, Vol. 33 (1/4), pp. 9–30, 2 plates, 3 text figures.

LIVING AND FOSSIL HOLOTHUROIDS.

Class Holothuroidea

The name of the class means "polyp-like." The popular name of the class "sea cucumber" derives from Pliny's *Cucumis marinus*, which referred to true holothuroids (Hyman, 1955). It is the oral-aboral elongate body ending in buccal tentacles that surficially resembles a polyp's body plan (Figure 12.10*A*, *B*).]

Diagnosis

Marine, generally benthonic, echinozoans that are mostly free-living. Cylindrical body with mouth encircled by tentacles at one end (oral end) and anus at the other (aboral); exceptions are forms like *Psolus* (Figure 12.10*E*). Radial symmetry indicated by disposition of ambulacra, nerve, and muscle systems, although, relative to the dorso-ventral plane, these are bilaterally symmetrical (Fell and Pawson, 1966). Reproduction commonly sexual (seasonal spawning) but asexual reproduction (fission) is known. Some species carry their young in the body cavity or special brood pouches (Deichmann, 1957). Hermaphroditism characterizes some species.

The skeleton is represented by five distinct elements: (1) Microscopic, calcareous sclerites embedded in the body wall (Figure 12.10*D*). (This is the most frequently fossilized holothuroid part and is the chief characteristic of the class.) These may form imbricating plates by coalescence. (2) Calcareous ring about pharynx (Figure 12.10*D*, *b*; *F*) composed of ten or more calcitic pieces—radials and interradials. (3) Anal plates (or teeth) occur in some forms. (4) Calcareous madreporite—internal and reduced. (5) In a single fossil order Arthrochirotida, there was an articulated axial skeleton in the tentacles.

Geologic range: (?) Ordovician, Lower Devonian to Recent.

Classification

Since sclerites are the most prominent holothuroid remains in the fossil record and the soft-part anatomy is chiefly used to categorize subclasses or orders of existing forms, problems arise in making accommodations to these two bodies of data. There are diverse opinions on the classification of fossil sclerites (Deflandre-Rigaud, 1953; Frizzell and Exline, 1955, 1966).

Pawson (1966) attempted to incorporate fossil material in his classification of subclasses and orders erected on the basis of living material. One can best grasp the difficulties by listing some major features used in the classification (see Figure 12.10): (for subclasses) presence or absence of tube feet, respiratory tree, and introvert (Figure 12.10*E*); shape of tentacles and number of digits in it; (for order) body shape, presence or absence of respiratory tree, and anal papillae, as well as skeletal deposits or test. On the level of species, sclerites characterize modern as well as fossil forms.

Despite limitations noted, each subclass and order can be briefly characterized solely in terms of sclerites, external tests, and so on, utilizing data in the literature. The families of Frizzell and Exline can be related to the classification that fits modern forms as well.

Subclass 1. Dendrochirotacea. All extant genera with plated skeletons are in this subclass. [Forms in this subclass are thought to display the basal plan from which other holothurians derived in post-Ordovician time (Fell and Pawson, 1966).] Two orders: Order Dendrochirotida. *Psolus*, Rec. (Figure 12.10*E*); *Calclamnella*, Eoc. (Figure 12.11*B*); Order Dactylochirotida, Recent forms only (Fell and Pawson, 1966).

Subclass 2. Aspidochirotacea. Test vestigial, original plates reduced to spicules (sclerites or ossicles) (Fell and Pawson, 1966; Pawson, 1966, Hyman, 1955). Two

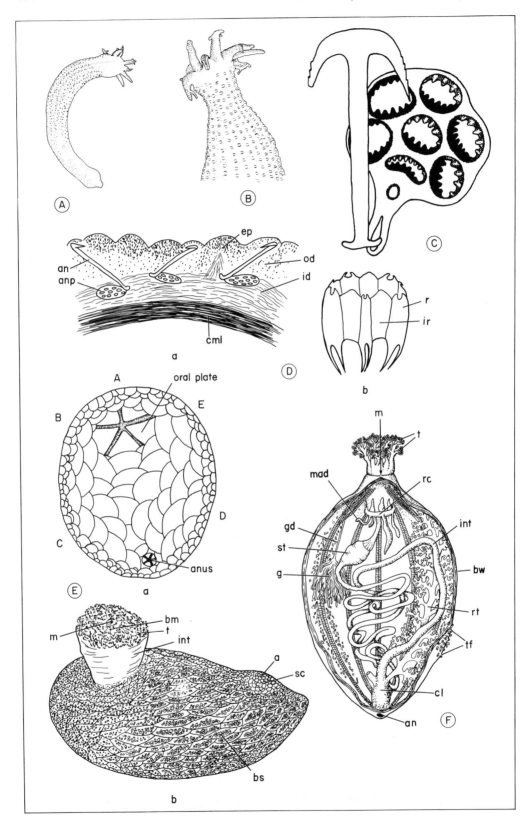

FIGURE 12.10 Information contained in living holothuroids: morphology.

(A) *Labidoplax buski.* Order Apodida, Rec., length 1–2 cm; note buccal tentacles or organs of feeding. Dashed lines perpendicular to body sides represent calcareous skeletal plates or sclerites.

(B) Detail of (A) with plates and buccal tentacles.

(C) Plate pair of (A) (used to classify species), greatly enlarged. These plates projecting through body wall seem to aid forward movements of the animal.

(D)(a) Location in body wall of skeletal plates similar to those in (C) in genus *Synapta* (Order Apodida); *ep*—epidermis; *od*—dermis-anterior connective tissue; *id*—dermis-inner dense layer: *cml*—circular muscular tissue; skeletal elements: *an*—anchor, *anp*—anchor plate, (b) *Thyone briareus,* calcareous ring (see Figure 12.10F, for position of such rings); *r*—radial, *ir*—interradial.

(E) *Psolus* (Order Dendrochirotida). Rec., (a) Dorsal view, schematic, comparable to an edrioasteroid in structure; (b) *Psolus fabricii.* Rec., *m*—mouth, *bm*—buccal membrane, *t*—tentacles, *int*—introvert, *a*—anus, *sc*—scales around anus, *bs*—body scales. Note that mouth and anus are dorsal.

(F) *Thyone briareus* (modified and redrawn from Guthrie and Anderson, 1957); *m*—mouth surrounded by tentacles (*t*); *mad*—madreporite; *rc*—ring canal; *int*—intestine; *bw*—body wall contains spicules; *rt*—respiratory tree, *tf*—tube feet (only a few shown; scattered over body); *cl*—cloaca; *an*—anus; *g*—gonad; *st*—stomach; *gd*—genital duct.

Illustrations after Fell and Pawson and others.

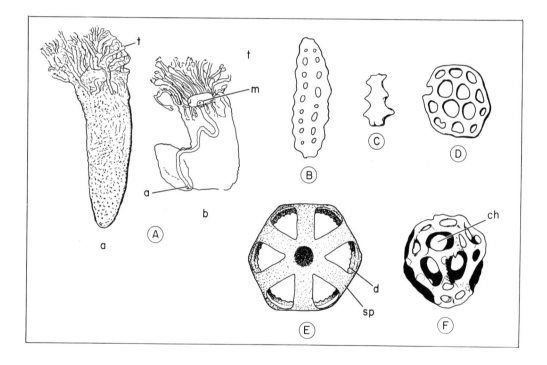

FIGURE 12.11 Fossil holothuroids.

(A) Order Arthrochirotida. *Palaeocucumaria hunsruckiana.* L. Dev., Germany, schematic (after Lehmann, 1958); anterior end bears some twisting, unbranched tentacles (*t*), each with an articulated axial skeleton (calcareous); (a) General configuration. (b) based on X-rays; *m*—mouth, *a*—anus (body length, exclusive of tentacles, 20–50 mm).

(B) Order Dendrochirotida. *Calclamnella irregularis.* Eoc., France, X 200. Note elongate plate with two rows of holes.

(C) Order Aspidochirotida. *Parvispina spinosa.* Florena shale, Perm., Kansas, enlarged.

(D) Order Elasipodida. *Protocandina hexoginaria.* Dev., X 65; wheel with large central part bearing four perforations.

(E) Order Apodida. *Thallocanthus consonus.* Penn., Oklahoma. X 125: *d*—denticulation, *sp*—spokes, six to ten.

(F) Order Molpadiida. *Exlinella frizzelli.* Jur., France. X 400; *ch*—three central holes (four on opposite surface).

Illustrations (B)–(F) redrawn after several authors.

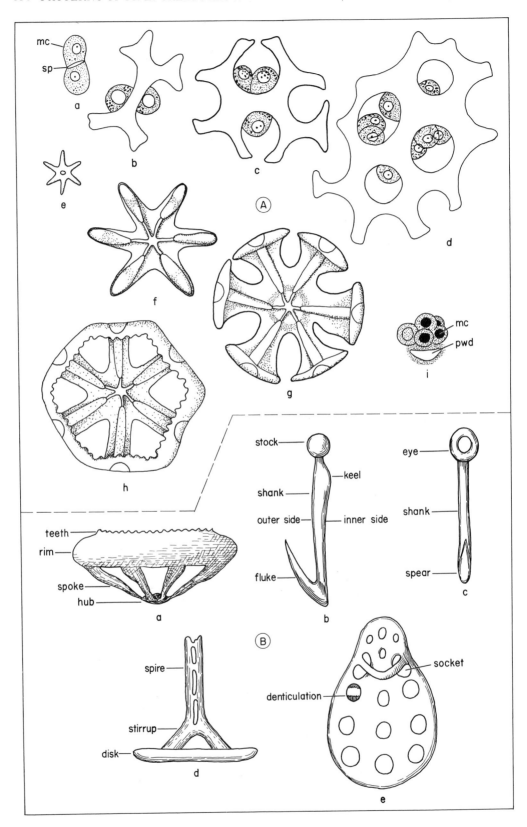

FIGURE 12.12 Holothurian sclerite development and essential terminology of fossil sclerites.
(A) Sclerite development (some stages omitted). *Cucumaria* sp. (a)–(d): Development of spinous sclerite to perforate ossicle (after Woodland, 1906); *mc* — mesenchyme cells, *sp* — initial rod; (e)–(i) Figure (Ai) depicts beginning of biomineralization (*pwd*) in a larval stage of a synaptid holothuroid; *mc* — mesenchymal cells; (e)–(h) *Chirodota venusta*, successive stages, wheel type of sclerite secretion. (Frizzell and Exline, 1955, after several authors).
(B) Types of fossil sclerites: (a) wheel, (b) hook, (c) anchor, (d) table, (e) anchor plate. All schematic after Frizzell and Exline, 1955.

orders: Order Aspidochirotida. *Parvispina*, Permian (Figure 12.11C). Order Elasipodida. *Protocaudina*, Dev. (Figure 12.11D).

Subclass 3. Apodacea. Test vestigial, skeletal spicules (or sclerites) may include anchor or anchor plates. Two orders: Order Apodida. *Labidoplax*, Rec. (Figure 12.10B, C). *Thallocanthus*, Penn. (Figure 12.11E). Order Molpadiida. *Exlinella*.

Subclass Undetermined. Order Arthrochirotida. Tentacles with articulated axial skeleton and stout imperforate body sclerite [Frizzell and Exline, 1966; not incorporated in Pawson's classification (1966, p. U645)]. *Palaeocucumaria*, L. Dev.

INFORMATION DERIVED FROM LIVING HOLOTHUROIDS

Skeletogenesis, Biomineralogy

Calcareous holothuroid sclerites are secreted by two or more mesenchymal cells. Commonly, a minute rod or spinous sclerite is secreted first, as Woodland (1906) showed for *Cucumaria* (Figure 12.12A, a). Growth continues until the sclerite fills almost the entire cell. As it grows, the sclerite will fork and branch, and the branches will then unite, leaving fenestrae (= windows or perforations) in the developing plate or ossicle (Figure 12.12A, b–d). A tiny disk-shaped calcite piece (Figure 12.12A, i) is the start of a spoked-wheel type of sclerite (Woodland, 1907).

There are several observations with reference to holothuroid perforate ossicles: (a) They are vestiges of a skeleton that was plated in remote ancestral stocks, as in the living Dendrochirotida. (b) Each sclerite (plate) reacts optically, as if composed of a single calcite crystal. (c) A general fenestration is characteristic of echinoderm skeletal elements. Nichols (1962) explained the advantages conferred by fenestrate skeletal elements: lightness and economy of materials; holes facilitating the binding together of elements by connective tissue that pervades them; holes conferring strength to resist whatever shearing and splitting forces might be applied to a given ossicle.

Chemical Composition of Sclerites

Hampton's analysis (1958) provided the first reliable data. The aspidochirote holothuroid studied was a species of *Callothuria*. The sclerites were carbonates (60.8 percent) with over 35 percent calcium, and 3+ percent magnesium, 0.29 percent phosphate radical, and trace quantities of copper, iron, lead, strontium, zinc, barium, bismuth, and sulphate. Some spicules in other types of holothuroids, are coated, or the calcium carbonate has been replaced by iron phosphate. Protein was absent.

Other researches (reviewed by Hyman, 1955) show that the body fluids of holothuroids are quite similar to that of seawater in composition but are less alkaline. Such fluids contain carbonates and phosphates and nonproteinaceous nitrogenous material. (In this regard, body wastes are deposited in the body wall and subsequently disposed of through any hollow organ, chiefly by amoebocyte transport.)

Pigments that are deposited in the body wall or elsewhere include fluorescent green pigments, black pigments (melanin), and fatty yellow pigments (carotenoids).

Sclerites in traction and body support. Sclerites (anchors, and so on) are used for clambering about or clinging to seaweeds (Hyman, 1955). For locomotion, anchors, wheels, and so on, are alternately extended and retracted aiding movement through soft substrates (Frizzell and Exline, 1955). The major function of sclerites, however, is that of a flexible skeletal frame or network (generally internal, but in forms like living *Psolus*, external).

Density of Sclerites

Hampton (1959) found that approximately 10 percent (2.2 grams) of the body weight of *Holothuria impatiens* consisted of calcareous spicules. That percentage of alcohol-preserved body weight was represented by an estimated 20,600,000 spicules. In turn, these data can explain the abundance of many fossil populations of sclerites. A population of hundreds or thousands of holothurians, even if distributed in groups of 5–20 individuals, would yield astronomical quantities of skeletal elements upon demise and settling into the substrate.

Substrate Ingestion

A study of holothuroids over some two square miles of surface in a sound (Bermuda area) revealed that in feeding the aspirochirote holothuroid (*Stichopus*) reworked 500 to 1000 tons of sand substrate a year (Crozier, 1918).

Populations of fossil sclerites (from burrowing forms) would carry equivalent implications and influence on the substrate. An individual, in another study, passed 6 to 8 grams of sediment per hour (cit. Hyman, 1955). Other holothuroids are plankton feeders.

Particularly at depths where they abound, holothuroids are considered to be "a major factor" (Bramlette and Bradley) in reworking and redistributing bottom sediments (red clay, *Globigerina* ooze, diatom ooze).

Bathymetry, Reefs

Bathymetry. Living holothuroids are distributed at all depths and occur in all latitudes. Nevertheless, it is in the deep waters of oceanic trenches—4000 to 8500 meters—that they constitute 50 to 90 percent of the total biomass (Zenkevitch, 1963; Bruun, 1956). (Since deep sea deposits are rare in the rock column, fossil sclerites from such substrates are unlikely to be encountered in most geological researches.) Generally, it is from the shallows—a few meters—to moderate depths (40 to 50 meters) that skeletal remains of living holothuroids can be expected to be fossilized (Pawson, 1966). One can assume the same for the geologic past (Frizzell and Exline, 1955).

Coral Reefs. Holothuroids are common around modern Pacific coral reefs (Yamanouchi, 1939, 1956). On the coral atoll Onotoa (Gilbert Island), the benthic fauna of the lagoonal sand flat includes a holothuroid density (*Holothuria atra*) of 5 to 15 individuals per square yard (Wiens, 1962). This density of individuals, upon demise, could yield up to 300 million individual sclerites per square yard.

Habit and Habitat. Holothuroids may be sessile in the benthic environment or they can move about in four different modes: by tube feet, by body contraction (in forms lacking tube feet), by tentacles that stick to bottom substrate and aid movement, or swimming by serpentine movements in a vertical plane that is reminiscent of an inverted medusa. This last mode of movement would apply to pelagic forms like *Pelagothuria* (Nichols, 1962, Figure 11g).

Sessile types attach to sandy substrates or seaweeds. Some forms burrow in sand or mud on the sea floor (*Holothuria* Figure 12.13D). Living *Leptosynapta inhaerens*, for example, is a burrower in sandy substrates (intertidal to deep water) and ejects a residue of ingested mud as coiled castings (Fenton and Fenton, 1934).

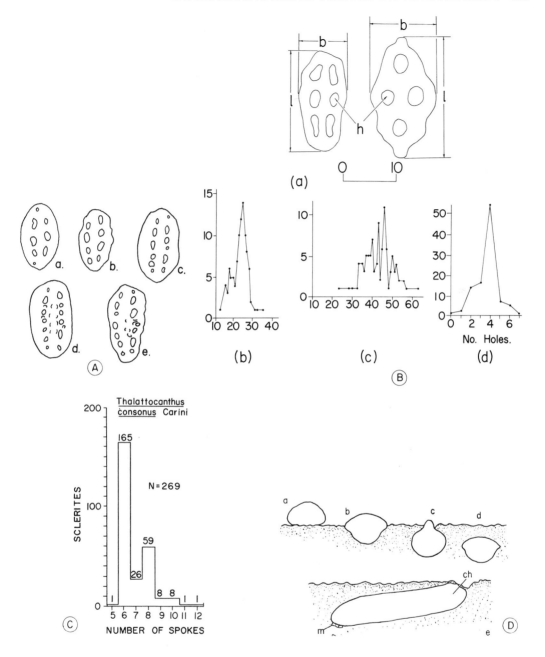

FIGURE 12.13 Information derived from living and fossil holothuroids.
(A) Intraspecific variation, living *Holothuria umbrina*.
(B)(a) Measurements of a holothuroid perforate plate. *Cucumaria saxicola: scale 0–10* = 0 to 0.02350 mm; *b* − breadth, *l* − length, *h* − holes. (b)–(d) Intraspecific variation in same species as (Ba); (b) distribution, breadth; (c) distribution, length; (d) distribution, holes. (Abscissa, 1 unit = 0.02350 mm.)
(C) Intraspecific variation in spoke number in a fossil holothuroid. *Thalattocanthus consonus.* Note that this is a bimodal distribution and hence is probably not a "coherent unit" in Hampton's sense (1959, and see text herein; also compare Carini, 1962 for unimodal distribution of same form).
(D) *Holothuria bivittata*. Rec., Burrowed in sandy substrate, coral reef. Palao Island, (a)–(d) Cross-sectional shape of body during burrowing: (e) burrowed position; *ch-cl* − cloaca, *m* − mouth. Burrowing head into sand first; muscular waves (peristaltic) aids penetration in (b) note lower part of body wedged in sand; in (c) note expanded circular shape that forces sand of burrow aside (Yamanouchi, 1939). [(A)–(C), after several authors].

Salinity, Temperature. Living holothuroids have small tolerance for waters of low salinity. They prefer seawater of normal salinity, although brackish water forms are known in the Philippine Islands.

Holothuroids can tolerate a wide range of temperatures. They are quite versatile and can survive harsh conditions such as freezing or desiccation (when tossed on a beach and subject to strong sunlight). Elevated temperatures are known to induce spawning in some instances (Pawson, 1966).

Evisceration, Autotomy, Regeneration. There is a triad of remarkable capacities in holothurians. Adverse environmental conditions lead to rupture of body wall or cloaca or other parts and ejection (*evisceration*) of the entire digestive tract, and so on. Subsequently, these parts grow back.

Another capacity is that of *autotomy*. The body is constricted and separates into two or more pieces. This is a mode of asexual reproduction in some forms (Deichmann, 1921, cit. Pawson, 1966).

Regeneration is a variable both in the extent and degree of its function in holothuroids. Essentially, body parts can regenerate but not all body parts and not in all species (Hyman, 1955). Body wall tears will heal readily and one can assume that biocrystalline skeletal elements are redeposited in such cases.

Ontogenetic and interpopulation modification of sclerites — various kinds of changes in sclerites occur in the life of the individual (ontogenetic) and under different environmental influences.

Ontogenetic Changes. (1) Change in plate type during life span of an individual. (Example: From large perforate plates to typically shaped small, thinlayered sclerites — Dendrochirotida.) (2) Distinct immature and mature sclerites. (Example: The genus *Molpadia* has racquets and anchors in immature stages, but in the adult these are lacking.) (3) Number and size increase or decrease in sclerites during ontogeny.

Interpopulation Variation. If the species population is geographically widespread, sclerites may vary from locality to locality (see Figure 12.13*A, C*).

Feeding

Holothurians are dominantly scavengers. They feed on living microorganisms and organic detritus. The former are caught on the mucoid coating of their tentacles, diatoms, for example. The detritus consists of protists, smaller worms, and crustaceans. Occasionally, some eat sponge flesh (Hyman, 1955; Frizzell and Exline, 1955). Yamanouchi (1939) studied the esophageal, rectal, and other content of 25 individuals of a species of *Holothuria* living in the shallow waters of a coral reef (Palao Islands). Sieve analysis showed that the major grade size ingested during feeding and/or burrowing consisted of coarse to fine silt. Such activity may be viewed as *bioturbation* of the respective substrate.

INFORMATION DERIVED FROM FOSSIL HOLOTHUROIDS

Fossil Record

Surprisingly, although millions of individual sclerites reach the sea floor upon the demise of a given holothuroid, the Lower Paleozoic record of such forms may be characterized as sparse. This may be an artifact of investigator emphasis on beds of Upper Paleozoic and younger ages.

Two workers (Gutchick, 1954; Reso and Wegner, 1964) have reported holothurian sieve plates in beds of Ordovician age (northern Illinois, southern Oklahoma). Both studies describe species of the genus *Thuroholia*. The problem arises whether these sieve plates are actually holothurian. Frizzell and Exline (1966) are uncertain since similar plates also formed in embryonic crinoids.

No Silurian holothurians have been reported (Pokorný, 1965), although it is plausible that such echinoderms might have thrived in seas of the time. Of course, proof of the occurrence of holothuroids in the Ordovician would make this last inference inevitable.

One of the rare finds of a complete individual holothurian — in this case with a unique articulated axial skeleton in the tentacles (*Paleocucumaria*, Figure 12.11*A*) — was reported in 1958. Lehmann found this form while X-raying Hunsrück shale slabs (L. Devonian of Germany). Sclerites have been described from the Devonian Cedar Valley Formation of Iowa (Martin, 1952) and the Devonian of Bohemia (Prantl, 1947).

Sclerites in the Permo-Carboniferous have a spotty, but not a narrow, distribution: Scotland, United States, Europe. Gutchick, Canis, and Brill (1967) characterized holothurian sclerites from the Kinderhookian of Montana and Missouri as "remarkably widespread." A growing number of papers treat the Permo-Carboniferous holothurian faunas: Kornicker and Imbrie, 1958 — Permian Florena shale of Kansas; Gutchick, 1959 — Lower Mississippi of northern Indiana; Langenheim and Epis, 1957 — Mississippian Escabrosa limestone of Arizona; Carini, 1962 — Middle Pennsylvanian Wewoka shale of Oklahoma; Summerson and Campbell, 1958 — Pennsylvanian Kendrick shale of Kentucky, among others.

Holothurian sclerites occur in the Triassic (Rioult, 1961; Kristan-Tollmann, 1963). Jurassic (Deflandre-Rigaud, 1953), and Cretaceous (Muller) of Europe. The Jurassic in particular have yielded many distinctive genera. Jurassic holothurian forms are also known from Egypt, the Dnieper-Donetz Basin, USSR, as well as from the Upper Oxfordian clays of Yorkshire, England (Frizzell and Exline, 1955, 1966; Fletcher, 1962, among others).

The record of the European Teritary and Pleistocene as far as pertains to fossil sclerites is not inconsiderable when compared to the Mesozoic. Many species described by Deflandre-Rigaud and Kristan-Tollmann occur in or range to the Tertiary. For example, Deflandre-Rigaud's genus *Cucumarites* ranges from the Jurassic to the Tertiary. The American Tertiary also yields sclerites. The family Calcancoridae Frizzell and Exline is based on a species from the Oligocene of Mississippi.

Migration, Shared Gene Pool

There is some evidence from the Paleozoic and Cenozoic that certain holothurian species described from fossil sclerites only occur in both Europe and Great Britain, as well as the United States: for example, Permo-Carboniferous-*Protocaudina traquarii*, and from the American Paleocene and the French Eocene, *Rigaudites cuvillieri*. These and other occurrences indicate a sharing of the same genetic code.

The same sort of broad geographic spread and unquestioned sharing of gene pools exists today for numerous species of holothurians. Living *Cucumaria frondosa* ranges across the North Atlantic (New England, northern Europe, Scandinavia). Some Mediterranean forms are distributed along western European shores and the British Isles. Some holothurians that live in the low-high tide zone (littoral) are found around the world at certain latitudes (circumtropical) (Hyman, 1955).

Stratigraphic Zonation

Limited use has been made of holothurian sclerites in zonation. In 1955 Frizzell and Exline observed that their use was paleobiological rather than stratigraphic. Nevertheless, more recent work seems to qualify this older judgment (Frizzel and Exline, 1966).

Rioult (1961 and this text, Figure 12.14) found that at least ten holothurian species had very limited vertical ranges in the French Liassic (Jurassic) and that the ranges of other species, while less restricted, were sufficiently limited to be of stratigraphic use also. This finding permitted zonation of Rhaetian to Toarcian stages. Gutchick et al. (1967, Table 1, and

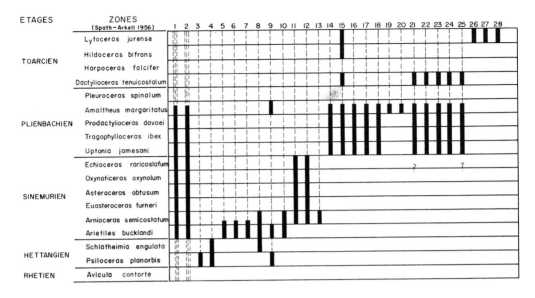

FIGURE 12.14 Information derived from fossil holothurian sclerites. Zonation of French Liassic (Lower Jurassic) based on 28 holothurian species. After Rioult, 1961.
The black vertical bars indicate ranges of a given holothurian species (numbers 1–28). These ranges are compared with standard Jurassic ammonite zones (see Spath and Arkell, 1956).
Many sclerite species have restricted ranges: compare species 3 and 7; 19 and 20.
Numbers 1 to 28 correspond to Liassic species of the following holothurian genera: *Achistrum* (Nos. 1, 2); *Binoculites* (Nos, 3, 5, 7, 8, 14, 16, 17); *Achistrum-Binoculites* aggregates (No. 6); *Calclamna* (Nos. 4, 9); *Staurocumites* (Nos. 10, 18); *Cucumarites* (No. 11); *Mortensites* (Nos. 12, 28); *Ambulacrites* (Nos. 15, 19, 20); *Chirodites* (Nos. 21–24); *Myriotrochites*(?) (No. 25); *Stichopites* (No. 26); *Calclamnoidea* (No. 27).

this text, Figure 12.13C) collected over 2500 specimens of sclerites from the Lower Mississippian strata in Missouri, Montana, Arizona, Texas, and Indiana. Some of the more familiar formations include the Lodgepole, Chouteau, and Redwall. These authors concluded that the holothurian sclerites they had retrieved from acetic acid residues appeared to have utility as stratigraphic markers for beds of Kinderhookian age in the American Mississippian at the places noted.

Substrates and Recovery of Fossil Sclerites

Sclerites are common in sedimentary substrates such as clays (Fletcher, 1962), marls, shales (Carini, 1962; Reso and Wegner, 1964), sandy shales. They occur less frequently in limestones and marly limestones (but see Gutchick et al., 1967). They are rarely encountered in arenaceous sediments, with some exceptions such as Oligocene greensands (Howe, 1942), or

in fine organic matrices (Frizzell and Exline, 1955, 1966).

Methods of Recovery

Sclerites can be retrieved from living holothurians by boiling in a weak solution of sodium hydroxide, lye, or caustic soda. Several decantings thereafter will yield clean sclerite residues. Fossil sclerites are released from carbonates by concentrated acetic acid, from shale or marl matrices by gentle soaking or boiling in water to which a suitable dispersing agent (sodium hydroxide or some other) has been added. (For fuller details, see Frizzell and Exline, 1955, pp. 46–47.)

Carini (1962) fluoridized sclerites released from the Wewoka shale to remove matrix and secondary calcite. He obtained specimens showing fine morphological details otherwise obscured.

Local Accumulation and Form Species.
Allowing for millions of discrete sclerites

embedded in the wall of individual holothurians, the problem arises: What happens to this skeletal material upon demise of the animal. Presumably they would ultimately be released to the sea floor. Predation, where it occurs, by fish, lobsters, crabs, sea gulls (intertidal area), may help to disperse these skeletal elements (Frizzell and Exline, 1955; Hyman, 1955). Bottom currents may further redistribute sclerites as sedimentary particles. Such particles would range from 0.05 mm to 1.0 mm and have a maximum dimension that would place them in the sand to coarse silt size grade. Thus two kinds of distribution occur: sclerites in local concentrations in a given rock (Gutchick et al., 1967) and normal dispersion through a given bed.

Local concentrations imply that predation and/or currents have not dispersed the skeletal elements and that the animals died *in situ*. Even thousands of sclerites dispersed in a given sediment may be attributed to a single individual (see Langenheim and Epis, 1957).

Assignment to two or more species of the sclerites found in a given residue may often mask the fact that a single biological species population gave rise to both types (Martin, 1952). Multiple types of sclerites are not uncommon in the same individual among living holothurians (Hyman, 1955). Similar sclerites may also occur in unrelated holothurian species (Hampton, 1959). Fletcher (1962) set up three new species of genus *Achistrum* from the Upper Jurassic of Yorkshire, England. He observed, following Hampton's study, that since morphological variation in fossil (as in Recent) sclerites did not necessarily denote biological species, his new species were to be regarded as form species (= parataxa).

Population Statistics and Interspecific Variability

Several problems arise in treating any sample of disjunct, discrete parts of an animal (conodonts, scolecodonts, holothurian sclerites, and so on). In the case of sclerites, Hampton (1959) observed that only under certain conditions could a given sample be considered as derived from a single biospecies population. Usually a sample of sclerites consisted of one or more *morphogroups* (that is, individual pieces that could be readily grouped on visible similarities). Essential criteria have to be satisfied by a given morphogroup before one can designate it as a distinct species: (1) It had to form a coherent unit. (2) Such a unit would be recognized by a unimodal distribution.

In Figure 12.13C, if the midpoint of each rectangle of the histogram were connected by a continuous curve, two distinct peaks would show. This is a bimodal distribution in contrast to a single peaked distribution. Bimodality would, in this instance, indicate the presence of more than one holothurian species population in the sample (see Hampton, 1959).

Carini (1962, Figure 7, and see Gutchick et al., 1967; this text, Figure 12.13C) found five morphologic variants (identified by different number of spokes) in the species *Thalattocanthus consonus*. This distribution was unimodal. By comparison, Gutchick et al. (1967) found a bimodal distribution. Interpreted in light of Hampton's study (1959), Carini's samples formed a "coherent unit"—a fossil morphogroup derived from a single biological species population. However, Gutchick et al. (1967), with their bimodal distribution, had more than one morphogroup, and apparently this sample derived from more than one biological species population.

Biotic Associates

Sclerites are often found with the fossils of other echinoderms, both macro- and microfossils. Gutchick et al. (1967) reported the following microfaunal associates of their Kinderhookian sclerites: conodonts, microcrinoids, agglutinate foraminifera, sponge spicules, fish teeth, worm tubes. Megafossils included skeletal debris of crinoids, echinoids, and blastoids; brachiopods, corals, mollusks (snails, clams, cephalopods), and trilobites. Frizzell and Exline (1955) also included bryozoans, ostracods, and ophiuroids.

Evolution

Very few observations can be made on holothurian evolution. Assuming that the Ordovician fossil genus *Thuroholia* (which was established on the basis of sieve plates) is a holothurian, Fell (Fell and Pawson, 1966, Figure 1-17*b*) envisioned the entire animal as resembling either *Helicoplacus* or *Eothuria* and showed the possible derivation of the holothurians from the *Helicoplacus* stock.

Pawson (1966) would place *Thuroholia* under the subclass Dendrochirotacea on the basis that the lattice plates of the former closely resemble those of some living dendrochirotacean genera. Furthermore, Fell and Pawson, on anatomical considerations, view the Dendrochirotacea as the ancestral and most primitive holothurian stock. All modern genera having skeletons of imbricating calcareous plates are assigned to this subclass (see *Psolus*, Figure 12.10*E*).

From this line of reasoning, it follows that (*a*) skeletal plates have been reduced to sclerites through geologic time, that is, from Ordovician time on, in all groups exclusive of the modern plated dendrochirotaceans; and (*b*) the holothurian calcareous ring (Figure 12.10*D*, *b*) formed from modified ambulacral plates. This last development explains the absence of external ambulacral plates in holothurians and the presence of relict radial water vessels and tube feet (Pawson, 1966).

Marked adaptive radiation into varied types of environments also characterize holothurian evolution: extant forms are benthic, pelagic; some burrow; they range from the shallows to deeper waters (Nichols, 1962, Figure 11).

It appears that both Carboniferous and Jurassic times were marked as times when "sclerites of new patterns were introduced" (Frizzell and Exline, 1955). Pawson (1966) remarked that several families (possibly genera also) were differentiated by Upper Paleozoic time. The range, for example, of the family Stichopitidae extends from Carboniferous to Pleistocene, suggesting major differentiation in this family occurred in Carboniferous time.

One order, known only from the Lower Devonian, the Arthrochirotida, if truly confined to such a narrow vertical span, could properly be termed an unsuccessful evolutionary experiment. Frizzell and Exline (1966) suggested that similar sclerites may have ranged up to Jurassic time. This does not mean, however, that the unique articulated axial skeleton in the tentacles also persisted.

There are a few families that are restricted to a single geological period: Carboniferous (family Paleochiridotidae Frizzell and Exline), Triassic (family Kaleobullitidae Kristan-Tollmann), Jurassic (family Schlumbergeritidae Deflandre-Rigaud), and Miocene (family Alexandritidae Kristan-Tollmann). Nonpersistence here may denote one or more inadaptive aspects of gross anatomy exclusive of skeletal elements.

Finally, anatomical homologies suggest relationships of the earliest holothuroids, edrioasteroids, and echinoids (Pawson, 1966; Fell and Pawson, 1966).

Needed Research

Frizzell and Exline (1966) emphasize that a search for Ordovician and Silurian holothurian sclerites could prove "critically important" in evolutionary relationships and taxonomic problems.

Working out of the food chain of the Paleozoic and Mesozoic seas would also be advanced by improved data on neglected groups such as the holothurians. In larval stages, holothurians are part of the plankton and serve as food for larger planktonic forms (Thorson, 1946) becoming scavengers in maturity.

REFERENCES

Broili, F., 1926. Eine Holothurie aus den oberen Jura von Franken. Bayerischen Akad. Wiss. zu München, Sitz., Math-Nat. Abt. Jahrg. 1926, pp. 341–351, 1 plate.

Bruun, A. F., 1956. Animals of the deep sea bottom, in The Galathea Deep Sea Expedition, 1950–1952, pp. 149–195.

Carini, G. F., 1962. A new holothurian sclerite from the Wewoka shale of Oklahoma: Micropaleontology, Vol. 8 (3), pp. 391–395, plate 1.

Crozier, W. J., 1918. The amount of bottom material ingested by holothurians: J. Exper. Zool., Vol. 26, pp. 379–389.

Deflandre-Rigaud, M. 1953. Classe des Holothurida, in Jean Piveteau, ed. Traité de Paléontologie, Vol. III, pp. 948–957.

Deichmann, E., 1957. Holothurians. Treat. Mar. Ecol. and Paleoecol., GSA Mem. 67, Vol. 1, pp. 1193–1195.

Fletcher, B. N., 1962. Some holothurian spicules from the Ampthill Clay of Melton, near Hull (Yorkshire): Geological Mag., Vol. XCIX (4), pp. 322–326, text figure 1.

Frizzell, D. L., and H. Exline, 1966. Holothuroidea—fossil record, in R. C. Moore, ed., Treat. Invert. Paleontology (U) Echinodermata 3², pp. U646–U672.

————, 1957. Holothurians. Treat. Mar. Ecol. and Paleoecol., GSA Mem. 67, Vol. II, pp. 983–986.

————, 1955. Monograph of fossil holothurian sclerites: Bull. Univ. Missouri, School of Mines and Metallurgy, Tech. Ser. #89. 164 pp., text figures 1–20, plates 1–11.

Gutschick, R. C., 1959. Lower Mississippian holothurian sclerites from the Rockford limestone of northern Indiana: J. Paleontology, Vol. 33 (1), pp. 130–137.

————, 1954. Holothurian sclerites from the Middle Ordovician of northern Illinois: J. Paleontology, Vol. 28 (6), pp. 827–829.

————, W. F. Canis and K. G. Brill, Jr., 1967. Kinderhook (Mississippian) holothurian sclerites from Montana and Missouri: J. Paleontology, Vol. 41 (6), pp. 1461–1480.

Hampton, J. S., 1959. Statistical analysis of holothurian sclerites: Micropaleontology, Vol. 5 (3), pp. 335–349.

————, 1958a. Frizzellus irregularis, a new holothurian sclerite from the Upper Bathonian of the Dorset coast, England. Micropaleontology, Vol. 4 (3), pp. 309–316.

————, 1958b. Chemical analysis of holothurian sclerites: Nature, Vol. 181, pp. 1608–1609.

Hyman, L. H., 1955. Class Holothuroida, in The Invertebrates. McGraw-Hill, Vol. 4, pp. 121–244.

Kornicker, L. S., and J. Imbrie, 1958. Holothurian sclerites from the Florena shale (Permian) of Kansas: Micropaleontology, Vol. 4 (1), pp. 93–96.

Kristan-Tollmann, E., 1963. Holothurien-Sklerite aus der Trias der Ostalpen: Östrr. Akad. Wiss. Math. Naturwiss. Kl., Sitzungsber., Abt. 1, Vol. 172, Part 6–8, pp. 351–380.

Langenheim, R. L., Jr., and R. C. Epis, 1957. Holothurian sclerites from the Mississippian Escabrosa limestone, Arizona. Micropaleontology, Vol. 3 (2), pp. 165–170.

Lehmann, W. M., 1958. Eine Holothurie zusammen mit Palaenectria devonica, etc.: Hess. Landesamt Bodenf. Notizbl., Vol. 86, pp. 81–86.

Martin, W. R., 1952. Holothuroidea from the Iowa Devonian: J. Paleontology, Vol. 26 (5), pp. 728–729.

Miksukuri, K., 1903. Notes on the habits and life history of Stichopus japonicus Selenka: Annot. Zool. Japan., Vol. 5 (1), pp. 1–21.

Nichols, D., 1962. Echinoderms. Hutchinson Univ. Library, 192 pp.

Pawson, D. L., 1966a. Phylogeny and evolution of holothuroids, in Treat. Invert. Paleontology, (U) Echinodermata, Vol. 3 (2), pp. U64–U646.

———, 1966b. Ecology of holothurians, in R. A. Boolootian, ed. Physiology of Echinodermata. Interscience, pp. 63–71.

——— and H. B. Fell, 1965. A revised classification of holothurians: Breviora, No. 214, pp. 1–7.

Pokorný, V., 1965. Principles of Zoological Micropaleontology. Pergamon Press, Vol. 2, p. 377.

Reso, A., and K. Wegner, 1964. Echinoderm (holothurian?) sclerites from the Branch Formation (Black Riverian) of southern Oklahoma: J. Paleontology, Vol. 38 (1), pp. 89–94.

Rioult, M., 1961. Les sclerites d'holothuries fossiles du Lias: France, Bur. Rech. Geol. et Min., Mem. No. 4, pp. 121–153.

Seilacher, A., 1961. Holothurien im Hunrückschiefer (Unter-Devon): Notizbl. hess. L. Amt. Bodenforsch., Vol. 89, pp. 66–72.

Wiens, H. J., 1962. Atoll environment and ecology. Yale Univ. Press, 484 pp.

Woodland, W., 1907. The scleroblastic development of the plate-and-anchor spicules of Synapta: Quart. J. Micros. Soc., Vol. 51.

———, 1906. The scleroblastic development of spicules in Cucumariidae: Quart. J. Micros. Soc., Vol. 49.

Yamanouchi, T., 1956. The daily activity rhythms of the holothurians in the coral reef of the Palao Islands. Publ. Seto mar. biol. Lab., Vol. 5, pp. 347–362.

———, 1939. Ecological and physiological studies on the holothurians in the coral reef of Palao Islands. Palao trop. biol. Stud., Vol. 4, pp. 603–636.

LIVING AND FOSSIL ECHINOIDS

Class Echinoidea

[After Hyman, 1955; Durham and Melville, 1957; Nichols, 1962; Fell and Pawson, 1966; Durham, 1966.] (The class name is derived from the Mediterranean sea urchin genus *Echinus.*) The class embraces sea urchins, heart urchins, and sand dollars. About 5800 species—fossil and living—with almost six times as many fossil echinoids as living, that is, 5000:850.

Diagnosis. Free-living pentamerous (Figure 12.15*A*), marine Echinozoa with a rigid endoskeletal test, generally of low relief relative to the substrate, and of variable shape—globose, ovate, discoidal, or attenuated (Figure 12.1*A*, part F). Test formed by interlocking calcareous plates organized in several systems. These include apical, coronal, peristomal, and periproctal. The last two consist of plated membranes about the mouth and anus, respectively (Figure 12.15*B*a, *C*a). The plates are arranged in columns (ambulacral and interambulacral) meridionally.

External surface bears variable types of *spines, pedicellariae,* and *spheridia* (= organs of equilibrium)—all of which are movable appendages. Tube feet extrude from pores in ambulacral plates. Jaws within an introvert (Aristotle's lantern) are present in primitive but not newly evolved forms. Sexes separate, hermaphroditism occasional.

Geologic range: Ordovician to Recent.

Skeletal Anatomy of Living Echinoids: Organs and Other Anatomical Systems

Body Wall. As schematically shown in Figure 12.15*C*, the echinoid body wall consists of an epidermis with ciliated cuticle and cuboidal- to columnar-shaped cells of a single-layered epithelium. The mesoderm can be studied only after decalcification since it bears skeletal plates. It consists of connective tissue and cells. The endoderm has flat, flagellate epithelial cells that line the coelom (= body

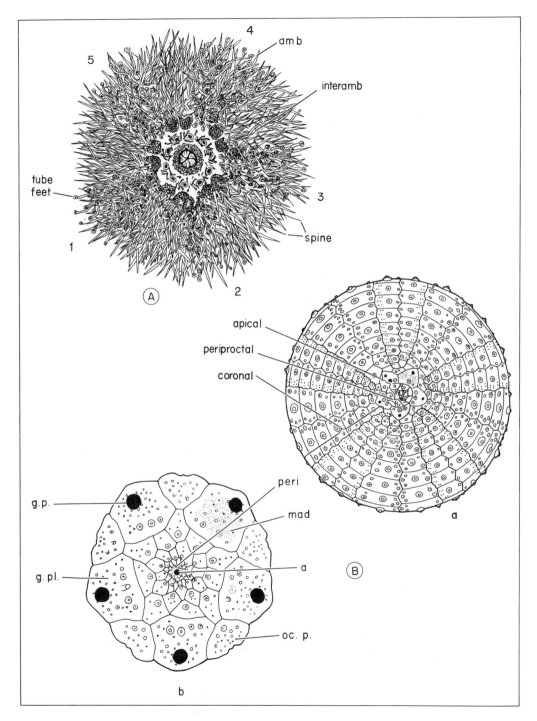

FIGURE 12.15 General morphology of echinoids—I. All figures schematic.
(A) *Echinus* sp., oral view, paired ambulacra (*amb*); interambulacra (*interamb*); numbers 1–5 show pentamerous symmetry (see *B*a).
(B)(a) Same as (A) but cleaned test, aboral surface; three plate systems: apical, periprocta, coronal (for fourth system, see *C*a). (b) *Goniocidaris,* apical system (that is, genital plus ocular plates); *gp* – genital pores (note five of these); *mad* – madreporite (or ''sieve pore,'' see *C*b); *gpl* – genital plate; *ocp* – ocular plate; *a* – anus; *peri* – periproct.

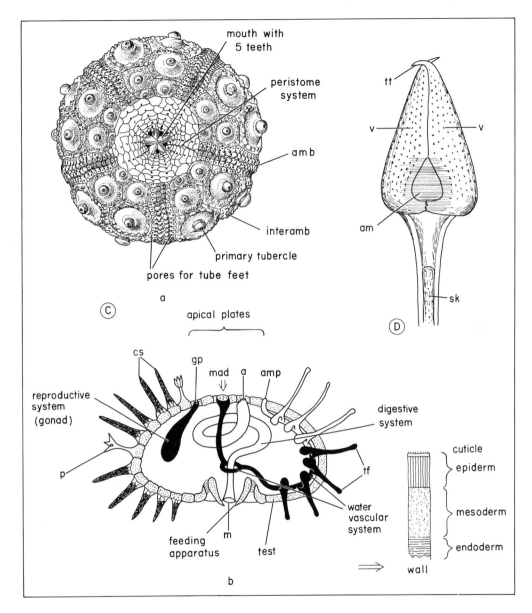

FIGURE 12.15 (cont.)
(C)(a) Peristomal system, and so on. *Stereocidaris mascaensis,* Rec., S.E. Pacific, oral view, a primary spine (radicle) articulates with each primary tubercle, (b) Body plan of a sea urchin. Note especially water vascular system, open area at madreporite (*mad*) for intake of sea water; *m* — mouth, *a* — anus. Observe apical plates; *amp* — ampulla, *tf* — tube feet, *p* — pedicellariae, *cs* — calcareous spines. Structure of echinoid body wall. Mesodermal layer bears endoskeletal elements, that is, plates.
(D) Pedicellariae, bidentate (*Echinarachnius parma*), *v* — calcareous valve, *am* — adductor muscle, *sk* — calcareous rod of muscular stem (a poison sac occurs in globiferous types at base or along wall of valves).
Illustrations redrawn and adapted from: (A) from G. A. Kerkut, ed. The Invertebrata, 3rd ed., 1959 by Borradaile and Potts. With permission of Cambridge Univ. Press. (Ba) MacBride, cit. Melville and Durham, 1966; (Bb) Fell and Pawson, 1966; (Ca) Allison et al., 1967; (Cb) Simpson et al., 1957; (D) from The Invertebrates, Vol. IV, by Libbie Hyman, Copyright © 1955 by McGraw-Hill, Inc. With permission of McGraw-Hill Book Co.

cavity enclosed by a rigid test and sub-divisible into a number of smaller spaces).

Although most organs are internal, some do attain externality (water vascular system, for example, through the tube feet, or the digestive system through Aristotle's lantern, and the respiratory system via the gills).

Digestive and Excretory Systems. Food, mainly vegetation, is taken in by mouth (Figure 12.15*C*, *b*) with or without a toothed masticatory apparatus (Aristotle's lantern); see Figure 12.16*A* for detail and Figure 12.15*C*, *b*. The food then passes the buccal cavity into the pharynx and then to the esophagus. It then follows a looped course through the intestine, and the waste is voided at the anal end.

Reproductive System. Gonads (ovaries and testes generally, or ovotestes occasion-ally) are five in number in "regular" echinoids and two in "irregular" echin-oids. These are connected to the genital pore (Figure 12.15*C*, *b*) on the genital plate (Figure 12.15*B*, *b*) by a brief gono-duct. There is one gonad in each inter-ambulacrum.

Eggs and sperm are shed directly into seawater, where fertilization takes place.

Water Vascular System. Through the hydropores in the madreporite or sieve plate, seawater enters the water vascular system of echinoids. These small open-ings, the hydropores, are present, with rare exceptions, and probably prevent blockage of the vital water passage (Nichols, 1966). Pores in these plates tend to increase in number with age of the in-dividual (Hyman, 1955). The water that has entered by these pores then passes along the stone canal to the ring canal and thence to tube feet (Figure 12.15*Cb*). It will be observed that the digestive tract (alimentary canal) is circled by the ring canal. (Not shown in Figure 12.15*C*, *b* are the five ambulacral water vessels or radial canals that arise from the ring canal and branch into the tube feet. Only one such vessel is shown. Likewise, not shown but coparallel and in contact with the stone canal, is the so-called "axial organ" that

moves coelomic fluids to and through the circulatory system.)

The main function of the water vascular system appears to be the maintenance of adequate hydraulic pressure in the tube feet—which are hydraulic organs—to facilitate their multifunctions (Nichols, 1962).

The stone canal has connective tissue of its wall heavily invested with calcite spicules. It is also lined with ciliated epithe-lial cells, and cilia movement maintains a current in the oral direction (Nichols, 1966; Hyman, 1955).

Nervous System. The essential parts are the main circumoral nerve ring, radial nerves that branch off the ring (see Figure 12.16), and the subepidermal skin plexus (Smith, 1966; Hyman, 1955).

The chief function of the nervous sys-tem is to coordinate, to the extent possible, that is, as patterns of movement, the main movable parts. These parts include spines, pedicellariae, spheridia, tube feet, and lantern (Smith, 1966).

Respiratory System. "Success" in a bio-logical sense in living echinoderms is at-tributed to their development of special-ized respiratory surfaces (Farmanfarma-ian, 1966). These include podia (for echin-oids), gills (sea urchins), and petaloids and podia (sand dollars, heart urchins). No known echinoid can tolerate anerobic conditions. Burrowing species respire the same way as nonburrowers but at a lower rate.

The largest percentage of oxygen up-take is through the body wall and appen-dages and by direct diffusion from sea-water. This seawater is essentially avail-able in the coelomic fluid.

Circulatory System. There is a direct connection between the circulatory (he-mal) system of echinoids and the water vascular system. Thus the so-called axial gland or organ (coparallel with the stone canal) regularly pulses along with the stone canal. The "gland" is considered a primitive heart (Boolootian and Camp-bell, 1964). Furthermore, it is the coelo-mic fluid (blood?) that moves into and out of the circulatory system of rings, vessels, and dendritic offshoots. Such fluids have the property of clotting and bear pigments, some of which may participate in oxygen transport (Endean, 1966).

Endoskeleton

Plate System, and So On. The phylum definition and accompanying illustrations reviewed the major plate system. The apical system (Figure 12.15*B,b*) consists of five smaller *oculars* and five or fewer *genitals*, one of which is larger than the other and modified to serve as a sieve plate (*madreporite*).

Ocular plates (in *regular* echinoids) can be classed as *insert* or *exsert* depending on which set (ocular or genital) is in contact with the margin of the *periproct*: oculars (= insert), or genitals only (= exsert). The apical system is referred to as being either monocyclic or dicyclic: if monocyclic, all oculars are insert; if dicyclic, all oculars are exsert. Obviously, *all* of either set of plates need not be insert or exsert, respectively. In Figure 12.15*B,b* the arrangement is monocyclic and all oculars are insert.

Genital plates of *irregular* echinoids may be reduced to four (tetrabasal) or, by fusion, can form a single large plate (monobasal) as in *Clypeaster* or be otherwise modified.

Echinus Skeletal Rudiment and Test Development.
Papers by Devanesen (1922), Isabella Gordon (hereafter referred to as "IG") (1926, 1927, 1929), Onada (1931), Okazaki (hereafter referred to as "OK") (1960), and Wolpert and Gustafson (hereafter referred to as "W & G") (1961) among others, have brought out many interesting observations on skeletogenesis of the echinus rudiment.

1. Three major steps precede spicule formation in the sea urchin larva (*Clypeaster japonicus* and *Psammechinus miliaris*) (OK and W & G):
(a) Primary mesenchymal cells bearing ray-like pseudopods arrange in an orderly pattern forming skeletal envelopes (Figure 12.16*G*), and dorsal and ventral paired longitudinal cell chains. All of these envelopes and chains may be viewed as *centers of calcification*, although usually it is the envelope only that shows continuous spicular growth.
(b) Fusion of pseudopods forms an organic matrix for biocrystallization (see Chapter 15). Separate matrices fuse subsequently.
(c) Small, calcareous granules (the spicular rudiment) are secreted in the center of the envelope. (The chains of 1(a) above often join the envelopes. If, because of certain stresses such as low calcium, the chains do not join the envelopes, they can compete with the envelopes as centers of calcification. In such an event their granules develop into spicules.)

2. Subsequently these spicules grow as triradiate objects which reorient in the embryo (W & G).

3. Given the echinopluteus (Figure 12.2) with all eight arms supported by fully developed calcareous rods (oral, midventral, body, and anal) (W & G), how then does growth proceed?

4. Part of the calcite is resorbed. Some larval spicules give rise to fenestrate plates (Figure 12.3) along branches, while their posterior ends undergo resorption (IG, and see Onada and this chapter Figure 12.16*D*). Growth of spicules will be straight unless bent by pseudopods of mesenchymal cells (see OK and herein Figure 12.16*G* for *Clypeaster japonicus*). These pseudopods determine bends, branches, and spine formation (W & G).

5. Ocular plates appear first (that is, they form by the method discussed in (4) above, Figure 12.16*C*) followed by a single plate on each interambulacrum (IG).

6. Other interambulacral plates appear subsequently (IG).

7. Spine rudiments appear as small calcareous granules, external to the ocular and interambulacral plates.

8. Small triradiate spicules next appear and develop into calcareous disks of the tube feet.

9. Rudiments of ambulacral and then buccal plates appear (Figure 12.16*C*).

Following step 9, Aristotle's lantern and further plate formation then develop.

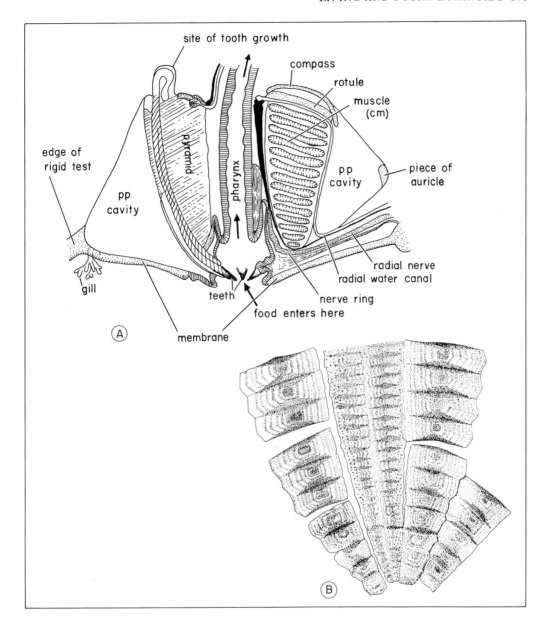

FIGURE 12.16 Gross morphology of Echinoids—II.

(A) Echinoid. *Paracentrotus lividus.* Rec., vertical slice through Aristotle's lantern and adjacent cavity. (Left pyramid piece removed to show muscle.) Parts of lantern include: compass, rotule (mere accessory pieces unrelated to feeding; see text), pyramids, teeth; *cm*—muscle that rocks pyramid during feeding; black = circulatory system (which plays no role in respiration). Adapted from Cuenot, 1891.

(B) *Echinus esculentus,* shell diameter = 100 mm. height = 78 mm; growth lines of test plate (adapted from photograph in Deutler, 1926. Plate 5, Figure 4). Ambulacra center on either side; interambulacra. Observe successive growth lines (increments) on plates. See text for use of such data.

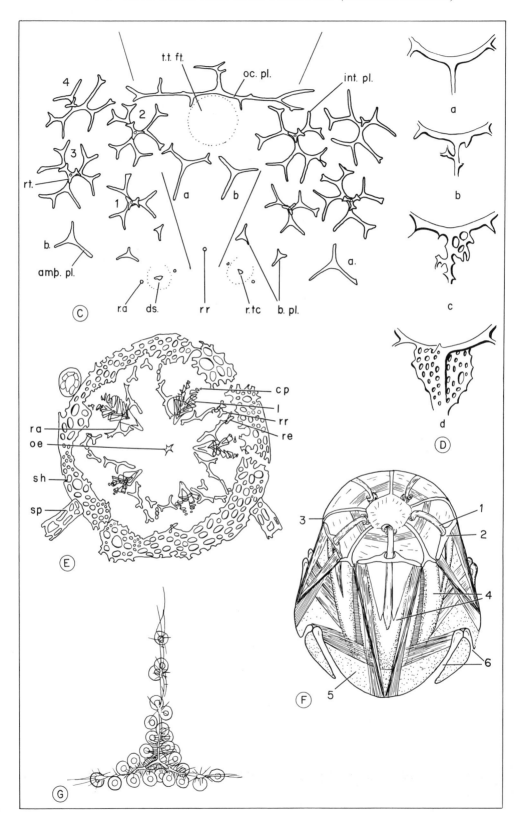

FIGURE 12.16 (cont.)

(C) Developing plates soon after calcification starts in Aristotle's lantern (all spines omitted). Ambulacrum labeled; *int, pl.* and numbers 1 to 4 = interambulacral plates; *amb,. pl.* = ambulacral plates (*a* and *b; oc, pl.* – ocular plate; *b, pl.* – buccal plate. Rudiments of Aristotle's lantern: *ds* – dental sac, *a* – alveolus, *rr* – rotula, *rtc* – first cone of tooth, *t. tft.* – portion of terminal tube foot (after Gordon, 1926).

(D) *Anthocidaris crassispina*, early developmental stage (*b–d*) in genital plate growth starting with a larval spicule (*a*). After Onoda, 1931, Figure 21.

(E) *Psammechinus miliaris*, Aristotle's lantern, rudiments of hard parts (after Devansen, 1922). Teeth: *cp* – calcareous palette, start of tooth lamella (*l*), which arrange in cone-and-cone structures during growth (see Figure 12.16F for fully formed teeth and lantern parts). Jaws: *ra* – rudiment of alveolus develops from triradiate spicule. Rotule: *rr* – rudiment of rotule; *re* – rudiment of epiphysis; *sh* – shell plate; *sp* – spine; *oe* – oesophagus.

(F) Aristotle's lantern (*Tripneustes esculantus*) in place, showing muscles and hard parts: (1) tooth, aboral end; (2) epiphysis; (3) compass; (4) pyramid; (5) peristomal membrane, see Figure 12.16A. (From The Invertebrates, Vol. IV, by Libbie Hyman. Copyright © 1955 by McGraw-Hill, Inc. With permission of McGraw-Hill Book Co.)

(G) *Clypeaster japonicus*, growth of triradiate spicule in triangular cluster of mesenchymal cells bearing pseudopods (after Okazaki).

The former is illustrated in Figure 12.16E. [Such lanterns consist of 40 skeletal elements: compasses (two portions each) -5×2; teeth $- 5$; half pyramid $- 10$; epiphyses $- 10$; rotulae $- 5$. These structures are taxonomically useful only in the study of living regular echinoids (Hyman, 1955). They are also noted in some paleontological studies (Spreng and Howe, 1963).]

Echinoid Skeletal Crystallography. See Chapter 15 and additional discussion in this chapter.

Mode of Plate Growth. Ontogeny of the echinoid skeleton is to some extent readable in individual plates: Growth bands below the surficial plate layer reveal the changing configuration of a given plate (Figure 12.16B) (Deutler, 1926). More recently, Raup (1967) studied growth line configuration in *Strongylocentrotus echinoides* and derived a general mathematical model for echinoid growth. This permitted plotting ideal plate patterns by computerization – digital computer. Slight shifts in constants in this model yielded a variety of echinoid plate patterns. Such results are possible because echinoid plate size and test shape correlate.

Plate shape and rate of plate growth are explained as governed respectively by size and position relative to neighboring plates (shape) and migration of plates away from the apical system (growth).

Migration itself is activated by appearance of newly formed apical plates.

Calcium Metabolism, Shell Repair. Aside from resorption of portions of calcite in larval spicules already mentioned, a sequence of resorption and redeposition (that is, recrystallization of calcite) occurs in the migration of pores for the tube feet and in the genital and ocular pores (Gordon, 1926; Moore, 1935; Swan, 1966).

Echinoids are known to repair broken spines and damaged tests. The former may also be completely replaced. [Resorption of the tubercle with which the spine articulates (Figure 12.17A,m) occurs during spine replacement, to reduce its size to fit the new thinner spine (Borig, 1933, cit. Hyman, 1955).]

Test repair has been observed by several workers (Cuenot, 1906; Oxner, 1917; Crozier, 1919, and cit. Hyman, 1955; Kindred, 1924; Okada, 1926; Swan, 1966). Essentially, if a piece of sea urchin test is surgically removed, it will be repaired in about a month's time.

In a natural setting, tests are injured by gull pecks at low tide, fish bites, or other means. Along the growing edge of the test calcareous deposition (concrescence) will occur and close such shell features as broken lunules (Figure 12.17B).

Spines (Adaptive Modification). On a size basis, large spines are *primary*, medium, *secondary*, and small, *miliary*. All

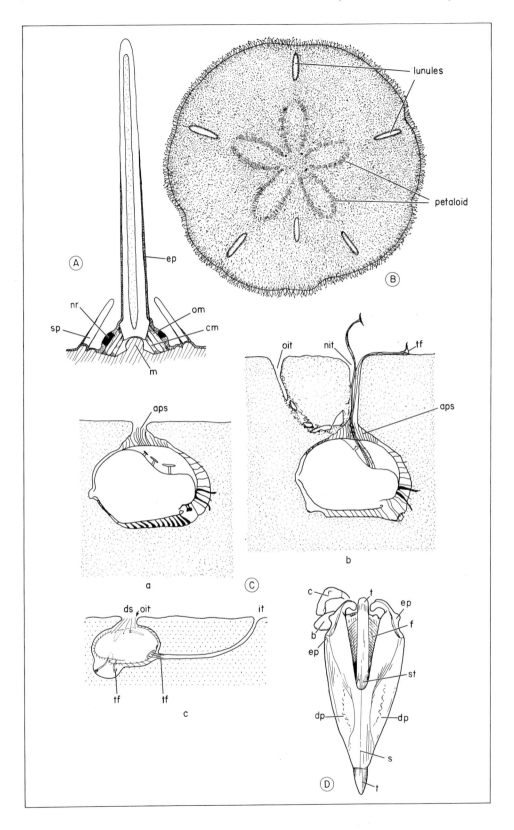

FIGURE 12.17 General morphology of echinoids—III.

(A) Echinoid spine and tubercle that bears it. Articulation of spine occurs on raised tip of tubercle (= mamelon, m); muscles (om, cm); sp—secondary spine; nr—nerve ring; ep—epithelium; stippled central area = dense calcite.

(B) Mellita sp., Rec., Bahamas, sand dollar; lunules and petaloids (gonopores between petaloids and tube feet).

Derivative data from living echinoids—I. (C)

(a)(b) Moira atropos, early and advanced burrowing stage; asp—apical spines; tf—as below, oit, nit—old and new intake tubes, the latter kept open by tube feet (after Chesher, 1963).

(c) Echinoid burrow, (Spatangus purpureus), after 15 minutes excavation in aquarium gravel; ds—dorsal spine, tf—tube feet, it—intake tube, oit—original intake tube. Schematic, after Nichols, 1960.

(D) Diadema sp., Pyramid of Aristotle's lantern (compare Figure 12.16 for details).

spines articulate with the thecal plate tubercles and are moved by muscles that receive impulses through nerve rings (Figure 12.17).

Adaptive modification of echinoid spines has led to their utility in a broad spectrum of activity: locomotion; protection and support; aid in digging and burrowing; regulatory function or external brood site (see Nichols, 1962, Figure 13 and text; Hyman, 1955).

The metachronal beat during movement can send the individual echinoid in transit over the sea floor. Such movements of echinoids on the sea floor and around an Indian Ocean atoll are depicted in a Cousteau film. The movement in this case was participated in by hundreds of individual echinoids. A predatory starfish that hovered among this echinoid community may have had something to do with the furious movement away from the area.

Other diadematoid forms (Plesiodiadema, for example) have developed unusually elongate, curved, and very thin spines with hooflike tips. The latter prevent sinking into the soft substrate. Spines may also aid the burrowing function, as is the case with the funnel-building spines in Spatangus and the apical spines in Moira (Figure 12.17C).

Spines may serve innocent or lethal functions; innocent function: brood site for larvae; lethal function: bearer of poison glands. They can also serve as a wave break or a generator of currents (during their metachronal beat).

Echinoid spines are often broken and are quite common as fossils. They break or fragment, and their articulation with the test is destroyed.

INFORMATION DERIVED FROM LIVING ECHINOIDS

Photosensitivity and its Significance

Millott (1954), in a study of the echinoid Diadema antillarium, detected photosensitivity. He found it to be related to pigment dispersion (melanin), and noted a direct effect of light on the nervous system.

Raup (1960, 1966) applied Millott's findings to studies on the crystallography of echinoid plates. He suggested: (a) photosensitivity and echinoid plate orientation are probably related (1960); (b) possibly echinoids utilized variation in polarization patterns as a crude navigational system, that is, adjacent test plates of different orientation observably transmit different amounts of light under polarized light (1966).

Burrowing

Use of teeth (see Aristotle's lantern, Figure 12.15C, a) and/or spines permits several intertidal echinoids to excavate into bedrock and softer substrates with new borings (burrows) or enlargements of existing ones. Such burrowing activity arises from a dual need: protection against wave action and against desiccation (Yonge, 1963). Umbgrove (1947) attributed erosional control of East Indian coral reefs to Echinomitra mathaii.

Among others, there are available excellent studies (Nichols, 1959; Chesher, 1963) on various burrowing sea urchins

(Figure 12.17*C*). Burrowing proceeds by a combined action of spines and tube feet; the latter secretes mucous to line the intake tubes (respiratory funnels) (Nichols, 1959). The funnel to the surface is kept open by extension of dorsal tube feet. The method of forward movement from inside a burrow is illustrated in Figure 12.17*C, b*.

Ionic Regulation

The carbonate ions occurring in body fluids are quantitatively less than those in seawater. Binyon (1966) thinks this difference may play a role in skeletogenesis. By contrast, potassium is enriched in the ambulacral fluid (water vascular system) up to as much as 90 percent over seawater. This may be related to tube feet movement (neuromuscular activity).

Ionic regulation is very elementary in echinoderms generally. Regulation governs toleration for various salinities. Poor control of internal osmotic pressure (relative density of fluids across body membranes) can explain why the phylum Echinodermata is marine and has been effectively excluded from the estuarine environment (Binyon, 1966). They cannot accommodate (generally) to reduced salinities, although brackish water penetration is known (20 percent salinity, for example, in Texas bays and elsewhere, Durham, 1966).

Biogeochemistry

A few exploratory studies have been made on echinoid skeletal concentration of various telltale elements, isotopes, or compounds: Sr, Ca, $MgCO_3$, C^{13}/C^{12}, O^{18}/O^{16} (Pilkey and Hower, 1960; Weber and Raup, 1968). The Sr/Ca ratio was found to be inversely related to the water temperature, and unaffected by salinity variation. $MgCO_3$ concentration, by contrast, was directly related to variation in water temperature and salinity.

Modern sand dollars and sea urchins were found to fractionate isotopic carbon

(C^{13}) and isotopic oxygen (O^{18}). Genetic control was indicated (Weber and Raup. 1966*b*, Figure 2; and this chapter, Figure 12.18). For O^{18}/O^{16} a temperature effect was noted. Different parts of a given Aristotle's lantern (Figure 12.16*F*) will differentially incorporate isotopic carbon and oxygen. Since these lantern structures are composed of calcite, such variability is thought to reflect carbonate ions in which the isotopic carbon and oxygen had derived from different sources (metabolic carbon dioxide and/or seawater bicarbonate).

Skeletal Pore Pattern and Tube Feet

Nichols (1959) showed that tube feet, and skeletal pores through which they project, vary according to function in burrowing forms. Thus funnel-building, oral food-gathering, burrow-building, and sensory tube feet each have distinctive ambulacral pore patterns. [Durham (1966) suggested that pore patterns in fossils might permit discrimination of kinds, functions, and numbers of tube feet.]

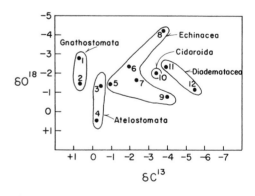

FIGURE 12.18 Information derived from living echinoids—II. Isotopic carbon (C^{13}) and oxygen (O^{18}) composition of Echinoidea orders grouped in superorders (after Weber and Raup, 1966*b*).
Orders: (1) Holectypoida. (2) Clypeasteroida. (3) Spatangoida. (4) Cassiduloida. (5) Temnopleuroida. (6) Arbacoida. (7) Echinoida. (8) Phymosomatoida. (9) Hemicidaroida. (10) Cidaroida. (11) Diadematoida. (12) Echinothurioida. Note: Cidaroida pattern uncertain.

Dimorphism, Spawning Cycles

Sexual dimorphism (that is, external differences in echinoid shell morphology) is generally lacking. Hyman (1955) cited the example of the deeper excavation of the *petaloids* in females of brooding spantangoids. Durham (1966) agrees that although sexual dimorphism is rare, it does occur — for example, in the clypeasteroids where the genital pore in females is larger than in the males.

In echinoids, spawning occurs according to extended breeding cycles in any given genus. Eggs and sperm are both shed into water, and fertilization and subsequent development ensues (Hyman, 1955; Boolootian, 1966).

Development and Metamorphosis

A microscopic creature, the larval echinopluteus (Figure 12.2) persists for four to six weeks. Temperature and available nutrients will affect the duration of this stage. Metamorphism from echinopluteus to juvenile stage is very rapid — about 60 minutes (Hyman, 1955); initially, mouth, anus, and periproct are lacking. Gordon (1926) designated completion of metamorphosis as the *imago* stage. Skeletal aspects of development have already been discussed.

Plates grow, resorb; soft-part anatomy alters, modifies, develops during postmetamorphosis — for example, gonad development or changes in the digestive tract. Juvenile and adult spines are distinguishable, and both may occur on the same test.

FOSSIL ECHINOIDS

Classification

The taxonomy of echinoids may be approached in several different ways. One can group characteristics as Philip has done (1965, and this text, Figure 12.20), establishing *morphological grades*. Characteristics include such items as whether the teeth of Aristotle's lantern are grooved or keeled; whether spines (radioles) are solid or hollow; whether the test is flexible or not; and so on.

To fit fossils and recent forms more closely into an evolutionary framework (that is, ancestor/descendant relationship) one can take the *phylogenetic approach*. This last method was used by Durham and Melville (1957), Durham (1966), and see Figure 12.19 of this chapter. All modern systems draw heavily on Mortensen's enormous contribution (1928–1952) in which recent and fossil forms were classified, described, and arranged into eleven orders and two subclasses: Regularia (that is, periproct within ring) and Irregularia (periproct outside the ring). (These are names for two types of tests defined as indicated. "The ring" is the oculogenital plate complex of the apical system.)

Raup (Chapter 15, and this chapter, Figure 12.18) applied both crystallographic data of the test and isotopic data (O^{18} plotted against C^{13}) to evaluate the coherence of the Durham-Melville classification. He found good agreement "on the whole" from crystallographic data (*C*-axis orientations) but noted some discrepancies on the family level in the order Echinoida. Also the isotopic data suggested that this order constituted an erroneous grouping, being polyphyletic. These crystallographic and isotopic approaches are more useful as a check on the limitations in classifications otherwise derived than a primary taxonomic mode.

Durham (1966) and Durham-Melville (1957) rejected Mortensen's subclasses. Their reasoning is of interest. (1) The *irregular* condition (periproct outside the ring) arose in three separate lineages. That is to say, the "subclass" Irregularia arose from more than one line of descent (polyphyletic). This point is disputed by Philip (1965, pp. 46–49). (2) Periproct migration was both geologically younger in time of appearance and in an evolutionary sense than a whole series of "more profound" morphological changes

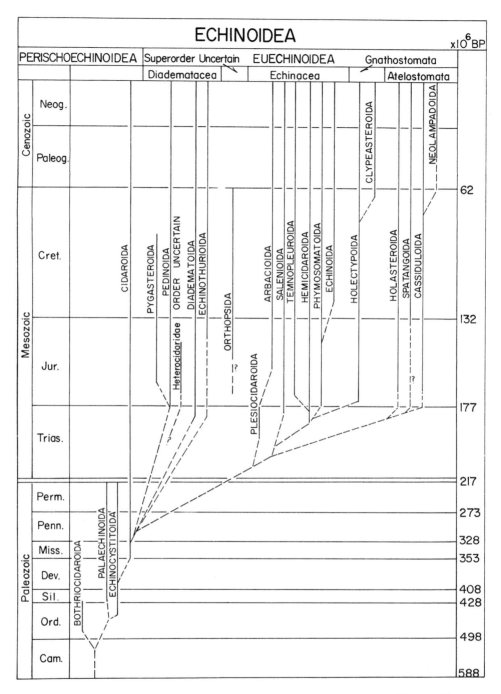

FIGURE 12.19 Derivation of echinoid orders through time (after Durham, 1966). (Compare Figure 15.12 and text, Chapter 15.) Note: 62, 132, and so on, are X $10^6 BP$—that is, millions of years before present. Scale used above Permian differs from that used for Paleozoic.

of the test (for example, condition of the perignathic girdle).

Having thus disposed of Mortensen's subclasses, the same authors set forth arguments for justifying their substitute subclasses: (1) *The late Paleozoic acquisition of external gills* in some echinoids, was a major evolutionary event. (2) A parallel and seemingly contemporaneous evolutionary event was ambulacral plate modi-

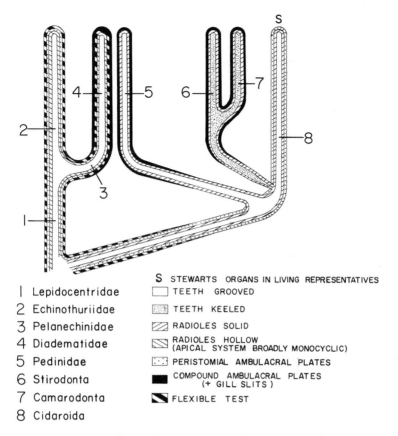

1	Lepidocentridae
2	Echinothuriidae
3	Pelanechinidae
4	Diadematidae
5	Pedinidae
6	Stirodonta
7	Camarodonta
8	Cidaroida

S STEWARTS ORGANS IN LIVING REPRESENTATIVES

☐ TEETH GROOVED

▦ TEETH KEELED

▨ RADIOLES SOLID

▧ RADIOLES HOLLOW (APICAL SYSTEM BROADLY MONOCYCLIC)

⋯ PERISTOMIAL AMBULACRAL PLATES

■ COMPOUND AMBULACRAL PLATES (+ GILL SLITS)

◣ FLEXIBLE TEST

FIGURE 12.20 Distribution of important morphological features in regular echinoids (after Philip, 1965).

S – Stewart's organs, seen from top view of Aristotle's lantern, there are five radially positioned, elongate sacs (expansion chambers for fluid), see Hyman, 1955, Figure 196A; radioles = spines; three structures of lantern teeth: *camarodont* – most regular urchins, teeth keeled longitudinally (keel = median crest along inner surface); *aulodont* – teeth lack keel, grow longitudinally; *stirodont* – teeth with keel, grow longitudinally.

Upper ends of paired demipyramids that enclose a tooth of Aristotle's lantern separated by a space (*foramen magnum*) that is "open" in the aulodont and stirodont condition but "closed" in camarodont condition. Closure of foramen magnum depends on whether epiphyses meet or not (see Figure 12.17D).

fication (= auricles develop) to allow supporting muscles of Aristotle's lantern a new attachment place.

As noted earlier, evolution of specialized respiratory surfaces (gills, and so on) in all the echinoderms has led to the *biological success* of modern forms. Any modification, such as muscle attachment site, pertains to the operation of the masticatory apparatus and hence involves feeding or, more probably, the efficiency of feeding. Thus since respiration and feeding are two critical elements in the life of an echinoid, there is a clear indication that a selective advantage en-

sues in certain kinds of skeletal modifications promoting such functions.

Actually, a broad radiation and diversification of post-Paleozoic noncidaroid echinoids is attributable, at least in part, to such skeletal changes (see Durham, 1966).

Following Treatise usage (with slight modifications), two new subclasses are recognized: Perischoechinoidea and Euechinoidea. Definition of these and of the 22 orders and illustrations of fossil representatives are given below.

Subclass 1. Perischoechinoidea. Flexible test usually (bothriocidaroids are

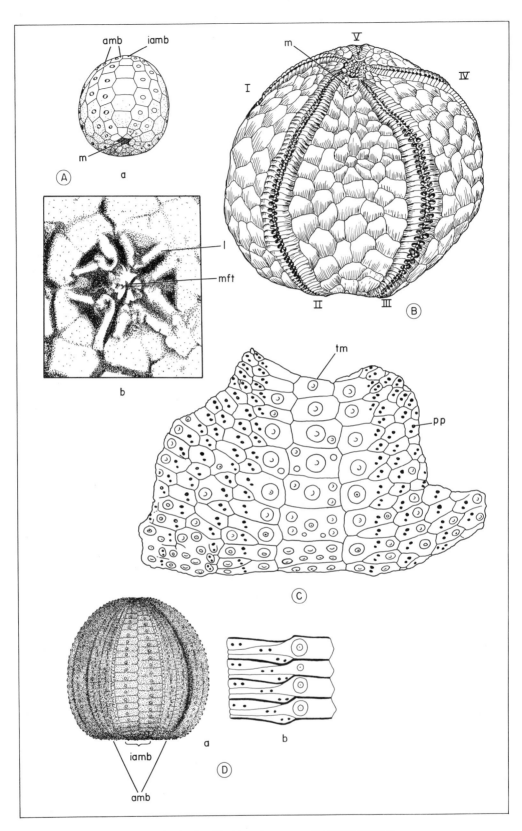

FIGURE 12.21 Representative fossil echinoids and some morphological features.
(1) Subclass Perischoechinoidea: (*A*) *Bothriocidaris* sp., M. Ord. (*a*) The oldest known echinoid.

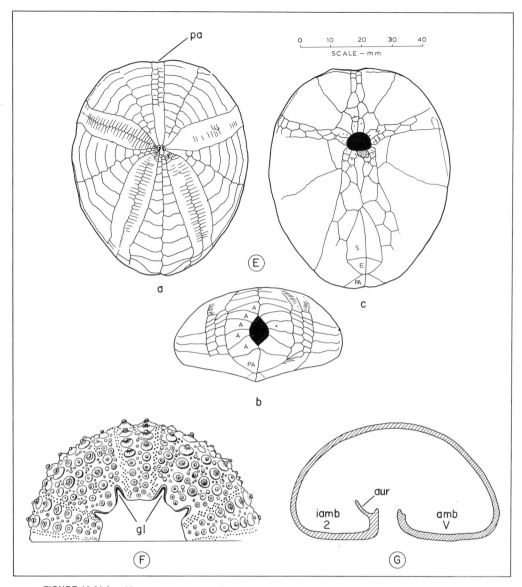

FIGURE 12.21 (cont.)

Note plate distribution; *amb*—two columns of ambulacral plates; a single column of inter-ambulacrals; *m*—mouth with teeth; schematic, (after Nichols); (*b*) lantern, internal view; *l*—laminae; *mft*—mouth with teeth; surrounding plates = peristome (after Myannil, 1962).

(*B*) *Aulechinus grayae*, lateral view, reconstructed; I–V, ambulacra; *m*—madreporite—genital. Note the many columns of interambulacral plates; see (*A*a) above, X 2.

(*C*) *Cravenechinus uniserialis*, Carb., Yorkshire, England, single column of interambulacrals; *tm*—median tubercle; *pp*—pore pairs (schematic).

(*D*) Subclass Euechinoidea (superorder Diadematacea), *Leiopedina tallevignesi*, Eoc., France. (*a*) Higher than broad; amb, interamb areas; (*b*) ambulacral plates, pore pairs in triads, enlarged.

(*E*) (Superorder Atelostomata); *Eupatagus (Gymnopatagus) Mooreanus*, U. Eoc., Fla. (*a*) Aboral view, ovoid test, ambulacral and interambulacral plates. Note: anterior ambulacrum is nonpetaloid, others show broad petals (a *petal* = differentiated adapical segment of ambulacrum with specialized tube feet); (*b*) anal view, plates surrounding anus (black) are *A*—anals, *PA*—postanals; (*c*) oral view, mouth (black) and surrounding peristomal plates; other plates include: *E*—episternals, *S*—sternals, *L*—labrum, *PA*—postanals.

(*F*) *Pseudodiadema* sp., gill slits (*gl*) (slits are indentations of peristomal margin of test through which stem of external gills passed); see Figure 12.16*A*.

(*G*) *Conoclypus aequidilatitus*, vertical section; peristomal invagination and auricle, schematic; *amb*—ambulacra; *iamb*—interambulacra; *aur*—auricle.

Illustrations redrawn and adapted from: (*A*a) Nichols, 1962; (*B*) MacBride and Spencer, 1939; (*C*) Hawkins, 1946; (*D*) Fell, 1966; (*E*) Fischer, 1951; (*F*) Forbes, cit. Durham, 1966; (*G*) Durham and Melville, 1957.

683

exceptions, having a rigid test). Endocyclic [*Endocyclic* = periproct located within oculo-genital ring; *exocyclic* = periproct located outside the ring.] (Fell and Pawson, 1966). Ambs of two to twenty columns; lantern present with grooved teeth; perignathic girdle absent or with only internal projection in the interambulacral plates (that is, no auricles); external gills absent; lack spheridia; possess specialized pedicellaria.

Geologic range: Ordovician to Recent.

The subclass consists of four orders (plus one uncertain order): Order 1. Bothriocidaroida (*Bothriocidaris*, Figure 12.21*A*). Order 2. Echinocystitoida (*Aulechinus*, Figure 12.21*B*). Order 3. Palaechinoida (*Cravenechinus*, Figure 12.21*C*). Order 4. Cidaroida (*Stereocidaris*, Figure 12.15*C, a*).

Subclass 2. Euechinoidea. Rigid test usually (Fell and Pawson, 1966). Endocyclic commonly but may be exocyclic secondarily; five bicolumnar ambs and interambulacral plates alternate. Plates are either held together or united by flexible integument or rigid sutures, or are imbricate. Lantern, gills, and gill slits may or may not be present. Auricles in perignathic girdle are either present or individuals derived from such forms (Durham, 1966). Spheridia and specialized pedicellariae present.

Geologic range: Carboniferous (?); Upper Triassic–Recent.

This subclass embraces 18 orders and one order uncertain. It is subdivided into four superorders and one uncertain superorder. A representative of each superorder is given below.

Superorder 1. Diadematacea. Range: L. Carb.–U. Trias.–Rec. (*Leiopedina*, Figure 12.21*D*).

Superorder 2. Echinacea. (*Echinus*, Plio.Rec., Figure 12.15*A*; *Tripneustes*, Plio.Rec., Figure 12.1*A*, part E).

Superorder 3. Gnathostomata. (*Mellita*. Mio.–Rec., Figure 12.17*B*).

Superorder 4. Atelostomata. (*Echinosigra*, Rec., Figure 12.1*A*, part F; *Eupatagus*, Eoc.–Rec., Figure 12.21*E*).

INFORMATION DERIVED FROM FOSSIL ECHINOIDS

Fossil Record and Evolution

Paleozoic Distribution. It has been observed that echinoids are among the rarest of Paleozoic fossils (MacBride and Spencer, 1939; Cooper, 1957). This holds for complete tests but not for spines and disarticulated plates, which are commonly encountered in Paleozoic well cuttings and on fossiliferous slabs. Although less frequently found than spines or isolated plates, pedicellariae are known from the Mississippian and Pennsylvanian of Texas, Illinois, and Missouri, and Aristotle's lanterns (jaws) are known from the Carboniferous of Missouri (Geis, 1936; Spreng and Howe, 1963).

In the older Paleozoic, MacBride and Spencer (1939) described some Scottish Upper Ordovician (Ashgillian) echinoids: *Aulechinus* (Figure 12.21*B*), *Ectinechinus*, and *Eothuria*—originally thought to be a holothurian. *Bothriocidaris* (Figure 12.21*A*) was described from the Ordovician outside the British Isles originally, although the type species, as now recognized, comes from the English Ordovician.

The Silurian of many parts of the world yield echinoid fossils: England (*Echinocystites*, for example), United States (*Koninckocidaris*), Sweden (*Gotlandechinus*). Devonian echinoids are widespread: in England, Europe, and America (*Lepidocentrus*, for example) and forms noted by Cooper (1931) and Thomas (1924).

By the Carboniferous (Mississippian–Pennsylvanian) in the United States, echinoids are frequently found as they are elsewhere [Great Britain, Europe (Belgium), USSR, and Asia]. Illustratively, one may cite a few Permian occurrences: Kansas (*Meekechinus*), Texas (*Xenechinus*).

Paleozoic Evolution. Figure 12.22-II graphically depicts the progress of echinoid speciation through the Paleozoic. These data are based on a world-wide sample ($n = 796$ fossils) assigned to a total of 125 species embraced in 36 genera.

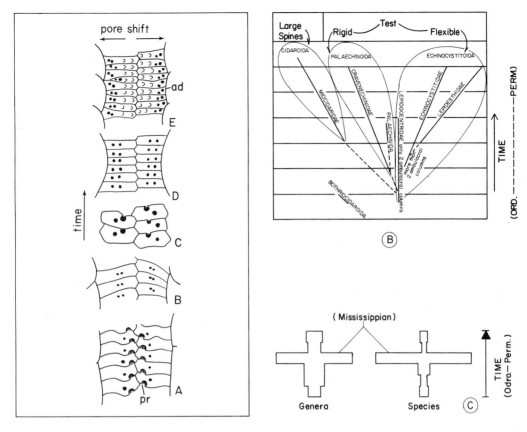

FIGURE 12.22 Information derived from Paleozoic echinoids—I (adapted from Kier, 1965).
(*A*) Ambulacral pore shift through geologic time. Pore shift from medial (perradial—*pr*) to adradial—*ad*. (*A*) *Ectinechinus;* (*B*) *Palaeodiscus;* (*C*) *Porechina;* (*D*) *Lepidocentrus;* (*E*) *Pholidechinus.* Of these, (*A*) is oldest, Ordovician; (*E*) is youngest, Mississippian; arrow indicates directions of pore shift, *pr* to *ad*.
Derivative data from Paleozoic echinoids—II.
(*B*) Possible phylogeny of Paleozoic echinoids. Compare Figure 12.19 where Bothriocidaroida, with rigid test, is now shown to be on the line from the same Cambrian common ancestor as Echinocystitoida, with flexible test.
(*C*) Echinoid speciation during Paleozoic time. (See text.) Observe the great radiation in Mississippian time.

(The Paleozoic species constitute about 2.5 percent of all described fossil echinoids. (See also Figure 12.22-I).

Within the indicated framework of echinoid speciation in the Paleozoic, one can place the evolutionary trends that it reflects. Kier (1965) identified 12 such trends. These are grouped below in order to facilitate relating these data to the molecular level of the genetic code.

*Ambulacra. Columns—*increase in number (some families); adoral expansion of ambulacra (Echinocystitoida); *plates—*pore shift on ambulacral plates (Figure 12.22); loss of perradial (median) groove along median suture (that is, where two ambulacral plates were in contact, see Figure 12.22); *interambulacra: plates—*trend to develop regular shape.

Test. Total number of plates increased in Echinocystitoida and Palaechinidae; *size—*trend to size increase in last-named orders; *shape—*tendency to flatten in Echinocystitoida; *spines—*differentiation of spines and development of tubercles.

Lantern. Trend to increase complexity (all 40 elements of the modern lantern were already present in Silurian forms).

Water Vascular System. Pore shift (see above); enclosure of tube feet; loss of internal enclosure of radial water vessel.

Many of the modifications and changes in skeletal components appear to have been related to *feeding, locomotion,* and probably *respiration.* Certain columnar, plate, pore, and related specifications conferred an advantage over others and were *selected for.* Data on modern forms suggest that two tendencies observed in Paleozoic echinoids allowed for a greater division of labor and probably greater efficiency: increase in the number of tube feet (a condition inferred from increase in number of ambulacral plates) and differentiated spines.

It is altogether likely that genetic modification affected more than one skeletal characteristic. The code for ambulacra as a test structure remained constant; this was a primary (stable) code. Gene loci affecting columns, plates, pores, and so on, in terms of their number, position, and shape either express one or another portion of a diverse potential or reflected establishment of new potential. This might be attributed to alteration on the molecular level of some parts of the secondary (variable) code.

On the other hand, Ernst Mayr (Chapter 14) reminds us that some loci may have been affected and morphological changes ensued, which did not confer any advantage.

Of great interest is the late Permian demise of all known Paleozoic echinoid genera except a cidaroid *Miocidaris* (Nichols, 1962; Fell, 1966, and Figure 12.22-II, this chapter). Apparently this stock was ancestral to all post-Paleozoic echinoids. All Paleozoic forms, as well as the cidaroids (which persist to the present day), appear to have lacked external gills. [The presence of external gills is recognized in fossilized tests by indentations (gill slits) around the peristomal margins, see Figure 12.21*F*).]

External gills are also referred to "outpouchings" of the peristomal membrane. In echinoderm evolution this was one of the modes of creating new specialized respiratory surfaces (Farmanfarmaian, 1966). Durham (1966) considered the acquisition of external gills in the Euechinoidea—which experienced a marked evolution and radiation (Figure 12.19)—as not just another event, but as a *major* happening in echinoid evolution. This judgment seems excessive in light of experimental evidence, unless an originally greater role in oxygen uptake was secondarily diminished or lost.

Modern *Strongylocentrotus purpura* (with five pairs of peristomal gills), for example, showed negligible diffusion of seawater oxygen across these gills. Oxygen uptake before and after excision of the gills was the same in the five test animals! By contrast, diffusion through the tube feet accounted for 40 percent of the oxygen uptake in the test animals.

Post-Paleozoic Distribution. Mesozoic time saw the greatest radiation of cidaroids (Fell, 1966). During the Triassic, *Cidaris,* for example, was part of the reef facies in the Alps. In the Jurassic, similarly, cidaroids lived in a coral reef environment of the Mediterranean. In Europe they were among characteristic Cretaceous fossils.

Mesozoic echinoids are abundant in places: Lower Cretaceous spatangoid marls (after the mud-eating *Spatangus* and related forms); Upper Cretaceous *Micraster* chalk (which includes various chalk echinoids such as *Holaster* and others as well) (Gignoux, 1955, and see Kermack, 1954; Nichols, 1960). The Lower and Upper Cretaceous of Texas has "vast numbers" of echinoids at several horizons, and along the Atlantic coast the New Jersey Cretaceous Vincentown Sand has a broad representation of echinoids (Clark and Twitchell, 1915).

The sand dollar *Dendraster excentricus* is the main fossil in certain Pleistocene deposits of California, but this biofacies is local and restricted in occurrence (Durham, 1966).

Tertiary to Recent echinoid faunas are extensive, and only a few selected illustrations can be cited. Genera belonging to families in the order Clypeasteroida

(*Clypeaster, Scutella, Mellita, Rotula*) range from the Tertiary (and a few genera, such as *Fibularia*, from the Cretaceous) to the Recent. In the Italian Nummulitic (Schio beds, Paleogene) *Clypeaster* and *Scutella* are common. These forms and others occur also in Mediterranean Neogene faunas (Miocene and Pliocene) (Gignoux, 1955). The Mediterranean and Caribbean areas both yield rich Tertiary faunas, among which are the spatangoids (Fischer, 1966). The New Zealand Tertiary also has several Indo-Pacific cidaroids, as well as several endemic genera, *Ogmocidaris*, and others. Fell (1954 and Figure 12.22*A*, this chapter) envisioned a shallow-water migration route from the north along the Malaysian-Indonesian arc to New Zealand for all the noted genera except *Goniocidaris*.

Some genera are worldwide in distribution as fossils, as, for example, *Echinocyamis*. Others are restricted geologically and geographically as *Periarchus* (Upper Eocene only, and from the Gulf of Mexico and southeast United States to Cuba) (Clark and Twitchell, 1915; Durham, 1966). Some genera are confined to

Europe or West Africa, and so on. Cassiduloid species range from Jurassic and Cretaceous through Tertiary and Quaternary (Kier, 1966).

Certain Tertiary forms are so abundant that their generic name designates the zone in which they occur as fossils as the *Periarchus* zone (*P. lyelli*, Eocene limestone, Ocala Formation of Florida) (Fischer, 1951; Richards, 1953).

Data on the echinoids of the California Tertiary were compiled by Grant and Hertler (1938). By Late Cenozoic time in the northeastern Pacific area, an estimated one-third to one-fourth of the echinoid populations (clypeasteroids chiefly) had evolved locally (= endemism). This markedly contrasted with few endemic genera previously. The changeover to enhanced local evolution yielding new genera and species is attributed to restriction of intermixing of Atlantic and Pacific populations via seaway connections across Mexico and Central America, cutoff and blocking of such seaway connections by Late Tertiary, and the progressive curtailment of latitudinal spread of tropical temperatures (Durham and Allison, 1960). Rise of local species and genera may be attributed to fractionation of echinoid gene pools by geographic isolation of the northeast Pacific fraction.

The Miocene of the Caliente Range yields *Echinarachnius* (Eaton et al., 1941); the Tertiary shallow-water facies of the Santa Maria district has *Dendraster* (Woodring et al., 1950), and the Lower Pliocene deep-water facies, *Araeosoma* (Woodring, 1938). As with molluscan and other faunas, many Tertiary echinoid genera continue through Pleistocene to Recent time (Durham, 1966, Figures 355, 356).

Post-Paleozoic Evolution. Cidaroid evolution is of extraordinary interest since the test pattern of Permo-Triassic cidaroids became the "fundamental" plan for all post-Paleozoic forms. The Devonian-Carboniferous cidaroids (*Archaeocidaris*, for example) had four or more interamb columns and narrow bicolum-

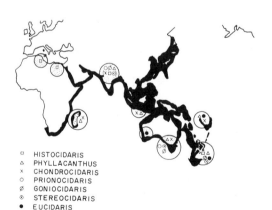

◻ HISTOCIDARIS
△ PHYLLACANTHUS
× CHONDROCIDARIS
○ PRIONOCIDARIS
⌀ GONIOCIDARIS
⊙ STEREOCIDARIS
● EUCIDARIS

FIGURE 12.22A Information derived from Mesozoic-Cenozoic echinoids—I. Probable distribution of seven genera (Cidaridae). Areas in heavy black indicate possible former shallow-water migration routes, mainly North to South along the continental shelf; these are parts of the sea floor presently above the 2000-meter contour.

Data for *Goniocidaris* suggest a possible migration in a reverse direction. *Pronocidaris* — during Cret. and Eoc. time, a European genus. Redrawn from Fell, 1954.

nar ambs. Permo-Triassic cidaroids had evolved bicolumnar intervals. The test now had 20 columns or 10 bicolumnars (five ambs, five intervals) — *Miocidaris* is illustrative (Fell and Pawson, 1966).

Genetically, the above-noted archetypal pattern denotes stability in that portion of the genetic code relevant to test architecture. Thus for the past $180 \pm$ million years despite other genetic modifications reflected in echinoid anatomy, no loci that could affect the 20-columnar pattern have been disrupted by recombinations, mutations, or other influences operative on the molecular level.

Two major genetic changes occurred in euechinoid evolution: plate fusion and anus migration. Several adjacent ambulacrals fused. The anus migrated out of the apical region (= endocyclic condition) to an interamb (= exocyclic condition).

Echinoid stocks in which anal migration did not occur (endocyclic) were genetically modified to give rise to compound ambulacral plates. By contrast, where the anus did migrate to an interamb, unfused single amb plates persisted. Fell (1966) observed that in the first case (fused amb plates) populations were and are reef-dwellers subject to strong wave action. The single amb plate types, with thin, delicate tests, constituted populations that generally, according to Fell, do not live in high energy zones and where they do, they bury themselves in the substrate and, hence, are protected. Accordingly, amb-plate fusion can be seen to give a selective advantage to endocyclic forms only.

Among primitive skeletal structures or components are less than 20-column plate pattern; endocyclic conditions; presence of jaws that have been secondarily lost in recently evolved groups.

Durham (1966) distinguished six principal tendencies in post-Ordovician evolution of echinoids. These include the following.

1. *Rigid Test*. This was general in post-Paleozoic forms except the deep-water Echinothurioida with reduced skeletal calcification.

2. *Antero-Posterior Orientation*. This tendency is associated with anal migration, discussed above (exocyclic condition) and follows it. Both test elongation (as in the cassiduloids — Kier, 1966, Figure 380) and preferred direction of locomotion followed the new antero-posterior axis. [This tendency led to the opening of new ecologic niches in the substrate into which burrowing types, such as the cassiduloids and spatangoids (Fischer, 1966) entered.]

3. *Water Vascular System Modification and Specialization*. (See earlier section on tube feet, and so on.) Accompanying the formation of compound amb plates noted above, tube feet multiplied. When the substrate niches opened up for burrowing types of echinoids (Figure 12.17*C*; Figure 12.23*B*), tube feet were modified and became specialized for respiration, food-gathering and sensing, as well as sanitary and other functions.

4. *Appendage Modification and Specialization*. Pedicellariae (Figure 12.15*C, D*) and spines, as with tube feet, are among the external appendages showing great variation in shape, size, and function. By Jurassic time, pendicellariae were highly specialized.

Echinoid spines display a broad spectrum of adaptive modification to serve varied functions: *protective* (shieldlike mosaic of flattened dorsal spines as in *Podophora* and related forms that effectively can resist wave action along a rocky coast); *lethal* (poison glands in spine tips, as in the family Echinothuriidae); *locomotor* (short, paddle-shaped spines); curved, flattened spines for *digging* or *water circulation* (by movement of small ciliated spines — clavicles that are borne on narrow bands or fascioles, Figure 12.23); or to serve an indirect, *respiratory* and *sanitary* function (elongated spines that aid in tube building).

Clearly, external appendages were one of the great genetic variables in echinoid evolution.

5. *Modification in Reproduction and Brooding*. (See earlier section on "Echinoderm Larval Types.") Development of

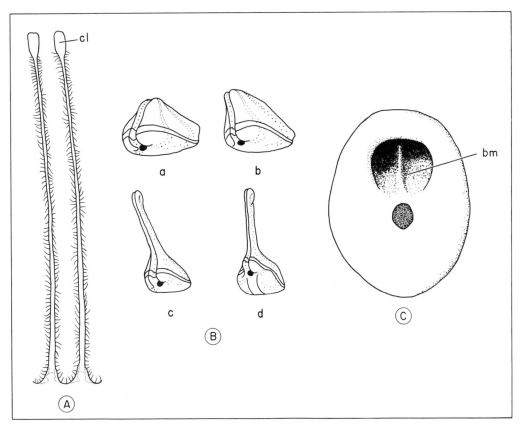

FIGURE 12.23 Information derived from Mesozoic-Cenozoic echinoids – II (Adaptive modifications).
(A) Spine modification: *Echinocardium cordatum*, spine; side view of two clavulae [= ciliated spines]; height of each clavulae (*cl*) = 2.0 mm.
(B) Modification of deeper burrowers. Rostrum lenthening; Cret., England. (A)–(D) *Infulaster – Hagenowia* Series: (A) *Infulaster excentricus;* (B) *Hagenowia infulasteroides;* (C) *H. blackmorei.* (A) Upper Middle Chalk; (B)–(D) Upper Chalk. (After Nichols, 1960.)
(C) Modification of test as brood chamber. *Fossulaster halli,* Pl. Mio., Australia; oral surface of female with bipartite marsupium (brood pouch, bm), X 6.

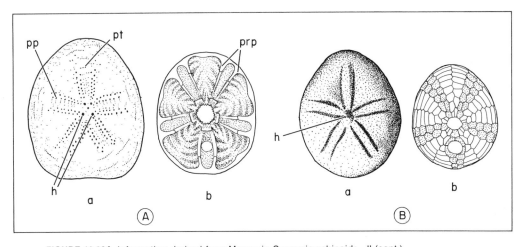

FIGURE 12.23A Information derived from Mesozoic-Cenozoic echinoids – II (cont.)
(A) *Echinocyamus pusillus,* Rec., Europe, (a) Aboral view, *pp* – pore pairs, *pt* – petals, *h* – hydropore; (b) oral view, interior, *prp* – five pairs of internal radiating partitions.
(B) *Mortonia australis,* Rec., Hawaii. (a) Aboral view, X 2.5; *h* – single hydropore; (b) oral surface, X 2, interamb stippled, showing central mouth and anus in posterior interamb. [Figures 12.23, 12.23A – after several authors].

echinoid lecithotrophic larvae (that is, that feed on egg yolks), alongside of planktotrophic types (that feed on other plankton), was related to shedding of fewer but larger eggs. This is deemed to have been a move toward greater efficiency since egg number is reduced.

A great waste occurs via the planktotrophic larval route (vast quantities of eggs and sperm are shed). Also, brooding on the surface of tests in special brood pouches or equivalents (curling of spines over a test depression, and so on) denotes a trend toward protection of the young as opposed to the more rigorous unprotected open water development. [The echinoid *Fossulaster* from the Australian Tertiary displays dimorphism, the female test has on its oral surface a brood chamber divided into two parts (Figure 12.23, *C*).]

6. Feeding—Specialization. This development relates to numbers 2 and 3 above. Among the feeding specializations are ingestion of organic-rich sediments (loss of lantern since no longer adaptive—spatangoids); in the clypeasteroids, food gathering along food grooves aided by accessory tube feet. Among such clypeasteroids as the sand dollar, *Dendraster*, a complex of food grooves occurs on the oral surface. Living in an inclined position in the substrate facilitated suspension feeding. A functional lantern is retained in clypeasteroids but does not serve in food gathering (lantern lacks compass and teeth have no lateral flanges but apparently were adequate for masticating food).

One can obtain some concept of the time involved in the evolution of such feeding specializations by realizing that the major development of a food groove complex on the oral surface in the clypeasteroids occurred from Oligocene to Pliocene time (Durham, 1966, Figure 347), although it extended through Pleistocene to Recent in some sand dollars. The major evolution of this feature spans some 23 million years. Oral food grooves were completely absent in the Upper Cretace-

ous through Paleocene ancestral condition. Hence their appearance represents a novel morphological feature.

The clypeasteroids underwent a marked radiation from Cretaceous time (two genera) to Miocene time (36 genera). Subsequently, about a dozen genera have become extinct. Spatangoids and cassiduloids, by contrast, attained a maximum in Eocene time, declining subsequently (Durham, 1966, Figure 355; Fischer, 1966, Figure 432; Kier, 1966, Figure 383).

Of a total of more than 8000 species of fossil and living echinoids, only about 10 percent inhabit modern seas. Despite this obvious decline, living forms display a broad spectrum of adaptive modifications and are distributed in almost all marine environments.

Durham (1966) holds that living echinoids are probably more successful now than at any time in the geologic past. Certainly, in the many adaptive modifications of external appendages and in their exploitation of substrate niches, this is true. However, whether echinoids that reached a peak of speciation in the Tertiary and subsequently declined, can be considered presently successful in the evolutionary sense, is doubtful. Data on new genera from Pleistocene to Recent time would be of interest in this respect.

Among the clypeasteroids only in the family Fibulariidae (suborder Laganina) did two new genera appear in the Recent in the Indo-Pacific area: *Fibulariella* and *Mortonia*. The first-mentioned of these genera is like *Fibularia*, which is from the same area but has a geologic range from Upper Cretaceous to Recent, and though of worldwide distribution in the fossil record, is restricted to the Indo-Pacific at present.

Mortonia (Figure 12.23*A*, part B) is like *Echinocyamus* (Figure 12.23*A*, part A), of similar size to *Fibularia* but not presently restricted to the Indo-Pacific area. The listing below gives several characteristics by which the new genera differ from their close relatives or parental stock.

Fibularia

1. Periproct: close to peristome with five large periproctal plates.
2. Tube feet: with calcareous disc.
3. Hydropores: in a groove.

Echinocyamus

1. Internal radiating partitions—five pairs.
2. Test moderately flattened.
3. Hydropores few; not in groove.
4. Pore pairs usually oblique.

Fibulariella

1. Periproct: elongated, with numerous small periproctal plates.
2. Tube feet: no calcareous disc.
3. Hydropores: no groove.

Mortonia

1. Single posterior of pair of partitions only.
2. Oral surface concave.
3. Single hydropore.
4. Radial ridge between members of pore pairs.

It can be seen that the following systems are involved: excretory (periproct—shape and plates), respiratory (arrangement of pore pairs), water vascular (tube feet, hydropores), skeletal [(test shape and strength (internal partitions)]. Isolates from the ancestral stock (*Fibularia, Echinocyamus*) apparently became distinct species and genera, essentially by a process of loss and reduction of structures: the loss of the disk in the tube feet, the loss of the groove for the hydropores, the loss of all but one of the internal partitions, reduction in the number of hydropores. In brief, that part of the parental code represented by such characteristics appears to have been selected against.

Much greater detailed information would be necessary to permit one to understand the relationship of the tendency noted to future speciation. Do loss and reduction of structures hold promise of a resurgence of new speciation in the Fibulariidae?

The fact that the indicated genetic events did occur in an ecologic niche and habitat shared by the parental stock, shows that some evolutionary potential probably exists in some other occupied niches. One wonders into what niches modern echinoids might radiate since they now occupy most suitable life zones.

As subsequent discussion of the holasteroids will indicate, additional migration into deeper waters may occur in the future. Even so, this is likely to be minor

for the echinoids if one bases his judgment on an analysis of tables of modern distribution (Durham, 1966, Figures 198–199). Thus 63 percent of all living echinoids occur in or on the littoral or sublittoral substrates, and below 500 meters species abundance progressively declines. There is also some evidence that population density of echinoids in deep sea substrates is relatively sparse at present (Bruun, 1956; Hessler and Sanders, 1967, Table 3).

Among the Holasteroida (Wagner and Durham, 1966), two families lacking a fossil record are apparently of Recent origin: the Calymnidae and the Pourtalesiidae (see genus *Echinosigra*, Figure 12.1*A*, part *F*). Another family of interest here is the Urechinidae that made its first appearance in Miocene time and are widespread today, and the family Holasteridae. The latter experienced great radiation and diversity from Cretaceous to Tertiary time. It had a large number of genera that became extinct by late Cretaceous time (some 30 genera). A single genus, *Stereopneustes*, known from the Recent, survives in the Japan-East Indian area (Indo-Pacific).

Many fossil holasteroids are found in Cretaceous chalks or in other fine sediments. Generally these represent shallow-water deposits. By contrast, most living forms that have brittle shells are deep-water dwellers. Thus urechinids, for example, range from 110 meters down to

4800+ meters, while the calyminids have been collected at depths of 4800+ meters. The range of pourtalesiids is from about 50 meters to some 7000 meters. [*Pourtalesia aurorae*, 7250 meters; *Echinosigra paradoxa*, 3000 and 4000 meters — Indian Ocean, Galathea Expedition (Bruun, 1956).]

A high percentage of Tertiary types constitute the modern abyssal faunas (Menzies and Imbrie, 1958). The study cited, however, referred only to crinoids among echinoderms. The urechinids among the echinoids appear to fit this general trend. The spread of the urechinids and pourtalesiids, for example, from the sublittoral to the abyssal, or hadal, depths indicates such migration. Similarly, the range of *Stereopneustes* from sublittoral to bathyal depths seems to be an equivalent indication. It seems clear that some Recent holasteroid genera and at least one Tertiary-derived genus have migrated to deeper waters during post-Tertiary and post-Pleistocene time, respectively.

Test Modification and Burrowing Habit

Numerous adaptations of skeletal configurations and structural development are observed in fossil as well as living echinoids. Many of these have to do with the burrowing capacity or existence (see Figure 12.23, *B*). The continuous vertical sequence of *Micraster* in the Cretaceous chalk of England is another excellent example (Figure 12.24). Nichols (1960, and see Neaverson, 1955, Figure 13) found a progressive increase in number of pore pairs in the main *Micraster* line. Deeper burrowing species replaced shallower burrowers after Middle Chalk time. Corresponding test changes suggested, by analogy to living forms, a gradual shift through time to reliance on ciliary feeding over employment of tube feet for this purpose.

Biostratigraphy

In addition to the use of *Micraster* in zonation of the Cretaceous chalks of

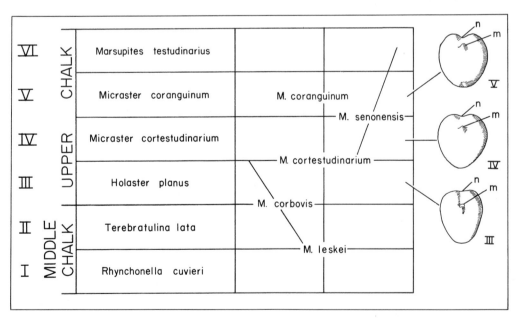

FIGURE 12.24 Evolution of echinoid genus *Micraster* (*M*) in Cretaceous Chalk Sea of England. Thickness, several hundred ft. Zones I–II, brachiopods; III–V, echinoids; VI, free-floating crinoids. * = shallow burrowers (*M. leskei, M. corbovis*); x = deep burrowers (*M. cortestudinarium, M. coranguinium*), + = partial burrowers, migrant stock (*M. senonensis*); n — interior notch; m — mouth (in IV and V, observe partial and then complete cover of mouth by labrum).
Illustration redrawn and modified from Kermack, 1954; Neaverson, 1955, including data from Nichols, 1960.

FIGURE 12.24A Zonation[a] of spatangoid Marls (L. Cret.) of France, by species of the echinoid, *Toxaster* (Data from Denizot, 1934).

Stages	Species of Toxaster
Aptien:	
Gargasien	
Beloulian—upper	*T. collegnoi*
lower	
Barrémien	
Barrémien	[b]
Baratélien	*T. ricondeaui*[c]
Cruasien	*T. amplus*
Néocomien	
Hautoverien—upper	*T. retusus*[d]
lower	*T. lorioli*
Valanginien	
Berrasien	*T. granosus*

[a]Ammonite and molluscan zones define these stages elsewhere.
[b]*Heteraster oblongus.*
[c]Also, echinoid *Heteraster couloni.*
[d]Illustrated species.

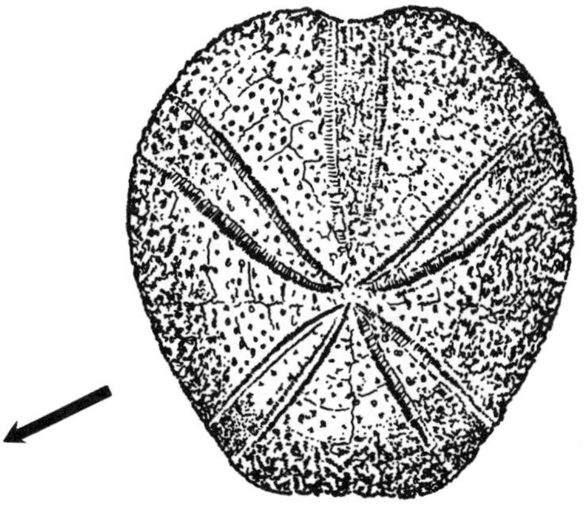

England (Figure 12.24), another interesting example is utilization of several distinct species of the echinoid genus *Toxaster* to zone the spatangoid marls (Lower Cretaceous) of France. (Denizot, 1934) (Figure 12.24*A*).

REFERENCES

Echinoids

Allison, E. C., J. W. Durham, L. W. Mintz, 1967. New Southeast Pacific echinoids: Occasional Papers, Calif. Acad. Sci. No. 62, 23 pp., 32 figures.

Borradaile, L. A., and F. A. Potts, 1958. "Class Echinoidea," in G. A. Kerkut, ed., The Invertebrata. Rev. 3rd ed., Cambridge Univ. Press, pp. 687–694.

Binyon, J., 1966. Salinity tolerance and ionic regulation, in R. A. Boolootian, ed., Physiology of Echinodermata (= PE). Interscience, pp. 359–373.

Boolootian, R. A., 1966. Reproductive physiology, in PE, pp. 561–606.

———— and J. L. Campbell, 1964. A primitive heart in the echinoid *Strongylocentrotus purpuratus*: Science, Vol. 145, pp. 173–175.

Bruun, A. F., 1956. Animal life of the deep sea bottom, in A. F. Bruun, et al., eds., The Galathea Deep Sea Expedition, 1950–1952, pp. 149–195.

Chesher, R. H., 1963. The morphology and function of the frontal ambulacrum of *Moira atropos* (Echinoidea: Spatangoida): Bull. Marine Sci. Gulf and Caribbean, Vol. 13 (4), pp. 550–573.

Clark, W. B., and M. W. Twitchell, 1915. The Mesozoic and Cenozoic Echinodermata of the United States: U. S. G. S. Monog. 54, pp. 1–131, plates 1–100.

Cooper, G. A., 1931. *Lepidechinoides* Olsson, a genus of Devonian echinoids: J. Paleontology, Vol. 5 (2), pp. 117–142.

Denizot, G., 1934. Description des massifs de Marseillevergre et de Puget: Ann. Musée d'hist. nat. Marseille, Vol. XXVI, Mem. V (pp. 138–142 esp.).

Deutler, F., 1926. Über das Wachstum des Seeigelskeletts: Zool. Jahrb. Abt. Anat. und Ontog. der Tiere, Vol. 48, pp. 119–200, plates 3–9.

Devanesen, D. W., 1922. The development of the calcareous parts of the lantern of Aristotle in *Echinus miliaris*: Roy. Soc. London, B., Vol. 93, pp. 468–485, plates 11–15.

Durham, J. W., 1966. "Anatomy," "Ecology and Paleoecology," "Phylogeny and Evolution," "Classification," in R. C. Moore, ed., Treat. Invert. Paleontology (= TIP), Part U, Echinodermata 3, pp. U214, U253, U266, U270, U297, U440, U450, U491.

————— et al., 1966. "Systematic Descriptions." TIP, p. U297.

————— and E. C. Allison, 1960. Cretaceous and Cenozoic history of Northeastern Pacific echinoid faunas: G. S. A. Program, Ann. Mtg., Denver, pp. 86–87.

————— and R. C. Melville, 1957. A classification of echinoids: J. Paleontology, Vol. 31 (1), pp. 242–272, 9 text figures.

Eaton, J. E., et al., 1941, Miocene of Caliente Range and environs, California: A.A.P.G. Bull., Vol. 25 (2), pp. 193–262, 9 plates.

Endean, R., 1966. The coelomocytes and coelomic fluids, PE, pp. 301–321.

Farmanfarmaian, A., 1966. The respiratory physiology of echinoderms, PE, pp. 245–261.

Fell, H. B., 1966. "Cidaroids," TIP, p. U312.

—————, 1954. Tertiary and Recent Echinoidea of New Zealand: Cidaridae: New Zealand Geol. Surv. Paleont. Bull., Vol. 23, 62 pp., 15 plates.

—————, et al., 1966a. "Euechinoids," TIP, p. U339.

————— and D. L. Pawson, 1966b. General Biology of Echinoderms, PE, pp. 1–45.

Fischer, A. G., 1966. "Spatangoids," TIP, pp. U543–U628.

—————, 1951. The echinoid fauna of the Inglis Member, Moody's Branch Formation: Part II: Florida Geol. Surv. Geol. Bull. No. 34. Paleont. Studies, 85 pp., plates, 1–17.

Geis, H. L., 1936. Recent and fossil pedicellarieae: J. Paleontology, Vol. 12 (6), pp. 427–448, plates 58–61, 1 text figure.

Gordon, Isabella, 1929. V. Skeletal development in *Arbacia, Echinarachnius*, and *Leptasterias*: Roy. Soc. London Philos. Trans. (= RST), Ser. B. Vol. 217, pp. 289–332.

—————, 1927. VI. The development of the calcareous test of *Echinocardium cordatum*: RST, Vol. 215: 255–312.

—————, 1926. VII. The development of the calcareous test of *Echinus miliaris*: RST, Vol. 214, pp. 259–311.

Hawkins, H. L., 1946. *Cravenechinus*, a new type of echinoid from the Carboniferous limestone: Geol. Mag., Vol. 83, pp. 192–197, plate 13.

Hessler, R. R., and H. L. Sanders, 1967. Faunal diversity in the deep sea: Deep Sea Research, Vol. 14, pp. 65–78.

Hyman, L. H., 1955. The Invertebrates: Echinodermata. McGraw-Hill, Vol. IV, 705 pp.

Kermack, K. A., 1954. A biometrical study of *Micraster coranguinium* and *M. (Isomicraster) senonensis*: RST, Vol. 237, pp. 375–428, plates 24–26.

Kier, P. M., 1966. "Non-cidaroid Paleozoic echinoids" and "Cassiduloids": TIP, pp. U298, U498.

—————, 1965. Evolutionary trends in Paleozoic echinoids: J. Paleontology, Vol. 39 (3), pp. 436–465, plates 55–60, 26 text figures.

MacBride, E. W., and W. K. Spencer, 1939. Two new Echinoidea, *Aulechinus* and *Ectinechinus*, and an adult plated holothurian, *Eothuria*, from the Upper Ordovician of Girvan, Scotland: RST, Vol. 29, pp. 91–134, plates, 10–17.

Melville, R. V., 1966. "Order Pygasteroida": TIP, p. U365.

————— and J. W. Durham, 1966. "Introduction" and "Skeletal Morphology": TIP, pp. U212, U220.

Menzies, R. J., and J. Imbrie, 1958. On the antiquity of the deep sea bottom fauna: Oikos, Vol. 9 (II), pp. 192–208.

Millott, N., 1954. Sensitivity to light and the reactions to changes in light intensity of the echinoid *Diadema antillarium* Philippi. RST, Ser. B, Vol. 238, p. 187–220.

Moore, H. B., 1966. Ecology of echinoids: PE, pp. 73–83.

Mortensen, T., 1928–1951. A monograph of the Echinoidea. Vol. I–V. C. A. Reitzel (Copenhagen) and Oxford Univ. Press (London).

Myannil (Männil), R. M., 1962. The taxonomy and morphology of *Bothriocidaris* (English summary, Russian text): Esti NSV Teaduste Akad. Geol. Inst. Uurimused, Vol. 9, pp. 144–188 (Russian text), 188–190 (English summary), 5 plates, 22 text figures.

Nichols, D., 1966. Functional morphology of the water vascular system: PE, pp. 219–239.

_____, 1962. Echinoderms. Hutchinson Univ. Libr., London, 200 pp.

_____, 1959. Changes in the Chalk heart-urchin *Micraster* interpreted in relation to living forms: RST, Vol. 242, pp. 342–437.

Okazaki, K., 1960. Skeletal formation of sea urchin larvae: II. Organic matrix of the spicule: Embryologia, Vol. 5 (3), pp. 283–320.

Onoda, K., 1931. Notes on the development of *Heliocidaris crassispina* with special reference to the structure of the larval body: Mem. Coll. Sci. Kyoto (b), Vol. 7, pp. 103–134.

Philip, G. M., 1965. Classification of echinoids: J. Paleontology, Vol. 39 (1), pp. 45–62.

Pilkey, O. H., and J. Hower, 1960. The effect of environment on the concentration of skeletal magnesium and strontium in *Dendraster*: J. Geology, Vol. 68 (2), pp. 203–214.

Raup, D. M., 1967. Echinoid growth: theoretical morphology: G. S. A. Program, Annual Mtg. (New Orleans): Abstract, pp. 181–182.

_____, 1966. The endoskeleton: PE, pp. 379–393.

_____, 1960. Ontogenetic variation in the crystallography of echinoid calcite: J. Paleontology, Vol. 34 (5), pp. 1041–1050.

Smith, J. E., 1966. The form and functions of the nervous system: PE, pp. 503–510.

Spreng, A. C., and W. B. Howe, 1963. Echinoid jaws from the Mississippian and Pennsylvanian of Missouri: J. Paleontology, Vol. 37 (4), pp. 931–938, 6 text figures.

Swan, E. F., 1966. Growth, autotomy, and regeneration: PE, pp. 397–425.

Wagner, C. D., and J. W. Durham, 1966. "Holectypoids," and "Holasteroids": TIP, pp. U440, U523.

Weber, J. N., and D. M. Raup, 1968. Comparison of C^{13}/C^{12} and O^{18}/O^{16} in the skeletal calcite of Recent and fossil echinoids: J. Paleontology, Vol. 42 (1), pp. 37–50, 5 text figures.

_____, 1966a. Fractionation of the stable isotopes of carbon and oxygen in marine calcareous organisms – the Echinoidea. Part 1. Variation of C^{13} and O^{18} content within individuals: Geochim. et Cosmochim. Acta, Vol. 30, pp. 681–703.

_____, 1966b. Same, Part 2. Environmental and genetic factors: same, Vol. 30, pp. 705–736.

Woodland, W., 1906. Studies in spicule formation, III. On a mode of formation of the spicular skeleton in the pluteus of *Echinus esculentus*: Quart. J. Micros. Sci., Vol. 49, n.s., pp. 305–322.

Wolpert, L., and T. Gustafson, 1961. Studies of the cellular basis of morphogenesis of the sea urchin embryo. Development of the skeletal pattern: Exper. Cell. Res., Vol. 25, pp. 311–325. Figures 1–5.

Woodring, W. P., 1938. Lower Pliocene mollusks and echinoids from the Los Angeles Basin, California: U.S.G.S. Prof. Paper 190, 67 pp., 9 plates.

_____ and M. N. Bramlette, 1950. Geology and Paleontology of the Santa Maria District, California: U.S.G.S. Prof. Paper 222, 142 pp., 23 plates.

FOSSIL CARPOIDS

Subphylum Homalozoa

Extinct echinoderms with body flattened dorso-ventrally; radial (pentamerous) symmetry absent (Caster, 1967); may be asymmetrical or display some degree of secondary bilateral symmetry (Caster, 1967; Fell and Moore, 1966). Geologic range: Middle Cambrian–L. Devonian. [Embraces old class Carpoidea (= Heterostelea), which is now distributed among three classes defined below.]

Class Stylophora. "Uniradiate" homalozoans with plated theca to which is attached a brachial process (= a jointed appendage, the "aulacophore" bearing the mouth at its base). The anus is located at or near the pole opposite the mouth. Middle Cambrian–Middle Devonian (*Enoploura*, Figure 12.25B). [This class has been assigned to a new subphylum of primitive chordates, the Calcichordata — R. P. S. Jefferies, 1968, British Mus. Nat. Hist. Geol. Bull., Vol. 16 (6), pp. 243–339. Caster (1971) disagrees.]

Class Homoiostelea. Biradiate homalozoans with multiplated theca, with single, biserial anterior arm bearing cover plates, and a long posterior heterostele ("tail") composed of differentiated plates. [Differentiation from four-part (tetra-merous) to two-part (dimerous) symmetry along the axis, that is, from the site of attachment of tail to base of theca (proximal) to site farthest from this base (distal). Anus lies in the right or left lobe of thecal base. Thecal plates near base of arm may bear hydropore and gonopore. Upper Cambrian–Lower Devonian (*Dendrocystites*, Figure 12.25A, b).

Class Homostelea. Triradiate homalozoans with skeleton of plated theca and tail. (Outer girdle of thecal plates, few, thick, and marginal; inner plates numerous and comparatively small.) Peduncular prolongation of theca — stele (= tail) composed of double series of primary, undifferentiated plates with some platelets intercalated; anus and mouth near one pole of the body. Middle Cambrian only (*Trochocystites*, Figure 12.25A, a).

Radial Homologs in Carpoids. If one equates "carpoid" tails in all classes with exothecal ambulacra (arms) in other free-living or attached echinoderms, then the problem arises of homologizing carpoid body appendages to the general radial symmetry of echinoderms. This is difficult since carpoids lack radial symmetry.

In this context, one can start with the postulated dipleurula ancestral type of the Echinodermata. It would have been attached to the substrate by one or another

---

FIGURE 12.25 Morphology, skeletal structure and other features of homalozoans (Orders Cornuta and Mitrata, the carpoid echinoderms, have been removed from echinoderms and classed as a vertebrate subphylum, Calcichordata (Jefferies, 1968).
(A) (a) *Trochocystites*, longitudinal section, Cambrian (after Jaekel, 1918); (b) *Dendrocystites*, Ord. (after Bather, 1925); m—mouth; an—anus; ip, mp, sp—interior, marginal, and stem plates respectively; br—food groove.
(B) *Enoploura popei* Caster, U. Ord., Cinncinnati, Ohio. Plate names: (a) plastron (= concave side); (b) carapace (= convex side); alm, mlm, plm—marginals; bm, lac, mac—adcolumnals; ax—sub-brachioles; br—brachioles; m_1–m_4—dorsal epibasals; ib—interbasals; A_1ms—irregular and median hypocentrals respectively; lam, cs—marginal and central somatics respectively; pp—peduncular or styloid process (= stylocone); M, mcm, am—adtegmenals; L—left side; R—right side (schematic, after Caster). [(C) reassigned, hence deleted here].

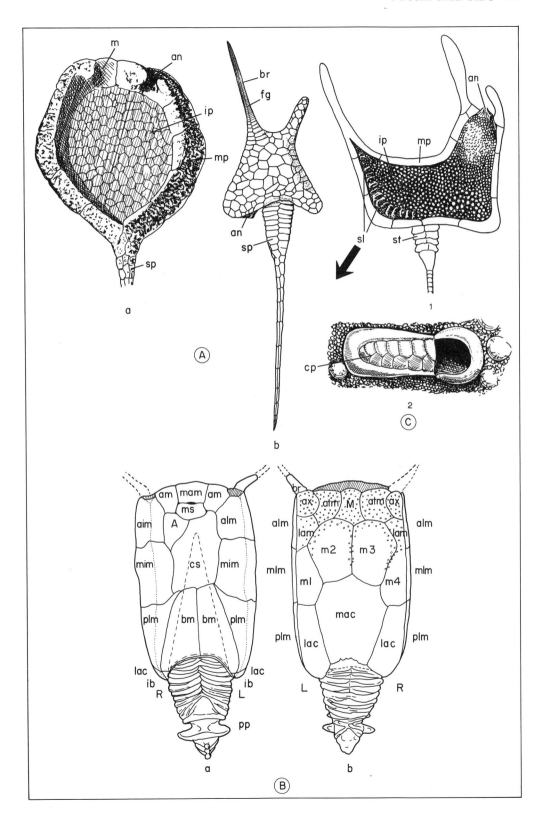

method (ventral surface or preoral lobe), that is, by the tail, and the opposite end with two lobes would yield mouth and anus. The tail at one and mouth/anus at the other would give the body a Y-shape (triradiate condition).

The Homostelea are thought to be close to the ancestral type of triradiate symmetry (Caster, 1967, Figure 380). By a sequence of genetic steps, one could envision ultimate convergence of mouth and anus and development of ambulacral grooves. That would yield a form like *Trochocystites*, according to Chauvel (1937–1941). The Homostelea may thus be referred to as the *triradiates*.

Similarly, the Stylophora may represent the *uniradiate* condition (tail-arm), whereas the Homoiostelea, with separate arm and tail, characterizes the *biradiate* condition. (All such interpretations, of course, are highly conjectural.)

Skeletal Structure. The flat plated theca may often be differentiated into a *plastron* on the concave underside and a carapace on the convex upperside (Caster, 1952). The crystalline calcitic plate system may vary although generally either side shows similarities. Frequently, there are a ring of marginal plates and an upper sequence of small, loosely articulated anterior plates, and a lower sequence of larger, and possibly better-fitting plates (Nichols, 1962).

The various kinds of plates, mostly irregular and polygonal in shape, are shown in Figure 12.25. In particular, one complete plate system is shown for *Enoploura*.

Caster (1952) remarked that the plates of the plastron were more stable (conservative), and hence any deviations observed might be of taxonomic significance. Specifically he found that in carpoid evolution there was a tendency toward fusion of somatic plates and toward asymmetry in their arrangement.

The theca (or calyx) ends in a posterior-plated stem or stalk composed of two or more columns of imbricated or abutted small skeletal elements. In some genera (*Corthurnocystis*) rows decrease to a single toward the free end. Skeletal appendages may or may not indicate one or more spines or brachioles (Figure 12.25 *A,b*).

The convex side of some forms, such as *Gyrocystis*, has one or two holes of different size defined by plates. The larger of these holes may have a covering plate. A series of slits occur on some theca and with them small, biserial skeletal plates. In some forms stem is fossilized at right angles to the theca. The underside of the theca of *Cothurnocystis* has knobs on its marginals.

These are the chief skeletal features recorded in homalozoans that are known only in the fossil state. Interpretation of these structures is given below.

Inferred Biology

Habit. How did these extinct presumed homalozoans live and feed? There are several pieces of evidence that suggest that at least some forms were moored or anchored to the bottom and capable of searching movements with the free body and protrusible set of tentacles (see Spencer, 1938, plate 1, Figure B).

The evidence bearing on the above interpretation is threefold: (1) The stem of some individuals occur in a position almost perpendicular to the body (2) Spencer (1938) reported many variants in body shape. (3) Two holes, one larger and one smaller, penetrate the marginal plate in some fossils. The larger has been interpreted to be a stomodial pouch that housed the protrusible set of tentacles when they were retracted in the non-feeding position. What appears to have been a movable plate is found adjacent to the larger hole and is thought to have functioned as an opercular lid or cover when the tentacles were retracted.

Feeding. There are various interpretations of the feeding mechanism. If, as Nichols and others indicate, all or almost all homalozoans fed by means of protrusible tentacles, then it is conceivable that anus and mouth were both situated in soft tissues and hence not preserved. If so, the holes in the marginal plate would denote where these soft parts were housed.

Ubaghs (1967) found an orifice in *Ceratocystis* (M. Cambrian, Czechoslovakia) where the aulacophore (that is, the "tail" or modified ambulacra) meets, or is attached to, the theca. He speculated that the mouth was probably internal at this site. Logically, the mouth toward which food is directed, might be expected to be related to the feeding arm.

The entire story of the feeding mechanism obviously is not encompassed solely by the protrusible set of tentacles, as can be seen by the food groove in *Dendrocystites*, which, in life, was probably ciliated and directed food to the mouth.

Sanitation. There is some indication of migration of the anus in evolution of homalozoans from Cambrian through Ordovician time. Cambrian forms such as *Trochocystites* (Figure 12.25*A*, *a*) and *Gyrocystis* have a smaller opening in the marginal plate that is interpreted to be the anus. It is situated close to the presumed mouth.

By contrast, *Dendrocystites* (Ordovician) may be cited. If the structure situated diametrically opposite to the food groove in this form (Figure 12.25*A*, *b*) is an anus, that would indicate a strong tendency in homalozoan evolution to achieve a separation between waste output and food input.

Slits. The multislits, grooves, or pores on a single side of the upper surface of some homalozoans have been given diverse interpretations: (*a*) gills to void water taken in by mouth and anus; (*b*) they are multiple gonopores; (*c*) they are ciliated components of a food-intake system; (*d*) respiratory organs that cover plates protected. Without more evidence it is difficult to choose.

Terminal Peduncle. In *Enoploura* (Figure 12.25*B*) the terminal peduncle is thought to have alternated its functions as anchor, a kind of rudder and counterbalance to thecal weight.

A quite different interpretation of the peduncle is given by Ubaghs (1961) for such forms as *Phyllocystites* and *Cothurnocystis*. Instead of a tail, he envisioned it as a crinozoan arm; instead of being the posterior end, he interpreted it to be the anterior end [an idea also expressed by Chauvel (1941)]. It also served as an endoskeletal support for a feeding mechanism.

Caster (1967) interpreted carpoid "tails" or peduncles in general as posterior in functional position and as having been derived from the exothecal ambulacra. Both anchorage and feeding function were performed by the same posterior tail.

Muscular and Nervous Systems. Associated with the varied functions and corresponding movements of the peduncle, was the need for muscular exertion. In *Mitrocystella*, Chauvel (1941) reported what he interpreted to be two nerve ganglia. These were located near, and with reference to, the peduncle, at the corners of the theca. Muscular and nerve control are also needed to explain a protrusible set of tentacles.

Carpoid Plate System/Ornamentation, and the Chordates. Several authors have remarked the similarity in plate ornament, a pebbled leather effect, that grades into labyrinthine pitting of plates of the carpoid calyx and plate arrangement found in carpoids and various armored early fishes (Gregory, Caster). Gislén interpreted both the asymmetrical calyx and slits in some Lower Silurian carpoids (*Cothurnocystis elizae*) as comparable to the asymmetry and larval gill apertures in "amphioxus," and he argued for a closer relationship between carpoids and chordates. Other workers rejected this view (McGregor, cit. Gregory, 1951, p. 97).

Plate ornament and arrangement may seem to be a stronger argument, according to Caster (1967). Even here, such resemblances, homologies, or convergences, may be quite superficial (Gregory, 1951, p. 96). On the other hand, with additional evidence, they may prove to be more important than previously thought. [Jefferies (1968) is the most recent worker to interpret the orders Cornuta and Mitrata (Class Stylophora) as primitive chordates ancestral to tunicates, amphioxus, and the vertebrates.]

Homalozoans and the Hypothetical Echinoderm Ancestor. Hyman (1955) was of the opinion that the construction of the hypothetical echinoderm ancestor seemed close to one of the homalozoans. (It was through modification of the tentacles that the water vascular system of the Echinodermata arose.)

REFERENCES

Carpoids

Bather, F. A., 1929. Echinoderms, in Encyclopedia Britannica, 14th Ed., Vol. 7, pp. 894–904.

————, 1925. *Cothurnocystis:* A study in adaptation: Paläont. Zeitschr., Vol. 7, pp. 1–15.

Caster, K. E., 1967. Possible triradiate background of some echinoderm and its bearing on phylogeny: Science, Vol. 158, p. 525.

————, 1952. Concerning *Enoploura* of the Upper Ordovician and its relation to other carpoid Echinodermata. Amer. Paleontology Bull., Vol. 34 (141), pp. 1–47.

Chauvel, J., 1937–1941. Recherches sur les Cystoides et les Carpoides armoricains: Mém. Soc. Géol. Mineral. Bretagne, Vol. 5.

Gill, E. D., and K. E. Caster, 1960. Carpoid echinoderms from the Silurian and Devonian of Australia: Amer. Paleontology Bull., Vol. 41 (185), 71 pp.

Gregory, W. K., 1951. Evolution Emerging. Macmillan Co., N.Y. 2 vols. Vol. 1, pp. 96–97.

Jaekel, O., 1918. Phylogenie und System der Pelmatozoen: Paläont. Zeitschr., Vol. 3, pp. 1–28.

Nichols, D., 1962. Echinoderms. Hutchinson Univ. Libr., London, 180 pp. (pp. 155–163).

Regnéll, G., 1945. Non-crinoid Pelmatozoa from the Paleozoic of Sweden: Medd. Lunds Geol.-Mineral. Inst., No. 108, pp. 1–255.

Spencer, W. K., 1938. Reconstruction of *Gyrocystis*, in G. R. de Beer, ed., Evolution, etc., Clarendon Press, Oxford (pp. 287–303). (Abbreviated Title.)

Ubaghs, G., 1967. Le genre *Ceratocystis* Jaekel (Echinodermata, Stylophora): Univ. Kansas Paleont. Contr. Paper No. 22, 16 pp.

————, 1963. *Cothurnocystis* Bather, *Phyllocystis* Thoral, and an undescribed member of the Order Soluta (Echinodermata Carpoidea) in the uppermost Cambrian of Nevada: J. Paleontology, Vol. 37, pp. 1133–1142.

————, 1961. Sur la nature de l'organe appelé tige ou pédoncle chez le Carpoides Cornuta et Mitrata: Acad. Sci. Paris, Comptes Rendus, Vol. 253, pp. 2738–2740, text figures 1–5.

———— and K. E. Caster, 1967. Homolozoans: Treat. Invert. Paleont., Part S, Echinodermata, Vol. 1 (2), pp. S495–S627.

FIGURE 12.26 Problematic echinoderms: haplozoans.
(A) *Peridionites navicula:* (a) lateral face, 4 of 5 thecal plates; apical plate, end plate, mediolateral plate, X 5; (b) ventral face, X 5.
(B) Haplozoan: *Cymbionites craticula*, M. Cambrian, Queensland, Australia. (a)–(d): (a) Complete specimen, oblique lateral view, cup-shaped theca arranged about a fluted crater-like calyx, X 3; (b) half a single radial, calcareous plate (1 of 5 plates), X 3; (c) microsection through calyx; a nearly vertical slice with sutures (s), growth lines, and calcitic fibrous arrangement; (d) honeycomb arrangement of stereom (characteristic of phylum Echinodermata); after Ubaghs, 1967; Whitehouse, 1941. (cf. Ubaghs, 1978, p. T359–T360).

PROBLEMATICAL ECHINODERMS

Middle Cambrian Haplozoans

Stratigraphic Setting. The lowest limit of the Middle Cambrian of Queensland, Australia, yielded a unique assemblage of unattached echinoderms (Whitehouse, 1941). One species of *Cymbionites* (Figure 12.26*A*) ranged through the basal 40 ft of a trilobite zone (*Xystridura*) occurring in "colossal numbers" and forming echinodermal limestones. The other species *Peridionites* (Figure 12.26*B*) occurred in a 5-ft-thick bed, some 24 ft above the youngest *Cymbionites* occurrence. The *Peridionites* bed was also dense with specimens. Between the two haplozoan horizons, there were unidentified echinoderms represented by ossicles that crowded the limestone.

Mineralogy of Plates. As can be seen in Ubaghs' photographs of the stereom (Figure 12.26*B*), the honeycomb structure is characteristically echinoderm. Whitehouse reported that the stereom of *Cymbionites* consisted of packed and intermittently impinging fibers between which there was a calcite mesh in optical continuity with the fibers (Figure 12.26*B*, *c*). The fibers of each thecal plate (diameter $= 12\mu$) united into a single calcite crystal. Thus each plate would show uniform extinction under crossed nicols.

Peridionites navicula has plates composed of nonfibrous calcite, and each plate behaves optically as a single crystal.

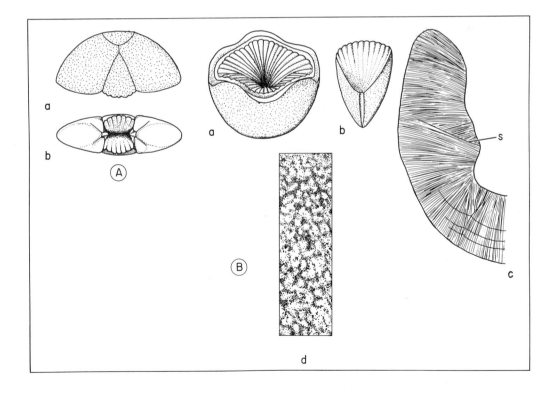

Affinities. Whitehouse erected a sub-phylum, Haplozoa, and two new classes as well as families, genera, and species. Following Ubaghs in the Treatise (1967), only the generic name will be recognized here and these will be considered un-classified echinoderms. But Whitehouse went further. He expressed the thought that his fossil genera represented the ancestral dipleurula.

It followed from the above assumption, according to Whitehouse, that asteroids, ophiuroids, and echinoids all derived from his haplozoan class Cycloidea and crinoids from his other class, Cyamoidea. (See earlier discussion, this chapter, on evolution and origin.) All of these con-cepts have been rejected by most echino-derm specialists.

Some other investigators have rejected the echinoderm affinities of haplozoans. These investigators favored an inter-mediate position for haplozoans between the coelenterate cnidarians and the ctenophores. Still others accepted echin-oderm affinities, yet regard *Cymbionites* as a cystoid or both genera as eocrinoids (Ubaghs, 1967; Termier and Termier, 1953). More recently Ubaghs (1978), based on newer data on plate microstruc-ture has rejected assignment of the hap-lozoans (class Machaeridia) to the echino-derms, and they are not included in the three Treatise volumes on the Echino-dermata. The placement of Whitehouse's genera and species is yet to be resolved.

TWO NEW ECHINODERM CLASSES

From specimens described in the litera-ture and some recent new material, two new classes of echinoderms have been established (Durham, 1966, 1967): the camptostromatoids and the lepidocys-toids. These classes are grouped together in this section for convenience only. They are not related. They share only a common bed; that is, they occur together as fossils in the same strata of the Kin-zers Formation of Pennsylvania, Lower Cambrian in age.

Camptostromatoids (Subphylum Echinozoa)

The medusaeform fossil, *Camptostroma*, was designated a coelenterate hydrozoan by Ruedemann (1933), and later, by sev-eral others, a scyphozoan (Harrington and Moore, 1956). How then is it possible to demonstrate, as Durham has, that it is, in fact, an echinoderm? Simply the find-ing of a better-preserved specimen with the original calcite plates. These showed typical echinoderm reticulate microstruc-tures and cleavage, and plates responded as single crystals.

In addition, other echinoderm struc-tures observed include a mouth sur-rounded by plates, plated anal pyramid, free plated arms, sutural pores that are thought to have contained tube feet, and indications via radial ridges of a well-defined radial symmetry.

Class Camptostromatoidea. Free-living medusaeform, radially symmetrical ech-inoderms, with plated body wall; mouth and anus at opposite poles; plated peri-pheral free arms. Range: Lower Cam-brian only (*Olenellus* Zone). (*Camptostroma*, Figure 12.27*A*).

Lepidocystoids (Subphylum Crinozoa)

Foerste (1938) originally described a *Lepidocystis* species, from the Kinzers Formation—*L. wanneri*. Durham (1967) found that Foerste's genus differed from the following classes as indicated:

1. *Eocrinoids.* Location of sutural pores, imbrication of plates, and circlet of arms, as well as mode of arm attachment.
2. *Edrioasteroids, Cystoids, Helicoplacoids.* Presence of sutural pores.
3. *Camptostromatoids.* Position of anus on oral surface and presence of a single plate type in any given area of the theca.

These distinctions led Durham (1967) to assign Foerste's genus to the new class.

Class Lepidocystoidea. Free-living plac-oid echinoderms with plates of oral sur-

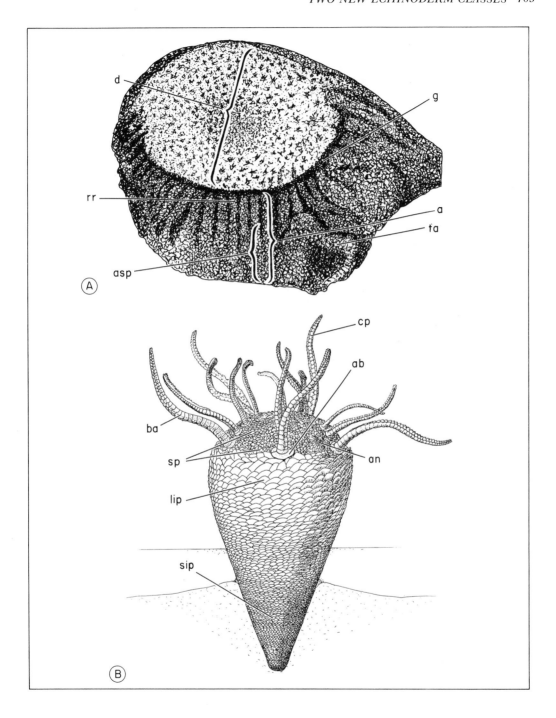

FIGURE 12.27 Two new classes and eocrinoids. Camptostromatoids: (A) *Camptostroma roddyi*, L. Cam., (*Olenellus* Zone), Kinzers Formation, Pennsylvania; aboral view; note medusaeform test, X 1.4; *d* — disk, *g* — groove, *a* — apron, *fa* — free arm, *rr* — radial rib, *asp* — area of sutural pores. Lepidocystoids: *Lepidocystis wanneri* (occurs with *Camptostroma*). Reconstruction of living position in substrate; schematic, *sip* — small, immature plate; *lip* — large, imbricating plates; *sp* — sutural pores; *ba* — biserial arms; *ab* — arm base; *cp* — cover plate; *an* — anal pyramid. (Illustrations redrawn after Durham).

face bearing sutural pores; free biserial arms arranged in circlets; anal pyramid marginal; aboral surface covered by imbricating plates that are small apically and enlarge up to the oral surface. Range: Lower Cambrian only (*Olenellus* Zone). (*Lepidocystis*, Figure 12.27B.)

REFERENCES

Problematical Echinoderms and Others

Calman, W. T., 1909. Crustacea, in A Treatise on Zoology. London, Adam and Black, p. 107, Figure 58.

Durham, J. W., 1967. "Camptostromatoids" and "Lepidocystoids," in Treat. Invert. Paleont. (= TIP) (S) Echinodermata, Vol. 1 (2), pp. S627–S634.

————, 1966. *Camptostroma*, an Early Cambrian supposed scyphozoan, referable to the Echinodermata: J. Paleontology, Vol. 40, pp. 1216–1220.

Foerste, A. F., 1938. "Echinodermata," in C. E. Resser & B. F. Howell, Lower Cambrian *Olenellus* Zone of the Appalachians:

G.S.A. Bull., Vol. 49, pp. 212–213, plate 2.

Ruedemann, R., 1933. *Camptostroma*, a Lower Cambrian floating hydrozoan: U.S. Nat. Mus. Proc., Vol. 82, art. 13, pp. 1–8, plates 1–4.

Termier, H., and G. Termier, 1953. Generalités sur les Echinodermes, in Jean Piveteau, ed., Traité de Paléontologie, Vol. III, pp. 588 ff.

Ubaghs, G., 1967. *Cymbionites* and *Peridionites* — unclassified Middle Cambrian Echinoderms. TIP, pp. S634–S637.

Whitehouse, F. W., 1941. The Cambrian faunas of Northeastern Australia: Part 4. Early Cambrian Echinoderms similar to the larval stage of Recent forms: Mem. Queensland Mus., Part 1, 288 pp.

SUBPHYLUM CRINOZOA

This subphylum (defined earlier) ranges from Lower Cambrian to Recent time and embraces eight classes: Eocrinoidea,* Cystoidea, Blastoidea,* Edrioblastoidea, Parablastoidea,* Lepidocystoidea (already discussed), Paracrinoidea, and Crinoidea. Of these, major treatment will be reserved for the three great classes — cystoids, blastoids, and crinoids. (*Subphylum Blastozoa Sprinkle, 1973.)

Eocrinoids

Introduction. Eocrinoids have numerous distinctions. They are the oldest (apparently first-appearing) of the Crinozoa, occurring in the lower Lower Cambrian, associated with helicoplacoids and edrioasteroids. Morphologically, they are a heterogeneous group and embrace widely varying forms, such as *Rhopalocystis* (Figure 12.28A) and *Lingulocystis* (Figure

12.28B), that differ markedly in thecal organization. They are thought to be ancestral to the camerate crinoids (Regnéll, 1945; 1960; Moore, 1954, Figures 1–8), and possible other echinoderms as well (Fell and Pawson, 1963, Figures 1–17a).

Essential Morphology

Gross Structure. Eocrinoids have a tripartite construction: plated theca (that enclosed the living animal's visceral mass), thecal appendages — brachioles (that served a food-transport function), and hollow column or stem (= *Hohlwurzel*) (for fixation to the substrate, either as anchor or ballast to orient the animal in a vertical position). It is conceivable that some internal organs were lodged in the hollow of the *Hohlwurzel* (Ehrenberg, 1929; Robison, 1965).

Thecal Construction. Eocrinoid theca come in varied shapes, ovoid to spheroid, as well as conical, subcylindrical, and so on. The generally rigid plated walls may bear from 20 to 500 plates. In turn, the plates are joined by sutures that may have sutural pores (Figure 12.28*A*, *D*).

Plates may be arranged in an orderly way, in alternating circlets (Figure 12.28*C*), and/or zones (Figure 12.28*A*), or irregularly (*Gogia*, Figure 12.28*D*, *a*). The compressed forms like *Lingulocystis* have theca framed by marginal plates (Figure 12.28*B*).

Inferred Biology: Systems

Digestive. Food grooves on the oral surface of the brachioles must have transported food particles to the mouth. Similarly, food grooves of the ambulacra.

Excretory. Presence of an anal pyramid.

Reproductive. There is one orifice of undetermined function on the oral plates of *Cryptocrinites* (Figure 12.28*E*) that is thought by some to be a gonopore and by others to have served an excretory function.

Respiratory. External ridges (Figure 12.28*C*) and sutural pores (Figure 12.28*D*, *b*) are thought to be related to the respiratory function, that is, exchange of gases between the eocrinoid and the seawater in which it lived. Other solutions must have also coexisted since all eocrinoid genera do not have such pores.

Muscular. The brachiolar facets with which the brachioles individually articulate, bear shallow hollows. Ubaghs (1967) interpreted these to be probably ligamentary. Brachioles must have moved in food-gathering, and the spiral twist of the brachioles in *Gogia* likewise implies a related muscularity. The brachiolars bear depressions where muscle fibers had been attached, and these apparently were capable of contraction and extension. Similarly, the column or stem were under some degree of muscular control.

Classification

Class Eocrinoidea. Extinct Blastozoa, having radial symmetry (often pentamerous), with or without stalk or stem; with theca composed of poreless plates (but sutural pores occur in many genera); biserial, unbranched exothecal brachioles.

Range: Lower Cambrian — Silurian. (Embraces ten families, five of which are illustrated here, and several genera unassigned to any family.) [Representative genera include *Rhopalocystis* (Figure 12.28*A*), *Lingulocystis* (Figure 12.28*B*), *Macrocystella* (Figure 12.28*C*), *Gogia* and *Lichenoides* (Figure 12.28*D*), and *Cryptocrinites* (Figure 12.28*E*).]

Information Derived from the Eocrinoids

Skeletal Features, Genetic Code, and Speciation. Plates, pores, brachioles, and general configuration are the major criteria for classification. The first three distinguish eocrinoid families, whereas overall thecal outline, hydropore position, plate and brachiole characteristics help to discriminate genera.

Features that distinguish the f.ve species of *Gogia* (Figure 12.29*A*), for example, are as follows: *G. spiralis* differs from the other four species by the helical twist (spiral construction) of its brachioles (Figure 12.28*D*, *a*). *G. longidactylus* differs from other species in its lack of sutural pores in the lower part of the theca. (All *Gogia* species have sutural pores.) The *Hohlwurzel* (stem), its shape, and plate structure distinguish *G. prolifica* from *G. longicaudatus* and *G. granulosis*. Other differences between species include presence or absence of ornamentation on thecal plates and thecal plate number.

In Middle Cambrian time where the trilobite *Glossopleura* moved over the sea floor of the Utah-Nevada-Arizona area, species of the eocrinoid *Gogia* (Figure 12.29*A*, *B*) were part of the contemporary biota. It is of interest that three different species of *Gogia* occur in each of these states. That suggests geographic and reproductive isolation from the parental gene pool during *Glossopleura* time.

A further clue to "isolation" can be gleaned from Robison (1965). If, as Clark (1921) postulated for comatulid crinoids, that number of arms correlated with depth and temperature, and if one can extrapolate from crinoids to eocrin-

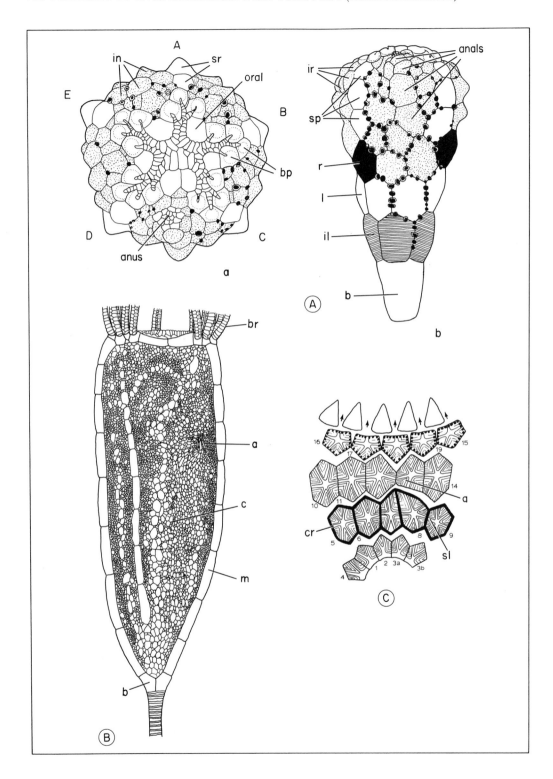

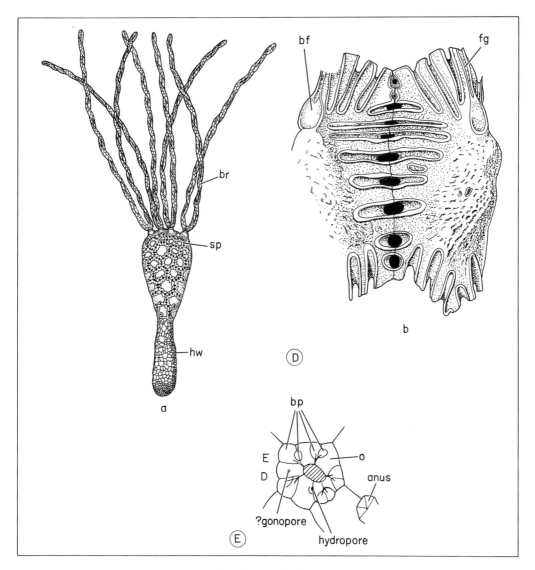

FIGURE 12.28 Information contained in various eocrinoids.

(A) Family Rhopalocystidae. *Rhopalocystis destombesi,* L. Ord., Morocco (after Ubaghs, 1967). Ambulacral rays marked (A) to (E) in clockwise progression. Interrays may be designated by letters of rays that bound them. Plates designated: *in = ir—*interradial, *sr—*supraradials, *bp—*brachioliferous plate (that is, bearing brachioles), *r—*radial, *l—*lateral, *il—*infralateral, *b—*basal. (a) Oral surface, brachioles lacking, slightly enlarged. (b) theca, side view, seen from *CD—*interray side.

(B) Family Lingulocystidae: *Lingulocystis elongata.* L. Ord., France; *br—*brachioles, *a—*anus; thecal plates: *c—*central, *m—*marginal, *b—*basal, enlarged.

(C) Family Macrocystellidae: *Macrocystella* sp., (?) U. Cam., England. *a* = position of anus, arrows denote brachioles that attach to radial plates. Individual plates numbered after Forbes.
In upward succession alternate circles of plates are lowermost circlet, basals; heavy outlined plates, laterals; third circlet, laterals; heavy-dotted plates, radials; and final circlet, orals. Except orals, other plates bear a ridge topography and across midpoints of sutures (*sl*) one ridge on each plate is confluent with that on adjacent plates (*cr*). After R. C. Moore.

(D)(a) Family Eocrinidae: *Gogia spiralis,* M. Cam., B.C., Canada, slightly larger than natural size; *sp—*sutural pores (black) on plate margins; *br—*brachioles with helical twist (in other *Gogia* species, twist is spiral); *hw—Hohlwurzel* or stem; (after Robison); (b) family Lichenoididae: *Lichenoides priscus,* M. Cam., Bohemia, enlarged; two lateral plates with epispires along sutural margins; *bf—*brachiole facet, *fg—*food groove (after Ubaghs, 1953).

(E) Family Cryptocrinitidae: *Cryptocrinites, laevis,* M. Ord., East Baltic; oral plates (*o*); brachioliferous plates (*bp*); interray *DE* bears orifice, possibly a gonopore. Note position of anus and hydropore.

oids, then certain inferences ensue. Each of these *Gogia* species (Figure 12.29*A*) in the *Glossopleura* zone lived at a different depth (based on brachiole count). *G. multibrachiata*, with 22–40 brachioles, presumably inhabited the warm shallows. *G. longidactylus*, with 8–16 brachioles, would then be a dweller of the intermediate depths. *G. granulosa*, with 8–12 brachioles, would have lived in the deeper cold water. This spatial "isolation" in distinct ecologic niches would insure fractionation of an original gene pool and the subsequent rise of new species.

Biomineralization and Growth. Skeletal elements are solid and composed of crystalline calcite.

One small specimen of *Gogia* suggested (Robison, 1965) that it was s young growth stage. The inferred ontogeny led to three major conclusions: (*a*) increase in thecal plate size and numbers with growth; (*b*) general lack of small plates in the central portion of the theca appeared to indicate that plate addition occurred at either end (adoral and aboral); (*c*) as the animal grew, there was a greater need for contact with seawater and a corresponding sutural pore number increase, as well as a deeper embayment of pores. Perforation of the theca along and across sutures may have involved crystal resorption.

Affinities. Eocrinoids are clearly cystoid-like in organization and structures (Ubaghs, 1967): in thecal form and plate construction, presence of brachioles instead of arms, and in comparable ambulacra. Ubaghs (1953) found that the sutural pores of *Lichenoides* were basically unlike the diplopores and rhombopores

MIDDLE CAMBRIAN TRILOBITE ASSEMBLAGE ZONES FOR NORTH AMERICA	Gogia prolifica Walcott	Gogia longidactlus (Walcott)	Gogia multibrachiatus (Kirk)	Gogia granulosa, n. sp.	Gogia spiralis n. sp.
BOLASPIDELLA					●
BATHYURISCUS– ELRATHINA	●				
GLOSSOPLEURA		●	●	●	
ALBERTELLA					
PLAGIURA– POLIELLA	●				

(A)   (B)

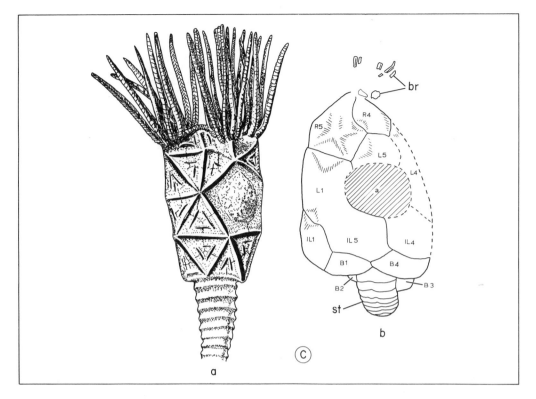

FIGURE 12.29 Information derived from fossil eocrinoids.
(A) Biostratigraphy: distribution of *Gogia* species relative to successively younger beds. M. Cam. Zone. *G. prolifica* has greatest vertical spread; three species of *Gogia* occur in the trilobite *Glossopleura* zone.
(B) Geographic spread: *Gogia* species in western North America. (Numbers denote localities).
(C) Affinities: between (a) eocrinoid *Mimocystites* and cystoid *Cheirocrinus* (see text), schematic.
Illustrations redrawn from (A)–(B) Robison; (C) Regnéll.

of the cystoids. A comparison of *Mimocystites* and *Cheirocrinus* (Figure 12.29C) discloses a close similarity in stem structures, and construction of theca (same number of plates and arrangement, including anal location, and so on. The hollow folds that transgress sutures in *Mimocystites* may be viewed as possibly homologous, intermediate or, at least, related to the pore rhombs of *Cheirocrinus*.

Genetically, such near identities may be a case of either convergent evolution or divergence from a common ancestral stock. Most investigators favor the second alternative.

Other widely recognized affinities are crinoidal. Moore (1954, Figures 1–8)

argued that the dorsal cup pattern of all types of camerate crinoids was "directly derivable" from an eocrinoid or a cystoid (rhombiferan) ancestor. There is no doubt that the encasement (dorsal capsule and ventral tegmen) of eocrinoids is crinoidal. However, eocrinoid "arms" are noncrinoidal but, in character, these exothecal brachioles are cystoid (Regnéll, 1945; 1960).

Ubaghs (1967) was of the opinion that the existence of these and still other differences between crinoid "arms" and eocrinoid brachioles did not support the hypothesis of crinoid descent from either eocrinoids or cystoids. This is one of the unresolved questions to be clarified by future research.

REFERENCES

Eocrinoids

Ehrenberg, K., 1929. Pelmatozoan root-forms (fixation): Amer. Mus. Nat. Hist. Bull. 59, art 1, pp. 1–76, 42 figures.

Moore, R. C., 1954. Status of Invertebrate Paleontology, 1953, IV, Echinodermata: Pelmatozoa: Mus. Comp. Zool. Bull. (Harvard), Vol. 112 (3), pp. 125–144.

Regnéll, G., 1960. "Intermediate" forms in early Paleozoic echinoderms. Inter. Geol. Congress Copenhagen, 1960. Pt. XXII–XXVI. I.P.U., pp. 71–80.

————, 1945. Non-crinoid Pelmatozoa from the Paleozoic of Sweden: Lunds Geo.-Mineral. Inst., Medd., No. 108, 255 pp., 15 plates.

Robison, R. A., 1965. Middle Cambrian Eocrinoids from Western North America: J. Paleontology, Vol. 39 (3), pp. 355–364, plates 50–52, 3 text figures.

Ubaghs, G., 1967. Eocrinoidea. Treat. Invert. Paleont., Pt. S, Echinodermata, Vol. 1 (2), pp. S455–S495.

————, 1953. Notes sur Lichenoides priscus Barrande, Eocrinoïde du Cambrien moyen de la Tchécoslovaquie: Inst. Royal Sci. Nat. Belgique Bull., Vol. 29 (34), pp. 1–24, 12 test figures.

Cystoids

Introduction. Originally regarded as organic objects (crystal apples), cystoids were first clearly recognized as being organic and as having an echinoderm organization in 1772. Not until three-fourths of a century later were they separated from the other major classes of echinoderms and put in a class of their own. Even so, this class carried a motley assortment of echinoderms.

During more than a century of further study, up to the year 1962, greater discrimination became possible among the several anomalies within the class Cystoidea. Each of the following groups was removed and placed in distinct classes: Edrioasteroidea (1858), Carpoidea (1900), Eocrinoidea (1918), Paracrinoidea (1945), Edrioblastoidea (1962) (Fay, 1962; Jaekel, 1899–1918; Kesling, 1967; Regnéll, 1945).

Classification
Class Cystoidea (after Bather, Jaekel, Kesling, and others).

1. Extinct, anchored or attached, crinozoans, with or without stem (Figure 12.30E).

2. With plated theca surrounding body plates that are calcareous; plates arranged in circlets and bearing pores (haplopores, diplopores, and subepithecal pore rhombs — Figure 12.30D); with ambulacra and biserial brachioles.

3. Systems include the following:

Water Vascular. Hydropore led to inferred stone canal. How this system functioned in cystoids is unknown. Bather (1900, p. 71) suggested that the water vascular system sent out branches from which the tube feet arose that presumably served the dual function of prehension of food and respiration.

Digestive. Food-collecting brachioles (biserially plated) arise from corners of exothecal ambulacra that bear a food groove and biserial floor plates. Food collected by brachioles was moved toward the food grooves, and thence toward the mouth. (See below for excretory function.)

Reproductive. Gonopore with separate opening on theca or combined with a hydropore. Presumably, via a duct, it discharged eggs and sperm into seawater. Such discharges probably occurred during definite spawning periods (see Boolootian, 1967, pp. 561 ff.)

Respiratory. Presumably the pore system(s) of the test were breathing organs.

Muscular. Aside from the muscles that move the brachioles either in feeding movements or as an aid to feeble propulsion along the bottom (Kirk, 1911), there must have been powerful muscles associated with the prehensile column in some forms (Figure 12.30*E*).

Excretory. Presence of an anus and valvular anal pyramid allows one to infer the probable course of the gut (Figure 12.30*A*) — the upper portion of which participated in digestion of food passing through the mouth. Wastes were apparently voided directly into seawater.

Skeletal. Plates of the theca and stem were composed of calcite crystals.

Geologic range: Lower Ordovician to Upper Devonian.

The class is divisible into two orders — Rhombifera and Diploporita — and these into seven superfamilies. [A representative genus in each superfamily is illustrated. Rhombifera*: Superfamily Glyptocystitida, *Cheirocrinus* (Figure 12.30*A, B*); Superfamily Hemicosmitida, *Caryocrinites* (Figure 12.31*A*); Superfamily Polycosmitida, *Stichocystis* (Figure 12.31*B*); Superfamily Caryocystitida, *Heliocrinites* (Figure 12.31*C*). Diploporita: Superfamily Glyptosphaerida, *Glyptosphaerites* (Figure 12.31*D*); Superfamily Sphaeronitida, *Triamara* (Figure 12.31*E*); Superfamily Asteroblastida, *Asteroblastus* (Figure 12.31*F*).] *(Now Blastozoa Sprinkle, 1973.)

Information Derived from Fossil Cystoids

Mineralogy, Plates. Secretion of the cystoid skeleton not only involved formation of calcite biocrystals in the lamellae of the plates, and equivalent exothecal secretion for the ambulacral plates and for the column or stem. All of the calcareous substance together constitutes the cystoid stereom. Limited data are available on biomineralogy details.

The epitheca (epistereom) or external layer of the plate is thin, nonporous, and may bear ornamentation. [In *Echinosphaerites*, this layer is composed of multiple fine laminae that bear a concentric pattern of growth lines (?) on their surface.] The mesotheca or mesostereom contains the characteristic cystoid pore canal system. The innermost integument

in noncalcareous and rarely preserved, being an organic continuation of the coelom (Bather, 1900; Cuénot, 1957).

Calcite biocrystals were secreted into the body of these tissues (lamellae). Kesling (1967) thought that thecal plates grew around pores and their canals during ontogenetic development. If so, the pore canal system must have been defined by soft tissues during the larval stages.

Additional indicators of biocrystallization include fixation by cementation, that is, secretion of aboral plate (Kirk, 1911); occasional fusion of thecal plates; resorption (along margins of peristomal and periproctal plates — Kesling, 1967); and secretion of a surficial layer (epithecal) that obscured canal openings to the surface and/or ambulacra (Figure 12.31*A*).

Thecal plates apparently grew in size by accretion of new laminae along the periphery, and secretion of biocrystals in these. Another way in which theca enlarged and changed configuration was secretion of smaller (secondary) plates interstitially, that is, between contacts of older plates (Kesling, 1967).

Some postdiagenetic deposits of calcite are possible and likely (Nichols, 1962, p. 137). Solution effects after burial were also common.

There also appears to have been genetically controlled deposition of excessive or reduced calcite during the life of certain species. Excessive deposition is seen in the case of *Pachycalix pachytheca* with very thick basal thecal plates (2 cm compared to the usual 2 mm!) (Kirk 1911).

Adaptive Modifications. Among the main modifications of cystoids were skeletal changes adaptive for feeding, attachment and/or movement, and respiration (Kesling, 1967; Nichols, 1962; Kirk, 1911). Food grooves progressively lengthened in the evolution of the Diploporita. Thinness or thickness of thecal plates noted above probably had selective value. Thus thickened plates could serve as ballast, and thin plates would have forwarded

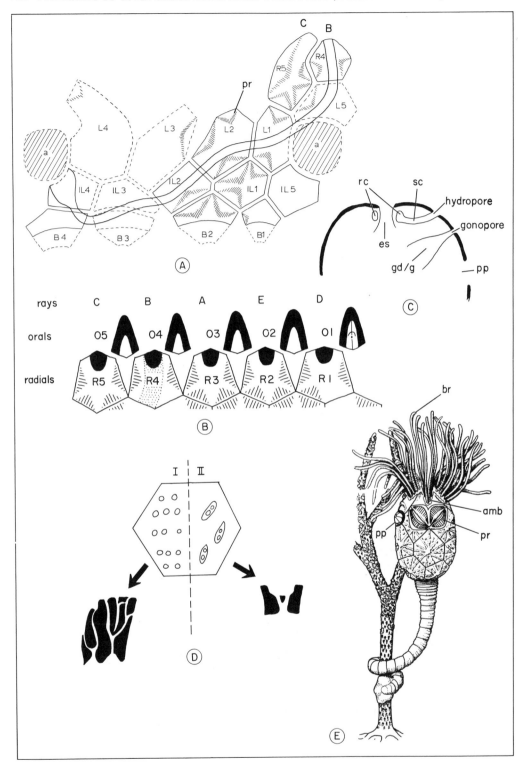

FIGURE 12.30 Information contained in fossil cystoids. Major morphological features.

(A) *Cheirocrinus hyperboreus* (superfamily Glyptocystitida), Ord., Norway. (A) Analysis of thecal plate pattern: B1–4, basals; *IL* – infralaterals; *L* – laterals; *R* – radials (compare Figure 12.29C, b, for same, restored); *pr* – pore rhombs; heavy line – inferred course of gut. After Regnéll, 1949.

(B) Generalized portion of a cystoid belonging to same superfamily as (A) above. Note row of orals above radials, also that ambulacra, where present, occur in both oral and underlying radials. Schematic. Rays of ambulacra, to right, (A)–(C); to left, (E)–(D). Portion of B-ray, shaded. To appreciate the system, a three-dimensional view needs to be envisioned, as in Figure 12.29C, b (schematic).

(C) *Glyptosphaerites sp.*, cross-section through peristome of reconstructed theca; *es* – esophagus; *sc* – stone canal; *pp* – periproct; *gd/g* – genital duct and gonad (hydropore and gonopore may be combined or separate); schematic.

(D) Pore types: exterior view of thecal plates bearing *haplopores* (1) consisting of unbranched canals to interior (1/2 plate shown), and *diplopores* (II) consisting of branched Y-shaped canals to interior. A third type of pore, is *subepithecal*. Pore rhombs shown in (A) above.

(E) *Lepidocystites moorei*, completely reconstructed individual, schematic; brachioles (*br*); pore rhombs (*pr*), mode of attachment to ramose bryozoa on an Ordovician substrate; *pp* – periproct; ambulacral system (*amb*).

Illustrations redrawn and adapted from (B)(C) Kesling, 1968; (E) Kesling and Mintz, 1961.

buoyancy in a complete or partial free-floating existence.

Fixation to the substrate took varied forms: cementation, a prehensile column as in *Macrocystella*, and columns, as well as brachioles, serving as organs of locomotion on the bottom Each of these tendencies had adaptive value. Cementation, for example, permitted the use of a gastropod shell as a substrate. *Aristocystis bohemicus* completely overgrew such shells. Attachment by cementation also occurred during ontogeny, and a subsequent free-living existence followed. Some forms had a prostrate habit as indicated by marked asymmetry of the theca (*Pleurocystis*, Kirk, 1911; see Sinclair, 1948).

Along with plate reduction, the actual surface coverage of the so-called "breathing organ" pore rhombs likewise diminished and became more restricted (Nichols).

Other changes: modified configurations, more globose form for a free-living existence, more flattened form for a bottom existence.

Larval Stages, Migration, Speciation.
Larval stages of cystoids are unknown except insofar as they may be inferred. Regnéll (1960) observed that for crinoids, a free-swimming larval stage lasted from 2 to 12 days and that that period sufficed to spread a given species over great distances. Reasoning by analogy, one could apply the yardstick to cystoid larvae in the early Paleozoic seas (Kesling, 1967). Kirk (1911) suggested that the globular theca of *Echinosphaerites* might denote a floating existence in the adult stage. Nevertheless, the likelihood is that most cystoids' horizontal spread may be attributed to free-swimming larvae.

Cystoids generally form restricted populations with some exceptions (reviewed by Kesling, 1967). Thus geographic isolation led to being cut off from the parental gene pool so that in the course of time new species and genera, as well as families, arose. Examples of restricted families are the Glyptocystitidae (Canada and United States only), and the Rhombiferidae known only from Bohemia.

The Echinosphaeritidae are of especial interest. The population of *E. ellipticus* clearly migrated between North America and Europe during upper Lower Ordovician time (Kesling, 1967, Figure 3, Esthonia B_3-zone). Russian, Esthonian, Swedish, Norwegian, and Welsh deposits of Middle Ordovician age yield *E. aurantium* fossils. It is clear in this case that at the time a common gene pool existed across much of Europe and parts of the British Isles.

Orviku (1927) studied variations in pore rhomb patterns in *Echinosphaerites aurantium* and the distribution of such variants

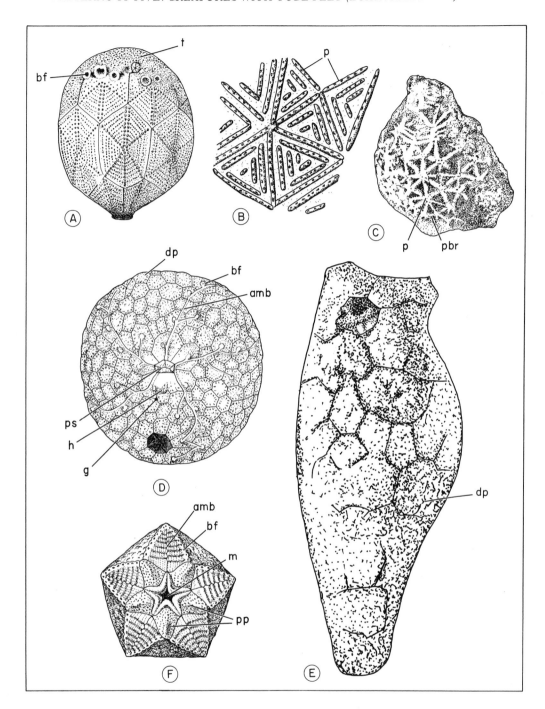

FIGURE 12.31 Representative cystoids.

(A) Superfamily Hemicosmitida: *Caryocrinites ornatus*, M. Sil., (Niagaran), United States; ovoid theca brachiole facets (*bf*) where biserial brachioles attached to tegmen (*t*) covered ambulacral grooves. Observe tubercle-type termination of pores, covered by epitheca, X 1.

(B) Superfamily Polycosmitida: *Stichocystis geometrica*, M. Ord., Germany. Ornamentation consists of pore-bearing straight ridges, X 3; *p* — pore.

(C) Superfamily Caryocystitida: *Heliocrinites rouvellei*, M. Ord.; (Caradocian), France, *pbr* — pore-bearing ridges weathered out; *p* — pores; slightly enlarged.

(D) Superfamily Glyptosphaeritida: *Glyptosphaerites leuchtenbergi*, L. Ord., U.S.S.R., *g* — gonopore; *h* — hydropore; *ps* — peristomal plate; *amb* — ambulacra; *bf* — brachiole facet; *dp* — diplopores. Oral view, X 1.

(E) Superfamily Sphaeronitida: *Triamara cutleri*, Sil., Indiana; anal surface on elongate, unsymmetrical calyx (not shown in this view are Y-arranged ambulacral grooves); periproct; calyx length, 78 mm; *dp* as in (D).

(F) Superfamily Asteroblastida: *Asteroblastus foveolatus*, L. Ord., U.S.S.R.; oral view; bud-shaped theca; five ambulacra (*amb*); diplopore-bearing pore plates (*pp*) restricted to interambulacral areas; *m* — mouth; *bf* — brachiole facets; enlarged.

Illustrations redrawn and adapted from (A) Bather, 1900; (B,D,F) Jaekel, 1899; (C) Dreyfuss, 1939; (E) Tillman, 1967.

from Leningrad across Esthonia to Scandinavia. This study yields some insight into the existing cystoid gene pool of the time. Orviku recognized numerous variants in the pore rhomb pattern that he called "mutations." Most important for this discussion is the increase of different variants between Leningrad and Scandinavia across Esthonia. This supports the view of a highly variable population, including geographic isolates on the way to speciation. The several possibilities in pore rhomb pattern in this ubiquitous species indicates the extreme plasticity of the genetic code governing this aspect of the respiratory system. In turn, local adaptations to new or changed respiratory requirements must have conferred a selective advantage.

In addition to shared gene pools, regional species of Middle Ordovician *Echinosphaerites* indicate fractionation of parental gene pools as a result of geographic isolation. Examples are *E. pogrebouvi*, China, and *E. granulatus* from Ireland and Scotland. Furthermore, some varieties have been named; *E. barrandei* (Czechoslovakia), having a variety, *E. b. belgicus* (Belgium), and *E. globosus* (Esthonia) with a variety, *E. g. anglicus* (Ireland). This also suggests migration and geographic isolation followed by subspeciation. If one could establish accurate regional and interregional correlations, it might be shown that *Echinosphaerites*

may have actually been a single polytypic species during Middle Ordovician time and many of the present species would then be subspecies, while the varieties would be "morphs" (see Chapter 14). Orviku's study lends support to this interpretation.

Population Density. Cystoids as fossils are relatively sparse, even though dense accumulations have been found in several parts of the rock column. *Echinosphaerites* abounds in limestones forming bank-type deposits near the Baltic. In the Esthonia Ordovician cystoid fossils have index value. There is a *Echinosphaerites aurantium* zone some 3.75 meters thick and an upper *Caryocystites* zone some 4 meters thick. In the former, one bed of three is denoted the *Echinosphaerites limestone*, because of the density of cystoids in it (Orviku, 1927, Table 1).

Stemmed *Cheirocrinus* species are packed on given bedding planes in the Canadian Ordovician. *Glyptocystites* species occur in crowds, about seven per sq ft of shale, in the Silurian of Canada. Dense cystoid populations of *Holocystites* have been recovered from the Indiana limestone quarries (Silurian Osgood Formation) (Tillman, 1967).

Biotic Associates. Fixation to gastropod shells or ramous bryozoans and encrustation by bryozoans, suggest that molluscan-bryozoan associates were probably common. Other biotic associates include brachiopods, auloporoid corals, paracrinoids (*Camarocystites*, Trentonian of Canada—Foerste, 1916). Kesling and Mintz (1961) provided a reconstruction of a Late Ordovician sea floor in southern Indiana (Figure 12.30*E*), and in addition to the bryozoans and brachiopods already noted, there were edrioasteroids.

Nutrition. A constant rain of microscopic organic debris was collected by the brachioles and transported to the mouth. There appears to have been a trend toward increase of the areal exposure of the ambulacra of the theca.

Aspects of Growth. Kesling (1967) spoke of a "physiological struggle" of two systems: pore rhombs (breathing apparatus) versus the ambulacral system (food collecting). He observed that (*a*) presence of one or more pore arms in the path of a developing ambulacrum "normally" appears to thwart further advance of the latter (for example, certain species of *Lepidocystites*); (*b*) however, sometimes a rhomb area is "smothered" by a developing ambulacrum.

The concept of "physiological struggle" during ontogenetic development tends to obscure more fundamental considerations. The given cystoid was genetically programmed for both ambulacra and rhombs. There are several possible explanations for the observed phenomenon. (1) The genetic coding for rhombs and amb position was controlled by the same loci on the twin DNA strands, and external selective factors determined which of the multiple potentials were to be realized in the phenotype. (2) Amb and rhomb position relative to the same thecal sites may have been governed by coexistence in the same population of alternate gene arrangements (Mayr, 1963). (3) There may have been a selective advantage in closer proximity of the two systems. Thus Nichols (1962) remarked on the proximity of diplopores to food grooves in forms like *Dactylocystis* (= *Proteroblastus*). He suggested that a beneficial effect might be the explanation; that is, brachiole currents could have aided respiration. If so, natural selection would favor tendencies toward such closeness of the two systems.

Ambulacra almost never occupy *precisely* the same position relative to ambulacral plates. In a sample of 41 specimens of *Jaeckelocystis hartleyi* (Kesling, 1967, Figure 43), a slight positional shift is observed from one specimen to another. What is denoted is a built-in leeway in phenotypic expression. Such variability in position on the theca of the ambulacra may have conferred a selective advantage relative to respiration and, in turn, been *selected for*.

Affinities. Cystoid ancestry of blastoids, echinoids, and affinities to eocrinoids and crinoids, have been discussed and argued against by many investigators (Cuénot, 1953; Kesling, 1967; Moore, 1954; Regnéll, 1960). Fell and Pawson (1963, Figure 1-17*a*) suggest derivation of cystoids, blastoids, edrioasteroids, and paracrinoids—all from a common eocrinoid stock. Obviously, even now the cystoids are a "polymorphous assemblage" (Regnéll, 1960).

REFERENCES

Cystoids

Bather, F. A., 1929. "Echinodermata," in Encyclopedia Britannica, 14th Ed., pp. 894–904.

————, et al., 1900. The Echinodermata, Pt. 2, in E. R. Lankester, A Treatise on Zoology. Adam and Charles Black, London, Ch. IX, Cystitidea.

Cuénot, L., 1953. Classe des Cystidés, in Jean Piveteau, ed., Traité de Paléontologie, Vol. III, pp. 607–628.

Chauvel, J., 1941. Recherches sur les cystoïdes et les carpoïdes armoricains: Soc. Géol. Minéral. Bretagne, Comptes Rendus, Mém., Vol. 5, pp. 1–286.

Foerste, A. F., 1916. *Camarocystites* and *Caryocrinites*, cystids with pinnuliferous free arms: Ottawa Naturalist, Vol. 30, pp. 69–79, 85–93, 101–113.

Jaekel, O., 1918. Phylogenie und System der Pelmatozoen: Paläont. Zeitschr., Vol. 3 (1), pp. 1–128.

————, 1899. Stammesgeschicte der Pelmatozoen. 1. Thecoidea und Cystoidea. Julius Springer (Berlin), 442 pp.

Kesling, R. V., 1967. Cystoids, in Treat. Invert. Paleont., Pt. S, Echinodermata, Vol. 1 (1), pp. S85–S267.

————, 1961. A new *Glyptocystites* from the Middle Ordovician strata of Michigan: Univ. Michigan Mus. Paleont. Contrib., Vol. 17 (2), pp. 59–76, 3 plates.

———— and L. W. Mintz, 1961. Notes on *Lepidocystis moorei* (Meek), an Upper Ordovician callocystitid cystoid: same, Vol. 17 (4), pp. 123–148, 7 plates.

Kirk, E., 1911. The structure and relationship of certain eleutherozoic Pelmatozoa: U.S. Nat. Mus. Proc., Vol. 41, 137 pp., 11 plates.

Moore, R. C., 1954. Status of Invertebrate Paleontology, 1953. IV. Echinodermata: Pelmatozoa: Mus. Comp. Zool. Bull. (Harvard), Vol. 112 (3), pp. 125–149, 8 text figures.

Orviku, K., 1927. Die Rassenvariationen bei *Echinosphaerites aurantium* Gyll., und ihre stratigraphische Verbreitung im estnischen Ordovidium: Publ. Inst. Geol. Univ. Tartu, No. 8.

Regnéll, G., 1960a. "Intermediate" forms in early Paleozoic echinoderms: Interntl. Geol. Congr. (Copenhagen, 1960), Rpt. XXI, Pt. XXV, pp. 71–80.

————, 1960b. The Lower Paleozoic Echinoderm Faunas of the British Isles and Balto-Scandia: Paleontology, Vol. 2, pp. 161–179.

———— 1949. Echinoderms (Hydrophoridea, Ophiocistia) from the Ordovician (Upper Skiddavian, 3cB) of the Oslo region: Norsk Geol. Tidsskr., Vol. 27, pp. 14–57, 2 plates, 6 text figures.

Sinclair, G. W., 1948. Three notes on Ordovician cystids: J. Paleontology, Vol. 22 (4), pp. 301–314, plates 42–44, 6 text figures.

Tillman, C. G., 1967. *Triamara cutleri*, a new cystoid from the Osgood Formation (Silurian) of Indiana: J. Paleontology, Vol. 41 (1), pp. 222–225, plate 25, text figure 1.

Paracrinoids

Introduction. Regnéll, in 1945, erected the class Paracrinoidea. Does this grouping embrace taxa that are true genetic affiliates? Fifteen years later, Regnéll recognized that the class may be artificial. Paracrinoids are neither true cystoids nor true crinoids. They appear to occupy a place somewhere intermediate between these classes.

Classification: Class Paracrinoidea. Crinozoans with a plated theca, undifferentiated into calyx and tegmen; calcareous thecal plates form irregular patterns and are variable in number (Figure 12.32*A*, *a*). Uniserial arms (no more than four) may be free or attached (Kesling, 1967, Table 3), bear uniserial pinnules — both constituting the ambulacral system; periproct with anal pyramid; hydropore and gonopore may be present. Pore system (Figure 12.32*A*, *b*) apparently distinctive. Geologic range: Middle Ordovician only.

The class may be divided into two orders on the basis of arm attachment to the theca: Order Varicata (arms attached) — genera include *Amygdalocystites*, *Sinclairocystis* (Strimple, 1952), Order Brachiata (arms free) — *Camarocystites* (Figure 12.32).

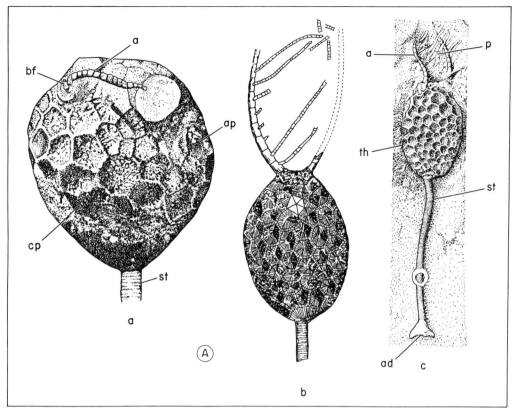

FIGURE 12.32 Data encoded in paracrinoids.
(A)(a) Order Brachiata. *Camarocystites punctatus,* posterior side; *bf* — brachial facet; *a* — arm; *ap* — anal pyramid; *st* — stem; *cp* — deeply concave exterior plate surface; note markings that are lunate pores; (*b*) thecal plate (*cp* in A*a*) with lunate pores in pairs on unweathered episteroem; (*c*) a nearly complete specimen, reduced; *st* — stem; *th* — theca; *p* — pinnules; *ad* — basal attachment disk (hold fast). After Foerste.

Information Derived from Fossil Paracrinoids

Affinities. Paracrinoid theca are cystoid-like, the arm-bearing pinnules are reminiscent of crinoids, whereas the column resembles that of blastoids. However, its uniserial appendages make assignment to the cystoids or to the eocrinoids difficult. The regular pattern of thecal plates (Foerste, 1916) ruled out assignment to the crinoids. Its lack of hydrospires and symmetry appears to eliminate blastoid affinities. Although paracrinoids were placed among the cystoids (Bather, 1900) and the carpoids (Jaekel, 1900), its theca is not flattened as that of the carpoid and its column is noncarpoid-like (Kesling, 1967).

At best, one should presently regard the class Paracrinoidea as tentative and subject to drastic revision with new discoveries.

Plate Structure. At least in the well-studied *Camarocystites* (Foerste, 1916), the calcitic thecal plates are known to consist of an outer, thin, nonporous epistereom (epitheca) and an underlying mesostereom (mesotheca). In this genus the mesostereom consists of vertical laminae that thicken toward the base and the interlamellar spaces into which the test pores lead (Figure 12.32*A, b*).

Feeding. Conceivably, the arms bearing multiple pinnules served as a dual food-gathering and food-straining device.

A downpour of microscopic organic debris seems to have been the main type of food (Kesling, 1967).

Substrate and Biotic Associates. *Camarocystites punctatus* was attached by holdfasts (Figure 12.32*A*, *c*) to a clay lime mud substrate. It was found associated in a Canadian quarry (Trentonian beds) with crinoids, edrioasteroids, cyclocystoids, and trilobites (*Isotelus*, *Amphilichas*). Other biotic associates are cystids and bryozoans (Foerste, 1916). *Sinclairocystis* (Bassler, 1950) occurred in green shales of the Oklahoma Bromide Formation. Clay mud and lime mud substrates are usual for paracrinoids.

REFERENCES

Paracrinoids

[Many of the same references under the cystoids include data on, or interpretation of, paracrinoids. These are not repeated here: Foerste, Regnéll, Jaekel, Bather, and some others.]

Bassler, R. S., 1950. New genera of American Middle Ordovician "Cystoidea": Washington Acad. Sci. J., Vol. 40 (9), pp. 273–277.

Kesling, R. V., 1967. Paracrinoids, in R. C. Moore, ed. Treat. Invert. Paleont.: S. Echinodermata 1 (1), pp. S268–S288.

Strimple, H. L., 1952. Two new species of *Sinclairocystis*: Washington Acad. Sci. J., Vol. 42 (5), pp. 158–160, Figures 1–4.

Edrioblastoids and Parablastoids

Introduction. The Edrioblastoidea was established on Whiteaves's two specimens of *Astrocystites* (Middle Ordovician, Cobourg limestone beds of Ottawa, Canada). Formerly, these specimens had been assigned to the Class Edrioasteroidea. Ubaghs (1967) commented that *Astrocystites* differed from all other Crinozoa in lacking arms and brachioles, possessing edrioasteroid-type ambulacral pores, among other characteristics.

Fay (1962, 1967) felt that edrioblastoids were sufficiently distinctive to merit establishment of a new class. They "came close" to the type that could have given rise to the fissiculate blastoid genus *Polydeltoideus* (Fay, 1967*b*).

Classification: Class Edrioblastoidea. Blastoid-like, stemmed crinozoans with regular pentameral symmetry; calcareous plated theca consisting of 20 main plates and four circlets. From the stem upward, plates include five basals, five radials, five deltoids, five orals, plus infradeltoids (Figure 12.33*A*). Ambulacra, subpetaloid, long and bearing row of close-spaced pores, covered by plates; anal orifice and hydropore present. Geologic range: Middle Ordovician, one locality only. (*Astrocystites*, Figure 12.33*A*.)

Information Derived from Edrioblastoids

Affinities. Bather (1900) commented on the cystid, blastoid, crinoid, and asteroid affinities of the genus *Steganoblastus* (= *Astrocystites*), which he had assigned to the edrioasteroids. He thought it was "simpler" to regard the genus as a *specialized* edrioasteroid.

Regnéll (1960) also regarded this genus as unique among the edrioasteroids and was unenthusiastic about Hudson's statement (1907) that it was a true blastoid. He did not agree that it was any closer to blastoids than to other stemmed echinoderms. Homologies to blastoids, Regnéll implied, might be due to convergence (that is, raising the theca above the sea floor by a stem). The chief genuine edrio-

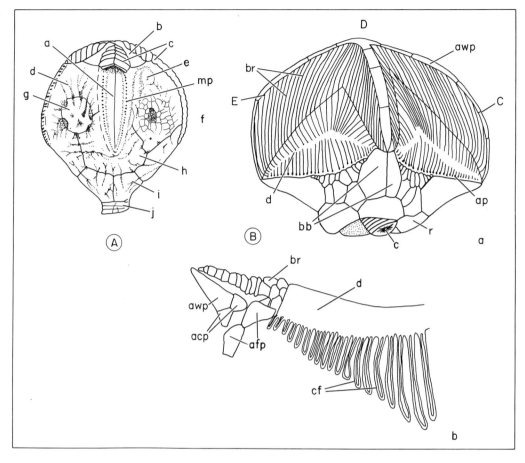

FIGURE 12.33 Information contained in edrioblastoids and parablastoids.
(*A*) Class Edrioblastoidea. *Astrocystites* (= *Steganoblastus*) *ottawaensis*, M. Ord., Canada, X 2. (After Bather, 1900). *a* — cover plate (*c*) removed from part of ambulacrum; pores (*mp* — about 26 on each side of each ambulacra along margins of floor plates); *b* — oral plate; *d* — deltoid; *e* — location of slit (hydropore?); *f* — anal orifice; *g* — infradeltoid plate; *h* — radial; *i* — basal plate; *j* — column.
(*B*) Class Parablastoidea: *Blastoidocrinus carchariaedens*, Ord. (Chazyian), United States. (*a*) Theca (viewed along *D*-ray), X 2 +; *D-E*, designate ambulacral rays; *c* — column; *r* — radial plates; *bb* — bibrachial plates; *d* — deltoid plates; *br* — brachioles; *awp* — ambulacral wing plates; *ap* — aboral pores. (*b*) Cross-sectional view across portion of deltoid and ambulacrum, X 8.
Observe cataspire folds (*cf*) and relationship of brachioles (*br*); *d* — deltoid plate; *afp* — ambulacral floor plate; ambulacral cover plates (*acp*). After R. C. Fay.

asteroid characteristic was the ambulacral structure with rows of marginal pores on floor plates (Figure 12.33).

Fay (1962) listed some anatomical features in which edrioblastoids differed from blastoids: (1) plates — five basal plates and interradially placed infradeltoids, (2) hydropore, (3) lack of brachioles (these may have been present as uncalcified structures and hence were not preserved). In fact, Fay (1967) thought edrioblastoids, having small intercalated plates between

radials and deltoids (Figure 12.33), were closer to the parablastoids than to the blastoids.

Water Vascular System. Bather (1914) and Fay (1962) referred to the tube feet of *Astrocystites* (Figure 12.33). The hydropore (water pore or madreporite) (Figure 12.33*A, e*) was also part of the water vascular system and served for intake of seawater.

Parablastoids

Introduction. In 1907 G. H. Hudson described some Chazyan echinoderms (Middle Ordovician of New York) and set up a new order of blastoids to incorporate them, the Parablastoidea. Hudson's order has now been elevated to the rank of a class.

Ubaghs (1967) stressed the fact that although the parablastoids are blastoid-like, they differ in the number of thecal plates and ambulacral structure. He thought that they were an aberrant, unsuccessful side branch of an early (that is, pre-Middle Ordovician) blastoid ancestral stock.

Classification. Blastoid-like stemmed blastozoans with strongly developed pentameral symmetry; calcareous theca of numerous plates arranged in aboral and oral regions as follows: aboral side—five basals and above them five radials in line with the last, five pairs of bibrachials, between any two of these, numerous interbrachials; oral side—five plated subpetaloid ambulacra and five interradials (= deltoids). Parallel infolds (= *cataspires*, Figure 12.33*B*, *b*) of deltoid plates connected to pores along ambulacral margins and also to aboral pores (Figure 12.33*B*, *a*, "ap"); parallel rows of biserial brachioles are borne by adambulacrals (Figure 12.33 *B,a*; *B,b*—"*br*"). Geologic range: Lower to Middle Ordovician.

[The class has a single family, Blastocystitidae, and it contains but two genera, *Blastocystis* and *Blastoidocrinus* (Figure 12.33*B*) from Leningrad, USSR and eastern United States, respectively.]

Information Derived from Parablastoids

Affinities. From a study of the essential anatomy of the parablastoids, their skeletal anatomy cannot be directly compared to either true blastoids or edrioblastoids.

Stable Genetic Configuration. There are many ways to explain blastoid-like thecae in nonblastoid creatures. Among others, one explanation could be evolutionary convergence in thecal configuration, ambulacral number and arrangement, and so on, of independent lines of descent. Or one could postulate a stable portion of the genetic code recurring in interrelated stocks, stable for the blastoid-like thecae, and so on.

Probably, equivalent environments for many stemmed echinoderms served as a selective factor for either of the above mechanisms to become effective repeatedly in distinct species populations.

Blastoids did not evolve until Silurian time and, as noted earlier, edrioblastoids might have been the precursor stock. Then, the blastoid-like characteristics could be due to certain stable portions of the genetic code. If parablastoids are an aberrant offshoot of an early blastoid stock (Ubaghs, 1967) (in Lower Ordovician time, rather, a blastoid-like stock), the same would apply. Thus, for edrioblastoids, blastoids, parablastoids, the stable portions of the genetic code could explain homologous thecae.

Restricted Populations and Extinction. Both parablastoid and edrioblastoid populations, as presently known, appear to have been restricted in both horizontal and vertical spread. The lowermost Middle Ordovician Chazyan beds of New York and Quebec both yield *Blastoidocrinus*. Fay (1967*c*) found, among many American specimens of this genus, a "morphological distinctness" suggestive also of the narrow range of intraspecific variability. The only other parablastoid genus known— and that poorly—is *Blastocystis* from the uppermost Lower Ordovician of Leningrad.

As for edrioblastoids, only a single genus occurred in the Middle Ordovician of Ottawa.

Clearly, the kind of populations that may be inferred from these sparse data appear to have been *local*. With such delimited gene pools, extinction was likely brought about by a continual decrease in the number of reproducing individuals. A further observation can be made on the vertical restriction of both classes; the unique qualities of each did not confer long enduring adaptive value in the Paleozoic seas. This is also shown by the lack of more than one genus in Chazyan time

that denotes a low to negligible fractionation of the existing gene pool.

It is conceivable that apparent restriction of spread and sparseness in occurrence is an artifact of collecting. More collections will be needed to resolve such negative possibilities.

REFERENCES

Edrioblastoids and Parablastoids

Edrioblastoids

Bather, F. A., 1914. I. Studies in Edrioasteroidea, *V. Steganoblastus*: Geol. Mag., n.s., Vol. 1, 205, pp. 193–203, plate 15, text figures 1–6.

_____, 1900. The Echinodermata. Pt. 3, in, E. R. Lankester, ed., A Treatise on Zoology. Adam and Charles Black, London, 332 pp. (pp. 209–210, Figure VII).

Fay, R. O., 1967a Edrioblastoids, in R. C. Moore, ed., Treat. Invert. Paleont. (S) Echinodermata 1 (1), pp. S289–S292.

_____, 1967b. Evolution of the Blastoidea, in Essays in Paleontology and Stratigraphy, Raymond C. Moore Commemorative Volume. Univ. Kansas Press, pp. 242–286.

_____, 1962. Edrioblastoidea, a new class of Echinodermata: J. Paleontology, Vol. 36 (2), pp. 201–205, plate 34, 3 text figures.

Regnéll, G., 1960. "Intermediate" forms in early Paleozoic Echinoderms: Interntl. Geol. Cong. XXI Session, Pt. XXII, Proc. I. P. U., pp. 71–80.

Ubaghs, G., 1967. General characteristics of Echinodermata: Treat. Invert. Paleont. (S) Echinodermata 1 (1), pp. S3–S60.

Parablastoids

Fay, R. O., 1967c. Parablastoids, in Treat. Invert. Paleont. (S) Echinodermata 1 (1), pp. S293–S296.

Hudson, G. H., 1907. On some Pelmatozoa from the Chazy Limestone of New York: N.Y. State Mus. Bull. 107, pp. 97–152, plates 1–10.

Blastoids

Introduction. As with other noncrinoid pelmatozoans (intermediate forms), blastoids have had a volatile history. Blastoid hydrospires (Figure 12.34D), especially, have inspired much speculation. Are the pore rhombs or diplopores of cystoids and blastoid hydrospires homologous structures? Are the genital slits of ophiuroids and blastoid hydrospires comparable? Can the dorsal plates of the crinoid arm be related to the lancet plate of the blastoid ambulacrum? Such notions have entered into the search for the ancestry,

affinity, and definition of the blastoids (Bergounioux, 1953; Fay, 1967a, 1967b; Regnéll, 1945, 1960).

The blastoids, now placed in a separate class by some authors, were previously treated as a subclass of the cystoids.

Classification: Class Blastoidea. Stemmed blastozoans bearing internal or external hydrospires (if internal, generally accompanied by pores and spiracles); with biserial brachioles attached at corner of ambulacra, and with a plated calcareous theca. Circlets of plates include (Figure

12.34*A*) three basals over which are five radials, and next higher and alternating with radials are the five deltoids (anal deltoids with up to six plates—Figure 12.34*F*). Other plates include five lancets radially disposed (see "Ambulacra," Figure 12.35*A*, *a*). Geologic range: Silurian—Permian.

(No Ordovician blastoids are presently known. As of this writing, the oldest known blastoid, *Decaschisma*, comes from the Middle Silurian of Indiana, the Waldron Shale.)

Hydrospire structures (slits, pores, spiracles) serve as criteria for subdivision of the blastoids into orders:

Order I—Fissiculata (with external hydrospire or spiracular slits) and Order II—Spiraculata (with pores and spiracles and internal hydrospire slits). The orders, in turn, are divisible into five and seven families, respectively. These 12 families embrace a total of some 80 described genera.

[Illustrations of nine of the 12 families are given below: Order Fissiculata—*Codaster* (Figure 12.34*E*, *b*), *Deltoschisma* (Figure 12.34*F*), *Orophocrinus* (Figure 12.34*E*, *a*), *Thaumatoblastus* (Figure 12.35*C*, *b*). Order Spiraculata—*Orbitremites* (see Chapter 14), *Deltoblastus* (Figure 12.35*D*, *b*), *Diploblastus* (Figure 12.35*D*, *a*), *Dentiblastus* (Figure 12.35*C*, *a*), *Pentremites* (Figure 12.34*B–C*; Figure 12.35*A*, *B*), *Pyramoblastus* (Figure 12.34*A*), *Auloblastus* (Figure 12.34*D*).]

Morphology and Systems

Plate Growth: Primary Deposition. Secretion of plates of the theca and brachioles, hydrospires and stem involve the major events in the skeletogenesis of blastoids.

Study of external growth lines on plates (Macurda, 1966) reveals that enlargement occurs along definite "growth fronts" in different directions. Three axial trends in the outward growth of radial plates in the species *Belocrinus cottaldi* were observed (Macurda, 1967). Secretion of biocrystals occurred in the mesodermal tissue. Plate

accretion or growth ensued from biocrystal formation in tissue between adjacent plate edges. Macurda (1967) observed differential deposition of calcite biocrystals along certain plate sutures (radiodeltoid) but uniform deposition in a given series of plates (radials).

Biocrystals. Other echinoderm data lead to the expectation that in blastoids each plate would optically react as a single crystal. It does. A distinct crystallographic orientation results in separate extinction for each plate (in thin section) when the stage of a polarizing microscope is rotated. This can be referred to as "optical continuity." If within a given plate separate extinctions take place in parts of it, this reaction would constitute an optical discontinuity. There is a possibility that the calcite latticework (reticulation) of the blastoid skeletal plates formed in two stages: (*a*) continuous secretion of discrete calcite fibers during growth, and later (*b*) recrystallization of these fibers to form individual crystals (Macurda, 1967*a*, and see Lucas, 1952, Joysey and Breimer, 1963).

In addition to the primary deposits there were other metabolic events involved in calcite deposition or transfer: Secondary deposits, resorption, and fusion.

Calcite Metabolism

Secondary Deposition of Calcite. Such deposits are not unusual in blastoid structures: extension of length of ambulacrum by secretion of crystals to form a prong on the radial plate (Figure 12.35*C*), overgrowths broadening the configuration of the basal plates where they are in contact with the stem, extension of the anus by secondary secretion of a ramp in *Orophocrinus stelliformis*.

Resorption. Diameter of spiracles enlarge in *Globoblastus*, among others, with the age of the individual. This must be attributed to resorption of existing regional calcite crystals of the deltoid plates that are pierced by the spiracles. Another example of resorption is the enlargement of the anal opening (Macurda, 1967).

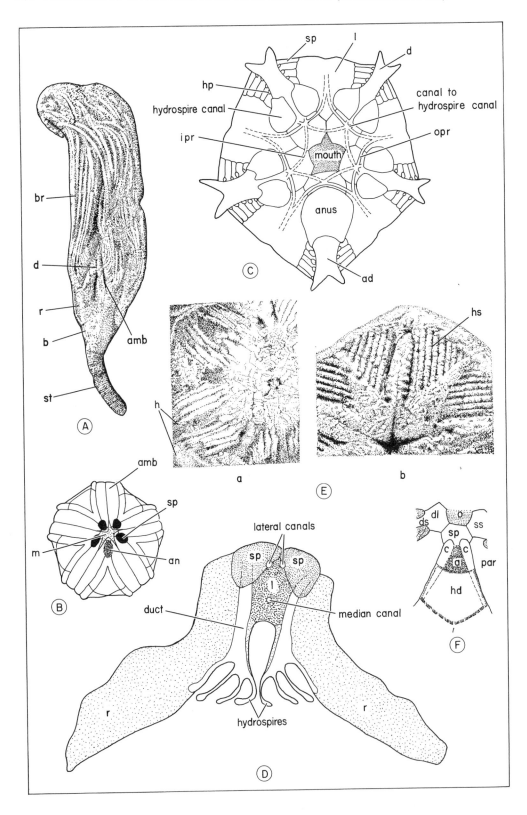

FIGURE 12.34 Information contained in blastoids. Morphology – I.

(A) *Pyramoblastus fusiformis.* L. Miss. (Iowa); lateral view, X 3; *st* – stem; plates; *b* – basal, *r* – radial, *d* – deltoid, *amb* – ambulacrum from corner of which, brachiole (*br*) arose. After Macurda.

(B) *Pentremites* sp., Miss.–Penn. Oral view, schematic, (after Fay); *sp* – spiracle, *an* – anispiracle (anus), *m* – mouth.

(C) *Pentremites godini,* U. Miss., Ill., thin section, view of oral ring canal and side canal to hydrospire. Plates: *ad* – anideltoid; *d* – deltoid; *sp* – side plate of ambulacrum; *l* – lancet (see text); *hp* – hydrospire pore; *opr, ipr* – outer and inner pentagonal ring. After Fay.

(D) *Auloblastus clinei,* Upper Burlington; thin section cut slightly lower than middle of theca; showing hydrospire relationship to various plates; *sp* – side plates; *r* – radials, ducts (hydrospire), and canals (lateral and medial). After Beaver.

(E) Hydrospire: internal and external types: (a) *Orophocrinus stelliformis,* L. Miss., Ill., internal view of hydrospire, X 2; (b) *Codaster acutus,* Carb., Eng., external hydrospire slits (*hs*), X 6.5. Adapted from photographs in Macurda, 1967.

(F) *Deltoschisma archiaci,* Dev., Spain; anal deltoid plate; *o* – oral opening; *a* – anus; *par* – posterior ambulacrum; *ad* – anal deltoids; *hd* – hypodeltoids; *sp* – superdeltoids; *c* – cryptodeltoids; *dl, ds* – deltoid plates with lip (*dl*) and septum (*ds*).

Fusion. Generally, there are three basal plates. Two of these are larger ones (zygous basal); one is smaller (azygous basal). Originally, these were five more-or-less uniform plates. Modified genetic programming led to fusion of two pairs of basals, each pair forming a larger zygous basal. All five plates may disappear as in a rare example of fusion found in *Acentrotremites* where a single basal plate occurs (Fay, 1967*a*).

Other possible indicators of fusion include multiple oral and anal plates present in early blastoids and apparently fused during subsequent evolution to form fewer plates. *Elaecrinus* has 18 to 21 orals, whereas a probable descendant, *Nucleocrinus* had five (Fay, 1967*b*) Fusion is also evident in some deltoid plates.

Other Systems

Water Vascular System. Hydrospires and thecal pores, slits and canals (Figure 12.34*C*) constitute the water vascular system. Seawater could enter the hydrospires in a given blastoid via spiracles or pores as in *Pentremites* (spiraculate blastoids, Figure 12.34*B*) and leave by the anispiracle carrying out wastes, or it could enter by slits (fissiculate blastoids) and exit through the anus.

Respiratory System. The calcitic hydrospire folds were probably encased in a saclike membrane and the fold area proper, covered by soft tissue (Nichols, 1962). The mineralized hydrospire folds

are likely to have been porous. It is doubtful that the calcitic hydrospire was "semipermeable" (see Fay, 1967*a*, p. S341), since this, more likely, would have been the condition of the soft tissue associated with it. Gaseous exchange could occur across such tissue in the hydrospire area, body fluids being aerated by it and releasing carbon dioxide to it (Nichols, 1962; Macurda, 1967).

Fay (1967, p. S393) raised the possibility that there may not have been a radial canal or water vascular system in blastoids. (The radial canal, being double, has been equated with the main nervous system; see below.) Regardless of alternate solutions, existence of pores, slits, and nonradial canals, all seem to be related to some type of a water-circulatory system.

In *Globoblastus*, a spiraculate form, it has been determined that given an arbitrary water-current velocity of 0.1 mm per second, the exchange of water volume in the hydrospires could take place in the following time intervals: 40, 75, and 100 seconds for small, medium, and large specimens, respectively. This volume of water exchange was deemed adequate to support a respiratory function (Macurda, 1965).

In fissiculate forms, with expanded hydrospire slits, the number of slits increased to as many as 600 per large individual (*Hadroblastus*). Such increase indicated a less efficient respiratory system

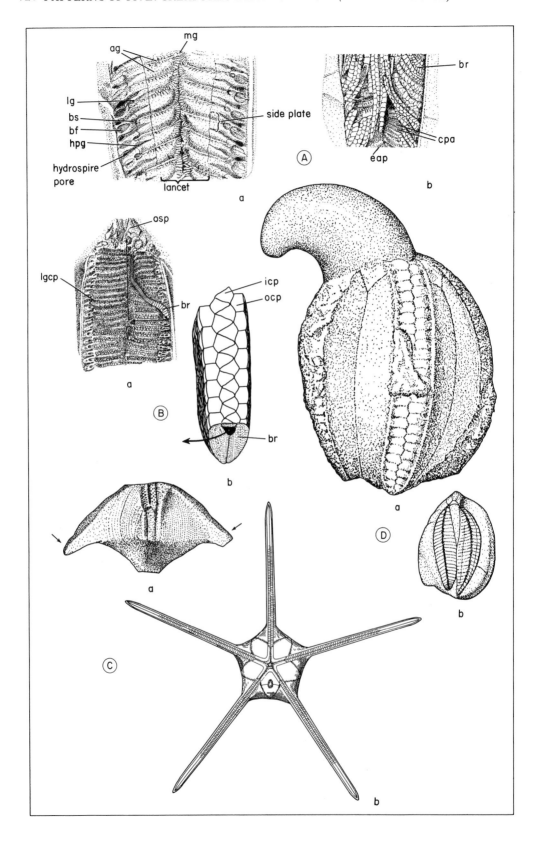

FIGURE 12.35 Information contained in blastoids. Morphology, II, and derivative data, I. (*Pentremites symmetrica*. (A)(a–b) Ambulacrum. (a) Detail, lacking brachioles and cover plates; *mg* — median groove; *lg* — lateral groove; *ag* — articulation grooves for ambulacral cover plates; *bs, bf* — brachiole socket and facet (*bf*). (See Figure 12.34, A for attached brachioles. (b) Same as "a" with brachioles (*br*); *cpa* — cover plates (amb.); *eap* — amb. plates with cover plates removed; *hpg* — pore groove (hydrospire).

(B)(a) Cover plates of lateral groove and medial groove; *br* — a single brachiole remaining attached; arrow shows current-borne food emerging from brachial groove; *icp-, ocp-* inner and outer cover plate; *br* — brachiolar.
Derivative Data, I.

(C) *Adaptive modification:* increase of ambulacral exposure to water. (a) *Dentiblastus sirius*. L. Miss., Missouri, X 6. After Macurda, 1967a. (b) *Thaumatoblastus longiramus*, E. Indies (Timor), Perm. After Wanner, 1924.

(D)(a) *Symbiosis:* Coprophagous *Platyceras* (*Platyceras*) sp. attached to anal plate of *Diploblastus kirkwoodensis* (see Gastropoda section). Miss., St. Louis limestone; (b) *Pathology:* abnormal development in *Deltoblastus permicus* (Timor, Perm.); apparent double ambulacra due to missing interradial plates (that is, radial limbs and deltoid plates). After Wanner, 1932.
Illustrations redrawn and adapted from: A, Ba — H. H. Beaver, 1967; Bb — Fay and Reimann, 1962.

than that found in spiraculate forms (Macurda, 1967) to accommodate a large volume of water per unit time.

Digestive and Excretory Systems. The food-gathering and transporting system of brachioles and ambulacra are known in the fossil state (Figures 12.34*A*; 12.35*A*, *B*). There are from 1000 to 10,000 individual plates to provide suitable protected channelways along brachioles that collect food and for the food's passage along radial and median grooves leading ultimately to the mouth and gut.

The beat of minute cilia along the entire tunnel route from brachiole to mouth was the propulsive force (Macurda, 1967). The food captured by the brachioles consisted, most probably, of live phytoplankton and organic detritus (Cline and Beaver, 1957).

One of the problems after digestion of these morsels in the gut was voiding of wastes. Fouling, or the danger of fouling, or consuming one's own metabolic wastes was a selective factor in at least several blastoids that developed special anal structures, such as ramps, plates, and so on.

In areas that were probably moderate in energy level, communities of blastoid species, dense with individuals, lived fixed to the substrate by a flexible stalk (Cline and Beaver, 1957). In crowded communities of this type, it is likely that there was active competition for available food and unfouled water.

Muscular and Nervous System. Macurda

(1965*a*, Figure 3 and plate 124, Figure 15) found several depressions on the underside of the deltoids surrounding the peristome (area of oral plates) in *Orophocrinus stelliformis*. He inferred that they were probably places of muscular attachment for the alimentary canal that could alternately expand and contract. If so, the reasoning went, this species would have been able to regulate the volumetric input of water.

Similarly, brachiole movements during food-gathering so as to present the maximum areal exposure to existing currents (Figure 12.36*B*), were under muscular control. The flexible stalk also presumably had some muscularity although incapable of major movements (Cline and Beaver, 1957).

Presence of an anal tube in some form suggests that the anus also was a muscular organ that could extend or retract (Macurda, 1967). If so, nutrition and excretion respectively were accompanied by a kind of peristalsis of the gut at either extremity.

Obviously, muscularity requires nerve control. Indicators of the nervous system are thought to be the "circumesophageal rings" in *Pentremites* (dashed lines about mouth and Figure 12.34*C*); offshoots to brachioles (Bather, 1900, nerve canal (?) in *Codaster*) and the stalk may also be inferred to have been parts of the nervous system.

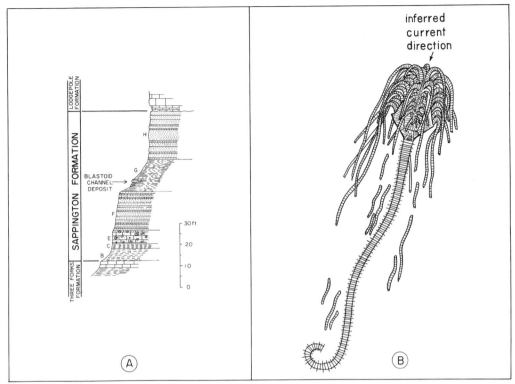

FIGURE 12.36 Information derived from fossil Blastoids—II. Biostratigraphy and paleocurrents (after Sprinkle and Gutschick, 1967). (*A*) Unit G is a laminated green shale, Sappington Formation, Montana. (*B*) Blastoid *Costablastus* specimens occur in a channel deposit, reconstruction, X 2.

Reproductive System and Larval Stages. The mode of reproduction is merely surmised by parallels with other echinoderms (see Crinoidea, this chapter). There has been some thought that hydrospires may have served a reproductive function, at least, in part. This last is more speculative than the postulated respiratory function after the hydrodynamic studies of Macurda. However, Fay (1967*b*) postulated that eggs and sperm formed in the hydrospire loops (assuming sexes were separate), whereas the sides of the hydrospires served for gas exchange.[4]

Very little is known of the larval stage(s) of blastoids. Once again, comparison to comulatid crinoids is usually made. Widespread dispersal of some fossil forms that are attached to the adult stages can only be explained by free-swimming larvae.

Such larvae were presumably planktonic, but some may have been nektonic (Cline and Beaver, 1957). Most fossils of adult blastoids represent individuals that had reached maturity or old age. However, some minute elongate tests also occur (a little over 0.8 mm), that are ranked as juveniles. In *Pentremites*, a specimen (less than 1.0 mm in maximum dimension) had a full complement of radials but only three basals (oral plates were not preserved). Did the larva have five basals at some still earlier stage?

Information Derived from Fossil Blastoids

Biostratigraphy. The occurrence of a restricted facies (channel deposit) of blastoids has been noted (Sprinkle and Gutschick, 1967 and Figure 12.36*A*, this text). Late Mississippian (Chesterian) limestones and calcareous shale interbeds are spoken of as "Pentremital" because of the abundance of the *Pentremites* blastoid popula-

[4]S. G. Katz and J. Sprinkle, 1976. Science, Vol. 192, pp. 1137–1139, modification of anal hydrospires for brooding eggs in female blastoids indicated by fossilized blastoid eggs (Pennsylvanian).

tions found in them in Illinois, Kentucky, and Alabama. Generally, blastoid fossils occur abundantly in the crinoidal facies, Devonian-Carboniferous (Cline and Beaver, 1957), and in the Permian of Timor and neighboring regions.

Adaptive Modification and Evolution

Ambulacra. Prolongation of the ambulacral area by vertical extension along theca and of the oral surface almost to the stem insertion site and of flange elongation laterally (Figure 12.35C) are among the adaptive modifications that related to improved efficiency in feeding. "Stretched ambulacra" appear to have had a strong selective value in some instances.

Hydrospire. In the spiraculate forms, a trend to reduction in the number of hydrospires on each side of the ambulacra

has been remarked; the numbers are 5, 4, 3, 2, or 1 in different genera, and reduction is attributed to atrophy.

What has been referred to as evolution by "atrophy" or "atrophy by disuse" (Fay, 1967a, 1967b) is really a nongenetic misnomer for genetic events: "selecting out" of inadaptive structures. Thus the structures referred to include blastoid stems, pores, hydrospires. During blastoid evolution, it is clear that several thecal structures were modified or lost. This particularly applies to structures related to respiration, feeding, and attachment.

Assuming that the Fissiculata were *primitive* and that this order gave rise to the Spiraculata, Fay (1961) envisioned the basic plan of a primitive blastoid contrasted to that of advanced forms as set forth below.

Structures	Primitive Blastoid	Advanced Blastoid
Hydrospire slits	Exposed	Hidden
Configuration	Steeply conical	Globular or winged
Anal Deltoid Plates	4	2 (or 1 as in *Pentremites*)
Deltoids/Radials	Radials overlap deltoids	Deltoids (elongate) overlap radials
Lancet	Elongate; approaches oral opening; covered by side plates	Short and wide; not covered; does not approach oral opening

Some additional primitive features have also been identified: paired spiracles; linear ambulacra (contrasted to "petaloid" ambulacra in advanced forms) (Fay, 1967a).

The interiorization of surface hydro-

spire slits was another trend. Presumably, following Macurda, it was a response toward greater efficiency in respiration.

Macurda (1965) defined two morphological adaptations to given environments in the species *Orophocrinus*.

Structures	Carbonate Shelves	vs.	Reef Knolls (Bioherms)
Hydrospire	Fewer (4–6)		Greater (7–9)
Brachiole Position	Close to center of main food grooves of ambulacrum		Away from center of main food grooves of ambulacrum
Species	*O. stelliformis*, etc.		*O. catactus*, etc.

Are the above-mentioned differences truly significant or artifacts of small sample, time difference, or other factors?

Speciation. Blastoid speciation is treated at length in Chapter 15.

Malformation. It will be apparent that in any large sample of blastoids or any other invertebrate fossil, one can expect to find evidence of some genetic errors, some abnormalities or pathologies. For the Timor blastoid species, *Deltoblastus* *permicus*, there was a total collection of $n = 24,000$. (Wanner, 1932). The tally of morphologically defective tests (ambulacra: abnormal plates) was $n = 133$.

Paleoecology

Lithology. Macurda's work on *Orophocrinus* species is an indication of some lithologies yielding blastoid fossils. Cline and Beaver (1957) noted the following types of lithologies in which blastoids (and frequently crinoids) occur: blue-gray marls, dark calcareous shales, argillaceous limestones, yellow gray to buff dolomites, crinoidal limestones, green shales, alternating calcareous shales and limestones.

Salinity. Normal salinity (35 parts per mille) is indicated for blastoids. There are no nonmarine blastoids. The recent find of channel deposit blastoids might denote some variation from normal seawater salinity.

Paleocurrents. Channel deposit blastoids showed a parting of brachioles on the bedding plane, a possible response to the prevailing channel current (Figure 12.36B). Macurda (1967b), in a study of Lower Carboniferous blastoids of Belgium, inferred paleocurrents from differential exposure of the hydrospires. The Tournasian forms had hydrospires partially exposed—denoting a low energy environment in which bottom current-induced turbulence was negligible. Cline and Beaver (1957) inferred at least moderate paleocurrents adequate to provide circulation of nutrient-bearing water wherever there were living blastoids.

Biotic Associates, Symbiosis, and Predation. Blastoid-bearing beds contain, at one or another sites, the following biotic associates: bryozoans (ectoprocts), worms, simple and compound corals, crinoids, brachiopods, pelecypods, gastropods, cephalopods, trilobites, ostracods, and fish. Phytoplankton also must be added as a probable food source (Cline and Beaver, 1957; Macurda, 1965).

The coprophagous gastropod *Platyceras* (*Platyceras*) (see Gastropoda, Chapter 8) not only attached to the anal aperture of crinoids, but to the anal outlet of some blastoids as well. In shale interbeds of the St. Louis limestone of Missouri, platycerids—12 different specimens attached to the anal outlet of *Diploblastus kirkwoodensis* (Levin and Fay, 1964).

Predatory gastropods (boring type) apparently sought to feed on blastoid soft parts; Macurda (1965, 1966) noted specimens of two species of *Orophocrinus* having such bore holes in their thecae.

Extinction. Extinction of blastoids by end of the Permian is rather curious, in a sense. There is no evidence of (a) gradual decline in speciation, (b) in population diversity, or (c) geographic spread. Contrariwise, 19 new genera made their apparent first appearance in the Permian. This last event would indicate a broad population diversity and vigorous speciation. Permian blastoids (mainly Fissiculata with 16 genera) were highly specialized. Furthermore, they have been found in Timor (chiefly), Australia, Sicily, and Russia—which certainly does not denote a small regional, or restricted terminal population. Rather, several aspects of evolving populations can be seen: interbreeding populations, marked diversity, species vigor, and migration.

Why, then, did the blastoids become extinct before Mesozoic time?

Since neither morphology nor distribution provide clues, it is plausible to assume, as Fay did (1967a, 1967b), that some external agency was responsible.

Were there some Late Paleozoic changes in seawater (temperature, salinity, or other) perhaps related to the Permo-Carboniferous glaciation of the continents of the southern hemisphere? Were there, perhaps, some other changes?

REFERENCES

Blastoids

Beaver, H. H., 1967. "Morphology," "Techniques," "Ontogeny," "Paleoecology," in Treat. Invert. Paleont., Echinodermata (= Ech), 1 (2), Blastoids, pp. S300, S350, S352, S382.

———, 1961a. Morphology of the blastoid *Globoblastus norwoodi:* J. Paleontology, Vol. 35 (6), pp. 1103–1112.

———, 1961b. Auloblastus, a new blastoid from the Mississippian Burlington limestone: same, Vol. 35 (6), pp. 1113–1116.

Bergounioux, F. M., 1953. Classe des Blastoïdes (Blastoidea Say, 1925): in Traité de Paléontologie, Jean Piveteau (ed.), Vol. 3, pp. 629–650. Masson et Cie (Paris). (With a section by Gabriel Lucas, on optical studies of the hydrospires of *Cryptoschisma* Schultzi, pp. 635–637; Figures 10–12.)

Cline, L. M., and H. Beaver, 1957. Blastoids, in Treat. on Mar. Ecol. and Paleoecol., H. S. Ladd, ed., GSA Mem. 67, pp. 955–960.

Fay, R. O., 1967a. "Introduction," "Classification," "Systematic Description" (with J. Wanner), in Treat. Invert. Paleont., Ech, pp. S298, S388, S392, S396.

———, 1967b. Evolution of Blastoidea, in Essays in Paleontology and Stratigraphy, Raymond C. Moore Commemorative Volume. Univ. Kansas Press, pp. 242–286.

———, 1961. Blastoid studies: Univ. Kansas, Paleon. Contrib. Echinodermata, Art. 3, pp. 1–147, 221 text figures, 54 plates.

——— and I. G. Reimann, 1962. Some brachiolar and ambulacral structures of blastoids: Okla. Geol. Surv., Okla. Geol. Notes, Vol. 23 (11), pp. 267–270, plate 1.

Joysey, K. A., and A. Breimer, 1963. The anatomical structure and systematic position of *Pentablastus* (Blastoidea) from the Carboniferous of Spain: Palaeontology, Vol. 6 (3), pp. 471–490, plates 66–69.

Levin, H. L., and R. O. Fay, 1964. The relationship between *Diploblastus kirkwoodensis* and *Platyceras* (*Platyceras*): Okla. Geol. Surv., Okla. Geol. Notes, Vol. 24 (2), pp. 22–29, text figure 1, plates 1–3.

Macurda, D. B., Jr., 1967a. "Development and hydrodynamics of blastoids," Stratigraphic and geographic distribution," in Treat. Invert. Paleont., Ech, pp. S356, S385.

———, 1967b. The Lower Carboniferous (Tournasian) blastoids of Belgium: J. Paleontology, Vol. 41 (2), pp. 455–486, plates 59–61, 6 text figures.

———, 1966. The ontogeny of the Mississippian blastoid *Orophocrinus:* J. Paleontology, Vol. 40 (1), pp. 92–124, plates 11–13, 10 text figures.

———, 1965a. The functional morphology and stratigraphic distribution of the Mississippian blastoid genus *Orophocrinus:* J. Paleontology (= JP), Vol. 39 (6), pp. 1045–1096, plates 121–126, 16 text figures.

———, 1965b. The hydrodynamics of the Mississippian blastoid genus *Globoblastus:* JP, pp. 1209–1217, 4 text figures.

Regnéll, G., 1960. "Intermediate" forms in early Paleozoic echinoderms: Interntl. Geol. Union, XXI Session (Copenhagen), Pt. XXII, I.P.U., pp. 71–80.

———, 1945. Non-crinoid Pelmatozoa from the Paleozoic of Sweden – A taxonomic study: Lunds Geol.-Mineral Inst., Medd., No. 108, 255 pp., 30 text figures, 15 plates.

Sprinkle, J., and R. C. Gutschick, 1967. *Costatoblastus*, a channel fill blastoid from the Sappington Formation of Montana: JP, Vol. 41 (2), pp. 385–402, plate 45, 6 text figures.

Wanner, J., 1932. Neue Beiträge zue Kenntnis der permischen Echinodermen von Timor. VII. Die Anomalieen der Schizoblasten (= *Deltoblastus*). Wetensch. Meded. no. 20, Dienst Mijnb. Nederland.-Oost-Indië, 46 pp., 4 plates, also Ech, p. S343, Figure 205.

———, 1924a. Die permischen Echinodermen von Timor, Tel II. Paläontologie von Timor, Lief. 14, Abhandl. 23, 81 pp., plates 199–206.

———, 1924b. Die permischen Blastoiden von Timor: Mijnwezen Nederland.-Oost-Indië, Jaarb., Verhandl. I, Jaarg. 51 for 1922, pp. 163–233, text figures 1–11, plates 1–5.

Living and Fossil Asterozoans.

Introduction. Anatomical studies made on a living form, *Platasterias latiradiata* Gray, presently in the British Museum collections, revealed a somasteroid thought to have been extinct in Ordovician time (Fell, 1962*a*).[5] A great windfall thus became available. Detailed anatomy of a "living fossil" could importantly supplement the data from Ordovician fossils (*Archegonaster*, *Chinianaster*, Figure 12.37*B*), and permit comparison with the living asteroids. It also placed in the hands of researchers, a near-duplicate of the ancestral type that gave rise to asteroids and ophiuroids.

Two different types of organization occur among the asteroids: those in which the arms are not distinguishable from the central body disk (asteroids), and those in which they are and capable of serpentine movement (hence, sometimes called "serpent stars").

Many interesting aspects distinguish the asteroids. Among other items, they are offshoots of crinoid ancestors; some forms have beautiful coloration; they have a capacity to learn; they have powerful muscular pull; can cast off arms and regenerate them; and some forms burrow. The subphylum also has a long geologic history but a sparse fossil record.

Classification. (After Spencer and Wright, 1966; Hyman, 1955; Bather, 1900). Subphylum Asterozoa (*aster* = star + *zoa* = animal). (Includes sea stars and serpent stars.)

1. Free-living echinoderms exclusively marine (but some occur in brackish coastal waters).

2. Body flattened or depressed along oral/aboral axis; stellate configuration

due to radiating *arms* (generally 5, but may be 4 to 50 in number).

3. *Central disk* continuous into arms; two gonads and paired digestive glands in each arm.

4. *Mouth* on underside of central disk (= oral surface); also open ambulacral groove with 2 to 4 rows of tube feet (= podia) that occur on lateral branches of the *radial canal* in each arm, lined with *spines*; with *pedicellariae* (pincer-like body appendages; compare echinoids) along grooves and margins of oral surface.

5. Aboral surface with *madreporite* (= inlet to water vascular system) and *anus*. (In asteroids only, since, in all ophiuroids known, there is a lack of an anus.)

6. *Systems: Skeletal*—endoskeleton of calcareous ossicles held intact by connective tissue. *Digestive*—micro- and macroscopic food intake via the mouth—ciliary mucous, detritus feeding or via everted stomach, as in the eating of bivalve meat (predatory, carnivorous feeding) (Reese, 1966). *Excretory*—phagocytic cells (amoebocytes) ingest wastes and, by way of *papulae*, release these to the outside (Hyman, 1955), or via slime secretions from tissues such as the disks of tube feet (Nichols, 1962). *Muscular*—all of the following structures and movements come under muscular control: skeletal flexing, arms, tube feet, narrowing and widening of ambulacral grooves, movement of mouth frame ossicles.

Nervous—a three-part system: circumoral nerve ring, radial nerve to each arm, and a general nerve plexus (subepidermal). *Reproductive*—sexes separate (hermaphroditism known); sexual and asexual reproduction; ovaries and testes; reproduction by copulation known, but spawning by shedding sex cells into the sea is common; brooding of young beneath disk occurs in several species.

Circulatory—haemal system enclosed chiefly in spaces of coelom; some movement of fluids; function of this system is

---

[5]Reassigned to Asteroidea, D. B. Blake, Science, 1972, 176:306–307.

uncertain. *Water vascular*—seawater enters through the sievelike madreporite, passes through numerous pores, down the stone canal to the ring canal, and thence to the radial canals and tube feet. *Respiratory*—gases exchange (O_2 in, CO_2 out) occurs across specialized respiratory surfaces (tissues) in asterozoans; podia, papulae, and genital bursae (pouches bearing internal gills in ophiuroids only) (Farmanfarmaian, 1966).

7. Geologic range: Lower Ordovician to Recent. (See Fig. 12.37E).

The subphylum includes three subclasses: I. Somasteroidea [*Chinianaster* (Figure 12.37B), *Villebrunaster* (Figure 12.37A)]; II. Asteroidea [*Astropecten* (Figure 12.37C)]; III. Ophiuroidea (*Archaeophiomusium*, Figure 12.37D).

Population Density and Brooding. Densities of modern asteroid populations are incredible (see Table 12.1, this chapter). In waters near Plymouth, England, uncountable numbers of individuals (*Ophiothrix*), one on top of another form a living mat, web, or sieve of external bodies (Vevers, 1952). This habit is thought to be functional: bottom currents bearing nutrients can thus be better sieved.

One reason for these densities, besides optimum conditions and the adaptiveness of asterozoan organization, is the enormous number of eggs per individual, actual or potential: from 2 million eggs at a single spawning (*Asterias rubens*), to 200 million eggs per female per season (*Luidia ciliaris*) (Thorson, 1946). All asteroids are not so prolific, only nonbrooders. Brooding types yield clutches of hundreds of eggs per event (Hyman, 1955).

Skeleton and Calcium Metabolism. Asterozoans have a flexible, calcareous endoskeleton. Changes in body shape are achieved by flexure of skeletal musculature (Nichols, 1962). Skeletal units or building blocks are varishaped: rods and plates that may be polygonal, rounded, or square. Dermal tissue embeds the skeleton but superficial calcareous projections (armature) (resting on, but not fused to, underlying ossicles) include tubercles, spines, and other structures.

Two patterns characterize ossicle arrangement: fenestrate and close-knit pavement; connective tissue fills the fenestra, and there are lesser interplatal spaces in the pavement types (Hyman, 1955).

Skeletogenesis begins with either rods or spicules. Later these units form perforate plates. Triradiate ossicles occur in the aboral skeleton of the fossil *Chinianaster* (Figure 12.37B). This is reminiscent of the triradiate spicules in echinoids as well as in the skeletogenesis of modern asteroids. During growth, new plates are interpolated between juveniles and the latter plates are moved apart.

Ossicles are grouped by their position on the test: axial, adaxial, and extraxial elements are recognized: (*a*) *Axial* elements comprise the ossicles bounding the midline on each arm and the mouth frame ossicle (Figure 12.37A); (*b*) *adaxial* elements are next to, and in series with, the ossicles of (*a*) ("virgals," Figure 12.37B); (*c*) *extraxial* elements include ossicles of the aboral surface; madreporite plate (Spencer and Wright, 1966). (In several asteroids, there are marginal plates and terminal arm plates.)

There is some evidence of fusion of plates: triangular ossicles of the mouthangle plate (Figure 12.37A); opposite ambulacral plates to form ophiuroid vertebrae (Figure 12.37D); and calcareous infilling of a template-like spicular arrangement in the scales of *Encrinaster's* aboral surface—which, even though each is a single plate, retains a trace of the original template.

Body Structures. Two body structures of unusual interest are: papulae and pedicellariae. *Papulae* are simply invaginations of the body wall that serve a respira-

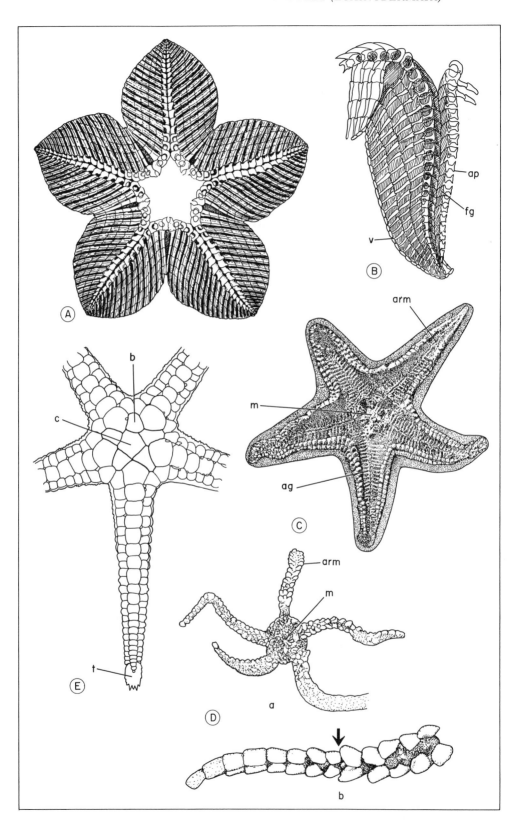

FIGURE 12.37 Information contained in living and fossil Asterozoans, and derivative data.

(A) *Villebrunaster thorali* (U. Cam.–L. Ord.), France, X 2 +; rays and interrays; ambulacrals; axial skeleton: calcareous ossicles of ambulacrals; adaxial skeleton: virgals and mouth angle plates. After H. B. Fell, 1962b.

(B) *Chinianaster levyi*, L. Ord., France, *v* — virgals; *fg* — food groove; *ap* — ambulacral plate, X 3+. After H. B. Fell, 1962a.

(C) *Astropecten matilijaensis*, Cret., Calif., aboral view; *m* — mouth; *ag* — ambulacral groove; numerous white dots — paxillae (= calcareous ossicles with shaft tufted by spines; an extraaxial structure), X 1.8. After Durham and Roberts, 1948.

(D)(a) *Archaeophiomusium burrisi*. L. Perm., Okla., oral side, X 4; *m* — mouth; arm showing individual segments (= vertebrae). *Information derived from fossil ophiuroids:* (b) arm, dorsal view; beyond arrow to left is a regenerated segment; observe change in vertebrae size, X 10. After Hatten.

(E) *Cnemidaster wyvillii*, Rec., plate system; *b* — basal; *c* — central; *t* — terminal, X 3. After W. P. Sladen, 1889.

tory function. They are spread over all or part of the asteroid body surface. *Pedicellariae* consist of two or more spines or fenestrate calcareous valves of great variety. These can be moved by appropriate muscles. They serve two main functions: (*a*) to keep the surface area clear, and (*b*) to protect the papulae (Hyman, 1955). Other protection for the papulae is provided by a network of paxillar spines when open. *Paxillae* (specialized ossicles bearing spinulets) also indirectly may serve a brood-chamber function (Spencer and Wright, 1966).

Information Derived from Living and Fossil Asteroids

Autotomy and Regeneration. Ophiuroids are capable of casting off their arms (autotomy) if injured, or if seized by a predator (Fell, 1966). A fossil example of such a regenerated arm is seen in Figure 12.37*D*, *b*.

Adaptive Modification. A variety of anatomical modifications have been found in asteroids (Hyman, 1955; Nichols, 1962; Spencer and Wright, 1966).

Ambulacral plates fused to form ophiuroid vertebrae. Arm ossicles articulated in a ball-socket joint: Such arrangements allowed only restricted movements, in the horizontal plane, and are thought to be the more primitive type; others, were unrestricted and permitted movement in all directions. Articulating vertebrae conferred greater capacity to twist and turn and so to attach to crinoids or seize prey.

Associated with burrowing habit, ophiuroids developed several structures (pockets or bursa to protect internal gills and a powerful movable mouth frame to excavate burrows). The latter capacity, in time, became adapted as tearing-crushing jaws.

Disk shape and arm arrangement were modified to facilitate different types of feeding. One such arrangement is found in the basket-arm ophiuroids that form a network on the substrate.

The metapinnules telescoped and its component ossicles (virgals, Figure 12.37*A*) became a new kind of ossicle; terminal ones, outer and inner respectively, became marginal plates or adambulacral plates (adjacent to ambulacral plates). Accompanying this development, as Nichols (1962, Figure 5) stressed, was a transference from exterior to interior of the ampullae (= bulbous part of the tube foot). This is seen in the Devonian form *Xenaster*, as well as in modern asteroids. One can detect the changeover in fossils by the absence or presence of pores to the interior of the ambulacral plates.

Other modifications included surface spines, pedicellariae, and marginal plates. Spines perform a sanitary role creating currents (by ciliary movement) that clean the respiratory areas of the body.

Biotic Associates, Predation, Parasitism. Asterozoans are natural predators on numerous mollusks and other echinoderms, as well as their own kind (cannibalism).

Asteroids may be expected to occur

(and do) in echinoderm-molluscan bottom communities. An excellent fossil example is in the Middle Devonian of New York State. Over 400 specimens of *Devonaster eucharis* were found fossilized in association with the pelecypod *Grammysia*. Reasonably, this has been interpreted as a starfish feeding ground in life (Schuchert, 1915).

A sea star can attach its tube feet on each valve of a bivalve and exert a pull of up to 5000 grams for five minutes or more, pulling the ventral edges apart, a distance of 0.1 mm. Once its pull has overcome the muscular resistance of the bivalve adductor, which keeps the valve shut, it can evert its stomach and feast on bivalve meat directly.

So ingrained are the predator-prey relationships of asteroids and certain mollusks and echinoderms, that even chemical extracts of asterozoan tissue invoke, in natural prey, an avoidance or negative response (Feder and Christensen, 1966, Table 5-3). The snail *Natica* draws a fold of tissue over its shell to prevent the tube feet of the sea star from attaching to its surface.

The biotic associates found with Permian ophiuroids of Oklahoma (Speiser shale) (Hattin, 1967), consisted of bryozoans, brachiopods, mollusks, trilobites, echinoderm fragments—including crinoid columnals. [The flexibility of the ophiuroid arm permitted coiling about and presumably attachment to crinoids in life.] Plant remains and substrate spoor are additional indicators of biotic associates of asteroids in the Cretaceous of California (Durham and Roberts, 1948).

The only likely Paleozoic predators on starfish were cephalopods (Schuchert).

Other biotic elements existing with, within, or upon asterozoans are several parasites: crustaceans (copepods, barnacles, amphipods), snails (*Thyca*), a protozoan ciliate. There is a commensal relationship between asterozoans and several polychaete worms (Polynoidae) that live in the ambulacral grooves (Hyman, 1955).

Evolutionary Affinities and Relationships. The Cambro-Ordovician somasteroids (Figure 12.37*A*) appear to have been ancestral to both asteroids and ophiuroids. The skeletal arm of the somasteroids resembled the biserial arms of crinoids. That resemblance suggests an original derivation from crinoidal stock. More precisely, what is the actual resemblance? Ossicles—*virgalia* in the more unspecialized somasteroids, had a pinnule-like arrangement found also in the crinoid biserial arm (see next section on crinoids) (Fell, 1962*b*). Other resemblances to crinoids include crinoid-like calyx plates in young asterozoans and gradual specialization of virgals (Fell, 1966).

Bryan (1967) found that discrete skeletal elements of asteroids in both fossil and Recent forms were useful in broad evolutionary studies. He observed that living *Platasterias* assigned to the class Somasteroidea by Fell has a plate morphology "very similar" to certain species of *Luidia* (class Asteroidea), and are in fact, asteroids Blake (1972).

Mode of Preservation. The way in which the arms are broken off in fossil ophiuroids probably resulted from their being in burrows at the time of demise. The buried disk and arm portions are preserved generally but not the free ends of the arm.

Durham and Roberts (1948) found a population of $n = 35$ of the genus *Astropecten* in the Cretaceous of Ventura County, California. These specimens were all molds. The Cottonwood limestone of Oklahoma and the Speiser shale have both yielded ophiuroid fossils. Schuchert (1915) provided some data on Paleozoic asterozoans: the occurrence of *Aganaster* molds commonly found in the crinoid beds of the Keokuk formation of Indiana.

Both Mesozoic and Tertiary beds contain dense accumulations of ophiuroid fossils: Middle Lias marls of France; Upper Miocene limestone, Santa Margarita formation of California. Such beds may be referred to as "ophiurite marls" and "ophiurite limestones." In another locality, the California formation yielded ophiuroid fossils preserved as limonite casts in a fine sandstone matrix (Merriam, 1931).

REFERENCES

Asterozoans

Bather, F. A., 1900. The Stelleroidea, in E. R. Lankester, ed. A Treatise on Zoology, Pt. III. The Echinodermata, London, Adam and Charles Black, Chapter XIV.

Bryan, B. D., 1967. Skeletal elements in asteroids. Program, 1967, Annual Mtg. GSA (New Orleans) Abstracts, pp. 15–16.

Durham, J. W. and W. A. Roberts, 1948. Cretaceous asteroids from California: J. Paleontology, Vol. 22 (4), pp. 432–439, plates 65–66.

Farmanfamaian, A., 1966. The respiratory physiology of Echinoderms: in R. A. Boolootian, ed., Physiology of Echinodermata (= PE). Interscience, New York, pp. 245 ff.

Feder, H. M., and A. M. Christensen, 1966. Aspects of asteroid biology: PE, pp. 87 ff.

Fell, H. B., 1966. The ecology of ophiuroids: PE, pp. 129 ff.

_____, 1962a. A surviving somasteroid from the eastern Pacific Ocean: Science, Vol. 136 (3516), pp. 633–636, 3 figures.

_____, 1962b. A living somasteroid *Platasterias latiradiata* Gray: Echinodermata. Art 6. Univ. Kansas. Paleont. Contrib., pp. 1–16, plates 1–4, text figures 1–8.

Hattin, D. E., 1967. Permian ophiuroids from northern Oklahoma: J. Paleontology, Vol. 41 (2), pp. 489–492, 3 text figures.

Hyman, L. H., 1955. The Invertebrates: Echinodermata. The coelomate Bilateria. McGraw-Hill, pp. 245–412.

Kesling, R. V., 1962. Notes on *Protopalaeaster* Narraway, Hudson: J. Paleontology, Vol. 36 (5), pp. 933–942, plates 133–134, 2 text figures.

Merriam, C. W., 1931. Notes on a brittlestar limestone from the Miocene of California: Am. J. Sci., Vol. 21, pp. 304–310.

Nichols, D., 1962. Echinoderms. Hutchinson Univ. Libr., London, pp. 35–64.

Reese, E. S., 1966. The complex behaviour of Echinoderms: PE, pp. 157 ff.

Schuchert, C., 1915. Revision of Paleozoic Stelleroidea with special reference to the North American Asteroidea: U.S.N. Mus. Bull. 88, 279 pp., 38 plates.

Sladen, W. P., and W. K. Spencer, 1891–1908. British Fossil Echinodermata from the Cretaceous formations: Vol. 2, Asteroidea and Ophiuroidea. Paleontographical Soc., London, Monog., Parts 1–5.

Spencer, W. K., and C. W. Wright, 1966. Asterozoans, in R. C. Moore, ed., Treat. Invert. Paleont., Pt. U, Echinodermata 3, Volume 1, pp. U4–U107.

LIVING CRINOIDS

Introduction

Although much of crinoid history was enacted during the Paleozoic, modern seas attest to the continued presence and the vigor of living forms: there are about 800-plus living species, with marked speciation among the free-swimming comatulids. Crinoids that are stalked resemble flowers and are popularly known as "sea lilies," whereas nonstalked forms are

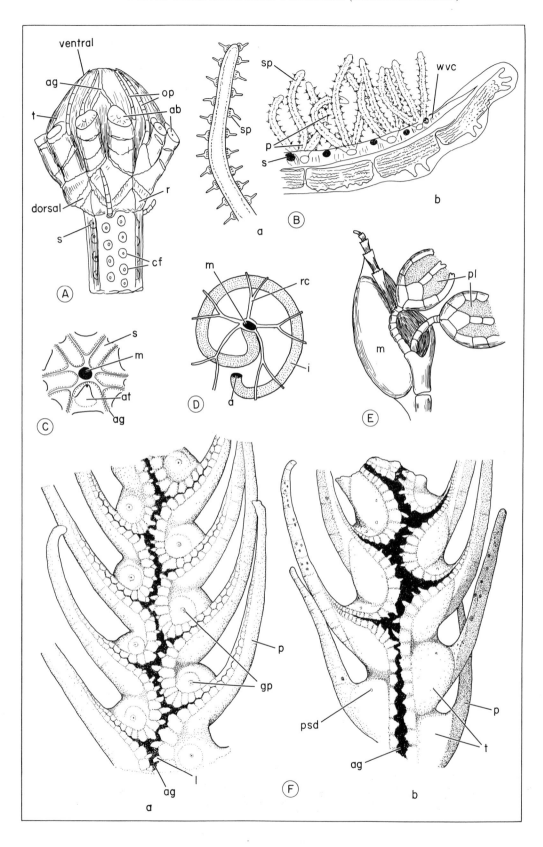

FIGURE 12.38 Information contained in living crinoids.

(A) Calyx, Tegmen, and so on: *Zenometra* sp., a comulatid (Rec.), schematic, with tegmen (*t*) and tegmenal structures (compare Figure 12.38*B*); *ag* – ambulacral groove; *op* – oral pinnules; *r* – radial plates of calyx; *ab* – arm bases cut to reveal tegmen; *s* – stem; *cf* – facets where cirri attach. Clark, 1915.

(B)(a) tube feet (podia) of *Antedon*, enlarged; *sp* – sensory papillae (Cf *Bb*). (b) Podia and saccules of pinnule end, schematic; *p* – groups of three podia; *s* – saccule; *wvc* – canal, water vascular system. Chadwick, 1907 and others.

(C) Tegmen (arms removed from disk). *Thysanometra tenelloides* (a comulatid), schematic; oral aspect of *ag*; see also "A"; *m* – mouth; *at* – anal tube; *s* – saccules; note pentamerous symmetry (Clark, 1921).

(D) Digestive (endocyclic) system, and water vascular system; *m* – mouth; *i* – intestine; *a* – anus; rc – radial canals (water vascular system). After several authors.

(E)–(F) Reproductive system. (E) Marsupium (*m*) of *Isometra vivipara* bearing pentacrinoid (*pl*) larvae; note larval stem, calyx plates, and arms. (F) Genitalia: *Notocrinus virilis*, Rec. (a) Female, *gp* – genital papillae; *p* – pinnules; *ag* – as in (*A*); *l* – lappets. (b) Male, *t* – testes; *psd* – sperm discharge pore. Clark, 1921.

called "feather stars." Crinoids are found in all oceans; at all depths – littoral to abyssal.

The ratio of extinct to living species is > 6:1 (5000:800+). Taking all known echinoderm species into account (ancient seas represented in the fossil record and species from modern seas), crinoids come to about one-third of the total.

Lacking economic value as a food for man (or in the fish food chain in the sea), crinoids found in fish nets are sold as colorful curios in the Orient and elsewhere; they are exquisitely pigmented – red, green, orange, brown, and other hues; in ultraviolet light, some forms show superb fluorescent effects. They are also remarkably facile and graceful in movement of their multiarms through the water, as seen in Catala's film of the New Caledonian reef fauna, "Carnival Under the Sea."

Some individuals with many branched arms have an impressive miniaturization in food-gathering apparatus: as much as one-quarter mile of food grooves; maximum arm spread is less than 20 ft (10 ft for an individual arm). Some stalked fossil crinoids had stems up to 27 yards long (compared to less than one yard in living types). But many small to minute crinoids occur as well, and are retrieved as larval stages attached to the pinnules of adult crinoids (Clark, 1921; Catala, 1964; Hyman, 1955; Ubaghs, 1953).

Classification

Class Crinoidea (sea lilies and feather stars). [Bather, 1900; Fell and Pawson, 1966; Hyman, 1955; Ubaghs, 1967].

1. Pentamerous crinozoans. (Primitively and normally, there are five brachial processes) (Figure 12.38*C*).

2. Commonly, with long stalk; secondarily may be stalkless and free-living, sessile, Figure 12.38*A*.

3. Cuplike calyx (theca) composed of a regular pattern of calcareous plates arranged in cycles; contains lower part of crinoid body.

4. Over the calyx there is a flexible ceiling or dome (tegmen) bearing the upper part of the body. This also extends to (5) below (Figure 12.38, part *A*).

5. The upper surface (adoral) contains mouth and anus (Figure 12.38*C*).

6. Whereas the calyx retains a meridional pattern of growth, the brachiation (multiple arms) that are well-developed, plated, and moveable, display radially divergent growth gradients (see Figure 12.4).

7. The arms are skeletal processes, evaginations of the body wall, continuous with thecal radial plates and bear extensions of all major body structures and systems: food grooves, coelom, and also reproductive, nervous, hemal, and water vascular system.

8. Food grooves from arms pass across

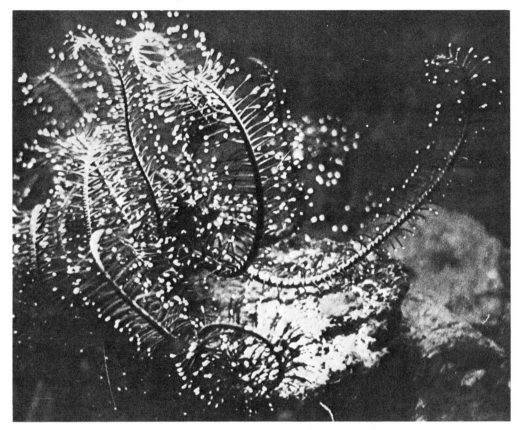

FIGURE 12.38A A living comatulid crinoid from New Caledonian waters where crinoids abound. Under ultraviolet light these crinoids show fluorescent effects. Multiple tips of pinnules are sensory papillae. Note the hooked cirri by which crinoid attaches to the solid substrate. Movement of arms brings food to the mouth (concealed). (Courtesy of Dr. René Catala.).

tegmen (Figure 12.38, part A) to mouth; tube feet (podia) margin the transit route from pinnules to arm groove, to mouth.

9. Sexes separate.

10. Systems: *Skeletal* — calcareous skeleton of typical porous (fenestrate) echinoderm plates includes plate cycles of calyx, stem disks, terminal and arm plates. *Reproductive* — gonads either on arms or in genital pinnules; swollen during maturity (Figure 12.38*F*); eggs develop in marsupium (brood chambers) in several species, in which case, eggs are fertilized before extrusion (Figure 12.38*E*). (Breeding at specific time — May to July, for example; sperm cells discharge first; extrusion of eggs leads to immediate fertilization in the sea.) *Digestive* — crinoids are ciliary-mucous feeders. Food gathered by pin-

nules and arms reaches the mouth by ciliary action along ambulacral grooves. From the mouth it passes to the stomach through a brief esophagus. If the mouth is approximately centered on the oral surface, the condition is "endocyclic"; if not, it is "exocyclic" (Figure 12.38*D*). *Excretory* — wastes apparently are voided in two ways: via coelomocytes through the *saccules* (Figure 12.38*C, s*), and through the anus. *Respiratory* — rhythmical contractions of the anal tube may participate in respiration, but the specialized respiratory surfaces in crinoids are those of the tentacles and podia (Farmanfamaian, 1966).

Water Vascular System. This follows the typical echinoderm plan but lacking an external sieve plate (madreporite), water must enter circuitously through ciliated funnel of the tegmen to outer cavity (coelom) to ciliated water pores and thence to the ring canal (Nichols, 1962). Components of the system include arms and tegmenal radial canals that enter the ring canal and, underlying the ambulacral groove, the system branches into podia; the diameter of the canals are under muscular control.

Other Systems: Nervous—Tripartite: oral, deeper oral, and aboral—the latter is critical, being the chief portion of the crinoid's nervous system; branched nerves regenerate when arms are lost or shed; free nerve endings may serve a sensory function (Hyman, 1955). *Muscular*—body wall lacks regular muscularity; skeletal plates including columnals are held together by an elastic ligament (fibrils) capable of some contractility. Arms and pinnules, however, do have regular muscle fibers, and, in swimming, the arms are capable of 100 synchronized strokes per minute. *Circulatory*—a rather poorly defined network of intercommunicating spaces with a proteinaceous colorless fluid (blood).

Growth Stages, Skeletogenesis, Calcium Metabolism

Growth Stages. (1) Spawning, as noted above. (2) Embryo develops in fertilized egg encased in a membrane (Figure 12.39*A*), (3) Embryo develops ciliated bands (60 to 84 hours after fertilization) (Figure 12.39*B*)—then body develops in the embryo before rupture of membrane, (4) Aided by enzymes the larvae (*Antedon bifida*) escapes the ruptured egg membrane and becomes a free-swimming doliolaria larva (84 to 120 hours after fertilization). Length is 0.40 mm. At 100 hours, embryos have calyx plates, oral and columnar plates present. (5) After a variable freeswimming larval stage (a few hours usually, but may extend from 2 to 3 days), the doliolaria larva (Lane, 1967) attaches to a substrate by means of an attachment disk (Figure 12.39*D*). (The ciliated bands, and so forth, degenerate.)

The process continues as follows: (6) The larva cannot feed itself because the oral end is closed, but a few days after attachment it breaks through (five grooves develop between the deltoid plates—at this time the larva is 1.0 mm long). (7) When the larva is about 3.3 mm long, having been attached for about six weeks, it puts out arms and becomes a pentacrinoidal larva (see *Pentacrinites*) (Figure 12.39*E*, *Fc*). (During the pentacrinus stages of living feather stars, all parts of the skeleton do not develop at the same rate: Growth rate of the stalk greatly exceeds that of the arms and crown.)

Crinoids may live a mean age of $20 \pm$ years. An annual growth increment is estimated to be about 10 mm. At the end of 2.5 years, a species of *Promechiocrinus* reached one-fourth of its maximum arm span. Living comulatids attain sexual maturity in one year (Fell, 1966a). Macurda (1968) found by using statistical analysis (correlation coefficients) that the entire ontogenetic development of a Paleozoic camerate crinoidal calyx was highly correlated: Thus a cup plate expanded to hold a larger viscera, the diameter of the stem increased to support larger theca, and so on.

Skeletogenesis and Calcium Metabolism
Skeletogenesis

1. The essential plate system is present in the embryo as observed above. Within mesenchymal cells, a spicule or needle of calcium carbonate is secreted.

2. Branches are added to these spicules at angles of 120° by joining other mesenchymal cells, each of which secretes a needlelike spicule of calcite.

3. A sievelike or fenestrate meshwork of calcite is formed constituting a given plate (radial, basal, oral, and so on).

[This fenestration corresponds to the organic template on which skeletogenesis proceeds. There are two types of cells in the organic base: a building unit and a secreting unit. The building unit consists of elongate spherical cells that lie at nodes and send out branching fibrillae.

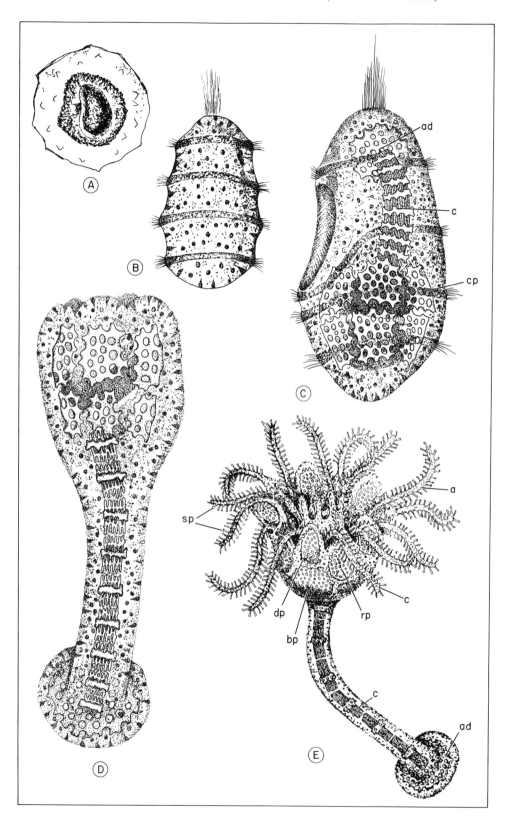

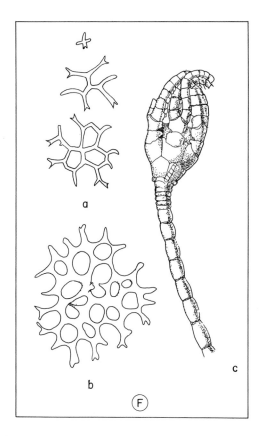

FIGURE 12.39 Growth stages and skeletogenesis of modern crinoids. Schematic. After Clark, 1921.

(*A*) *Antedon bifidia* (Recent), young embryo seen through vitelline membrane.

(*B*) Same as (*A*) shortly after rupture of membrane, observe ciliated tuft and bands.

(*C*) Same as (*A*) developing fenestrate skeletal plate; *ad* — attachment disk; *c* — columnals (stem); *cp* — calyx plate.

(*D*) Same as (*A*), young larva recently attached; note development of skeletal plates.

(*E*) Same as (*A*), pentacrinoid larva with expanded calyx and more advanced plate development; *bp, rp, dp* — back, radial, and deltoid plates, respectively; *c* — calyx; *a* — arm; *sp* — sensory papilla of podia.

(*F*)(*a*) Skeletal plate growth; from spicule to fenestrate network (Seeliger, 1892, and see Hyman, 1955). (*b*) Basal plate, 10-day-old larva (*Antedon mediterranea*). Bury, cit. Clark, 1921. (*c*) Pentacrinoid larva more advanced skeletal development, the comulatid, *Heliometra glacialis* (P. H. Carpenter, cit. Clark, 1921).

The intervals with equivalent branching fibrillae of neighboring cells thus create the organic mesh or net. In the interspaces of the porous or fenestrate mat, presumably the ovoid/spherical secreting mesenchymal cells deposit biocrystals of calcite (Clark, 1921, p. 313, and compare Chapter 15, this text). See Figure 12.39*F*.

4. Fusion of all branches yield the juvenile plated endoskeleton (Figure 12.39*E*).

5. Increase in size and fusion of more branches lead to thickening of plates.

6. Larval columnals, the individual components formed, as noted above (Figure 12.39*C*) eventually become hollow cylinders. New columnals are added at the posterior end.

Calcium Metabolism. Certain plates are resorbed as the large pavement plate of the disk of young *Thaumatocrinus*. (Generally, larval deltoid plates atrophy, that is, are resorbed.) Other plates are fused into a single row of calcite pieces (forming five separate rows) and give rise to the crinoid stalk. The basal plates of *Antedon* fuse to form a rosette. (Resorption, fusion, thickening, and secretion of intercalated plates are the subsequent modes of skeletal growth.)

Organs and tissues also are variably calcified exclusive of the general endoskeleton: There is a calcified wall of the digestive tube, of the mesenteries, and of the ventral body covering. Clark (1915) estimated that in a ten-armed large comulatid in which both side and covering plates developed, there were as many as 600,000 distinct skeletal elements, each of which had its own center of ossification, that is, mesenchymal cell. Recent pentacrinids have up to 2.5 million ossicles. Comulatids discard the columnal to lead a free-living adult existence, indicating that the calcium carbonate locked up in columnar biocrystals is neither available nor needed for reuse.

Mineralogy and Biogeochemistry

The high magnesium content of crinoids was discussed earlier (Chapter 1). Biocrystals secreted to form the crinoid endoskeleton are mineral calcite. Other chemical components of the crinoid skeleton are tabulated in Chapter 15.

Regeneration. Crinoids can shed or cast off arms naturally (autotomy), and regenerate new arms. Arms will break at points (*syzygies*) along their length where flexibility is lacking (that is, joints between skeletal units). Pinnules, cirri, and even the visceral mass will regenerate. Smaller size and lighter color often distinguish a regenerated crinoid arm.

Causes for casting off arms are not only predation (being grasped) but unstable environmental conditions such as decreased oxygen or enhanced temperature.

Crinoid spines as well as arms may also regenerate. Hattin (1958) described an example of a regenerated brachial plate spine in a Pennsylvanian crinoid.

Adaptive Modification. Among other adaptive changes in skeletal morphology of modern crinoids are those affecting cirri, skeletal plates, free-swimming habit in adult stages. Clark (1915, Figures 306–309) illustrated some variations in cirri of some living crinoids which are tabulated below.

Genus and Species	Modification of Cirri	Inferred Adaptive Value
Comactinia echinoptera	Short, strong	Grasping-arborescent organisms
Pentametacrinus tuberculatus	Short, numerous	Grasping marine organisms
Pentametacrinus varians	Numerous, long	Attachment to soft ooze substrates
Asterometra macrops	Few, long, stout, spinose dorsally	Attachment to rough, hard substrates

The multibrachiate condition in modern comulatid crinoids is related to the whole process of food-gathering. Clark (1921, p. 84) thought it was due to some ecological factor in the tropical littoral marine environment. Thus an enlarged digestive system (and visceral mass) might follow if food intake had an excess of phytoplankton and silt. Multiplication of arms might be adaptive to such a condition. Species with 40 or more arms occur only in the littoral facies; those with 15 to 30 arms inhabit intermediate depths where moderate temperatures prevail; and deep-sea, cold-water species rarely have more than ten arms.

In living crinoids, the infrabasal skeletal plates are reduced or lost (fused to the aboral calyx surface of juvenile comulatids); basals in comulatids are reduced.

Another associated tendency is the changeover from a pelmatozoan (attached) to a eleutherozoan (free-living) existence. Modern comulatids attach at early stages; during ontogeny they break away from their stems (at syzygies) and thereafter lead a free-moving life.

Ecology

Population Density. A single haul of *Antedon* is reported to have contained about 10,000 individuals. Crinoid communities often consist of closely spaced individuals on reefs or other substrates. Off northern Spain, for example, at a depth of 650 meters, an unknown comulatid species occurs in a density of 6 individuals per sq ft over an area of 1000 sq ft (Figure 12.1). Lesser densities also occur as in Antarctic waters (Ross Sea bottom) (Fell, 1966; Reese, 1966).

Reef comulatids occur in abundance in the Australian-Malayan region and the western oceans (Clark, 1921).

Aggregations of crinoids in closely packed communities are attributed to the feeble swimming potential of the larvae. The resultant is that larvae tend to settle close to parents (Hyman, 1955). Temperature may also exert control on bottom distribution in shallow waters.

Diet. Generally, the food of modern crinoids consists of varied phytoplankton and zooplankton: copepods and diatoms, dinoflagellates, radiolarians, foraminiferans, crustacean larvae, and so on.

Biotic Associates. In addition to its normal planktonic food intake, modern crinoids are associated with coral reef faunas. Mollusks and certain echinoderms (asterozoans) prey on crinoid communities. Other associates include polychaete worms, crustaceans and, in the Torres Strait fauna, other echinoderms (H. L. Clark, 1921). Bryozoan associates are also known as are a mixed flora.

Parasites, Commensals, and Predators. Clark (1921) reported on parasitism and commensalism in exhaustive detail. Polychaete worms (many different species of myzostomes) either suck up food particles streaming along the crinoid ambulacral grooves or actually enter the digestive tube for a while and feed there. Myzostome parasites cause galls, swellings, cysts, and so on, to rise on crinoids. Other animal parasites include various gastropods.

Genera of the gastropod family Melanellidae occur on all echinoderms including crinoids. Some species are nonparasitic, whereas others are parasitic. The skeletons of melanellids attached to crinoids are delicate and fixed to the calyx plates, cirri, brachials, and pinnules.

Other parasites and commensals include various crustaceans (shrimp, prawn, crab—which show protective coloration so as to be indistinguishable from the crinoid host, and also parasitic barnacles and copepods); serpulid worms (tubes are common on cirri and stems of pentacrinitids); hydroids (attached to stems and cirri); coelenterates (corals—attached to columnals of stalked crinoids); sponges (attached to stem and cirri); protists (foraminiferans that attach to column and cirri); internal parasites (dinoflagellates and ciliate protozoans); and others.

Crinoids (comulatids) live as commensals in cavities of large sponges feeding on incurrent streams containing minute organisms. Other crinoids cling to gorgonians or live symbiotically in crevices of corals.

One effect of an abundance of parasites in an overspecialized giant crinoid (large species of the family Comasteridae of the Indo-Pacific and North Australian reefs) may be ultimate extinction. This seems indicated because such giant species are helpless in warding off parasites.

Hybridization. Experiments in the fertilization of echinoid eggs (*Echinus*, and so forth) with crinoid spermatozoan (*Antedon*) (Godlewski, cit. Clark, 1921, pp. 590–592) led, in a few instances (5 percent of the cases), to the plutei stage but normally reached only the gastrula stage. The plutei, which lived only a few days, were invariably echinoids with only echinoid characters and the beginning of the typical five-rayed skeleton. The crinoid sperm merely serving as the inducer of fertilization.

Such experiments are of value as indicators of the *degree of evolutionary divergence* of two echinoderm lineages (Echinoidea, Crinoidea) that are traceable to a remote common ancestor.

Attachment. A root system proceeds from the crinoidal stem ending and ramifies in soft mud bottoms. In the stalked pentacrinus stage of free-living comulatids, carbonate cement attaches the stem to the hard bottom (Clark, 1921, plate 37). Many existing and extinct crinoids, however, were moored to the bottom by an attachment disk. Attachment to a given rock or coral fragment may endure for years (Catala, 1964).

In the free-swimming stage, as noted under "Adaptive Modification," temporary attachment to substrates is achieved by clawlike cirri.

Eelgrasses, algal, rock and mud substrates, shell and skeletal debris, sponges and corals are among the attachment sites for modern crinoids.

Bathymetry, Salinity, and Temperature

Bathymetry. Most living crinoids prefer shallower water. At a depth of 3 kilometers, 19 crinoid species were found and species decreased thereafter to depths of 9 kilometers (Zenkevitch et al., 1955, 1959, cit. Fell, 1966).

An insight into the variety of shallow-water environments can be gleaned from Clark's review of littoral species (1921, pp. 593 ff): species found between tide marks, coral reef species, species living among seaweeds, on half-submerged cans, beneath stones and pilings, mangrove roots, or on mud. H. L. Clark (1921) found that most reef crinoids at Maei Island, Torres Strait, lived in different parts of reef flats in water from a few inches to 3 ft deep.

Salinity. Associated with the preferred shallow-water environments is the salinity tolerance of modern crinoids. There are, however, major factors that might explain fluctuating salinities in crinoid-inhabited shallows: rain water runoff on reef flats, lagoons, and so on, which could lower salinity or where ice melts, as in the Ross Sea, during austral summer (December–February), and is refrozen again in the winter.

The range for tolerable salinities for crinoids is 24 ppt (parts per thousand) to 36 ppt. Normal seawater is 35 ppt (H. L. Clark, 1915).

Temperature. Arm number varies with temperature belt: warm tropical/subtropical—crinoids generally with many arms; cold deeps or surface waters as, for example, the Antarctic—crinoids with few arms (Fell, 1966).

Clark (1921, p. 743) could find no temperature dependence of crinoid coloration (pigmentation) but did, nevertheless, list a range of temperatures (and related colors of crinoids) in the tropical littoral zone: 28+°C to 43+°C.

At Maei (H. L. Clark, 1915), an optimum temperature recorded was 26°C–27°C; maturity set in at 29°C and withdrawal at 31°C. The actual restricted distribution on the reef flat may be temperature-related since parts of the flat at low tide have temperatures far above 31°C.

Predation. According to Mortensen, there are no regular animal predators on crinoids except their own kind. Fish and sea stars tend to avoid them. In Antarctic shelf waters, adults of the crinoid *Isometra* cannibalize their own young (larval stages) (Fell, 1960). However, in other opinion, carnivorous snails (Melanellidae) discussed earlier, are viewed as natural predators. These animals bore through hard skeletal plates of crinoids to find the soft tissue and eat it (Hyman, 1955).

Despite relative freedom from natural animal predators, the quantity of parasitic infestation of a given crinoid is very high.

FOSSIL CRINOIDS

Morphology and Plate Terminology and Diagrams

A necessary preliminary of this section is a review of the gross skeletal morphology used in classification, symbols for plates, and utilization of plate diagrams.

Skeletal Morphology. [After Bather, 1900; Moore and Laudon, 1941, 1943; Ubaghs, 1953; and others.]

The entire crinoid skeleton above the stem is the *crown*. It embraces the body in a plated skeleton; calyx and arms. The *calyx* has two parts: dorsal cup and ventral tegmen (oral surface).

A constant plate arrangement occurs in the dorsal cup of any given genus. There are three major types of plates: *Radial* (the symbol is *R* or its plural, *RR*), *Basal* (*B* or *BB*), *Infrabasal* (*IB* or *IBB*). Radials support the arms; basals form the base and attach to the stem in many crinoids, while the infrabasals may occur as a circlet of plates below the basals (or be incorporated into a centrodorsal plate).

Dorsal cup plate arrangements may be simple or complex. Simpler arrangements are *monocyclic* and *dicyclic*. Monocyclic arrangements have radial plate (*R*) and basals (*B*, plural *BB*) (Figure 12.40*A, a*); dicyclic arrangements include, in addition, an infrabasal series (*IB* or *IBB*) (Figure 12.40*A, b*). The number of plates in any circlet is five (but see below for variants) and subsequent circlets of five plates alternate with those of the row above (Figure 12.40*A*).

More complex plate arrangements involve the introduction of additions such as *interradials* (*IRR*) or interbrachials. In the course of evolution, plate number and size may be reduced by *resorption*, and plate number itself may be decreased by *fusion*. Such, for example, is the case where

an original circlet of five infrabasals reduce to three plates, and ultimately yield a single fused plate.

Bather and others related monocyclic/dicyclic plate arrangement to two distinct courses for the axial nerve cord. These relationships were derived from embryological studies of the living crinoid *Antedon* (Figure 12.40*B*). Both mono- and dicyclic calices occur in the two subclasses Inadunata and Camerata; dicyclic alone in the Flexibilia, whereas in the Articulata, the dicyclic calyx appears monocyclic.

Tegmen. The oral surface of the calyx (tegmen) has a pentamerous arrangement of oral plates to cover or guard the mouth. In the simplest condition, the tegmen has mouth, open ambulacral grooves (that is, lacking cover plates) in an interradial position, and the anus. Through time cover plates evolve to conceal food grooves, which are ultimately partly or completely submerged below the outer surface. The mouth too, is covered by peristomal plates. A plated valvular pyramid or similar structure often bore the anus (Figure 12.38*C*).

Arms and Pinnules. Food grooves, usually five, radiate from the mouth, across the tegmen to each of five arms. Arms most often have multiple bifurcations; these may be from 10 to 100 branches. Arms may also be of unequal length usually due to the asymmetry of the mouth.

Arms have miniature extensions called *pinnules*. These occur along the borders of each side. There are three types of pinnules with distinct morphology and functional differences. Ray pinnules bear food grooves. Some are highly specialized and lack grooves and podia [*oral* pinnules— Figure 12.38*A, op*) which serve a sensory function, and being spinelike, probably have some protective function as well]. The next set of pinnules are the genital *pinnules*. These bear the gonads (Figure 12.38*F*). Genital pinnules are followed by *distal* pinnules.

The skeletal support for arms and pin-

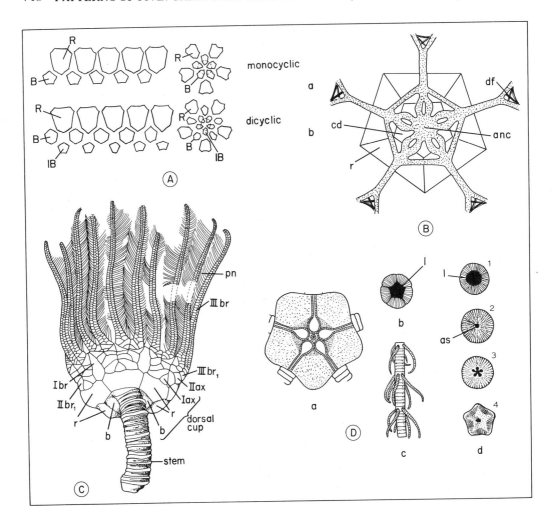

FIGURE 12.40 Morphology and body systems. Schematic.
(A)(a) Monocyclic crinoids (2 circlets *B* and *R*); *(b)* same, dicyclic (3 circlets, *IB, B, R*); *R* — radial plate; *B* — basal plate; *IB* — infrabasal.
(B) Antedon, aboral nervous system; *cd* — centrodorsal plate; *r* — radial plate; *anc* — aboral nervous center; *df* — decussating fibers.
(C) Marsupiocrinus. Sil. (Gotland and N.A.); 2 circlets of basals (*B*) and radials (*R*), axillary plate (*Ax*), supports two arm branchs, and biserial brachials (*Br* I, II, III, *Br* — 3 bifurcations or branchings of arms); *pn* — pinnules.
(D) Stem Type and Development: *(a)* pentamerous arrangement of primitive stem that, during evolution, fused into a single columnal; *(b)*–*(c) Isocrinus* sp.: *(b) nodal* (= columnal-bearing cirri); section across stem shows extension of central tubular cavity bearing coelum and nerves which continue into cirri; *(c) I. decorus*, cirri in whorls of 5 nodals, columns between these are internodals; *(d)* types of columnals; (1) single fused columnal with large lumen (1); (2)(3) note articulating surface (*as*) by which successive columnals mesh; (4) observe pentacrinoid outline in single fused columnal reminiscent of primitive pentamerous structure of columnals (compare *D(a)*.

nules are a series of calcareous ossicles. Brachials are a continuation of the radial plates. On the brachials and across the tegmen, food grooves are protected by calcareous cover plates. Each pinnule ossicle (= pinnulars) arises from a single brachial ossicle and alternates from one side to the other along the length of the arm.

Lappets and Saccules. Other structures or bodies associated with the food grooves are the *lappets* and *saccules* (Figure 12.38*F*, *a* and 12.38*C, s*).

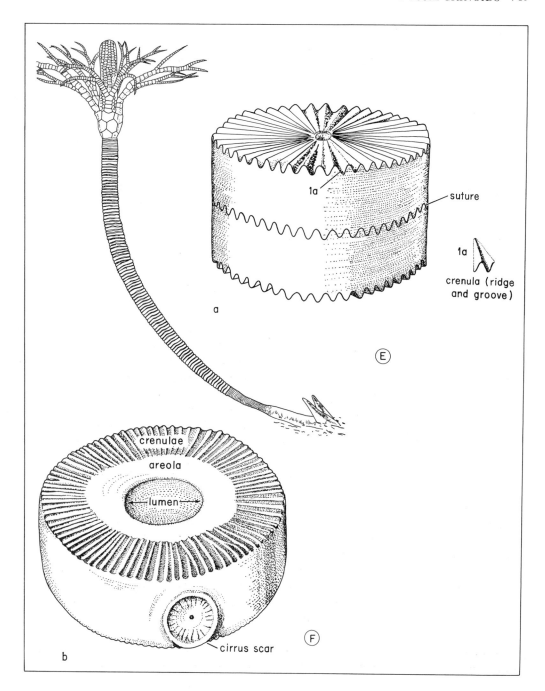

FIGURE 12.40 (cont.)

(*E*) *Ancryocrinus bulbosus.* Dev., reconstruction (based on Hall and Goldring, by Ubaghs); holdfast anchored to substrate.

(*F*)(*a*) (upper right) Articulating calcite columnals; crenulae composed of ridges and grooves; crenulate suture where crenulae intermesh; (*b*) columnals have structures—cirri scars where cirri emerge. Illustrations redrawn and adapted from (*A–C, D*) Bather, 1900; (*F*) Moore and Jeffords, 1968; Jeffords and Miller, 1968.

Stem, Cirri, and Holdfast

Stem. The stem or stalk of crinoids is composed of successive rounded or pentamerous calcite disks called *columnals*. (Circular cross sections of crinoid stems are the dominant configuration.) These intervals are where *crenulae* interlock and form a suture. A given stalk may bear differing shaped or sized disks in different portions (Figures 12.40F and 12.40D,d, 2–4). *Botryocrinus*, a Silurian form, is an example. Disks are often invariant in a given stem.

Even so, disks from which cirri ensue (Figure 12.40F, b) are called *nodals*; those in between, without cirri, *internodals*.

The central lumen bears extension of the coelom and nervous system (Figure 12.40E, b). In life, columnals possessed flexibility due to the elastic bundle of fibrils (2 to 5) that bound them together.

Two or more attached columnals are called *pleuricolumnals*. Both columnals and pleuricolumnals may be grouped in four clearcut morphological categories (discussed later). The ratio of occurrence in the rock record of disarticulated crinoid skeletal parts to whole crinoid skeletons is one hundred million to one (Moore et al., 1968). Crinoidal limestones often are dense with columnals, joined or disarticulated. Obviously, study of columnals, cirrals (segments of cirri) or pleuricolumnals, in lieu of more complete material, are of stratigraphic and paleoecologic value.

Disarticulated skeletal parts of crinoids, particularly the parts mentioned above and holdfasts, also play a significant role in taxonomy. Most post-Paleozoic stalked species are described on stemmed features alone: (131 out of 167 of the Jurassic species of *Pentacrinites* or 90 percent of all *Balanocrinus* species are examples). All in all, about 600 species covering both Mesozoic and Cenozoic as well as Paleozoic, have been described solely on either stem parts or holdfasts (Moore et al., 1968, Table 4).

The length of a stem increases by secretion of biocrystals to form new columnals just below the calyx. New columnals are also secreted between existing internodes. A genetic control appears to limit the number of columnals for each species. This control probably arose by a long selective process.

Modern comatulids have an early *attached* stage (pentacrinid larva). Before breaking away from their stem and an attached existence, cirrals are secreted and then where the centrodorsal plate meets the uppermost stem columnals, the separation (breakaway) occurs. Thereafter, the crinoid is free-living.

The genetic program involved in the above transformation from a pelmatozoan (attached) to an eleutherozoan (free-living) existence illustrates how more recent gene modification at specific loci may be carried along with the older evolutionary information. Thus the whole development of stem and holdfasts required for a fixed existence on varied substrates (the older evolution) was superseded in certain crinoids by genetic changes that permitted entrance into a new free-living life with its greater possibilities.

Natural selection that had operated on stem and holdfast problems at one time then proceeded to favor arm and cirral development that facilitated feeding, locomotion, and temporary attachment — all commensurate with a free-living existence.

Cirri. Cirri constitute organs that are outgrowths of the body wall (Clark, 1915, p. 258). They ensue from nodes arising from columnals. At the place of attachment, a scar is left when cirri are broken off (Figure 12.40F, b). The skeletal segments of a given cirri are called cirrals. Each cirrus may be composed of up to fifty cirrals but typically, from fifteen to twenty. The central tubular cavity (lumen) has extensions (equal to one-fifth the diameter of cirrals in modern forms) into each cirri (Figure 12.40D, b–c). In life nerve impulses and nutrients also were brought to cirral tips along these perforations. Elastic fibers bind the skeletal calcite cirrals and in life conferred contractility that allowed cirrals to move and to serve a prehensile function. Cirri length is inversely proportional to the amount of motion they had to accommodate.

Cirri may occur in young stages only and be suppressed in the adult; they may persist in adulthood; or, they may never occur at all in a given crinoid's life history. Generally, the more specialized and later types of modern and fossil crinoids bore cirri. Clark (1915) saw resemblances between crinoid cirri and the arthropod limb in that both were specialized anteriorly. The extreme tip of cirri often terminates in a hooklike process (terminal claw) suitable for holdfast function in different substrates. The dorsal side of a cirrus often bore spines. Besides functioning as a holdfast to some object or substrate, cirri served as a broad base to prevent a given crinoid from being submerged in soft bottom oozes and in keeping the animal upright.

Holdfast. The role of cirri as holdfasts has already been noted. Other holdfast adaptations included: *anchor* type, as in *Ancryocrinus* (Figure 12.40*E*) *discoid* type, as in *Calamocrinus*, and a *rootlike* holdfast (= *radix*), as in *Dictenocrinus* of the Silurian (Bather, 1900, p. 98). Another way of remaining moored to the substrate was by the development of a heavy *basal bulb*.

Orientation

Crinoids have no defined head or tail. They do have a mouth and an anus. One need but remember that the food grooves of the arm lead to the mouth in order to find the natural living position for any given specimen. In this position, the anus occurs at an interradius (= the posterior side of the crinoid), and the radius opposite the anal interradius, denotes the *anterior* side. So oriented, right and left in the crinoid skeleton conforms to the right and left side of the observer. The side bearing cup and stem attachment or, in attached forms, cirri only, is *dorsal*. The side in the opposite direction to that of stem attachment and cup, that is, the upper surface, is *ventral*.

Plates may be identified as left posterior, right anterior, and so on — and hence orientation is important.

Crinoid Plate Diagrams

Symbols. Before discussing the plate diagrams, a review of certain symbols and terminology is necessary. The discussion that follows is abbreviated, and symbols are grouped to facilitate recognition.

There are three large groups of crinoid plates (Bather, 1900; Moore and Laudon, 1944): those of the *dorsal cup* (radials, basals, infrabasals, and see others to be discussed below); those of the *arms* (brachials); and those of the *food grooves* (ambulacrals). For radials, brachials, and ambulacrals, other plates may be interposed and will then bear the symbol "inter," for example, interradial (*IRR*), interbrachial (*IBrBr*), interambulacrals (*IAmb-Amb*).

The branching of the arms results in names such as primibranch (*PBrBr*), secundibranch (*IIBrBr*), and so on); brachials may also be fixed (*fBrBr*). Where a ray plate supports two branches, it is an axillary plate (*Ax*, or plural, *AxAx*, see Figure 12.41*D*).

The first anal plate is designated *X* (all anals together are designated *XX*); left and right anal plates are *LX* or *RX*, respectively.

There are also plates designated by duel names such as the radianal (*RA*). The "radi" part refers to the portion of the plate relative to the right posterior radial (*RPR*). It generally occurs below it; the anal part of the name arises because the left side of the plate supports the *X* (anal plate) (Figure 12.41*E*). *RA* may be absent entirely or exist in a primitive condition of *iRA* (infraradianal) and superradianal (*sRA*). Another plate that supports the anal plate (*X*) (Order Disparida) is the aniradial (*AR*) (see Figure 12.41*C, b*).

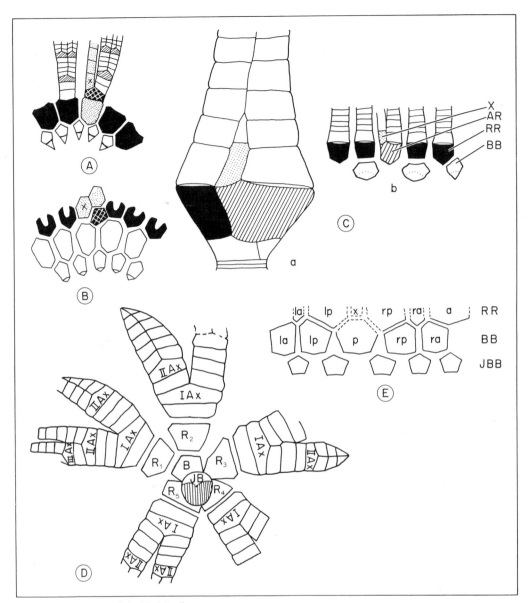

FIGURE 12.41 Crinoid plate diagrams.
(A) Poteriocrinites. (B) Iocrinus. (C) Synbathocrinus. (a) Camera lucida drawing, dorsal cup and portion of arms; *(b)* plate diagram. *(D) Plagiocrinus torynocrinoides.*
(E) Sundacrinus sp.; *RR* — radials, *BB* — basals; *IBB* or *JBB* (singular, *JB*) infrabasals; *x* — first anal plate; *Rx* — right second anal plate; *RA* — radianal plate; *Ax* — axillary plate; *Ar* — aniradial plate; portion of *RA* to *RPR* (right posterior radial and *x*).
Illustrations redrawn and adapted from: *(A)–(B)* Moore and Laudon; *(C)* Van Sant and Lane; *(D)* Wanner; *(E)* Bather.

Plates surrounding the mouth are called "orals" (*O*; plural, *OO*); those of the tegmen, *T*.

Plate Diagrams. To reduce the three-dimensional dorsal cup and brachials to appear in space as a two-dimensional diagram on paper requires expanding the plates to a flat layout (Figure 12.41). The layout must be envisioned in the round to restore the original spatial relationship and missing dimension. These diagrams facilitate comparison of plate systems between genera and species.

Generally, crinoid plate diagrams show the other plates in relation to the anal X and related plates (radianal, RA). (See discussion of "Orientation." Consult Figure 12.41A, which shows radials, basals, and infrabasals, as well as the XX plates and RA.)

Classification

The Class Crinoidea defined earlier is generally subdivided into four subclasses: Inadunata, Camerata, Flexibilia, Articulata (Bather, 1900; Moore and Laudon, 1943, 1944; Fell and Pawson, 1966; Ubaghs, 1953).

Subclass 1. Inadunata. Mono- or dicyclic; calyx plates sutured together firmly; tegmen plates; arms generally not incorporated into calyx. Inadunates are referred to as "the free-armed crinoids." Geologic range: Lower Ordovician – Upper Permian. There are three orders.

Order **Disparida.** Includes many very small forms with calyx height or diameter measuring a few millimeters. Common plate pattern of dorsal cup – RR, BB, XX (fixed Br sometimes). (Variants occur.) Middle Ordovician – Upper Permian. (*Atyphocrinus*, Figure 12.42A, part A.)

Order **Hybocrinida.** Plate pattern of dorsal cup – RR, BB, XX, RA (development of Br restricted). Middle Ordovician – Lower Silurian (*Hybocystites*, Figure 12.42A, part B.)

Order **Cladida.** Plate pattern of dorsal cup – RR, BB, IBB, XX [radial may be multiple, that is, radianal (RA) + radial (R)]. Lower Ordovician – Upper Permian. (*Cupulocrinus*, Figure 12.42A, part C.)

Subclass 2. Camerata (*camera* = box). Monocyclic or dicyclic. Calyx rigid. Dorsal cup plate pattern: lower $BrBr$, interbrachials (*IBrBr*), interradials (*IRR* – plates between any two rays), RR, BB; tegmen plated (mouth and ambulacral grooves covered). Middle Ordovician – Upper Permian. There are two orders:

Order **Diplobathrida.** Dicyclic; infrabasals (*IBB*) present. Middle Ordovician – Upper Mississippian. (*Eudimerocrinus*, Figure 12.42A, part F.)

Order **Monobathrida.** Monocyclic; infrabasals (*IBB*) not present. Middle Ordovician – Upper Permian (*Periechocrinites*, Figure 12.42B, part A.)

Subclass 3. Flexibilia. Dicyclic. Dorsal cup plate pattern: lower $BrBr$ (loosely incorporated), RR, BB, 3 *IBB*, and XX; plates not rigidly sutured together; tegmen flexible; plated uncurved arms form globular crown. Middle Ordovician – Upper Permian. There are two orders.

Order **Taxocrinida.** Crown elongate; calyx plate articulation comparatively weak. Middle Ordovician – Upper Pennsylvanian (*Taxocrinus*, Figure 12.42A, part D).

Order **Sagenacrinida.** Crown subglobular; plate articulation strong. Middle Silurian – Upper Permian. (*Forbesiocrinus*, Figure 12.42A, part E.)

Subclass 4. Articulata. Generally dicyclic. Tendency to fusion and reduction of dorsal cup plates; *IBB* number five (five-to-one); tegmen flexible. Mouth and food grooves exposed; lower brachial element flexibly articulates with radial (R). Triassic – Recent.

Orders with Stalks
1. *Order* Isocrinida. Dicyclic; stem with true cirri; radial facets. (These are articulating surfaces or radial plates formed by the fusion of radial and smaller intercalated plates] wide; arms dominantly

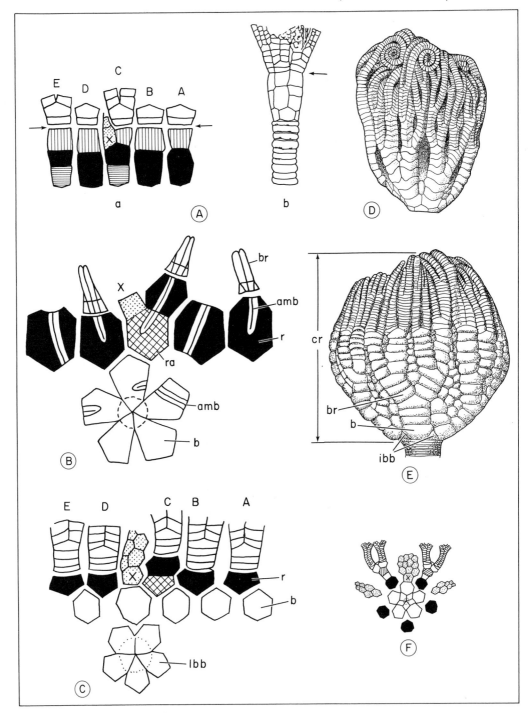

FIGURE 12.42A Representative genera. Schematic.

(A) Order Disparida: *Atyphocrinus*, U. Ord., Ohio. (a) Plate diagram: (A)–(E) letter symbols for rays (A = anterior ray); black = radials; close-spaced horizontals = infraradials; close-spaced parallels below arms = fixed brachials; above arrow–brachial; x–anals, (b) Crown and large stem; arrow points to top of crown, compare (a).

(B) Order Hybocrinida *Hybocystites*, M. Ord., Ky.-Ohio. Basal plates (stellate pattern = basals) (b), ambulacral grooves (amb); r–radials, ra–radianal; x–anal; br–brachial. Note: only 3 radials bear appendages (br).

(C) Order Cladida. *Cupulocrinus* sp. M. Ord.–U. Ord., United States, Canada; hatched plate = radi-anal directly below radial; r–radial, b–basal, ibb–infrabasal.

(D) Order Taxocrinida, *Taxocrinus ornatus*, Miss., Iowa.

(E) Order Sagenocrinida. *Forbesiocrinus* sp., Miss., *cr* — subglobular crown.

(F) Subclass Camerata. *Eudimerocrinus multibrachiatus*, Sil., Tenn., plate pattern, dorsal cup; dicyclic; black — *rr*; *x* — anals; label *br* — brachials; *ibr* — interbrachials.

Illustrations redrawn and adapted from several authors including R. C. Moore; Springer, Laudon.

uniserial). Lower Triassic to Recent. (*Isocrinus*, Figure 12.42*B*, part *B*).

2. *Order* Millercrinida. Monocyclic or cryptodicyclic; stem lacks true cirri; arms and radial facets as in Isocrinida. Middle Triassic to Eocene. (*Dunnicrinus*, Figure 12.42*B*, part *C*.)

3. *Order* Cyrtocrinida. Small articulates. Monocyclic; generally short stem without cirri or, may lack stem; radial facets small and separated by arm processes; ten arms. Jurassic-Recent. (*Eugeniacrinites*, Figure 12.42*B*, part *D*.)

Orders without Stalks

4. *Order* Roveacrinida. Small articulates. *IBB* atrophied radial facets narrow. Upper Triassic-Upper Cretaceous. (*Saccocoma*, Figure 12.42*B*, part *E*.)

5. *Order* Comatulida. *BB* present as rosette; *IBB* absent in adult because these plates fuse with other plates to form a single ossicle — the centrodorsal. The latter usually bears cirri. Jurassic to Recent. [*Zenometra*, Recent (Figure 12.38*A*); *Antedon*, Recent (Figure 12.1*A*, part *B*).]

6. *Order* Uintacrinida. Besides the presence of radials and basals, there is a centrale and, in addition, *IBB* occurs (in *Uintacrinus*, *IBB* is obsolete); globular cup, thin-plated, includes fixed *IRR*, proximal brachials (*fBrBr*) and pinnulars. Upper Cretaceous. (*Uintacrinus*, Figure 12.42*B*, part *F*.)

INFORMATION DERIVED FROM FOSSIL CRINOIDS

Reef, Interreef, and Other Crinoidal Facies

Some prefer the phrase "bioaccumulated crinoidal limestones" rather than "reefal or biohermal limestones" because crinoids are not the *framework* — building structure of such wave-resistant mounds or reefs as are corals, stromato-

poroids, or algae. Whatever term is employed, as set forth in Figure 12.43, moundlike structures are often encountered in Mississippian beds, among others.

The construction of the crinoidal mounds of Indiana proceeded in the following manner (Carozzi and Suderman, 1962):

1. Crinoid communities grew in shallow depressions in the sea floor.

2. Upon demise, the skeletons fragmented and accumulated on the substrate to form (what would later become after consolidation) bioaccumulated crinoidal limestones.

3. The carbon dioxide released by the crinoid community during normal metabolism apparently sponsored phytoplankton blooms.

4. Algae precipitated calcium carbonate (algal dust) that covered the bioaccumulated crinoidal limestones. Subsequently these consolidated into calcilutite.

5. Another generation of crinoid larvae settled on the calcilutite substrate and a new crinoid community formed.

The process from items 2 to 4 was repeated several times.

Crinoids are often components of the interreefal facies of stromatoporoid, coral, and other reefs (Langton and Chin, 1968). Particular crinoid genera with almost intact skeletons may be represented in great density on a given bedding plane. Such forms are often planktonic or floating types — for example, *Uintacrinus* of the Kansas Cretaceous.

Structure and Habit

Structure

Attachment/Protection. A variety of coiled stems developed to fit a pseudoplanktonic habit (Figure 12.44*A*) or protection of the crown under adverse conditions (Figure 12.44*B*). Dendritic-type

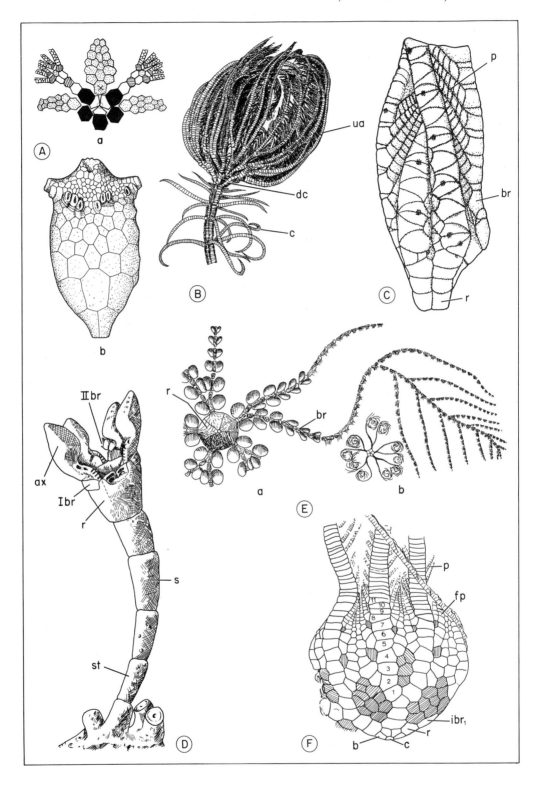

FIGURE 12.42B Representative genera (Camerata, continued).
(A) *Periechocrinites.* (a) Plate diagram; brachials; interbrachials; (b) *P. tennesseensis,* Sil., Tenn,
X 1; anal tube; arm bases; radials; basals.
Articulata: (B) *Isocrinus asteria,* schematic; only a portion of stem shown; *ua* — uniserial arm; *dc* —
dorsal cup; *c* — cirri.
(C) *Dunnicrinus mississippiensis,* U. Cret., Miss, X 3.3. Black dots indicate division of brachials; *br* —
brachials; *p* — pinnules; *r* — radial plate of dorsal cup; *bb* — missing.
(D) *Eugeniacrinites caryophyllatus,* Jur., Europe, *ibr,* II *br* — primibranch and secundibranch; *ax* — ax-
illary plate; *r* — radial plate, *st* — stem.
(E) *Saccocoma,* U. Jur., Germany. (a) *S. tenella,* aboral surface, X 7; *r* — radial plate, *br* — brachial;
(b) *S. pectinata,* arm branches coiled as in swimming movements in life, X 0.5.
(F) *Uintacrinus sociale,* U. Cret., Kansas, reduced; *c* — centrale; *r* — radial; *b* — basal, *ibr* — first primi-
branch; *p* — pinnule; *fp* — fixed pinnulars; shaded plates = intercalated plates, *irr* and *ibr.*
Illustrations redrawn and adapted from Moore and Laudon; R. C. Moore, 1967; Bather.

branched systems helped moor some forms to the shifting substrate (Figure 12.44*F*).

Sanitation. Sanitation requirements, to prevent fouling of intake with the creature's own waste, led to the development of elongate anal tubes or some with anal openings variably positioned (Figure 12.44*D*). Sometimes the tegmen became conical and ended in an elongate tube with the anal opening at the end.

Habit. Crinoids adapted for bottom crawling (Figure 12.44*E*), pseudoplanktonic (Figure 12.44*A*), and sessile existence (Figure 12.44*F*).

Feeding. Requirements for better food collection or more efficiency in intake led to such modifications as spatulate wing processes bearing ambulacra (Figure 12.44*C*) as well as others (Figure 12.44*C, b*). Enhanced efficiency in food gathering is indicated by ontogenetic study of primitive Ordovician camerates. This tendency apparently had high selective value in many crinoid stocks. Structural strengthening and greater arm-sweep of the food-collecting domain is inferred for fixed brachials and from the intricate spreading apart of the arm of a single ray (Brower, 1967).

Dissociated Columnals: Classification, Stratigraphic Utility

Disarticulated skeletal parts such as stems and columnals have been infrequently used in fossil and stratigraphic studies. Conceivably, they have been bypassed by the mere recording in faunal records of "crinoid stems." Moore and Jeffords (1968) feel this is inadequate utilization of potentially valuable data. Accordingly, they have written a monographic study covering the morphology and taxonomy of such disarticulated crinoid parts.

The essential morphology of pluricolumnals is shown in Figure 12.45. Most terms are self-evident from the illustrations or are defined in the legend of the cited figure. [Further details are given in Moore and Jeffords (1968).] *Noditaxis* (*NT*) is a nodal (*N*) combined with any related internodals (*IN*) (Figure 12.45*A*). The axial canal (Figure 12.45*B*) may be simple or complex, lack or possess an intracolumnal passage (canalicula). The lumen may be circular, pentagonal, stellate, and so on (Figure 12.40*Fb*).

Stem parts can be classified into four distinct assemblages. To do this, one must consider the following features: transverse shape, articula (surface of articulation between columnals), size and outline of lumen and the longitudinal structural division of its parts. The four assemblages or groups are:

Group I. Pentameri. Fivefold columnals and pluricolumnals. (Any type of division of stem or parts into five segments or five-ray pattern). (Included are crinoid genera assignable to Inadunata, Camerata and Articulata.

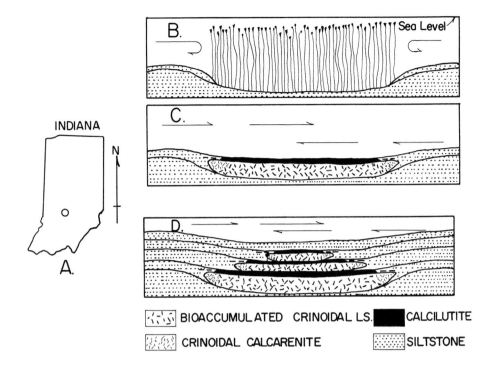

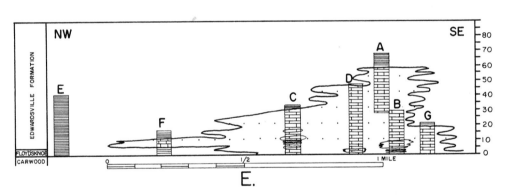

FIGURE 12.43 Information derived from fossil crinoids—I. Crinoidal limestone of Indiana (Mississippian).

(*A*)–(*E*): (*A*) Location map; (*E*) cross-section of crinoidal accumulation, Stobo, Indiana (two-dimensional view of structure completely delineated by peripheral lines; SE–NW.

(*B*)–(*D*): How structure in (*E*) accumulated: (*B*) living crinoids in shallow depression on sea floor (observe position of sea level); (*C*) calcilutite is then precipitated as carbonate "algal dust" (originally sponsored by CO_2 discharge of crinoidal metabolism) and covers bioaccumulated crinoidal lime mud. Three successive (*B*)–(*C*) pairs of events follow each other as shown in (*D*).
Figures redrawn and adapted from Carozzi and Suderman, 1962.

Group II. Elliptici. Elliptical columnals. (Even if quadrangular in outline, the column belongs in this assemblage if the articula facets are elliptical). (Included are genera belonging to the Camerata and Articulata.)

Group III. Cyclici. Cylindrical stem. (Included are genera from all crinoid classes.)

Group IV. Varii. All stems or parts not included in Groups I–III. (Included are genera belonging to the Inadunata, Camerata, and Flexibilia.)

It will be seen that this classification is nongeneric, an operational or practical arrangement to permit separation of stems and columnals into distinctive categories. In turn, these can have stratigraphic utility.

Within these groupings, one can distinguish variation in crenulae, areolae, lumen size and shape, and so on. Moore and Jeffords (1968) suggested that after one has separated fossil stems into Groups I–IV, then a further segregation into seven subgroups can be made: (1) Both genera and species can be identified. (2) Genus identity is certain; species identity uncertain. (3) Genus is certain but the species is undescribed – but can usually be described if stem parts are distinctive. (4) Both genus and species is doubtful. (5) Species undescribed and genus doubtful. (6) Genus and species are both undescribed. (7) Poor preservation preventing meaningful taxonomic designation.

The very fact that one studies disarticulated crinoidal skeletal parts, in itself can yield important paleoecological data: energy levels, sediment transport, and the formation of bioclastic limestones.

Jeffords and Miller (1968) studied the ontogenetic development of stem and other parts found in Late Pennsylvanian shales and limestones of Texas. The number of ridges (Figure 12.45B) were counted and found to increase as the columnal diameter increased. These data suggest that ontogenetic development characterized each species with some distinction, while it was more or less persistent in a particular way in a given species (or genus).

Such data could have index value in stratigraphic correlation.

Major Events in Crinoid Evolution

Origin. The origin of crinoids has not been resolved. Adding to the puzzle of crinoid origin is nonhomology of crinoid arms and pinnules to noncrinoid Crinozoa appendages. A thesis has been advanced for cystoid or eocrinoid ancestry (Moore, 1954). If one shifts thecal plates of either of the last-named echinoderms in a longitudinal direction, the resultant dorsal cup pattern would correspond to that of the camerate crinoids. Furthermore, one may interpret the fold and ridge pattern on any camerate's cup plates as a reflection or modification of the pore arms of an ancestral cystoid or eocrinoid.

Even if camerate crinoid origin is linked to an ancestral eocrinoid or cystoid stock, that still leaves the other subclasses to be accounted for. Crinoids may have had a polyphyletic origin. If so, that would imply convergent evolution in several distinct lineages.

Moore and Laudon (1943) observed some identical evolutionary tendencies in *all* Paleozoic crinoid stocks: shape and structure of dorsal cup, organization of the rays, and evolutionary end form of a pentamerous crown. Such equivalences in adunates, flexibles, and camerates seem to denote transmission of *certain stable patterns* in genetic coding on the molecular level.

Convergence in shape is more likely to arise in *distinct* lines of descent than similar ray organization, or dorsal cup structure.

The rise of Articulata in Mesozoic time implies descent from nonarticulates, thus linking at least two crinoid subclasses. Other kinds of genetic linkages between Paleozoic and Mesozoic stocks rest on a chain of observations and consequent inferences. There is evidence, for example, that certain evolutionary trends in Paleozoic stemmed inadunates may have been preadaptive for a completely different function in unstemmed Mesozoic articulates (Lane, 1968). Muscular articulation originally selected for in conjunction with improved balance (Paleozoic stemmed forms) is thought to have been preadaptive for swimming in Mesozoic unstemmed forms. Even the direction of the arms' swimming strokes in modern comulatids, aboral, out-and-down, is

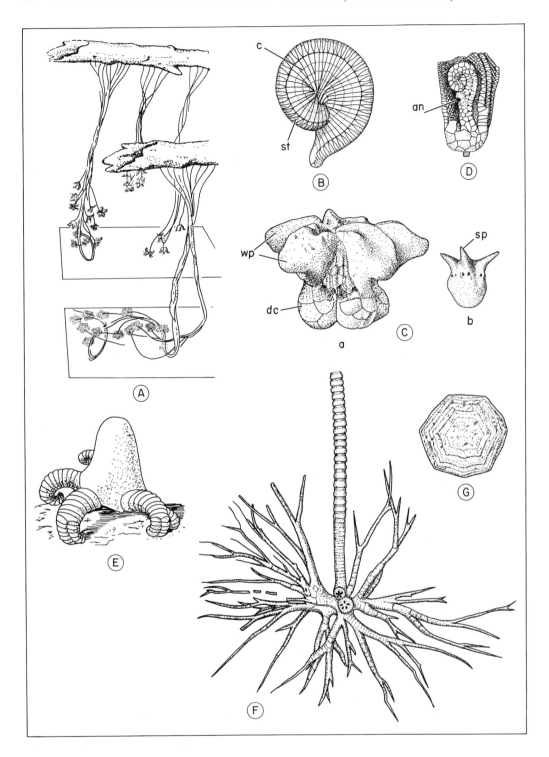

FIGURE 12.44 Information derived from fossil crinoids—II. Schematic.

(A) *Pseudoplanktonic habit. Seirocrinus subangularis.* Lias., Germany, schematic, in life stem coiled around floating logs with calyx free to move over substrate (after Seilacher et al., 1968).

(B) *Coiled stems. Myelodactylus ammonis,* Sil., Gotland; crown completely concealed by closely coiled stem (st) with cirri (c) emergent from each interlocking columnal, pseudo-radiate.

(C) *Spatulate "wing" process.* (a) *Pterotocrinus coronarius,* Miss., Ky. Observe huge, massive spatulate process (wp) that bear ambulacrals; (b) *Talarocrinus cornigerus,* Miss., Ky., same as (Ca) but development spinose (sp), a precursor condition of wp.

(D) *Elongate and coiled anal tube. Pachylocrinus arboreus,* Miss., Alabama, slightly enlarged. Anal opening to left (an). (B)–(D) after Springer, 1926.

(E) *Crawling habit. Edriocrinus* sp., L. Dev., interpretive reconstruction to emphasize crawling habit. After Cuenot, 1948, cit. Moore and Jeffords, 1968.

(F) *Bottom attachment. Eucalyptocrinites ovalis,* Sil., N.Y., root system of stem. After Wachsmuth.

(G) *Seasonal growth rings.* Genus and species unknown; isolated basal, X 1; with concentric growth lines ranging from 7 to 9 in number, and probably seasonal. After Lane and Webster, 1966.

viewed as a Paleozoic inheritance. Similar strokes are attributed to Paleozoic stalked ancestors, but for a different function—maintenance of upright position.

Study of the doliolaria larval stage (Figure 12.39D) of living crinoids suggests that the extinct inadunates and flexibles larvae were similar in respect to the plane of symmetry. In fact, the modern condition is thought to be a relic of Paleozoic precursors (Lane, 1967).

The apparent first appearance of a crinoid occurs in the Lower Ordovician of England (Ubaghs, 1953, 1967). Fully developed, it resembled a monocyclic inadunate. It was not an intermediate form. It was not a primitive link with older Crinozoa ancestors. The lack of a sequence of transitional types leading back to the ancestral stock is, of course, the chief reason for the uncertainty about origin of the class.

Paleozoic Expansion and Evolution. The inadunate order Hybocrinida shows a stratigraphically-based succession of Ordovician crinoids that undergo arm reduction: five (*Hybocrinus*) to three (*Baerocrinus*) to complete atrophy of free arms (*Cornucrinus*) (Ubaghs, 1953). This is clearly indicative of a high selective advantage conferred by a more compact (fixed) food-gathering apparatus in the seas of the time. It points to the variable genetic potential for this feature available in the respective gene pools. Geographic spread of the cited forms range from

North America to Sweden and the Baltic area (a circumpolar distribution).

(Genetic coding for the food-gathering apparatus is still highly plastic and variable as witnessed by the ramification of arms and pinnules in modern forms. The last tendency is the reverse of that in the Ordovician seas insofar as it applies to stalkless forms.)

By middle Ordovician time, the oldest known fossils of the Flexibilia and Camerata occur. Their exact evolutionary relationship to the inadunates is unclear.

In terms of number of species, the Flexibilia, all during the Paleozoic always had the lowest count of the three subclasses. Camerata species assumed the dominant role during the Silurian through Mississippian time. Paleozoic evolution of crinoids can be summarized as a tendency toward skeletal modification (Bather, Moore and Laudon, Swinnerton, Ubaghs). The essential skeletal unit is the individual calcite plate of the dorsal cup, brachials, interbrachials, and so on. Many of the plates were *fused, moved upward, reduced,* or *lost* in the course of evolution. A symmetrical pentamerous crown follows such simplification in the evolution of several subclasses.

Ray development was important as a part of the food-gathering apparatus. Arms may be divided or undivided. The atomous (undivided) condition occurred primitively and also as a result of skeletal simplification. Many Paleozoic crinoids had bifurcate arms (isotomous, that is,

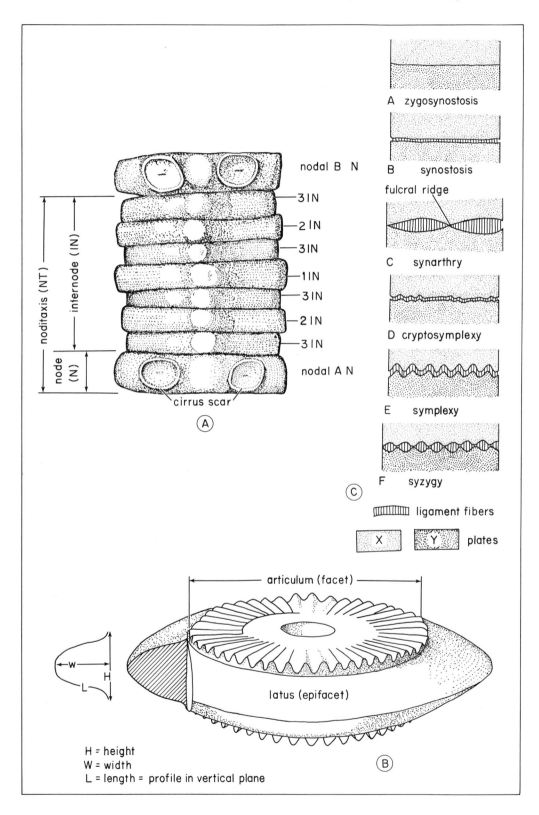

A   zygosynostosis

B   synostosis

fulcral ridge

C   synarthry

D   cryptosymplexy

E   symplexy

F   syzygy

ligament fibers

X   Y   plates

nodal B N

3 IN

2 IN

3 IN

1 IN

3 IN

2 IN

3 IN

nodal A N

noditaxis (NT)

internode (IN)

node (N)

cirrus scar

Ⓐ

Ⓒ

articulum (facet)

latus (epifacet)

W

H

L

H = height
W = width
L = length = profile in vertical plane

Ⓑ

FIGURE 12.45 Information derived from fossil crinoids—III. Details of crinoid columnals.
(A) *Isocrinus*, pluricolumnals showing nodals (N) bearing cirrus scar (where cirri attached in life) and internodals of three orders based on sequence of intercalations between nodals: 1N—first order, 2N—second order, and so on.
(B) Nonstraight-sided columnals with apron-like extension (= *latus*), articulum (facet) = the entire articulating surface, and mode of measuring epifacet; H, height from ridge to ridge; W, width from rim of facet (see arrow); L, length of profile in vertical plane (see Figure 12.40 Fa; for mode of articulation).
(C) Types of articulation (A–F) of successive columnals in crinoid stem; ligament fibers occupied spaces shown; x, y = columnals. Illustrations redrawn and adapted from Moore and Jeffords, 1968.

equal branches, or heteronomous, that is, unequal branches). The heteronomous tendency is more advanced. Pinnulation, or the formation of arm offshoots or pinnules, developed in the course of arm evolution.

The individual calcite segments (ossicles) comprising rays may be varishaped —polygonal to cuneate or wedge-shaped —and be arranged in a uniserial or biserial fashion. The polygonal/uniserial pair is considered primitive but nevertheless, this arrangement is retained in the flexibles and some camerates. The cuneate/biserial pair appears as free arms evolve. The changeover led to alignment, as with the others, of pinnules on opposite sides (Swinnerton, 1950) and hence enhanced efficiency in the food-gathering apparatus.

Increase in surface area of the arms by heteronomy, pinnulation, and biseriality did more than favor food-gathering of greater efficiency. It also affected balancing movements. An upright crinoid is hydrodynamically unstable. It lacks a center of buoyancy. If, for Paleozoic stemmed forms, the drag of gravity went uncompensated, the tendency would have been to bend the stem until the crown was horizontal to the substrate. One consequence would be less effective food-gathering, as well as, sanitation. A chief evolutionary factor in Paleozoic stemmed crinoids is thought to have been this tendency toward maintenance of an upright position (Lane, 1968).

An essential compensation to counteract gravitational drag would be an appropriate movement of arms and stem. Given this prerequisite, there are a whole series of consequences: form; more rapid, powerful arm movement; a greater number of articulations between radial plates and brachial plates; added stress thereby imposed on the dorsal cup. Thus a selective advantage would accrue if dorsal cups were lower and stronger and, therefore, better able to withstand new stresses.

Other remarkable variability was displayed in the broad shape diversity of holdfasts or attachment mechanisms—from anchors to root and bulb types.

Some early Permian inadunate crinoids (Battleship Wash fauna, Nevada) showed a reversal of the general evolutionary trends observed in Pennsylvanian inadunates: these had more uniserial than biserial arms and a dorsal cup with flat or convex, rather than concave, base. These reversals of ongoing evolutionary trends are attributed to (*a*) dwindling of advanced Pennsylvanian types by early Permian time, (*b*) greater numbers of Permian inadunates descended from more conservative stocks of older crinoids—that is, stocks that were unspecialized in arm and dorsal cup plate arrangements (Lane and Webster, 1966).

Paleozoic Extinction. Before one can meaningfully discuss the Paleozoic mass extinctions of crinoid species, certain clarifying items must be reviewed.

1. Suborder Encrinina (genus *Encrinus* of the European Triassic) shares skeletal characteristics with both inadunates and articulates. Articulate features include open ambulacral grooves and mouth. Inadunate features include: dicyclic dorsal cup that bears a stem. In arm structure they differ from both subclasses

having uniserial-becoming-biserial arms, while articulates generally have uniserial arms and inadunates, to which the Ecrinidae have been assigned by some authors, have uniserial or biserial arms.

2. The subclasses can be listed by geological period and ranked from I to III indicating relative dominance in terms of gross numbers of species.

Harrell Strimple (1974, personal communication) sees the evolutionary route from inadunate to articulate as follows: take the living stalked articulate *Pentacrinus* as typical of the majority. Next, find a Paleozoic inadunate morphologically closest to *Pentacrinus*. The candidate is the Pennsylvanian inadunate *Chlidonocrinus*. What changes were necessary

Ordovician	Silurian through Mississippian	Post-Mississippian through Permian
I Inadunata	I Camerata	I Inadunata
II Camerata	II Inadunata	II Flexibilia
III Flexibilia	III Flexibilia	III Camerata

The above listing makes clear that speciation among the camerates exceeded that of the inadunates and flexibles for a period of about 195 million years. By Mississippian time, it reached a peak— some 2+ camerate species for every one inadunate species. Yet by the beginning of Pennsylvanian time, camerate populations practically disappeared from the seas of the world. Only a few relict species populations remained and even during the advance of Permian time, the slight resurgence in camerate speciation failed to guarantee survival into the Mesozoic. So impoverished were the camerate populations in number of species by post-Mississippian time, that even the species-poor flexible crinoids surpassed them.

All Paleozoic crinoids diminished in number of species at the end of Mississippian time. That time can be regarded as a prelude to their extinction. The precise explanation is not available.

3. Since the dicyclic inadunates are thought to be progenitors of the oldest articulates, one cannot speak unqualifiedly of inadunate extinction. Some inadunates disappeared by *evolving* into articulates.

4. Final demise by end of Permian time corresponds to demise or marked decline of many other Paleozoic forms—blastoids, brachiopods, fusilinaceans, goniatites, trilobites.

to modify the latter into the former? Strimple lists three: (1) reduction in size, (2) eventual resorption of infrabasals, and (3) elimination of the single anal plate. Finally, what other Mesozoic crinoid is morphologically closest to *Pentacrinus*? The Jurassic articulate crinoid *Isocrinus* (Figure 12.42*B*, *B*) fits this prescription with a single qualification: it still has infrabasals. Accordingly, Strimple concluded that the most probable Paleozoic inadunate progenitor of the modern *Pentacrinus*-like articulate crinoids, was a type like the inadunate *Chlidonocrinus*. (For details about this genus, see Strimple and Moore, 1971;) H. Wienberg Rasmussen (p. T302); H. L. Strimple (p. 298) Treatise Invertebrate Paleontology, Echinodermata, 1978, 3 volumes.

5. A new subclass, Articulata, arose in Mesozoic time after the demise of all Paleozoic populations (see Figures 12.42*A* and 12.42*B* for representative types) (see Bather, 1909; Sieverts-Doreck, 1939; Moore, 1967). The encrinid crinoids have been assigned to the oldest Mesozoic order—Isocrinida, Lower Triassic. By Middle Triassic time there must have been dense populations of *Encrinus liliformis* in the Muschelkalk seas of Europe, as evidenced by the rock record of that sea floor. Other Triassic forms (Middle Triassic) belong to a second articulate

order—the genus *Millercrinus*, for example, of the order Millercrinida, which is also present in the European Upper Jurassic. Isocrinids (*Pentacrinus*) are prominent in the Lower Jurassic, while millercrinids abound in the Middle Jurassic (genus *Apiocrinus*, for example), and certain Upper Jurassic limestones of Europe (Gignoux, 1955).

Among the 200+ different Cretaceous crinoids, both stemmed and unstemmed forms occur. Among the stalked forms known in European faunas, there are twenty-eight species. Crinoids attached to the bottom have species restricted to a single continent compared to the broad geographic spread of planktonic forms like *Marsupites* represented in India, Europe, North America, Australia, and Africa (see Moore, 1967).

Plate and arm segments are dense in thin sections of certain Alpine crinoidal limestones. Not recognized as crinoids, called "Lombardia" after the original investigator, they were later restudied and identified as *Saccocoma* (Figure 12.42 B, part E)—a pelagic crinoid (Verniory, 1954).

A remarkable example of a free-swimming (planktonic) crinoid is the genus *Uintacrinus* (Figure 12.42B, part F) with ten simple arms bearing pinnules that had a spread of 2.5 meters (H. W. Miller, 1968). It abounds in restricted areas of certain horizons in the Upper Cretaceous of Kansas, the western interior of North America (Uinta Mountains), and is known from England and Germany. Some Kansas slabs (Niobrara Chalk) have as many as 250 complete individuals with arms interlocked to form a kind of floating mat of about thirty-five square feet. Juveniles or stunted individuals range from 14 to 35 mm in diameter, while other calices run to 55 mm or greater.

In the Niobrara, beds with crinoid calices are effectively crinoid coquinas of *Uintacrinus*. This genus did not extend beyond the Upper Cretaceous.

Modern Comatulids. The modern articulates, Order Comatulida, have evolved so that, at an early stage, the stem is shed and subsequently, individuals lead a free-swimming existence. Development of cirri as attachment appendages thus replaced the stem. Further, multibrachiation became a selective factor for creatures that swam from place to place.

Comatulids in modern seas can be grouped into families according to the number of arms they possess and the relative size and depth of the centrodorsal cavity in the calyx. Thus oligophreates have ten plus arms and small, shallow cavities; macrophreates have ten arms and large deep cavities. Most comatulids are oligophreates. There are about twice the number of families of oligophreates as there are macrophreates. The well-known genus *Comatula* is an oligophreate, while genus *Antedon* is a macrophreate.

The 700+ species of articulates living in modern seas together with the abundant population densities in some areas, such as the Indo-Pacific, attest to genetic viability. The array of adaptations for life at all depths from shallows to deeps, for attachment, for swimming, for feeding suggests that strong selective factors will continue.

The heyday of crinoid speciation was obviously Paleozoic time. Articulates of the Mesozoic seas showed the strong operation of natural selection among alleles favoring planktonic existence. (In adaptive evolution, certain alleles in the genetic code are substituted, that is, selected out, for others more fit.) The greatest radiation and expansion into diverse ecologic niches transpired among the Cenozoic to Recent comatulids as already remarked.

Several Mesozoic articulates became extinct (Uintacrinids, Roseacrinids, and so on) or diminished in species number in Cretaceous and post-Cretaceous seas of the world—with the exception of the Comatulida. This can be taken as a prelude to further extinctions in the future.

The extent of the vigor in modern

comatulid diversification and distribution could possibly lead to a new resurgence in crinoid evolution in the future. However, the history of the multiple extinctions reviewed above as well as the abundance of parasites, among other factors, may be an ominous augury of comatulid, and finally crinoid extinction in the seas of the world.

Biochemistry

Blumer (1960) isolated several pigments — derivations of aromatic hydrocarbons — from Jurassic crinoids. It should be possible to reconstruct the chemical steps leading to the formation of such pigments. In turn, this could yield derivative data on the reduction-oxidation potential and variations in dispersed organic matter in the sedimentary substrate of the Jurassic sea.

Paleopathology

The observation was made that worms and other parasites infest modern forms. Similarly, worm parasites may be inferred from cystlike swellings on stem and brachials of Carboniferous and younger sea lilies (Tasnádi-Kubacska, 1962).

Examples of skeletal injury and subsequent repair — injury brought on by predators or rough water — were detected in an extensive Permian fauna (Battle Ship Wash crinoids of Nevada). Biserial arm arrangements, for example, were disrupted in some, but not all, arms at the same level. Repairs resulted in an abrupt change of shape and irregularities (Lane and Webster, 1966).

There is a considerable literature on crinoid regeneration or shell repair and rebuilding after injury (Hattin, 1958). However, repair of discrete skeletal elements has seldom been studied.

Seasonal Growth

Under "Echinoids," successive growth lines in ambulacral plates were described (see Raup, 1968). Here an equivalent pos-

sibility of detecting growth lines in crinoids is illustrated (Figure 12.44G). Lane and Webster (1966) found many isolated cup plates in a rich Permian collection of crinoids from Nevada, that had from 7 to 9 successive growth lines. These were laid down concentrically and could be counted on the interior of the basal and radial plates.

These crinoid growth lines were more widely spaced toward the center of the plate's interior and more tightly spaced near their more lateral edges. This was interpreted to denote initial rapid growth and decelerated growth in maturity. Possibly seven or more such growth increments corresponded to years in the creature's life cycle.

Decoding the seasonal growth increment in crinoidal plates may offer a fruitful field of research.

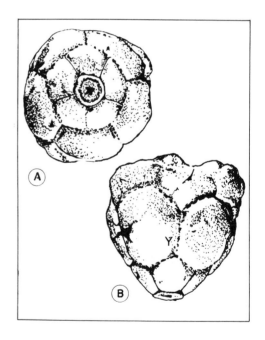

FIGURE 12.46 *Ulocrinus fistulosus* Strimple and Moore, 1971, holotype, dorsal cup. (Schematic). (*A*) Basal view; note bowl-shaped infrabasals and sharply impressed stem attachment area. (*B*) Anterior view; note slight constriction of dorsal cup at summit and deeply impressed suture areas between plates that appear as a series of pits.

Identification of a Crinoid to Genus and Species

(Based on data in Strimple and Moore, 1971, pp. 24–26, plate 15, figures 2*a*–*d*; plate 16, figures 1*a*–*c* 2*a*–*c*.)

Raw Data. $N = 4$. The holotype cup has two infrabasal circlets and one crown referred to as "excellent."

Condition of Preservation. Generally critical plate characteristics are very clear.

Description of the Study Material. The chief characteristics are the dorsal cup is slightly constricted at the summit and has an elongate (or high) bowl shape; with two anal plates. (No data are given for rest of the crown); infrabasals upflared (Figure 12.46).

Family and Genus. The above characteristics occur in two genera of the Cromyocrinidae: *Ulocrinus* and *Metacromyocrinus*. These genera share many features in common: attitude of infrabasals, surface ornamentation, dorsal cup constricted near summit, and similar branching arms. However, the dorsal cup is distinctly lower in *Metacromyocrinus*. From these considerations, the specimens were assigned to Family Cromyocrinidae, genus *Ulocrinus*.

Species. The material at hand from Illinois, Kansas, and Oklahoma was closest to *Ulocrinus convexus* and *Ulocrinus zeschi*. It differed from *convexus* in having a cup-shaped infrabasal circlet of plates (more evenly upflared in *convexus*), thicker dorsal cup-plates, and more deeply impressed sutures. *U. zeschi* is closer to *U. convexus* in several features: cup-shape, lack of tumidity of plates, etc.

Another species, *Goleocrinus confossus*, shares the characteristics of "tumid" (swollen or protruding) plates and "well-defined holes" on the impressed sutures (see Figure 12.46) with the material being studied. However, its dorsal cup is distinctly different—it has a low, truncate bowl shape.

Measurements. A table of measurements (in millimeters) of the holotype was included: height and width of dorsal cup and infrabasal circlet of plates; length and width of basals and radial plates; diameter of stem attachment area.

Species Designation. The above data led to establishment of a new species named *Ulocrinus fistulosus* Strimple and Moore, 1971. (The species name applies to the suture area.)

QUESTIONS

Echinodermata

General

1. Discuss the diagnostic characteristics of echinoderms, living and fossil.
2. What was the significance of hydrocoel growth gradient in echinoderm evolution? (*Hint.* See Figure 12.4.)

Helicoplacoids

1. How does a paleobiologist go about establishing a new class?
2. Cite the diagnostic features of the class Helicoplacoidea, and indicate what specifically is considered "primitive" about its skeletal anatomy.

Edrioasteroids, Ophiscistioids, Cyclocystoids

1. Given an array of fossil echinoderm skeletal material, how would you convince yourself that you had specimens assignable to each of the three classes listed above?
2. What do the three classes have in common?

Holothuroids

1. Discuss skeletogenesis, mineralogy, and chemistry of holothurian skeletal parts.
2. What derivative data is obtainable from fossil sclerites on: population density, bioturbated sediments, bathymetry, ontogenetic and interpopulation skeletal modification.

3. Study Figure 12.14 and indicate information on zonation obtainable from fossil slcerite data.

4. How can population statistics help to clarify intraspecific variability in fossil holothurian sclerites.

Echinoids

1. (a) Outline the organization of the echinoid skeletal plate.
   (b) Draw and fully label an echinoid.

2. Distinguish: insert/exsert, monocyclic/dicyclic. Identify: echinus rudiment, madreporite, Aristotle's lantern.

3. Discuss the various approaches to echinoid classification and give details of the Durham and Melville system.

4. (a) What is the significance of the modfication and changes in skeletal components of Paleozoic echinoids?
   (b) What major anatomical changes occurred in post-Paleozoic echinoids?

5. Review steps in biomineralization of the echinoid skeleton. (See Chapter 15.)

Crinozoans — Exclusive of Crinoids

1. How do edrioblastoids differ from parablastoids and blastoids?

2. What major changes in thecal structures characterize blastoid evolution?

3. Illustrate derived information from blastoides using paleoccurents paleosalinity, biotic associates and predators, paleopathology, biostratigraphy.

4. Distinguish between the subclasses of the phylum Asterozoa in terms of ancestral and descendant lineages.

Crinozoans — Crinoids

1. Review essential information contained in a fossil crinoid skeleton. Discuss the steps in skeletogenesis of a crinoid.

2. Study Figure 12.43 and review steps leading to accumulation of a crinoidal limestone.

3. Define the role of natural selection in the evolution of Paleozoic crinoids, the rise of Mesozoic planktonic crinoids.

4. What are the possible factors that can explain Paleozoic extinction of various crinoid classes. From what lineage did the Articulates arise? Review steps according to Strimple.

5. Discuss Moore and Jeffords' use of disarticulated crinoidal parts in classification and biostratigraphy.

6. Prepare a labeled plate diagram of any fossil crinoid.

REFERENCES

Crinoids

Bather, F. A., 1900. The Echinoderma, in E. R. Lankester, ed., A Treatise on Zoology, Pt. III. London, Charles Black, 332 pp. (Crinoids, Chapter 11.)

————, 1909. Triassic echinoderms of Bakony. Resultate des wiss. Erforsch. des Balaton-Sees. 1 Bd, 1 Teil. Anhang Palaeontologie, Wien.

Blumer, Max, 1960. Pigments of a fossil echinoderm. G.S.A. Ann. Mtg. (Denver) Abstracts, p. 61.

Brower, J. C., 1967. Growth patterns in primitive Ordovician camerate crinoids. G.S.A. Ann. Mtg. (New Orleans) Abstracts, pp. 25–26.

Carozzi, A. V., and J. G. W. Suderman, 1962. Petrography of Mississippian (Borden) crinoidal limestones at Stobo, Indiana: J. Sed. Petrology, Vol. 32, pp. 397–414.

Clark, A. H., 1915. Monograph of the existing crinoids. The Comatulids. Smithsonian Inst., U.S.N.M. Bull. 82, Vol. 1 (Pt. 1), 382 pp., 17 plates.

————, 1921. Same, Vol. 1 (Pt. 2), 755 pp., 57 plates.

Clark, A. M., 1966. Some crinoids from New Zealand waters: N.Z. J. Science, Vol. 9 (3), pp. 684–705.

Fell, H. B., 1948. Echinoderm embryology and the origin of chordates. Biol. Rev., Cambridge Philos. Soc., Vol. 23, pp. 81–107.

————, 1966. Ecology of crinoids, in R. A. Boolootian, ed. Physiology of Echinodermata: Interscience, pp. 49–71.

———— and D. L. Pawson, 1966. General biology of echinoderms. Same, pp. 1–48.

Hattin, D. E., 1958. Regeneration in a Pennsylvanian crinoid spine: J. Paleontology, Vol. 32 (4), pp. 701–702, plate 98.

Hyman, L. H., 1955. The Invertebrata. Echinodermata. The coelomate Bilateria. McGraw-Hill, Vol. IV, 705 pp.

Jeffords, R. M., and Theo. H. Miller, 1968. Ontogenetic development of fossil crinoids, etc.: U. Kansas Paleont. Contributions (= UKPC), Echinodermata, Art. 10, pp. 1–14, Figures 1–5, plates 1–4.

Lane, N. Gary, 1967. Larval-adult orientation of fossil and living crinoids. G.S.A. Ann. Mtg. (New Orleans) Abstracts, pp. 127–128.

_____, 1968. The adaptive significance of balance in Paleozoic stalked crinoids. G.S.A. Ann. Mtg. (Mexico City) Abstracts, p. 169.

_____ and G. D. Webster, 1966. New Permian crinoid fauna from Southern Nevada: U. Cal. Publ. Geol. Sci. U. Calif. Press, Vol. 63, 60 pp., 13 plates.

Langton, T. R., and G. E. Chin, 1968. Rainbow Lake facies and related reservoir properties, Rainbow Lake, Alberta: A.A.P.G. Bull., Vol. 52 (10), 1925–1955.

Macurda, D. B., Jr., 1968. Ontogeny of the crinoid *Eucalyptocrinus:* J. Paleontology, Vol. 42 Suppl. to N. 5, Pt. II of II. The Paleont. Soc. Mem. 2, pp. 99–117.

Miller, H. W., 1968. Invertebrate fauna and environment of deposition of the Niobrara Formation (Cretaceous) of Kansas: Ft. Hays Studies. New Series. Science Ser. No. 8, 66 pp.

Moore, R. C., 1954. Echinodermata: Pelmatozoa, in B. Kummel, ed., Status of Invertebrate Paleontology, 1953, Pt. IV. Bull. M.C.Z. (Harvard) Vol. 112 (3), pp. 125–144.

_____, 1962. Ray structures in some inadunate crinoids: UKPC, Art. 5, pp. 1–47, plates 1–4, text figures 1–17.

_____, 1967. Unique stalked crinoid from Upper Cretaceous of Mississippi: UKPC, Paper 17, U. Kansas Paleont. Inst., 35 pp.

_____ and L. R. Laudon, 1943. Evolution and classification of Paleozoic crinoids: G.S.A. Sp. Paper #46, 117 pp., 14 plates.

_____ and L. R. Laudon, 1944. Class Crinoidea, in H. W. Shimer and R. S. Shrock, Index Fossils of North America: John Wiley, pp. 137–209.

_____ and R. Jeffords, 1968. Classification and nomenclature of fossil crinoids based on studies of dissociated parts of their columns: UKPC, Echinodermata; Art. 10, pp. 1–86, Figures 1–6, plates 1–28.

Raup, D. M., 1968. Theoretical morphology of echinoid growth: J. Paleontology: Paleontol. Soc. Mem. #2, pp. 50–63.

Sieverts-Doreck, Hertha, 1939, Jura-und-Kreide-crinoideen aus Deutsch-Ostafrika.

Palaeontographia. Suppl. VII, II, Reihe Teil II, Lief. 3.

Seilacher, A., G. Drozdzewski, and R. Haude, 1968. Form and function of the stem in a pseudoplankton crinoid (*Seirocrinus*): Paleontology, Vol. 11, Pt. 2, pp. 275–282, plates 48.

Springer, Frank, 1917. On the crinoid genus *Scyphocrinus* and its bulbous root *Camarocrinus:* Smithsonian Instit. Publ. 2440, 55 pp., 9 plates.

_____, 1920. The Crinoidea Flexibilia: Smithsonian Inst. Publ. 2501, 76 plates.

_____, 1926. American Silurian crinoids. Smithsonian Inst. Publ. 2851, 166 pp., 33 plates.

Swinnerton, H. H., 1950. Outlines of Paleontology. 3rd ed., London, Edw. Arnold & Co., 370 pp. (Crinoids, pp. 121–136.)

Tasnádi-Kubacska, A., 1962. Paläopathologie. Gustav Fischer, Jena, 240 pp. (Crinoids, pp. 80–85.)

Teichert, Curt, 1949. Permian crinoid *Calceolispongia*, G.S.A. Mem. #34, 97 pp., plates 1–26.

Ubaghs, Georges, 1953. Classe des Crinoides, in Jean Piveteau, ed. Traité de Paleontologie, Vol. III, pp. 658–773. (Also, 1967, Treat. Invert Paleont. Pt. S).

Van Sant, J. F., and N. G. Lane, 1964. Crawfordville (Indiana) crinoid studies. Echinodermata: UKPC, Art. 7, pp. 1–136, plates 1–8.

Verniory, R., 1954. *Eothrix alpina* Lombard Algue ou Crinoide? Arch. Sci., Vol. 7 (4), pp. 327–330, 1 plate, Geneva.

Wachsmuth, C., and Frank Springer, 1897. The North American Crinoidea Camerata: M.C.Z. (Harvard) Mem., Vol. XX, pp. 1–359; Vol. XXI, pp. 361–810; Atlas, Vol. XX–XXI, plates I–LXXXIII.

Wanner, Joh., 1924. Die Permischen krinoiden von Timor, II, 2ᵉ Nederlandische Timor-Exped., 1916, pp. 1–330.

Webster, G. D., and N. G. Lane, 1967. Additional Permian crinoids from Southern Nevada: UKPC, Paper 27, 32 pp.

[H. L., Clark, 1915, 1921. Carnegie Instit. Washington. Mar. Biol. 8, 10. H. L. Strimple, and R. C. Moore, 1971. Crinoids of the La Salle Limestone (Pennsylvanian) of Illinois. U. Kansas Paleont. Contrib. Article 55 (Echinodermata 11), p. 29, pl. 6, figs. 1*a–c* (*Chlidonocrinus*). Consult the Treatise on Invertebrate Paleontology. (T) Echinodermata, volumes 2(1) and 2(2), Crinoidea.]

chapter thirteen

PALEOCREATURES IN SEARCH OF RELATIVES (GRAPTOLITHINES, CONODONTS, AND CHITINOZOA)

GENERAL VIEWPOINT

Although graptolithines, conodonts, and chitinozoans, are considered in this chapter, they are completely unrelated; what they share in common is the uncertain taxonomic position of each group.

PART I. GRAPTOLITHINES

Introduction

Graptolithine biology must be inferred from morphology. Comparative study of forms somewhat homologous can supplement morphological indications.

The class of graptolithines endured for about two hundred million years (Cambrian–Lower Carboniferous). Graptolite fossils are important index forms of the Ordovician-Silurian. They have been monographed repeatedly—collections from North America (Ruedemann, 1947), South America (Bulman, 1931), England, Europe (Bohemia, Germany, Sweden, Poland, and so on), Australia, and elsewhere. Still other reports treat graptolites from Asia (for example, Kobayashi and Kimura, 1942; Hsü, 1934), North Syria (Flügel, 1963) and elsewhere.

Two different types of fossil material are available: the bulk of all fossils are compressed, flattened carbonized remains; limited exceptional cases have yielded excellently preserved three-dimensional specimens (Kozlowski, Urba-

nek, Thorsteinsson, Whittington, among others). From the exceptional three dimension forms and based on Kozlowski's particular researches, details of ontogeny, biology, and so on, of graptolithines have become known. However, Bouček (1957) remarked in his monograph on dendroid graptolites of Bohemian Silurian, internal structures "cannot unfortunately be used in practice" in most cases because carbonized, flattened specimens tend to dominate given collections.

What, then, is the data relevance from three-dimensional specimens to study of relatively poor two-dimensional material? Detailed structures of three-dimensional graptolithines give the key to equivalence in two-dimensional specimens. This critical realization provides graptolite specimens meaning as biological objects.

To what phylum do the graptolithines belong? Many excellent researchers have been engaged in finding an answer. Assignments that have been favored include: Bryozoa (Ruedemann, 1931); Coelenterata (Bohlen, 1950; Decker, 1956; Hyman, 1959); Hemichordates (Kozlowski, 1938, 1947, 1948, 1966; Bulman, 1955, 1966, 1970).

Fossil Graptolithines

Essential Graptolithine Structure
Megascopic Observations. Examine a fossil representative of the five grapto-

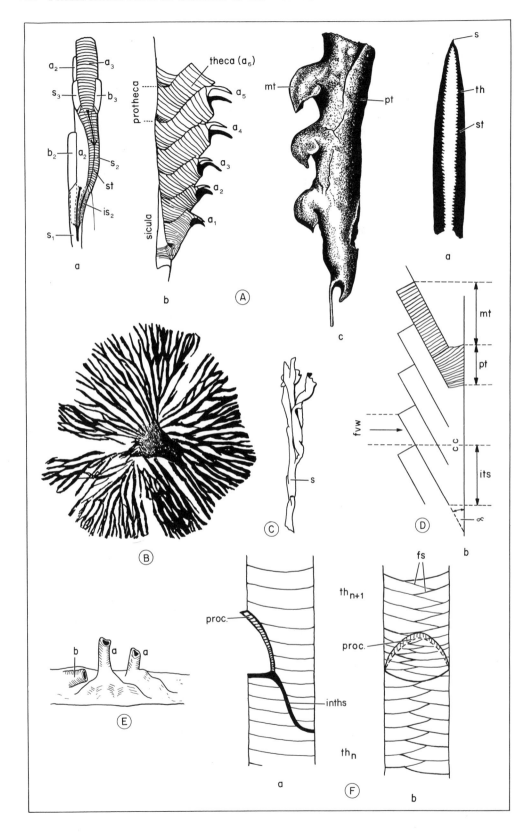

FIGURE 13.1 The Five graptolithine orders and graptolite morphology.
(A)(a) Order Dendroidea. Note three types of thecae; s — stolotheca, b — bithecae, a — autothecae (is$_2$ = internal portion). (b) Order Graptoloidea. Shown for comparison with (A)a. Note that there is only one type of theca. [For further Graptoloidea, see (A)c, (D), (F).] (c) Order Graptoidea. *Monograptus* (*Monograptus*) *uncinatus*. Erratic boulders, Silurian, Poland (X 7); *mt* — metatheca, *pt* — protheca [compare *pt* with *s* and *mt* with "*a*" in (A)b].
(B) Order Dendroidea. (?) *Calyptograptus cyathiformis*, Middle Silurian, Canada.
(C) Order Stolonoidea, *Stolonodendrum uniramosum*, Lower Ordovician, Poland.
(D) Graptoloidea. (a) *Didymograptus murchisoni*, Lower Ordovician, South Wales; s — sicula, st — stipe, th — thecae. (Terminology of graptolite stipe and theca; *its* — interthecal septae, *alpha* — angle of inclination, *fvw* — free ventral wall, *cc* — common canal, as well as *mt* and *pt* as defined above.)
(E) Order Camaroidea. *Bithecocamara platicellata*, Tremadocian, Poland (X 35); a — autotheca, b — bitheca.
(F)(a–b) Graptoidea: *Monograptus* showing additional features: *inths* — interthecal septum, *proc* — apertural process, *thn + 1* — distal theca, *thn* — proximal theca, *fs* — fusellar structure of theca. All figures X 1 unless otherwise stated. Illustrations adapted from Bulman, 1955; Kozlowski, 1938; Urbanek, 1958.

lithine orders (Figure 13.1); only a few features are immediately obvious. Generally fossils are flattened on bedding planes. More rarely, acid digestion releases three-dimensional specimens from rock matrices. Most apparent is a branching structure. Along the branches one or more rows of cups or tubes (= thecae) can be readily identified.

From this first-level observation, it is apparent that graptolites were *colonial* organisms (multiple individuals corresponding to multiple thecae); they were *thecate* and *branching*.

Details of Thecae. The next inquiry concerns the nature of the thecae. Are they all alike in size, position, and inferred function? Three types of graptolite thecae have been distinguished (Figure 13.1).

Stolotheca. This is the proximal immature portion of the daughter autotheca; the dendroid stolotheca is equal to the *protheca* of graptoloids (Figure 13.1A, c). (A stolon is the internal thread from which thecae originate.)

[*Note: Proximal* — nearest to point of origin; first formed portion of theca, and so on. *Distal* — last formed portion of theca, and so on; farthest away from point of origin (Figure 13.1F, b).]

Autotheca. The cup or tube is thought to have housed the female zooid; the dendroid autotheca equals the graptoloid metatheca (Figure 13.1A, c).

Bitheca. The cup or tube is thought to have housed the male zooid; lacks aper-

tural processes (Figure 13.1A, a).

Details of the Periderm. The periderm is the wall of the graptolithine skeleton (= rhabdosome). Largely owing to the researches of Kozlowski (1938, 1947, 1949), details have been made clear. Figure 13.2A–B indicates the two-part structure: an external laminated layer of cortical tissue, and an inner layer of fusellar tissue. The fusellar layer consists of short growth segments intercalated to form a vertical zigzag suture.

Ontogenetic Stages

Prosicula. The initial stage of graptolite ontogeny began with a prosicula secretion (Figure 13.2C, a). This resembles an elongate, narrow belljar about 400 μ long, with a basal opening (= aperture) and an apical closure (= neck or nema). Bulman (1955) equated the prosicula with the larval skeleton developed from a fertilized egg.

According to Kraft (1926) and subsequently illustrated by Kozlowski (1953), by Urbanek (1958) and others, a helicoidal line is often visible on the initial prosicula (Figure 13.2C, a, nl). Gradually, two additional prosicula stages followed: appearance of several longitudinal fibers, and later, secondary fibers.

Metasicula. The next stage gave rise to a metasicula; it is markedly distinguished from the prosicula by intercalated growth bands (= fuselli) extending on the ventral side to form a short spine (= *virgella*). When growth of this stage is almost

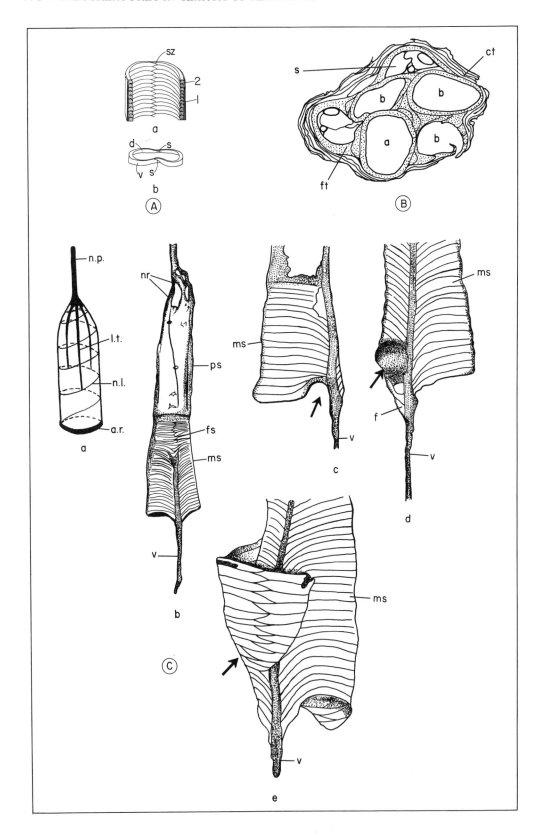

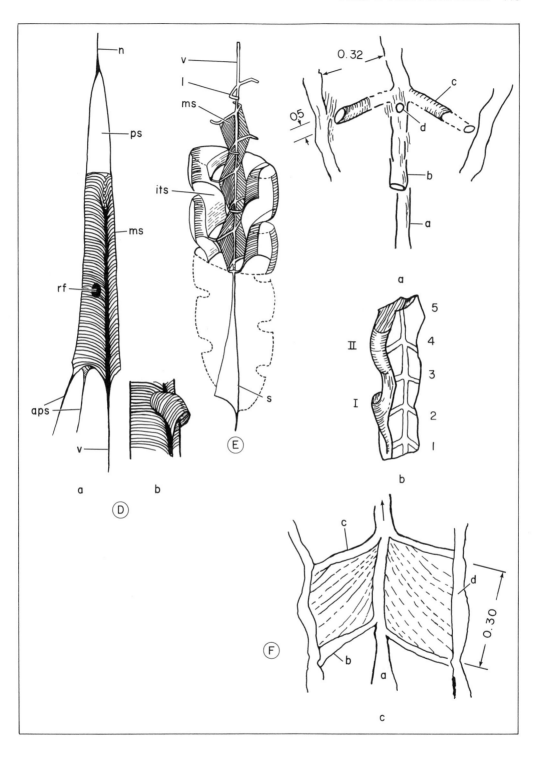

FIGURE 13.2 Structural graptolite features. (A)(a–b) Periderm; diagram represents two periderm layers: 1. cortical tissue (cortex), 2. fusellar tissue; *sz*—zigzag suture of short growth segments of fusellar tissue. (b) fusellar tissue connects two semiannular fuselli: *d*—dorsal, *v*—ventral, *s*—oblique suture. Note bilateral segment symmetry.
(B) *Koremagraptus* stipe (= one branch of branched rhabdosome), transverse section; *s*—stolotheca, *b*—bitheca, *a*—autotheca, *ft*—fusellar tissue, *ct*—cortical tissue.
(C) Ontogenetic stages: Monograptidae: (a) prosicula (*ps*), primary nema (*np*) (this structure often destroyed, regenerates), longitudinal threads (*lt*), helicoidal line (*nl* = hl), apertural ring (*ar*). All figures schematic. (b) Prosicula (*ps*) and metasicula (*ms*) stages show a regenerated nema (*nr*) after damage, virgula (*v*) (note fusellar structure); (c) (d) (e) *Monograptus colonus*: (c) *sinus stage*, arrow points to apertural notch; (d) *lacuna stage*, note fuselli band closes sinus notch on right side; (e) *prothecal stage*, first theca budding (see arrow). All figures schematic and enlarged.
(D) Ontogeny resorption occurs: (a) Notice the resorption foramen (*RF*) near completion of the metasicula (*ms*) growth; *aps*—apertural dorsal side spines, *v*—virgella; (b) initial bud—first theca passes through pore to exterior. Schematic.
(E) *Climacograptus* showing sicula (*s*), virgula (*v*), lists (*l*), median septum (*ms*), interhecal septum (*its*).
(F) Graptolite genus uncertain, Maquoketa shale. (a) Virgula and one list pair; (b) virgula detail and list system (1–5), I–II = thecae (not ontogenetic but as positioned in fossil fragment); (c) median septum (see *E*) [dimensions for (*F*) in mm].
Redrawn, adapted illustrations from: (A) Kozlowski, 1947; (B, D, E) Bulman, 1955; (C) Urbanek, 1958; (F) Tasch, 1958.

completed, a pair of apertural spines appear (Figure 13.2*D, a*).

The prosicula and metasicula, taken together, constitute the "sicula."

Foramen or Primary Notch. A hole (foramen) occurs generally in the graptolite metasicula by resorption. In dendroids and some dichograptids, this foramen occurs in the prosicula. The initial bud then passes to the exterior through this opening. In the Monograptidae, a primary notch forms instead. Eisenack (1942) recognized three stages in the appearance of an initial bud (from the metasicula) (Figure 13.2*C, c–e*): the sinus stage, in which the apertural notch forms to the right of the virgella; the lacuna stage, in which the notches close by a band of fuselli; the prothecal stage, in which the first theca subsequently develops.

Rhabdosome. The exoskeleton of a colony (= rhabdosome) is composed of multiple thecae. A pseudobranch may develop attached to the parental rhabdosome as in the case of *Cryptograptus*. In a pseudobranch the term "cladium" is used. One can derive all subsequent buds starting with an initial bud that may be referred to as "1'." If diverse modes of branching had not developed, all graptolites would have had similar architecture. However, modes of branching are a variable grapto-

lite feature. Branching accounts for the unique aspect presented by a dicellograptid, a diplograptid, or a dichograptid, among others. (Figure 13.4*C*).

Internal Supportive and Strengthening Structures. Three related internal structures are the *virgula*, *lists*, and *median septum*.

Virgula. An extension of the prosicula apex, that is hollow, and about 0.03 to 0.05 mm thick (Tasch, 1958, Table 1) (Figure 13.2*E*). Thorsteinsson (1955, p. 42 and text figures) showed that in *Cryptograptus*, the thecal spines ("apertural spines," Figure 13.2*D*) of the mother theca extended to form the virgula.

Lists. These are hollow supports attached to the virgula at one end, and to the rhabdosome wall at the other.

Median Septum. This structure consists of a diagonal band series in echelon arrangement on either side of the virgula between two successive lists. The virgula is embedded in it as are the lists where present. Aseptate forms also occur in which the virgula is free in the cavity (Figure 13.2*E, ms; F, c*).

Serial Thecal Arrangement. A rhabdosome or a branch of one containing a single chain or more of thecae is *uniserial*. *Biserial* forms have two rows of thecae in contact side by side or back to back, while *quadriserial* forms show four rows of thecae.

Skeletal Composition.

Skeletal Composition. Almost universal reference is made to the chitinous exoskeleton of graptolites.

M. F. Foucart et al. (1966) seem to have the latest and most authoritative resolution. They reported absence of chitin in graptolite and pterobranch tests and a suite of amino acids as present. That is to say, the test is not chitinous but proteinaceous.

Taxonomic Placement and Inferred Biology. One goes from the initial recognition of equivalences between graptolites and some other known animal group, (Bryozoa, Coelenterata, or Hemichordata) and applies to all pertinent structures or to each and every morphological feature of graptolites, the stamp of equivalence.

Thesis I. Graptolites are an extinct side branch of Bryozoa. The "muscle scar" impressions noted by Ulrich and Ruedemann, and by Haberfelner (1933) were thought to be artifacts by Kozlowski (1947, p. 99); probably due to accidental compression of different parts of the thecae. The presence of ovicells (gonothecae) in dendroid graptolites, are problematical, according to Kozlowski, and Bulman (1955). Bilateral symmetry in graptolites and the growth form of the colony (thought bryozoan-like by Ulrich and Ruedemann) can be explained by convergence during evolution in animals of similar habit (Sinclair, 1948; Kozlowski, 1947, p. 98).

Thesis II. Graptolites are coelenterate hydrozoans or a specialized coelenterate class. Decker and Gold (1957), from a study of Upper Cambrian dendroid graptolites and various genera of the Graptoloidea, concluded that gonotheca and numerous nemotothecae (see Chapter 5) were definitely present. Furthermore, bisexuality existed in the bithecae (male) and gonothecae (female). One carbonized impression of a polyp preserved

what was interpreted to be a single row of tentacles. All of these are coelenterate features.

Almost a decade before, Kozlowski (1948) had argued against coelenterate affinities based on several observations. These included the following: (*a*) The "common canal" of graptolites (that is actually formed by the common portions of all thecae) is not homologous with the coelenterate hydrocoel. (*b*) Hydrozoans have radial symmetry, graptolites have bilateral symmetry—and this symmetry is manifest in the larval stages of each group, showing its fundamental character. Graptolite bilateral symmetry could not, therefore, be a primitive secondary acquisition, as Ulrich and Ruedemann had suggested. (*c*) Presence of intrathecal stolons (the most important argument, according to Kozlowski). (*d*) External thickening of the thecal walls which implied extrathecal tissues—unknown in Hydrozoa.

Decker and Gold (1957) reopened the case for coelenterate affinities. Other workers (Bohlen, 1950; Hyman, 1959) also support this argument.

Thesis III. Graptolites have hemichordate (pterobranch) affinities. Kozlowski (1938, 1947, 1948, 1966) and others (Bulman, 1955, and so on) have advanced major arguments in favor of pterobranch affinities. (1) Graptolites and pterobranchs (*Rhabdopleura*) both secrete proteinaceous tubes. (2) The fusellar structure of such tubes (Figure 13.2*A*) are found only in graptolites and *Rhabdopleura*. (3) Both pterobranchs and graptolites have a stolon system that connects zooids and functions in asexual multiplication.

Hyman (1959), Decker (1956), and Bohlen (1950) argued against this thesis for the following reasons:

(*a*) Graptolite exoskeleton composition is unknown and is probably chitinoid. The coenoecium of *Rhabdopleura* is, however, known (see Chapter 15) and it is not chitinous. [Foucart (1966), already referred to, found equivalent amino acid suites in graptolites and *Cephalodiscus*—this eliminates the argument over comparable composition.]

(b) The coenoecium of pterobranchs and the periderm of graptolites were not secreted either in the same way or from equivalent tissues. *Rhabdopleura* tubes consist of a single layer, but the graptolithine periderm has an inner fusellar and an outer organic layer.

(c) Fusellar structure is found in several different kinds of worms (Bohlen, 1950, and Bohlen, in Decker, 1956, p. 1073). Hence this type of structure is not restricted, as Kozlowski had said [Kozlowski (1966) challenged this statement].

(d) Stolon and stolon systems are common in the animal kingdom. Accordingly, no unusual weight can be given to the fact that in *Rhabdopleura* and in graptolites, stolons connect zooids (Hyman, 1959). [Kozlowski (1966) challenged this statement.]

(e) *Rhabdopleura* lacks graptolite structures such as nema, sicula, virgula.

(f) Decker listed a series of further arguments in addition to those already cited under Thesis II: There is a notable difference in budding—irregular in *Rhabdopleura*, regular in graptolites; graptolites uniquely originated from a tiny, cone-shaped sicula (= prosicula), but *Rhabdopleura* did not.

Barrington (1965) allowed the possibility of "some relationship" to the pterobranchs.

Kozlowski (1966) considered the Decker/Bohlen/Hyman interpretation erroneous, fantastic and reaffirmed the hemichordate affinities of graptolites. Bulman, in a personal communication (1966) agreed with Kozlowski's 1966 reevaluation of the evidence bearing on this matter and favored pterobranch affinities. At the present writing, the dual opinion of Kozlowski and Bulman, backed by further evidence, seems to tip the scale in favor of hemichordate affinities as the likely interpretation of known evidence.

Inferred Biology. If we accept hemichordate affinities, then the autotheca housed the female zooid, and the bitheca, the smaller males. In the Graptoloidea, the loss of the bitheca implies transition to hermaphroditism. With the elimination of the male, the female became hermaphroditic. The stolotheca, Bulman inferred, was secreted by an immature zooid, and the bitheca, by a reduced male.

The inferred embryology envisions (a) a bisexual generation arising from male and female zooids; (b) following fertilization of the egg, a larval stage—whether or not it was free-living, is undetermined, although it seems likely; (c) the prosicula is equated with the larval skeleton; (d) it may have been attached to the substrate by a fleshy peduncle or primary nema; (e) the development then ensued from prosicula to metasicula to the initial bud (thecae) (Figure 13.2*C*, *D*).

Three modes of life have been postulated for graptolites: *sessile*—living in shallow water attached to pebbles, shells, or seaweeds; *epiplanktonic*—as in *Dictyonema flabelliforme*, attached to floating seaweeds, or *truly planktonic*—several rhabdosomes (= synrhabdosome) were linked by nemata to a common central float (see Figure 13.4); or flotation by buoyant tissue.

Classification. Class Graptolithina (after Bulman, 1955, 1970).

1. Colonial marine organisms.
2. Exoskeleton proteinaceous (see Foucart et al., 1966). [Cf Towe and Urbanek (1972).]
3. Periderm with characteristic growth bands and growth lines (Figure 13.2*A*).
4. Theca, uniserial or biserial, in rows along stipes (that is, a branch of a branched rhabdosome or a whole colony of unbranched rhabdosomes).
5. Some orders show polymorphism, that is, thecae: modified, different sizes.
6. Polymorphic thecae in some orders are related to the stolonal system (Figure 13.2*B*).
7. Conical sicula gave rise to rhabdosome.
8. Attachment was by nema or virgula (Figures 13.2 and 13.4).

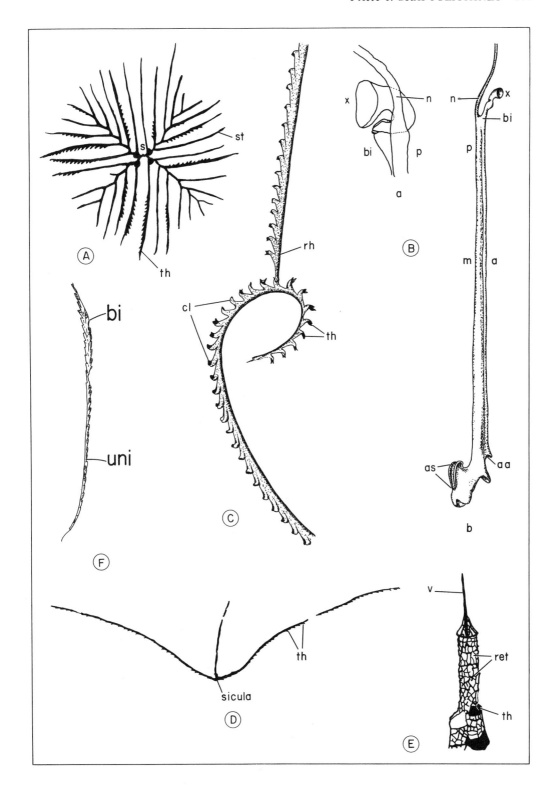

9. Stipes generally free and rarely encrusting.

10. Geologic range: Middle Cambrian to Carboniferous.

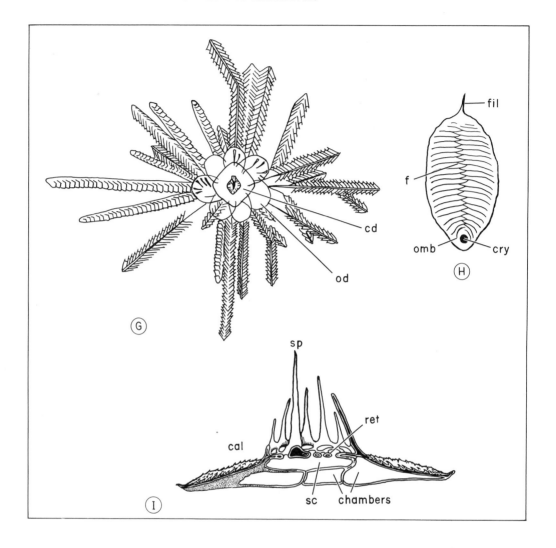

FIGURE 13.3 and 13.4 Representative graptolite genera, and others.

(A) Family Dichograptidae: (multiramous form) *Goniograptus thureaui,* L. Ord., Australia (also known from Deepkill and Normanskill shale, United States), X 1; *s*—sicula, *st*—stipes, *th*—theca, dots = origination point of subsequent stipes.

(B)(a) Family Corynoididae: *Corynoides divnoviensis,* L. Caradocian, Poland, X 25; *a*—theca, *aa*—thecal aperture, *as*—sicula aperture, *bi*—initial bud, *n*—nema, *m*—metasicula, *p*—prosicula, *x*—rudimentary thecae, last formed.

(B)(b) theca detail, X 1 [labels as in (B)a].

(C) Family Monograptidae: *Cyrtograptus rigidus,* schematic; *rh*—rhabdosome, *th*—thecae, *cl*—cladium.

(D) *Leptograptus (flaccidus) flaccidus,* M. Ord., S. Scotland, X 1. Note sicula and central branch position arisen from it. This is the "centribrachiate mutation" since other forms of genus are biramous.

(E) *Gothograptus nassa,* U. Sil., Baltic, X 12; *th*—theca, *v*—virgula, *ret*—reticula.

(F) *Dimorphograptus elongatus,* L. Sil., S. Scotland, X 2; *bi*—biserial, *uni*—uniserial.

(G) *Orthograptus,* synrhabdosome, X 1; *od*—several oral/ circular discs (contain bundles of siculae); *cd*—central disc (immature forms attached here by virgula).

(H)–(I) Problematical graptolithines: (H) *Graptoblastus planus,* Tremadocian, Poland, X 40; *f*—fusellar segments; *fil*—filum; *cry*—cryptopyle; *omb* (= *umb*) umbilicus. (I) *Acanthastus luniewski,* Tremadocian, Poland, X 15; *cal*—calotte; *sp*—spinarium; *ret*—reticulum; *sc*—subreticular cavity. Illustrations adapted from: (A, d–I) Bulman, Kozlowski and others; B—Kozlowski, 1953; (C) Thorsteinsson, 1955.

The class is divided into six orders: Dendroidea, Tuboidea, Camaroidea, Stolonoidea, and the most numerous order, Graptoloidea. (Also, Crustoidea).[1]

***Order 1*. Dendroidea.** Sessile graptolithines. Polymorphic thecae [autothecae, bithecae, and stolothecae (Figure 13.1*A*) according to Kozlowski, Bulman et al.]. Geologic range: Middle Cambrian–Carboniferous.

(Growth by alternating triads: stolotheca gives rise to bi-, auto-, and stolotheca. This last stolotheca repeats the triad, and so on. Such development is known as the *Wiman triads*. Specifically, it refers to the three theca types repeatedly arising from the stolotheca.)

Representative genus: *Dictyonema* (Figure 13.1*A*).

***Order 2*. Tuboidea:** Sessile graptolithines. Budding by diads or without fixed rule such as Wiman triads in Dendroidea. Autothecae and bithecae present; stolothecae limited, or feebly expressed. Geologic range: Lower Ordovician (Tremadocian)–Silurian.

Representative genus: (?) *Calyptograptus* (Figure 13.1*B*) (reassigned, Dendroidea).

***Order 3*. Camaroidea:** Encrusting graptolithines. Autotheca consists of two parts: *camara*—an inflated basal vesicle—and *collum*—the free tubular distal portion; stolotheca indistinct, forms bifurcating network above camarae, or in extracamaral tissue surrounding stolons; bithecae tubular. Geologic range: Ordovician.

Representative genus: *Bithecocamara* (Figure 13.1*E*).

***Order 4*. Stolonoidea.** Sessile or encrusting graptolithines. Irregular development of stolons; stolotheca and possibly autotheca present. Geologic range: Lower Ordovician (Tremadocian).

Representative genus: *Stolonodendrum* (Figure 13.1*C*).

***Order 5*. Graptoloidea.** Planktonic or epiplanktonic graptolithines. Sicula pendant from nema (virgula); stipes range from parallel to nema (=scandent) to

[1]Bulman, 1970, p. V 51.

perpendicular to nema (=horizontal) and may have a reclined, reflexed, deflexed, declined or pendant attitude; uniserial or biserial (rare forms are quadriserial); few stipes on rhabdosome. Only the autotheca is present, according to Bulman (1955) and others. (Following Bulman, the bitheca loss equates with elimination of the male and the transition to hermaphroditism.) Geologic range: Lower Ordovician to Silurian.

One or more representative genera in each family of this order are: Family Dichograptidae [Multiramous (= many branches) forms]: *Goniograptus* (Figure 13.4*A*); Pauciramous (= few branches) forms: *Didymograptus* (Figure 13.1*D, a*). Family Corynoididae: *Corynoides* (Figure 13.4*B*). Family Cryptograptidae: *Glossograptus* (not figured). Family Leptograptidae: *Leptograptus* (Figure 13.4*D*). Family Dicranograptidae: *Dicellograptus* Diplograptidae: *Climacograptus* (13.2*E*); *Orthograptus* (13.4*G*). Family Lasiograptidae: *Lasiograptus*. Family Retiolitidae: *Gothograptus* (13.4*E*). Family Dimorphograptidae: *Dimorphograptus* (13.4*F*). Family Monograptidae, subfamily Monograptinae: *Monograptus* (13.2*C*), *Rastrites* (Figure 1.6); subfamily Cyrtograptinae: *Cyrtograptus* (Figures 1.6 and 13.4*C*).]

Problematical Graptolithines. Evidence of much still to be learned about animal evolution and affinities, shows in Kozlowski's problematical graptolithines. With one exception, they are all from Poland's Lower Ordovician (Tremadocian). Three groups erected embrace problematical forms: *Graptovermis* (Group Graptovermida) is an irregular coil of tubes attached at one end. The tubes have a fusellar structure reminiscent of graptolithines. *Graptoblastus* (Group Graptoblasti) (Figure 13.4*H*), small and ovoid, was attached by its lower surface. It also has fusellar segments. Curiously, structures, unknown in other graptolithines, occur. These include: *filium*, an anterior spine; *ombilic*, a rounded protuberance bears an apparent circular pore (= *cryptostyle*). The body divides by a transverse partition into anterior and posterior chambers. These chambers are quite unlike any known graptolithine thecal arrangement.

Even more unusual complex structures are found in *Acanthastus* (Group Acanthastida). The colonial organization with subdivisions arranges a number of radially noncommunicating chambers. Attachment of the discoidal colony was by its flattened lower surface. A central perforated area (= reticulum) bears a ring of long spines. The reticulum, the underlying subreticular cavity, and the spines constitute the *spinarium*. A surficial, rugose/spinose surface (= *calotte*) surrounds the spinarium. The chambers, referred to above, underlie the calotte and spinarium (Figure 13.4*I*).

Reticula, an attenuated meshwork periderm, characterizes the graptolite family Retiolidae (see Figure 13.4*E*). This structure is graptolithine. The extensive spinosity and segments of *Acanthastus* is extraordinary however common moderate apertural spine development is in graptolites. Strong spinosity is not unknown in such forms as *Tetragraptus* (Ruedemann, 1947, plate 51, Figures 14–17). The noncommunicating chambers and their basal arrangement is quite unusual in graptolithines.

Information Derived from Graptolithines

Graptolithine biology is known solely from derivative data obtained from fossils and homologies drawn from living forms with comparable anatomical structures.

Substrates/Graptolite Facies. Black, carbonaceous (up to 13 percent carbon), noncalcareous shales are the common substrates on which graptolites occur. These probably accumulated in shallow, coastal lagoons, landlocked embayments, periodically flooded coastal plains, and so on under anaerobic conditions. Evidence for syngenetic deposition of sulphides is seen in pyrite infillings of graptolite rhabdosomes in full relief (Bulman, 1957).

As much as 7 percent sulphur has been recorded. (Since hydrogen sulphide is generated in the presence of decaying organic matter, graptolite tissue could have filled this function.)

However, as many workers have realised, graptolites have been found in all sorts of sedimentary environments and contexts. Bouček (1957) described several such lithofacies in the Silurian of Bohemia bearing dendroid graptolites. These include (*a*) dark-gray to black graptolitic shales to claystones with abundant disseminated or concretionary pyrite; (*b*) marly or tuffitic shales containing carbonized graptolite; (*c*) dark, marly, platey limestones alternating with shales, or compact calcareous shales to shaley limestones (platey limestones are bituminous and in these graptolite remains are fairly abundant); the final lithofacies consists of organodetritus or platey mud limestones and calcareous concretions.

Berry (1960) provided excellent lithologic as well as faunal data (Figures 13.6, 13.7) on the Ordovician graptolite faunas of the Marathon Region, West Texas. His measured sections shows graptolites in the following type lithologies: black shale; thin-bedded, fine-grained, cross-laminated limestones; dark-gray, sublithographic limestone; limestone interbedded with conglomeratic lenses containing flat chips and slabs of limestone; coarse-grained, subgraywacke lenses; black shale interbedded dolomitic limestones; buff, calcareous shales; black chert interbedded with fine-grained black limestone; chert pebble conglomerate; flaggy, buff, argillaceous limestone interbedded with calcareous shale.

Three-dimensional specimens have been recovered from cherts and chalcedonic lenses by acid digestion (*HF*) and from limestones by HCl digestion. In some instances, noncompressed graptolites have been preserved in a sedimentary matrix permitting study of usually obscured details (*Cryptograptus*, for example, Cornwallis Island, Canadian Arctic Archipelago).

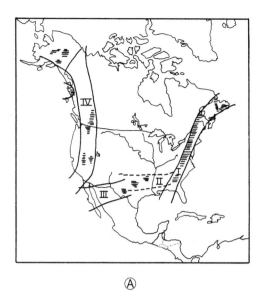

Ⓐ

UNITED STATES	ZONE	AUSTRALIA	ZONE	BRITISH ISLES	ZONE
Richmond	15	Bolindian	25	Ashgill	15
					14
Maysville			24		13
	14				12
Eden		Eastonian	23	Caradoc	
Trenton	13		22		11
Wilderness	12	Gisbornian	21		10
Porterfield	11		20		9
Ashby	10	Middle Ordovician	19	Llandeilo	8
			18		
Marmor			17		
	9		16		
Whiterock	8	Yapeenian	15	Llanvirn	7
			14		
		Castlemainian	13		
			12		6
			11		
	7	Chewtonian	10		
			9		
			8		5
	6	Bendigonian	7	Arenig	4
Canadian series	5		6		
	4		5		
			4		3
	3		3		
	2	Lancefieldian	2		2
	1		1	Tremadoc	1

Ⓑ

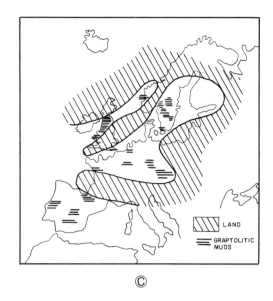

©

FIGURE 13.5 Information derived from graptolites—I.

(*A*) Graptolite-bearing North American rocks. Paleozoic geosyncline: I. Appalachian geosyncline, II. Arbuckle-Wichita geosyncline. III. Pacific embayment. IV. Cordilleran geosyncline (after Ruedemann, 1947).

(*B*) Ordovician sequences correlation, United States (Marathon Area), Australia (Victoria), and British Isles based on graptolite zones. Graptolites in respective zone numbers, figures 13.6, 13.7. After W. B. N. Berry, 1960.

(*C*) Graptolites: England and Europe. Map shows Stockdale Shales (Silurian, Llandovery) of Northern England and its European equivalents. After J. C. Marr, 1925. (Basal 50 feet of the Stockdale Shales contains 7 graptolite zones, upper 200 feet, 3 zones. Silurian, a time of great *Monograptus* abundance. G. M. Bennison and A. E. Wright, 1969. The Geological History of the British Isles. St. Martin's Press).

Distribution and Stratigraphic Zonation. Figure 13.5*A, B* graphically depicts North American and European distributions. The bulk of graptolite facies in North America occur in geosynclinal belts and in black carbonaceous shales (Athens, Canajoharie, Utica, Normanskill, Deepkill, and so on). These occurrences have been exhaustively reviewed by Ruedemann (1947), supplemented by Decker from the Athens shale (1952) and Berry's study of Marathon Region graptolite faunas (1960).

Texas graptolite zones correlate closely with those of the British Isles and Australia (Figure 13.5*B*). Decker (1952) showed that equivalents of the Athens fauna were distributed in a shallow, marginal North Atlantic Sea, that extended from Newfoundland to Great Britain and Europe. Equivalents also occurred southwestward from Alaska to China, Korea, Australia, and New Zealand (Benson et al., 1936)

and presumably via Antarctica with South America. Thus, from Trentonian time, graptolites formed world-wide reference zones.

In light of the above details, the superb utility of graptolites as index fossils will be apparent.

Paleozoic Sargasso Seas, Biotic Associates of Graptolites. Lapworth (1897) first suggested that the modern Sargasso Seas (Atlantic, Pacific, and Indian Oceans) might be viewed as the type in which Paleozoic graptolites lived, attached by rhabdosomes to masses of floating seaweeds. Ruedemann (1934), among others, examined this concept in great detail and agreed. Some 20 million tons of floating seaweeds occur in the Atlantic oceanic meadow alone. Associated biota include: fishes, crustaceans, worms, hydroids, bryozoans.

When Ruedemann compared the biotic

associates of Paleozoic graptolites to those of modern Sargasso Seas, he found "an astonishingly similar complex" with a few modifications. Differences included such items as: fishes had not yet appeared; there were graptolites instead of bryozoans; and small brachiopods were present. To sustain this view, one need but cite evidence for a pseudoplanktonic or epiplanktonic existence (various float-like structures) for juvenile graptolites.

Bulman (1957) decided that Lapworth's Sargasso Seas theory explained both the significance of the *nema* (a thread attaching the graptolite rhabdosome to floating seaweeds) and also the worldwide geographical spread of graptolites.

Anomalous graptolite occurrences in a given bed can be better understood when they are attributed to inswept plankton among which were graptolithines. It is also clear that a planktonic existence would not have been interfered with when the bottom fouled; upon demise, graptolite exoskeletons would sink downwards into this fouled bottom to be incorporated into the black mud. A variety of observations point to the demise of epiplanktonic graptolites in the area where they sank to the bottom (rather than considering them inswept from other sea areas). These include rhabdosomes occurred quite frequently in swarms aligned in parallel—suggesting gentle sinking and settling on an undisturbed bottom; occurrences in quiet waters evidenced by graptolites in sublithographic limestones; association of rainprints, suncracks and ripple marks on the very same bedding planes as graptolites—indicators of emerged substrates later covered by the eustatic rise in sea level, or of shallow, relatively low energy zones.

Some Adaptive Modifications.

Adaptive modifications in graptolites will be considered: flotation, buoyancy, and spinosity. These are, of course, meant to be only illustrative. Bulman (1970) carefully qualified buoyant mechanisms and need

to attach to floating seaweed: many forms lacked nemata; others (many biserial forms) had buoyant tissues; chiefly juveniles attached to discs. A planktonic rather than an epiplanktonic (that is, attached) existence is presently favored.

Flotation. The attachment thread (= *nema*) is one such adaptation. So-called *secondary air bladders* or *floats* that have been recorded for several graptolites, is clearly another (Ruedemann, 1934, plate 15). *Recurvature* of rhabdosome branches (stipes) have been interpreted as accommodations for attachment to a float or seaweed, as in *Didymograptus similis*.

Buoyancy. Graptolites of the family Retiolitidae seem to have lightened their exoskeletons. Whorls of horizontal spines formed a lacy network in place of a solid continuous periderm.

Spinosity. Graptolite spines arise at different times, at different sites, from different selective pressures: protective spines, apertural spines, lateral spines, and so on: the *virgella* occurring at the mouth of the sicula; bent branches with spines in their most convex parts; degenerate or modified thecae assume the form of spines in some species of *Monograptus*, and so on; development of large spines in "dwarfed" and gerontic forms; spines bear extra-thecal structures such as a network of fibers (*Lasiograptus*).

Numerous adaptive modifications not examined here include: *seriality* (uni-, bi-, quadriserial), *branching*, *attachment organ* variation, and *thecal polymorphism*.

Malformation and Body Repair.

Malformations are expected in the progeny of any breeding population due to genetic mistakes and also due to injuries. Thus, in graptolites, buds or thecae may be abnormal; one or another end of a stipe may be misformed; all or part of a stipe may be suppressed; the prosicula may be lost; the nema is often torn and regenerated (Figure 13.2C, *b*, *nr*); the periderm may be damaged and reconstructed (Kraft, 1926; Kozlowski, 1948; Bulman, 1955; Urbanek, 1958).

STAGE	ZONE
RICHMOND	15. Dicellograptus complanatus
MAYSVILLE	
	14. Orthograptus quadrimucronatus
EDEN	
TRENTON	13. Orthograptus truncatus var intermedius
WILDERNESS Black River in Part	12. Climacograptus bicornis
PORTERFIELD	11. Nemagraptus gracillis
ASHBY	10. Glyptograptus cf. G. teretiusculus
MARMOR Chazy	9. Hallograptus etheridgei
WHITEROCK	8. Isograptus caduceus
	7. Didymograptus bifidus
	6. Didymograptus protobifidus
	5. Tetragraptus fruticosus (3-and 4-branched)
CANADIAN SERIES	4. Tetragraptus fruticosus (4-branched)
	3. Tetragraptus approximatus
	2. Clonograptus
	1. Anisograptus

FIGURE 13.6 Graptolite derivative data—II. Graptolite zones—principal species. After W. B. N. Berry, 1960.

	STAGES	GRAPTOLITE ZONES	MARATHON	OUACHITAS	ARBUCKLES	NEW YORK	QUEBEC
UPPER ORDOVICIAN	Richmond	15. D. complanatus		Polk Creek	Sylvan	Queenston Oswego	
	Maysville Eden	14. O. quadrimucronatus	Maravillas	Bigfork	Fernvale Viola	Lorraine group Upper Utica	Magog
MIDDLE ORDOVICIAN	Trenton	13. O. truncatus var. intermedius	Woods Hollow	Womble	Bromide	Loyal Creek Nowadaga Canajoharie	
	Wilderness	12. C. bicornis				Normanskill	
	Porterfield	11. N. gracilis			Tulip Creek		
	Ashby	10. G. cf. G. teretiusculus	Fort Peña		McLish		
	Marmor	9. H. etheridgei		Blakely	Oil Creek	Deepkill (Zone 3)	Levis (Zone D)
	Whiterock	8. I. caduceus	Alsate				
		7. D. bifidus			Joins		
LOWER ORDOVICIAN	Stages not yet estab- lished	6. D. protobifidus	Marathon	Mazarn	West Spring Creek	Deepkill (Zones 1–2)	Levis (Zones A–C)
		5. T. fruticosus (3–br. and 4–br.)					
		4. T. fruticosus (4–br.)					
		3. T. approximatus		Crystal Mountain	Kindblade		
		2. Clonograptus			Cool Creek	Schaghticoke	
		1. Anisograptus		Collier	Mc Kenzie Hill		Matane

FIGURE 13.7 Graptolite-bearing correlation sequences in North America. For graptolites characteristic zones, Fig. 13.6. After W. B. N. Berry, 1960.

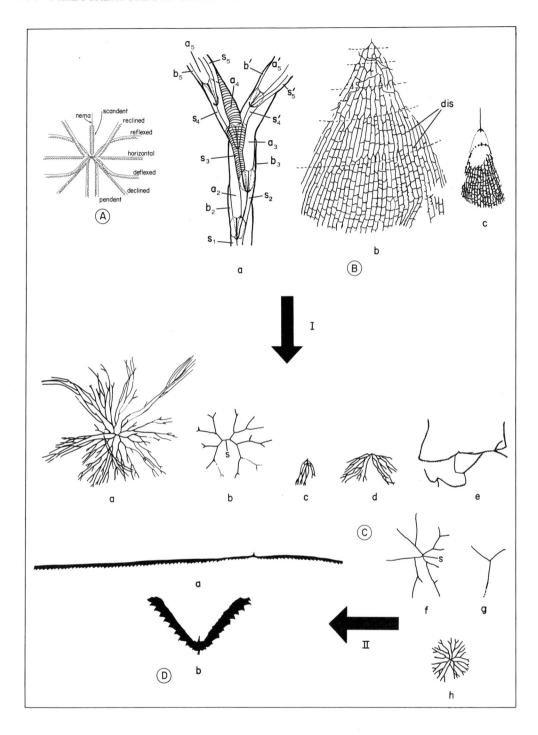

Colony Size. Table 13.1 shows the comparable population density on one rhabdosome (colony).

Evolution. Graptoloidea geological history was marked by three major evolutionary events: their origination (late Tremadocian), first biserial modification, scandent rhabdosomes (early Ordovician) and later into uniserial modification scandent rhabdosomes (Silurian time). For terminology, see Figure 13.8*A* (Bulman, 1954).

Some steps in sequence were the following.

I(*A*) 1. Ancestral dendroid stock (*Dictyonema flabelliforme*) gave rise to Graptoloidea through a transitional group, the Anisograptidae (see Figure 13.8*B, C*).

2. This transition involved (*a*) change in habit from sessile (attached to bottom substrate) to epiplanktonic (attached by nema to seaweed), (*b*) loss of dissepiments (see Figure 13.8*B, C*), (*c*) changed attitudes of branches (pendant to reclined), that is, change in the growth direction of stipe relative to the nema.

While most Anisograptidae retained the dendroid branch pattern, a marked change occurred in the unbranched *Triograptus* (Figure 13.8*C, g*).

These changes denote broad-band genetic variability first within the *Dictyonema flabelliforme* populations, and second, within the Anisograptidae populations —

its immediate descendants. Most Dendrograptidae genera to which *Dictyonema* belongs had a thickened nonthecate stem that ended in a rootlike process or disk for substrate attachment. However, *Dictyonema flabelliforme* from the Lower Ordovician of England was attached by a nema; that is, it was epiplanktonic and not sessile (Figure 13.8*B, c*). This indicates the change to disk from nema conferred a selective advantage because it came to dominate in the Anisograptidae. Further it suggests that changed sea conditions may have favored structures suitable for seaweed attachment.

The Anisograptidae clearly illustrate the seaweed attachment point (Figure 13.8*C*) for, even though they display great variability in the several genera as well as species in such items as attitude of stipes (pendant to reclined), and also vary in number of primary stipes, they all had evolved to the nema stage of attachment (epiplanktonic habit).

What environmental changes could exert selective pressures favorable to a genotype encoded for nema instead of disk? The Sargasso Sea meadows may have extended to new areas; shifting bottom substrates may have caused high mortality among attached dictyonemas. Some such factor(s) exerted a kind of selection pressure. However, Ernst Mayr reminds us (Chapter 12), the phenotype is a compromise, a compromise between

FIGURE 13.8 Further morphology and evolution data.

(*A*) Graptolite stipes relationship to nema. Terminology self-explanatory; after Lapworth, cit. Ruedemann, 1947.

(*B*)(*a–c*) Ancestral dendroid—*Dictyonema flabelliforme*, L. Ord., England: (*a*) Alternating Wiman triads (see text), *s*—stolotheca, *a*—autotheca, *b*—bitheca; note branching. (*b*) Dashes indicate approximate branching zones, dissepiments (= periderm strands connect adjacent branches in dendroid graptolites). (*c*) Same, attached by nema, L. Ord., England, X 1.

(*C*) Order Dendroidea, family Anisograptidae: (*a–b*) *Clonograptus* (*flexilis* and *tenellus* species); (*c–d*) *Bryograptus* (*kjerulfi* and *patens* species); (*e*) *Adelograptus hunnebergensis*; (*f*) *Anisograptus matanensis*; (*g*) *Triograptus canadensis*, (*h*) *Staurograptus dichotomas* (X 1/2 approx.).

(*D*) Order Graptoloidea, Didymograptidae: (*a*) *Didymograptus extensus*, two-stiped horizontal form, L. Ord., Quebec, X 1; (*b*) *Meandrograptus schmalensee*, two-stiped reclined form, L. Ord., Sweden, X 5.

Large arrows (I and II) indicate an evolutionary sequence from ancestral dendroid, D. *flabelliforme* figured in (*B*) to immediate descendants, the Anisograptidae, figured in (*C*), to the true dichographtidae, figured in (*D*). See text on "Evolution." Data from Bulman, 1954.

TABLE 13.1. Comparable Population Density in Graptolite Colonies (Per Rhabdosome)

	Number of Individual Zooids or Polyps/Colony	*Comments*
Dendroidea:		
Dictyonema flabelliforme (See Figure 13.8B, b)	35,000	Autothecal, bithecal; large rhabdosome; up to 64 stipes (Bulman, 1933)
Graptoidea:		
Dichograptid	3,000	All autothecae; large rhabdosome
Leptograptid Dicellograptid Diplograptid	100–200[a]	
Monograptus (Silurian)	50	
Monograptus leintwardinensis	10–17	

[a]Usually closer to lower limit.
Source. Data from Bulman, 1954.

contrary selective pressures.

Bulman (1933) reported the nema underwent modification in diplograptids and monograptids to develop a terminal float for truly planktonic existence. This contrasted with the epiplanktonic habit, nema attached to seaweed. Nema evolution, begun in *Dictyonema* populations, does not seem to have been an isolated event.

Among the Dendrograptidae (*Dendrograptus, Callograptus*), in some species there is a loss of dissepiments or a lack of anastomosed stipes. This tendency toward free stipes departing from the united condition reached full expression in the descendant Anisograptidae. It must have allowed a wider spread of the stipes (from the original clustered conditions). In turn, this would insure a greater volume of water around each moving tentacle of a zooid housed on any given stipe. Therefore greater oxygen/food supply would be available for each zooid.

However, in the Anisograptidae populations, the zooid selective advantage gained by living in freed stipes may have created new selective pressures by increasing the survival span of an individual zooid. This would be akin to a population explosion. Natural selection would then tend to favor reduction in colonial populations, that is, the number of zooids.

That would be reflected in the reduction of the number of branches (see also *Triograptus*, Figure 13.8*C, g*, and Table 13.1).

I. *A.* More than three decades ago, Bulman (1933) expressed the view that there was little evidence that any trends in graptolite evolution were adaptive responses to external conditions. He included such items as stipe reduction, scandent direction of growth trends, thecal elaboration. Rather than recourse to an external factor, he favored some internal factor. He did acknowledge that the manner of stipe reduction favored the symmetrical rhabdosome and this, in turn, related to the balance or equilibrium of the colony—an external factor. However, he doubted that "reduction itself" was a selective modification by which decrease in zooid density per colony would insure adequate food supply for the remaining zooids.

A simple answer is unlikely to subtle questions raised by Bulman. However, the internal factors to which he referred interpreted in light of modern knowledge (Mayr, 1963) as change in genotype, apply to further concepts: (*a*) given a species of $n \times 10^6$ individuals and on lowest estimates it "is bound to have a couple of mutations per (gene) locus in every generation except at most inert loci" (Mayr, 1963, p. 173); it follows that modified genotypes repeatedly become available in a given gene pool; (*b*) the differential perpetuation of genotypes is the work of natural selection (Mayr, 1963, p. 183); (*c*) the phenotype (rhabdosome with reduced branches, for example), is a resolution, a compromise among all selection pressures often contrary in their effects (Mayr, 1963, p. 194).

Clearly, internal (modified genotypes) plus external factors (selection pressures toward greater food/oxygen supply per zooid, or the need to maintain colony structure balance, and so on), both were responsible. There is, in other words, no one-to-one correspondence between a single factor and a trend in graptolite evolution. Rather, correspondence exists between a complex of factors and trends.

I.B. The dominant graptolite (Graptoloidea) of the Middle Lower Ordovician was the two-stiped, horizontal or reclined didymograptid (Figure 13.8*D, a, b*).

Selective pressures begun in the Anisograptidae population, reflected in the *Triograptus* genotype favored reducing the number of branches, therefore, population density. (Forms, like *Clonograptus*, nevertheless retained the primitive dendroid rhabdosome form.) The genotype, reflected in didymograptids, established branch reduction.

[An equally significant development, according to both Kozlowski and Bulman (1954) was the bithecae loss, the male zooids, in transition from anisograptids to graptoloid dichograptids.]

II. Origin of biserial scandent rhabdosome (early Ordovician).

[The oldest diplograptid, *Glyptograptus* with a novel structure—sigmoid thecae curvature—may have been the ancestral stock leading to true diplograptids such as *Climacograptus* (Figure 13.2*E*).]

III. Origin of monograptids with uniserial, scandent rhabdosome (somewhat above the base of the Silurian).

Bulman (1954), said, that once the monograptids arose (that is, after about half of graptoloid history had passed), for the next 30 million years, the uniserial, scandent forms dominated and remained so until the extinction of graptolites.

What kinds of changes were involved in going from II to III? Not only the loss of branches, that is, decrease in polyp/zooid population per stipe, but thecal rearrangements, thecal reduction and loss, and modification also occurred. To appreciate genetically theca change in succeeding graptolite populations, natural selection operation, the compromises between contrary selective pressures represented in given genotypes, one need only compare Figure 13.8*B, b* and Figure

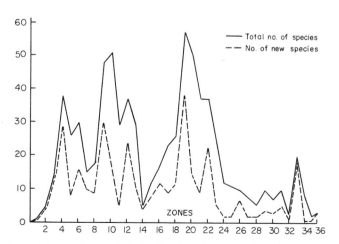

FIGURE 13.9 Graptolite species zonal distribution throughout Ordovician-Silurian time, British Isles. Continuous line = total number of species; broken line = number of new species appearing. Data on zonal species from Ellis and Wood (1902, 1908, Table 1). Zone 4—*Didymograptus extensus;* zone 9—*Nemograptus gracilis;* zone 12—*Dicranograptus clingani.* After Bulman, 1933.

1.6, a dendroid and monograptid graptolite. (See also *Dictyonema* and *Monograptus*, Table 13.1.)

It should be realised that scandent inclination persisted from II to III, only thecae seriality changed.

Unfortunately, graptolite specialists have yet to resolve many details of graptolite evolution. Despite limitations, the three major events in graptolite evolution, as viewed by leading students of these forms, delineates in broadest terms, the main trends.

Graptolite Species—Maxima/Minima.
Bulman (1933, Figure 7-9, and this chapter, Figure 13.9) found a regular tendency to periodic increase (maxima) followed by sharp decline minima) from analysis of plots of graptolite species. G. G. Simpson (1953, p. 155) explained this phenomenon in modern terms as follows: the number of graptolite species at any given time appears to have been a function of *relative adaptation*. The times of species bursts (maxima) were times of great fluidity in adaptive types. This was the *less* adaptive phase and was characterized by fractionation into sub-gene pools. By contrast, the times of species decline (minima) were times of restricted fluidity in adaptive types—that is, while few, the species populations were *more* adaptive. In the transition from maxima to minima, there was a weeding out of the less adaptive types by natural selection operative on sub-gene pools. Result: extinction of less "successful" groups or subpopulations.

QUESTIONS

Graptolithines

1. (a) Study Figure 13.1 and indicate features the five orders of graptolithines share in common; then ways in which they differ.
   (b) Place problematical graptolithines in the same context as (a).
2. Discuss the evidence pro/con graptolithine affinity to Bryozoa, Coelenterata, Hemichordata.
3. Identify: rhabdosome, cladium, stipe, virgula, nema, median septum, lists, reticulum, virgella, stolon, biseriality, polymorphism in thecal types, sicula.
4. Discuss sexuality in graptolites as viewed by Kozlowski/Bulman.
   (See Jenkins, p. 833.)
5. (a) Relate Lapworth's Sargasso Sea concept to worldwide graptolite zones. How did Bulman (1970) qualify this interpretation?
   (b) Discuss graptolite extinction.
   (c) Upon the extinction of the graptolites, what other marine creatures could have filled the ecologic niches they vacated?
6. Discuss the three major events in the Graptoloidea evolution.

REFERENCES

Graptolithines

Barrington, E. T. W., 1965. The biology of Hemichordata and Protochordata. Univ. Rev. Biol., W. H. Freeman & Co., 167 pp.

Benson, W. N., R. A. Keble, L. C. King, and J. T. McKee, 1936. The Ordovician graptolites of North-West Nelson, N.Z. Second Paper; with notes on other Ordovician fossils: Trans. & Proc. Roy. Soc., New Zealand, Vol. 65 (4), pp. 357–382.

Berry, W. B. N., 1965. Graptolites, in Handbook of Paleontological Techniques. W. H. Freeman & Co., pp. 103–109.

_____, 1964. Early Ludlow Graptolites from Presque Isle Quadrangle, Maine: J. Paleontology, Vol. 38 (3), pp. 587–599, 3 text figures.

_____, 1960. Graptolite Faunas of the Marathon Region, West Texas: Publ. 6005, Bur. Eco. Geol., 103 pp., plates 1–20.

Bohlen, Birger, 1950. The affinities of the graptolites: Bull. Geol. Univ. Uppsala, Vol. 34, pp. 107–113.

Bouček, Bedřich, 1957. The dendroid graptolite of the Silurian of Bohemia. Rozpravy. Ústředniho Ústavo Geol. Svazek XXIII, 294 pp., 39 plates.

Bulman, O. M. B., 1957. Graptolites, in Treat. Ecol. and Paleocol., Vol. 2, G.S.A. Mem. 67, pp. 987–991.

—————, 1955. Graptolithines, in Treat. Invert. Paleont., Pt. V (Graptolithina), pp. V3–V98; 1970, V (Revised).

—————, 1954a. Graptolite fauna of the *Dictyonema* shales of the Oslo Region: Norsk Geol. Tiddskr., Vol. 33, pp. 1–40, plates 1–8, Figures 1–13.

—————, 1954b. Graptolithina, in Status of Invert. Paleont., 1953, Bull. M.C.Z. (Harvard), pp. 201–215.

—————, 1933. Programme evolution in graptolites: Biol. Rev. Cambridge Philos. Soc., Vol. 8, pp. 311–334.

Bulman, O. M. B., 1931. South American graptolites: Arkiv. f. Zool., Vol. 22A (3), pp. 1–111, plates 1–12, Figures 1–41.

Decker, C. E., 1957. Bithecae, Gonothecae, and Nematothecae on Graptoloidea: J. Paleontology, Vol. 31 (6), pp. 1154–1158, plates 145–148.

—————, 1952. Stratigraphic significance of graptolites of Athens Shale: A.A.P.G. Bull., Vol. 36 (1), pp. 1–145.

Eisenack, A., 1952. Uber einige Funde von Graptolithen aus ostpreussischen Silurgeschieben: Zeitsch. Geschiebeforsch., Vol. 18, pp. 29–42.

Ellis, G. L., and E. M. E. Wood, 1914–1918. Monograph of British Graptolites. Pts. I–XI. Palaeontgr. Soc., London, 539 pp., 52 plates, 359 figures.

Flügel, Helmut, 1963. Graptolithen aus dem mittleren Ordovicium von Nord-Syrien: N. Jb. Geol. Paläont. Abh., Vol. 118 (1), pp. 21–26, Table 1.

Hsü, S. C., 1934. The graptolites of the Lower Yangtze Valley: Mon. Nat. Research Instit.: Geol. Ser. A, Vol. 4, pp. 1–106, plates 1–7, Figures 1–36.

Kobayashi, T., and T. Kimura, 1942. A discovery of a few Lower Ordovician Graptolites in South Chosen with a brief note on the Graptolite Zones in Eastern Asia; Japan. J. Geol. Geogr., Vol. 18 (4), pp. 307–312, plate 29.

Kozlowski, Roman, 1966. On the structure and relationships of graptolites: J. Paleontology, Vol. 40 (3), pp. 589–501.

—————, 1953. Étude d'une nouvelle espèce du genre *Corynoides* (Graptolithina): Acta Geol. Polonica, Vol. 3, pp. 193–209 (Polish), Conspectus: pp. 68–81 (French).

—————, 1948(1949). Les graptolithes et quelques nouveaux groupes d'animaux du Tremadoc de la Pologne: Palaeont. Polonica, Vol. 3, pp. 1–235, plates 1–42, Figures 1–66.

—————, 1947. Les Affinités des Graptolithes: Biol. Rev. Cambridge Philos. Soc., Vol. 22, pp. 93–108.

Kraft, Paul, 1926. Ontogenetische Entwicklung und Biologie von *Diplograptus* und *Monograptus*: Paleont. Zeitschr., Vol. 7, pp. 207–249.

Marr, J. E., 1925. Conditions of deposition of the Stockdale shales: Quart. J. Geol. Soc. London, Vol. 81 (2), pp. 113–133, Figures 1–3.

Rigby, J. K., 1958. Lower Ordovician graptolite faunas of Western Utah: J. Paleontology, Vol. 32 (5), pp. 907–917, plate 118, 1 text figure.

Ruedemann, Rudolf, 1947. Graptolites of North America: G.S.A. Mem. 19, 498 pp., plates 1–92.

—————, 1934. Paleozoic plankton of North America: G.S.A. Mem. 2, 97 pp., 26 plates, 6 text figures.

Thorsteinsson, Raymond, 1955. The mode of cladial generation in *Cyrtograptus*: Geol. Mag., Vol. 92, pp. 37–49, plates 3–4, Figures 1–4.

Urbanek, Adam, 1958. Monograptidae from erratic boulders of Poland: Palaeont. Polonica, #9, 96 pp., 68 text figures, 7 text plates, 5 plates.

Whittington, H. B., 1954. A new Ordovician graptolite from Oklahoma: J. Paleontology, Vol. 28 (5), pp. 613–621, plate 63, 13 text figures. (Three-dimensional specimen of *Orthoretiolites.*)

[Foucart, M. F., and Ch. Jeuniaux, 1966. Paleóbiochimie et position systématique des graptolithes: Ann. Soc. Royal Zool. Belgique, Vol. 95, pp. 39–45. Towe, K. M., and A. Urbanek, 1972. Collagen-like structures in Ordovician graptolite periderm: Nature, Vol. 237, pp. 443–445.]

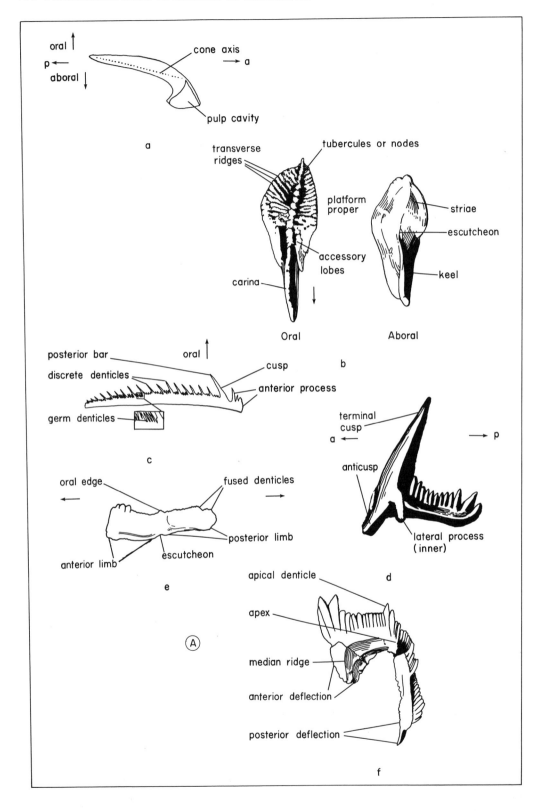

FIGURE 13.10 Essential Megascopic Structure and Orientation of Conodonts. [After R. O. Fay, 1952]. Escutcheon = a basal attachment scar, navel or pit. The preferred term is "pulp cavity" (Hass, 1962).

PART II. CONODONTS

Introduction

More than a century ago (1856), Pander described samples from the Lower Ordovician glauconitic sand that outcropped south of St. Petersburg, Russia; a curious group of microscopic objects that were toothlike and platelike. He concluded that they "closely resembled fish teeth in external form" (Pander, cit. Hass, 1962; Fay, 1952).

Since Pander's discovery, conodonts have been widely reported, particularly from the Paleozoic but also Triassic and Cretaceous. They are known from North America, England, and Europe (Germany, Sweden, Czechoslovakia, Poland, Russia, Belgium, Denmark). Available are a few published notices of conodonts in Australia and Africa (Egypt, West Africa). It is likely that conodonts occur in the Paleozoic and possibly too in parts of the Mesozoic in all continents.

Are conodonts really fish teeth? Since Pander's time, the following conodont affinities have been suggested: mollusks (gastropods, cephalopods), arthropods (crustaceans), worms (annelids), algae, and chordates (chiefly fish). Probably a thousand published papers discuss conodonts. Of these, six hundred had been published by 1952 (Fay, 1952). The reason for great productive conodont research arises from their excellent utility in biostratigraphic studies (zonation), plus the continuing dilemma of their true affinities.

Essential Structure of Fossil Objects Called Conodonts

Megascopic Structure. Figure 13.10 illustrates simple cone, bar (bowed and arched), blade (straight and arched), and platform-type conodonts. Most of these objects are toothlike, have either a pulp cavity or attachment scar (that is, they had been attached to some mooring or supportive anatomical animal part and are visibly varishaped. Essential terminology and orientation (oral, aboral, anterior, posterior) is shown. Cusps, tubercles, and denticles face the oral side.

Lindström (1964) pointed to certain conodont tendencies: (*a*) *increased complexity* (simple cones to compound and then to platform types), (*b*) *expansion of base*: accompanying (*a*), the expanding base permitted increasing numbers of denticles or tubercles to be secreted.

Do all these different conodont types belong to the same animal? Pander expressed doubts. Could different groupings of some or all of such conodonts represent different parts of the same animal (that is, have served functions or have been components of one functioning apparatus)? Subsequent discussion of natural assemblages, and consideration of some concepts and reconstructions (Hinde, 1879; Schmidt, 1934; Lindström (1964), show this is possible in some instances.

Could some objects be unrelated to the others, that is, belonging to taxonomically separate groups within the same phylum? Different phylum? This inquiry will be further explored with more data reviewed.

Notice (Figure 13.10) that some denticles are fused, others sharply defined, some peg-like, others fang-like (= cusp); in platform types, tubercles or nodes are present instead of so-called teeth or denticles. If we view several of these objects as probable parts of a single functional apparatus, then the variation and denticulation (size, kind, number, thickness, fused or free) reflect apparent adaptation for a partial function.

Microscopic Structure. Conodonts fine structure can be studied by transmitted light either in thin section (Hass, 1941; Schwab, 1965) or in three-dimensional specimens rendered translucent by chemical treatment (clorox, and so on).

Lamellae. All conodonts presently known (including the so-called "fibrous" conodonts) have a laminated structure (Hass, 1941, 1962; Lindström, 1964; Schwab, 1965). Accretion of lamellae (secreted by surrounding cells) build their structure about the apex of the pulp cavity (Figure 13.11*A–C*). The aboral side is the termination zone of lamellae free edges (Figure 13.11*G*).

Discussed elsewhere, a transition series of species with slightly variant dental plans often links up two apparent extreme end member species (Lindström, 1964). Diversity of denticulation for distinct growth stages within a given species is also marked (Scott, Collinson, Rexroad, 1960).

Interlamellar Layers and Thickened Lamellae. The *interlamellar area* has an apparent tubelike appearance (Figure 13.11*A, F, G*) but in simple cones, will appear cone-shaped (Figure 13.11*H*). These areas open to the exterior, generally along the aboral midline, in lateral blade, bar, or platelike forms.

If separation of adjacent lamellae had been localized, then distinctive conodont features would have resulted, that is, pus-

tules, ridges, or nodes (Hass, 1941, 1962). One might speak of "rapid growth" under the following circumstance: when a succession of interlamellar spaces regularly occurred between growth lamellae along the growth axis of a denticle/cusp (Figure 13.11*A, B,* and so on). Effectively, this situation may indicate growth leaps over spaces.

Apparently involved in such rapid growth was the arching of surrounding cells *above* the last lamella. This would, to some extent, spatially redispose the three-dimensional organic molecules with their several loci; at these sites biocrystals formed. The conical shape would have been determined in this way.

Individual laminae probably represent discrete time intervals of ontogenetic growth (portions of a year) and interlamellar spaces, possibly denote single year spans or smaller units. Although these suggestions are highly speculative, other fossil material with analogous kinds of annular and periodic markings, lead the writer to infer that conodonts are little geochrons and that they carry a decipherable record of their life span.

Lamellae thickening is thought to be the mode of differential growth in many simple, as well as, compound conodonts. However, both successive interlamellar spaces and thickened lamellae may occur in different parts of a given conodont (carina might form by thickened lamellae; blade might form a sequence of interlamellar spaces). The polygnathids are an example (Lindström, 1964, p. 17).

Nonlaminated Layer. Schwab (1965) reported that several Middle Ordovician forms (*Zygognathus, Cordylodus,* etc.) revealed a two-unit construction: outer laminated (ca. 100 μ thick) and inner nonlaminated lining (20 to 80 μ thick) in thin sections set parallel to the growth axis (Figure 13.11*B*). A similar construction was seen in other genera (*Microcoelodus*) viewed in transmitted light. Is this nonlaminated layer primary or the result of postdeposition? Schwab thought it was primary.

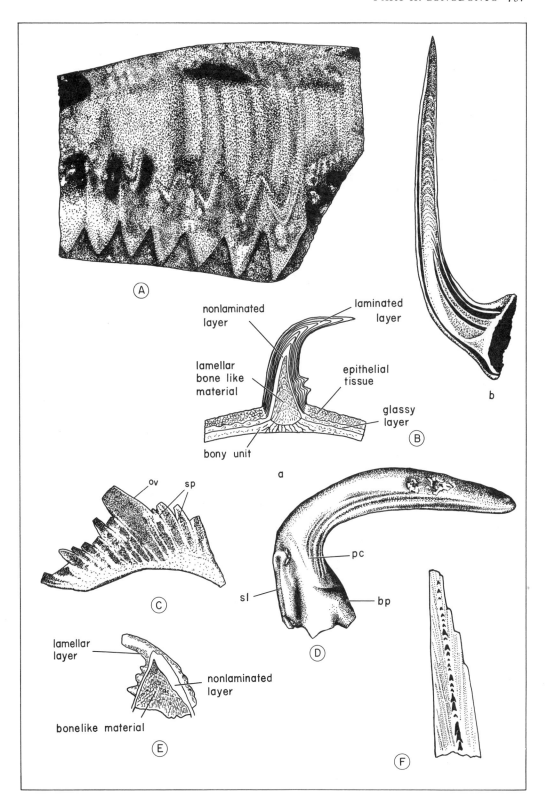

nonlaminated layer

laminated layer

lamellar bone like material

epithelial tissue

glassy layer

bony unit

a

b

B

ov sp

C

pc

sl

bp

a

D

lamellar layer

nonlaminated layer

bonelike material

E

F

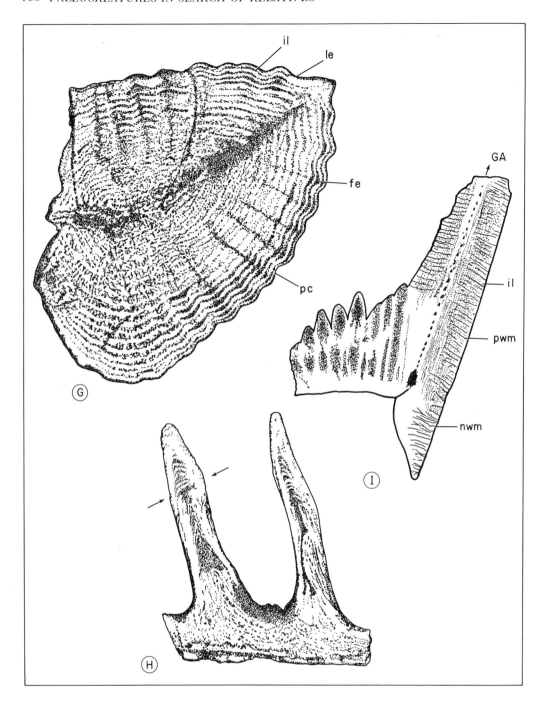

FIGURE 13.11 Fine Structure and Other Morphological Features of Conodonts. (A) *Polygnathus* sp., Hushpuckney shale, Kansas, longitudinal view of complete specimen, cloroxed, laminar structure (echelon pattern in denticles; interlaminar spaces are black in denticles and represent longitudinal gray areas in the blade) X 18. (B)(a) simple cone attached to a dermal (?) plate, showing laminated and non-laminated layers (other labels as interpreted by K. W. Schwab) X 30; (b) *Oistodus lanceolatus* showing lamellar structure, Ord., X 100. (C) Lateral view of compound conodont showing suppression of parts: *sp* — suppressed denticles, overgrowth by "*ov*" (*Subbryantodus sp.*) X 55. (D) Basal plate of Scolopodus, *sp.*, lateral view, X 20. Note slit (*sl*) along interior side of basal plate (*bp*). *pc* — basal plate fills pulp cavity (*pc*), cf. this figure to "*B*". (E) *Zygognathus* showing laminated, non-laminated areas and what Schwab has interpreted to be "bone-like material" in the pulp cavity, X 25. (F.) *Lonchodina sp.*, same beds as "*A*", tip of cusp, X 200. Black = interlaminar space along growth axis; Stipple = rays of "white matter" (see text).

(G) *Siphongnathus sp.*, Miss., Okla., X 60. horizontal view, *pc* — pulp cavity, *fe* — oblique lines are free edges of lamellae on aboral side, *le* — lamellae, *il* — interlaminar space (black) (H) *Lonchodus*. Longitudinal view. Same horizon as "*G*", X 100. Arrow indicates injury site. Repair (rejuvenation) indicated by different kind of growth lamellae above arrow-zone, X 100. (*I.*) *Neoprioniodus sp.*, Miss., Chappel Limestone, Texas, X 65, showing cancellate structure ("white matter" in megascopic view), *nwm* — "white matter" — rays or their canals normal to growth axis (GA), *pwm* — "white matter" or rays (stippled) almost parallel to growth axix (GA), *il* — interlamellar spaces.
Illustrations adapted from Schwab, Hass and others.

"White Matter" or Cancellate Structure.

Megascopic inspection of any conodont collection will reveal that numerous, although not all, cusps and denticles appear shiny, opaque white in reflected light. Lindström (1964) aptly termed this condition "white matter." With conodonts, sectioned and studied in transmitted light, the "white matter" is not seen as a homogeneous color mass. Rather, it appears to by an array of fine structures. Pander and Hass called this the *cellular* or *cancellate* structure. The cancellate denticle appearance in transmitted light arises because the vari-sized and shaped tubules (fine canals or hollow spaces) occur in layers whose long axes have two orientations, roughly parallel and normal to the long axis of the denticles (Figure 13.11*I*) (Hass, 1941; Beckmann, 1949).

These so-called hollow tubules, cellules, fine canals, may have a dual origin. Some may be minute hollows but most may include "particles of unknown composition" (!) Lindström argued that if "white matter" were a mesh of voids, there would be normal channelways for fluids and hence, one would expect the denticles also to be mineralized. However, they rarely are. "White matter" is apparently impermeable to water-soluble colors.

Lindström thought the "ghost conodont" (that is, what remains after acid digestion) might be largely organic. It probably is and it might very well enclose miniscule translucent particles. (Many workers have recovered amino acids from conodonts.) The evidence points to all possibilities rather than preferentially to one. Perhaps voids occur where original translucent particles, held in an organic mesh, had dissolved.

Other aspects pertain to the "white matter": (*a*) Evidence points to the formation of "white matter" by resorption. (*b*) Evidence indicates its formation during growth. (*c*) It is present in the oldest known conodonts (Müller, 1959). (*d*) Blades of platform conodonts — an expansion of the basal anterior edge with well-developed interlamellar spaces — tend to contain little "white matter". [Lindström (1964) considered this last point significant.]

Composition

Chemical. Composition given in percent: CaO — 48.05, P_2O_5 — 34.96, insoluble — 3.96, remainder [(undifferentiated iron oxide (Fe_2O_3), fluorine (F_2), CO_2, organic, and other matter)] — 13.03 (Ellison, 1944; Hass, 1962). Calcium phosphate is the chief component of conodonts.

Fluorine content also is of interest. It is well established that phosphates (apatites) in fossil bones absorb fluorine, and that the fluorine content in the cited instance increases with the geological age of the bones. Similarly in conodonts. The fluorine amount incorporated in a conodont after it reached the seafloor and was held in a bottom sediment matrix, is a function of its local abundance and of rock porosity (Roy Phillips, cit. Rhodes, 1954, p. 429; see also Branson and Mankin, 1964). However, trace quantities of fluorine could be primary.

Mineralogy. X-ray diffraction and spectrographic studies both confirm the chief conodont mineral is in the apatite group (that is, the apatite isomorphous series, hydroxy-carbonate-fluor apatite).

Hass and Lindberg (1946) placed conodont composition in the dahllite-francolite isomorphous series, an assignment that Roy Phillips contested. Schwab (1965) attributed slight differences in reported mineral composition of conodonts (for example, dahllite, fluorapatite, francolite, lewistonite, and so on) to possible differences in the composition of laminated and non-laminated layers.

From X-ray diffraction studies of several conodonts, Branson and Mankin (1964) concluded that the ratios obtained (c/a) corresponded to that given for the carbonate-fluorapatite of fossil vertebrates (Osmund and Sawin, 1959). They favored a closer affinity to francolite in the dahllite-francolite isomorphous series — which confirmed Hass and Lindberg (1946).

X-ray diffraction studies have shown that conodont apatite is in crystallite form ranging from 0.1 to 0.01 mm according to Roy Phillips. Crystallites in each lamella were found to be oriented in the main direction of conodont growth (that is, the c-axis of each crystallite in its location in a given lamella, is so oriented). Thus, if thousands of c-axes were plotted, accounting for each site, the general orientation would be discerned. Although crystallites are superimposed and align-

ment is inexact, Hass and Lindberg inferred from the low birefringence that it could be explained if the crystallites were "actually" parallel to the growth direction (that is, crystallites at right angles to the lamellae). (For extinction figures, see Hass, 1962, Figure 17, or Hass and Lindberg, 1946.)

Histology. A histological study (microtome sections) by Gross (1954) led him to conclude that conodonts grew by *outer* instead of inner deposition.

Although Figure 13.12 shows a continuum of cells surrounding the growing conodont, one must return to sites numbered 4 and 5 and surround these areas by cells in order to initiate apatite crystallite nucleation. Even then, probably, restricted and specific loci in organic molecules comprised secreting cells whereby crystallite nuclei formed preferentially. In this way, cells built up the initial lamellae while another suite of such sites in the organic molecules of the cells (CIII in Figure 13.12) secreted the basal filling.

Basal Filling or Plates. The aboral side of many (but not all) conodonts bears a basal filling in the pulp cavity. This is called the basal plate, cone filling, basal cone. The continuity of cells (Figure 13.12 from CIII to CI) suggests conodont secretion took place from specialized cells that in phylogeny were solely devoted once to secretion of the basal filling. However, a large body of negative evidence indicates that basal plate/filling and conodont are *not* of the same composition and structure, in some instances, (Figure 13.11*D*). [Gross (1954) concluded that conodonts were the only preservable hard parts of the conodont-bearing animal. However, the nature of some basal fillings as, for example, the anterior slit in the basal plate (cone) of *Scolopodus* sp., suggests downward continuity with still other structures.]

McConnell (cit. Stewart and Sweet, 1956) reported such marked difference in composition: the basal plate's X-ray pow-

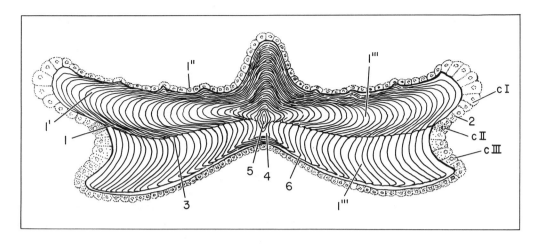

FIGURE 13.12 Mode of cellular secretion of the conodont *Palmatolepis*. Schematic reconstruction shown in cross-section. Adapted from W. Gross, 1960. CI–III = cells; CI = conodont-secreting cell (that is, biocrystallites of the apatite series secreted by such cells build up individual lamellae); CII = postulated resorbing cell – note counterpart on opposite side; CIII = cells secreting basal filling. 1, 1′, 1″, 1‴ = growth lamellae each composed of thousands of crystallites, 0.1 − 0.01 mm, with *c*-axis at right angles to given lamellae (see text). 2 = lower surface of platform. 3 = basal attachment surface. 4 = initial portion of basal filling (see text). 5 = initial portion of conodont, that is, initial lamellae. 6 = sharply refolded lamella of basal filling.

der-diffraction pattern was found to be identical with collophane, that is, crypto-crystalline carbonate-hydroxyl-fluorine apatite. Lindström (1954) found two different basal plate compositions of Swedish Ordovician conodonts: identical to conodonts to which it was attached; or possibly chitinous. Rhodes (1954), following Phillips's X-ray diffraction studies of Silurian conodonts, reached the conclusion that conodonts and basal plate material were identical – in agreement with Lindström, in part.

Hass (1962) presumed that basal plate growth and conodont proper synchronized, that is, occurred simultaneously. This is certainly plausible in light of the cellular reconstruction by Gross. However, Gross had postulated that conodont material-apatite group crystallites – was resorbed *before* basal cone secretion. Lindström modified this by suggesting alternates. (*a*) Resorption entirely predated conodont lamellae formation; (*b*) or it was simultaneous in the outer zone while conodont lamellae were deposited in the inner zone. He ruled out (*a*) but allowed for (*b*) – if the existence of theoretical resorbing cells could be proved. Alternative (*b*) would also accommodate the Hass concept noted above.

Lindström (1964) found that his basal plates yielded a greater residue than the conodonts. This consisted of black/brown probably organic material, which, in 1954, he thought might be chitinous. (See Chapter 15. Alleged chitinous material, in conodont basal plates subjected to tests, can settle the matter.)

Analysis of Harding sandstone conodonts led Schwab (1965) to conclude that basal filling was cartilagelike, had areas of cell-like structure as well as lamellae. He referred to this basal filling as "bone-like" – a statement that prejudges the case for affinities! [Vertebrate paleontologists and anatomists almost unanimously reject conodonts as parts of any known or inferred vertebrate. Melton and Scott (1973) favor protochordate affinites – also questionable.]

Repair and Suppression. Much light is afforded on conodont growth by the available evidence on repair (regeneration by secretion or fusion), and suppression of "germ denticles" (Figure 13.11*C*, *H*). Thin section studies have shown numerous germ denticles growing along a given growth axis were abandoned due to injury, some pathological condition, or an error in the genetic code. These denticles were overgrown by new ones that often were larger and followed a different growth axis. Localization of biocrystal-forming sites in the organic molecules of individual cells seems indicated.

Hass (1962) was of the opinion that conodont growth represented a balance between suppression of parts (germ denticles, for example) and restoration of parts. The trend, as he saw it, maintained the maximum number of growth axes along the growing edge.

There are alternate explanations. Rhodes (1954) thought that regeneration was post-functioning, that the so-called fine canals (Figure 13.11*I*) might have brought mineralizing fluids for bio-crystallization to repair sites. Further, regeneration was viewed as abnormal deformation during growth.

Suppressed parts (possible primitive characteristics) are probably, in some instances, clues to phylogeny and especially occur early in ontogeny. It might be valuable to reconstruct the appearance of several conodonts with their germ denticles as contrasted to their so-called re-generated mature states. The suppressed denticles might very well be atavistic to an earlier evolutionary stage. If true, that raises the question: are all so-called suppressed parts really injuries? Do they represent damaged or malformed structures?

Natural Assemblages. Conodonts are known both as disarticulated, discrete objects or, as apparently complete, articulated, or naturally arranged (although sometimes disarranged) assemblages of the conodont-bearing animal (Figure 13.13). Table 13.2 shows which components (described from isolated pieces, the so-called "form genera") comprise selected natural assemblages.

Important generalities arise from study of all the known natural assemblages (Rhodes, 1954, 1962). Component genera, for example, occur as *paired pieces* (mirror images, right and left forms). As shown in Table 13.2, the same form genus *Hindeodella*, may be present in several distinct natural assemblages. Where this is not so, "a similar genus of same structural type" is sometimes present. (See "Transitional Types" under "Derivative Data," this section.) The number of individual component parts or pieces in a natural assemblage varies from 12 to 22; extensive "specific variation" has been

TABLE 13.2. Types of Conodonts and Form Genera that Comprise Selected Natural Conodont Assemblages

Component Conodont Type	Lochriea[a] Scott	Lewistownella Scott	Westfalicus Schmidt	Scottognathus Rhodes
Elongated Blades	4 pairs *Hindeodella* ———————————————————————————————————————→			
Arched Blades	*Prioniodella*	*Ozarkodina* ————————————————————————————→		
Pick-shaped Blades	*Neoprioniodus* ————————————→ *Synprioniodina* ————————————→			
Platform Blades	*Spathognathodus*	*Cavusgnathus*	*Gnathodus*	*Streptognathodus* or *Idiognathodus*

[a]The largest natural assemblage, 9.0 mm± length, 2.0 to 3.0 mm width.
[b]Arrows indicate recurrence of same component form-genus in respective natural assemblages. Thus, four pairs of *Hindeodella* appear in all assemblages shown; *Ozarkodina* in three; and so on.
Source. After Rhodes, 1954, 1962.

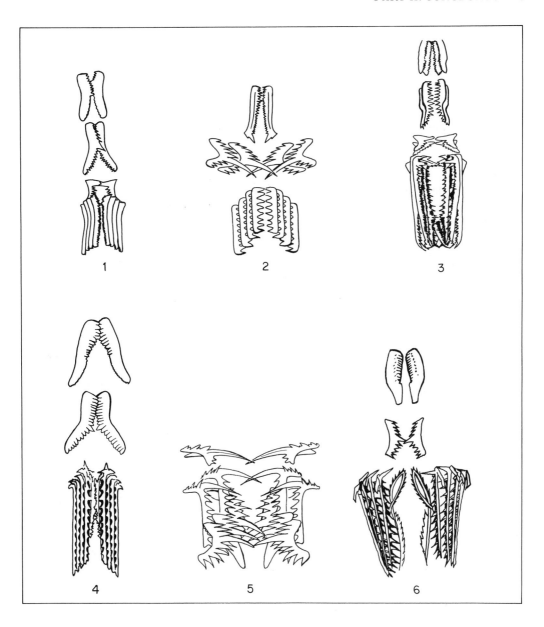

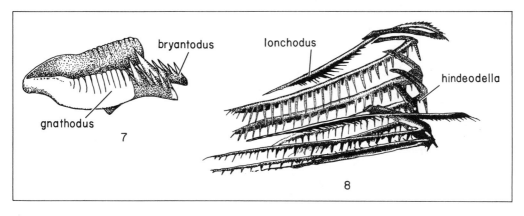

FIGURE 13.13 Natural conodont assemblages. (1) *Lewistownella* Scott, Miss., ca. X 20. (2) *Illinella* Rhodes, Penn. (3) *Lochriea* Scott, Miss. (4) *Scottognathus* Rhodes, Penn. (5) *Duboisella* Rhodes, Penn. (2–5 ca. X 10). (6) *Westfalicus* Schmidt. (7–8) Reconstruction of (6) by Schmidt (1934); (7) interpreted to be mouth and gill arch structures of a placoderm (*Pisces*); *Gnathodus,* mandibles; *Bryantodus,* teeth on hyoid arch; (8) interpreted to be ceratobranchial, epibranchial gill and arch structures of *Lonchodus* and *Hindeodella,* X 30. Note the actual find is shown in Figure 6. Illustrations after Rhodes, and other authors.

observed in components of several form genera (Rhodes, 1952). As found on bedding planes, both arrangement and alignment are such that a given natural assemblage was apparently elongate and positioned antero-posterior in the living conodont-bearing animal (thought by many investigators to be vermiform).

No basal filling or any bonelike substance yet has been reported attached to one or more components in a natural assemblage. Rhodes considered this observation singular in light of the apparently undisturbed condition of preservation of such assemblages.

Clearly, the problem, of affinities, as well as modern concepts of classification, have been importantly influenced by these data. Many discrete (disjunct) conodonts belong to assemblages, either known or inferred, as a component of a pair (right or left piece) or as an isolated piece. Others may have been located at different sites within the animal and not have been part of any assemblage. Such isolated forms may have served a function distinct from that of the assemblage.

Affinities

A confusing, often contradictory welter of hypotheses have been put forth to explain or to decipher conodont affinities. Were conodonts part of some known or unknown invertebrate? Vertebrate? Possibly plant? Various investigators have surveyed the total evidence (Rhodes, 1954; Hass, 1962; Lindström, 1964) without anything conclusive forthcoming. However, since the whole debate may contribute to future discoveries, it is briefly reviewed here. (Many years ago, in teaching a course on "Comparative Anatomy of Vertebrates," all the available postulates on conodont affinities

were explored by the writer. Specimens were dissected from every phylum, including the invertebrates, in search of possible affinities of selected hard parts or accessory structures for conodonts. The results were negative. No such affinities were discovered.)

Postulate 1. The conodont-bearing animal was a mollusk. Some workers have compared the teeth of gastropod radulae (Chapter 8) or spines of other mollusks to conodonts. It is true that a few conodonts do bear resemblances but reference to natural assemblages and isolated pieces that do not resemble radula teeth eliminates the comparison's validity. The composition, calcium phosphate for conodonts, vs. chitin for radulae, supports rejection. Against the mollusk theory is the general absence of molluscan remains in many conodont-bearing beds.

Postulate 2. The conodont-bearing animal was an arthropod. This view, held at the end of the nineteenth century, has not been heard since. Conodonts, it was suggested, were the tips of segments of the trilobite exoskeleton, possibly claws or some external part of crustaceans, or spines of an undetermined arthropod. Relative to trilobites, extinct by the end of Permian time, the occurrence of Triassic and Cretaceous conodonts of the same type as in the Paleozoic in itself rules out this postulate. Study of indicated arthropod structures and comparison with conodonts or conodont assemblages reveals no detectable affinities.

Postulate 3. The conodonts were not animal in origin but rather were secretions of algae. (Fahlbusch, 1963; and see Lindström, 1964, p. 121). This postulate set forth in a note is inadequately documented. A specialist on conodonts, Ziegler, studied Fahlbusch's material and, as reported in Lindström, found the conclusions untenable. Natural assemblages (Figure 13.13) alone seem adequate to refute this postulate.

Postulate 4. The conodont-bearing animal was an annelid worm.

(See Chapter 9) Rhodes (1954), after a review of arguments previously set forth for annelid affinities, was disposed favorably to this interpretation because of "general lack of wear, form, size, and assemblage arrangement of conodonts."

Other than those already given, the chief arguments for annelid affinities are: (a) the lack of wear suggests that conodonts did not function as teeth but as grasping components as in polychaete worm jaws; (b) conodont attachment by posterior or medial portion is wormlike rather than fishlike; (c) certain polychaete worms secrete calcium phosphate tubes. According to Du Bois (1943), this would place annelid worm jaws as "physiologic homologues" of such tubes. It also satisfies the necessity that the conodont-bearing animal must have been a calcium phosphate secreter. (d) Natural assemblages indicate a vermiform, soft-bodied conodont-bearing animal. (This analogy is misleading, however, since worm-like types are known in almost all phyla. For example, see *Pogonophora*, a hemichordate.)

Proponents of this view include: Owen (1860), Zittel and Rohen (1886), Grabau and Shimer (1912), Scott (1934), Du Bois (1943), and Rhodes (1954), among others. On the mistaken view that conodonts were chitinous, Denham (1944) suggested conodont affinity to the copulatory structures in worms such as Nematoda and Turbellaria. The similarities are negligible; the different size ranges, composition, and irrelevant arrangement of natural assemblages rule against this postulate which has gained no independent support.

The writer's own researches along with others indicate a very large gap between scolecodonts and conodonts. The gap covers differences on every level: composition, shape, natural assemblage arrangement and components, detailed maxilla morphology, microstructure (Tasch et al., 1961; Schwab 1965), absence of basal fillings, upper and lower jaw components in scolecodonts, that is, mandible vs. upper jaw apparatus with no equivalents in conodont assemblages. This concept also fails to account for "cartilaginous" or other basal fillings which have no homologues in annelid or other worms.

Postulate 5. The conodont-bearing animal was a lophophorate.

The natural conodont assemblage, according to this view, was arranged in the tentacles in a way to serve, aided by ciliary movements, a food straining and water current-channelling function (Figure 13.14). (1) The postulate does not account for basal fillings, particularly the bonelike or cartilaginous type—known in disjunct conodonts but not in natural assemblages; (2) it does not show how natural assemblages could have been preserved undisturbed with not a single instance of a tentacle impression or trace; (3) the model lacks economy in that multiple tentacles each bearing a natural assemblage, are visualized. (By contrast, but with no better supportive evidence, Du Bois' view of a given natural assemblage with buccal and pharyngeal apparatus of an individual worm is simpler.)

S. C. Morris [Paleontology, 1976, Vol. 19 (2), p. 99] noted elements in a newly described Burgess shale lophophorate that he thought were the equivalents of certain Cambrian conodonts.

Postulate 6. The conodont-bearing animal was a chordate

(fish, or some undetermined vertebrate). Conodonts have been referred to hagfish, sharks, placoderms (Schmidt, 1934). Disjunct conodonts, arranged in multiple ways much like a jigsaw puzzle, resemble the hyoid

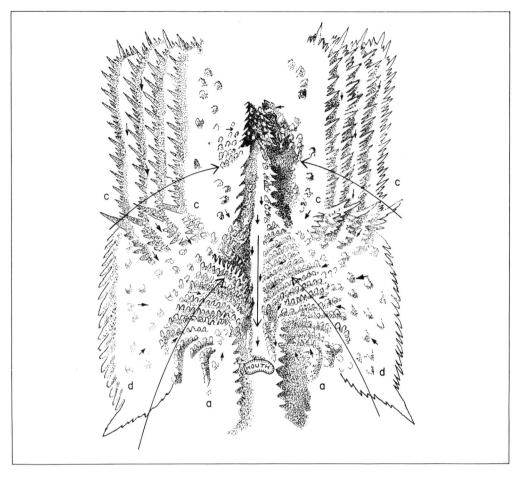

FIGURE 13.14 The conodont-bearing animal as a soft-bodied lophophorate with conodont assemblage in functioning position in a tentacle (Schematic). After Lindstrom, 1964; *c* — position of *Hindeodella* element, *a* — position of *Prionodus-Spathognathodus-Polygnathus-Gnathodus* elements, *b* — location of *Ozarkodina* elements along groove, *d* — *Neoprionodus-Synprioniodina* elements. Small arrows: direction of ciliary movements. Long arrows: water circulation. Follow passage through *Hindeodella* food straining elements aided by ciliary movements to *Ozarkodina* elements along groove to mouth surrounded by *a*-elements. Expelled water might be partially recirculated by mouth many times.

arch, brachial gill apparatus, spines, scales, or other parts of one or another fish. Chemical composition, noted earlier, of conodonts and vertebrates have been compared and found to show more than a chance relationship. Branson and Mankin, without presenting "new" evidence, expressed a strong conviction that conodonts probably belong to some fish such as the Chondrichthys. (If this last interpretation were true, one should be able to find independent verification in one or another fish. However, this has not as yet been forthcoming.)

Schmidt (1934), Beckmann (1949),

Gross (1954), Schwab (1965), among others, hold that conodonts have clearcut chordate affinities. The cartilaginous or lamellar bonelike basal fillings usually cited support such views. Suppose one tentatively agreed that conodonts do have affinities for, say the Placodermi (Geologic range: Upper Ordovician to Upper Permian), then how can one account for conodonts older and younger than any placoderm fish fossil? The same applies to the Elasmobranchs (Geologic range: Upper Silurian to Recent). Similar problems come from postulated affinities to jawless fish (Agnatha) (Rhodes, 1954).

In 1935, (Romer and Grove), and in 1945, Romer rejected vertebrate affinities. These investigators could find no satisfactory fit of conodonts into any fish type — ostracoderm, lamprey, shark, or higher bony fish. Gross (1954), an histologist, however, considered the conodont-bearing animal as a possible distinctive offshoot of the chordates or agnathid fishes.

Affinities. A conodont-bearing animal (70 mm long, carbonaceous impression with disarranged conodont assemblages overlaid — *Lochreia, Scottognathus*), was compared to the living protochordate *Amphioxus* (Melton and Scott, 1973). Detailed study of *Amphioxus* in the writer's laboratory did not support this analogy. Lindström thought the animal represented a conodont eater. The problem of affinity remains unresolved.

Classification

Without positive knowledge on affinities, it is not feasible to assign conodonts to a phylum or class. The state of taxonomy has been reviewed by several workers (Müller, 1962; Hass, 1962; Lindström, 1964; Webers et al., 1965). The last-named group of authors recognized *multi-element conodont species* (that is, several distinctive disjunct conodonts were assigned to a given paleobiological species). Details will be reviewed later. Lindström (1964) provided a valuable key to the conodont genera that he considered valid. In light of the study by Webers et al., it now appears that the multi-element species concept will necessitate redefinition of genera.

The present state of flux in conodont classification and lack of agreement among specialists deems prudent excluding familial assignments and restricting illustrations to those for natural assemblages (Figure 13.13), a few disjunct conodonts (Figure 13.11), and transition series within a single species (Figure 13.15).

Order Conodontophorida

Microscopic toothlike and platelike laminated objects — *conodonts*, composed chiefly of calcium phosphate; occur as disjunct pieces or in natural assemblages. Varisized pulp cavity, site of attachment may or may not bear basal filling or plate. Characteristically denticulate with denticles fused or free; fanglike structure with or without denticulation. May occur as bars, blades, or platformed platelike structures.

The conodont-bearing animal is known only through fossil objects — "conodonts" and basal fillings (ranging from cartilaginous- to collophane-type material). natural assemblages with right-handed–left-handed conodont pieces suggest bilateral symmetry. Affinities unknown. Thought to be monophyletic (but may prove to be polyphyletic, that is, vertebrate origins for some, invertebrate for others). Geologic range: Upper Cambrian to Upper Jurassic.

[*Polygnathus, Oistodus, Subbryantodus, Scolopodus, Lonchodus, Lonchodina, Neoprioniodus, Siphonognathus, Zygognathus* (Figure 13.11). Natural assemblages: *Illinella, Lochriea, Scottognathus, Duboisella, Westfalicus* (Figure 13.13). *Cordylodus, Ligonodina, Cladognathodus,* and *Roundya* are shown as "elements" of the species *Periodon flabellum* (Figure 13.15).]

Information derived from Conodonts

Inferred Biology. There seems to be widespread agreement on some aspects of the conodont-bearing animal. Except for conodonts arranged in natural assemblages and possibly some disjunct conodonts and basal fillings, the animal appears to have been soft-bodied. Several workers have suggested that it was vermiform. (This, of course, allows for either invertebrate or hemichordate affinities.) From the size of disjunct conodonts and complete natural assemblages, it is apparent the animal was quite small — possibly 3 mm wide by 3 to 5mm in length (Du Bois, 1943).

The role of hindeodellids as food particle (plankton) strainers seems to have

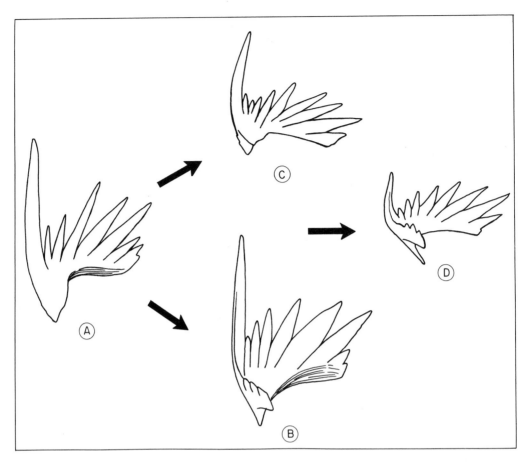

FIGURE 13.15 Transition series of conodonts. (A) *Cordylodus-Roundya* transition series. (A)–(D) *Periodon flabellum* (Lindström), Ord., Sweden. (A) *Cordylodus* element, (C) *Ligonodina* element, (B) *Cladognathodus* element, to (D) *Roundya* element, X 57. Elements bear the name of the respective genera and represent transitions from one conodont plan to another within the same conodont species. After Lindström. 1964.

the agreement of several workers. The conodont-bearing animal secreted calcium phosphate. Only marine forms are known. A pelagic existence is inferred for most conodonts (their worldwide distribution and occurrence in black shales indicates foul bottom environments—hence, conodont-bearing animals presumably lived high above such bottoms). Lindström (1964) suggested some conodont-bearing animals may have been passive floaters living in colonies. Müller (1962) thought that certain "near reef" form genera may have adapted to a benthonic existence or lived near the bottom (*Icriodus,* "Belodus" from Silurian beds).[2]

That the animal lacked upper and lower jaws and was bilaterally symmetrical seems indicated by right and left-handed elements with no corresponding underlying elements. Some asymmetrical disjunct conodonts are known (Lindström, 1964). Were these unpaired pieces in an original symmetrical assemblage, as is the case in scolecodonts? Were they positioned elsewhere in the animal, that is, outside the site of the natural assemblage? Or, were they, as Lindström suggests, truly part of an asymmetrical assemblage? If this be true, bilaterally symmetrical, as well as asymmetrical forms had to coexist!

How to account that no other parts than the conodonts themselves (disjunct

[2] Cf G. Klapper and J. E. Barrick (1977).

or natural assemblages) have ever been found? Occurrence of conodonts in some fish coprolites and certain evidences of denticle breakage (injury?) and repair point to predators. In the former instance, they must have consumed the fleshy parts and disgorged or voided the indigestible conodonts. In turn, this could explain several reports of fish plates, scales, etc. in conodont-bearing beds, as well as, some anomalous assemblages (Rhodes, 1954, p. 433) and the abundance of some disjunct conodonts. If correct, disjunct conodonts would be testimonials to certain aspects of the food chain in Paleozoic-Mesozoic seas.

Although conodont terminology has built-in a semantic prejudgment of affinities ("jaw," "bonelike," "teeth," "denticles," "oral/aboral," and so on), there is no acceptable evidence that any conodont served as biting or tearing teeth. Rather, lack of wear of denticles points to a more passive function.

Without some breakthrough pointing more clearly to actual rather than inferred affinity, little more can be said on the biology of the conodont-bearing animal. Of course, several workers have said "more" (Lindström, Gross, Schmidt, Melton and Scott, etc.), but their extrapolations, while admittedly provocative, are still pure speculation (see Figures 13.12, 13.13, 13.14).

Biotic Associates. Conodonts have been found, at one time or another, in some instances, frequently, with almost every known invertebrate (including scolecodonts) and several vertebrates (chiefly fish fossils, as the ostracoderm plates in the Harding Formation — Kirk, 1929). They are not found in every fish bone bed, however, nor are they invariably associated with scolecodonts.

They are found with cephalopods in limestones of the German Devonian (Müller, 1962, Figure 47); in trilobite beds — as in the Keisley limestone in England (Rhodes, 1955). Rarely do they occur in biohermal or biostromal facies composed dominantly of corals, sponges, calcareous algae, stromatoporoids. They are effectively absent in the fusulinid phase of Pennsylvanian-Permian beds. In the Maquoketa shale depauperate zone they are associated with numerous brachiopods, mollusks and other invertebrates (Tasch, unpublished field and laboratory notes). In the American Upper Ordovician (Cincinnatian Series in S. W. Ohio, Kentucky, and Indiana) conodonts abound in certain formations, such as the Fairview, associated with numerous articulate brachiopods, bryozoans, and some common gastropods (*Cyclonema*, and so on). Association with fish remains and ostracods is also encountered. Conodonts have been fossilized with graptolites in black graptolitic shales (Lindström, 1957), with radiolaria (Lindström, 1964), and are commonly found with plant remains (Youngquist et al., 1951).

Du Bois (1943, plate 25) collected three hundred pounds of black, fissile shale just below the Pennsylvanian La Salle limestone and directly above a thin coal seam near La Salle, Illinois. From this sample he retrieved over 75 conodont assemblages and 500 complete disjunct conodonts.

One of his figures (Figure 16, plate 25) is indecipherable. It is reported to contain impressions associated with conodonts. They occurred as part of a dark, carbonaceous film. Not shown was a hindeodellid "sandwiched" between two layers of membrane (?). Conical structures associated with this film were thought to be parapodia or cirri. Rhodes (1954, p. 437) reexamined this specimen and others and found the interpretation reasonable.

Hass (1962) suggested a search in fine-grained sedimentaries representing quiet marine or lagoonal deposits; in concretions developed in reducing environments. A search of numerous black shale slabs from the Kansas Pennsylvanian by the present writer and students, has not yet yielded significant clues; bedding planes had only disjunct conodonts which may be the explanation since these could represent voided indigestibles.

Substrates. Conodonts, not restricted to particular lithofacies, abound in black carbonaceous shale (Devonian, Mississippian, Pennsylvanian). They occur in calcareous facies: limestones of the Marathon Basin, Texas (Graves and Ellison, 1941), siltstones, quartz sandstones, glauconitic sands, and cherty shales. Lindström (1964) understood the greater likelihood of finding disjunct conodonts in limestones and fine-grained shales than in coarser sediments, such as, sandy shales, sandstones, and conglomerates.

Phosphorus (as nodules or pellets— Tasch, 1959), glauconite and iron oxides are frequent minerals in conodont-bearing beds (Phosphoria and Maquoketa Formations, for example).

Bathymetry. No one who has collected conodonts in black shales of a Pennsylvanian cyclothem, can doubt that they were deposited in relatively shallow water. Ordovician and Silurian brachiopod-rich limestones, plus coral and molluscan-rich carbonates are, according to Lindström (1964), among the best conodont sources. These, of course, denote shallow, well-oxygenated waters. By contrast, graptolite shales bearing conodonts with associations of conodonts and radiolarians denote deep water (Lindström, 1957). Müller (1956, 1962), however, excluded certain shallow-water environments as

generally unfavorable for conodonts (biostromes and bioherms, for example). If most conodont-bearing animals were pelagic, one could expect to find disjunct conodonts in every kind of lithofacies representing both shallow and deep-water facies. That actually does happen. Conodonts, obviously not restricted to particular lithofacies and except for some forms secondarily adapted to a benthonic existence, swam or floated in the upper water levels of the sea.

Population Density. The density of disjunct conodonts per weight of sediment. Table 13.3 obliquely reflects times of relatively high and/or low population density in the given area. This seems to hold even if the disjunct conodonts were voided indigestibles. It would also appear to be most usable for the shallow water facies. Transport could have been involved in deeper water facies.

Biostratigraphy, Zonal Correlation, and Mixed Fauna. Conodonts (suites, or assemblages of disjunct pieces) are often excellent index fossils. This is particularly true where other biota are scarce or absent. The Devonian-Mississippian black shale sequence in interior United States and Canada, is a good example. Hass (1962) stressed the value of conodonts in the subdivision of otherwise fossil-poor beds. Collinson et al. (1962) have effectively used conodonts in zonation of the Devonian-Mississippian of the Upper Mississippi Valley. Bischoff and Ziegler (1957)

TABLE 13.3. Relative Density of Disjunct Conodonts per Kilogram of Sediments

Age	Locality	Relative Density Number of Conodonts/kg	Source
L. Ord.	Sweden	1000–10,000	Lindström, 1964
U. Ord.	Wales	350	Lindström, 1964
U. Dev.	U. Miss. Valley, United States	75–100	Collinson et al., 1959
U. Dev.	Germany (Cheiloceras Bed)	23,000+[a]	Bischoff and Ziegler, 1957

[a]In three kilograms, 70,000 disjunct conodonts occurred and were assigned to 80 species.

and others (Müller, 1962) have demonstrated the superb stratigraphic value of the conodont genus *Palmatolepsis* in subdivisions of the Upper Devonian deposits of Germany. Walliser (1957) traced conodonts through the complete Silurian sequence of the European Carnic Alps.

Middle and Upper Ordovician beds of Europe and America can be correlated (Lindström, 1964) and conodonts are widespread in the American Ordovician (Bighorn dolomite, Maquoketa shale, Marathon Basin limestone, etc.).

Müller found conodonts in the Upper Cambrian of the Black Hills of South Dakota (Deadwood Formation) and A. R. Palmer has others from the Upper Cambrian of Wyoming. Surely, Cambrian beds are most readily dated and correlated by trilobite faunas. The same applies to other Paleozoic beds which can be zoned by graptolites, cephalopods, and so on. Pennsylvanian beds, readily correlated by other megafossils, often contain facies of certain conodont genera: *Cavusgnathus, Idiognathus, Streptognathodus, Polygnathus*, and so on.

Following Lower Carboniferous time, conodonts diminish and then extend into the Permian, Triassic, and questionably, Cretaceous time (Müller, 1956a; Diebel, 1956 — the upper Cretaceous of the Cameroons).

Special qualities of conodonts are resistivity to acids that digest most rock types; minute size which permits retrieval from cuttings and cores as well as small samples; utility for regional correlations. Pulse and Sweet (1960, text figure 1), for example, collected 10,000 disjunct conodonts referrable to 29 species and 18 genera in the Upper Ordovician of Ohio, Kentucky, and Indiana. They were able to show a correlation in successive formations across the area sampled, and to link this up with three other North American sections as well as the Welsh and English sections of Rhodes.

There are drawbacks in the use of conodonts in stratigraphy as well as strengths. Anomalous, mixed, or reworked assemblages of disjunct conodonts are not uncommon. As Figure 13.16 shows, *Palmatolepsis* in the Glen Park (Kinderhookian) is apparently reworked from Upper Devonian beds. Numerous examples are on record of similar occurrences.

The resolution of such anomalies come about by (*a*) recognition in the field of reworked beds, (*b*) comparison of stratigraphic ranges of disjunct conodonts occurring in undisturbed sections, (*c*) determination of the phylogenetic relationships of disjunct conodonts, and (*d*) notation of differences in physical appearance — color, luster, preservation — between elements of a given fauna (Hass, 1962; Lindström, 1964).

Evolution, Biological Species, Transition Types

Evolution. Evolutionary trends have been deciphered by several specialists: Müller (1962), Lindström (1964). Simple cones gave rise to bar and blade types; blades gave rise to platform types in different evolutionary lines. In the latter, particularly for *Palmatolepsis*, a rather complete picture of morphogenetic development through time has been worked out (Müller, 1962, Figure 47; 1956b).

It may seem that sequences, based on incomplete knowledge of the conodont-bearing animal, could hardly indicate true evolutionary events. However, such sequences, relying only on disjunct conodonts, do show vertical ranges, first appearances, and apparent extinctions of different genera inferred from changing morphology through time.

Transitional Types. There are certain difficulties in working out the evolution of a "phantom" animal. Homeomorphy is not uncommon; the same tendencies recur in different lines. Transitional series of conodonts (Figure 13.15) due to changing trends from one conodont plan to another, are known.

Lindström (1964) observed that recur-

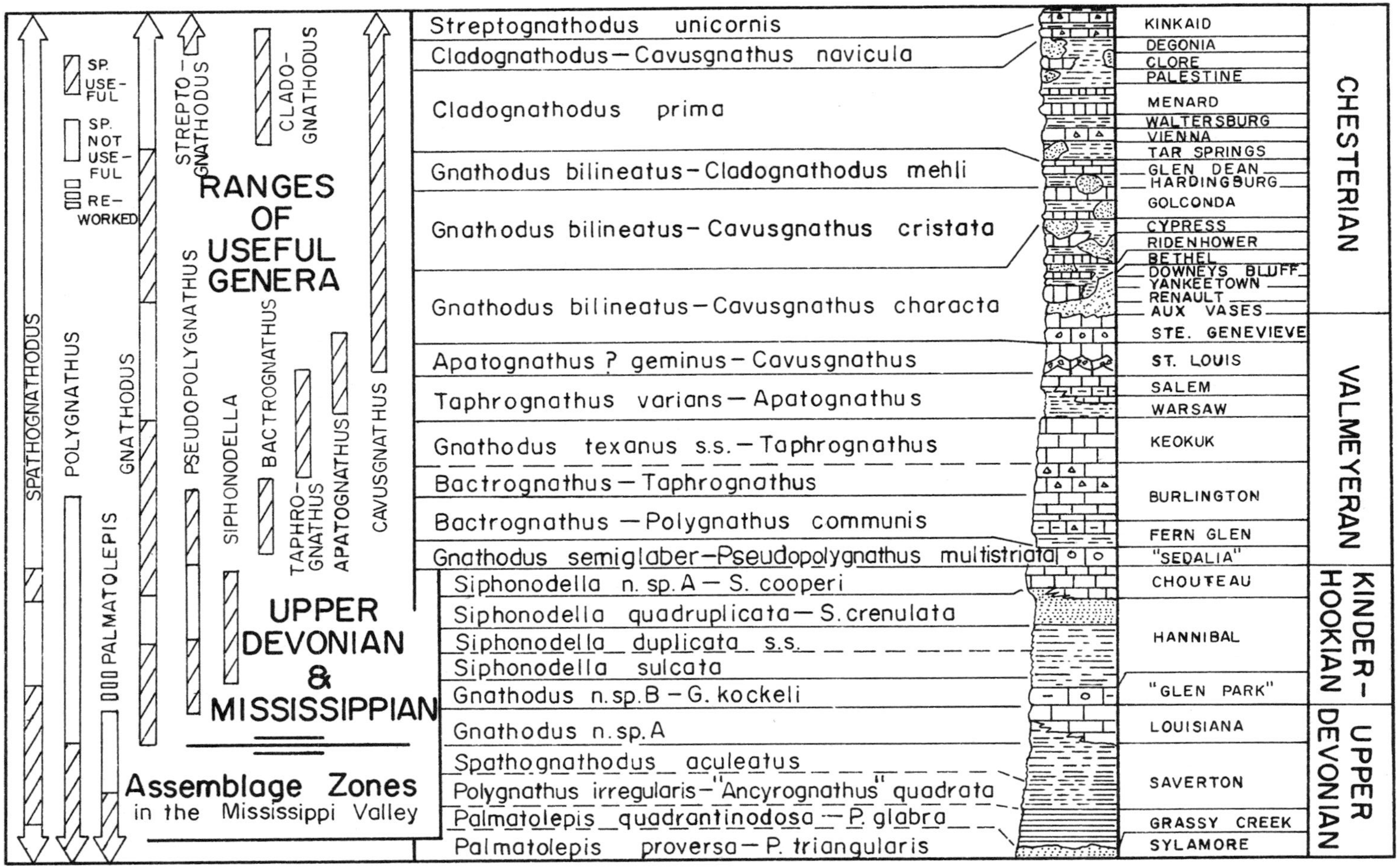

FIGURE 13.16 Information derived from conodonts: utility of conodonts in stratigraphic zonation. After Collinson et al. 1962. Check three classes of species: very useful for zonation, not particularly useful, reworked from older beds.

rences of a given conodont plan cannot be directly attributed to evolution. Simply, indirectly such plans must have been favored by natural selection for the advantage they conferred in meeting respective or similar selection pressures. Viewed as right and left-handed parts of an apparatus associated with the ingestion of food, it is clear that efficiency as a partial component depend on the whole apparatus function. In turn, this could have created important selection pressure.

Biological Species. In the current taxonomy, disjunct conodonts are assigned to a given species; in a transitional series different "elements" are identified in a single species. However, Webers et al. (1965) collected three hundred thousand discrete conodont elements from the Middle Ordovician of Minnesota, Ohio-Kentucky, Indiana, and the New York-Ontario area. Their findings are of unusual interest: (*a*) The faunas include some 75 single-element species (*b*) but these could be reduced to 25 biological species—such species defined as composed of multielements (that is, as natural assemblages).

Sweet (1970) referred 21,000 conodont elements to 28 species of six conodont genera. Multi-element taxonomy is covered in Lindström and Ziegler (1972, Sympos. on Conodont Taxonomy, Geol. et Palaeont., Marburg, Germany).

How can a natural assemblage—that is, a multi-element biological species—be recognized by discrete elements? (1) By similar size; (2) by common stratigraphic range; (3) from similarities in secondary structural features, such as dentition, basal cavity, and so on; (4) from occurrence of discrete elements together at different localities with relatively uniform abundance ratios.

Effectively, the reasoning goes as follows: even in the absence of natural assemblages, obviously disjunct conodonts did, in fact, belong to such assemblages and hence, by the cited criteria, can be so assigned. A similar approach to the taxonomy of scolecodonts has been applied (Szaniawski and Gazdicht, 1978, ref. p. 470).

QUESTIONS

Conodonts

1. (a) Give the significance of the following: pulp cavity, fused vs. free denticles, cone-bar-blade-platform types, laminae, white matter.

   (b) What does composition denote about chemical oceanography during the Paleozoic?

2. (a) W. H. Hass (1959) presented a utilitarian rather than a biological classification of conodonts. Discuss both strengths and weaknesses of this approach. Include the work of Webers et al. (1965), and contrast both concepts of conodont taxonomy.

   (b) Contrast the multielement (inferred) biological species and the natural assemblage species. Do they differ? If yes, how? What difficulties arise from the former? Why does it represent a breakthrough?

3. Account for the predominance of disjunct conodonts (discrete elements).

4. Discuss disjunct conodont suites in stratigraphic correlation and zonation (refer to Upper Mississippi Valley, Carnic Alps, Devonian-Mississippian black shale sequences).

5. Devise an exploration program on the theme "Search for the conodont-bearing animal" by a carefully argued review of telltale evidence, best search prospects, and proposed procedures. (See Melton and Scott, 1973, *in* F.H.T. Rhodes ed. Conodont Palaeozoology, *in* G.S.A. Special Paper 141.)

REFERENCES

Conodonts

Beckmann, H., 1949. Conodonten aus dem Iberger Kalk (Ober-Devon) des Bergischen Landes and ihr Fenibau. Senckenbergiana Lethaia, Vol. 30, pp. 153–168.

Bischoff, G., and W. Ziegler, 1957. Die Conodontenchronologie des Mitteldevons und des tiefsten Oberdevons: Abhandl. Hess. Landesamtes Bodenforsch., Vol. 22, 136 pp.

Branson, C. C., and C. J. Mankin, 1964. Composition of conodonts: Okla. Geol. Notes, Vol. 24 (12), pp. 296–302.

Branson, E. B., and M. G. Mehl, 1933–1934. Conodont Studies 1–4. Univ. Missouri Studies 8, 343 pp.

Collinson, C. W., 1963. Collection and preparation of conodonts through mass production techniques: Ill. St. Geol. Surv. Circ. 343, 16 pp.

————, 1965. Conodonts, in Handbook of Paleontological Techniques. W. H. Freeman & Co., pp. 94–102.

————, A. J. Scott and C. B. Rexroad, 1962. Six charts showing biostratigraphic zones and correlations based on conodonts from the Devonian and Mississippian rocks of the Upper Mississippi Valley: Ill. St. Geol. Surv. Circ. 328, 32 pp.

Diebel, K., 1956. Conodonten in der Oberkreide von Kamerun: Geologie, Vol. 5, pp. 424–450.

Du Bois, E. P., 1943. Evidence on the nature of conodonts: J. Paleontology, Vol. 17, pp. 155–159, plate 25.

Fay, R. O., 1952. Catalog of conodonts: Univ. Kansas Paleont. Contr. Vertebrata, Vol. 3, 206 pp.

Graves, R. W., and S. P. Ellison, Jr., 1941. Ordovician conodonts of the Marathon Basin, Texas: Missouri Univ. School of Mines and Metallurgy, Tech. Ser. 14 (2), 26 pp.

Gross, W., 1960. Über der Basis bei den Gattungen *Palmatolepsis* und *Polygnathus* (Conodontida): Paläontol. Z., Vol. 34, pp. 40–58.

Hass, W. H., 1962. Conodonts: Treat. Invert. Paleont., W. Miscellanea, pp. W3–W69.

————, 1959. Conodonts from the Chappel Limestone of Texas: U.S. Geol. Surv. Prof. Paper 294-J, pp. 365–399.

————, 1941. Morphology of conodonts: J. Paleontology, Vol. 15, pp. 71–81.

————, and M. L. Lindberg, 1946. Orientation of the crystal units of conodonts: J. Paleontology, Vol. 20, pp. 501–504.

Kirk, S. R., 1929. Conodonts associated with the Ordovician fish fauna of Colorado—a preliminary note: Am. J. Sci., Ser. 5, Vol. 18 (No. 108), pp. 493–496, Figures 1–14.

Lindström, M., 1964. Conodonts. Elsevier Publ. Co., 176 pp.

————, 1957. Two Ordovician conodont forms found with zonal graptolites: Geol. Fören. Stockholm, Förh., Vol. 79, pp. 161–178.

————, 1954. Conodonts from the lowermost Ordovician strata of south-central Sweden: Geol. Fören. Stockholm, Förh., Vol. 76, pp. 517–603, plates 1–10. [G. Klapper, and J. E. Barrick, 1978. Conodont ecology: pelagic versus benthic: Lethaia, Vol. 11 (1), pp. 15–23, prefer a neritic pelagic or nektobenthic mode of life for the conodont-bearing animal].

Müller, K. J., 1962. Taxonomy, evolution, and ecology of conodonts: Treat. Invert. Paleont., W-Miscellanea, pp. W246–W249.

————, 1959. Kambrische Conodonten: Z. Deut. Geol. Ges., Vol. 111, pp. 434–485.

————, 1956a. Triassic conodonts from Nevada: J. Paleontology, Vol. 30, pp. 818–830.

————, 1956b. Die Gattung *Palmatolepis:* Abhandl. Senckenberg. Naturforsch. Ges., Vol. 494, 70 pp.

Osmund, J. K., and H. J. Sawin, 1959. Unit cell dimensions of recent and fossil tooth apatite: G.S.A. Bull. Abstracts, Vol. 70, p. 1653.

Pulse, R. R., and W. C. Sweet, 1960. The Upper Ordovician standard. III. Conodonts from the Fairview and McMillan Formations of Ohio, Kentucky and Indiana: J. Paleontology, Vol. 34, pp. 237–264.

Rexroad, C. B., and C. W. Collinson, 1963. Conodonts from the St. Louis Formation (Valmayeran Ser.) of Illinois, Indiana, and Missouri: Ill. St. Geol. Surv. Circ. 355, 28 pp.

Rhodes, F. H. T., 1963. Conodonts from the topmost Tensleep Sandstone of the eastern Big Horn Mountains, Wyoming: J. Paleontology, Vol. 37, pp. 401–408.

———, 1962. Recognition, interpretation and taxonomic positions of conodont assemblages: Treat. Invert. Paleont., W. Miscellanea, pp. W70–W83.

———, 1955. The conodont fauna of the Keisley Limestone: Quart. J. Geol. Soc. London, Vol. 111, pp. 117–142.

———, 1954. The zoological affinities of the conodonts: Biol. Rev., Cambridge Philos. Soc., Vol. 29, pp. 419–452.

———, and P. Wingard, 1957. Chemical composition, microstructure and affinities of the Neurodontiformes: J. Paleontology, Vol. 31, pp. 448–454.

Schwab, K. W., 1965. Microstructure of some Middle Ordovician conodonts: J. Paleontology, Vol. 39 (4), pp. 590–593.

Scott, A. J., and C. W. Collinson, 1959. Intraspecific variability in conodonts: *Palmatolepis glabra* Ulrich and Bassler: J. Paleontology, Vol. 33, pp. 550–565, plates 75–76.

——— and C. B. Rexroad, 1960. Interpretation of growth stages and specific variation in Paleozoic conodonts: A.A.P.G.-S.E.P.M. Program (Atlantic City) Abstracts, pp. 81–82.

Schmidt, H., 1934. Conodonten-Funde in ursprünglichen Zusammenhang: Paläont. Zeitschr., Vol. 16, No. 1/2, pp. 76–85, plate 6.

Stewart, G. A., and W. C. Sweet, 1956. Conodonts from the Middle Devonian bone beds of central and west-central Ohio: J. Paleontology, Vol. 30, pp. 261–273, plates 33–34.

Stone, G. L., and W. M. Furnish, 1959. Bighorn conodonts from Wyoming: J. Paleontology, Vol. 33 (2).

Sweet, W. C., 1955. Conodonts from the Harding Formation (Middle Ordovician) of Colorado: J. Paleontology, Vol. 29, pp. 226–262, plates 27–29.

Tasch, P., 1958. Significance of conodont control of pellet formation in the basal Maquoketa: Micropaleontology, Vol. 4 (2), pp. 187–191, text Figures 1–15.

Thompson, T. L., and E. D. Goebel, 1963. Preliminary report on conodonts of the Meramecian Stage (Upper Mississippian) from the subsurface of western Kansas: State Geol. Surv. Kans. Bull., Vol. 165 (1), 16 pp.

Youngquist, W. L., R. W. Hawley, and A. K. Miller, 1951. Phosphoria conodonts from southeastern Idaho: J. Paleontology, Vol. 25, pp. 356–364, plate 54.

Ziegler, Willi, 1956. Unterdevonische Conodonten insbesondere aus dem Schonauer und dem Zorgensis-Kalk: Hess. Landesamt. Bodenf., Notizbl., Vol. 84, pp. 93–106, plates 6–7. (W. C. Sweet, and S. M. Bergström, Eds., 1971. Symposium on conodont biostratigraphy. G.S.A. Memoir 127. C. R. Barnes, Ed., 1976. Conodont paleoecology. Geol. Assn. Canada, Special Paper).

Walliser, O. H., 1957. Conodonten aus dem oberen Gotlandium Deutschlands und der Karnischen Alpen: Notizbl. Hess. Landesamtes Bodenforsch. Weisbaden, Vol. 85, pp. 28–52.

Webers, G. F., J. M. Schopf, and W. C. Sweet, 1965. Multi-element Ordovician conodont species: G.S.A. Program (Kansas City, Missouri, 1965), Abstracts, pp. 180–181.

PART III. CHITINOZOA

Introduction

In 1931 Eisenack discovered a new group of extinct microfossils in the Ordovician/Silurian sedimentaries of the Baltic region, East Prussia. He described these, devised a classificatory system in general use at present, and erected a new order, Chitinozoa, to embrace such objects. (Subsequent papers by Eisenack on this subject appeared in 1932, 1934, 1938, 1955, 1959, 1962.)

Certain generalities are apparent: these objects have a relatively simple flask/bottle shape, may occur free or in chains, all specimens are rather small, generally 150 μ to 300 μ (although some range from 50 μ to 1000 μ), they occur in abundance through a restricted Paleozoic range, and appear to be cosmopolitan in distribution.

Affinities

As with graptolites and conodonts, affinities of chitinozoans are uncertain and unknown. There are no lack of suggested affinities, however. These include affinities to various protozoans (thecamoebaea testaceans, flagellates, ciliates—tintinnids—as well as affinities to metazoans [unidentified eggs and egg capsules (Kozlowski, 1963) or the gonothecae of Hydrozoa (Eisenack), see Figure 13.18, *Desmochitina minor*].

Two prominent investigators have expressed diverse opinions on the matter. Deflandre (1944–1945) raised the question as to why it is necessary to view such objects as being completely unlike and unrelated to any living protozoan or metazoan. Kozlowski (1963) responded that for a large part of the Paleozoic, particularly Ordovician-Devonian time,

numerous organisms were of a type completely unknown in contemporary marine faunas.

The problem of affinities is clearly not resolved. (See Jenkins, p. 833).

Fossil Chitinozoans

Essential Structure. Chitinozoa are represented by small, hollow, varishaped cysts or tests (= vesicles), individual or in chains (Figure 13.17*B*, *b*). Tests, axially symmetrical, (Figure 13.17*B*) consist of a two- or three-layered wall (see discussion below) that encloses a relatively large interior body chamber (Figure 13.17*A*). (If vesicles are "egg cysts," then the term "chamber" would be preferable.) The chamber narrows to a long or short neck at one end (the open or oral end) and is closed at the opposite, broader end (aboral).

[The aboral end may have a central perforation, basal pit or scar that served for attachment to a holdfast if chitinozoans are viewed as protozoans (Collinson and Schwalb, 1955), or if one does not consider probable affinities, it may have served to couple individual vesicles in a chain-like colony or arrangement.]

The neck terminates in a *lip*. Sunk within an elongate neck, is the *operculum* with a membranous fringe. It may also be external as in the desmochitinids. Very much like a flask stopper, this circular outline of the operculum may have a sleevelike appendage called the *prosome*, the appendage develops into a separate organ or plug) (see Figures 13.17*A* and 13.18*A*). Apparent prosome properties include the capacity to contract and retract within the vesicular neck. Such capacity, however, may be unreal—an

FIGURE 13.17 Essential chitinozoan structure, terminology, and ranges.
(*A*) Labeled diagram (after Jansonius, 1964). (*B*)*a–b*. Diagram, after Collinson and Schwalb, 1955.
(*C*) Suggested relationships and approximate ranges of genera (Jansonius, 1964).

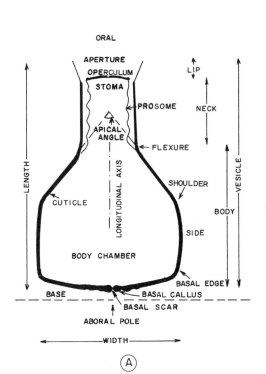

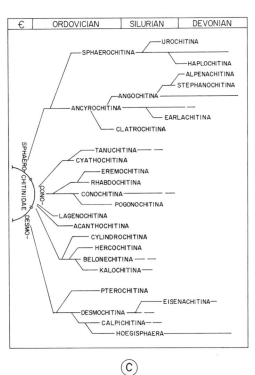

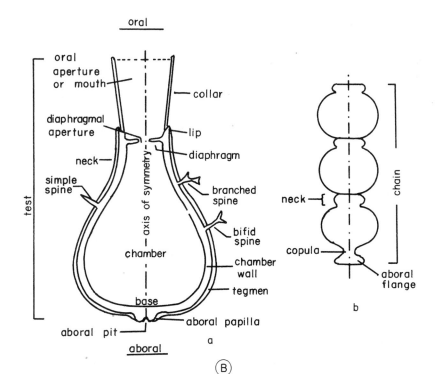

artifact due to processing. Staplin (1961) figured the plug is less resistant than the body in the chitinozoans.

It will be understood that if one follows the protozoan affinity concept, a structure at the oral end inserted at the neck is seen as a "collar" (see collared flagellates, Chapter 3) (Figure 13.17*B, a*). If one rejects protozoans (Jansonius, 1964), then the same structure becomes a "prosome" (Figure 13.17*A*).

The vesicle exterior may have a surface texture, smooth, tuberculate, or hispid (= hairy; with stiff hairs or bristles). Any of three texture kinds may also bear simple, branched, or bifid spines, spines may be lacking entirely). Other ornamentation or surficial structures include skirt-like flanges (Figure 13.17*B, b*) and near the shoulders, longitudinal striae. Jansonius (1964) insisted that ornamentation can involve either outer layer. In the case of *Acanthochitina*, a mat of intertwined, intergrown spines form a rough cocoon around the vesicle, the outer cuticle layer. In some species of this genus the cocoon is lost. Jansonius reported that conceivably smooth vesicles in other genera might denote a lost spinous cocoon.

Composition and Wall Structure. It is quite important to establish whether or not the test of chitinozoans is composed of true chitin or pseudochitin. These tests, as any student can confirm for himself in the laboratory, resist heating in concentrated acids—HCl, HF, H_2SO_4— and in concentrated potash lye. Hot (160° to 180°) concentrated alkali solutions will remove some of the acetyl groups $(CH_3—CO)$ from chitin yielding a group of chitin derivatives called "chitinosans" (see Chapter 15, "Chitin").

Involved in chitin degradation to chitinosans is the shortening of the chitin chain. Eisenack's experiments (1931) indicated some difference in composition between modern chitin and the chitinozoan test. If first treated with concentrated HCl or fifty per cent H_2SO_4 (hydrolization) at 100°C—modern chitin will react to caustic soda. The differences in composition that Eisenack saw, could have been those of degraded chitin (chitinosans) compared to undigested chitin.

Independent evidence that chitinozoans might be composed of chitin comes from W. F. Bradley (see, Collinson and Schwalb, 1955). X-ray studies of the wall of *Angochitina* yielded different haloes indicative of amorphous structure, similar to Kesling's data on the cladoceran *Daphnia longispina* a known chitinous test.

It has been assumed, reasoning by analogy to the protozoan *Gromia*, that chitinozoans may be composed of pseudochitin (Collinson and Schwalb, 1955 and see Chapter 3, this text). However, true chitin composition has not been eliminated until proof becomes available that the differences Eisenack recorded were not due to the degradation of true chitin.

Chitinozoans did not stain brown when given the chlorzinc-iodide test (Chapter 3). However, this cannot be considered conclusive since this chemical test was devised for Recent biological materials. Fossil chitinozoans cannot be expected to have adequate chitin purity to make a test valid. Accordingly, the true composition of Chitinozoa is still open.

Growth, Reproduction. If one assumes protozoan affinities (and it is important to admit that this assumption must precede the elucidation of growth, and so on), then one can further infer protozoan type of reproduction such as binary fission, multiple fission, and so on. One can interpret the vesicle chains as in *Desmochitina*, equivalent to those in dinoflagellates. In that event, they would be formed by repeated binary fission (Collinson and Schwalb).

By contrast, assuming metazoan affinities of chitinozoans and conceiving of them as eggs or egg cysts, then analogies can be made to gastropod or polychaete worm chains of egg capsules or eggs and, hence, to equivalent reproductive cycles (Kozlowski, 1963).

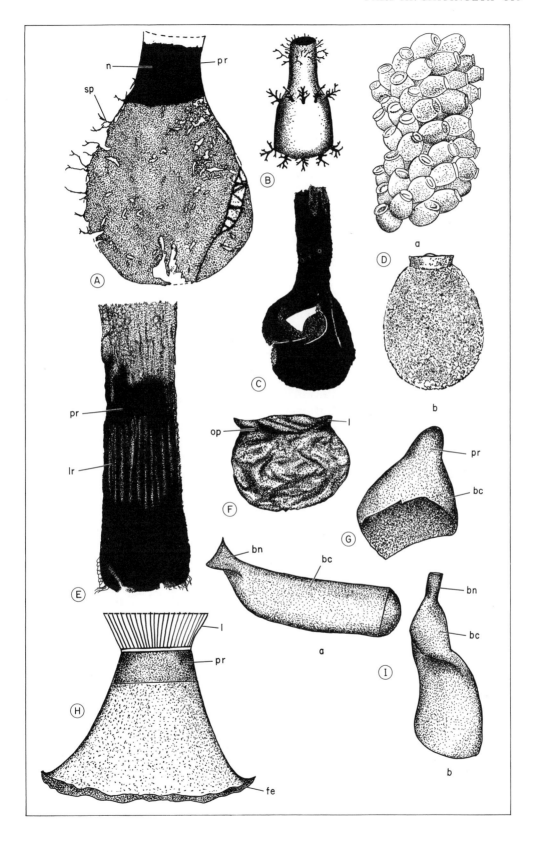

With present limited knowledge, it is prudent to suspend judgment on elusive subjects requiring greater details not now available.

Jansonius (1964) made a few observations on growth and development. He allowed for propagation by budding but was unclear whether new individuals arose orally or aborally. Did the operculum operate in the budding process? Staplin (1961) decided that a recent chrysophycean *Ochromonas* has a similar "plug" — sealed cellulose cyst (Fritsch, 1935, Figure 164). During germination in *Ochromonas*, the plug dissolves allowing cyst contents to escape through the pore, that is, orally, in the form of motile cells. Archeomonad cysts have a cytoplasmic plug (Chapter 3). Probably Jansonius had this in mind when he inquired about the "operculum" in propagation.

[Observations that both operculum (family Desmochitinidae) and base of the chitinozoan vesicle may have concentric structure, and that forms like *Tanuchitina* show annular thickening of the prosome, appear promising for further research in deciphering growth mode of test or cyst.]

Classification

(Assignment to phylum and class seems unwarranted with conflicting evidence — see Collinson and Schwalb, 1955.)

Order **Chitinozoa Eisenack, *1931***

1. Varishaped vesicles (tests or cysts included in spinous cocoons) representing a part of an unknown extinct metazoan (as represented by an egg cyst, for example) or the test of some unknown extinct protist.

2. Vesicles are single or joined in chains or other colony-like association (Kozlowski, 1963, Figure 3*B*-2).

3. Size ranges from 50 μ to 1000 μ (1 micron = 0.001 mm).

4. Vesicles are symmetrical about a longitudinal axis.

5. Composition is uncertain; may be chitinous or pseudochitinous.

6. Vesicles, retrieved by acid digestion of the containing sedimentary rock, are black, opaque, and structureless. Some tests, however, are translucent, brown, and amber. [Many, but not all, chitinozoans can be rendered partially or almost completely translucent by processing insoluble residues with potassium chlorate and nitric acid; fine structures can be seen by using infra-red photography (L. R. Wilson, 1959).]

7. Vesicle is open at the narrow end that defines a neck which may end in a lip. Inside the neck there may be an operculum or its membranous elaboration (= prosome). The operculum, sometimes spinous, may be external.

8. The broader base at the aboral end may bear a basal pit or scar. (Conceivably,

this scar functioned to attach a stalk or articulation to other vesicles.)

9. Wall, single or double-layered, in which the outer layer may be a spinous cocoon that is not preserved. Cuticular surfaces range from smooth to tuberculate and bristly.

10. Ornamentation includes spines and longitudinal striations.

11. Geologic range: Ordovician–Devonian: Pennsylvanian,[3] Permian.

Following Jansonius' modified classification (1964) three families are recognized: I—Sphaerochitinidae, II—Desmochitinidae, III—Conochitinidae. [Representative genera in each of these families include: Family I, *Angochitina* (Figure 13.18*A, G*); *Alpenachitina* (Figure 13.18*B*). Family II, *Desmochitina* (Figure 13.18*D*); *Calpichitina* (Figure 13.18*F*). Family III, *Lagenochitina* (Figure 13.18*C*); *Herco-chitina* (Figure 13.18*E*).]

Permian Chitinozoa. Since there had been no reports of Permian chitinozoa heretofore, Figure 13.18*G–I* is of surprise interest. These chitinozoa were recovered after acid digestion of samples, from the Ft. Riley Limestone, Augusta Quarry, Butler County, Kansas. Previously Stude, and Tasch and Stude (1966a) had recovered and described an abundant scolecodont fauna from these same beds.

Two students in a course in micropaleontology, E. L. Gafford, Jr., and E. J. Kidson, retrieved this unexpected fauna associated with additional assemblages of scolecodont jaws from the Augusta Quarry. [Another student, C. R. Young, subsequently restudied these beds at several quarries and secured further chitinozoans from undisturbed beds. Descriptions of this fauna and field observations will be published at a later date.]

Figure 13.19 gives a columnar section from the Augusta Quarry in which asterisks indicate precisely where chitinozoans were retrieved. All sampling levels were determined in feet above or below

the inarticulate brachiopod *Orbiculoides* zone.

The nature of the fauna indicates the evolution of new types of chitinozoa yet, with characteristic chitinozoan organization and morphology (plug, neck, broad base, chains). Figure 13.17*C*, with the known ranges of presently described genera, may need extension to accommodate this new material.

Discovery of Pennsylvanian chitinozoans helps to establish the Permian chitinozoans more firmly. Vertical ranges of fossil biota can be extended forwards or backwards in geologic time whenever the evidence so indicates.

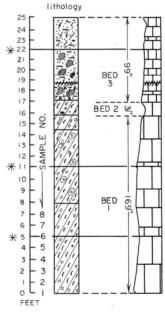

SE Corner of
Augusta Quarry
N ¼ Sec. 9 T28S,
R4E, Butler Co.

Ⓐ

KEY:

Sparite		Biogenic	
Micrite		Clayey (10–50%)	
Dismicrite		Osagia Algae	
Intraclastic		Fossiliferous (1–10%)	
Orbiculoidae Zone		Chitinozoa	

NOTE: Oketo sh. & Florence ls. directly below Fort Riley.

---

[3]P. Tasch, and T. Hutter, 1978.

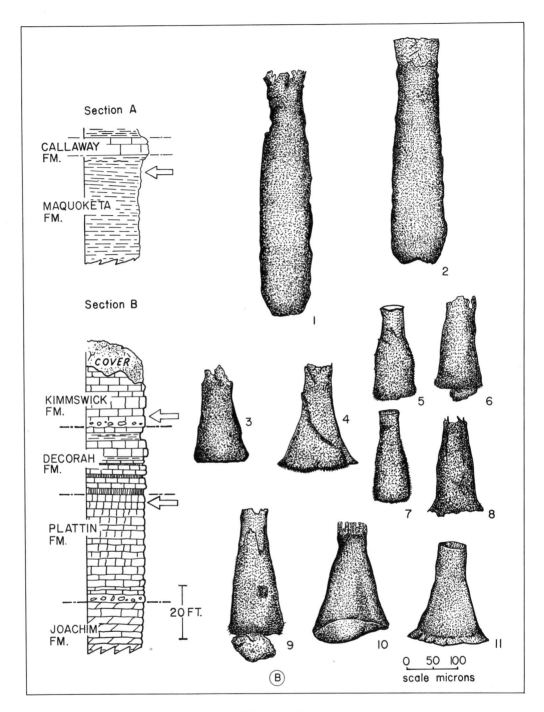

FIGURE 13.19 Information derived from chitinozoans—I.

(*A*) Columnar section, Ft. Riley Limestone, S.E. corner Augusta Quarry. N/Sec. 9, T 28S. R4E, Butler County, Kansas. Asterisks indicate where chitinozoans were retrieved. All chitinozoan-bearing beds were located above or below the *Orbiculoides* zone (misspelled in legend) and the elevation above the Florence Limestone. Data from J. R. Stude, 1964; Tasch and Stude, 1966; faunal data from E. L. Gafford, Jr., 1962, unpublished report.

(*B*) Composite columnar section of Ordovician outcrop in N.E. Missouri (Sec. A, SE, NW, Sec. 13, T54N, R5W, Pike County, Sec. B, SE, SE, Sec. 21, T55N, R4W, Ralls County). Chitinozoan fauna include (1) (2) *Rhabdochitina*; (3) *Conochitina*, (4) (9) (10) *Cyathochitina*; (5–7) (9) forms assignable to either *Conochitina* or *Belonechitina*; (11) Another species of Cyathochitina (after Echols and Levin, 1966).

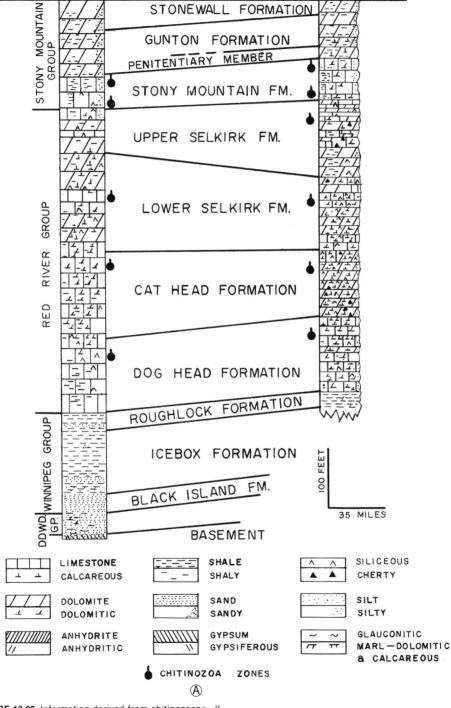

SUBSURFACE   SURFACE

TOP ORDOVICIAN

FIGURE 13.20 Information derived from chitinozoans—II.
(*A*) Subsurface and surface correlations by chitinozoans (Interlake Area, Manitoba, Can., surface section, and subsurface section in southwest Manitoba). After Jodry and Campau, 1961.

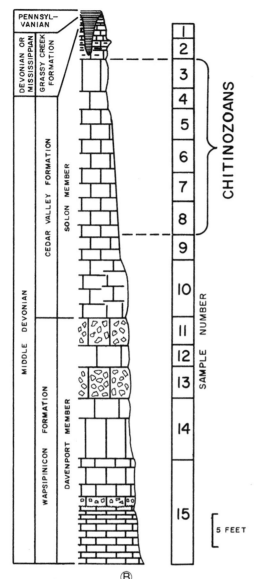

FIGURE 13.20 (cont.)
(*B*) Cedar Valley chitinozoan faunule (Dev., Illinois; E1/2 NW1/4 sec. 25 T17N, R2W, Rock Island County, Ill). (Collinson and Scott, 1958).

There is no field evidence in the portion of the Ft. Riley Limestone from which chitinozoans were retrieved of any of the following: disturbed beds, stratigraphic leaks, mixed faunas, anomalous exotic faunal elements. This tends to support the concept that the chitinozoans referred to could be of Permian age.]

Lithologies

Chitinozoans are most often recovered from acid digested samples of limestone.

Occurrences in chert, dolomite, and shale are also known. In the Permian Ft. Riley limestone, they were found in biosparite, algal biosparite, and argillaceous biomicrite facies. They are known from erratic boulders in the Polish Ordovician, in a glauconitic zone (Clear Creek, Illinois), in a Devonian off-reef calcareous facies (Alberta, Canada) in subsurface limestones and shales. Other occurrences include: cherty argillaceous limestone (Sexton Creek Formation, Illinois), siliceous dolomites, argillaceous, cherty, and arenaceous dolomites of Manitoba, Canada. A variety of calcareous facies in the Baltic contain chitinozoan faunules: Ordovician, Ostseekalk, Silurian, Krinoidenkalk, for example.

Chitinozoans are also known from Lower and Middle Devonian Brazilian shales, in various lithologies on many continents and in numerous localities surface and subsurface.

Biotic Associates

Reports of chitinozoans found in association with scolecodonts are common in many places: Devonian of Illinois, Canada, Ordovician of the Williston Basin and Oklahoma, the Permian of Kansas, and so on. Other biota include: ostracodes, sponges (spicules), bryozoa, algae (*Tasmanites*, acritarchs), plant spores, graptolites (Whittington, 1955), echinoderms, and conodonts.

Certain common occurrences of other forms with chitinozoans can be explained because they are all found in the insoluble residues, acid-resistant forms: scolecodonts, graptolites, acritarchs, and so on.

In outcrop many usual Paleozoic megafossils occur in calcareous beds whose insoluble residues yield those acid-resistant suites including chitinozoans. Common megafossils are brachiopods, trilobites, mollusks, coelenterates, and so forth.

Zonation and Stratigraphic Correlation

A number of studies are available on using chitinozoans in zonation and stratigraphic correlation. Some data are schematically summarized in Figures 13.19 and 13.20. A search for acid-resistant microfossil suites is needed in all Paleozoic beds yielding megafossils as well as in beds barren of megafossils. Clearly, in otherwise barren beds, both in surface and subsurface, chitinozoan faunas can be valuable zonal fossils and permit stratigraphic correlations otherwise highly speculative (Figures 13.19*B*, 13.20*A*; also see Dunn, 1959, Figures 1, 2).

Taugourdeau and Jekhowsky (1960) studied chitinozoan faunules from almost 300 cores (12 borings) in the Saharan Lower Paleozoic. They found 84 species (49 were new) distributed in 12 genera—12 were new. Chitinozoans allowed the Silurian-Devonian of the Saharan subsurface to be divided into 10 faunal zones traceable over 1000 kilometers.

Paleoecology

Collinson and Schwalb (1955) suggested that at least for the Illinois Devonian, chitinozoans may have lived in an environment suitable for glauconite formation.

From megafossil associations, one infers an environment favorable to chitinozoans was shallow marine, with good circulation, limited detritus (although the arenaceous content of some dolomites must be taken into account in some situations). Tasch and Stude (1966*a*) showed that the environment indicated by scolecodont remains in the Permian Ft. Riley Limestone of Kansas was inshore, shallow water. Further support for well-lighted, clear water seemed to follow from the association with algal reef facies (*Osagia*). Apparently indigenous, chitinozoans of these beds must have lived under similar conditions.

The chitinozoan *Calpichitina* occurred in graptolitic, greenish-gray, calcareous shale of Oklahoma (Sylvan shale). Staplin's off-reef fauna in the Devonian of Alberta, Canada was found in facies interfingered with reef carbonates. Such association points to clear, shallow, well-

lighted waters. Dunn's Cedar Valley (Devonian of Iowa) fauna occurred in argillaceous, fine-grained limestones and were absent from coralline biostromes and sublithographic to brecciated lithographic limestones (Dunn, 1959, Figure 2).

Further Notes on Scolecodont/Chitinozoan Associations

Polychaete worm jaws (scolecodonts), as already remarked, are frequent associates of chitinozoan faunules. Such finds are particularly marked in the Ordovician, Devonian, and Permian. Kozlowski (1963), persuaded by such Ordovician associations, investigated possible relationships between chitinozoans and the egg capsules of some annelid polychaete worms. He saw that egg capsules or cocoons are frequently produced by sedentary polychaetes (*Spirorbis*, among others). These are greatly varied. The eggs form an agglutinated mass enveloped by a common outer membrane. Such capsules are free or attached. Sometimes they form chains as in *Polydora ciliata*. A dozen eggs may be inside a single capsule, quadrangular in shape, and joined in chains to others. Such chains are analogous to the chitinozoan.

Despite the above observations, Kozlowski reported that the very thin capsules, lack distinctive morphological differentiation, structure, and ornamentation. Accordingly he rejected the notion that chitinozoans might be related as their egg capsules to annelid polychaetes.

While it is abundantly clear that chitinozoans cannot be shown to have annelid affinities via the egg capsule-chain analogy, one can still persist in trying to understand a common association with polychaete worm jaws. Employing the multiple-hypotheses approach, one can say that (*a*) such occurrences are fortuitous or (*b*) they are controlled by some relationship.

Pursuing hypothesis (*b*), we ask what this relationship might have been. (1) Some metazoans are known to prey on polychaete larvae (Thorson, 1944, pp.

145–146) although the latter yield a small part in the food chain in modern seas. Chitinozoans viewed as metazoan egg capsules from which larval forms hatch, could include a predator-prey relationship in the Paleozoic seas. Supporting this view are coelenterates in association with chitinozoans and scolecodonts. (2) Polychaetes are common reef faunal elements (crevice dwellers) and, as in Staplin's Devonian off-reef fauna, some chitinozoans and polychaetes could have become associated after demise by bottom mixing or interfingering of facies. (3) Viewed as protozoans, the chitinozoans with a basal pit were probably attached by a stalk or holdfast. Such holdfasts may have been attached to the substrate inhabited by polychaete worms. (4) Again, viewed as metazoan egg capsules, these capsules might have been part of the plankton during the same time that the polychaete larvae passed through a planktonic stage.

Any possibility reviewed or more than one may help explain the common errant polychaete-chitinozoan association. Ultimate clarification must await decisive evidence on chitinozoan affinities.

QUESTIONS

Chitinozoans

1. (a) Sketch the essential structure of an individual chitinozoan; of a chain of vesicles. Label all parts and distinguish oral/aboral, operculum/prosome, basal flange.
   (b) Discuss existing data on wall structure and ornament; composition.
2. Compare the chitinozoan vesicle chain to the gastropod and annelid egg chain.
3. Cite several examples of "loaded word descriptive terms" for the chitinozoan cyst, test or cocoon. Show in each case how words prejudge probable affinities.
4. Kozlowski (1963) stated that no known protozoan was organized as is *Desmochitina* (Figure 13.18*D*). Refer to Chapter 3 and cite evidence for your agreement or disagreement with this evaluation.
5. What light, if any, do the biotic associates of Chitinozoa provide pertinent to
   (a) substrates,
   (b) water depths,
   (c) location relative to shore,
   (d) ecosystem at the given time in the Paleozoic sea?
6. Study Figure 13.20*A* to discuss the value of chitinozoans, and related microfossils as aids in subsurface stratigraphy and in correlation within the Williston Basin Paleozoic.
7. Outline a program for chitinozoan retrieval from various types of lithologies and in beds of pre-Cambrian through Permian age. How would you proceed?

Outline steps and justification for selection of localities?

REFERENCES

Chitinozoans

Collinson, C. W., and H. Schwalb, 1955. North American Paleozoic Chitinozoa. Ill. St. Geol. Surv. Rpt. of Inv. 186, 33 pp., plates 1–2.

————, A. J. Scott, 1958. Chitinozoan faunule of the Devonian Cedar Valley Formation. Ill. St. Geol. Surv. Circ. 247, 34 pp., plates 1–3.

Combaz, A., and C. Poumont, 1962. Observations sur la structure des Chitinozoaires: Rev. Micropal., Vol. 5 (3), pp. 147–160, plates 1–5, Figures 1–2.

Deflandre, G., 1952. Groupe des Chitinozoaires, in J. Piveteau, ed., Traité de Paléontologie (Paris), Vol. 1, pp. 327–329, Figures 1–17.

————, 1944–1945. Microfossils des Calcaires Siluriens de la Montagne Noire: Ann. de Paleontologie, Vol. 31, pp. 41–75, 41 Figures, 3 plates (Ch. III, I. Chitinozoaires).

Dunn, D. L., 1959. Devonian chitinozoans from the Cedar Valley Formation of Iowa: J. Paleontology, Vol. 33 (6): pp. 1001–1017, plates 125–127, 2 text-figures.

———— and T. H. Miller, 1964. A distinctive chitinozoan from the Alpena limestone (Middle Devonian) of Michigan: J. Paleontology, Vol. 38 (4), pp. 725–728, plate 119.

Echols, D. J., and H. L. Levin, 1966. Ordovician chitinozoa from Missouri: Okla. Geol.

Notes, Vol. 26 (5), pp. 134–139, plates 1–2, 1 text figure.

Eisenack, A., 1962. Neotypen baltischen Silur-Chitinozoen und neue Arten (Fortsetzung): N. Jb. Geol. Paläont. Abh., Vol. 114, No. 3, pp. 291–316, plates 14–17, Figures 1–8, 1 table.

―――――, 1959. Neotypen baltischen Silur-Chitinozoen und neue Arten: N. Jb. Geol. Paläont. Abh., Vol. 108, No. 1, pp. 1–20, plates 1–3, Figures 1–4.

―――――, 1955a. Chitinozoen, Hystrichosphaeren und andere Mikrofossilien aus dem Beyrichia-Kalk: Senck, leth., Vol. 36, No. 1–2, pp. 157–188, plates 1–5, Figures 1–13.

―――――, 1955b. Neue Chitinozoen aus dem Silur des Baltikums und dem Devon der Eifel: Senck, leth., Vol. 36, No. 5–6, pp. 311–319, plate 1, Figures 1–3.

―――――, 1938 (1937). Neue Mikrofossilien des baltischen Silurs, IV: Pal. Zeitschr., Vol. 19, pp. 217–244, plates 15–16, Figures 1–22.

―――――, 1934. Neue Mikrofossilien des baltischen Silurs, III; und: Neue Mikrofossilien des böhmischen Silurs. I: Pal. Zeitschr., Vol. 16, pp. 52–76, plates 4–5, Figures 1–35.

―――――, 1932. Neue Mikrofossilien des baltischen Silurs, II. (Foraminiferen, Hydrozoen, Chitinozoen u.a.): Pal. Zeitschr., Vol. 14, pp. 257–277, plates 1–12, Figures 1–13.

―――――, 1931. Neue Mikrofossilien des baltischen Silurs, I: Pal. Zeitschr., Vol. 13, pp. 74–118, plates 1–5, Figures 1–5.

Fritsch, F. E., 1935. See references Chapter 3.

Grignani, D., and M. P. Mantovani, 1964. Les Chitinozoaires du sondage Oum Doul I (Maroc): Rev. Micropal., Vol. 6 (4), pp. 243–258, plates 1–4.

Jansonius, J., 1964. Morphology and classification of some Chitinozoa: Canad. Petrol. Geol. Bull., Vol. 12, pp. 901–918, 2 plates, 2 figures.

Jodry, R. L., and D. E. Campau, 1961. Small pseudochitinous and resinous microfossils: A.A.P.G. Bull., Vol. 45 (8), pp. 1378–1391, 3 plates, 5 figures.

Kozlowski, R., 1963. Sur la nature des Chitinozoaires: Acta Paleont. Polonica, Vol. 8 (4), pp. 427–445, Figures 1–11.

Schallreuter, R., 1963. Neue Chitinozoen aus ordovizischen Geschieben und Bemerkungen zur Gattung *Illichitina*: Paläontologische Abhandlungen Geol. Gesellsch. Deutsch. Dem. Rep., Vol. 1, No. 4, pp. 391–405, plates 1–2.

Staplin, F. L., 1961. Reef-controlled distribution of Devonian microplankton in Alberta: Palaeontology, Vol. 4 (3), pp. 392–424, plates 48–51.

Tasch, P., and J. R. Stude, 1966. Permian scolecodonts from the Ft. Riley Limestone of Southeastern Kansas: Wichita State Univ. Bull., Univ. Studies, Vol. 68 (3), 35 pp., 3 plates.

Taugourdeau, P., 1962a. Associations de Chitinozoaires dans quelques sondages de la region d'Edjelé (Sahara): Rev. Micropal., Vol. 4 (4), pp. 229–236, 1 plate, 1 table.

―――――, 1962b. Association de Chitinozoaires sahariens du Gothlandien supérieur (Ludlowien): Bull. Soc. Géol. de France, Ser. 7, Vol. 4 (6), pp. 806–808, Figures 1–8.

―――――, 1961. Chitinozoaires due Silurien d'Aquitaine: Rev. Micropal., Vol. 4 (3), pp. 135–154, plates 1–6.

――――― and B. de Jekhowsky, 1960. Répartition et description des Chitinozoaires Siluro-Dévoniens de quelques sondages de la C.R.E.P.S., de la C.F.P.A. et de la S.N. Repal au Sahara: Rev. Inst. Franc. Pétr., Vol. 15 (9), pp. 1199–1260, plates 1–12, Figures 1–21.

Wilson, L. R., 1958. A chitinozoan faunule from the Sylvan Shale of Oklahoma: Okla. Geol. Notes, Vol. 18 (4), pp. 67–71, 1 plate.

――――― and R. W. Hedlund, 1964. *Calpichitina scabiosa*, a new chitinozoan from the Sylvan Shale (Ordovician) of Oklahoma: Okla. Geol. Notes, Vol. 24 (7), pp. 161–164, 1 plate.

Jenkins, W. A. M., 1970. Chitinozoa, *In* Geoscience and Man (A.A.S.P. Proc., L.S.U.,) vol. 1, p. 1–20, suggested that chitinozoans resemble (and hence might be) reproductive bodies of graptolites. If so, two of the paleocreatures of this chapter would be related after all. One difficulty with Jenkins' suggestion is that chitinozoans are not invariably found in graptolite beds and vice versa. (J. Jansonius, 1969. Classification and stratigraphic application of Chitinozoa. *In*, North Amer. Paleont. Convention, Pt. G, pp. 789–808. P. Tasch and T. J. Hutter, 1978. Chitinozoa in the Pennsylvanian Leavenworth Formation of Kansas. Palinologia, núm. extraord. 1, pp. 443–452.)

chapter fourteen
GENETICS OF SPECIATION: MODERN CONCEPTS APPLIED TO FOSSIL POPULATIONS

INTRODUCTION

Speciation and evolution were briefly considered in Chapter 1. In Chapter 2, genetic code and natural selection were examined. Further, observations were made earlier on speciation among the several invertebrate phyla. Here speciation and variation with fossil populations will be more closely studied.

CRITICAL PRELIMINARIES

Basic Genetic Alphabet

The older concept of the gene as a particulate entity of the chromosome that functions, mutates, and is the smallest unit of recombination, has been replaced (Mayr, 1963). The picture emerging from molecular genetics is more complex (Waddington, 1962; Mayr, 1963). To follow contemporary thought, review Figure 2.7, Chapter 2.

A segment of the DNA molecule is schematically shown in Figure 2.7. The entire configuration illustrated in part constitutes one given gene (that is, two linear chains of nucleotides twisted or coiled about each other in a spiral). Many such genes occur in the linear array of varisized units comprising any given chromosome. To distinguish one DNA segment (gene) from another on a given chromosome, the place where it occurs is called a given "locus," example, C-locus.

At each such gene locus are sub-units (sites) called nucleotides. The essential alphabet of a nucleotide (Figure 2.7) is A, C, G, T—corresponding to four bases: A—adenine, C—cytosine, G—guanine, T—thymine. Experiments have shown that only certain arrangements are possible in ordering nucleotides along the double spiral. Thus in Waddington's illustration (Figure 2.7) the code along one strand is $TCAAGATG$. Since thymine can only fit with adenine, cytosine with guanine, the code on the other strand must be $AGTTCGTAC$.

Cistron

Given the gene and its subunits (that is, a gene locus such as the C-locus and its individual sites, TA, CG, AT, and so on, in Figure 2.7), then the *cistron* is the summation or totality of both. The cistron is considered the functionary gene controlling a given function, such as enzyme production (Mayr, 1963). More speculatively, one or more cistrons might also control visible traits such as pelecypod dental formula. This last type of control has yet to be experimentally demonstrated.

Unit of Mutation

A mutation can affect a single site (nucleotide pair—TA or CG, and so on) on a given gene locus. Or the mutation may affect an entire chromosome section (with numerous cistrons). Mayr (1963) estimated the probable mutation rate in a theoretical species. Given: a population of $n = 3 - 4 \times 10^6$ individuals; assume several hundred to thousands of available gene loci, and these with several hundred thousand nucleotide sites, then it follows that *several mutations per gene locus per generation* are likely on probability considerations alone.

These mutations actually have small effects and changes so induced in skeletal or shell features are rare to absent. How then, do morphological changes observed in skeletal material come about? (See "Natural Selection" below.)

Genotype and Phenotype

The *genotype* of any particular animal, a bryozoan, for example, is the sum of genetic factors transmitted by heredity (from parents, grandparents, and so on). The genotype constitutes the individual's genetic makeup; it is relatively constant for an individual. The *phenotype* corresponds to the individual's appearance; it is the sum of an individual's characteristics arising from genotype interaction with the environment. Throughout an individual's lifetime, the phenotype varies (Dobzhansky, 1955). Comparison of larval, postlarval, and adult morphology of a fossil brachiopod reveals a sequence of changing phenotypes.

Genetic Basis of Individual Traits

Given the set of characteristics found in an orthocone cephalopod, shape, septation, cameral deposits, can any or all of these be ascribed to a given gene? The answer is no. Dobzhansky (1955) stressed that observation and/or recording or enumerating kinds of traits in an animal is subjective, a convenience for discourse; it arises from relative intensity of study applied to a given object.

The cistron itself, or large groups of cistrons, may act together as units to control the total morphological pattern of a growing organism (Pontecorvo, cit. Waddington, 1962, p. 51). Exact structure of the unit controlling a suite of characters

in any living thing still eludes researchers (see Chapter 2, Hoyer et al., 1964).

Gene Flow and Recombination

In any given local population, Mayr (1963) estimated that 99 percent of "new" genes in a local gene pool were introduced by immigration (that is, migrant species subpopulations entered a local area; by interbreeding with local residents, they contributed their genes to the local gene pool). This is "gene flow."

Recombination is now judged the main contributing source of genetic variation within populations. How does it work? A genotype is a combination of genes. Given: one parental pair whose genetic makeup or genotype differs only at 100 gene loci. Mayr estimates that thereafter this difference can give rise to 3^{100} *genetically different descendants*. The time involved will vary depending upon the particular reproductive cycle. Surely, many progeny will die young, others will be nonviable or nonreproductive. However, in sexual individuals, recombinations of genes can yield in time a vast array of actual or potential variability. Furthermore, new genotypes in reality do arise in this way.

Although mutation primarily causes change at a specific nucleotide site, it is *gene flow* of normal plus mutated genes plus *recombinations* of these by interbreeding that account largely for genotypic variation within a population (Mayr, 1963, p. 171).

Natural Selection

Additional notations on natural selection are necessary (Mayr, 1963, Chapter 8):

1. Chance does enter into natural selection. It is therefore a statistical phenomenon. Accordingly, its exact effect cannot be predicted. Future evolutionary history may be surmised but not forecast. Modern coleoid cephalopods (Chapter 8) are a splendid example. In this group an unexpected expansion "into mollusks of superior power and activity to all others"

(Morton, 1960) followed shell reduction. Selection pressures at work.

2. What does natural selection (N.S.) accomplish? How does it work? N.S. brings together (selects) favorable gene combinations (genotypes). N.S. effects frequency of given genes both mutated and nonmutated in a particular gene pool. Those that confer an adaptive advantage are favored; others are selected out.

Natural selection, in some cases (Mayr, 1960), not bringing together individual units operates only after they have been combined into a new character complex (genotype).

3. Mayr's definition of the phenotype is illuminating. The phenotype is a compromise of *all* selection pressures. Some pressures operate in one direction, others in the opposite. Skeletal characteristics of an invertebrate fossil, or the sum of these (that is, its appearance or phenotype), is a balance or compromise between contrary selection pressures.

In a fossil shell or carapace, a structure might have conferred a selective advantage. It might have served a protective function. Contrarily, this might increase individual longevity and hence, population density on a given marine substrate. High population density could, in turn, diminish available food supply (assuming sessile animals) in an area of the sea floor. Opposing pressures are thus at work on the gene pool of this population. If gene combinations favor protective structures and are selected, in time survival per individual declines. If other gene combinations are selected instead, the individual's longevity will still create a survival threat through the food supply. The solution arrived at (that is, the phenotype) presumably will be a balance between some such indicated selection pressures.

Subsystems that Affect Gene Frequency Changes

Waddington (1959, 1960) suggested that the evolutionary system involved four major subsystems: (1) genetic, (2) natural

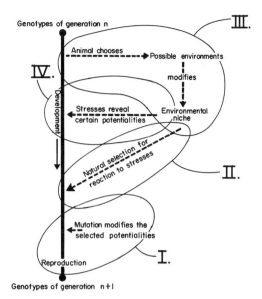

FIGURE 14.1 Subsystems that affect gene frequency change. Adapted from Waddington, 1959, 1960. Subsystems: I—Genetic. II—Natural Selection. III—Exploitive. IV—Epigenetic.

selection, (3) exploitive, (4) epigenetic (Figure 14.1).

1. Genetic Subsystem. This involves the complete chromosomal-genic mechanism of hereditary transmission.

2. Natural Selection Subsystem. Subsystems 1, 3, and 4, as well as natural selection affect gene frequencies.

3. Exploitive Subsystem. A given environment has *n* open and available niches. Among these possibilities, an individual animal can choose to exploit or modify niche 1 or niche *x*. Waddington sees the complex of processes in such choice as the exploitive subsystem.

4. Epigenetic Subsystem. This sequence of causal processes advances development from fertilized egg to fertile adult. (Referring to some evolving organism, a coral, it is important to understand that each subsystem partially determines the others.)

Polymorphism and Pleiotrophy

Polymorphism. Between individuals of an invertebrate population, existing differences are commonly small and intergrading. In some species one can readily separate the sample into groups sharing one or another characteristic—maybe shell color or pattern as in African Pleistocene land snails (Owen, 1966). Within the same land snail population, forms coiled right or left occur; or, among insects, venational patterns may vary in parts only. Grouping reflects discontinuous individual variation within a species population. It is known as *genetic polymorphism*—and may be sex- or nonsex-related (Mayr et al., 1953).

Variation observed in such instances appears to be controlled by a single gene. Within a given gene pool, so-called *morph genes* coexist with others and affect morphological and/or other characters (physiological, and so on) of the phenotype (Mayr, 1963).

Other variations within a population may not have a genetic basis such as age, seasonal, social, ecological or traumatic (Mayr et al., 1953).

What neo- and paleobiologists have called "varieties" of a given species are actually, in many cases, *polymorphs*. The term "variety," ambiguous and misleading, should be abandoned. Dobzhansky and others demonstrated that "several alternate gene arrangements" can coexist in a given population. Selection controls which of the available "gene arrangements" will appear in a given sector of a species geographic spread.

Pleiotrophy. Every gene (gene and its subunits, as defined earlier) apparently has a greater or lesser capacity to promote multiple effects in the phenotype — a condition known as *pleiotrophy*. A given gene might affect pigmentation, chemical constitution, and biological features—reproduction, ontogenetic development, oxygen consumption, and so on (Mayr, 1963, Table 7-2).

Many characters of fossil shells, carapaces, and so on, were probably governed, controlled, or affected by pleiotrophic gene action.

POPULATION VARIATION AND SPECIATION

In addition to "Critical Preliminaries," several biometrical studies of fossil invertebrates and vertebrates contribute insights and analytical approaches to population studies—Olson and Miller (1958), Imbrie (1956), among others. These would take us too far afield and hence are not examined in detail here. Statistics as used in these studies are more adequately and fittingly studied in courses offered by Departments of Mathematics. Serious students should avail themselves of this valuable background early in their careers.

Researches considered here, in whole or part, include three on the Coelenterata (Wells, 1937; Oliver, 1958; Stumm, 1964), four on Brachiopoda (Corroy, 1927; Fage, 1934; Alexander, 1949; Grinnell and Andrews, 1964), one on Mollusca (Stenzel, 1949), and one on the Echinodermata (Joysey, 1960).

Coelenterata

Varieties or Forma. Wells (1937) resampled the coral bed of the Hamilton shale (M. Devonian) of New York State (Onondaga County). He was curious about individual variation in a Devonian rugose coral population living under the same environmental conditions.

In 1876, James Hall had figured Hamilton corals including five new species, one old species, and two new varieties of the genus *Heliophyllum* (see Chapter 4). Hall's species were *H. arachne*, *H. confluens*, *H. degener*, *H. irregulare*, *H. proliferum*. The old species named by Edwards and Haime was *H. halli*. Hall's two varieties were *H. halli* var. *obconicum*, and *H. halli* var. *reflexum*.

Wells, comparing a few specimens collected at two localities with those figures by Hall, found no difficulty in identification of species and varieties. However, he realized that after accumulating several hundred specimens and trying to relate them to the fewer number identified from Hall's figures, difficulties arose (Figure 14.2).

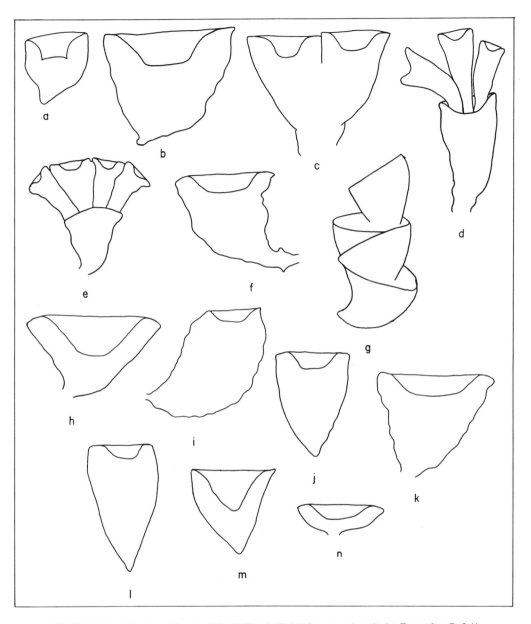

FIGURE 14.2 Individual variation in *Heliophyllum halli*, (all figures reduced). (*a*) Type after E. & H., 1851. (*b*) Specimen from Ludlowville Formation. (*c*) *H. confluens.* (*d*) *H. proliferum.* (*e*) *H. halli* forma *praecoquus.* (*f*) *H. halli reflexum.* (*g*) *H. halli* forma *pravum.* (*h*) *H. arachne.* (*i*) *H. degener.* (*j*) *H. halli obconicum.* (*k*) *H. halli-reflexum-obconicum.* (*l*) *H. halli* "*degener-reflexum obconicum.*" (*m*) *H. halli* forma *infundibulum.* (*n*) *H. halli* forma *aplatum.*

Analysis of his collections and comparison with Hall's figures led Wells to conclude: because a large number of specimens showed intermediate stages between "species" and "varieties" (Figure 14.2, *k*, *l*), only one species, *Heliophyllum halli*, and only one variety, *H. halli confluens*, were present. Instead of multiple species

and varieties, Wells included these in eleven formae (that is, morphological forms) of a single species.

Causes of Local Variation. What factors produced a broad spectrum of variation in growth forms within the species population *H. halli*? Table 14.1 summar-

TABLE 14.1. Causes of Local Variation and Forma[a] of *Heliophyllum halli*

1. Reproductive precocity	*H. halli,* forma *praecoquus*
2. Stability (relative to substrate)	
(a) Stable	
aa. Straight conical corallum	*H. halli,* forma *obconicum*
bb. Curved corallum	*H. halli,* forma *typicum*
(b) Unstable	*H. halli,* forma *irregulare*
3. Gerontism (old age)	
(a) Senility	*H. halli,* forma *degener*
(b) Gerontic budding	*H. halli,* forma *proliferum*
4. Irregular rejuvenescence	*H. halli,* forma *proliferum*
5. Differential growth rate	
(a) Growth rate of polyp faster than development of corallum	
aa. Corallum broadly conical	*H. halli,* forma *arachne*
bb. Corallum patelliform	*H. halli,* forma *aplatum*
(b) Variation in growth rate of parts of polyp (see Chapter 5)	
aa. Rate of lumen faster than periphery	*H. halli,* forma *reflexum*
bb. Rate of periphery faster than lumen	*H. halli,* forma *infundibulum*[b]

[a]*H. halli* var. *confluens* is excluded from this table (see text).
[b]Stumm (1964) recognized Hall's species, *H. infundibulum,* as a distinct species from *H. halli* and not as a *forma*
Wells (1937) did not provide new data in this instance. *H. halli* and *H. infundibulum* occurred together in the
Beechwood limestone (Figure 14.3).
Source. Data from Wells, 1937.

izes such factors for each forma (compare with Figure 14.2).

There were at least five possible causes for local variation in this species population. Another study (Oliver, 1958) of the Onondaga rugose corals (these strata underlie the Hamilton) was noted in Chapter 4. Certain aspects will be discussed here that were not previously considered. Oliver's sample ($n = 250$) came from a single horizon and locality, and represented a single species, *Metriophyllum (Aemulophyllum) exiguum*. Six formae were recognized. These formed a completely intergradational system. The basis for distinguishing formae is indicated in Table 14.2 (see Chapter 5).

Oliver avoided attributing the observed

TABLE 14.2. Analytical Key to the Formae of the Onondaga Rugose *Coral Metriophyllum (Aemulophyllum) exiguum*

	Forma
I. Patellate corallum	
A. Apical angle $> 80°$	a_1
B. Apical angle $< 80°$	a_2
II. Elongate corallum	
A. Corallum patellate becoming subcylindrical	
1. Cylindrical stage $< 1/2$ of length	B_1
2. Cylindrical stage $> 1/2$ of length	B_2
(a) Stages sharply defined	
(i) Diameter > 12 mm	γ
(ii) Diameter < 12 mm	ϵ
B. Corallum subconical	
1. Apical angle $70°–85°$, diameter > 12 mm	ζ
2. Apical angle $60°–70°$, diameter > 12 mm	ζ

Source. Data from Oliver, 1958.

variation to a particular cause for each *forma*. He qualifiedly favored a probable genetic basis for both differential growth rate and stage development (Table 14.2).

The use of *forma*—morphological growth forms—has no formal taxonomic standing such as a species. Oliver's use of Greek letters to designate "form variance" within a species is preferable to bestowing distinct names. Wells' variety, *H. halli* var. *confluens* does not differ from *H. halli* in any critical features; as Wells noted, a colonial form of *H. halli* results from peripheral increase.

Morph Genes. In both studies, apparently morph genes can account for many so-called forma (for example, Table 14.1, "Differential growth rate"). "Stability" derives from coralline form (Oliver, 1958) and is attributable to genetics. Irregular rejuvenescence may be due to temperature or to other external factors as are forma due to gerontism (age variation). Reproductive precocity, bringing early maturation, may be sponsored by temperature or other external factor.

The condition of genetic polymorphism may occur with nongenetic variance within a given species population as in the examples given.

Favosites. Stumm (1964) discussed and figured Silurian and Devonian corals from the Falls of the Ohio River. Because he studied some two dozen species of one coralline genus, *Favosites*, some comparisons are possible. How do species from the same locality and formation (zone) differ? What new aspect do species have that replace older species in the same area (Figure 14.3)?

Favosites species recovered from the Louisville limestone displayed the following characteristics.

1. ***Configuration.*** *F. discoideus* coralla vary from low hemispherical to discoid. *F. favosus* and *F. niagarensis* are both subhemispherical and hemispherical, and *F. discus* is flat and discoid. There is an overlap of all shape variants.

2. ***Corallites—shape and diameter.*** All corallites are polygonal in shape (except *F. discus* whose thick walls give apertures a rounded appearance). *F. discoideus* and *F. discus* both have corallites of two sizes: 3–4 mm or 0.8–1.0 mm and 1.5 mm or 0.2–0.6 mm. *F. favosus* corallites are 4–5 mm, while those of *F. niagarensis* are 1.5 mm in diameter. Corallite size for the last two species range within either larger or smaller corallite size for species with dimorphic corallites.

3. ***Tabulae.*** In all four cited species, tabulae are *complete*. Distance apart of tabulae for three species is less than 1.5 mm and for *F. discoideus*, 1.0–3.0 mm apart. All tabulae are horizontal except *F. favosus* whose tabulae are distally convex.

4. ***Mural Pores.*** Mural pores are present in three species. They have not been observed in *F. discus*. Pores are uni/biserial in two species; triserial (that is, three rows) in one species—*F. niagarensis*.

5. ***Corallite Walls.*** Corallite walls are thin in one species, thick in another, and in *F. favosus* are divided into twelve vertically fluted segments.

All commonly shared characteristics of these four species were encoded in stable gene combinations within the gene pool. These are the generic characters. Features such as fluted segment of walls and distally convex tabulae of *F. favosus* may be viewed as phenotypic expressions of new gene combinations within the common gene pool. Creation of novelty is the work of natural selection.

Original Setting. During Middle Silurian time, a common seaway spread over the areas where *Favosites* species occur in the rock record (Indiana, Kentucky, New York, and Tennessee). Within this geographic spread and in the shallow epicontinental sea of the time, ecological niches occupied by these species were probably similar to those now occupied by modern lagoonal reef corals (see Chapter 5; Vaughan, 1911; Wells, 1957).

In these niches the common gene pool became fractionated, genes mutated, then

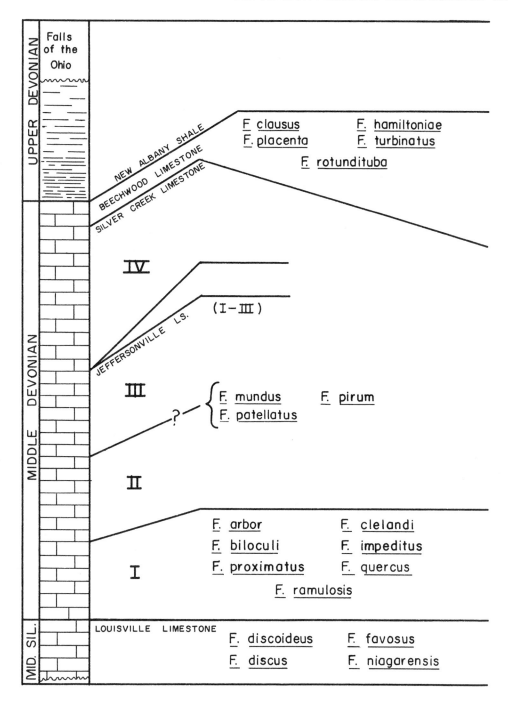

FIGURE 14.3 Species of the tabulate coral *Favosites* from the Falls of the Ohio, Louisville, Kentucky. I—Coral Zone. II—*Brevispirifer* Zone. III—*Paraspirifer* Zone. IV—*Platyrachella oweni* Zone. F = *Favosites*. Data from Stumm, 1964. Figure 2; Table 1. Note that no *Favosites* species were present in this area from Upper Silurian, Lower Devonian, or during much of Upper Middle Devonian time. The unconformity between Louisville and Jeffersonville limestones represents an estimated 20 + million years.

both mutated genes as well as stable genes were selected and formed new combinations. But all the characters of the *Favosites* species reviewed above need not have been genetically based, for example, spacing of tabulae. Duncan (1956; Chapter 5 this text) thought tabulae spacing indicated seasonal growth in one tabulate coral where tabulae were periodically more closely spaced. What of the variable spacing of tabulae between the four species? Was this also a non-inherited character attributable to external factors, temperature or some other?

Significance of Certain Coralline Structures.

Dimorphic corallites in some individuals are probably due to inherited factors. Jones (1936; Chapter 5 this text) attributed the unequal sizes of corallites in given tabulates (*Favosites*) to differential growth rates: equi-sized corallites indicated normal growth with a moderate budding rate in two of the above-mentioned species; unequal corallites represented slow growth with relatively rapid budding rate. Wells (Table 14.1) also suggested different forma of *Heliophyllum halli* could be attributed to differential growth rate. Both examples may be explicable in terms of morph genes (genetic polymorphism).

Can more be said about mural pore relationship to intercorallite communication? While two, or more rarely, three rows of pores provided greater communication between adjacent corallites than one row, pores have been interpreted to be *aborted buds* occurring at sites where new corallites would normally appear (Zittel, and others). If this version is correct, mural pores are related to the reproductive system. The total number of such pores in a given corallite would then be an index of the total number of aborted buds. In turn, this probably relates to one or more lethal mutant loci in the genotype (within the functional gene — cistron). Occurrence and number of mural pores may be the result of weeding out unfavorable genotype components.

A by-product of the above-postulated event has interest — an intercorallite communication system. In time, greater selection pressure would exist for more aborted buds, if this system provided an adaptive advantage, and if it was favored by natural selection. This may explain the occasional tendency for the triserial condition. The mural pore example illustrates very well what Mayr wished to denote in defining the phenotype as compromise between all selection pressures.

Allowing for genetic polymorphism within a species population, for external factors, for seasonal and other changes reflected in the genotype, what actual differences remain among the four *Favosites* species? Differences include: coralla configuration, relative thickness and/or fluting of corallite walls, presence or absence of dimorphic corallites, and number of rows of mural pores. The last two items may be attributed to different morph genes occurring in some genotypes and not in others.

Individual corallite shape can be explained: given a system of cylinders or spheres so associated that their walls are in complete surface contact (that is, compressed together), then in section or surface view there will be an hexagonal (or other polygonal) pattern. Subsequent thickening by calcite deposition (secondary deposits as in *F. discus*) will restore the original circular cross-section. Why? It must follow the law of minimal area achieved by filling up the angles of the primary cell (see D'Arcy Thompson, 1942, pp. 509 ff). Briefly, a circle is the limiting condition for an inscribed n-sided polygon; under compression, the circle reverts to the polygonal configuration. Accordingly, corallite shape in *Favosites*, among other tabulates, can be attributed to external mechanical factors. This feature, therefore, can hardly be diagnostic of any species.

Another question concerns the suite of *Favosites* species in Zone 1 of the Middle Devonian Jeffersonville limestone (Figure 14.3) that succeeded the Middle Silurian Louisville limestone assemblage. The time lapse between the Louisville and Jeffersonville limestones was an estimated 20+ million years. What *new* aspect(s) did this suite of Middle Devonian *Favosites* species display?

Because some seven new species of the same genus, *Favosites*, appear in Zone 1, stable gene combinations governing generic characters must have persisted. In the Middle Silurian suite of *Favosites* species, four variables distinguished one species from another. Did these variables retain their variable qualities into Middle Devonian time?

Coralla Configuration. The dendroid, staghorn growth form not seen in the Louisville limestone species was new. Of seven species, only one, *F. clelandi* was "discoid to low hemispherical." Another, *F. biloculi*, had a range of growth forms: "discoid, palmate, hemispherical or irregular branching." The remaining species were all dendroid, either with staghorn growth form, irregularly palmate, or with ramose branching.

Relative Thickness and/or Fluting of Corallite Walls. Generally, corallite walls in the seven species were thin; exception—*F. impeditus*, with relatively thick walls. Thin walls of *F. biloculi* were either smooth or wrinkled (this last denotes possible obsolescence of the fluted condition). Thin walls were not new.

Presence or Absence of Dimorphic Corallites. Three of seven species (Figure 14.3) had dimorphic corallites—*F. arbor*, *F. impeditus*, *F. proximatus*; both larger and smaller were in the size range of the Middle Silurian species reviewed earlier. The other species, with corallites of equal size, had diameters within size range of Middle Silurian forms. Thus either presence or absence of dimorphism was not new.

Number of Rows of Mural Pores. Mural pores were either uniserial (in four species), biserial (in one species), biserial-

triserial (in *F. biloculi*). In *F. ramulosis*, pores occur in one vertical row on each side of each corallite. Except for *F. ramulosis*, this feature was not new.

Tabulae-spacing can be regarded as probably caused by seasonal changes, that is, nongenetic factors But changed aspects in the very tabulae themselves remain for Middle Devonian *Favosites* species.

Tabulae. All Middle Silurian species had complete or horizontal tabulae. In several Middle Devonian species, tabulae are both complete and incomplete (*F. arbor*, *F. clelandi*, *F. proximatus*). In the first and last-named species, further variations were observed: *F. arbor* had cystose tabulae in part; *F. proximatus* displayed a dual condition of "complete-incomplete tabulae" producing an "anastomosing network" in parts of some corallites. *These tabulae variations are a new aspect* not observed in the Middle Silurian species.

In summary, after a twenty plus million year lapse, the new suite of *Favosites* species found in exposures along the Falls of the Ohio, retained virtually unchanged, many characteristics of Middle Silurian species. In addition, a few new aspects did appear. These changes may be attributed to mutated gene effects, pleiotrophic genes, morph genes, and the work of natural selection controlling favored recombinations of these. Phenotypic expressions of changed genotypes were reflected primarily in *tabulae* and *coralla configuration* (ecological changes may have enhanced selective pressures favoring dendroid-staghorn growth forms). Because tabulae partition the corallite when the polyp vacates a lower level during growth, such variation must have been part of the skeletal demands of the growing process.

If incomplete, cystose, or anastomosing tabulae were modifications of the "complete" condition, they may have arisen as mutated genes in several genotypes. However, it is conceivable that they were incidental by-products—one of multiple phenotypic effects of pleiotrophic genes.

Brachiopoda

Three studies describing so-called varieties illustrate one kind of problem arising in the study of variation in fossil brachiopod populations. These examples cover Paleozoic and Mesozoic articulates (Corroy, 1927; Fage, 1934; Alexander, 1948).

Atrypa. Data from the Alexander study on *Atrypa reticularis*, are summarized in Table 14.3. This investigator examined many hundreds of *Atrypa* specimens from America and Europe, including the British Isles.

Major findings of the Alexander study were the following.

(a) *Atrypa reticularis sensu stricto* (that is, in the restricted sense) does not appear in Silurian beds of the British Isles.

(b) There are eight varieties in the species, two being local and found only at Dudley (see Table 14.3 — var. *lonsdalei*; var. *sowerbyi*).

(c) Exclusive of the Dudley varieties, each of the other six varieties dominate in one of the main Silurian subdivisions (Table 14.3) — although they all do occur subordinately in other subdivisions. Extremely common, such varieties can serve for rough zonal work in England. Kozlowski (1929, pp. 169–173, table, p. 24) utilized the variable external characters of the Polish *Atrypa reticularis* for partial correlations by localities and horizons. He recognized three varieties of this species: *A. dzwinogrodensis, A. tajnensis, A. nieczlawiensis.* The last-named variety was restricted to the Borszcow Stage (Polish Gotlandian).

Recall that *Atrypa* species are cosmopolitan and range from Silurian to Devonian. Alexander referred to the six varieties of *Atrypa reticularis* (Table 14.3) as an "evolutionary series" and the two Dudley varieties as "offshoots of the main line." Dudley forms were thus geographic isolates from *Atrypa reticularis* populations during Wenlock limestone time and subsequently during Aymestry limestone time. Nevertheless, implied in Alexander's explanation, is the concept that so-

called varieties are effectively subspecies. Are they?

Additional evolutionary concepts need review for the answer.

Polytypic and Monotypic Species. A polytypic species is defined as a species containing two or more subspecies (see Chapter 2). A monotypic species is not divided into sub-populations (subspecies).

Are vertical polytypic species possible? In response to my question Dr. Ernst Mayr explained that once you accept the notion of vertical subspecies, you must, of necessity, accept vertical polytypic species. The answer is yes, you can define a vertical polytypic species. It will be based on vertical distribution of its subspecies (see "Subspecies" below).

Allopatric and Sympatric Populations. Allopatric populations occupy *mutually exclusive* (usually adjacent) geographical areas. Sympatric populations occupy the *same* area (Mayr, 1942, 1963).

Subspecies. A subspecies consists of geographically isolated subpopulations of a parent species that differ from other subpopulations of the same species in one or more characters. The term "subspecies" may be used in paleobiology for "any subspecific category" whether or not it occurs on the same time plane, that is, is contemporaneous with the parent species (Mayr et al., 1953, p. 37). Thus both vertical subspecies (geologic — through time), as well as, horizontal subspecies (geographic — at the same time) are possible.

Lateral Variation in Atrypa Reticularis. The next step examines the nature of variation in Alexander's varieties. From Table 14.3 and her discussion, apparently *Atrypa reticularis* extends through *all* the British subdivisions of the Silurian. What constants, what variations are displayed by successive populations of this species from the same general region?

1. The hinge, general plan of the brachidium (see Figure 14.4), and ornament are relatively constant. This denotes cer-

tain stable gene combinations in successive gene pools (see below—Duration, 10–15 ± million years).

2. Shape (elongate, circular, globular, and so on), size of shell (average length, 18–27 mm; average width, 16–31 mm), costation (coarse to fine; four to eight costae in 5 mm), and shell thickness (thin to very thick)—all are variables.

Alexander interpreted the constant items in (1) to indicate one species—*Atrypa reticularis*—represented in both the type material from Gotland and the specimens from Britain (Table 14.3). The variable items in (2) were attributed to two factors: a time function and an environmental function. Thus, through time (top of Llandovery to *Dayia navicula* beds—Table 14.3), these changes occurred: Average adult size steadily increased, brachial valve convexity increased greatly, while pedicle valve decreased slightly; shell outline elongated at both ends of the time series shown in Table 14.3; wide or circular outlines occurred in the middle of this series; shell depth (pertaining to interior volume) also increased; inner layer of shell steadily thickened.

A survey of the oldest atrypas in this series (Llandovery) shows that they had small, almost equiconvex, shells. After a 10 to 15 million year lapse, large, thick-shelled atrypas appeared (*Dayia navicula* beds). These had a gibbous (= hunched, shaped almost like a half-moon) brachial valve and a nearly plane pedicle valve. Throughout this time span British atrypas showed "a steady evolution." Further, environment apparently influenced costation, which was either coarse or fine (see Chapter 7). Finally, costate forms dominated in the calcareous seas (Wenlock, Aymestry limestone).

Variation within a given "variety"—var. *davidsoni*—also has interest. Specimens displayed variable shell size and thickness. Size and thickness were greater in calcareous seas than in seas with detrital (sandy) substrates.

Varieties or Subspecies (?). From the concepts reviewed, it is clear that *Atrypa reticularis* varieties did show notable dif-

ferences from the type species where they had some degree of geographic isolation. This implied parental gene pool fractionation. Dudley "varieties" which occur nowhere else, and the Gotlandian type species (*sensu stricto*), absent in the British Isles, illustrate such isolation.

Envision successive *Atrypa reticularis* populations during the British Silurian in which numerous subpopulations developed by migrating to different sectors in the species range. Articulates provide a plausible mechanism for this migration through the free-living larval stage. At least one probable articulate brachiopod polytypic species, is on record. Helmcke (1940, cit. Mayr, 1942) defined three geographic races of living genus *Cryptopora* (Eocene-to-Recent) in modern seas. These include a North Atlantic race—*C. gnomon*, a South African race—*C. boettgeri*, and a South Australian race—*C. bratzieri*. The existence of these races—geographic isolates—indicates how polytypic species of brachiopods could have arisen in the geologic past.

Another clue to geographic isolation among the *Atrypa* populations in Silurian seas is in the patchy distribution of articulates in modern seas (see Chapter 7). Increased distances between individual brachiopod patches or clusters could have preceded more extensive geographic isolation.

Several of Alexander's varieties coexisted at a given time. Nevertheless, only one became predominant. Presumably, the dominant variety for each Silurian subdivision increased at the expense of competing with other "varieties." At first, populations could have been sympatric, later becoming allopatric. Effective but incomplete reproductive isolation of subpopulations from the parental population would prevail. In turn, this accounted for Alexander's "varietal"—really subspecific differences. Complete reproductive isolation was never achieved even over a span of 10 to 15 million years. Therefore new species never arose in the study area. Subspecies, however, did persist in time as ancestors and/or descendants.

TABLE 14.3. Summary of the Characters of the British Varieties of the Articulate Brachiopod *Atrypa reticularis* in the on the bottom. The *Dayia* beds are in the Mocktree shale. The entire sequence shown is part of the calcareous (shell)

Variety	Horizon	Average Length	External Characters Average Breadth	Average Depth	Shape	Costation	Growth-Lines
Davidsoni	Top of Lland-overy and Wool-hope Limestone	Complete specimens rare about 18 mm	About 18 mm	About 10 mm	Elongate, Both valves moderately convex	Coarse, 4 in 5 mm	Weak, Lamellae thin, feeble
Harknessi	Wenlock Shales	17 mm	16 mm	11 mm	Circular or elongate. Convexity of brachial valve moderate, slightly greater than in pedicle valve	Coarse, some fine, 5 or 6 in 5 mm	Rather weak. Lamellae thin, short
Lapworthi	Wenlock Limestone	18 mm	18 mm	11 mm	Wide. Convexity of brachial valve moderate, slightly greater than in pedicle valve	Coarse or fine, 5 to 7 in 5 mm	Well marked. Lamellae large, thin
Lonsdalei	Wenlock Limestone of Dudley	16 mm	18 mm	10 mm	Wide to circular. Brachial valve moderately, pedicle valve slightly convex	Fine, 7 or 8 in 5 mm	Strong. Lamellae large, thin
Murchisoni	Lower Ludlow Shales	17 mm	17 mm	14 mm	Globular. Brachial valve strongly, pedicle valve slightly convex	Coarse or fine, 5 to 7 in 5 mm	Strong. Lamellae short. thick
Sedgwicki	Aymestry Limestone	27 mm	30 mm	18 mm	Circular. Brachial valve very strongly curved, pedicle valve flat. Margin recurved.	Coarse or fine 4 to 7 in 5 mm	Strong. Lamellae large, thick
Sowerbyi	Aymestry Limestone of Dudley	23 mm	31 mm	12 mm	Wide to circular. Brachial valve strongly curved, pedicle valve flat. Margin recurved	Fine, 7 or 8 in 5 mm	Strong. Lamellae large, thin
Woodwardi	Davia beds	Complete specimens rare about 24 mm	About 21 mm	About 14 mm	Elongate. Brachial valve very strongly curved, pedicle valve flat. Margin re-curved.	Coarse, 4 in 5 mm	Strong. Lamellae large, thick

Source. After Alexander, 1949.

Silurian (Gotlandian) of England. [Formations shown represent older strata on the top and younger facies.]

Flange	Brachial Valve Impressions of Soft Parts	Thickness of Shell	Turns Per Spire	Internal Characters Pedicle Valve Impressions of Soft Parts	Muscle Callosity	Thickness of Shell	Groove for First Ascending Lamella
Feeble	Weak	Thin	Few	Weak	Absent	Thin	Absent
Feeble	Weak	Rather thin	10–12	Moderate	Absent	Rather thin	Very weak
Large, thin	Moderate	Rather thick	14	Fairly strong	Very weak	Rather thick	Weak
Short, thick	Moderate	Rather thick	12–14	Fairly strong	Weak	Rather thick	Moderate
Short, thick	Strong	Thick	10	Strong	Strong	Thick	Fairly strong
Large, thick	Very strong	Very thick	17	Very strong	Strong	Very thick	Very strong
Short, thick	Very strong	Very thick	17	Very strong	Strong	Very thick	Very strong
Large, thick	Very strong	Very thick	17	Very strong	Very strong	Very thick	Very strong

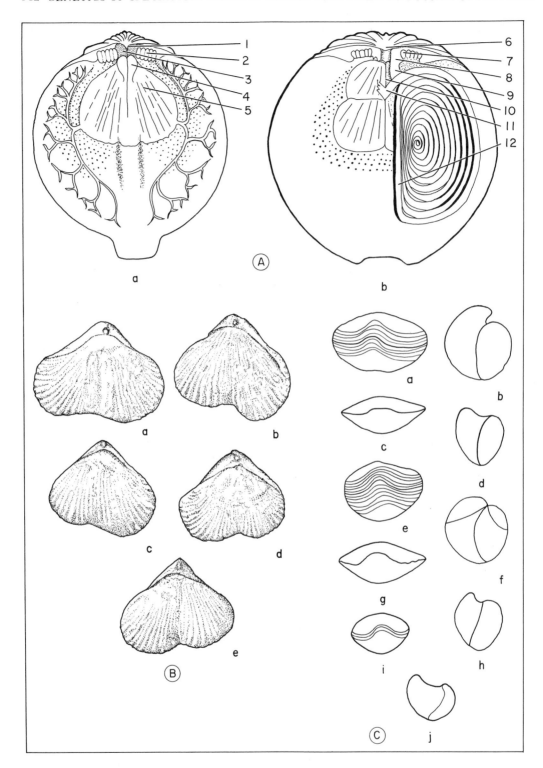

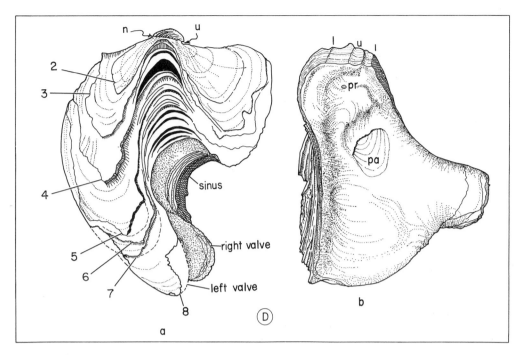

FIGURE 14.4 Articulate brachiopods to illustrate gross morphology and/or intraspecific and interspecific variations.

(A) *Atrypa reticularis*, reconstruction based on serial sections, free valves and internal molds (enlarged). (A) a Pedicle valve (1, umbo; 2, tooth; 3, pseudoseptum; 4, adductor-muscle impression; 5, diductor-muscle impression). (A)b Brachial valve; 6, umbo; 7, socket; 8, callosity for didactor muscle attachment; 9, fused crural base and inner socket walls; 10, pseudoseptum; 11, jugum; 12; first ascending lamella. (Only one spire shown).

(B) *Rhynchonella* (= *Cyclothyris*) *difformis*, U. Cret., Charentes, France. (a) Type of species; (b–e) intermediate forms.

(C) *Spiriferina tumida* and its varieties, L. Jur. (Lias), Alsace and Lorraine, France. (a, b) Type species; (c, d) var. *ascendens*; (e, f) var. *sicula*; (g, h) var. *acuta*; (i, j) *rupestris*. X 0.25.

(D) The Oyster *Cubitostrea sellaeformis*, Archusa Marl, M. Eoc., Clarke County, Miss. (a) Senile shell showing both valves; nos. 2–8 indicate growth stages (maximum age = 8 years); the ribbed neanic stage ends at N, U = umbo and small attachment scar. (b) Senile right valve, same individual as in (a); oblique anterior view of inside of right valve (viewed at about 45° to the hinge line). Note warping of the valve; U—umbo, L—ligament, PA—posterior adductor muscle imprint, PR—pedal retractor muscle imprint. Both figures (a) and (b) are reduced.

Illustrations redrawn and adapted from Alexander, 1948 (Figure A); Fage, 1934 (Figure B); Corroy, 1927 (Figure C); Stenzel, 1949 (Figure D).

In summary, Alexander's varieties representing an evolutionary series through time, are subspecies; *Atrypa reticularis* and all the varieties (= subspecies) shown in Table 14.3 constitute a vertical polytypic species. (The same conclusion, of course, applies to Kozlowski's Polish varieties of the same species.)

Cretaceous Rhynchonellids. Another example of vertical subspecies and polytypic species are two species and their varieties described by Genevieve Fage (1934) from Upper Cretaceous rhynchon-

ellids of Charentes, France. *Rhynchonella* (= *Cyclothyris*) *vespertilio* and *R. difformis* were both studied. The first species is symmetrical, the latter, dissymmetrical (Figure 14.4).

Species previously named, were found to be varieties of the *Rhynchonella* species in Fage's relatively large sample. Sample size was $n = 500$. A set of intermediate forms (Figure 14.4) was found that linked the type species to the form designated as a variety. Only one species, *R.* (= *Cyclothyris*) *difformis*, is discussed here. Fage found that *R. rudis* was indistinguishable from the type *R. globata* (Figure 14.4), with a rounder, thicker shell; otherwise, it was similar to the type. *R. vesticularis* was also found to be like the type except for ornamentation of many fine striae and acuminate (acute) posterior sector of the ventral valve.

As with British Silurian atrypas, both rhynchonellid species—the types and its varieties (= subspecies) showed a vertical distribution through Upper Cretaceous time in the area studied (Table 14.4). Each species constitutes a vertical polytypic species. Since they occurred together in various stages of the French Upper Cretaceous, their respective populations were probably sympatric during life.

Jurassic Spiriferinids. Corroy (1927) also found suites of intermediate forms linking known species of *Spiriferina* (Lower Jurassic—Lias—Alsace-Lorainne, France); end members of the series were designated as "varieties." Stages of the Lower Jurassic (Lias) are from older to younger: Rhaetien, Hettangien, Sinemurien, Charmouthien, Toarcien, and Aalenien. Particular attention was given to *Spiriferina tumida* and its varieties (Figure 14.4) which, along with other spiriferinids, abound in the calcareous facies.

The type species ranges from Upper Hettangien to Lower Toarcien, while the five varieties (*S. ascendens*, *S. sicula*, *S. acuta*, *S. haueri*, and *S. rupestris*) range from Lower Sinemurien to Lower Toarcien. This species varied particularly in general form and in posterior sector of the ventral valve (Figure 14.4).

Costation showed a progressive then subsequently a regressive evolution. The parent population of *Spiriferina tumida* at its first appearance had very fine costae. Successive populations then reached a maximum development of costation in the Sinemurien. In the Charmouthien, this species regressed to weakly expressed costation. Noncostate forms of this species (var. *haueri*) and a few other spiriferinids did continue to the Toarcian.

Corroy (1927, table, pp. 34–35) grouped

spiriferinid species and varieties into four categories: multicostate, markedly costate, weakly costate, noncostate. All *Spiriferina tumida* forms were weakly costate or noncostate.

At the beginning of the Sinemurien, four of the five "varieties" co-existed with the parent species, *S. tumida*. Only one variety made its apparent first appearance in the younger Charmouthien (var. *haueri*). Were these "varieties" *forma* or *subspecies*? Was *S. tumida* a monotypic species displaying polymorphism or a polytypic species with subspecies persisting through time?

Since costation responded to environmental influence, this particular shell character (also of stratigraphic utility) may be disregarded at present. Costation cannot illuminate the evolutionary history. However, as mentioned, marked changes in overall configuration may be attributed to genetic modification mediated by natural selection. These changes can be traced through intermediate forms coextant with both end members of a given series (that is, the type species at one end, and the so-called "variety" at the other).

Additional observations support the interpretation that Corroy's varieties were true subspecies persistent in time. It follows that *S. tumida* was a polytypic species: (1) The parental species did not differentiate into distinctive variants at its first appearance; (2) variants (Corroy's varieties) arose subsequently—implying a degree of geographic isolation; (3) a variant arose after a lapse of time represented by a whole stage—indicative of belated fractionation of part of the parental gene pool.

Composita: Speciation and Phylogeny. Another study is a morphological examination of the long-ranging Paleozoic brachiopod genus *Composita*, common fossils in the midcontinent Pennsylvanian-Permian (Grinnel and Andrews, 1964).

Background Data. The genus ranges from Late Devonian or Early Mississip-

TABLE 14.4. Vertical Distribution of Two Polytypic Species of *Cyclothyris* (= formerly *Rhynchonella*) through Stages of the Upper Cretaceous, Charentes, France[a]

Étages	Notations d'Arnaud	*Rhynchonella vespertilio*						*Rhynchonella difformis*			
		Var. Octoplicata	*Var. expansa*	*Var. Bangasi*	*Var. Eudesi*	Typique	Typique	*Var. Lamarckiana*	*Var. rudis*	*Var. globata*	*Var. vesicularis*
Dordonien	R								+		+
Campanien	Q						+			+	+
	P	+			+		+			+	
Santonien	M, N		+		+		+			+	
Coniacien	K, L			+		+	+		+		
Turonien	H, I	+									
	E, F, G							+			
Cénomanien	C, D										
	A, B										

[a]Instead of "variety" substitute "subspecies"; letters A through R represent zones.
Source. After Fage, 1934.

pian to Late Permian. Forty-two species and "varieties" are embraced by the North American record of this genus.

Actual samples studied by the investigators represented largely midcontinental and eastern interior regions. These were supplemented by published data. The study investigated "the relationships of associated synchronous forms." The collected sample ($n = 1649$) was found on visual inspection to be subdivisible as follows: *C. subtilita*—642 specimens, *C. argentea*—566, *C. ovata*—280, *C. magna*—105, *C. elongata*—56 (Figure 14.5).

Five sampling localities were in the following formations (from older to younger): Pennsylvanian of Kansas—Drum limestone, Plattsburg limestone, Oread limestone, La Salle limestone of Illinois (nonsynchronous with, but separated over a short time interval from the Drum limestone), and the Permian of Kansas—Red Eagle limestone.

Samples were studied by graphical means—frequency polygons and triangular graphs—and by the method of reduced major axis (Imbrie, 1954, and in the cited study, pp. 242–243). Data are not given for sample size for each species per horizon. Investigators noted that *C. subtilita* was present in each of the five localities.

Analysis and Findings. The authors first asked, "Can plotting length or depth against frequency of occurrence (yielding simple frequency polygons) distinguish between five *Composita* species separated visually?" No. Species discrimination was too blurred by this method.

Next, they wanted to know, "Can the triangular graph method (that is, plotting percentage width, length, and depth on separate axes) provide discrimination on the subspecific level?" Although subspecies could be detected, the authors found the results not only varied, but even reversed, depending upon number of points plotted (each point representing an individual in the population).

The same investigators compared the Drum limestone and La Salle limestone

(nearly synchronous) samples. Drum contained many specimens of *C. subtilita* and *C. argentea* and a few of *C. elongata*; La Salle also had an abundance of the first two named species. Statistically significant differences were found between Drum and La Salle samples and also between Oread and Red Eagle samples (nonsynchronous)—both contained representatives of *C. subtilita*, *C. ovata*, and *C. elongata*.

It is unclear what this last finding denotes since species are combined in each sample; each mixed species sample had some of the same species as the compared sample. The authors, however, felt that it denoted significant differences.

Inter- and Intraspecific Variation. With reference to Figures 14.5 and 14.6, it is now possible to examine further aspects of the two types of variation identified: interpopulation (that is, between chronodemes, through time, and intrapopulation (that is, between topodemes, at the same time—see Chapter 2).

Varieties, Conspecific Forms, Forma, Subspecies.

1. Several uses of the term "variety" are employed (Figure 14.5). These bear scrutiny. It is technically impossible to have *C. madisonensis* var. *pusilla* precede in time its parent population as shown. If one regards *madisonensis* as a subspecies, then it must be a subspecies arising by geographic isolation from parental stock. If, as indicated, *C. humilis* gave rise to it, then *pusilla* was one of its subspecies that persisted vertically in the Kinderhookian. *C. athabaskensis* var. *esplanadensis* is obviously a subspecies of the type species. Other varieties correctly shown in Figure 14.5 are either conspecific (that is, indistinguishable from type species) or examples of polymorphism as in *C. subtilita* var. *angusta*, and hence, *forma*.

2. Among species virtually identical in morphology (Figures 14.5 and 14.6) and probably conspecific or polymorphs of the same species, are: *C. corpulenta*, *C. claytoni* (conspecific), *C. subquadrata*, and *C. sulcata*—the latter a *forma* of *subquad-*

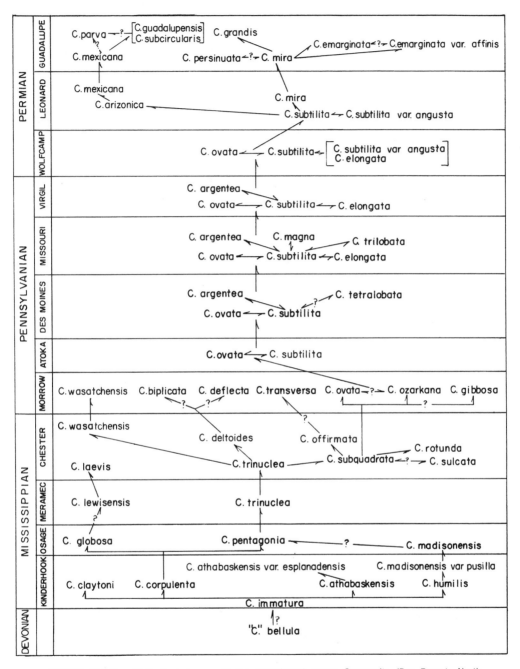

FIGURE 14.5 Hypothetical phylogeny of the brachiopod genus *Composita* (Dev.-Perm.). North American species and "varieties" only. Single-headed arrows indicate developmental trends; double-headed arrows, intrapopulation variations; brackets enclose possible conspecific forms (after Grinnell and Andrews, 1964).

rata; C. ovata and *C. ozarkana*—the last-named species probably a *forma* of *ovata; C. ovata* and *C. wasatchensis*—*wasatchensis* probably a juvenile form of *ovata.*

3. Various *Composita* species apparently represent subspecies as indicated by the intergradation between *C. elongata, C. subtilita,* and *C. ovata* (Figure 14.6). Grinnel and Andrews acknowledge that "a case could be made for suppression of all species names except *Composita subtilita* (for Pennsylvanian and Permian forms)

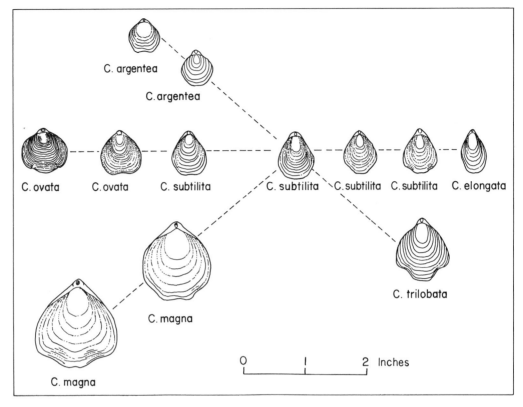

FIGURE 14.6 Main morphological trends in *Composita* species (dorsal view). The basic, most variable form is *C. subtilita*. Variant shapes of *C. ovata* and *C. subtilita* may be regarded as *forma*. After Grinnel and Andrews, 1964.

and a few others." However, they retained the names even though technically doubtful.

4. Evidence presented favors regarding *C. subtilita* as a polytypic species embracing numerous, vertically persistent subspecies such as *ovata, argentea, elongata,* and a few short-lived subspecies such as *magna* and *trilobata*.

A problem remains because *C. ovata,* presumably a subspecies of *subtilita* precedes it in time. One must therefore assume *subtilita* presence in the basal Pennsylvanian (Morrowan) even though it was not found below the Atokan (Figure 14.5).

Mollusca

Successional Speciation (Oysters). Although Chapter 8 treated oyster evolution, a particular oyster—the *sellaeformis*

stock—is considered here (Stenzel, 1949).

What is successional speciation? Assume the following model: A topodeme—an interbreeding parental population in a given geographic area (Gulf Coast); let this population undergo genetic modification; let such modification give rise to one or more successor species through geologic time spans; allow all of this to transpire without either geographic, ecological, or adaptive segregation or isolation. Then it follows from this model that vertical speciation observed in this study may be called *successional speciation* after Julian Huxley. Stenzel intentionally illustrated this mode of speciation in an oyster stock from the Gulf Coast Tertiary.

Geologic and Distributional Setting. Figure 14.7 outlines essential geologic data. Three successive *Cubitostrea* species of the same lineage, that is, same sub-

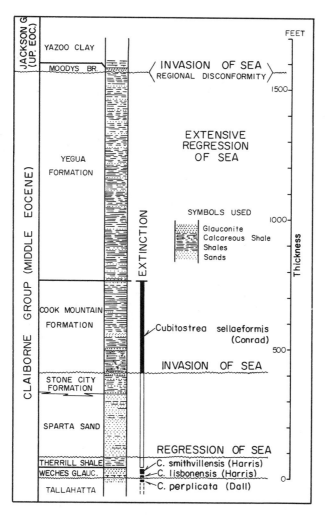

FIGURE 14.7 Stratigraphic ranges of the sequence of species composing *Cubitostrea sellaeformis* (Conrad) stock in the Gulf Coastal Plain. After Stenzel, 1949.

genus (unnamed) are shown: *Cubitostrea lisbonensis* (Figure 14.8*B*), *C. smithvillensis* (Figure 14.8*C*), and *C. sellaeformis* (Figure 14.4*D*, also see Figure 14.8*D*). These forms belong to subgenus II.

These species represent an offbranch of the typical branch to which the oldest species in the sequence belongs, that is, *C. perplicata*. The last-named species belongs to another subgenus of *Cubitostrea* (that is, subgenus I). This subgenus embraces two other species correlated to the problem. Note should be taken of the disconformities on top of the Tallahatta, Therill Shale, Stone City Formations; and the depositional hiatus within the Weches.

Between the Weches and Cook Mountains Formation, deposits are largely fluvatile and nonmarine. While some oysters occur here, none of the *sellaeformis* stock has been reported.

The first appearance of a species that would culminate in *C. sellaeformis* was in Middle Eocene beds of Alabama (that species being *C. perplicata*). The succeeding species, *C. lisbonensis*, is widespread from Alabama to Mexico, while *C. smithvillensis*, the next successor species, is restricted to Texas and Alabama. *C. sellaeformis*, the Cook Mt. Formation species, is widespread in equivalent beds along the Atlantic shore north to Virginia-Maryland, and south to Mexico.

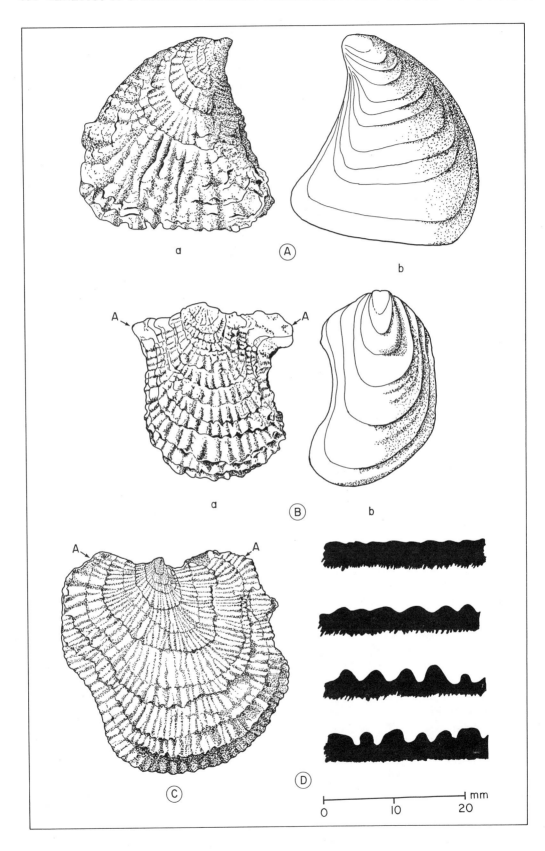

FIGURE 14.8 *Cubitostrea* species (M. Eocene, Gulf Coastal Plain), left valve only.
(*A*) *C. perplicata*, Alabama, (*a*) costate form, (*b*) smooth form, both reduced.
(*B*) *C. lisbonensis*, Texas, (*a*) costate form, (*b*) smooth form, reduced.
(*C*) *C. smithvillensis*, Alabama, reduced.
(*D*) Sections across radial costae of left valve, mature stage, distance from umbo six centimeters, X 2. From top to bottom; *C. sellaeformis. C. smithvillensis, C. lisbonensis, C. perplicta A–A* = auricles).
(After Stenzel, 1949).

Absence of Intergrades and Migrants. No intergrading populations were found as fossils between any of these four species. Thus, there is an apparently abrupt replacement of faunas. Disconformities and more local depositional breaks shown on Figure 14.7 might partly explain absence of intergrades. A second factor might be that population density declined markedly between levels occupied by *C. lisbonensis* and *C. smithvillensis*.

One cannot attribute the chain of species shown (Figure 14.7) to migration into or from, regions outside of the Tertiary Gulf of Mexico for several reasons: (1) exclusive of *C. sellaeformis*, all other species were restricted to the Tertiary Gulf of Mexico, (2) Tertiary deposits of England, France, and South America have not yielded any equivalent fossils that closely compare to the Gulf Coast species of the *sellaeformis* stock.

Sympatric Populations. An important consideration is the contemporaneity of *Cubitostrea sensu stricto* (subgenus I) to which *C. perplicata* belonged. Species of this subgenus are common both in the European and Gulf Tertiary. They lived side by side with the species shown in Figure 14.7. This denotes that species of subgenera I and II were sympatric.

Some morphologic changes through time reflect genetic modifications at specific gene loci in the common gene pool arising from a branch of the older subgenus I. Other changes responded to environmental conditions (for example, thick shell in high energy areas, thin shell in low energy areas). Figure 14.7 should be read with the understanding that *C. perplicata*, *C. lisbonensis*, *C. smithvillensis*, and *C. sellaeformis* respectively shared certain gene combinations in common.

Hence these must have been transmitted from generation to generation without evidence of intergradational populations in the fossil record (due to depositional or erosional breaks).

Duration. To appreciate the time involved (allowing for lost intervals in the rock record), estimate duration of the entire Middle Eocene (Claibornean Group) at some $7 \pm \times 10^6$ (Kulp, 1961, Figure 1). In addition, take into account Nicol's figures on duration of some Tertiary pelecypod species (Chapter 8) — an average of 6.5×10^6 years, with a low of 1×10^6 years. Furthermore, include the fact that all of Claiborne time is not represented by the *sellaeformis* chain of species time span (Figure 14.7). With these data more clarification becomes possible:

1. The portion of the Claiborne represented spans several million years less than 7×10^6 (subtract the time to deposit the 700+ ft of the Yegua sand and shale, plus beds below the Tallahatta). (The concerned portion ranges from time of *C. perplicata* to extinction of *C. sellaeformis*.)

2. Accordingly, some lower duration figure than Nicol's average is indicated — one closer to the $1 \pm \times 10^6$ year figure. Obviously, this is highly speculative since depositional breaks indicated must be evaluated in terms of years. Still, it does provide a temporal frame of reference. Actual time involved is unlikely to have extended over five million years, but may have occurred in three to four million years.

Changing Morphology. Chief morphological changes through time included: (1) Increase in shell size. Maximum sizes (in mm) that were attained (measured from umbo to opposite margin of left valve) follow: *perplicata*—64, *lisbonensis*—86, *smithvillensis*—147, *sellae-formis*—182. (2) Increase in proportions of auricles (see Figures 14.4 and 14.8). The proportional height of the posterior auricle to valve height (measured on left valve of specimens with large auricles): *perplicata*—0.00, *lisbonensis*—0.33, *smith-villensis*—0.41, *sellaeformis*—0.51 (3) Decrease in strength of costae (Figure 14.8*D*). (4) Disappearance of triangular shell outline (Figure 14.8*A–C* and Figure 14.4). (5) Appearance of twist in shell (Figure 14.4*D*).

Primitive Features. In the oldest of the *sellaeformis* chain of species, *C. per-plicata*, Stenzel considered primitive shell characteristics to be: small shell size, lack of auricles at any growth stage, strength of costae, triangular shell outline, complete lack of arching of the valves.

Phyletic Size Increase. Newell (1949) wrote that a progressive size increase through time characterizes many invertebrate faunas. He attributed it to evolution, that is, large size gave a selective advantage. Stenzel (1948) thought size increase might be due to intraspecific competition. Competition favors features advantageous to the individual, not necessarily to the whole species. If so, he reasoned, that explains why large size attained by *C. sellaeformis* "predisposed" the species to extinction. Progressive size increase might be transmitted from generation to generation (that is, encoded in the genotype) and if so, size increase in an evolving stock could be readily explained.

Some Genetical Considerations. The expression "primitive features" lacks biological meaning. Rather, such features in older populations disappear through time. All that can be said is that they were inadaptive. Why did such features as small triangular shell, disappear to be replaced by larger, more ovate configurations? Mayr's definition of the phenotype—a compromise between all selection pressures—provides an answer.

What were some diverse selection pressures on the Eocene Gulf sea floor? Oysters of the time did not form reefs nor oyster beds. Their spat settled in areas already populated by corals, echinoderms, and other forms. Thus selection pressure arose from this context.

Many oysters were unattached in adult stages. Selection pressure for auricle development, shell shape, and size would thus have arisen. Heavy costation might serve as sediment traps in areas of high energy wave action and ultimately lead to oyster burial. Thus selection pressure to diminish coarseness of costation could have grown.

Thicker shells lend ponderosity and attendant stability in high energy areas of the sea floor. In this way, greater calcium secretion may have been a selection factor.

Multiple pressures probably coexisted and simultaneously operated in the gene pool of this oyster population. Morphological trends through time really mark off actual resolutions of these several pressures.

Some morphological changes may have resulted from pleiotrophic gene action. If so, some changes may have had no "trend" significance being merely by-products. Without detailed population statistics, it is impossible to distinguish adequately a "trend" from other conditions such as polymorphism (large and small size variants of same age group, and so on).

Other Factors. A whole array of un-documented factors may have been at work. Example, oyster spat will not settle in the absence of certain elements in trace amounts (copper). Spectrographic study of the fossil oyster beds might reveal presence or absence of these critical elements. Stenzel observed a high selection factor operative on the fringes of the geographic range of *C. lisbonensis*—where thin shell forms lived. Few reached maturity; some were buried under sediment; demise of others was assigned to natural causes. However, "natural causes" may have been generated by anything from lower salinity than that of normal sea water due to heavy rains, to the absence of adequate food supply, parasitic infection, contagious diseases, and so on (Korringa, 1952).

Echinodermata

Details of blastoid morphology, taxonomy, etc. were given in Chapter 12. This section treats variation and relative growth in the blastoid *Orbitremites* (Joysey, 1955).

Background. Before this investigation started, previous workers had described five species of *Orbitremites* from the British Lower Carboniferous: *O. derbiensis*, *O. ellipticus*, *O. orbicularis*, *O. campanulatus*, and *O. mccoyi* (see Figure 14.9 for the first two species).

In rocks of Lower Visean age (C₂ subzone), these species were collected at three localities: *O. ellipticus*, *O. orbicularis*, *O. campanulatus* and *O. mccoyi*. At one locality, Clitheroe, all four species were found at the same horizon. Higher up in the rock column at four additional places (Upper Visean, D₂ subzone), only specimens of *O. derbiensis* were found.

Study Sample. The actual collections Joysey studied represented two samples: *O. derbiensis* from the Middle Limestone of Grassington, Yorkshire, population density—$n = 581$ (See Figures 14.10, 14.11)—and another from the Upper Clitheroe Limestone of Clitheroe, Lancashire—$n = 112$ specimens including four other species besides *O. derbiensis*.

After examination, the *O. derbiensis* sample from Grassington, was reduced by one-third; 225 specimens showed shape distortion due to crushing; 186 specimens were fractured or badly weathered. These were rejected. The Clitheroe sample also was reduced to $n = 55$ (although only 40 specimens had ambulacral plates adequately preserved to allow counts to be made).

Questions to be Answered. Joysey set forth three objectives formulated here as

questions: (1) What kinds of variations occur in a blastoid community? (2) How does shape change during ontogenetic growth? (3) What differences occurred between species of same genus through geologic time? (This last question could not be answered because four of the original five species proved to be synonyms of the same species.)

Blastoid Communities and Genetics. One *O. derbiensis* collection had several hundred blastoids obtained from soil derived from a weathered 2-ft-thick limestone at Grassington. The exposure was four feet long. A few yards away another collection again yielded several hundred blastoids from 4-in.-thick limestone covering an area not more than 3 sq ft.

Apparently dense populations existed on the Visean (D₂ subzone) sea floor. All these blastoids occurred in a highly crinoidal limestone matrix in restricted areas. The lateral extent of the *Orbitremites* band in the Grassington area was 2+ miles (Figure 14.11). It is noteworthy that this community shared a common gene pool. Even though minor spatial separation may have existed between segments of the population, there was only one true species.

Furthermore, the reduced Grassington sample yielded several abnormal specimens. The most unusual specimen among the abnormals lacked one plate in each of the basal, radial, and deltoid circlets, but retained all five ambulacra, two of which lay side by side. These data provide more insights into genetic events.

If the cistron is envisioned as containing multiple loci capable of being mutated, then abnormalities probably reflect mutated loci. Available in the blastoid gene pool at Grassington at that time were, besides normal genes, these mutated genes. The fact that abnormal phenotypes were relatively few—seven abnormal to 188 normal—indicates that normal plate count and arrangement must have bestowed adaptive advantage.

Intergrades and Young Forms. Joysey found the Clitheroe sample had inter-

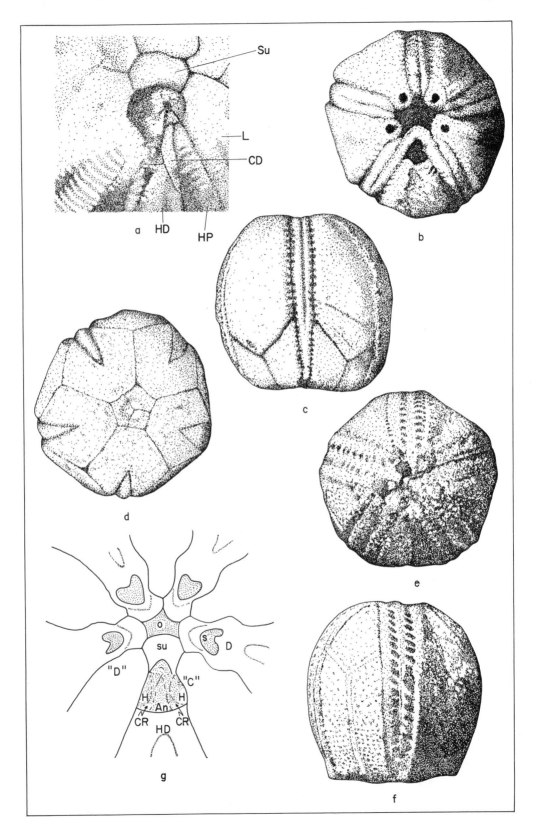

FIGURE 14.9 Blastoids, Lower Carboniferous, Britian, *a–d Orbitremites, derbiensis* (Sowerby), plesiotype. L. Carb., Cockleshell Beds. Grassington, Yorkshire, Eng. (X 6.3); (*a*) detailed oral view, *CD*—cryptodeltoids, *HD*—hypodeltoids, *HP*—hydrospire plate, *L*—lancet plate, *Su*—superdeltoid plate (X 15.3); (*b*) Oral view; (*c*) *D* ambulacral; (*d*) aboral view.

e–g Orbitremites ellipticus (Sowerby), L. Carb., Yorkshire, Eng., plesiotype (X 7.2): (*e*) oral view; (*f*) *D* ambulacral; (*g*) same species from Clitheroe Ls., Yorkshire, summit area (X 15); *O*—oral opening, *D*—deltoid, *H*—hydrospire, *An*—anal opening, *S*—spiracle, *D* and *C*—ambulacra; other labels as in (*A*) above. After Fay, 1961.

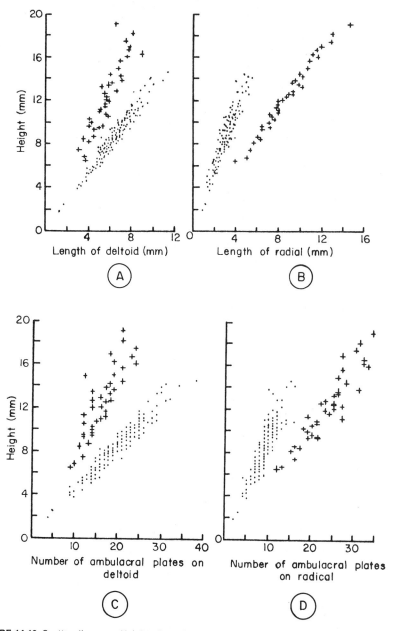

FIGURE 14.10 Scatter diagrams. Height of specimen plotted against length of (*a*) deltoid, (*b*) radial, (*c*) number of ambulacral plates on deltoid, (*d*) number of ambulacral plates on radial. Black dots = *Orbitremites derbiensis* individuals from Grassington; + = *O. ellipticus* from Clitheroe. Lack of overlap of areas covered by scatter points is due to marked distinction in the relative size of component plates in the two samples. After Joysey, 1959.

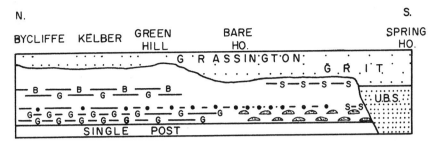

FIGURE 14.11 Reconstructed lateral distribution of the blastoid *Orbitremites* and associated bio-facies. Middle Limestone, Carboniferous, Grassington, Yorkshire, England. Vertical scale, 1″ = 200′; horizontal scale, 1.5″ = 1 mile; *Orbitremites* band = ·; G−*Gigantoproductus* or Cockleshell Bed; B−Bryozoan band; S−*Spirifer* bed; UBS−Upper Bowland Shale. Knoll-forming banks of calcite-mudstone are shown diagrammatically and not to scale. Lateral extent of *Orbitremites* band is 2 + miles. (Joysey, 1955).

grades of a single species instead of four distinct species. One, *O. orbicularis*, is merely a shape variant or *forma*; *O. mccoyi*, a young form of *O. ellipticus*, and *O. companulatus*, a young form of *O. orbicularis*.

Clearly, *O. ellipticus* was characterized by polymorphism.

Statistics. All usable specimens were measured for the following characters: height, diameter, height of deltoid plate, height of radial plate, total number of ambulacral plates, number of ambulacral plates on the deltoid region, and on the radial region.

These data were treated statistically. Since statistics would lead too far afield, refer to Joysey's paper with its several explanatory pages and cited references, and other standard texts.

Problems Pertinent to Variation and Relative Growth Studies
Errors of Measurement. Three groups of errors must be taken into account in any such study: (1) *Instrument error*−in this case, measurements made by a traversing microscope fitted with crosswire and a vernier scale. Such errors are constant and affect all measurements equally. (2) *Accidental or random observational error*− these can hardly be anticipated. (3) *Secular error*−due to investigator. Measurements made on the same day tend to be closer together than those on different days (Joysey, 1961, p. 109). Nevertheless,

statistical tests of repeated measurements made at different times can be used. Also, the degree of reliability of measurements by different investigators (operator error) can be determined.

Sampling Bias. Joysey observed that during actual field collections, larger blastoids are more likely to protrude from the rock matrix and thus more readily seen by the collector than smaller, imbedded ones. Smaller specimens, freed from the matrix by chemical erosion, are likely to fall into crevices between larger fragments and be lost. Thus, in one of his samples, Joysey found a complete absence of very young individuals. Was this real?

Age Determinations. Blastoid samples have no acceptable criterion for age determination; individuals of different ages are represented in such samples. Are the observed variations due to a mixture of varied age groups? Joysey found that they were.

Four larval stages are known in the ontogeny of two Late Mississippian blastoids (*Mesoblastus* and *Pentremites*−Croneis and Geis, 1940). The smallest individual had a maximum dimension of 0.09–0.14 mm (= first stage); the largest individual (representing the fourth stage), for the same dimension, had a length of 0.4 mm. In any sample, size differences of blastoids will reflect differences in ontogenetic age. Joysey remarked that one of the smallest individuals in the Grassington sample had a height of 2.1 mm and seven ambulacral plates. This individual was about twenty times larger than the size of the first larval stage of the forms studied by Croneis and Geis, and about five times the size at which the adult plate arrangement becomes established.

Where it is possible to separate individuals in a sample that are of same age, different shell characters in such individuals can be compared. Only in a crude inferential sense could insights into comparative ages be ascertained in Joysey's samples.

Phylogeny. Can the two species of *Orbitremites* discussed above be related—one stratigraphically younger and morphologically distinguishable from the other? Is the older ancestral to the younger?

Joysey showed the two species had much in common. That means they shared certain stable gene combinations. Common characters included: similar shape, similar relationship between gross size and number of ambulacral plates. They also inhabited similar environments, and occurred in the British Isles.

The two species are, nevertheless, quite distinct. In *O. ellipticus*, radial plate is always larger than deltoid plate. The converse is true for *O. derbiensis*. What could be the adaptive significance of this difference? Joysey speculated that since the hydrospires bear a different relationship to each of these plates, possibly hydrospire-efficiency was modified. It is unclear, however, whether modification was directed toward greater or lesser efficiency.

Whatever the significance, difference in relative plate size required new gene combinations through multiple generations. Between the time represented by the two species in the British Lower Carboniferous, no fossils of *Orbitremites* are known. One possibility Joysey entertained was that the two species might have evolved independently outside the area of the British Isles.

To explore this possibility further, more data are needed:

1. *Orbitremites* species outside of the British Isles. Mississippian beds of Illinois, Iowa, and Missouri contain one Kinderhookian species and in the succeeding Osagian, 14 species occur. No recorded *Orbitremites* are presently known outside of North America during the Mississippian. There are no records of *Orbitremites* beyond England during Lower Visean time. A doubtful form has been reported from the Australian Permo-Carboniferous, *Orbitremites* (?) *wachsmuthi*. A single species also comes from the Middle Permian of Timor (with deltoid plate larger than radial plate as in *O. derbiensis*).

2. All the older American species of *Orbitremites* and the British *O. ellipticus* never have the deltoid plate larger than the radial. The converse holds for *O. derbiensis* and, as noted above, for the Timor species. (Did this represent a trend toward relative size increase of deltoid plate and relative size decrease of radial plate through time? Joysey cautiously observed that such a general trend was apparent although only more data on North American species could check it out.)

3. According to Fay (1961), both the genera *Ellipticoblastus* and *Orbitremites* are confined to the English Lower Carboniferous. *Ellipticoblastus* gave rise to *Orbitremites*. In turn, the genus *Ellipticoblastus* was thought to be a derivative of *Globoblastus* (confined to the North American Mississippian). (Further details on evolution in Chapter 12.)

4. Vertical separation between the older *O. ellipticus* (C_2 subzone) and younger *O. derbiensis* (D_2 subzone) has a time value of x years. About one thousand plus feet separates C_2 and D_2 subzones elsewhere in England (Sibley, 1920, cit. Stamp, 1953, Figure 36).

With these factors in mind, implications of Joysey's suggested origin of two British *Orbitremites* species can now be discussed. Restricted geographic spread of the Mississippian *Globoblastus* stock must have been extended as Fay held. It gave rise to the *Ellipticoblastus* stock of the British Lower Carboniferous. Thus a geographic isolate of the parental *Globoblas-*

tus population became reproductively isolated during time. In this separated gene pool, genetic loci governing hydrospire size and lancet-plate position were modified; new gene combinations, favored by natural selection, gave rise to the characteristic *Ellipticoblastus* phenotype. [In this genus, the two hydrospires on each side of the hydrospire in *Globoblastus* were reduced to one and the lancet-plate moved outward. (Fay, 1961, p. 72).] *Ellipticoblastus orbicularis* individuals migrated to Great Britain.

However, if Fay is correct, the derivation of *Orbitremites* from *Ellipticoblastus* might not have required, as Joysey suspected, any new migration to the British area. The gene pool of the British *Ellipticoblastus* stock could now undergo fractionation repeatedly, mutations would occur at specific gene loci governing radial-deltoid plate arrangements. Such mutated genes could eventually enter into new combinations phenotypically expressed in *Orbitremites ellipticus*. (In the *Orbitremites* species, deltoids overlap radials; in *Ellipticoblastus* species, radials overlap deltoids.)

Throughout a time span of *x* years, no *Orbitremites* survivors or long-persistent subspecies occurred. Instead, a new species, *O. derbiensis*, a very distinct form (Figure 14.9), appeared at the end of this interval. Despite distinctiveness, it shared certain stable gene combinations with *O. ellipticus*, namely, generic characters.

The record of *Orbitremites* outside the British Carboniferous is very sparse. Nevertheless, the few existing indicators point to several contemporaneous populations elsewhere. The same trend, radial plate larger than deltoids, is found in at least two populations (*O. derbiensis*, Carboniferous of England, and the Timor species, *O. malaianus*, Middle Permian). This occurrence in widely separated populations suggests some shared gene combinations *other than* those characteristic of the genus. Possibly, the *O. ellipticus* stock had migrated from the British locale, subdivided into geographic races; ultimately, where reproductive isolation followed, giving rise to British, and later, Australian and Timor species. Migration to British waters from this foreign population could then have happened during D_2 time. At any rate, this might explain the appearance of *O. derbiensis* having greater affinities for populations outside England than for the older British *O. ellipticus*.

QUESTIONS

1. (a) Compare older and newer concepts of the gene.
    (b) Define a cistron, gene locus, nucleotide.
    (c) Contrast genotype and phenotype with reference to a fossil arthropod, bryozoan, and crinoid.
    (d) What is the paleobiological significance of the genetic concept of "gene flow"?
    (e) How do new genotypes arise? (Use a fossil population to illustrate.)
2. (a) Outline the work of natural selection in any given fossil population.
    (b) Describe a Paleozoic gene pool with reference to any given fossil population.
    (c) How does natural selection create evolutionary novelty?

    (d) How is one gene pool kept distinct from another in closely related populations?
3. (a) Can pleiotrophy be detected in fossil populations? Explain.
    (b) Can the effects of recombination, as distinct from mutation, be recognized in fossil populations? Discuss.
4. Distinguish between genetic variation and nongenetic variation in the following fossils: rugose corals and articulate brachiopods.
5. Distinguish among polymorphism, subspecies, conspecific forms, and monotypic species in a fossil assemblage of compositas or atrypas.
7. (a) What is the difference between horizontal and vertical subspecies?
    (b) Referring to Carruthers' work on *Zaphrentoides* (Chapter 5), define the kind of speciation involved.

(c) Interpret the chart on graptolite speciation in Chapter 2 in light of genetic factors.

(d) Can a living population of invertebrates persist into the indefinite future without any phenotypic change? Explain.

8. L. F. Simpson (G.S.A. Bull., 1966, Vol. 77 (2), pp. 197–204) found a good correlation between times of magnetic polarity reversal and evolutionary spurts through geologic time. For present purposes, assume such a factor. What bearing might increases in radiation have had on: (a) Evolution of archeocyathids, (b) trilobite speciation in the Cambro-Ordovician, (c) rise of the scleratinian corals? (d) Might this have had a reverse effect accounting for some extinctions? (e) Could it relate to the abrupt appearance of *Orbitremites derbiensis*? (f) Can enhanced mutation rate of itself exert selection pressure? (g) Is the mutation rate more important than the rate at which mutations distribute through a population over a time span?

(An abridged edition of Ernst Mayr's *Animal Species and Evolution* is available under the title *Populations, Species and Evolution*. Harvard Univ. Press, 1970.)

REFERENCES

Speciation

Alexander, F. E. S., 1949. (See Chapter 7, references).

Corroy, G., 1927. Les Spiriféridés du Lias Européen et Principalement Du Lias de Lorraine et D'Alsace. Ann. de Paléontologie, Vol. 16, pp. 2–36, plates 1–4, 5 text figures.

Croneis, C., and H. L. Geis, 1940. Microscopic Pelmatozoa: Part I, Ontogeny of the Blastoidea: J. Paleontology, Vol. 14 (4), pp. 345–355.

Dobzhansky, T., 1955. Evolution, Genetics, and Man. John Wiley and Sons, 379 pp.

Fage, Genviéve, 1934. Les Rhynchonelles des Cretácé Supérieur des Charentes: Bull. Soc. Géol. France, 5e Ser., Vol. IV, pp. 433–441, plate 23.

Fay, R. O., 1961. Echinodermata: Blastoid Studies. U. Kansas Paleont. Contrib. Art. 3, pp. 1–147, plates 1–54, text figures 1–221.

Grinnel, R. S., Jr., and G. W. Andrews, 1964. Morphologic studies of the brachiopod genus *Composita:* J. Paleontology, Vol. 38 (2), pp. 227–248, plates 37–39, 9 text figures.

Imbrie, J., 1956. Biometrical methods in the study of invertebrate fossils: Amer. Mus. Nat. Hist. Bull., Vol. 108, pp. 217 ff.

Joysey, K. A., 1955. On the geological distribution of Carboniferous blastoids in the Craven area, based on a study of their occurrence in the Yoredale Series of Grassington, Yorkshire: Quart. J. Geol. Soc., London, Vol. III, p. 209 ff.

———, 1959. A study of variation and relative growth in the blastoid *Orbitremites*: Phil. Trans. Roy. Soc., London, Ser. B, Vol. 242 (Biological Sciences), pp. 99–125.

Kermack, K. A., 1954. A biometrical study of *Micraster coranguinum* and *M. (Isomicraster) senonensis*: Phil. Trans. Roy. Soc., London, Ser. B., Vol. 237, pp. 375 ff.

Korringa, P., 1952. Recent advances in oyster biology: Quart. Rev. Biol., Vol. 27, pp. 266–308, 339–365.

Kozlowski, R., 1929. Les brachiopodes Gothlandiens de la Podolie Polonaise: Paleontologia Polonica, Vol. 1, 245 pp.

Kulp, J. L., 1961. Geologic time scale: Science, Vol. 133 (No. 3459), pp. 1105–1114.

Mayr, E., 1960. The emergence of evolutionary novelties, in Sol Tax, Ed., The Evolution of Life. Univ. of Chicago Press, pp. 349–380.

———, 1963. Animal Species and Evolution. Belknap Press, Harvard Univ. Press, 662 pp.

Newell, N. D., 1949. Phyletic size increase, an important trend illustrated by fossil invertebrates: Evolution, Vol. 3 (2), pp. 103–124.

Olson, E. C., and R. L. Miller, 1958. Morphological Integration. Univ. Chicago Press, 308 pp.

Owen, D. F., 1966. Polymorphism in Pleistocene land snails: Science, Vol. 152, pp. 71–72, Table 1.

Simpson, G. G., 1953. The Major Features of Evolution. Columbia Univ. Press, 393 pp.

Stamp, L. D., 1950. An Introduction to Stratigraphy, British Isles. Thom. Murby and Co., London, 356 pp. (see Figure 36, p. 140).

Stenzel, H. B., 1949. Successional speciation in paleontology: the case of the oysters of the *Sellaeformis* stock: Evolution, Vol. 3, pp. 34–50.

Stumm, E. C., 1964. Silurian and Devonian corals of the Falls of the Ohio: G.S.A. Mem. 93, 85 pp., 80 plates, 2 text figures.

Waddington, C. H., 1959. Canalization of development and genetic assimilation of acquired characters: Nature, Vol. 183, pp. 1654–1655.

————, 1960. Evolutionary adaptation, in Sol Tax, Ed., The Evolution of Life. Univ. of Chicago Press, 6 pp. 381–402.

Wells, J. W., 1937. Individual variation in the rugose coral species *Heliophyllum halli* E. & H., Palaeontographica Americana, Vol. 2, pp. 5–20, plate 1, 30 text figures.

chapter fifteen

MINERALOGY, BIOCHEMISTRY, AND SKELETOGENESIS OF LIVING AND FOSSIL INVERTEBRATES

INTRODUCTION

Study of the paleobiology of invertebrates must consider the general theme of mineralized tissues as skeletal building materials (Lowenstam, 1954). This field of investigation has rapidly expanded, having overflowed the boundaries of narrow specialities.

SOME ESSENTIAL PRELIMINARIES

Glimcher (1960) outlined some physical-chemical principles pertaining to crystallizing systems:

1. Phase Transformation

A change of state occurs whether inorganic crystals form from solutions inside, outside, or unrelated to biological systems. (A change from liquid to solid crystalline is a change of state.) This change is a "phase transformation." Therefore biocrystallization is a particular case of a more general phenomenon, phase transformation.

2. Stability of Phases

How stable is the crystalline solid phase in a given biological system? That depends on the equilibrium properties of the system.

3. Four Kinds of Equilibrium

Of the four kinds of equilibrium—

stable, neutral, unstable, and metastable—the last is of importance in biocrystallization even though unstable equilibrium may be involved.

Body fluids (= a solution) that bathe living cells or tissues in an unstable equilibrium can undergo a phase change to a crystalline solid. Fluids can do so without drastic alteration of any given system's properties.

4. Mechanisms of Phase Transformation

(See Figure 15.1.)

(a) *Crystal nucleation.* Initially, minute fragments that form from the metastable solution of body fluids, are a step in the direction of a new, more stable phase, that is, the crystalline solid.

(b) *Crystal growth.* When clearly defined particles appear (crystals begin to grow), the new, more stable, solid phase replaces the metastable liquid phase (Figure 15.1).

(c) *Recrystallization.* The growth of large crystals by cannibalism of smaller ones can occur even after the solid phase has come into being.

Physical-chemical considerations for mineralization do not explain the mechanism that operates to favor specific sites on tissues and cells over other possible sites. The following items do help explain

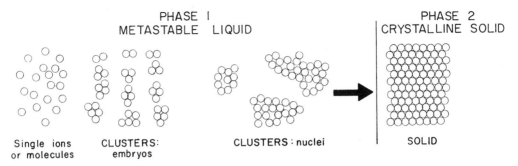

FIGURE 15.1 Common mechanism of nucleation in most biological systems (adapted from Glimcher, 1960).

site preference in crystal formation:

(a) Of all tissues bathed in body fluids (fluids being in a metastable equilibrium), only those will be mineralized that possess organic components with a particular molecular structure and configuration. (Example: The conchiolin matrix in molluscan shells.)

(b) Organic components possessing such properties are not distributed throughout the tissue but are restricted to specific loci within it.

(c) These specific loci act as sites for the formation of crystal nuclei (Figure 15.1).

UNICELLULAR FORMS

A few unicellular forms will be considered in this section—the coccolithophores, foraminifera, and radiolaria.

Coccolithogenesis

[Data from Wilbur and Watabe (1963).]

1. *Precrystal Formation*. The organic matrix in a given region of the cell (= reticular network) takes the shape of the coccolith to be (Figure 15.2*a–c*). Probably, in this way, specific loci for nucleation are provided (see Glimcher above and Isenberg et al., 1965).

2. *Calcification*. Calcification proceeds from several centers in the organic matrix (Figure 15.2*c*). A thin membrane marks the location of central extensions of the

base of the coccolith-to-be (Figure 15.2*a–c*).

3. *Growth of Coccolith Elements*. In cross-section (Figure 15.2*c*), a coccolith shows peripheral upper and lower outer elements, and a central basal portion composed of inner elements. Figure 15.2*c, d* illustrates growth stages of these elements.

4. *Crystalline Solid State*. More than one crystal may compose the main coccolith cylinder as well as upper and lower elements.

5. *Interlocking of Coccoliths*. (Figure 15.2). By interlocking, the coccoliths form a spherical shell at the cell's surface. Only external coccoliths are interlocked. At present, details are unknown on the extrusion from inside the cell to the periphery of coccoliths.

6. *Multiple Shells of Coccoliths*. Multiple shells envelope the cell in concentric layers as new coccoliths form within the cell and are extruded.

7. *Duration of Change of State*. Paasche (1962) reported that decalcified cells of *C. huxleyi*, on the average, yielded new coccoliths in less than one hour. Wilbur and Watabe (1963) periodically examined decalcified living cells of this species and found new coccoliths had formed after 30 hours. Apparently the processes begin much earlier since, in polarized light, internal crystals appeared in decalcified specimens after six hours.

8. *Mineralogy*. X-ray diffraction powder patterns of extracellular coccoliths, compared to those of synthetic calcite (Isenberg et al., 1963), clearly demonstrated that the biocalcite of the marine coccolithophorid *Hymenomonas* sp. compares closely with mineral calcite. However, a dark central portion signals presence of organic material. This evidence hints 'at how the organic material could have been incorporated when the coccolith formed within a "mold" made by the organic matrix.

9. *Calcium Chemistry*. Isenberg et al., (1963) found that cultures required at least $10^{-5} M$ of calcium for coccolithogenesis. Paasche (1962) determined the calcium content of one specimen of C. *huxleyi* — 5.5×10^{-13} grams — and noted that coccolithogenesis required the bicarbonate as the principal carbon source (1963, cit. Isenberg et al., 1965).

Foraminifera

Mineralogy. By X-ray diffraction methods, Blackman and Todd (1959) studied 131 genera in 29 families of Recent foraminifers with calcareous shells. These were widely representative (Tropics to the Arctic; depths, 27–494 fathoms). The authors found mineralogy of the shell (that is, presence of calcite or aragonite) to be a *genetic* character. Mineralogical composition could therefore distinguish between families (Table 15.1).

Wood's systematic study (1949) of foraminifers examined under polarized light concluded that the hyaline or glassy appearance (*Lagena* and *Globigerina*) can be attributed to radial wall structure.

Silica in Foraminifers. Chemical analysis of the tests of calcareous foraminiferida (*Globigerina, Amphistegina, Orbitolites*, and so on) indicates a silicon content (in terms of percent ash) generally from a few tenths of a percent, to as much as 15 percent in rare instances (Loeblich, 1964, Table 1).

Vinogradov observed that siliceous foraminifers had skeletal parts composed of pseudoquartz and opal. These had been biochemically formed, in contrast to siliceous particles, from external sources that some foraminifers incorporated into their tests by agglutination. (Compare family Rzehazinidae, formerly the Silicinidae.)

Spectrographic and X-Ray Analysis. Various studies have shown that strontium is frequently substituted in the calcite structure of planktonic foraminifers (Said, 1951; Emiliani, 1955; Krinsley, 1960; and see Table 2, p. C-98, Loeblich, 1964). Said noted that among marine invertebrates, some Recent foraminifers have a higher percentage of strontium than any other forms except acantharian radiolarians.

Excluding calcium and magnesium (see Table 15.1), strontium and silicon, the element composition of foraminiferal tests includes Mn, Fe, Al, (from −1 to 0.1 percent), Pb, Cr, Sn, Ag, Ba, B, V, Cu (from 0.01 to 0.0001 percent). Pb, Ag, Cr occur in amounts of less than 0.0001 percent. Titanium and aluminum were probable sediment contaminants and presence of nickel was inexplicable (Krinsley, 1960).

Hedley (1963) investigated cement and iron content of arenaceous foraminifers. He found that iron was organically bound and occurred as ferric iron in positive granules within the cytoplasm. The iron content (Fe_2O_3) ranged from 0.6 to 2.0 percent in tests of the arenaceous foraminifers analyzed.

Amino Acid Content of Organic Cement. In the same study, Hedley (1963, Table 4) indicated the amino acids found in the organic cement of four genera, determined by chromatographic analysis. All had small amounts of leucine, valine, alanine, glycine, serine, glutamic acid, aspartic acid, B-phenyalanine, proline, threonine, argine, lysine, and cystine.

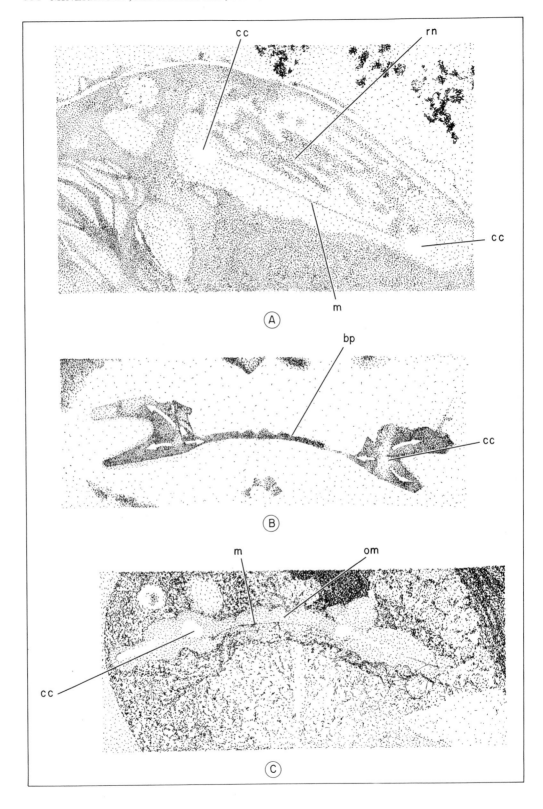

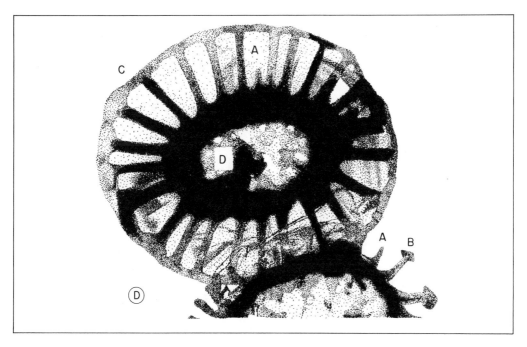

FIGURE 15.2 Coccolith formation. (a) *Coccolithus huxleyi* cell; *rn* — reticular network or strands; dashed area — *cc* — centers of calcification; *m* — membrane connecting the two centers. (b) Cross-section of coccolith showing basal plate (*bp*) and portions of both lower and upper outer elements; directions of growth (emanating from centers of calcification — *cc*) are shown by white arrows. (c) *om* — organic matrix and *m* as in (a) above; outline of *om* has general form of the coccolith that will form within it. (d) Stages in growth of upper elements; *A* — elements with "heads" not yet formed, *B* — "heads" present, *C* — growth of "heads" in contact with each forming continuous outer ring, *D* — central extensions of elements of the base. Data from Wilbur and Watabe, 1963. Figures reduced from electron micrographs taken at X 15,400 to X 55,500.

Foraminiferal Calcite Secretion at Depths. Bé and Ericson (1963, pp. 65 ff.) showed that the living planktonic foraminifer *Globoratalia truncatulinoides* sank to deeper waters seasonally (500 to 1000 meter depths), at which time they were virtually absent in the western North Atlantic epipelagic zone.

When these foraminifers left the zone of light penetration, there was an increase in wall thickness although overall test growth was retarded (Figure 15.3*A*). Of interest here is calcite secretion at great depths and in relatively low temperatures (4° to 17°C) in the form of an outer crust.

Radiolaria

Spumellaria have skeletons of protein plus hydrated silica ($SiO_2 \cdot nH_2O$) — opaline silica. Acantharia have skeletons of protein plus strontium sulphate — celestite ($SrSO_4$).

The most accurate analysis of an acantharian radiolarian (*Acanthometra pellucidium*) was given by Odum (1951): in percentage of ash, strontium, 28.1; calcium content, 0.8; calcium and strontium content undifferentiated, 19.4; silica content, 8.7; sulphate ran to 36.9, and R_2O_3 (= either Fe_2O_3 or Al_2O_3), 1.1.

Most analyses are of radiolarian silts or oozes rather than of clean, radiolarian scleracomas: boron, selenium, and yttrium; barium oxide; and fluorine content — each hundredths of one percent. (Compare Sverdrup et al., 1946, and Vinogradov, 1953.)

Revelle (cit. Sverdrup et al., 1946, p. 1035) assumed that the radium content of radiolarian and diatom oozes was bio-

TABLE 15.1. Mineralogical Classification of Foraminifera by Families

Imperforate	Perforate				
High Magnesium Calcite (10 to 10+ mol %)[a]		Low Magnesium Calcite (0 to 5 mol %)			Aragonite[b]
Single Crystal	Single Crystal	Radial Microstructure		Granular Microstructure	
Alveolinellidae	Spirillinidae	Calcarinidae	Anomalinidae ←→	←→ Anomalinidae	Ceratobuliminidae
Fischerinidae		Camerinidae	Elphidiidae	Cassidulinidae	Robertinidae
Miliolidae		Discorbidae	Globigerinidae	Chilostomellidae	
Opthalmididae		Heterohelicidae	Lagenidae	Nonionidae	
Peneroplidae		Homotremidae	Polymorphinidae		
		Planorbolinidae	Rupertiidae		
		Buliminidae ←→	Buliminidae		
		Pegidiidae ←→	Pegidiidae		
		Rotaliidae ←→	Rotaliidae ←→	Rotaliidae	
			Amphisteginidae[c] ←→		
			Cymbaloporidae[c]		

[a]Chave's analysis (1954a) suggested that the high magnesium calcite in Recent foraminifers taken from areas with known water temperature indicated temperatures of 18° to 29°C, whereas low magnesium calcite indicated temperatures of 0° to 10°C.

[b]A Recent *Textularia* is reported to contain both calcite and aragonite. A normally aragonitic foraminiferal genera that had inverted to calcite has never been observed. The two families (listed under "Aragonite" above) are embraced by the superfamily Robertinacea, which is perforate radial in structure (Loeblich and Tappan, 1961, 1964).

[c]Intermediate to low range of magnesium substitution in calcitic shells of Recent foraminifers.

←→ Indicates families that display both high and low magnesium or two different microstructures.

Source. Data from Blackman and Todd, 1956.

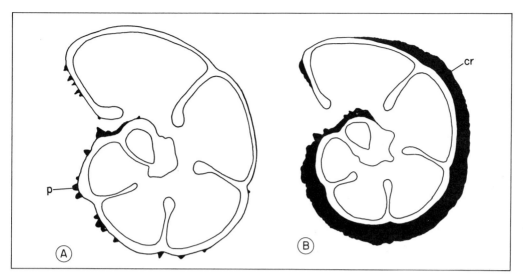

FIGURE 15.3A *Globorotalia truncatulinoides*, Recent. Horizontal Sections: (*A*) Surface plankton tow specimen, Bermuda waters; *p*—punctae (early stage of calcite crust). (*B*) Empty test from bottom sediment, central N. Atlantic (35°45′N, 47°05′W); *Cr*—calcite crust secreted at depth. After Bé and Ericson, 1961.

genic in origin; individuals extracted it from seawater.

PORIFERA

Introduction

Data on the biochemistry and mineralogy of different sponges are limited, yet shed light both on mineralized tissues and on the process of mineralization in fossils.

Composition and Mineralogy

Table 15.2 is a compilation of essential data. A few living sponges lack skeletons.

In terms of percentage weight of living sponges, water is the largest component, ranging from 70 to some 93 percent. Siliceous spicules also contain 5 to 13 percent bound water (hydrated silica = opal).

Element and Trace Element Composition

Elements found in living sponges include boron (1550 grams/ton in ash of siliceous forms) (Rankama and Sahama, 1950, p. 489); iron in sponge pigments; traces of manganese, copper, and zinc.

Still other elements may be present in negligible quantities: titanium, chromium, cobalt, silver, gold, cadmium, gallium, germanium, tin, lead, arsenic, antimony, bismuth, molybdenum, and vanadium. Vinogradov observed that aluminum as Al_2O_3 is "found consistently" in sponge ash. (The sponge is charred or burned and the ash chemically studied.)

Bowen and Sutton (1951) found that some marine sponge species accumulated Fe, K, Ti; Cu and Na; or Mn, Sn, and V. High nickel content in one species is attributed to microfloral inhabitants. In the axial canals of some sponge spicules (Black Sea muds), there was 0.3 to 3.0 percent pyrite (Breger, 1963, p. 561).

Organic Compounds

Compounds isolated from sponges include ten different sterols (= high molecular weight alcohols having a benzene complex); numerous unsaturated fatty acids; nucleic acids with common nucleosides; the usual suite of amino acids; phosphoarginine and phosphocreatine. Noteworthy is the conspicuous absence of the complex polysaccharide, chitin, in sponges (Bergmann, 1949, Table 3 and

TABLE 15.2. Major Mineral and Chemical Components of Porifera

Chemical Compound	Mineral	Organism	Remarks
$SiO_2 \cdot nH_2O$ (with or without spongin)	Opal	Demospongea Hyalospongea[a]	
$CaCO_3 + MgCO_3$	Calcite[b]	Calcarea[c]	$MgCO_3$[d] ranges from 4.0 to 14.0 percent by weight
$CaCO_3$	Aragonite	Living *Astrosilera wellyana*, and fossil pharetronids (Dev.–Tert.)	
Spongin[e] (supporting tissues, spicules)	Organic	Demospongea (Order Keratosa)[f]	In *Euspongea*, spongin consists of C–48.51, H–6.30, N–14.79, S–0.73, I–1.5 (percent).

[a]A ton of seawater must enter the inhalent pores to yield in the fleshy parts, one ounce of siliceous spicules. Since there are 32×10^3 ounces in a ton, the ratio is 32,000 : 1.
[b]In thin section, a calcareous spicule optically responds as a single calcite crystal.
[c]Deformed spicules composed only of organic matter develop in calcisponges raised in carbonate-free media.
[d]Chave (1954a) found $MgCO_3$ in three calcareous sponges ranged from 5.5 to 14.1 percent.
[e]Spongin is an almost insoluble organic compound; nitrogenous; containing iodine and often bromine; composed of very fine silklike threads that branch and anastomose; chemically related to horn and silk. (For details see Gross and Piez, 1960).
[f]The bulk of protein in the Keratosa consists of an extracellular fibrous protein, collagen (Gross and Piez, 1960).
Source. Data from Vinogradov (1935–1944), and others.

pp. 146–156; Bergmann et al., 1957; Inskip and Cassidy, 1955; Mark et al., 1949; Roche and Robin, 1954; Rudall, 1955, cit. Hyman, 1959, pp. 715–716). The close chemical affinity of spongin and mammalian collagen is established (Gross and Piez, 1960). Bergmann and others isolated liquid and solid hydrocarbons from species of living sponges (Breger, 1963).

Living demosponges' organic content (in percent of living weight) varies from 7 to 13; for calcisponges, one determination indicates 3.61.

Halogens

All sponge groups, except the Hexactinellida, some 46 species, had dry sponge iodine that varied from 0.05 to 1.21 percent; bromine, from 0.01 to 2.66 percent. Halogens (iodine, bromine) accumulate in sponges in the absence of spongin.

Silica Content

A certain silica proportion in siliceous sponges may come from diatoms that sometimes fill openings of modern sponges (Vinogradov). *Euplectella speciosa*, the well-known Philippine Islands siliceous sponge, is composed of 99.18 percent silica.

Spicular Skeleton

For the mode of formation of the spicular skeleton, see Chapter 3.

COELENTERATA

Composition

Coelenterate skeletogenesis was reviewed in Chapter 5. Here, these data are supplemented. Table 15.3 compares composition of a hydrozoan milleporid, scler-

TABLE 15.3 Comparative Composition of the Coelenterata (Numbers expressed as percentage of ash)

	$CaCO_3$	$MgCO_3$	P_2O_5	Al_2O_3 $+Fe_2O_3$	SiO_2	$CaSO_4$
Hydrozoa (Milleporida)						
Millepora alcicornis[a]	98.22	0.95	Trace	0.11	0.24	0.48
Anthozoa (Scleratinia)						
Porites furcata[b]	99.95	0.82	Trace	0.11	0.12	?
Anthozoa (Octocorallia)						
Helioporacoerulea[c]	98.93	0.35	Trace	0.07	0.15	0.50

[a]*M. braziliensis* has 2.14 percent $MgCO_3$; other hydrocorals contain less than 1 percent; mineral aragonite composes its skeleton.

[b]Skeleton of aragonite. Scleratinians studied by Chave showed 0.12 to 0.76 percent $MgCO_3$ (1954*a*, Table 1). (Compare Figure 15.4.)

[c]Octocorals, exclusive of *Heliopora* (compare Chapter 5) have spicules of mineral calcite. *Heliopora's* skeleton is aragonite (Chave, 1954*a*). Octocorals generally have 6.0 to 17.0 percent $MgCO_3$ (Vinogradov); 6.0 to 14.9 percent (Chave).

Source. Data from Vinogradov (1935–1944) and others.

atinian coral, *Porites*, and octocoral, *Heliopora*. Calcium carbonate content and other constituents are quite similar in all of these species. However, important variations occur: One milleporid species has twice the magnesium carbonate as another (*Millepora braziliensis*)

Mineral calcite spicules characterize all octocorals except *Heliopora*. Scleratinian corals invariably have aragonite skeletons; hydrozoan milleporids also have such skeletons.

Greater substitution of magnesium occurs in the octocoralline gorgonacean calcite lattice structure. Compared to other octocorals, they have about 10 percent less calcium carbonate but contain up to 14 percent magnesium carbonate.

Trace Elements

Spectrographic analysis of sea anemone *Metridium dianthus* (Noddack and Noddack, 1939, cit. Vinogradov, 1935–1944) showed trace amounts of Ti, V, Mo, Mn, Fe, Co, Ni, Cu, Ag, Au, Zn, Cd, Ga, Th, Ge, Sn, Pb, As, Sb, Bi.

Mollusks and crustaceans, among other invertebrates, contain more zinc than coelenterates. Zinc is several times greater than copper in coelenterates, as in all other invertebrates. Corals have about the same cobalt/nickel ratio as seawater. *Porites* contains 0.42 percent strontium (in dry ash assays).

Octocorals are unusual among coelenterates since they concentrate the halogens (iodine, bromine). In octocorals (exclusive of gorgonaceans) bromine content exceeds iodine by a factor of 2 to 3. A reverse ratio holds for gorgonaceans; iodine exceeds bromine. Iodine ranges from 0.06 to 9.3 percent dry organic matter.

Ionium/uranium ratio was studied in Pacific coral reef limestone drill cuttings taken to depths of several hundred feet. The uranium is known in newly formed corals; ionium (an isotope of [230]Thorium, a decay product of uranium) is a later addition (Barnes et al., 1956).

Organic Components

Bergmann (1963) found organic matter in reef-building corals in significant quantities (2.0 to 8.0 percent). The most conspicuous component was lipids, including, waxes. These were found in both the living tissues and the deeper portions.

Trapped reef lipids that could no longer circulate might, in the geologic past, have been a petroleum source, according

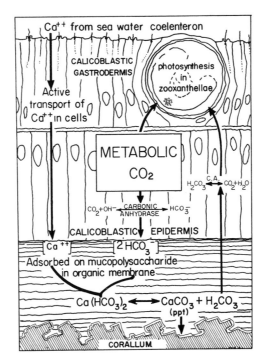

FIGURE 15.3B Skeletogenesis in reef-building corals. Schematic representation to show possible pathways of calcium and carbonate during calcification. Top of figure, coelenteron and flagellated gastrodermis; middle of figure, calico-blastic epidermis; bottom of figure, organic membrane with crystals of calcareous matter. Direction of growth is upward; calcium deposition in downward direction. After Goreau, 1959. See Chapter 5 and text of Chapter 15 for further details.

to Bergmann and co-workers. This excludes extraction of lipids by reef-boring worms, mollusks, and so on—a situation hard to envision (Yonge, 1963; Goreau and Hartmann, 1963).

Skeletogenesis

Goreau (1959) subdivided the calcification process in scleratinian corals into separate chemical pathways by which Ca^{++} (= one pathway) and $CO_3^=$ (= another distinct pathway) are brought to the calcification centers. Study of thin slices of polyp tissues helped to isolate the site of the mineralization process, outside of the calico-blastic epidermis (see Figure 15.3 B). (Steps diagrammatically shown in Figure 15.3B).

Pathway I (Calcium Ca⁺⁺)

1. Seawater enters coelenteron; calcium ions are extracted.

2. Body wall cells transport Ca^{++} ions to the calicoblast external surface.

3. At this site, organic matrix characteristics come into play.

4. The lattice structure provided by the mucopolysaccharide in the organic matrix serves as a control. By ion exchange, the Ca^{++} ions are adsorbed onto this structure.

Pathway II (HCO_3^-)

1. An end product of metabolism in polyp cells is carbon dioxide (CO_2).

2. Metabolic CO_2 goes partly to photosynthesizing zooxanthellae.

3. Another part remains in the epidermal tissue and in presence of the enzyme, carbonic anhydrase combines

$$CO_2 + OH^- \xrightarrow[\text{Anhydrase}]{\text{Carbonic}} HCO_3^-$$

4. Carbonate is then adsorbed on the organic matrix, as was calcium.

5. After adsorption, the following reaction occurs: the Ca^{++} and HCO_3^- combine to

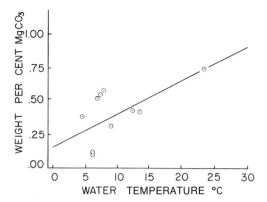

FIGURE 15.4 Relationship of water temperature and magnesium content in stony corals. Genera include *Flabellum, Madrepora, Dendrophyllia,* and others. All stony corals are 100 percent aragonite; range of $MgCO_3$ in percent, 0.12–0.76. After Chave, 1954a, Table 1, and Figure 4.

form the unstable compound, $Ca(HCO_3)_2$. This compound then yields calcium carbonate and carbonic acid:

$$Ca(HCO_3)_2 \rightleftharpoons CaCO_3 + H_2CO_3$$

6. The carbonate is precipitated as mineral aragonite (in a downward direction, that is, opposite to direction of coral growth, which is upward).

7. Carbonic acid is transported from the site of calcification into the epidermal tissue where, influenced by the enzyme, carbonic anhydrase, it yields CO_2 and H_2O.

8. This CO_2 is also fixed by photosynthesizing zooxanthellae.

At what rate will the calcium carbonate form? Goreau (1959) postulated that if the quantity of available calcium is adequate, calcium carbonate formation will depend on the rate of removal of carbonic acid from the calcification site. This would depend on how rapidly carbon dioxide is fixed (taken up) by photosynthesizing zooxanthellae. Calcification is slowed down if the coral is kept in the dark.

Yet Goreau's experiments disclosed that the coral kept in darkness, as well as enzyme inhibition, did not stop but merely greatly decreased calcification rate. Accordingly, neither zooxanthellae nor enzyme constitute part of the essential "message" for calcification in corals. They merely mediate or control the rate. Components of the essential "message" appear to be supply of Ca^{++}, metabolic CO_2, and the presence of suitable molecular structures and configurations in the organic matrix, which will determine sites of crystallization.

(At this point, review the steps in skeletal formation outlined by Olgilvie; Bryan and Hill, discussed in Chapter 5.)

Another valuable finding of Goreau's researches concerns calcification rates in different parts of a colony of *Acropora conferta.* Branching corals usually calcify faster in the apical region by a factor of 4 to 8 than in the lateral or basal regions (Goreau, 1959, Figure 4 and Table III).

ECTOPROCTS (BRYOZOA)

Introduction

Calcification in the cheilostome *Escharioides coccinea,* and body wall development in cyclostomes, were treated in Chapter 6. Limited supplementary data are added here.

Mineralogy

Schmidt (1924) and others (cit. Vinogradov, 1953) found only mineral calcite present in analyses of cyclostome *Crisia* and cheilostomes *Bugula* and *Flustra.*

Borg (1926) demonstrated that *Crisia's* calcareous layer is deposited between a surficial chitinous cuticle and the epidermis (Figure 6:3; *D,* Chapter 6). Actual deposition of calcareous salts (as calcite crystals) occurs on chitinous fibrils that occupy the space between the outer cuticle and inner epidermis. Thus *sites on these chitinous strands appear to control biomineralization.*

Since chitin is essential as the mineralization substrate in calcareous ectoprocts, Loppens's findings (1920) are useful. Chitin content (in percent of dry matter) was found to vary according to growth habit and habitat. A littoral zone *Membranipora membranipora* contained 13.55 percent chitin compared to 31.6 to 56.0 percent chitin in other representatives of this species living beyond the low tide–high tide zone.

Chemical Composition

Besides calcium carbonate, the major ectoproct component, magnesium carbonate, is usually present in quantities up to 11.08 percent. Clarke and Wheeler (1922) found that the percent $MgCO_3$ in less compact ectoproct structures was twice as great as in more compact structures. This could be related to greater or lesser readiness for Mg^{++} ion substitution in the calcite lattice provided by the actual surface area exposed to seawater.

Phosphorus also appears to be relatively high in some living and fossil ectoprocts — to 3.0 percent. Various older Paleozoic ectoprocts (family Ceramiporidae) from North Wales and Gotland contained phosphate in the form of dahllite crystals. This phosphate may have been a replacement of calcite after burial. Another possibility is that it was actually concentrated during life as with some living cheilostomes. Vinogradov (1953, p. 223) speculated that ectoprocts with phosphate skeletons (that is, with phosphate the dominant component) may have existed in the geologic past.

Other compounds reported (as percent of the ash residue) include $CaSO_4$ — 1.32 to 8.47 percent (*Eschara foliacea*); Fe_2O_3 plus Al_2O_3 — 0.12 to 4.82 percent (*Flustra*); and SiO_2 — 0.35 to 5.53 percent (*Lepralia*). Higher silica, up to 16.7 percent that has been reported, undoubtedly includes the contribution of associated sand grains.

Trace Elements

Trace elements of Cu, Mn, Mb, Va, Pb, K, Ba, and the halides (Cl, I) are known (Vinogradov, 1953).

Biochemistry

Although data on ectoproct biochemistry are practically nil, Bergmann (in Breger, 1963, p. 533) mentioned bryozoan reefs, among others, as places where lipids might accumulate. Various ectoproct species contain pigments (carotenes and xanthophylls) that give larvae yellow, orange, and red coloration and lend a reddish color to adults of several species (Hyman, 1959). A complex polysaccharide, chitin, ranks as an important constituent.

BRACHIOPODA

Introduction

In Chapter 7 investigations by Jope, Williams, and others were examined. Here shell mineralogy, chemical composition, element, and trace element content are especially emphasized.

Shell Layers

Shell may be of three types in inarticulates (Cooper, 1944):

1. Unlayered. All chitin — *Acrothele*
2. Unlayered. Calcium phosphate uniformly distributed through the chitin — *Discinisca*
3. Layered. Alternate layers or bands of chitinous material and calcium phosphate — *Lingula* (Figure 15.5C).

Modern articulates have a triple-layered shell (Williams, 1956) (Figure 15.5A, B):

1. A thin, external organic skin with honeycombed structure — the *periostracum* (not preserved in fossils but sometimes represented by a mosaic of convex, minute polygons).
2. Two carbonate layers of lamellar calcite:
   (A) Primary layer — cryptocrystalline; thickness relatively constant.
   (B_1) Inner secondary layer — obliquely disposed coarse fibers enclosed in cytoplasmic sheaths. (The boundary between 2A and B_1 is sharply defined.)
   (B_2) Alexander (1948, p. 157) described a third or "prismatic" layer. It comprises a relatively thick layer. Williams (1956) recognizes the prismatic layer as a modification of the B_1 inner, secondary layer characteristic of the Pentameracea (suborder) and some of the Spiriferida (order). For the association

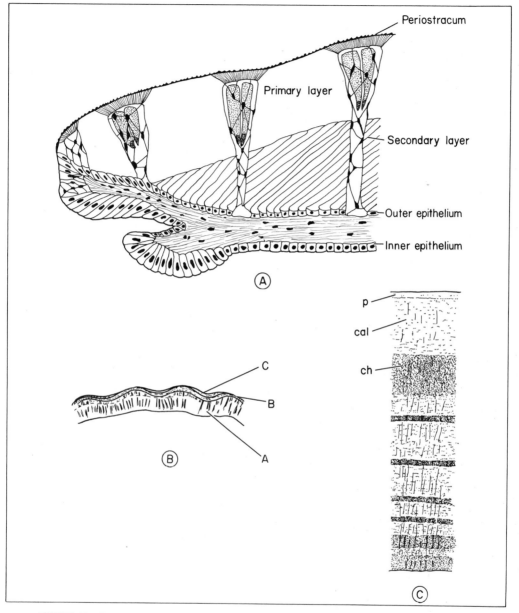

FIGURE 15.5 Skeletogenesis and shell layering in articulate and inarticulate brachiopods.
(A) Terebratuloid brachiopod, based on section of shell and mantle and serial sections of decalcified specimens. Further details in Figure 7.1C. Adapted from Williams, 1956.
(B) *Conchidium knightii,* transverse section through pedicle valve; A – prismatic layer; B – inner shell layer, C – outer shell layer. After Alexander, 1947.
(C) *Lingula* sp., section through valve, p – periostracum, *cal* – calcareous layer, *ch* – chitinous layers. Note alternation of *cal* and *ch*. After Blochman, 1900, cit. Hyman, 1959.

of these layers and shell structures, see Chapter 7.

The B_2 prismatic layer may (for example, *Conchidium*) consist of two or three layers laid down one over the other. Alexander (1948, p. 153) commented that the thickness of the prismatic layer may be a function of the individual's age.

Skeletogenesis

(See Figure 15.5A, B.) Each articulate valve starts as a minute chitinous plate (the *protegulum*, see Chapter 7). Along its edges shell enlargement occurs. Only a portion or extension of the mantle, the advancing mantle edge, is the deposition site of the periostracum and the primary carbonate layer. These do not thicken or receive additions, once deposited. Both layers are extracellular secretions.

By contrast, an intracellular secretion accounts for the secondary carbonate layer. Each obliquely dispersed fiber in it is secreted by a cell of the mantle's outer epithelium. The enclosing cytoplasmic sheath of each fiber is thought to be the stretched membranous wall of one epithelial cell. Hence secretion is intracellular.

However, the prismatic layer is in continuity with these underlying fibers of the secondary layer—probably an extracellular secretion of the outer epithelium (Williams, 1956). This explains why the prismatic layer can be thickened subsequently and may consist of multiple thin layers. It appears to result from excess carbonate in solution in body fluids bathing the mantle's epithelial cells.

The relationship of punctae and shell layers—various types—was covered in Chapter 7.

Biochemistry

Chitin. With few exceptions, a considerable amount of chitin is present in inarticulate shells. Hyman (1958) confirmed this older finding but could find no chitin in articulate shells. Chitin is present in the cuticle of the articulate pedicle as in the linguloids.

Periostracum. Data on the biochemistry of this organic layer are given in Chapter 7 (Table 7.1).

Respiratory Pigments. The respiratory pigment, haemerythrin, in the coelomic fluids contains iron but not copper (cit. Hyman, 1959).

Organic Content. (in percent of dry matter). Living *Terebratulina* and *Wald-*

heimia contain between 1.0 and 4.73 percent organic matter. Inarticulates, by contrast, have from 25.0 to 51.1 percent organic matter. Vinogradov (1953) reported proteinaceous organic matter impregnated with phosphate in *Lingula*, *Obolela*, and *Obolus*. Paper chromatographic study and other chemical analyses should help elucidate details of organic content in both articulates and inarticulates.

Mineralogy. The Paleozoic articulate brachiopods Stehli (1956) studied by X-ray and chemical methods were exclusively calcite—Spiriferida and Strophomenida. McConnell (1963) found, by X-ray powder diffraction analysis, that the microcrystalline substance in living *Lingula* was identical with the mineral francolite—a carbonate fluorapatite.

Element and Trace Element Content. A *Lingula* shell from Ceylon was analyzed: organic matter, 51.1 percent; in percent ash, P_2O_5–42.99, CaO–50.66, CaF_2–3.92, MgO–0.30, Mn_3O_4–0.46, Fe_2O_3–0.22. X-ray analysis indicated presence of apatite and fluorapatite (Vinogradov, 1953, p. 228)—francolite (McConnell, 1963).

McConnell also assayed fluorine content. A dried central portion of the shell had 1.6 percent *F* and the ashed inorganic material, 2.44 percent *F*. He assumed that *Lingula* probably had a special tolerance for fluorine. Opik (1929, cit. Vinogradov, 1953) also reported high fluorine in the inarticulate *Obolus*.

Various inarticulates have high $MgCO_3$ (*Lingula*, *Discinisca*, *Crania*). Clarke and Wheeler (1922) analyzed various phosphatic brachiopods and found shell $Ca_3P_2O_8$ ranged from 74.73 to 91.74 percent. $MgCO_3$ content is generally less in articulates (0.47 to 1.37 percent) than in inarticulates. *Crania*, for example, had 8.63 percent and *Discinisca*, 6.68 percent $MgCO_3$.

Articulate shells contain from 94.6 to 98.6 percent calcium carbonate.

High sulphate ($CaSO_4$) from 2.93 to

8.37 percent was recorded by Clarke and Wheeler (1922) for inarticulates, whereas it was 0.36 to 1.18 percent in articulate shells.

Trace elements (in percentage of ash) include Cu, Ba, Al, Ti, and Zr (in *Lingula*) and boron as B_2O_3 in the ash of *Lingula* and *Terebratula*.

Kulp et al. (1952) published valuable data on strontium in carbonate rocks and contained fossils. Fossils, on the average, contained twice as much strontium as the surrounding carbonate matrix. This finding applied to articulate brachiopods as well as other forms. Included were articulate valves of Silurian through Permian age, and such common genera as *Atrypa*, *Leptaena*, *Composita*, *Marginifera*, *Derbya*.

A typical suite of Pennsylvanian brachiopods (*Composita*, *Chonetes*, *Derbya*, *Dictyoclostus*, *Neospirifer*, and *Linoproductus*) from a single formation (Lower Francis Formation, Stonewall Quadrangle, Oklahoma) was studied. Strontium of the carbonate matrix compared to that of the named brachiopods was almost 1:1 except for *Chonetes* and *Derbya* where the ratio was as elsewhere, about 1:2.

One of the highest strontium assays — 4.75 (Sr/1000 Ca) — occurred in shells of the inarticulate *Lingulepsis*. High phosphate content of this form may explain high strontium. Among randomly selected articulates, strontium varied as follows: *Pentamerus*, 0.44; spiriferids, 0.66 to 0.75; *Terebratula*, 0.78.

MOLLUSCA

Introduction

Together with material in Chapter 8, the data of this section can introduce a broad spectrum of exciting researches.

Each of the five molluscan classes has its own distinctive skeletal characteristics. Nevertheless, certain generalities in composition apply to all. Despite these generalities, chemical composition varies between species of a given genus, between different genera, and even within a species population.

Composition

With a few exceptions (Table 15.4 and *Loligo*), some 97.0 to 99.0+ percent of min-

TABLE 15.4. Comparative Composition of Molluscan Shells (in Percent)

Compounds	Pelecypods		Gastropods		Scaphopods	Amphineurans	Cephalopods	
	1	*2*	*3*	*4*	*5*	*6*	*7*	*8*
SiO_2	0.32	0.36	0.15	0.26	0.40	0.61	0.19	0.09
$(Al, Fe)_2O_3$	0.08	0.05	0.16	1.89	0.27	0.22	0.15	0.13
$MgCO_3$	1.00	Trace	0.41	0.44	0.20	0.45	0.16	6.02
$CaCO_3$	98.60	98.74	99.28	97.21	99.13	98.37	99.50	93.76
$Ca_3P_2O_8$	Trace	0.40	Trace	Trace	Trace	Trace	Trace	Trace
$CaSO_4$	—	—	?	0.20	?	0.35	—	—

1. *Pecten dislocatus* Say, Charlotte Harbor, Florida.
2. *Nucula expanser* Hancock, north of Bering Strait.
3. *Parpura lapillus* Linné, Eastport, Maine.
4. *Turritella gonostoma* Valenciennes, Mulege, Gulf of California.
5. *Dentalium solidum* Verrill, Off Georges Bank, East of Cape Cod, Mass., depth of water, 2361 meters, temperature 4.5°C.
6. *Mopalia muscosa* Gould, Santa Barbara, California. A chiton.
7. *Nautilus pompilius* Linné, Mindinao, Philippine Islands.
8. *Argonauta argo* Linné, High Seas, Pacific Ocean.
Source. After Clarke and Wheeler, 1922.

eral residues in mollusks, regardless of class, is calcium carbonate.

Magnesium carbonate varies from a trace to as much as 38.1 percent in *Loligo vulgaris* (Turek, 1933, cit. Vinogradov, 1953, Table 177).

The amount of calcite present in the shell controls quantity of magnesium in pelecypods, gastropods, and cephalopods (Chave, 1954). Those forms with only aragonite or high aragonitic shells had least magnesium. Temperature of water in which several mollusks lived was one controlling factor in amount of magnesium present. Earlier, Chave (1952) found that in high-magnesium calcitic skeletons, magnesium formed a solid solution with calcium.

Silica generally occurs in tenths, sometimes hundredths, of 1 percent, although exceptions are known. *Oncidium*, a genus of gastropods, has about 10 percent of its total weight composed of siliceous spicules. Spicules are present as amorphous silica (Labbe, 1933; Kahane, 1935, cit. Vinogradov, 1953, pp. 295, 296). This gastropod genus is of unusual interest in that it is siliceous and has an active silicon metabolism. The liver of *Oncidium planatum* contains 11.3 percent silica.

Radulae (see Ch. 8), characteristic of large numbers of gastropods, may also contain high silica (and iron). One analysis of the ash in *Patella athletica* showed 32.85 percent SiO_2.

Phosphate is present as a trace to a few tenths of 1 percent (see Table 15.4) (see Watabe, 1956).

Calcium Metabolism

Calcium, along with water content in pelecypod tissues, varies with different growth stages and in different organs of the same individual. Molluscan blood generally carries more calcium than magnesium. Calcium, stored in the liver as amorphous carbonate, may be available for shell repairs in *Helix*.

Other Element and Trace Element Content

Some mollusks concentrate mangan-ese, copper, iron, and zinc. The copper is present as haemocyanin — a respiratory pigment (see Chapter 8). Vinogradov (1953) thought that in a sense cephalopods, concentrating several times more copper in the blood than other mollusks, are copper organisms. He also speculated that Paleozoic ammonites and Mesozoic belemnites probably contained the same pigment.

Fresh-water molluscan shells have greater manganese content than marine. Fresh-water mussels *Anodonta* and *Unio*, in particular, have an exceptionally large manganese content (4.80 to 5.02 percent of ash respectively). Manganese is present in molluscan blood as a metallo-organic complex (Mn + protein), but it serves no respiratory function as was once thought.

An analysis of the pelecypod *Crassostrea virginica* recorded, among other elements, Cu, Mn, Zn, and Ni present in trace amounts. Three genera of pelecypods had a sulfur content of 0.1 to 0.2 percent.

Cephalopods and pelecypods in the family Ostreidae tend to concentrate iron or zinc (and copper, as already noted).

Many trace elements are found in mollusks: Sn, Cd, Ti, Li, B, Pb, Mb, Au, Ag, Ni, Co, Sr, Rb, As. *Mytilus edulis*, a pelecypod, contained barium (Ba).

Halogens (fluorine, iodine, and bromine) occur in mollusks. Fluorine is present in all molluscan shells. The pelecypod *Ostrea* has a fluorine content of 0.003 percent. Fossil mollusks have a higher fluorine content. This undoubtedly is due to secondary enrichment, which may be used as an age-dating method for Pleistocene to post-Pleistocene fossils. Shells of *Crassostrea virginica* contain iodine. A dye used in antiquity, the famous Tyrian purple, is derived from a bromine-rich substance present as a bromi-organic compound in both the skin and shell of the snail *Murex brandaris*. Exposed to the air, this compound turns purple.

Kulp et al. (1952) studied strontium content in randomly selected fossils. In a single formation (Calvert Formation, Miocene of Maryland) they found these Sr/ Ca ratios, using different genera of gastropods and pelecypods and a single genus of scaphopods (*Dentalium*):

Class	Sr(atoms)/1000 Ca(atoms) (Average)
Gastropoda	2.12–3.21
Pelecypoda	2.05–6.57
Scaphopoda	3.88

Assessment of these results indicated the following: (*a*) There is a relatively constant strontium content in a given genus. (*b*) Certain animals are selective in their strontium uptake. (*c*) Low or high strontium content appears to be related to the crystalline lattice structure (the calcium lattice being less receptive to strontium than the aragonite lattice).

Both belemnite (*Atractites pusillus*) and pelecypod (*Glycymeris rubicus*) were sectioned to determine the strontium content by growth zones. Would individuals show greater or lesser strontium content as they grew? The conclusion reached: no appreciable variation in strontium content with an individual's age.

Mineralogy

As seen in Table 15.4, the general composition of mollusks is calcium carbonate. Three possibilities exist in biological systems for crystalline carbonates—calcite, aragonite and, more infrequently, vaterite (Wilbur and Watabe, 1963). Vinogradov (1953) recorded increasing stability from amorphous $CaCO_3$ to vaterite to aragonite to calcite.

Wilbur and Watabe studied calcification in mollusks (1963) by means of a shell regeneration study. Holes were bored in shells, or the edges were removed. Both shell matrix and crystal type varied in regenerated, compared to original, shell. *Crassostrea virginica*, which is calcitic, regenerated its shell with both calcite and aragonite. *Elliptio complanatus*, which is aragonitic, repaired its shell with calcite,

aragonite, and vaterite. Aragonitic *Mercenaria mercenaria* showed no change in crystal type, depositing only aragonite. The organic matrix appeared to be a factor in type of carbonate crystal formed. Temperature also played an important role (see Lowenstam, 1954).

Further complications in mineral type must be considered, especially for larval shells (prodissoconchs) of pelecypods. Watabe (1956) found the mineral dahllite (hydroxyapatite), previously unreported in invertebrate skeletons, in the prodissoconch of *Pinctada martensii*.

Bøggild (1930) observed that the majority of fresh-water forms have aragonitic shells; among the lamellibranchs (see Chapter 8), some are partly or entirely of calcite, whereas others are entirely of aragonite. He found that gastropods were characteristically aragonitic even though both calcite and aragonite occurred in some forms.

In Figure 15.6 comparative shell structures in pelecypods, scaphopods, gastropods, and cephalopods are shown. Further details will be reviewed under "Skeletogenesis." The general plan of biomineralization is confined to a few zones: a fibrous, prismatic, or porcelaneous outer layer and an inner nacreous or lamellar layer. In addition there is a thin, external film, or periostracum, as in *Nautilus*, usually lacking in fossils.

In many, but not in all Archeogastropoda, there is a pearly or nacreous inner layer (not shown in Figure 15.6C). Similar nacreous layers occur in fossil and Recent cephalopods and pelecypods. The crossed lamellar structure, generally aragonitic, is most common in layers of gastropod shells (Figure 15.6E). First order or primary lamellae are perpendicular to shell surface. They are 0.02 to 0.04 mm thick and are composed of second-order lamellae that are less than 0.001 mm thick. The latter are transverse to the primaries (Cox, 1960).

Figure 15.6C shows an outer pseudoprismatic layer. Bøggild observed an outer prismatic calcitic layer in a number of gastropod families. There are variants.

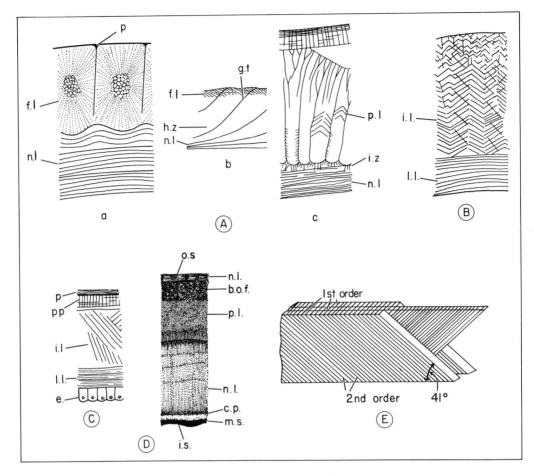

FIGURE 15.6 Comparative shell structure in Mollusca (schematic).
(A) Pelecypoda: (a) *Nucula* (Recent), transverse section, 500 μ long, composition aragonite; (b) fossil, *Nucula strigillata* (Trias., Saint-Cassian), radial cut along shell margin, 900μ long; *p* — periostracum; *f.l.* — fibrous layer; *n.l.* — nacreous layer; *hz* — homogeneous zone crossed by growth surface traces (*gt*); (c) *Avicula* (*Meleagrina*), Rec., detail of radial section of shell margin (1000μ) aragonite, *p.l.* — prismatic layer; *n.l.*, as in (a) above; *i.z.* — intermediate zone.
(B) Scaphopoda: *Dentalium*, Rec. — prismatic layer; transverse section (1000μ long), composition aragonite; *l.l.* — lamellar layer; *i.l.* — intercrossed layer.
(C) Gastropoda: Generalized transverse section of shell; generally, composition aragonite (see text), *i.l., l.l.*, as in (B) above; *p* — periostracum, *e* — pallial epithelium.
(D) Cephalopoda: *Nautilus*, Rec., X 30; *o.s.* — *i.s.* — outer and inner surfaces of shell; *n.l.* — nacreous layer; *b.o.f.* — black organic film (tyrosine derivative); *p.l.* — porcelaneous layer (= aragonite in conchiolin matrix); *c.p.* — periphract conchiolin; *m.s.* — mural part of septum (schematic).
(E) Crossed lamellar structure in gastropods (compare Figure C). Portion of three first-order laminae and thin lamellae of second order are shown. Note alternate direction. First-order laminae (primary) are parallel to shell surface (schematic).
Illustrations modified, adapted and redrawn from: (A)(B) from Jean Piveteau, ed., Traité de Paléontologie, 1952, Vol. II, Colette Deschaseaux, "Classe des Lamellibranches", in Traité, Vol. II; (C) after several authors; (D) Appelhöf, 1893, cit. Stenzel, 1964; (E) Cox, 1964.

Buccinum undatum has an outer layer of prismatic aragonite, whereas shells of *Haliotis* (Haliotidae) have a prismatic layer of calcite between aragonite layers.

Shell microstructure, after Bøggild, was tabulated by Newell (1965, Table 1) for 36 extant superfamilies of Bivalvia (= Pelecypoda = Lamellibranchiata) in six subclasses.

All extant superfamilies, such as Myacea or Tridacneacea, in the subclass Heterodonta have only cross-lamellar microstructures; all in subclass Anomalodesmata and subclass Palaeoheterodonta show only nacreous shell microstructure. In subclass Pteriomorphia, superfamilies vary: cross-lamellar microstructure only (Arcacea), nacreous microstructure only (Pinnacea), both of these (Pteriacea), or both of these plus some variant such as Bøggild's "foliate" structure consisting of parallel flakes of calcite (Pectinacea). Furthermore, only foliate structure is present in some forms (Ostreacea). Only a "homogenous" structure occurs in subclass Cryptodonta and Paleotaxodonta. (The homogenous structure is characterized by $CaCO_3$ which appears structureless in ordinary transmitted light.) The last-named subclass contains another superfamily, Nuculacea (Figure 15.6A, a, b), with forms having both nacreous and cross-lamellar shell microstructure.

Biochemistry and Submicroscopic Shell Structure

The conchiolin matrix of molluscan shells has been shown to consist of three fractions: a protein, a sclero-protein, and a polypeptide (Grégoire et al., 1955).

Some 18 amino acids have been found in conchiolin (Grégoire, et al., 1955; Roche et al., 1951). Tanaka, Hatano, and Itasaka (1960) studied the amino acid residues in the conchiolin of nacre of pearl, the nacreous substance of molluscan shells, and the prismatic substance of such shells (Table 15.5). These investigators found a large amount of cystine in conchiolin, suggesting a fibrous structure for conchiolin reminiscent of keratin in hair and skin. Some "remarkable differences" were found in phenylalanine, alanine, and proline of conchiolin obtained from nacreous substance as compared with prismatic shell substance (Table 15.5).

Beedham (1954) showed that different concentrations of amino acids occurred in the outer layer compared to the inner layer of valves of *Anodonta cygnea*. Variation in amino acid concentration between different species has also been observed.

Electron microscopic studies of the nacreous layer fine structure, particularly by Grégoire (1955–1962) and by other workers (Tsuji et al., 1958), further advanced biochemical researches.

The essential findings of numerous researches on conchiolin fine structure indicate that at magnifications of $\times$ 18,000, it occurs as thin lamellae-fenestrate, lace-like sheets (Figure 15.7A), whereas at magnifications of $\times$ 173,400, these sheets are seen to be composed of *trabeculae* that have the appearance of a knotted cord rug (Figure 15.7B). Grégoire has noted that the fenestrae or pores embraced by this corded matrix vary in shape, size, and frequency distribution in nautiloids, pelecypods, and gastropods. His studies of many fossil cephalopods also demonstrated persistence through geologic time of conchiolin patterns (fenestration or pore characteristics). Thus Recent and Pennsylvanian nacreous conchiolin membranes have a great similarity of structure (Grégoire, 1959). It follows that even where only minute fragments of shell are available in fossil collections, the conchiolin microarchitecture in such fragments can permit class identification.

As indicated earlier, biomineralization will partly be controlled by "organic components having particular molecular structure and configuration." Conchiolin's fenestrate structure, with its knotted trabeculae, appears to provide just such an organic matrix. Variation in fenestration, thickness of conchiolin sheets, and protuberances on trabeculae – all seem related to particular three-dimensional molecular configurations that create crystallization sites.

The distinct conchiolin patterns that Grégoire found among the molluscan classes must therefore be, at least, a partial determinant of the individuality in shell structures. (Study Figure 15.6 from this viewpoint.)

Skeletogenesis

(After Wilbur, 1960.) As diagrammed in Figure 15.8, biomineralization in mollusks requires several items: source of carbon dioxide, presence of an enzyme – carbonic anhydrase (see Figure 15.3B) –

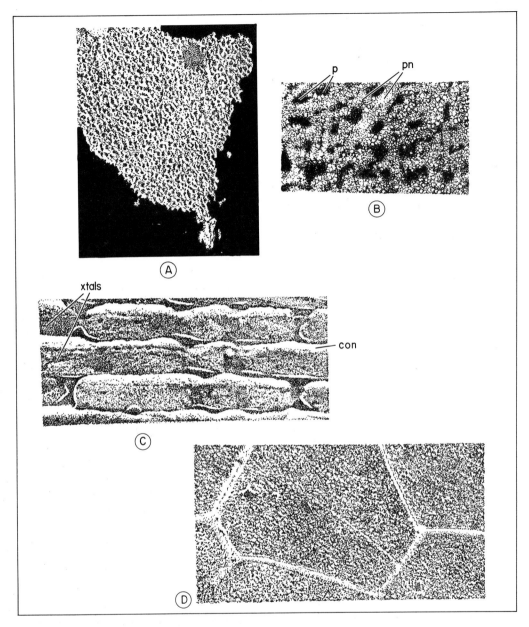

FIGURE 15.7 Submicroscopic structure of conchiolin. After Grégoire et al., 1955, Grégoire, 1957.
(A) The gastropod, *Astrea olivacea,* note trabeculae with pore pattern, X ca. 18,000;
(B) The gastropod, *Turbo* sp., cordlike trabeculae, showing protuberances (*pn*) and pore (*p*) characteristics, X ca. 173,400.
(C) *Turbo* sp., nacreous region. Note bricklike arrangement of tabular aragonite crystals (*xtal*). Conchiolin (white)—*con*—occurs as one or more sheets between crystalline laminae; note conchiolin bridges in intercrystalline spaces, X 19,000.
(D) The pelecypod, *Pinctada martensii,* prismatic region. Note the large polygonal chambers bounded by relatively thick layers of conchioliu. Adapted from Watabe and Wada, 1956, in Wilbur, 1960.

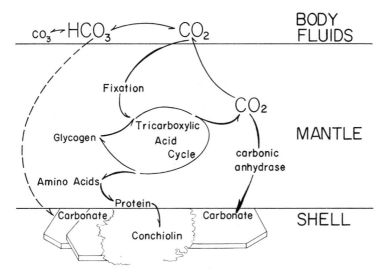

FIGURE 15.8 Schematic diagram indicating possible steps in the biomineralization of a molluscan shell (after Hammen, in Wilbur, 1960). For details and elementary discussion of the TCA Cycle, see M. J. Guthrie and J. M. Anderson, 1957, Zoology (John Wiley and Sons), Figure 2.9 and pp. 35–38. TCA Cycle = Krebs Cycle = Citric Acid Cycle. (See Wilbur, 1972).

synthesis of a conchiolin matrix, uptake of calcium from seawater as well as the operation of the tricarboxylic acid cycle (TCA).

1. Figure 15.8 indicates the aerobic phase of carbohydrate metabolism—the TCA cycle. Out of this cycle one of the waste products of metabolism, carbon dioxide, is fixed in molluscan tissue. The CO_2 occurs in mantle and body fluids. There is thus a source of CO_2.

2. Carbonic anhydrase is present in mollusks. (Chemical inhibition of this enzyme retards calcification rate.)

3. CO_2 fixation contributes to amino acid formation and probably glycogen (a stored polysaccharide carbohydrate). In turn, amino acids (see Table 15.5) formed in this way in the molluscan mantle contribute to conchiolin matrix synthesis.

4. Seawater calcium enters the mantle directly. A small part of the total calcium in the mantle is turned over rapidly and deposited as shell. [There are not two but three interrelated processes in shell calcification. The third, a non-continuous process, involves shell growth by small increments (individual crystal layers) (Wilbur, 1972).]

Given the conditions set forth in items 1–4 above, then CO_2 is converted to carbonate in the body fluids by either (1) or (2):

$$(1) \quad CO_2 + H_2O \longrightarrow H_2CO_3,$$

or

$$(2) \quad CO_3 + OH^- \longrightarrow HCO_3^-.$$

There are now two possible pathways by which crystalline carbonate can form the shell layers in the conchiolin matrix:

Pathway I: HCO_3^- dissociates in body fluids, yielding the CO_3 that combines with calcium uptake in the mantle and directly deposits crystals in the several sites of the conchiolin trabeculae.

Pathway II: The CO_2 from the TCA cycle is converted to carbonate by combining with calcium in the mantle. There, under the influence of an enzyme, carbonic anhydrase, the process is accelerated and crystals form in the organic substrate.

Glycogen, highly concentrated in the mantle's outer epithelium, probably allows a given mollusk to continue normal metabolism for long periods without food intake. The mantle's outer epithelium is the plate where shell deposition occurs.

Although details of the process are still under investigation by specialists, and the

TABLE 15.5. **Amino Acid Composition of Conchiolin**

| | | Conchiolin[a] | |
Amino Acid Residues	Nacre of Pearl	Nacreous Substance of Shell	Prismatic Substance of Shell
Leucine	9.2	13.6	9.0
Phenylalanine	1.1	0.0	16.9
Valine	2.1	0.5	1.0
Tyrosine	2.7	7.2	3.0
Methionine	0.4	0.4	0.0
Proline	0.0	0.0	7.9
Alanine	14.0	16.3	4.6
Glutamic acid	3.1	1.5	1.5
Threonine	0.6	9.3	0.3
Aspartic acid	6.2	2.2	3.7
Serine	5.4	2.1	3.5
Glycine	24.3	12.8	16.8
Ammonia	0.7	0.5	0.1
Arginine	7.2	15.3	10.2
Lysine	7.4	3.3	1.5
Histidine	0.5	0.0	1.0
Cystine/2	12.2	11.8	14.7

[a]Figures represent grams of amino acid residue per 100 g. of protein.
Source. After Tanaka, Hatano, and Itasaka, 1960.

cycle is much simplified in Figure 15.8, modification of certain chemical steps, rather than the generalities outlined, can be expected.

The sequence from the initial biochemical steps to the layered crystalline carbonate in molluscan shells can best be followed if the illustrations are reviewed once more serially in the following order: Figure 15.8, Table 15.5, Figure 15.7, Fig-15.8 again, and then Figure 15.6.

ARTHROPODA

Introduction

Literature is rather spotty on the mineralogy and biochemistry of shell or carapace. Nevertheless, data available are excellent (Richards, 1951, 1953; Travis, 1960, 1963; Dennell, 1947, 1958, 1960; Passano, 1960; Lafon, 1948; Vinogradov, 1953; Clarke and Wheeler, 1922; among others). In Chapter 9, details of molting as elaborated by Passano (1960) were studied; many aspects of encoded data in arthropod shells and carapaces were re-

viewed. Here, additional details and factors will be considered including chemical determinations on fossils such as trilobites.

Composition

Major inorganic chemical components of crustaceans, particularly of barnacles (Cirripedia) and malacostracans are given in Table 15.6. Malacostracans generally have a high magnesium carbonate content. Their calcium carbonate content ranges from 70 to 80 percent or less. Similarly, they also have relatively high phosphate and from negligible to high sulphate content. Barnacles, by contrast, like mollusks, have high calcium carbonate content and little else. In the mantis shrimp (Table 15.6, column 4) phosphate content exceeds total carbonates ($MgCO_3$ + $CaCO_3$).

Calcium, magnesium, and phosphate content are variables in decapods. Quantitatively, these elements vary with age, size, sex, and part of skeleton. The chelae increase in phosphate content with age

TABLE 15.6. **Percent Composition of Various Crustaceans (after Clarke and Wheeler, 1922)**

Compound	Class Cirripedia	Class Malacostraca			
	1	2	3	4	5
SiO_2	0.03	0.06	0.00	0.00	—
$(Al, Fe)_2O_3$	0.15	0.06	0.34	0.50	—
$MgCO_3$	0.75	6.69	8.02	15.99	2.42
$CaCO_3$	97.47	78.14	79.50	28.56	83.40
$Ca_3P_2O_8$	Trace	14.45	10.91	49.56	11.88
$CaSO_4$	—	0.60	1.23	5.33	2.30

1. Barnacle, *Balanus hameri*, Georges Bank, east of Cape Cod, Mass.
2. Blue Crab, *Callinectes sapidus*, ranges from Cape Cod to Florida.
3. Common Lobster, *Homarus americanus*, Vineyard Sound, Mass.
4. Mantis Shrimp, *Chloridella empusa*, ranges from Cape Cod to Florida.
5. Fresh-water Crayfish, *Astacus fluviatilis* (analysis by O. Bütschli, 1908).
Source. After Clarke and Wheeler, 1922.

and have generally more magnesium and phosphate than has the carapace (Clarke and Wheeler, 1922). However, Koller (1930, cit. Vinogradov, 1953) found the ratio of carapace weight to calcium carbonate constant despite size and age differences in specimens.

Crustacean gastroliths (Figure 15.9*E*), sometimes referred to as "krebsaugen," often have high phosphate content—to 18 percent. Travis (1963, p. 181) observed that total phosphorus steadily increases in gastroliths from growth to maturity and that the calcium carbonate constitutes about 76.37 percent of the mineral in the mature gastrolith.

Phosphates in decapods and other crustaceans are apparently present as amorphous Mg and Ca salts, and not as the mineral apatite (Vinogradov, 1953).

Phosphate is present in fossil and living xiphosurans (*Limulus*) and in Middle and Lower Cambrian trilobites from Wales (*Paradoxides davidii*). Sedimentary layers containing many carapaces of the last-named trilobite were found to be enriched in phosphate. Phosphate, as P_2O_5 in such carapaces, was present in amounts of 17.0 to 20.0 percent compared to the P_2O_5 in neighboring sedimentary layers (0.3 to 4.0 percent). *Olenellus* carapaces from the Cambro-Silurian of Sweden, had up to 19.45 percent P_2O_5. Other trilobites, such

as *Belfura*, contained up to 36.0 percent phosphate. The literature remains unclear as to what extent some or all of this phosphate is secondary, that is, enrichment after burial. Nevertheless, greater or lesser phosphate was part of the chemical content of several lower Paleozoic arthropods (Vinogradov, 1953, pp. 414–417). In turn, this suggests a comparable role for the enzyme, alkaline phosphatase in the mineralization of Paleozoic arthropod skeletons and in modern skeletogenesis.

Calcium Metabolism

Different crustaceans draw the calcium for their carapaces from various sources. Calcium can be stored in the body in (*a*) ionized form in the blood, (*b*) hepatopancreas glands (see Figure 15.9*E*), (*c*) gastroliths (see Figure 15.9*E*), or elsewhere. It is resorbed from old exoskeletons (Passano, 1960, p. 511); or the calcium salts that harden the new exoskeleton, after ecdysis, may be obtained from the external medium. Passano (1960) stressed this last source as the "major" supply for integument calcification. However, Scott (1961, p. Q-21) observed that ostracods will secrete a new shell after molting in a calcium-free environment; this points up the role of internal stockpiling.

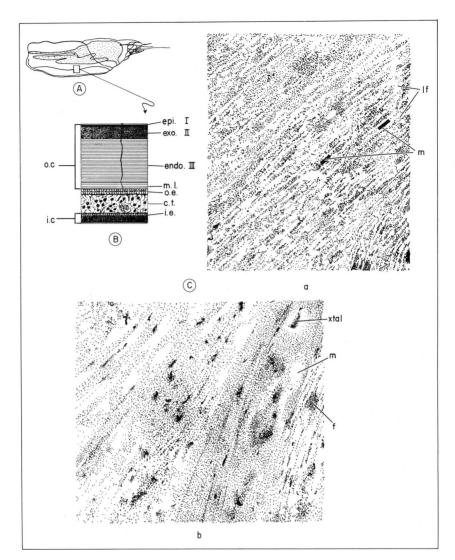

FIGURE 15.9 Crustacean mineralization. (*A*) Branchial exoskeleton (carapace) Of a crayfish (Order Decapoda), schematic.
(*B*) Differentiated layers of crayfish carapace, detail of segment from (*A*) above, see arrow; *Epi*—epicuticle, hardened by compounding of organic components (proteins, and so on): *Exo*—exocuticle, hardened by compounding of organic components and by calcification; *Endo*—endocuticle, hardened only by calcification; *m.l.*—membranous layer, not calcified; *o.e.* and *i.e.*—outer and inner epidermis; *c.t.*—connective tissue.
(*C*)(*a*) Portion of completely calcified fiber, characteristic of the entire lamellae fiber system. Note mineral distribution as rectangular bricks (shown in black) along longitudinal axis of fiber, X 133,000 (drawn from electron micrograph); *m*—mineral, *lf*—longitudinal axis of fiber; (*b*) Surface section (drawn from electron micrograph) showing mineral particles along fibers of lamella and large crystals (calcite) deposited in chitin-protein matrix (see arrows), X 31,500; *xtal*—crystals, *f*—fibers, *m*—matrix.

Element and Trace Element Content

Potassium (K_2O) is present in many decapods. It ranges in percent of dry matter from 0.0 to 2.74 percent. Decapod tissues have a relatively high sulfur content (see Table 15.6).

The liver of arthropods, such as *Limulus*, tends to concentrate Cu, Mn, Fe, and Zn. Quantities range from tenths of 1 percent for Fe, Cu, and Zn, up to hundredths and hundred-thousandths of 1 percent for Mn (Vinogradov, 1953, Tables 249–252). For crustaceans generally, Mn, in

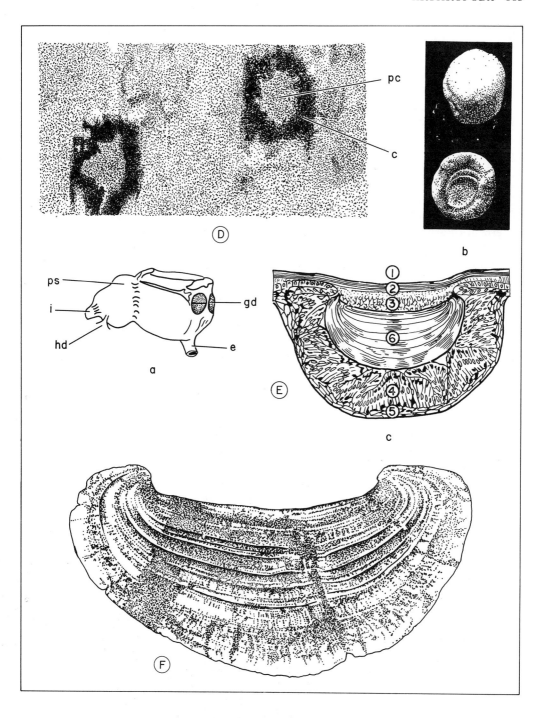

FIGURE 15.9 (cont.) (*D*) Surface section showing heavy mineralization (in black) in walls of the pore canal; *pc* — pore wall, *c* — calcite.

(*E*) (*a*) Sketch of stomach showing position of gastrolith disks; *ps* — pyloric stomach, *cs* — cardiac stomach, *HP* — hepatopancreas, *e* — esophagus, *in* — intestine, *gd* — gastrolith discs; (*b*) pair of mature gastroliths (see fossil gastroliths, Figure 1.7, X 2; (*c*) stage D_4 (oldest mineralized gastrolith stage), showing relationship of gastroliths to cuticles and so on; (1) epicuticle, (2) exocuticle, (3) endocuticle, (4) epidermis, (5) connective tissue, (6) gastrolith (mature stage), schematic.

(*F*) mature gastrolith (vertical section) showing high degree of ordering in the mineral; note alternation of high-density mineral lamellae (dark), and low-density mineral lamellae; note also dense vertical, compact mineral prisms; X 12. Adapted from Travis, 1963.

percent of living matter, occurs in quantities that may be expressed as $n \times 10^{-3}$. The notostracan, *Apus* (= *Triops*), a branchiopod, has an exceptionally high Mn concentration even in the carapace chitin.

Other trace element content includes Ni, Co, Ti (in many crustacea, including the planktonic copepod, *Calanus finmarchius*), and B. In the tissues of crabs and in the blood of other crustaceans, silver has been spectrographically detected. The barnacle *Balanus* contains 0.02 percent barium and 0.2 percent strontium (percent of ash); tin; lead; molybdenum; vanadium; and chromium. Radium is also present in the ash of some crustaceans (10^{-13} percent).

The halogens (I, Br, and possibly F) have also been recorded in various crustaceans. Crustacean carapaces are richer in iodine than are other parts. Bromine content exceeds iodine in crustaceans. However, even though the magnitude of Br in crustaceans is about equal to that of seawater, zooplankton, such as the copepods, contain as much as $n \times 10$ times the amount of iodine as seawater.

Arsenic, which occurs in high percentages on ocean bottoms, has also been reported in crustaceans. This appears to follow logically since benthonic crustaceans are bottom scavengers. Apparently, the arsenic intake migrates into the areas of fat storage.

Mineralogy

The calcium carbonate of the shells of living barnacles occurs as mineral calcite, although among fossil genera the skeleton may be all chitin. Planktonic crustaceans (Mysidae) have otoliths of aragonite; some show the presence of the mineral dahllite (Vinogradov, 1953, p. 399). Only calcite has been reported from trilobite carapaces (Bøggild, 1930).

In decapods (Travis, 1963), calcium carbonate occurs in (*a*) a poorly crystalline state in gastroliths, and (*b*) as calcite in the pore-canal system and matrix located between fibers in the exocuticle-epicuticle

junction (see Figure 15.9*C–E*). Crystalline calcite occurs in cladoceran carapaces. Amorphous carbonate is also a component. Crystalline calcite has been found in large numbers of fossil and living ostracods (Schmidt, 1924, cit. Vinogradov, 1953). In addition, amorphous calcium carbonate and vaterite have been reported.

Aside from calcite and its polymorphs, the important components of the arthropod exo- and endoskeleton are chitin and other organic complexes.

Biochemistry

(Dennell, 1960; Travis, 1960, 1963; Goodwin, 1960.)

Chitin. Chitin of the arthropod skeleton is of the *alpha* (α) type in contrast to that of coelenterates, annelid worms, mollusks, and brachiopods, which have the *beta* (β) type.

In the decapod cuticle, chitin constitutes 60 to 80 percent of the dry weight of the organic fraction. This type of chitin is "arthropod chitin."

Protein. Amino acids reported in arthropod cuticular protein include serine, alanine, threonine, tyrosine, and tryptophane. Dennell suggests that other amino acids are present. Skeletal proteins of crustaceans can be removed in two fractions: water-soluble arthropodin and alkali-soluble, water-insoluble sclerotin (= cuticulin). The arthropod skeleton is composed of mixed polymers of protein complexed with the polysaccharide, chitin.

A sterol-protein complex is present in both insect and crustacean cuticles, that is "tanned" when the cuticle is hardened (see Chapter 9). Surface waxes and cholesterol are known components of the cuticle. Cholesterol also occurs in the calcifying gastroliths.

Enzymes. Among the enzymes important in the calcification of crustacean carapaces are alkaline phosphatase, which may play a role in phosphate resorption from the old exoskeleton and the transport of resorbed calcium, carbonic anhydrase, and polyphenoloxidase (this plays a role in the tanning process).

Citric Acid. Citric acid may act in forming weakly ionized salts with calcium and thus enter into the mineral complex (Travis, 1960).

Pigments. Hemoglobin occurs in the Branchiopoda, Ostracoda, Copepoda, and Cirripedia. Hemocyanin, a copper-containing respiratory pigment, occurs in arthropod and molluscan blood. In arthropods, hemocyanins occur in xiphosurans, arachnids, and certain crustaceans (namely, decapods and stomatopods). Carotenoids are present chiefly as astaxanthin in both fresh-water and marine crustaceans. This last substance accounts for some crustacean coloration, for example, dark blue of decapods, green of lobster eggs, brown of amphipods, and so on.

Riboflavin and derivatives, although true pigments, play a minor part in crustacean coloration. Other pigments, such as the melanins, occur in crustaceans (Goodwin, 1960).

Skeletogenesis

Problem. A living decapod has a rigid skeleton that it must shed to grow. What are the steps in making a new, larger skeleton and discarding the old?

Steps. Before a new exoskeleton can be secreted, existing substances must be resorbed, redistributed, and stockpiled. Resorbed calcium carbonate is deposited in gastrolith disks (Figure 15.9E). A molting fluid secreted by the epidermis aids resorption of both mineral and organic matter. It contains the enzyme, chitinase, which breaks down chitin and alkaline phosphatase. The resorbed constituents are carried to the blood and redistributed (Passano, 1960). Whatever is not resorbed

is removed from the system during ecdysis.

1. *Premolt Stages.* Essentially two related processes go on: (*a*) resorption of the old exoskeleton and (*b*) new exoskeletal synthesis.

(a) *Resorption.* First, the membranous layer begins to be resorbed. This causes the enlarging epidermal cells to separate from the old exoskeleton.

(b) *New Epicuticle.* A new epicuticle is deposited under the old skeleton. Composition of this cuticle is lipoprotein. Pore canals (Figure 15.9D) and tegumental ducts cross this cuticle, which lacks lamellae. Compounding of organic components, protein plus sterols, and calcification hardens the new epicuticle.

(c) *New Exocuticle.* A new exocuticle is deposited under the freshly developed epicuticle. It is composed of chitin, protein, lipid, and so on. Pore canals cross closely spaced lamellae. Protoplasmic extensions of the epidermis pass through these canals.

[(b)–(c) During (b)–(c) time resorbed mineral calcium carbonate is conveyed for stockpiling in the gastroliths. Glycogen and other organic reserves concentrate in the epidermis, connective tissues, reserve cells, and the matrix of the newly developed epicuticle. The reserve cells contain calcium and are sites of phosphatase activity.]

2. *Ecdysis.* Shedding the old Skeleton.

3. *Postmolt Stages.* Endocuticle deposition begins. It is deposited under the exocuticle. During these stages the gastrolith stockpile gradually breaks down and some of the constituents are conveyed by the blood to the epidermis. In this way, an internal calcium source is provided.

(a) *Endocuticle.* The endocuticle has widely spaced lamellae and is crossed by pore canals and tegumental ducts. Its composition is chitin-protein, which occur in thin layers. Lipids are also present. It is hardened by calcification. The endocuticle is called the "principal layer" or "calcified zone."

(b) *Calcification.* Calcium uptake from the external media occurs. According to some authors, the calcium uptake is primarily via internal recycling. Calcification goes on simultaneously with the laying down of the endocuticle organic matrix. Viewed microscopically, granules appear to be extruded from the epidermis.

(c) *Calcification of the New Exocuticle.* Calcification occurs by way of the epidermal cells that extrude strings of granules, including calcium, into the exocuticle through the pore canals (Figure 15.9*D*).

(d) *Formation of a Membranous Layer.* This layer, an endocuticle part, immediately contacts the epidermis. Composed principally of lipids in the region of the pore canals, it has very closely spaced lamellae. The layer is not calcified.

4. Intermolt Conditions. After a period of rest or stability, another premolt period begins (Travis, 1963; Passano, 1960).

Further Details on Crustacean Mineralization

Mineral particles are deposited in the chitinous, fibrous matrix of the exo- and endocuticle. These fibers consist of filaments 50 to 120 Å long (Travis, 1963) (1 Å = 10^{-8} cm or 3.937×10^{-9} in.). Such microfibers visible on electron micrographs may represent molecular chains of chitin (Dennell, 1960). Calcite particles have no preferred orientation. They are 400 to 500 Å long and about 150 to 200 Å wide.

Decapod carapace mineralization has two major crystal systems: pore canal calcite crystal system (Figure 15.9*D*) and calcite crystal system of the matrix. The latter occurs between widely spaced fibers of the lamellae system near the exo-epicuticle junction. Very poorly crystalline to amorphous $CaCO_3$ is deposited in the lamellae fiber system forming rectangular bricks (see Figure 15.9*C* and *D*.

Even though initially deposited mineral particles are most likely controlled by a close association with the chitinous fila-

ments, it is in the interfilamentous spaces themselves that major crystal growth occurs.

Shell carapace and integument characteristics of trilobites, branchiopods, ostracods, insects, and so on, are given in Chapter 9. Details reviewed for decapod crustaceans have analogies with those pertinent to the insect integument (exclusive of calcification) and with other crustaceans. Since growth requires ecdysis in arthropods, it is understandable that many of the factors considered above apply more widely. Of course, details may vary. All steps will not be duplicated.

ECHINODERMATA

Skeletal development (Gordon, 1926–1929), mineralogy, and chemical composition of echinoderms were reviewed in Chapter 12. Here these data are supplemented and/or expanded.

Chemical Composition

Tables 15.7 and 15.8 summarize and compare the composition of living and fossil crinoids and representative living echinoderms.

In Table 15.7, column 6, high silica is certainly due to silicification after burial. Low $MgCO_3$ in fossil crinoids has already been considered (see Chapter 2), although a fossil *Encrinus* stem is an exception.

Echinoderm $CaCO_3$ content ranges from 80.0 to 90.0+ percent, but phosphate is consistently present as a trace or tenths of 1 percent. Presence of $MnCO_3$ and $FeCO_3$ in fossil crinoid plates may be accounted for by calcium replacement in the carbonate lattice structure after burial.

In embryogenesis of the sand dollar raised in seawater, high calcium content yielded multiple skeletal *anlage* (that is, spicules) (Bevelander and Nakahara, 1960).

Strontium content of echinoderm skeletons varies from 0.005 to 0.5 percent of dry matter. Barium is present in lesser quantities.

TABLE 15.7. Composition of Selected Living and Fossil Crinoids

Compound	1	2	3	4	5	6	7
SiO_2	0.08	0.42	0.02	1.57	6.94	29.30	0.24
$(Al, Fe)_2O_3$	0.10	0.33	0.57	2.64	0.64	1.74	0.44
$MgCO_3$	10.16	10.09	7.86	1.79	0.80	1.23	20.23
$CaCO_3$	89.66	87.16	95.55	93.80	91.62	67.24	79.09
$Ca_3P_2O_8$	Trace	Trace	Trace	0.20	Trace	0.00	Trace
$CaSO_4$	—	—	?	—	—	—	—
$MnCO_3$	—	—	—	—	Trace	0.06	Trace
$FeCO_3$	—	—	—	—	0.00	0.43	0.00

1. *Hypalocrinus naresianus* Carpenter, Philippine Islands; depth, 612 meters, bottom temperature, 10.22°C.
2. *Bythocrinus robustus* Clark, Gulf of Mexico; depth, 255 meters, no temperature recorded.
3. *Promachocrinus kerguelensis* Carpenter, Shores of Antarctica (near Gaussberg); depth 350 to 400 meters, bottom temperature 1.85°C; salinity, 3.3 percent.
4. *Pentacrinus basaltiformis* Miller, stem, Middle Lias (L. Jr.), Germany.
5. *Dorycrinus unicornis* Owen and Shumard, calyx and stem, lower part of Burlington limestone, Mississippian (L. Carb.), Iowa.
6. *Eucalyptocrinus crassus*, plates, Silurian, Western Tennessee.
7. *Encrinus liliiformis* Lamarck, stem, Triassic, Germany.
Source. After Clarke and Wheeler, 1922.

TABLE 15.8. Comparative Composition of Selected Echinoderms

Compound	Crinoid	Echinoid		Asteroid	Ophiurian	Holothurian
	1	2	3	4	5	6
SiO_2	2.01	0.13	3.25	0.27	2.39	0.15
$(Al, Fe)_2O_3$	1.31	0.37	2.38	0.20	0.84	0.34
$MgCO_3$	7.91[a]	5.99[a]	5.41	13.33	9.53	13.84
$CaCO_3$	88.48	93.13	88.96	85.99	86.60	83.29
$Ca_3P_2O_8$	0.29	Trace	Trace	0.21	0.64	Trace
$CaSO_4$	—	0.38	Trace	?	?	2.38

1. *Ptilocrinus pinnatus* A. H. Clark. Off Queen Charlotte Islands, British Columbia; depths 2858 meters, bottom temperature 1.83°C (crinoid).
2. *Strongylocentrotus drobachiensis* O. F. Müller. Upernivik, Greenland (sea urchin).
3. *Echinus affinis* Mortensen. Between Hatteras and Nantucket; depths, 1919 meters, bottom temperature, 3.6°C (sea urchin).
4. *Acanthaster planci* Linné. Palymyra Island, Pacific Ocean (starfish).
5. *Gorgonocephalus arcticus* Gray. Off Cape Cod, Mass. (brittle star).
6. *Holothuria floridana* Pourtalès, Fajardo, Porto Rico (sea cucumber).
[a]Compare Figure 15.10.
Source. After Clarke and Wheeler, 1922.

Shell trace elements of echinoid *Brissopsis lyrifera* (order Spatangoida) included Ti, V, Cr, Mo, Mn, Fe, Co, Ni, Cu, Ag, Au (7×10^{-7} percent dry weight), and Zn (I. and W. Noddack, 1939, cit. Vinogradov, 1953, Table 158). Radium occurs in sea stars and sea urchins. It is reckoned in 10^{-13} percent of living matter.

Absence of lithium and potassium should be noted. Experimental studies of echinoderms indicate that addition of lithium to seawater results in abnormal skeletal formation. Multiple skeletal starts result with addition of potassium chloride, lithium bromide, or sulphuric acid (cit. Bevelander and Nakahara, 1960).

Goldschmidt and Peters (1932) reported boron in the skeleton of *Echinus esculentus* (0.01 percent B_2O_3).

Some echinoderms have trace quantities of the halogens, iodine and bromine, plus arsenic. In sea cucumbers, iodine forms an organic compound in the body wall. Sea urchin shells and spines contain about 0.27 percent sulphur and the starfish *Asterias*, 0.87 percent.

Biochemistry

Collagen (a major fibrous protein found in tissues of most invertebrates) occurs in the body wall of the sea cucumber *Thyone*. Amino acid composition of *Thyone* collagen proved very similar to that of mammalian tissues but was very low in lysine (Gross and Piez, 1960). In this regard, creatinine and creatine have been reported from the coelomic fluids of echinoids. Although these are typical nitrogenous products of vertebrates, they are unknown in invertebrates other than echinoderms (Myers, 1920; Baldwin, 1947, cit. Hyman, 1955).

Invertebrates generally have substantial quantities of unsaponifiable material in contrast to vertebrates. (Conversion of a fat or oil into a soap by the action of an alkali is called "saponification." Some fractions of lipids are "unsaponifiable.") Bergmann (1963, Table 2; 1949) recorded that Protozoa, Porifera, and Coelenterata have a high percent (35.0, 37.0, and 35.0 respectively) reckoned in terms of unsaponifiable percent of total lipids. Annelida has an intermediate percent (22.0); Echinodermata, Crustacea, and marine Mollusca, a comparatively low percent of unsaponifiable fractions. These figures considerably exceed the vertebrate percent, which averages less than 2.0.

Among other pigments, echinoderms contain echinochrome ($C_{12}H_{8-10}O_{7-8}$) (Dunning, 1963, Table 17). Blumer (1951) reported a mixture of fossil dyes in fossil crinoids (*Millericrinus*, U. Jur.), and he attributed the original source to echinochrome. Iron-bearing respiratory pigments (erythrocruorin) are known in echinoderms, annelid worms, crustaceans, insects, and mollusks (Dunning, 1963).

Mineralogy

In echinoderms the entire skeleton is a rhombohedral crystal system (Vinogradov, 1953) and consists of calcite crystals with definite distribution and orientation. Raup's work (1959, 1960, 1962a, 1962b) explored echinoid calcite crystallography.

To introduce these researches, some preliminary observations are necessary:

1. *Elements of the Skeleton.* Each element in the echinoderm skeleton is a single calcite crystal. Elements such as echinoid plates, spines, and crinoidal columnals, respond optically as a single crystallographic unit.

2. *Essential Calcite Crystallography.* For convenience, crystal faces can be located in space using imaginary lines or directions (= crystal axes). Six groupings of such axes embrace all crystals. Calcite falls in the hexagonal system. This system has four axes, three of which are in the same plane, intersect at 60° and 120° angles, and are of equal length. These are called a_1, a_2, a_3. The fourth axis, the *C*-axis, is perpendicular to the plane of the *a* axes and is either longer or shorter (Wahlstrom, 1943, pp. 1–3, Figure 1).

Raup's investigations concern only the principal optic axis, the *C*-axis.

3. *Volume of Calcite in Echinoderm Skeletal Elements.* West (1937) showed that the true volume of solid matter in an echinoid spine was about 38 percent. The solid matter, of course, is calcite. Raup (1962b) found that the typical echinoid plate is about 40 percent calcite by volume. Thus skeletal elements are spongelike and in life form a porous latticework of interconnected rods filled with organic

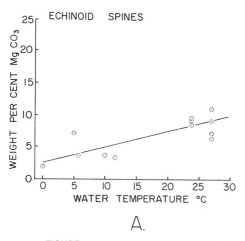

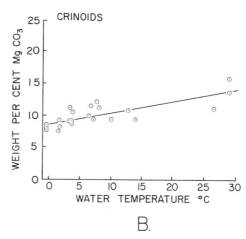

FIGURE 15.10 Relationship of water temperature and magnesium content in echinoderms. (A) Echinoid spines. (B) Crinoids. After Chave, 1954.

tissue. Following fossilization, secondary calcite filled these open spaces. This calcite is in optical continuity with the original calcite rods and has good calcite cleavage. Skeletal elements of living forms do not show good cleavage because of spatial discontinuities between calcite subunits (Raup, 1959; 1962a and 1962b).

4. Lattice Warping and Crystal Curvature. X-ray analysis demonstrated that lattice orientation changes slightly across individual hexagonal echinoid plates and appears to follow its curvature (Donnay, 1956, p. 206); compare curved surfaces in Figure 15.11. Thus calcite lattice warping appears to conform to morphological curvature of a given plate (Raup, 1962a). Similar warping also occurs in the structure called "Aristotle's lantern" (see Chapter 12). The bony parts consist of calcite fiber bundles each, with rare exceptions, a curved, single crystal.

5. Larval Forms (Plutei). In *Strongylocentrotus* plutei are 0.1 mm long. They consist of single crystals of calcite. The skeletal length coincides with the C-axis of the crystal (Donnay, 1956). This larval skeleton is completely lost in metamorphosis. Postmetamorphic echinoids have fewer plates than the adult (Gordon, 1929; Hyman, 1955).

6. Ontogenetic Indicators. In echinoids, plates nearest the peristome are oldest and they decrease in relative age as the apical system is approached (see Figure 15.11). Thus, in Raup's studies, thin sections were prepared from interambulacral and ambulacral plates along various points of the plate column. This made possible the detection of any variation in C-axis orientation between older and younger plates.

Raup's Studies

Thin sections cut from the skeleton, a polarizing microscope, and a four-axis universal stage permitted positioning the C-axis with respect to the surface of a given echinoid plate within one or two degrees (Raup, 1962b).

Major Findings. (1960, 1962a, 1962b)
1. Each echinoid studied exhibited a remarkably preferred orientation of the C-axis. This could relate to both individual plate morphology and to whole skeleton symmetry.

2. Five types of orientation pattern of C-axis were found. Of these, the two most common were Types 1 and 2:

Type 1. All C-axes nearly perpendicular to the plate surface (Figure 15.11c) (73 species).

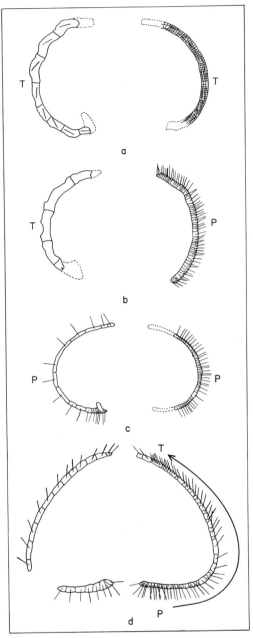

a

b

c

d

Electron microscopic studies (Travis, 1970) of the echinoderm skeleton, among others, have led to clarification and correction of the conclusions reached by polarizing microscope studies (Raup). At the ultrastructural level, individual calcite crystals are oriented with their long or crystallographic C-axes parallel to the fiber axes of the collagen fibrils. Raup's conclusions that these crystals are oriented with crystallographic C-axes either tangential or perpendicular to the plate surface, thus reflect an apparent rather than a real situation. (Travis, D. F., 1970, in Harold Schraer, ed., Biological Calcification. Appleton-Century-Croft, Chapter 5.)

Type 2. All C-axes nearly tangent to plate surface and aligned longitudinally (dorso-ventrally) in the skeleton (Figure 15.11a) (35 species).

Type 3. Ambulacral plates with perpendicular C-axes and interambulacral plates with tangential C-axes—a combination of Types 1 and 2. Type 3 is apparently restricted to the primitive family Cidaridae

FIGURE 15.11 Echinoid crystallography (adapted from Raup, 1962*a,* 1962*b*). Variant *C*-axes orientations seen in cross-section (schematic). *C*-axes—straight lines; in each figure, apical system is upper open portion; peristome, lower open portion. Oldest plates adjacent to peristome. Ambulacra on the right, interambulacra on the left, in each figure. *T*—*C*-axes uniformly tangential to plate surface, both in amb and interamb plates (shown as *T–T*); *P*–*C*-axes perpendicular to plate surface both in amb and interamb plates (shown as *P–P*); *P–T*—amb and interamb plates (shown as *P–P*); *P–T*—variation in *C*-axes orientation, perpendicular in amb plates and tangential in interambs, P-to-T in a single plate column—a 0° to 90° variation during ontogeny, that is, perpendicular near peristome and tangential adjacent to apical system.
(*a*) *Eucidaris thouarsii* (Rec.).
(*b*) *Plegiocidaris florigemma* (Jur.).
(*c*) *Pedina sublaevis* (Jur.).
(*d*) *Conulus albogalerus* (Cret.). Periproct opening to far right.

(Triassic to Recent) (Figure 15.11*b*) (6 species).

Type 4. Systematic variation in orientation amounting to 45° or less occurred during growth of the organism. This was observed in a single column of an adult individual such as *Hemicentrotus pulcherrimus* (Recent, Japan) (2 species).

Type 5. In this type, ontogenetic variation was recorded. It approached 90° in a single plate column [from perpendicular near the peristome (oldest), to tangential adjacent to the apical system (younger) (Figure 15.11*d*) (3 species).

3. Phylogenetic Affinity. When Raup compared orientation of the *C*-axes in species of given families and orders, he found that within a given order, defined solely on morphological criteria, these orientations remained relatively constant and/or stable. Thus, from species to species assigned to a given order, marked divergences in *C*-axes orientation were lacking except in two orders: Cidaroida and Holectypoida (Figure 15.12).

Calcite crystallography of echinoid plates, while generally in agreement with Durham-Melville's morphological classification of orders (1957), revealed the following: (1) Family Cidaridae—lack of crystallographic homogeneity within the family may require placing certain cidarid genera in a separate family. (2) Order Echinoida—this order is considered unnatural by several workers. Two families in it have Type 1 *C*-axes (Echinometridae and Stronglyocentrotidae) and two families have Type 2 *C*-axes (Echinidae and Parasileneidae). These data suggest a different ancestry for each type. (3) In the Arbaciidae, a Cretaceous form (*Goniopygus menardi*) is anomalous since it has Type 1 *C*-axes. Crystallographic evidence partly sustains the possibility (based on morphology) that the Arbaciidae is of polyphyletic origin.

Thus calcite crystallography has been applied variously: an independent check on classificatory schemes, on evolutionary lineages, as an indicator of anomalies within morphologically homogenous groupings, or homologies in morphologically uncertain groupings. [The Travis findings (1970) on *C*-axis orientation will require reexamination of many of these interpretations.]

4. Evolution of Preferred Orientation. The oldest orientations known are perpendicular (U. Sil.–Perm), and those of eight out of 14 extant orders, with one exception, are also perpendicular.

The perpendicular orientation (Type 1) has apparently two adaptive attributes, according to Raup (1962, 1962*b*): Such plates allow less light to pass than those with tangential *C*-axes; such plates tend to be more highly curved. The first attribute confirmed in the laboratory may be

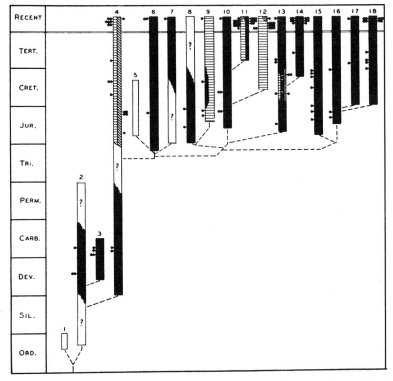

FIGURE 15.12 Echinoidea—phylogeny according to Durham-Melville (1957) and distribution by orders of C-axes orientations in echinoid calcite according to Raup (1962b). Each dot = a studied species within a given order. C-axes orientations: black—perpendicular; parallel lines—tangential; right-to-left oblique lines—ontogenetic variation; left-to-right oblique lines—ambulacra, perpendicular; interambulacra, tangential; blank—no data.

Echinoid orders: (1) Bothriocidaroida. (2) Echinocystitoida. (3) Palaechinoida. (4) Cidaroida. (5) Pygasteroida. (6) Echinothurioida. (7) Diadematoida. (8) Hemicidaroida. (9) Arbacioida. (10) Phymosomatoida. (11) Echinoida. (12) Temnopleuroida. (13) Holectypoida. (14) Clypeasteroida. (15) Holasteroida. (16) Nucleolitoida. (17) Cassiduloida. (18) Spatangoida. (19) Megalopoda—*incertae sedis.*

an adaptation to a sunlit, shallow-water environment.

Tangential orientation that confers *functional disadvantages* relative to perpendicular orientation was present in two extant orders of Jurassic and Cretaceous origin.

Since natural selection weeds out inadaptive portions of the genetic message through time, how can the retention of the functionally disadvantageous (and hence inadaptive) tangential orientation be explained? Raup (1960) held that tangential orientation "evolved in spite of functional disadvantages." If so, orientation of C-axes relative to plate surfaces must lack adaptive value (a point that Raup allowed). Furthermore, they perhaps represent not a primary, but a secondary effect of a more fundamental cause that had adaptive value.

At any rate, calcite crystallography in echinoids has been applied to taxonomy, paleoecology, and evolution.

Skeletogenesis. This section will survey the Bevelander-Nakahara findings (1960) on skeletogenesis of sand dollar *Echinarachnius parma* supplemented by Gordon's data on the same species (1929).

Fertilized eggs of *E. parma* were systematically examined at the blastula stage. [After fertilization, cell cleavage occurs, resulting in a *blastula* which is oval, hollow, and ciliated all over (Nichols, 1962, p. 120 and Chapter 12).]

Stage 1. Embryonic mesenchymal cells proliferate. A group of cells adjacent to the basal or baso-lateral ectoderm results (Figure 15.13a).

Stage 2. First micromineral crystals appear. Eighteen to 20 hours after Stage 1, three events occur: (*a*) One cell in each lateral cell cluster markedly differentiates, assuming an elongate, irregular shape; (*b*) numerous large, cytoplasmic vacuoles appear; and (*c*) some vacuoles contain birefringent bodies (viewed in polarizing light) that can be identified as micromineral calcite crystals (Figure 15.13*b–d*).

Stage 3. Formation of birefringent granules and resorption of microcrystals take place. After one to two hours lapse, the microcrystals formed in Stage 2 yield birefringent granules in a single vacuole. Such growth of granules apparently arises from resorption of the microcrystals in other vacuoles. This last is indicated because the original birefringent bodies of these other vacuoles disappear from them (Figure 15.13*e*).

Stage 4. Transformation of granule to early triradiate crystal occurs. The single granule within the cell continues growth. At 26 hours past Stage 1, a triangular-shaped crystal first appears in the cytoplasm of the cell (Figure 15.13*f*).

Stage 5. (a) Continued crystal growth is observed. A crystal continues to grow, assuming a more definite triangular appearance (in surface view). This transpires at 28 hours and 30 hours, respectively, past Stage 1 (Figure 15.13*g*).

(b) Subsequent steps in skeletal formation occur. After 32 hours and 34 hours, respectively, a structure develops—a *triradiate spicule* (Figure 15.13*h*).

(c) Once the skeleton is initiated by a triradiate spicule, largely by intracellular events, development becomes extracellular. It is also carried out by the mesenchymal cells. Further skeletal development has two components: (1) elaboration of a protein matrix, (2) transfer of mineral salts from seawater to this matrix.

Use of tagged calcium (Ca^{45}) as calcium chloride enabled the investigators to locate concentrations of Ca^{45} in the primary mesenchyme cells (*gastrula* stage) associated with triradiate formation. Thus calcium from seawater, mobilized in these cells, contributed to the growing skeleton.

Figure 15.13*a–h* refers to the blastula

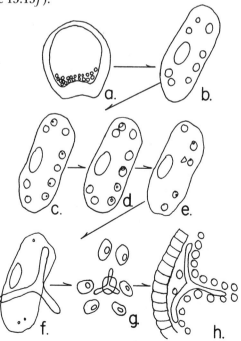

FIGURE 15.13 Skeletogenesis of the living sand dollar, *Echinoarachnius parma* (after Bevelander and Nakahara). Steps: (*a*) mesenchymal cells proliferate; (*b, c*) vacuoles form; (*d*) microcrystals grow (black dots); (*e*) granules form; (*f*) granules transform to beginning triradiate crystals; (*g–h*) continued growth of triradiate crystals. Repetition of this process yields the sand dollar skeleton.

through gastrula stages in the sand dollar. The last-named stage alters into the larval pluteus (Figure 12.2). The pluteus, after four to six weeks, undergoes rapid metamorphosis into an "imago." Subsequent skeletal development occurs following metamorphosis. Individual echinoderm skeletal plates develop in this general sequence: (1) formation of triradiate spicules, (2) dichotomous branching of spicular rays, (3) repeated branching resulting in an open meshwork (Gordon, 1926, 1929). (The solid rods are calcite; organic matter fills the intervening spaces.) Resorption of larval spicules and certain plates, to a degree, accompanies skeletal development.

CHITIN

Introduction

The complex polysaccharide, chitin, and a fibrous protein, collagen, are the chief supporting tissues that keep animal cellular structures intact. There are thus two skeletons: the collagenous skeleton and the chitinous skeleton (Rudall, 1955; Gross and Piez, 1960).

Occurrence of Chitin

Chitin is found in coelenterates (hydrozoan polyps, millepores, and pneumatophores of the Siphonophora—see Chapter 5). It is absent in hydromedusae, scyphozoans, and anthozoans (Rudall, 1955). The arthropod integument is composed of a chitin-protein complex (Richards, 1953). Only one-quarter to one-half of the cuticle's dry weight is chitin (Richards, 1951, Table 10).

Hyman (1958) studied chitin in the lophophorate phyla (see Chapter 6). She found chitin in secreted tubes of phoronoid worms but not in their bodies. Hyman's other findings include the fact that the entire exoskeleton of ctenostome bryozoans and noncalcareous cheilostomes consisted of chitin, whereas in calcareous cheilostomes, avicularia, opercula, and frontal membranes (plus other parts) were generally chitinous. In inarticulate brachiopods (except family Craniidae), chitin is abundantly present in shell, mantle setae, and pedicle cuticle (see Chapter 7). Articulate valves are devoid of chitin.

In Mollusca (see Chapter 8), localized chitin is commonly present. It occurs, for example, in the gastropod radula and buccal cuticle (*Patella*) as alpha chitin (for chitin types, see discussion below). In cephalopods, there are two chitin types— alpha and beta; alpha chitin is found in beaks, radula and gut linings, whereas beta chitin has been located in *Loligo's* pen. Furthermore, beta chitin was reported in a decalcified skeleton of *Sepia* (Rudall, 1955).

Annelid worms have beta chitin in chaetae and gizzard lining. Rudall (1955) noticed that chitins of annelids differ in their X-ray diffraction patterns from arthropods. So far, chitin is apparently absent in sponges (Porifera) and the Echinodermata.

Presence or absence of chitin in invertebrates has additional aspects. Collagen, the fibrous protein, is of mesodermal (that is, internal) origin; chitin is ectodermal (that is, external) in origin. Collagen and chitin systems can replace each other in supporting structures (that is, cuticles). In fact, with chitin absent (sponges and echinoderms, for example), an abundance of collagen is present. Thus, among sponges, the mesogleal protein is typical collagen, and most of the connective tissue in echinoderms is collagen. Similarly, whereas collagen is abundant in mollusks, the chitin is localized. Also note the chitin-protein complex in arthropods. Protein of this complex is noncollagenous. (Nevertheless, some collagen is present—barnacle peduncle of *Lepas*, subcuticular tissue of the lobster, among others.)

Chemistry and Structure of Chitin

Chitin is a polymer; it forms a long, several-hundred-unit molecular chain (Figure 15.14*a*). Figure 15.14*d* shows one

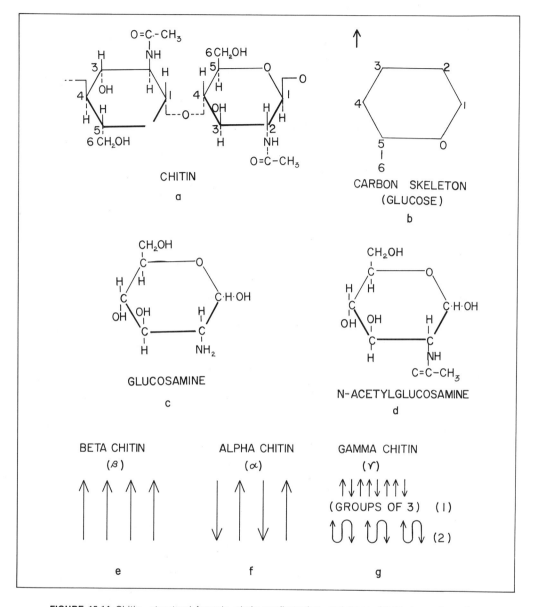

FIGURE 15.14 Chitin—structural formula, chain configuration, and types. (a) Steric configuration; compare numbered carbon position to (b) (Richards, 1953). (b) Carbon skeleton of glucose with carbon positions numbered (c) Glucosamine. (d) N-acetylglucosamine—multiple units of which form the long, molecular chain of chitin (C=C–CH₃ should read O=C–CH₃). The name can be broken down into its components: acetyl group = $CH_3 \cdot CO$; glucose: amine = NH_2. (e) Types of chitin showing chitin chain arrangements: groups of 1 = Beta chitin; groups of 2 = Alpha chitin; groups of 3 = Gamma chitin; folding of chitin chain upon itself forms three parallel segments (Rudall, 1963).

unit of *N*-acetylglucosamine. The length of a chain may range from 0.1 to 1.0 micron.

Purified, chitin consists of highly ordered molecular chains, "micelles." Micelles have a diameter of 0.01 to 0.03 micron.

A crystal lattice is formed by precise arrangement of chitin chains in any given micelle. Chitin fibrils observed by plain light microscopy show subdivisions into smaller micellar fibrils when scanned by the electron microscope.

Types of Chitin

Different chitins are identifiable in X-ray diffraction patterns. Crystallographically three distinct chitin types are recognized. They may all occur in the same animal but in different tissues (Rudall, 1963): Alpha and beta chitin are found in cephalopods. The three types have the same chemical composition and differ only in chain arrangement (Figure 15.14*e–g*.

Long molecular chains are grouped into sets of one (beta chitin), sets of two (alpha chitin), or sets of three (gamma chitin). In gamma chitin, a single chain yields three parallel segments by folding on itself (see Figure 15.14*g*). The "bent" chain is the only one structurally plausible that fits X-ray diffraction data, according to Carlström (1962).

Chitin-Protein

Covalent links to protein holding the chitin chains probably determine the orientation and/or folding of such chains (Figure 15.14*e*). There appears to be a rather exact fit of protein molecules on the pattern made by small groups of chitin chains (Rudall, 1963). Richards (1953) stressed that chitin never occurs as a separate entity in the arthropod cuticle but is always only one component of the chitin-protein complex. Accordingly, chitin cannot be considered a "natural" compound any more than its degradation products.

In dry weight percent of organic fraction of the arthropod cuticle, chitin will be about one-half of the extractable protein, and so on, in arachnids; in calcified and uncalcified crustacean tissues, chitin is relatively high compared to the protein fraction, with rare exceptions; in insects, the situation is reversed—comparatively high protein and lower chitin generally. In dipterid insects, chitin is higher than the protein fraction (Richards, 1951, Table 10).

Chitin Degradation

Chitin is insoluble in water, alcohol, ether, dilute acids, alkalis, and most other solvents. Hot (160° to 180°C) concentrated alkali solutions will not dissolve it. However, some acetyl groups (CH_3-CO) thus removed (see Figure 15.14) yield a chitin-derivative group—the *chitosans*. Chitosans are soluble in dilute acids.

In concentrated mineral acids, chitin dissolves. In sodium hypochlorite, it decomposes. Enzymes degrade it.

Essentially, chitin degradation yields shorter chains. Breakdown products include *N*-acetylglucosamine, glucosamine, and acetic acid (Richards, 1951, 1953).

Chitosan-Iodine Color Test

Place a clean cuticle fragment in a test tube with a few millimeters of KOH saturated in distilled water at room temperature. Seal tube, place in beaker of glycerine. Heat glycerine bath slowly for 15 to 20 minutes to 160°C. Hold at this temperature for about 15 minutes. Remove test tube, and cool to room temperature before examining. Any residue present in the alkali after heating may contain chitin (Hyman, 1958). Wash out alkali in tap water and again in distilled water.

Place one piece of residue in 3 percent acetic acid. If complete solution occurs, add one drop of 1 percent H_2SO_4, which will yield a white precipitate. This indicates a chitinous residue.

Place another piece of residue in 0.2 percent potassium iodide solution. If chitinous, this residue will turn brown. Remove iodide solution and add 1 percent H_2SO_4. If chitinous, the piece will immediately turn reddish-violet. (If thick, the piece may turn black.)

Remove dilute sulphuric acid and replace by 75.0 percent sulphuric acid. The color should disappear and the object go slowly into solution.

A brown-purple color change is a positive color test for the presence of chitin (see Richards, 1951, Chapter 4).

Chitin Accumulation and Decomposition in Nature

Of unusual interest for the study of invertebrate paleobiology is the chitin inventory in nature. Copepod crustaceans alone yield an estimated *several billion tons* annually (Johnson, 1908, cit. Zobell and Rittenberg, 1937). Richards (1951) calculated marine crustacea and terrestrial arthropods probably yield several times that amount. Obviously, the total annual yield of all arthropods must be enormously high. Why, then, are there no significantly large accumulations?

Richards found that chitin will not spontaneously decompose, hence should accumulate, yet it does not. Why? Further data are needed to explore this question.

Arthropod cuticles will decompose in water in the natural state. However, that is a very slow and protracted process. Foster and Benson (1958) estimated that ostracod chitin lost its identity in some 5000 to 15,000 years after bottom mud deposition. Zobell and Rittenberg (1937) designated any bacterium that alters chitin molecules "chitinoclastic." Such chitin oclasts, mostly aerobic, abound in all environments tested — fresh, marine, hypersaline muds, and soils — and the guts of animals from all habitats. In an animal's gut, chitinoclasts may aid chitin digestion. Zobell and Rittenberg identified 31 types of chitinoclasts presumably representing distinct but unnamed species. Rod, coccus, vibrio, and myxobacterial forms are known among chitinoclasts.

Marine chitinoclastic species constitute from 0.1 to 1.0 percent of all marine bacteria: *Bacillus chitinovorus*, *B. chitinophilum*, *B. chitinochroma*, among others. *Cytophaga johnsonae* inhabit soils and muds; *Serratia salinaria* (?) and *Sarcina littoralis* (?) are cosmopolitan, occurring worldwide in salts accumulated by solar evaporation.

Widespread chitinoclastic bacteria, common in decomposing arthropods (crabs, lobsters, and other crustaceans), produce chitin-digesting enzymes (Johnson, 1932; Hess, 1937, cit. Zobell and Rittenberg, 1937). These enzymes probably are the most important among possible factors in the failure of the enormous annual chitin production to accumulate. Serious depletion of available carbon and nitrogen reserves would result from such accumulation (Richards, 1951, Chapter 6, Table 3).

In addition, the snail *Helix pomata* produces chitin-digesting enzymes (chitinases), and its digestive juices can attack lobster cuticle. Possibly, in this case, gut-dwelling symbiont bacteria produce these enzymes. Among the fungi, *Penicillium* also appears to have chitin-decomposing capacity (Richards, 1951, Table 3).

Fossil Chitin

Chitin has been identified in Tertiary insects, the coleopterans (Abderhalden, 1933); in Silurian eurypterids, *Pterygotus osiliensis* (Rosenheim, 1905), among others. (Foucart et al., 1965, found only scleroproteins in graptolites.) Longevity of chitinous skeletal material after burial may be considerable in some instances. Modern chemical tests should be used before declaring any fossil chitinous.

Carbonaceous residues of original chitinous structures are common for many fossil arthropods.

By extrapolation from equivalent and/or related living forms, chitin presence during life in fossil bryozoans, coelenterates, arthropods, mollusks, and inarticulate brachiopods may be inferred.

Disappearance of chitin in almost all fossils that during life had chitinous structures (cuticles, body walls, and so on) should be attributed to either decomposing chitinoclastic bacteria, water action over extended time periods, or volatilization. This last would lead to carbonaceous residues commonly encountered in the rock record.

QUESTIONS

1. (a) Outline the mechanism of crystal nucleation common in most biological systems.
   (b) What physical-chemical considerations importantly affect crystallization in such systems?

2. Indicate significant differences and/or similarities in skeletogenesis of a coral, mollusk, and echinoderm.

3. Evaluate the comparative significance of collagen, chitin, and conchiolin.

4. Compare skeletogenesis in a brachiopod and a calcareous bryozoan.

5. What is the role of chitinoclastic bacteria in the sea? Relate their activity to biogeochemical cycles and the fossil record.

6. Discuss why and how Travis's electron microscope studies might require reevaluation of echinoderm calcite crystallography (Raup) applied to phylogeny and taxonomy.

7. Carefully review Travis' studies of decapod Crustacea, then attempt to apply relevant items to inferred trilobite skeletogenesis.

8. Apply known skeletogensis in the living sand dollar to a fossil sand dollar. How does it differ from crinoid skeletogenesis? (See Chapter 12.)

9. Review Grégoire's chief findings and explain their relevance to the molluscan fossil record.

10. (a) Select a variety of Tertiary fossils that have skeleton chitin and perform the chitosan-iodine test. Tabulate your results. (*Hint*: First decalcify specimen by immersion in HCL.)
    (b) Decalcify the following fossils: a fusulinid, a fragment of a pearly *Nautilus*, a rugose coral, a trilobite cephalon or pygidium; then report the resulting residue, if any. (If residues result, how can they be explained?)

REFERENCES

Abderhalden, E., and K. Hegns, 1931. Nachweis vom Chitin, in Flugelresten von Coeleopteren des oberan Mitteleocäne: Biochem. Z., Vol. 259, pp. 320–321.

Barnes, J. W., E. J. Lang, and H. A. Potratz, 1956. Ratio of Ionium to Uranium in coral limestone: Science, Vol. 124, pp. 175–176, Table 1.

Bé, H., and D. B. Ericson, 1961. Aspects of calcification in planktonic foraminifera (Sarcodina), in M. L. Moss, ed., Comparative biology of calcified tissue: Ann. N.Y. Acad. Sci., Vol. 109, pp. 65–81.

Beedham, G. E., 1954. Properties of non-calcareous material in the shell of *Anodonta cygnea*: Nature, Vol. 194, pp. 750.

Bergmann, Werner, 1949. Comparative biochemical study on the lipids of marine invertebrates with special reference to the sterols: J. Marine Res., Vol. 8, pp. 137–176.

_____, 1963. Geochemistry of lipids, in I. A. Breger, ed., Organic Geochemistry. Macmillan, pp. 503–542.

Bevelander, G., and Hiroshi Nakahara, 1960. Development of the sand dollar (*Echinarachnius parma*), in Calcification in biological systems: A.A.A.S. Publ. 64, pp. 41–56.

Blackman, P. D., and R. Todd, 1959. Mineralogy of some foraminifera as related to their classification and ecology: J. Paleontology, Vol. 33 (1), pp. 1–15.

Blumer, M., 1951. Fossile Kohlenwasserstoffe und Farbstoffe in Kalksteinen: Mikrochemie, 36/37, pp. 1048–1055.

Bøggild, O. B., 1930. The shell structure of the mollusks: K. Dansk Vidensk. K. Skr. Naturvidensk. K. Math. afd. Raekke 2, pp. 232–326.

Bowen, V. T., and D. Sutton, 1951. Comparative studies of mineral constituents of marine sponges. The genera *Dysidea, Chondrilla, Terpios*: J. Marine Res., Vol. 10 (2), pp. 153–165.

Carlström, Diego., 1962. The polysaccharide chain of chitin: Biochim. Biophys. Acta, Vol. 59, pp. 361–364.

Chave, K. E., 1954a, Aspects of the biogeochemistry of magnesium. 1. Calcareous marine organisms: J. Geology, Vol. 62 (3), pp. 266–283 (see Chapter 2 for other references).

Clarke, F. W., and W. C. Wheeler, 1922. The inorganic constituents of marine invertebrates: U.S.G.S. Prof. Paper 124, 62 pp.

Donnay, G., 1955. Crystallography: Carn. Instit. Washington Yearbook No. 55, pp. 203–206.

Dunning, H. N., 1963. Geochemistry of

organic pigments, in I. A. Breger, ed., Organic Geochemistry, Chapter 9.

Emiliani, C., 1955. Mineralogical and chemical composition of the tests of certain pelagic Foraminifera: Micropaleontology, Vol. 1, pp. 377–380, text figures 1–3, Tables 1–4.

Foster, G. L., and R. H. Benson, 1958. Constituents and structural arrangement in ostracode valves: G.S.A. Program, Annual Meetings (St. Louis), abstr., p. 61.

Glimcher, M. J., 1960. Specificity of the organic matrices in mineralization, in Calcification in biological systems: A.A.A.S. Publ. 64, pp. 421–487.

Goodwin, T. W., 1960. Biochemistry of pigments, in T. H. Waterman, ed., The Physiology of Crustacea. Vol. 1, pp. 101–135.

Gordon, Isabella, 1929. Skeletal development in *Arbacia, Echinarachinius and Leptasterias:* Roy. Soc. London, Philos. Trans. Ser. B, Vol. 217, pp. 289–334.

Goreau, T. F., 1959. The physiology of skeleton formation in corals: Biol. Bull., Vol. 116, pp. 59–75, 5 text figures.

———— and W. O. Hartmann, 1963. Boring sponges as controlling factors in the formation and maintenance of coral reefs, in R. F. Sognnaes, ed., Mechanisms of hard tissue destruction: A.A.A.S. Publ. 75, pp. 25–54.

Grégoire, C., 1957. Topography of the organic component in mother-of-pearl: J. Biophys. Biochem. Cytol., Vol. 3, p. 797.

————, 1959a. Conchiolin remnants in mother-of-pearl from fossil Cephalopoda: Nature, Vol. 184, pp. 1157–1158.

————, 1959b. A study of the remains of organic components in fossil mother-of-pearl: Instit. Roy. Sciences Nat. Belg. Bull., Vol. 37 (3), 27 pp., 10 plates.

————, 1960. Further studies in the structure of the organic components in mother-of-pearl, especially in pelecypods, Pt. I: same, Vol. 36 (23), 18 pp., 5 plates.

————, 1961a. Sur la structure submicroscopique de la conchioline associee aux prismes des coquilles de mollusques: same, Vol. 37 (3), 27 pp., 10 plates.

Gross, J., and K. A. Piez, 1960. The nature of collagen. 1. Invertebrate collagens, in Calcification of biological systems: A.A.A.S. Publ. 64, pp. 395–409.

Hedley, R. H., 1963. Cement and iron in the arenaceous foraminifera: Micropaleontology, Vol. 9 (4), pp. 433–441, plate 1.

Hammen, C. S., and K. M. Wilbur, 1959. Carbon dioxide fixation in marine invertebrates. The main pathway in the oyster: J. Biol. Chemistry, Vol. 234, pp. 1268–1271.

Hay, W. H., K. M. Towe, and R. C. Wright, 1963. Ultramicrostructure of some selected foraminiferal tests: Micropaleontology, Vol. 9 (2), pp. 171–195, plates 1–16.

Hyman, L. H., 1958. The occurrence of chitin in the lophophorate phyla: Biol. Bull., Vol. 114, pp. 106–112.

Isenberg, H. D., et al., 1963. Calcification in a marine coccolithophorid, in Comparative biology of calcified tissues: Ann. N.Y. Acad. Sci., Vol. 109, pp. 49–64.

Kulp, J. L., K. Turekian, and D. W. Boyd, 1952. Strontium content of limestones and fossils: G.S.A. Bull., Vol. 63, pp. 701–716, 4 figures.

Krinsley, David, 1960. Trace elements in the tests of planktonic foraminifera: Micropaleontology, Vol. 6 (3), pp. 297–300, Tables 1–2.

Loppens, K., 1920. Influence du milieu sur la composition chimique des Zoécies des Bryozoaires marins: Ann. Soc. Zool. malac. Belg., Vol. 51, pp. 91 ff.

Low, E. M., 1951. Halogenated amino acids of the bath sponge: J. Mar. Res., Vol. 10 (2), pp. 239–245.

Lowenstam, H. A., 1954. Systematic, paleoecologic and evolutionary aspects of skeletal building materials: Bull. Mus. Comp. Zool. (Harvard), Vol. 112 (3), pp. 287–317.

McConnell, Duncan, 1963. Inorganic constituents in the shell of the living brachiopod *Lingula:* G.S.A. Bull., Vol. 74, pp. 363–364, 1 plate.

Odum, H. T., 1951. Notes on the strontium content of sea water, celestite Radiolaria and strontianite snail shells: Science, Vol. 114, pp. 211–213.

Paasche, E., 1962. Coccolith formation: Nature, Vol. 193, p. 1094.

Piez, K. A., 1963. The amino acid chemistry of some calcified tissues, in Comparative biology of calcified tissues: Ann. N.Y. Acad. Sci., Vol. 109, pp. 256–268.

Passano, L. M., 1960. Molting and its control, in T. H. Waterman, ed., The Physiology of Crustacea, Vol. 1, pp. 473–536.

Raup, David, 1959. Crystallography of echinoid calcite: J. Geology, Vol. 67, pp. 661–674.

————, 1960. Calcite crystallography in sea urchins: Yearbook, The Amer. Philos. Soc., 1960, pp. 267–270.

————, 1962a. The phylogeny of calcite crystallography in echinoids: J. Paleontology, Vol. 36 (4), pp. 793–810, 4 text figures.

————, 1962b. Crystallographic data in echinoderm classification: Systematic Zool., Vol. 11 (3), pp. 97–108.

Richards, A. G., 1951. The Integument of Arthropods. Univ. Minnesota Press, 319 pp.

————, 1953. "Structure and development of the integument" and "Chemical and physical properties of chitin," in K. D. Roeder, ed., Insect Physiology: John Wiley and Sons, Chapter 1, 2.

Roche, J., G. Ranson, and M. Eysseric-Lafon, 1951. Sur la composition des scléroprotéines des coquilles des mollusques (conchiolines): Compt. redn. soc. biol., Vol. 145, p. 1474.

Rosenheim, D., 1905. Chitin in the carapace of *Pterygotus osiliensus* from the Silurian rocks of Oesil: Proc. Roy. Soc. London, Ser. B, Vol. 76, pp. 398–400.

Rudall, K. M., 1955. The distribution of collagen and chitin, in Fibrous proteins and their biological significance: S.E.B. Symp. No. IX, Academic Press, pp. 49–71.

————, 1963. The chitin/protein complexes of insect cuticle, in Advances in Insect Physiology. Academic Press, Vol. 1, pp. 257–311.

Schmidt, W. J., 1924/1925. Bau und Bildung der Prismen in den Muschelschalen: Mikrokosmus, Vol. 12, p. 49.

Stehli, F. G., 1956. Shell mineralogy in Paleozoic invertebrates: Science, Vol. 123, pp. 1031, 1032.

Tanaka, S., H. Hatano, and O. Itasaka, 1960. Biochemical studies on pearl. IX. Amino acid composition of conchiolin in shell and pearl: Chem. Soc. Japan Bull., Vol. 33, pp. 543–545.

Travis, D. F., 1960. Matrix and mineral deposition in skeletal structures of the decapod Crustacea, in R. F. Sognnaes, ed., Calcification in biological systems: A.A.A.S. Publ. 64, pp. 57–116.

————, 1963. Structural features of mineralization from tissue and macromolecular levels of organization in the decapod Crustacea, in M. L. Moss, consulting ed.,

Comparative biology of calcified tissue: Ann. N.Y. Acad. Sci., Vol. 109, pp. 177–245.

Tsujii, Tadashi, 1960–1962. Studies in the mechanism of shell and pearl formation in Mollusca: J. Fac. Fisheries Prefectural Univ. of Mie, Vol. 5, pp. 2–70, 21 text figures, 13 plates.

————, D. G. Sharp, and K. M. Wilbur, 1958. Studies on shell formation. VII. The submicroscopic structure of the shell of the oyster: Biophys. Biochem. Cytol., Vol. 4 (3), pp. 275–279, plates 148–151.

Vinogradov, A. P. (1935–1944). The elementary chemical composition of marine organisms: Sears Found. Marine Res. (Yale Univ.) Mem. No. 2 (1953), 647 pp.

Watabe, N., 1956. Dahllite identified as a constituent of prodissoconch I of *Pinctada martensii*: Science, Vol. 124, p. 630.

West, C. D., 1937. Note on the crystallography of the echinoderms: J. Paleontology, Vol. 11, pp. 458–459.

Wilbur, K. M., 1960. Shell structure and mineralization in molluscs, in Calcification in biological systems: A.A.A.S. Publ. 64, pp. 15–36, refs.

———— and Norimitsu Watabe, 1963. Experimental studies on calcification in molluscs and the alga *Coccolithus huxleyi*, in Comparative biology of calcified tissue: Ann. N.Y. Acad. Sci., Vol. 109, pp. 82–112.

Wood, Alan, 1949. The structure of the wall of the test in the Foraminifera: its value in classification: Geol. Soc. London Quart. J., Vol. 104, pp. 229 ff., plates 13–15.

ZoBell, C. E., 1963. Organic geochemistry of sulfur, in I. A. Breger, ed., Organic Geochemistry. Macmillan, Chapter 13.

———— and S. C. Rittenberg, 1937. The occurrence and characteristics of chitinoclastic bacteria in the sea: J. Bacteriology, Vol. 35, pp. 275–287.

(Some additional useful bio- and geochemical data: Ting Yan Ho, 1966, G.S.A. Bull., Vol. 77, pp. 375–392; Hare, P. E., and R. M. Mitterer, 1967, Carnegie Inst. Ann. Rpt., 1965–1966, pp. 362–364; Mook, W. G., and T. C. Vogel 1968, Science, Vol. 159 pp. 874–875; Swain, F. M., 1970, Nonmarine Organic Geochemistry. Cambridge Univ. Press). Wilbur, K. M., 1972, in, M. Florkin and B. T. Schneer, Chem. Zool. Vol. 7, Academic Press, pp. 103–145.

appendix one
SELECTED CASE HISTORIES IN PALEOBIOLOGY

INTRODUCTION

Four case histories are considered. These cover a few of the many exciting new possibilities in paleobiology.

The object of this section's discussion is to italicize the important bearing paleobiologic data have on major scientific problems ranging from molecular biology to astronomy.

I. GEOCHRONOMETRY

[References: Wells, J. W., 1963. Coral growth and geochronometry: Nature, Vol. 197, pp. 948–950. Scrutton, C. T., 1965. Periodicity in Devonian coral growth: Paleontology, Vol. 7 (4), pp. 552–558; Aveni, A. F., 1966. Middle Devonian lunar month: Science, Vol. 151, pp. 1221–1222; Clark, G. R., II, 1968. Mollusk shell: daily growth lines: Science, Vol. 161, pp. 800–802; Pannella, G., C. MacClintock, M. N. Thompson, 1968. Paleontological evidence of variation in length of synodic month since Late Cambrian: Science, Vol. 162, pp. 792–796.]

Problem. (a) Can one count daily (= circadian) or monthly (= synodic) growth intervals in selected fossilized inverte-brate skeletons?

(b) If so, how can these data shed new light on geophysical and astronomical events?

Evidence and Findings

Wells (1963) assumed that the major annulations observed in the epitheca of fossil Paleozoic corals represented annual growth increments. This was tested by study of Devonian, Pennsylvanian, and scleratinian corals. (See below).

Scrutton (1965) checked Wells results in a study of M. Devonian rugose corals. He found that fine growth ridges were grouped into regular bands of about 30 ridges each. The length of the year (M. Dev.), using astronomical data, was taken to be 399 ± days. When 399 was divided by the average number of ridges per band, a figure was obtained of 13.04 for 10 specimens. The M. Devonian lunar month lasted 30.5 days; during the M. Devonian year, the moon circled the earth 13 times. Banding in the rugose corals studied (that included some of the same forms studied by Wells) was probably related to a lunar breeding periodicity (see Chapter 5 for modern corals).

Wells and Scrutton's results were ad-

M. Dev.	*Heliophyllum* *Eridophyllum* *Favosites*	400 ± Growth Lines/annum
Penn.	*Lophophylidium*	385–390 Growth Lines/annum
Rec.	*Manicina areolata*	360 Growth Lines/annum

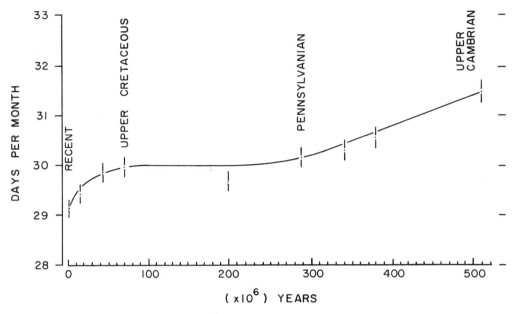

FIGURE A.1 Variations in length of the synodic month through geologic time. Note two major changes in curve slope: Pennsylvanian and Upper Cretaceous. (See text for interpretation.) The bar beyond each plotted point represents the standard error for each point. Adapted from Pannella, MacClintock, and Thompson, 1968.

vanced by a study of variation in the synodic month from U. Cambrian to Recent time (Pannella et al., 1968). This investigation determined growth increments from fossil mollusks (chiefly bivalves, but including one L. Pennsylvanian cephalopod) and a stromatolite from the U. Cambrian (Figure A.1). Good agreement was found with Scrutton's corals for the Devonian bivalve, *Conocardium*.

This study suggests (see Figure A.1, change of curve slope) that the slowing down of earth's rotation has not transpired uniformly: Pennsylvanian to Cretaceous time, slow down was negligible; Post-Cretaceous to Recent, marked slow down. Is this last observation an indication of rapid continental drifting apart by Late Cretaceous time? The authors allowed for this possibility. What would control slowdown of earth's rotation? Tidal dissipation in shallow seas. From this line of reasoning it follows that post-Drift, the seaways broadened (proto-Atlantic Ocean) and hence increased tidal effects on earth's rotation.

Geophysical data indicate a uniform

slowing down, about two milliseconds per century, in earth's rotation due to tidal effects. The paleontological data reviewed support the interpretation of tidal effect influence. It goes further and suggests a nonuniform slowdown.

What is needed now is comparable data for the same geologic period (L, M, or Upper Devonian, or Pennsylvanian, for example) for a half-dozen *different* fossils: mollusks, stromatolites, corals, and so on.

The existing evidence appears to establish that certain well-preserved fossil invertebrates can serve as geochronometers.

II. THEORETICAL MORPHOLOGY AND MATHEMATICAL MODELS

[References: Raup, D. M., 1961. The geometry of coiling in gastropods: Nat. Acad. Sci. Proc., Vol. 47 (4), pp. 602–609. Raup, D. M., 1962. Computer as an aid in describing form in gastropod shells: Science, Vol. 138, pp. 150–152. Raup, D. M., 1966. Geometric analysis of shell coiling: general problems: J. Paleontology, Vol. 40 (5), pp. 1178–1190. Raup, D. M., 1967. Geometric analysis of shell coiling: coiling in ammonoids: J. Paleontology, Vol. 41 (1), pp. 43–65. Raup, D. M., and A. Michelson, 1965. Theoretical morphology of the coiled shell: Science, Vol. 147, pp. 1294–1295. Thompson, D'Arcy W., 1942. On Growth and Form. Cambridge Univ. Press, 1116 pp.]

Given

(a) Most coiled forms are variants of a simple model. (b) A mathematical model based on the logarithmic spiral to be established for one or another morphological feature (see Thompson, 1942, and Chapter 8, this text). (c) With (b), to predict the total spectrum of possible forms. (d) To use (c) to interpret existing forms or those known as fossils.

Problem. To establish (b) and to determine (c) and (d).

Solution. Raup and Michelson (1965) used a logarithmic model that required four parameters to describe the general form of a coiled shell. These parameters include the following (see Figure A.2C):

S—shape of generating curve. For snails, it is a cross-section of a hollow tube; for ammonoids, width/height ratio of whorl cross-section. The generating curve progressively enlarges after each complete whorl about the axis of coiling.

D—relative distance between generating curve and axis of coiling.

W—a constant factor by which any linear dimension of the generating curve is enlarged

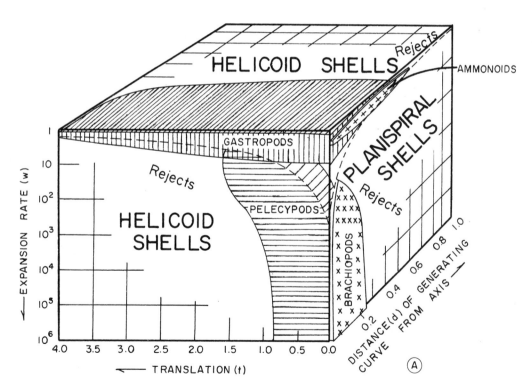

FIGURE A.2 Theoretical spectrum of actual and possible shell forms in mollusks and brachiopods. Modified from Raup, 1966, 1967.

(A) Geometric model, block diagram. Shape of generating curve taken as a constant. Note region of diagram occupied by majority of mollusks and brachiopods. Blank spaces represent the more voluminous potential sparsely or not occupied by any known living or fossil forms (natural selection's "rejects"). These spaces can be filled by computer snails, ammonoids, and so on. See (B) below.

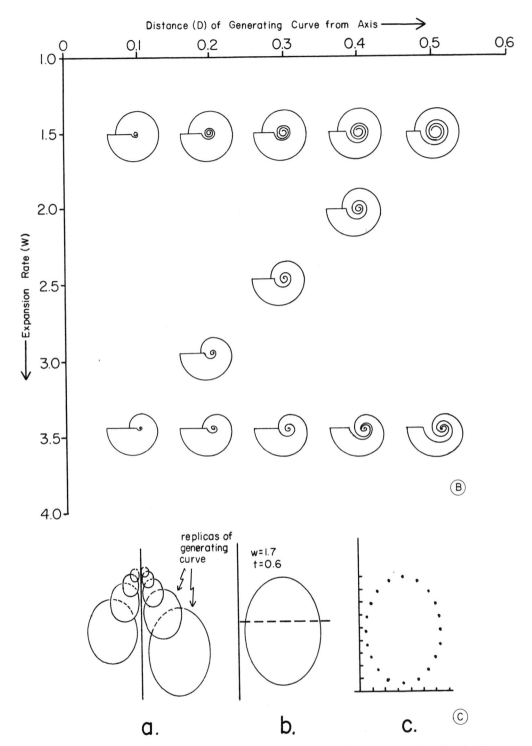

(B) Computer ammonoids. (Generating curve taken as a perfect circle.) Compare parameters W and D, and relative space occupied in the block diagram by ammonoids. (All spaces on diagram can be filled. Only a few are shown.)

(C)(a, b): (a) Shape of generating curve for a computer snail. (b) Four parameters shown: oval = shape of generating curve (S); horizontal dashed line through it = translation; rate of increase; vertical line = axis of coiling — note position of generating curve relative to it; w = a constant factor by which any linear dimension of curve is enlarged during one revolution; compare (C)a for enlarging generating curve per revolution; (c) selected pairs of coordinates (x,y) are taken from a given snail (y = axis of coiling) that define shape (S), and D = distance of curve from axis; W and t are constants.

during a full revolution about axis of coiling.

t—a constant that indicates rate of whorl translation along the axis. It is the portion of the height of a generating curve (S) measured parallel to its axis of coiling, which is covered by the succeeding generating curve on the same side (see Figure A.2Ca, Cb).

To fix the position of the generating curve, an x-y coordinate system was used, where the y-axis coincided with the axis of coiling. These data are measured on a given fossil or Recent specimen. Thus S and D are defined. The constants w and t are also determined for a given species. The coordinate points plus the constants are the input data for the computer.

In order to obtain theoretical coiled forms, a computer program can be written so that any other generating curve on a given cross-section (Figure A.2Ca) can yield x and y values. Variation of w and t can also be programmed. In this way a whole spectrum of possible coiled forms can be obtained.

Raup and Michelson (1965), using a digital and analog computer, obtained both printed outputs and oscilloscope photographs of computer snails. Raup subsequently (1967) obtained computer planispiral ammonoids (Figure A.2B) and produced a geometric model of the total possible spectrum of shell forms (Figure A.2A) (1966).

Significance and Application

Natural Selection. As pointed out, the blank spaces on the geometric model (Figure A.2A) are negative choices made by natural selection's operation on the existing gene pool through time (Tasch, 1965). Thus, of n possible shapes potential in the genetic code of a given mollusk or other coiled form, only a limited range was *selected for* and most others selected against.

By reviewing the rejects of natural selection (computer snails, and so on), we can get a glimpse of how it works. It may prove possible, using block diagrams (Figure A.2A), to determine volumetrically (or areally) the ratio of actual to potential

shell forms per taxonomic group for a given formation and geologic period or its subdivisions. Then a comparative index would become available of the *rate of natural selection in a given group/per period of time.* For mollusks, this could be done readily starting with the Tertiary. Year values obtained by radioactive dating would be available. From this, the selective rate per thousand or million years might be determined and used for comparative purposes.

Function. A further consequence of observing the end results of natural selection is to probe the adaptive significance or advantage conferred by the successful configurations that would not have been available if the rejects had prevailed. Enhanced shell strength, greater buoyancy, improved feeding, and protection are among the kinds of advantages that might ensue.

There is also the possibility that every change in shape may not be related to function directly but could be a meaningless or chance genetic event (see Chapter 14, this text). The genetic code has a great deal of redundancy (repetition) and nonsense syllables as well as sensible signals or messages.

Raup (1967) recognized the "primitive" nature of the analysis used for ammonoids. Nevertheless, he ventured an explanation of why ammonoids are restricted to the planispiral face of the block diagram (Figure A.2A). This was attributed to ammonoid requirements relative to orientation in water and streamlining.

Clearly, studies are needed of all aspects of theoretical morphology so that the range of its possibilities as a research tool can be determined.

III. BIOMINERALIZATION AND PHYLOGENY

[References: Williams, Alwyn, 1968a. A history of skeletal secretion among articulate brachiopods: Lethaia, Vol. 1(3), pp. 268–287. Williams, Alwyn, 1968b.

Evolution of the shell structure of articulate brachiopods: Sp. Papers in Paleontology 2, Paleont. Assn. London, 55 pp., and see Chapters 7 and 15, this text.]

Axioms

(a) The *process* of secretion of the living articulate brachiopod shell has been a constant through geologic time (that is, without any basic change on the molecular level). (b) The *pattern* of mantle secretion (biomineralization) has changed through geologic time. (c) Hence (b) should permit inferred phylogenies of skeletal successions since secretory regimens have evolved.

Problem. To decipher shell microfabric (ultrastructure) and, from this information, to deduce cellular morphology of the outer epithelial tissue that secreted the shell.

Procedure. Ultrastructure was determined from prepared shell sections of the living articulate brachiopod, *Notosaria nigricans.* Comparative studies of secretory regimen were made in other articulates such as the terebratulid *Waltonia inconspicua.* Prepared sections were studied by electron microscope and scanning electron microscope.

Evidence and Findings

Ultrastructure in *Notosaria nigricans* revealed six steps in biomineralization of the shell. These are considered standard for articulate brachiopods. The six secretory steps produce the following shell layers (Figure A.3):

1. Mucopolysaccharide.*
2. Outer fibrillar triple-layered membrane* (upper portion a series of protein rods arranged in regular pattern; whole proteinaceous membrane is about 140 Å thick).
3. Mucoprotein (= chief constituent of periostracum; up to 1 μm thick).

4. Inner fibrillar triple-layered membrane.
5. Calcareous primary layer.*
6. Calcareous-organic secondary layer.*

[*Only the starred four steps are fundamental, as indicated by comparison with terebratulids. Actually, the four steps can be reduced to three basic operations that provided the following sequence: (1) mucopolysaccharide, (2) fibrillar triple-layered membrane, (3) calcareous shell.]

The review of skeletal succession in articulate brachiopods through time (Figure A.3) discloses the following events: (*a*) by early Paleozoic, introduction of a crystalline primary layer (orthids), which persisted through subsequent time in the Pentamerida-Spiriferida lines of descent; (*b*) introduction of a primary layer in other offshoots of the orthids during the Paleozoic, with loss of the fibrous secondary layer in most strophomenids as well as some Mesozoic forms; (*c*) introduction of the second triple-layered membrane in Rhynchonellida; (*d*) loss of (*c*) in terebratulids and introduction of a spicular endoskeleton; (*e*) introduction of a tertiary prismatic layer as in the Mesozoic koninckinaceans.

What occurred at the gene loci in light of the given information? Coding for fibrous shell layer was clearly *stable* in most lines of descent, as was the outer triple-layered membrane. However, the fibrous shell layer was *selected out* twice (strophomenids and thecideidineans). The crystalline primary layer was intercalated between unchanged existing layers in the orthids. That would suggest addition of genetic information, possibly by genetic recombination. In the plectambonitacean line, a primary laminar layer *replaced* the crystalline primary layer. Here it is probable that the concerned gene loci were modified by mutation.

In the Paleozoic articulate brachiopod gene pools, genetic potential existed for fibrous secondary layer, crystalline primary layer, primary laminar layer, and endospicular skeleton. Which of these were to be selected for can only be deter-

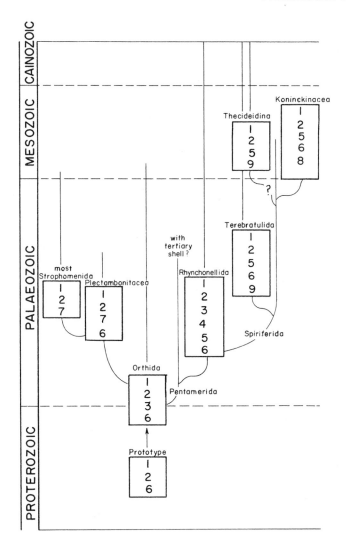

FIGURE A.3 Evolution of shell ultrastructure in articulate brachiopods. Data from Williams, 1968a. 1. Mucopolysaccharide layer; 2. Fibrillar rods with outer bounding membrane; 3. Mucoprotein layer; 4. Fibrils with inner bounding membrane; 5. Crystalline primary layer; 6. Fibrous secondary layer; 7. Primary laminar layer; 8. Prismatic tertiary layer; 9. Spicules. (1–8 are exoskeletal; 9 is endoskeletal). Under Orthida replace 3 by 5.

mined by the end result in given lines of descent.

The full significance of such researches may well extend beyond inferred phylogenies (which can be equally inferred from morphologies), to such items as selection pressure of the environment, daily accretion (banding), seasonal effects, and many other factors.

IV. NUMERICAL TAXONOMY

[References: Mayr, Ernst, 1965. Numerical phenetics and taxonomic theory: Systematic Zoology, Vol. 14 (2), pp. 73–97; Kaesler, R. L., 1967. Numerical taxonomy in invertebrate paleontology, in Curt Teichert and E. L. Yochelson, eds., Essays in Paleontology and Stratigraphy, Univ. Press Kansas, pp. 63–81; Rowell, A. J., 1967a. A numerical taxonomic study of the chonetacean brachiopods. Same: pp. 113–140; Rowell, A. J., 1967b. Raw data from above study, unpublished, supplied by Dr. Rowell and reviewed in this section; Sokal, R. P., and P. H. A. Sneath, 1963,

Principles of Numerical Taxonomy. Freeman, 359 pp.; Sneath, P. H. A., 1957. The application of computers to taxonomy: J. Gen. Microbiol., Vol. 175, pp. 201–206. (See appendix for "Details of Analysis of Example"—a given genus *Chromobacterium*.]

Major Concepts and Terminology

1. (a) To establish repeatable and "objective" methodology for classifying and comparing species of living things, numerical taxonomy was devised (Sokal and Sneath, 1963).

(b) Where numerical taxonomy differs from classical taxonomy is precisely indicated in the term "numerical." Multiple unit characters or features (according to Sokal, not less than 60) are measured or qualitatively evaluated. Such measurements or evaluations are then treated by statistical techniques, such as multivariate analysis. (Sneath, 1957, used 105 unit characters.)

(c) The purpose of (b) is to establish similarities and differences between taxa (species, for example—referred to as OTU's, operational taxonomic units). This is done in order to arrange these species, genera, families, and so on, by affinities.

2. Some major concepts in numerical taxonomy relate to genes. It is assumed that most genes have multiple effects, affecting more than one unit character. Furthermore, no large class of genes can affect exclusively a given class of unit characters, for example, skeleton only, or circulation only (see Chapter 14, "Pleiotrophy," this text). By use of sampling theory, Sneath illustrated what seems almost self-evident, that any species being studied can be comprehended by a certain minimum of information, and that no amount of measuring or evaluating unit characters thereafter adds anything new.

Each character can have *n* states. Each character and each state provide one piece of new information. Thus 60 characters yield 60 pieces and each character may have, say, five states. For example, consider a trilobite: The pygidium may bear the unit character *spinosity*, and this may be strongly, moderately, or weakly expressed, lacking or be present in a pseudo-condition. This would add up to five states. Thus $60 \times 5 = 300$ pieces of information about the given species if each character has five states. A similar number of pieces can be determined for another species, and so on, and statistical comparison made by cluster analysis.

A graphic representation (phenogram) will show a branching diagram of the clusters, for example, which clusters have closer affinities than others. Then one can draw lines of equal similarity (= phenon lines) across the phenogram, forming phenons that correspond to taxonomic categories attained by classical taxonomy (Figure A.4, A). (See Sokal and Sneath, 1963, appendix for "Computational Methods in Numerical Taxonomy.)

Numerical Taxonomy Applied to Fossils

The best available study of numerical taxonomy applied to a fossil taxa is that by Rowell (1967).

Problem. Given: two classifications of the superfamily Chonetacea (Silurian through Permian) each by a specialist of the group:

Sokolskaya (1960) recognized three families, three subfamilies, embracing 21 genera. Muir-Wood (1962) independently recognized four families, 11 subfamilies, embracing 29 genera. (Of these seven were new and one not included by Sokolskaya.) In 1965 Muir-Wood incorporated four more genera raising the total to 33— Figure A.4*B*.)

Both classifications, which differed markedly, were regarded as tentative by their authors. What was the relative merit of each? Could one determine by a non-subjective approach (numerical taxonomic analysis) which of the two was preferable and if it had any flaws in it? Rowell's study was designed to resolve these

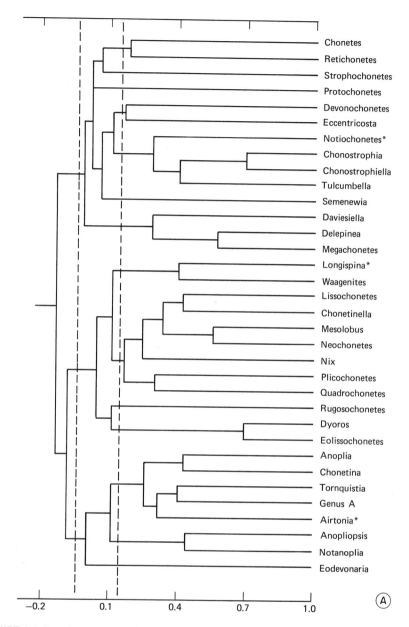

FIGURE A.4 Superfamily Chonetacea (Sil.–Perm.). Classification by two methods: numerical and classical taxonomy.
(A) Numerical taxonomy: Phenogram. 0.14 and –0.05 phenon lines, shown by dashed lines. These are lines of equal similarity. Where they intersect the phenogram, they mark off groupings that should ideally be equivalent to those obtained by classical taxonomy.

questions as well as to test Sokal and Sneath's numerical taxonomic techniques.

Methodology

Sixty unit characters were recognized These pertained to ornament, gross mor-

phology, interarea features, pedicle, and brachial valve–internal features, and cardinalia. Most characters chosen were *qualitative* (observed but not measured) and were divided into a number of states. Each state bore a corresponding number depending on degree of development of the given character. *Chonetes*, State 2,

FIGURE A.4 (cont.)

(B) Classical Taxonomy: Diagrammatic layout of Muir-Wood's classification of the chonetaceans; left column—family; middle column—subfamily; right column—genus. This classification agrees with that marked off in (A) by the 0.14 phenon except for the starred genera. These, Rowell suggested, should be assigned as indicated in (A).

means the brachial valve is weakly concave, whereas for *Tulcumbella*, State 4, it is convex.

    Tabulation of Characters and States in

Chonetacean Genera. (Raw data by Rowell, arrangement and added notations by Tasch.) (Only two genera are treated for illustration.)

Unit Characters (Cu)	States of Unit Characters (Scu in pieces)
1. Brachial valve	Strongly concave – 1, weakly concave – 2, plane – 3, convex – 4, strongly convex – 5, globose – 6. These numbers are what is tallied for each genus. *Chonetes** – 2, *Eccentricosta** – 1, and so on.

[*Notation: none of the 34 OTU's (= genera) are in States 5 or 6, 18 are in State 2. In this example, $Cu \times Scu = 1 \times 6 = 6$ pieces (pcs) of information (= info).*]

2. Pedicle valve	6 States as in Brachial valve; *Chonetes* – 4, *Eccentricosta* – 4. [6 pcs of info.]

3. Commissure

Straight line—1, weakly concave—2, strongly concave—3, double looped, low-lying alate, W—4. *Chonetes*—1, *Eccentricosta*—1.

[Notation: 25 OTU's are in State 1 . . . 4 pcs of info.]

4. Ornament

Fasciculate, yes—1, no—0. *Chonetes*—0, *Eccentricosta*—0.

[Notation: 33 OTU's are 0 . . . 2 pcs of info.]

5. Radial ornament

Smooth—1, capillate—2, costellate—3, costate—4. *Chonetes*—2, *Eccentricosta*—4.

[Notation: 15 OTU's are in State 2 . . . 4 pcs of info.]

6. Concentric ornament

Rugae present—1, absent—0. *Chonetes*—0, *Eccentricosta*—0.

[Notation: 33 OTU's are in State 0 . . . 2 pcs of info.]

7. Growth lines

Absent to rare—0, common, anterior—1, common—2. *Chonetes*—1, *Eccentricosta*—0.

[Notation: 24 OTU's are in State 2 . . . 3 pcs of info.]

8. Pedicle valve interarea

Absent—0, linear—1, modest—2, high—3. *Chonetes*—2, *Eccentricosta*—2.

[Notation: 27 OTU's are in State 2 . . . 4 pcs of info.]

9. Brachial valve interarea

4 States same as "8" above. *Chonetes*—2, *Eccentricosta*—1.

[Notation: 24 OTU's are in State 2 . . . 4 pcs of info.]

10. Brachial valve interarea reflexed

Yes—1, no—0. *Chonetes*—0, *Eccentricosta*—0.

[Notation: 24 OTU's are in State 0 . . . 2 pcs of info.]

11. Delthyrial cover

Open—1, Apical pseudo-deltid—2, large pseudo-deltid—3. *Chonetes*—2, *Eccentricosta*—2.

[Notation: 26 OTU's in State 2 . . . 2 pcs of info.]

12. Calliot plugging delthyrium

Yes—1, no—0. *Chonetes*—0, *Eccentricosta*—1.

[Notation: 22 OTU's in State 0 . . . 2 pcs of info.]

13. Chilidium

Yes—1, no—0. *Chonetes*—0, *Eccentricosta*—0.

[Notation: 19 OTU's in State 0 . . . 2 pcs of info.]

The other 47 unit characters and states (listed as pieces only) are given below without detail.

14. Chilidial plates—2 pcs. 15. Posterior/lateral margins—3 pcs. 16. Umbo incurvature—2 pcs. 17. Brachial valve, posterior sulcus—2 pcs. 18. Brachial valve, median valve—3 pcs. 19. Pedicle valve, median ridge bifurcating posteriorly—2 pcs. 20. Pedicle valve, median ridge extending to umbo—2 pcs. 21. Accessory ridge, pedicle valve—2 pcs. 22. Outer socket ridge—2 pcs. 23. Alveolus—2 pcs.

24. Cardinal process, exterior face—5 pcs. 25. Cardinal process, internal face—4 pcs. 26. Cardinal process projecting—2 pcs. 27. Denticulation on hinge—2 pcs. 28. Anderidia—2 pcs. 29. Anderidia fused with cardinalia—2 pcs. 30. Anderidia fused with inner socket ridge—2 pcs. 31. Anderidia, parallel—subparallel—2 pcs. 32. Brachial valve median septum—3 pcs. 33. Brachial valve median septum elevated anteriorly—2 pcs. 34. Brachial valve median septum elevated posteriorly—2 pcs. 35. Brachial valve median septum spinose

anteriorly – 2 pcs. 36. Internal periphery pustulose – 2 pcs. 37. Pustules – 2 pcs. 38. Accessory ridges, number – 4 pcs. 39. Accessory ridges, type – 3 pcs. 40. Median accessory ridges greater than lateral and enlarged – 2 pcs. 41. Pedicle valve, mantle canals forming ridges – 2 pcs. 42. Pedicle valve muscle bounding ridge – 2 pcs. 43. Teeth, kind – 2 pcs. 44. Teeth, size – 3 pcs. 45. "Spinules" – 2 pcs. 46. Brachial ridges – 2 pcs. 47. Shell of pedicle valve strongly thickened – 2 pcs. 48. Enlarged median capillus – 2 pcs. 49. Outline – 3 pcs.

50. "Ears" – 2 pcs. 51. Pedicle valve, diductor field, longitudinal ridges – 2 pcs. 52. Pustulose lateral-to-median septum – 2 pcs. 53. Brachial valve, adductors, dendritic – 2 pcs. 54. Accessory adductor scars – 2 pcs. 55. Inner socket ridges – 2 pcs. 56. Median septum – 2 pcs. 57. Maximum number spines for exemplar (each side) – 1 pc. 58. Spine angle for exemplar – 1 pc. 59. Maximum width cited of exemplar – 1 pc. 60. All ribs radial from umbo – 2 pcs.

The character states above yield 149 pieces of information for 60 characters. This comes to many thousands of pieces. The figure would be reduced if one deducted from the 149 pieces figure the empty character states that do not characterize any of the 34 OTU's. Nevertheless, a very large number of pieces were available for statistical treatment (cluster analysis).

Following various analyses, the computer printed out a graphic representation (phenogram) of clusters of similar genera (Figure A.4). These were then compared with the classical taxonomic grouping. Rowell found that the Muir-Wood classification was very close to that derived by numerical taxonomy with a few anomalies (see starred genera, Figure A.4).

Conclusion

Rowell concluded that the placing of the three anomalous genera *Airtonia*, *Notiochonetes*, and *Longispina*, as indicated by numerical analysis (Figure A.4*A*), should be followed as superior to both classifications. Furthermore, he found the numerical classification closer to Muir-Wood's than to Sokolskaya's. He suggested that numerical taxonomic methods could provide clarification for groups *where the lines of descent are vague*, as in the Chonetacea.

Comment

For the best critical comment on numerical taxonomy, see Mayr (1965); for the most supportive view by a paleontologist, see Kaesler (1967). It is significant that the classical methods used by Muir-Wood could not be faulted except in placement of 3 out of 34 genera.

What is the deeper meaning, if any, of this low percent of disagreement between numerical and classical taxonomic treatment of the same subfamily? Rowell, as we know, used literally thousands more pieces of information.

Although all characters are given equal weight, they do form a natural hierarchy of primary and secondary and/or auxiliary pieces of information. From this it follows that even if fewer variables were treated, five or ten instead of the larger number, say 60 (plus States), the result would be *the equivalent of the larger number in informational content.*

Nevertheless, any statistical tool that brings to light closer relationships than can be obtained by other methods – even if partial – is an important and usable approach.

appendix two

MINIMUM SCHEDULE FOR USABLE COLLECTIONS OF FOSSIL SAMPLES

[See Kummel and Raup, 1965, *Handbook of Paleontological Techniques*, for further details of sampling, preparatory techniques, and many useful hints and guides for collecting.]

1. GEOGRAPHIC

(a) Locate site or outcrop on a road map, topographic or geologic map.

(b) If more than one site, repeat (a).

(c) If unknown locality, prepare field sketch map with nontransient reference point and compass directions from this point. Subsequently, relate this map to a standard map.

2. GEOLOGIC

Stratigraphic

(a) If site is at a roadcut, pace out length or tape distance from one end to the other. If a quarry, do the same for each exposed face to be sampled.

(b) Determine overall thickness of entire sequence of beds to be sampled by standard methods (tape and compass).

(c) Before systematic sampling, explore the accessible portions of the outcrop (if not too extensive or forbidding) and sample for fossils. This will provide an early indication of the potential. Leave fossils on collecting bag where they were released from the rock matrix. These can be collected later during systematic sampling.

Systematic Sampling

(a) Fossils should be carefully collected with regard to time, that is, from oldest to youngest beds (or vice versa, from top of outcrop down to base, or the reverse). Fossils from each bed should be numbered by locality and bed, for example, from Bed I to Bed X.

(b) If a given bed is a few feet thick, it can be divided into three or more parts. Thus Bed 9 collections, for example, would be labelled "9a, 9b, 9c," and so on, or equivalents, from older to younger or vice versa. (Customarily, the oldest bed is labelled "1," the next older "2," and so on. However, for field convenience, it may be necessary or desirable to collect from the top down to the base. In this event, numbers can be reversed and corrected subsequently.)

(c) If multiple sacks are used for a given fossiliferous bed, each should bear the same locality and bed label plus a unique sack number.

(d) Beds found to be nonfossiliferous in the field should also be sampled for subsequent laboratory examination and acid digestion; they should be searched for microfossils in the residues.

(e) Sampling may be in a single columnar (vertical) section at an outcrop or at several such sections. If the latter, the distance between each section should be determined. Collecting sacks would then all have the same locality number but

915

distinctive section numbers. Thus Locality Enco, 80 paces from north end of outcrop; Section A, 300 paces from Section A; Section B, and so on.

(f) Sampling of talus slopes, isolated rock debris, morainal train debris, or infills, and so on, require special steps. All of these situations involve rocks not in place.

(1) Try to locate the bed from which fossiliferous slab or piece was derived. Put these data in your notes.

(2) Draw a sketch of talus, rock pile, and so on, relative to both the outcrop being sampled and specific measured sections.

(3) Sack these "float" fossils separately and refer to notes (1) and (2), above, on the labels.

(g) If an isolated fossil is located in the field where there are no visible outcrops, or in chert pebbles in a creek bed or road gravel, precise location will prove more difficult. In such cases subsequent reference to maps or further field exploration in the vicinity — upstream direction, and so on — may locate the source bed.

Beds of Dubious Age

Fossils may be collected from beds that have been displaced, disturbed, and/or subsequently deeply dissected by erosional processes. In such cases, these data (attitude of beds, structure) should accompany other notes on the fossil site since they will bear importantly on the relative age of the given fossils.

STUDENT'S GUIDE TO SUBJECT INDEX

This index is alphabetically arranged. The respective alphabet letters are printed in bold type and centered above the first column in which relevant entries occur. The A's or S's, for example, may cover several columns and, in some instances, may extend to more than one page. However, it will become evident as one uses the index, how many columns and pages are devoted to a single alphabet letter, since the next alphabet letter will follow and also be centered and printed in bold type.

There are three kinds of entries in this index:

1. MAJOR ENTRIES are printed in all capital letters (e.g., ACRITARCHS, ALGAE, AMINO ACIDS...).

2. SECONDARY ENTRIES are indented two spaces from each major entry on the line below the major entry generally. However, the first occurrence of a secondary entry will appear in whole or part on the same line as the major entry.

3. TERTIARY ENTRIES are indented two spaces from the secondary entry.

A student wishing to look up a specific subject, such as "Evolution of Paleozoic Echinoids," can proceed as follows: find the letter "E"; find the main entry, "ECHINOID(S)"; indented two spaces from this main entry, but in the next column, find the secondary entry, "Paleozoic"; indented two spaces from this secondary entry, find the tertiary entry "evolution" and page numbers.

ECHINOID(S) — — Main Entry
Paleozoic — — Secondary Entry
evolution — — Tertiary Entry

ALPHABETICAL GUIDE TO INDEX OF SCIENTIFIC NAMES

[*Name endings*: -idea = subclass; -ida = order; -ina = suborder; -acea = superfamily; -idae = family].

To find the pages on which any scientific name from phylum to species or subspecies occurs in this text, proceed as follows: find the respective alphabet letter (A, B, C. . .) for your particular entry (e.g., *Pachydicta acuta*, in this case "P"), then locate the genus name, "*Pachydicta*"; to find the species name, look on the same line or some line below the genus name. Genus and species names are always in italics.

Since this index is arranged alphabetically, it is not necessary to know if the desired entry is a bryozoan, cephalopod, or other form to find it. In the event that only the phylum name is known (Bryozoa, Cephalopoda, etc.), refer to the subject index for more details on all other aspects exclusive of the name of a given taxa.